Moments of Inertia of Common Geometric Shapes

Rectangle

$$\bar{I}_{x'} = \tfrac{1}{12}bh^3$$
$$\bar{I}_{y'} = \tfrac{1}{12}b^3h$$
$$I_x = \tfrac{1}{3}bh^3$$
$$I_y = \tfrac{1}{3}b^3h$$
$$J_C = \tfrac{1}{12}bh(b^2 + h^2)$$

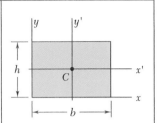

Triangle

$$\bar{I}_{x'} = \tfrac{1}{36}bh^3$$
$$I_x = \tfrac{1}{12}bh^3$$

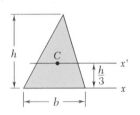

Circle

$$\bar{I}_x = \bar{I}_y = \tfrac{1}{4}\pi r^4$$
$$J_O = \tfrac{1}{2}\pi r^4$$

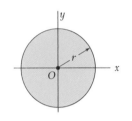

Semicircle

$$I_x = I_y = \tfrac{1}{8}\pi r^4$$
$$J_O = \tfrac{1}{4}\pi r^4$$

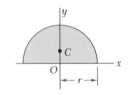

Quarter circle

$$I_x = I_y = \tfrac{1}{16}\pi r^4$$
$$J_O = \tfrac{1}{8}\pi r^4$$

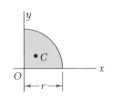

Ellipse

$$\bar{I}_x = \tfrac{1}{4}\pi ab^3$$
$$\bar{I}_y = \tfrac{1}{4}\pi a^3b$$
$$J_O = \tfrac{1}{4}\pi ab(a^2 + b^2)$$

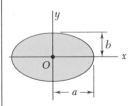

Mass Moments of Inertia of Common Geometric Shapes

Slender rod

$$I_y = I_z = \tfrac{1}{12}mL^2$$

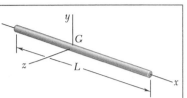

Thin rectangular plate

$$I_x = \tfrac{1}{12}m(b^2 + c^2)$$
$$I_y = \tfrac{1}{12}mc^2$$
$$I_z = \tfrac{1}{12}mb^2$$

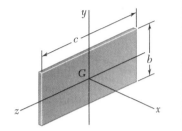

Rectangular prism

$$I_x = \tfrac{1}{12}m(b^2 + c^2)$$
$$I_y = \tfrac{1}{12}m(c^2 + a^2)$$
$$I_z = \tfrac{1}{12}m(a^2 + b^2)$$

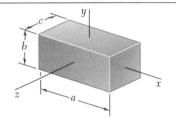

Thin disk

$$I_x = \tfrac{1}{2}mr^2$$
$$I_y = I_z = \tfrac{1}{4}mr^2$$

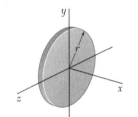

Circular cylinder

$$I_x = \tfrac{1}{2}ma^2$$
$$I_y = I_z = \tfrac{1}{12}m(3a^2 + L^2)$$

Circular cone

$$I_x = \tfrac{3}{10}ma^2$$
$$I_y = I_z = \tfrac{3}{5}m(\tfrac{1}{4}a^2 + h^2)$$

Sphere

$$I_x = I_y = I_z = \tfrac{2}{5}ma^2$$

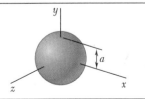

Vector Mechanics
For Engineers

Dynamics

Eleventh Edition

Vector Mechanics For Engineers

Dynamics

Ferdinand P. Beer

Late of Lehigh University

E. Russell Johnston, Jr.

Late of University of Connecticut

Phillip J. Cornwell

Rose-Hulman Institute of Technology

Brian P. Self

California Polytechnic State University—San Luis Obispo

Mc
Graw
Hill
Education

VECTOR MECHANICS FOR ENGINEERS: DYNAMICS, ELEVENTH EDITION

Published by McGraw-Hill Education, 2 Penn Plaza, New York, NY 10121. Copyright © 2016 by McGraw-Hill Education. All rights reserved. Printed in the United States of America. Previous editions © 2013, 2010, and 2007. No part of this publication may be reproduced or distributed in any form or by any means, or stored in a database or retrieval system, without the prior written consent of McGraw-Hill Education, including, but not limited to, in any network or other electronic storage or transmission, or broadcast for distance learning.

Some ancillaries, including electronic and print components, may not be available to customers outside the United States.

This book is printed on acid-free paper.

1 2 3 4 5 6 7 8 9 0 DOW/DOW 1 0 9 8 7 6 5

ISBN 978-0-07-768734-2
MHID 0-07-768734-5

Managing Director: *Thomas Timp*
Global Brand Manager: *Raghothaman Srinivasan*
Director of Development: *Rose Koos*
Product Developer: *Robin Reed*
Brand Manager: *Thomas Scaife, Ph.D.*
Digital Product Analyst: *Dan Wallace*
Editorial Coordinator: *Samantha Donisi-Hamm*
Marketing Manager: *Nick McFadden*
LearnSmart Product Developer: *Joan Weber*
Content Production Manager: *Linda Avenarius*
Content Project Managers: *Jolynn Kilburg and Lora Neyens*
Buyer: *Laura Fuller*
Designer: *Matthew Backhaus*
Content Licensing Specialist (Image): *Carrie Burger*
Typeface: *10/12 Times LT Std*
Printer: *R. R. Donnelley*

All credits appearing on page or at the end of the book are considered to be an extension of the copyright page.

Library of Congress Cataloging-in-Publication

Data on File

The Internet addresses listed in the text were accurate at the time of publication. The inclusion of a website does not indicate an endorsement by the authors or McGraw-Hill Education, and McGraw-Hill Education does not guarantee the accuracy of the information presented at these sites.

www.mhhe.com

About the Authors

Ferdinand P. Beer. Born in France and educated in France and Switzerland, Ferd received an M.S. degree from the Sorbonne and an Sc. D. degree in theoretical mechanics from the University of Geneva. He came to the United States after serving in the French army during the early part of World War II and taught for four years at Williams College in the Williams-MIT joint arts and engineering program. Following his service at Williams College, Ferd joined the faculty of Lehigh University where he taught for thirty-seven years. He held several positions, including University Distinguished Professor and chairman of the Department of Mechanical Engineering and Mechanics, and in 1995 Ferd was awarded an honorary Doctor of Engineering degree by Lehigh University.

E. Russell Johnston, Jr. Born in Philadelphia, Russ received a B.S. degree in civil engineering from the University of Delaware and an Sc. D. degree in the field of structural engineering from the Massachusetts Institute of Technology. He taught at Lehigh University and Worcester Polytechnic Institute before joining the faculty of the University of Connecticut where he held the position of chairman of the Civil Engineering Department and taught for twenty-six years. In 1991 Russ received the Outstanding Civil Engineer Award from the Connecticut Section of the American Society of Civil Engineers.

Phillip J. Cornwell. Phil holds a B.S. degree in mechanical engineering from Texas Tech University and M.A. and Ph.D. degrees in mechanical and aerospace engineering from Princeton University. He is currently a professor of mechanical engineering and Vice President of Academic Affairs at Rose-Hulman Institute of Technology where he has taught since 1989. Phil received an SAE Ralph R. Teetor Educational Award in 1992, the Dean's Outstanding Teacher Award at Rose-Hulman in 2000, and the Board of Trustees' Outstanding Scholar Award at Rose-Hulman in 2001. Phil was one of the developers of the Dynamics Concept Inventory.

Brian P. Self. Brian obtained his B.S. and M.S. degrees in Engineering Mechanics from Virginia Tech, and his Ph.D. in Bioengineering from the University of Utah. He worked in the Air Force Research Laboratories before teaching at the U.S. Air Force Academy for seven years. Brian has taught in the Mechanical Engineering Department at Cal Poly, San Luis Obispo since 2006. He has been very active in the American Society of Engineering Education, serving on its Board from 2008–2010. With a team of five, Brian developed the Dynamics Concept Inventory to help assess student conceptual understanding. His professional interests include educational research, aviation physiology, and biomechanics.

Brief Contents

Contents

11 Kinematics of Particles   615

12 Kinetics of Particles: Newton's Second Law   718

13 Kinetics of Particles: Energy and Momentum Methods   795

*Advanced or specialty topics

14 Systems of Particles 915

15 Kinematics of Rigid Bodies 977

16 Plane Motion of Rigid Bodies: Forces and Accelerations 1107

17 Plane Motion of Rigid Bodies: Energy and Momentum Methods 1181

18 Kinetics of Rigid Bodies in Three Dimensions 1264

19 Mechanical Vibrations 1332

Preface

Objectives

A primary objective in a first course in mechanics is to help develop a student's ability first to analyze problems in a simple and logical manner, and then to apply basic principles to their solutions. A strong conceptual understanding of these basic mechanics principles is essential for successfully solving mechanics problems. We hope that this text, as well as the preceding volume, *Vector Mechanics for Engineers: Statics*, will help instructors achieve these goals.[†]

General Approach

Vector algebra was introduced at the beginning of the first volume and is used in the presentation of the basic principles of statics, as well as in the solution of many problems, particularly three-dimensional problems. Similarly, the concept of vector differentiation will be introduced early in this volume, and vector analysis will be used throughout the presentation of dynamics. This approach leads to more concise derivations of the fundamental principles of mechanics. It also makes it possible to analyze many problems in kinematics and kinetics which could not be solved by scalar methods. The emphasis in this text, however, remains on the correct understanding of the principles of mechanics and on their application to the solution of engineering problems, and vector analysis is presented chiefly as a convenient tool.[‡]

Practical Applications Are Introduced Early. One of the characteristics of the approach used in this book is that mechanics of *particles* is clearly separated from the mechanics of *rigid bodies*. This approach makes it possible to consider simple practical applications at an early stage and to postpone the introduction of the more difficult concepts. For example:

- In *Statics,* the statics of particles is treated first, and the principle of equilibrium of a particle was immediately applied to practical situations involving only concurrent forces. The statics of rigid bodies is considered later, at which time the vector and scalar products of two vectors were introduced and used to define the moment of a force about a point and about an axis.
- In *Dynamics,* the same division is observed. The basic concepts of force, mass, and acceleration, of work and energy, and of impulse and momentum are introduced and first applied to problems involv-

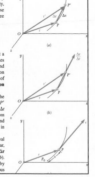

11.4 CURVILINEAR MOTION OF PARTICLES

When a particle moves along a curve other than a straight line, we say that the particle is in **curvilinear motion**. We can use position, velocity, and acceleration to describe the motion, but now we must treat these quantities as vectors because they can have directions in two or three dimensions.

11.4A Position, Velocity, and Acceleration Vectors

To define the position P occupied by a particle in curvilinear motion at a given time t, we select a fixed reference system, such as the x, y, z axes shown in Fig. 11.12a, and draw the vector $\mathbf{r}$ joining the origin O and point P. The vector $\mathbf{r}$ is characterized by its magnitude r and its direction with respect to the reference axes, so it completely defines the position of the particle with respect to those axes. We refer to vector $\mathbf{r}$ as the **position vector** of the particle at time t.

Consider now the vector $\mathbf{r}'$ defining the position P' occupied by the same particle at a later time $t + \Delta t$. The vector $\Delta \mathbf{r}$ joining P and P' represents the change in the position vector during the time interval Δt and is called the **displacement vector**. We can check this directly from Fig. 11.12a, where we obtain the vector $\mathbf{r}'$ by adding the vectors $\mathbf{r}$ and $\Delta \mathbf{r}$ according to the triangle rule. Note that $\Delta \mathbf{r}$ represents a change in *direction* as well as a change in *magnitude* of the position vector $\mathbf{r}$.

We define the **average velocity** of the particle over the time interval Δt as the quotient of $\Delta \mathbf{r}$ and Δt. Since $\Delta \mathbf{r}$ is a vector and Δt is a scalar, the quotient $\Delta \mathbf{r}/\Delta t$ is a vector attached at P with the same direction as $\Delta \mathbf{r}$ and a magnitude equal to the magnitude of $\Delta \mathbf{r}$ divided by Δt (Fig. 11.12b).

We obtain the **instantaneous velocity** of the particle at time t by taking the limit as the time interval Δt approaches zero. The instantaneous

---

[†]Both texts also are available in a single volume, *Vector Mechanics for Engineers: Statics and Dynamics,* eleventh edition.

[‡]In a parallel text, *Mechanics for Engineers: Dynamics,* fifth edition, the use of vector algebra is limited to the addition and subtraction of vectors, and vector differentiation is omitted.

ing only particles. Thus, students can familiarize themselves with the three basic methods used in dynamics and learn their respective advantages before facing the difficulties associated with the motion of rigid bodies.

New Concepts Are Introduced in Simple Terms.

Since this text is designed for the first course in dynamics, new concepts are presented in simple terms and every step is explained in detail. On the other hand, by discussing the broader aspects of the problems considered, and by stressing methods of general applicability, a definite maturity of approach has been achieved. For example, the concept of potential energy is discussed in the general case of a conservative force. Also, the study of the plane motion of rigid bodies is designed to lead naturally to the study of their general motion in space. This is true in kinematics as well as in kinetics, where the principle of equivalence of external and effective forces is applied directly to the analysis of plane motion, thus facilitating the transition to the study of three-dimensional motion.

Fundamental Principles Are Placed in the Context of Simple Applications.

The fact that mechanics is essentially a *deductive* science based on a few fundamental principles is stressed. Derivations have been presented in their logical sequence and with all the rigor warranted at this level. However, the learning process being largely *inductive,* simple applications are considered first. For example:

- The kinematics of particles (Chap. 11) precedes the kinematics of rigid bodies (Chap. 15).
- The fundamental principles of the kinetics of rigid bodies are first applied to the solution of two-dimensional problems (Chaps. 16 and 17), which can be more easily visualized by the student, while three-dimensional problems are postponed until Chap. 18.

The Presentation of the Principles of Kinetics Is Unified.

The eleventh edition of *Vector Mechanics for Engineers* retains the unified presentation of the principles of kinetics which characterized the previous ten editions. The concepts of linear and angular momentum are introduced in Chap. 12 so that Newton's second law of motion can be presented not only in its conventional form $\mathbf{F} = m\mathbf{a}$, but also as a law relating, respectively, the sum of the forces acting on a particle and the sum of their moments to the rates of change of the linear and angular momentum of the particle. This makes possible an earlier introduction of the principle of conservation of angular momentum and a more meaningful discussion of the motion of a particle under a central force (Sec. 12.3A). More importantly, this approach can be readily extended to the study of the motion of a system of particles (Chap. 14) and leads to a more concise and unified treatment of the kinetics of rigid bodies in two and three dimensions (Chaps. 16 through 18).

17.1 ENERGY METHODS FOR A RIGID BODY

We now use the principle of work and energy to analyze the plane motion of rigid bodies. As we pointed out in Chap. 13, the method of work and energy is particularly well adapted to solving problems involving velocities and displacements. Its main advantage is that the work of forces and the kinetic energy of particles are scalar quantities.

17.1A Principle of Work and Energy

To apply the principle of work and energy to the motion of a rigid body, we again assume that the rigid body is made up of a large number n of particles of mass Δm_i. From Eq. (14.30) of Sec. 14.2B, we have

Principle of work and energy, rigid body

$$T_1 + U_{1\to 2} = T_2 \tag{17.1}$$

where T_1, T_2 = the initial and final values of total kinetic energy of particles forming the rigid body

$U_{1\to 2}$ = work of all forces acting on various particles of the body

Just as we did in Chap. 13, we can express the work done by nonconservative forces as $U_{1\to 2}^{NC}$, and we can define potential energy terms for conservative forces. Then we can express Eq. (17.1) as

$$T_1 + V_{g_1} + V_{e_1} + U_{1\to 2}^{NC} = T_2 + V_{g_2} + V_{e_2} \tag{17.1'}$$

where V_{g_1} and V_{g_2} are the initial and final gravitational potential energy of the center of mass of the rigid body with respect to a reference point or datum, and V_{e_1} and V_{e_2} are the initial and final values of the elastic energy associated with springs in the system.

We obtain the total kinetic energy

$$T = \frac{1}{2}\sum_{i=1}^{n} \Delta m_i\, v_i^2 \tag{17.2}$$

by adding positive scalar quantities, so it is itself a positive scalar quantity. You will see later how to determine T for various types of motion of a rigid body.

The expression $U_{1\to 2}$ in Eq. (17.1) represents the work of all the forces acting on the various particles of the body whether these forces are internal or external. However, the total work of the internal forces holding together the particles of a rigid body is zero. To see this, consider two particles A and B of a rigid body and the two equal and opposite forces $\mathbf{F}$ and $-\mathbf{F}$ they exert on each other (Fig. 17.1). Although, in general, small displacements $d\mathbf{r}$ and $d\mathbf{r}'$ of the two particles are different, the components of these displacements along AB must be equal; otherwise, the particles would not remain at the same distance from each other and the body would not be rigid. Therefore, the work of $\mathbf{F}$ is equal in magnitude and

Systematic Problem-Solving Approach.   New to this edition of the text, all the sample problems are solved using the steps of *S*trategy, *M*odeling, *A*nalysis, and *R*eflect & *T*hink, or the "SMART" approach. This methodology is intended to give students confidence when approaching new problems, and students are encouraged to apply this approach in the solution of all assigned problems.

Free-Body Diagrams Are Used Both to Solve Equilibrium Problems and to Express the Equivalence of Force Systems.   Free-body diagrams were introduced early in statics, and their importance was emphasized throughout. They were used not only to solve equilibrium problems but also to express the equivalence of two systems of forces or, more generally, of two systems of vectors. In dynamics we will introduce a kinetic diagram, which is a pictorial representation of inertia terms. The advantage of this approach becomes apparent in the study of the dynamics of rigid bodies, where it is used to solve three-dimensional as well as two-dimensional problems. By placing the emphasis on the free-body diagram and kinetic diagram, rather than on the standard algebraic equations of motion, a more intuitive and more complete understanding of the fundamental principles of dynamics can be achieved. This approach, which was first introduced in 1962 in the first edition of *Vector Mechanics for Engineers,* has now gained wide acceptance among mechanics teachers in this country. It is, therefore, used in preference to the method of dynamic equilibrium and to the equations of motion in the solution of all sample problems in this book.

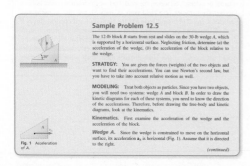

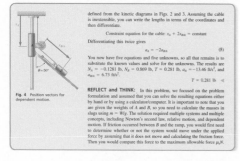

A Careful Balance between SI and U.S. Customary Units Is Consistently Maintained.   Because of the current trend in the American government and industry to adopt the international system of units (SI metric units), the SI units most frequently used in mechanics are introduced in Chap. 1 and are used throughout the text. Approximately half of the sample problems and 60 percent of the homework problems are stated in these units, while the remainder are in U.S. customary units. The authors believe that this approach will best serve the need of the students, who, as engineers, will have to be conversant with both systems of units.

It also should be recognized that using both SI and U.S. customary units entails more than the use of conversion factors. Since the SI system of units is an absolute system based on the units of time, length, and mass, whereas the U.S. customary system is a gravitational system based on the units of time, length, and force, different approaches are required for the solution of many problems. For example, when SI units are used, a body is generally specified by its mass expressed in kilograms; in most problems of statics it will be necessary to determine the weight of the body in newtons, and an additional calculation will be required for this purpose. On the other hand, when U.S. customary units are used, a body is specified by its weight in pounds and, in dynamics problems, an additional calculation will be required to determine its mass in slugs (or $lb \cdot s^2/ft$). The authors, therefore, believe that problem assignments should include both systems of units.

The *Instructor's and Solutions Manual* provides six different lists of assignments so that an equal number of problems stated in SI units and in U.S. customary units can be selected. If so desired, two complete lists of assignments can also be selected with up to 75 percent of the problems stated in SI units.

Optional Sections Offer Advanced or Specialty Topics.

A large number of optional sections have been included. These sections are indicated by asterisks and thus are easily distinguished from those which form the core of the basic dynamics course. They can be omitted without prejudice to the understanding of the rest of the text.

The topics covered in the optional sections include graphical methods for the solution of rectilinear-motion problems, the trajectory of a particle under a central force, the deflection of fluid streams, problems involving jet and rocket propulsion, the kinematics and kinetics of rigid bodies in three dimensions, damped mechanical vibrations, and electrical analogues. These topics will be found of particular interest when dynamics is taught in the junior year.

The material presented in the text and most of the problems require no previous mathematical knowledge beyond algebra, trigonometry, elementary calculus, and the elements of vector algebra presented in Chaps. 2 and 3 of the volume on statics.[†] However, special problems are included, which make use of a more advanced knowledge of calculus, and certain sections, such as Secs. 19.5A and 19.5B on damped vibrations, should be assigned only if students possess the proper mathematical background. In portions of the text using elementary calculus, a greater emphasis is placed on the correct understanding and application of the concepts of differentiation and integration, than on the nimble manipulation of mathematical formulas. In this connection, it should be mentioned that the determination of the centroids of composite areas precedes the calculation of centroids by integration, thus making it possible to establish the concept of moment of area firmly before introducing the use of integration.

[†]Some useful definitions and properties of vector algebra have been summarized in Appendix A at the end of this volume for the convenience of the reader. Also, Secs. 9.5 and 9.6 of the volume on statics, which deal with the moments of inertia of masses, have been reproduced in Appendix B.

Guided Tour

11

Kinematics of Particles

The motion of the paraglider can be described in terms of its position, velocity, and acceleration. When landing, the pilot of the paraglider needs to consider the wind velocity and the relative motion of the glider with respect to the wind. The study of motion is known as kinematics and is the subject of this chapter.

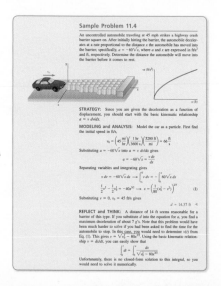

Chapter Introduction. Each chapter begins with a list of learning objectives and an outline that previews chapter topics. An introductory section describes the material to be covered in simple terms, and how it will be applied to the solution of engineering problems.

Chapter Lessons. The body of the text is divided into sections, each consisting of one or more sub-sections, several sample problems, and a large number of end-of-section problems for students to solve. Each section corresponds to a well-defined topic and generally can be covered in one lesson. In a number of cases, however, the instructor will find it desirable to devote more than one lesson to a given topic. *The Instructor's and Solutions Manual* contains suggestions on the coverage of each lesson.

Sample Problems. The Sample Problems are set up in much the same form that students will use when solving assigned problems, and they employ the SMART problem-solving methodology that students are encouraged to use in the solution of their assigned problems. They thus serve the double purpose of reinforcing the text and demonstrating the type of neat and orderly work that students should cultivate in their own solutions. In addition, in-problem references and captions have been added to the sample problem figures for contextual linkage to the step-by-step solution.

Solving Problems on Your Own. A section entitled *Solving Problems on Your Own* is included for each lesson, between the sample problems and the problems to be assigned. The purpose of these sections is to help students organize in their own minds the preceding theory of the text and the solution methods of the sample problems so that they can more successfully solve the homework problems. Also included in these sections are specific suggestions and strategies that will enable the students to more efficiently attack any assigned problems.

Homework Problem Sets. Most of the problems are of a practical nature and should appeal to engineering students. They are primarily designed, however, to illustrate the material presented in the text and to help students understand the principles of mechanics. The problems are grouped according to the portions of material they illustrate and, in general, are arranged in order of increasing difficulty. Problems requiring special attention are indicated by asterisks. Answers to 70 percent of the problems are given at the end of the book. Problems for which the answers are given are set in straight type in the text, while problems for which no answer is given are set in italic and red font color.

Chapter Review and Summary. Each chapter ends with a review and summary of the material covered in that chapter. Marginal notes are used to help students organize their review work, and cross-references have been included to help them find the portions of material requiring their special attention.

Review Problems. A set of review problems is included at the end of each chapter. These problems provide students further opportunity to apply the most important concepts introduced in the chapter.

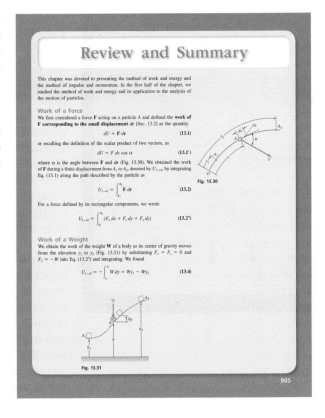

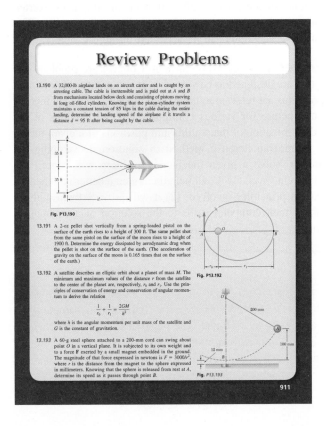

Concept Questions. Educational research has shown that students can often choose appropriate equations and solve algorithmic problems without having a strong conceptual understanding of mechanics principles.[†] To help assess and develop student conceptual understanding, we have included Concept Questions, which are multiple choice problems that require few, if any, calculations. Each possible incorrect answer typically represents a common misconception (e.g., students often think that a vehicle moving in a curved path at constant speed has zero acceleration). Students are encouraged to solve these problems using the principles and techniques discussed in the text and to use these principles to help them develop their intuition. Mastery and discussion of these Concept Questions will deepen students' conceptual understanding and help them to solve dynamics problems.

[†]Hestenes, D., Wells, M., and Swakhamer, G (1992). The force concept inventory. *The Physics Teacher,* 30: 141–158.
Streveler, R. A., Litzinger, T. A., Miller, R. L., and Steif, P. S. (2008). Learning conceptual knowledge in the engineering sciences: Overview and future research directions, *JEE,* 279–294.

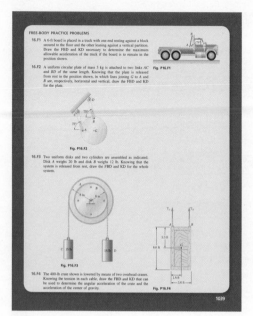

Free Body and Impulse-Momentum Diagram Practice Problems.
Drawing diagrams correctly is a critical step in solving kinetics problems in dynamics. A new type of problem has been added to the text to emphasize the importance of drawing these diagrams. In Chaps. 12 and 16 the Free Body Practice Problems require students to draw a free-body diagram (FBD) showing the applied forces and an equivalent diagram called a "kinetic diagram" (KD) showing $m\mathbf{a}$ or its components and $\bar{I}\alpha$. These diagrams provide students with a pictorial representation of Newton's second law and are critical in helping students to correctly solve kinetic problems. In Chaps. 13 and 17 the Impulse-Momentum Diagram Practice Problems require students to draw diagrams showing the momenta of the bodies before impact, the impulses exerted on the body during impact, and the final momenta of the bodies. The answers to all of these questions can be accessed through Connect.

Computer Problems.
Accessible through Connect are problem sets for each chapter that are designed to be solved with computational software. Many of these problems are relevant to the design process; they may involve the analysis of a structure for various configurations and loadings of the structure, or the determination of the equilibrium positions of a given mechanism that may require an iterative method of solution. Developing the algorithm required to solve a given mechanics problem will benefit the students in two different ways: (1) it will help them gain a better understanding of the mechanics principles involved; (2) it will provide them with an opportunity to apply their computer skills to the solution of a meaningful engineering problem.

Digital Resources

 **Connect**® Engineering provides online presentation, assignment, and assessment solutions. It connects your students with the tools and resources they'll need to achieve success. With Connect Engineering you can deliver assignments, quizzes, and tests online. A robust set of questions and activities are presented and aligned with the textbook's learning outcomes. As an instructor, you can edit existing questions and author entirely new problems. Integrate grade reports easily with Learning Management Systems (LMS), such as WebCT and Blackboard—and much more. Connect Engineering also provides students with 24/7 online access to a media-rich eBook, allowing seamless integration of text, media, and assessments. **To learn more, visit connect.mheducation.com**

Find the following instructor resources available through Connect:

- **Instructor's and Solutions Manual.** *The Instructor's and Solutions Manual* that accompanies the eleventh edition features solutions to all end of chapter problems. This manual also features a number of tables designed to assist instructors in creating a schedule of assignments for their course. The various topics covered in the text have been listed in Table I and a suggested number of periods to be spent on each topic has been indicated. Table II prepares a brief description of all groups of problems and a classification of the problems in each group according to the units used. Sample lesson schedules are shown in Tables III, IV, and V, together with various alternative lists of assigned homework problems.
- **Lecture PowerPoint Slides** for each chapter that can be modified. These generally have an introductory application slide, animated worked-out problems that you can do in class with your students, concept questions, and "what-if?" questions at the end of the units.
- **Textbook images**
- **Computer Problem sets** for each chapter that are designed to be solved with computational software.
- **C.O.S.M.O.S.**, the Complete Online Solutions Manual Organization System that allows instructors to create custom homework, quizzes, and tests using end-of-chapter problems from the text.

LEARNSMART® LearnSmart is available as an integrated feature of McGraw-Hill Connect. It is an adaptive learning system designed to help students learn faster, study more efficiently, and retain more knowledge for greater success. LearnSmart assesses a student's knowledge of course content through a series of adaptive questions. It pinpoints concepts the student does not understand and maps out a personalized study plan for success. This innovative study tool also has features that allow instructors to see exactly what students have accomplished and a built-in assessment tool for graded assignments.

NEW!

NEW!

▌ SMARTBOOK™

SmartBook™ is the first and only adaptive reading experience available for the higher education market. Powered by an intelligent diagnostic and adaptive engine, SmartBook facilitates the reading process by identifying what content a student knows and doesn't know through adaptive assessments. As the student reads, the reading material constantly adapts to ensure the student is focused on the content he or she needs the most to close any knowledge gaps.

Visit the following site for a demonstration of LearnSmart or SmartBook: www.learnsmartadvantage.com

CourseSmart.   This text is offered through CourseSmart for both instructors and students. CourseSmart is an online browser where students can purchase access to this and other McGraw-Hill textbooks in a digital format. Through their browser, students can access the complete text online at almost half the cost of a traditional text. Purchasing the eTextbook also allows students to take advantage of CourseSmart's web tools for learning, which include full text search, notes and highlighting, and e-mail tools for sharing notes among classmates. To learn more about CourseSmart options, contact your sales representative or visit **www.coursesmart.com**.

Acknowledgments

A special thanks to Jim Widmann of California Polytechnic State University, who thoroughly checked the solutions and answers of all problems in this edition and then prepared the solutions for the accompanying Instructor's and Solutions Manual. The authors would also like to thank Baheej Saoud, who helped develop and solve several of the new problems in this edition.

We are pleased to acknowledge David Chelton, who carefully reviewed the entire text and provided many helpful suggestions for revising this edition.

The authors thank the many companies that provided photographs for this edition. We also wish to recognize the determined efforts and patience of our photo researcher Danny Meldung.

The authors also thank the members of the staff at McGraw-Hill for their support and dedication during the preparation of this new edition.

Phillip J. Cornwell
Brian P. Self

The authors gratefully acknowledge the many helpful comments and suggestions offered by focus group attendees and by users of the previous editions of *Vector Mechanics for Engineers:*

George Adams
Northeastern University

William Altenhof
University of Windsor

Sean B. Anderson
Boston University

Manohar Arora
Colorado School of Mines

Gilbert Baladi
Michigan State University

Francois Barthelat
McGill University

Oscar Barton, Jr.
U.S. Naval Academy

M. Asghar Bhatti
University of Iowa

Shaohong Cheng
University of Windsor

Philip Datseris
University of Rhode Island

Timothy A. Doughty
University of Portland

Howard Epstein
University of Connecticut

Asad Esmaeily
Kansas State University,
Civil Engineering Department

David Fleming
Florida Institute of Technology

Jeff Hanson
Texas Tech University

David A. Jenkins
University of Florida

Shaofan Li
University of California, Berkeley

William R. Murray
Cal Poly State University

Eric Musslman
University of Minnesota, Duluth

Masoud Olia
Wentworth Institute of
Technology

Renee K. B. Petersen
Washington State University

Amir G Rezaei
California State Polytechnic
University, Pomona

Martin Sadd
University of Rhode Island

Stefan Seelecke
North Carolina State University

Yixin Shao
McGill University

Muhammad Sharif
The University of Alabama

Anthony Sinclair
University of Toronto

Lizhi Sun
University of California, lrvine

Jeffrey Thomas
Northwestern University

Jiashi Yang
University of Nebraska

Xiangwa Zeng
Case Western Reserve University

List of Symbols

$\mathbf{a}, a$	Acceleration
a	Constant; radius; distance; semimajor axis of ellipse
$\bar{\mathbf{a}}, \bar{a}$	Acceleration of mass center
$\mathbf{a}_{B/A}$	Acceleration of B relative to frame in translation with A
$\mathbf{a}_{P/\mathscr{F}}$	Acceleration of P relative to rotating frame $\mathscr{F}$
$\mathbf{a}_c$	Coriolis acceleration
$\mathbf{A}, \mathbf{B}, \mathbf{C}, \dots$	Reactions at supports and connections
$A, B, C, \dots$	Points
A	Area
b	Width; distance; semiminor axis of ellipse
c	Constant; coefficient of viscous damping
C	Centroid; instantaneous center of rotation; capacitance
d	Distance
$\mathbf{e}_n, \mathbf{e}_t$	Unit vectors along normal and tangent
$\mathbf{e}_r, \mathbf{e}_\theta$	Unit vectors in radial and transverse directions
e	Coefficient of restitution; base of natural logarithms
E	Total mechanical energy; voltage
f	Scalar function
f_f	Frequency of forced vibration
f_n	Natural frequency
$\mathbf{F}$	Force; friction force
g	Acceleration of gravity
G	Center of gravity; mass center; constant of gravitation
h	Angular momentum per unit mass
$\mathbf{H}_O$	Angular momentum about point O
$\dot{\mathbf{H}}_G$	Rate of change of angular momentum $\mathbf{H}_G$ with respect to frame of fixed orientation
$(\dot{\mathbf{H}}_G)_{Gxyz}$	Rate of change of angular momentum $\mathbf{H}_G$ with respect to rotating frame $Gxyz$
$\mathbf{i}, \mathbf{j}, \mathbf{k}$	Unit vectors along coordinate axes
i	Current
$I, I_x, \dots$	Moments of inertia
$\bar{I}$	Centroidal moment of inertia
$I_{xy}, \dots$	Products of inertia
J	Polar moment of inertia
k	Spring constant
k_x, k_y, k_O	Radii of gyration

$\bar{k}$	Centroidal radius of gyration
l	Length
$\mathbf{L}$	Linear momentum
L	Length; inductance
m	Mass
m'	Mass per unit length
$\mathbf{M}$	Couple; moment
$\mathbf{M}_O$	Moment about point O
$\mathbf{M}_O^R$	Moment resultant about point O
M	Magnitude of couple or moment; mass of earth
M_{OL}	Moment about axis OL
n	Normal direction
$\mathbf{N}$	Normal component of reaction
O	Origin of coordinates
$\mathbf{P}$	Force; vector
$\dot{\mathbf{P}}$	Rate of change of vector $\mathbf{P}$ with respect to frame of fixed orientation
q	Mass rate of flow; electric charge
$\mathbf{Q}$	Force; vector
$\dot{\mathbf{Q}}$	Rate of change of vector $\mathbf{Q}$ with respect to frame of fixed orientation
$(\dot{\mathbf{Q}})_{Oxyz}$	Rate of change of vector $\mathbf{Q}$ with respect to frame $Oxyz$
$\mathbf{r}$	Position vector
$\mathbf{r}_{B/A}$	Position vector of B relative to A
r	Radius; distance; polar coordinate
$\mathbf{R}$	Resultant force; resultant vector; reaction
R	Radius of earth; resistance
$\mathbf{s}$	Position vector
s	Length of arc
t	Time; thickness; tangential direction
$\mathbf{T}$	Force
T	Tension; kinetic energy
$\mathbf{u}$	Velocity
u	Variable
U	Work
U_{1-2}^{NC}	work done by non-conservative forces
$\mathbf{v}, v$	Velocity
v	Speed
$\bar{\mathbf{v}}, \bar{v}$	Velocity of mass center
$\mathbf{v}_{B/A}$	Velocity of B relative to frame in translation with A
$\mathbf{v}_{P/\mathscr{F}}$	Velocity of P relative to rotating frame $\mathscr{F}$

$\mathbf{V}$	Vector product	θ	Angular coordinate; Eulerian angle; angle; polar coordinate
V	Volume; potential energy		
w	Load per unit length	μ	Coefficient of friction
$\mathbf{W}, W$	Weight; load	ρ	Density; radius of curvature
x, y, z	Rectangular coordinates; distances	τ	Periodic time
$\dot{x}, \dot{y}, \dot{z}$	Time derivatives of coordinates x, y, z	τ_n	Period of free vibration
$\bar{x}, \bar{y}, \bar{z}$	Rectangular coordinates of centroid, center of gravity, or mass center	ϕ	Angle of friction; Eulerian angle; phase angle; angle
$\boldsymbol{\alpha}, \alpha$	Angular acceleration	φ	Phase difference
α, β, γ	Angles	ψ	Eulerian angle
γ	Specific weight	$\boldsymbol{\omega}, \omega$	Angular velocity
δ	Elongation	ω_f	Circular frequency of forced vibration
ε	Eccentricity of conic section or of orbit	ω_n	Natural circular frequency
$\boldsymbol{\lambda}$	Unit vector along a line	Ω	Angular velocity of frame of reference
η	Efficiency		

11

Kinematics of Particles

The motion of the paraglider can be described in terms of its *position*, *velocity*, and *acceleration*. When landing, the pilot of the paraglider needs to consider the wind velocity and the *relative motion* of the glider with respect to the wind. The study of motion is known as *kinematics* and is the subject of this chapter.

Introduction

Objectives

- **Describe** the basic kinematic relationships between position, velocity, acceleration, and time.
- **Solve** problems using these basic kinematic relationships and calculus or graphical methods.
- **Define** position, velocity, and acceleration in terms of Cartesian, tangential and normal, and radial and transverse coordinates.
- **Analyze** the relative motion of multiple particles by using a translating coordinate system.
- **Determine** the motion of a particle that depends on the motion of another particle.
- **Determine** which coordinate system is most appropriate for solving a curvilinear kinematics problem.
- **Calculate** the position, velocity, and acceleration of a particle undergoing curvilinear motion using Cartesian, tangential and normal, and radial and transverse coordinates.

Introduction

Chapters 1 to 10 were devoted to **statics**, i.e., to the analysis of bodies at rest. We now begin the study of **dynamics**, which is the part of mechanics that deals with the analysis of bodies in motion.

Although the study of statics goes back to the time of the Greek philosophers, the first significant contribution to dynamics was made by Galileo (1564–1642). Galileo's experiments on uniformly accelerated bodies led Newton (1642–1727) to formulate his fundamental laws of motion.

Dynamics includes two broad areas of study:

1. **Kinematics**, which is the study of the geometry of motion. The principles of kinematics relate the displacement, velocity, acceleration, and time of a body's motion, without reference to the cause of the motion.
2. **Kinetics**, which is the study of the relation between the forces acting on a body, the mass of the body, and the motion of the body. We use kinetics to predict the motion caused by given forces or to determine the forces required to produce a given motion.

Chapters 11 through 14 describe the **dynamics of particles**; in Chap. 11, we consider the **kinematics of particles**. The use of the word *particles* does not mean that our study is restricted to small objects; rather, it indicates that in these first chapters we study the motion of bodies—possibly as large as cars, rockets, or airplanes—without regard to their size or shape. By saying that we analyze the bodies as particles, we mean that we consider only their motion as an entire unit; we neglect any rotation about their own centers of mass. In some cases, however, such a rotation is not negligible, and we cannot treat the bodies as particles. Such motions are analyzed in later chapters dealing with the **dynamics of rigid bodies**.

In the first part of Chap. 11, we describe the rectilinear motion of a particle; that is, we determine the position, velocity, and acceleration of a particle at every instant as it moves along a straight line. We first use general methods of analysis to study the motion of a particle; we then consider two important particular cases, namely, the uniform motion and the uniformly accelerated motion of a particle (Sec. 11.2). We then discuss the simultaneous motion of several particles and introduce the concept of the relative motion of one particle with respect to another. The first part of this chapter concludes with a study of graphical methods of analysis and their application to the solution of problems involving the rectilinear motion of particles.

In the second part of this chapter, we analyze the motion of a particle as it moves along a curved path. We define the position, velocity, and acceleration of a particle as vector quantities and introduce the derivative of a vector function to add to our mathematical tools. We consider applications in which we define the motion of a particle by the rectangular components of its velocity and acceleration; at this point, we analyze the motion of a projectile (Sec. 11.4C). Then we examine the motion of a particle relative to a reference frame in translation. Finally, we analyze the curvilinear motion of a particle in terms of components other than rectangular. In Sec. 11.5, we introduce the tangential and normal components of an object's velocity and acceleration and then examine the radial and transverse components.

11.1 RECTILINEAR MOTION OF PARTICLES

A particle moving along a straight line is said to be in **rectilinear motion**. The only variables we need to describe this motion are the time, t, and the distance along the line, x, as a function of time. With these variables, we can define the particle's position, velocity, and acceleration, which completely describe the particle's motion. When we study the motion of a particle moving in a plane (two dimensions) or in space (three dimensions), we will use a more general position vector rather than simply the distance along a line.

11.1A Position, Velocity, and Acceleration

At any given instant t, a particle in rectilinear motion occupies some position on the straight line. To define the particle's position P, we choose a fixed origin O on the straight line and a positive direction along the line. We measure the distance x from O to P and record it with a plus or minus sign, according to whether we reach P from O by moving along the line in the positive or negative direction. The distance x, with the appropriate sign, completely defines the position of the particle; it is called the **position coordinate** of the particle. For example, the position coordinate corresponding to P in Fig. 11.1a is $x = +5$ m; the coordinate corresponding to P' in Fig. 11.1b is $x' = -2$ m.

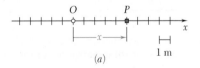

(a)

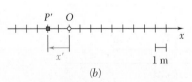

(b)

Fig. 11.1 Position is measured from a fixed origin. (a) A positive position coordinate; (b) a negative position coordinate.

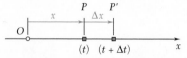

Fig. 11.2 A small displacement Δx from time t to time $t + \Delta t$.

Photo 11.1 The motion of this solar car can be described by its position, velocity, and acceleration.

When we know the position coordinate x of a particle for every value of time t, we say that the motion of the particle is known. We can provide a "timetable" of the motion in the form of an equation in x and t, such as $x = 6t^2 - t^3$, or in the form of a graph of x versus t, as shown in Fig. 11.6. The units most often used to measure the position coordinate x are the meter (m) in the SI system of units[†] and the foot (ft) in the U.S. customary system of units. Time t is usually measured in seconds (s).

Now consider the position P occupied by the particle at time t and the corresponding coordinate x (Fig. 11.2). Consider also the position P' occupied by the particle at a later time $t + \Delta t$. We can obtain the position coordinate of P' by adding the small displacement Δx to the coordinate x of P. This displacement is positive or negative according to whether P' is to the right or to the left of P. We define the **average velocity** of the particle over the time interval Δt as the quotient of the displacement Δx and the time interval Δt as

$$\text{Average velocity} = \frac{\Delta x}{\Delta t}$$

If we use SI units, Δx is expressed in meters and Δt in seconds; the average velocity is then expressed in meters per second (m/s). If we use U.S. customary units, Δx is expressed in feet and Δt in seconds; the average velocity is then expressed in feet per second (ft/s).

We can determine the **instantaneous velocity** v of a particle at the instant t by allowing the time interval Δt to become infinitesimally small. Thus,

$$\text{Instantaneous velocity} = v = \lim_{\Delta t \to 0} \frac{\Delta x}{\Delta t}$$

The instantaneous velocity is also expressed in m/s or ft/s. Observing that the limit of the quotient is equal, by definition, to the derivative of x with respect to t, we have

Velocity of a particle along a line

$$v = \frac{dx}{dt} \qquad (11.1)$$

We represent the velocity v by an algebraic number that can be positive or negative.[‡] A positive value of v indicates that x increases, i.e., that the particle moves in the positive direction (Fig. 11.3a). A negative value of v indicates that x decreases, i.e., that the particle moves in the negative direction (Fig. 11.3b). The magnitude of v is known as the **speed** of the particle.

Consider the velocity v of the particle at time t and also its velocity $v + \Delta v$ at a later time $t + \Delta t$ (Fig. 11.4). We define the **average acceleration** of the particle over the time interval Δt as the quotient of Δv and Δt as

$$\text{Average acceleration} = \frac{\Delta v}{\Delta t}$$

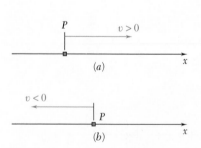

Fig. 11.3 In rectilinear motion, velocity can be only (a) positive or (b) negative along the line.

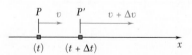

Fig. 11.4 A change in velocity from v to $v + \Delta v$ corresponding to a change in time from t to $t + \Delta t$.

[†]See Sec. 1.3.
[‡]As you will see in Sec. 11.4A, velocity is actually a vector quantity. However, since we are considering here the rectilinear motion of a particle where the velocity has a known and fixed direction, we need only specify its sense and magnitude. We can do this conveniently by using a scalar quantity with a plus or minus sign. This is also true of the acceleration of a particle in rectilinear motion.

If we use SI units, Δv is expressed in m/s and Δt in seconds; the average acceleration is then expressed in m/s². If we use U.S. customary units, Δv is expressed in ft/s and Δt in seconds; the average acceleration is then expressed in ft/s².

We obtain the **instantaneous acceleration** a of the particle at the instant t by again allowing the time interval Δt to approach zero. Thus,

$$\text{Instantaneous acceleration} = a = \lim_{\Delta t \to 0} \frac{\Delta v}{\Delta t}$$

The instantaneous acceleration is also expressed in m/s² or ft/s². The limit of the quotient, which is by definition the derivative of v with respect to t, measures the rate of change of the velocity. We have

**Acceleration of a
particle along a line**

$$a = \frac{dv}{dt} \tag{11.2}$$

or substituting for v from Eq. (11.1),

$$a = \frac{d^2x}{dt^2} \tag{11.3}$$

We represent the acceleration a by an algebraic number that can be positive or negative (see the footnote on the preceding page). A positive value of a indicates that the velocity (i.e., the algebraic number v) increases. This may mean that the particle is moving faster in the positive direction (Fig. 11.5a) or that it is moving more slowly in the negative direction (Fig. 11.5b); in both cases, Δv is positive. A negative value of a indicates that the velocity decreases; either the particle is moving more slowly in the positive direction (Fig. 11.5c), or it is moving faster in the negative direction (Fig. 11.5d).

Sometimes we use the term *deceleration* to refer to a when the speed of the particle (i.e., the magnitude of v) decreases; the particle is then moving more slowly. For example, the particle of Fig. 11.5 is decelerating in parts b and c; it is truly accelerating (i.e., moving faster) in parts a and d.

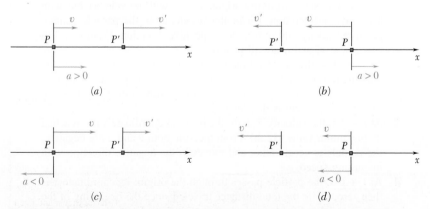

Fig. 11.5 Velocity and acceleration can be in the same or different directions. (a, d) When a and v are in the same direction, the particle speeds up; (b, c) when a and v are in opposite directions, the particle slows down.

We can obtain another expression for the acceleration by eliminating the differential dt in Eqs. (11.1) and (11.2). Solving Eq. (11.1) for dt, we have $dt = dx/v$; substituting into Eq. (11.2) gives us

$$a = v \frac{dv}{dx}$$

(11.4)

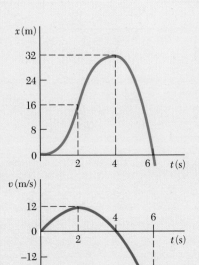

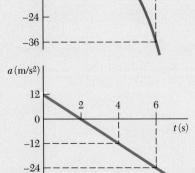

Fig. 11.6 Graphs of position, velocity, and acceleration as functions of time for Concept Application 11.1.

Concept Application 11.1

Consider a particle moving in a straight line, and assume that its position is defined by

$$x = 6t^2 - t^3$$

where t is in seconds and x in meters. We can obtain the velocity v at any time t by differentiating x with respect to t as

$$v = \frac{dx}{dt} = 12t - 3t^2$$

We can obtain the acceleration a by differentiating again with respect to t. Hence,

$$a = \frac{dv}{dt} = 12 - 6t$$

In Fig. 11.6, we have plotted the position coordinate, the velocity, and the acceleration. These curves are known as *motion curves*. Keep in mind, however, that the particle does not move along any of these curves; the particle moves in a straight line.

Since the derivative of a function measures the slope of the corresponding curve, the slope of the x–t curve at any given time is equal to the value of v at that time. Similarly, the slope of the v–t curve is equal to the value of a. Since $a = 0$ at $t = 2$ s, the slope of the v–t curve must be zero at $t = 2$ s; the velocity reaches a maximum at this instant. Also, since $v = 0$ at $t = 0$ and at $t = 4$ s, the tangent to the x–t curve must be horizontal for both of these values of t.

A study of the three motion curves of Fig. 11.6 shows that the motion of the particle from $t = 0$ to $t = \infty$ can be divided into four phases:

1. The particle starts from the origin, $x = 0$, with no velocity but with a positive acceleration. Under this acceleration, the particle gains a positive velocity and moves in the positive direction. From $t = 0$ to $t = 2$ s, x, v, and a are all positive.

2. At $t = 2$ s, the acceleration is zero; the velocity has reached its maximum value. From $t = 2$ s to $t = 4$ s, v is positive, but a is negative. The particle still moves in the positive direction but more slowly; the particle is decelerating.

3. At $t = 4$ s, the velocity is zero; the position coordinate x has reached its maximum value (32 m). From then on, both v and a are negative; the particle is accelerating and moves in the negative direction with increasing speed.

4. At $t = 6$ s, the particle passes through the origin; its coordinate x is then zero, while the total distance traveled since the beginning of the motion is 64 m (i.e., twice its maximum value). For values of t larger than 6 s, x, v, and a are all negative. The particle keeps moving in the negative direction—away from O—faster and faster. ∎

11.1B Determining the Motion of a Particle

We have just seen that the motion of a particle is said to be known if we know its position for every value of the time t. In practice, however, a motion is seldom defined by a relation between x and t. More often, the conditions of the motion are specified by the type of acceleration that the particle possesses. For example, a freely falling body has a constant acceleration that is directed downward and equal to 9.81 m/s^2 or 32.2 ft/s^2, a mass attached to a stretched spring has an acceleration proportional to the instantaneous elongation of the spring measured from its equilibrium position, etc. In general, we can express the acceleration of the particle as a function of one or more of the variables x, v, and t. Thus, in order to determine the position coordinate x in terms of t, we need to perform two successive integrations.

Let us consider three common classes of motion.

1. $a = f(t)$. **The Acceleration Is a Given Function of** t. Solving Eq. (11.2) for dv and substituting $f(t)$ for a, we have

$$dv = a\,dt$$
$$dv = f(t)\,dt$$

Integrating both sides of the equation, we obtain

$$\int dv = \int f(t)\,dt$$

This equation defines v in terms of t. Note, however, that an arbitrary constant is introduced after the integration is performed. This is due to the fact that many motions correspond to the given acceleration $a = f(t)$. In order to define the motion of the particle uniquely, it is necessary to specify the **initial conditions** of the motion, i.e., the value v_0 of the velocity and the value x_0 of the position coordinate at $t = 0$. Rather than use an arbitrary constant that is determined by the initial conditions, it is often more convenient to replace the indefinite integrals with **definite integrals**. Definite integrals have lower limits corresponding to the initial conditions $t = 0$ and $v = v_0$ and upper limits corresponding to $t = t$ and $v = v$. This gives us

$$\int_{v_0}^{v} dv = \int_{0}^{t} f(t)\,dt$$

$$v - v_0 = \int_{0}^{t} f(t)\,dt$$

which yields v in terms of t.

We can now solve Eq. (11.1) for dx as

$$dx = v\,dt$$

and substitute for v the expression obtained from the first integration. Then we integrate both sides of this equation via the left-hand side with respect to x from $x = x_0$ to $x = x$ and the right-hand side with respect to t from $t = 0$ to $t = t$. In this way, we obtain the position coordinate x in terms of t; the motion is completely determined.

We will study two important cases in greater detail in Sec. 11.2: the case when $a = 0$, corresponding to a *uniform motion*, and the case when a = constant, corresponding to a *uniformly accelerated motion*.

2. $a = f(x)$. **The Acceleration Is a Given Function of** x. Rearranging Eq. (11.4) and substituting $f(x)$ for a, we have

$$v \, dv = a \, dx$$
$$v \, dv = f(x) \, dx$$

Since each side contains only one variable, we can integrate the equation. Denoting again the initial values of the velocity and of the position coordinate by v_0 and x_0, respectively, we obtain

$$\int_{v_0}^{v} v \, dv = \int_{x_0}^{x} f(x) \, dx$$

$$\tfrac{1}{2}v^2 - \tfrac{1}{2}v_0^2 = \int_{x_0}^{x} f(x) \, dx$$

which yields v in terms of x. We now solve Eq. (11.1) for dt, giving

$$dt = \frac{dx}{v}$$

and substitute for v the expression just obtained. We can then integrate both sides to obtain the desired relation between x and t. However, in most cases, this last integration cannot be performed analytically, and we must resort to a numerical method of integration.

3. $a = f(v)$. **The Acceleration Is a Given Function of** v. We can now substitute $f(v)$ for a in either Eqs. (11.2) or (11.4) to obtain either

$$f(v) = \frac{dv}{dt} \qquad f(v) = v\frac{dv}{dx}$$

$$dt = \frac{dv}{f(v)} \qquad dx = \frac{v \, dv}{f(v)}$$

Integration of the first equation yields a relation between v and t; integration of the second equation yields a relation between v and x. Either of these relations can be used in conjunction with Eq. (11.1) to obtain the relation between x and t that characterizes the motion of the particle.

Sample Problem 11.1

The position of a particle moving along a straight line is defined by the relation $x = t^3 - 6t^2 - 15t + 40$, where x is expressed in feet and t in seconds. Determine (a) the time at which the velocity is zero, (b) the position and distance traveled by the particle at that time, (c) the acceleration of the particle at that time, (d) the distance traveled by the particle from $t = 4$ s to $t = 6$ s.

STRATEGY: You need to use the basic kinematic relationships between position, velocity, and acceleration. Because the position is given as a function of time, you can differentiate it to find equations for the velocity and acceleration. Once you have these equations, you can solve the problem.

MODELING and ANALYSIS: Taking the derivative of position, you obtain

$$x = t^3 - 6t^2 - 15t + 40 \tag{1}$$

$$v = \frac{dx}{dt} = 3t^2 - 12t - 15 \tag{2}$$

$$a = \frac{dv}{dt} = 6t - 12 \tag{3}$$

These equations are graphed in Fig. 1.

a. Time at Which v = 0. Set $v = 0$ in Eq. (2) for

$$3t^2 - 12t - 15 = 0 \qquad t = -1 \text{ s} \qquad \text{and} \qquad t = +5 \text{ s} \blacktriangleleft$$

Only the root $t = +5$ s corresponds to a time after the motion has begun: for $t < 5$ s, $v < 0$ and the particle moves in the negative direction; for $t > 5$ s, $v > 0$ and the particle moves in the positive direction.

b. Position and Distance Traveled When v = 0. Substitute $t = +5$ s into Eq. (1), yielding

$$x_5 = (5)^3 - 6(5)^2 - 15(5) + 40 \qquad x_5 = -60 \text{ ft} \blacktriangleleft$$

The initial position at $t = 0$ was $x_0 = +40$ ft. Since $v \neq 0$ during the interval $t = 0$ to $t = 5$ s, you have

$$\text{Distance traveled} = x_5 - x_0 = -60 \text{ ft} - 40 \text{ ft} = -100 \text{ ft}$$

$$\text{Distance traveled} = 100 \text{ ft in the negative direction} \blacktriangleleft$$

c. Acceleration When v = 0. Substitute $t = +5$ s into Eq. (3) for

$$a_5 = 6(5) - 12 \qquad a_5 = +18 \text{ ft/s}^2 \blacktriangleleft$$

d. Distance Traveled from t = 4 s to t = 6 s. The particle moves in the negative direction from $t = 4$ s to $t = 5$ s and in the positive direction from $t = 5$ s to $t = 6$ s; therefore, the distance traveled during each of these time intervals must be computed separately.

From $t = 4$ s to $t = 5$ s: $\qquad x_5 = -60$ ft

$$x_4 = (4)^3 - 6(4)^2 - 15(4) + 40 = -52 \text{ ft}$$

(continued)

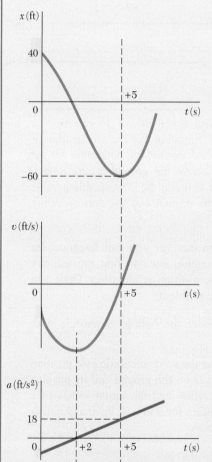

Fig. 1 Motion curves for the particle.

$$\text{Distance traveled} = x_5 - x_4 = -60 \text{ ft} - (-52 \text{ ft}) = -8 \text{ ft}$$
$$= 8 \text{ ft in the negative direction}$$

From $t = 5$ s to $t = 6$ s:　　$x_5 = -60$ ft

$$x_6 = (6)^3 - 6(6)^2 - 15(6) + 40 = -50 \text{ ft}$$
$$\text{Distance traveled} = x_6 - x_5 = -50 \text{ ft} - (-60 \text{ ft}) = +10 \text{ ft}$$
$$= 10 \text{ ft in the positive direction}$$

Total distance traveled from $t = 4$ s to $t = 6$ s is 8 ft + 10 ft　　$= 18$ ft

REFLECT and THINK: The total distance traveled by the particle in the 2-second interval is 18 ft, but because one distance is positive and one is negative, the net change in position is only 2 ft (in the positive direction). This illustrates the difference between total distance traveled and net change in position. Note that the maximum displacement occurs at $t = 5$ s, when the velocity is zero.

Sample Problem 11.2

You throw a ball vertically upward with a velocity of 10 m/s from a window located 20 m above the ground. Knowing that the acceleration of the ball is constant and equal to 9.81 m/s² downward, determine (*a*) the velocity v and elevation y of the ball above the ground at any time t, (*b*) the highest elevation reached by the ball and the corresponding value of t, (*c*) the time when the ball hits the ground and the corresponding velocity. Draw the v–t and y–t curves.

STRATEGY: The acceleration is constant, so you can integrate the defining kinematic equation for acceleration once to find the velocity equation and a second time to find the position relationship. Once you have these equations, you can solve the problem.

MODELING and ANALYSIS: Model the ball as a particle with negligible drag.

a. Velocity and Elevation. Choose the y axis measuring the position coordinate (or elevation) with its origin O on the ground and its positive sense upward. The value of the acceleration and the initial values of v and y are as indicated in Fig. 1. Substituting for a in $a = dv/dt$ and noting that, when $t = 0$, $v_0 = +10$ m/s, you have

$$\frac{dv}{dt} = a = -9.81 \, \text{m/s}^2$$

$$\int_{v_0 = 10}^{v} dv = -\int_{0}^{t} 9.81 \, dt$$

$$[v]_{10}^{v} = -[9.81t]_{0}^{t}$$

$$v - 10 = -9.81t$$

$$v = 10 - 9.81t \quad \text{(1)} \ \blacktriangleleft$$

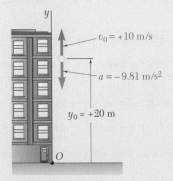

Fig. 1 Acceleration, initial velocity, and initial position of the ball.

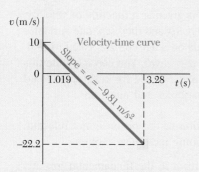

Fig. 2 Velocity of the ball as a function of time.

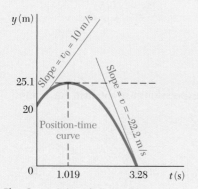

Fig. 3 Height of the ball as a function of time.

Substituting for v in $v = dy/dt$ and noting that when $t = 0$, $y_0 = 20$ m, you have

$$\frac{dy}{dt} = v = 10 - 9.81t$$

$$\int_{y_0=20}^{y} dy = \int_{0}^{t} (10 - 9.81t)\, dt$$

$$[y]_{20}^{y} = [10t - 4.905t^2]_0^t$$

$$y - 20 = 10t - 4.905t^2$$

$$y = 20 + 10t - 4.905t^2 \quad \text{(2)} \blacktriangleleft$$

Graphs of these equations are shown in Figs. 2 and 3.

b. Highest Elevation. The ball reaches its highest elevation when $v = 0$. Substituting into Eq. (1), you obtain

$$10 - 9.81t = 0 \qquad\qquad t = 1.019 \text{ s} \blacktriangleleft$$

Substituting $t = 1.019$ s into Eq. (2), you find

$$y = 20 + 10(1.019) - 4.905(1.019)^2 \quad y = 25.1 \text{ m} \blacktriangleleft$$

c. Ball Hits the Ground. The ball hits the ground when $y = 0$. Substituting into Eq. (2), you obtain

$$20 + 10t - 4.905t^2 = 0 \qquad t = -1.243 \text{ s} \qquad \text{and} \qquad t = +3.28 \text{ s} \blacktriangleleft$$

Only the root $t = +3.28$ s corresponds to a time after the motion has begun. Carrying this value of t into Eq. (1), you find

$$v = 10 - 9.81(3.28) = -22.2 \text{ m/s} \qquad v = 22.2 \text{ m/s} \downarrow \blacktriangleleft$$

REFLECT and THINK: When the acceleration is constant, the velocity changes linearly, and the position is a quadratic function of time. You will see in Sec. 11.2 that the motion in this problem is an example of free fall, where the acceleration in the vertical direction is constant and equal to $-g$.

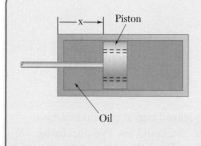

Sample Problem 11.3

Many mountain bike shocks utilize a piston that travels in an oil-filled cylinder to provide shock absorption; this system is shown schematically. When the front tire goes over a bump, the cylinder is given an initial velocity v_0. The piston, which is attached to the fork, then moves with respect to the cylinder, and oil is forced through orifices in the piston. This causes the piston to decelerate at a rate proportional to the velocity at $a = -kv$. At time $t = 0$, the position of the piston is $x = 0$. Express (a) the velocity v in terms of t, (b) the position x in terms of t, (c) the velocity v in terms of x. Draw the corresponding motion curves.

(continued)

STRATEGY: Because the acceleration is given as a function of velocity, you need to use either $a = dv/dt$ or $a = v\, dv/dx$ and then separate variables and integrate. Which one you use depends on what you are asked to find. Since part a asks for v in terms of t, use $a = dv/dt$. You can integrate this again using $v = dx/dt$ for part b. Since part c asked for $v(x)$, you should use $a = v\, dv/dx$ and then separate the variables and integrate.

MODELING and ANALYSIS: Rotation of the piston is not relevant, so you can model it as a particle undergoing rectilinear motion.

a. v in Terms of t. Substitute $-kv$ for a in the fundamental formula defining acceleration, $a = dv/dt$. You obtain

$$-kv = \frac{dv}{dt} \qquad \frac{dv}{v} = -k\, dt \qquad \int_{v_0}^{v} \frac{dv}{v} = -k \int_{0}^{t} dt$$

$$\ln \frac{v}{v_0} = -kt \qquad\qquad v = v_0 e^{-kt} \quad \blacktriangleleft$$

b. x in Terms of t. Substitute the expression just obtained for v into $v = dx/dt$. You get

$$v_0 e^{-kt} = \frac{dx}{dt}$$

$$\int_{0}^{x} dx = v_0 \int_{0}^{t} e^{-kt}\, dt$$

$$x = -\frac{v_0}{k}[e^{-kt}]_0^t = -\frac{v_0}{k}(e^{-kt} - 1)$$

$$x = \frac{v_0}{k}(1 - e^{-kt}) \quad \blacktriangleleft$$

c. v in Terms of x. Substitute $-kv$ for a in $a = v\, dv/dx$. You have

$$-kv = v\frac{dv}{dx}$$

$$dv = -k\, dx$$

$$\int_{v_0}^{v} dv = -k \int_{0}^{x} dx$$

$$v - v_0 = -kx \qquad\qquad v = v_0 - kx \quad \blacktriangleleft$$

The motion curves are shown in Fig. 1.

REFLECT and THINK: You could have solved part c by eliminating t from the answers obtained for parts a and b. You could use this alternative method as a check. From part a, you obtain $e^{-kt} = v/v_0$; substituting into the answer of part b, you have

$$x = \frac{v_0}{k}(1 - e^{-kt}) = \frac{v_0}{k}\left(1 - \frac{v}{v_0}\right) \qquad v = v_0 - kx \qquad \text{(checks)}$$

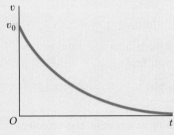

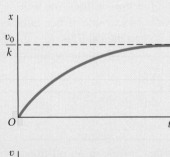

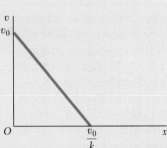

Fig. 1 Motion curves for the piston

Sample Problem 11.4

An uncontrolled automobile traveling at 45 mph strikes a highway crash barrier square on. After initially hitting the barrier, the automobile decelerates at a rate proportional to the distance x the automobile has moved into the barrier; specifically, $a = -60\sqrt{x}$, where a and x are expressed in ft/s^2 and ft, respectively. Determine the distance the automobile will move into the barrier before it comes to rest.

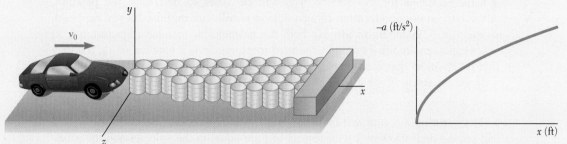

STRATEGY: Since you are given the deceleration as a function of displacement, you should start with the basic kinematic relationship $a = v \, dv/dx$.

MODELING and ANALYSIS: Model the car as a particle. First find the initial speed in ft/s,

$$v_0 = \left(45 \frac{\text{mi}}{\text{hr}}\right)\left(\frac{1 \text{ hr}}{3600 \text{ s}}\right)\left(\frac{5280 \text{ ft}}{\text{mi}}\right) = 66 \frac{\text{ft}}{\text{s}}$$

Substituting $a = -60\sqrt{x}$ into $a = v \, dv/dx$ gives

$$a = -60\sqrt{x} = \frac{v \, dv}{dx}$$

Separating variables and integrating gives

$$v \, dv = -60\sqrt{x} \, dx \longrightarrow \int_{v_0}^{0} v \, dv = -\int_{0}^{x} 60\sqrt{x} \, dx$$

$$\frac{1}{2}v^2 - \frac{1}{2}v_0^2 = -40x^{3/2} \longrightarrow x = \left(\frac{1}{80}(v_0^2 - v^2)\right)^{2/3} \quad \textbf{(1)}$$

Substituting $v = 0$, $v_0 = 45$ ft/s gives

$$d = 14.37 \text{ ft} \quad \blacktriangleleft$$

REFLECT and THINK: A distance of 14 ft seems reasonable for a barrier of this type. If you substitute d into the equation for a, you find a maximum deceleration of about 7 g's. Note that this problem would have been much harder to solve if you had been asked to find the time for the automobile to stop. In this case, you would need to determine $v(t)$ from Eq. (1). This gives $v = \sqrt{v_0^2 - 80x^{3/2}}$. Using the basic kinematic relationship $v = dx/dt$, you can easily show that

$$\int_{0}^{t} dt = \int_{0}^{x} \frac{dx}{\sqrt{v_0^2 - 80x^{3/2}}}$$

Unfortunately, there is no closed-form solution to this integral, so you would need to solve it numerically.

SOLVING PROBLEMS
ON YOUR OWN

In the problems for this section, you will be asked to determine the **position, velocity,** and/or **acceleration** of a particle in **rectilinear motion**. As you read each problem, it is important to identify both the independent variable (typically t or x) and what is required (for example, the need to express v as a function of x). You may find it helpful to start each problem by writing down both the given information and a simple statement of what is to be determined.

1. Determining $v(t)$ and $a(t)$ for a given $x(t)$. As explained in Sec. 11.1A, the first and second derivatives of x with respect to t are equal to the velocity and the acceleration, respectively, of the particle [Eqs. (11.1) and (11.2)]. If the velocity and acceleration have opposite signs, the particle can come to rest and then move in the opposite direction [Sample Prob. 11.1]. Thus, when computing the total distance traveled by a particle, you should first determine if the particle comes to rest during the specified interval of time. Constructing a diagram similar to that of Sample Prob. 11.1, which shows the position and the velocity of the particle at each critical instant ($v = v_{max}$, $v = 0$, etc.), will help you to visualize the motion.

2. Determining $v(t)$ and $x(t)$ for a given $a(t)$. We discussed the solution of problems of this type in the first part of Sec. 11.1B. We used the initial conditions, $t = 0$ and $v = v_0$, for the lower limits of the integrals in t and v, but any other known state (for example, $t = t_1$ and $v = v_1$) could be used instead. Also, if the given function $a(t)$ contains an unknown constant (for example, the constant k if $a = kt$), you will first have to determine that constant by substituting a set of known values of t and a in the equation defining $a(t)$.

3. Determining $v(x)$ and $x(t)$ for a given $a(x)$. This is the second case considered in Sec. 11.1B. We again note that the lower limits of integration can be any known state (for example, $x = x_1$ and $v = v_1$). In addition, since $v = v_{max}$ when $a = 0$, you can determine the positions where the maximum or minimum values of the velocity occur by setting $a(x) = 0$ and solving for x.

4. Determining $v(x)$, $v(t)$, and $x(t)$ for a given $a(v)$. This is the last case treated in Sec. 11.1B; the appropriate solution techniques for problems of this type are illustrated in Sample Probs. 11.3 and 11.4. All of the general comments for the preceding cases once again apply. Note that Sample Prob. 11.3 provides a summary of how and when to use the equations $v = dx/dt$, $a = dv/dt$, and $a = v\,dv/dx$.

We can summarize these relationships in Table 11.1.

Table 11.1

If....	Kinematic relationship	Integrate
$a = a(t)$	$\dfrac{dv}{dt} = a(t)$	$\displaystyle\int_{v_0}^{v} dv = \int_{0}^{t} a(t)dt$
$a = a(x)$	$v\dfrac{dv}{dx} = a(x)$	$\displaystyle\int_{v_0}^{v} v\, dv = \int_{x_0}^{x} a(x)dx$
$a = a(v)$	$\dfrac{dv}{dt} = a(v)$	$\displaystyle\int_{v_0}^{v} \dfrac{dv}{a(v)} = \int_{0}^{t} dt$
	$v\dfrac{dv}{dx} = a(v)$	$\displaystyle\int_{x_0}^{x} dx = \int_{v_0}^{v} \dfrac{v\, dv}{a(v)}$

Problems[†]

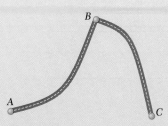

Fig. P11.CQ1

CONCEPT QUESTIONS

11.CQ1 A bus travels the 100 miles between A and B at 50 mi/h and then another 100 miles between B and C at 70 mi/h. The average speed of the bus for the entire 200-mile trip is:
a. More than 60 mi/h.
b. Equal to 60 mi/h.
c. Less than 60 mi/h.

11.CQ2 Two cars A and B race each other down a straight road. The position of each car as a function of time is shown. Which of the following statements are true (more than one answer can be correct)?
a. At time t_2 both cars have traveled the same distance.
b. At time t_1 both cars have the same speed.
c. Both cars have the same speed at some time $t < t_1$.
d. Both cars have the same acceleration at some time $t < t_1$.
e. Both cars have the same acceleration at some time $t_1 < t < t_2$.

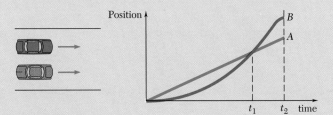

Fig. P11.CQ2

END-OF-SECTION PROBLEMS

11.1 A snowboarder starts from rest at the top of a double black diamond hill. As she rides down the slope, GPS coordinates are used to determine her displacement as a function of time: $x = 0.5t^3 + t^2 + 2t$, where x and t are expressed in feet and seconds, respectively. Determine the position, velocity, and acceleration of the boarder when $t = 5$ seconds.

11.2 The motion of a particle is defined by the relation $x = 2t^3 - 9t^2 + 12t + 10$, where x and t are expressed in feet and seconds, respectively. Determine the time, the position, and the acceleration of the particle when $v = 0$.

11.3 The vertical motion of mass A is defined by the relation $x = 10 \sin 2t + 15 \cos 2t + 100$, where x and t are expressed in millimeters and seconds, respectively. Determine (a) the position, velocity, and acceleration of A when $t = 1$ s, (b) the maximum velocity and acceleration of A.

Fig. P11.3

†Answers to all problems set in straight type (such as **11.1**) are given at the end of the book. Answers to problems with a number set in italic type (such as *11.6*) are not given.

11.4 A loaded railroad car is rolling at a constant velocity when it couples with a spring and dashpot bumper system. After the coupling, the motion of the car is defined by the relation $x = 60e^{-4.8t} \sin 16t$, where x and t are expressed in millimeters and seconds, respectively. Determine the position, the velocity, and the acceleration of the railroad car when (a) $t = 0$, (b) $t = 0.3$ s.

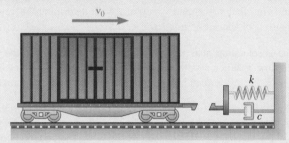

Fig. P11.4

11.5 The motion of a particle is defined by the relation $x = 6t^4 - 2t^3 - 12t^2 + 3t + 3$, where x and t are expressed in meters and seconds, respectively. Determine the time, the position, and the velocity when $a = 0$.

11.6 The motion of a particle is defined by the relation $x = t^3 - 9t^2 + 24t - 8$, where x and t are expressed in inches and seconds, respectively. Determine (a) when the velocity is zero, (b) the position and the total distance traveled when the acceleration is zero.

11.7 A girl operates a radio-controlled model car in a vacant parking lot. The girl's position is at the origin of the xy coordinate axes, and the surface of the parking lot lies in the x-y plane. She drives the car in a straight line so that the x coordinate is defined by the relation $x(t) = 0.5t^3 - 3t^2 + 3t + 2$, where x and t are expressed in meters and seconds, respectively. Determine (a) when the velocity is zero, (b) the position and total distance travelled when the acceleration is zero.

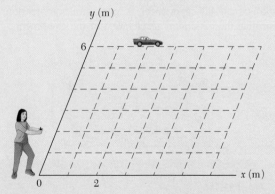

Fig. P11.7

11.8 The motion of a particle is defined by the relation $x = t^2 - (t - 2)^3$, where x and t are expressed in feet and seconds, respectively. Determine (a) the two positions at which the velocity is zero (b) the total distance traveled by the particle from $t = 0$ to $t = 4$ s.

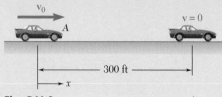

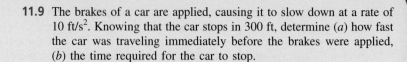

11.9 The brakes of a car are applied, causing it to slow down at a rate of 10 ft/s². Knowing that the car stops in 300 ft, determine (a) how fast the car was traveling immediately before the brakes were applied, (b) the time required for the car to stop.

Fig. P11.9

11.10 The acceleration of a particle is defined by the relation $a = 3e^{-0.2t}$, where a and t are expressed in ft/s² and seconds, respectively. Knowing that $x = 0$ and $v = 0$ at $t = 0$, determine the velocity and position of the particle when $t = 0.5$ s.

11.11 The acceleration of a particle is directly proportional to the square of the time t. When $t = 0$, the particle is at $x = 24$ m. Knowing that at $t = 6$ s, $x = 96$ m and $v = 18$ m/s, express x and v in terms of t.

11.12 The acceleration of a particle is defined by the relation $a = kt^2$. (a) Knowing that $v = -8$ m/s when $t = 0$ and that $v = +8$ m/s when $t = 2$ s, determine the constant k. (b) Write the equations of motion, knowing also that $x = 0$ when $t = 2$ s.

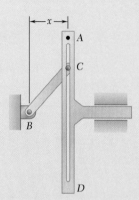

11.13 A Scotch yoke is a mechanism that transforms the circular motion of a crank into the reciprocating motion of a shaft (or vice versa). It has been used in a number of different internal combustion engines and in control valves. In the Scotch yoke shown, the acceleration of point A is defined by the relation $a = -1.8 \sin kt$, where a and t are expressed in m/s² and seconds, respectively, and $k = 3$ rad/s. Knowing that $x = 0$ and $v = 0.6$ m/s when $t = 0$, determine the velocity and position of point A when $t = 0.5$ s.

11.14 For the Scotch yoke mechanism shown, the acceleration of point A is defined by the relation $a = -1.08 \sin kt - 1.44 \cos kt$, where a and t are expressed in m/s² and seconds, respectively, and $k = 3$ rad/s. Knowing that $x = 0.16$ m and $v = 0.36$ m/s when $t = 0$, determine the velocity and position of point A when $t = 0.5$ s.

Fig. P11.13 and P11.14

11.15 A piece of electronic equipment that is surrounded by packing material is dropped so that it hits the ground with a speed of 4 m/s. After contact the equipment experiences an acceleration of $a = -kx$, where k is a constant and x is the compression of the packing material. If the packing material experiences a maximum compression of 20 mm, determine the maximum acceleration of the equipment.

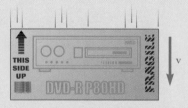

Fig. P11.15

11.16 A projectile enters a resisting medium at $x = 0$ with an initial velocity $v_0 = 900$ ft/s and travels 4 in. before coming to rest. Assuming that the velocity of the projectile is defined by the relation $v = v_0 - kx$, where v is expressed in ft/s and x is in feet, determine (a) the initial acceleration of the projectile, (b) the time required for the projectile to penetrate 3.9 in. into the resisting medium.

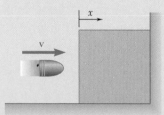

Fig. P11.16

11.17 The acceleration of a particle is defined by the relation $a = -k/x$. It has been experimentally determined that $v = 15$ ft/s when $x = 0.6$ ft and that $v = 9$ ft/s when $x = 1.2$ ft. Determine (a) the velocity of the particle when $x = 1.5$ ft, (b) the position of the particle at which its velocity is zero.

11.18 A brass (nonmagnetic) block A and a steel magnet B are in equilibrium in a brass tube under the magnetic repelling force of another steel magnet C located at a distance $x = 0.004$ m from B. The force is inversely proportional to the square of the distance between B and C. If block A is suddenly removed, the acceleration of block B is $a = -9.81 + k/x^2$, where a and x are expressed in m/s^2 and meters, respectively, and $k = 4 \times 10^{-4}$ m^3/s^2. Determine the maximum velocity and acceleration of B.

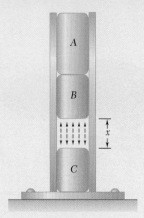

Fig. P11.18

11.19 Based on experimental observations, the acceleration of a particle is defined by the relation $a = -(0.1 + \sin x/b)$, where a and x are expressed in m/s^2 and meters, respectively. Knowing that $b = 0.8$ m and that $v = 1$ m/s when $x = 0$, determine (a) the velocity of the particle when $x = -1$ m, (b) the position where the velocity is maximum, (c) the maximum velocity.

11.20 A spring AB is attached to a support at A and to a collar. The unstretched length of the spring is l. Knowing that the collar is released from rest at $x = x_0$ and has an acceleration defined by the relation $a = -100(x - lx/\sqrt{l^2 + x^2})$, determine the velocity of the collar as it passes through point C.

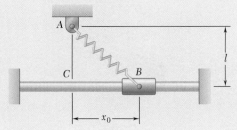

Fig. *P11.20*

11.21 The acceleration of a particle is defined by the relation $a = k(1 - e^{-x})$, where k is a constant. Knowing that the velocity of the particle is $v = +9$ m/s when $x = -3$ m and that the particle comes to rest at the origin, determine (a) the value of k, (b) the velocity of the particle when $x = -2$ m.

11.22 Starting from $x = 0$ with no initial velocity, a particle is given an acceleration $a = 0.1\sqrt{v^2 + 16}$, where a and v are expressed in ft/s^2 and ft/s, respectively. Determine (a) the position of the particle when $v = 3$ ft/s, (b) the speed and acceleration of the particle when $x = 4$ ft.

11.23 A ball is dropped from a boat so that it strikes the surface of a lake with a speed of 16.5 ft/s. While in the water the ball experiences an acceleration of $a = 10 - 0.8v$, where a and v are expressed in ft/s^2 and ft/s, respectively. Knowing the ball takes 3 s to reach the bottom of the lake, determine (a) the depth of the lake, (b) the speed of the ball when it hits the bottom of the lake.

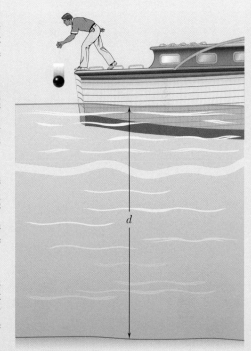

11.24 The acceleration of a particle is defined by the relation $a = -k\sqrt{v}$, where k is a constant. Knowing that $x = 0$ and $v = 81$ m/s at $t = 0$ and that $v = 36$ m/s when $x = 18$ m, determine (a) the velocity of the particle when $x = 20$ m, (b) the time required for the particle to come to rest.

11.25 The acceleration of a particle is defined by the relation $a = -kv^{2.5}$, where k is a constant. The particle starts at $x = 0$ with a velocity of 16 mm/s, and when $x = 6$ mm, the velocity is observed to be 4 mm/s. Determine (a) the velocity of the particle when $x = 5$ mm, (b) the time at which the velocity of the particle is 9 mm/s.

Fig. P11.23

Fig. P11.26

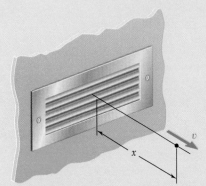

Fig. P11.27

Fig. P11.28

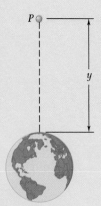

Fig. P11.29   **Fig. P11.30**

11.26 A human-powered vehicle (HPV) team wants to model the acceleration during the 260-m sprint race (the first 60 m is called a flying start) using $a = A - Cv^2$, where a is acceleration in m/s^2 and v is the velocity in m/s. From wind tunnel testing, they found that $C = 0.0012$ m^{-1}. Knowing that the cyclist is going 100 km/h at the 260-meter mark, what is the value of A?

11.27 Experimental data indicate that in a region downstream of a given louvered supply vent the velocity of the emitted air is defined by $v = 0.18v_0/x$, where v and x are expressed in m/s and meters, respectively, and v_0 is the initial discharge velocity of the air. For $v_0 = 3.6$ m/s, determine (*a*) the acceleration of the air at $x = 2$ m, (*b*) the time required for the air to flow from $x = 1$ to $x = 3$ m.

11.28 Based on observations, the speed of a jogger can be approximated by the relation $v = 7.5(1 - 0.04x)^{0.3}$, where v and x are expressed in mi/h and miles, respectively. Knowing that $x = 0$ at $t = 0$, determine (*a*) the distance the jogger has run when $t = 1$ h, (*b*) the jogger's acceleration in ft/s^2 at $t = 0$, (*c*) the time required for the jogger to run 6 mi.

11.29 The acceleration due to gravity at an altitude y above the surface of the earth can be expressed as

$$a = \frac{-32.2}{[1 + (y/20.9 \times 10^6)]^2}$$

where a and y are expressed in ft/s^2 and feet, respectively. Using this expression, compute the height reached by a projectile fired vertically upward from the surface of the earth if its initial velocity is (*a*) 1800 ft/s, (*b*) 3000 ft/s, (*c*) 36,700 ft/s.

11.30 The acceleration due to gravity of a particle falling toward the earth is $a = -gR^2/r^2$, where r is the distance from the *center* of the earth to the particle, R is the radius of the earth, and g is the acceleration due to gravity at the surface of the earth. If $R = 3960$ mi, calculate the *escape velocity*, that is, the minimum velocity with which a particle must be projected vertically upward from the surface of the earth if it is not to return to the earth. (*Hint:* $v = 0$ for $r = \infty$.)

11.31 The velocity of a particle is $v = v_0[1 - \sin(\pi t/T)]$. Knowing that the particle starts from the origin with an initial velocity v_0, determine (*a*) its position and its acceleration at $t = 3T$, (*b*) its average velocity during the interval $t = 0$ to $t = T$.

11.32 An eccentric circular cam, which serves a similar function as the Scotch yoke mechanism in Problem 11.13, is used in conjunction with a flat face follower to control motion in pumps and in steam engine valves. Knowing that the eccentricity is denoted by e, the maximum range of the displacement of the follower is d_{max} and the maximum velocity of the follower is v_{max}, determine the displacement, velocity, and acceleration of the follower.

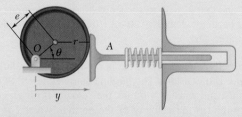

Fig. P11.32

11.2 SPECIAL CASES AND RELATIVE MOTION

In this section, we derive the equations that describe uniform rectilinear motion and uniformly accelerated rectilinear motion. We also introduce the concept of relative motion, which is of fundamental importance whenever we consider the motion of more than one particle at the same time.

11.2A Uniform Rectilinear Motion

Uniform rectilinear motion is a type of straight-line motion that is frequently encountered in practical applications. In this motion, the acceleration a of the particle is zero for every value of t. The velocity v is therefore constant, and Eq. (11.1) becomes

$$\frac{dx}{dt} = v = \text{constant}$$

We can obtain the position coordinate x by integrating this equation. Denoting the initial value of x by x_0, we have

Distance in uniform rectilinear motion

$$\int_{x_0}^{x} dx = v \int_{0}^{t} dt$$
$$x - x_0 = vt$$
$$\boxed{x = x_0 + vt} \tag{11.5}$$

This equation can be used *only if the velocity of the particle is known to be constant*. For example, this would be true for an airplane in steady flight or a car cruising along a highway at a constant speed.

11.2B Uniformly Accelerated Rectilinear Motion

Uniformly accelerated rectilinear motion is another common type of motion. In this case, the acceleration a of the particle is constant, and Eq. (11.2) becomes

$$\frac{dv}{dt} = a = \text{constant}$$

We obtain the velocity v of the particle by integrating this equation as

$$\int_{v_0}^{v} dv = a \int_{0}^{t} dt$$
$$v - v_0 = at$$
$$\boxed{v = v_0 + at} \tag{11.6}$$

where v_0 is the initial velocity. Substituting for v in Eq, (11.1), we have

$$\frac{dx}{dt} = v_0 + at$$

Denoting by x_0 the initial value of x and integrating, we have

$$\int_{x_0}^{x} dx = \int_{0}^{t} (v_0 + at)dt$$

$$x - x_0 = v_0 t + \tfrac{1}{2}at^2$$

$$x = x_0 + v_0 t + \tfrac{1}{2}at^2 \tag{11.7}$$

We can also use Eq. (11.4) and write

$$v\frac{dv}{dx} = a = \text{constant}$$

$$v \, dv = a \, dx$$

Integrating both sides, we obtain

$$\int_{v_0}^{v} v \, dv = a \int_{x_0}^{x} dx$$

$$\tfrac{1}{2}(v^2 - v_0^2) = a(x - x_0)$$

$$v^2 = v_0^2 + 2a(x - x_0) \tag{11.8}$$

The three equations we have derived provide useful relations among position, velocity, and time in the case of constant acceleration, once you have provided appropriate values for a, v_0, and x_0. You first need to define the origin O of the x axis and choose a positive direction along the axis; this direction determines the signs of a, v_0, and x_0. Equation (11.6) relates v and t and should be used when the value of v corresponding to a given value of t is desired, or inversely. Equation (11.7) relates x and t; Eq. (11.8) relates v and x. An important application of uniformly accelerated motion is the motion of a body in **free fall**. The acceleration of a body in free fall (usually denoted by g) is equal to 9.81 m/s^2 or 32.2 ft/s^2 (we ignore air resistance in this case).

It is important to keep in mind that the three equations can be used *only when the acceleration of the particle is known to be constant*. If the acceleration of the particle is variable, you need to determine its motion from the fundamental Eqs. (11.1) through (11.4) according to the methods outlined in Sec. 11.1B.

11.2C Motion of Several Particles

When several particles move independently along the same line, you can write independent equations of motion for each particle. Whenever possible, you should record time from the same initial instant for all particles and measure displacements from the same origin and in the same direction. In other words, use a single clock and a single measuring tape.

Relative Motion of Two Particles. Consider two particles A and B moving along the same straight line (Fig. 11.7). If we measure the position coordinates x_A and x_B from the same origin, the difference $x_B - x_A$ defines the **relative position coordinate of B with respect to A**, which is denoted by $x_{B/A}$. We have

**Relative position
of two particles**

$$x_{B/A} = x_B - x_A \qquad \text{or} \qquad x_B = x_A + x_{B/A} \tag{11.9}$$

Regardless of the positions of A and B with respect to the origin, a positive sign for $x_{B/A}$ means that B is to the right of A, and a negative sign means that B is to the left of A.

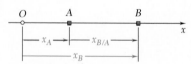

Fig. 11.7 Two particles A and B in motion along the same straight line.

The rate of change of $x_{B/A}$ is known as the **relative velocity of B with respect to A** and is denoted by $v_{B/A}$. Differentiating Eq. (11.9), we obtain

Relative velocity of two particles $v_{B/A} = v_B - v_A$ or $v_B = v_A + v_{B/A}$ **(11.10)**

A positive sign for $v_{B/A}$ means that B is *observed from A* to move in the positive direction; a negative sign means that it is observed to move in the negative direction.

The rate of change of $v_{B/A}$ is known as the **relative acceleration of B with respect to A** and is denoted by $a_{B/A}$. Differentiating Eq. (11.10), we obtain[†]

Relative acceleration of two particles $a_{B/A} = a_B - a_A$ or $a_B = a_A + a_{B/A}$ **(11.11)**

Dependent Motion of Particles.   Sometimes, the position of a particle depends upon the position of another particle or of several other particles. These motions are called **dependent**. For example, the position of block B in Fig. 11.8 depends upon the position of block A. Since the rope $ACDEFG$ is of constant length, and since the lengths of the portions of rope CD and EF wrapped around the pulleys remain constant, it follows that the sum of the lengths of the segments AC, DE, and FG is constant. Observing that the length of the segment AC differs from x_A only by a constant and that, similarly, the lengths of the segments DE and FG differ from x_B only by a constant, we have

$$x_A + 2x_B = \text{constant}$$

Since only one of the two coordinates x_A and x_B can be chosen arbitrarily, we say that the system shown in Fig. 11.8 has **one degree of freedom**. From the relation between the position coordinates x_A and x_B, it follows that if x_A is given an increment Δx_A—that is, if block A is lowered by an amount Δx_A—the coordinate x_B receives an increment $\Delta x_B = -\frac{1}{2}\Delta x_A$. In other words, block B rises by half the same amount. You can check this directly from Fig. 11.8.

In the case of the three blocks of Fig. 11.9, we can again observe that the length of the rope that passes over the pulleys is constant. Thus, the following relation must be satisfied by the position coordinates of the three blocks:

$$2x_A + 2x_B + x_C = \text{constant}$$

Since two of the coordinates can be chosen arbitrarily, we say that the system shown in Fig. 11.9 has **two degrees of freedom**.

When the relation existing between the position coordinates of several particles is *linear*, a similar relation holds between the velocities and between the accelerations of the particles. In the case of the blocks of Fig. 11.9, for instance, we can differentiate the position equation twice and obtain

$$2\frac{dx_A}{dt} + 2\frac{dx_B}{dt} + \frac{dx_C}{dt} = 0 \quad \text{or} \quad 2v_A + 2v_B + v_C = 0$$

$$2\frac{dv_A}{dt} + 2\frac{dv_B}{dt} + \frac{dv_C}{dt} = 0 \quad \text{or} \quad 2a_A + 2a_B + a_C = 0$$

[†]Note that the product of the subscripts A and B/A used in the right-hand sides of Eqs. (11.9), (11.10), and (11.11) is equal to the subscript B that appears in the left-hand sides. This may help you remember the correct order of subscripts in various situations.

Photo 11.2   Multiple cables and pulleys are used by this shipyard crane.

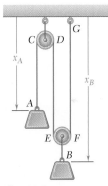

Fig. 11.8   A system of blocks and pulleys with one degree of freedom.

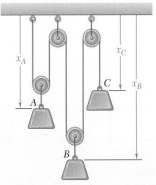

Fig. 11.9   A system of blocks and pulleys with two degrees of freedom.

Sample Problem 11.5

In an elevator shaft, a ball is thrown vertically upward with an initial velocity of 18 m/s from a height of 12 m above ground. At the same instant, an open-platform elevator passes the 5-m level, moving upward with a constant velocity of 2 m/s. Determine (a) when and where the ball hits the elevator (b) the relative velocity of the ball with respect to the elevator when the ball hits the elevator.

STRATEGY: The ball has a constant acceleration, so its motion is *uniformly accelerated*. The elevator has a constant velocity, so its motion is *uniform*. You can write equations to describe each motion and then set the position coordinates equal to each other to find when the particles meet. The relative velocity is determined from the calculated motion of each particle.

MODELING and ANALYSIS:

Motion of Ball. Place the origin O of the y axis at ground level and choose its positive direction upward (Fig. 1). Then the initial position of the ball is $y_0 = +12$ m, its initial velocity is $v_0 = +18$ m/s, and its acceleration is $a = -9.81$ m/s^2. Substituting these values in the equations for uniformly accelerated motion, you get

$$v_B = v_0 + at \qquad v_B = 18 - 9.81t \tag{1}$$

$$y_B = y_0 + v_0 t + \frac{1}{2}at^2 \qquad y_B = 12 + 18t - 4.905t^2 \tag{2}$$

Motion of Elevator. Again place the origin O at ground level and choose the positive direction upward (Fig. 2). Noting that $y_0 = +5$ m, you have

$$v_E = +2 \text{ m/s} \tag{3}$$

$$y_E = y_0 + v_E t \qquad y_E = 5 + 2t \tag{4}$$

Ball Hits Elevator. First note that you used the same time t and the same origin O in writing the equations of motion for both the ball and the elevator. From Fig. 3, when the ball hits the elevator,

$$y_E = y_B \tag{5}$$

Substituting for y_E and y_B from Eqs. (2) and (4) into Eq. (5), you have

$$5 + 2t = 12 + 18t - 4.905t^2$$
$$t = -0.39 \text{ s} \qquad \text{and} \qquad t = 3.65 \text{ s} \quad \blacktriangleleft$$

Only the root $t = 3.65$ s corresponds to a time after the motion has begun. Substituting this value into Eq. (4), you obtain

$$y_E = 5 + 2(3.65) = 12.30 \text{ m}$$

Elevation from ground = 12.30 m $\quad \blacktriangleleft$

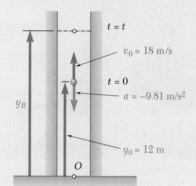

Fig. 1 Acceleration, initial velocity, and initial position of the ball.

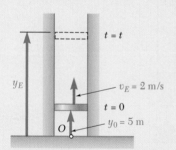

Fig. 2 Initial velocity and initial position of the elevator.

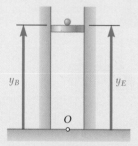

Fig. 3 Position of ball and elevator at time t.

Relative Velocity. The relative velocity of the ball with respect to the elevator is

$$v_{B/E} = v_B - v_E = (18 - 9.81t) - 2 = 16 - 9.81t$$

When the ball hits the elevator at time $t = 3.65$ s, you have

$$v_{B/E} = 16 - 9.81(3.65) \qquad v_{B/E} = -19.81 \text{ m/s} \blacktriangleleft$$

The negative sign means that if you are riding on the elevator, it will appear as if the ball is moving downward.

REFLECT and THINK: The key insight is that, when two particles collide, their position coordinates must be equal. Also, although you can use the basic kinematic relationships in this problem, you may find it easier to use the equations relating a, v, x, and t when the acceleration is constant or zero.

Sample Problem 11.6

Car A is travelling at a constant 90 mi/h when she passes a parked police officer B, who gives chase when the car passes her. The officer accelerates at a constant rate until she reaches the speed of 105 mi/h. Thereafter, her speed remains constant. The police officer catches the car 3 mi from her starting point. Determine the initial acceleration of the police officer.

STRATEGY: One car is traveling at a constant speed and the other has a constant acceleration, so you can start with the algebraic relationships found in Sec. 11.2 rather than separating and integrating the basic kinematic relationships.

MODELING and ANALYSIS: A clearly labeled picture will help you understand the problem better (Fig. 1). The position, x, is defined from the point the car passes the officer.

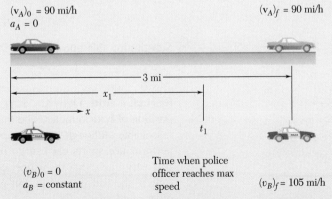

Fig. 1 Velocities and accelerations of the cars at various times.

(continued)

Unit Conversions.   First you should convert everything to units of feet and seconds. Use the subscript A for the car and B for the officer

$$v_A = \left(90\,\frac{\text{mi}}{\text{hr}}\right)\left(\frac{1\ \text{hr}}{3600\ \text{s}}\right)\left(\frac{5280\ \text{ft}}{\text{mi}}\right) = 132\,\frac{\text{ft}}{\text{s}}$$

$$v_B = \left(105\,\frac{\text{mi}}{\text{hr}}\right)\left(\frac{1\ \text{hr}}{3600\ \text{s}}\right)\left(\frac{5280\ \text{ft}}{\text{mi}}\right) = 154\,\frac{\text{ft}}{\text{s}}$$

Motion of the Speeding Car A.   Since the car has a constant speed,

$$x_A = v_A t = 132t \tag{1}$$

Motion of the Officer B.   The officer has a constant acceleration until she reaches a final speed of 105 mph. This time is labeled t_1 in Fig. 1. Therefore, from time $0 < t < t_1$, the officer has a velocity of

$$v_B = a_B t \quad \text{for } 0 < t < t_1$$

or at time $t = t_1$, it is

$$154 = a_B t_1 \tag{2}$$

The distance the officer travels is going to be the distance from 0 to t_1 and then from t_1 to t_f. Hence,

$$x_B = \frac{1}{2}a_B t_1^2 + v_B(t - t_1) \quad \text{for } t > t_1 \tag{3}$$

The officer catches the speeder when $x_A = x_B = 3$ mi $= 15{,}840$ ft. From Eq. (1), you can solve for the time $t_f = (15{,}840\ \text{ft})/(132\ \text{ft/s}) = 120$ s. Therefore, you have two equations: Eq. (2) and

$$15{,}840 = \frac{1}{2}a_B t_1^2 + 154(120 - t_1) \tag{4}$$

Substituting Eq. (2) into Eq. (4) allows you to solve for t_1:

$$t_1 = 34.39\ \text{s}$$

Substituting this into Eq. (2) gives

$$a_B = 4.49\ \text{ft/s} \quad \blacktriangleleft$$

REFLECT and THINK: It is important to use the same origin for the position of both vehicles. The time to accelerate from 0 to 105 mph seems reasonable, although it is perhaps longer than you would expect. A high-performance sports car can go from 0 to 60 mph in less than 5 seconds. It is very likely that the officer could have accelerated to 105 mph in less time if she had wanted to, but perhaps she had to consider the safety of other motorists.

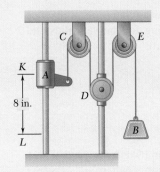

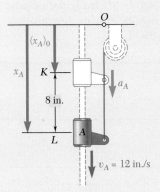

Fig. 1 Position, velocity, and acceleration of collar A.

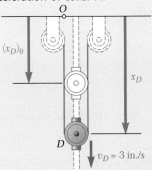

Fig. 2 Position and velocity of pulley D.

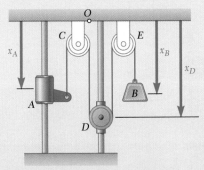

Fig. 3 Position of A, B, and D.

Sample Problem 11.7

Collar A and block B are connected by a cable passing over three pulleys C, D, and E as shown. Pulleys C and E are fixed, but D is attached to a collar that is pulled downward with a constant velocity of 3 in./s. At $t = 0$, collar A starts moving downward from position K with a constant acceleration and no initial velocity. Knowing that the velocity of collar A is 12 in./s as it passes through point L, determine the change in elevation, the velocity, and the acceleration of block B when collar A passes through L.

STRATEGY: You have multiple objects connected by cables, so this is a problem in *dependent motion*. Use the given data to write a single equation relating the changes in position coordinates of collar A, pulley D, and block B. Based on the given information, you will also need to use the algebraic relationships we found for uniformly accelerated motion.

MODELING and ANALYSIS:

Motion of Collar A. Place the origin O at the upper horizontal surface and choose the positive direction downward. Then when $t = 0$, collar A is at position K and $(v_A)_0 = 0$ (Fig. 1). Since $v_A = 12$ in./s and $x_A - (x_A)_0 = 8$ in. when the collar passes through L, you have

$$v_A^2 = (v_A)_0^2 + 2a_A[x_A - (x_A)_0] \qquad (12)^2 = 0 + 2a_A(8)$$
$$a_A = 9 \text{ in./s}^2$$

To find the time at which collar A reaches point L, use the equation for velocity as a function of time with uniform acceleration. Thus,

$$v_A = (v_A)_0 + a_A t \qquad 12 = 0 + 9t \qquad t = 1.333 \text{ s}$$

Motion of Pulley D. Since the positive direction is downward, you have (Fig. 2)

$$a_D = 0 \qquad v_D = 3 \text{ in./s} \qquad x_D = (x_D)_0 + v_D t = (x_D)_0 + 3t$$

When collar A reaches L at $t = 1.333$ s, the position of pulley D is

$$x_D = (x_D)_0 + 3(1.333) = (x_D)_0 + 4$$

Thus, $$x_D - (x_D)_0 = 4 \text{ in.}$$

Motion of Block B. Note that the total length of cable ACDEB differs from the quantity $(x_A + 2x_D + x_B)$ only by a constant. Since the cable length is constant during the motion, this quantity must also remain constant. Thus, considering the times $t = 0$ and $t = 1.333$ s, you can write

$$x_A + 2x_D + x_B = (x_A)_0 + 2(x_D)_0 + (x_B)_0 \qquad \textbf{(1)}$$

$$[x_A - (x_A)_0] + 2[x_D - (x_D)_0] + [x_B - (x_B)_0] = 0 \qquad \textbf{(2)}$$

But you know that $x_A - (x_A)_0 = 8$ in. and $x_D - (x_D)_0 = 4$ in. Substituting these values in Eq. (2), you find

$$8 + 2(4) + [x_B - (x_B)_0] = 0 \qquad x_B - (x_B)_0 = -16 \text{ in.}$$

Thus, Change in elevation of B = 16 in. ↑ ◀

(continued)

Differentiating Eq. (1) twice, you obtain equations relating the velocities and the accelerations of A, B, and D. Substituting for the velocities and accelerations of A and D at $t = 1.333$ s, you have

$$v_A + 2v_D + v_B = 0: \qquad 12 + 2(3) + v_B = 0$$
$$v_B = -18 \text{ in./s} \qquad v_B = 18 \text{ in./s} \uparrow \quad \blacktriangleleft$$

$$a_A + 2a_D + a_B = 0: \qquad 9 + 2(0) + a_B = 0$$
$$a_B = -9 \text{ in./s}^2 \qquad a_B = 9 \text{ in./s}^2 \uparrow \quad \blacktriangleleft$$

REFLECT and THINK: In this case, the relationship we needed was not between position coordinates, but between changes in position coordinates at two different times. The key step is to clearly define your position vectors. This is a two-degree-of-freedom system, because two coordinates are required to completely describe it.

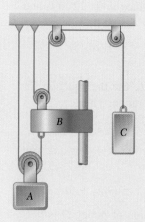

Sample Problem 11.8

Block C starts from rest and moves down with a constant acceleration. Knowing that after block A has moved 1.5 ft its velocity is 0.6 ft/s, determine (a) the acceleration of A and C, (b) the change in velocity and the change in position of block B after 2.5 seconds.

STRATEGY: Since you have blocks connected by cables, this is a dependent-motion problem. You should define coordinates for each mass and write constraint equations for both cables.

MODELING and ANALYSIS: Define position vectors as shown in Fig. 1, where positive is defined to be down.

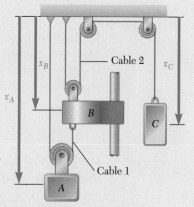

Fig. 1 Position of A, B, and C.

Constraint Equations. Assuming the cables are inextensible, you can write the lengths in terms of the defined coordinates and then differentiate.

Cable 1: $$x_A + (x_A - x_B) = \text{constant}$$

Differentiating this, you find

$$2v_A = v_B \quad \text{and} \quad 2a_A = a_B \tag{1}$$

Cable 2: $$2x_B + x_C = \text{constant}$$

Differentiating this, you find

$$v_C = -2v_B \quad \text{and} \quad a_C = -2a_B \tag{2}$$

Substituting Eq. (1) into Eq. (2) gives

$$v_C = -4v_A \quad \text{and} \quad a_C = -4a_A \tag{3}$$

Motion of A. You can use the constant-acceleration equations for block A:, as

$$v_A^2 - v_{A_0}^2 = 2a_A[x_A - (x_A)_0] \quad \text{or} \quad a_A = \frac{v_A^2 - (v_A)_0^2}{2[x_A - (x_A)_0]} \tag{4}$$

a. Acceleration of A and C. You know v_C and a_C are down, so from Eq. (3), you also know v_A and a_A are up. Substituting the given values into Eq. (4), you find

$$a_A = \frac{(0.6 \text{ ft/s})^2 - 0}{2(-1.5 \text{ ft})} = -0.12 \text{ ft/s}^2 \quad \mathbf{a_A} = 0.120 \text{ ft/s}^2 \uparrow \quad \blacktriangleleft$$

Substituting this value into $a_C = -4a_A$, you obtain

$$\mathbf{a_C} = 0.480 \text{ ft/s}^2 \downarrow \quad \blacktriangleleft$$

b. Velocity and change in position of B after 2.5 s. Substituting a_A in $a_B = 2a_A$ gives

$$a_B = 2(-0.2 \text{ ft/s}^2) = -0.24 \text{ ft/s}^2$$

You can use the equations of constant acceleration to find

$$\Delta v_B = a_B t = (-0.24 \text{ ft/s}^2)(2.5 \text{ s}) = -0.600 \text{ ft/s} \quad \Delta v_B = 0.600 \text{ ft/s} \uparrow \quad \blacktriangleleft$$

$$\Delta x_B = \tfrac{1}{2}a_B t = \tfrac{1}{2}(-0.24 \text{ ft/s}^2)(2.5 \text{ s})^2 = -0.750 \text{ ft} \quad \Delta x_B = 0.750 \text{ ft} \uparrow \quad \blacktriangleleft$$

REFLECT and THINK: One of the keys to solving this problem is recognizing that since there are two cables, you need to write two constraint equations. The directions of the answers also make sense. If block C is accelerating downward, you would expect A and B to accelerate upward.

SOLVING PROBLEMS
ON YOUR OWN

In this section, we derived the equations that describe **uniform rectilinear motion** (constant velocity) and **uniformly accelerated rectilinear motion** (constant acceleration). We also introduced the concept of **relative motion**. We can apply the equations for relative motion [Eqs. (11.9) through (11.11)] to the independent or dependent motions of any two particles moving along the same straight line.

A. Independent motion of one or more particles. Organize the solution of problems of this type as follows.

1. Begin your solution by listing the given information, sketching the system, and selecting the origin and the positive direction of the coordinate axis [Sample Prob. 11.5]. It is always advantageous to have a visual representation of problems of this type.

2. Write the equations that describe the motions of the various particles as well as those that describe how these motions are related [Eq. (5) of Sample Prob. 11.5].

3. Define the initial conditions, i.e., specify the state of the system corresponding to $t = 0$. This is especially important if the motions of the particles begin at different times. In such cases, either of two approaches can be used.

    a. Let $t = 0$ be the time when the last particle begins to move. You must then determine the initial position x_0 and the initial velocity v_0 of each of the other particles.

    b. Let $t = 0$ be the time when the first particle begins to move. You must then, in each of the equations describing the motion of another particle, replace t with $t - t_0$, where t_0 is the time at which that specific particle begins to move. It is important to recognize that the equations obtained in this way are valid only for $t \geq t_0$.

B. Dependent motion of two or more particles. In problems of this type, the particles of the system are connected to each other, typically by ropes or cables. The method of solution of these problems is similar to that of the preceding group of problems, except that it is now necessary to describe the *physical connections* between the particles. In the following problems, the connection is provided by one or more cables. For each cable, you will have to write equations similar to the last three equations of Sec. 11.2C. We suggest that you use the following procedure.

1. Draw a sketch of the system and select a coordinate system, indicating clearly a positive sense for each of the coordinate axes. For example, in Sample Probs. 11.7 and 11.8, we measured lengths downward from the upper horizontal support. It thus follows that those displacements, velocities, and accelerations that have positive values are directed downward.

2. Write the equation describing the constraint imposed by each cable on the motion of the particles involved. Differentiating this equation twice, you obtain the corresponding relations among velocities and accelerations.

3. If several directions of motion are involved, you must select a coordinate axis and a positive sense for each of these directions. You should also try to locate the origins of your coordinate axes so that the equations of constraints are as simple as possible. For example, in Sample Prob. 11.7, it is easier to define the various coordinates by measuring them downward from the upper support than by measuring them upward from the bottom support.

Finally, keep in mind that the method of analysis described in this section and the corresponding equations can be used only for particles moving with *uniform* or *uniformly accelerated rectilinear motion.*

Problems

11.33 An airplane begins its take-off run at A with zero velocity and a constant acceleration a. Knowing that it becomes airborne 30 s later at B and that the distance AB is 900 m, determine (a) the acceleration a (b) the take-off velocity v_B.

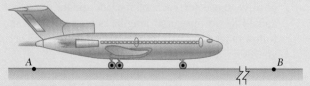

Fig. P11.33

11.34 A motorist is traveling at 54 km/h when she observes that a traffic light 240 m ahead of her turns red. The traffic light is timed to stay red for 24 s. If the motorist wishes to pass the light without stopping just as it turns green again, determine (a) the required uniform deceleration of the car, (b) the speed of the car as it passes the light.

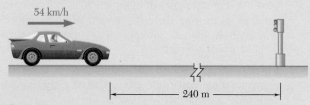

Fig. P11.34

11.35 Steep safety ramps are built beside mountain highways to enable vehicles with defective brakes to stop safely. A truck enters a 750-ft ramp at a high speed v_0 and travels 540 ft in 6 s at constant deceleration before its speed is reduced to $v_0/2$. Assuming the same constant deceleration, determine (a) the additional time required for the truck to stop (b) the additional distance traveled by the truck.

Fig. P11.35

11.36 A group of students launches a model rocket in the vertical direction. Based on tracking data, they determine that the altitude of the rocket was 89.6 ft at the end of the powered portion of the flight and that the rocket landed 16 s later. Knowing that the descent parachute failed to deploy so that the rocket fell freely to the ground after reaching its maximum altitude and assuming that $g = 32.2$ ft/s^2, determine (a) the speed v_1 of the rocket at the end of powered flight, (b) the maximum altitude reached by the rocket.

89.6 ft

Fig. P11.36

11.37 A small package is released from rest at A and moves along the skate wheel conveyor $ABCD$. The package has a uniform acceleration of 4.8 m/s^2 as it moves down sections AB and CD, and its velocity is constant between B and C. If the velocity of the package at D is 7.2 m/s, determine (a) the distance d between C and D, (b) the time required for the package to reach D.

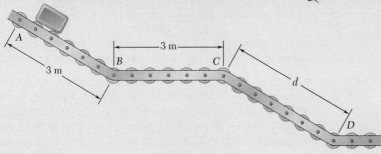

Fig. P11.37

11.38 A sprinter in a 100-m race accelerates uniformly for the first 35 m and then runs with constant velocity. If the sprinter's time for the first 35 m is 5.4 s, determine (a) his acceleration, (b) his final velocity, (c) his time for the race.

Fig. P11.38

11.39 Automobile A starts from O and accelerates at the constant rate of 0.75 m/s^2. A short time later it is passed by bus B which is traveling in the opposite direction at a constant speed of 6 m/s. Knowing that bus B passes point O 20 s after automobile A started from there, determine when and where the vehicles passed each other.

Fig. P11.39

11.40 In a boat race, boat A is leading boat B by 50 m and both boats are traveling at a constant speed of 180 km/h. At $t = 0$, the boats accelerate at constant rates. Knowing that when B passes A, $t = 8$ s and $v_A = 225$ km/h, determine (a) the acceleration of A, (b) the acceleration of B.

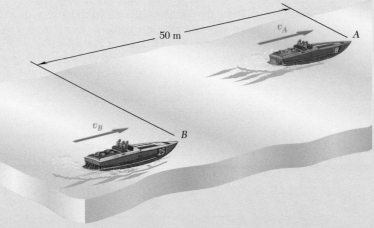

Fig. P11.40

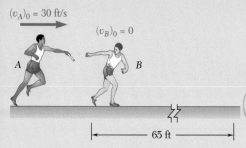

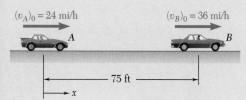

Fig. P11.41

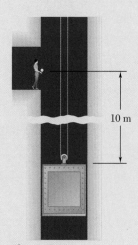

Fig. P11.42

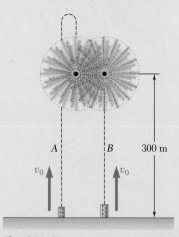

Fig. P11.44

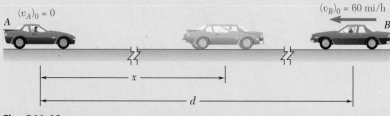

Fig. P11.45

11.41 As relay runner A enters the 65-ft-long exchange zone with a speed of 30 ft/s, he begins to slow down. He hands the baton to runner B 2.5 s later as they leave the exchange zone with the same velocity. Determine (a) the uniform acceleration of each of the runners, (b) when runner B should begin to run.

11.42 Automobiles A and B are traveling in adjacent highway lanes and at $t = 0$ have the positions and speeds shown. Knowing that automobile A has a constant acceleration of 1.8 ft/s^2 and that B has a constant deceleration of 1.2 ft/s^2, determine (a) when and where A will overtake B, (b) the speed of each automobile at that time.

11.43 Two automobiles A and B are approaching each other in adjacent highway lanes. At $t = 0$, A and B are 3200 ft apart, their speeds are $v_A = 65$ mi/h and $v_B = 40$ mi/h, and they are at points P and Q, respectively. Knowing that A passes point Q 40 s after B was there and that B passes point P 42 s after A was there, determine (a) the uniform accelerations of A and B, (b) when the vehicles pass each other, (c) the speed of B at that time.

Fig. P11.43

11.44 An elevator is moving upward at a constant speed of 4 m/s. A man standing 10 m above the top of the elevator throws a ball upward with a speed of 3 m/s. Determine (a) when the ball will hit the elevator, (b) where the ball will hit the elevator with respect to the location of the man.

11.45 Two rockets are launched at a fireworks display. Rocket A is launched with an initial velocity $v_0 = 100$ m/s and rocket B is launched t_1 seconds later with the same initial velocity. The two rockets are timed to explode simultaneously at a height of 300 m as A is falling and B is rising. Assuming a constant acceleration $g = 9.81$ m/s^2, determine (a) the time t_1, (b) the velocity of B relative to A at the time of the explosion.

11.46 Car A is parked along the northbound lane of a highway, and car B is traveling in the southbound lane at a constant speed of 60 mi/h. At $t = 0$, A starts and accelerates at a constant rate a_A, while at $t = 5$ s, B begins to slow down with a constant deceleration of magnitude $a_A/6$. Knowing that when the cars pass each other $x = 294$ ft and $v_A = v_B$, determine (a) the acceleration a_A, (b) when the vehicles pass each other, (c) the distance d between the vehicles at $t = 0$.

Fig. P11.46

11.47 The elevator E shown in the figure moves downward with a constant velocity of 4 m/s. Determine (*a*) the velocity of the cable C, (*b*) the velocity of the counterweight W, (*c*) the relative velocity of the cable C with respect to the elevator, (*d*) the relative velocity of the counterweight W with respect to the elevator.

11.48 The elevator E shown starts from rest and moves upward with a constant acceleration. If the counterweight W moves through 30 ft in 5 s, determine (*a*) the acceleration of the elevator and the cable C, (*b*) the velocity of the elevator after 5 s.

11.49 An athlete pulls handle A to the left with a constant velocity of 0.5 m/s. Determine (*a*) the velocity of the weight B, (*b*) the relative velocity of weight B with respect to the handle A.

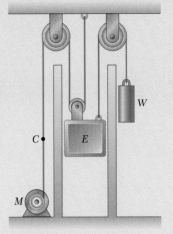

Fig. P11.47 and P11.48

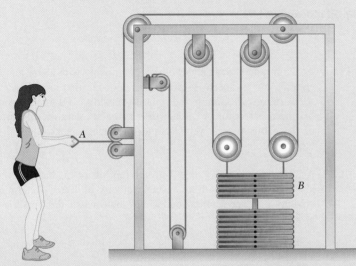

Fig. P11.49

11.50 An athlete pulls handle A to the left with a constant acceleration. Knowing that after the weight B has been lifted 4 in. its velocity is 2 ft/s, determine (*a*) the accelerations of handle A and weight B, (*b*) the velocity and change in position of handle A after 0.5 sec.

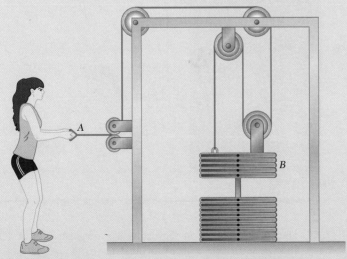

Fig. P11.50

11.51 Slider block B moves to the right with a constant velocity of 300 mm/s. Determine (*a*) the velocity of slider block A, (*b*) the velocity of portion C of the cable, (*c*) the velocity of portion D of the cable, (*d*) the relative velocity of portion C of the cable with respect to slider block A.

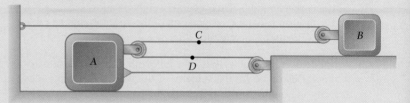

Fig. P11.51 and P11.52

11.52 At the instant shown, slider block B is moving with a constant acceleration, and its speed is 150 mm/s. Knowing that after slider block A has moved 240 mm to the right its velocity is 60 mm/s, determine (*a*) the accelerations of A and B, (*b*) the acceleration of portion D of the cable, (*c*) the velocity and the change in position of slider block B after 4 s.

11.53 A farmer lifts his hay bales into the top loft of his barn by walking his horse forward with a constant velocity of 1 ft/s. Determine the velocity and acceleration of the hay bale when the horse is 10 ft away from the barn.

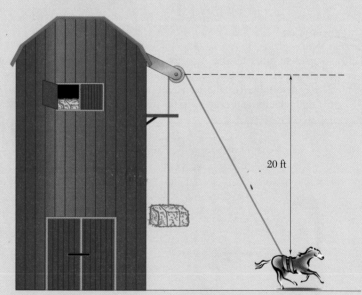

Fig. P11.53

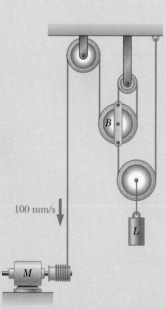

Fig. P11.54

11.54 The motor M reels in the cable at a constant rate of 100 mm/s. Determine (*a*) the velocity of load L, (*b*) the velocity of pulley B with respect to load L.

11.55 Collar A starts from rest at $t = 0$ and moves upward with a constant acceleration of 3.6 in./s^2. Knowing that collar B moves downward with a constant velocity of 18 in./s, determine (*a*) the time at which the velocity of block C is zero, (*b*) the corresponding position of block C.

11.56 Block A starts from rest at $t = 0$ and moves downward with a constant acceleration of 6 in./s^2. Knowing that block B moves up with a constant velocity of 3 in./s, determine (*a*) the time when the velocity of block C is zero, (*b*) the corresponding position of block C.

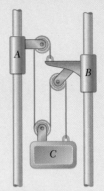

Fig. P11.55

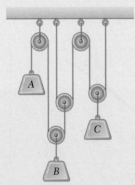

Fig. P11.56

11.57 Block B starts from rest, block A moves with a constant acceleration, and slider block C moves to the right with a constant acceleration of 75 mm/s^2. Knowing that at $t = 2$ s the velocities of B and C are 480 mm/s downward and 280 mm/s to the right, respectively, determine (*a*) the accelerations of A and B, (*b*) the initial velocities of A and C, (*c*) the change in position of slider block C after 3 s.

11.58 Block B moves downward with a constant velocity of 20 mm/s. At $t = 0$, block A is moving upward with a constant acceleration, and its velocity is 30 mm/s. Knowing that at $t = 3$ s slider block C has moved 57 mm to the right, determine (*a*) the velocity of slider block C at $t = 0$, (*b*) the accelerations of A and C, (*c*) the change in position of block A after 5 s.

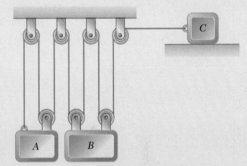

Fig. P11.57 and P11.58

11.59 The system shown starts from rest, and each component moves with a constant acceleration. If the relative acceleration of block C with respect to collar B is 60 mm/s^2 upward and the relative acceleration of block D with respect to block A is 110 mm/s^2 downward, determine (*a*) the velocity of block C after 3 s, (*b*) the change in position of block D after 5 s.

11.60 The system shown starts from rest, and the length of the upper cord is adjusted so that A, B, and C are initially at the same level. Each component moves with a constant acceleration, and after 2 s the relative change in position of block C with respect to block A is 280 mm upward. Knowing that when the relative velocity of collar B with respect to block A is 80 mm/s downward, the displacements of A and B are 160 mm downward and 320 mm downward, respectively, determine (*a*) the accelerations of A and B if $a_B > 10$ mm/s^2, (*b*) the change in position of block D when the velocity of block C is 600 mm/s upward.

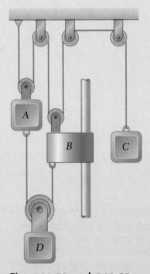

Fig. *P11.59* and *P11.60*

*11.3 GRAPHICAL SOLUTIONS

In analyzing problems in rectilinear motion, it is often useful to draw graphs of position, velocity, or acceleration versus time. Sometimes these graphs can provide insight into the situation by indicating when quantities increase, decrease, or stay the same. In other cases, the graphs can provide numerical solutions when analytical methods are not available. In many experimental situations, data are collected as a function of time, and the methods of this section are very useful for the analysis.

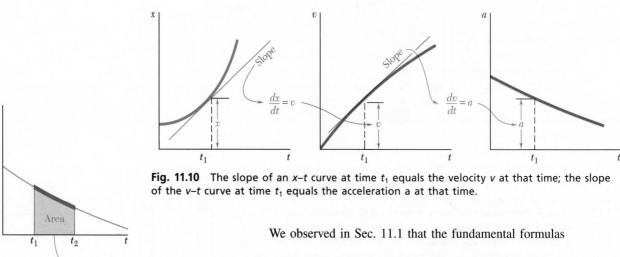

Fig. 11.10 The slope of an x–t curve at time t_1 equals the velocity v at that time; the slope of the v–t curve at time t_1 equals the acceleration a at that time.

We observed in Sec. 11.1 that the fundamental formulas

$$v = \frac{dx}{dt} \quad \text{and} \quad a = \frac{dv}{dt}$$

have a geometrical significance. The first formula says that the velocity at any instant is equal to the slope of the x–t curve at that instant (Fig. 11.10). The second formula states that the acceleration is equal to the slope of the v–t curve. We can use these two properties to determine graphically the v–t and a–t curves of a motion when the x–t curve is known.

Integrating the two fundamental formulas from a time t_1 to a time t_2, we have

$$x_2 - x_1 = \int_{t_1}^{t_2} v\,dt \quad \text{and} \quad v_2 - v_1 = \int_{t_1}^{t_2} a\,dt \qquad \textbf{(11.12)}$$

The first formula says that the area measured under the v–t curve from t_1 to t_2 is equal to the change in x during that time interval (Fig. 11.11). Similarly, the second formula states that the area measured under the a–t curve from t_1 to t_2 is equal to the change in v during that time interval. We can use these two properties to determine graphically the x–t curve of a motion when its v–t curve or its a–t curve is known (see Sample Prob. 11.9).

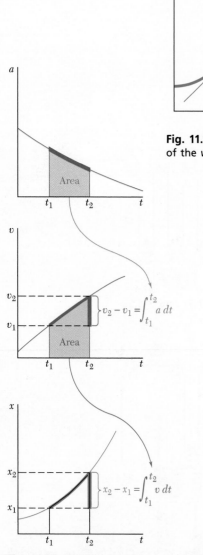

Fig. 11.11 The area under an a–t curve equals the change in velocity during that time interval; the area under the v–t curve equals the change in position during that time interval.

Graphical solutions are particularly useful when the motion considered is defined from experimental data and when x, v, and a are not analytical functions of t. They also can be used to advantage when the motion consists of distinct parts and when its analysis requires writing a different equation for each of its parts. When using a graphical solution, however, be careful to note that (1) the area under the v–t curve measures the *change in x*—not x itself—and similarly, that the area under the a–t curve measures the change in v; (2) an area above the t axis corresponds to an *increase* in x or v, whereas an area located below the t axis measures a *decrease* in x or v.

In drawing motion curves, it is useful to remember that, if the velocity is constant, it is represented by a horizontal straight line; the position coordinate x is then a linear function of t and is represented by an oblique straight line. If the acceleration is constant and different from zero, it is represented by a horizontal straight line; v is then a linear function of t and is represented by an oblique straight line, and x is a second-degree polynomial in t and is represented by a parabola. If the acceleration is a linear function of t, the velocity and the position coordinate are equal, respectively, to second-degree and third-degree polynomials; a is then represented by an oblique straight line, v by a parabola, and x by a cubic. In general, if the acceleration is a polynomial of degree n in t, the velocity is a polynomial of degree $n + 1$, and the position coordinate is a polynomial of degree $n + 2$. These polynomials are represented by motion curves of a corresponding degree.

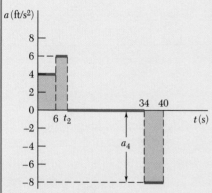

Fig. 1 Acceleration of the subway car as a function of time.

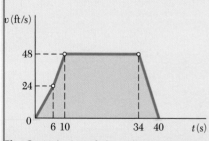

Fig. 2 Velocity of the subway car as a function of time.

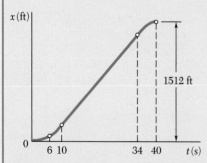

Fig. 3 Position of the subway car as a function of time.

Sample Problem 11.9

A subway car leaves station A; it gains speed at the rate of 4 ft/s^2 for 6 s and then at the rate of 6 ft/s^2 until it has reached the speed of 48 ft/s. The car maintains the same speed until it approaches station B; then the driver applies the brakes, giving the car a constant deceleration and bringing it to a stop in 6 s. The total running time from A to B is 40 s. Draw the a–t, v–t, and x–t curves, and determine the distance between stations A and B.

STRATEGY: You are given acceleration data, so first draw the graph of a versus t. You can calculate areas under the curve to determine the v–t curve and calculate areas under the v–t curve to determine the x–t curve.

MODELING and ANALYSIS: You can model the subway car as a particle without drag.

Acceleration–Time Curve. Since the acceleration is either constant or zero, the a–t curve consists of horizontal straight-line segments. Determine the values of t_2 and a_4 as

$0 < t < 6$:   Change in v = area under a–t curve
$$v_6 - 0 = (6 \text{ s})(4 \text{ ft/s}^2) = 24 \text{ ft/s}$$

$6 < t < t_2$:   Since the velocity increases from 24 to 48 ft/s,
     Change in v = area under a–t curve
$$48 \text{ ft/s} - 24 \text{ ft/s} = (t_2 - 6)(6 \text{ ft/s}^2) \qquad t_2 = 10 \text{ s}$$

$t_2 < t < 34$:   Since the velocity is constant, the acceleration is zero.

$34 < t < 40$:   Change in v = area under a–t curve
$$0 - 48 \text{ ft/s} = (6 \text{ s})a_4 \qquad a_4 = -8 \text{ ft/s}^2$$

The acceleration is negative, so the corresponding area is below the t axis; this area represents a decrease in velocity (Fig. 1).

Velocity–Time Curve. Since the acceleration is either constant or zero, the v–t curve consists of straight-line segments connecting the points determined previously (Fig. 2).

      Change in x area under v–t curve
$0 < t < 6$: $x_6 - 0 = \frac{1}{2}(6)(24) = 72$ ft
$6 < t < 10$: $x_{10} - x_6 = \frac{1}{2}(4)(24 + 48) = 144$ ft
$10 < t < 34$: $x_{34} - x_{10} = (24)(48) = 1152$ ft
$34 < t < 40$: $x_{40} - x_{34} = \frac{1}{2}(6)(48) = 144$ ft

Adding the changes in x gives you the distance from A to B:
$$d = x_{40} - 0 = 1512 \text{ ft}$$

$$d = 1512 \text{ ft} \blacktriangleleft$$

Position–Time Curve. The points determined previously should be joined by three parabolic arcs and one straight-line segment (Fig. 3). In constructing the x–t curve, keep in mind that for any value of t, the slope of the tangent to the x–t curve is equal to the value of v at that instant.

REFLECT and THINK: This problem also could have been solved using the uniform motion equations for each interval of time that has a different acceleration, but it would have been much more difficult and time consuming. For a real subway car, the acceleration does not instantaneously change from one value to another.

SOLVING PROBLEMS
ON YOUR OWN

In this section, we reviewed and developed several **graphical techniques** for the solution of problems involving rectilinear motion. These techniques can be used to solve problems directly or to complement analytical methods of solution by providing a visual description, and thus a better understanding, of the motion of a given body. We suggest that you sketch one or more motion curves for several of the problems in this section, even if these problems are not part of your homework assignment.

1. **Drawing x–t, v–t, and a–t curves and applying graphical methods.** We described the following properties in Sec. 11.3, and they should be kept in mind as you use a graphical method of solution.

    a. The slopes of the x–t and v–t curves at a time t_1 are equal to the velocity and the acceleration at time t_1, respectively.

    b. The areas under the a–t and v–t curves between the times t_1 and t_2 are equal to the change Δv in the velocity and to the change Δx in the position coordinate, respectively, during that time interval.

    c. If you know one of the motion curves, the fundamental properties we have summarized in paragraphs a and b will enable you to construct the other two curves. However, when using the properties of paragraph b, you must know the velocity and the position coordinate at time t_1 in order to determine the velocity and the position coordinate at time t_2. Thus, in Sample Prob. 11.9, knowing that the initial value of the velocity was zero allowed us to find the velocity at $t = 6$ s: $v_6 = v_0 + \Delta v = 0 + 24$ ft/s $= 24$ ft/s.

If you have studied the shear and bending-moment diagrams for a beam previously, you should recognize the analogy between the three motion curves and the three diagrams representing, respectively, the distributed load, the shear, and the bending moment in the beam. Thus, any techniques that you have learned regarding the construction of these diagrams can be applied when drawing the motion curves.

2. **Using approximate methods.** When the a–t and v–t curves are not represented by analytical functions or when they are based on experimental data, it is often necessary to use approximate methods to calculate the areas under these curves. In those cases, the given area is approximated by a series of rectangles of width Δt. The smaller the value of Δt, the smaller is the error introduced by the approximation. You can obtain the velocity and the position coordinate from

$$v = v_0 + \Sigma a_{\text{ave}}\, \Delta t \qquad x = x_0 + \Sigma v_{\text{ave}}\, \Delta t$$

where a_{ave} and v_{ave} are the heights of an acceleration rectangle and a velocity rectangle, respectively.

Problems

11.61 A particle moves in a straight line with a constant acceleration of -4 ft/s^2 for 6 s, zero acceleration for the next 4 s, and a constant acceleration of $+4$ ft/s^2 for the next 4 s. Knowing that the particle starts from the origin and that its velocity is -8 ft/s during the zero acceleration time interval, (*a*) construct the v–t and x–t curves for $0 \leq t \leq 14$ s, (*b*) determine the position and the velocity of the particle and the total distance traveled when $t = 14$ s.

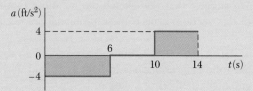

Fig. P11.61 and P11.62

11.62 A particle moves in a straight line with a constant acceleration of -4 ft/s^2 for 6 s, zero acceleration for the next 4 s, and a constant acceleration of $+4$ ft/s^2 for the next 4 s. Knowing that the particle starts from the origin with $v_0 = 16$ ft/s, (*a*) construct the v–t and x–t curves for $0 \leq t \leq 14$ s, (*b*) determine the amount of time during which the particle is further than 16 ft from the origin.

11.63 A particle moves in a straight line with the velocity shown in the figure. Knowing that $x = -540$ m at $t = 0$, (*a*) construct the a–t and x–t curves for $0 < t < 50$ s, and determine (*b*) the total distance traveled by the particle when $t = 50$ s, (*c*) the two times at which $x = 0$.

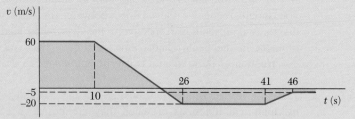

Fig. P11.63 and P11.64

11.64 A particle moves in a straight line with the velocity shown in the figure. Knowing that $x = -540$ m at $t = 0$, (*a*) construct the a–t and x–t curves for $0 < t < 50$ s, and determine (*b*) the maximum value of the position coordinate of the particle, (*c*) the values of t for which the particle is at $x = 100$ m.

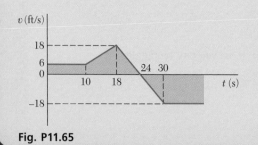

Fig. P11.65

11.65 A particle moves in a straight line with the velocity shown in the figure. Knowing that $x = -48$ ft at $t = 0$, draw the a–t and x–t curves for $0 < t < 40$ s and determine (*a*) the maximum value of the position coordinate of the particle, (*b*) the values of t for which the particle is at a distance of 108 ft from the origin.

11.66 A parachutist is in free fall at a rate of 200 km/h when he opens his parachute at an altitude of 600 m. Following a rapid and constant deceleration, he then descends at a constant rate of 50 km/h from 586 m to 30 m, where he maneuvers the parachute into the wind to further slow his descent. Knowing that the parachutist lands with a negligible downward velocity, determine (*a*) the time required for the parachutist to land after opening his parachute, (*b*) the initial deceleration.

11.67 A commuter train traveling at 40 mi/h is 3 mi from a station. The train then decelerates so that its speed is 20 mi/h when it is 0.5 mi from the station. Knowing that the train arrives at the station 7.5 min after beginning to decelerate and assuming constant decelerations, determine (*a*) the time required for the train to travel the first 2.5 mi, (*b*) the speed of the train as it arrives at the station, (*c*) the final constant deceleration of the train.

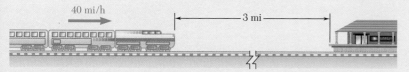

Fig. P11.67

11.68 A temperature sensor is attached to slider *AB* which moves back and forth through 60 in. The maximum velocities of the slider are 12 in./s to the right and 30 in./s to the left. When the slider is moving to the right, it accelerates and decelerates at a constant rate of 6 in./s^2; when moving to the left, the slider accelerates and decelerates at a constant rate of 20 in./s^2. Determine the time required for the slider to complete a full cycle, and construct the *v–t* and *x–t* curves of its motion.

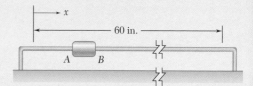

Fig. P11.68

11.69 In a water-tank test involving the launching of a small model boat, the model's initial horizontal velocity is 6 m/s and its horizontal acceleration varies linearly from -12 m/s^2 at $t = 0$ to -2 m/s^2 at $t = t_1$ and then remains equal to -2 m/s^2 until $t = 1.4$ s. Knowing that $v = 1.8$ m/s when $t = t_1$, determine (*a*) the value of t_1, (*b*) the velocity and the position of the model at $t = 1.4$ s.

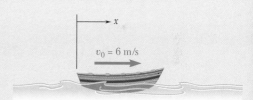

Fig. P11.69

11.70 The acceleration record shown was obtained for a small airplane traveling along a straight course. Knowing that $x = 0$ and $v = 60$ m/s when $t = 0$, determine (*a*) the velocity and position of the plane at $t = 20$ s, (*b*) its average velocity during the interval $6\text{ s} < t < 14\text{ s}$.

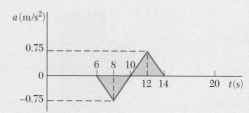

Fig. P11.70

11.71 In a 400-m race, runner *A* reaches her maximum velocity v_A in 4 s with constant acceleration and maintains that velocity until she reaches the halfway point with a split time of 25 s. Runner *B* reaches her maximum velocity v_B in 5 s with constant acceleration and maintains that velocity until she reaches the halfway point with a split time of 25.2 s. Both runners then run the second half of the race with the same constant deceleration of 0.1 m/s^2. Determine (*a*) the race times for both runners, (*b*) the position of the winner relative to the loser when the winner reaches the finish line.

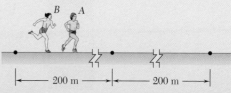

Fig. P11.71

11.72 A car and a truck are both traveling at the constant speed of 35 mi/h; the car is 40 ft behind the truck. The driver of the car wants to pass the truck, i.e., he wishes to place his car at B, 40 ft in front of the truck, and then resume the speed of 35 mi/h. The maximum acceleration of the car is 5 ft/s^2 and the maximum deceleration obtained by applying the brakes is 20 ft/s^2. What is the shortest time in which the driver of the car can complete the passing operation if he does not at any time exceed a speed of 50 mi/h? Draw the v–t curve.

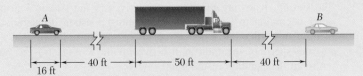

16 ft 40 ft 50 ft 40 ft

Fig. P11.72

11.73 Solve Prob. 11.72, assuming that the driver of the car does not pay any attention to the speed limit while passing and concentrates on reaching position B and resuming a speed of 35 mi/h in the shortest possible time. What is the maximum speed reached? Draw the v–t curve.

11.74 Car A is traveling on a highway at a constant speed $(v_A)_0 = 60$ mi/h and is 380 ft from the entrance of an access ramp when car B enters the acceleration lane at that point at a speed $(v_B)_0 = 15$ mi/h. Car B accelerates uniformly and enters the main traffic lane after traveling 200 ft in 5 s. It then continues to accelerate at the same rate until it reaches a speed of 60 mi/h, which it then maintains. Determine the final distance between the two cars.

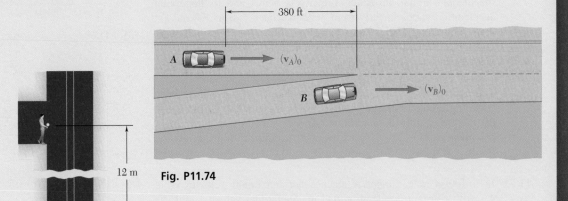

380 ft

A $(v_A)_0$

B $(v_B)_0$

Fig. P11.74

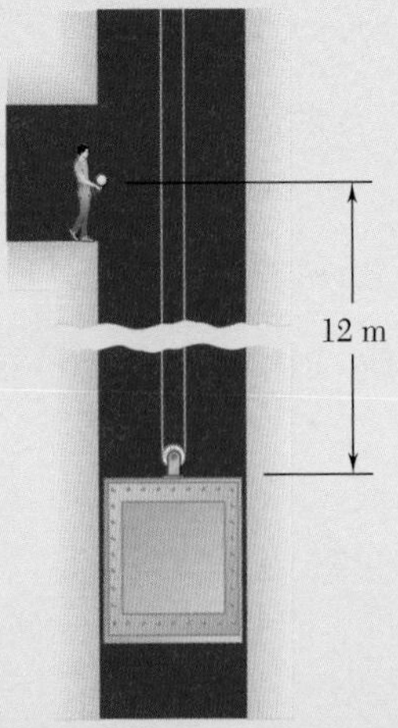

12 m

Fig. P11.75

11.75 An elevator starts from rest and moves upward, accelerating at a rate of 1.2 m/s^2 until it reaches a speed of 7.8 m/s, which it then maintains. Two seconds after the elevator begins to move, a man standing 12 m above the initial position of the top of the elevator throws a ball upward with an initial velocity of 20 m/s. Determine when the ball will hit the elevator.

11.76 Car A is traveling at 40 mi/h when it enters a 30 mi/h speed zone. The driver of car A decelerates at a rate of 16 ft/s² until reaching a speed of 30 mi/h, which she then maintains. When car B, which was initially 60 ft behind car A and traveling at a constant speed of 45 mi/h, enters the speed zone, its driver decelerates at a rate of 20 ft/s² until reaching a speed of 28 mi/h. Knowing that the driver of car B maintains a speed of 28 mi/h, determine (*a*) the closest that car B comes to car A, (*b*) the time at which car A is 70 ft in front of car B.

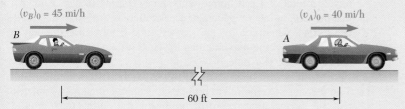

$(v_B)_0 = 45$ mi/h $(v_A)_0 = 40$ mi/h

B A

60 ft

Fig. P11.76

11.77 An accelerometer record for the motion of a given part of a mechanism is approximated by an arc of a parabola for 0.2 s and a straight line for the next 0.2 s as shown in the figure. Knowing that $v = 0$ when $t = 0$ and $x = 0.8$ ft when $t = 0.4$ s, (*a*) construct the v–t curve for $0 \leq t \leq 0.4$ s, (*b*) determine the position of the part at $t = 0.3$ s and $t = 0.2$ s.

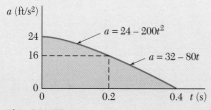

a (ft/s²)

24

16

0

$a = 24 - 200t^2$

$a = 32 - 80t$

0   0.2   0.4 t (s)

Fig. P11.77

11.78 A car is traveling at a constant speed of 54 km/h when its driver sees a child run into the road. The driver applies her brakes until the child returns to the sidewalk and then accelerates to resume her original speed of 54 km/h; the acceleration record of the car is shown in the figure. Assuming $x = 0$ when $t = 0$, determine (*a*) the time t_1 at which the velocity is again 54 km/h, (*b*) the position of the car at that time, (*c*) the average velocity of the car during the interval $1\ s \leq t \leq t_1$.

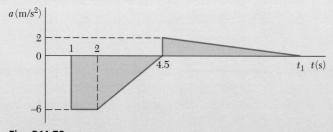

a (m/s²)

2

0

1   2

4.5

t_1 t (s)

−6

Fig. P11.78

11.79 An airport shuttle train travels between two terminals that are 1.6 mi apart. To maintain passenger comfort, the acceleration of the train is limited to ± 4 ft/s^2, and the jerk, or rate of change of acceleration, is limited to ± 0.8 ft/s^2 per second. If the shuttle has a maximum speed of 20 mi/h, determine (*a*) the shortest time for the shuttle to travel between the two terminals, (*b*) the corresponding average velocity of the shuttle.

11.80 During a manufacturing process, a conveyor belt starts from rest and travels a total of 1.2 ft before temporarily coming to rest. Knowing that the jerk, or rate of change of acceleration, is limited to ± 4.8 ft/s^2 per second, determine (*a*) the shortest time required for the belt to move 1.2 ft, (*b*) the maximum and average values of the velocity of the belt during that time.

11.81 Two seconds are required to bring the piston rod of an air cylinder to rest; the acceleration record of the piston rod during the 2 s is as shown. Determine by approximate means (*a*) the initial velocity of the piston rod, (*b*) the distance traveled by the piston rod as it is brought to rest.

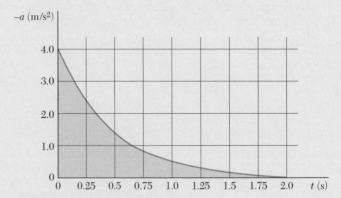

Fig. P11.81

11.82 The acceleration record shown was obtained during the speed trials of a sports car. Knowing that the car starts from rest, determine by approximate means (*a*) the velocity of the car at $t = 8$ s, (*b*) the distance the car has traveled at $t = 20$ s.

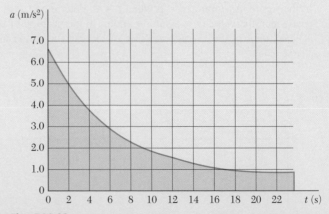

Fig. P11.82

11.83 A training airplane has a velocity of 126 ft/s when it lands on an aircraft carrier. As the arresting gear of the carrier brings the airplane to rest, the velocity and the acceleration of the airplane are recorded; the results are shown (solid curve) in the figure. Determine by approximate means (*a*) the time required for the airplane to come to rest, (*b*) the distance traveled in that time.

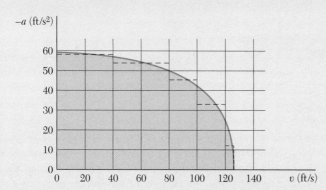

Fig. P11.83

11.84 Shown in the figure is a portion of the experimentally determined *v–x* curve for a shuttle cart. Determine by approximate means the acceleration of the cart when (*a*) $x = 10$ in., (*b*) $v = 80$ in./s.

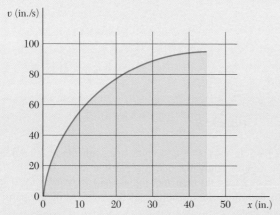

Fig. P11.84

11.85 An elevator starts from rest and rises 40 m to its maximum velocity in *T* s with the acceleration record shown in the figure. Determine (*a*) the required time *T*, (*b*) the maximum velocity, (*c*) the velocity and position of the elevator at $t = T/2$.

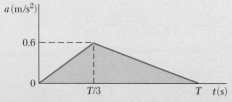

Fig. P11.85

11.86 Two road rally checkpoints A and B are located on the same highway and are 8 mi apart. The speed limits for the first 5 mi and the last 3 mi are 60 mi/h and 35 mi/h, respectively. Drivers must stop at each checkpoint, and the specified time between points A and B is 10 min 20 s. Knowing that the driver accelerates and decelerates at the same constant rate, determine the magnitude of her acceleration if she travels at the speed limit as much as possible.

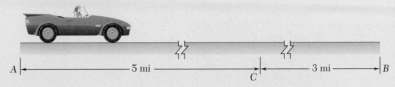

Fig. P11.86

11.87 As shown in the figure, from $t = 0$ to $t = 4$ s, the acceleration of a given particle is represented by a parabola. Knowing that $x = 0$ and $v = 8$ m/s when $t = 0$, (a) construct the v–t and x–t curves for $0 < t < 4$ s, (b) determine the position of the particle at $t = 3$ s. (*Hint:* Use table inside the front cover.)

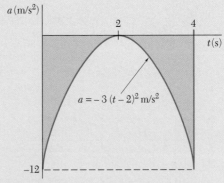

Fig. P11.87

11.88 A particle moves in a straight line with the acceleration shown in the figure. Knowing that the particle starts from the origin with $v_0 = -2$ m/s, (a) construct the v–t and x–t curves for $0 < t < 18$ s, (b) determine the position and the velocity of the particle and the total distance traveled when $t = 18$ s.

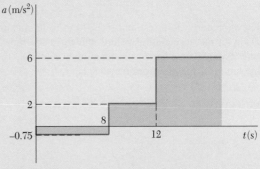

Fig. P11.88

11.4 CURVILINEAR MOTION OF PARTICLES

When a particle moves along a curve other than a straight line, we say that the particle is in **curvilinear motion**. We can use position, velocity, and acceleration to describe the motion, but now we must treat these quantities as vectors because they can have directions in two or three dimensions.

11.4A Position, Velocity, and Acceleration Vectors

To define the position P occupied by a particle in curvilinear motion at a given time t, we select a fixed reference system, such as the x, y, z axes shown in Fig. 11.12a, and draw the vector **r** joining the origin O and point P. The vector **r** is characterized by its magnitude r and its direction with respect to the reference axes, so it completely defines the position of the particle with respect to those axes. We refer to vector **r** as the **position vector** of the particle at time t.

Consider now the vector **r**′ defining the position P' occupied by the same particle at a later time $t + \Delta t$. The vector Δ**r** joining P and P' represents the change in the position vector during the time interval Δt and is called the **displacement vector**. We can check this directly from Fig. 11.12a, where we obtain the vector **r**′ by adding the vectors **r** and Δ**r** according to the triangle rule. Note that Δ**r** represents a change in *direction* as well as a change in *magnitude* of the position vector **r**.

We define the **average velocity** of the particle over the time interval Δt as the quotient of Δ**r** and Δt. Since Δ**r** is a vector and Δt is a scalar, the quotient Δ**r**/Δt is a vector attached at P with the same direction as Δ**r** and a magnitude equal to the magnitude of Δ**r** divided by Δt (Fig. 11.12b).

We obtain the **instantaneous velocity** of the particle at time t by taking the limit as the time interval Δt approaches zero. The instantaneous velocity is thus represented by the vector

$$\mathbf{v} = \lim_{\Delta t \to 0} \frac{\Delta \mathbf{r}}{\Delta t} \qquad (11.13)$$

As Δt and Δ**r** become shorter, the points P and P' get closer together. Thus, the vector **v** obtained in the limit must be tangent to the path of the particle (Fig. 11.12c).

Because the position vector **r** depends upon the time t, we can refer to it as a **vector function** of the scalar variable t and denote it by **r**(t). Extending the concept of the derivative of a scalar function introduced in elementary calculus, we refer to the limit of the quotient Δ**r**/Δt as the **derivative** of the vector function **r**(t). We have

Velocity vector $$\mathbf{v} = \frac{d\mathbf{r}}{dt} \qquad (11.14)$$

The magnitude v of the vector **v** is called the **speed** of the particle. We can obtain the speed by substituting the magnitude of this vector,

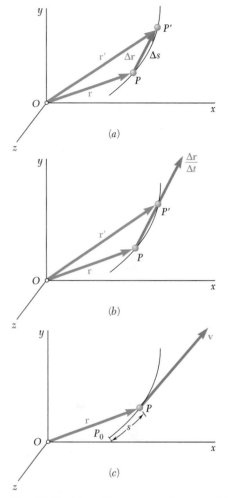

Fig. 11.12 (a) Position vectors for a particle moving along a curve from P to P'; (b) the average velocity vector is the quotient of the change in position to the elapsed time interval; (c) the instantaneous velocity vector is tangent to the particle's path.

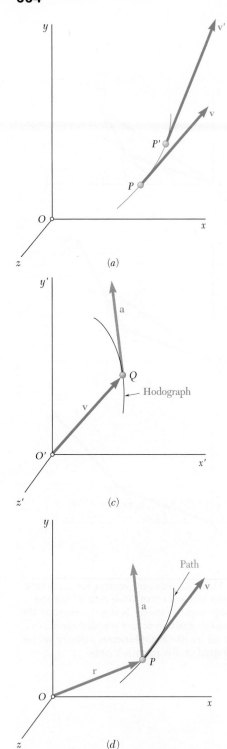

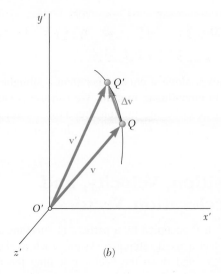

which is represented by the straight-line segment PP', for the vector $\Delta\mathbf{r}$ in formula (11.13). However, the length of segment PP' approaches the length Δs of arc PP' as Δt decreases (Fig. 11.12a). Therefore, we can write

$$v = \lim_{\Delta t \to 0} \frac{PP'}{\Delta t} = \lim_{\Delta t \to 0} \frac{\Delta s}{\Delta t} \qquad \boxed{v = \frac{ds}{dt}} \qquad \textbf{(11.15)}$$

Thus, we obtain the speed v by finding the length s of the arc described by the particle and differentiating it with respect to t.

Now let's consider the velocity $\mathbf{v}$ of the particle at time t and its velocity $\mathbf{v}'$ at a later time $t + \Delta t$ (Fig. 11.13a). Let us draw both vectors $\mathbf{v}$ and $\mathbf{v}'$ from the same origin O' (Fig. 11.13b). The vector $\Delta\mathbf{v}$ joining Q and Q' represents the change in the velocity of the particle during the time interval Δt, since we can obtain the vector $\mathbf{v}'$ by adding the vectors $\mathbf{v}$ and $\Delta\mathbf{v}$. Again, note that $\Delta\mathbf{v}$ represents a change in the *direction* of the velocity as well as a change in *speed*. We define the **average acceleration** of the particle over the time interval Δt as the quotient of $\Delta\mathbf{v}$ and Δt. Since $\Delta\mathbf{v}$ is a vector and Δt is a scalar, the quotient $\Delta\mathbf{v}/\Delta t$ is a vector in the same direction as $\Delta\mathbf{v}$.

We obtain the **instantaneous acceleration** of the particle at time t by choosing increasingly smaller values for Δt and $\Delta\mathbf{v}$. The instantaneous acceleration is thus represented by the vector

$$\mathbf{a} = \lim_{\Delta t \to 0} \frac{\Delta\mathbf{v}}{\Delta t} \qquad \textbf{(11.16)}$$

Noting that the velocity $\mathbf{v}$ is a vector function $\mathbf{v}(t)$ of the time t, we can refer to the limit of the quotient $\Delta\mathbf{v}/\Delta t$ as the derivative of $\mathbf{v}$ with respect to t. We have

Acceleration vector $\qquad\qquad \mathbf{a} = \dfrac{d\mathbf{v}}{dt} \qquad\qquad$ **(11.17)**

Fig. 11.13 (a) Velocities v and v' of a particle at two different times; (b) the vector change in the particle's velocity during the time interval; (c) the instantaneous acceleration vector is tangent to the hodograph; (d) in general, the acceleration vector is not tangent to the particle's path.

Observe that the acceleration **a** is tangent to the curve described by the tip Q of the vector **v** when we draw **v** from a fixed origin O' (Fig. 11.13c). However, in general, the acceleration is *not* tangent to the path of the particle (Fig. 11.13d). The curve described by the tip of **v** and shown in Fig. 11.13c is called the *hodograph* of the motion.

11.4B Derivatives of Vector Functions

We have just seen that we can represent the velocity **v** of a particle in curvilinear motion by the derivative of the vector function **r**(t) characterizing the position of the particle. Similarly, we can represent the acceleration **a** of the particle by the derivative of the vector function **v**(t). Here we give a formal definition of the derivative of a vector function and establish a few rules governing the differentiation of sums and products of vector functions.

Let **P**(u) be a vector function of the scalar variable u. By that, we mean that the scalar u completely defines the magnitude and direction of the vector **P**. If the vector **P** is drawn from a fixed origin O and the scalar u is allowed to vary, the tip of **P** describes a given curve in space. Consider the vectors **P** corresponding, respectively, to the values u and $u + \Delta u$ of the scalar variable (Fig. 11.14a). Let Δ**P** be the vector joining the tips of the two given vectors. Then we have

$$\Delta\mathbf{P} = \mathbf{P}(u + \Delta u) - \mathbf{P}(u)$$

Dividing through by Δu and letting Δu approach zero, we define the derivative of the vector function **P**(u) as

$$\frac{d\mathbf{P}}{du} = \lim_{\Delta u \to 0} \frac{\Delta\mathbf{P}}{\Delta u} = \lim_{\Delta u \to 0} \frac{\mathbf{P}(u + \Delta u) - \mathbf{P}(u)}{\Delta u} \quad \textbf{(11.18)}$$

As Δu approaches zero, the line of action of Δ**P** becomes tangent to the curve of Fig. 11.14a. Thus, the derivative $d\mathbf{P}/du$ of the vector function **P**(u) *is tangent to the curve described by the tip of* **P**(u) (Fig. 11.14b).

The standard rules for the differentiation of the sums and products of scalar functions extend to vector functions. Consider first the **sum of two vector functions** **P**(u) and **Q**(u) of the same scalar variable u. According to the definition given in Eq. (11.18), the derivative of the vector **P** + **Q** is

$$\frac{d(\mathbf{P} + \mathbf{Q})}{du} = \lim_{\Delta u \to 0} \frac{\Delta(\mathbf{P} + \mathbf{Q})}{\Delta u} = \lim_{\Delta u \to 0} \left(\frac{\Delta\mathbf{P}}{\Delta u} + \frac{\Delta\mathbf{Q}}{\Delta u} \right)$$

or since the limit of a sum is equal to the sum of the limits of its terms,

$$\frac{d(\mathbf{P} + \mathbf{Q})}{du} = \lim_{\Delta u \to 0} \frac{\Delta\mathbf{P}}{\Delta u} + \lim_{\Delta u \to 0} \frac{\Delta\mathbf{Q}}{\Delta u}$$

$$\frac{d(\mathbf{P} + \mathbf{Q})}{du} = \frac{d\mathbf{P}}{du} + \frac{d\mathbf{Q}}{du} \quad \textbf{(11.19)}$$

That is, the derivative of a sum of vector functions equals the sum of the derivative of each function separately.

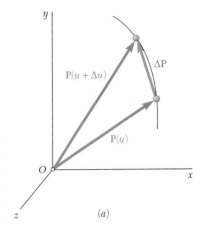

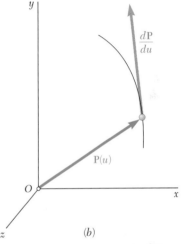

Fig. 11.14 (a) The change in vector function for a particle moving along a curvilinear path; (b) the derivative of the vector function is tangent to the path described by the tip of the function.

We now consider the **product of a scalar function** $f(u)$ **and a vector function** $\mathbf{P}(u)$ of the same scalar variable u. The derivative of the vector $f\mathbf{P}$ is

$$\frac{d(f\mathbf{P})}{du} = \lim_{\Delta u \to 0} \frac{(f + \Delta f)(\mathbf{P} + \Delta \mathbf{P}) - f\mathbf{P}}{\Delta u} = \lim_{\Delta u \to 0} \left(\frac{\Delta f}{\Delta u}\mathbf{P} + f\frac{\Delta \mathbf{P}}{\Delta u} \right)$$

or recalling the properties of the limits of sums and products,

$$\frac{d(f\mathbf{P})}{du} = \frac{df}{du}\mathbf{P} + f\frac{d\mathbf{P}}{du} \tag{11.20}$$

In a similar way, we can obtain the derivatives of the **scalar product** and the **vector product** of two vector functions $\mathbf{P}(u)$ and $\mathbf{Q}(u)$. Thus,

$$\frac{d(\mathbf{P} \cdot \mathbf{Q})}{du} = \frac{d\mathbf{P}}{du} \cdot \mathbf{Q} + \mathbf{P} \cdot \frac{d\mathbf{Q}}{du} \tag{11.21}$$

$$\frac{d(\mathbf{P} \times \mathbf{Q})}{du} = \frac{d\mathbf{P}}{du} \times \mathbf{Q} + \mathbf{P} \times \frac{d\mathbf{Q}}{du} \tag{11.22}^\dagger$$

We can use the properties just established to determine the **rectangular components of the derivative of a vector function** $\mathbf{P}(u)$. Resolving $\mathbf{P}$ into components along fixed rectangular axes x, y, and z, we have

$$\mathbf{P} = P_x\mathbf{i} + P_y\mathbf{j} + P_z\mathbf{k} \tag{11.23}$$

where P_x, P_y, and P_z are the rectangular scalar components of the vector $\mathbf{P}$, and $\mathbf{i}$, $\mathbf{j}$, and $\mathbf{k}$ are the unit vectors corresponding, respectively, to the x, y, and z axes (Sec. 2.12 or Appendix A). From Eq. (11.19), the derivative of $\mathbf{P}$ is equal to the sum of the derivatives of the terms in the right-hand side. Since each of these terms is the product of a scalar and a vector function, we should use Eq. (11.20). However, the unit vectors $\mathbf{i}$, $\mathbf{j}$, and $\mathbf{k}$ have a constant magnitude (equal to 1) and fixed directions. Their derivatives are therefore zero, and we obtain

$$\frac{d\mathbf{P}}{du} = \frac{dP_x}{du}\mathbf{i} + \frac{dP_y}{du}\mathbf{j} + \frac{dP_z}{du}\mathbf{k} \tag{11.24}$$

Note that the coefficients of the unit vectors are, by definition, the scalar components of the vector $d\mathbf{P}/du$. We conclude that we can obtain the rectangular scalar components of the derivative $d\mathbf{P}/du$ of the vector function $\mathbf{P}(u)$ by differentiating the corresponding scalar components of $\mathbf{P}$.

Rate of Change of a Vector. When the vector $\mathbf{P}$ is a function of the time t, its derivative $d\mathbf{P}/dt$ represents the **rate of change** of $\mathbf{P}$ with respect to the frame $Oxyz$. Resolving $\mathbf{P}$ into rectangular components and using Eq. (11.24), we have

$$\frac{d\mathbf{P}}{dt} = \frac{dP_x}{dt}\mathbf{i} + \frac{dP_y}{dt}\mathbf{j} + \frac{dP_z}{dt}\mathbf{k}$$

†Since the vector product is not commutative (see Sec. 3.4), the order of the factors in Eq. (11.22) must be maintained.

Alternatively, using dots to indicate differentiation with respect to t gives

$$\dot{\mathbf{P}} = \dot{P}_x\mathbf{i} + \dot{P}_y\mathbf{j} + \dot{P}_z\mathbf{k} \qquad (11.24')$$

As you will see in Sec. 15.5, the rate of change of a vector as observed from a *moving frame of reference* is, in general, different from its rate of change as observed from a fixed frame of reference. However, if the moving frame $O'x'y'z'$ is in *translation*, i.e., if its axes remain parallel to the corresponding axes of the fixed frame $Oxyz$ (Fig. 11.15), we can use the same unit vectors $\mathbf{i}$, $\mathbf{j}$, and $\mathbf{k}$ in both frames, and at any given instant, the vector $\mathbf{P}$ has the same components P_x, P_y, and P_z in both frames. It follows from Eq. (11.24') that the rate of change $\dot{\mathbf{P}}$ is the same with respect to the frames $Oxyz$ and $O'x'y'z'$. Therefore,

> **The rate of change of a vector is the same with respect to a fixed frame and with respect to a frame in translation.**

This property will greatly simplify our work, since we will be concerned mainly with frames in translation.

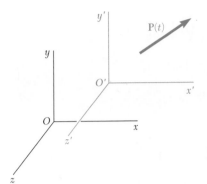

Fig. 11.15 The rate of change of a vector is the same with respect to a fixed frame of reference and with respect to a frame in translation.

11.4C Rectangular Components of Velocity and Acceleration

Suppose the position of a particle P is defined at any instant by its rectangular coordinates x, y, and z. In this case, it is often convenient to resolve the velocity $\mathbf{v}$ and the acceleration $\mathbf{a}$ of the particle into rectangular components (Fig. 11.16).

To resolve the position vector $\mathbf{r}$ of the particle into rectangular components, we write

$$\mathbf{r} = x\mathbf{i} + y\mathbf{j} + z\mathbf{k} \qquad (11.25)$$

Here the coordinates x, y, and z are functions of t. Differentiating twice, we obtain

Velocity and acceleration in rectangular components

$$\mathbf{v} = \frac{d\mathbf{r}}{dt} = \dot{x}\mathbf{i} + \dot{y}\mathbf{j} + \dot{z}\mathbf{k} \qquad (11.26)$$

$$\mathbf{a} = \frac{d\mathbf{v}}{dt} = \ddot{x}\mathbf{i} + \ddot{y}\mathbf{j} + \ddot{z}\mathbf{k} \qquad (11.27)$$

where $\dot{x}$, $\dot{y}$, and $\dot{z}$ and $\ddot{x}$, $\ddot{y}$, and $\ddot{z}$ represent, respectively, the first and second derivatives of x, y, and z with respect to t. It follows from Eqs. (11.26) and (11.27) that the scalar components of the velocity and acceleration are

$$v_x = \dot{x} \qquad v_y = \dot{y} \qquad v_z = \dot{z} \qquad (11.28)$$
$$a_x = \ddot{x} \qquad a_y = \ddot{y} \qquad a_z = \ddot{z} \qquad (11.29)$$

A positive value for v_x indicates that the vector component $\mathbf{v}_x$ is directed to the right, and a negative value indicates that it is directed to the left. The sense of each of the other vector components is determined in a similar way from the sign of the corresponding scalar component. If desired, we can obtain the magnitudes and directions of the velocity and acceleration from their scalar components using the methods of Secs. 2.2A and 2.4A (or Appendix A).

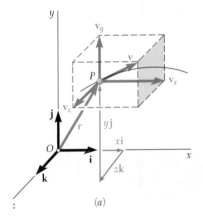

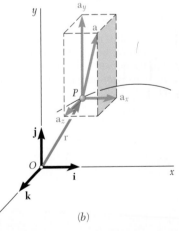

Fig. 11.16 (a) Rectangular components of position and velocity for a particle P; (b) rectangular components of acceleration for particle P.

Photo 11.3 The motion of this snowboarder in the air is a parabola, assuming we can neglect air resistance.

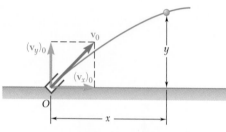

(a) Motion of a projectile

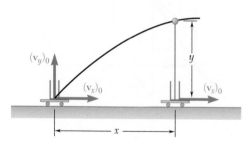

(b) Equivalent rectilinear motions

Fig. 11.17 The motion of a projectile (a) consists of uniform horizontal motion and uniformly accelerated vertical motion and (b) is equivalent to two independent rectilinear motions.

The use of rectangular components to describe the position, velocity, and acceleration of a particle is particularly effective when the component a_x of the acceleration depends only upon t, x, and/or v_x, and similarly when a_y depends only upon t, y, and/or v_y, and when a_z depends upon t, z, and/or v_z. In this case, we can integrate Equations (11.28) and (11.29) independently. In other words, the motion of the particle in the x direction, its motion in the y direction, and its motion in the z direction can be treated separately.

In the case of the **motion of a projectile**, we can show (see Sec. 12.1D) that the components of the acceleration are

$$a_x = \ddot{x} = 0 \qquad a_y = \ddot{y} = -g \qquad a_z = \ddot{z} = 0$$

if the resistance of the air is neglected. Denoting the coordinates of a gun by x_0, y_0, and z_0 and the components of the initial velocity $\mathbf{v}_0$ of the projectile by $(v_x)_0$, $(v_y)_0$, and $(v_z)_0$, we can integrate twice in t and obtain

$$v_x = \dot{x} = (v_x)_0 \qquad v_y = \dot{y} = (v_y)_0 - gt \qquad v_z = \dot{z} = (v_z)_0$$
$$x = x_0 + (v_x)_0 t \qquad y = y_0 + (v_y)_0 t - \tfrac{1}{2}gt^2 \qquad z = z_0 + (v_z)_0 t$$

If the projectile is fired in the xy plane from the origin O, we have $x_0 = y_0 = z_0 = 0$ and $(v_z)_0 = 0$, so the equations of motion reduce to

$$v_x = (v_x)_0 \qquad v_y = (v_y)_0 - gt \qquad v_z = 0$$
$$x = (v_x)_0 t \qquad y = (v_y)_0 t - \tfrac{1}{2}gt^2 \qquad z = 0$$

These equations show that the projectile remains in the xy plane, that its motion in the horizontal direction is uniform, and that its motion in the vertical direction is uniformly accelerated. Thus, we can replace the motion of a projectile by two independent rectilinear motions, which are easily visualized if we assume that the projectile is fired vertically with an initial velocity $(\mathbf{v}_y)_0$ from a platform moving with a constant horizontal velocity $(\mathbf{v}_x)_0$ (Fig. 11.17). The coordinate x of the projectile is equal at any instant to the distance traveled by the platform, and we can compute its coordinate y as if the projectile were moving along a vertical line. Additionally, because the $(\mathbf{v}_x)_0$ values are the same, the projectile will land on the platform regardless of the value of $(\mathbf{v}_y)_0$.

Note that the equations defining the coordinates x and y of a projectile at any instant are the parametric equations of a parabola. Thus, the trajectory of a projectile is *parabolic*. This result, however, ceases to be valid if we take into account the resistance of the air or the variation with altitude of the acceleration due to gravity.

11.4D   Motion Relative to a Frame in Translation

We have just seen how to describe the motion of a particle by using a single frame of reference. In most cases, this frame was attached to the earth and was considered to be fixed. Now we want to analyze situations in which it is convenient to use several frames of reference simultaneously. If one of the frames is attached to the earth, it is called a **fixed frame of reference**, and the other frames are referred to as **moving frames of reference**. You should recognize, however, that the selection of a fixed frame of reference is purely arbitrary. Any frame can be designated as "fixed"; all other frames not rigidly attached to this frame are then described as "moving."

Consider two particles A and B moving in space (Fig. 11.18). The vectors $\mathbf{r}_A$ and $\mathbf{r}_B$ define their positions at any given instant with respect to the fixed frame of reference $Oxyz$. Consider now a system of axes x', y', and z' centered at A and parallel to the x, y, and z axes. Suppose that, while the origin of these axes moves, their orientation remains the same; then the frame of reference $Ax'y'z'$ is in *translation* with respect to $Oxyz$. The vector $\mathbf{r}_{B/A}$ joining A and B defines **the position of B relative to the moving frame** $Ax'y'z'$ (or for short, **the position of B relative to A**).

Figure 11.18 shows that the position vector $\mathbf{r}_B$ of particle B is the sum of the position vector $\mathbf{r}_A$ of particle A and of the position vector $\mathbf{r}_{B/A}$ of B relative to A; that is,

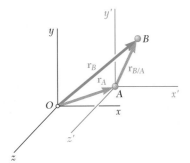

Fig. 11.18 The vector $\mathbf{r}_{B/A}$ defines the position of B with respect to moving frame A.

Relative position

$$\mathbf{r}_B = \mathbf{r}_A + \mathbf{r}_{B/A} \tag{11.30}$$

Differentiating Eq. (11.30) with respect to t within the fixed frame of reference, and using dots to indicate time derivatives, we have

$$\dot{\mathbf{r}}_B = \dot{\mathbf{r}}_A + \dot{\mathbf{r}}_{B/A} \tag{11.31}$$

The derivatives $\dot{\mathbf{r}}_A$ and $\dot{\mathbf{r}}_B$ represent, respectively, the velocities $\mathbf{v}_A$ and $\mathbf{v}_B$ of the particles A and B. Since $Ax'y'z'$ is in translation, the derivative $\dot{\mathbf{r}}_{B/A}$ represents the rate of change of $\mathbf{r}_{B/A}$ with respect to the frame $Ax'y'z'$ as well as with respect to the fixed frame (Sec. 11.4B). This derivative, therefore, defines **the velocity $\mathbf{v}_{B/A}$ of B relative to the frame** $Ax'y'z'$ (or for short, **the velocity $\mathbf{v}_{B/A}$ of B relative to A**). We have

Photo 11.4 The pilot of a helicopter landing on a moving carrier must take into account the relative motion of the ship.

Relative velocity

$$\mathbf{v}_B = \mathbf{v}_A + \mathbf{v}_{B/A} \tag{11.32}$$

Differentiating Eq. (11.32) with respect to t, and using the derivative $\dot{\mathbf{v}}_{B/A}$ to define **the acceleration $\mathbf{a}_{B/A}$ of B relative to the frame** $Ax'y'z'$ (or for short, **the acceleration $\mathbf{a}_{B/A}$ of B relative to A**), we obtain

Relative acceleration

$$\mathbf{a}_B = \mathbf{a}_A + \mathbf{a}_{B/A} \tag{11.33}$$

We refer to the motion of B with respect to the fixed frame $Oxyz$ as the **absolute motion of** B. The equations derived in this section show that **we can obtain the absolute motion of B by combining the motion of A and the relative motion of B with respect to the moving frame attached to A.** Equation (11.32), for example, expresses that the absolute velocity $\mathbf{v}_B$ of particle B can be obtained by vectorially adding the velocity of A and the velocity of B relative to the frame $Ax'y'z'$. Equation (11.33) expresses a similar property in terms of the accelerations. (Note that the product of the subscripts A and B/A used in the right-hand sides of Eqs. (11.30) through (11.33) is equal to the subscript B used in their left-hand sides.) Keep in mind, however, that the frame $Ax'y'z'$ is in *translation;* that is, while it moves with A, it maintains the same orientation. As you will see later (Sec. 15.7), you must use different relations in the case of a rotating frame of reference.

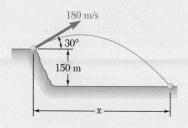

Fig. 1 Acceleration and initial velocity of the projectile in the y-direction.

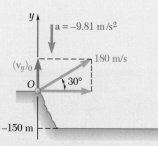

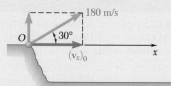

Fig. 2 Initial velocity of the projectile in the x-direction.

Sample Problem 11.10

A projectile is fired from the edge of a 150-m cliff with an initial velocity of 180 m/s at an angle of 30° with the horizontal. Neglecting air resistance, find (a) the horizontal distance from the gun to the point where the projectile strikes the ground, (b) the greatest elevation above the ground reached by the projectile.

STRATEGY: This is a projectile motion problem, so you can consider the vertical and horizontal motions separately. First determine the equations governing each direction, and then use them to find the distances.

MODELING and ANALYSIS: Model the projectile as a particle and neglect the effects of air resistance. The vertical motion has a constant acceleration. Choosing the positive sense of the y axis upward and placing the origin O at the gun (Fig. 1), you have

$$(v_y)_0 = (180 \text{ m/s}) \sin 30° = +90 \text{ m/s}$$
$$a = -9.81 \text{ m/s}^2$$

Substitute these values into the equations for motion with constant acceleration. Thus,

$$v_y = (v_y)_0 + at \qquad v_y = 90 - 9.81t \qquad (1)$$
$$y = (v_y)_0 t + \tfrac{1}{2}at^2 \qquad y = 90t - 4.90t^2 \qquad (2)$$
$$v_y^2 = (v_y)_0^2 + 2ay \qquad v_y^2 = 8100 - 19.62y \qquad (3)$$

The horizontal motion has zero acceleration. Choose the positive sense of the x axis to the right (Fig. 2), which gives you

$$(v_x)_0 = (180 \text{ m/s}) \cos 30° = +155.9 \text{ m/s}$$

Substituting into the equation for constant acceleration, you obtain

$$x = (v_x)_0 t \qquad x = 155.9t \qquad (4)$$

a. Horizontal Distance. When the projectile strikes the ground,

$$y = -150 \text{ m}$$

Substituting this value into Eq. (2) for the vertical motion, you have

$$-150 = 90t - 4.90t^2 \qquad t^2 - 18.37t - 30.6 = 0 \qquad t = 19.91 \text{ s}$$

Substituting $t = 19.91$ s into Eq. (4) for the horizontal motion, you obtain

$$x = 155.9(19.91) \qquad x = 3100 \text{ m} \blacktriangleleft$$

b. Greatest Elevation. When the projectile reaches its greatest elevation, $v_y = 0$; substituting this value into Eq. (3) for the vertical motion, you have

$$0 = 8100 - 19.62y \qquad y = 413 \text{ m}$$
$$\text{Greatest elevation above ground} = 150 \text{ m} + 413 \text{ m} = 563 \text{ m} \blacktriangleleft$$

REFLECT and THINK: Because there is no air resistance, you can treat the vertical and horizontal motions separately and can immediately write down the algebraic equations of motion. If you did want to include air resistance, you must know the acceleration as a function of the speed (you will see how to derive this in Chapter 12), and then you need to use the basic kinematic relationships, separate variables, and integrate.

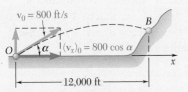

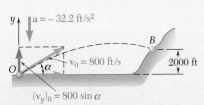

Fig. 1 Initial velocity of the projectile in the x-direction.

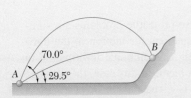

Fig. 2 Acceleration and initial velocity of the projectile in the y-direction.

Sample Problem 11.11

A projectile is fired with an initial velocity of 800 ft/s at a target B located 2000 ft above the gun A and at a horizontal distance of 12,000 ft. Neglecting air resistance, determine the value of the firing angle α needed to hit the target.

STRATEGY: This is a projectile motion problem, so you can consider the vertical and horizontal motions separately. First determine the equations governing the motion in each direction, and then use them to find the firing angle.

MODELING and ANALYSIS:

Horizontal Motion. Place the origin of the coordinate axes at the gun (Fig. 1). Then

$$(v_x)_0 = 800 \cos \alpha$$

Substituting into the equation of uniform horizontal motion, you obtain

$$x = (v_x)_0 t \qquad x = (800 \cos \alpha)t$$

Obtain the time required for the projectile to move through a horizontal distance of 12,000 ft by setting x equal to 12,000 ft.

$$12,000 = (800 \cos \alpha)t$$
$$t = \frac{12,000}{800 \cos \alpha} = \frac{15}{\cos \alpha}$$

Vertical Motion. Again, place the origin at the gun (Fig. 2).

$$(v_y)_0 = 800 \sin \alpha \qquad a = -32.2 \text{ ft/s}^2$$

Substituting into the equation for constant acceleration in the vertical direction, you obtain

$$y = (v_y)_0 t + \tfrac{1}{2}at^2 \qquad y = (800 \sin \alpha)t - 16.1t^2$$

Projectile Hits Target. When $x = 12,000$ ft, you want $y = 2000$ ft. Substituting for y and setting t equal to the value found previously, you have

$$2000 = 800 \sin \alpha \frac{15}{\cos \alpha} - 16.1\left(\frac{15}{\cos \alpha}\right)^2 \qquad (1)$$

Since $1/\cos^2 \alpha = \sec^2 \alpha = 1 + \tan^2 \alpha$, you have

$$2000 = 800(15) \tan \alpha - 16.1(15^2)(1 + \tan^2 \alpha)$$
$$3622 \tan^2 \alpha - 12,000 \tan \alpha + 5622 = 0$$

Solving this quadratic equation for $\tan \alpha$ gives you

$$\tan \alpha = 0.565 \qquad \text{and} \qquad \tan \alpha = 2.75$$
$$\alpha = 29.5° \qquad \text{and} \qquad \alpha = 70.0° \blacktriangleleft$$

The target will be hit if either of these two firing angles is used (Fig. 3).

REFLECT and THINK: It is a well-known characteristic of projectile motion that you can hit the same target by using either of two firing angles. We used trigonometry to write the equation in terms of $\tan \alpha$, but most calculators or computer programs like Maple, Matlab, or Mathematica also can be used to solve (1) for α. You must be careful when using these tools, however, to make sure that you find both angles.

Fig. 3 Firing angles that will hit target B.

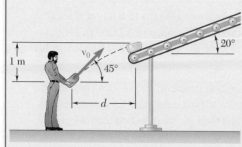

Sample Problem 11.12

A conveyor belt at an angle of 20° with the horizontal is used to transfer small packages to other parts of an industrial plant. A worker tosses a package with an initial velocity $\mathbf{v}_0$ at an angle of 45° so that its velocity is parallel to the belt as it lands 1 m above the release point. Determine (a) the magnitude of v_0, (b) the horizontal distance d.

STRATEGY: This is a projectile motion problem, so you can consider the vertical and the horizontal motions separately. First determine the equations governing the motion in each direction, then use them to determine the unknown quantities.

MODELING and ANALYSIS:

Horizontal Motion. Placing the axes of your origin at the location where the package leaves the workers hands (Fig. 1), you can write

Horizontal: $v_x = v_0 \cos 45°$ and $x = (v_0 \cos 45°)t$

Vertical: $v_y = v_0 \sin 45° - gt$ and $y = (v_0 \sin 45°)t - \dfrac{1}{2}gt^2$

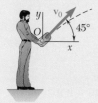

Fig. 1 Initial velocity of the package.

Landing on the Belt. The problem statement indicates that when the package lands on the belt, its velocity vector will be in the same direction as the belt is moving. If this happens when $t = t_1$, you can write

$$\frac{v_y}{v_x} = \tan 20° = \frac{v_0 \sin 45° - gt_1}{v_0 \cos 45°} = 1 - \frac{gt_1}{v_0 \cos 45°} \quad (1)$$

This equation has two unknown quantities: t_1 and v_0. Therefore, you need more equations. Substituting $t = t_1$ into the remaining projectile motion equations gives

$$d = (v_0 \cos 45°)t \quad (2)$$

$$1 \text{ m} = (v_0 \sin 45°)t_1 - \frac{1}{2}gt_1^2 \quad (3)$$

You now have three equations (1), (2), and (3) and three unknowns t_1, v_0, and d. Using $g = 9.81 \text{ m/s}^2$ and solving these three equations give $t_1 = 0.3083$ s and

$$v_0 = 6.73 \text{ m/s} \blacktriangleleft$$
$$d = 1.466 \text{ m} \blacktriangleleft$$

REFLECT and THINK: All of these projectile problems are similar. You write down the governing equations for motion in the horizontal and vertical directions and then use additional information in the problem statement to solve the problem. In this case, the distance is just less than 1.5 meters, which is a reasonable distance for a worker to toss a package.

Sample Problem 11.13

Airplane B, which is travelling at a constant 560 km/h, is pursuing airplane A, which is travelling northeast at a constant 800 km/hr. At time $t = 0$, airplane A is 640 km east of airplane B. Determine (*a*) the direction of the course airplane B should follow (measured from the east) to intercept plane A, (*b*) the rate at which the distance between the airplanes is decreasing, (*c*) how long it takes for airplane B to catch airplane A.

STRATEGY: To find when B intercepts A, you just need to find out when the two planes are at the same location. The rate at which the distance is decreasing is the magnitude of $v_{B/A}$, so you can use the relative velocity equation.

MODELING and ANALYSIS: Choose x to be east, y to be north, and place the origin of your coordinate system at B (Fig. 1).

Positions of the Planes: You know that each plane has a constant speed, so you can write a position vector for each plane. Thus,

$$\mathbf{r}_A = [(v_A \cos 45°)\,t + 640 \text{ km}]\mathbf{i} + [(v_A \sin 45°)\,t]\mathbf{j} \qquad (1)$$
$$\mathbf{r}_B = [(v_B \cos \theta)\,t]\mathbf{i} + [(v_B \sin \theta)\,t]\mathbf{j} \qquad (2)$$

a. Direction of B. Plane B will catch up when they are at the same location, that is, $\mathbf{r}_A = \mathbf{r}_B$. You can equate components in the $\mathbf{j}$ direction to find

$$v_A \sin 45° t_1 = v_B \sin \theta\, t_1$$

After you substitute in values,

$$\sin \theta = \frac{(v_A \sin 45°)t_1}{v_B t_1} = \frac{(560 \text{ km/hr})\sin 45°}{800 \text{ km/hr}} = 0.4950$$

$$\theta = \sin^{-1} 0.4950 = 29.67° \qquad\qquad \theta = 29.7° \quad \blacktriangleleft$$

b. Rate. The rate at which the distance is decreasing is the magnitude of $\mathbf{v}_{B/A}$, so

$$\mathbf{v}_{B/A} = \mathbf{v}_B - \mathbf{v}_A = (v_B \cos \theta\, \mathbf{i} + v_B \sin \theta\, \mathbf{j}) - (v_A \cos 45°\, \mathbf{i} + v_A \sin 45°\, \mathbf{j})$$
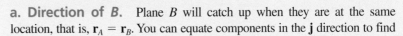
$$= [(800 \text{ km/h})\cos 29.668° - (560 \text{ km/h})\cos 45°]\mathbf{i}$$
$$+ [(800 \text{ km/h})\sin 29.668° - (560 \text{ km/h})\sin 45°]\mathbf{j}$$
$$= 299.15 \text{ km/h } \mathbf{i} \qquad\qquad |\mathbf{v}_{B/A}| = 299 \text{ km/h} \quad \blacktriangleleft$$

c. Time for B to catch up with A. To find the time, you equate the $\mathbf{i}$ components of each position vector, giving

$$(v_A \cos 45°)\,t_1 + 640 \text{ km} = (v_B \cos \theta)\,t_1$$

Solve this for t_1. Thus,

$$t_1 = \frac{640 \text{ km}}{v_B \cos \theta - v_A \cos 45°}$$

$$= \frac{640 \text{ km}}{(800 \text{ km/h})\cos 29.67° - (560 \text{ km/h})\cos 45°} = 2.139 \text{ h}$$

$$t_1 = 2.14 \text{ h} \quad \blacktriangleleft$$

REFLECT and THINK: The relative velocity is only in the horizontal (eastern) direction. This makes sense, because the vertical (northern) components have to be equal in order for the two planes to intersect.

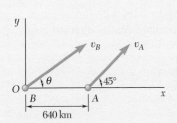

Fig. 1 Initial velocity of airplanes A and B.

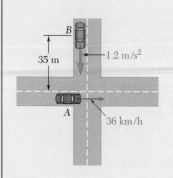

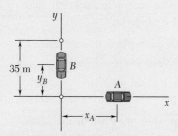

Fig. 1 Initial positions of car A and B.

Sample Problem 11.14

Automobile A is traveling east at the constant speed of 36 km/h. As automobile A crosses the intersection shown, automobile B starts from rest 35 m north of the intersection and moves south with a constant acceleration of 1.2 m/s^2. Determine the position, velocity, and acceleration of B relative to A 5 s after A crosses the intersection.

STRATEGY: This is a relative motion problem. Determine the motion of each vehicle independently, and then use the definition of relative motion to determine the desired quantities.

MODELING and ANALYSIS:

Motion of Automobile A. Choose x and y axes with the origin at the intersection of the two streets and with positive senses directed east and north, respectively. First express the speed in m/s, as

$$v_A = \left(36 \frac{km}{h}\right)\left(\frac{1000 \text{ m}}{1 \text{ km}}\right)\left(\frac{1 \text{ h}}{3600 \text{ s}}\right) = 10 \text{ m/s}$$

The motion of A is uniform, so for any time t

$$a_A = 0$$
$$v_A = +10 \text{ m/s}$$
$$x_A = (x_A)_0 + v_A t = 0 + 10t$$

For $t = 5$ s, you have (Fig. 1)

$$a_A = 0 \qquad\qquad \mathbf{a}_A = 0$$
$$v_A = +10 \text{ m/s} \qquad\qquad \mathbf{v}_A = 10 \text{ m/s} \rightarrow$$
$$x_A = +(10 \text{ m/s})(5 \text{ s}) = +50 \text{ m} \qquad \mathbf{r}_A = 50 \text{ m} \rightarrow$$

Motion of Automobile B. The motion of B is uniformly accelerated, so

$$a_B = -1.2 \text{ m/s}^2$$
$$v_B = (v_B)_0 + at = 0 - 1.2t$$
$$y_B = (y_B)_0 + (v_B)_0 t + \tfrac{1}{2}a_B t^2 = 35 + 0 - \tfrac{1}{2}(1.2)t^2$$

For $t = 5$ s, you have (Fig. 1)

$$a_B = -1.2 \text{ m/s}^2 \qquad\qquad \mathbf{a}_B = 1.2 \text{ m/s}^2 \downarrow$$
$$v_B = -(1.2 \text{ m/s}^2)(5 \text{ s}) = -6 \text{ m/s} \qquad \mathbf{v}_B = 6 \text{ m/s} \downarrow$$
$$y_B = 35 - \tfrac{1}{2}(1.2 \text{ m/s}^2)(5 \text{ s})^2 = +20 \text{ m} \qquad \mathbf{r}_B = 20 \text{ m} \uparrow$$

Motion of B Relative to A. Draw the triangle corresponding to the vector equation $\mathbf{r}_B = \mathbf{r}_A + \mathbf{r}_{B/A}$ (Fig. 2) and obtain the magnitude and direction of the position vector of B relative to A.

$$r_{B/A} = 53.9 \text{ m} \qquad \alpha = 21.8° \qquad \mathbf{r}_{B/A} = 53.9 \text{ m} \ \text{\reflectbox{$\searrow$}}\ 21.8° \quad \blacktriangleleft$$

Proceeding in a similar fashion (Fig. 2), find the velocity and acceleration of B relative to A. Hence,

$$\mathbf{v}_B = \mathbf{v}_A + \mathbf{v}_{B/A}$$
$$v_{B/A} = 11.66 \text{ m/s} \qquad \beta = 31.0° \qquad \mathbf{v}_{B/A} = 11.66 \text{ m/s} \ \nearrow\ 31.0° \quad \blacktriangleleft$$
$$\mathbf{a}_B = \mathbf{a}_A + \mathbf{a}_{B/A}$$
$$\mathbf{a}_{B/A} = 1.2 \text{ m/s}^2 \downarrow \quad \blacktriangleleft$$

Fig. 2 Vector triangles for position, velocity, and acceleration.

REFLECT and THINK: Note that the relative position and velocity of B relative to A change with time; the values given here are only for the moment $t = 5$ s. Rather than drawing triangles, you could have also used vector algebra. When the vectors are at right angles, as in this problem, drawing vector triangles is usually easiest.

Sample Problem 11.15

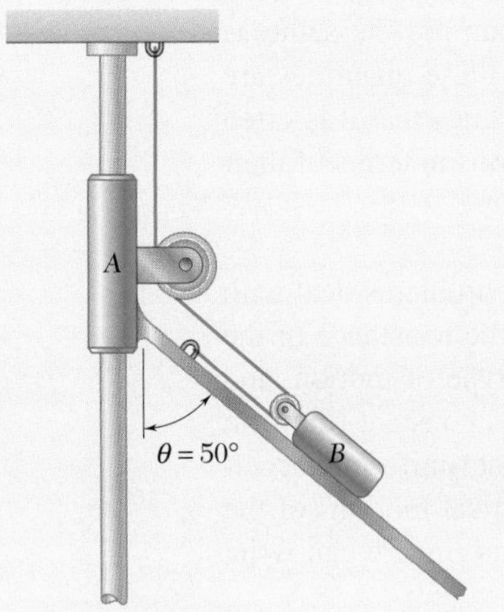

Knowing that at the instant shown cylinder/ramp A has a velocity of 8 in./s directed down, determine the velocity of block B.

STRATEGY: You have objects connected by cables, so this is a dependent-motion problem. You should define coordinates for each block-object and write a constraint equation for the cable. You will also need to use relative motion, since B slides on A.

MODELING and ANALYSIS: Define position vectors, as shown in Fig. 1.

Constraint Equations. Assuming the cable is inextensible, you can write the length in terms of the coordinates and then differentiate.

The constraint equation for the cable is

$$x_A + 2x_{B/A} = \text{constant}$$

Differentiating this gives

$$v_A = -2v_{B/A} \tag{1}$$

Substituting for v_A gives $v_{B/A} = -4$ in./s or 4 in./s up the incline.

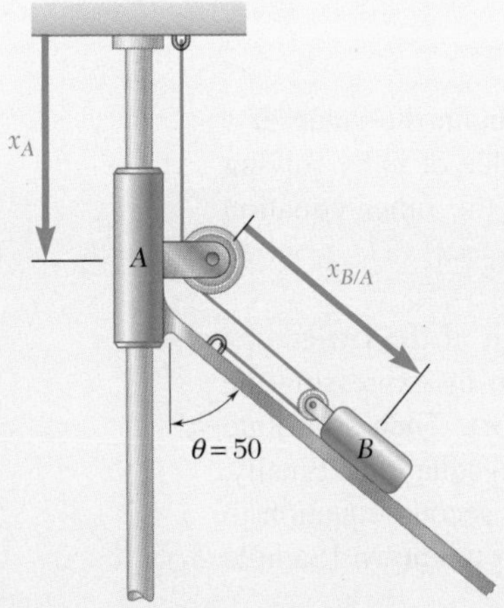

Dependent Motion: You know that the direction of $v_{B/A}$ is directed up the incline. Therefore, the relative motion equation relating the velocities of blocks A and B is $\mathbf{v}_B = \mathbf{v}_A + \mathbf{v}_{B/A}$. You could either draw a vector triangle or use vector algebra. Let's use vector algebra. Using the coordinate system shown in Fig. 2 and substituting in the magnitudes gives

$$(v_B)_x \mathbf{i} + (v_B)_y \mathbf{j} = (-8 \text{ in./s})\mathbf{j} + (-4 \text{ in./s}) \sin 50° \mathbf{i} + (4 \text{ in./s}) \cos 50° \mathbf{j}$$

Equating components gives

i: $(v_B)_x = -(4 \text{ in./s})\sin 50°$ $\qquad \rightarrow v_{B_x} = -3.064$ in./s

j: $(v_B)_y = (-8 \text{ in./s}) + (4 \text{ in./s})\cos 50°$ $\qquad \rightarrow v_{B_y} = -5.429$ in./s

Finding the magnitude and direction gives

$$\mathbf{v}_B = 6.23 \text{ in./s} \; \nearrow 60.6° \blacktriangleleft$$

REFLECT and THINK: Rather than using vector algebra, you could have also drawn a vector triangle, as shown in Fig. 3. To use this vector triangle, you need to use the law of cosines and the law of sines. Looking at the mechanism, block B should move up the incline if block A moves downward; our mathematical result is consistent with this. It is also interesting to note that, even though B moves up the incline relative to A, block B is actually moving down and to the left, as shown in the calculation here. This occurs because block A is also moving down.

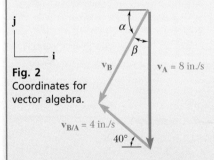

Fig. 1 Position vectors to A and B.

Fig. 2
Coordinates for vector algebra.

Fig. 3 Vector triangle for velocity of blocks A and B.

SOLVING PROBLEMS
ON YOUR OWN

In the problems for this section, you will analyze the **curvilinear motion** of a particle. The physical interpretations of velocity and acceleration are the same as in the first sections of the chapter, but you should remember that these quantities are vectors. In addition, recall from your experience with vectors in statics that it is often advantageous to express position vectors, velocities, and accelerations in terms of their rectangular scalar components [Eqs. (11.25) through (11.27)].

A. Analyzing the motion of a projectile. Many of the following problems deal with the two-dimensional motion of a projectile where we can neglect the resistance of the air. In Sec. 11.4C, we developed the equations that describe this type of motion, and we observed that the horizontal component of the velocity remains constant (uniform motion), while the vertical component of the acceleration is constant (uniformly accelerated motion). We are able to consider the horizontal and the vertical motions of the particle separately. Assuming that the projectile is fired from the origin, we can write the two equations as

$$x = (v_x)_0 t \qquad y = (v_y)_0 t - \tfrac{1}{2} g t^2$$

1. If you know the initial velocity and firing angle, you can obtain the value of y corresponding to any given value of x (or the value of x for any value of y) by solving one of the previous equations for t and substituting for t into the other equation [Sample Prob. 11.10].

2. If you know the initial velocity and the coordinates of a point of the trajectory and you wish to determine the firing angle α, begin your solution by expressing the components $(v_x)_0$ and $(v_y)_0$ of the initial velocity as functions of α. Then substitute these expressions and the known values of x and y into the previous equations. Finally, solve the first equation for t and substitute that value of t into the second equation to obtain a trigonometric equation in α, which you can solve for that unknown [Sample Prob. 11.11].

B. Solving translational two-dimensional relative-motion problems. You saw in Sec. 11.4D that you can obtain the absolute motion of a particle B by combining the motion of a particle A and the **relative motion** of B with respect to a frame attached to A that is in *translation* [Sample Probs. 11.12 and 11.13]. You can then express the velocity and acceleration of B as shown in Eqs. (11.32) and (11.33), respectively.

1. To visualize the relative motion of B with respect to A, imagine that you are attached to particle A as you observe the motion of particle B. For example, to a passenger in automobile A of Sample Prob. 11.14, automobile B appears to be heading in a southwesterly direction (*south* should be obvious; *west* is due to the fact that automobile A is moving to the east—automobile B then appears to travel to the west). Note that this conclusion is consistent with the direction of $\mathbf{v}_{B/A}$.

2. To solve a relative-motion problem, first write the vector equations (11.30), (11.32), and (11.33), which relate the motions of particles A and B. You may then use either of the following methods.

    a. Construct the corresponding vector triangles and solve them for the desired position vector, velocity, and acceleration [Sample Prob. 11.14].

    b. Express all vectors in terms of their rectangular components and solve the resulting two independent sets of scalar equations [Sample Prob. 11.15]. If you choose this approach, be sure to select the same positive direction for the displacement, velocity, and acceleration of each particle.

Problems

11.CQ3 Two model rockets are fired simultaneously from a ledge and follow the trajectories shown. Neglecting air resistance, which of the rockets will hit the ground first?
- **a.** *A*.
- **b.** *B*.
- **c.** They hit at the same time.
- **d.** The answer depends on *h*.

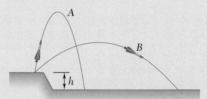

Fig. P6.CQ3

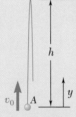

Fig. P6.CQ4

11.CQ4 Ball *A* is thrown straight up. Which of the following statements about the ball are true at the highest point in its path?
- **a.** The velocity and acceleration are both zero.
- **b.** The velocity is zero, but the acceleration is not zero.
- **c.** The velocity is not zero, but the acceleration is zero.
- **d.** Neither the velocity nor the acceleration is zero.

11.CQ5 Ball *A* is thrown straight up with an initial speed v_0 and reaches a maximum elevation *h* before falling back down. When *A* reaches its maximum elevation, a second ball is thrown straight upward with the same initial speed v_0. At what height, *y*, will the balls cross paths?
- **a.** $y = h$
- **b.** $y > h/2$
- **c.** $y = h/2$
- **d.** $y < h/2$
- **e.** $y = 0$

11.CQ6 Two cars are approaching an intersection at constant speeds as shown. What velocity will car *B* appear to have to an observer in car *A*?
- **a.** → **b.** ↘ **c.** ↖ **d.** ↗ **e.** ↙

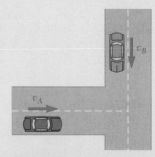
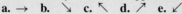

Fig. P6.CQ6

11.CQ7 Blocks *A* and *B* are released from rest in the positions shown. Neglecting friction between all surfaces, which figure best indicates the direction α of the acceleration of block *B*?

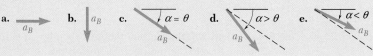

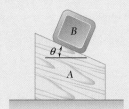

Fig. P6.CQ7

END-OF-SECTION PROBLEMS

11.89 A ball is thrown so that the motion is defined by the equations $x = 5t$ and $y = 2 + 6t - 4.9t^2$, where x and y are expressed in meters and t is expressed in seconds. Determine (*a*) the velocity at $t = 1$ s, (*b*) the horizontal distance the ball travels before hitting the ground.

Fig. P11.89

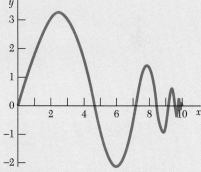

Fig. P11.90

11.90 The motion of a vibrating particle is defined by the position vector $\mathbf{r} = 10(1 - e^{-3t})\mathbf{i} + (4e^{-2t} \sin 15t)\mathbf{j}$, where $\mathbf{r}$ and t are expressed in millimeters and seconds, respectively. Determine the velocity and acceleration when (*a*) $t = 0$, (*b*) $t = 0.5$ s.

11.91 The motion of a vibrating particle is defined by the position vector $\mathbf{r} = (4 \sin \pi t)\mathbf{i} - (\cos 2\pi t)\mathbf{j}$, where r is expressed in inches and t in seconds. (*a*) Determine the velocity and acceleration when $t = 1$ s. (*b*) Show that the path of the particle is parabolic.

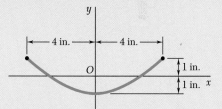

Fig. P11.91

11.92 The motion of a particle is defined by the equations $x = 10t - 5 \sin t$ and $y = 10 - 5 \cos t$, where x and y are expressed in feet and t is expressed in seconds. Sketch the path of the particle for the time interval $0 \leq t \leq 2\pi$, and determine (*a*) the magnitudes of the smallest and largest velocities reached by the particle, (*b*) the corresponding times, positions, and directions of the velocities.

11.93 The damped motion of a vibrating particle is defined by the position vector $\mathbf{r} = x_1[1 - 1/(t + 1)]\mathbf{i} + (y_1 e^{-\pi t/2} \cos 2\pi t)\mathbf{j}$, where t is expressed in seconds. For $x_1 = 30$ mm and $y_1 = 20$ mm, determine the position, the velocity, and the acceleration of the particle when (*a*) $t = 0$, (*b*) $t = 1.5$ s.

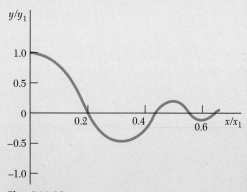

Fig. P11.93

11.94 A girl operates a radio-controlled model car in a vacant parking lot. The girl's position is at the origin of the xy coordinate axes, and the surface of the parking lot lies in the x–y plane. The motion of the car is defined by the position vector $\mathbf{r} = (2 + 2t^2)\mathbf{i} + (6 + t^3)\mathbf{j}$ where $\mathbf{r}$ and t are expressed in meters and seconds, respectively. Determine (a) the distance between the car and the girl when $t = 2$ s, (b) the distance the car traveled in the interval from $t = 0$ to $t = 2$ s, (c) the speed and direction of the car's velocity at $t = 2$ s, (d) the magnitude of the car's acceleration at $t = 2$ s.

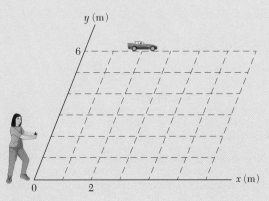

Fig. P11.94

11.95 The three-dimensional motion of a particle is defined by the position vector $\mathbf{r} = (Rt \cos \omega_n t)\mathbf{i} + ct\mathbf{j} + (Rt \sin \omega_n t)\mathbf{k}$. Determine the magnitudes of the velocity and acceleration of the particle. (The space curve described by the particle is a conic helix.)

*11.96 The three-dimensional motion of a particle is defined by the position vector $\mathbf{r} = (At \cos t)\mathbf{i} + (A\sqrt{t^2 + 1})\mathbf{j} + (Bt \sin t)\mathbf{k}$, where r and t are expressed in feet and seconds, respectively. Show that the curve described by the particle lies on the hyperboloid $(y/A)^2 - (x/A)^2 - (z/B)^2 = 1$. For $A = 3$ and $B = 1$, determine (a) the magnitudes of the velocity and acceleration when $t = 0$, (b) the smallest nonzero value of t for which the position vector and the velocity are perpendicular to each other.

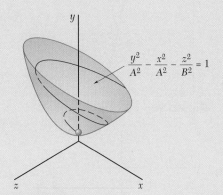

$$\frac{y^2}{A^2} - \frac{x^2}{A^2} - \frac{z^2}{B^2} = 1$$

Fig. P11.96

11.97 An airplane used to drop water on brushfires is flying horizontally in a straight line at 180 mi/h at an altitude of 300 ft. Determine the distance d at which the pilot should release the water so that it will hit the fire at B.

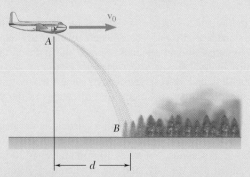

Fig. P11.97

11.98 A ski jumper starts with a horizontal take-off velocity of 25 m/s and lands on a straight landing hill inclined at 30°. Determine (*a*) the time between take-off and landing, (*b*) the length *d* of the jump, (*c*) the maximum vertical distance between the jumper and the landing hill.

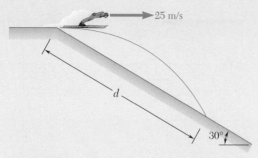

Fig. P11.98

11.99 A baseball pitching machine "throws" baseballs with a horizontal velocity $\mathbf{v}_0$. Knowing that height *h* varies between 788 mm and 1068 mm, determine (*a*) the range of values of v_0, (*b*) the values of α corresponding to $h = 788$ mm and $h = 1068$ mm.

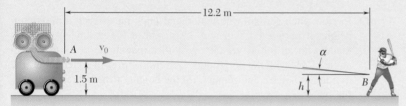

Fig. P11.99

11.100 While delivering newspapers, a girl throws a newspaper with a horizontal velocity $\mathbf{v}_0$. Determine the range of values of v_0 if the newspaper is to land between points *B* and *C*.

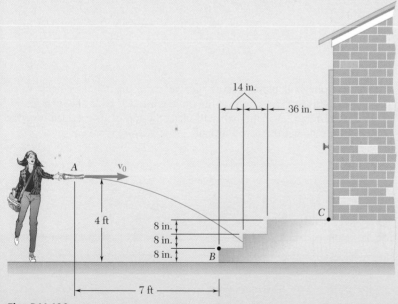

Fig. P11.100

11.101 Water flows from a drain spout with an initial velocity of 2.5 ft/s at an angle of 15° with the horizontal. Determine the range of values of the distance d for which the water will enter the trough BC.

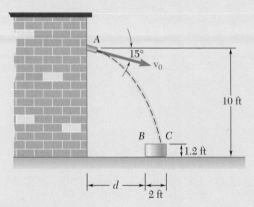

Fig. P11.101

11.102 In slow pitch softball, the underhand pitch must reach a maximum height of between 1.8 m and 3.7 m above the ground. A pitch is made with an initial velocity $\mathbf{v}_0$ with a magnitude of 13 m/s at an angle of 33° with the horizontal. Determine (*a*) if the pitch meets the maximum height requirement, (*b*) the height of the ball as it reaches the batter.

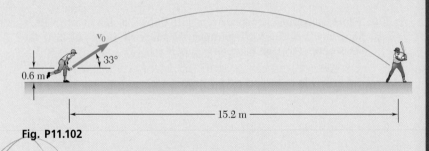

Fig. P11.102

11.103 A volleyball player serves the ball with an initial velocity $\mathbf{v}_0$ of magnitude 13.40 m/s at an angle of 20° with the horizontal. Determine (*a*) if the ball will clear the top of the net, (*b*) how far from the net the ball will land.

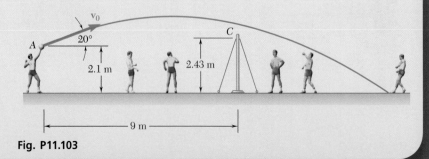

Fig. P11.103

11.104 A golfer hits a golf ball with an initial velocity of 160 ft/s at an angle of 25° with the horizontal. Knowing that the fairway slopes downward at an average angle of 5°, determine the distance d between the golfer and point B where the ball first lands.

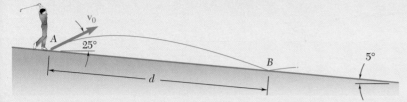

Fig. P11.104

11.105 A homeowner uses a snowblower to clear his driveway. Knowing that the snow is discharged at an average angle of 40° with the horizontal, determine the initial velocity v_0 of the snow.

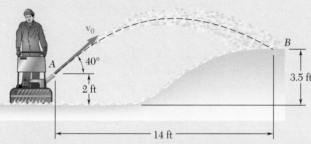

Fig. P11.105

11.106 At halftime of a football game souvenir balls are thrown to the spectators with a velocity v_0. Determine the range of values of v_0 if the balls are to land between points B and C.

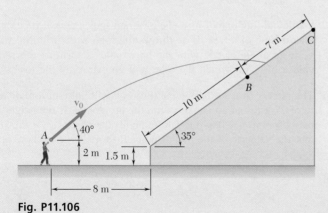

Fig. P11.106

11.107 A basketball player shoots when she is 16 ft from the backboard. Knowing that the ball has an initial velocity v_0 at an angle of 30° with the horizontal, determine the value of v_0 when d is equal to (*a*) 9 in., (*b*) 17 in.

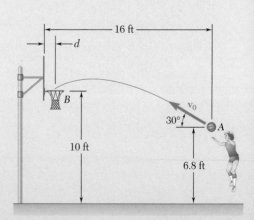

Fig. P11.107

11.108 A tennis player serves the ball at a height $h = 2.5$ m with an initial velocity of $\mathbf{v_0}$ at an angle of 5° with the horizontal. Determine the range of v_0 for which the ball will land in the service area that extends to 6.4 m beyond the net.

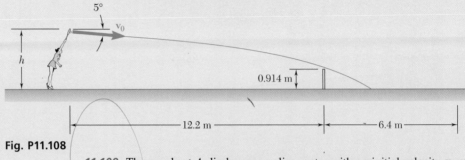

Fig. P11.108

11.109 The nozzle at A discharges cooling water with an initial velocity $\mathbf{v_0}$ at an angle of 6° with the horizontal onto a grinding wheel 350 mm in diameter. Determine the range of values of the initial velocity for which the water will land on the grinding wheel between points B and C.

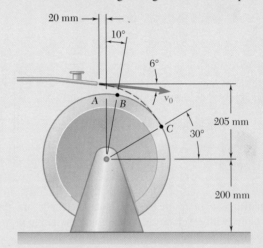

Fig. *P11.109*

11.110 While holding one of its ends, a worker lobs a coil of rope over the lowest limb of a tree. If he throws the rope with an initial velocity $\mathbf{v_0}$ at an angle of 65° with the horizontal, determine the range of values of v_0 for which the rope will go over only the lowest limb.

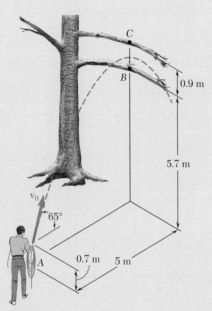

Fig. P11.110

11.111 The pitcher in a softball game throws a ball with an initial velocity $\mathbf{v_0}$ of 72 km/h at an angle α with the horizontal. If the height of the ball at point B is 0.68 m, determine (a) the angle α, (b) the angle θ that the velocity of the ball at point B forms with the horizontal.

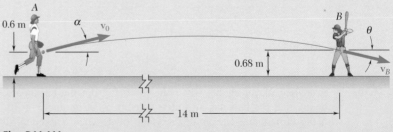

Fig. P11.111

11.112 A model rocket is launched from point A with an initial velocity $\mathbf{v}_0$ of 75 m/s. If the rocket's descent parachute does not deploy and the rocket lands a distance $d = 100$ m from A, determine (a) the angle α that $\mathbf{v}_0$ forms with the vertical, (b) the maximum height above point A reached by the rocket, (c) the duration of the flight.

11.113 The initial velocity $\mathbf{v}_0$ of a hockey puck is 105 mi/h. Determine (a) the largest value (less than 45°) of the angle α for which the puck will enter the net, (b) the corresponding time required for the puck to reach the net.

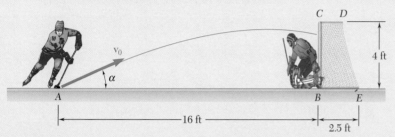

Fig. P11.113

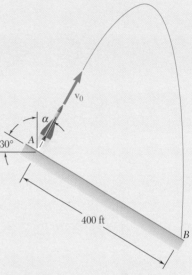

Fig. P11.112

11.114 A worker uses high-pressure water to clean the inside of a long drainpipe. If the water is discharged with an initial velocity $\mathbf{v}_0$ of 11.5 m/s, determine (a) the distance d to the farthest point B on the top of the pipe that the worker can wash from his position at A, (b) the corresponding angle α.

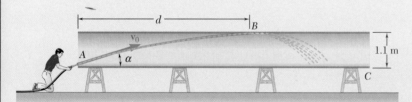

Fig. P11.114

11.115 An oscillating garden sprinkler which discharges water with an initial velocity $\mathbf{v}_0$ of 8 m/s is used to water a vegetable garden. Determine the distance d to the farthest point B that will be watered and the corresponding angle α when (a) the vegetables are just beginning to grow, (b) the height h of the corn is 1.8 m.

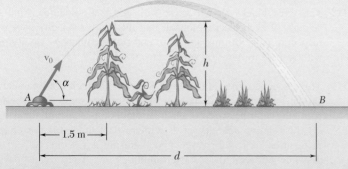

Fig. P11.115

*11.116 A nozzle at *A* discharges water with an initial velocity of 36 ft/s at an angle α with the horizontal. Determine (*a*) the distance *d* to the farthest point *B* on the roof that the water can reach, (*b*) the corresponding angle α. Check that the stream will clear the edge of the roof.

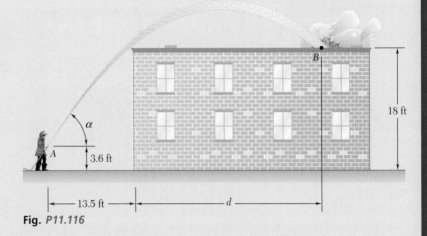

Fig. P11.116

11.117 The velocities of skiers *A* and *B* are as shown. Determine the velocity of *A* with respect to *B*.

Fig. P11.117

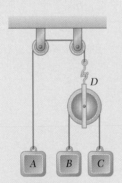

Fig. P11.118

11.118 The three blocks shown move with constant velocities. Find the velocity of each block, knowing that the relative velocity of *A* with respect to *C* is 300 mm/s upward and that the relative velocity of *B* with respect to *A* is 200 mm/s downward.

11.119 Three seconds after automobile *B* passes through the intersection shown, automobile *A* passes through the same intersection. Knowing that the speed of each automobile is constant, determine (*a*) the relative velocity of *B* with respect to *A*, (*b*) the change in position of *B* with respect to *A* during a 4-s interval, (*c*) the distance between the two automobiles 2 s after *A* has passed through the intersection.

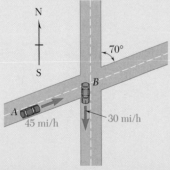

Fig. P11.119

11.120 Shore-based radar indicates that a ferry leaves its slip with a velocity $\mathbf{v} = 18$ km/h ⬈ 70°, while instruments aboard the ferry indicate a speed of 18.4 km/h and a heading of 30° west of south relative to the river. Determine the velocity of the river.

11.121 Airplanes A and B are flying at the same altitude and are tracking the eye of hurricane C. The relative velocity of C with respect to A is $\mathbf{v}_{C/A} = 350$ km/h ⬈ 75°, and the relative velocity of C with respect to B is $\mathbf{v}_{C/B} = 400$ km/h ⬊ 40°. Determine (a) the relative velocity of B with respect to A, (b) the velocity of A if ground-based radar indicates that the hurricane is moving at a speed of 30 km/h due north, (c) the change in position of C with respect to B during a 15-min interval.

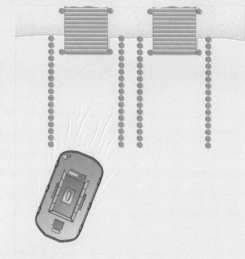

Fig. P11.120

Fig. P11.121

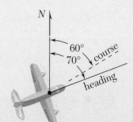

Fig. P11.122

11.122 Instruments in an airplane which is in level flight indicate that the velocity relative to the air (airspeed) is 120 km/h and the direction of the relative velocity vector (heading) is 70° east of north. Instruments on the ground indicate that the velocity of the airplane (ground speed) is 110 km/h and the direction of flight (course) is 60° east of north. Determine the wind speed and direction.

11.123 Knowing that at the instant shown block B has a velocity of 2 ft/s to the right and an acceleration of 3 ft/s² to the left, determine (a) the velocity of block A, (b) the acceleration of block A.

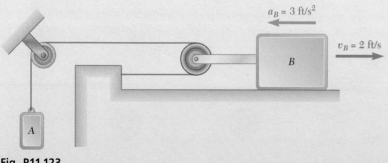

Fig. P11.123

11.124 Knowing that at the instant shown block A has a velocity of 8 in./s and an acceleration of 6 in./s² both directed down the incline, determine (a) the velocity of block B, (b) the acceleration of block B.

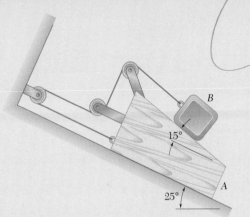

Fig. P11.124

11.125 A boat is moving to the right with a constant deceleration of 0.3 m/s² when a boy standing on the deck D throws a ball with an initial velocity relative to the deck which is vertical. The ball rises to a maximum height of 8 m above the release point and the boy must step forward a distance d to catch it at the same height as the release point. Determine (a) the distance d, (b) the relative velocity of the ball with respect to the deck when the ball is caught.

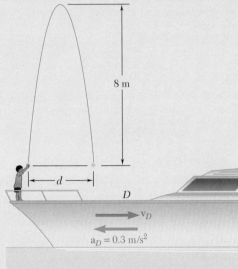

Fig. P11.125

11.126 The assembly of rod A and wedge B starts from rest and moves to the right with a constant acceleration of 2 mm/s². Determine (a) the acceleration of wedge C, (b) the velocity of wedge C when $t = 10$ s.

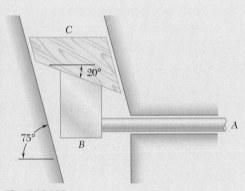

Fig. P11.126

11.127 Determine the required velocity of the belt B if the relative velocity with which the sand hits belt B is to be (a) vertical, (b) as small as possible.

11.128 Conveyor belt A, which forms a 20° angle with the horizontal, moves at a constant speed of 4 ft/s and is used to load an airplane. Knowing that a worker tosses duffel bag B with an initial velocity of 2.5 ft/s at an angle of 30° with the horizontal, determine the velocity of the bag relative to the belt as it lands on the belt.

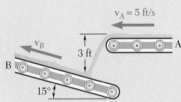

Fig. P11.127

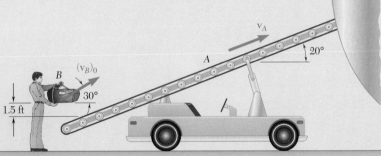

Fig. P11.128

11.129 During a rainstorm the paths of the raindrops appear to form an angle of 30° with the vertical and to be directed to the left when observed from a side window of a train moving at a speed of 15 km/h. A short time later, after the speed of the train has increased to 24 km/h, the angle between the vertical and the paths of the drops appears to be 45°. If the train were stopped, at what angle and with what velocity would the drops be observed to fall?

11.130 Instruments in airplane A indicate that, with respect to the air, the plane is headed 30° north of east with an air speed of 300 mi/h. At the same time, radar on ship B indicates that the relative velocity of the plane with respect to the ship is 280 mi/h in the direction 33° north of east. Knowing that the ship is steaming due south at 12 mi/h, determine (a) the velocity of the airplane, (b) the wind speed and direction.

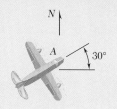
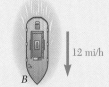

Fig. P11.130

11.131 When a small boat travels north at 5 km/h, a flag mounted on its stern forms an angle $\theta = 50°$ with the centerline of the boat as shown. A short time later, when the boat travels east at 20 km/h, angle θ is again 50°. Determine the speed and the direction of the wind.

11.132 As part of a department store display, a model train D runs on a slight incline between the store's up and down escalators. When the train and shoppers pass point A, the train appears to a shopper on the up escalator B to move downward at an angle of 22° with the horizontal, and to a shopper on the down escalator C to move upward at an angle of 23° with the horizontal and to travel to the left. Knowing that the speed of the escalators is 3 ft/s, determine the speed and the direction of the train.

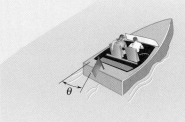

Fig. P11.131

Fig. P11.132

11.5 NON-RECTANGULAR COMPONENTS

Sometimes it is useful to analyze the motion of a particle in a coordinate system that is not rectangular. In this section, we introduce two common and important systems. The first system is based on the path of the particle; the second system is based on the radial distance and angular displacement of the particle.

11.5A Tangential and Normal Components

We saw in Sec. 11.4 that the velocity of a particle is a vector tangent to the path of the particle, but in general, the acceleration is not tangent to the path. It is sometimes convenient to resolve the acceleration into components directed, respectively, along the tangent and the normal to the path of the particle. We will refer to this reference frame as tangential and normal coordinates, which are sometimes called path coordinates.

Planar Motion of a Particle. First we consider a particle that moves along a curve contained in a plane. Let P be the position of the particle at a given instant. We attach at P a unit vector $\mathbf{e}_t$ tangent to the path of the particle and pointing in the direction of motion (Fig. 11.19a). Let $\mathbf{e}_t'$ be the unit vector corresponding to the position P' of the particle at a later instant. Drawing both vectors from the same origin O', we define the vector $\Delta\mathbf{e}_t = \mathbf{e}_t' - \mathbf{e}_t$ (Fig. 11.19b). Since $\mathbf{e}_t$ and $\mathbf{e}_t'$ are of unit length, their tips lie on a circle with a radius of 1. Denote the angle formed by $\mathbf{e}_t$ and $\mathbf{e}_t'$ by $\Delta\theta$. Then the magnitude of $\Delta\mathbf{e}_t$ is $2\sin(\Delta\theta/2)$. Considering now the vector $\Delta\mathbf{e}_t/\Delta\theta$, we note that, as $\Delta\theta$ approaches zero, this vector becomes tangent to the unit circle of Fig. 11.19b, i.e., perpendicular to $\mathbf{e}_t$, and that its magnitude approaches

$$\lim_{\Delta\theta \to 0} \frac{2\sin(\Delta\theta/2)}{\Delta\theta} = \lim_{\Delta\theta \to 0} \frac{\sin(\Delta\theta/2)}{\Delta\theta/2} = 1$$

Thus, the vector obtained in the limit is a unit vector along the normal to the path of the particle in the direction toward which $\mathbf{e}_t$ turns. Denoting this vector by $\mathbf{e}_n$, we have

$$\mathbf{e}_n = \lim_{\Delta\theta \to 0} \frac{\Delta\mathbf{e}_t}{\Delta\theta}$$

$$\mathbf{e}_n = \frac{d\mathbf{e}_t}{d\theta} \tag{11.34}$$

Now, since the velocity $\mathbf{v}$ of the particle is tangent to the path, we can express it as the product of the scalar v and the unit vector $\mathbf{e}_t$. We have

$$\mathbf{v} = v\mathbf{e}_t \tag{11.35}$$

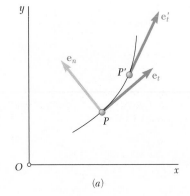

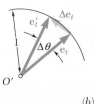

Fig. 11.19 (a) Unit tangent vectors for two positions of particle P; (b) the angle between the unit tangent vectors and their difference $\Delta\mathbf{e}_t$.

To obtain the acceleration of the particle, we differentiate Eq. (11.35) with respect to t. Applying the rule for the differentiation of the product of a scalar and a vector function (Sec. 11.4B), we have

$$\mathbf{a} = \frac{d\mathbf{v}}{dt} = \frac{dv}{dt}\mathbf{e}_t + v\frac{d\mathbf{e}_t}{dt} \qquad (11.36)$$

However,

$$\frac{d\mathbf{e}_t}{dt} = \frac{d\mathbf{e}_t}{d\theta}\frac{d\theta}{ds}\frac{ds}{dt}$$

Recall from Eq. (11.15) that $ds/dt = v$, from Eq. (11.34) that $d\mathbf{e}_t/d\theta = \mathbf{e}_n$, and from elementary calculus that $d\theta/ds$ is equal to $1/\rho$, where ρ is the radius of curvature of the path at P (Fig. 11.20). Then we have

$$\frac{d\mathbf{e}_t}{dt} = \frac{v}{\rho}\mathbf{e}_n \qquad (11.37)$$

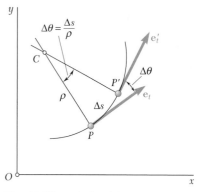

Fig. 11.20 Relationship among $\Delta\theta$, Δs, and ρ. Recall that for a circle, the arc length is equal to the radius multiplied by the angle.

Substituting into Eq. (11.36), we obtain

Acceleration in normal and tangential components

$$\mathbf{a} = \frac{dv}{dt}\mathbf{e}_t + \frac{v^2}{\rho}\mathbf{e}_n \qquad (11.38)$$

Thus, the scalar components of the acceleration are

$$a_t = \frac{dv}{dt} \qquad a_n = \frac{v^2}{\rho} \qquad (11.39)$$

These relations state that the **tangential component** of the acceleration is equal to the **rate of change of the speed of the particle**, whereas the **normal component** is equal to the **square of the speed divided by the radius of curvature of the path at P**. For a given speed, the normal acceleration increases as the radius of curvature decreases. If the particle travels in a straight line, then ρ is infinite, and the normal acceleration is zero. If the speed of the particle increases, a_t is positive, and the vector component $\mathbf{a}_t$ points in the direction of motion. If the speed of the particle decreases, a_t is negative, and $\mathbf{a}_t$ points against the direction of motion. The vector component $\mathbf{a}_n$, on the other hand, **is always directed toward the center of curvature C of the path** (Fig. 11.21).

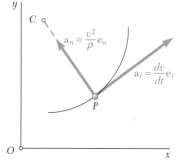

Fig. 11.21 Acceleration components in normal and tangential coordinates; the normal component always points toward the center of curvature of the path.

We conclude from this discussion that the tangential component of the acceleration reflects a change in the speed of the particle, whereas its normal component reflects a change in the direction of motion of the particle. The acceleration of a particle is zero only if both of its components are zero. Thus, the acceleration of a particle moving with constant speed along a curve is not zero unless the particle happens to pass through a point of inflection of the curve (where the radius of curvature is infinite) or unless the curve is a straight line.

The fact that the normal component of acceleration depends upon the radius of curvature of the particle's path is taken into account in the design of structures or mechanisms as widely different as airplane wings, railroad tracks, and cams. In order to avoid sudden changes in the acceleration of the air particles flowing past a wing, wing profiles are designed without any sudden change in curvature. Similar care is taken in designing

Photo 11.5 The passengers in a train traveling around a curve experience a normal acceleration toward the center of curvature of the path.

railroad curves to avoid sudden changes in the acceleration of the cars (which would be hard on the equipment and unpleasant for the passengers). A straight section of track, for instance, is never directly followed by a circular section. Special transition sections are used to help pass smoothly from the infinite radius of curvature of the straight section to the finite radius of the circular track. Likewise, in the design of high-speed cams (that can be used to transform rotary motion into translational motion), abrupt changes in acceleration are avoided by using transition curves that produce a continuous change in acceleration.

Motion of a Particle in Space. The relations in Eqs. (11.38) and (11.39) still hold in the case of a particle moving along a space curve. However, since an infinite number of straight lines are perpendicular to the tangent at a given point P of a space curve, it is necessary to define more precisely the direction of the unit vector $\mathbf{e}_n$.

Let us consider again the unit vectors $\mathbf{e}_t$ and $\mathbf{e}'_t$ tangent to the path of the particle at two neighboring points P and P' (Fig. 11.22a). Again the vector $\Delta\mathbf{e}_t$ represents the difference between $\mathbf{e}_t$ and $\mathbf{e}'_t$ (Fig. 11.22b). Let us now imagine a plane through P (Fig. 11.22c) parallel to the plane defined by the vectors $\mathbf{e}_t$, $\mathbf{e}'_t$, and $\Delta\mathbf{e}_t$ (Fig. 11.22b). This plane contains the tangent to the curve at P and is parallel to the tangent at P'. If we let P' approach P, we obtain in the limit the plane that fits the curve most closely in the neighborhood of P. This plane is called the **osculating plane** at P (from the Latin *osculari*, to kiss*)*. It follows from this definition that the osculating plane contains the unit vector $\mathbf{e}_n$, since this vector represents the limit of the vector $\Delta\mathbf{e}_t/\Delta\theta$. The normal defined by $\mathbf{e}_n$ is thus contained in the osculating plane; it is called the **principal normal** at P. The unit vector $\mathbf{e}_b = \mathbf{e}_t \times \mathbf{e}_n$ that completes the right-handed triad $\mathbf{e}_t$, $\mathbf{e}_n$, and $\mathbf{e}_b$ (Fig. 11.22c) defines the **binormal** at P. The binormal is thus perpendicular to the osculating plane. We conclude that the acceleration of the particle at P can be resolved into two components: one along the tangent and the other along the principal normal at P, as indicated in Eq. (11.38). Note that the acceleration has no component along the binormal.

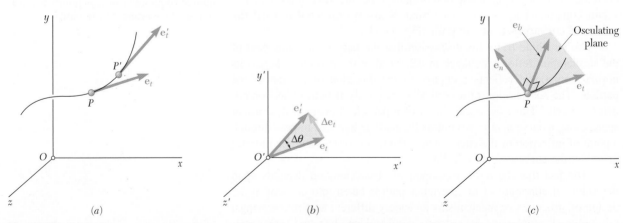

Fig. 11.22 (a) Unit tangent vectors for a particle moving in space; (b) the plane defined by the unit vectors and the vector difference $\Delta\mathbf{e}_t$; (c) the osculating plane contains the unit tangent and principal normal vectors and is perpendicular to the unit binormal vector.

11.5B Radial and Transverse Components

In some situations in planar motion, the position of particle P is defined by its polar coordinates r and θ (Fig. 11.23a). It is then convenient to resolve the velocity and acceleration of the particle into components parallel and perpendicular to the radial line OP. These components are called **radial and transverse components**.

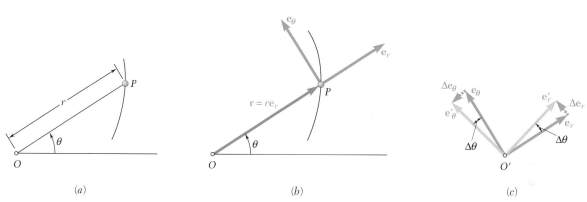

Fig. 11.23 (a) Polar coordinates r and θ of a particle at P; (b) radial and transverse unit vectors; (c) changes of the radial and transverse unit vectors resulting from a change in angle $\Delta\theta$.

We attach two unit vectors, $\mathbf{e}_r$ and $\mathbf{e}_\theta$, at P (Fig. 11.23b). The vector $\mathbf{e}_r$ is directed along OP and the vector $\mathbf{e}_\theta$ is obtained by rotating $\mathbf{e}_r$ through 90° counterclockwise. The unit vector $\mathbf{e}_r$ defines the **radial** direction, i.e., the direction in which P would move if r were increased and θ were kept constant. The unit vector $\mathbf{e}_\theta$ defines the **transverse** direction, i.e., the direction in which P would move if θ were increased and r were kept constant. A derivation similar to the one we used in the preceding section to determine the unit vector $\mathbf{e}_t$ leads to the relations

$$\frac{d\mathbf{e}_r}{d\theta} = \mathbf{e}_\theta \qquad \frac{d\mathbf{e}_\theta}{d\theta} = -\mathbf{e}_r \qquad \textbf{(11.40)}$$

Here $-\mathbf{e}_r$ denotes a unit vector with a sense opposite to that of $\mathbf{e}_r$ (Fig. 11.23c). Using the chain rule of differentiation, we express the time derivatives of the unit vectors $\mathbf{e}_r$ and $\mathbf{e}_\theta$ as

$$\frac{d\mathbf{e}_r}{dt} = \frac{d\mathbf{e}_r}{d\theta}\frac{d\theta}{dt} = \mathbf{e}_\theta\frac{d\theta}{dt} \qquad \frac{d\mathbf{e}_\theta}{dt} = \frac{d\mathbf{e}_\theta}{d\theta}\frac{d\theta}{dt} = -\mathbf{e}_r\frac{d\theta}{dt}$$

or using dots to indicate differentiation with respect to t as

$$\dot{\mathbf{e}}_r = \dot{\theta}\mathbf{e}_\theta \qquad \dot{\mathbf{e}}_\theta = -\dot{\theta}\mathbf{e}_r \qquad \textbf{(11.41)}$$

To obtain the velocity $\mathbf{v}$ of particle P, we express the position vector $\mathbf{r}$ of P as the product of the scalar r and the unit vector $\mathbf{e}_r$ and then differentiate with respect to t for

$$\mathbf{v} = \frac{d}{dt}(r\mathbf{e}_r) = \dot{r}\mathbf{e}_r + r\dot{\mathbf{e}}_r$$

Photo 11.6 The foot pedals on an elliptical trainer undergo curvilinear motion.

Using the first of the relations of Eq. (11.41), we can rewrite this as

**Velocity in radial and
transverse components**

$$\mathbf{v} = \dot{r}\mathbf{e}_r + r\dot{\theta}\mathbf{e}_\theta \qquad (11.42)$$

Differentiating again with respect to t to obtain the acceleration, we have

$$\mathbf{a} = \frac{d\mathbf{v}}{dt} = \ddot{r}\mathbf{e}_r + \dot{r}\dot{\mathbf{e}}_r + \dot{r}\dot{\theta}\mathbf{e}_\theta + r\ddot{\theta}\mathbf{e}_\theta + r\dot{\theta}\dot{\mathbf{e}}_\theta$$

Substituting for $\dot{\mathbf{e}}_r$ and $\dot{\mathbf{e}}_\theta$ from Eq. (11.41) and factoring $\mathbf{e}_r$ and $\mathbf{e}_\theta$, we obtain

**Acceleration in radial and
transverse components**

$$\mathbf{a} = (\ddot{r} - r\dot{\theta}^2)\mathbf{e}_r + (r\ddot{\theta} + 2\dot{r}\dot{\theta})\mathbf{e}_\theta \qquad (11.43)$$

The scalar components of the velocity and the acceleration in the radial and transverse directions are

$$v_r = \dot{r} \qquad\qquad v_\theta = r\dot{\theta} \qquad (11.44)$$

$$a_r = \ddot{r} - r\dot{\theta}^2 \qquad a_\theta = r\ddot{\theta} + 2\dot{r}\dot{\theta} \qquad (11.45)$$

It is important to note that a_r is *not* equal to the time derivative of v_r and that a_θ is *not* equal to the time derivative of v_θ.

In the case of a particle moving along a circle with a center O, we have $r = \text{constant}$ and $\dot{r} = \ddot{r} = 0$, so the formulas (11.42) and (11.43) reduce, respectively, to

$$\mathbf{v} = r\dot{\theta}\mathbf{e}_\theta \qquad \mathbf{a} = -r\dot{\theta}^2\mathbf{e}_r + r\ddot{\theta}\mathbf{e}_\theta \qquad (11.46)$$

Compare this to using tangential and normal coordinates for a particle in a circular path. In this case, the radius of curvature ρ is equal to the radius of the circle r, and we have $\mathbf{v} = v\mathbf{e}_t$ and $\mathbf{a} = \dot{v}\mathbf{e}_t + (v^2/r)\mathbf{e}_n$. Note that $\mathbf{e}_r$ and $\mathbf{e}_n$ point in opposite directions ($\mathbf{e}_n$ inward and $\mathbf{e}_r$ outward).

Extension to the Motion of a Particle in Space: Cylindrical Coordinates.

Sometimes it is convenient to define the position of a particle P in space by its cylindrical coordinates R, θ, and z (Fig. 11.24a). We can then use the unit vectors $\mathbf{e}_R$, $\mathbf{e}_\theta$, and $\mathbf{k}$ shown in Fig. 11.24b. Resolving the position vector $\mathbf{r}$ of particle P into components along the unit vectors, we have

$$\mathbf{r} = R\mathbf{e}_R + z\mathbf{k} \qquad (11.47)$$

Observe that $\mathbf{e}_R$ and $\mathbf{e}_\theta$ define the radial and transverse directions in the horizontal xy plane, respectively, and that the vector $\mathbf{k}$, which defines the **axial** direction, is constant in direction as well as in magnitude. Then we can verify that

$$\mathbf{v} = \frac{d\mathbf{r}}{dt} = \dot{R}\mathbf{e}_R + R\dot{\theta}\mathbf{e}_\theta + \dot{z}\mathbf{k} \qquad (11.48)$$

$$\mathbf{a} = \frac{d\mathbf{v}}{dt} = (\ddot{R} - R\dot{\theta}^2)\mathbf{e}_R + (R\ddot{\theta} + 2\dot{R}\dot{\theta})\mathbf{e}_\theta + \ddot{z}\mathbf{k} \qquad (11.49)$$

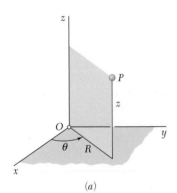

(a)

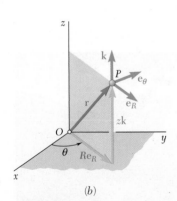

(b)

Fig. 11.24 (a) Cylindrical coordinates R, θ, and z; (b) unit vectors in cylindrical coordinates for a particle in space.

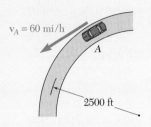

Sample Problem 11.16

A motorist is traveling on a curved section of highway with a radius of 2500 ft at a speed of 60 mi/h. The motorist suddenly applies the brakes, causing the automobile to slow down at a constant rate. If the speed has been reduced to 45 mi/h after 8 s, determine the acceleration of the automobile immediately after the brakes have been applied.

STRATEGY: You know the path of the motion, and that the forward speed of the vehicle defines the direction of $\mathbf{e}_t$. Therefore, you can use tangential and normal components.

MODELING and ANALYSIS:

Tangential Component of Acceleration. First express the speeds in ft/s.

$$60 \text{ mi/h} = \left(60 \frac{\text{mi}}{\text{h}}\right)\left(\frac{5280 \text{ ft}}{1 \text{ mi}}\right)\left(\frac{1 \text{ h}}{3600 \text{ s}}\right) = 88 \text{ ft/s}$$

$$45 \text{ mi/h} = 66 \text{ ft/s}$$

Since the automobile slows down at a constant rate, you have the tangential acceleration of

$$a_t = \text{average } a_t = \frac{\Delta v}{\Delta t} = \frac{66 \text{ ft/s} - 88 \text{ ft/s}}{8 \text{ s}} = -2.75 \text{ ft/s}^2$$

Normal Component of Acceleration. Immediately after the brakes have been applied, the speed is still 88 ft/s. Therefore, you have

$$a_n = \frac{v^2}{\rho} = \frac{(88 \text{ ft/s})^2}{2500 \text{ ft}} = 3.10 \text{ ft/s}^2$$

Magnitude and Direction of Acceleration. The magnitude and direction of the resultant $\mathbf{a}$ of the components $\mathbf{a}_n$ and $\mathbf{a}_t$ are (Fig. 1)

$$\tan \alpha = \frac{a_n}{a_t} = \frac{3.10 \text{ ft/s}^2}{2.75 \text{ ft/s}^2} \qquad \alpha = 48.4° \blacktriangleleft$$

$$a = \frac{a_n}{\sin \alpha} = \frac{3.10 \text{ ft/s}^2}{\sin 48.4°} \qquad a = 4.14 \text{ ft/s}^2 \blacktriangleleft$$

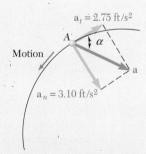

Fig. 1 Acceleration of the car.

REFLECT and THINK: The tangential component of acceleration is opposite the direction of motion, and the normal component of acceleration points to the center of curvature, which is what you would expect for slowing down on a curved path. Attempting to do this problem in Cartesian coordinates is quite difficult.

Sample Problem 11.17

Determine the minimum radius of curvature of the trajectory described by the projectile considered in Sample Prob. 11.10.

STRATEGY: You are asked to find the radius of curvature, so you should use normal and tangential coordinates.

MODELING and ANALYSIS: Since $a_n = v^2/\rho$, you have $\rho = v^2/a_n$. Therefore, the radius is small when v is small or when a_n is large. The speed v is minimum at the top of the trajectory, since $v_y = 0$ at that point; a_n is maximum at that same point, since the direction of the vertical coincides with the direction of the normal (Fig. 1). Therefore, the minimum radius of curvature occurs at the top of the trajectory. At this point, you have

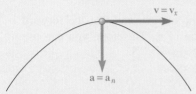

Fig. 1 Acceleration and velocity of the projectile.

$$v = v_x = 155.9 \text{ m/s} \qquad a_n = a = 9.81 \text{ m/s}^2$$

$$\rho = \frac{v^2}{a_n} = \frac{(155.9 \text{ m/s})^2}{9.81 \text{ m/s}^2} \qquad \rho = 2480 \text{ m} \blacktriangleleft$$

REFLECT and THINK: The top of the trajectory is the easiest point to determine the radius of curvature. At any other point in the trajectory, you need to find the normal component of acceleration. You can do this easily at the top, because you know that the total acceleration is pointed vertically downward and the normal component is simply the component perpendicular to the tangent to the path. Once you have the normal acceleration, it is straightforward to find the radius of curvature if you know the speed.

Sample Problem 11.18

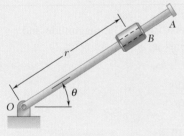

The rotation of the 0.9-m arm OA about O is defined by the relation $\theta = 0.15t^2$, where θ is expressed in radians and t in seconds. Collar B slides along the arm in such a way that its distance from O is $r = 0.9 - 0.12t^2$, where r is expressed in meters and t in seconds. After the arm OA has rotated through 30°, determine (a) the total velocity of the collar, (b) the total acceleration of the collar, (c) the relative acceleration of the collar with respect to the arm.

STRATEGY: You are given information in terms of r and θ, so you should use polar coordinates.

MODELING and ANALYSIS: Model the collar as a particle.

Time t at which $\theta = 30°$. Substitute $\theta = 30° = 0.524$ rad into the expression for θ. You obtain

$$\theta = 0.15t^2 \qquad 0.524 = 0.15t^2 \qquad t = 1.869 \text{ s}$$

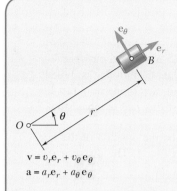

$$\mathbf{v} = v_r\mathbf{e}_r + v_\theta\mathbf{e}_\theta$$
$$\mathbf{a} = a_r\mathbf{e}_r + a_\theta\mathbf{e}_\theta$$

Fig. 1 Radial and transverse coordinates for collar *B*.

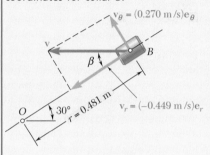

Fig. 2 Velocity of collar *B*.

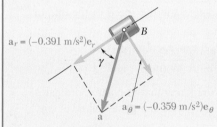

Fig. 3 Acceleration of collar *B*.

Equations of Motion. Substituting $t = 1.869$ s in the expressions for r, θ, and their first and second derivatives, you have

$$r = 0.9 - 0.12t^2 = 0.481 \text{ m} \qquad \theta = 0.15t^2 = 0.524 \text{ rad}$$
$$\dot{r} = -0.24t = -0.449 \text{ m/s} \qquad \dot{\theta} = 0.30t = 0.561 \text{ rad/s}$$
$$\ddot{r} = -0.24 = -0.240 \text{ m/s}^2 \qquad \ddot{\theta} = 0.30 = 0.300 \text{ rad/s}^2$$

a. Velocity of B. Using Eqs. (11.44), you can obtain the values of v_r and v_θ when $t = 1.869$ s (Fig. 1).

$$v_r = \dot{r} = -0.449 \text{ m/s}$$
$$v_\theta = r\dot{\theta} = 0.481(0.561) = 0.270 \text{ m/s}$$

Solve the right triangle shown in Fig. 2 to obtain the magnitude and direction of the velocity,

$$v = 0.524 \text{ m/s} \qquad \beta = 31.0° \qquad \blacktriangleleft$$

b. Acceleration of B. Using Eqs. (11.45), you obtain (Fig. 3)

$$a_r = \ddot{r} - r\dot{\theta}^2$$
$$= -0.240 - 0.481(0.561)^2 = -0.391 \text{ m/s}^2$$
$$a_\theta = r\ddot{\theta} + 2\dot{r}\dot{\theta}$$
$$= 0.481(0.300) + 2(-0.449)(0.561) = -0.359 \text{ m/s}^2$$
$$a = 0.531 \text{ m/s}^2 \qquad \gamma = 42.6° \qquad \blacktriangleleft$$

c. Acceleration of B with Respect to Arm OA. Note that the motion of the collar with respect to the arm is rectilinear and defined by the coordinate r (Fig. 4). You have

$$a_{B/OA} = \ddot{r} = -0.240 \text{ m/s}^2$$
$$a_{B/OA} = 0.240 \text{ m/s}^2 \text{ toward } O. \qquad \blacktriangleleft$$

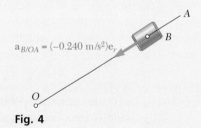

Fig. 4

REFLECT and THINK: You should consider polar coordinates for any kind of rotational motion. They turn this problem into a straightforward solution, whereas any other coordinate system would make this problem much more difficult. One way to make this problem harder would be to ask you to find the radius of curvature in addition to the velocity and acceleration. To do this, you would have to find the normal component of the acceleration; that is, the component of acceleration that is perpendicular to the tangential direction defined by the velocity vector.

Sample Problem 11.19

A boy is flying a kite that is 60 m high with 75 m of cord out. The kite moves horizontally from this position at a constant 6 km/h that is directly away from the boy. Ignoring the sag in the cord, determine how fast the cord is being let out at this instant and how fast this rate is increasing.

STRATEGY: The most natural way to describe the position of the kite is using a radial vector and angle, as shown in Fig. 1. The distance r is changing, so use polar coordinates.

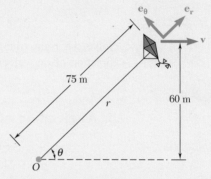

Fig. 1 Radial and transverse coordinates for the kite.

MODELING and ANALYSIS: The angle and the speed of the kite in m/s are found by

$$\theta = \sin^{-1}\left(\frac{60}{75}\right) = 53.13° \quad \text{and} \quad v = 6\left(\frac{km}{hr}\right)\left(\frac{hr}{3600\ s}\right)\left(\frac{1000\ m}{km}\right) = \frac{5}{3}\ \text{m/s}$$

Velocity in Polar Coordinates: You know that in polar coordinates the velocity is $\mathbf{v} = \dot{r}\mathbf{e}_r + r\dot{\theta}\mathbf{e}_r$. Using Fig. 1, you can resolve the velocity vector into polar coordinates, giving

$$\dot{r} = v\cos\theta = \left(\frac{5}{3}\ \text{m/s}\right)\cos 53.13° \quad \dot{r} = 1.000\ \text{m/s} \quad \blacktriangleleft$$

$$r\dot{\theta} = -v\sin\theta \quad \dot{\theta} = -\frac{v\sin\theta}{r} = -\frac{(5/3\ \text{m/s})\sin 53.13°}{75\ \text{m}} = 0.01778\ \text{rad/s}$$

Acceleration in Polar Coordinates: You know that the acceleration is zero, because the kite is traveling at a constant speed. This means that both components of the acceleration need to be zero. You know the radial component is $a_r = \ddot{r} - r\dot{\theta}^2 = 0$. So

$$\ddot{r} = r\dot{\theta}^2 = (75\ \text{m})(-0.01778\ \text{rad/s})^2 \quad \ddot{r} = 0.0237\ \text{m/s}^2 \quad \blacktriangleleft$$

REFLECT and THINK: When the angle is 90°, then $\dot{r}$ will be zero. When the angle is very small—that is, when the kite is far away—you would expect the cord to increase at a rate of 6 m/s, which is the speed of the kite. Our answer is reasonable since it is between these two limits.

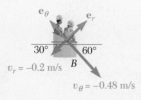

Sample Problem 11.20

At the instant shown, the length of the boom *AB* is being *decreased* at the constant rate of 0.2 m/s, and the boom is being lowered at the constant rate of 0.08 rad/s. Determine (*a*) the velocity of point *B*, (*b*) the acceleration of point *B*.

STRATEGY: Use polar coordinates, since that is the most natural way to describe the position of point *B*.

MODELING and ANALYSIS: From the problem statement, you know

$$\dot{r} = -0.2 \text{ m/s} \qquad \ddot{r} = 0 \qquad \dot{\theta} = -0.08 \text{ rad/s} \qquad \ddot{\theta} = 0$$

a. Velocity of B. Using Eqs.(11.44), you can determine the values of v_r and v_θ at this instant to be

$$v_r = \dot{r} = -0.2 \text{ m/s}$$
$$v_\theta = r\dot{\theta} = (6 \text{ m})(-0.08 \text{ rad/s}) = -0.48 \text{ m/s}$$

Therefore, you can write the velocity vector as

$$\mathbf{v} = (-0.200 \text{ m/s})\mathbf{e}_r + (-0.480 \text{ m/s})\mathbf{e}_t \quad \blacktriangleleft$$

b. Acceleration of B. Using Eqs. (11.45), you find

$$a_r = \ddot{r} - r\dot{\theta}^2 = 0 - (6 \text{ m})(-0.08 \text{ rad/s})^2 = -0.0384 \text{ m/s}^2$$
$$a_\theta = r\ddot{\theta} + 2\dot{r}\dot{\theta} = 0 + 2(-0.02 \text{ m/s})(-0.08 \text{ rad/s}) = 0.00320 \text{ m/s}^2$$

or

$$\mathbf{a} = (-0.0384 \text{ m/s}^2)\mathbf{e}_r + (0.00320 \text{ m/s}^2)\mathbf{e}_\theta \quad \blacktriangleleft$$

REFLECT and THINK: Once you identify what you are given in the problem statement, this problem is quite straightforward. Sometimes you will be asked to express your answer in terms of a magnitude and direction. The easiest way is to first determine the *x* and *y* components and then to find the magnitude and direction. From Fig. 1,

$$\xrightarrow{\;+\;}: (v_B)_x = 0.48 \cos 60° - 0.2 \cos 30° = 0.06680 \text{ m/s}$$

$$+\uparrow: (v_B)_y = -0.48 \sin 60° - 0.2 \sin 30° = -0.5157 \text{ m/s}$$

So the magnitude and direction are

$$v_B = \sqrt{0.06680^2 + 0.5157^2}$$

$$= 0.520 \text{ m/s} \qquad \tan\beta = \frac{0.51569}{0.06680}, \; \beta = 82.6°$$

So, an alternative way of expressing the velocity of *B* is $\mathbf{v}_B = 0.520 \text{ m/s} \; \diagdown 82.6°$

You could also find the magnitude and direction of the acceleration if you needed it expressed in this way. It is important to note that no matter what coordinate system we choose, the resultant velocity vector is the same. You can choose to express this vector in whatever coordinate system is most useful. Figure 2 shows the velocity vector $\mathbf{v}_B$ resolved into *x* and *y* components and *r* and θ coordinates.

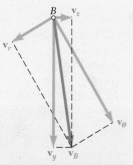

Fig. 1 Velocity of *B*.

Fig. 2 Resultant velocity of collar *B* in Cartesian and in radial and transverse coordinates.

SOLVING PROBLEMS
ON YOUR OWN

In the following problems, you will be asked to express the velocity and the acceleration of particles in terms of either their **tangential and normal components** or their **radial and transverse components**. Although these components may not be as familiar to you as rectangular components, you will find that they can simplify the solution of many problems and that certain types of motion are more easily described when they are used.

1. Using tangential and normal components. These components are most often used when the particle of interest travels along a known curvilinear path or when the radius of curvature of the path is to be determined [Sample Prob. 11.16]. Remember that the unit vector $\mathbf{e}_t$ is tangent to the path of the particle (and thus aligned with the velocity), whereas the unit vector $\mathbf{e}_n$ is directed along the normal to the path and always points toward its center of curvature. It follows that the directions of the two unit vectors are constantly changing as the particle moves.

2. Acceleration in terms of tangential and normal components. We derived in Sec. 11.5A the following equation, which is applicable to both the two-dimensional and the three-dimensional motion of a particle:

$$\mathbf{a} = \frac{dv}{dt}\mathbf{e}_t + \frac{v^2}{\rho}\mathbf{e}_n \tag{11.38}$$

The following observations may help you in solving the problems of this section.

    a. The tangential component of the acceleration measures the rate of change of the speed as $a_t = dv/dt$. It follows that, when a_t is constant, you can use the equations for uniformly accelerated motion with the acceleration equal to a_t. Furthermore, when a particle moves at a constant speed, we have $a_t = 0$, and the acceleration of the particle reduces to its normal component.

    b. The normal component of the acceleration is always directed toward the center of curvature of the path of the particle, and its magnitude is $a_n = v^2/\rho$. Thus, you can determine the normal component if you know the speed of the particle and the radius of curvature ρ of the path. Conversely, if you know the speed and normal acceleration of the particle, you can find the radius of curvature of the path by solving this equation for ρ [Sample Prob. 11.17].

3. Using radial and transverse components. These components are used to analyze the planar motion of a particle P when the position of P is defined by its polar coordinates r and θ. As shown in Fig. 11.23, the unit vector $\mathbf{e}_r$, which defines the **radial** direction, is attached to P and points away from the fixed point O, whereas the unit vector $\mathbf{e}_\theta$, which defines the **transverse** direction, is obtained by rotating $\mathbf{e}_r$ *counterclockwise* through 90°. The velocity and acceleration of a particle are expressed in terms of their radial and transverse components in Eqs. (11.42) and (11.43), respectively. Note that the expressions obtained contain the first and second derivatives with respect to t of both coordinates r and θ.

In the problems of this section, you will encounter the following types of problems involving radial and transverse components.

    a. Both r and θ are known functions of t. In this case, you compute the first and second derivatives of r and θ and substitute the resulting expressions into Eqs. (11.42) and (11.43).

    b. A certain relationship exists between r and θ. First, you should determine this relationship from the geometry of the given system and use it to express r as a function of θ. Once you know the function $r = f(\theta)$, you can apply the chain rule to determine $\dot{r}$ in terms of θ and $\dot{\theta}$, and $\ddot{r}$ in terms of θ, $\dot{\theta}$, and $\ddot{\theta}$:

$$\dot{r} = f'(\theta)\dot{\theta}$$

$$\ddot{r} = f''(\theta)\dot{\theta}^2 + f'(\theta)\ddot{\theta}$$

You can then substitute these expressions into Eqs. (11.42) and (11.43).

    c. The three-dimensional motion of a particle, as indicated at the end of Sec. 11.5B, often can be described effectively in terms of the **cylindrical coordinates** R, θ, and z (Fig. 11.24). The unit vectors then should consist of $\mathbf{e}_R$, $\mathbf{e}_\theta$, and $\mathbf{k}$. The corresponding components of the velocity and the acceleration are given in Eqs. (11.48) and (11.49). Note that the radial distance R is always measured in a plane parallel to the xy plane, and be careful not to confuse the position vector $\mathbf{r}$ with its radial component $R\mathbf{e}_R$.

Problems

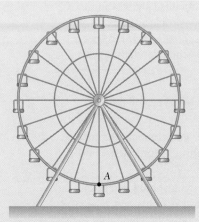

Fig. P11.CQ8

CONCEPT QUESTIONS

11.CQ8 The Ferris wheel is rotating with a constant angular velocity ω. What is the direction of the acceleration of point A?
a. → **b.** ↑ **c.** ↓ **d.** ← **e.** The acceleration is zero.

11.CQ9 A race car travels around the track shown at a constant speed. At which point will the race car have the largest acceleration?
a. A. **b.** B. **c.** C. **d.** D. **e.** The acceleration will be zero at all the points.

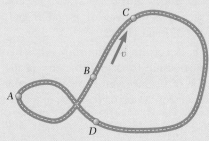

Fig. P11.CQ9

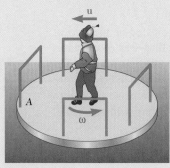

Fig. P11.CQ10

11.CQ10 A child walks across merry-go-round A with a constant speed u relative to A. The merry-go-round undergoes fixed-axis rotation about its center with a constant angular velocity ω counterclockwise. When the child is at the center of A, as shown, what is the direction of his acceleration when viewed from above?
a. → **b.** ← **c.** ↑ **d.** ↓ **e.** The acceleration is zero.

END-OF-SECTION PROBLEMS

11.133 Determine the smallest radius that should be used for a highway if the normal component of the acceleration of a car traveling at 72 km/h is not to exceed 0.8 m/s².

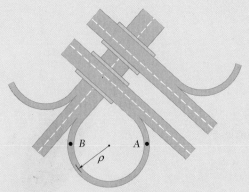

Fig. P11.133

11.134 Determine the maximum speed that the cars of the roller-coaster can reach along the circular portion AB of the track if $\rho = 25$ m and the normal component of their acceleration cannot exceed 3g.

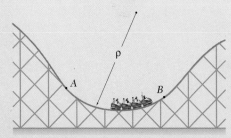

Fig. P11.134

11.135 Human centrifuges are often used to simulate different acceleration levels for pilots and astronauts. Space shuttle pilots typically face inwards towards the center of the gondola in order to experience a simulated 3-g forward acceleration. Knowing that the astronaut sits 5 m from the axis of rotation and experiences 3 g's inward, determine her velocity.

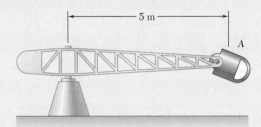

Fig. P11.135

11.136 The diameter of the eye of a stationary hurricane is 20 mi and the maximum wind speed is 100 mi/h at the eye wall with $r = 10$ mi. Assuming that the wind speed is constant for constant r and decreases uniformly with increasing r to 40 mi/h at $r = 110$ mi, determine the magnitude of the acceleration of the air at (a) $r = 10$ mi, (b) $r = 60$ mi, (c) $r = 110$ mi.

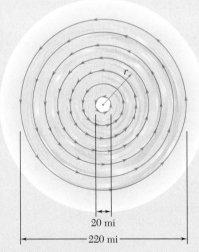

20 mi

220 mi

Fig. P11.136

11.137 The peripheral speed of the tooth of a 10-in.-diameter circular saw blade is 150 ft/s when the power to the saw is turned off. The speed of the tooth decreases at a constant rate, and the blade comes to rest in 9 s. Determine the time at which the total acceleration of the tooth is 130 ft/s^2.

11.138 A robot arm moves so that P travels in a circle about point B, which is not moving. Knowing that P starts from rest, and its speed increases at a constant rate of 10 mm/s^2, determine (a) the magnitude of the acceleration when $t = 4$ s, (b) the time for the magnitude of the acceleration to be 80 mm/s^2.

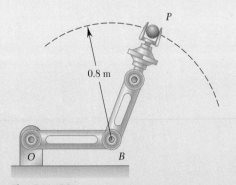

Fig. P11.138

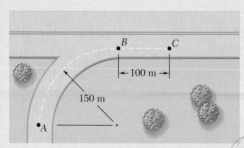

Fig. P11.140

11.139 A monorail train starts from rest on a curve of radius 400 m and accelerates at the constant rate a_t. If the maximum total acceleration of the train must not exceed 1.5 m/s^2, determine (a) the shortest distance in which the train can reach a speed of 72 km/h, (b) the corresponding constant rate of acceleration a_t.

11.140 A motorist starts from rest at point A on a circular entrance ramp when $t = 0$, increases the speed of her automobile at a constant rate and enters the highway at point B. Knowing that her speed continues to increase at the same rate until it reaches 100 km/h at point C, determine (a) the speed at point B, (b) the magnitude of the total acceleration when $t = 20$ s.

11.141 Race car A is traveling on a straight portion of the track while race car B is traveling on a circular portion of the track. At the instant shown, the speed of A is increasing at the rate of 10 m/s^2, and the speed of B is decreasing at the rate of 6 m/s^2. For the position shown, determine (a) the velocity of B relative to A, (b) the acceleration of B relative to A.

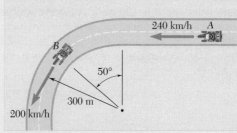

Fig. P11.141

11.142 At a given instant in an airplane race, airplane A is flying horizontally in a straight line, and its speed is being increased at the rate of 8 m/s^2. Airplane B is flying at the same altitude as airplane A and, as it rounds a pylon, is following a circular path of 300-m radius. Knowing that at the given instant the speed of B is being decreased at the rate of 3 m/s^2, determine, for the positions shown, (a) the velocity of B relative to A, (b) the acceleration of B relative to A.

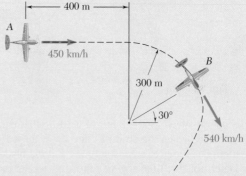

Fig. P11.142

11.143 A race car enters the circular portion of a track that has a radius of 70 m. When the car enters the curve at point P, it is travelling with a speed of 120 km/h that is increasing at 5 m/s^2. Three seconds later, determine the x and y components of velocity and acceleration of the car.

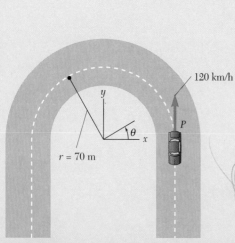

Fig. P11.143

11.144 An airplane flying at a constant speed of 240 m/s makes a banked horizontal turn. What is the minimum allowable radius of the turn if the structural specifications require that the acceleration of the airplane shall never exceed 4 g?

11.145 A golfer hits a golf ball from point A with an initial velocity of 50 m/s at an angle of 25° with the horizontal. Determine the radius of curvature of the trajectory described by the ball (a) at point A, (b) at the highest point of the trajectory.

11.146 Three children are throwing snowballs at each other. Child A throws a snowball with a horizontal velocity $\mathbf{v}_0$. If the snowball just passes over the head of child B and hits child C, determine the radius of curvature of the trajectory described by the snowball (a) at point B, (b) at point C.

Fig. P11.144

Fig. P11.145

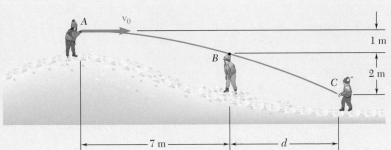

Fig. P11.146

11.147 Coal is discharged from the tailgate A of a dump truck with an initial velocity $\mathbf{v}_A = 2$ m/s ∠ 50°. Determine the radius of curvature of the trajectory described by the coal (a) at point A, (b) at the point of the trajectory 1 m below point A.

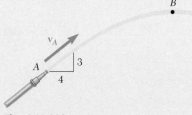

Fig. P11.147

11.148 From measurements of a photograph, it has been found that as the stream of water shown left the nozzle at A, it had a radius of curvature of 25 m. Determine (a) the initial velocity $\mathbf{v}_A$ of the stream, (b) the radius of curvature of the stream as it reaches its maximum height at B.

11.149 A child throws a ball from point A with an initial velocity $\mathbf{v}_0$ at an angle of 3° with the horizontal. Knowing that the ball hits a wall at point B, determine (a) the magnitude of the initial velocity, (b) the minimum radius of curvature of the trajectory.

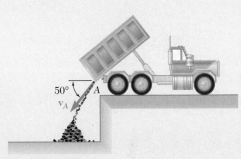

Fig. P11.148

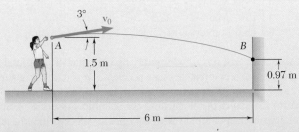

Fig. P11.149

11.150 A projectile is fired from point A with an initial velocity $\mathbf{v}_0$. (*a*) Show that the radius of curvature of the trajectory of the projectile reaches its minimum value at the highest point B of the trajectory. (*b*) Denoting by θ the angle formed by the trajectory and the horizontal at a given point C, show that the radius of curvature of the trajectory at C is $\rho = \rho_{min}/\cos^3\theta$.

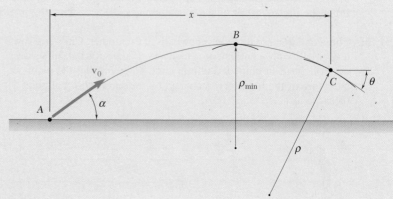

Fig. P11.150

*11.151 Determine the radius of curvature of the path described by the particle of Prob. 11.95 when $t = 0$.

*11.152 Determine the radius of curvature of the path described by the particle of Prob. 11.96 when $t = 0$, $A = 3$, and $B = 1$.

11.153 and 11.154 A satellite will travel indefinitely in a circular orbit around a planet if the normal component of the acceleration of the satellite is equal to $g(R/r)^2$, where g is the acceleration of gravity at the surface of the planet, R is the radius of the planet, and r is the distance from the center of the planet to the satellite. Knowing that the diameter of the sun is 1.39 Gm and that the acceleration of gravity at its surface is 274 m/s^2, determine the radius of the orbit of the indicated planet around the sun assuming that the orbit is circular.

  11.153 Earth: $(v_{mean})_{orbit} = 107$ Mm/h.
  11.154 Saturn: $(v_{mean})_{orbit} = 34.7$ Mm/h.

11.155 through 11.157 Determine the speed of a satellite relative to the indicated planet if the satellite is to travel indefinitely in a circular orbit 100 mi above the surface of the planet. (See information given in Probs. 11.153–11.154.)

  11.155 Venus: $g = 29.20$ ft/s^2, $R = 3761$ mi.
  11.156 Mars: $g = 12.17$ ft/s^2, $R = 2102$ mi.
  11.157 Jupiter: $g = 75.35$ ft/s^2, $R = 44,432$ mi.

11.158 A satellite will travel indefinitely in a circular orbit around the earth if the normal component of its acceleration is equal to $g(R/r)^2$, where $g = 9.81$ m/s^2, R = radius of the earth = 6370 km, and r = distance from the center of the earth to the satellite. Assuming that the orbit of the moon is a circle with a radius of 384×10^3 km, determine the speed of the moon relative to the earth.

11.159 Knowing that the radius of the earth is 6370 km, determine the time of one orbit of the Hubble Space Telescope if the telescope travels in a circular orbit 590 km above the surface of the earth. (See information given in Probs. 11.153–11.154.)

11.160 Satellites A and B are traveling in the same plane in circular orbits around the earth at altitudes of 120 and 200 mi, respectively. If at $t = 0$ the satellites are aligned as shown and knowing that the radius of the earth is $R = 3960$ mi, determine when the satellites will next be radially aligned. (See information given in Probs. 11.153–11.154.)

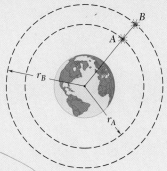

Fig. P11.160

11.161 The oscillation of rod OA about O is defined by the relation $\theta = (3/\pi)(\sin \pi t)$, where θ and t are expressed in radians and seconds, respectively. Collar B slides along the rod so that its distance from O is $r = 6(1 - e^{-2t})$ where r and t are expressed in inches and seconds, respectively. When $t = 1$ s, determine (a) the velocity of the collar, (b) the acceleration of the collar, (c) the acceleration of the collar relative to the rod.

11.162 The path of a particle P is a limaçon. The motion of the particle is defined by the relations $r = b(2 + \cos \pi t)$ and $\theta = \pi t$ where t and θ are expressed in seconds and radians, respectively. Determine (a) the velocity and the acceleration of the particle when $t = 2$ s, (b) the value of θ for which the magnitude of the velocity is maximum.

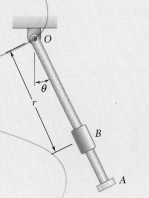

Fig. P11.161

11.163 During a parasailing ride, the boat is traveling at a constant 30 km/hr with a 200-m long tow line. At the instant shown, the angle between the line and the water is 30° and is increasing at a constant rate of 2°/s. Determine the velocity and acceleration of the parasailer at this instant.

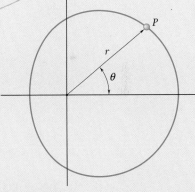

Fig. P11.162

Fig. P11.163

11.164 Some parasailing systems use a winch to pull the rider back to the boat. During the interval when θ is between 20° and 40° (where $t = 0$ at $\theta = 20°$), the angle increases at the constant rate of 2°/s. During this time, the length of the rope is defined by the relationship $r = 600 - \frac{1}{8}t^{5/2}$, where r and t are expressed in feet and seconds, respectively. Knowing that the boat is travelling at a constant rate of 15 knots (where 1 knot = 1.15 mi/h), (a) plot the magnitude of the velocity of the parasailer as a function of time, (b) determine the magnitude of the acceleration of the parasailer when $t = 5$ s.

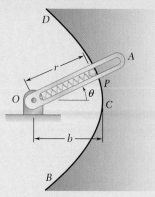

Fig. P11.165

11.165 As rod OA rotates, pin P moves along the parabola BCD. Knowing that the equation of this parabola is $r = 2b/(1 + \cos\theta)$ and that $\theta = kt$, determine the velocity and acceleration of P when (a) $\theta = 0$, (b) $\theta = 90°$.

11.166 The pin at B is free to slide along the circular slot DE and along the rotating rod OC. Assuming that the rod OC rotates at a constant rate $\dot\theta$, (a) show that the acceleration of pin B is of constant magnitude, (b) determine the direction of the acceleration of pin B.

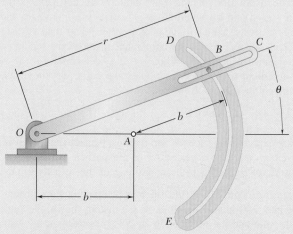

Fig. P11.166

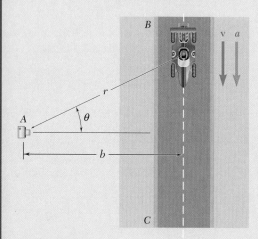

Fig. P11.167

11.167 To study the performance of a race car, a high-speed camera is positioned at point A. The camera is mounted on a mechanism which permits it to record the motion of the car as the car travels on straightaway BC. Determine (a) the speed of the car in terms of b, θ, and $\dot\theta$, (b) the magnitude of the acceleration in terms of b, θ, $\dot\theta$, and $\ddot\theta$.

11.168 After taking off, a helicopter climbs in a straight line at a constant angle β. Its flight is tracked by radar from point A. Determine the speed of the helicopter in terms of d, β, θ, and $\dot\theta$.

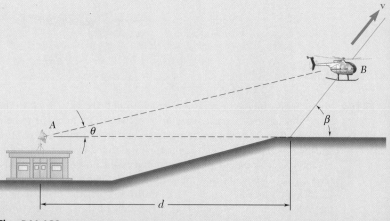

Fig. P11.168

11.169 At the bottom of a loop in the vertical plane an airplane has a horizontal velocity of 315 mi/h and is speeding up at a rate of 10 ft/s². The radius of curvature of the loop is 1 mi. The plane is being tracked by radar at O. What are the recorded values of $\dot{r}$, $\ddot{r}$, $\dot{\theta}$, and $\ddot{\theta}$ for this instant?

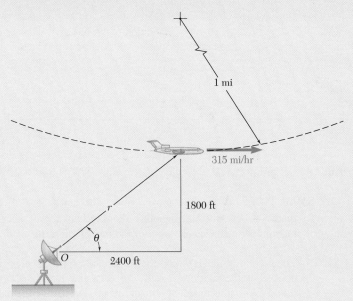

Fig. P11.169

11.170 Pin C is attached to rod BC and slides freely in the slot of rod OA which rotates at the constant rate ω. At the instant when $\beta = 60°$, determine (a) $\dot{r}$ and $\dot{\theta}$, (b) $\ddot{r}$ and $\ddot{\theta}$. Express your answers in terms of d and ω.

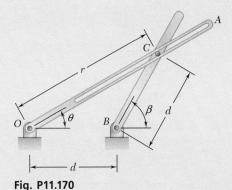

Fig. P11.170

11.171 For the race car of Prob. 11.167, it was found that it took 0.5 s for the car to travel from the position $\theta = 60°$ to the position $\theta = 35°$. Knowing that $b = 25$ m, determine the average speed of the car during the 0.5-s interval.

11.172 For the helicopter of Prob. 11.168, it was found that when the helicopter was at B, the distance and the angle of elevation of the helicopter were $r = 3000$ ft and $\theta = 20°$, respectively. Four seconds later, the radar station sighted the helicopter at $r = 3320$ ft and $\theta = 23.1°$. Determine the average speed and the angle of climb β of the helicopter during the 4-s interval.

11.173 and 11.174 A particle moves along the spiral shown. Determine the magnitude of the velocity of the particle in terms of b, θ, and $\dot{\theta}$.

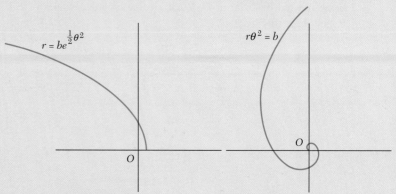

$r = be^{\frac{1}{2}\theta^2}$

$r\theta^2 = b$

Fig. P11.173 and P11.175   **Fig. P11.174 and P11.176**

11.175 and 11.176 A particle moves along the spiral shown. Knowing that $\dot{\theta}$ is constant and denoting this constant by ω, determine the magnitude of the acceleration of the particle in terms of b, θ, and $\dot{\theta}$.

11.177 The motion of a particle on the surface of a right circular cylinder is defined by the relations $R = A$, $\theta = 2\pi t$, and $z = B\sin 2\pi nt$, where A and B are constants and n is an integer. Determine the magnitudes of the velocity and acceleration of the particle at any time t.

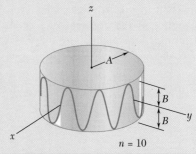

$n = 10$

Fig. P11.177

11.178 Show that $\dot{r} = h\dot{\phi}\sin\theta$ knowing that at the instant shown, step AB of the step exerciser is rotating counterclockwise at a constant rate $\dot{\phi}$.

11.179 The three-dimensional motion of a particle is defined by the relations $R = A(1 - e^{-t})$, $\theta = 2\pi t$, and $z = B(1 - e^{-t})$. Determine the magnitudes of the velocity and acceleration when (a) $t = 0$, (b) $t = \infty$.

*11.180 For the conic helix of Prob. 11.95, determine the angle that the osculating plane forms with the y axis.

*11.181 Determine the direction of the binormal of the path described by the particle of Prob. 11.96 when (a) $t = 0$, (b) $t = \pi/2$ s.

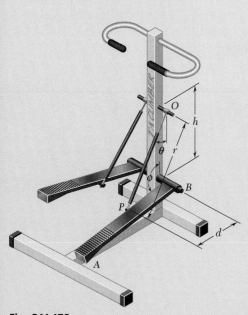

Fig. P11.178

Review and Summary

Position Coordinate of a Particle in Rectilinear Motion

In the first half of this chapter, we analyzed the **rectilinear motion of a particle**, i.e., the motion of a particle along a straight line. To define the position P of the particle on that line, we chose a fixed origin O and a positive direction (Fig. 11.25). The distance x from O to P, with the appropriate sign, completely defines the position of the particle on the line and is called the **position coordinate** of the particle [Sec. 11.1A].

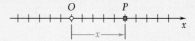

Fig. 11.25

Velocity and Acceleration in Rectilinear Motion

The **velocity** v of the particle was shown to be equal to the time derivative of the position coordinate x, so

$$v = \frac{dx}{dt} \tag{11.1}$$

And we obtained the **acceleration** a by differentiating v with respect to t, as

$$a = \frac{dv}{dt} \tag{11.2}$$

or

$$a = \frac{d^2x}{dt^2} \tag{11.3}$$

We also noted that a could be expressed as

$$a = v\frac{dv}{dx} \tag{11.4}$$

We observed that the velocity v and the acceleration a are represented by algebraic numbers that can be positive or negative. A positive value for v indicates that the particle moves in the positive direction, and a negative value shows that it moves in the negative direction. A positive value for a, however, may mean that the particle is truly accelerated (i.e., moves faster) in the positive direction or that it is decelerated (i.e., moves more slowly) in the negative direction. A negative value for a is subject to a similar interpretation [Sample Prob. 11.1].

Determination of the Velocity and Acceleration by Integration

In most problems, the conditions of motion of a particle are defined by the type of acceleration that the particle possesses and by the initial conditions [Sec. 11.1B]. Then we can obtain the velocity and position of the particle by integrating two of the equations (11.1) to (11.4). The selection of these equations depends upon the type of acceleration involved [Sample Probs. 11.2 through 11.4].

Uniform Rectilinear Motion

Two types of motion are frequently encountered. **Uniform rectilinear motion** [Sec. 11.2A], in which the velocity v of the particle is constant, is described by

$$x = x_0 + vt \tag{11.5}$$

Uniformly Accelerated Rectilinear Motion

Uniformly accelerated rectilinear motion [Sec. 11.2B], in which the acceleration a of the particle is constant, is described by

$$v = v_0 + at \tag{11.6}$$
$$x = x_0 + v_0 t + \tfrac{1}{2}at^2 \tag{11.7}$$
$$v^2 = v_0^2 + 2a(x - x_0) \tag{11.8}$$

Relative Motion of Two Particles

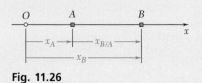

Fig. 11.26

When two particles A and B (such as two aircraft) move, we may wish to consider the **relative motion** of B with respect to A [Sec. 11.2C]. Denoting the **relative position coordinate** of B with respect to A by $x_{B/A}$ (Fig. 11.26), we have

$$x_B = x_A + x_{B/A} \tag{11.9}$$

Differentiating Eq. (11.9) twice with respect to t, we obtained successively

$$v_B = v_A + v_{B/A} \tag{11.10}$$
$$a_B = a_A + a_{B/A} \tag{11.11}$$

where $v_{B/A}$ and $a_{B/A}$ represent, respectively, the **relative velocity** and the **relative acceleration** of B with respect to A.

Dependent Motion

When several blocks are **connected by inextensible cords**, it is possible to write a linear relation between their position coordinates. We can then write similar relations between their velocities and between their accelerations, which we can use to analyze their motion [Sample Probs. 11.7 and 11.8].

Graphical Solutions

It is sometimes convenient to use a **graphical solution** for problems involving the rectilinear motion of a particle [Sec. 11.3]. The graphical solution most commonly used involves the x–t, v–t, and a–t curves [Sample Prob. 11.10]. It was shown at any given time t that

$$v = \text{slope of } x\text{–}t \text{ curve}$$
$$a = \text{slope of } v\text{–}t \text{ curve}$$

Also, over any given time interval from t_1 to t_2, we have

$$v_2 - v_1 = \text{area under } a\text{–}t \text{ curve}$$
$$x_2 - x_1 = \text{area under } v\text{–}t \text{ curve}$$

Position Vector and Velocity in Curvilinear Motion

In the second half of this chapter, we analyzed the **curvilinear motion of a particle**, i.e., the motion of a particle along a curved path. We defined the position P of the particle at a given time [Sec. 11.4A] by the **position vector r** joining the O of the coordinates and point P (Fig. 11.27). We defined the **velocity v** of the particle by the relation

$$\mathbf{v} = \frac{d\mathbf{r}}{dt} \qquad (11.14)$$

The velocity is a **vector tangent to the path of the particle** with a magnitude v (called the **speed** of the particle) equal to the time derivative of the length s of the arc described by the particle. Thus,

$$v = \frac{ds}{dt} \qquad (11.15)$$

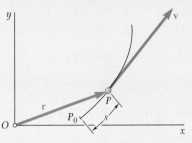

Fig. 11.27

Acceleration in Curvilinear Motion

We defined the **acceleration a** of the particle by the relation

$$\mathbf{a} = \frac{d\mathbf{v}}{dt} \qquad (11.17)$$

and we noted that, in general, *the acceleration is not tangent to the path of the particle.*

Derivative of a Vector Function

Before proceeding to the consideration of the components of velocity and acceleration, we reviewed the formal definition of the derivative of a vector function and established a few rules governing the differentiation of sums and products of vector functions. We then showed that the rate of change of a vector is the same with respect both to a fixed frame and to a frame in translation [Sec. 11.4B].

Rectangular Components of Velocity and Acceleration

Denoting the rectangular coordinates of a particle P by x, y, and z, we found that the rectangular components of the velocity and acceleration of P equal, respectively, the first and second derivatives with respect to t of the corresponding coordinates. Thus,

$$v_x = \dot{x} \qquad v_y = \dot{y} \qquad v_z = \dot{z} \qquad (11.28)$$
$$a_x = \ddot{x} \qquad a_y = \ddot{y} \qquad a_z = \ddot{z} \qquad (11.29)$$

Component Motions

When the component a_x of the acceleration depends only upon t, x, and/or v_x; when, similarly, a_y depends only upon t, y, and/or v_y; and a_z upon t, z, and/or v_z, Eq. (11.29) can be integrated independently. The analysis of the given curvilinear motion then reduces to the analysis of three independent rectilinear component motions [Sec. 11.4C]. This approach is particularly effective in the study of the motion of projectiles [Sample Probs. 11.10 and 11.11].

Relative Motion of Two Particles

For two particles A and B moving in space (Fig. 11.28), we considered the relative motion of B with respect to A, or more precisely, with respect to a moving frame attached to A and in translation with A [Sec. 11.4D]. Denoting the **relative position vector** of B with respect to A by $\mathbf{r}_{B/A}$ (Fig. 11.28), we have

$$\mathbf{r}_B = \mathbf{r}_A + \mathbf{r}_{B/A} \qquad (11.30)$$

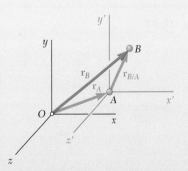

Fig. 11.28

Denoting the **relative velocity** and the **relative acceleration** of B with respect to A by $\mathbf{v}_{B/A}$ and $\mathbf{a}_{B/A}$, respectively, we also showed that

$$\mathbf{v}_B = \mathbf{v}_A + \mathbf{v}_{B/A} \tag{11.32}$$

and

$$\mathbf{a}_B = \mathbf{a}_A + \mathbf{a}_{B/A} \tag{11.33}$$

Tangential and Normal Components

It is sometimes convenient to resolve the velocity and acceleration of a particle P into components other than the rectangular x, y, and z components. For a particle P moving along a path contained in a plane, we attached to P unit vectors $\mathbf{e}_t$ tangent to the path and $\mathbf{e}_n$ normal to the path and directed toward the center of curvature of the path [Sec. 11.5A]. We then express the velocity and acceleration of the particle in terms of tangential and normal components. We have

$$\mathbf{v} = v\mathbf{e}_t \tag{11.35}$$

and

$$\mathbf{a} = \frac{dv}{dt}\mathbf{e}_t + \frac{v^2}{\rho}\mathbf{e}_n \tag{11.38}$$

where v is the speed of the particle and ρ is the radius of curvature of its path [Sample Probs. 11.16, ,and 11.17]. We observed that, while the velocity $\mathbf{v}$ is directed along the tangent to the path, the acceleration $\mathbf{a}$ consists of a component $\mathbf{a}_t$ directed along the tangent to the path and a component $\mathbf{a}_n$ directed toward the center of curvature of the path (Fig. 11.29).

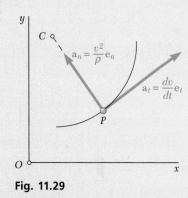

Fig. 11.29

Motion Along a Space Curve

For a particle P moving along a space curve, we defined the plane that most closely fits the curve in the neighborhood of P as the **osculating plane**. This plane contains the unit vectors $\mathbf{e}_t$ and $\mathbf{e}_n$ that define the tangent and principal normal to the curve, respectively. The unit vector $\mathbf{e}_b$, which is perpendicular to the osculating plane, defines the **binormal**.

Radial and Transverse Components

When the position of a particle P moving in a plane is defined by its polar coordinates r and θ, it is convenient to use radial and transverse components directed, respectively, along the position vector $\mathbf{r}$ of the particle and in the direction obtained by rotating $\mathbf{r}$ through $90°$ counterclockwise [Sec. 11.5B]. We attached to P unit vectors $\mathbf{e}_r$ and $\mathbf{e}_\theta$ directed in the radial and transverse directions, respectively (Fig. 11.30). We then expressed the velocity and acceleration of the particle in terms of radial and transverse components as

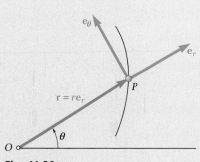

Fig. 11.30

$$\mathbf{v} = \dot{r}\mathbf{e}_r + r\dot{\theta}\mathbf{e}_\theta \tag{11.42}$$
$$\mathbf{a} = (\ddot{r} - r\dot{\theta}^2)\mathbf{e}_r + (r\ddot{\theta} + 2\dot{r}\dot{\theta})\mathbf{e}_\theta \tag{11.43}$$

where dots are used to indicate differentiation with respect to time. The scalar components of the velocity and acceleration in the radial and transverse directions are therefore

$$v_r = \dot{r} \qquad v_\theta = r\dot{\theta} \tag{11.44}$$
$$a_r = \ddot{r} - r\dot{\theta}^2 \qquad a_\theta = r\ddot{\theta} + 2\dot{r}\dot{\theta} \tag{11.45}$$

It is important to note that a_r is *not* equal to the time derivative of v_r and that a_θ is *not* equal to the time derivative of v_θ [Sample Probs. 11.18, 11.19, and 11.20].

This chapter ended with a discussion of the use of cylindrical coordinates to define the position and motion of a particle in space.

Review Problems

11.182 The motion of a particle is defined by the relation $x = 2t^3 - 15t^2 + 24t + 4$, where x and t are expressed in meters and seconds, respectively. Determine (*a*) when the velocity is zero, (*b*) the position and the total distance traveled when the acceleration is zero.

11.183 A drag car starts from rest and moves down the racetrack with an acceleration defined by $a = 50 - 10t$, where a and t are in m/s² and seconds, respectively. After reaching a speed of 125 m/s, a parachute is deployed to help slow down the dragster. Knowing that this deceleration is defined by the relationship $a = -0.02v^2$, where v is the velocity in m/s, determine (*a*) the total time from the beginning of the race until the car slows back down to 10 m/s, (*b*) the total distance the car travels during this time.

11.184 A particle moves in a straight line with the acceleration shown in the figure. Knowing that the particle starts from the origin with $v_0 = -2$ m/s, (*a*) construct the v–t and x–t curves for $0 < t < 18$ s, (*b*) determine the position and the velocity of the particle and the total distance traveled when $t = 18$ s.

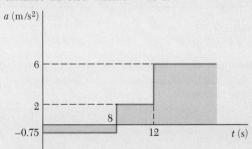

Fig. P11.184

11.185 The velocities of commuter trains A and B are as shown. Knowing that the speed of each train is constant and that B reaches the crossing 10 min after A passed through the same crossing, determine (*a*) the relative velocity of B with respect to A, (*b*) the distance between the fronts of the engines 3 min after A passed through the crossing.

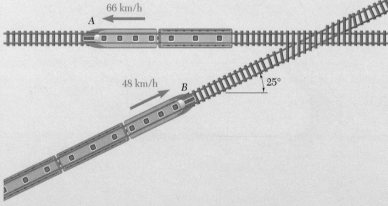

Fig. P11.185

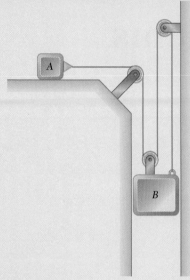

Fig. P11.186

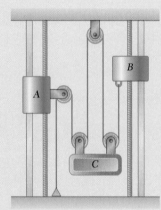

Fig. P11.187

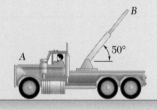

Fig. P11.189

11.186 Knowing that slider block *A* starts from rest and moves to the left with a constant acceleration of 1 ft/s², determine (*a*) the relative acceleration of block *A* with respect to block *B*, (*b*) the velocity of block *B* after 2 s.

11.187 Collar *A* starts from rest at *t* = 0 and moves downward with a constant acceleration of 7 in./s². Collar *B* moves upward with a constant acceleration, and its initial velocity is 8 in./s. Knowing that collar *B* moves through 20 in. between *t* = 0 and *t* = 2 s, determine (*a*) the accelerations of collar *B* and block *C*, (*b*) the time at which the velocity of block *C* is zero, (*c*) the distance through which block *C* will have moved at that time.

11.188 A golfer hits a ball with an initial velocity of magnitude v_0 at an angle α with the horizontal. Knowing that the ball must clear the tops of two trees and land as close as possible to the flag, determine v_0 and the distance *d* when the golfer uses (*a*) a six-iron with $\alpha = 31°$, (*b*) a five-iron with $\alpha = 27°$.

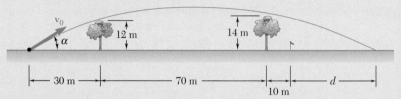

Fig. P11.188

11.189 As the truck shown begins to back up with a constant acceleration of 4 ft/s², the outer section *B* of its boom starts to retract with a constant acceleration of 1.6 ft/s² relative to the truck. Determine (*a*) the acceleration of section *B*, (*b*) the velocity of section *B* when *t* = 2 s.

11.190 A velodrome is a specially designed track used in bicycle racing that has constant radius curves at each end. Knowing that a rider starts from rest $a_t = (11.46 - 0.01878v^2)$ m/s², determine her acceleration at point *B*.

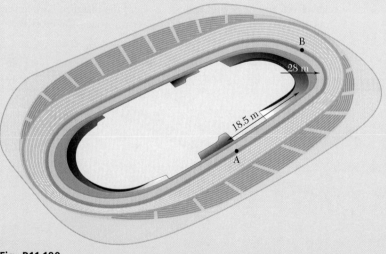

Fig. P11.190

716

11.191 Sand is discharged at A from a conveyor belt and falls onto the top of a stockpile at B. Knowing that the conveyor belt forms an angle $\alpha = 25°$ with the horizontal, determine (a) the speed v_0 of the belt, (b) the radius of curvature of the trajectory described by the sand at point B.

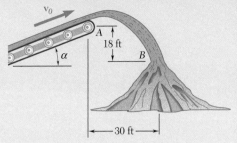

Fig. P11.191

11.192 The end point B of a boom is originally 5 m from fixed point A when the driver starts to retract the boom with a constant radial acceleration of $\ddot{r} = -1.0$ m/s^2 and lower it with a constant angular acceleration $\ddot{\theta} = -0.5$ rad/s^2. At $t = 2$ s, determine (a) the velocity of point B, (b) the acceleration of point B, (c) the radius of curvature of the path.

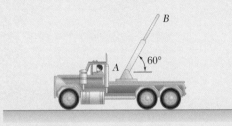

Fig. P11.192

11.193 A telemetry system is used to quantify kinematic values of a ski jumper immediately before she leaves the ramp. According to the system $r = 500$ ft, $\dot{r} = -105$ ft/s, $\ddot{r} = -10$ ft/s^2, $\theta = 25°$, $\dot{\theta} = 0.07$ rad/s, $\ddot{\theta} = 0.06$ rad/s^2. Determine (a) the velocity of the skier immediately before she leaves the jump, (b) the acceleration of the skier at this instant, (c) the distance of the jump d neglecting lift and air resistance.

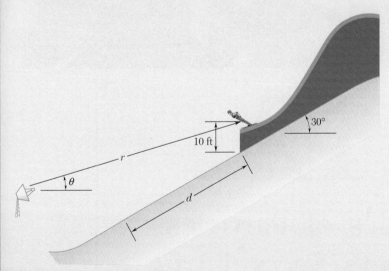

Fig. P11.193

12

Kinetics of Particles:
Newton's Second Law

The forces experienced by the passengers on a roller coaster will depend on whether the roller-coaster car is traveling up a hill or down a hill, in a straight line, or along a horizontal or vertical curved path. The relation existing among force, mass, and acceleration will be studied in this chapter.

Objectives

- **Explain** the relationships between mass, force, and acceleration.
- **Model** physical systems by drawing complete free-body diagrams and kinetic diagrams.
- **Apply** Newton's second law of motion to solve particle kinetics problems using different coordinate systems.
- **Analyze** central force motion problems using principles of angular momentum and Newton's law of gravitation.

Introduction

In statics, we used Newton's first and third laws of motion extensively to study bodies at rest and the forces acting upon them. We also use these two laws in dynamics; in fact, they are sufficient for analyzing the motion of bodies that have no acceleration. However, when a body is accelerated—that is, when the magnitude or the direction of its velocity changes—it is necessary to use Newton's second law of motion to relate the motion of the body to the forces acting on it.

In this chapter, we discuss Newton's second law and apply it to analyzing the motion of particles. According to the second law, if the resultant of the forces acting on a particle is not zero, the particle has an acceleration proportional to the magnitude of the resultant and in the direction of this resultant force. Moreover, we use the ratio of the magnitudes of the resultant force and of the acceleration to define the *mass* of the particle. In Sec. 12.1B, we define the *linear momentum* of a particle as the product $\mathbf{L} = m\mathbf{v}$ of the mass m and velocity $\mathbf{v}$ of the particle. Then we can express Newton's second law in an alternative form, relating the rate of change of the linear momentum to the resultant of the forces acting on that particle.

In the Sample Problems, we apply Newton's second law to the solution of engineering problems using either rectangular components, tangential and normal components, or radial and transverse coordinates of the forces and accelerations involved. Recall that we can consider an actual body—including bodies as large as a car, rocket, or airplane—as a particle for the purpose of analyzing its motion, as long as the effect of a rotation of the body about its center of mass can be ignored. We stress the need for consistent units in solving these problems, briefly reviewing the International System of Units (SI units) and the system of U.S. customary units.

The second part of this chapter is devoted to the study of the motion of particles under central forces. We define the *angular momentum* $\mathbf{H}_O$ of a particle about a point O as the moment about O of the linear momentum of the particle: $\mathbf{H}_O = \mathbf{r} \times m\mathbf{v}$. It then follows from Newton's second law that the rate of change of the angular momentum $\mathbf{H}_O$ of a particle is equal to the sum of the moments about O of the forces acting on that particle.

We can use this form of the second law to deal with the motion of a particle under a *central force,* i.e., under a force directed toward or away from a fixed point O. Since such a force has zero moment about O, it follows that the angular momentum of the particle about O is conserved. This property greatly simplifies the analysis of the motion, as we show by solving problems involving the orbital motion of bodies under gravitational attraction.

In Sec.12.3, which is optional, we present a more extensive discussion of orbital motion, including several problems related to space mechanics.

12.1 NEWTON'S SECOND LAW AND LINEAR MOMENTUM

In statics, we dealt with forces acting on particles that led to a state of equilibrium. Now we study forces acting on particles that lead to a state of motion. The key relationship connecting force and motion is Newton's second law.

12.1A Newton's Second Law of Motion

We can state Newton's second law as follows:

If the resultant force acting on a particle is not zero, the particle has an acceleration proportional to the magnitude of the resultant and in the direction of this resultant force.

Newton's second law of motion is best understood by imagining the following experiment: A particle is subjected to a force $\mathbf{F}_1$ of constant direction and constant magnitude F_1. Under the action of that force, the particle moves in a straight line and *in the direction of the force* (Fig. 12.1a). By determining the position of the particle at various instants, we find that its acceleration has a constant magnitude a_1. If we repeat the experiment with forces $\mathbf{F}_2$, $\mathbf{F}_3$, . . . of a different magnitude or direction (Fig. 12.1b and c), we find each time that the particle moves in the direction of the force acting on it and that the magnitudes $a_1, a_2, a_3, . . .$ of the accelerations are proportional to the magnitudes $F_1, F_2, F_3, . . .$ of the corresponding forces. Thus,

$$\frac{F_1}{a_1} = \frac{F_2}{a_2} = \frac{F_3}{a_3} = \cdots = \text{constant}$$

The constant value obtained for the ratio of the magnitudes of the forces and accelerations is a characteristic of the particle under consideration; it is called the **mass** of the particle and is denoted by m. When a particle of mass m is acted upon by a force $\mathbf{F}$, the force $\mathbf{F}$ and the acceleration $\mathbf{a}$ of the particle must therefore satisfy the relation

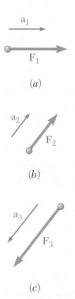

Fig. 12.1 Experiments show that a force applied to a particle gives the particle an acceleration proportional to the magnitude of the force and in the same direction as the force.

Newton's second law $\mathbf{F} = m\mathbf{a}$ (12.1)

This relation provides a complete formulation of Newton's second law; it states not only that the magnitudes of **F** and **a** are proportional, but also (since m is a positive scalar) that the vectors **F** and **a** have the same direction (Fig. 12.2). Note that Eq. (12.1) still holds when **F** is not constant, but varies with time in magnitude or direction. The magnitudes of **F** and **a** remain proportional, and the two vectors have the same direction at any given instant. However, they are not, in general, tangent to the path of the particle.

When a particle is subjected simultaneously to several forces, Eq. (12.1) should be replaced by

Newton's second law, multiple forces

$$\Sigma \mathbf{F} = m\mathbf{a} \tag{12.2}$$

where $\Sigma \mathbf{F}$ represents the sum or resultant of all the forces acting on the particle.

Note that the system of axes with respect to which we determine the acceleration **a** is not arbitrary. These axes must have a constant orientation with respect to the stars, and their origin either must be attached to the sun (more accurately, to the center of mass of the solar system) or move with a constant velocity with respect to the sun. Such a system of axes is called a **newtonian frame of reference**[†]. A system of axes attached to the earth does *not* constitute a newtonian frame of reference, since the earth rotates with respect to the stars and is accelerated with respect to the sun. However, in most engineering applications, we can determine the accele-ration **a** with respect to axes attached to the earth and use Eqs. (12.1) and (12.2) without any appreciable error. However, these equations do not hold if **a** represents a relative acceleration measured with respect to moving axes, such as axes attached to an accelerated car or to a rotating piece of machinery.

If the resultant $\Sigma \mathbf{F}$ of the forces acting on the particle is zero, it follows from Eq. (12.2) that the acceleration **a** of the particle is also zero. If the particle is initially at rest ($\mathbf{v}_0 = 0$) with respect to the newtonian frame of reference used, it will thus remain at rest ($\mathbf{v} = 0$). If originally moving with a velocity $\mathbf{v}_0$, the particle will maintain a constant velocity $\mathbf{v} = \mathbf{v}_0$; that is, it will move with the constant speed v_0 in a straight line. This, we recall, is the statement of Newton's first law (Sec. 2.3B); thus, Newton's first law is a particular case of Newton's second law.

12.1B Linear Momentum of a Particle and its Rate of Change

Suppose we replace the acceleration **a** in Eq. (12.2) by the derivative $d\mathbf{v}/dt$. We have

$$\Sigma \mathbf{F} = m\frac{d\mathbf{v}}{dt}$$

Since the mass m of the particle is constant, we can write this as

$$\Sigma \mathbf{F} = \frac{d}{dt}(m\mathbf{v}) \tag{12.3}$$

[†]Stars are not actually fixed, so a more rigorous definition of a newtonian frame of reference (also called an *inertial system*) is one with respect to which Eq. (12.2) holds.

Fig. 12.2 By Newton's second law, the proportionality constant between an applied force and the resulting acceleration is the particle's mass m.

Photo 12.1 When the racecar accelerates forward, the rear tires have a friction force acting on them in the direction the car is moving.

Fig. 12.3 Linear momentum is the product of the mass m and the velocity **v** of a particle. It is a vector in the same direction as the velocity.

The product $m\mathbf{v}$ is called the **linear momentum**, or simply the **momentum**, of the particle. It has the same direction as the velocity of the particle, and its magnitude is equal to the product of the mass m and the speed v of the particle (Fig. 12.3). Equation (12.3) says:

> **The resultant of the forces acting on the particle is equal to the rate of change of the linear momentum of the particle.**

The second law of motion was originally stated by Newton in this form. Denoting the linear momentum of the particle by **L**, we have

Linear momentum
$$\mathbf{L} = m\mathbf{v} \tag{12.4}$$

If we denote its derivative with respect to t as $\dot{\mathbf{L}}$, we can write Eq. (12.3) in the alternative form as

Newton's second law, momentum form
$$\Sigma\mathbf{F} = \dot{\mathbf{L}} \tag{12.5}$$

We assumed that the mass m of the particle is constant in Eqs. (12.3) through (12.5). Therefore, you should not use Equation (12.3) or (12.5) to solve problems involving the motion of bodies, such as rockets, that gain or lose mass. We will consider problems of that type in Sec. 14.3B.[†]

It follows from Eq. (12.3) that the rate of change of the linear momentum $m\mathbf{v}$ is zero when $\Sigma\mathbf{F} = 0$. Thus, we have the statement:

> **If the resultant force acting on a particle is zero, the linear momentum of the particle remains constant in both magnitude and direction.**

This is the principle of **conservation of linear momentum** for a particle.

12.1C Systems of Units

In using the fundamental equation $\mathbf{F} = m\mathbf{a}$, the units of force, mass, length, and time cannot be chosen arbitrarily. If they are, the magnitude of the force **F** required to give an acceleration **a** to the mass m will *not* be numerically equal to the product ma; it will only be proportional to this product. Thus, we can choose three of the four units arbitrarily, but we must choose the fourth unit so that the equation $\mathbf{F} = m\mathbf{a}$ is satisfied. The units are then said to form a system of consistent kinetic units.

Two systems of consistent kinetic units are currently used by American engineers: the International System of Units (SI units[‡]) and the system of U.S. customary units. Both systems were discussed in detail in Sec. 1.3, so we describe them only briefly in this section.

International System of Units (SI Units).   In this system, the base units are the units of length, mass, and time and are called, respectively, the *meter (m)*, the *kilogram* (kg), and the *second* (s). All three are

---

[†]Note that Eqs. (12.3) and (12.5) do hold in relativistic mechanics, where the mass m of the particle is assumed to vary with the speed of the particle.

[‡]SI stands for *Système International d'Unités* (French).

arbitrarily defined (Sec. 1.3). The unit of force is a derived unit. It is called the *newton* (N) and is defined as the force that gives an acceleration of 1 m/s^2 to a mass of 1 kg (Fig. 12.4). From Eq. (12.1), we have

$$1 \text{ N} = (1 \text{ kg})(1 \text{ m/s}^2) = 1 \text{ kg·m/s}^2$$

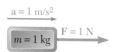

Fig. 12.4 A force of 1 newton gives a 1-kilogram mass an acceleration of 1 m/s^2.

The SI units are said to form an *absolute* system of units. This means that the three base units chosen are independent of the location where measurements are made. The meter, the kilogram, and the second may be used anywhere on the earth; they may even be used on another planet. They always have the same meaning.

The *weight* **W** of a body, or the *force of gravity* exerted on that body, should, like any other force, be expressed in newtons. A body subjected only to its own weight acquires an acceleration equal to the acceleration due to gravity g. (Be careful using the term *acceleration due to gravity,* since the only time an object accelerates with a magnitude g is during free-fall in the absence of drag.) It follows from Newton's second law that the magnitude W of the weight of a body of mass m is

$$W = mg \tag{12.6}$$

Recall that $g = 9.81$ m/s^2, so the weight of a body of mass 1 kg (Fig. 12.5) is

$$W = (1 \text{ kg})(9.81 \text{ m/s}^2) = 9.81 \text{ N}$$

This value would be much less on the moon, where the acceleration due to gravity is 1.6249 m/s^2.

Multiples and submultiples of the units of length, mass, and force are frequently used in engineering practice. They are, respectively, the *kilometer* (km) and the *millimeter* (mm); the *megagram* (Mg, which is also called the metric ton) and the *gram* (g); and the *kilonewton* (kN). By definition,

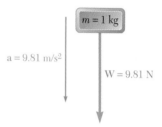

Fig. 12.5 In the SI system, a block with mass 1 kg has a weight of 9.81 N.

$$1 \text{ km} = 1000 \text{ m} \qquad 1 \text{ mm} = 0.001 \text{ m}$$
$$1 \text{ Mg} = 1000 \text{ kg} \qquad 1 \text{ g} = 0.001 \text{ kg}$$
$$1 \text{ kN} = 1000 \text{ N}$$

You can convert these units to meters, kilograms, and newtons, respectively, simply by moving the decimal point three places to the right or to the left.

Units other than those of mass, length, and time all can be expressed in terms of these three base units. For example, we can obtain the unit of linear momentum by recalling the definition and writing

$$mv = (\text{kg})(\text{m/s}) = \text{kg·m/s}$$

U.S. Customary Units.   Most practicing American engineers still commonly use a system in which the base units are those of length, force, and time. These units are, respectively, the *foot* (ft), the *pound* (lb), and the *second* (s). The second is the same as the corresponding SI unit. The foot is equal to 0.3048 m. The pound is defined as the *weight* of a platinum standard, called the *standard pound,* which is kept at the National Institute of Standards and Technology outside Washington, D.C. The mass of this standard is 0.453 592 43 kg. Since the weight of a body depends upon the gravitational attraction of the earth, which varies with location, the standard

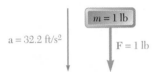

Fig. 12.6 In the U.S. customary system, a block with a weight of 1 lb in free fall has an acceleration of 32.2 ft/s².

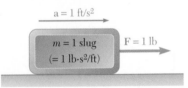

Fig. 12.7 In the U.S. customary system, a force of 1 lb applied to a block with a mass of 1 slug produces an acceleration of 1 ft/s².

pound should be placed at sea level and at a latitude of 45° to properly define a force of 1 lb. Clearly, the U.S. customary units do not form an absolute system of units. Because of their dependence upon the earth's gravitational attraction, they are said to form a *gravitational* system of units.

Although the standard pound also serves as the unit of mass in commercial transactions in the United States, it cannot be used that way in engineering computations because such a unit would not be consistent with the base units defined in this system. Indeed, when acted upon by a force of 1 lb—that is, when subjected to its own weight—the standard pound receives the acceleration of gravity, $g = 32.2$ ft/s² (Fig. 12.6), and not the unit acceleration required by Eq. (12.1). The unit of mass consistent with the foot, the pound, and the second is the mass that receives an acceleration of 1 ft/s² when a force of 1 lb is applied to it (Fig. 12.7). This unit, sometimes called a *slug*, can be derived from the equation $F = ma$ after substituting 1 lb and 1 ft/s² for F and a, respectively. We have

$$F = ma \qquad 1\text{ lb} = (1\text{ slug})(1\text{ ft/s}^2)$$

From this, we obtain

$$1\text{ slug} = \frac{1\text{ lb}}{1\text{ ft/s}^2} = 1\text{ lb·s}^2/\text{ft}$$

Comparing Figs. 12.6 and 12.7, we conclude that the slug is a mass 32.2 times larger than the mass of the standard pound. (On a horizontal surface, when acted on by a force of 1 pound, the motion of the larger mass is relatively "sluggish.")

The fact that bodies are characterized in the U.S. customary system of units by their weight in pounds rather than by their mass in slugs was convenient in the study of statics, where we were dealing (for the most part) with weights and other forces and only seldom with masses. However, in the study of kinetics, which involves forces, masses, and accelerations, we will often have to express the mass m of a body in slugs, the weight W of which is given in pounds. Recalling Eq. (12.6), we have

$$m = \frac{W}{g} \tag{12.7}$$

where g is the acceleration due to gravity ($g = 32.2$ ft/s²).

Units other than the units of force, length, and time all can be expressed in terms of these three base units. For example, we can obtain the unit of linear momentum from its definition as

$$mv = (\text{slug})(\text{ft/s}) = (\text{lb·s}^2/\text{ft})(\text{ft/s}) = \text{lb·s}$$

Conversion from One System of Units to Another.   The conversion from U.S. customary units to SI units, and vice versa, was discussed in Sec. 1.4. Recall that the conversion factors obtained for the units of length, force, and mass are, respectively,

Length:	1 ft = 0.3048 m
Force:	1 lb = 4.448 N
Mass:	1 slug = 1 lb·s²/ft = 14.59 kg

Thermodynamicists often use a unit called the pound-mass (lbm); this is not a unit that is consistent with Newton's second law, and whenever we use pounds in dynamics, it will refer to pounds-force (lbf).

Although it cannot be used as a consistent unit of mass, the mass of the standard pound is, by definition,

$$1 \text{ pound-mass} = 0.4536 \text{ kg}$$

This constant can be used to determine the *mass* in SI units (kilograms) of a body that has been characterized by its *weight* in U.S. customary units (pounds).

12.1D Equations of Motion

Consider a particle of mass m acted upon by several forces. Recall that we can express Newton's second law by the equation

$$\Sigma \mathbf{F} = m\mathbf{a} \qquad \textbf{(12.2)}$$

which relates the forces acting on the particle to the vector $m\mathbf{a}$ (Fig. 12.8).[†] Two of the most important tools you will use in solving dynamics problems, particularly those involving Newton's second law, are the free-body diagram and the kinetic diagram. These diagrams will help you to model dynamic systems and apply appropriate equations of motion. The free-body diagram shown on the left side of Fig. 12.9 is no different from what you did in statics in Chapter 4 and consists of the following steps:

Fig. 12.8 The sum of forces applied to a particle of mass m produces a vector $m\mathbf{a}$ in the direction of the resultant force.

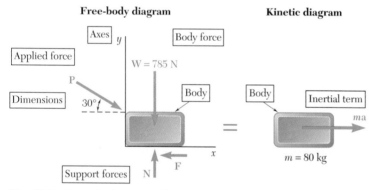

Fig. 12.9 Steps in drawing a free-body diagram and a kinetic diagram for solving dynamics problems.

Body: Define your system by isolating the body (or bodies) of interest. If a problem has multiple bodies (such as in Sample Problems 12.3 through 12.5), you may have to draw multiple free-body diagrams and kinetic diagrams.

Axes: Draw an appropriate coordinate system (e.g., Cartesian, normal and tangential, or radial and transverse).

Support Forces: Replace supports or constraints with appropriate forces (e.g., two perpendicular forces for a pin, normal forces, friction forces).

[†]In the 1700s, Jean-Baptiste le Rond d'Alembert expressed Newton's second law as $\Sigma \mathbf{F} - m\mathbf{a} = 0$ so he could solve dynamics problems using the principles of statics. The $-m\mathbf{a}$ term has been called a fictitious *inertial force*, but it is important for you to realize that there is no such thing as inertial forces (or centrifugal forces that "push" you outward when going around a curve). D'Alembert's principle (also called dynamic equilibrium) is seldom used in modern engineering.

Applied Forces and Body Forces: Draw any applied forces and body forces (also sometimes called field forces) on your diagram (e.g., weight, magnetic forces, a known pulling force).

Dimensions: Add any angles or distances that are important for solving the problem.

In statics problems, we deal with bodies in equilibrium, and the inertial term in Newton's second law is zero. For dynamics problems, this is not the case. We utilize the kinetic diagram to visualize this term.

Body: This is the same body as in the free-body diagram; place this beside the free-body diagram.

Inertial terms: Draw the $m\mathbf{a}$ term to be consistent with the coordinate system. Generally, draw this term in different components (e.g., ma_x and ma_y or ma_n and ma_t). If they are unknown quantities, it is best to draw them in the positive directions as defined by your coordinates.

Drawing these two diagrams clarifies how to develop your equations of motion. The free-body diagram is a visual representation of the $\Sigma \mathbf{F}$ term, and the kinetic diagram is a visual representation of the $m\mathbf{a}$ term. Since Newton's second law is a vector equation, you can use the free-body diagram and kinetic diagram to write $\Sigma \mathbf{F} = m\mathbf{a}$ directly in component form. Examples of using these diagrams to help you write your equations of motion are shown in the Sample Problems, and you can get extra practice by solving the Free-Body Problems 12.F1 through 12.F12.

As mentioned, it is usually more convenient to replace Eq. (12.2) with equivalent equations involving scalar quantities. As we saw in Chapter 11, we can resolve these vectors into components using several different coordinate systems (e.g., Cartesian, tangential and normal, or radial and transverse), depending on the type of problem we are solving.

Rectangular Components.   Resolving each force $\mathbf{F}$ and the acceleration $\mathbf{a}$ into rectangular components, we have

$$\Sigma(F_x\mathbf{i} + F_y\mathbf{j} + F_z\mathbf{k}) = m(a_x\mathbf{i} + a_y\mathbf{j} + a_z\mathbf{k})$$

It follows from this equation that

$$\Sigma F_x = ma_x \qquad \Sigma F_y = ma_y \qquad \Sigma F_z = ma_z \qquad \textbf{(12.8)}$$

Recall from Sec. 11.4C that the components of the acceleration are equal to the second derivatives of the coordinates of the particle. This gives us

$$\Sigma F_x = m\ddot{x} \qquad \Sigma F_y = m\ddot{y} \qquad \Sigma F_z = m\ddot{z} \qquad \textbf{(12.8')}$$

Consider, as an example, the motion of a projectile. If we neglect air resistance, the only force acting on the projectile after it has been fired is its weight $\mathbf{W} = -W\mathbf{j}$. The equations defining the motion of the projectile are therefore

$$m\ddot{x} = 0 \qquad m\ddot{y} = -W \qquad m\ddot{z} = 0$$

and the components of the acceleration of the projectile are

$$\ddot{x} = 0 \quad \ddot{y} = -\frac{W}{m} = -g \quad \ddot{z} = 0$$

Photo 12.2 Biomechanics researchers use video analysis and force plate measurements in Cartesian coordinates to analyze human motion.

where g is 9.81 m/s^2 or 32.2 ft/s^2. You can integrate these equations independently, as shown in Sec. 11.4C, to obtain the velocity and displacement of the projectile at any instant.

When a problem involves two or more bodies, you should write equations of motion for each of the bodies (see Sample Probs. 12.3 through 12.5). Recall from Sec. 12.1A that all accelerations should be measured with respect to a newtonian frame of reference. In most engineering applications, you can determine accelerations with respect to axes attached to the earth, but relative accelerations measured with respect to moving axes, such as axes attached to an accelerated body, cannot be substituted for **a** in the equations of motion.

Tangential and Normal Components.   We can also resolve the forces and the acceleration of the particle into components along the tangent to the path (in the direction of motion) and the normal (toward the inside of the path) (Fig. 12.10). Substituting into Eq. (12.2), we obtain the two scalar equations of

$$\Sigma F_t = ma_t \qquad \Sigma F_n = ma_n \qquad \textbf{(12.9)}$$

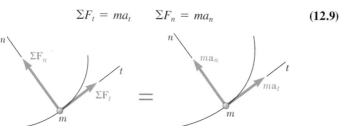

Fig. 12.10 The net force acting on a particle moving in a curvilinear path can be resolved into components tangent to the path and normal to the path, producing tangential and normal components of acceleration.

Now substituting for a_t and a_n from Eqs. (11.39), we have

$$\Sigma F_t = m\frac{dv}{dt} \qquad \Sigma F_n = m\frac{v^2}{\rho} \qquad \textbf{(12.9}')$$

We can solve these equations for two unknowns.

Radial and Transverse Components.   Consider a particle P, with polar coordinates r and θ, that moves in a plane under the action of several forces. Resolving the forces and the acceleration of the particle into radial and transverse components (Fig. 12.11) and substituting into Eq. (12.2), we obtain the two scalar equations of

$$\Sigma F_r = ma_r \qquad \Sigma F_\theta = ma_\theta \qquad \textbf{(12.10)}$$

Substituting for a_r and a_θ from Eqs. (11.45), we have

$$\Sigma F_r = m(\ddot{r} - r\dot{\theta}^2) \qquad \textbf{(12.11)}$$
$$\Sigma F_\theta = m(r\ddot{\theta} + 2\dot{r}\dot{\theta}) \qquad \textbf{(12.12)}$$

We can solve these equations for two unknowns.

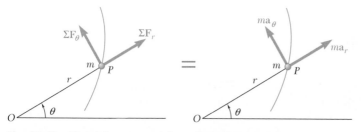

Fig. 12.11 Pictorial representation of Newton's second law in radial and transverse components.

Photo 12.3 A fighter jet making a sharp turn has a large normal component of acceleration, often equal to several *g*. As a result, the pilot experiences a large normal force, which in extreme cases, can cause blackouts.

Photo 12.4 The forces on the specimens used in a high speed centrifuge can be described in terms of radial and transverse components.

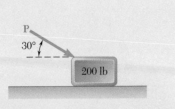

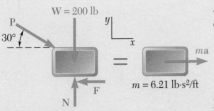

Fig. 1 Free-body diagram and kinetic diagram for the block.

Sample Problem 12.1

A 200-lb block rests on a horizontal plane. Find the magnitude of the force **P** required to give the block an acceleration of 10 ft/s² to the right. The coefficient of kinetic friction between the block and the plane is $\mu_k = 0.25$.

STRATEGY: You are given an acceleration and want to find the applied force. Therefore, you need to use Newton's second law.

MODELING: Pick the block as your system and model it as a particle. Drawing its free-body and kinetic diagrams, you obtain Fig. 1.

ANALYSIS: Before using Fig. 1, it is convenient to determine the mass of the object.

$$m = \frac{W}{g} = \frac{200 \text{ lb}}{32.2 \text{ ft/s}^2} = 6.21 \text{ lb·s}^2/\text{ft}$$

From Fig. 1, it is clear that the forces acting on the block shown in the free-body diagram need to be equal to the vector *m***a**, as shown in the kinetic diagram. Using these diagrams, you can write

$\xrightarrow{+} \Sigma F_x = ma$: $P \cos 30° - 0.25N = (6.21 \text{ lb·s}^2/\text{ft})(10 \text{ ft/s}^2)$
                $P \cos 30° - 0.25N = 62.1 \text{ lb}$ **(1)**

$+\uparrow \Sigma F_y = 0$: $N - P \sin 30° - 200 \text{ lb} = 0$ **(2)**

Solving Eq. (2) for N and substituting the result into Eq. (1), you obtain

$$N = P \sin 30° + 200 \text{ lb}$$

$$P \cos 30° - 0.25(P \sin 30° + 200 \text{ lb}) = 62.1 \text{ lb} \qquad P = 151 \text{ lb} \blacktriangleleft$$

REFLECT and THINK: When you begin pushing on an object, you first have to overcome the static friction force ($F = \mu_s N$) before the object will move. Also note that the downward component of force **P** increases the normal force **N**, which in turn increases the friction force **F** that you must overcome.

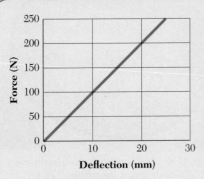

Sample Problem 12.2

A 0.5-kg fragile glass vase is dropped onto a thick pad that has a force-deflection relationship as shown. Knowing that the vase has a speed of 3 m/s when it first contacts the pad, determine the maximum downward displacement of the vase.

STRATEGY: Use Newton's second law to find the acceleration of the vase and then integrate it to find the displacement.

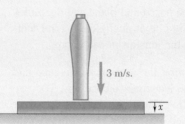

MODELING: Choose the vase to be your system and model it as a particle. Because the force is a linear function of displacement, you can write the force acting on the vase as

$$F_P = \frac{200 \text{ N}}{0.02 \text{ m}} y = (10\,000 \text{ N/m})x$$

Draw its free-body diagram and kinetic diagram (Fig. 1).

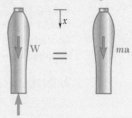

$F_P = 10\,000x$

Fig. 1 Free-body diagram and kinetic diagram for the vase.

ANALYSIS: You can obtain a scalar equation by applying Newton's second law in the vertical direction. Thus,

$$+\downarrow \Sigma F_x = ma \qquad W - (10\,000)x = ma$$

Substituting in values and solving for a gives

$$a = 9.81 - 20\,000x$$

Maximum Displacement. Now that you have the acceleration as a function of displacement, you need to use the basic kinematic relationships to find the maximum compression of the pad. Substituting $a = 9.81 - 20\,000x$ into $a = v\,dv/dx$ gives

$$a = 9.81 - 20\,000x = \frac{v\,dv}{dx}$$

Separating variables and integrating, you find

$$v\,dv = (9.81 - 20\,000x)dx \longrightarrow \int_{v_0}^{0} v\,dv = \int_{0}^{x_{\max}} (9.81 - 20\,000x)dx$$

$$0 - \frac{1}{2}v_0^2 = 9.81x_{\max} - 10\,000x_{\max}^2 \tag{1}$$

Substituting $v_0 = 3$ m/s into Eq. (1) and solving for $x_{\max}$ using the quadratic formula gives $x_{\max} = 0.0217$ m.

$$x_{\max} = 21.7 \text{ mm} \quad \blacktriangleleft$$

REFLECT and THINK: A distance of 21.7 mm indicates that the pad must be relatively thick. For a real pad, the assumption that it acts as a linear spring may not be an accurate model. For the numbers given in this problem, the maximum acceleration the vase experiences is

$$a = 9.81 - (20\,000)(0.0217) = -207.3 \text{ m/s}^2 \text{ or about 21 g's}$$

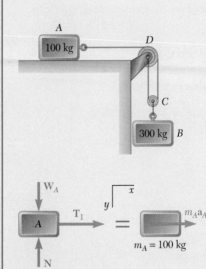

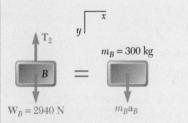

Fig. 1 Free-body diagram and kinetic diagram for A.

Fig. 2 Free-body diagram and kinetic diagram for B.

Fig. 3 Free-body diagram and kinetic diagram for the pulley.

Sample Problem 12.3

The two blocks shown start from rest. The horizontal plane and the pulley are frictionless, and the pulley is assumed to be of negligible mass. Determine the acceleration of each block and the tension in each cord.

STRATEGY: You are interested in finding the tension in the rope and the acceleration of the two blocks, so use Newton's second law. The two blocks are connected by a cable, indicating that you need to relate their accelerations using the techniques discussed in Chapter 11 for objects with dependent motion.

MODELING: Treat both blocks as particles and assume that the pulley is massless and frictionless. Since there are two masses, you need two systems: block A by itself and block B by itself. The free-body and kinetic diagrams for these objects are shown in Figs. 1 and 2. To help determine the forces acting on block B, you can also isolate the massless pulley C as a system (Fig. 3).

ANALYSIS: You can start with either kinetics or kinematics. The key is to make sure you keep track of your equations and unknowns.

Kinetics. Apply Newton's second law successively to block A, block B, and pulley C.

Block A. Denote the tension in cord ACD by T_1 (Fig. 1). Then you have

$$\xrightarrow{+} \Sigma F_x = m_A a_A: \qquad T_1 = 100a_A \qquad (1)$$

Block B. Observe that the weight of block B is

$$W_B = m_B g = (300 \text{ kg})(9.81 \text{ m/s}^2) = 2940 \text{ N}$$

Denote the tension in cord BC by T_2 (Fig. 2). Then

$$+\downarrow \Sigma F_y = m_B a_B: \qquad 2940 - T_2 = 300a_B \qquad (2)$$

Pulley C. Assuming m_C is zero, you have (Fig. 3)

$$+\downarrow \Sigma F_y = m_C a_C = 0: \qquad T_2 - 2T_1 = 0 \qquad (3)$$

At this point, you have three equations, (1), (2), and (3), and four unknowns, T_1, T_2, a_B, and a_A. Therefore, you need one more equation, which you can get from kinematics.

Kinematics. It is important to make sure that the directions you assumed for the kinetic diagrams are consistent with the kinematic analysis. Note that if block A moves through a distance x_A to the right, block B moves down through a distance

$$x_B = \frac{1}{2}x_A$$

Differentiating twice with respect to t, you have

$$a_B = \frac{1}{2}a_A \qquad (4)$$

You now have four equations and four unknowns, so you can solve this problem. You can do this using a computer, a calculator, or by hand. To solve these equations by hand, you can substitute for a_B from Eq. (4) into Eq. (2) for

$$2940 - T_2 = 300(\tfrac{1}{2}a_A)$$
$$T_2 = 2940 - 150a_A \qquad (5)$$

Now substitute for T_1 and T_2 from Eqs. (1) and (5), respectively, into Eq. (3).

$$2940 - 150a_A - 2(100a_A) = 0$$
$$2940 - 350a_A = 0 \qquad a_A = 8.40 \text{ m/s}^2 \quad \blacktriangleleft$$

Then substitute the value obtained for a_A into Eqs. (4) and (1).

$$a_B = \tfrac{1}{2}a_A = \tfrac{1}{2}(8.40 \text{ m/s}^2) \qquad a_B = 4.20 \text{ m/s}^2 \quad \blacktriangleleft$$
$$T_1 = 100a_A = (100 \text{ kg})(8.40 \text{ m/s}^2) \qquad T_1 = 840 \text{ N} \quad \blacktriangleleft$$

Recalling Eq. (3), you have

$$T_2 = 2T_1 \qquad T_2 = 2(840 \text{ N}) \qquad T_2 = 1680 \text{ N} \quad \blacktriangleleft$$

REFLECT and THINK: Note that the value obtained for T_2 is *not* equal to the weight of block B. Rather than choosing B and the pulley as separate systems, you could have chosen the system to be B *and* the pulley. In this case, T_2 would have been an internal force.

Sample Problem 12.4

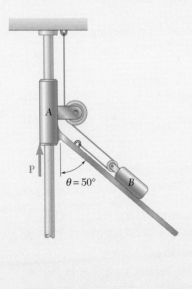

Collar A has a ramp that is welded to it and a force $P = 5$ lb applied as shown. Collar A and the ramp weigh 3 lb, and block B weighs 0.8 lb. Neglecting friction, determine the tension in the cable.

STRATEGY: The principle you need to use is Newton's second law. Since a block is sliding down an incline and a cable is connecting A and B, you also need to use relative motion and dependent motion.

MODELING: Model A and B as particles and assume all surfaces are smooth. As usual, start by choosing a system and then drawing a free-body diagram and a kinetic diagram. This problem has two systems, and you need to be careful with how you define them. The easiest systems to use are (*a*) collar A with its pulley and the ramp welded to it (system 1) and (*b*) block B and the pulley attached to it (system 2), as shown in Fig. 1. The free-body and kinetic diagrams for system 1 are shown in Fig. 2. The free-body and kinetic diagrams for B are a little trickier, because you don't know the direction of the acceleration of B.

(continued)

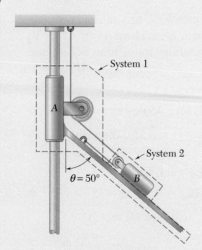

System 1

System 2

$\theta = 50°$

B

A

Fig. 1 System boundaries.

Kinematics for Block *B*. Express the acceleration $\mathbf{a}_B$ of block *B* as the sum of the acceleration of *A* and the acceleration of *B* relative to *A*. Hence,

$$\mathbf{a}_B = \mathbf{a}_A + \mathbf{a}_{B/A}$$

Here $\mathbf{a}_{B/A}$ is directed along the inclined surface of the wedge. Now you can draw the appropriate diagrams (Fig. 3). Note that you do not need to use the same *x–y* coordinate system for each mass, since these directions are simply used for obtaining the scalar equations.

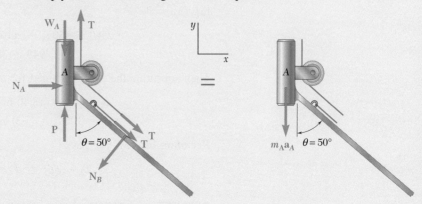

Fig. 2 Free-body diagram and kinetic diagram for system 1.

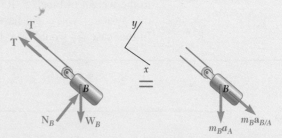

Fig. 3 Free-body diagram and kinetic diagram for *B*.

ANALYSIS: You can obtain a scalar equation by applying Newton's second law to each of these systems.

a. System 1:

$$\xrightarrow{+} \Sigma F_x = m_A a_{A_x} \quad N_A - N_B \cos 50° + 2T \cos 40° = 0 \tag{1}$$

$$+\uparrow \Sigma F_y = m_A a_{A_y} \quad -W_A + P + T - 2T \sin 40° - N_B \sin 50° = -m_A a_A \tag{2}$$

b. Block *B*:

$$+\searrow \Sigma F_x = m_B a_{B_x} \quad -2T + W_B \sin 40° = m_B a_{B/A} + m_B a_A \sin 40° \tag{3}$$

$$+\nearrow \Sigma F_y = m_B a_{B_y} \quad N_B - W_B \cos 40° = -m_B a_A \cos 40° \tag{4}$$

You now have four equations and five unknowns (T, N_A, N_B, a_A, and $a_{B/A}$), so you need one more equation. The motion of *A* and *B* are related because they are connected by a cable.

Constraint Equations. Define position vectors as shown in Fig. 4. Note that the positive directions for the position vectors for *A* and *B* are

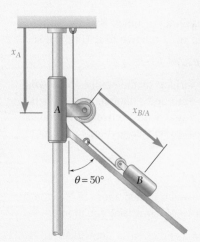

Fig. 4 Position vectors for dependent motion.

defined from the kinetic diagrams in Figs. 2 and 3. Assuming the cable is inextensible, you can write the lengths in terms of the coordinates and then differentiate.

Constraint equation for the cable: $x_A + 2x_{B/A} =$ constant

Differentiating this twice gives

$$a_A = -2a_{B/A} \tag{5}$$

You now have five equations and five unknowns, so all that remains is to substitute the known values and solve for the unknowns. The results are $N_A = -0.1281$ lb, $N_B = 0.869$ lb, $T = 0.281$ lb, $a_A = -13.46$ ft/s^2, and $a_{B/A} = 6.73$ ft/s^2.

$$T = 0.281 \text{ lb} \blacktriangleleft$$

REFLECT and THINK: In this problem, we focused on the problem formulation and assumed that you can solve the resulting equations either by hand or by using a calculator/computer. It is important to note that you are given the weights of A and B, so you need to calculate the masses in slugs using $m = W/g$. The solution required multiple systems and multiple concepts, including Newton's second law, relative motion, and dependent motion. If friction occurred between B and the ramp, you would first need to determine whether or not the system would move under the applied force by assuming that it does not move and calculating the friction force. Then you would compare this force to the maximum allowable force $\mu_s N$.

Sample Problem 12.5

The 12-lb block B starts from rest and slides on the 30-lb wedge A, which is supported by a horizontal surface. Neglecting friction, determine (*a*) the acceleration of the wedge, (*b*) the acceleration of the block relative to the wedge.

STRATEGY: You are given the forces (weights) of the two objects and want to find their accelerations. You can use Newton's second law, but you have to take into account relative motion as well.

MODELING: Treat both objects as particles. Since you have two objects, you will need two systems: wedge A and block B. In order to draw the kinetic diagrams for each of these systems, you need to know the direction of the accelerations. Therefore, before drawing the free-body and kinetic diagrams, look at the kinematics.

Kinematics. First examine the acceleration of the wedge and the acceleration of the block.

Wedge A. Since the wedge is constrained to move on the horizontal surface, its acceleration $\mathbf{a}_A$ is horizontal (Fig. 1). Assume that it is directed to the right.

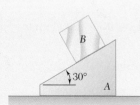

Fig. 1 Acceleration of *A*.

(continued)

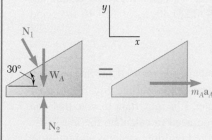

Fig. 2   Acceleration of B.

Fig. 3   Free-body diagram and kinetic diagram for A.

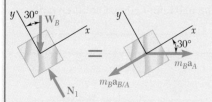

Fig. 4   Free-body diagram and kinetic diagram for B.

Block B. You can express the acceleration $\mathbf{a}_B$ of block B as the sum of the acceleration of A and the acceleration of B relative to A (Fig. 2), so

$$\mathbf{a}_B = \mathbf{a}_A + \mathbf{a}_{B/A}$$

Here $\mathbf{a}_{B/A}$ is directed along the inclined surface of the wedge. Now you can draw the appropriate diagrams. The free-body diagrams and kinetic diagrams for A and B are shown in Figs. 3 and 4, respectively. The forces exerted by the block and the horizontal surface on wedge A are represented by $\mathbf{N}_1$ and $\mathbf{N}_2$, respectively.

ANALYSIS:

Kinetics.   Recall that Figs. 3 and 4 are visual representations of Newton's second law. Therefore, you can use them to obtain scalar equations.

Wedge A. For Wedge A, the positive x-direction is defined to be to the right. Applying Newton's second law in the x-direction gives

$$\xrightarrow{+} \Sigma F_x = m_A a_A: \qquad N_1 \sin 30° = m_A a_A$$
$$0.5 N_1 = (W_A/g) a_A \tag{1}$$

Block B. Using the coordinate axes shown in Fig. 4 and resolving $\mathbf{a}_B$ into its components $\mathbf{a}_A$ and $\mathbf{a}_{B/A}$, you have

$$+\nearrow \Sigma F_x = m_B a_x: \qquad -W_B \sin 30° = m_B a_A \cos 30° - m_B a_{B/A}$$
$$-W_B \sin 30° = (W_B/g)(a_A \cos 30° - a_{B/A})$$
$$a_{B/A} = a_A \cos 30° + g \sin 30° \tag{2}$$
$$+\nwarrow \Sigma F_y = m_B a_y: \qquad N_1 - W_B \cos 30° = -m_B a_A \sin 30°$$
$$N_1 - W_B \cos 30° = -(W_B/g) a_A \sin 30° \tag{3}$$

You now have three equations, (1), (2), and (3), and three unknowns, N_1, a_A, and $a_{B/A}$, so you can solve these with your calculator or by hand as shown here.

a. Acceleration of Wedge A.   Substitute for N_1 from Eq. (1) into Eq. (3).

$$2(W_A/g) a_A - W_B \cos 30° = -(W_B/g) a_A \sin 30°$$

Then solve for a_A and substitute the numerical data.

$$a_A = \frac{W_B \cos 30°}{2W_A + W_B \sin 30°} g = \frac{(12 \text{ lb}) \cos 30°}{2(30 \text{ lb}) + (12 \text{ lb}) \sin 30°}(32.2 \text{ ft/s}^2)$$
$$a_A = +5.07 \text{ ft/s}^2 \qquad \mathbf{a}_A = 5.07 \text{ ft/s}^2 \rightarrow \ \blacktriangleleft$$

b. Acceleration of Block B Relative to A.   Now substitute the value obtained for a_A into Eq. (2).

$$a_{B/A} = (5.07 \text{ ft/s}^2) \cos 30° + (32.2 \text{ ft/s}^2) \sin 30°$$
$$a_{B/A} = +20.5 \text{ ft/s}^2 \qquad \mathbf{a}_{B/A} = 20.5 \text{ ft/s}^2 \ \measuredangle 30° \ \blacktriangleleft$$

REFLECT and THINK: Many students are tempted to draw the acceleration of block B down the incline in the kinetic diagram. It is important to recognize that this is the direction of the *relative* acceleration. Rather than the kinetic diagram you used for block B, you could have simply put unknown accelerations in the x and y directions and then used your relative motion equation to obtain more scalar equations.

Sample Problem 12.6

The bob of a 2-m pendulum describes an arc of a circle in a vertical plane. If the tension in the cord is 2.5 times the weight of the bob for the position shown, find the velocity and the acceleration of the bob in that position.

STRATEGY: The most direct approach is to use Newton's law with tangential and normal components.

MODELING: Choose the bob as your system; if its radius is small, you can model it as a particle. Draw the free-body and kinetic diagrams for the bob knowing that the weight of the bob is $W = mg$; the tension in the cord is $2.5mg$. The normal acceleration $\mathbf{a}_n$ is directed toward O, and you can assume that $\mathbf{a}_t$ is in the direction shown in Fig. 1.

ANALYSIS: You can obtain scalar equations by applying Newton's second law in the normal and tangential directions. Hence,

$$+\swarrow \Sigma F_t = ma_t: \qquad mg \sin 30° = ma_t$$
$$a_t = g \sin 30° = +4.90 \text{ m/s}^2 \quad \mathbf{a}_t = 4.90 \text{ m/s}^2 \swarrow \quad \blacktriangleleft$$

$$+\nwarrow \Sigma F_n = ma_n: \qquad 2.5mg - mg \cos 30° = ma_n$$
$$a_n = 1.634g = +16.03 \text{ m/s}^2 \quad \mathbf{a}_n = 16.03 \text{ m/s}^2 \nwarrow \quad \blacktriangleleft$$

Since $a_n = v^2/\rho$, you have $v^2 = \rho a_n = (2 \text{ m})(16.03 \text{ m/s}^2)$. Thus,

$$v = \pm 5.66 \text{ m/s} \qquad \mathbf{v} = 5.66 \text{ m/s} \nearrow \text{(up or down)} \quad \blacktriangleleft$$

REFLECT and THINK: If you look at these equations for an angle of zero instead of 30°, you will see that when the bob is straight below point O, the tangential acceleration is zero, and the velocity is a maximum. The normal acceleration is not zero because the bob has a velocity at this point.

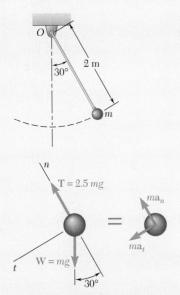

Fig. 1 Free-body diagram and kinetic diagram for the bob.

Sample Problem 12.7

Determine the rated speed of a highway curve with a radius of $\rho = 400$ ft banked through an angle $\theta = 18°$. The *rated speed* of a banked highway curve is the speed at which a car should travel to have no lateral friction force exerted on its wheels.

STRATEGY: You are given information about the lateral friction force—that is, it is equal to zero—so use Newton's second law. Use normal and tangential components, since the car is traveling in a curved path and the problem involves speed and a radius of curvature.

MODELING: Choose the car to be the system. Assuming you can neglect the rotation of the car about its center of mass, treat it as a particle.

(continued)

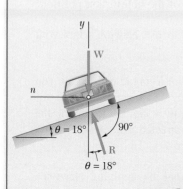

Fig. 1 Free-body diagram and kinetic diagram of the car.

The car travels in a *horizontal* circular path with a radius of ρ. The normal component $\mathbf{a}_n$ of the acceleration is directed toward the center of the path, as shown in the kinetic diagram (Fig. 1); its magnitude is $a_n = v^2/\rho$, where v is the speed of the car in ft/s. The mass m of the car is W/g, where W is the weight of the car. Since no lateral friction force is exerted on the car, the reaction $\mathbf{R}$ of the road is perpendicular to the roadway, as shown in the free-body diagram (Fig. 1).

ANALYSIS: You can obtain scalar equations by applying Newton's second law in the vertical and normal directions. Thus,

$$+\uparrow \Sigma F_y = 0: \qquad R \cos \theta - W = 0 \qquad R = \frac{W}{\cos \theta} \tag{1}$$

$$\xrightarrow{+} \Sigma F_n = ma_n: \qquad R \sin \theta = \frac{W}{g} a_n \tag{2}$$

Substituting R from Eq. (1) into Eq. (2), and recalling that $a_n = v^2/\rho$, you obtain

$$\frac{W}{\cos \theta} \sin \theta = \frac{W}{g} \frac{v^2}{\rho} \qquad v^2 = g\rho \tan \theta$$

Finally, substituting $\rho = 400$ ft and $\theta = 18°$ into this equation, you get $v^2 = (32.2 \text{ ft/s}^2)(400 \text{ ft}) \tan 18°$. Hence,

$$v = 64.7 \text{ ft/s} \qquad\qquad v = 44.1 \text{ mi/h} \blacktriangleleft$$

REFLECT and THINK: For a highway curve, this seems like a reasonable speed for avoiding a spin-out. For this problem, the tangential direction is into the page; since you were not asked about forces or accelerations in this direction, you did not need to analyze motion in the tangential direction. If the roadway were banked at a larger angle, would the rated speed be larger or smaller than this calculated value?

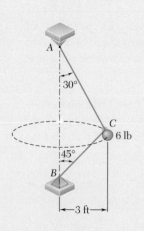

Sample Problem 12.8

Two wires AC and BC are tied at C to a sphere that revolves at the constant speed v in the horizontal circle shown. Knowing that the wires will break if their tension exceeds 15 lb, determine the range of values of v for which both wires remain taut and the wires do not break.

STRATEGY: You are given information about the forces in the wires, so use Newton's second law. The sphere is moving along a curved path, so use normal and tangential coordinates.

MODELING: Choose the sphere for the system and assume you can treat it as a particle. Draw the free-body and kinetic diagrams as shown in Fig. 1. The tensions act in the direction of the wires, and the normal direction is toward the center of the circular path.

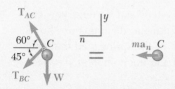

Fig. 1 Free-body diagram and kinetic diagram for the sphere.

ANALYSIS: You can obtain a scalar equation by applying Newton's second law in the normal and vertical directions. Thus,

$$+ \leftarrow \Sigma F_n = ma_n \qquad T_{AC} \cos 60° + T_{BC} \cos 45° = ma_n = m\frac{v^2}{\rho} \qquad (1)$$

$$+ \uparrow \Sigma F_y = ma_y \qquad -W + T_{AC} \sin 60° - T_{BC} \sin 45° = 0 \qquad (2)$$

where $m = W/g = 6 \text{ lb}/(32.2 \text{ ft/s}^2) = 0.1863 \text{ lb·s}^2/\text{ft}$ and $\rho = 3$ ft. In these two equations, you have three unknowns, T_{AC}, T_{BC}, and v, so you need a third equation. The problem statement indicates that you want the range of speeds when the both wires remain taut (that is, the tension is positive) and that this tension must be less than 15 lb. To find this range, first set each tension equal to zero and solve the resulting set of equations.

For $T_{AC} = 0$, you find $v = 9.83$ ft/s and $T_{BC} = -8.485$ lb, which is impossible for a wire.

For $T_{BC} = 0$, you find $v = 7.468$ ft/s and $T_{AC} = 6.928$ lb.

Thus, the minimum speed is 7.47 ft/s. Now set the tensions equal to 15 lb to find the maximum speed.

For $T_{AC} = 15$ lb, you find $v = 15.29$ ft/s and $T_{BC} = 9.886$ lb.

For $T_{BC} = 15$ lb, you find $v = 18.03$ ft/s and $T_{AC} = 19.18$ lb.

Therefore, the maximum speed is 15.29 ft/s. Combining these results gives you

$$7.47 \text{ ft/s} \leq v \leq 15.29 \text{ ft/s} \quad \blacktriangleleft$$

REFLECT and THINK: In this problem, you needed to use the information in the problem statement to obtain additional equations so that you could determine the range of speeds. Another way to look at the solution is to solve Eqs. (1) and (2) for T_{AC} and T_{BC} in terms of v and to plot these as shown in Fig. 2. It is easy to see from this graph that T_{AC} determines the maximum speed and T_{BC} determines the minimum speed if both wires are to remain taut and also have tensions less than 15 lb.

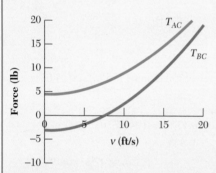

Fig. 2 Tension in cables as a function of speed.

Sample Problem 12.9

A 0.5-kg collar is attached to a spring and slides without friction along a circular rod in a *vertical* plane. The spring has an undeformed length of 150 mm and a constant $k = 200$ N/m. Knowing that the collar has a speed of 3 m/s as it passes through point B, determine the tangential acceleration of the collar and the force of the rod on the collar at this instant.

STRATEGY: This problem deals with forces and accelerations, so you need to use Newton's second law. The collar moves along a curved path, so you should use normal and tangential coordinates.

(continued)

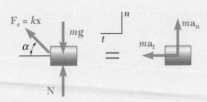

Fig. 1 Free-body diagram and kinetic diagram for the collar.

MODELING: Choose the collar as your system and assume you can treat it as a particle. Draw the free-body and kinetic diagrams as shown in Fig. 1. The spring force acts in the direction of the spring, and the force is drawn assuming that the spring is stretched and not compressed. Check this using geometry.

$$\sin \alpha = \frac{125 \text{ mm}}{300 \text{ mm}} = 0.4167 \longrightarrow \alpha = 24.62°$$

$$L_{BD} = \sqrt{(300 \text{ mm})^2 + (125 \text{ mm})^2} = 325 \text{ mm}$$

Thus, when the collar is at B, the spring is extended as $x = L_{BD} - L_0 = 325$ mm $- 150$ mm $= 175$ mm.

ANALYSIS: You can obtain scalar equations by applying Newton's second law in the normal and tangential directions. Hence,

$$+\uparrow \Sigma F_n = ma_n \quad kx \sin \alpha + N - mg = ma_n = m\frac{v^2}{\rho} \tag{1}$$

$$\xleftarrow{+} \Sigma F_t = ma_t \qquad Fx \cos \alpha = ma_t \tag{2}$$

You now have two equations (1) and (2) and two unknown a_t and N. You can solve for these by hand or using your calculator/computer. You can solve for the normal force in Eq. (1) as

$$N = mg + m\frac{v^2}{\rho} - kx \cos \alpha$$

Substituting values gives

$$N = (0.5 \text{ kg})(9.81 \text{ m/s}^2) + (0.5 \text{ kg})\frac{(3 \text{ m/s})^2}{0.125 \text{ m}} - (200 \text{ N/m})(0.175 \text{ m})\sin(24.62°)$$

$$N = 26.3 \text{ N} \blacktriangleleft$$

$$a_t = \frac{Fx \cos \alpha}{m} = \frac{(200 \text{ N/m})(0.175)\cos(24.62°)}{0.5 \text{ kg}}$$

$$a_t = 63.6 \text{ m/s}^2 \blacktriangleleft$$

REFLECT and THINK: How would this problem have changed if you had been told friction was acting between the rod and the collar? You would have had one additional term in your free-body diagram, $\mu_k N$, in the direction opposite to the velocity. Thus, you would need to be told the direction the collar was moving as well as the coefficient of kinetic friction.

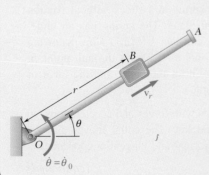

Sample Problem 12.10

A block B with a mass m can slide freely on a frictionless arm OA that rotates in a horizontal plane at a constant rate $\dot{\theta}_0$. Knowing that B is released at a distance r_0 from O, express as a function of r, *(a)* the component v_r of the velocity of B along OA, *(b)* the magnitude of the horizontal force **F** exerted on B by the arm OA.

STRATEGY: You want to find a force, so use Newton's second law. The radial distance r of the mass is changing, as is the angular displacement θ, so use radial and transverse coordinates.

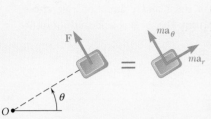

Fig. 1 Free-body diagram and kinetic diagram for the block.

MODELING: Choose block B as your system and assume you can model it as a particle. Since all other forces are perpendicular to the plane of the figure, the only force shown acting on B is the force **F** perpendicular to OA. Draw free-body and kinetic diagrams for block B as shown in Fig. 1.

ANALYSIS:

Equations of Motion. You can obtain scalar equations by applying Newton's second law in the radial and transverse directions. Hence,

$$+\nearrow\Sigma F_r = ma_r: \qquad\qquad 0 = m(\ddot{r} - r\dot{\theta}^2) \qquad\qquad (1)$$
$$+\nwarrow\Sigma F_\theta = ma_\theta: \qquad\qquad F = m(r\ddot{\theta} + 2\dot{r}\dot{\theta}) \qquad\qquad (2)$$

a. Component v_r of Velocity. Since $v_r = \dot{r}$, you have

$$\ddot{r} = \dot{v}_r = \frac{dv_r}{dt} = \frac{dv_r}{dr}\frac{dr}{dt} = v_r\frac{dv_r}{dr}$$

After using Eq. (1) to obtain $\ddot{r} = r\dot{\theta}^2$ and recalling that $\dot{\theta} = \dot{\theta}_0$, you can separate the variables to obtain

$$v_r\,dv_r = \dot{\theta}_0^2 r\,dr$$

Multiply by 2 and integrate from 0 to v_r and from r_0 to r. The result is

$$v_r^2 = \dot{\theta}_0^2(r^2 - r_0^2) \qquad v_r = \dot{\theta}_0(r^2 - r_0^2)^{1/2} \quad \blacktriangleleft$$

b. Horizontal Force F. Set $\dot{\theta} = \dot{\theta}_0$, $\ddot{\theta} = 0$, and $\dot{r} = v_r$ in Eq. (2). Then substitute for v_r the expression obtained in part a. The result is

$$F = 2m\dot{\theta}_0(r^2 - r_0^2)^{1/2}\dot{\theta}_0 \qquad F = 2m\dot{\theta}_0^2(r^2 - r_0^2)^{1/2} \quad \blacktriangleleft$$

REFLECT and THINK: Introducing radial and transverse components of force and acceleration involves using components of velocity as well in the computations. But this is still much simpler and more direct than trying to use other coordinate systems. Even though the radial acceleration is zero, the block accelerates relative to the rod with acceleration $\ddot{r}$.

Sample Problem 12.11

NASA flies a reduced-gravity aircraft (affectionately known as the Vomit Comet) in an elliptic flight to train astronauts in a microgravity environment. The plane is being tracked by radar located at O. When the plane is near the bottom of its trajectory, as shown, values from the radar tracking station are $\dot{r} = 120$ m/s, $\dot{\theta} = -0.900$ rad/s, $\ddot{r} = 34.8$ m/s^2, and $\ddot{\theta} = 0.0156$ rad/s^2. At the instant shown, determine the force exerted on the 80-kg pilot by his seat.

STRATEGY: You want to find the force the pilot experiences at this instant and you can calculate the accelerations, so you should use Newton's second law. Since you know that the radial distance and the angle are changing with time, use radial and transverse components.

(continued)

MODELING: Choosing the pilot as the system, draw the free-body and kinetic diagrams as shown in Fig. 1. You could choose to put the forces and the pilot in the r and θ direction or the x and y direction (we chose F_x and F_y to represent the forces from the seat back and bottom, respectively).

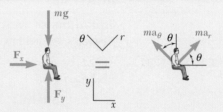

Fig. 1

ANALYSIS: Before you apply Newton's second law, determine r and θ from the geometry.

$$r = \sqrt{800^2 + 600^2} = 1000 \text{ m} \qquad \theta = \tan^{-1}(600/800) = 36.87°$$

Kinematics. Determine the components of the accelerations as

$$a_r = \ddot{r} - r\dot{\theta}^2 = 34.8 \text{ m/s}^2 - (100 \text{ m})(-0.090 \text{ rad/s})^2 = 26.7 \text{ m/s}^2$$

$$a_\theta = r\ddot{\theta} + 2\dot{r}\dot{\theta} = (1000 \text{ m})(0.0156 \text{ rad/s}^2) + 2(120 \text{ m/s})(-0.090 \text{ rad/s})$$
$$= -6.00 \text{ m/s}^2$$

Kinetics. Obtain scalar equations by applying Newton's second law in the horizontal and vertical directions. Thus,

$$\xrightarrow{+} \Sigma F_x = ma_x \qquad\qquad F_x = ma_r\cos\theta - ma_\theta\sin\theta \qquad\qquad (1)$$

$$+\uparrow\Sigma F_y = ma_y \qquad\qquad F_y - mg = ma_r\sin\theta + ma_\theta\cos\theta \qquad\qquad (2)$$

You have two equations, (1) and (2), and two unknowns, F_x and F_y. Substituting the known values into Eqs. (1) and (2) gives

$$F_x = (80 \text{ kg})(26.7 \text{ m/s}^2)\cos 36.87° - (80 \text{ kg})(-6.00 \text{ m/s}^2)\sin 36.87°$$

$$F_x = 1997 \text{ N} \rightarrow \ \blacktriangleleft$$

$$F_y = (80 \text{ kg})(9.81 \text{ m/s}^2) + (80\text{kg})(26.7 \text{ m/s}^2)\sin 36.87° +$$

$$(80 \text{ kg})(-6.00 \text{ m/s}^2)\cos 36.87°$$

$$F_y = 1682 \text{ N} \uparrow \ \blacktriangleleft$$

REFLECT and THINK: These forces correspond to a forward acceleration of 2.54 g and a vertical acceleration of 2.14 g. Although this is a bit high for a passenger aircraft, it is within the flight characteristics for the Vomit Comet. If you had been asked to determine whether the plane was speeding up or slowing down, you would need to find the component of the acceleration in the tangential direction, which is defined by the direction of the velocity vector.

SOLVING PROBLEMS
ON YOUR OWN

In the problems for this section, you will apply **Newton's second law of motion**, $\Sigma \mathbf{F} = m\mathbf{a}$, to relate the forces acting on a particle to its motion.

1. Writing the equations of motion. When applying Newton's second law to the types of motion discussed in this section, you will find it most convenient to express the vectors **F** and **a** in terms of either their rectangular components, their tangential and normal components, or their radial and transverse components.

   a. When using rectangular components [Sample Probs. 12.1 through 12.5], recall from Sec. 11.4C the expressions found for a_x, a_y, and a_z. Then you can write

$$\Sigma F_x = m\ddot{x} \qquad \Sigma F_y = m\ddot{y} \qquad \Sigma F_z = m\ddot{z}$$

   b. When using tangential and normal components [Sample Probs. 12.6 and 12.9], recall from Sec. 11.5A the expressions found for a_t and a_n. Then you can write

$$\Sigma F_t = m\frac{dv}{dt} \qquad \Sigma F_n = m\frac{v^2}{\rho}$$

   c. When using radial and transverse components [Sample Probs. 12.10 and 12.11], recall from Sec. 11.5B the expressions found for a_r and a_θ. Then you can write

$$\Sigma F_r = m(\ddot{r} - r\dot{\theta}^2) \qquad \Sigma F_\theta = m(r\ddot{\theta} + 2\dot{r}\dot{\theta})$$

2. Drawing a free-body diagram and a kinetic diagram. Drawing a free-body diagram showing the applied forces and a kinetic diagram showing the vector $m\mathbf{a}$ or its components will provide you with a pictorial representation of Newton's second law [Sample Probs. 12.1 through 12.11]. These diagrams will be of great help to you when writing the equations of motion. Note that when a problem involves two or more bodies, it is usually best to consider each body separately.

3. Applying Newton's second law. As we observed in Sec. 12.1A, the acceleration used in the equation $\Sigma \mathbf{F} = m\mathbf{a}$ always should be the absolute acceleration of the particle (that is, it should be measured with respect to a newtonian frame of reference). Also, if the sense of the acceleration **a** is unknown or is not easily deduced, assume an arbitrary sense for **a** (usually the positive direction of a coordinate axis), and then let the solution provide the correct sense. Finally, note how the solutions of Sample Probs. 12.3 through 12.5 were divided into a *kinematics* portion and a *kinetics* portion, and how in Sample Probs. 12.4 and 12.5 we used two systems of coordinate axes to simplify the equations of motion.

4. When a problem involves dry friction, be sure to review the relevant section of *Statics* [Sec. 8.1] before attempting to solve that problem. In particular, you should know when to use each of the equations $F = \mu_s N$ and $F = \mu_k N$. You should also recognize that if the motion of a system is not specified, it is necessary first to assume a possible motion and then to check the validity of that assumption. For example, you can assume that the motion is impending, then check to see if the friction force is greater than $\mu_s N$ (if it is, then your assumption was wrong and the particle is moving).

5. Solving problems involving relative motion. When a body B moves with respect to a body A, as in Sample Probs. 12.4 and 12.5, it is often convenient to express the acceleration of B as

$$\mathbf{a}_B = \mathbf{a}_A + \mathbf{a}_{B/A}$$

where $\mathbf{a}_{B/A}$ is the acceleration of B relative to A, that is, the acceleration of B as observed from a frame of reference attached to A and in translation. If B is observed to move in a straight line, $\mathbf{a}_{B/A}$ is directed along that line. On the other hand, if B is observed to move along a circular path, you should resolve the relative acceleration $\mathbf{a}_{B/A}$ into components tangential and normal to that path.

6. Finally, always consider the implications of any assumption you make. Thus, in a problem involving two cords, if you assume that the tension in one of the cords is equal to its maximum allowable value, check whether any requirements set for the other cord will be satisfied. For instance, will the tension T in that cord satisfy the relation $0 \leq T \leq T_{max}$? That is, will the cord remain taut and will its tension be less than its maximum allowable value?

Problems

12.CQ1 A 1000-lb boulder *B* is resting on a 200-lb platform *A* when truck *C* accelerates to the left with a constant acceleration. Which of the following statements are true (more than one may be true)?

a. The tension in the cord connected to the truck is 200 lb.
b. The tension in the cord connected to the truck is 1200 lb.
c. The tension in the cord connected to the truck is greater than 1200 lb.
d. The normal force between *A* and *B* is 1000 lb.
e. The normal force between *A* and *B* is 1200 lb.
f. None of the above are true.

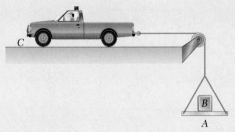

Fig. P12.CQ1

12.CQ2 Marble *A* is placed in a hollow tube, and the tube is swung in a horizontal plane causing the marble to be thrown out. As viewed from the top, which of the following choices best describes the path of the marble after leaving the tube?

a. 1 b. 2 c. 3 d. 4 e. 5

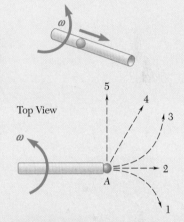

Fig. P12.CQ2

12.CQ3 The two systems shown start from rest. On the left, two 40-lb weights are connected by an inextensible cord, and on the right, a constant 40-lb force pulls on the cord. Neglecting all frictional forces, which of the following statements is true?

a. Blocks *A* and *C* will have the same acceleration.
b. Block *C* will have a larger acceleration than block *A*.
c. Block *A* will have a larger acceleration than block *C*.
d. Block *A* will not move.
e. None of the above are true.

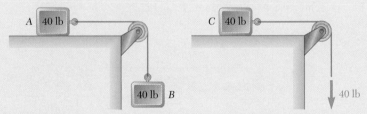

Fig. P12.CQ3

12.CQ4 Blocks *A* and *B* are released from rest in the position shown. Neglecting friction, the normal force between block *A* and the ground is:

a. Less than the weight of *A* plus the weight of *B*.
b. Equal to the weight of *A* plus the weight of *B*.
c. Greater than the weight of *A* plus the weight of *B*.

Fig. P12.CQ4

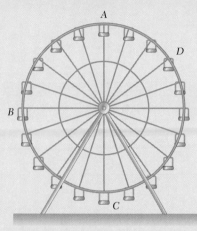

Fig. P12.CQ5

12.CQ5 People sit on a Ferris wheel at points *A*, *B*, *C,* and *D*. The Ferris wheel travels at a constant angular velocity. At the instant shown, which person experiences the largest force from his or her chair (back and seat)? Assume you can neglect the size of the chairs— that is, the people are located the same distance from the axis of rotation.

a. *A*
b. *B*
c. *C*
d. *D*
e. The force is the same for all the passengers.

FREE-BODY PRACTICE PROBLEMS

12.F1 Crate *A* is gently placed with zero initial velocity onto a moving conveyor belt. The coefficient of kinetic friction between the crate and the belt is μ_k. Draw the FBD and KD for *A* immediately after it contacts the belt.

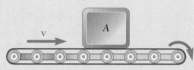

Fig. P12.F1

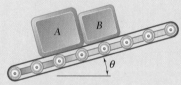

Fig. P12.F2

12.F2 Two blocks weighing W_A and W_B are at rest on a conveyor that is initially at rest. The belt is suddenly started in an upward direction so that slipping occurs between the belt and the boxes. Assuming the coefficient of friction between the boxes and the belt is μ_k, draw the FBDs and KDs for blocks *A* and *B*. How would you determine if *A* and *B* remain in contact?

12.F3 Objects *A*, *B*, and *C* have masses m_A, m_B, and m_C, respectively. The coefficient of kinetic friction between *A* and *B* is μ_k, and the friction between *A* and the ground is negligible and the pulleys are massless and frictionless. Assuming *B* slides on *A*, draw the FBD and KD for each of the three masses *A*, *B*, and *C*.

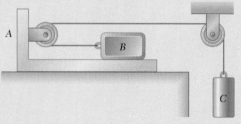

Fig. P12.F3

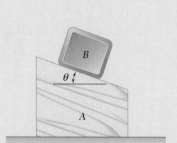

Fig. P12.F4

12.F4 Blocks *A* and *B* have masses m_A and m_B, respectively. Neglecting friction between all surfaces, draw the FBD and KD for each mass.

12.F5 Blocks A and B have masses m_A and m_B, respectively. Neglecting friction between all surfaces, draw the FBD and KD for the two systems shown.

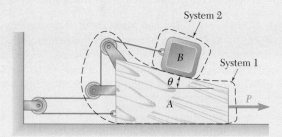

Fig. P12.F5

12.F6 A pilot of mass m flies a jet in a half-vertical loop of radius R so that the speed of the jet, v, remains constant. Draw a FBD and KD of the pilot at points A, B, and C.

12.F7 Wires AC and BC are attached to a sphere which revolves at a constant speed v in the horizontal circle of radius r as shown. Draw a FBD and KD of C.

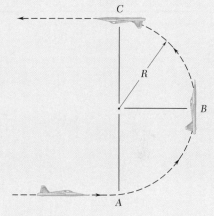

Fig. P12.F6

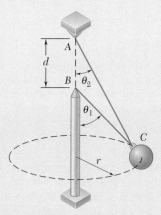

Fig. P12.F7

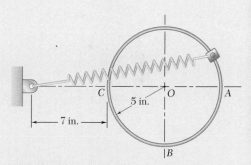

Fig. P12.F8

12.F8 A collar of mass m is attached to a spring and slides without friction along a circular rod in a vertical plane. The spring has an undeformed length of 5 in. and a constant k. Knowing that the collar has a speed v at point C, draw the FBD and KD of the collar at this point.

12.F9 Four pins slide in four separate slots cut in a horizontal circular plate as shown. When the plate is at rest, each pin has a velocity directed as shown and of the same constant magnitude u. Each pin has a mass m and maintains the same velocity relative to the plate when the plate rotates about O with a constant counterclockwise angular velocity ω. Draw the FBDs and KDs to determine the forces on pins P_1 and P_2.

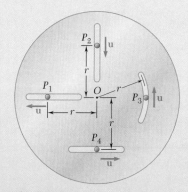

Fig. P12.F9

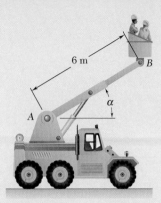

6 m

B

α

A

Fig. P12.F10

12.F10 At the instant shown, the length of the boom AB is being *decreased* at the constant rate of 0.2 m/s, and the boom is being lowered at the constant rate of 0.08 rad/s. If the mass of the men and lift connected to the boom at point B is m, draw the FBD and KD that could be used to determine the horizontal and vertical forces at B.

12.F11 Disk A rotates in a horizontal plane about a vertical axis at the constant rate $\dot{\theta}_0$. Slider B has a mass m and moves in a frictionless slot cut in the disk. The slider is attached to a spring of constant k, which is undeformed when $r = 0$. Knowing that the slider is released with no radial velocity in the position $r = r_0$, draw a FBD and KD at an arbitrary distance r from O.

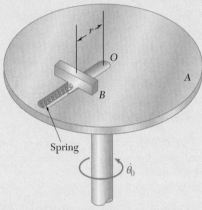

r

O

A

B

Spring

$\dot{\theta}_0$

Fig. P12.F11

12.F12 Pin B has a mass m and slides along the slot in the rotating arm OC and along the slot DE which is cut in a fixed horizontal plate. Neglecting friction and knowing that rod OC rotates at the constant rate $\dot{\theta}_0$, draw a FBD and KD that can be used to determine the forces $\mathbf{P}$ and $\mathbf{Q}$ exerted on pin B by rod OC and the wall of slot DE, respectively.

END-OF-SECTION PROBLEMS

12.1 Astronauts who landed on the moon during the Apollo 15, 16, and 17 missions brought back a large collection of rocks to the earth. Knowing the rocks weighed 139 lb when they were on the moon, determine (*a*) the weight of the rocks on the earth, (*b*) the mass of the rocks in slugs. The acceleration due to gravity on the moon is 5.30 ft/s^2.

12.2 The value of g at any latitude ϕ may be obtained from the formula

$$g = 32.09(1 + 0.0053 \sin^2 \phi)\text{ft/s}^2$$

which takes into account the effect of the rotation of the earth, as well as the fact that the earth is not truly spherical. Knowing that the weight of a silver bar has been officially designated as 5 lb, determine to four significant figures, (*a*) the mass in slugs, (*b*) the weight in pounds at the latitudes of 0°, 45°, and 60°.

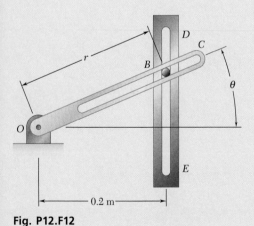

D

C

r

B

θ

O

E

0.2 m

Fig. P12.F12

12.3 A 400-kg satellite has been placed in a circular orbit 1500 km above the surface of the earth. The acceleration of gravity at this elevation is 6.43 m/s^2. Determine the linear momentum of the satellite, knowing that its orbital speed is 25.6×10^3 km/h.

12.4 A spring scale A and a lever scale B having equal lever arms are fastened to the roof of an elevator, and identical packages are attached to the scales as shown. Knowing that when the elevator moves downward with an acceleration of 1 m/s² the spring scale indicates a load of 60 N, determine (*a*) the weight of the packages, (*b*) the load indicated by the spring scale and the mass needed to balance the lever scale when the elevator moves upward with an acceleration of 1 m/s².

12.5 In anticipation of a long 7° upgrade, a bus driver accelerates at a constant rate of 3 ft/s² while still on a level section of the highway. Knowing that the speed of the bus is 60 mi/h as it begins to climb the grade and that the driver does not change the setting of his throttle or shift gears, determine the distance traveled by the bus up the grade when its speed has decreased to 50 mi/h.

12.6 A 0.2-lb model rocket is launched vertically from rest at time $t = 0$ with a constant thrust of 2 lb for one second and no thrust for $t > 1$ s. Neglecting air resistance and the decrease in mass of the rocket, determine (*a*) the maximum height h reached by the rocket, (*b*) the time required to reach this maximum height.

12.7 A tugboat pulls a small barge through a harbor. The propeller thrust minus the drag produces a net thrust that varies linearly with speed. Knowing that the combined weight of the tug and barge is 3600 kN, determine (*a*) the time required to increase the speed from an initial value $v_1 = 1.0$ m/s to a final value $v_2 = 2.5$ m/s, (*b*) the distance traveled during this time interval.

12.8 Determine the maximum theoretical speed that may be achieved over a distance of 60 m by a car starting from rest, knowing that the coefficient of static friction is 0.80 between the tires and the pavement and that 60 percent of the weight of the car is distributed over its front wheels and 40 percent over its rear wheels. Assume (*a*) four-wheel drive, (*b*) front-wheel drive, (*c*) rear-wheel drive.

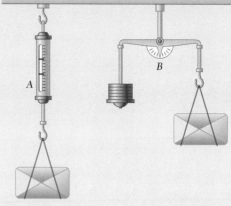

Fig. P12.4

Fig. P12.6

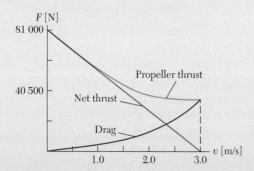

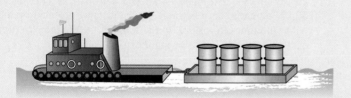

Fig. P12.7

12.9 If an automobile's braking distance from 90 km/h is 45 m on level pavement, determine the automobile's braking distance from 90 km/h when it is (*a*) going up a 5° incline, (*b*) going down a 3-percent incline. Assume the braking force is independent of grade.

12.10 A mother and her child are skiing together, and the mother is holding the end of a rope tied to the child's waist. They are moving at a speed of 7.2 km/h on a gently sloping portion of the ski slope when the mother observes that they are approaching a steep descent. She pulls on the rope with an average force of 7 N. Knowing the coefficient of friction between the child and the ground is 0.1 and the angle of the rope does not change, determine (*a*) the time required for the child's speed to be cut in half, (*b*) the distance traveled in this time.

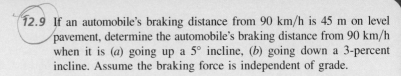

Fig. P12.10

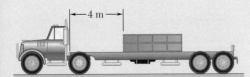

Fig. P12.11

12.11 The coefficients of friction between the load and the flatbed trailer shown are $\mu_s = 0.40$ and $\mu_k = 0.30$. Knowing that the speed of the rig is 72 km/h, determine the shortest distance in which the rig can be brought to a stop if the load is not to shift.

12.12 A light train made up of two cars is traveling at 90 km/h when the brakes are applied to both cars. Knowing that car *A* has a mass of 25 Mg and car *B* a mass of 20 Mg, and that the braking force is 30 kN on each car, determine (*a*) the distance traveled by the train before it comes to a stop, (*b*) the force in the coupling between the cars while the train is slowing down.

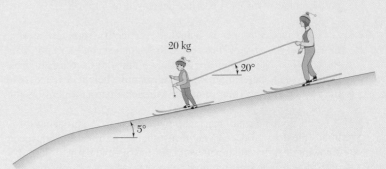

Fig. P12.12

12.13 The two blocks shown are originally at rest. Neglecting the masses of the pulleys and the effect of friction in the pulleys and between block *A* and the incline, determine (*a*) the acceleration of each block, (*b*) the tension in the cable.

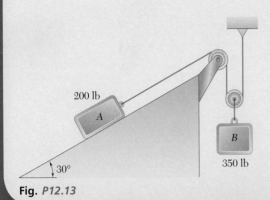

Fig. P12.13

12.14 Solve Prob. 12.13, assuming that the coefficients of friction between block *A* and the incline are $\mu_s = 0.25$ and $\mu_k = 0.20$.

12.15 Each of the systems shown is initially at rest. Neglecting axle friction and the masses of the pulleys, determine for each system (*a*) the acceleration of block *A*, (*b*) the velocity of block *A* after it has moved through 10 ft, (*c*) the time required for block *A* to reach a velocity of 20 ft/s.

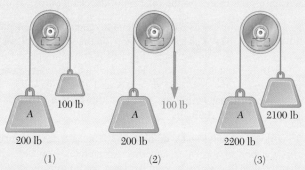

(1)   (2)   (3)

Fig. P12.15

12.16 Boxes *A* and *B* are at rest on a conveyor belt that is initially at rest. The belt is suddenly started in an upward direction so that slipping occurs between the belt and the boxes. Knowing that the coefficients of kinetic friction between the belt and the boxes are $(\mu_k)_A = 0.30$ and $(\mu_k)_B = 0.32$, determine the initial acceleration of each box.

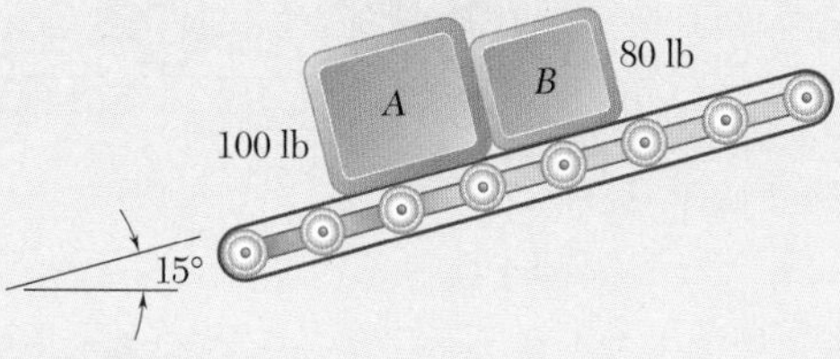

12.17 A 5000-lb truck is being used to lift a 1000-lb boulder *B* that is on a 200-lb pallet *A*. Knowing the acceleration of the truck is 1 ft/s^2, determine (*a*) the horizontal force between the tires and the ground, (*b*) the force between the boulder and the pallet.

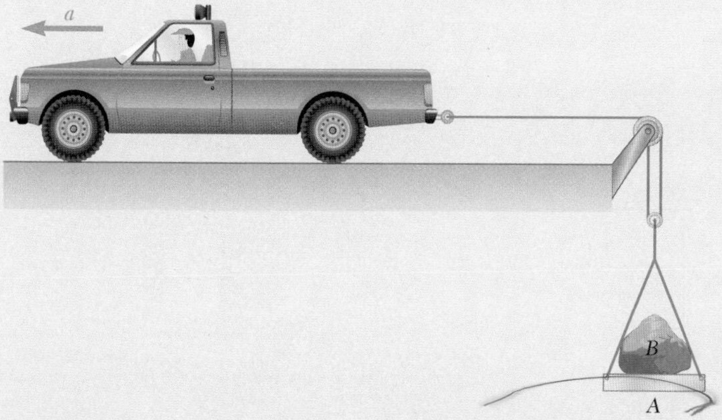

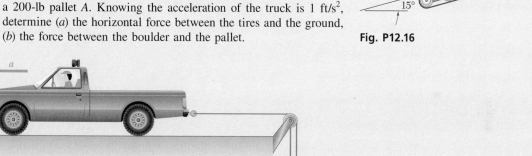

Fig. P12.16

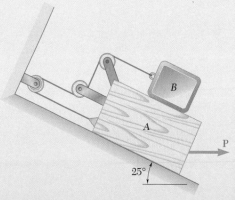

Fig. P12.17

12.18 Block *A* has a mass of 40 kg, and block *B* has a mass of 8 kg. The coefficients of friction between all surfaces of contact are $\mu_s = 0.20$ and $\mu_k = 0.15$. If $P = 0$, determine (*a*) the acceleration of block *B*, (*b*) the tension in the cord.

12.19 Block *A* has a mass of 40 kg, and block *B* has a mass of 8 kg. The coefficients of friction between all surfaces of contact are $\mu_s = 0.20$ and $\mu_k = 0.15$. If $P = 40$ N, determine (*a*) the acceleration of block *B*, (*b*) the tension in the cord.

Fig. P12.18 and P12.19

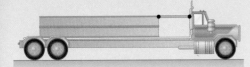

Fig. P12.20

Fig. P12.21

Fig. P12.22

12.20 The flat-bed trailer carries two 1500-kg beams with the upper beam secured by a cable. The coefficients of static friction between the two beams and between the lower beam and the bed of the trailer are 0.25 and 0.30, respectively. Knowing that the load does not shift, determine (*a*) the maximum acceleration of the trailer and the corresponding tension in the cable, (*b*) the maximum deceleration of the trailer.

12.21 A baggage conveyor is used to unload luggage from an airplane. The 10-kg duffel bag *A* is sitting on top of the 20-kg suitcase *B*. The conveyor is moving the bags down at a constant speed of 0.5 m/s when the belt suddenly stops. Knowing that the coefficient of friction between the belt and *B* is 0.3 and that bag *A* does not slip on suitcase *B*, determine the smallest allowable coefficient of static friction between the bags.

12.22 To unload a bound stack of plywood from a truck, the driver first tilts the bed of the truck and then accelerates from rest. Knowing that the coefficients of friction between the bottom sheet of plywood and the bed are $\mu_s = 0.40$ and $\mu_k = 0.30$, determine (*a*) the smallest acceleration of the truck which will cause the stack of plywood to slide, (*b*) the acceleration of the truck which causes corner *A* of the stack to reach the end of the bed in 0.9 s.

12.23 To transport a series of bundles of shingles *A* to a roof, a contractor uses a motor-driven lift consisting of a horizontal platform *BC* which rides on rails attached to the sides of a ladder. The lift starts from rest and initially moves with a constant acceleration $\mathbf{a}_1$ as shown. The lift then decelerates at a constant rate $\mathbf{a}_2$ and comes to rest at *D*, near the top of the ladder. Knowing that the coefficient of static friction between a bundle of shingles and the horizontal platform is 0.30, determine the largest allowable acceleration $\mathbf{a}_1$ and the largest allowable deceleration $\mathbf{a}_2$ if the bundle is not to slide on the platform.

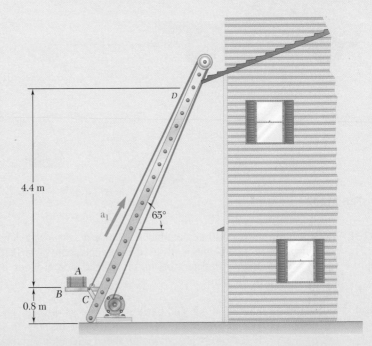

Fig. P12.23

12.24 An airplane has a mass of 25 Mg and its engines develop a total thrust of 40 kN during take-off. If the drag **D** exerted on the plane has a magnitude $D = 2.25\ v^2$, where v is expressed in meters per second and D in newtons, and if the plane becomes airborne at a speed of 240 km/h, determine the length of runway required for the plane to take off.

12.25 A 4-kg projectile is fired vertically with an initial velocity of 90 m/s, reaches a maximum height, and falls to the ground. The aerodynamic drag **D** has a magnitude $D = 0.0024\ v^2$ where D and v are expressed in newtons and m/s, respectively. Knowing that the direction of the drag is always opposite to the direction of the velocity, determine (*a*) the maximum height of the trajectory, (*b*) the speed of the projectile when it reaches the ground.

12.26 A constant force **P** is applied to a piston and rod of total mass m to make them move in a cylinder filled with oil. As the piston moves, the oil is forced through orifices in the piston and exerts on the piston a force of magnitude kv in a direction opposite to the motion of the piston. Knowing that the piston starts from rest at $t = 0$ and $x = 0$, show that the equation relating x, v, and t, where x is the distance traveled by the piston and v is the speed of the piston, is linear in each of these variables.

Fig. P12.26

12.27 A spring *AB* of constant k is attached to a support at *A* and to a collar of mass m. The unstretched length of the spring is l. Knowing that the collar is released from rest at $x = x_0$ and neglecting friction between the collar and the horizontal rod, determine the magnitude of the velocity of the collar as it passes through point *C*.

12.28 Block *A* has a mass of 10 kg, and blocks *B* and *C* have masses of 5 kg each. Knowing that the blocks are initially at rest and that *B* moves through 3 m in 2 s, determine (*a*) the magnitude of the force **P**, (*b*) the tension in the cord *AD*. Neglect the masses of the pulleys and axle friction.

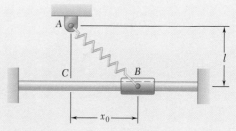

Fig. P12.27

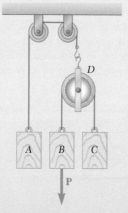

Fig. P12.28

12.29 A 40-lb sliding panel is supported by rollers at B and C. A 25-lb counterweight A is attached to a cable as shown and, in cases a and c, is initially in contact with a vertical edge of the panel. Neglecting friction, determine in each case shown the acceleration of the panel and the tension in the cord immediately after the system is released from rest.

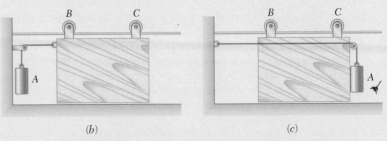

(a)   (b)   (c)

Fig. P12.29

12.30 An athlete pulls handle A to the left with a constant force of $P = 100$ N. Knowing that after the handle A has been pulled 30 cm and its velocity is 3 m/s, determine the mass of the weight stack B.

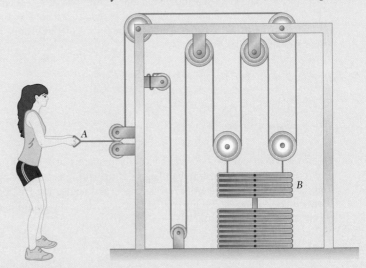

Fig. P12.30

12.31 A 10-lb block B rests as shown on a 20-lb bracket A. The coefficients of friction are $\mu_s = 0.30$ and $\mu_k = 0.25$ between block B and bracket A, and there is no friction in the pulley or between the bracket and the horizontal surface. (a) Determine the maximum weight of block C if block B is not to slide on bracket A. (b) If the weight of block C is 10 percent larger than the answer found in a, determine the accelerations of A, B, and C.

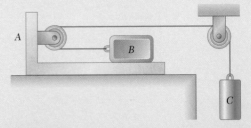

Fig. P12.31

12.32 Knowing that $\mu_k = 0.30$, determine the acceleration of each block when $m_A = m_B = m_C$.

12.33 Knowing that $\mu_k = 0.30$, determine the acceleration of each block when $m_A = 5$ kg, $m_B = 30$ kg, and $m_C = 15$ kg.

12.34 A 25-kg block A rests on an inclined surface, and a 15-kg counterweight B is attached to a cable as shown. Neglecting friction, determine the acceleration of A and the tension in the cable immediately after the system is released from rest.

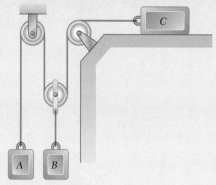

Fig. P12.32 and P12.33

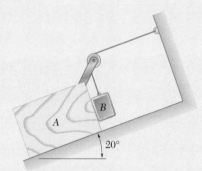

Fig. P12.34

12.35 Block B of mass 10 kg rests as shown on the upper surface of a 22-kg wedge A. Knowing that the system is released from rest and neglecting friction, determine (a) the acceleration of B, (b) the velocity of B relative to A at $t = 0.5$ s.

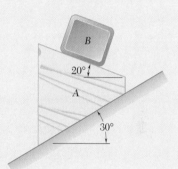

Fig. P12.35

12.36 A 450-g tetherball A is moving along a horizontal circular path at a constant speed of 4 m/s. Determine (a) the angle θ that the cord forms with pole BC, (b) the tension in the cord.

12.37 During a hammer thrower's practice swings, the 7.1-kg head A of the hammer revolves at a constant speed v in a horizontal circle as shown. If $\rho = 0.93$ m and $\theta = 60°$, determine (a) the tension in wire BC, (b) the speed of the hammer's head.

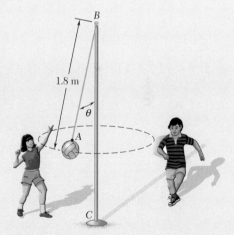

Fig. P12.36

Fig. P12.37

12.38 Human centrifuges are often used to simulate different acceleration levels for pilots. When aerospace physiologists say that a pilot is pulling 9 *g's*, they mean that the resultant normal force on the pilot from the bottom of the seat is nine times their weight. Knowing that the centrifuge starts from rest and has a constant angular acceleration of 1.5 RPM per second until the pilot is pulling 9 *g's* and then continues with a constant angular velocity, determine (*a*) how long it will take for the pilot to reach 9 *g's* (*b*) the angle θ of the normal force once the pilot reaches 9 *g's*. Assume that the force parallel to the seat is zero.

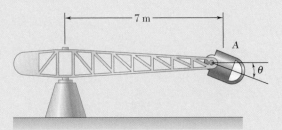

Fig. P12.38

12.39 A single wire *ACB* passes through a ring at *C* attached to a sphere that revolves at a constant speed *v* in the horizontal circle shown. Knowing that the tension is the same in both portions of the wire, determine the speed *v*.

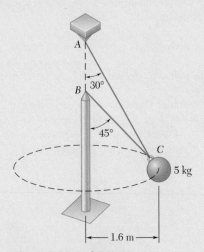

Fig. P12.39 and P12.40

*12.40 Two wires *AC* and *BC* are tied at *C* to a sphere that revolves at a constant speed *v* in the horizontal circle shown. Determine the range of the allowable values of *v* if both wires are to remain taut and if the tension in either of the wires is not to exceed 60 N.

12.41 A 1-kg sphere is at rest relative to a parabolic dish that rotates at a constant rate about a vertical axis. Neglecting friction and knowing that $r = 1$ m, determine (*a*) the speed *v* of the sphere, (*b*) the magnitude of the normal force exerted by the sphere on the inclined surface of the dish.

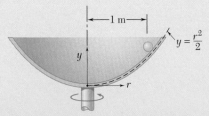

Fig. P12.41

*12.42 As part of an outdoor display, a 12-lb model *C* of the earth is attached to wires *AC* and *BC* and revolves at a constant speed *v* in the horizontal circle shown. Determine the range of the allowable values of *v* if both wires are to remain taut and if the tension in either of the wires is not to exceed 26 lb.

*12.43 The 1.2-lb flyballs of a centrifugal governor revolve at a constant speed *v* in the horizontal circle of 6-in. radius shown. Neglecting the weights of links *AB*, *BC*, *AD*, and *DE* and requiring that the links support only tensile forces, determine the range of the allowable values of *v* so that the magnitudes of the forces in the links do not exceed 17 lb.

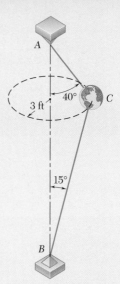

Fig. P12.42

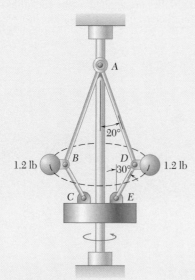

Fig. P12.43

12.44 A 130-lb wrecking ball *B* is attached to a 45-ft-long steel cable *AB* and swings in the vertical arc shown. Determine the tension in the cable (*a*) at the top *C* of the swing, (*b*) at the bottom *D* of the swing, where the speed of *B* is 13.2 ft/s.

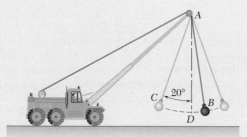

Fig. P12.44

12.45 During a high-speed chase, a 2400-lb sports car traveling at a speed of 100 mi/h just loses contact with the road as it reaches the crest *A* of a hill. (*a*) Determine the radius of curvature ρ of the vertical profile of the road at *A*. (*b*) Using the value of ρ found in part *a*, determine the force exerted on a 160-lb driver by the seat of his 3100-lb car as the car, traveling at a constant speed of 50 mi/h, passes through *A*.

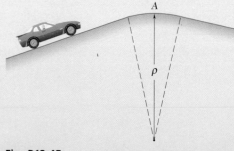

Fig. P12.45

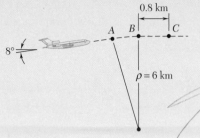

0.8 km

A B C

8°

$\rho = 6$ km

Fig. P12.46

12.46 An airline pilot climbs to a new flight level along the path shown. Knowing that the speed of the airplane decreases at a constant rate from 180 m/s at point A to 160 m/s at point C, determine the magnitude of the abrupt change in the force exerted on a 90-kg passenger as the airplane passes point B.

12.47 The roller-coaster track shown is contained in a vertical plane. The portion of track between A and B is straight and horizontal, while the portions to the left of A and to the right of B have radii of curvature as indicated. A car is traveling at a speed of 72 km/h when the brakes are suddenly applied, causing the wheels of the car to slide on the track ($\mu_k = 0.20$). Determine the initial deceleration of the car if the brakes are applied as the car (a) has almost reached A, (b) is traveling between A and B, (c) has just passed B.

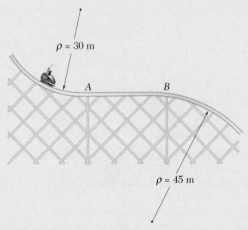

$\rho = 30$ m

A B

$\rho = 45$ m

Fig. P12.47

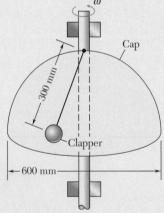

ω

Cap

300 mm

Clapper

←600 mm→

Fig. P12.48

12.48 A spherical-cap governor is fixed to a vertical shaft that rotates with angular velocity ω. When the string-supported clapper of mass m touches the cap, a cutoff switch is operated electrically to reduce the speed of the shaft. Knowing that the radius of the clapper is small relative to the cap, determine the minimum angular speed at which the cutoff switch operates.

12.49 A series of small packages, each with a mass of 0.5 kg, are discharged from a conveyor belt as shown. Knowing that the coefficient of static friction between each package and the conveyor belt is 0.4, determine (a) the force exerted by the belt on the package just after it has passed point A, (b) the angle θ defining the point B where the packages first *slip* relative to the belt.

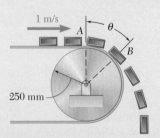

1 m/s

A θ

B

250 mm

Fig. P12.49

12.50 A 54-kg pilot flies a jet trainer in a half-vertical loop of 1200-m radius so that the speed of the trainer decreases at a constant rate. Knowing that the pilot's apparent weights at points A and C are 1680 N and 350 N, respectively, determine the force exerted on her by the seat of the trainer when the trainer is at point B.

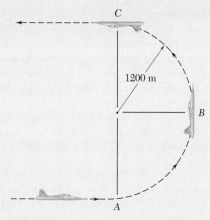

Fig. P12.50

12.51 A carnival ride is designed to allow the general public to experience high-acceleration motion. The ride rotates about point O in a horizontal circle such that the rider has a speed v_0. The rider reclines on a platform A which rides on rollers such that friction is negligible. A mechanical stop prevents the platform from rolling down the incline. Determine (a) the speed v_0 at which the platform A begins to roll upward, (b) the normal force experienced by an 80-kg rider at this speed.

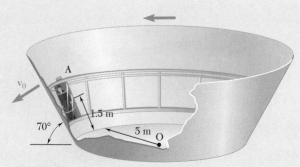

Fig. P12.51

12.52 A curve in a speed track has a radius of 1000 ft and a rated speed of 120 mi/h. (See Sample Prob. 12.7 for the definition of rated speed.) Knowing that a racing car starts skidding on the curve when traveling at a speed of 180 mi/h, determine (a) the banking angle θ, (b) the coefficient of static friction between the tires and the track under the prevailing conditions, (c) the minimum speed at which the same car could negotiate the curve.

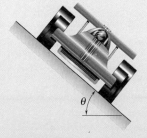

Fig. P12.52

Fig. P12.53 and *P12.54*

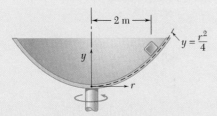

Fig. P12.55

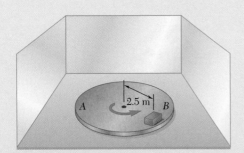

Fig. P12.57

12.53 Tilting trains, such as the *American Flyer* which will run from Washington to New York and Boston, are designed to travel safely at high speeds on curved sections of track which were built for slower, conventional trains. As it enters a curve, each car is tilted by hydraulic actuators mounted on its trucks. The tilting feature of the cars also increases passenger comfort by eliminating or greatly reducing the side force $\mathbf{F}_s$ (parallel to the floor of the car) to which passengers feel subjected. For a train traveling at 100 mi/h on a curved section of track banked through an angle $\theta = 6°$ and with a rated speed of 60 mi/h, determine (*a*) the magnitude of the side force felt by a passenger of weight W in a standard car with no tilt ($\phi = 0$), (*b*) the required angle of tilt ϕ if the passenger is to feel no side force. (See Sample Prob. 12.7 for the definition of rated speed.)

12.54 Tests carried out with the tilting trains described in Prob. 12.53 revealed that passengers feel queasy when they see through the car windows that the train is rounding a curve at high speed, yet do not feel any side force. Designers, therefore, prefer to reduce, but not eliminate that force. For the train of Prob. 12.53, determine the required angle of tilt ϕ if passengers are to feel side forces equal to 10 percent of their weights.

12.55 A 3-kg block is at rest relative to a parabolic dish which rotates at a constant rate about a vertical axis. Knowing that the coefficient of static friction is 0.5 and that $r = 2$ m, determine the maximum allowable velocity v of the block.

12.56 A polisher is started so that the fleece along the circumference undergoes a constant tangential acceleration of 4 m/s². Three seconds after it is started, small tufts of fleece from along the circumference of the 225-mm-diameter polishing pad are observed to fly free of the pad. At this instant, determine (*a*) the speed v of a tuft as it leaves the pad, (*b*) the magnitude of the force required to free a tuft if the average mass of a tuft is 1.6 mg.

Fig. P12.56

12.57 A turntable *A* is built into a stage for use in a theatrical production. It is observed during a rehearsal that a trunk *B* starts to slide on the turntable 10 s after the turntable begins to rotate. Knowing that the trunk undergoes a constant tangential acceleration of 0.24 m/s², determine the coefficient of static friction between the trunk and the turntable.

12.58 The carnival ride from Prob. 12.51 is modified so that the 80-kg riders can move up and down the inclined wall as the speed of the ride increases. Assuming that the friction between the wall and the carriage is negligible, determine the position h of the rider if the speed $v_0 = 13$ m/s.

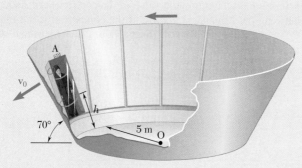

Fig. P12.58 and *P12.59*

12.59 The carnival ride from Prob 12.51 is modified so that the 80-kg riders can move up and down the inclined wall as the speed of the ride increases. Knowing that the coefficient of static friction between the wall and the platform is 0.2, determine the range of values of the constant speed v_0 for which the platform will remain at $h = 1.5$ m.

12.60 A semicircular slot of 10-in. radius is cut in a flat plate that rotates about the vertical AD at a constant rate of 14 rad/s. A small, 0.8-lb block E is designed to slide in the slot as the plate rotates. Knowing that the coefficients of friction are $\mu_s = 0.35$ and $\mu_k = 0.25$, determine whether the block will slide in the slot if it is released in the position corresponding to (a) $\theta = 80°$, (b) $\theta = 40°$. Also determine the magnitude and the direction of the friction force exerted on the block immediately after it is released.

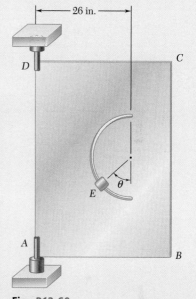

Fig. *P12.60*

12.61 A small block B fits inside a slot cut in arm OA that rotates in a vertical plane at a constant rate. The block remains in contact with the end of the slot closest to A and its speed is 1.4 m/s for $0 \leq \theta \leq 150°$. Knowing that the block begins to slide when $\theta = 150°$, determine the coefficient of static friction between the block and the slot.

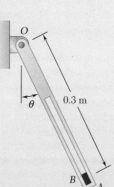

Fig. P12.61

12.62 The parallel-link mechanism *ABCD* is used to transport a component *I* between manufacturing processes at stations *E*, *F*, and *G* by picking it up at a station when $\theta = 0$ and depositing it at the next station when $\theta = 180°$. Knowing that member *BC* remains horizontal throughout its motion and that links *AB* and *CD* rotate at a constant rate in a vertical plane in such a way that $v_B = 2.2$ ft/s, determine (*a*) the minimum value of the coefficient of static friction between the component and *BC* if the component is not to slide on *BC* while being transferred, (*b*) the values of θ for which sliding is impending.

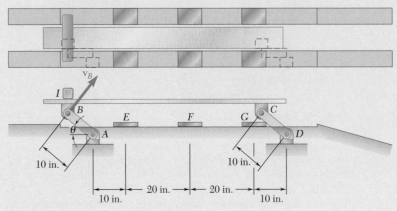

Fig. P12.62

12.63 Knowing that the coefficients of friction between the component *I* and member *BC* of the mechanism of Prob. 12.62 are $\mu_s = 0.35$ and $\mu_k = 0.25$, determine (*a*) the maximum allowable constant speed v_B if the component is not to slide on *BC* while being transferred, (*b*) the values of θ for which sliding is impending.

12.64 A small 250-g collar *C* can slide on a semicircular rod which is made to rotate about the vertical *AB* at a constant rate of 7.5 rad/s. Determine the three values of θ for which the collar will not slide on the rod, assuming no friction between the collar and the rod.

12.65 A small 250-g collar *C* can slide on a semicircular rod which is made to rotate about the vertical *AB* at a constant rate of 7.5 rad/s. Knowing that the coefficients of friction are $\mu_s = 0.25$ and $\mu_k = 0.20$, indicate whether the collar will slide on the rod if it is released in the position corresponding to (*a*) $\theta = 75°$, (*b*) $\theta = 40°$. Also, determine the magnitude and direction of the friction force exerted on the collar immediately after release.

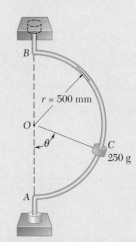

Fig. P12.64 and P12.65

12.66 An advanced spatial disorientation trainer allows the cab to rotate around multiple axes as well as to extend inwards and outwards. It can be used to simulate driving, fixed-wing aircraft flying, and helicopter maneuvering. In one training scenario, the trainer rotates and translates in the horizontal plane where the location of the pilot is defined by the relationships $r = 10 + 2\cos\left(\frac{\pi}{3}t\right)$ and $\theta = 0.1(2t^2 - t)$, where r, θ, and t are expressed in feet, radians, and seconds, respectively. Knowing that the pilot has a weight of 175 lbs, (*a*) determine the magnitude of the resulting force acting on the pilot at $t = 5$ s, (*b*) plot the magnitudes of the radial and transverse components of the force exerted on the pilot from 0 to 10 seconds.

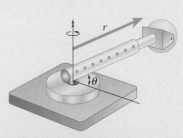

Fig. P12.66 and P12.67

12.67 An advanced spatial disorientation trainer is programmed to only rotate and translate in the horizontal plane. The pilot's location is defined by the relationships $r = 8(1 - e^{-t})$ and $\theta = 2/\pi\left(\sin\frac{\pi}{2}t\right)$, where r, θ, and t are expressed in feet, radians, and seconds, respectively. Determine the radial and transverse components of the force exerted on the 175-lb pilot at $t = 3$ s.

12.68 The 3-kg collar B slides on the frictionless arm AA'. The arm is attached to drum D and rotates about O in a horizontal plane at the rate $\dot\theta = 0.75t$, where $\dot\theta$ and t are expressed in rad/s and seconds, respectively. As the arm-drum assembly rotates, a mechanism within the drum releases cord so that the collar moves outward from O with a constant speed of 0.5 m/s. Knowing that at $t = 0$, $r = 0$, determine the time at which the tension in the cord is equal to the magnitude of the horizontal force exerted on B by arm AA'.

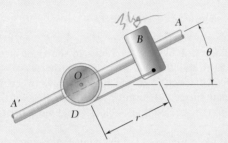

Fig. P12.68

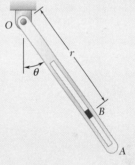

12.69 A 0.5-kg block B slides without friction inside a slot cut in arm OA that rotates in a vertical plane. The rod has a constant angular acceleration $\ddot\theta = 10$ rad/s^2. Knowing that when $\theta = 45°$ and $r = 0.8$ m the velocity of the block is zero, determine at this instant, (a) the force exerted on the block by the arm, (b) the relative acceleration of the block with respect to the arm.

Fig. P12.69

12.70 Pin B weighs 4 oz and is free to slide in a horizontal plane along the rotating arm OC and along the fixed circular slot DE of radius $b = 20$ in. Neglecting friction and assuming that $\dot\theta = 15$ rad/s and $\ddot\theta = 250$ rad/s^2 for the position $\theta = 20°$, determine for that position (a) the radial and transverse components of the resultant force exerted on pin B, (b) the forces **P** and **Q** exerted on pin B, respectively, by rod OC and the wall of slot DE.

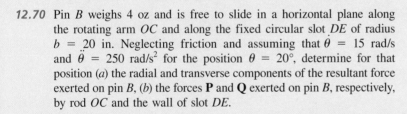

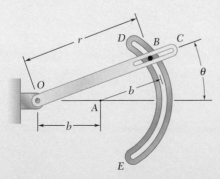

Fig. P12.70

12.71 The two blocks are released from rest when $r = 0.8$ m and $\theta = 30°$. Neglecting the mass of the pulley and the effect of friction in the pulley and between block A and the horizontal surface, determine (a) the initial tension in the cable, (b) the initial acceleration of block A, (c) the initial acceleration of block B.

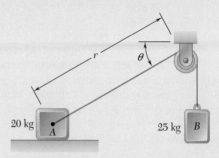

Fig. P12.71 and P12.72

12.72 The velocity of block A is 2 m/s to the right at the instant when $r = 0.8$ m and $\theta = 30°$. Neglecting the mass of the pulley and the effect of friction in the pulley and between block A and the horizontal surface, determine, at this instant, (a) the tension in the cable, (b) the acceleration of block A, (c) the acceleration of block B.

*12.73 Slider C has a weight of 0.5 lb and may move in a slot cut in arm AB, which rotates at the constant rate $\dot\theta_0 = 10$ rad/s in a horizontal plane. The slider is attached to a spring of constant $k = 2.5$ lb/ft, which is unstretched when $r = 0$. Knowing that the slider is released from rest with no radial velocity in the position $r = 18$ in. and neglecting friction, determine for the position $r = 12$ in. (a) the radial and transverse components of the velocity of the slider, (b) the radial and transverse components of its acceleration, (c) the horizontal force exerted on the slider by arm AB.

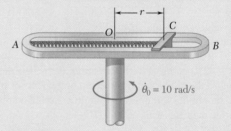

Fig. P12.73

12.2 ANGULAR MOMENTUM AND ORBITAL MOTION

In Sec. 12.1, we introduced the idea of linear momentum and showed how Newton's second law could be expressed as the rate of change of linear momentum. Angular momentum, or the moment of linear momentum, is another useful quantity. In this section, we define angular momentum for a particle and discuss the motion of a particle under a central force, which is applicable to many types of orbital motion.

12.2A Angular Momentum of a Particle and Its Rate of Change

Consider a particle P with a mass m moving with respect to a newtonian frame of reference $Oxyz$. As we saw in Sec. 12.1B, the linear momentum of the particle at a given instant is defined as the vector $m\mathbf{v}$ that is obtained by multiplying the velocity $\mathbf{v}$ of the particle by its mass m. The moment about O of the vector $m\mathbf{v}$ is called the *moment of momentum,* or the **angular momentum,** of the particle about O at that instant and is denoted by $\mathbf{H}_O$. Recall the definition of the moment of a vector (Sec. 3.1E) and denote the position vector of P by $\mathbf{r}$. Then we have

Angular momentum of a particle
$$\mathbf{H}_O = \mathbf{r} \times m\mathbf{v} \qquad \textbf{(12.13)}$$

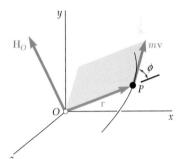

Fig. 12.12 The angular momentum vector of a particle is the vector product of the position vector **r** and the linear momentum vector *m*v.

Note that $\mathbf{H}_O$ is a vector perpendicular to the plane containing $\mathbf{r}$ and $m\mathbf{v}$ and has a magnitude

$$H_O = rmv \sin \phi \qquad \textbf{(12.14)}$$

where ϕ is the angle between $\mathbf{r}$ and $m\mathbf{v}$ (Fig. 12.12). We can determine the sense of $\mathbf{H}_O$ from the sense of $m\mathbf{v}$ by applying the right-hand rule. The unit of angular momentum is obtained by multiplying the units of length and of linear momentum (Sec. 12.1C). In SI units, we have

$$(\text{m})(\text{kg·m/s}) = \text{kg·m}^2/\text{s}$$

In U.S. customary units, we have

$$(\text{ft})(\text{slug})(\text{ft/s}) = (\text{ft})(\text{lb·s}) = \text{ft·lb·s}$$

Resolving the vectors $\mathbf{r}$ and $m\mathbf{v}$ into components and applying formula (3.10), we obtain

$$\mathbf{H}_O = \begin{vmatrix} \mathbf{i} & \mathbf{j} & \mathbf{k} \\ x & y & z \\ mv_x & mv_y & mv_z \end{vmatrix} \qquad \textbf{(12.15)}$$

The components of $\mathbf{H}_O$, which also represent the moments of the linear momentum $m\mathbf{v}$ about the coordinate axes, can be obtained by expanding the determinant in Eq. (12.15). The results are

$$\begin{aligned} H_x &= m(yv_z - zv_y) \\ H_y &= m(zv_x - xv_z) \\ H_z &= m(xv_y - yv_x) \end{aligned} \qquad \textbf{(12.16)}$$

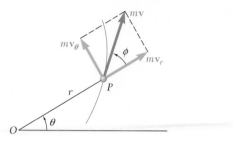

Fig. 12.13 In polar coordinates, angular momentum of a particle is the product of the position r and the transverse component of linear momentum.

In the case of a particle moving in the xy plane, we have $z = v_z = 0$ and the components H_x and H_y reduce to zero. The angular momentum is thus perpendicular to the xy plane; it is then completely defined by the scalar

$$H_O = H_z = m(xv_y - yv_x) \tag{12.17}$$

This value can be positive or negative, according to the sense in which the particle is observed to move from O. If we use polar coordinates, we resolve the linear momentum of the particle into radial and transverse components (Fig. 12.13), which gives us

$$H_O = rmv \sin \phi = rmv_\theta \tag{12.18}$$

Alternatively, recalling from Eq. (11.44) that $v_\theta = r\dot{\theta}$ we have

Angular momentum in polar coordinates

$$H_O = mr^2\dot{\theta} \tag{12.19}$$

Let us now compute the derivative with respect to t of the angular momentum $\mathbf{H}_O$ of a particle P moving in space. Differentiating both sides of Eq. (12.13) and recalling the rule for the differentiation of a vector product (Sec. 11.4B), we have

$$\dot{\mathbf{H}}_O = \dot{\mathbf{r}} \times m\mathbf{v} + \mathbf{r} \times m\dot{\mathbf{v}} = \mathbf{v} \times m\mathbf{v} + \mathbf{r} \times m\mathbf{a}$$

Since the vectors $\mathbf{v}$ and $m\mathbf{v}$ are collinear, the first term of this expression is zero; by Newton's second law, $m\mathbf{a}$ is equal to the sum $\Sigma\mathbf{F}$ of the forces acting on P. Noting that $\mathbf{r} \times \Sigma\mathbf{F}$ represents the sum $\Sigma\mathbf{M}_O$ of the moments about O of these forces, we obtain

$$\Sigma\mathbf{M}_O = \dot{\mathbf{H}}_O \tag{12.20}$$

Equation (12.20), which results directly from Newton's second law, states:

> **The sum of the moments about O of the forces acting on the particle is equal to the rate of change of angular momentum (or moment of momentum) of the particle about O.**

12.2B Motion Under a Central Force and Conservation of Angular Momentum

When the only force acting on a particle P is a force $\mathbf{F}$ directed toward or away from a fixed point O, the particle is said to be moving under a **central force**, and the point O is referred to as the **center of force** (Fig. 12.14). Since the line of action of $\mathbf{F}$ passes through O, we must have $\Sigma\mathbf{M}_O = 0$ at any given instant. Substituting into Eq. (12.20), we obtain

$$\dot{\mathbf{H}}_O = 0$$

for all values of t and, integrating in t,

$$\mathbf{H}_O = \text{constant} \tag{12.21}$$

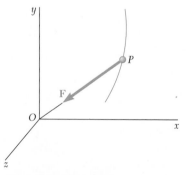

Fig. 12.14 The central force $\mathbf{F}$ acts towards the center of force O.

We thus conclude that

> **The angular momentum of a particle moving under a central force is constant in both magnitude and direction.**

Recall the definition of the angular momentum of a particle (Sec. 12.2A). From that, we have

Conservation of angular momentum
$$\mathbf{r} \times m\mathbf{v} = \mathbf{H}_O = \text{constant} \qquad \text{(12.22)}$$

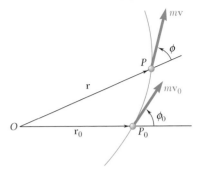

Fig. 12.15 Angular momentum of a particle moving in a fixed plane under the action of a central force.

It follows that the position vector $\mathbf{r}$ of the particle P must be perpendicular to the constant vector $\mathbf{H}_O$. Thus, a particle under a central force moves in a fixed plane perpendicular to $\mathbf{H}_O$. The vector $\mathbf{H}_O$ and the fixed plane are defined by the initial position vector $\mathbf{r}_0$ and the initial velocity $\mathbf{v}_0$ of the particle. For convenience, let us assume that the plane of the figure coincides with the fixed plane of motion (Fig. 12.15).

Since the magnitude H_O of the angular momentum of the particle P is constant, the right-hand side in Eq. (12.14) must be constant. Therefore, we have

$$rmv \sin \phi = r_0 m v_0 \sin \phi_0 \qquad \text{(12.23)}$$

This is another way to express the conservation of angular momentum; this relation applies to the motion of any particle under a central force. Since the gravitational force exerted by the sun on a planet is a central force directed toward the center of the sun, Eq. (12.23) is fundamental to the study of planetary motion. For a similar reason, it is also fundamental to studying the motion of space vehicles in orbit about the earth.

Alternatively, from Eq. (12.19), we can express the fact that the magnitude H_O of the angular momentum of the particle P is constant by writing

$$mr^2\dot{\theta} = H_O = \text{constant} \qquad \text{(12.24)}$$

Dividing by m and using h to denote the angular momentum per unit mass H_O/m, we have

$$r^2\dot{\theta} = h \qquad \text{(12.25)}$$

Equation (12.25) has an interesting geometric interpretation. Note from Fig. 12.16 that the radius vector OP sweeps across an infinitesimal area $dA = \frac{1}{2}r^2 d\theta$ as it rotates through an angle $d\theta$. Then, defining the **areal velocity** of the particle as the quotient dA/dt, we see that the left-hand side of Eq. (12.25) represents twice the areal velocity of the particle. We thus conclude that

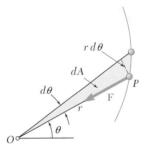

> **When a particle moves under a central force, its areal velocity is constant.**

Fig. 12.16 When a particle moves under a central force, its areal velocity is constant.

12.2C Newton's Law of Gravitation

As you saw in the preceding section, the gravitational force exerted by the sun on a planet or by the earth on an orbiting satellite is an important example of a central force. In this section, you will learn how to determine the magnitude of a gravitational force.

Newton's **law of universal gravitation** states that two particles of masses M and m at a distance r from each other have a mutual attraction

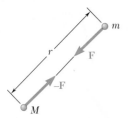

Fig. 12.17 By Newton's law of gravitation, two masses attract each other with equal force.

of equal and opposite forces **F** and $-$**F** directed along the line joining the particles (Fig. 12.17). The common magnitude F of the two forces is

Newton's law of universal gravitation

$$F = G \frac{Mm}{r^2} \tag{12.26}$$

where G is a universal constant, called the **constant of gravitation**. Experiments show that the value of G is $(66.73 \pm 0.03) \times 10^{-12}$ m³/kg·s² in SI units or approximately 34.4×10^{-9} ft⁴/lb·s⁴ in U.S. customary units. Gravitational forces exist between any pair of bodies, but their effect is appreciable only when one of the bodies has a very large mass. The effect of gravitational forces is apparent in the cases of the motion of a planet about the sun, of satellites orbiting about the earth, or of bodies falling on the surface of the earth.

Since the force exerted by the earth on a body of mass m located on or near its surface is defined as the weight **W** of the body, we can substitute the magnitude $W = mg$ of the weight for F, and the earth's radius R for r in Eq. (12.26). We obtain

$$W = mg = \frac{GM}{R^2} m \quad \text{or} \quad g = \frac{GM}{R^2} \tag{12.27}$$

where M is the mass of the earth. Since the earth is not truly spherical, the distance R from the center of the earth depends upon the point selected on its surface. Thus, the values of W and g vary with the altitude and latitude of the point considered. Another reason for the variation of W and g with latitude is that a system of axes attached to the earth does not constitute a newtonian frame of reference (see Sec. 12.1A). A more accurate definition of the weight of a body should therefore include a component representing the effects of this centripetal acceleration due to the earth's rotation. Values of g at sea level vary from 9.781 m/s² (or 32.09 ft/s²) at the equator to 9.833 m/s² (or 32.26 ft/s²) at the poles.[†]

We can use Eq. (12.26) to find the force exerted by the earth on a body of mass m located in space at a distance r from its center. The computations are somewhat simplified by noting that, according to Eq. (12.27), we can express the product of the constant of gravitation G and the mass M of the earth as

$$GM = gR^2 \tag{12.28}$$

Here we give g and the earth's radius R as their average values $g = 9.81$ m/s² and $R = 6.37 \times 10^6$ m in SI units[‡] and $g = 32.2$ ft/s² and $R = (3960 \text{ mi})(5280 \text{ ft/mi})$ in U.S. customary units.

The discovery of the law of universal gravitation often has been attributed to the belief that, after observing an apple falling from a tree, Newton realized that the earth must attract an apple in much the same way as the moon. It is doubtful that this incident actually took place, but we can say that Newton would not have formulated his law if he had not first perceived that the acceleration of a falling body must have the same cause as the acceleration that keeps the moon in its orbit.

[†]A formula expressing g in terms of the latitude ϕ was given in Prob. 12.2.

[‡]You can find the value of R simply by relating the earth's circumference to its radius as $2\pi r = 40 \times 10^6$ m.

Sample Problem 12.12

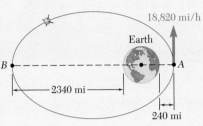

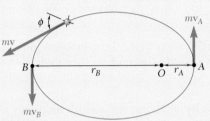

Fig. 1 The satellite at various positions.

A satellite is launched in a direction parallel to the surface of the earth with a velocity of 18,820 mi/h from an altitude of 240 mi. Determine the velocity of the satellite as it reaches its maximum altitude of 2340 mi. Recall that the earth's radius is 3960 mi.

STRATEGY: The satellite is acted on by a central force, so angular momentum is conserved. You can use the principle of conservation of angular momentum to determine the velocity of the satellite.

MODELING and ANALYSIS: Since the satellite is moving under a central force directed toward the center O of the earth, its angular momentum $\mathbf{H}_O$ is constant. From Eq. (12.14), you have

$$rmv \sin \phi = H_O = \text{constant}$$

This equation shows that v is at a minimum at B, where both r and $\sin \phi$ are maximum. Expressing the conservation of angular momentum between A and B, we have

$$r_A m v_A = r_B m v_B$$

Hence,

$$v_B = v_A \frac{r_A}{r_B} = (18{,}820 \text{ mi/h}) \frac{3960 \text{ mi} + 240 \text{ mi}}{3960 \text{ mi} + 2340 \text{ mi}}$$
$$v_B = 12{,}550 \text{ mi/h} \quad \blacktriangleleft$$

REFLECT and THINK: Note that in order to increase velocity, you could choose to apply thrusters pushing the spacecraft closer to the earth. Since this is a central force, the spacecraft's angular momentum remains constant. Therefore, its speed v increases as the radial distance r decreases.

Sample Problem 12.13

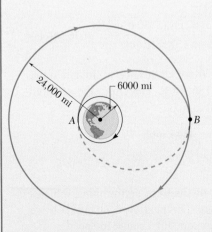

A space tug travels a circular orbit with a 6000-mi radius around the earth. In order to transfer it to a larger orbit with a 24,000-mi radius, the tug is first placed on an elliptical path AB by firing its engines as it passes through A, thus increasing its velocity by 3810 mi/h. Determine how much the tug's velocity should be increased as it reaches B to insert it into the larger circular orbit.

STRATEGY: Use Newton's second law and conservation of angular momentum.

MODELING: Choose the space tug as the system, and assume you can treat it as a particle. Draw free-body and kinetic diagrams of the system at A as shown in Fig. 1.

(continued)

ANALYSIS:

Circular Orbit through A. Applying Newton's second law in the normal direction when the tug is at A gives

$$\xrightarrow{+}\Sigma F_n = ma_n \qquad\qquad \frac{GMm}{r_A^2} = \frac{mv_A^2}{r_A} \qquad\qquad (1)$$

Solve Eq. (1) for v_A and substitute in numbers to find

$$v_A = \sqrt{\frac{GM}{r_A}} = \sqrt{\frac{gR^2}{r_A}} = \sqrt{\frac{(32.2\ \text{ft/s}^2)((3960\ \text{mi})(5280\ \text{ft/mi}))^2}{(600\ \text{mi})(5280\ \text{ft/mi})}} = 21{,}080\ \text{ft/s}$$

Converting this to mi/h gives $v_A = 14{,}370$ mi/h. Thus, the velocity to put the space tug into an elliptic orbit is $(v_A)_{\text{ell}} = 14{,}370$ mi/h $+ 3810$ mi/h $= 18{,}180$ mi/h.

Elliptic Path AB. To find the velocity at B, use the conservation of angular momentum between A and B. The velocity is perpendicular to r at both A and B, so you have

$$H_O = r_A m v_A = r_B m v_B \qquad\qquad (2)$$

Solving Eq. (2) for v_B and substituting in numbers give

$$(v_B)_{\text{ell}} = \frac{r_A}{r_B}(v_A)_{\text{ell}} = \frac{6000\ \text{mi}}{24{,}000\ \text{mi}}(18{,}180\ \text{mi/h}) = 4545\ \text{mi/h}$$

Circular Orbit through B. Applying Newton's second law in the normal direction when the tug is at B gives

$$\xleftarrow{+}\Sigma F_n = ma_n \qquad\qquad \frac{GMm}{r_B^2} = \frac{mv_B^2}{r_B} \qquad\qquad (3)$$

By solving Eq. (3) for v_B and substituting in numbers, you find

$$v_B = \sqrt{\frac{GM}{r_B}} = \sqrt{\frac{gR^2}{r_B}} = \sqrt{\frac{(32.2\ \text{ft/s}^2)((3960\ \text{mi})(5280\ \text{ft/mi}))^2}{(24{,}000\ \text{mi})(5280\ \text{ft/mi})}} = 10{,}540\ \text{ft/s}$$

This is the speed of the space tug at B for it to have a circular orbit. Converting this to mi/h gives $v_B = 7186$ mi/h. Therefore, the required increase in velocity is

$$\Delta v_B = 7186\ \text{mi/h} - 4545\ \text{mi/h}$$

$$\Delta v_B = 2640\ \text{mi/h} \quad \blacktriangleleft$$

REFLECT and THINK: The speeds of satellites and orbiting vehicles are quite large, as seen in this problem. The next type of question we could ask is what force is required to impart this change in speed.

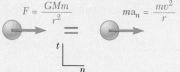

Fig. 1 Free-body diagram and kinetic diagram of satellite at point A.

SOLVING PROBLEMS
ON YOUR OWN

In this section, we introduced the *angular momentum* or the *moment of the momentum*, $\mathbf{H}_O$, of a particle about O as

$$\mathbf{H}_O = r \times m\mathbf{v} \tag{12.13}$$

and found that $\mathbf{H}_O$ is constant when the particle moves under a **central force** with its center located at O.

1. **Solving problems involving the motion of a particle under a central force.** In problems of this type, the angular momentum $\mathbf{H}_O$ of the particle about the center of force O is conserved. Therefore, we can express the conservation of angular momentum of particle P about O by $rmv \sin \phi = r_0 mv_0 \sin \phi_0$.

2. **In space mechanics problems** involving the orbital motion of a planet about the sun or of a satellite about the earth, the moon, or some other planet, the central force $\mathbf{F}$ is the force of gravitational attraction. This force is directed *toward* the center of force O and has the magnitude

$$F = G\frac{Mm}{r^2} \tag{12.26}$$

Note that in the particular case of the gravitational force exerted by the earth, the product GM can be replaced by gR^2, where R is the earth's radius [Eq. 12.28].

The following two cases of orbital motion are frequently encountered:

  a. **For a satellite in a circular orbit,** the force $\mathbf{F}$ is normal to the orbit and you can write $F = ma_n$ [Sample Prob. 12.13]. Substituting for F from Eq. (12.26) and observing that $a_n = v^2/\rho = v^2/r$, you obtain

$$G\frac{Mm}{r^2} = m\frac{v^2}{r} \quad \text{or} \quad v^2 = \frac{GM}{r}$$

  b. **For a satellite in an elliptical orbit,** the radius vector $\mathbf{r}$ and the velocity $\mathbf{v}$ of the satellite are perpendicular to each other at points A and B, which are closest and farthest to the center of force O, respectively [Sample Prob. 12.12]. Thus, the conservation of angular momentum of the satellite between these two points can be expressed as

$$r_A mv_A = r_B mv_B$$

Problems

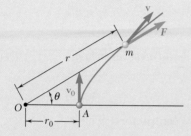

Fig. P12.74

12.74 A particle of mass m is projected from point A with an initial velocity $\mathbf{v}_0$ perpendicular to line OA and moves under a central force $\mathbf{F}$ directed away from the center of force O. Knowing that the particle follows a path defined by the equation $r = r_0/\sqrt{\cos 2\theta}$ and using Eq. (12.25), express the radial and transverse components of the velocity $\mathbf{v}$ of the particle as functions of θ.

12.75 For the particle of Prob. 12.74, show (a) that the velocity of the particle and the central force $\mathbf{F}$ are proportional to the distance r from the particle to the center of force O, (b) that the radius of curvature of the path is proportional to r^3.

12.76 A particle of mass m is projected from point A with an initial velocity $\mathbf{v}_0$ perpendicular to line OA and moves under a central force $\mathbf{F}$ along a semicircular path of diameter OA. Observing that $r = r_0 \cos \theta$ and using Eq. (12.25), show that the speed of the particle is $v = v_0/\cos^2 \theta$.

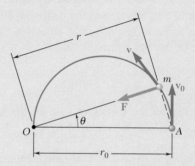

Fig. P12.76

12.77 For the particle of Prob. 12.76, determine the tangential component F_t of the central force $\mathbf{F}$ along the tangent to the path of the particle for (a) $\theta = 0$, (b) $\theta = 45°$.

12.78 Determine the mass of the earth knowing that the mean radius of the moon's orbit about the earth is 238,910 mi and that the moon requires 27.32 days to complete one full revolution about the earth.

12.79 Show that the radius r of the moon's orbit can be determined from the radius R of the earth, the acceleration of gravity g at the surface of the earth, and the time τ required for the moon to complete one full revolution about the earth. Compute r knowing that $\tau = 27.3$ days, giving the answer in both SI and U.S. customary units.

12.80 Communication satellites are placed in a geosynchronous orbit, i.e., in a circular orbit such that they complete one full revolution about the earth in one sidereal day (23.934 h), and thus appear stationary with respect to the ground. Determine (a) the altitude of these satellites above the surface of the earth, (b) the velocity with which they describe their orbit. Give the answers in both SI and U.S. customary units.

12.81 Show that the radius r of the orbit of a moon of a given planet can be determined from the radius R of the planet, the acceleration of gravity at the surface of the planet, and the time τ required by the moon to complete one full revolution about the planet. Determine the acceleration of gravity at the surface of the planet Jupiter knowing that $R = 71\,492$ km and that $\tau = 3.551$ days and $r = 670.9 \times 10^3$ km for its moon Europa.

12.82 The orbit of the planet Venus is nearly circular with an orbital velocity of 126.5×10^3 km/h. Knowing that the mean distance from the center of the sun to the center of Venus is 108×10^6 km and that the radius of the sun is 695.5×10^3 km, determine (*a*) the mass of the sun, (*b*) the acceleration of gravity at the surface of the sun.

12.83 A satellite is placed into a circular orbit about the planet Saturn at an altitude of 2100 mi. The satellite describes its orbit with a velocity of 54.7×10^3 mi/h. Knowing that the radius of the orbit about Saturn and the periodic time of Atlas, one of Saturn's moons, are 85.54×10^3 mi and 0.6017 days, respectively, determine (*a*) the radius of Saturn, (*b*) the mass of Saturn. (The *periodic time* of a satellite is the time it requires to complete one full revolution about the planet.)

12.84 The periodic time (see Prob. 12.83) of an earth satellite in a circular polar orbit is 120 minutes. Determine (*a*) the altitude h of the satellite, (*b*) the time during which the satellite is above the horizon for an observer located at the north pole.

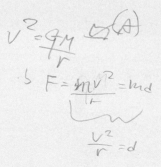
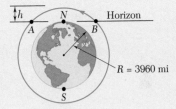

Fig. P12.84

12.85 A 500-kg spacecraft first is placed into a circular orbit about the earth at an altitude of 4500 km and then is transferred to a circular orbit about the moon. Knowing that the mass of the moon is 0.01230 times the mass of the earth and that the radius of the moon is 1737 km, determine (*a*) the gravitational force exerted on the spacecraft as it was orbiting the earth, (*b*) the required radius of the orbit of the spacecraft about the moon if the periodic times (see Prob. 12.83) of the two orbits are to be equal, (*c*) the acceleration of gravity at the surface of the moon.

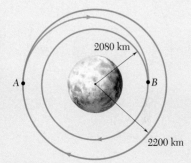

Fig. P12.86

12.86 A space vehicle is in a circular orbit of 2200-km radius around the moon. To transfer it to a smaller circular orbit of 2080-km radius, the vehicle is first placed on an elliptic path AB by reducing its speed by 26.3 m/s as it passes through A. Knowing that the mass of the moon is 73.49×10^{21} kg, determine (*a*) the speed of the vehicle as it approaches B on the elliptic path, (*b*) the amount by which its speed should be reduced as it approaches B to insert it into the smaller circular orbit.

12.87 As a first approximation to the analysis of a space flight from the earth to the planet Mars, assume the orbits of the earth and Mars are circular and coplanar. The mean distances from the sun to the earth and to Mars are 149.6×10^6 km and 227.8×10^6 km, respectively. To place the spacecraft into an elliptical transfer orbit at point A, its speed is increased over a short interval of time to v_A, which is 2.94 km/s faster than the earth's orbital speed. When the spacecraft reaches point B on the elliptical transfer orbit, its speed v_B is increased to the orbital speed of Mars. Knowing that the mass of the sun is 332.8×10^3 times the mass of the earth, determine the increase in speed required at B.

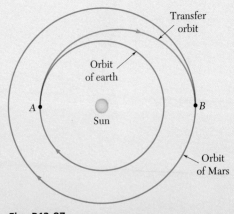

Fig. P12.87

12.88 To place a communications satellite into a geosynchronous orbit (see Prob. 12.80) at an altitude of 22,240 mi above the surface of the earth, the satellite first is released from a space shuttle, which is in a circular orbit at an altitude of 185 mi, and then is propelled by an upper-stage booster to its final altitude. As the satellite passes through A, the booster's motor is fired to insert the satellite into an elliptic transfer orbit. The booster is again fired at B to insert the satellite into a geosynchronous orbit. Knowing that the second firing increases the speed of the satellite by 4810 ft/s, determine (a) the speed of the satellite as it approaches B on the elliptic transfer orbit, (b) the increase in speed resulting from the first firing at A.

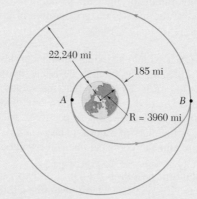

Fig. P12.88

12.89 A space vehicle is in a circular orbit with a 1400-mi radius around the moon. To transfer to a smaller orbit with a 1300-mi radius, the vehicle is first placed in an elliptic path AB by reducing its speed by 86 ft/s as it passes through A. Knowing that the mass of the moon is 5.03×10^{21} lb·s²/ft, determine (a) the speed of the vehicle as it approaches B on the elliptic path, (b) the amount by which its speed should be reduced as it approaches B to insert it into the smaller circular orbit.

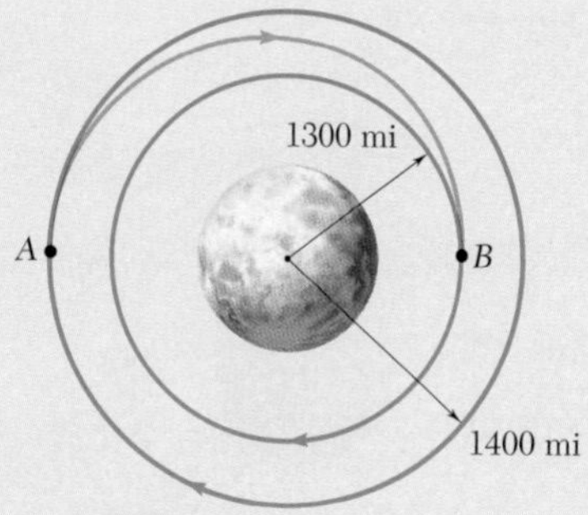

12.90 A 1-kg collar can slide on a horizontal rod that is free to rotate about a vertical shaft. The collar is initially held at A by a cord attached to the shaft. A spring of constant 30 N/m is attached to the collar and to the shaft and is undeformed when the collar is at A. As the rod rotates at the rate $\dot\theta = 16$ rad/s, the cord is cut and the collar moves out along the rod. Neglecting friction and the mass of the rod, determine (a) the radial and transverse components of the acceleration of the collar at A, (b) the acceleration of the collar relative to the rod at A, (c) the transverse component of the velocity of the collar at B.

Fig. P12.89

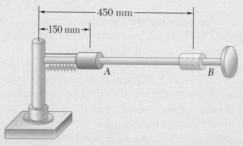

Fig. P12.90

12.91 A 1-lb ball A and a 2-lb ball B are mounted on a horizontal rod that rotates freely about a vertical shaft. The balls are held in the positions shown by pins. The pin holding B is suddenly removed and the ball moves to position C as the rod rotates. Neglecting friction and the mass of the rod and knowing that the initial speed of A is $v_A =$ 8 ft/s, determine (a) the radial and transverse components of the acceleration of ball B immediately after the pin is removed, (b) the acceleration of ball B relative to the rod at that instant, (c) the speed of ball A after ball B has reached the stop at C.

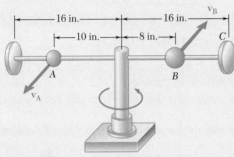

Fig. P12.91

12.92 Two 2.6-lb collars A and B can slide without friction on a frame, consisting of the horizontal rod OE and the vertical rod CD, which is free to rotate about CD. The two collars are connected by a cord running over a pulley that is attached to the frame at O and a stop prevents collar B from moving. The frame is rotating at the rate $\dot{\theta} = 12$ rad/s and $r = 0.6$ ft when the stop is removed allowing collar A to move out along rod OE. Neglecting friction and the mass of the frame, determine, for the position $r = 1.2$ ft, (a) the transverse component of the velocity of collar A, (b) the tension in the cord and the acceleration of collar A relative to the rod OE.

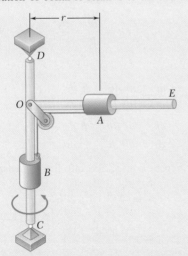

Fig. P12.92

12.93 A small ball swings in a horizontal circle at the end of a cord of length l_1, which forms an angle θ_1 with the vertical. The cord is then slowly drawn through the support at O until the length of the free end is l_2. (a) Derive a relation among l_1, l_2, θ_1, and θ_2. (b) If the ball is set in motion so that initially $l_1 = 0.8$ m and $\theta_1 = 35°$, determine the angle θ_2 when $l_2 = 0.6$ m.

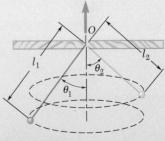

Fig. P12.93

*12.3   APPLICATIONS OF CENTRAL-FORCE MOTION

The most important examples of a particle moving under the action of a central force occur in space mechanics, where gravity is the central force. In this section, we examine some of the basic ideas of this motion, concentrating on the motions of satellites around the earth and planets around a star.

12.3A   Trajectory of a Particle Under a Central Force

Consider a particle P moving under a central force $\mathbf{F}$. In order to fully characterize the motion of particle P (which could represent a satellite, a moon, etc.), we must develop a differential equation that defines its trajectory.

Assuming that the force $\mathbf{F}$ is directed toward the center of force O, we note that ΣF_r and ΣF_θ reduce, respectively, to $-F$ and zero in Eqs. (12.11) and (12.12). Therefore, we have

$$m(\ddot{r} - r\dot{\theta}^2) = -F \tag{12.29}$$

$$m(r\ddot{\theta} + 2\dot{r}\dot{\theta}) = 0 \tag{12.30}$$

These equations define the motion of P. We can also use Eq. (12.25) to analyze the motion of P, obtaining

$$r^2\dot{\theta} = h \quad \text{or} \quad r^2\frac{d\theta}{dt} = h \tag{12.31}$$

We can use Eq. (12.31) to eliminate the independent variable t from Eq. (12.29). Solving Eq. (12.31) for $\dot{\theta}$, or $d\theta/dt$, we have

$$\dot{\theta} = \frac{d\theta}{dt} = \frac{h}{r^2} \tag{12.32}$$

It follows that

$$\dot{r} = \frac{dr}{dt} = \frac{dr}{d\theta}\frac{d\theta}{dt} = \frac{h}{r^2}\frac{dr}{d\theta} = -h\frac{d}{d\theta}\left(\frac{1}{r}\right) \tag{12.33}$$

$$\ddot{r} = \frac{d\dot{r}}{dt} = \frac{d\dot{r}}{d\theta}\frac{d\theta}{dt} = \frac{h}{r^2}\frac{d\dot{r}}{d\theta}$$

If we substitute for $\dot{r}$ from Eq. (12.33) into the expression for $\ddot{r}$, we obtain

$$\ddot{r} = \frac{h}{r^2}\frac{d}{d\theta}\left[-h\frac{d}{d\theta}\left(\frac{1}{r}\right)\right]$$

$$\ddot{r} = -\frac{h^2}{r^2}\frac{d^2}{d\theta^2}\left(\frac{1}{r}\right) \tag{12.34}$$

Now, substituting for $\dot{\theta}$ and $\ddot{r}$ from Eqs. (12.32) and (12.34), respectively, in Eq. (12.29) and introducing the function $u = 1/r$, we obtain, after reductions,

$$\frac{d^2u}{d\theta^2} + u = \frac{F}{mh^2u^2} \tag{12.35}$$

In deriving Eq. (12.35), we assumed force **F** to be directed toward O. The magnitude F therefore should be positive if **F** is actually directed toward O (attractive force) and negative if **F** is directed away from O (repulsive force). If F is a known function of r and thus of u, Eq. (12.35) is a differential equation in u and θ. This differential equation defines the trajectory followed by the particle under the central force **F**. We can obtain the equation of the trajectory by solving the differential equation (12.35) for u as a function of θ and determining the constants of integration from the initial conditions.

*12.3B Application to Space Mechanics

After the last stages of their launching rockets have burned out, earth satellites and other space vehicles are subject to only the gravitational pull of the earth. We can therefore determine their motion from Eqs. (12.31) and (12.35), which govern the motion of a particle under a central force, after replacing F by the expression for the force of gravitational attraction.[†] We set F in Eq. (12.35) as

$$F = \frac{GMm}{r^2} = GMmu^2$$

where M = mass of the earth
m = mass of space vehicle
r = distance from center of the earth to vehicle
$u = 1/r$

Then we obtain the differential equation

$$\frac{d^2u}{d\theta^2} + u = \frac{GM}{h^2} \qquad \textbf{(12.36)}$$

Note that the right-hand side is a constant.

To solve the differential equation (12.36), we add the particular solution $u = GM/h^2$ to the general solution $u = C \cos(\theta - \theta_0)$ of the corresponding homogeneous equation (i.e., the equation obtained by setting the right-hand side equal to zero). Choosing the polar axis so that $\theta_0 = 0$, we have

$$\frac{1}{r} = u = \frac{GM}{h^2} + C \cos \theta \qquad \textbf{(12.37)}$$

Equation (12.37) is the equation of a *conic section* (ellipse, parabola, or hyperbola) in the polar coordinates r and θ. The origin O of the coordinates, which is located at the center of the earth, is a *focus* of this conic section, and the polar axis is one of its axes of symmetry (Fig. 12.18).

The ratio of the constants C and GM/h^2 defines the **eccentricity** ε of the conic section. If we set

$$\varepsilon = \frac{C}{GM/h^2} = \frac{Ch^2}{GM} \qquad \textbf{(12.38)}$$

[†]We assume that the space vehicles considered here are attracted by the earth only and that their masses are negligible compared to the mass of the earth. If a vehicle travels very far from the earth, its path may be affected by the gravitational attraction of the sun, the moon, or another planet.

Photo 12.5 The Hubble telescope was carried into orbit by the space shuttle in 1990.

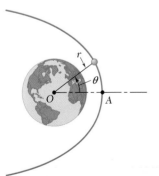

Fig. 12.18 The trajectory of an earth satellite is a conic section with the center of the earth as one of its foci.

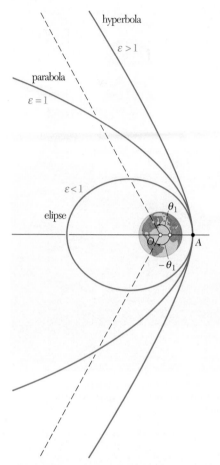

Fig. 12.19 Depending on the eccentricity, the orbit of an earth satellite can be a hyperbola, a parabola, or an ellipse.

we can write Eq. (12.37) in the form

$$\frac{1}{r} = \frac{GM}{h^2}(1 + \varepsilon \cos \theta) \qquad (12.37')$$

This equation represents three possible trajectories.

1. $\varepsilon > 1$, or $C > GM/h^2$: There are two values θ_1 and $-\theta_1$ of the polar angle, defined by $\cos \theta_1 = -GM/Ch^2$, for which the right-hand side of Eq. (12.37) becomes zero. For both these values, the radius vector r becomes infinite; the conic section is a *hyperbola* (Fig. 12.19).
2. $\varepsilon = 1$, or $C = GM/h^2$: The radius vector becomes infinite for $\theta = 180°$; the conic section is a *parabola*.
3. $\varepsilon < 1$, or $C < GM/h^2$: The radius vector remains finite for every value of θ; the conic section is an *ellipse*. In the particular case when $\varepsilon = C = 0$, the length of the radius vector is constant; the conic section is a circle.

Let's now see how we can determine the constants C and GM/h^2, which characterize the trajectory of a space vehicle, from the vehicle's position and velocity at the beginning of its free flight. We assume that, as is generally the case, the powered phase of its flight has been programmed in such a way that as the last stage of the launching rocket burns out, the vehicle has a velocity parallel to the surface of the earth (Fig. 12.20). In other words, we assume that the space vehicle begins its free flight at the vertex A of its trajectory. (In Sec. 13.2D, we consider problems involving oblique launchings.)

Denoting the radius and speed of the vehicle at the beginning of its free flight by r_0 and v_0, respectively, we observe that the velocity reduces to its transverse component. Thus, $v_0 = r_0\dot{\theta}_0$. Recalling Eq. (12.25), we express the angular momentum per unit mass h as

$$h = r_0^2\dot{\theta}_0 = r_0v_0 \qquad (12.39)$$

The value obtained for h can be used to determine the constant GM/h^2. We also note that the computation of this constant is simplified if we use the relation obtained in Sec. 12.2C.

$$GM = gR^2 \qquad (12.28)$$

where R is the radius of the earth ($R = 6.37 \times 10^6$ m or 3960 mi) and g is the acceleration due to gravity at the earth's surface.

We obtain the constant C by setting $\theta = 0$, $r = r_0$ in Eq. (12.37). Hence,

$$C = \frac{1}{r_0} - \frac{GM}{h^2} \qquad (12.40)$$

Substituting for h from Eq. (12.39), we can easily express C in terms of r_0 and v_0.

Initial Conditions. Now we can determine the initial conditions corresponding to each of the three fundamental trajectories indicated. Considering first the parabolic trajectory, we set C equal to GM/h^2 in Eq. (12.40) and eliminate h between Eqs. (12.39) and (12.40). Solving for v_0, we obtain

$$v_0 = \sqrt{\frac{2GM}{r_0}}$$

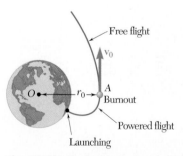

Fig. 12.20 Typically, a space vehicle has a velocity parallel to the surface of the earth after the powered portion of its flight.

We can check that a larger value of the initial velocity corresponds to a hyperbolic trajectory and a smaller value corresponds to an elliptic orbit. Since the value of v_0 obtained for the parabolic trajectory is the smallest value for which the space vehicle does not return to its starting point, it is called the **escape velocity**. Therefore, making use of Eq. (12.28), we have

$$v_{esc} = \sqrt{\frac{2GM}{r_0}} \quad \text{or} \quad v_{esc} = \sqrt{\frac{2gR^2}{r_0}} \qquad \textbf{(12.41)}$$

Note that the trajectory is (1) hyperbolic if $v_0 > v_{esc}$, (2) parabolic if $v_0 = v_{esc}$, and (3) elliptic if $v_0 < v_{esc}$.

Among the various possible elliptic orbits, the one obtained when $C = 0$, the *circular orbit,* is of special interest. Taking into account Eq. (12.28), the value of the initial velocity corresponding to a circular orbit is

$$v_{circ} = \sqrt{\frac{GM}{r_0}} \quad \text{or} \quad v_{circ} = \sqrt{\frac{gR^2}{r_0}} \qquad \textbf{(12.42)}$$

Note from Fig. 12.21 that, for values of v_0 larger than v_{circ} but smaller than v_{esc}, point A is the point of the orbit closest to the earth where free flight begins. This point is called the *perigee,* whereas point A', which is farthest away from the earth, is known as the *apogee.* For values of v_0 smaller than v_{circ}, point A is the apogee and point A'', which is on the other side of the orbit, is the perigee. For values of v_0 much smaller than v_{circ}, the trajectory of the space vehicle intersects the surface of the earth; in such a case, the vehicle does not go into orbit.

Ballistic missiles, which were designed to hit the surface of the earth, also travel along elliptic trajectories. In fact, you should now realize that any object projected in vacuum with an initial velocity v_0 smaller than v_{esc} moves along an elliptic path. Only when the distances involved are small enough that we can assume the gravitational field of the earth is uniform can we approximate the elliptic path by a parabolic path, as we did earlier (Sec. 11.4C) in the case of conventional projectiles.

Periodic Time.   An important characteristic of the motion of an earth satellite is the time required by the satellite to travel through one complete orbit. This time, which is known as the satellite's **periodic time**, is denoted by τ. We first observe, in view of the definition of areal velocity (Sec. 12.2B), that we can obtain τ by dividing the area inside the orbit by the areal velocity. The area of an ellipse is equal to πab, where a and b denote the semimajor and semiminor axes, respectively. Since the areal velocity is equal to $h/2$, we have

$$\tau = \frac{2\pi ab}{h} \qquad \textbf{(12.43)}$$

Although we can readily determine h from r_0 and v_0 in the case of a satellite launched in a direction parallel to the earth's surface, the semiaxes a and b are not directly related to the initial conditions. However, the values r_0 and r_1 of r corresponding to the perigee and apogee of the orbit can be determined from Eq. (12.37), so we can express the semiaxes a and b in terms of r_0 and r_1.

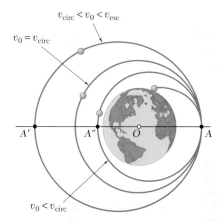

Fig. 12.21 Various elliptic orbits are possible for earth satellites, depending on the initial velocity.

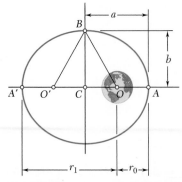

Fig. 12.22 For an elliptic orbit, the distances to apogee (A') and perigee (A) are related to the semimajor and semiminor axes.

Consider the elliptic orbit shown in Fig. 12.22. The earth's center is located at O and coincides with one of the two foci of the ellipse, and the points A and A' represent, respectively, the perigee and apogee of the orbit. We easily check that

$$r_0 + r_1 = 2a$$

and thus

$$a = \tfrac{1}{2}(r_0 + r_1) \tag{12.44}$$

Recall that the sum of the distances from each of the foci to any point of the ellipse is constant, so we have

$$O'B + BO = O'A + OA = 2a \qquad \text{or} \qquad BO = a$$

On the other hand, we have $CO = a - r_0$. We can therefore write

$$b^2 = (BC)^2 = (BO)^2 - (CO)^2 = a^2 - (a - r_0)^2$$
$$b^2 = r_0(2a - r_0) = r_0 r_1$$

and thus

$$b = \sqrt{r_0 r_1} \tag{12.45}$$

Formulas (12.44) and (12.45) indicate that the semimajor and semiminor axes of the orbit are equal, respectively, to the arithmetic and geometric means of the maximum and minimum values of the radius vector. Once you have determined r_0 and r_1, you can compute the lengths of the semiaxes and substitute for a and b in Eq. (12.43).

*12.3C Kepler's Laws of Planetary Motion

We can use the equations governing the motion of an earth satellite to describe the motion of the moon around the earth. In that case, however, the mass of the moon is not negligible compared with the earth's mass, and the results are not entirely accurate.

We can also apply the theory developed in the preceding sections to the study of the motion of the planets around the sun. Although another error is introduced by neglecting the forces exerted by the planets on one another, the approximation obtained is excellent. Indeed, even before Newton had formulated his fundamental theory, the properties expressed by Eq. (12.37), where M now represents the mass of the sun, and by Eq. (12.31) had been discovered by the German astronomer Johannes Kepler (1571–1630) from astronomical observations of the motion of the planets.

Kepler's three **laws of planetary motion** can be stated as follows.

1. The path of each planet describes an ellipse, with the sun located at one of its foci.
2. The radius vector drawn from the sun to a planet sweeps equal areas in equal times.
3. The squares of the periodic times of the planets are proportional to the cubes of the semimajor axes of their orbits.

The first law states a particular case of the result established in Sec. 12.3B, and the second law expresses that the areal velocity of each planet is constant (see Sec. 12.2B). Kepler's third law also can be derived from the results obtained in Sec. 12.3B. (See also Prob. 12.120.)

Sample Problem 12.14

A satellite is launched in a direction parallel to the earth's surface with a velocity of 36 900 km/h from an altitude of 500 km. Determine (a) the maximum altitude reached by the satellite, (b) the periodic time of the satellite.

STRATEGY: After the satellite is launched, it is subjected to the earth's gravitational attraction only and undergoes central-force motion. Knowing this, you can determine the satellite's trajectory, maximum altitude, and periodic time.

MODELING and ANALYSIS: The satellite can be modeled as a particle.

a. Maximum Altitude. After the satellite is launched, it is subject only to the earth's gravitational attraction. Thus, its motion is governed by Eq. (12.37), so

$$\frac{1}{r} = \frac{GM}{h^2} + C \cos \theta \tag{1}$$

Since the radial component of the velocity is zero at the point of launching A, you have $h = r_0 v_0$. Recalling that for the earth, $R = 6370$ km, you can compute

$$r_0 = 6370 \text{ km} + 500 \text{ km} = 6870 \text{ km} = 6.87 \times 10^6 \text{ m}$$

$$v_0 = 36\,900 \text{ km/h} = \frac{36.9 \times 10^6 \text{ m}}{3.6 \times 10^3 \text{ s}} = 10.25 \times 10^3 \text{ m/s}$$

$$h = r_0 v_0 = (6.87 \times 10^6 \text{ m})(10.25 \times 10^3 \text{ m/s}) = 70.4 \times 10^9 \text{ m}^2/\text{s}$$

$$h^2 = 4.96 \times 10^{21} \text{ m}^4/\text{s}^2$$

Since $GM = gR^2$, where R is the radius of the earth, you also have

$$GM = gR^2 = (9.81 \text{ m/s}^2)(6.37 \times 10^6 \text{ m})^2 = 398 \times 10^{12} \text{ m}^3/\text{s}^2$$

$$\frac{GM}{h^2} = \frac{398 \times 10^{12} \text{ m}^3/\text{s}^2}{4.96 \times 10^{21} \text{ m}^4/\text{s}^2} = 80.3 \times 10^{-9} \text{ m}^{-1}$$

Substituting this value into Eq. (1) gives

$$\frac{1}{r} = 80.3 \times 10^{-9} \text{ m}^{-1} + C \cos \theta \tag{2}$$

Note that at point A, $\theta = 0$ and $r = r_0 = 6.87 \times 10^6$ m (Fig. 1). From this, you can compute the constant C as

$$\frac{1}{6.87 \times 10^6 \text{ m}} = 80.3 \times 10^{-9} \text{ m}^{-1} + C \cos 0° \qquad C = 65.3 \times 10^{-9} \text{ m}^{-1}$$

At A', which is the point on the orbit farthest from the earth, you have $\theta = 180°$ (Fig. 1). Using Eq. (2), you can compute the corresponding distance r_1 to be

$$\frac{1}{r_1} = 80.3 \times 10^{-9} \text{ m}^{-1} + (65.3 \times 10^{-9} \text{ m}^{-1}) \cos 180°$$

$$r_1 = 66.7 \times 10^6 \text{ m} = 66\,700 \text{ km}$$

Maximum altitude $= 66\,700 \text{ km} - 6370 \text{ km} = 60\,300 \text{ km}$ ◄

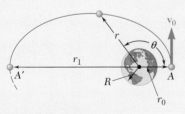

Fig. 1 Satellite orbit after launch velocity v_o.

(continued)

b. Periodic Time. Since A and A' are the perigee and apogee, respectively, of the elliptic orbit, use Eqs. (12.44) and (12.45) to compute the semimajor and semiminor axes of the orbit (Fig. 2):

$$a = \tfrac{1}{2}(r_0 + r_1) = \tfrac{1}{2}(6.87 + 66.7)(10^6) \text{ m} = 36.8 \times 10^6 \text{ m}$$

$$b = \sqrt{r_0 r_1} = \sqrt{(6.87)(66.7)} \times 10^6 \text{ m} = 21.4 \times 10^6 \text{ m}$$

$$\tau = \frac{2\pi ab}{h} = \frac{2\pi(36.8 \times 10^6 \text{m})(21.4 \times 10^6\text{m})}{70.4 \times 10^9 \text{ m}^2/\text{s}}$$

$$\tau = 70.3 \times 10^3 \text{ s} = 1171 \text{ min} = 19 \text{ h } 31 \text{ min} \quad \blacktriangleleft$$

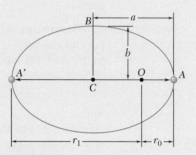

Fig. 2 Semimajor and semiminor axes of the orbit.

REFLECT and THINK: The satellite takes less than one day to travel over 60 000 km around the earth. In this problem, you started with Eq. 12.37, but it is important to remember that this formula was the solution to a differential equation that was derived using Newton's second law.

SOLVING PROBLEMS
ON YOUR OWN

In this section, we continued our study of the motion of a particle under a central force and applied the results to problems in space mechanics. We found that the trajectory of a particle under a central force is defined by the differential equation

$$\frac{d^2u}{d\theta^2} + u = \frac{F}{mh^2u^2} \tag{12.35}$$

where u is the reciprocal of the distance r of the particle from the center of force ($u = 1/r$), F is the magnitude of the central force $\mathbf{F}$, and h is a constant equal to the angular momentum per unit mass of the particle. In space mechanics problems, $\mathbf{F}$ is the force of gravitational attraction exerted on the satellite or spacecraft by the sun, the earth, or other planet about which it travels. Substituting $F = GMm/r^2 = GMmu^2$ into Eq. (12.35), we obtain for that case

$$\frac{d^2u}{d\theta^2} + u = \frac{GM}{h^2} \tag{12.36}$$

where the right-hand side is a constant.

1. Analyzing the motion of satellites and spacecraft. The solution of the differential equation (12.36) defines the trajectory of a satellite or spacecraft. We obtained it in Sec. 12.3B in the alternative forms

$$\frac{1}{r} = \frac{GM}{h^2} + C \cos \theta \quad \text{or} \quad \frac{1}{r} = \frac{GM}{h^2}(1 + \varepsilon \cos \theta) \tag{12.37, 12.37'}$$

Remember when applying these equations that $\theta = 0$ always corresponds to the perigee (the point of closest approach) of the trajectory (Fig. 12.18) and that h is a constant for a given trajectory. Depending on the value of the eccentricity ε, the trajectory is either a hyperbola, a parabola, or an ellipse.

   a. $\varepsilon > 1$: **The trajectory is a hyperbola.** For this case, the spacecraft never returns to its starting point.

   b. $\varepsilon = 1$: **The trajectory is a parabola.** This is the limiting case between open (hyperbolic) and closed (elliptic) trajectories. We had observed for this case that the velocity v_0 at the perigee is equal to the escape velocity v_{esc}. Hence,

$$v_0 = v_{\text{esc}} = \sqrt{\frac{2GM}{r_0}} \tag{12.41}$$

Note that the escape velocity is the smallest velocity for which the spacecraft does not return to its starting point.

   c. $\varepsilon < 1$: **The trajectory is an elliptic orbit.** For problems involving elliptic orbits, you may find that the relation derived in Prob. 12.102

$$\frac{1}{r_0} + \frac{1}{r_1} = \frac{2GM}{h^2}$$

is useful in the solution of subsequent problems. When you apply this equation, remember that r_0 and r_1 are the distances from the center of force to the perigee ($\theta = 0$) and apogee ($\theta = 180°$), respectively; that $h = r_0v_0 = r_1v_1$; and that, for a satellite orbiting the earth, $GM_{earth} = gR^2$, where R is the radius of the earth. Also recall that the trajectory is a circle when $\varepsilon = 0$.

2. Determining the point of impact of a descending spacecraft. For problems of this type, you may assume that the trajectory is elliptic and that the initial point of the descent trajectory is the apogee of the path (Fig. 12.21). Note that at the point of impact, the distance r in Eqs. (12.37) and (12.37′) is equal to the radius R of the body on which the spacecraft lands or crashes. In addition, we have $h = Rv_I \sin \phi_I$, where v_I is the speed of the spacecraft at impact and ϕ_I is the angle that its path forms with the vertical at the point of impact.

3. Calculating the time to travel between two points on a trajectory. For central-force motion, you can determine the time t required for a particle to travel along a portion of its trajectory by recalling from Sec. 12.2B that the rate at which area is swept per unit time by the position vector $\mathbf{r}$ is equal to one-half of the angular momentum per unit mass h of the particle: $dA/dt = h/2$. Since h is a constant for a given trajectory, it follows that

$$t = \frac{2A}{h}$$

where A is the total area swept in the time t.

    a. **In the case of an elliptic trajectory,** the time required to complete one orbit is called the **periodic time** and is expressed as

$$\tau = \frac{2(\pi ab)}{h} \tag{12.43}$$

where a and b are the semimajor and semiminor axes, respectively, of the ellipse and are related to the distances r_0 and r_1 by

$$a = \tfrac{1}{2}(r_0 + r_1) \qquad \text{and} \qquad b = \sqrt{r_0 r_1} \tag{12.44, 12.45}$$

    b. **Kepler's third law** provides a convenient relation between the periodic times of two satellites describing elliptic orbits about the same body [Sec. 12.3C]. Denoting the semimajor axes of the two orbits by a_1 and a_2, respectively, and the corresponding periodic times by τ_1 and τ_2, we have

$$\frac{\tau_1^2}{\tau_2^2} = \frac{a_1^3}{a_2^3}$$

    c. **In the case of a parabolic trajectory,** you may be able to use the expression given on the inside of the front cover of this book for a parabolic or a semiparabolic area to calculate the time required to travel between two points of the trajectory.

Problems

12.CQ6 A uniform crate C with mass m is being transported to the left by a forklift with a constant speed v_1. What is the magnitude of the angular momentum of the crate about point D, that is, the upper left corner of the crate?

    a. 0
    b. mv_1a
    c. mv_1b
    d. $mv_1\sqrt{a^2 + b^2}$

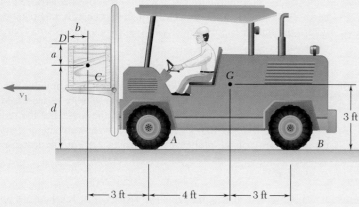

Fig. P12.CQ6 and P12.CQ7

12.CQ7 A uniform crate C with mass m is being transported to the left by a forklift with a constant speed v_1. What is the magnitude of the angular momentum of the crate about point A, that is, the point of contact between the front tire of the forklift and the ground?

    a. 0
    b. mv_1d
    c. $3mv_1$
    d. $mv_1\sqrt{3^2 + d^2}$

END-OF-SECTION PROBLEMS

12.94 A particle of mass m is projected from point A with an initial velocity $\mathbf{v}_0$ perpendicular to OA and moves under a central force $\mathbf{F}$ along an elliptic path defined by the equation $r = r_0/(2 - \cos\theta)$. Using Eq. (12.35), show that $\mathbf{F}$ is inversely proportional to the square of the distance r from the particle to the center of force O.

12.95 A particle of mass m describes the logarithmic spiral $r = r_0\,e^{b\theta}$ under a central force $\mathbf{F}$ directed toward the center of force O. Using Eq. (12.35), show that $\mathbf{F}$ is inversely proportional to the cube of the distance r from the particle to O.

12.96 A particle with a mass m describes the path defined by the equation $r = r_0/(6\cos\theta - 5)$ under a central force $\mathbf{F}$ directed away from the center of force O. Using Eq. (12.35), show that $\mathbf{F}$ is inversely proportional to the square of the distance r from the particle to O.

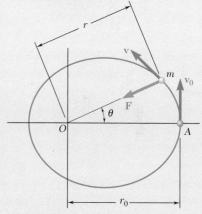

Fig. P12.94

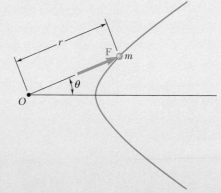

Fig. P12.96

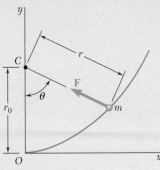

Fig. P12.97

12.97 A particle of mass m describes the parabola $y = x^2/4r_0$ under a central force $\mathbf{F}$ directed toward the center of force C. Using Eq. (12.35) and Eq. (12.37′) with $\varepsilon = 1$, show that $\mathbf{F}$ is inversely proportional to the square of the distance r from the particle to the center of force and that the angular momentum per unit mass $h = \sqrt{2GMr_0}$.

12.98 It was observed that during its second flyby of the earth, the Galileo spacecraft had a velocity of 14.1 km/s as it reached its minimum altitude of 303 km above the surface of the earth. Determine the eccentricity of the trajectory of the spacecraft during this portion of its flight.

12.99 It was observed that during the Galileo spacecraft's first flyby of the earth, its minimum altitude was 600 mi above the surface of the earth. Assuming that the trajectory of the spacecraft was parabolic, determine the maximum velocity of Galileo during its first flyby of the earth.

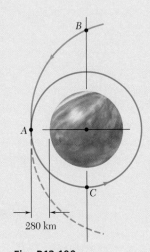

Fig. P12.100

12.100 As a space probe approaching the planet Venus on a parabolic trajectory reaches point A closest to the planet, its velocity is decreased to insert it into a circular orbit. Knowing that the mass and the radius of Venus are 4.87×10^{24} kg and 6052 km, respectively, determine (*a*) the velocity of the probe as it approaches A, (*b*) the decrease in velocity required to insert it into the circular orbit.

12.101 It was observed that as the Voyager I spacecraft reached the point of its trajectory closest to the planet Saturn, it was at a distance of 185×10^3 km from the center of the planet and had a velocity of 21.0 km/s. Knowing that Tethys, one of Saturn's moons, describes a circular orbit of radius 295×10^3 km at a speed of 11.35 km/s, determine the eccentricity of the trajectory of Voyager I on its approach to Saturn.

12.102 A satellite describes an elliptic orbit about a planet of mass M. Denoting by r_0 and r_1, respectively, the minimum and maximum values of the distance r from the satellite to the center of the planet, derive the relation

$$\frac{1}{r_0} + \frac{1}{r_1} = \frac{2GM}{h^2}$$

where h is the angular momentum per unit mass of the satellite.

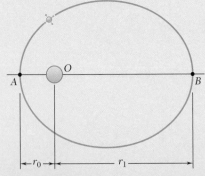

Fig. P12.102

12.103 A space probe is describing a circular orbit about a planet of radius R. The altitude of the probe above the surface of the planet is αR and its speed is v_0. To place the probe in an elliptic orbit which will bring it closer to the planet, its speed is reduced from v_0 to βv_0, where $\beta < 1$, by firing its engine for a short interval of time. Determine the smallest permissible value of β if the probe is not to crash on the surface of the planet.

12.104 A satellite describes a circular orbit at an altitude of 19 110 km above the surface of the earth. Determine (*a*) the increase in speed required at point *A* for the satellite to achieve the escape velocity and enter a parabolic orbit, (*b*) the decrease in speed required at point *A* for the satellite to enter an elliptic orbit with a minimum altitude of 6370 km, (*c*) the eccentricity ε of the elliptic orbit.

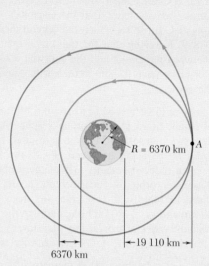

Fig. P12.104

12.105 A space probe is to be placed in a circular orbit of 5600-mi radius about the planet Venus in a specified plane. As the probe reaches *A*, the point of its original trajectory closest to Venus, it is inserted in a first elliptic transfer orbit by reducing its speed by Δv_A. This orbit brings it to point *B* with a much reduced velocity. There the probe is inserted in a second transfer orbit located in the specified plane by changing the direction of its velocity and further reducing its speed by Δv_B. Finally, as the probe reaches point *C*, it is inserted in the desired circular orbit by reducing its speed by Δv_C. Knowing that the mass of Venus is 0.82 times the mass of the earth, that $r_A = 9.3 \times 10^3$ mi and $r_B = 190 \times 10^3$ mi, and that the probe approaches *A* on a parabolic trajectory, determine by how much the velocity of the probe should be reduced (*a*) at *A*, (*b*) at *B*, (*c*) at *C*.

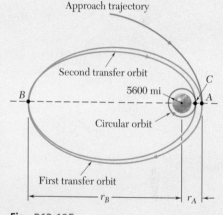

Fig. P12.105

12.106 For the space probe of Prob. 12.105, it is known that $r_A = 9.3 \times 10^3$ mi and that the velocity of the probe is reduced to 20,000 ft/s as it passes through *A*. Determine (*a*) the distance from the center of Venus to point *B*, (*b*) the amounts by which the velocity of the probe should be reduced at *B* and *C*, respectively.

12.107 As it describes an elliptic orbit about the sun, a spacecraft reaches a maximum distance of 202×10^6 mi from the center of the sun at point *A* (called the aphelion) and a minimum distance of 92×10^6 mi at point *B* (called the perihelion). To place the spacecraft in a smaller elliptic orbit with aphelion at *A'* and perihelion at *B'*, where *A'* and *B'* are located 164.5×10^6 mi and 85.5×10^6 mi, respectively, from the center of the sun, the speed of the spacecraft is first reduced as it passes through *A* and then is further reduced as it passes through *B'*. Knowing that the mass of the sun is 332.8×10^3 times the mass of the earth, determine (*a*) the speed of the spacecraft at *A*, (*b*) the amounts by which the speed of the spacecraft should be reduced at *A* and *B'* to insert it into the desired elliptic orbit.

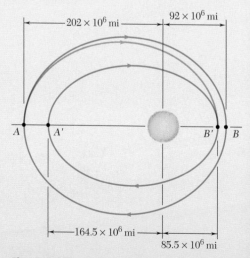

Fig. P12.107

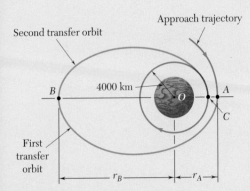

Second transfer orbit

Approach trajectory

4000 km

First transfer orbit

r_B r_A

Fig. P12.110

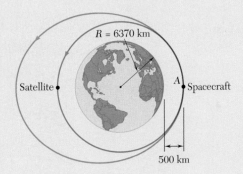

$R = 6370$ km

Satellite

A Spacecraft

500 km

Fig. P12.111

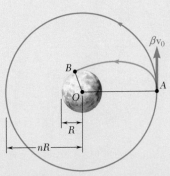

βv_0

B

O

R

nR

Fig. P12.114

786

12.108 Halley's comet travels in an elongated elliptic orbit for which the minimum distance from the sun is approximately $\frac{1}{2} r_E$, where $r_E = 150 \times 10^6$ km is the mean distance from the sun to the earth. Knowing that the periodic time of Halley's comet is about 76 years, determine the maximum distance from the sun reached by the comet.

12.109 Based on observations made during the 1996 sighting of comet Hyakutake, it was concluded that the trajectory of the comet is a highly elongated ellipse for which the eccentricity is approximately $\varepsilon = 0.999887$. Knowing that for the 1996 sighting the minimum distance between the comet and the sun was $0.230R_E$, where R_E is the mean distance from the sun to the earth, determine the periodic time of the comet.

12.110 A space probe is to be placed in a circular orbit of radius 4000 km about the planet Mars. As the probe reaches A, the point of its original trajectory closest to Mars, it is inserted into a first elliptic transfer orbit by reducing its speed. This orbit brings it to point B with a much-reduced velocity. There the probe is inserted into a second transfer orbit by further reducing its speed. Knowing that the mass of Mars is 0.1074 times the mass of the earth, that $r_A = 9000$ km and $r_B = 180\,000$ km, and that the probe approaches A on a parabolic trajectory, determine the time needed for the space probe to travel from A to B on its first transfer orbit.

12.111 A spacecraft and a satellite are at diametrically opposite positions in the same circular orbit of altitude 500 km above the earth. As it passes through point A, the spacecraft fires its engine for a short interval of time to increase its speed and enter an elliptic orbit. Knowing that the spacecraft returns to A at the same time the satellite reaches A after completing one and a half orbits, determine (a) the increase in speed required, (b) the periodic time for the elliptic orbit.

12.112 The Clementine spacecraft described an elliptic orbit of minimum altitude $h_A = 400$ km and maximum altitude $h_B = 2940$ km above the surface of the moon. Knowing that the radius of the moon is 1737 km and that the mass of the moon is 0.01230 times the mass of the earth, determine the periodic time of the spacecraft.

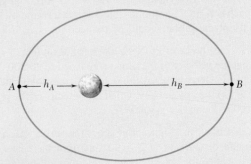

A h_A h_B B

Fig. P12.112

12.113 Determine the time needed for the space probe of Prob. 12.100 to travel from B to C.

12.114 A space probe is describing a circular orbit of radius nR with a velocity v_0 about a planet of radius R and center O. As the probe passes through point A, its velocity is reduced from v_0 to βv_0, where $\beta < 1$, to place the probe on a crash trajectory. Express in terms of n and β the angle AOB, where B denotes the point of impact of the probe on the planet.

12.115 A long-range ballistic trajectory between points A and B on the earth's surface consists of a portion of an ellipse with the apogee at point C. Knowing that point C is 1500 km above the surface of the earth and the range $R\phi$ of the trajectory is 6000 km, determine (a) the velocity of the projectile at C, (b) the eccentricity ε of the trajectory.

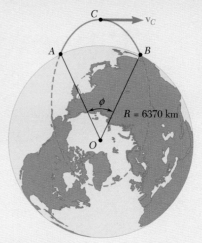

Fig. P12.115

12.116 A space shuttle is describing a circular orbit at an altitude of 563 km above the surface of the earth. As it passes through point A, it fires its engine for a short interval of time to reduce its speed by 152 m/s and begin its descent toward the earth. Determine the angle AOB so that the altitude of the shuttle at point B is 121 km. (*Hint:* Point A is the apogee of the elliptic descent trajectory.)

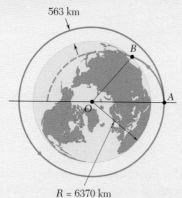

Fig. *P12.116*

12.117 As a spacecraft approaches the planet Jupiter, it releases a probe which is to enter the planet's atmosphere at point B at an altitude of 280 mi above the surface of the planet. The trajectory of the probe is a hyperbola of eccentricity $\varepsilon = 1.031$. Knowing that the radius and the mass of Jupiter are 44,423 mi and 1.30×10^{26} slug, respectively, and that the velocity $\mathbf{v}_B$ of the probe at B forms an angle of $82.9°$ with the direction of OA, determine (a) the angle AOB, (b) the speed v_B of the probe at B.

Fig. *P12.117*

12.118 A satellite describes an elliptic orbit about a planet. Denoting by r_0 and r_1 the distances corresponding, respectively, to the perigee and apogee of the orbit, show that the curvature of the orbit at each of these two points can be expressed as

$$\frac{1}{\rho} = \frac{1}{2}\left(\frac{1}{r_0} + \frac{1}{r_1}\right)$$

12.119 (a) Express the eccentricity ε of the elliptic orbit described by a satellite about a planet in terms of the distances r_0 and r_1 corresponding, respectively, to the perigee and apogee of the orbit. (b) Use the result obtained in part *a* and the data given in Prob. 12.109, where $R_E = 149.6 \times 10^6$ km, to determine the approximate maximum distance from the sun reached by comet Hyakutake.

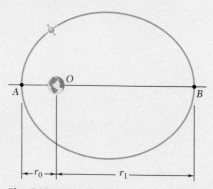

Fig. P12.118 and *P12.119*

12.120 Derive Kepler's third law of planetary motion from Eqs 12.37 and 12.43.

12.121 Show that the angular momentum per unit mass h of a satellite describing an elliptic orbit of semimajor axis a and eccentricity ε about a planet of mass M can be expressed as

$$h = \sqrt{GMa(1 - \varepsilon^2)}$$

Review and Summary

This chapter was devoted to Newton's second law and its application to analyzing the motion of particles.

Newton's Second Law

Denote the mass of a particle by m, the sum (or resultant) of the forces acting on the particle by $\Sigma\mathbf{F}$, and the acceleration of the particle relative to a newtonian frame of reference by $\mathbf{a}$ [Sec. 12.1A]. Then we have

$$\Sigma\mathbf{F} = m\mathbf{a} \tag{12.2}$$

Linear Momentum

Introducing the **linear momentum** of a particle, $\mathbf{L} = m\mathbf{v}$ [Sec. 12.1B], we saw that Newton's second law also can be written in the form

$$\Sigma\mathbf{F} = \dot{\mathbf{L}} \tag{12.5}$$

This equation states that **the resultant of the forces acting on a particle is equal to the rate of change of the linear momentum of the particle**.

Consistent Systems of Units

Equation (12.2) holds only if we use a consistent system of units. With SI units, the forces should be expressed in newtons, the masses in kilograms, and the accelerations in m/s²; with U.S. customary units, the forces should be expressed in pounds, the masses in lb·s²/ft (also referred to as *slugs*), and the accelerations in ft/s² [Sec. 12.1C].

Free-Body Diagram and Kinetic Diagram

A **free-body diagram** for a system shows the applied forces and a **kinetic diagram** shows the vector $m\mathbf{a}$ or its components. These diagrams provide a pictorial representation of Newton's second law. Drawing them will be of great help to you when writing the equations of motion. Note that when a problem involves two or more bodies, it is usually best to consider each body separately.

Equations of Motion for a Particle

To solve a problem involving the motion of a particle, we should first draw the free-body diagram and kinetic diagram for each particle in the system. Then we can use these diagrams to help us write equations containing scalar quantities (Sec. 12.1D). Using **rectangular components** of $\mathbf{F}$ and $\mathbf{a}$, we have

$$\Sigma F_x = ma_x \qquad \Sigma F_y = ma_y \qquad \Sigma F_z = ma_z \tag{12.8}$$

Using **tangential and normal components**, we have

$$\Sigma F_t = m\frac{dv}{dt} \qquad \Sigma F_n = m\frac{v^2}{\rho} \tag{12.9'}$$

Using **radial and transverse components**, we have

$$\Sigma F_r = m(\ddot{r} - r\dot{\theta}^2) \tag{12.11}$$

$$\Sigma F_\theta = m(r\ddot{\theta} + 2\dot{r}\dot{\theta}) \tag{12.12}$$

Sample Probs. 12.1 through 12.5 used rectangular components, Sample Probs. 12.6 through 12.9 used tangential and normal coordinates, and Sample Probs. 12.10 and 12.11 used radial and transverse coordinates.

Angular Momentum

In the second part of this chapter, we defined the **angular momentum H_O** of a particle about a point O as the moment about O of the linear momentum $m\mathbf{v}$ of that particle [Sec. 12.2A]. Thus,

$$\mathbf{H}_O = \mathbf{r} \times m\mathbf{v} \tag{12.13}$$

We noted that $\mathbf{H}_O$ is a vector perpendicular to the plane containing $\mathbf{r}$ and $m\mathbf{v}$ (Fig. 12.23) and has a magnitude of

$$H_O = rmv \sin \phi \tag{12.14}$$

Resolving the vectors $\mathbf{r}$ and $m\mathbf{v}$ into rectangular components, we expressed the angular momentum $\mathbf{H}_O$ in the determinant form

$$\mathbf{H}_O = \begin{vmatrix} \mathbf{i} & \mathbf{j} & \mathbf{k} \\ x & y & z \\ mv_x & mv_y & mv_z \end{vmatrix} \tag{12.15}$$

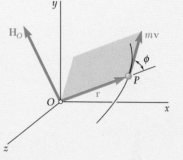

Fig. 12.23

In the case of a particle moving in the xy plane we have $z = v_z = 0$. The angular momentum is perpendicular to the xy plane and is completely defined by its magnitude. We have

$$H_O = H_z = m(xv_y - yv_x) \tag{12.17}$$

Rate of Change of Angular Momentum

Computing the rate of change $\dot{\mathbf{H}}_O$ of the angular momentum $\mathbf{H}_O$ and applying Newton's second law, we obtain the equation

$$\Sigma \mathbf{M}_O = \dot{\mathbf{H}}_O \tag{12.20}$$

This equation states that **the sum of the moments about O of the forces acting on a particle is equal to the rate of change of the angular momentum of the particle about O.**

Motion Under a Central Force

When the only force acting on a particle P is a force $\mathbf{F}$ directed toward or away from a fixed point O, the particle is said to be moving **under a central force** [Sec. 12.2B]. Since $\Sigma \mathbf{M}_O = 0$ at any given instant, it follows from Eq. (12.20) that $\dot{\mathbf{H}}_O = 0$ for all values of t and thus

$$\mathbf{H}_O = \text{constant} \tag{12.21}$$

We concluded that **the angular momentum of a particle moving under a central force is constant, both in magnitude and direction**, and that the particle moves in a plane perpendicular to the vector $\mathbf{H}_O$.

Recalling Eq. (12.14), we wrote the relation

$$rmv \sin \phi = r_0 mv_0 \sin \phi_0 \tag{12.23}$$

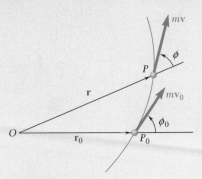

Fig. 12.24

for the motion of any particle under a central force (Fig. 12.24). Using polar coordinates and recalling Eq. (12.19), we also had

$$r^2\dot{\theta} = h \tag{12.25}$$

where h is a constant representing the angular momentum per unit mass, H_O/m, of the particle. We observed (Fig. 12.25) that the infinitesimal area dA swept by the radius vector OP as it rotates through $d\theta$ is equal to $dA = \frac{1}{2}r^2 d\theta$ and, thus, that the left-hand side of Eq. (12.25) represents twice the **areal velocity** dA/dt of the particle. Therefore, **the areal velocity of a particle moving under a central force is constant**.

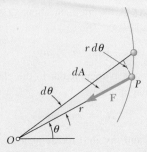

Fig. 12.25

Newton's Law of Universal Gravitation

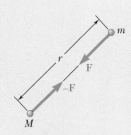

Fig. 12.26

An important application of the motion under a central force is provided by the orbital motion of bodies under gravitational attraction [Sec. 12.2C]. According to **Newton's law of universal gravitation**, two particles at a distance r from each other and of masses M and m, respectively, attract each other with equal and opposite forces $\mathbf{F}$ and $-\mathbf{F}$ directed along the line joining the particles (Fig. 12.26). The common magnitude F of the two forces is

$$F = G\frac{Mm}{r^2} \tag{12.26}$$

where G is the **constant of gravitation**. In the case of a body of mass m subjected to the gravitational attraction of the earth, we can express the product GM, where M is the mass of the earth, as

$$GM = gR^2 \tag{12.28}$$

where $g = 9.81$ m/s^2 = 32.2 ft/s^2 and R is the radius of the earth.

Orbital Motion

We showed in Sec. 12.3A that a particle moving under a central force describes a trajectory defined by the differential equation

$$\frac{d^2u}{d\theta^2} + u = \frac{F}{mh^2u^2} \tag{12.35}$$

where $F > 0$ corresponds to an attractive force and $u = 1/r$. In the case of a particle moving under a force of gravitational attraction [Sec. 12.12C], we substituted for F the expression given in Eq. (12.26). Measuring θ from the axis OA joining the focus O to the point A of the trajectory closest to O (Fig. 12.27), we found that the solution to Eq. (12.35) is

$$\frac{1}{r} = u = \frac{GM}{h^2} + C\cos\theta \tag{12.37}$$

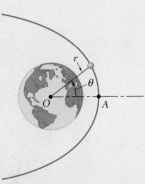

Fig. 12.27

This is the equation of a conic of eccentricity $\varepsilon = Ch^2/GM$. The conic is an **ellipse** if $\varepsilon < 1$, a **parabola** if $\varepsilon = 1$, and a **hyperbola** if $\varepsilon > 1$. We can determine the constants C and h from the initial conditions; if the particle is projected from point A ($\theta = 0$, $r = r_0$) with an initial velocity $\mathbf{v}_0$ that is perpendicular to OA, we have $h = r_0 v_0$ [Sample Prob. 12.14].

Escape Velocity

We also showed that the values of the initial velocity corresponding, respectively, to a parabolic and a circular trajectory are

$$v_{esc} = \sqrt{\frac{2GM}{r_0}} \tag{12.41}$$

$$v_{circ} = \sqrt{\frac{GM}{r_0}} \tag{12.42}$$

The first of these values, called the **escape velocity**, is the smallest value of v_0 for which the particle will not return to its starting point.

Periodic Time

The **periodic time** τ of a planet or satellite is defined as the time required by that body to describe its orbit. We showed that

$$\tau = \frac{2\pi ab}{h} \tag{12.43}$$

where $h = r_0 v_0$ and where a and b represent the semimajor and semiminor axes of the orbit. We further showed that these semiaxes are respectively equal to the arithmetic and geometric means of the maximum and minimum values of the radius r.

Kepler's Laws

The last part of the chapter [Sec. 12.3C] presented **Kepler's laws of planetary motion** and showed that these empirical laws, obtained from early astronomical observations, confirm Newton's laws of motion as well as his law of gravitation.

Review Problems

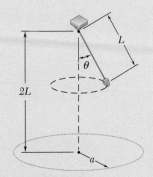

Fig. P12.123

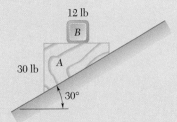

Fig. P12.124

12.122 In the braking test of a sports car its velocity is reduced from 70 mi/h to zero in a distance of 170 ft with slipping impending. Knowing that the coefficient of kinetic friction is 80 percent of the coefficient of static friction, determine (*a*) the coefficient of static friction, (*b*) the stopping distance for the same initial velocity if the car skids. Ignore air resistance and rolling resistance.

12.123 A bucket is attached to a rope of length $L = 1.2$ m and is made to revolve in a horizontal circle. Drops of water leaking from the bucket fall and strike the floor along the perimeter of a circle of radius a. Determine the radius a when $\theta = 30°$.

12.124 A 12-lb block B rests as shown on the upper surface of a 30-lb wedge A. Neglecting friction, determine immediately after the system is released from rest (*a*) the acceleration of A, (*b*) the acceleration of B relative to A.

12.125 A 500-lb crate B is suspended from a cable attached to a 40-lb trolley A which rides on an inclined I-beam as shown. Knowing that at the instant shown the trolley has an acceleration of 1.2 ft/s² up and to the right, determine (*a*) the acceleration of B relative to A, (*b*) the tension in cable CD.

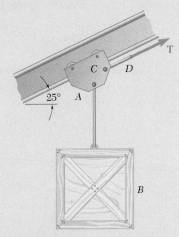

Fig. P12.125

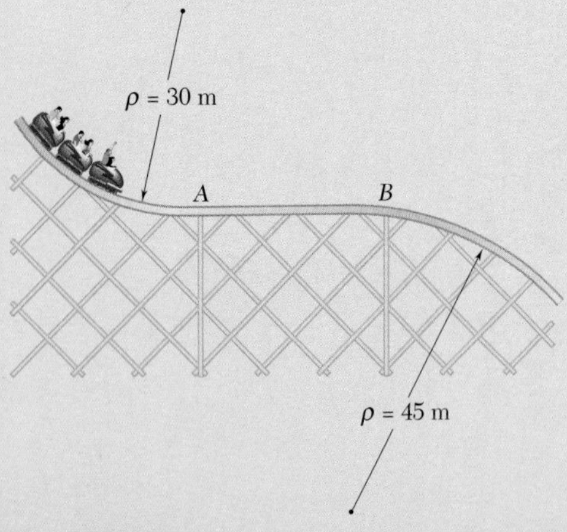

12.126 The roller-coaster track shown is contained in a vertical plane. The portion of track between A and B is straight and horizontal, while the portions to the left of A and to the right of B have radii of curvature as indicated. A car is traveling at a speed of 72 km/h when the brakes are suddenly applied, causing the wheels of the car to slide on the track ($\mu_k = 0.25$). Determine the initial deceleration of the car if the brakes are applied as the car (*a*) has almost reached A, (*b*) is traveling between A and B, (*c*) has just passed B.

Fig. P12.126

12.127 The parasailing system shown uses a winch to pull the rider in towards the boat, which is travelling with a constant velocity. During the interval when θ is between $20°$ and $40°$ (where $t = 0$ at $\theta = 20°$), the angle increases at the constant rate of $2°/s$. During this time, the length of the rope is defined by the relationship $r = 125 - \frac{1}{3}t^{3/2}$, where r and t are expressed in meters and seconds, respectively. At the instant when the rope makes a $30°$ angle with the water, the tension in the rope is 18 kN. At this instant, what is the magnitude and direction of the force of the parasail on the 75 kg parasailor?

Fig. P12.127

12.128 A small 200-g collar C can slide on a semicircular rod that is made to rotate about the vertical AB at the constant rate of 6 rad/s. Determine the minimum required value of the coefficient of static friction between the collar and the rod if the collar is not to slide when (a) $\theta = 90°$, (b) $\theta = 75°$, (c) $\theta = 45°$. Indicate in each case the direction of the impending motion.

12.129 Telemetry technology is used to quantify kinematic values of a 200-kg roller-coaster cart as it passes overhead. According to the system, $r = 25$ m, $\dot{r} = -10$ m/s, $\ddot{r} = -2$ m/s², $\theta = 90°$, $\dot{\theta} = -0.4$ rad/s, $\ddot{\theta} = -0.32$ rad/s². At this instant, determine (a) the normal force between the cart and the track, (b) the radius of curvature of the track.

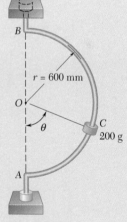

Fig. P12.128

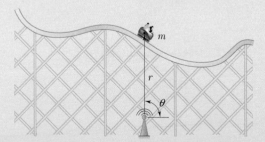

Fig. P12.129

12.130 The radius of the orbit of a moon of a given planet is equal to twice the radius of that planet. Denoting by ρ the mean density of the planet, show that the time required by the moon to complete one full revolution about the planet is $(24\pi/G\rho)^{1/2}$, where G is the constant of gravitation.

12.131 At engine burnout on a mission, a shuttle had reached point A at an altitude of 40 mi above the surface of the earth and had a horizontal velocity $\mathbf{v}_0$. Knowing that its first orbit was elliptic and that the shuttle was transferred to a circular orbit as it passed through point B at an altitude of 170 mi, determine (a) the time needed for the shuttle to travel from A to B on its original elliptic orbit, (b) the periodic time of the shuttle on its final circular orbit.

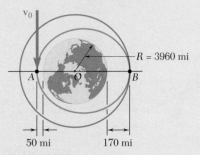

Fig. P12.131

12.132 A space probe in a low earth orbit is inserted into an elliptic transfer orbit to the planet Venus. Knowing that the mass of the sun is 332.8×10^3 times the mass of the earth and assuming that the probe is subjected only to the gravitational attraction of the sun, determine the value of ϕ, which defines the relative position of Venus with respect to the earth at the time the probe is inserted into the transfer orbit.

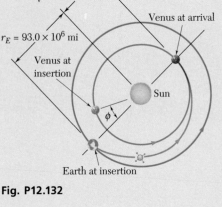

$r_V = 67.2 \times 10^6$ mi

$r_E = 93.0 \times 10^6$ mi

Fig. P12.132

*12.133 Disk A rotates in a horizontal plane about a vertical axis at the constant rate $\dot{\theta}_0 = 10$ rad/s. Slider B has mass 1 kg and moves in a frictionless slot cut in the disk. The slider is attached to a spring of constant k, which is undeformed when $r = 0$. Knowing that the slider is released with no radial velocity in the position $r = 500$ mm, determine the position of the slider and the horizontal force exerted on it by the disk at $t = 0.1$ s for (a) $k = 100$ N/m, (b) $k = 200$ N/m.

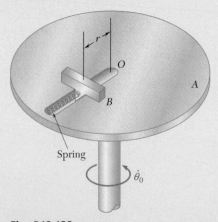

Fig. P12.133

13

Kinetics of Particles: Energy and Momentum Methods

A golf ball will deform upon impact as shown by this high-speed photo. The maximum deformation will occur when the club head velocity and the ball velocity are the same. In this chapter impacts will be analyzed using the coefficient of restitution and conservation of linear momentum. The kinetics of particles using energy and momentum methods is the subject of this chapter.

Introduction

Objectives

- **Calculate** the work done by a force.
- **Calculate** the kinetic energy of a particle.
- **Calculate** the gravitational and elastic potential energy of a system.
- **Solve** particle kinetics problems using the principle of work and energy.
- **Calculate** the power and efficiency of a mechanical system.
- **Solve** particle kinetics problems using conservation of energy.
- **Solve** particle kinetic problems involving conservative central forces.
- **Draw** complete and accurate impulse-momentum diagrams.
- **Solve** particle kinetics problems using the principle of impulse and momentum.
- **Solve** particle kinetics problems using conservation of linear momentum.
- **Solve** impact problems using the principle of impact and momentum and the coefficient of restitution.
- **Determine** the appropriate principle(s) to apply when solving a particle dynamics problem.
- **Solve** multi-step dynamics problems using multiple kinetics principles.

Introduction

In the preceding chapter, we solved most problems dealing with the motion of particles through the use of the fundamental equation of motion $\mathbf{F} = m\mathbf{a}$. Given a particle acted upon by a force $\mathbf{F}$, we could solve this equation for the acceleration $\mathbf{a}$; then by applying the principles of kinematics, we could determine from $\mathbf{a}$ the velocity and position of the particle at any time.

However, using the general equation $\mathbf{F} = m\mathbf{a}$ together with kinematics allows us to obtain two additional concepts: the **principle of work and energy** and the **principle of impulse and momentum**. The advantage of these ideas lies in the fact that they make the determination of the acceleration unnecessary. Indeed, the principle of work and energy directly relates force, mass, velocity, and displacement, whereas the principle of impulse and momentum relates force, mass, velocity, and time.

We present work and energy first. In Sec. 13.1, we define the *work of a force* and the *kinetic energy of a particle.* Then we apply the principle of work and energy to the solution of engineering problems. We also introduce the concepts of *power* and *efficiency* of a machine, which are important in engineering applications such as motors and hydraulic actuators.

In Sec. 13.2, we examine the concept of *potential energy* of a conservative force, and we apply the principle of conservation of energy to various problems of practical interest. In Sec. 13.2D, we use the principles of conservation of energy and of conservation of angular momentum jointly to solve problems of space mechanics.

The second part of this chapter deals with the principle of impulse and momentum and its application to the study of the motion of a particle. You will see in Sec. 13.3B that this principle is particularly effective in the study of the *impulsive motion* of a particle, where very large forces act for a very short time interval—like hitting a nail with a hammer.

We also consider the *central impact* of two bodies. We will show that a relation exists between the relative velocities of the two colliding bodies before and after impact. We can use this relation, together with the fact that the total momentum of the two bodies is conserved, to solve several types of practical problems.

Finally, we will discuss how to choose the best principle for solving a given problem from among Newton's second law, work and energy, or impulse and momentum. You may even need to apply multiple principles in order to solve some dynamics problems.

13.1 WORK AND ENERGY

Work and energy have very specific meanings in science and engineering. In everyday speech, you might say that holding up a concrete block is a lot of work, but in science, if the block doesn't move, you don't do any work at all while holding it. Similarly, people talk about energy all the time, from how you feel on a particular day ("I don't seem to have much energy today") to national and international policy ("The high cost of energy is affecting our trade balance with other countries."). In science and engineering, work and energy have very specific definitions that involve forces, displacements, masses, and velocities. These two concepts are of great value in analyzing a wide range of engineering problems.

13.1A Work of a Force

We first define the terms *displacement* and *work* as they are used in mechanics.[†] Consider a particle that moves from a point A to a neighboring point A' (Fig. 13.1). If $\mathbf{r}$ denotes the position vector corresponding to point A, we can denote the small vector joining A and A' by the differential $d\mathbf{r}$; the vector $d\mathbf{r}$ is called the **displacement** of the particle. Now, let us assume that a force $\mathbf{F}$ is acting on the particle. We define the **work of the force F corresponding to the displacement** $d\mathbf{r}$ as the quantity

$$dU = \mathbf{F} \cdot d\mathbf{r} \tag{13.1}$$

We obtain dU by taking the scalar product of the force $\mathbf{F}$ and the displacement $d\mathbf{r}$. We denote the magnitudes of the force and of the displacement by F and ds, respectively, and the angle formed by $\mathbf{F}$ and

[†]We defined work in Sec. 10.1A and outlined its basic properties in Secs. 10.1A and 10.2A. For convenience, we repeat here the portions of this material that relate to the kinetics of particles.

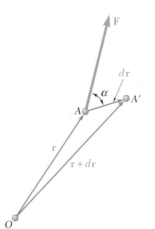

Fig. 13.1 The work of a force acting on a particle is the scalar product of the force **F** and the displacement $d\mathbf{r}$ of the particle.

$d\mathbf{r}$ by α. Then from the definition of the scalar product of two vectors (Sec. 3.2A), we have

$$dU = F\,ds\,\cos\alpha \qquad\qquad \textbf{(13.1')}$$

Using Eq. (3.30), we can also express the work dU in terms of the rectangular components of the force and of the displacement:

$$dU = F_x\,dx + F_y\,dy + F_z\,dz \qquad\qquad \textbf{(13.1'')}$$

Work is a scalar quantity, so it has a magnitude and a sign but no direction.

Note that work is expressed in units obtained by multiplying units of length by units of force. Thus, if we use U.S. customary units, work is expressed in ft·lb or in·lb. If we use SI units, work is expressed in N·m. The unit of work N·m is called a **joule** (J).[†] Recalling the conversion factors indicated in Sec. 12.1C, we have

$$1\text{ ft·lb} = (1\text{ ft})(1\text{ lb}) = (0.3048\text{ m})(4.448\text{ N}) = 1.356\text{ J}$$

It follows from Eq. (13.1′) that the work dU is positive if angle α is acute and negative if α is obtuse. Three particular cases are of special interest. If the force $\mathbf{F}$ has the same direction as $d\mathbf{r}$, the work dU reduces to $F\,ds$. If $\mathbf{F}$ has a direction opposite to that of $d\mathbf{r}$, the work is $dU = -F\,ds$. Finally, if $\mathbf{F}$ is perpendicular to $d\mathbf{r}$, the work dU is zero.

We can obtain the work of $\mathbf{F}$ during a *finite* displacement of the particle from A_1 to A_2 (Fig. 13.2a) by integrating Eq. (13.1) along the path described by the particle. This work, denoted by $U_{1\to2}$, is

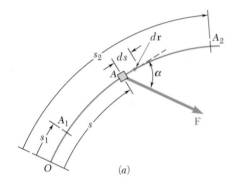

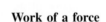

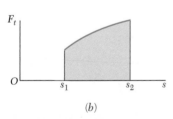

Work of a force	$$U_{1\to2} = \int_{A_1}^{A_2} \mathbf{F}\cdot d\mathbf{r} \qquad\qquad \textbf{(13.2)}$$

Using the alternative expression of Eq. (13.1′) for the elementary work dU, and observing that $F\cos\alpha$ represents the tangential component F_t of the force, we can also express the work $U_{1\to2}$ as

$$U_{1\to2} = \int_{s_1}^{s_2} (F\cos\alpha)\,ds = \int_{s_1}^{s_2} F_t\,ds \qquad\qquad \textbf{(13.2')}$$

where the variable of integration s measures the distance traveled by the particle along the path. The work $U_{1\to2}$ is represented by the area under the curve obtained by plotting $F_t = F\cos\alpha$ against s (Fig. 13.2b).

When the force $\mathbf{F}$ is defined by its rectangular components, we can use the expression of Eq. (13.1″) for the elementary work. We have

$$U_{1\to2} = \int_{A_1}^{A_2} (F_x\,dx + F_y\,dy + F_z\,dz) \qquad\qquad \textbf{(13.2'')}$$

where the integration is performed along the path described by the particle.

Fig. 13.2 (a) The work of force **F** over a finite displacement is the integral of Eq. (13.1) from point A_1 to point A_2. (b) The work is represented by the area under the graph of F_t versus s from s_1 to s_2.

[†]The joule (J) is the SI unit of energy, whether in mechanical form (work, potential energy, or kinetic energy) or in chemical, electrical, or thermal form. Note that even though N·m = J, the moment of a force must be expressed in N·m and not in joules, since the moment of a force is not a form of energy.

We can use these equations to derive formulas for the work done by a force in several common and important situations, as we now show. These formulas can simplify the calculations needed to solve many common problems. For other situations, you can return to the basic Equations (13.1) and (13.2) and their variants.

Work of a Constant Force in Rectilinear Motion.

When a particle moving in a straight line is acted upon by a force $\mathbf{F}$ of constant magnitude and of constant direction (Fig. 13.3), formula (13.2') yields

$$U_{1\rightarrow2} = (F \cos \alpha)\, \Delta x \tag{13.3}$$

where α = angle the force forms with direction of motion
Δx = displacement from A_1 to A_2

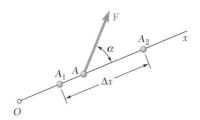

Fig. 13.3 For a constant force in rectilinear motion, the work equals the displacement times the component of force in the direction of the displacement.

Work of the Force of Gravity.

We can obtain the work of the weight $\mathbf{W}$ of a body—i.e., of the force of gravity exerted on that body—by substituting the components of $\mathbf{W}$ into Eqs. (13.1″) and (13.2″). Choosing the y axis upward (Fig. 13.4), we have $F_x = 0$, $F_y = -W$, and $F_z = 0$. This gives us

$$dU = -W\,dy$$

$$U_{1\rightarrow2} = -\int_{y_1}^{y_2} W\,dy = Wy_1 - Wy_2 \tag{13.4}$$

or

$$U_{1\rightarrow2} = -W(y_2 - y_1) = -W\,\Delta y \tag{13.4'}$$

where Δy is the vertical displacement from A_1 to A_2. The work of the weight $\mathbf{W}$ is thus equal to **the product of W and the vertical displacement of the center of gravity of the body**. The work is positive when $\Delta y < 0$, that is, when the body moves down. When the body moves up (and $\Delta y > 0$), the force and displacement are in opposite directions, and the work is negative.

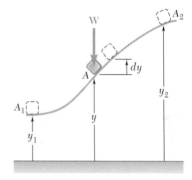

Fig. 13.4 The work done by the force of gravity is the product of the weight and the vertical displacement of the object's center of gravity. If the object moves up, the work done by gravity is negative.

Work of the Force Exerted by a Spring.

Consider a body A attached to a fixed point B by a spring; we assume that the spring is undeformed when the body is at A_0 (Fig. 13.5a). For a linear spring, the magnitude of the force $\mathbf{F}$ exerted by the spring on body A is proportional to the deflection x of the spring measured from the unstretched position A_0 (i.e., $x = L_{\text{stretched}} - L_{\text{unstretched}}$). We have

$$F = kx \tag{13.5}$$

where k is the **spring constant** expressed in N/m or kN/m if SI units are used and in lb/ft or lb/in. if U.S. customary units are used.[†]

[†]The relation $F = kx$ is correct under static conditions only. Under dynamic conditions, Eq. (13.5) should be modified to take into account the inertia of the spring. However, the error introduced by using $F = kx$ in the solution of kinetics problems is small if the mass of the spring is small compared with the other masses in motion.

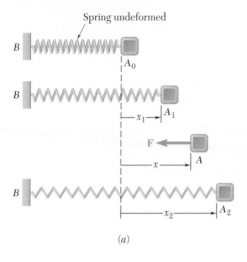

Spring undeformed

(a)

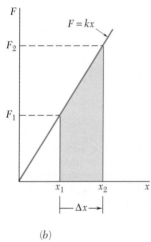

(b)

Fig. 13.5 (a) The work of a force exerted by a spring depends on the spring constant and the initial and final positions of the spring. (b) The work is represented by the area under the graph of force versus position.

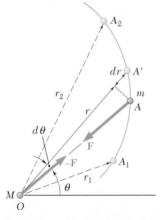

Fig. 13.6 The work of a gravitational force depends on the gravitational constant, the masses of the interacting bodies, and the radial distance between them.

We can obtain the work of force **F** exerted by the spring during a finite displacement of the body from $A_1(x = x_1)$ to $A_2(x = x_2)$ by writing

$$dU = -F\,dx = -kx\,dx$$

$$U_{1\to2} = -\int_{x_1}^{x_2} kx\,dx = \tfrac{1}{2}kx_1^2 - \tfrac{1}{2}kx_2^2 \tag{13.6}$$

You need to be careful in expressing k and x in consistent units. For example, if you use U.S. customary units, k should be expressed in lb/ft and x in feet, or k should be given in lb/in. and x in inches. In the first case, the work will have units of ft·lb; in the second case, it will have units of in·lb. Note that the work of force **F** exerted by the spring on the body is positive when $x_2 < x_1$; that is, when the spring is returning to its undeformed position. When the body is moved from x_1 to x_2, the work of the force is negative, since the displacement and force are in opposite directions.

Since Eq. (13.5) is the equation of a straight line of slope k passing through the origin, we can also obtain the work $U_{1\to2}$ of **F** during the displacement from A_1 to A_2 by evaluating the area of the trapezoid shown in Fig. 13.5b. We can do this by computing F_1 and F_2 and multiplying the base Δx of the trapezoid by its mean height $\tfrac{1}{2}(F_1 + F_2)$. Since the work of the force **F** exerted by the spring is positive for a negative value of Δx, we have

$$U_{1\to2} = -\tfrac{1}{2}(F_1 + F_2)\,\Delta x \tag{13.6'}$$

Work of a Gravitational Force. We saw in Sec. 12.2C that two particles of mass M and m separated by a distance r attract each other with equal and opposite forces **F** and $-\mathbf{F}$, directed along the line joining the particles and of magnitude as

$$F = G\frac{Mm}{r^2}$$

Let us assume that particle M occupies a fixed position O while particle m moves along the path shown in Fig. 13.6. We can obtain the work of force **F** exerted on particle m during an infinitesimal displacement of the particle from A to A' by multiplying the magnitude F of the force by the radial component dr of the displacement. Since **F** is directed toward O and dr is directed away from O, the work is negative, and we have

$$dU = -F\,dr = -G\frac{Mm}{r^2}\,dr$$

The work of the gravitational force **F** during a finite displacement from $A_1(r = r_1)$ to $A_2(r = r_2)$ is therefore

$$U_{1\to2} = -\int_{r_1}^{r_2} \frac{GMm}{r^2}\,dr = \frac{GMm}{r_2} - \frac{GMm}{r_1} \tag{13.7}$$

where M is the mass of the earth. We can use this formula to determine the work of the force exerted by the earth on a body of mass m at a distance r from the earth's center when r is larger than the radius R of the earth. Recalling the first of the relations in Eq. (12.27), we can replace the product GMm in Eq. (13.7) by WR^2, where R is the earth's radius

($R = 6.37 \times 10^6$ m or 3960 mi) and W is the weight of the body at the earth's surface.

Some forces frequently encountered in kinetics problems *do no work*. They are forces applied to fixed points ($ds = 0$) or acting in a direction perpendicular to the displacement ($\cos \alpha = 0$). Forces that do no work include the reaction at a frictionless pin when the body supported rotates about the pin; the normal force at a frictionless fixed surface when the body in contact moves along the surface; the reaction at a roller moving along its track; and the weight of a body when its center of gravity moves horizontally.

13.1B Principle of Work and Energy

Consider a particle of mass m acted upon by a force $\mathbf{F}$ and moving along a path that is either rectilinear or curved (Fig. 13.7). Expressing Newton's second law in terms of the tangential components of the force and of the acceleration (see Sec. 12.1D), we have

$$F_t = ma_t \quad \text{or} \quad F_t = m\frac{dv}{dt}$$

where v is the speed of the particle. Recalling from Sec. 11.4A that $v = ds/dt$, we obtain

$$F_t = m\frac{dv}{ds}\frac{ds}{dt} = mv\frac{dv}{ds}$$
$$F_t\, ds = mv\, dv$$

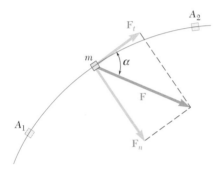

Fig. 13.7 A particle m acted upon by a force **F**.

Integrating from A_1, where $s = s_1$ and $v = v_1$, to A_2, where $s = s_2$ and $v = v_2$, we have

$$\int_{s_1}^{s_2} F_t\, ds = m\int_{v_1}^{v_2} v\, dv = \tfrac{1}{2}mv_2^2 - \tfrac{1}{2}mv_1^2 \tag{13.8}$$

The left-hand side of Eq. (13.8) represents the work $U_{1\to2}$ of the force $\mathbf{F}$ exerted on the particle during the displacement from A_1 to A_2; as indicated earlier, the work $U_{1\to2}$ is a scalar quantity. Thus, the expression $\tfrac{1}{2}mv^2$ is also a scalar quantity. We define it as the kinetic energy of the particle, denoted by T. That is,

Kinetic energy of a particle
$$T = \frac{1}{2}mv^2 \tag{13.9}$$

Substituting into Eq. (13.8), we have

Principle of work and energy
$$U_{1\to2} = T_2 - T_1 \tag{13.10}$$

This equation states that when a particle moves from A_1 to A_2 under the action of a force $\mathbf{F}$, **the work of the force F is equal to the change in kinetic energy of the particle**. This is known as the **principle of work and energy**. Rearranging the terms in Eq. (13.10) gives

$$T_1 + U_{1\to2} = T_2 \tag{13.11}$$

Like Newton's second law from which it is derived, the principle of work and energy applies only with respect to a newtonian frame of reference (Sec. 12.1A). The speed v used to determine the kinetic energy T therefore should be measured with respect to a newtonian frame of reference.

Since both work and kinetic energy are scalar quantities, we can compute their sum as an ordinary algebraic sum with the work $U_{1\to2}$ being positive or negative according to the direction of $\mathbf{F}$. When several forces act on the particle, the expression $U_{1\to2}$ represents the total work of the forces acting on the particle; it is obtained by adding algebraically the work of the various forces.

As just noted, the kinetic energy of a particle is a scalar quantity. It further appears from the definition $T = \frac{1}{2}mv^2$ that, regardless of the particle's direction of motion, the kinetic energy is always positive. Considering the particular case when $v_1 = 0$ and $v_2 = v$, and substituting $T_1 = 0$ and $T_2 = T$ into Eq. (13.10), we observe that the work done by the forces acting on the particle is equal to T. Thus, the kinetic energy of a particle moving with a speed v represents the work that must be done to bring the particle from rest to the speed v. Substituting $T_1 = T$ and $T_2 = 0$ into Eq. (13.10), we also note that when a particle moving with a speed v is brought to rest, the work done by the forces acting on the particle is $-T$. Assuming that no energy is dissipated into heat, we conclude that the work done by the forces exerted *by the particle* on the bodies that cause it to come to rest is equal to T. Thus, the kinetic energy of a particle also represents **the capacity to do work associated with the speed of the particle**.

The kinetic energy is measured in the same units as work, i.e., in joules if we use SI units and in ft·lb if we use U.S. customary units. We check that, in SI units,

$$T = \tfrac{1}{2}mv^2 = \text{kg(m/s)}^2 = (\text{kg·m/s}^2)\text{m} = \text{N·m} = \text{J}$$

whereas in customary units,

$$T = \tfrac{1}{2}mv^2 = (\text{slug})(\text{ft/s})^2 = (\text{lb·s}^2/\text{ft})(\text{ft/s})^2 = \text{ft·lb}$$

13.1C   Applications of the Principle of Work and Energy

Using the principle of work and energy greatly simplifies the solution of many problems involving forces, displacements, and velocities. Consider, for example, the pendulum OA consisting of a bob A of weight W attached to a cord of length l (Fig. 13.8a). The pendulum is released with no initial velocity from a horizontal position OA_1 and allowed to swing in a vertical plane. We wish to determine the speed of the bob as it passes through A_2, directly under O.

We first determine the work done during the displacement from A_1 to A_2 by the forces acting on the bob. We draw a free-body diagram of the bob, showing all the *actual* forces acting on it; i.e., the weight $\mathbf{W}$ and the force $\mathbf{P}$ exerted by the cord (Fig. 13.8b). (Recall that an inertia vector is not an actual force and *should not* be included in the free-body diagram.) Note that force $\mathbf{P}$ does no work, since it is normal to the path; the only force that does work is thus the weight $\mathbf{W}$. We obtain the work of $\mathbf{W}$ by

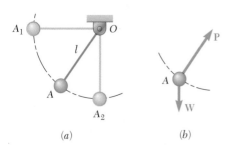

Fig. 13.8   (a) A bob of weight W swings from an initial position A_1 to a final position A_2; (b) free-body diagram of the bob at position A.

multiplying its magnitude W by the vertical displacement l (Sec. 13.1A); since the displacement is downward, the work is positive. We therefore have $U_{1\rightarrow2} = Wl$.

Now consider the kinetic energy of the bob. We have $T_1 = 0$ at A_1 and $T_2 = \frac{1}{2}(W/g)v_2^2$ at A_2. We can now apply the principle of work and energy. From Eq. (13.11), we have

$$T_1 + U_{1\rightarrow2} = T_2 \quad 0 + Wl = \frac{1}{2}\frac{W}{g}v_2^2$$

Solving for v_2, we find $v_2 = \sqrt{2gl}$. Note that this speed is also that of a body falling freely from a height l.

This example illustrates the following advantages of the method of work and energy:

1. In order to find the speed at A_2, there is no need to determine the acceleration in an intermediate position A and to integrate the acceleration expression from A_1 to A_2.
2. All quantities involved are scalars and can be added directly, without using x and y components.
3. Forces that do no work are eliminated from the solution of the problem.

What is an advantage in one problem, however, may be a disadvantage in another. It is evident, for instance, that the method of work and energy cannot be used to directly determine an acceleration. It is also evident that to determine a force that is normal to the path of the particle (i.e., a force that does no work) we must supplement the method of work and energy by the direct application of Newton's second law. Suppose, for example, that we wish to determine the tension in the cord of the pendulum of Fig. 13.8a as the bob passes through A_2. We draw a free-body diagram and kinetic diagram of the bob in that position (Fig. 13.9) and express Newton's second law in terms of tangential and normal components. The equations $\Sigma F_t = ma_t$ and $\Sigma F_n = ma_n$ yield, respectively, $a_t = 0$ and

$$P - W = ma_n = \frac{W}{g}\frac{v_2^2}{l}$$

But earlier, we determined the speed at A_2 by the method of work and energy. Substituting $v_2^2 = 2gl$ and solving for P, we have

$$P = W + \frac{W}{g}\frac{2gl}{l} = 3W$$

If we used only statics principles and designed the cord to hold the weight of the bob (or even twice the weight of the bob), the cord would have failed.

When a problem involves two particles or more, we can apply the principle of work and energy to each particle separately. Adding the kinetic energies of the various particles and considering the work of all the forces acting on them, we can also write a single equation of work and energy for all the particles involved. We have

$$T_1 + U_{1\rightarrow2} = T_2 \tag{13.11}$$

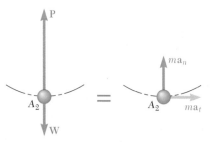

Fig. 13.9 Free-body diagram and kinetic diagram for determining the force on a pendulum bob.

where T_1 represents the arithmetic sum of the kinetic energies of the particles involved at position 1, T_2 represents the arithmetic sum of the kinetic energies of the particles involved at position 2, and $U_{1\rightarrow2}$ is the work of all the forces acting on the particles, including the forces of action and reaction exerted by the particles on each other. In problems involving bodies connected by inextensible cords or links, however, the work of the forces exerted by a given cord or link on the two bodies it connects cancels out, since the points of application of these forces move through equal distances (see Sample Prob. 13.2). (In Chapter 14 we discuss how to apply the method of work and energy to a system of particles.)

Friction forces have a direction opposite of that of the displacement of the body on which they act, so **the work of friction forces is always negative**. This work represents energy dissipated into heat and always results in a decrease in the kinetic energy of the body involved (see Sample Prob. 13.3).

13.1D   Power and Efficiency

We define **power** as the time rate at which work is done. In the selection of a motor or engine, power is a much more important criterion than is the actual amount of work to be performed. Either a small motor or a large power plant can be used to do a given amount of work, but the small motor may require a month to do the work done by the power plant in a matter of minutes. If ΔU is the work done during the time interval Δt, the average power during that time interval is

$$\text{Average power} = \frac{\Delta U}{\Delta t}$$

Letting Δt approach zero, we obtain in the limit

Power
$$\text{Power} = \frac{dU}{dt} \tag{13.12}$$

Substituting the scalar product $\mathbf{F}\cdot d\mathbf{r}$ for dU, we can also write

$$\text{Power} = \frac{dU}{dt} = \frac{\mathbf{F}\cdot d\mathbf{r}}{dt}$$

Then, recalling that $d\mathbf{r}/dt$ represents the velocity $\mathbf{v}$ of the point of application of $\mathbf{F}$, we have

$$\text{Power} = \mathbf{F}\cdot\mathbf{v} \tag{13.13}$$

Photo 13.1 The power used to operate a chair lift at a ski resort is the product of the force applied and the speed of the lift.

Since we defined power as the time rate at which work is done, we obtain its units by dividing units of work by the unit of time. Thus, if we use SI units, power is expressed in J/s; this unit is called a *watt* (W). We have

$$1 \text{ W} = 1 \text{ J/s} = 1 \text{ N·m/s}$$

If we use U.S. customary units, power is expressed in ft·lb/s or in *horsepower* (hp), where one horsepower is defined as

$$1 \text{ hp} = 550 \text{ ft·lb/s}$$

Recall from Sec. 13.1A that 1 ft·lb = 1.356 J, so we can verify that

$$1 \text{ ft·lb/s} = 1.356 \text{ J/s} = 1.356 \text{ W}$$

$$1 \text{ hp} = 550(1.356 \text{ W}) = 746 \text{ W} = 0.746 \text{ kW}$$

We defined the **mechanical efficiency** of a machine in Sec. 10.1D as the ratio of the output work to the input work:

$$\eta = \frac{\text{output work}}{\text{input work}} \qquad \textbf{(13.14)}$$

This definition is based on the assumption that work is done at a constant rate. The ratio of the output to the input work is therefore equal to the ratio of the rates at which output and input work are done, and we have

Mechanical efficiency

$$\eta = \frac{\text{power output}}{\text{power input}} \qquad \textbf{(13.15)}$$

Because of energy losses due to friction, the output work is always smaller than the input work, and consequently, the power output is always smaller than the power input. The mechanical efficiency of a machine is therefore always less than 1.

When we use a machine to transform mechanical energy into electrical energy or thermal energy into mechanical energy, we can obtain its *overall efficiency* from Eq. (13.15). The overall efficiency of a machine is always less than 1; it provides a measure of all the various energy losses involved (losses of electric or thermal energy as well as frictional losses). Note that you have to express the power output and the power input in the same units before using Eq. (13.15).

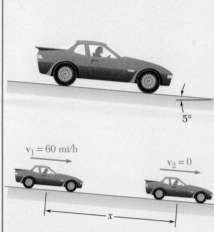

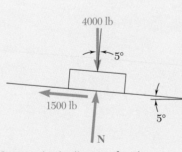

Fig. 1 Car at the two positions of interest.

Sample Problem 13.1

An automobile weighing 4000 lb is driven down a 5° incline at a speed of 60 mi/h when the brakes are applied, causing a constant total braking force (applied by the road on the tires) of 1500 lb. Determine the distance traveled by the automobile as it comes to a stop.

STRATEGY: You are given the velocity of the car at two positions along the road and need to determine the distance x between them (Fig. 1), so use the principle of work and energy.

MODELING: Choose the car as the system and assume it can be modeled as a particle.

ANALYSIS: To apply the principle of work and energy, you find the kinetic energy at each position of the car. The difference between the kinetic energies will be equal to the work done by the braking force.

Principle of Work and Energy.

$$T_1 + U_{1 \rightarrow 2} = T_2 \tag{1}$$

Therefore, you need to calculate each term in this equation.

Kinetic Energy.

Position 1.
$$v_1 = \left(60\,\frac{\text{mi}}{\text{h}}\right)\left(\frac{5280\,\text{ft}}{1\,\text{mi}}\right)\left(\frac{1\,\text{h}}{3600\,\text{s}}\right) = 88\,\text{ft/s}$$

$$T_1 = \tfrac{1}{2}mv_1^2 = \tfrac{1}{2}(4000/32.2)(88)^2 = 481{,}000\,\text{ft·lb}$$

Position 2. $\qquad v_2 = 0 \qquad T_2 = 0$

Work. The best way to identify which forces do work is to draw a free-body diagram, as shown in Fig. 2. It is clear that the only external forces that do work are the total braking force and the weight. The normal force does no work because it is perpendicular to the motion. Using the definition of work gives

$$U_{1 \rightarrow 2} = -1500x + (4000\sin 5°)x = -1151x$$

Note that the work of the gravitational force is positive since the automobile is moving down. Substituting into Eq. (1) gives

$$481{,}000 - 1151x = 0 \qquad x = 418\,\text{ft} \blacktriangleleft$$

Fig. 2 Free-body diagram for the car.

REFLECT and THINK: Solving this problem using Newton's second law would require determining the car's deceleration from the free-body diagram (Fig. 2) and then integrating this using the given velocity information. Using the principle of work and energy allows you to avoid that calculation.

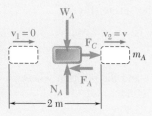

Sample Problem 13.2

Two blocks are joined by an inextensible cable as shown. If the system is released from rest, determine the velocity of block A after it has moved 2 m. Assume that the coefficient of kinetic friction between block A and the plane is $\mu_k = 0.25$ and that the pulley is weightless and frictionless.

STRATEGY: You are interested in determining the velocity and are given two locations in space, so use the principle of work and energy. You can apply this principle to each block and combine the resulting equations, or you can choose your system to be both blocks and the cable, thereby avoiding the need to determine the work of internal forces.

MODELING: Define two separate systems, one for each block, and model them as particles. As stated in the problem, assume the pulley is weightless and frictionless.

ANALYSIS:

Work and Energy for Block A. Denote the friction force by $\mathbf{F}_A$ and the force exerted by the cable by $\mathbf{F}_C$. Then you have (Fig. 1)

Fig. 1 Free-body diagram and two positions for block A.

$$m_A = 200 \text{ kg} \qquad W_A = (200 \text{ kg})(9.81 \text{ m/s}^2) = 1962 \text{ N}$$

$$F_A = \mu_k N_A = \mu_k W_A = 0.25(1962 \text{ N}) = 490 \text{ N}$$

$$T_1 + U_{1\to2} = T_2: \qquad 0 + F_C(2 \text{ m}) - F_A(2 \text{ m}) = \tfrac{1}{2}m_A v^2$$

$$F_C(2 \text{ m}) - (490 \text{ N})(2 \text{ m}) = \tfrac{1}{2}(200 \text{ kg})v^2 \tag{1}$$

Work and Energy for Block B. From the free-body diagram for Block B (Fig. 2), you have

$$m_B = 300 \text{ kg} \qquad W_B = (300 \text{ kg})(9.81 \text{ m/s}^2) = 2940 \text{ N}$$

$$T_1 + U_{1\to2} = T_2: \qquad 0 + W_B(2 \text{ m}) - F_C(2 \text{ m}) = \tfrac{1}{2}m_B v^2$$

$$(2940 \text{ N})(2 \text{ m}) - F_C(2 \text{ m}) = \tfrac{1}{2}(300 \text{ kg})v^2 \tag{2}$$

Now add the left-hand and right-hand sides of Eqs. (1) and (2). The work of the forces exerted by the cable on A and B cancels out. This is why when solving problems using work and energy, it is usually best to choose your system to include all the objects of interest, so you don't need to worry about the work of internal forces. Therefore, after combining Eqs. (1) and (2) or by choosing your system to be block A, block B, and the cable, you get

$$(2940 \text{ N})(2 \text{ m}) - (490 \text{ N})(2 \text{ m}) = \tfrac{1}{2}(200 \text{ kg} + 300 \text{ kg})v^2$$

$$4900 \text{ J} = \tfrac{1}{2}(500 \text{ kg})v^2 \qquad v = 4.43 \text{ m/s} \quad \blacktriangleleft$$

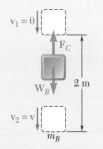

Fig. 2 Free-body diagram and two positions for block B.

REFLECT and THINK: When using the principle of work and energy, it usually saves time to choose your system to be everything that moves. Now that you know the velocity of the block, you could use Eq. (1) to determine the force in the cable. Only when you need to determine an internal force would you need to isolate part of a system.

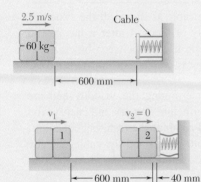

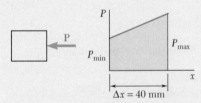

Fig. 1 The package at position 1 and position 2.

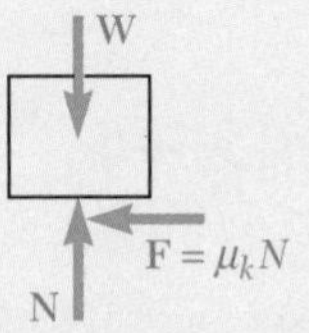

Fig. 2 Free-body diagram before spring is engaged.

Fig. 3 Force P on the block after it hits the spring.

Sample Problem 13.3

A spring is used to stop a 60-kg package that is sliding on a horizontal surface. The spring has a constant $k = 20$ kN/m and is held by cables so that it is initially compressed 120 mm. The package has a velocity of 2.5 m/s in the position shown, and the maximum additional deflection of the spring is 40 mm. Determine (a) the coefficient of kinetic friction between the package and the surface, (b) the velocity of the package as it passes again through the position shown.

STRATEGY: You have velocity information and specific locations in space, so use the principle of work and energy. Break the motion into two segments: segment 1 is the initial position to the point where the spring has a maximum deflection (Fig. 1), and segment 2 is from the point the spring has a maximum deflection back to the original position.

MODELING: The system is the crate, which you can model as a particle. A free-body diagram for the crate when it is not in contact with the spring is shown in Fig. 2. After it hits the spring, it has an additional force P acting on it due to the compression of the spring (Fig. 3).

ANALYSIS: The principle of work and energy is

$$T_1 + U_{1\rightarrow2} = T_2 \qquad (1)$$

Call the initial position of the package position 1 and the position where maximum spring deflection occurs position 2 (Fig. 1)

a. Motion from Position 1 to Position 2

Kinetic Energy. *Position 1.* $v_1 = 2.5$ m/s

$$T_1 = \tfrac{1}{2}mv_1^2 = \tfrac{1}{2}(60 \text{ kg})(2.5 \text{ m/s})^2 = 187.5 \text{ N·m} = 187.5 \text{ J}$$

Position 2. (maximum spring deflection): $v_2 = 0 \qquad T_2 = 0$

Work. *Friction Force* F. You have (Fig. 2)

$$F = \mu_k N = \mu_k W = \mu_k mg = \mu_k(60 \text{ kg})(9.81 \text{ m/s}^2) = (588.6 \text{ N})\mu_k$$

The work of F is negative and equal to

$$(U_{1\rightarrow2})_f = -Fx = -(588.6 \text{ N})\mu_k(0.600 \text{ m} + 0.040 \text{ m}) = -(377 \text{ J})\mu_k$$

Spring Force P. The variable force P exerted by the spring does an amount of negative work equal to the area under the force-deflection curve of the spring force. You have

$$P_{\text{min}} = kx_0 = (20 \text{ kN/m})(120 \text{ mm}) = (20\,000 \text{ N/m})(0.120 \text{ m}) = 2400 \text{ N}$$
$$P_{\text{max}} = P_{\text{min}} + k\,\Delta x = 2400 \text{ N} + (20 \text{ kN/m})(40 \text{ mm}) = 3200 \text{ N}$$
$$(U_{1\rightarrow2})_e = -\tfrac{1}{2}(P_{\text{min}} + P_{\text{max}})\,\Delta x = -\tfrac{1}{2}(2400 \text{ N} + 3200 \text{ N})(0.040 \text{ m}) = -112.0 \text{ J}$$

The total work between positions 1 and 2 is thus

$$U_{1\rightarrow2} = (U_{1\rightarrow2})_f + (U_{1\rightarrow2})_e = -(377 \text{ J})\mu_k - 112.0 \text{ J}$$

Principle of Work and Energy. You can determine the coefficient of kinetic friction from the expression for the principle of work and energy in this segment of the motion.

$$T_1 + U_{1\rightarrow2} = T_2: \qquad 187.5 \text{ J} - (377 \text{ J})\mu_k - 112.0 \text{ J} = 0 \qquad \mu_k = 0.20 \quad \blacktriangleleft$$

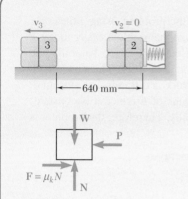

Fig. 4 Free-body diagram when the package is moving to the left.

b. Motion from Position 2 to Position 3. Call the position where the package returns to its initial position as position 3 (Fig. 4).

Kinetic Energy. *Position 2.*

$$v_2 = 0 \qquad T_2 = 0$$

Position 3. $\qquad T_3 = \frac{1}{2}mv_3^2 = \frac{1}{2}(60 \text{ kg})v_3^2$

Work. Since the distances involved are the same, the numerical values of the work of the friction force **F** and of the spring force **P** are the same as before. However, the work of **F** is still negative, whereas the work of **P** is now positive.

$$U_{2\to3} = -(377 \text{ J})\mu_k + 112.0 \text{ J} = -75.5 \text{ J} + 112.0 \text{ J} = +36.5 \text{ J}$$

Principle of Work and Energy.

$$T_2 + U_{2\to3} = T_3: \qquad 0 + 36.5 \text{ J} = \frac{1}{2}(60 \text{ kg})v_3^2$$
$$v_3 = 1.103 \text{ m/s} \qquad \mathbf{v}_3 = 1.103 \text{ m/s} \leftarrow \quad \blacktriangleleft$$

REFLECT and THINK: You needed to break this problem into two segments. From the first segment you were able to determine the coefficient of friction. Then you could use the principle of work and energy to determine the velocity of the package at any other location. Note that the system does not lose any energy due to the spring; it returns all of its energy back to the package. You would need to design something that could absorb the kinetic energy of the package in order to bring it to rest.

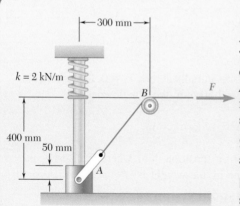

Sample Problem 13.4

The 2-kg collar *A* starts from rest in the position shown when a constant force $F = 100$ N is applied to the cable, causing the collar *A* to move up the smooth vertical shaft. Neglecting the mass of the frictionless pulley and the spring, determine the speed of *A* when the spring is compressed 50 mm.

STRATEGY: You have information about two positions and are asked to find a speed, so use the principle of work and energy.

MODELING: You have several choices of systems. Two possible systems are shown in Fig. 1.

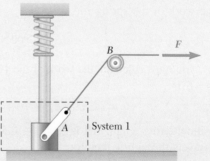 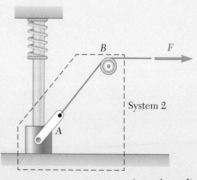

Fig. 1 Possible systems for this problem.

(continued)

Which one should you use? You can solve the problem using either one, but it turns out that some system choices make the problem easier to solve than others. For system 1, the tension in the rope is F, but only the component of F in the direction of the motion does work. This component is continually changing, so calculating the work is difficult. For system 2, the work the force F does is just the magnitude of F (since it is constant) times the distance the force travels horizontally. Therefore, the problem is easiest to solve using system 2.

ANALYSIS: The principle of work and energy is

$$T_1 + U_{1 \rightarrow 2} = T_2 \tag{1}$$

To start, draw the system in the two positions shown in Fig. 2. Since the figure will be very cluttered if you draw the two positions on the same figure, you should draw them side by side.

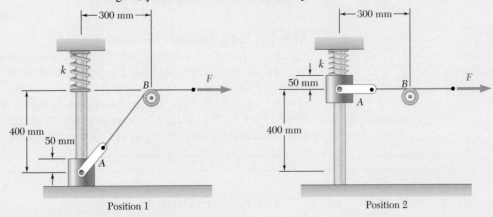

Position 1   Position 2

Fig. 2 System in the two positions of interest.

Kinetic Energy.   Since the collar is initially at rest, $T_1 = 0$. In position 2, when the upper spring is compressed 50 mm, the kinetic energy is

$$T_2 = \frac{1}{2}mv_2^2 = \frac{1}{2}(2 \text{ kg})v_2^2 = v_2^2$$

Work.   As the collar is raised from position 1 to where the spring is compressed 50 mm, the work done by the weight is

$$(U_{1 \rightarrow 2})_g = -mgy_2 = -(2 \text{ kg})(9.81 \text{ m/s}^2)(0.4 \text{ m}) = -7.848 \text{ J}$$

and the work of the spring force is

$$(U_{1 \rightarrow 2})_s = \frac{1}{2}kx_1^2 - \frac{1}{2}kx_2^2 = 0 - \frac{1}{2}(2000 \text{ N/m})(0.05 \text{ m})^2 = -2.50 \text{ J}$$

Finally, you must calculate the work of the 100-N force. In position 1, the length AB is

$$(l_{AB})_1 = \sqrt{(0.4)^2 + (0.3)^2} = 0.5 \text{ m}$$

In position 2, the length AB is $(l_{AB})_2 = 0.3$ m. The distance the 100-N force travels through is therefore

$$d = (l_{AB})_1 - (l_{AB})_2 = 0.5 \text{ m} - 0.3 \text{ m} = 0.2 \text{ m}$$

The work done by the 100-N force F is

$$(U_{1 \rightarrow 2})_F = Fd = (100 \text{ N})(0.5 \text{ m} - 0.3 \text{ m}) = 20 \text{ J}$$

Thus, the total work is

$$U_{1 \rightarrow 2} = (U_{1 \rightarrow 2})_g + (U_{1 \rightarrow 2})_s + (U_{1 \rightarrow 2})_F = -7.848 \text{ J} - 2.50 \text{ J} + 20 \text{ J} = 9.652 \text{ J}$$

Substituting these values in the principle of work and energy gives

$$T_1 + U_{1 \rightarrow 2} = T_2$$
$$0 + 9.652 = v_2^2$$

$$v_2 = 3.11 \text{ m/s} \quad \blacktriangleleft$$

REFLECT and THINK: What if the force had been only 10 N instead of 100 N? The work would have been a factor of 10 smaller (that is, 2 J), and you would have had $v_2^2 = -8.348$, which obviously makes no sense. What does this mean? It means the assumption that the mass will actually reach position 2 is incorrect.

Sample Problem 13.5

The 650-kg hammer of a drop-hammer pile driver falls onto the top of a 140-kg pile. After the impact, the hammer and the pile stick together and have a velocity of 3 m/s. The vertical force exerted on the pile by the ground after the impact is given by $F = 0.02x^2$, where x and F are expressed in mm and kN, respectively. Determine the velocity of the system after it has penetrated 80 mm into the ground.

STRATEGY: You are given a force as a function of displacement and are interested in two positions; therefore, use the principle of work and energy.

MODELING: The system is the hammer and the pile together after the impact. They can be modeled as a single particle. A free-body diagram for this system (Fig. 1) shows that the only two forces that do work are the weight and the force from the ground.

ANALYSIS: The principle of work and energy is

$$T_1 + U_{1 \rightarrow 2} = T_2 \tag{1}$$

Kinetic Energy. The two positions being considered are immediately after the impact and after the system has moved down 50 mm. Since the system is initially traveling at 3 m/s, the initial kinetic energy is

$$T_1 = \frac{1}{2}mv_1^2 = \frac{1}{2}(650 \text{ kg} + 140 \text{ kg})(3 \text{ m/s})^2 = 3555 \text{ J}$$

In position 2, the kinetic energy is

$$T_2 = \frac{1}{2}mv_2^2 = \frac{1}{2}(650 \text{ kg} + 140 \text{ kg})v_2^2 = 395v_2^2$$

Work. As the system moves into the ground, the weight and the resisting force, F, do work. The work the weight does is

$$(U_{1 \rightarrow 2})_g = mgy = (790 \text{ kg})(9.81 \text{ m/s}^2)(0.08 \text{ m}) = 620.0 \text{ J}$$

(continued)

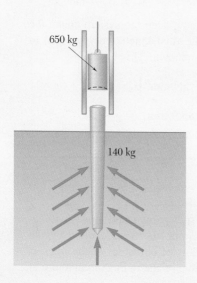
650 kg

140 kg

F(kN)
250
200
150
100
50
0
0   20   40   60   80   100 x(mm)

mg

F

Fig. 1 Free-body diagram after the impact.

The given equation for the force is such that F is in kN when x is expressed in mm. This means that the number in front (that is, the 0.02) has to have the units of kN/mm^2 for the units to work out. The work of the resisting force is

$$(U_{1\rightarrow2})_F = \int_{x_1}^{x_2} F_x\, dx$$

$$= \int_0^{0.05} -(0.02 \text{ kN/mm}^2)x^2\, dx = -\left(\frac{0.02}{3} \text{ kN/mm}^2\right)x^3\Big|_0^{80}$$

$$= -3413 \text{ kN·mm} = -3413 \text{ J}$$

Thus, the total work is

$$U_{1\rightarrow2} = (U_{1\rightarrow2})_g + (U_{1\rightarrow2})_F = 620.0 \text{ J} - 3413 \text{ J} = -2793 \text{ J}$$

Substituting the kinetic energies and total work in the principle of work and energy gives

$$T_1 + U_{1\rightarrow2} = T_2$$
$$3555 - 2793 = 395v_2^2$$

$$v_2 = 1.389 \text{ m/s} \downarrow \blacktriangleleft$$

REFLECT and THINK: To determine how deep the system enters the ground before it stops, you need to set the final kinetic energy equal to zero and make the maximum depth, x_m, unknown. This gives

$$3555 + 790(9.81)x_m - \left(\frac{0.02}{3} \text{ kN/mm}^2\right)x_m^3 = 0$$

Solving this, you find $x_m = 0.0859$ m or 85.9 mm.

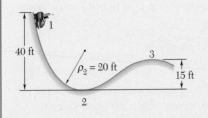

40 ft

1

$\rho_2 = 20$ ft

3

15 ft

2

Sample Problem 13.6

A 2000-lb roller coaster car starts from rest at point 1 and moves without friction down the track shown. (*a*) Determine the force exerted by the track on the car at point 2, where the radius of curvature of the track is 20 ft. (*b*) Determine the minimum safe value of the radius of curvature at point 3.

STRATEGY: Use the principle of work and energy to determine the speed of the car at any location along the track. To determine the force exerted by the track, you need to use Newton's second law. You will need to draw a free-body diagram and kinetic diagram of the car at each position.

MODELING: Choose the car as the system and assume it can be modeled as a particle.

ANALYSIS: Apply the principle of work and energy

$$T_1 + U_{1\rightarrow2} = T_2 \tag{1}$$

a. Force Exerted by the Track at Point 2. Use the principle of work and energy to determine the velocity of the car as it passes through point 2.

Kinetic Energy. $\quad T_1 = 0 \qquad T_2 = \frac{1}{2}mv_2^2 = \frac{1}{2}\frac{W}{g}v_2^2$

Work. The only force that does work is the weight **W**. Since the vertical displacement from point 1 to point 2 is 40 ft downward, the work of the weight is

$$U_{1 \to 2} = +W(40 \text{ ft})$$

Principle of Work and Energy. Substituting these values into Eq. (1) gives

$$T_1 + U_{1 \to 2} = T_2 \qquad 0 + W(40 \text{ ft}) = \frac{1}{2}\frac{W}{g}v_2^2$$

$$v_2^2 = 80g = 80(32.2) \qquad v_2 = 50.8 \text{ ft/s}$$

Newton's Second Law at Point 2. The acceleration $\mathbf{a}_n$ of the car at point 2 has a magnitude of $a_n = v_2^2/\rho$ and is directed upward. Since the external forces acting on the car are **W** and **N** (Fig. 1), you have

$$+\uparrow \Sigma F_n = ma_n: \qquad -W + N = ma_n$$

$$= \frac{W}{g}\frac{v_2^2}{\rho}$$

$$= \frac{W}{g}\frac{80g}{20}$$

$$N = 5W \qquad \mathbf{N} = 10{,}000 \text{ lb} \uparrow \quad \blacktriangleleft$$

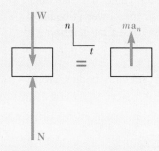

Fig. 1 Free-body diagram and kinetic diagram at point 2.

b. Minimum Value of ρ at Point 3.

Principle of Work and Energy. Applying the principle of work and energy between point 1 and point 3, you obtain

$$T_1 + U_{1 \to 3} = T_3 \qquad 0 + W(25 \text{ ft}) = \frac{1}{2}\frac{W}{g}v_3^2$$

$$v_3^2 = 50g = 50(32.2) \qquad v_3 = 40.1 \text{ ft/s}$$

Newton's Second Law at Point 3. The minimum safe value of ρ occurs when **N** = 0. In this case, the acceleration $\mathbf{a}_n$ with a magnitude of $a_n = v_3^2/\rho$, is directed downward (Fig. 2), and you have

$$+\downarrow \Sigma F_n = ma_n: \qquad W = \frac{W}{g}\frac{v_3^2}{\rho}$$

$$= \frac{W}{g}\frac{50g}{\rho} \qquad \rho = 50 \text{ ft} \quad \blacktriangleleft$$

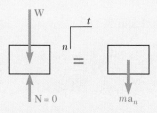

Fig. 2 Free-body diagram and kinetic diagram at point 3.

REFLECT and THINK: This is an example where you need both Newton's second law and the principle of work and energy. Work–energy is used to determine the speed of the car, and Newton's second law is used to determine the normal force. A normal force of 5W is equivalent to a fighter pilot pulling 5g's and should only be experienced for a very short time. For safety, you would also want to make sure your radius of curvature was quite a bit larger than 50 ft.

Sample Problem 13.7

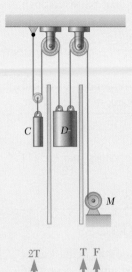

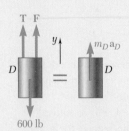

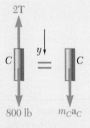

2T   T   F

v_C | C   D | v_D

800 lb   600 lb

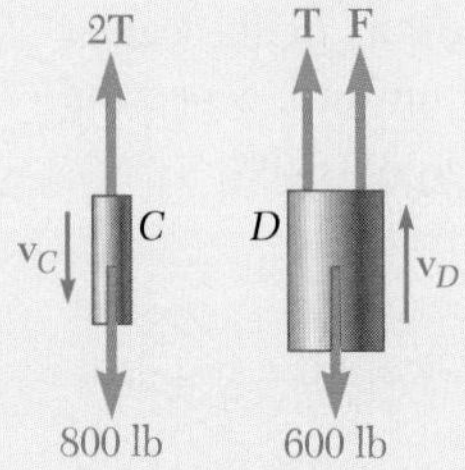

Fig. 1 Free-body diagrams for C and D.

2T

C = C

$y\downarrow$

800 lb $m_C a_C$

Fig. 2 Free-body diagram and kinetic diagram for C.

T   F

D = D

$y\uparrow$ $m_D a_D$

600 lb

Fig. 3 Free-body diagram and kinetic diagram for D.

The dumbwaiter D and its load have a combined weight of 600 lb, whereas the counterweight C weighs 800 lb. Determine the power delivered by the electric motor M when the dumbwaiter (a) is moving up at a constant speed of 8 ft/s, (b) has an instantaneous velocity of 8 ft/s and an acceleration of 2.5 ft/s^2, where both are directed upward.

STRATEGY: This problem requires you to use the definition of power. You will need to use Newton's second law to determine the tensions in the two cables.

MODELING: Define two separate systems, one for body C and one for body D, and model them as particles. Assume the pulley is weightless and frictionless.

ANALYSIS: The force **F** exerted by the motor cable has the same direction as the velocity $\mathbf{v}_D$ of the dumbwaiter, so the power is equal to Fv_D, where $v_D = 8$ ft/s. To obtain the power, you must first determine **F** in each of the two given situations.

a. Uniform Motion. You have $\mathbf{a}_C = \mathbf{a}_D = 0$; both bodies are in equilibrium (Fig. 1).

Body C: $+\uparrow \Sigma F_y = 0:$ $2T - 800 \text{ lb} = 0$ $T = 400 \text{ lb}$
Body D: $+\uparrow \Sigma F_y = 0:$ $F + T - 600 \text{ lb} = 0$

$$F = 600 \text{ lb} - T = 600 \text{ lb} - 400 \text{ lb} = 200 \text{ lb}$$
$$Fv_D = (200 \text{ lb})(8 \text{ ft/s}) = 1600 \text{ ft·lb/s}$$

$$\text{Power} = (1600 \text{ ft·lb/s})\frac{1 \text{ hp}}{550 \text{ ft·lb/s}} = 2.91 \text{ hp} \blacktriangleleft$$

b. Accelerated Motion. You have

$$\mathbf{a}_D = 2.5 \text{ ft/s}^2 \uparrow \qquad \mathbf{a}_C = -\frac{1}{2}\mathbf{a}_D = 1.25 \text{ ft/s}^2 \downarrow$$

The equations of motion are obtained using Figs 2 and 3.

Body C: $+\downarrow \Sigma F_y = m_C a_C:$ $800 - 2T = \dfrac{800}{32.2}(1.25)$ $T = 384.5 \text{ lb}$

Body D: $+\uparrow \Sigma F_y = m_D a_D:$ $F + T - 600 = \dfrac{600}{32.2}(2.5)$

$$F + 384.5 - 600 = 46.6 \qquad F = 262.1 \text{ lb}$$
$$Fv_D = (262.1 \text{ lb})(8 \text{ ft/s}) = 2097 \text{ ft·lb/s}$$

$$\text{Power} = (2097 \text{ ft·lb/s})\frac{1 \text{ hp}}{550 \text{ ft·lb/s}} = 3.81 \text{ hp} \blacktriangleleft$$

REFLECT and THINK: As you might expect, the motor needs to deliver more power to produce accelerated motion than to produce motion at constant velocity.

SOLVING PROBLEMS
ON YOUR OWN

In the preceding chapter, you solved problems dealing with the motion of a particle by using the fundamental equation $\mathbf{F} = m\mathbf{a}$ to determine the acceleration $\mathbf{a}$. By applying the principles of kinematics, you could then use $\mathbf{a}$ to determine the velocity and displacement of the particle at any time. In this section, we combined $\mathbf{F} = m\mathbf{a}$ and kinematic relationships to obtain an additional principle called the **principle of work and energy**. This eliminates the need to calculate the acceleration and enables you to relate the velocities of the particle at two points along its path of motion. To solve a problem using work and energy, you need to follow these steps:

1. Compute the work of each of the external forces. The work $U_{1 \rightarrow 2}$ of a given force $\mathbf{F}$ during the finite displacement of a particle from A_1 to A_2 is defined as

$$U_{1 \rightarrow 2} = \int \mathbf{F} \cdot d\mathbf{r} \quad \text{or} \quad U_{1 \rightarrow 2} = \int (F \cos \alpha)\, ds \qquad \textbf{(13.2, 13.2')}$$

where α is the angle between $\mathbf{F}$ and the displacement $d\mathbf{r}$. The work $U_{1 \rightarrow 2}$ is a scalar quantity and is expressed in ft·lb or in·lb in the U.S. customary system of units and in N·m or joules (J) in the SI system of units. Note that the work done is zero for a force perpendicular to the displacement ($\alpha = 90°$). Negative work is done for $90° < \alpha < 180°$ and in particular for a friction force, which is always opposite in direction to the displacement ($\alpha = 180°$).

The work $U_{1 \rightarrow 2}$ can be easily evaluated in the following cases that you will encounter.

    a. Work of a constant force in rectilinear motion [Sample Prob. 13.1]

$$U_{1 \rightarrow 2} = (F \cos \alpha)\, \Delta x \qquad \textbf{(13.3)}$$

    where α = angle the force forms with the direction of motion
          Δx = displacement from A_1 to A_2 (Fig. 13.3)

    b. Work of the force of gravity [Sample Probs. 13.2 and 13.6]

$$U_{1 \rightarrow 2} = -W \Delta y \qquad \textbf{(13.4')}$$

where Δy is the vertical displacement of the center of gravity of the body of weight W. Note that the work is positive when Δy is negative, that is, when the body moves down (Fig. 13.4).

    c. Work of the force exerted by a linear spring [Sample Probs. 13.3 and 13.4]

$$U_{1 \rightarrow 2} = \tfrac{1}{2}kx_1^2 - \tfrac{1}{2}kx_2^2 \qquad \textbf{(13.6)}$$

where k is the spring constant and x_1 and x_2 are the elongations of the spring corresponding to the positions A_1 and A_2 (Fig. 13.5).

d. Work of a gravitational force

$$U_{1 \to 2} = \frac{GMm}{r_2} - \frac{GMm}{r_1} \tag{13.7}$$

for a displacement of the body from $A_1(r = r_1)$ to $A_2(r = r_2)$ (Fig. 13.6).

2. **Calculate the kinetic energy at A_1 and A_2.** The kinetic energy T is

$$T = \tfrac{1}{2}mv^2 \tag{13.9}$$

where m is the mass of the particle and v is the magnitude of its velocity. The units of kinetic energy are the same as the units of work, that is, ft·lb or in·lb if you use U.S. customary units and N·m or joules (J) if you use SI units.

3. **Substitute the values for the work done $U_{1 \to 2}$ and the kinetic energies T_1 and T_2 into the equation**

$$T_1 + U_{1 \to 2} = T_2 \tag{13.11}$$

You will now have one scalar equation that you can solve for one unknown. Note that this equation does not yield the time of travel or the acceleration directly. However, if you know the radius of curvature ρ of the path of the particle at a point where you have obtained the velocity v, you can express the normal component of the acceleration as $a_n = v^2/\rho$ and obtain the normal component of the force exerted on the particle by using Newton's second law.

4. **We introduced power in this section as the time rate at which work is done as $P = dU/dt$.** Power is measured in ft·lb/s or *horsepower* (hp) in U.S. customary units and in J/s or *watts* (W) in the SI system of units. To calculate the power, you can use the equivalent formula

$$P = \mathbf{F} \cdot \mathbf{v} \tag{13.13}$$

where $\mathbf{F}$ and $\mathbf{v}$ denote the force and the velocity, respectively, at a given time [Sample Prob. 13.7]. In some problems [see, e.g., Prob. 13.47], you will be asked for the *average power* that you can obtain by dividing the total work by the time interval during which the work is done.

Problems

CONCEPT QUESTIONS

13.CQ1 Block A is traveling with a speed v_0 on a smooth surface when the surface suddenly becomes rough with a coefficient of friction of μ causing the block to stop after a distance d. If block A were traveling twice as fast, that is, at a speed $2v_0$, how far will it travel on the rough surface before stopping?

   a. $d/2$
   b. d
   c. $\sqrt{2}d$
   d. $2d$
   e. $4d$

Fig. P13.CQ1

END-OF-SECTION PROBLEMS

13.1 A 400-kg satellite is placed in a circular orbit 6394 km above the surface of the earth. At this elevation, the acceleration of gravity is 4.09 m/s². Knowing that its orbital speed is 20 000 km/h, determine the kinetic energy of the satellite.

13.2 A 1-lb stone is dropped down the "bottomless pit" at Carlsbad Caverns and strikes the ground with a speed of 95 ft/s. Neglecting air resistance, (*a*) determine the kinetic energy of the stone as it strikes the ground and the height h from which it was dropped. (*b*) Solve part *a* assuming that the same stone is dropped down a hole on the moon. (Acceleration of gravity on the moon = 5.31 ft/s².)

Fig. P13.2

13.3 A baseball player hits a 5.1-oz baseball with an initial velocity of 140 ft/s at an angle of 40° with the horizontal as shown. Determine (*a*) the kinetic energy of the ball immediately after it is hit, (*b*) the kinetic energy of the ball when it reaches its maximum height, (*c*) the maximum height above the ground reached by the ball.

13.4 A 500-kg communications satellite is in a circular geosynchronous orbit and completes one revolution about the earth in 23 h and 56 min at an altitude of 35 800 km above the surface of the earth. Knowing that the radius of the earth is 6370 km, determine the kinetic energy of the satellite.

13.5 In an ore-mixing operation, a bucket full of ore is suspended from a traveling crane which moves along a stationary bridge. The bucket is to swing no more than 10 ft horizontally when the crane is brought to a sudden stop. Determine the maximum allowable speed v of the crane.

13.6 In an ore-mixing operation, a bucket full of ore is suspended from a traveling crane which moves along a stationary bridge. The crane is traveling at a speed of 10 ft/s when it is brought to a sudden stop. Determine the maximum horizontal distance through which the bucket will swing.

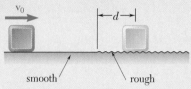

Fig. P13.3

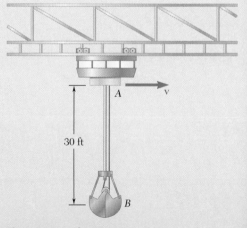

Fig. P13.5 and P13.6

13.7 Determine the maximum theoretical speed that may be achieved over a distance of 110 m by a car starting from rest assuming there is no slipping. The coefficient of static friction between the tires and pavement is 0.75, and 60 percent of the weight of the car is distributed over its front wheels and 40 percent over its rear wheels. Assume (a) front-wheel drive, (b) rear-wheel drive.

13.8 A 2000-kg automobile starts from rest at point *A* on a 6° incline and coasts through a distance of 150 m to point *B*. The brakes are then applied, causing the automobile to come to a stop at point *C*, which is 20 m from *B*. Knowing that slipping is impending during the braking period and neglecting air resistance and rolling resistance, determine (a) the speed of the automobile at point *B*, (b) the coefficient of static friction between the tires and the road.

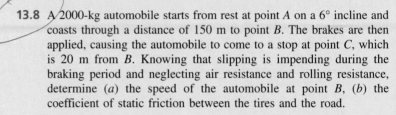

Fig. P13.8

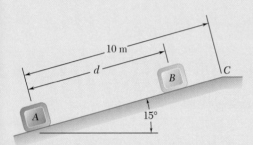

Fig. P13.9

13.9 A package is projected up a 15° incline at *A* with an initial velocity of 8 m/s. Knowing that the coefficient of kinetic friction between the package and the incline is 0.12, determine (a) the maximum distance *d* that the package will move up the incline, (b) the velocity of the package as it returns to its original position.

13.10 A 1.4-kg model rocket is launched vertically from rest with a constant thrust of 25 N until the rocket reaches an altitude of 15 m and the thrust ends. Neglecting air resistance, determine (a) the speed of the rocket when the thrust ends, (b) the maximum height reached by the rocket, (c) the speed of the rocket when it returns to the ground.

13.11 Packages are thrown down an incline at *A* with a velocity of 1 m/s. The packages slide along the surface *ABC* to a conveyor belt which moves with a velocity of 2 m/s. Knowing that $\mu_k = 0.25$ between the packages and the surface *ABC*, determine the distance *d* if the packages are to arrive at *C* with a velocity of 2 m/s.

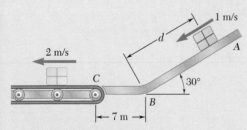

Fig. P13.11 and P13.12

13.12 Packages are thrown down an incline at *A* with a velocity of 1 m/s. The packages slide along the surface *ABC* to a conveyor belt which moves with a velocity of 2 m/s. Knowing that $d = 7.5$ m and $\mu_k = 0.25$ between the packages and all surfaces, determine (a) the speed of the package at *C*, (b) the distance a package will slide on the conveyor belt before it comes to rest relative to the belt.

13.13 Boxes are transported by a conveyor belt with a velocity $\mathbf{v}_0$ to a fixed incline at *A* where they slide and eventually fall off at *B*. Knowing that $\mu_k = 0.40$, determine the velocity of the conveyor belt if the boxes leave the incline at *B* with a velocity of 8 ft/s.

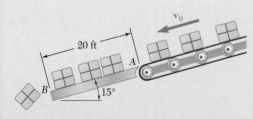

Fig. P13.13 and P13.14

13.14 Boxes are transported by a conveyor belt with a velocity $\mathbf{v}_0$ to a fixed incline at *A* where they slide and eventually fall off at *B*. Knowing that $\mu_k = 0.40$, determine the velocity of the conveyor belt if the boxes are to have zero velocity at *B*.

13.15 A 1200-kg trailer is hitched to a 1400-kg car. The car and trailer are traveling at 72 km/h when the driver applies the brakes on both the car and the trailer. Knowing that the braking forces exerted on the car and the trailer are 5000 N and 4000 N, respectively, determine (*a*) the distance traveled by the car and trailer before they come to a stop, (*b*) the horizontal component of the force exerted by the trailer hitch on the car.

Fig. P13.15

13.16 A trailer truck enters a 2 percent uphill grade traveling at 72 km/h and reaches a speed of 108 km/h in 300 m. The cab has a mass of 1800 kg and the trailer 5400 kg. Determine (*a*) the average force at the wheels of the cab, (*b*) the average force in the coupling between the cab and the trailer.

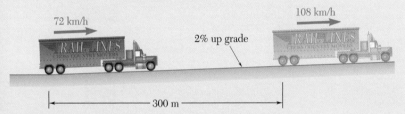

Fig. P13.16

13.17 The subway train shown is traveling at a speed of 30 mi/h when the brakes are fully applied on the wheels of cars *B* and *C*, causing them to slide on the track, but are not applied on the wheels of car *A*. Knowing that the coefficient of kinetic friction is 0.35 between the wheels and the track, determine (*a*) the distance required to bring the train to a stop, (*b*) the force in each coupling.

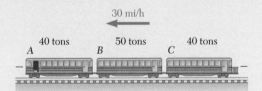

Fig. P13.17 and P13.18

13.18 The subway train shown is traveling at a speed of 30 mi/h when the brakes are fully applied on the wheels of car *A,* causing it to slide on the track, but are not applied on the wheels of cars *B* or *C*. Knowing that the coefficient of kinetic friction is 0.35 between the wheels and the track, determine (*a*) the distance required to bring the train to a stop, (*b*) the force in each coupling.

13.19 Blocks *A* and *B* weigh 25 lb and 10 lb, respectively, and they are both at a height 6 ft above the ground when the system is released from rest. Just before hitting the ground, block *A* is moving at a speed of 9 ft/s. Determine (*a*) the amount of energy dissipated in friction by the pulley, (*b*) the tension in each portion of the cord during the motion.

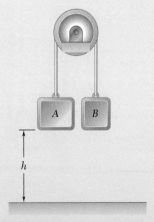

Fig. P13.19

13.20 The system shown is at rest when a constant 30-lb force is applied to collar B. (a) If the force acts through the entire motion, determine the speed of collar B as it strikes the support at C. (b) After what distance d should the 30-lb force be removed if the collar is to reach support C with zero velocity?

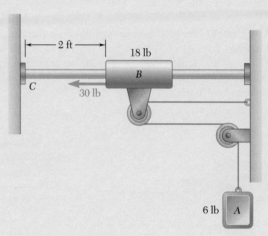

Fig. P13.20

13.21 Car B is towing car A at a constant speed of 10 m/s on an uphill grade when the brakes of car A are fully applied causing all four wheels to skid. The driver of car B does not change the throttle setting or change gears. The masses of the cars A and B are 1400 kg and 1200 kg, respectively, and the coefficient of kinetic friction is 0.8. Neglecting air resistance and rolling resistance, determine (a) the distance traveled by the cars before they come to a stop, (b) the tension in the cable.

Fig. P13.21

13.22 The system shown is at rest when a constant 250-N force is applied to block A. Neglecting the masses of the pulleys and the effect of friction in the pulleys and between block A and the horizontal surface, determine (a) the velocity of block B after block A has moved 2 m, (b) the tension in the cable.

13.23 The system shown is at rest when a constant 250-N force is applied to block A. Neglecting the masses of the pulleys and the effect of friction in the pulleys and assuming that the coefficients of friction between block A and the horizontal surface are $\mu_s = 0.25$ and $\mu_k = 0.20$, determine (a) the velocity of block B after block A has moved 2 m, (b) the tension in the cable.

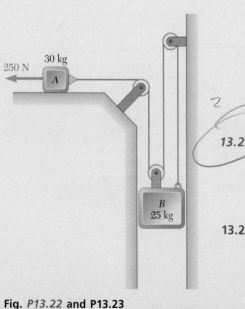

Fig. *P13.22* and P13.23

13.24 Two blocks A and B, of mass 4 kg and 5 kg, respectively, are connected by a cord that passes over pulleys as shown. A 3-kg collar C is placed on block A and the system is released from rest. After the blocks have moved 0.9 m, collar C is removed and blocks A and B continue to move. Determine the speed of block A just before it strikes the ground.

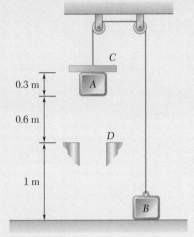

Fig. P13.24

13.25 Four 3-kg packages are held in place by friction on a conveyor which is disengaged from its drive motor. When the system is released from rest, package 1 leaves the belt at A just as package 4 comes onto the inclined portion of the belt at B. Determine (a) the velocity of package 2 as it leaves the belt at A, (b) the velocity of package 3 as it leaves the belt at A. Neglect the mass of the belt and rollers.

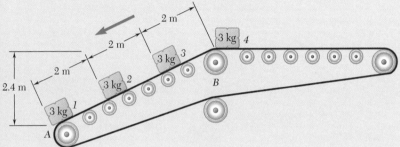

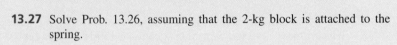

Fig. P13.25

13.26 A 3-kg block rests on top of a 2-kg block supported by, but not attached to, a spring of constant 40 N/m. The upper block is suddenly removed. Determine (a) the maximum speed reached by the 2-kg block, (b) the maximum height reached by the 2-kg block.

13.27 Solve Prob. 13.26, assuming that the 2-kg block is attached to the spring.

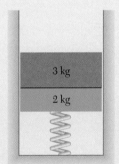

Fig. P13.26

13.28 People with mobility impairments can gain great health and social benefits from participating in different recreational activities. You are tasked with designing an adaptive spring-powered shuffleboard attachment that can be utilized by people who use wheelchairs. Knowing that the coefficient of kinetic friction between the 15 ounce puck A and the wooden surface is 0.3, the maximum spring displacement you desire is 6 inches, and that you want the puck to travel at least 30 ft/s, determine (a) the spring constant k, (b) how far the athlete should pull back the spring to make the puck come to rest after 34 ft.

Fig. P13.28

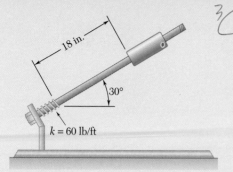

Fig. P13.29

13.29 A 7.5-lb collar is released from rest in the position shown, slides down the inclined rod, and compresses the spring. The direction of motion is reversed and the collar slides up the rod. Knowing that the maximum deflection of the spring is 5 in., determine (*a*) the coefficient of kinetic friction between the collar and the rod, (*b*) the maximum speed of the collar.

13.30 A 10-kg block is attached to spring *A* and connected to spring *B* by a cord and pulley. The block is held in the position shown with both springs unstretched when the support is removed and the block is released with no initial velocity. Knowing that the constant of each spring is 2 kN/m, determine (*a*) the velocity of the block after it has moved down 50 mm, (*b*) the maximum velocity achieved by the block.

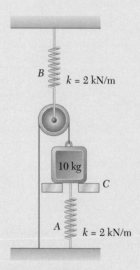

Fig. P13.30

13.31 A 5-kg collar *A* is at rest on top of, but not attached to, a spring with stiffness $k_1 = 400$ N/m when a constant 150-N force is applied to the cable. Knowing *A* has a speed of 1 m/s when the upper spring is compressed 75 mm, determine the spring stiffness k_2. Ignore friction and the mass of the pulley.

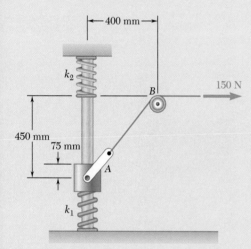

Fig. P13.31

13.32 A piston of mass *m* and cross-sectional area *A* is in equilibrium under the pressure *p* at the center of a cylinder closed at both ends. Assuming that the piston is moved to the left a distance *a*/2 and released, and knowing that the pressure on each side of the piston varies inversely with the volume, determine the velocity of the piston as it again reaches the center of the cylinder. Neglect friction between the piston and the cylinder and express your answer in terms of *m*, *a*, *p*, and *A*.

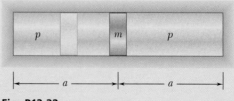

Fig. P13.32

13.33 An uncontrolled automobile traveling at 65 mph strikes squarely a highway crash cushion of the type shown in which the automobile is brought to rest by successively crushing steel barrels. The magnitude F of the force required to crush the barrels is shown as a function of the distance x the automobile has moved into the cushion. Knowing that the weight of the automobile is 2250 lb and neglecting the effect of friction, determine (a) the distance the automobile will move into the cushion before it comes to rest, (b) the maximum deceleration of the automobile.

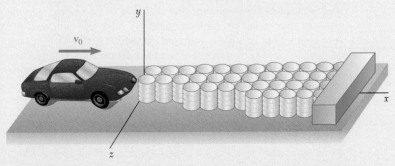

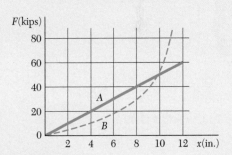

Fig. P13.33

13.34 Two types of energy-absorbing fenders designed to be used on a pier are statically loaded. The force-deflection curve for each type of fender is given in the graph. Determine the maximum deflection of each fender when a 90-ton ship moving at 1 mi/h strikes the fender and is brought to rest.

Fig. P13.34

13.35 Nonlinear springs are classified as hard or soft, depending upon the curvature of their force-deflection curve (see figure). If a delicate instrument having a mass of 5 kg is placed on a spring of length l so that its base is just touching the undeformed spring and then inadvertently released from that position, determine the maximum deflection x_m of the spring and the maximum force F_m exerted by the spring, assuming (a) a linear spring of constant $k = 3$ kN/m, (b) a hard, nonlinear spring, for which $F = (3$ kN/m$)(x + 160x^3)$.

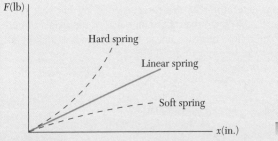

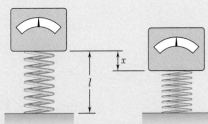

Fig. P13.35

13.36 A meteor starts from rest at a very great distance from the earth. Knowing that the radius of the earth is 6370 km and neglecting all forces except the gravitational attraction of the earth, determine the speed of the meteor (a) when it enters the ionosphere at an altitude of 1000 km, (b) when it enters the stratosphere at an altitude of 50 km, (c) when it strikes the earth's surface.

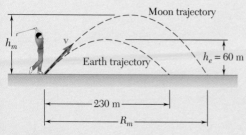

Fig. P13.38

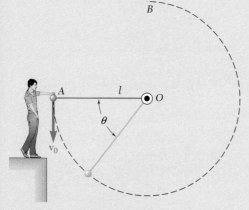

Fig. P13.39 and P13.40

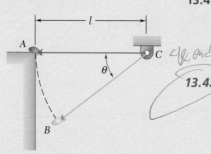

Fig. P13.41

13.37 Express the acceleration of gravity g_h at an altitude h above the surface of the earth in terms of the acceleration of gravity g_0 at the surface of the earth, the altitude h, and the radius R of the earth. Determine the percent error if the weight that an object has on the surface of the earth is used as its weight at an altitude of (*a*) 0.625 mi, (*b*) 625 mi.

13.38 A golf ball struck on earth rises to a maximum height of 60 m and hits the ground 230 m away. How high will the same golf ball travel on the moon if the magnitude and direction of its velocity are the same as they were on earth immediately after the ball was hit? Assume that the ball is hit and lands at the same elevation in both cases and that the effect of the atmosphere on the earth is neglected, so that the trajectory in both cases is a parabola. The acceleration of gravity on the moon is 0.165 times that on earth.

13.39 The sphere at A is given a downward velocity $\mathbf{v}_0$ of magnitude 5 m/s and swings in a vertical plane at the end of a rope of length $l = 2$ m attached to a support at O. Determine the angle θ at which the rope will break, knowing that it can withstand a maximum tension equal to twice the weight of the sphere.

13.40 The sphere at A is given a downward velocity $\mathbf{v}_0$ and swings in a vertical circle of radius l and center O. Determine the smallest velocity $\mathbf{v}_0$ for which the sphere will reach point B as it swings about point O (*a*) if AO is a rope, (*b*) if AO is a slender rod of negligible mass.

13.41 A bag is gently pushed off the top of a wall at A and swings in a vertical plane at the end of a rope of length l. Determine the angle θ for which the rope will break, knowing that it can withstand a maximum tension equal to twice the weight of the bag.

13.42 A roller coaster starts from rest at A, rolls down the track to B, describes a circular loop of 40-ft diameter, and moves up and down past point E. Knowing that $h = 60$ ft and assuming no energy loss due to friction, determine (*a*) the force exerted by his seat on a 160-lb rider at B and D, (*b*) the minimum value of the radius of curvature at E if the roller coaster is not to leave the track at that point.

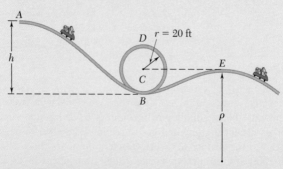

Fig. *P13.42*

13.43 In Prob. 13.42, determine the range of values of h for which the roller coaster will not leave the track at D or E, knowing that the radius of curvature at E is $\rho = 75$ ft. Assume no energy loss due to friction.

13.44 A small block slides at a speed v on a horizontal surface. Knowing that $h = 0.9$ m, determine the required speed of the block if it is to leave the cylindrical surface BCD when $\theta = 30°$.

13.45 A small block slides at a speed $v = 8$ ft/s on a horizontal surface at a height $h = 3$ ft above the ground. Determine (a) the angle θ at which it will leave the cylindrical surface BCD, (b) the distance x at which it will hit the ground. Neglect friction and air resistance.

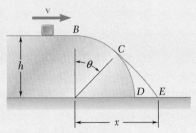

Fig. P13.44 and P13.45

13.46 A chair-lift is designed to transport 1000 skiers per hour from the base A to the summit B. The average mass of a skier is 70 kg and the average speed of the lift is 75 m/min. Determine (a) the average power required, (b) the required capacity of the motor if the mechanical efficiency is 85 percent and if a 300-percent overload is to be allowed.

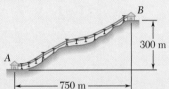

Fig. P13.46

13.47 It takes 15 s to raise a 1200-kg car and the supporting 300-kg hydraulic car-lift platform to a height of 2.8 m. Determine (a) the average output power delivered by the hydraulic pump to lift the system, (b) the average electric power required, knowing that the overall conversion efficiency from electric to mechanical power for the system is 82 percent.

Fig. P13.47

13.48 The velocity of the lift of Prob. 13.47 increases uniformly from zero to its maximum value at mid-height in 7.5 s and then decreases uniformly to zero in 7.5 s. Knowing that the peak power output of the hydraulic pump is 6 kW when the velocity is maximum, determine the maximum lift force provided by the pump.

13.49 (a) A 120-lb woman rides a 15-lb bicycle up a 3-percent slope at a constant speed of 5 ft/s. How much power must be developed by the woman? (b) A 180-lb man on an 18-lb bicycle starts down the same slope and maintains a constant speed of 20 ft/s by braking. How much power is dissipated by the brakes? Ignore air resistance and rolling resistance.

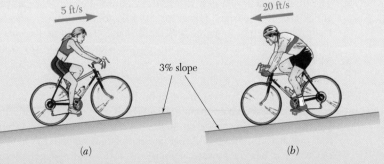

Fig. P13.49

13.50 A power specification formula is to be derived for electric motors which drive conveyor belts moving solid material at different rates to different heights and distances. Denoting the efficiency of a motor by η and neglecting the power needed to drive the belt itself, derive a formula (a) in the SI system of units for the power P in kW, in terms of the mass flow rate m in kg/h, the height b and horizontal distance l in meters and (b) in U.S. customary units, for the power in hp, in terms of the material flow rate w in tons/h, and the height b and horizontal distance l in feet.

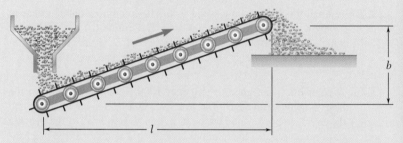

Fig. P13.50

13.51 A 1400-kg automobile starts from rest and travels 400 m during a performance test. The motion of the automobile is defined by the relation $x = 4000 \ln(\cosh 0.03t)$, where x and t are expressed in meters and seconds, respectively. The magnitude of the aerodynamic drag is $D = 0.35v^2$, where D and v are expressed in newtons and m/s, respectively. Determine the power dissipated by the aerodynamic drag when (a) $t = 10$ s, (b) $t = 15$ s.

Fig. P13.51 and P13.52

13.52 A 1400-kg automobile starts from rest and travels 400 m during a performance test. The motion of the automobile is defined by the relation $a = 3.6e^{-0.0005x}$, where a and x are expressed in m/s² and meters, respectively. The magnitude of the aerodynamic drag is $D = 0.35v^2$, where D and v are expressed in newtons and m/s, respectively. Determine the power dissipated by the aerodynamic drag when (a) $x = 200$ m, (b) $x = 400$ m.

13.53 The fluid transmission of a 15-Mg truck allows the engine to deliver an essentially constant power of 50 kW to the driving wheels. Determine the time required and the distance traveled as the speed of the truck is increased (a) from 36 km/h to 54 km/h, (b) from 54 km/h to 72 km/h.

13.54 The elevator E has a weight of 6600 lb when fully loaded and is connected as shown to a counterweight W of weight of 2200 lb. Determine the power in hp delivered by the motor (a) when the elevator is moving down at a constant speed of 1 ft/s, (b) when it has an upward velocity of 1 ft/s and a deceleration of 0.18 ft/s².

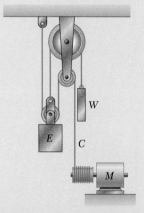

Fig. P13.54

13.2 CONSERVATION OF ENERGY

The principle of work and energy is useful for solving many different types of engineering problems. However, in many engineering applications, the total mechanical energy remains constant, although it may be transformed from one form into another. This is known as the principle of conservation of energy. To formulate this principle, we must first define a quantity known as potential energy. (Some of the material in this section was considered in Sec. 10.2B.)

13.2A Potential Energy

Let's consider again a body of weight **W** that moves along a curved path from a point A_1 of elevation y_1 to a point A_2 of elevation y_2 (Fig. 13.4). Recall from Sec. 13.1A that the work done by the force of gravity **W** during this displacement is

$$U_{1\to2} = -(Wy_2 - Wy_1) = Wy_1 - Wy_2 \qquad \textbf{(13.4)}$$

That is, we obtain the work done by **W** by subtracting the value of the function Wy corresponding to the second position of the body from its value corresponding to the first position. The work of **W** is independent of the actual path followed; it depends only upon the initial and final values of the function Wy. This function is called the **potential energy** of the body with respect to the **force of gravity W** and is denoted by V_g. We have

Gravitational potential energy on earth

$$U_{1\to2} = (V_g)_1 - (V_g)_2 \qquad \text{where } V_g = Wy \qquad \textbf{(13.16)}$$

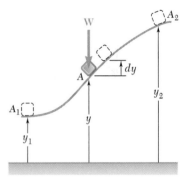

Fig. 13.4 *(repeated)*

where y is measured from an arbitrary horizontal datum where the potential energy is zero by definition. Note that if $(V_g)_2 > (V_g)_1$, that is, **if the potential energy increases** during the displacement (as in the case considered here), **the work $U_{1\to2}$ is negative.** On the other hand, if the work of **W** is positive, the potential energy decreases. Therefore, the potential energy V_g of the body provides a measure of the work that can be done by its weight **W**. Also note that the *change* in potential energy—not the actual value of V_g—is involved in formula (13.16). For this reason, the level, or datum, from which we measure the elevation y can be chosen arbitrarily. Finally, note that potential energy is expressed in the same units as work, i.e., in joules we use if SI units and in ft·lb or in·lb if we use U.S. customary units.

This expression for the potential energy of a body with respect to gravity is valid only as long as we can assume the weight **W** of the body remains constant, i.e., as long as the displacements of the body are small compared with the radius of the earth. In the case of a space vehicle, however, we need to take into consideration the variation of the force of gravity with the distance r from the center of the earth. Using the expression obtained in Sec. 13.1A for the work of a gravitational force, we have (Fig. 13.6)

$$U_{1\to2} = \frac{GMm}{r_2} - \frac{GMm}{r_1} \qquad \textbf{(13.7)}$$

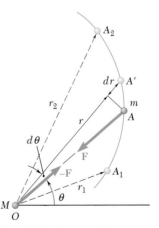

Fig. 13.6 *(repeated)*

Therefore, we can obtain the work of the force of gravity by subtracting the value of the function $-GMm/r$ corresponding to the second position of the body from its value corresponding to the first position. Thus, the expression that we use for the potential energy V_g when the variation in the force of gravity cannot be neglected is

Gravitational potential energy in space

$$V_g = -\frac{GMm}{r} \tag{13.17}$$

Taking the first of the relations of Eq. (12.27) into account, we can write V_g in the alternative form

$$V_g = -\frac{WR^2}{r} \tag{13.17'}$$

where R is the radius of the earth and W is the value of the weight of the body at the surface of the earth. When using either of the relations in Eqs. (13.17) or (13.17′) to express V_g, the distance r should, of course, be measured from the center of the earth.[†] Note that V_g is always negative and that it approaches zero for very large values of r.

Consider now a body attached to a spring and moving from a position A_1, corresponding to a deflection x_1 of the spring, to a position A_2, corresponding to a deflection x_2 of the spring (Fig. 13.5). Recall from Sec. 13.1A that the work of the force **F** exerted by the spring on the body is

$$U_{1 \to 2} = \tfrac{1}{2}kx_1^2 - \tfrac{1}{2}kx_2^2 \tag{13.6}$$

That is, we obtain the work of the elastic force by subtracting the value of the function $\tfrac{1}{2}kx^2$ corresponding to the second position of the body from its value corresponding to the first position. This function is denoted by V_e and is called the **potential energy** of the body with respect to the **elastic force F**. We have

Elastic potential energy

$$U_{1 \to 2} = (V_e)_1 - (V_e)_2 \quad \text{with } V_e = \tfrac{1}{2}kx^2 \tag{13.18}$$

where $x = L_{\text{stretched}} - L_{\text{unstretched}}$, or the deflection of the spring from its undeformed position. Note that, during the displacement from A_1 to A_2, the work of the force **F** exerted by the spring on the body is negative and that the potential energy V_e increases. We can use formula (13.18) even when the spring is rotated about its fixed end (Fig. 13.10a). The work of the elastic force depends only upon the initial and final deflections of the spring (Fig. 13.10b).

We can use the concept of potential energy when forces other than gravity forces and elastic forces are involved. Indeed, it remains valid as long as the work of the force considered is independent of the path followed by its point of application, as this point moves from a given position A_1 to a given position A_2. Such forces are said to be **conservative forces** or **path-independent forces**. We next consider their general properties.

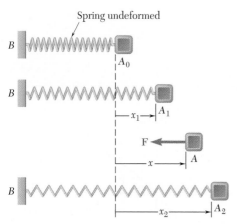

Spring undeformed

Fig. 13.5 (repeated)

[†]The expressions for V_g in Eqs. (13.17) and (13.17′) are valid only when $r \geq R$; that is, when the body considered is above the surface of the earth.

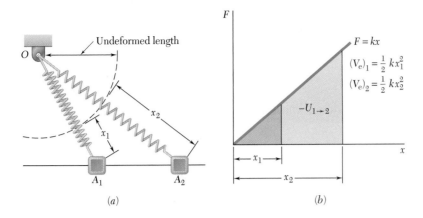

Fig. 13.10 (a) The equation for potential energy of a spring force is valid if the spring stretches when rotated about a fixed end; (b) the work of the elastic force depends only on the initial and final deflections of the spring.

*13.2B   Conservative Forces

As indicated in the preceding section, a force **F** acting on a particle A is said to be **conservative if its work** $U_{1\to2}$ **is independent of the path followed by the particle** A **as it moves from** A_1 **to** A_2 (Fig. 13.11a). We then have

$$U_{1\to2} = -(V(x_2, y_2, z_2) - V(x_1, y_1, z_1)) = V(x_1, y_1, z_1) - V(x_2, y_2, z_2) \quad \textbf{(13.19)}$$

or for short,

$$U_{1\to2} = V_1 - V_2 \quad \textbf{(13.19')}$$

The function $V(x, y, z)$ is called the potential energy, or **potential function**, of **F**.

Note that if A_2 is chosen to coincide with A_1—that is, if the particle describes a closed path (Fig. 13.11b)—we have $V_1 = V_2$ and the work is zero. Thus for any conservative force **F**, we can write

$$\oint \mathbf{F} \cdot d\mathbf{r} = 0 \quad \textbf{(13.20)}$$

where the circle on the integral sign indicates that the path is closed.

Let us now apply Eq. (13.19) between two neighboring points $A(x, y, z)$ and $A'(x + dx, y + dy, z + dz)$. The elementary work dU corresponding to the displacement $d\mathbf{r}$ from A to A' is

$$dU = V(x, y, z) - V(x + dx, y + dy, z + dz)$$

or

$$dU = -dV(x, y, z) \quad \textbf{(13.21)}$$

Thus, the elementary work of a conservative force is an **exact differential**.

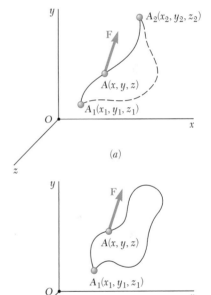

Fig. 13.11 (a) The work of a conservative force acting on a particle is independent of the path of the particle; (b) if the particle travels a closed path, the work of a conservative force is zero.

If we substitute the expression obtained in Eq. 13.1″ for dU in Eq. (13.21) and recall the definition of the differential of a function of several variables, we have

$$F_x \, dx + F_y \, dy + F_z \, dz = -\left(\frac{\partial V}{\partial x}dx + \frac{\partial V}{\partial y}dy + \frac{\partial V}{\partial z}dz\right)$$

from which it follows that

$$F_x = -\frac{\partial V}{\partial x} \qquad F_y = -\frac{\partial V}{\partial y} \qquad F_z = -\frac{\partial V}{\partial z} \qquad \textbf{(13.22)}$$

It is clear that the components of $\mathbf{F}$ must be functions of the coordinates x, y, and z. Thus, a necessary condition for a conservative force is that it depends only upon the position of its point of application. The relations in Eq. (13.22) can be expressed more concisely if we write

$$\mathbf{F} = F_x\mathbf{i} + F_y\mathbf{j} + F_z\mathbf{k} = -\left(\frac{\partial V}{\partial x}\mathbf{i} + \frac{\partial V}{\partial y}\mathbf{j} + \frac{\partial V}{\partial z}\mathbf{k}\right)$$

The vector in parentheses is known as the **gradient of the scalar function** V and is denoted by **grad** V. We thus have for any conservative force

$$\mathbf{F} = -\textbf{grad } V \qquad \textbf{(13.23)}$$

The relations in Eqs. (13.19) to (13.23) are satisfied by any conservative force. It can also be shown that if a force $\mathbf{F}$ satisfies one of these relations, $\mathbf{F}$ must be a conservative force.

13.2C   The Principle of Conservation of Energy

We saw in the preceding two sections that we can express the work of a conservative force, such as the weight of a particle or the force exerted by a spring, as a change in potential energy. When a particle moves under the action of conservative forces, the principle of work and energy stated in Sec. 13.B can be expressed in a modified form. Substituting for $U_{1\to2}$ from Eq. (13.19′) into Eq. (13.10), we have

$$V_1 - V_2 = T_2 - T_1$$

or

Conservation of energy
$$T_1 + V_1 = T_2 + V_2 \qquad \textbf{(13.24)}$$

Formula (13.24) indicates that when a particle moves under the action of conservative forces, **the sum of the kinetic energy and of the potential energy of the particle remains constant**. The sum $T + V$ is called the **total mechanical energy** of the particle and is denoted by E. So far, we have discussed two types of potential energy: gravitational potential energy, V_g, and elastic potential energy, V_e. Therefore, another way to write Eq. (13.24) is

$$T_1 + V_{g_1} + V_{e_1} = T_2 + V_{g_2} + V_{e_2} \qquad \textbf{(13.24′)}$$

Consider, for example, the pendulum analyzed in Sec. 13.1C that is released with no velocity from A_1 and allowed to swing in a vertical plane (Fig. 13.12). Measuring the potential energy from the level of A_2, that is, placing our datum at A_2, we have at A_1

$$T_1 = 0 \qquad V_1 = Wl \qquad T_1 + V_1 = Wl$$

Recalling that at A_2 the speed of the pendulum is $v_2 = \sqrt{2gl}$, we have

$$T_2 = \tfrac{1}{2}mv_2^2 = \frac{1}{2}\frac{W}{g}(2gl) = Wl \quad V_2 = 0$$

$$T_2 + V_2 = Wl$$

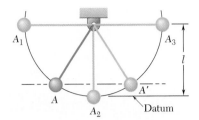

Fig. 13.12 The motion of a pendulum is easily analyzed using conservation of energy.

Thus, we can check that the total mechanical energy $E = T + V$ of the pendulum is the same at A_1 and A_2. Whereas the energy is entirely potential at A_1, it becomes entirely kinetic at A_2, and as the pendulum keeps swinging to the right past A_2, the kinetic energy is transformed back into potential energy. At A_3, $T_3 = 0$ and $V_3 = Wl$.

Because the total mechanical energy of the pendulum remains constant and its potential energy depends only upon its elevation, the kinetic energy of the pendulum must have the same value at any two points located at the same height. Thus, the speed of the pendulum is the same at A and at A' (Fig. 13.12). We can extend this result to the case of a particle moving along any given path, regardless of the shape of the path, as long as the only forces acting on the particle are its weight and the normal reaction of the path. The particle of Fig. 13.13, for example, which slides in a vertical plane along a frictionless track, has the same speed at A, A', and A''.

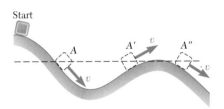

Fig. 13.13 A particle moving along a frictionless track has the same speed every time it passes the same elevation.

The weight of a particle and the force exerted by a spring are conservative forces, but **friction forces are nonconservative, or path-dependent, forces**. In other words, *the work of a friction force cannot be expressed as a change in potential energy*. The work of a friction force depends upon the path followed by its point of application; and whereas the work $U_{1\to2}$ defined by Eq. (13.19) is positive or negative according to the sense of motion, the work of a friction force, as we noted in Sec. 13.1C, is always negative. It follows that when a mechanical system involves friction, its total mechanical energy does not remain constant but decreases. The energy of the system, however, is not lost; it is transformed into heat, and the sum of the *mechanical energy* and of the *thermal energy* of the system remains constant.

Other forms of energy also can be involved in a system. For instance, a generator converts mechanical energy into *electrical energy*; a gasoline engine converts *chemical energy* into mechanical energy; a nuclear reactor converts *mass* into thermal energy. If all forms of energy are considered, the energy of any system can be considered as constant, and the principle of conservation of energy remains valid under all conditions.

If we express the work done by non-conservative forces as $U_{1\to2}^{NC}$, we can express Eq. (13.2) as

$$T_1 + V_{g_1} + V_{e_1} + U_{1\to2}^{NC} = T_2 + V_{g_2} + V_{e_2} \qquad \textbf{(13.24'')}$$

Note that if $U_{1\to2}^{NC}$ is zero, then the expression reduces to the conservation of energy equation of Eq. (13.24').

Photo 13.2 The potential energy of the roller coaster car is converted into kinetic energy as it descends the track.

Photo 13.3 Once in orbit, Earth satellites move under the action of gravity, which acts as a central force.

13.2D Application to Space Mechanics: Motion Under a Conservative Central Force

We saw in Sec. 12.2B that when a particle P moves under a central force $\mathbf{F}$, the angular momentum $\mathbf{H}_O$ of the particle about the center of force O is constant. If the force $\mathbf{F}$ is also conservative, there exists a potential energy V associated with $\mathbf{F}$, and the total energy $E = T + V$ of the particle is constant. Thus, when a particle moves under a conservative central force, we can use both the principle of conservation of angular momentum and the principle of conservation of energy to study its motion.

Consider, for example, a space vehicle of mass m moving under the earth's gravitational force. Let us assume that it begins its free flight at point P_0 at a distance r_0 from the center of the earth with a velocity $\mathbf{v}_0$ forming an angle ϕ_0 with the radius vector OP_0 (Fig. 13.14). Let P be a point of the trajectory described by the vehicle; we denote by r the distance from O to P, by $\mathbf{v}$ the velocity of the vehicle at P, and by ϕ the angle formed by $\mathbf{v}$ and the radius vector OP. Applying the principle of conservation of angular momentum about O between P_0 and P (Sec. 12.2B), we have

$$r_0 m v_0 \sin \phi_0 = r m v \sin \phi \qquad (13.25)$$

Recalling the expression in Eq. (13.17) for the potential energy due to a gravitational force, we apply the principle of conservation of energy between P_0 and P, obtaining

$$T_0 + V_0 = T + V$$
$$\tfrac{1}{2}mv_0^2 - \frac{GMm}{r_0} = \tfrac{1}{2}mv^2 - \frac{GMm}{r} \qquad (13.26)$$

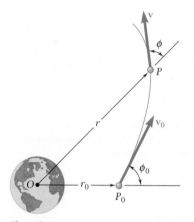

Fig. 13.14 A space vehicle moving from P_0 to P under the earth's gravitational force.

where M is the mass of the earth.

We can solve Eq. (13.26) for the magnitude v of the velocity of the vehicle at P when we know the distance r from O to P. Then we can use Eq. (13.25) to determine the angle ϕ that the velocity forms with the radius vector OP.

We can also use Eqs. (13.25) and (13.26) to determine the maximum and minimum values of r in the case of a satellite launched from P_0 in a direction forming an angle ϕ_0 with the vertical OP_0 (Fig. 13.15). We obtain the desired values of r by making $\phi = 90°$ in Eq. (13.25) and eliminating v between Eqs. (13.25) and (13.26).

Note that applying the principles of conservation of energy and of conservation of angular momentum leads to a more fundamental formulation of the problems of space mechanics than does the method indicated in Sec. 12.3B. It also results in much simpler computations in all cases involving oblique launchings. Although you must use the method of Sec. 12.3B when the actual trajectory or the periodic time of a space vehicle is to be determined, the calculations will be simplified if you first use the conservation principles to compute the maximum and minimum values of the radius vector r.

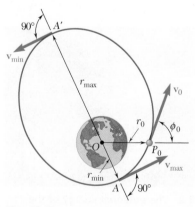

Fig. 13.15 A space vehicle launched from point P_0 into an orbit around the earth.

Sample Problem 13.8

A 20-lb collar slides without friction along a vertical rod as shown. The spring attached to the collar has an undeformed length of 4 in. and a spring constant of 3 lb/in. If the collar is released from rest in position 1, determine its velocity after it has moved 6 in. to position 2.

STRATEGY: You are given two positions and want to determine the velocity of the collar. No non-conservative forces are involved, so use the conservation of energy.

MODELING: For your system, choose the collar and the spring. You can treat the collar as a particle.

ANALYSIS: Conservation of Energy. Applying the principle of conservation of energy between positions 1 and 2 gives

$$T_1 + V_{g_1} + V_{e_1} = T_2 + V_{g_2} + V_{e_2} \tag{1}$$

You need to determine the kinetic and potential energy at these positions.

Position 1. *Potential Energy.* The elongation of the linear spring (Fig. 1) is

$$x_1 = 8 \text{ in.} - 4 \text{ in.} = 4 \text{ in.}$$

This gives

$$V_{e_1} = \tfrac{1}{2}kx_1^2 = \tfrac{1}{2}(3 \text{ lb/in.})(4 \text{ in.})^2 = 24 \text{ in·lb} = 2 \text{ ft·lb}$$

Choosing the datum as shown, you have $V_{g_1} = 0$.

Kinetic Energy. Since the velocity at position 1 is zero, $T_1 = 0$.

Position 2. *Potential Energy.* The elongation of the spring is

$$x_2 = 10 \text{ in.} - 4 \text{ in.} = 6 \text{ in.}$$

so you have

$$V_{e_2} = \tfrac{1}{2}kx_2^2 = \tfrac{1}{2}(3 \text{ lb/in.})(6 \text{ in.})^2 = 54 \text{ in·lb} = 4.5 \text{ ft·lb}$$
$$V_{g_2} = Wy_2 = (20 \text{ lb})(-6 \text{ in.}) = -120 \text{ in·lb} = -10 \text{ ft·lb}$$

Kinetic Energy.

$$T_2 = \tfrac{1}{2}mv_2^2 = \frac{1}{2}\frac{20}{32.2}v_2^2 = 0.311v_2^2$$

Conservation of Energy. Substituting into Eq. (1) gives

$$T_1 + V_{g_1} + V_{e_1} = T_2 + V_{g_2} + V_{e_2}$$
$$0 + 0 + 2 \text{ ft·lb} = 0.311v_2^2 + (-10 \text{ ft·lb}) + (4.5 \text{ ft·lb})$$
$$v_2 = \pm 4.91 \text{ ft/s}$$

$$\mathbf{v_2} = 4.91 \text{ ft/s} \downarrow \ \blacktriangleleft$$

REFLECT and THINK: If you had not included the spring in your system, you would have needed to treat it as an external force; therefore, you would have needed to determine the work. Similarly, if there was friction acting on the collar, you would have needed to use the more general work–energy principle to solve this problem. It turns out that the work done by friction is not very easy to calculate because the normal force depends on the spring force.

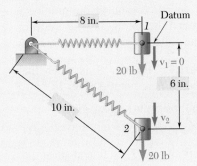

Fig. 1 The system in position 1 and position 2.

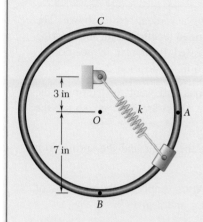

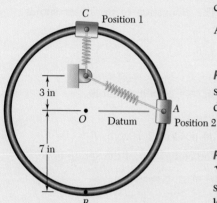

Fig. 1 The system in the two positions of interest.

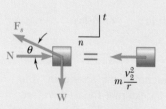

Fig. 2 Free-body diagram and kinetic diagram for the collar at point A.

Sample Problem 13.9

A 2.5-lb collar is attached to a spring and slides along a smooth circular rod in a vertical plane. The spring has an undeformed length of 4 in. and a spring constant k. The collar is at rest at point C and is given a slight push to the right. Knowing that the maximum velocity of the collar is achieved as it passes through point A, determine (a) the spring constant k, (b) the force exerted by the rod on the collar at point A.

STRATEGY: Since you have two positions and are given information about the speed, use the conservation of energy. To find the force, you need to use Newton's second law.

MODELING: For the conservation of energy portion of the problem, model the collar as a particle and use it and the spring as your system. When using Newton's second law, use the collar as your system.

ANALYSIS: Conservation of Energy. Position 1 is when the collar is at point C, and position 2 is when it is at point A (Fig. 1).

Applying conservation of energy between positions 1 and 2 gives

$$T_1 + V_{g_1} + V_{e_1} = T_2 + V_{g_2} + V_{e_2} \tag{1}$$

Position 1. Because the system starts from rest, $T_1 = 0$, and since the spring has an unstretched length of 4 in., you know $V_{e_1} = 0$. Putting the datum at A gives.

$$V_{g_1} = (2.5 \text{ lb})(7/12 \text{ ft}) = 1.4583 \text{ ft·lb}$$

Position 2. From geometry, the distance from the pin to A is $\sqrt{(3 \text{ in.})^2 + (7 \text{ in.})^2} = 7.616$ in. Therefore, the elongation of the linear spring (Fig. 1) is $x_2 = 7.616$ in. $-$ 4 in. $=$ 3.616 in. $=$ 0.3013 ft. You know $V_{g_2} = 0$ because the datum is at position 2. You also know

$$V_{e_2} = \tfrac{1}{2}kx_2^2 = \frac{1}{2}k(0.3013 \text{ ft})^2 = 0.04539k$$

$$T_2 = \tfrac{1}{2}mv_2^2 = \frac{1}{2}\left(\frac{2.5 \text{ lb}}{32.2 \text{ ft/s}^2}\right)v_2^2 = 0.03882v_2^2$$

Substituting these expressions into Eq. (1) gives

$$0 + 1.4585 + 0 = 0.03882v_2^2 + 0 + 0.04539k \tag{2}$$

You have two unknowns in this equation, so you need another equation. In the problem statement, you are also given that the collar has a maximum velocity at point A. Therefore, the tangential acceleration must be zero at A, and you should use Newton's second law to get additional equations. The system now includes only the collar; the spring applies an external force to the system. A free-body diagram and kinetic diagram for the collar at position 2 are shown in Fig. 2. Applying Newton's second law in the t-direction gives

$$+\uparrow\Sigma F_t = 0 = kx_2 \sin\theta - W \quad \text{or} \quad k(0.3013 \text{ ft})(3/7.652) - 2 \text{ lb} = 0$$

Solving for k,

$$k = 21.06 \text{ lb/ft} \blacktriangleleft$$

Force Exerted by the Rod: Substituting this value of k into Eq. (2) gives $v_2 = 3.597$ ft/s. Applying Newton's second law in the n-direction gives

$$\xleftarrow{+} \Sigma F_n = m\frac{v_2^2}{r} \qquad kx_2 \cos\theta - N = \frac{mv_2^2}{r}$$

Solving for N and substituting in values provides

$$N = kx_2 \cos\theta - \frac{mv_2^2}{r}$$
$$= (21.06 \text{ lb/ft})(0.3013 \text{ ft})(7/7.616)$$
$$- \frac{(2.5 \text{ lb/}32.2 \text{ ft/s}^2)(3.597 \text{ ft/s}^2)}{(7/12 \text{ ft})}$$

$$N = 4.11 \text{ lb} \quad \blacktriangleleft$$

REFLECT and THINK: When the collar is pushed to the right, its speed increases until it reaches point A, and then it begins to decrease. The minimum speed occurs when the collar is at B, since the only forces are in the normal direction; that is, no forces act in the tangential direction. Therefore, the acceleration in the tangential direction is zero, indicating a minimum speed.

Sample Problem 13.10

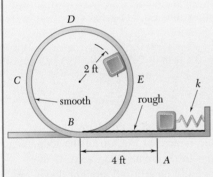

A 0.5-lb pellet is pushed against the spring at A and released from rest. It moves 4 ft along a rough horizontal surface until it reaches a smooth loop. The coefficient of kinetic friction along the rough horizontal surface is $\mu_k = 0.3$, and the spring is initially compressed 0.25 ft. Determine the minimum spring constant k for which the pellet will travel around $BCDE$ and always remain in contact with the loop.

STRATEGY: You are given two positions and a non-conservative force is present, so use the work-energy principle. Also, for the pellet to remain in contact with the loop, the force $\mathbf{N}$ exerted on the pellet by the loop must be equal to or greater than zero. Therefore, you also need to use Newton's second law.

MODELING: Choose the pellet as your system and model it as a particle. A free-body diagram and kinetic diagram for the pellet when it is at point D are shown in Fig. 1.

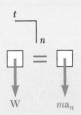

Fig. 1 Free-body diagram and kinetic diagram for the pellet at point D.

ANALYSIS:

Newton's Second Law. Applying Newton's second law in the normal direction and setting $N = 0$ gives you

$$+\downarrow\Sigma F_n = ma_n: \qquad W = ma_n \qquad mg = ma_n \qquad a_n = g$$
$$a_n = \frac{v_D^2}{r}: \qquad v_D^2 = ra_n = rg = (2 \text{ ft})(32.2 \text{ ft/s}^2) = 64.4 \text{ ft}^2/\text{s}^2$$

This is the minimum speed of the pellet at D in order for it to remain in contact with the path.

Work–Energy. Choose the system to be the pellet and the spring. Apply the principle of work–energy between positions 1 and 2 (Fig. 2)

$$T_1 + V_{g_1} + V_{e_1} + U^{NC}_{1\to2} = T_2 + V_{g_2} + V_{e_2} \tag{1}$$

You need to determine the kinetic and potential energy at positions 1 and 2 and the work done by friction.

Position 1. *Potential Energy.* The elastic potential energy is

$$V_{e_1} = \tfrac{1}{2}kx^2 = \tfrac{1}{2}(k)(0.25\text{ ft})^2 = 0.03125k$$

Choosing the datum at A, you have $V_{g_1} = 0$.

Kinetic Energy. Since the pellet is released from rest, $v_A = 0$ and $T_1 = 0$.

Position 2. *Potential Energy.* The spring is now undeformed; thus $V_{e_2} = 0$. Since the pellet is 4 ft above the datum, you have

$$V_{g_2} = Wy_2 = (0.5\text{ lb})(4\text{ ft}) = 2\text{ ft·lb}$$

Kinetic Energy. Using the value of v_D^2 obtained above, you have

$$T_2 = \tfrac{1}{2}mv_D^2 = \frac{1}{2}\frac{0.5\text{ lb}}{32.2\text{ ft/s}^2}(64.4\text{ ft}^2/\text{s}^2) = 0.5\text{ ft·lb}$$

Work. Since the normal force is equal to the weight on a horizontal surface, you find the work that friction does to be

$$U^{NC}_{1\to2} = -\mu_k Nd = -0.3(0.5\text{ lb})(4\text{ ft}) = -0.6\text{ ft·lb}$$

Work–Energy. Substituting these values into Eq. (1) gives you

$$T_1 + V_{g_1} + V_{e_1} + U^{NC}_{1\to2} = T_2 + V_{g_2} + V_{e_2}$$
$$0 + 0 + 0.3125k - 0.6\text{ ft·lb} = 0.5\text{ ft·lb} + 2\text{ ft·lb} + 0$$

You can solve this for k.

$$k = 99.2\text{ lb/ft} \blacktriangleleft$$

REFLECT and THINK: A common misconception in problems like this is assuming that the speed of the particle is zero at the top of the loop, rather than that the normal force is equal to or greater than zero. If the pellet had a speed of zero at the top, it would clearly fall straight down, which is impossible.

Fig. 2 The system at the positions of interest.

Sample Problem 13.11

A sphere of mass $m = 0.6$ kg is attached to an elastic cord of constant $k = 100$ N/m, which is undeformed when the sphere is located at the origin O. The sphere may slide without friction on the horizontal surface and in the position shown its velocity $\mathbf{v}_A$ has a magnitude of 20 m/s. Determine (*a*) the maximum and minimum distances from the sphere to the origin O, (*b*) the corresponding values of its speed.

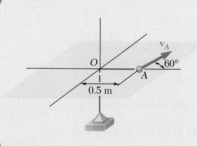

STRATEGY: The force exerted by the cord on the sphere passes through the fixed point O, so use conservation of angular momentum. Also, you are interested in the speed at two locations, and no non-conservative forces act on the sphere. You can therefore use conservation of energy.

MODELING: Choose the sphere, which can be modeled as a particle, as your system.

ANALYSIS:

Conservation of Angular Momentum About O. At point B, where the distance from O is maximum (Fig. 1), the velocity of the sphere is perpendicular to OB and the angular momentum is $r_m m v_m$. A similar property holds at point C, where the distance from O is minimum. Expressing conservation of angular momentum between A and B, you have

$$r_A m v_A \sin 60° = r_m m v_m$$
$$(0.5 \text{ m})(0.6 \text{ kg})(20 \text{ m/s}) \sin 60° = r_m(0.6 \text{ kg})v_m$$
$$v_m = \frac{8.66}{r_m} \qquad (1)$$

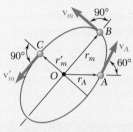

Fig. 1 The particle at locations A, B, and C.

You have one equation and two unknowns, v_m and r_m. Therefore, you need to use conservation of energy to get a second equation.

Conservation of Energy.

At Point A. $\qquad T_A = \frac{1}{2}mv_A^2 = \frac{1}{2}(0.6 \text{ kg})(20 \text{ m/s})^2 = 120 \text{ J}$
$\qquad\qquad\qquad V_A = \frac{1}{2}kr_A^2 \ = \frac{1}{2}(100 \text{ N/m})(0.5 \text{ m})^2 = 12.5 \text{ J}$

At Point B. $\qquad T_B = \frac{1}{2}mv_m^2 = \frac{1}{2}(0.6 \text{ kg})v_m^2 = 0.3v_m^2$
$\qquad\qquad\qquad V_B = \frac{1}{2}kr_m^2 \ = \frac{1}{2}(100 \text{ N/m})r_m^2 = 50r_m^2$

Apply the principle of conservation of energy between points A and B:

$$T_A + V_A = T_B + V_B$$
$$120 + 12.5 = 0.3v_m^2 + 50r_m^2 \qquad (2)$$

a. Maximum and Minimum Values of Distance. Substituting for v_m from Eq. (1) into Eq. (2) and solving for r_m^2, you obtain

$$r_m^2 = 2.468 \text{ or } 0.1824 \qquad r_m = 1.571 \text{ m}, r_m' = 0.427 \text{ m} \quad \blacktriangleleft$$

b. Corresponding Values of Speed. Substituting the values obtained for r_m and r_m' into Eq. (1), you have

$$v_m = \frac{8.66}{1.571} \qquad\qquad v_m = 5.51 \text{ m/s} \quad \blacktriangleleft$$
$$v_m' = \frac{8.66}{0.427} \qquad\qquad v_m' = 20.3 \text{ m/s} \quad \blacktriangleleft$$

REFLECT and THINK: This problem is similar to problems dealing with space mechanics; instead of the gravitational central force acting on an orbiting body, you have the spring force acting on the sphere. It can be shown that the path of the sphere is an ellipse with center O.

Sample Problem 13.12

A satellite is launched from an altitude of 500 km in a direction parallel to the surface of the earth with a velocity of 36 900 km/h. Determine (a) the maximum altitude reached by the satellite, (b) the maximum allowable error in the direction of launching if the satellite is to go into orbit and come no closer than 200 km to the surface of the earth.

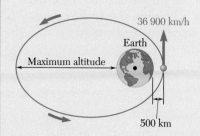

STRATEGY: Since the only force acting on the satellite is the force of gravity, which is a central force, and you are interested in two positions (the position of the satellite at launch and at its maximum altitude), you can use conservation of angular momentum and conservation of energy.

MODELING: Choose the satellite as your system and model it as a particle.

ANALYSIS:

a. Maximum Altitude. Denote the point of the orbit farthest from the earth by A' and the corresponding distance from the center of the earth by r_1 (Fig. 1). Since the satellite is in free flight between A and A', you can apply the principle of conservation of energy as

$$T_A + V_A = T_{A'} + V_{A'}$$

$$\tfrac{1}{2}mv_0^2 - \frac{GMm}{r_0} = \tfrac{1}{2}mv_1^2 - \frac{GMm}{r_1} \tag{1}$$

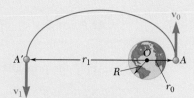

Fig. 1 The system in the two positions of interest.

Now apply the principle of conservation of angular momentum of the satellite about O. Considering points A and A', you have

$$r_0mv_0 = r_1mv_1 \qquad v_1 = v_0\frac{r_0}{r_1} \tag{2}$$

Substitute this expression for v_1 into Eq. (1), divide each term by the mass m, and rearrange the terms. The result is

$$\tfrac{1}{2}v_0^2\left(1 - \frac{r_0^2}{r_1^2}\right) = \frac{GM}{r_0}\left(1 - \frac{r_0}{r_1}\right) \qquad 1 + \frac{r_0}{r_1} = \frac{2GM}{r_0v_0^2} \tag{3}$$

Recall that the radius of the earth is $R = 6370$ km. This gives you

$r_0 = 6370$ km $+ 500$ km $= 6870$ km $= 6.87 \times 10^6$ m
$v_0 = 36\,900$ km/h $= (36.9 \times 10^6$ m$)/(3.6 \times 10^3$ s$) = 10.25 \times 10^3$ m/s
$GM = gR^2 = (9.81$ m/s$^2)(6.37 \times 10^6$ m$)^2 = 398 \times 10^{12}$ m^3/s^2

Substituting these values into Eq. (3), you obtain $r_1 = 66.8 \times 10^6$ m.

Maximum altitude $= 66.8 \times 10^6$ m $- 6.37 \times 10^6$ m $= 60.4 \times 10^6$ m $=$

60 400 km ◀

b. Allowable Error in Direction of Launch.

The satellite is launched from P_0 in a direction forming an angle ϕ_0 with the vertical OP_0 (Fig. 2). You obtain the value of ϕ_0 corresponding to $r_{min} = 6370$ km $+$ 200 km $= 6570$ km by applying the principles of conservation of energy and of conservation of angular momentum between P_0 and A:

$$\tfrac{1}{2}mv_0^2 - \frac{GMm}{r_0} = \tfrac{1}{2}mv_{max}^2 - \frac{GMm}{r_{min}} \tag{4}$$

$$r_0 mv_0 \sin \phi_0 = r_{min} mv_{max} \tag{5}$$

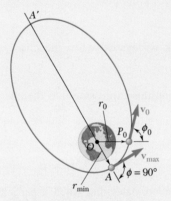

Fig. 2 Two locations used to determine maximum allowable error in direction.

Solving (5) for v_{max} and then substituting for v_{max} into (4), you can solve (4) for $\sin \phi_0$. Finally, using the values of v_0 and GM computed in part a and noting that $r_0/r_{min} = 6870/6570 = 1.0457$, you find

$\sin \phi_0 = 0.9801$ $\phi_0 = 90° \pm 11.5°$ Allowable error $= \pm 11.5°$ ◀

REFLECT and THINK: Space probes and other long-distance vehicles are designed with small rockets to allow for mid-course corrections. Satellites launched from the Space Station usually do not need this kind of fine-tuning.

SOLVING PROBLEMS
ON YOUR OWN

In this section you learned that when the work done by a force $\mathbf{F}$ acting on a particle A is independent of the path followed by the particle as it moves from a given position A_1 to a given position A_2 (Fig. 13.11*a*), then we can define a function V, called **potential energy**, for the force $\mathbf{F}$. Such forces are said to be **conservative forces**, and you can write

$$U_{1\to2} = -(V(x_2, y_2, z_2) - V(x_1, y_1, z_1)) = V(x_1, y_1, z_1) - V(x_2, y_2, z_2) \quad \textbf{(13.19)}$$

or for short,

$$U_{1\to2} = V_1 - V_2 \quad \textbf{(13.19′)}$$

The work is negative when the change in potential energy is positive, i.e., when $V_2 > V_1$.

Substituting this expression into the equation for work and energy, you can write

$$T_1 + V_1 = T_2 + V_2 \quad \textbf{(13.24)}$$

or

$$T_1 + V_{g_1} + V_{e_1} = T_2 + V_{g_2} + V_{e_2} \quad \textbf{(13.24′)}$$

This equation states that when a particle moves under the action of a conservative force, **the sum of the kinetic and potential energies of the particle remains constant**. We expanded this equation for cases when there are non-conservative forces present:

$$T_1 + V_{g_1} + V_{e_1} + U^{\text{NC}}_{1\to2} = T_2 + V_{g_2} + V_{e_2} \quad \textbf{(13.24″)}$$

Your solutions of problems using the above formulas will consist of the following steps.

1. Determine whether all the forces involved are conservative. If some of the forces are not conservative—for example, if friction is involved—you must use the second equation (13.24″), since the work done by such forces depends upon the path followed by the particle and a potential function does not exist for these non-conservative forces. You can then determine the work done by non-conservative forces as:

$$U^{\text{NC}}_{1\to2} = \int_1^2 \mathbf{F}^{\text{NC}} \cdot d\mathbf{s}$$

2. Determine the kinetic energy $T = \frac{1}{2}mv^2$ **at each end of the path.**

3. Compute the potential energy for all the forces involved at each end of the path. Recall the following expressions for potential energy derived in this section.

a. The potential energy of a weight W close to the surface of Earth and at a height y above a given datum:

$$V_g = Wy \tag{13.16}$$

b. The potential energy of a mass m located at a distance r from the center of the earth, large enough so that the variation of the force of gravity must be taken into account:

$$V_g = -\frac{GMm}{r} \tag{13.17}$$

where the distance r is measured from the center of the earth and V_g is equal to zero at $r = \infty$.

c. The potential energy of a body with respect to an elastic force $F = kx$:

$$V_e = \tfrac{1}{2}kx^2 \tag{13.18}$$

where the distance x is the deflection of the elastic spring measured from its *undeformed* position and k is the spring constant. Note that V_e depends only upon the deflection x and not upon the path of the body attached to the spring. Also, V_e is always positive, whether the spring is compressed or elongated.

4. Substitute your expressions for the non-conservative work and the kinetic and potential energies into Eq. (13.24″). You will be able to solve this equation for one unknown—for example, for a velocity [Sample Prob. 13.8]. If more than one unknown is involved, you will have to search for another condition or equation, such as Newton's second law [Sample Prob. 13.10], the maximum speed [Sample Prob. 13.9], minimum speed [Sample Prob. 13.10], or the minimum potential energy of the particle. For problems involving a central force, you can obtain a second equation by using conservation of angular momentum [Sample Prob. 13.11]. This is especially useful in space mechanics applications [Sec. 13.2D].

Problems

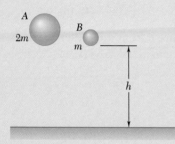

Fig. P13.CQ2

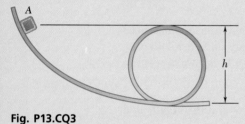

Fig. P13.CQ3

13.CQ2 Two small balls A and B with masses $2m$ and m, respectively, are released from rest at a height h above the ground. Neglecting air resistance, which of the following statements is true when the two balls hit the ground?

a. The kinetic energy of A is the same as the kinetic energy of B.
b. The kinetic energy of A is half the kinetic energy of B.
c. The kinetic energy of A is twice the kinetic energy of B.
d. The kinetic energy of A is four times the kinetic energy of B.

13.CQ3 A small block A is released from rest and slides down the friction-less ramp to the loop. The maximum height h of the loop is the same as the initial height of the block. Will A make it completely around the loop without losing contact with the track?

a. Yes
b. No
c. Need more information

END-OF-SECTION PROBLEMS

13.55 A force **P** is slowly applied to a plate that is attached to two springs and causes a deflection x_0. In each of the two cases shown, derive an expression for the constant k_e, in terms of k_1 and k_2, of the single spring equivalent to the given system, that is, of the single spring which will undergo the same deflection x_0 when subjected to the same force **P**.

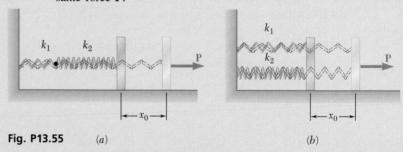

Fig. P13.55   (a)   (b)

13.56 A loaded railroad car of mass m is rolling at a constant velocity $\mathbf{v_0}$ when it couples with a massless bumper system. Determine the maximum deflection of the bumper assuming the two springs are (a) in series (as shown), (b) in parallel.

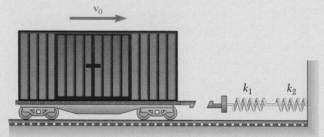

Fig. *P13.56*

842

13.57 A 750-g collar can slide along the horizontal rod shown. It is attached to an elastic cord with an undeformed length of 300 mm and a spring constant of 150 N/m. Knowing that the collar is released from rest at A and neglecting friction, determine the speed of the collar (a) at B, (b) at E.

13.58 A 4-lb collar can slide without friction along a horizontal rod and is in equilibrium at A when it is pushed 1 in. to the right and released from rest. The springs are undeformed when the collar is at A and the constant of each spring is 2800 lb/in. Determine the maximum velocity of the collar.

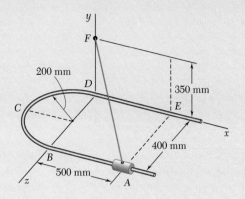

Fig. P13.57

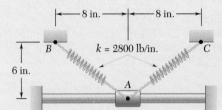

Fig. P13.58 and P13.59

13.59 A 4-lb collar can slide without friction along a horizontal rod and is released from rest at A. The undeformed lengths of springs BA and CA are 10 in. and 9 in., respectively, and the constant of each spring is 2800 lb/in. Determine the velocity of the collar when it has moved 1 in. to the right.

13.60 A 500-g collar can slide without friction on the curved rod BC in a *horizontal* plane. Knowing that the undeformed length of the spring is 80 mm and that $k = 400$ kN/m, determine (a) the velocity that the collar should be given at A to reach B with zero velocity, (b) the velocity of the collar when it eventually reaches C.

13.61 For the adapted shuffleboard device in Prob 13.28, you decide to utilize an elastic cord instead of a compression spring to propel the puck forward. When the cord is stretched directly between points A and B, the tension is 20 N. The 425-gram puck is placed in the center and pulled back through a distance of 400 mm; a force of 100 N is required to hold it at this location. Knowing that the coefficient of friction is 0.3, determine how far the puck will travel.

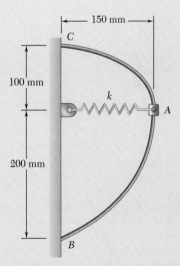

Fig. P13.60

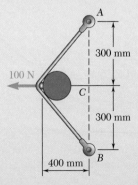

Fig. P13.61

13.62 An elastic cable is to be designed for bungee jumping from a tower 130 ft high. The specifications call for the cable to be 85 ft long when unstretched, and to stretch to a total length of 100 ft when a 600-lb weight is attached to it and dropped from the tower. Determine (a) the required spring constant k of the cable, (b) how close to the ground a 186-lb man will come if he uses this cable to jump from the tower.

13.63 It is shown in mechanics of materials that the stiffness of an elastic cable is $k = AE/L$, where A is the cross-sectional area of the cable, E is the modulus of elasticity, and L is the length of the cable. A winch is lowering a 4000-lb piece of machinery using a constant speed of 3 ft/s when the winch suddenly stops. Knowing that the steel cable has a diameter of 0.4 in., $E = 29 \times 10^6$ lb/in^2, and when the winch stops $L = 30$ ft, determine the maximum downward displacement of the piece of machinery from the point it was when the winch stopped.

Fig. P13.62

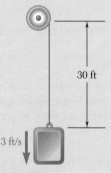

3 ft/s

30 ft

Fig. P13.63

13.64 A 2-kg collar is attached to a spring and slides without friction in a vertical plane along the curved rod ABC. The spring is undeformed when the collar is at C and its constant is 600 N/m. If the collar is released at A with no initial velocity, determine its velocity (a) as it passes through B, (b) as it reaches C.

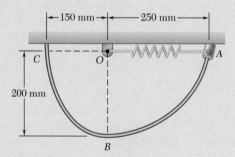

150 mm · 250 mm

C · O · A

200 mm

B

Fig. P13.64

13.65 A 500-g collar can slide without friction along the semicircular rod BCD. The spring is of constant 320 N/m and its undeformed length is 200 mm. Knowing that the collar is released from rest at B, determine (a) the speed of the collar as it passes through C, (b) the force exerted by the rod on the collar at C.

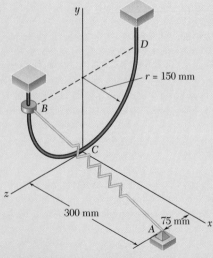

y

D

r = 150 mm

B

C

z

300 mm

75 mm *x*

A

Fig. P13.65

13.66 A thin circular rod is supported in a *vertical plane* by a bracket at A. Attached to the bracket and loosely wound around the rod is a spring of constant $k = 3$ lb/ft and undeformed length equal to the arc of circle AB. An 8-oz collar C, not attached to the spring, can slide without friction along the rod. Knowing that the collar is released from rest at an angle θ with the vertical, determine (*a*) the smallest value of θ for which the collar will pass through D and reach point A, (*b*) the velocity of the collar as it reaches point A.

13.67 Cornhole is a game that requires you to toss beanbags through a hole in a wooden board. People with limited arm mobility often have difficulty enjoying this favorite tailgating activity. An adapted launching device attaches to a wheelchair so that points O and A are fixed. The device mimics an underhand throw by utilizing an elastic band to power the arm OC, which rotates about pin O. The elastic cord has an unstretched length of 1 ft and is attached to the fixed point A and to point B on the arm. The combined weight of the beanbag and holder at C is 4 lbs, and you can neglect the weight of the rod OB. Knowing that the starting position is 30° from the horizontal, as shown in the figure, determine the spring constant if the velocity of the bean bag is 31 ft/s when the bag is released at an angle of $\theta = 45°$.

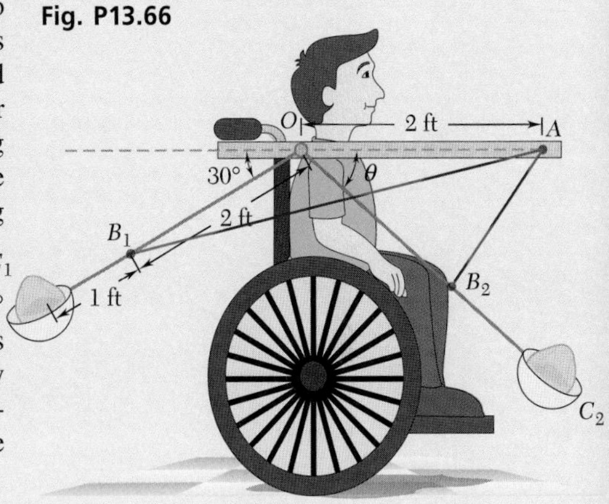

13.68 A spring is used to stop a 50-kg package that is moving down a 20° incline. The spring has a constant k = 30 kN/m and is held by cables so that it is initially compressed 50 mm. Knowing that the velocity of the package is 2 m/s when it is 8 m from the spring and neglecting friction, determine the maximum additional deformation of the spring in bringing the package to rest.

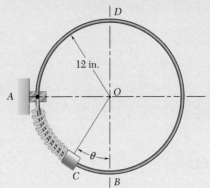

Fig. P13.66

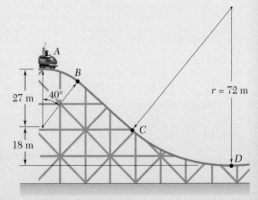

Fig. P13.67

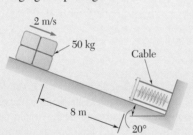

Fig. P13.68

13.69 Solve Prob. 13.68 assuming the kinetic coefficient of friction between the package and the incline is 0.2.

13.70 A section of track for a roller coaster consists of two circular arcs AB and CD joined by a straight portion BC. The radius of AB is 27 m and the radius of CD is 72 m. The car and its occupants, of total mass 250 kg, reach point A with practically no velocity and then drop freely along the track. Determine the normal force exerted by the track on the car as the car reaches point B. Ignore air resistance and rolling resistance.

13.71 A section of track for a roller coaster consists of two circular arcs AB and CD joined by a straight portion BC. The radius of AB is 27 m and the radius of CD is 72 m. The car and its occupants, of total mass 250 kg, reach point A with practically no velocity and then drop freely along the track. Determine the maximum and minimum values of the normal force exerted by the track on the car as the car travels from A to D. Ignore air resistance and rolling resistance.

Fig. P13.70 and P13.71

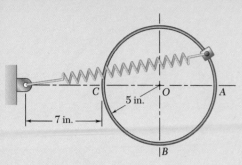

Fig. P13.72

13.72 A 1-lb collar is attached to a spring and slides without friction along a circular rod in a *vertical* plane. The spring has an undeformed length of 5 in. and a constant $k = 10$ lb/ft. Knowing that the collar is released from being held at A, determine the speed of the collar and the normal force between the collar and the rod as the collar passes through B.

13.73 A 10-lb collar is attached to a spring and slides without friction along a fixed rod in a vertical plane. The spring has an undeformed length of 14 in. and a constant $k = 4$ lb/in. Knowing that the collar is released from rest in the position shown, determine the force exerted by the rod on the collar at (*a*) point A, (*b*) point B. Both these points are on the curved portion of the rod.

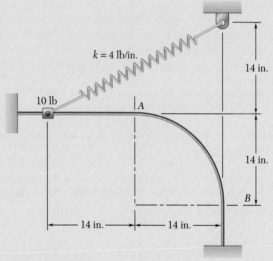

Fig. *P13.73*

13.74 An 8-oz package is projected upward with a velocity v_0 by a spring at A; it moves around a frictionless loop and is deposited at C. For each of the two loops shown, determine (*a*) the smallest velocity v_0 for which the package will reach C, (*b*) the corresponding force exerted by the package on the loop just before the package leaves the loop at C.

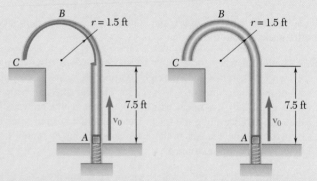

Fig. P13.74 and *P13.75*

13.75 If the package of Prob. 13.74 is not to hit the horizontal surface at C with a speed greater than 10 ft/s, (*a*) show that this requirement can be satisfied only by the second loop, (*b*) determine the largest allowable initial velocity v_0 when the second loop is used.

13.76 A small package of weight W is projected into a vertical return loop at A with a velocity $\mathbf{v}_0$. The package travels without friction along a circle of radius r and is deposited on a horizontal surface at C. For each of the two loops shown, determine (a) the smallest velocity $\mathbf{v}_0$ for which the package will reach the horizontal surface at C, (b) the corresponding force exerted by the loop on the package as it passes point B.

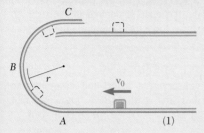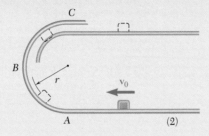

Fig. P13.76

13.77 The 1-kg ball at A is suspended by an inextensible cord and given an initial horizontal velocity of 5 m/s. If $l = 0.6$ m and $x_B = 0$, determine y_B so that the ball will enter the basket.

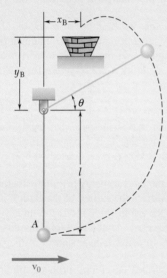

Fig. P13.77

13.78 The pendulum shown is released from rest at A and swings through $90°$ before the cord touches the fixed peg B. Determine the smallest value of a for which the pendulum bob will describe a circle about the peg.

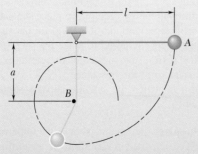

Fig. P13.78

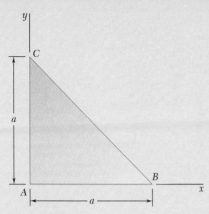

Fig. P13.81

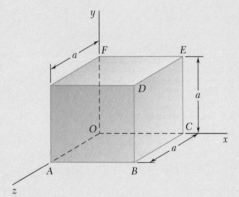

Fig. P13.82

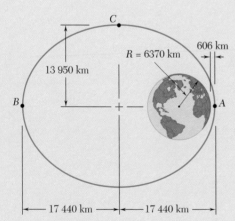

Fig. P13.86

*13.79 Prove that a force $F(x, y, z)$ is conservative if, and only if, the following relations are satisfied:

$$\frac{\partial F_x}{\partial y} = \frac{\partial F_y}{\partial x} \qquad \frac{\partial F_y}{\partial z} = \frac{\partial F_z}{\partial y} \qquad \frac{\partial F_z}{\partial x} = \frac{\partial F_x}{\partial z}$$

13.80 The force $\mathbf{F} = (yz\mathbf{i} + zx\mathbf{j} + xy\mathbf{k})/xyz$ acts on the particle $P(x, y, z)$ which moves in space. (a) Using the relation derived in Prob. 13.79, show that this force is a conservative force. (b) Determine the potential function associated with $\mathbf{F}$.

*13.81 A force $\mathbf{F}$ acts on a particle $P(x, y)$ which moves in the xy plane. Determine whether $\mathbf{F}$ is a conservative force and compute the work of $\mathbf{F}$ when P describes the path $ABCA$ knowing that (a) $\mathbf{F} = (kx + y)\mathbf{i}, + (kx + y)\mathbf{j}$, (b) $\mathbf{F} = (kx + y)\mathbf{i} + (x + ky)\mathbf{j}$.

*13.82 The potential function associated with a force $\mathbf{P}$ in space is known to be $V(x, y, z) = -(x^2 + y^2 + z^2)^{1/2}$. (a) Determine the x, y, and z components of $\mathbf{P}$. (b) Calculate the work done by $\mathbf{P}$ from O to D by integrating along the path $OABD$, and show that it is equal to the negative of the change in potential from O to D.

*13.83 (a) Calculate the work done from D to O by the force $\mathbf{P}$ of Prob. 13.82 by integrating along the diagonal of the cube. (b) Using the result obtained and the answer to part b of Prob. 13.82, verify that the work done by a conservative force around the closed path $OABDO$ is zero.

*13.84 The force $\mathbf{F} = (x\mathbf{i} + y\mathbf{j} + z\mathbf{k})/(x^2 + y^2 + z^2)^{3/2}$ acts on the particle $P(x, y, z)$ which moves in space. (a) Using the relations derived in Prob. 13.79, prove that $\mathbf{F}$ is a conservative force. (b) Determine the potential function $V(x, y, z)$ associated with $\mathbf{F}$.

13.85 (a) Determine the kinetic energy per unit mass that a missile must have after being fired from the surface of the earth if it is to reach an infinite distance from the earth. (b) What is the initial velocity of the missile (called the *escape velocity*)? Give your answers in SI units and show that the answer to part b is independent of the firing angle.

13.86 A satellite describes an elliptic orbit of minimum altitude 606 km above the surface of the earth. The semimajor and semiminor axes are 17 440 km and 13 950 km, respectively. Knowing that the speed of the satellite at point C is 4.78 km/s, determine (a) the speed at point A, the perigee, (b) the speed at point B, the apogee.

13.87 While describing a circular orbit 200 mi above the earth, a space vehicle launches a 6000-lb communications satellite. Determine (a) the additional energy required to place the satellite in a geosynchronous orbit at an altitude of 22,000 mi above the surface of the earth, (b) the energy required to place the satellite in the same orbit by launching it from the surface of the earth, excluding the energy needed to overcome air resistance. (A *geosynchronous orbit* is a circular orbit in which the satellite appears stationary with respect to the ground.)

13.88 How much energy per pound should be imparted to a satellite in order to place it in a circular orbit at an altitude of (a) 400 mi, (b) 4000 mi?

13.89 Knowing that the velocity of an experimental space probe fired from the earth has a magnitude $v_A = 32.5$ Mm/h at point A, determine the speed of the probe as it passes through point B.

13.90 A spacecraft is describing a circular orbit at an altitude of 1500 km above the surface of the earth. As it passes through point A, its speed is reduced by 40 percent and it enters an elliptic crash trajectory with the apogee at point A. Neglecting air resistance, determine the speed of the spacecraft when it reaches the earth's surface at point B.

13.91 Observations show that a celestial body traveling at 1.2×10^6 mi/h appears to be describing about point B a circle of radius equal to 60 light years. Point B is suspected of being a very dense concentration of mass called a black hole. Determine the ratio M_B/M_S of the mass at B to the mass of the sun. (The mass of the sun is 330,000 times the mass of the earth, and a light year is the distance traveled by light in 1 year at 186,300 mi/s.)

13.92 (a) Show that, by setting $r = R + y$ in the right-hand member of Eq. (13.17') and expanding that member in a power series in y/R, the expression in Eq. (13.16) for the potential energy V_g due to gravity is a first-order approximation for the expression given in Eq. (13.17'). (b) Using the same expansion, derive a second-order approximation for V_g.

13.93 Collar A has a mass of 3 kg and is attached to a spring of constant 1200 N/m and of undeformed length equal to 0.5 m. The system is set in motion with $r = 0.3$ m, $v_\theta = 2$ m/s, and $v_r = 0$. Neglecting the mass of the rod and the effect of friction, determine the radial and transverse components of the velocity of the collar when $r = 0.6$ m.

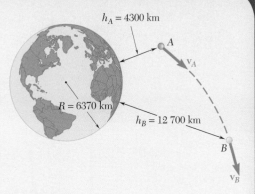

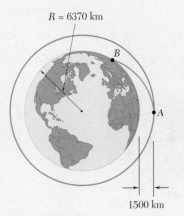

Fig. P13.89

Fig. P13.90

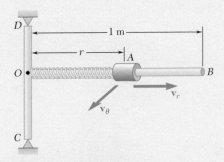

Fig. P13.93 and P13.94

13.94 Collar A has a mass of 3 kg and is attached to a spring of constant 1200 N/m and of undeformed length equal to 0.5 m. The system is set in motion with $r = 0.3$ m, $v_\theta = 2$ m/s, and $v_r = 0$. Neglecting the mass of the rod and the effect of friction, determine (a) the maximum distance between the origin and the collar, (b) the corresponding speed. (*Hint:* Solve the equation obtained for r by trial and error.)

13.95 A governor is designed so that the valve of negligible mass at D will open once a vertical force greater than 20 lbs is exerted on it. In initial testing of the device, the two 1-lb masses are at $x = 1$ in. and are prevented from sliding along the rod by stops. Each mass is connected to the valve by 10 lb/in. springs that are both unstretched at $x = 1$ in. The governor rotates so that $v_1 = 30$ ft/s when the stops are removed. When the valve opens, determine the position and velocity of the masses.

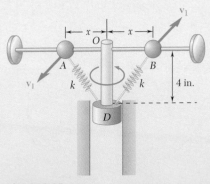

Fig. P13.95

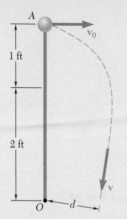

Fig. P13.96 and P13.97

13.96 A 1.5-lb ball that can slide on a *horizontal* frictionless surface is attached to a fixed point O by means of an elastic cord of constant $k = 1$ lb/in. and undeformed length 2 ft. The ball is placed at point A, 3 ft from O, and given an initial velocity $\mathbf{v}_0$ perpendicular to OA. Determine (*a*) the smallest allowable value of the initial speed v_0 if the cord is not to become slack, (*b*) the closest distance d that the ball will come to point O if it is given half the initial speed found in part *a*.

13.97 A 1.5-lb ball that can slide on a *horizontal* frictionless surface is attached to a fixed point O by means of an elastic cord of constant $k = 1$ lb/in. and undeformed length 2 ft. The ball is placed at point A, 3 ft from O, and given an initial velocity $\mathbf{v}_0$ perpendicular to OA, allowing the ball to come within a distance $d = 9$ in. of point O after the cord has become slack. Determine (*a*) the initial speed v_0 of the ball, (*b*) its maximum speed.

13.98 Using the principles of conservation of energy and conservation of angular momentum, solve part *a* of Sample Prob. 12.14.

13.99 Solve Sample Prob. 13.11, assuming that the elastic cord is replaced by a central force $\mathbf{F}$ with a magnitude of $(80/r^2)$ N directed toward O.

13.100 A spacecraft is describing an elliptic orbit of minimum altitude $h_A = 2400$ km and maximum altitude $h_B = 9600$ km above the surface of the earth. Determine the speed of the spacecraft at A.

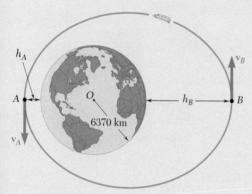

Fig. P13.100

13.101 While describing a circular orbit, 185 mi above the surface of the earth, a space shuttle ejects at point A an inertial upper stage (IUS) carrying a communications satellite to be placed in a geosynchronous orbit (see Prob. 13.87) at an altitude of 22,230 mi above the surface of the earth. Determine (*a*) the velocity of the IUS relative to the shuttle after its engine has been fired at A, (*b*) the increase in velocity required at B to place the satellite in its final orbit.

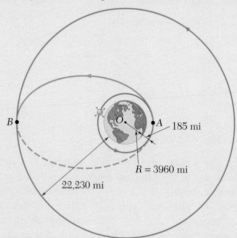

Fig. P13.101

13.102 A spacecraft approaching the planet Saturn reaches point A with a velocity $\mathbf{v}_A$ of magnitude 68.8×10^3 ft/s. It is to be placed in an elliptic orbit about Saturn so that it will be able to periodically examine Tethys, one of Saturn's moons. Tethys is in a circular orbit of radius 183×10^3 mi about the center of Saturn, traveling at a speed of 37.2×10^3 ft/s. Determine (*a*) the decrease in speed required by the spacecraft at A to achieve the desired orbit, (*b*) the speed of the spacecraft when it reaches the orbit of Tethys at B.

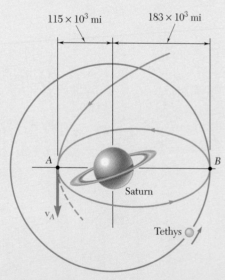

Fig. P13.102

13.103 A spacecraft traveling along a parabolic path toward the planet Jupiter is expected to reach point A with a velocity $\mathbf{v}_A$ of magnitude 26.9 km/s. Its engines will then be fired to slow it down, placing it into an elliptic orbit which will bring it to within 100×10^3 km of Jupiter. Determine the decrease in speed Δv at point A which will place the spacecraft into the required orbit. The mass of Jupiter is 319 times the mass of the earth.

13.104 As a first approximation to the analysis of a space flight from the earth to Mars, it is assumed that the orbits of the earth and Mars are circular and coplanar. The mean distances from the sun to the earth and to Mars are 149.6×10^6 km and 227.8×10^6 km, respectively. To place the spacecraft into an elliptical transfer orbit at point A, its speed is increased over a short interval of time to v_A, which is faster than the earth's orbital speed. When the spacecraft reaches point B on the elliptical transfer orbit, its speed v_B is increased to the orbital speed of Mars. Knowing that the mass of the sun is 332.8×10^3 times the mass of the earth, determine the increase in velocity required (a) at A, (b) at B.

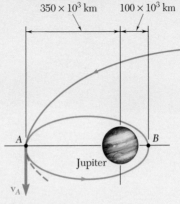

Fig. P13.103

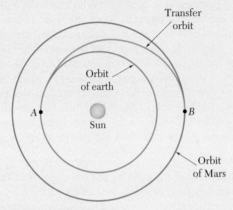

Fig. P13.104

13.105 The optimal way of transferring a space vehicle from an inner circular orbit to an outer coplanar circular orbit is to fire its engines as it passes through A to increase its speed and place it in an elliptic transfer orbit. Another increase in speed as it passes through B will place it in the desired circular orbit. For a vehicle in a circular orbit about the earth at an altitude $h_1 = 200$ mi, which is to be transferred to a circular orbit at an altitude $h_2 = 500$ mi, determine (a) the required increases in speed at A and at B, (b) the total energy per unit mass required to execute the transfer.

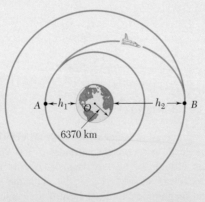

Fig. P13.105

13.106 During a flyby of the earth, the velocity of a spacecraft is 10.4 km/s as it reaches its minimum altitude of 990 km above the surface at point *A*. At point *B* the spacecraft is observed to have an altitude of 8350 km. Determine (*a*) the magnitude of the velocity at point *B*, (*b*) the angle ϕ_B.

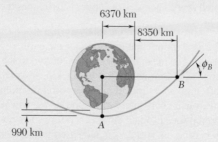

Fig. P13.106

13.107 A space platform is in a circular orbit about the earth at an altitude of 300 km. As the platform passes through *A*, a rocket carrying a communications satellite is launched from the platform with a relative velocity of magnitude 3.44 km/s in a direction tangent to the orbit of the platform. This was intended to place the rocket in an elliptic transfer orbit bringing it to point *B*, where the rocket would again be fired to place the satellite in a geosynchronous orbit of radius 42 140 km. After launching, it was discovered that the relative velocity imparted to the rocket was too large. Determine the angle γ at which the rocket will cross the intended orbit at point *C*.

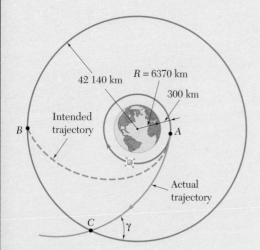

Fig. P13.107

13.108 A satellite is projected into space with a velocity $\mathbf{v}_0$ at a distance r_0 from the center of the earth by the last stage of its launching rocket. The velocity $\mathbf{v}_0$ was designed to send the satellite into a circular orbit of radius r_0. However, owing to a malfunction of control, the satellite is not projected horizontally but at an angle α with the horizontal and, as a result, is propelled into an elliptic orbit. Determine the maximum and minimum values of the distance from the center of the earth to the satellite.

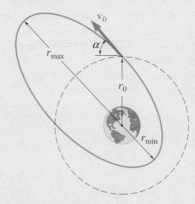

Fig. P13.108

13.109 A space vehicle is to rendezvous with an orbiting laboratory that circles the earth at a constant altitude of 360 km. The vehicle has reached an altitude of 60 km when its engine is shut off, and its velocity $\mathbf{v}_0$ forms an angle $\phi_0 = 50°$ with the vertical *OB* at that time. What magnitude should $\mathbf{v}_0$ have if the vehicle's trajectory is to be tangent at *A* to the orbit of the laboratory?

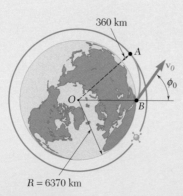

Fig. P13.109

13.110 A space vehicle is in a circular orbit at an altitude of 225 mi above the earth. To return to earth, it decreases its speed as it passes through A by firing its engine for a short interval of time in a direction opposite to the direction of its motion. Knowing that the velocity of the space vehicle should form an angle $\phi_B = 60°$ with the vertical as it reaches point B at an altitude of 40 mi, determine (a) the required speed of the vehicle as it leaves its circular orbit at A, (b) its speed at point B.

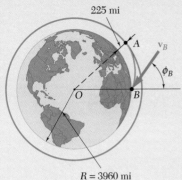

Fig. P13.110

*13.111 In Prob. 13.110, the speed of the space vehicle was decreased as it passed through A by firing its engine in a direction opposite to the direction of motion. An alternative strategy for taking the space vehicle out of its circular orbit would be to turn it around so that its engine would point away from the earth and then give it an incremental velocity $\Delta\mathbf{v}_A$ toward the center O of the earth. This would likely require a smaller expenditure of energy when firing the engine at A, but might result in too fast a descent at B. Assuming this strategy is used with only 50 percent of the energy expenditure used in Prob. 13.110, determine the resulting values of ϕ_B and v_B.

13.112 Show that the values v_A and v_P of the speed of an earth satellite at the apogee A and the perigee P of an elliptic orbit are defined by the relations

$$v_A^2 = \frac{2GM}{r_A + r_P}\frac{r_P}{r_A} \qquad v_P^2 = \frac{2GM}{r_A + r_P}\frac{r_A}{r_P}$$

where M is the mass of the earth, and r_A and r_P represent, respectively, the maximum and minimum distances of the orbit to the center of the earth.

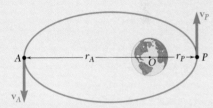

Fig. P13.112 and P13.113

13.113 Show that the total energy E of an earth satellite of mass m describing an elliptic orbit is $E = -GMm/(r_A + r_P)$, where M is the mass of the earth, and r_A and r_P represent, respectively, the maximum and minimum distances of the orbit to the center of the earth. (Recall that the gravitational potential energy of a satellite was defined as being zero at an infinite distance from the earth.)

*13.114 A space probe describes a circular orbit of radius nR with a velocity $\mathbf{v}_0$ about a planet of radius R and center O. Show that (a) in order for the probe to leave its orbit and hit the planet at an angle θ with the vertical, its velocity must be reduced to $\alpha\mathbf{v}_0$, where

$$\alpha = \sin\theta\sqrt{\frac{2(n-1)}{n^2 - \sin^2\theta}}$$

(b) the probe will not hit the planet if α is larger than $\sqrt{2/(1+n)}$.

13.115 A missile is fired from the ground with an initial velocity $\mathbf{v}_0$ forming an angle ϕ_0 with the vertical. If the missile is to reach a maximum altitude equal to αR, where R is the radius of the earth, (a) show that the required angle ϕ_0 is defined by the relation

$$\sin\phi_0 = (1+\alpha)\sqrt{1 - \frac{\alpha}{1+\alpha}\left(\frac{v_{esc}}{v_0}\right)^2}$$

where v_{esc} is the escape velocity, (b) determine the range of allowable values of v_0.

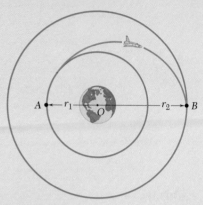

Fig. P13.116

13.116 A spacecraft of mass m describes a circular orbit of radius r_1 around the earth. (*a*) Show that the additional energy ΔE that must be imparted to the spacecraft to transfer it to a circular orbit of larger radius r_2 is

$$\Delta E = \frac{GMm(r_2 - r_1)}{2r_1 r_2}$$

where M is the mass of the earth. (*b*) Further show that if the transfer from one circular orbit to the other is executed by placing the space-craft on a transitional semielliptic path AB, the amounts of energy ΔE_A and ΔE_B which must be imparted at A and B are, respectively, proportional to r_2 and r_1:

$$\Delta E_A = \frac{r_2}{r_1 + r_2} \Delta E \qquad \Delta E_B = \frac{r_1}{r_1 + r_2} \Delta E$$

*13.117 Using the answers obtained in Prob. 13.108, show that the intended circular orbit and the resulting elliptic orbit intersect at the ends of the minor axis of the elliptic orbit.

*13.118 (*a*) Express in terms of $r_{\min}$ and $v_{\max}$ the angular momentum per unit mass, h, and the total energy per unit mass, E/m, of a space vehicle moving under the gravitational attraction of a planet of mass M (Fig. 13.15). (*b*) Eliminating $v_{\max}$ between the equations obtained, derive the formula

$$\frac{1}{r_{\min}} = \frac{GM}{h^2} \left[1 + \sqrt{1 + \frac{2E}{m}\left(\frac{h}{GM}\right)^2} \right]$$

(*c*) Show that the eccentricity £ of the trajectory of the vehicle can be expressed as

$$\varepsilon = \sqrt{1 + \frac{2E}{m}\left(\frac{h}{GM}\right)^2}$$

(*d*) Further show that the trajectory of the vehicle is a hyperbola, an ellipse, or a parabola, depending on whether E is positive, negative, or zero.

13.3 IMPULSE AND MOMENTUM

We now consider a third basic method for the solution of problems dealing with the motion of particles. This method is based on the principle of impulse and momentum and can be used to solve problems involving force, mass, velocity, and time. It is of particular interest in the solution of problems involving impulsive motion and problems involving impacts (Secs. 13.3B and 13.4).

13.3A Principle of Impulse and Momentum

Consider a particle of mass m acted upon by a force $\mathbf{F}$. As we saw in Sec. 12.1B, we can express Newton's second law in the form

$$\mathbf{F} = \frac{d}{dt}(m\mathbf{v}) \tag{13.27}$$

where $m\mathbf{v}$ is the linear momentum of the particle. Multiplying both sides of Eq. (13.27) by dt and integrating from a time t_1 to a time t_2, we have

$$\mathbf{F}\, dt = d(m\mathbf{v})$$

$$\int_{t_1}^{t_2} \mathbf{F}\, dt = m\mathbf{v}_2 - m\mathbf{v}_1$$

Moving $m\mathbf{v}_1$ to the left side of this equation gives us

$$m\mathbf{v}_1 + \int_{t_1}^{t_2} \mathbf{F}\, dt = m\mathbf{v}_2 \tag{13.28}$$

The integral in Eq. (13.28) is a vector known as the **linear impulse**, or simply the **impulse**, of the force $\mathbf{F}$ during the interval of time considered. Resolving $\mathbf{F}$ into rectangular components, we have

$$\begin{aligned}
\mathbf{Imp}_{1\to2} &= \int_{t_1}^{t_2} \mathbf{F}\, dt \\
&= \mathbf{i}\int_{t_1}^{t_2} F_x\, dt + \mathbf{j}\int_{t_1}^{t_2} F_y\, dt + \mathbf{k}\int_{t_1}^{t_2} F_z\, dt
\end{aligned} \tag{13.29}$$

Note that the components of the impulse of force $\mathbf{F}$ are, respectively, equal to the areas under the curves obtained by plotting the components F_x, F_y, and F_z against t (Fig. 13.16). In the case of a force $\mathbf{F}$ of constant magnitude and direction, the impulse is represented by the vector $\mathbf{F}(t_2 - t_1)$, which has the same direction as $\mathbf{F}$.

If we use SI units, the magnitude of the impulse of a force is expressed in N·s. However, recalling the definition of the newton, we have

$$\text{N·s} = (\text{kg·m/s}^2)\text{·s} = \text{kg·m/s}$$

which is the unit obtained in Sec. 12.1C for the linear momentum of a particle. This verifies that Eq. (13.28) is dimensionally correct. If we use U.S. customary units, the impulse of a force is expressed in lb·s, which is also the unit obtained in Sec. 12.1C for the linear momentum of a particle.

Photo 13.4 This impact test between an F-4 Phantom and a rigid reinforced target was to determine the impact force as a function of time.

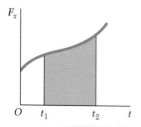

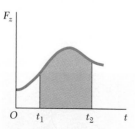

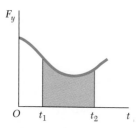

Fig. 13.16 Components of the impulse of a force **F** acting from times t_1 to t_2.

Equation (13.28) states that when a particle is acted upon by a force **F** during a given time interval, **we can obtain the final momentum $m\mathbf{v}_2$ of the particle by adding vectorially its initial momentum $m\mathbf{v}_1$ and the impulse of the force F during the time interval considered.** This can be expressed as:

Principle of impulse and momentum

$$m\mathbf{v}_1 + \mathbf{Imp}_{1\to2} = m\mathbf{v}_2 \qquad (13.30)$$

Figure 13.17 is a pictorial representation of this principle and is called an *impulse-momentum diagram*. To obtain an analytic solution, it is thus necessary to replace Eq. (13.30) with the corresponding component equations. Note that whereas kinetic energy and work are scalar quantities, momentum and impulse are vector quantities.

$$(mv_x)_1 + \int_{t_1}^{t_2} F_x \, dt = (mv_x)_2$$

$$(mv_y)_1 + \int_{t_1}^{t_2} F_y \, dt = (mv_y)_2 \qquad (13.31)$$

$$(mv_z)_1 + \int_{t_1}^{t_2} F_z \, dt = (mv_z)_2$$

When several forces act on a particle, we must consider the impulse of each of the forces. We have

$$m\mathbf{v}_1 + \Sigma\mathbf{Imp}_{1\to2} = m\mathbf{v}_2 \qquad (13.32)$$

Again, this equation represents a relation between vector quantities; in the actual solution of a problem, it should be replaced by the corresponding component equations.

When a problem involves two or more particles, we can consider each particle separately and write Eq. (13.32) for each particle. We can also add vectorially the momenta of all the particles and the impulses of all the forces involved. We then have

$$\Sigma m\mathbf{v}_1 + \Sigma\mathbf{Imp}_{1\to2} = \Sigma m\mathbf{v}_2 \qquad (13.33)$$

Since the forces of action and reaction exerted by the particles on each other form pairs of equal and opposite forces, and since the time interval from t_1 to t_2 is common to all of the forces involved, the impulses of the forces of action and reaction cancel out. Thus, we need consider only the impulses of the external forces.[†]

[†]Note the difference between this statement and the corresponding statement in Sect. 13.1C regarding the work of the forces of action and reaction between several particles. Although the sum of the impulses of these forces is always zero, the sum of their work is zero only under special circumstances, e.g., when the particles involved are connected by inextensible cords or links and thus are constrained to move through equal distances.

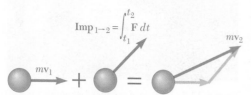

Fig. 13.17 Impulse-momentum diagram. Initial momentum plus impulse of a force **F** equals final momentum.

If no external force is exerted on the particles or, more generally, if the sum of the external forces is zero, the second term in Eq. (13.33) vanishes and the equation reduces to

**Conservation of
linear momentum**

$$\Sigma m\mathbf{v}_1 = \Sigma m\mathbf{v}_2 \qquad \textbf{(13.34)}$$

For two particles A and B, this is

$$m_A\mathbf{v}_A + m_B\mathbf{v}_B = m_A\mathbf{v}_A' + m_B\mathbf{v}_B' \qquad \textbf{(13.34$'$)}$$

where $\mathbf{v}_A'$ and $\mathbf{v}_B'$ represent the velocities of the bodies at the second time. This equation says that **the total momentum of the particles is conserved**. Consider, for example, two boats with masses of m_A and m_B, initially at rest, that are being pulled together (Fig. 13.18). If we neglect the resistance of the water, the only external forces acting on the boats are their weights and the buoyant forces exerted on them. Since these forces are balanced, we have

$$\Sigma m\mathbf{v}_1 = \Sigma m\mathbf{v}_2$$
$$0 = m_A\mathbf{v}_A' + m_B\mathbf{v}_B'$$

where $\mathbf{v}_A'$ and $\mathbf{v}_B'$ represent the velocities of the boats after a finite interval of time. This equation indicates that the boats move in opposite directions (toward each other) with velocities inversely proportional to their masses.[†]

Fig. 13.18 Neglecting the resistance of the water, linear momentum is conserved for two boats being pulled together.

13.3B Impulsive Motion

A force acting on a particle during a very short time interval but large enough to produce a definite change in momentum is called an **impulsive force**. The resulting motion is called an **impulsive motion**. For example, when a baseball is struck, the contact between bat and ball takes place during a very short time interval Δt. But the average value of the force $\mathbf{F}_{\text{avg}}$ exerted by the bat on the ball is very large, and the resulting impulse $\mathbf{F}_{\text{avg}}\,\Delta t$ is large enough to change the sense of motion of the ball (Fig. 13.19).

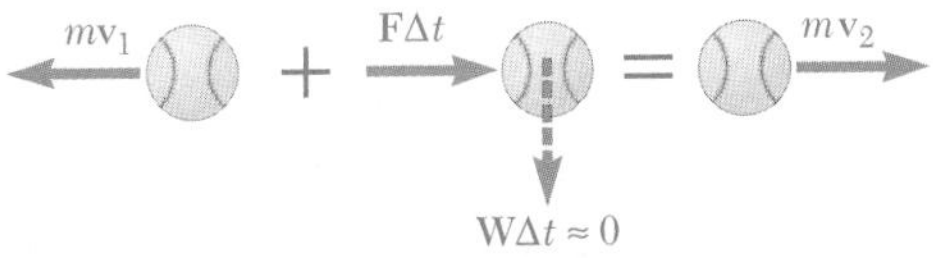

[†]We use blue equals signs in Fig. 13.18 and throughout the remainder of this chapter to indicate that two systems of vectors are *equipollent*; i.e., that they have the same resultant and moment resultant (cf. Sec. 3.4B). We continue to use red equals signs to indicate that two systems of vectors are *equivalent*; i.e., they have the same effect. We will discuss this and the concept of the conservation of momentum for a system of particles in greater detail in Chap. 14.

Fig. 13.19 When an impulsive force (i.e., a large force that acts over a short time) acts on a system, we can often neglect non-impulsive forces, such as weight.

When impulsive forces act on a particle, Eq. (13.32) becomes

**Impulse-momentum principle
for impulsive motion**

$$m\mathbf{v}_1 + \Sigma\mathbf{F}_{avg} \, \Delta t = m\mathbf{v}_2 \qquad (13.35)$$

We can neglect any force that is not an impulsive force because the corresponding impulse $\mathbf{F}_{avg}\Delta t$ is very small. Non-impulsive forces include the weight of the body, the force exerted by a spring, or any other force that is known to be small compared with an impulsive force. Unknown reactions may or may not be impulsive; their impulses therefore should be included in Eq. (13.35) as long as they have not been proved negligible. For example, we may neglect the impulse of the weight of the baseball considered previously. If we analyze the motion of the bat, we can neglect the impulse of the weight of the bat. The impulses of the reactions of the player's hands on the bat, however, should be included; these impulses are not negligible if the ball is incorrectly hit.

Note that the method of impulse and momentum is particularly effective in analyzing the impulsive motion of a particle, since it involves only the initial and final velocities of the particle and the impulses of the forces exerted on the particle. The direct application of Newton's second law, on the other hand, would require determining the forces as functions of time and integrating the equations of motion over the time interval Δt.

In the case of the impulsive motion of several particles, we can use Eq. (13.33). It reduces to

$$\Sigma m\mathbf{v}_1 + \Sigma\mathbf{F}_{avg} \, \Delta t = \Sigma m\mathbf{v}_2 \qquad (13.36)$$

where the second term involves only impulsive, external forces. If all of the external forces acting on the various particles are non-impulsive, the second term in Eq. (13.36) vanishes, and this equation reduces to Eq. (13.34):

$$\Sigma m\mathbf{v}_1 = \Sigma m\mathbf{v}_2 \qquad (13.34)$$

As before, for two particles, this reduces to

$$m_A\mathbf{v}_A + m_B\mathbf{v}_B = m_A\mathbf{v}'_A + m_B\mathbf{v}'_B \qquad (13.34')$$

In other words, the total momentum of the particles is conserved. This situation occurs, for example, when two freely moving particles collide with one another. We should note, however, that although the total momentum of the particles is conserved, their total energy is generally *not* conserved. Problems involving the collision or *impact* of two particles are discussed in detail in Sec. 13.4.

Sample Problem 13.13

An automobile weighing 4000 lb is moving down a 5° incline at a speed of 60 mi/h when the brakes are applied, causing a constant total braking force (applied by the road on the tires) of 1500 lb. Determine the time required for the automobile to come to a stop.

STRATEGY: Since you are given velocities at two different times, use the principle of impulse and momentum.

MODELING: Choose the automobile to be your system and assume you can model it as a particle. The impulse-momentum diagram for this system is shown in Fig. 1.

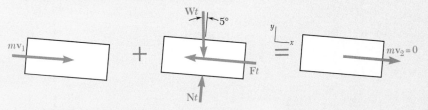

Fig. 1 Impulse-momentum diagram for the car.

ANALYSIS: The general impulse-momentum principle is

$$m\mathbf{v}_1 + \Sigma\mathbf{Imp}_{1\rightarrow2} = m\mathbf{v}_2$$

This is a vector equation, and since the impulsive force is constant, the impulse is simply equal to the force multiplied by its time duration. You can obtain scalar equations by using Fig. 1. In the direction down the incline, you get

$+\searrow$ components: $mv_1 + (W \sin 5°)t - Ft = 0$

$(4000/32.2)(88 \text{ ft/s}) + (4000 \sin 5°)t - 1500t = 0$ $\qquad t = 9.49 \text{ s}$ ◀

REFLECT and THINK: You could use Newton's second law to solve this problem. First, you would determine the car's deceleration, separate variables, and then integrate $a = dv/dt$ to relate the velocity, deceleration, and time. You could not use conservation of energy to solve this problem, because this principle does not involve time.

Sample Problem 13.14

In order to determine the weight of a freight train of 40 identical boxcars, an engineer attaches a dynamometer between the train and the locomotive. The train starts from rest, travels over a straight, level track, and reaches a speed of 30 mi/h after three minutes. During this time interval, the average reading of the dynamometer is 120 tons. Knowing that the effective coefficient of friction in the system is 0.03 and air resistance is negligible, determine (a) the weight of the train (in tons), (b) the coupling force between boxcars A and B.

(continued)

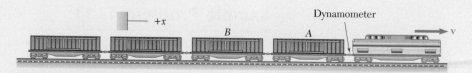

STRATEGY: This problem could be solved using Newton's second law and kinematic relationships, but since you are given velocities at two times and asked to find the force, you can also use impulse and momentum.

MODELING: Choose the system to be the 40 boxcars behind the engine. An impulse-momentum diagram for this system is shown in Fig. 1, where **F** is the dynamometer force.

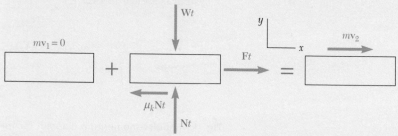

Fig. 1 Impulse-momentum diagram for the 40 boxcars.

ANALYSIS: Apply the principle of impulse and momentum

$$m\mathbf{v}_1 + \Sigma\mathbf{Imp}_{1\to2} = m\mathbf{v}_2$$

You can obtain scalar equations by using Fig. 1 and looking at the x and y directions.

$+\uparrow y$ components: $\qquad Nt - Wt = 0 \qquad N = W$

$\xrightarrow{+} x$ components: $\qquad 0 + Ft - \mu_k Nt = mv_2$

$0 + (120 \text{ ton})(2000 \text{ lb/ton})(180 \text{ s}) - 0.03(W)(180 \text{ s})$

$$= \left(\frac{W}{32.2 \text{ ft/s}^2}\right)(30 \text{ mi/h})\left(\frac{1 \text{ h}}{3600 \text{ s}}\right)\left(\frac{5280 \text{ ft}}{\text{mi}}\right)$$

Solving for W, you obtain

$$W = 6.384 \times 10^6 \text{ lb} = 3190 \text{ tons} \blacktriangleleft$$

Coupling Force Between Cars A and B. You need to define a new system where the force of interest is an external force. Therefore, choose car A to be your system and define F_A as the coupling force between cars A and B. The impulse-momentum diagram for this system is shown in Fig. 2.

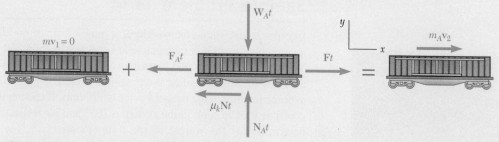

Fig. 2 Impulse-momentum diagram for car A.

Since all the cars weigh the same amount, the weight of A is $W_A = W/40 = 159{,}600$ lb. Applying impulse-momentum in the y-direction gives you $N_A = W_A$. Considering the x-direction,

$\xrightarrow{+} x$ components: $0 + Ft - \mu_k N_A t - F_A t = m_A v_2$

Substituting in numbers and solving for F_A gives

$$F_A = 117.0 \text{ tons} \quad \blacktriangleleft$$

REFLECT and THINK: Rather than using A as your system, you could have chosen the remaining 39 cars to be your system. In this case, you would find

$$0 - \mu_k N_{39} t + F_A t = m_{39} v_2$$

where N_{39} and m_{39} are the normal force and the mass, respectively, for the remaining 39 cars. The answer, as you would expect, is the same.

Sample Problem 13.15

A hammer and punch is used by a surgeon when inserting a hip implant. To better understand this process, an instrumented implant is inserted into a fixed replicate femur. The upward resisting force from the replicate femur on the hip implant can be neglected during the impact, and the impact force from the punch can be approximated by a half sine wave. Determine the speed of the 0.3-kg implant immediately after impact.

STRATEGY: Since you are relating force, time, and velocities, you should use the principle of impulse and momentum.

MODELING: Choose the system to be the implant. An impulse-momentum diagram for this system is shown in Fig. 1. The resisting force is left off Fig.1 since it is assumed to be negligible.

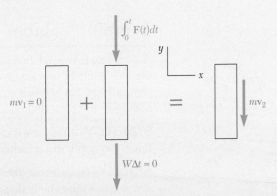

Fig. 1   Impulse-momentum diagram for the implant.

F (kN)

35

2   *t* (ms)

(continued)

ANALYSIS: Apply the principle of impulse and momentum

$$m\mathbf{v}_1 + \Sigma\mathbf{Imp}_{1\rightarrow 2} = m\mathbf{v}_2$$

You can obtain scalar equations by looking at the vertical components.

$\downarrow + y$ components:
$$0 + \int_0^t F(t)\,dt = mv_2 \tag{1}$$

where

$$\int_0^t F(t)\,dt = \int_0^{0.002} 35\,000\,\sin\left(\frac{2\pi}{0.004}t\right)dt = -35\,000\frac{0.004}{2\pi}\cos\left(\frac{2\pi}{0.004}t\right)_0^{0.002}$$

$$= 45.56\ \text{N·s}$$

Substituting this into Eq. (1) and solving for v_2 gives

$$v_2 = \frac{\int_0^t F(t)\,dt}{m} = \frac{45.36\ \text{N·s}}{0.3\ \text{kg}}$$

$$v_2 = 148.5\ \text{m/s} \blacktriangleleft$$

REFLECT and THINK: This problem is similar to Sample Prob. 13.5, where the drop-hammer pile driver hits the pile, then the hammer and pile move down, and the earth resists the motion. In that problem, you analyzed the motion *after* the impact; in this problem, you are analyzing the motion *during* the impact. In reality, you would need to do some experimental measurements to determine if the resisting force really is negligible during the impact. If you knew the force relationship of the femur on the implant, you could solve this as a two-part problem to first find the velocity of the implant immediately after the impact using impulse and momentum, and then determine how far the implant moves down into the femur using work and energy.

Sample Problem 13.16

A 4-oz baseball is pitched with a velocity of 80 ft/s toward a batter. After the ball is hit by the bat B, it has a velocity of 120 ft/s in the direction shown. If the bat and ball are in contact for 0.015 s, determine the average impulsive force exerted on the ball during the impact.

STRATEGY: This situation features an impact, and therefore impulsive forces, so apply the principle of impulse and momentum to the ball.

MODELING: Choose the ball as your system and treat it as a particle. The impulse-momentum diagram for this system is shown in Fig. 1. Because the weight of the ball is a non-impulsive force that is typically much smaller than the impulsive force, you can neglect it.

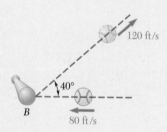

120 ft/s

40°

B

80 ft/s

ANALYSIS: Apply the principle of impulse and momentum

$$m\mathbf{v}_1 + \Sigma\mathbf{Imp}_{1\to2} = m\mathbf{v}_2$$

Applying this in the x and y directions gives

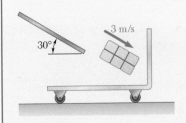

$\xrightarrow{+}x$ components: $\qquad -mv_1 + F_x\Delta t = mv_2\cos 40°$

$$-\frac{\frac{4}{16}}{32.2}(80\text{ ft/s}) + F_x(0.015\text{ s}) = \frac{\frac{4}{16}}{32.2}(120\text{ ft/s})\cos 40°$$

$$F_x = +89.0\text{ lb}$$

$+\uparrow y$ components: $\qquad 0 + F_y\Delta t = mv_2\sin 40°$

$$F_y(0.015\text{ s}) = \frac{\frac{4}{16}}{32.2}(120\text{ ft/s})\sin 40°$$

$$F_y = +39.9\text{ lb}$$

From its components F_x and F_y you can determine the magnitude and direction of the average impulsive force $\mathbf{F}$ as

$$\mathbf{F} = 97.5\text{ lb} \measuredangle 24.2° \quad \blacktriangleleft$$

Fig. 1 Impulse-momentum diagram for the ball.

REFLECT and THINK: In this problem, we neglected the impulse due to the weight. This would have had a magnitude of $(4/16\text{ lb})(0.015\text{ s})$ $= 0.00288$ lb·s. This indeed is much smaller than the impulse exerted on the ball by the bat, which is $(97.4\text{ lb})(0.015\text{ s}) = 1.463$ lb·s.

Sample Problem 13.17

A 10-kg package drops from a chute into a 25-kg cart with a velocity of 3 m/s. The cart is initially at rest and can roll freely. Determine (*a*) the final velocity of the cart, (*b*) the impulse exerted by the cart on the package, (*c*) the fraction of the initial energy lost in the impact.

STRATEGY: Since you have an impact, and therefore impulsive forces, use the principle of impulse and momentum.

MODELING: Choose the package and the cart to be your system, and assume that both can be treated as particles. The impulse-momentum diagram for this system is shown in Fig. 1. Note that a vertical impulse occurs between the cart and the ground, because the cart is constrained to move horizontally.

ANALYSIS: Apply the principle of impulse and momentum

$$m\mathbf{v}_1 + \Sigma\mathbf{Imp}_{1\to2} = m\mathbf{v}_2$$

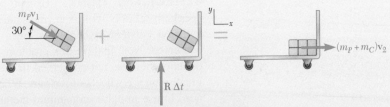

Fig. Impulse-momentum diagram for the system.

(continued)

a. Package and Cart. Applying this principle in the x-direction gives

$\xrightarrow{+}x$ components:
$$m_P v_1 \cos 30° + 0 = (m_P + m_C)v_2$$
$$(10 \text{ kg})(3 \text{ m/s}) \cos 30° = (10 \text{ kg} + 25 \text{ kg})v_2$$
$$v_2 = 0.742 \text{ m/s}\rightarrow \blacktriangleleft$$

In Fig. 1, the force between the package and the cart is not shown because it is internal to the defined system. To determine this force, you need a new system; that is, just the package by itself. The impulse-momentum diagram for the package alone is shown in Fig. 2.

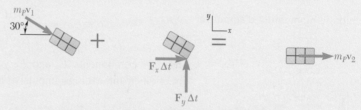

Fig. 2 Impulse-momentum diagram for the package.

b. Impulse-Momentum Principle: Package. The package moves in both x and y directions, so write the conservation of momentum equation for each component of the motion.

$\xrightarrow{+}x$ components:
$$-mv_1 + F_x\Delta t = mv_2 \cos 40°$$
$$(10 \text{ kg})(3 \text{ m/s}) \cos 30° + F_x \Delta t = (10 \text{ kg})(0.742 \text{ m/s})$$
$$F_x \Delta t = -18.56 \text{ N·s}$$
$+\uparrow y$ components:
$$-m_P v_1 \sin 30° + F_y \Delta t = 0$$
$$-(10 \text{ kg})(3 \text{ m/s}) \sin 30° + F_y \Delta t = 0$$
$$F_y \Delta t = +15 \text{ N·s}$$

The impulse exerted on the package is

$$\mathbf{F} \Delta t = 23.9 \text{ N·s} \; \text{⦩} \; 38.9° \blacktriangleleft$$

c. Fraction of Energy Lost. The initial and final energies are
$$T_1 = \tfrac{1}{2}m_P v_1^2 = \tfrac{1}{2}(10 \text{ kg})(3 \text{ m/s})^2 = 45 \text{ J}$$
$$T_2 = \tfrac{1}{2}(m_P + m_C)v_2^2 = \tfrac{1}{2}(10 \text{ kg} + 25 \text{ kg})(0.742 \text{ m/s})^2 = 9.63 \text{ J}$$

The fraction of energy lost is
$$\frac{T_1 - T_2}{T_1} = \frac{45\text{J} - 9.63\text{J}}{45 \text{ J}} = 0.786 \blacktriangleleft$$

REFLECT and THINK: Except in the purely theoretical case of a "perfectly elastic" collision, mechanical energy is never conserved in a collision between two objects, even though linear momentum may be conserved. Note that, in this problem, momentum was conserved in the x direction but was not conserved in the y direction because of the vertical impulse on the wheels of the cart. Whenever you deal with an impact, you need to use impulse-momentum methods.

SOLVING PROBLEMS ON YOUR OWN

In this section, we integrated Newton's second law to derive the **principle of impulse and momentum** for a particle. Recalling that we defined the *linear momentum* of a particle as the product of its mass m and its velocity $\mathbf{v}$ [Sec. 12.1B], we have

$$m\mathbf{v}_1 + \Sigma\mathbf{Imp}_{1\rightarrow2} = m\mathbf{v}_2 \tag{13.32}$$

This equation states that we can obtain the linear momentum $m\mathbf{v}_2$ of a particle at time t_2 by adding its linear momentum $m\mathbf{v}_1$ at time t_1 to the **impulses** of the forces exerted on the particle during the time interval t_1 to t_2. For computing purposes, we can express the momenta and impulses in terms of their rectangular components and replace Eq. (13.32) by the equivalent scalar equations. The units of momentum and impulse are N·s in the SI system of units and lb·s in U.S. customary units. To solve problems using this equation you can follow these steps.

1. Draw an impulse-momentum diagram showing the particle, its momentum at t_1 and at t_2, and the impulses of the forces exerted on the particle during the time interval t_1 to t_2.

2. Calculate the impulse of each force, expressing it in terms of its rectangular components if more than one direction is involved. You may encounter the following cases:

    a. The time interval is finite and the force is constant.

$$\mathbf{Imp}_{1\rightarrow2} = \mathbf{F}(t_2 - t_1)$$

    b. The time interval is finite and the force is a function of t.

$$\mathbf{Imp}_{1\rightarrow2} = \int_{t_1}^{t_2} \mathbf{F}(t)\,dt$$

    c. The time interval is very small and the force is very large. The force is called an **impulsive force**, and its impulse over the time interval $t_2 - t_1 = \Delta t$ is

$$\mathbf{Imp}_{1\rightarrow2} = \mathbf{F}_{avg}\,\Delta t$$

Note that this impulse is assumed to be zero for a non-impulsive force such as the weight of a body, the force exerted by a spring, or any other force that is known to be small by comparison with the impulsive forces. However, we *cannot* assume unknown reactions are non-impulsive, and you should take their impulses into account.

3. Substitute the values obtained for the impulses into Eq. (13.32) or into the equivalent scalar equations. You will find that the forces and velocities in the problems of this section are contained in a plane. Therefore, you can write two scalar equations and solve these equations for two unknowns. These unknowns may be a *time* [Sample Prob. 13.13], a *force* [Sample Prob. 13.14], a *velocity* [Sample Prob. 13.15], an *average impulsive force* [Sample Prob. 13.16], or an *impulse* [Sample Prob. 13.17].

4. When several particles are involved, it is often necessary to draw a separate diagram for each particle showing the initial and final momentum of the particle as well as the impulses of the forces exerted on the particle.

  a. It is usually convenient, however, to first consider a system that includes all of the particles. This system leads to

$$\Sigma m\mathbf{v}_1 + \Sigma \mathbf{Imp}_{1\rightarrow 2} = \Sigma m\mathbf{v}_2 \tag{13.33}$$

where you need to consider the impulses of only the forces external to the system. Therefore, the two equivalent scalar equations will not contain any of the impulses of the unknown internal forces.

  b. If the sum of the impulses of the external forces is zero, Eq. (13.33) reduces to

$$\Sigma m\mathbf{v}_1 = \Sigma m\mathbf{v}_2 \tag{13.34}$$

or for two particles as

$$m_A\mathbf{v}_A + m_B\mathbf{v}_B = m_A\mathbf{v}'_A + m_B\mathbf{v}'_B \tag{13.34'}$$

which says that *the total linear momentum of the particles is conserved*. This occurs when the time interval is very short and the external forces are negligible compared to the impulsive forces. Keep in mind, however, that the total momentum may be conserved in one direction, but not in another [Sample Prob. 13.17].

Problems

CONCEPT QUESTIONS

13.CQ4 A large insect impacts the front windshield of a sports car traveling down a road. Which of the following statements is true during the collision?
  a. The car exerts a greater force on the insect than the insect exerts on the car.
  b. The insect exerts a greater force on the car than the car exerts on the insect.
  c. The car exerts a force on the insect, but the insect does not exert a force on the car.
  d. The car exerts the same force on the insect as the insect exerts on the car.
  e. Neither exerts a force on the other; the insect gets smashed simply because it gets in the way of the car.

Case 1

13.CQ5 The expected damages associated with two types of perfectly plastic collisions are to be compared. In the first case, two identical cars traveling at the same speed impact each other head-on. In the second case, the car impacts a massive concrete wall. In which case would you expect the car to be more damaged?
  a. Case 1
  b. Case 2
  c. The same damage in each case

Case 2

Fig. P13.CQ5

IMPULSE-MOMENTUM DIAGRAM PRACTICE PROBLEMS

13.F1 The initial velocity of the block in position A is 30 ft/s. The coefficient of kinetic friction between the block and the plane is $\mu_k = 0.30$. Draw the impulse-momentum diagram that can be used to determine the time it takes for the block to reach B with zero velocity, if $\theta = 20°$.

13.F2 A 4-lb collar which can slide on a frictionless vertical rod is acted upon by a force **P** which varies in magnitude as shown. Knowing that the collar is initially at rest, draw the impulse-momentum diagram that can be used to determine its velocity at $t = 3$ s.

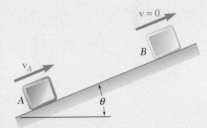

Fig. P13.F1

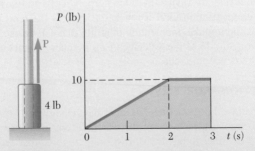

Fig. P13.F2

13.F3 The 15-kg suitcase A has been propped up against one end of a 40-kg luggage carrier B and is prevented from sliding down by other luggage. When the luggage is unloaded and the last heavy trunk is removed from the carrier, the suitcase is free to slide down, causing the 40-kg carrier to move to the left with a velocity v_B of magnitude 0.8 m/s. Neglecting friction, draw the impulse-momentum diagrams that can be used to determine (*a*) the velocity of A as it rolls on the carrier, (*b*) the velocity of the carrier after the suitcase hits the right side of the carrier without bouncing back.

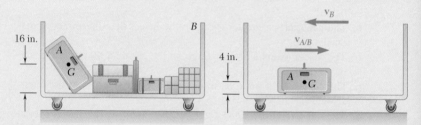

Fig. P13.F3

13.F4 Car A was traveling west at a speed of 15 m/s and car B was traveling north at an unknown speed when they slammed into each other at an intersection. Upon investigation it was found that after the crash the two cars got stuck and skidded off at an angle of 50° north of east. Knowing the masses of A and B are m_A and m_B, respectively, draw the impulse-momentum diagram that can be used to determine the velocity of B before impact.

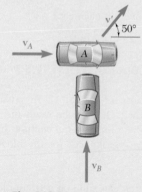

Fig. P13.F4

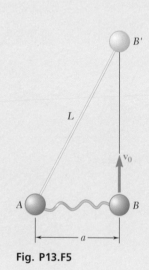

Fig. P13.F5

13.F5 Two identical spheres A and B, each of mass m, are attached to an inextensible inelastic cord of length L and are resting at a distance a from each other on a frictionless horizontal surface. Sphere B is given a velocity $\mathbf{v}_0$ in a direction perpendicular to line AB and moves it without friction until it reaches B' where the cord becomes taut. Draw the impulse-momentum diagram that can be used to determine the magnitude of the velocity of each sphere immediately after the cord has become taut.

END-OF-SECTION PROBLEMS

13.119 A 35 000-Mg ocean liner has an initial velocity of 4 km/h. Neglecting the frictional resistance of the water, determine the time required to bring the liner to rest by using a single tugboat that exerts a constant force of 150 kN.

13.120 A 2500-lb automobile is moving at a speed of 60 mi/h when the brakes are fully applied, causing all four wheels to skid. Determine the time required to stop the automobile (*a*) on dry pavement ($\mu_k = 0.75$), (*b*) on an icy road ($\mu_k = 0.10$).

13.121 A sailboat weighing 980 lb with its occupants is running downwind at 8 mi/h when its spinnaker is raised to increase its speed. Determine the net force provided by the spinnaker over the 10-s interval that it takes for the boat to reach a speed of 12 mi/h.

Fig. P13.121

13.122 A truck is hauling a 300-kg log out of a ditch using a winch attached to the back of the truck. Knowing the winch applies a constant force of 2500 N and the coefficient of kinetic friction between the ground and the log is 0.45, determine the time for the log to reach a speed of 0.5 m/s.

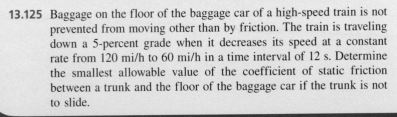

Fig. *P13.122*

13.123 The coefficients of friction between the load and the flatbed trailer shown are $\mu_s = 0.40$ and $\mu_k = 0.35$. Knowing that the speed of the rig is 55 mi/h, determine the shortest time in which the rig can be brought to a stop if the load is not to shift.

Fig. P13.123

13.124 Steep safety ramps are built beside mountain highways to enable vehicles with defective brakes to stop. A 10-ton truck enters a 15° ramp at a high speed $v_0 = 108$ ft/s and travels for 6 s before its speed is reduced to 36 ft/s. Assuming constant deceleration, determine (*a*) the magnitude of the braking force, (*b*) the additional time required for the truck to stop. Neglect air resistance and rolling resistance.

Fig. P13.124

13.125 Baggage on the floor of the baggage car of a high-speed train is not prevented from moving other than by friction. The train is traveling down a 5-percent grade when it decreases its speed at a constant rate from 120 mi/h to 60 mi/h in a time interval of 12 s. Determine the smallest allowable value of the coefficient of static friction between a trunk and the floor of the baggage car if the trunk is not to slide.

13.126 The 18 000-kg F-35B uses thrust vectoring to allow it to take off vertically. In one maneuver, the pilot reaches the top of her static hover at 200 m. The combined thrust and lift force on the airplane applied at the end of the static hover can be expressed as $\mathbf{F} = (44t + 2500t^2)\mathbf{i} + (250t^2 + t + 176\ 580)\mathbf{j}$, where $\mathbf{F}$ and t are expressed in newtons and seconds, respectively. Determine (a) how long it will take the airplane to reach a cruising speed of 1000 km/hr (cruising speed is defined to be in the x-direction only), (b) the altitude of the plane at this time.

Fig. P13.126

13.127 A truck is traveling down a road with a 4-percent grade at a speed of 60 mi/h when its brakes are applied to slow it down to 20 mi/h. An antiskid braking system limits the braking force to a value at which the wheels of the truck are just about to slide. Knowing that the coefficient of static friction between the road and the wheels is 0.60, determine the shortest time needed for the truck to slow down.

13.128 In anticipation of a long 6° upgrade, a bus driver accelerates at a constant rate from 80 km/h to 100 km/h in 8 s while still on a level section of the highway. Knowing that the speed of the bus is 100 km/h as it begins to climb the grade at time $t = 0$ and that the driver does not change the setting of the throttle or shift gears, determine (a) the speed of the bus when $t = 10$ s, (b) the time when the speed is 60 km/h.

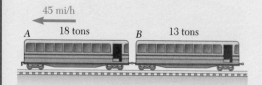

45 mi/h

A   18 tons   *B*   13 tons

Fig. P13.129

13.129 A light train made of two cars travels at 45 mi/h. Car A weighs 18 tons, and car B weighs 13 tons. When the brakes are applied, a constant braking force of 4300 lb is applied to each car. Determine (a) the time required for the train to stop after the brakes are applied, (b) the force in the coupling between the cars while the train is slowing down.

13.130 Solve Problem 13.129, assuming that a constant braking force of 4300 lb is applied to car B but that the brakes on car A are not applied.

13.131 A tractor-trailer rig with a 2000-kg tractor, a 4500-kg trailer, and a 3600-kg trailer is traveling on a level road at 90 km/h. The brakes on the rear trailer fail, and the antiskid system of the tractor and front trailer provide the largest possible force that will not cause the wheels to slide. Knowing that the coefficient of static friction is 0.75, determine (*a*) the shortest time for the rig to a come to a stop, (*b*) the force in the coupling between the two trailers during that time. Assume that the force exerted by the coupling on each of the two trailers is horizontal.

Fig. P13.131

13.132 The system shown is at rest when a constant 150-N force is applied to collar *B*. Neglecting the effect of friction, determine (*a*) the time at which the velocity of collar *B* will be 2.5 m/s to the left, (*b*) the corresponding tension in the cable.

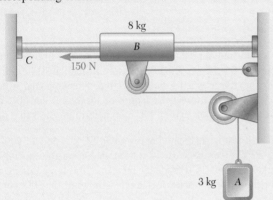

Fig. P13.132

13.133 An 8-kg cylinder *C* rests on a 4-kg platform *A* supported by a cord that passes over the pulleys *D* and *E* and is attached to a 4-kg block *B*. Knowing that the system is released from rest, determine (*a*) the velocity of block *B* after 0.8 s, (*b*) the force exerted by the cylinder on the platform.

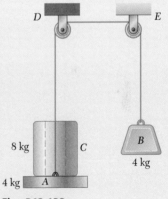

Fig. P13.133

13.134 An estimate of the expected load on over-the-shoulder seat belts is to be made before designing prototype belts that will be evaluated in automobile crash tests. Assuming that an automobile traveling at 45 mi/h is brought to a stop in 110 ms, determine (*a*) the average impulsive force exerted by a 200-lb man on the belt, (*b*) the maximum force F_m exerted on the belt if the force-time diagram has the shape shown.

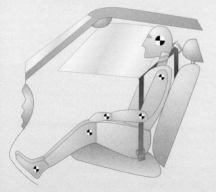

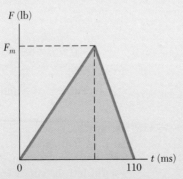

Fig. P13.134

13.135 A 60-g model rocket is fired vertically. The engine applies a thrust **P** which varies in magnitude as shown. Neglecting air resistance and the change in mass of the rocket, determine (*a*) the maximum speed of the rocket as it goes up, (*b*) the time for the rocket to reach its maximum elevation.

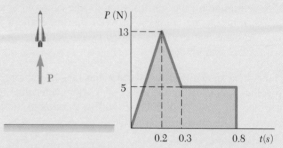

Fig. P13.135

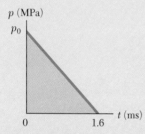

Fig. P13.136

13.136 A simplified model consisting of a single straight line is to be obtained for the variation of pressure inside the 10-mm-diameter barrel of a rifle as a 20-g bullet is fired. Knowing that it takes 1.6 ms for the bullet to travel the length of the barrel and that the velocity of the bullet upon exit is 700 m/s, determine the value of p_0.

13.137 A crash test is performed between an SUV *A* and a 2500-lb compact car *B*. The compact car is stationary before the impact and has its brakes applied. A transducer measures the force during the impact, and the force **P** varies as shown. Knowing that the coefficients of friction between the tires and road are $\mu_s = 0.9$ and $\mu_k = 0.7$, determine (*a*) the time at which the compact car will start moving, (*b*) the maximum speed of the car, (*c*) the time at which the car will come to a stop.

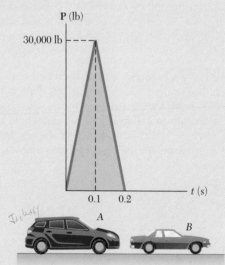

Fig. *P13.137* and P13.138

13.138 A crash test is performed between a 4500 lb SUV *A* and a compact car *B*. A transducer measures the force during the impact, and the force **P** varies as shown. Knowing that the SUV is travelling 30 mph when it hits the car, determine the speed of the SUV immediately after the impact.

13.139 A baseball player catching a ball can soften the impact by pulling his hand back. Assuming that a 5-oz ball reaches his glove at 90 mi/h and that the player pulls his hand back during the impact at an average speed of 30 ft/s over a distance of 6 in., bringing the ball to a stop, determine the average impulsive force exerted on the player's hand.

Fig. P13.139

13.140 A 1.62-oz golf ball is hit with a golf club and leaves it with a velocity of 100 mi/h. We assume that for $0 \leq t \leq t_0$, where t_0 is the duration of the impact, the magnitude F of the force exerted on the ball can be expressed as $F = F_m \sin (\pi t/t_0)$. Knowing that $t_0 = 0.5$ ms, determine the maximum value F_m of the force exerted on the ball.

13.141 The triple jump is a track-and-field event in which an athlete gets a running start and tries to leap as far as he can with a hop, step, and jump. Shown in the figure is the initial hop of the athlete. Assuming that he approaches the takeoff line from the left with a horizontal velocity of 10 m/s, remains in contact with the ground for 0.18 s, and takes off at a 50° angle with a velocity of 12 m/s, determine the vertical component of the average impulsive force exerted by the ground on his foot. Give your answer in terms of the weight W of the athlete.

Fig. P13.141

13.142 The last segment of the triple jump track-and-field event is the jump, in which the athlete makes a final leap, landing in a sand-filled pit. Assuming that the velocity of a 80-kg athlete just before landing is 9 m/s at an angle of 35° with the horizontal and that the athlete comes to a complete stop in 0.22 s after landing, determine the horizontal component of the average impulsive force exerted on his feet during landing.

13.143 The design for a new cementless hip implant is to be studied using an instrumented implant and a fixed simulated femur. Assuming the punch applies an average force of 2 kN over a time of 2 ms to the 200-g implant, determine (*a*) the velocity of the implant immediately after impact, (*b*) the average resistance of the implant to penetration if the implant moves 1 mm before coming to rest.

Landing pit

Fig. P13.142

Fig. *P13.143*

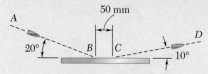

Fig. P13.144

13.144 A 28-g steel-jacketed bullet is fired with a velocity of 650 m/s toward a steel plate and ricochets along path *CD* with a velocity of 500 m/s. Knowing that the bullet leaves a 50-mm scratch on the surface of the plate and assuming that it has an average speed of 600 m/s while in contact with the plate, determine the magnitude and direction of the impulsive force exerted by the plate on the bullet.

13.145 A 25-ton railroad car moving at 2.5 mi/h is to be coupled to a 50-ton car that is at rest with locked wheels ($\mu_k = 0.30$). Determine (*a*) the velocity of both cars after the coupling is completed, (*b*) the time it takes for both cars to come to rest.

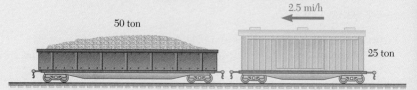

Fig. P13.145

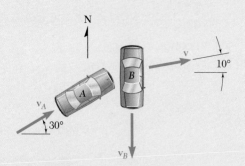

Fig. P13.146

13.146 At an intersection, car *B* was traveling south and car *A* was traveling 30° north of east when they slammed into each other. Upon investigation it was found that after the crash the two cars got stuck and skidded off at an angle of 10° north of east. Each driver claimed that he was going at the speed limit of 50 km/h and that he tried to slow down but couldn't avoid the crash because the other driver was going a lot faster. Knowing that the masses of cars *A* and *B* were 1500 kg and 1200 kg, respectively, determine (*a*) which car was going faster, (*b*) the speed of the faster of the two cars if the slower car was traveling at the speed limit.

13.147 The 650-kg hammer of a drop-hammer pile driver falls from a height of 1.2 m onto the top of a 140-kg pile, driving it 110 mm into the ground. Assuming perfectly plastic impact ($e = 0$), determine the average resistance of the ground to penetration.

13.148 A small rivet connecting two pieces of sheet metal is being clinched by hammering. Determine the impulse exerted on the rivet and the energy absorbed by the rivet under each blow, knowing that the head of the hammer has a weight of 1.5 lb and that it strikes the rivet with a velocity of 20 ft/s. Assume that the hammer does not rebound and that the anvil is supported by springs and (*a*) has an infinite mass (rigid support), (*b*) has a weight of 9 lb.

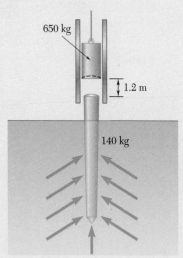

Fig. P13.147

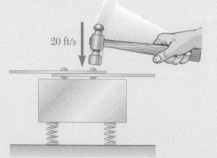

Fig. P13.148

13.149 Bullet B weighs 0.5 oz and blocks A and C both weigh 3 lb. The coefficient of friction between the blocks and the plane is $\mu_k = 0.25$. Initially the bullet is moving at v_0 and blocks A and C are at rest (Fig. 1). After the bullet passes through A it becomes embedded in block C and all three objects come to stop in the positions shown (Fig. 2). Determine the initial speed of the bullet v_0.

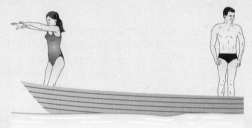

Fig. P13.149

13.150 A 180-lb man and a 120-lb woman stand at opposite ends of a 300-lb boat, ready to dive, each with a 16-ft/s velocity relative to the boat. Determine the velocity of the boat after they have both dived, if (a) the woman dives first, (b) the man dives first.

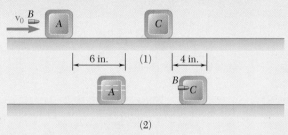

Fig. P13.150

13.151 A 75-g ball is projected from a height of 1.6 m with a horizontal velocity of 2 m/s and bounces from a 400-g smooth plate supported by springs. Knowing that the height of the rebound is 0.6 m, determine (a) the velocity of the plate immediately after the impact, (b) the energy lost due to the impact.

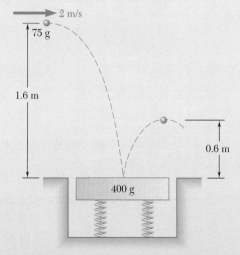

Fig. P13.151

13.152 A ballistic pendulum is used to measure the speed of high-speed projectiles. A 6-g bullet A is fired into a 1-kg wood block B suspended by a cord with a length of $l = 2.2$ m. The block then swings through a maximum angle of $\theta = 60°$. Determine (a) the initial speed of the bullet v_0, (b) the impulse imparted by the bullet on the block, (c) the force on the cord immediately after the impact.

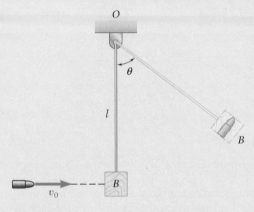

Fig. P13.152

13.153 A 1-oz bullet is traveling with a velocity of 1400 ft/s when it impacts and becomes embedded in a 5-lb wooden block. The block can move vertically without friction. Determine (a) the velocity of the bullet and block immediately after the impact, (b) the horizontal and vertical components of the impulse exerted by the block on the bullet.

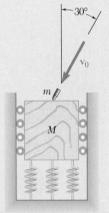

Fig. P13.153

13.154 In order to test the resistance of a chain to impact, the chain is suspended from a 240-lb rigid beam supported by two columns. A rod attached to the last link is then hit by a 60-lb block dropped from a 5-ft height. Determine the initial impulse exerted on the chain and the energy absorbed by the chain, assuming that the block does not rebound from the rod and that the columns supporting the beam are (a) perfectly rigid, (b) equivalent to two perfectly elastic springs.

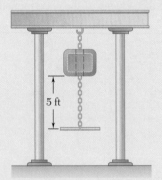

Fig. P13.154

13.4 IMPACTS

A collision between two bodies that occurs in a very small interval of time, and during which the two bodies exert relatively large forces on each other, is called an **impact**. The common normal to the surfaces in contact during the impact is called the **line of impact**. If the mass centers of the two colliding bodies are located on this line, the impact is called a **central impact**. Otherwise, the impact is said to be **eccentric**. Our present study is limited to the central impact of two particles. In Chapter 17, we consider the analysis of the eccentric impact of two rigid bodies.

If the velocities of the two particles are directed along the line of impact, the impact is said to be a **direct impact** (Fig. 13.20*a*). If either or both particles move along a line other than the line of impact, the impact is said to be an **oblique impac***t* (Fig. 13.20*b*).

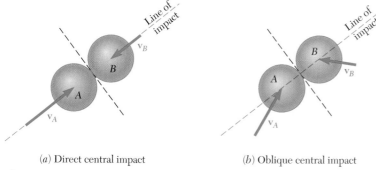

(*a*) Direct central impact   (*b*) Oblique central impact

Fig. 13.20   Central impacts can be (*a*) direct (or "head-on") or (*b*) oblique.

13.4A   Direct Central Impact

Consider two particles *A* and *B* with mass m_A and m_B that are moving in the same straight line and to the right with known velocities $\mathbf{v}_A$ and $\mathbf{v}_B$ (Fig. 13.21*a*). If $\mathbf{v}_A$ is larger than $\mathbf{v}_B$, particle *A* eventually strikes particle *B*. Under the impact, the two particles **deform**, and at the end of the period of deformation, they have the same velocity **u** (Fig. 13.21*b*). A period of **restitution** then takes place. At the end of this period, depending upon the magnitude of the impact forces and upon the materials involved, the two particles either have regained their original shape or will stay permanently deformed. Our purpose here is to determine the velocities $\mathbf{v}'_A$ and $\mathbf{v}'_B$ of the particles at the end of the period of restitution (Fig. 13.21*c*).

Considering first the two particles as a single system, we note that there is no impulsive, external force. Thus, the total linear momentum of the two particles is conserved, and we have

$$m_A\mathbf{v}_A + m_B\mathbf{v}_B = m_A\mathbf{v}'_A + m_B\mathbf{v}'_B \qquad \textbf{(13.34′)}$$

Since all of the velocities considered are directed along the same axis, we can replace this equation by the following relation involving only scalar components, as

$$m_A v_A + m_B v_B = m_A v'_A + m_B v'_B \qquad \textbf{(13.37)}$$

A positive value for any of the scalar quantities v_A, v_B, v'_A, or v'_B means that the corresponding vector is directed to the right; a negative value indicates that the corresponding vector is directed to the left.

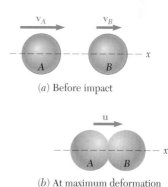

(*a*) Before impact

(*b*) At maximum deformation

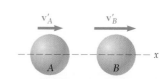

(*c*) After impact

Fig. 13.21   Every impact has three stages: (*a*) before the impact, (*b*) a maximum deformation when the particles have the same velocity, and (*c*) after the impact.

To obtain the velocities $\mathbf{v}'_A$ and $\mathbf{v}'_B$, it is necessary to establish a second relation between the scalars $\mathbf{v}'_A$ and $\mathbf{v}'_B$. For this purpose, let us now consider the motion of particle A during the period of deformation and apply the principle of impulse and momentum. Since the only impulsive force acting on A during this period is the force $\mathbf{P}$ exerted by B (Fig. 13.22a), we have, again using scalar components,

$$m_A v_A - \int P \, dt = m_A u \tag{13.38}$$

where the integral extends over the period of deformation. Considering now the motion of A during the period of restitution and denoting the force exerted by B on A during this period by $\mathbf{R}$ (Fig. 13.22b), we have

$$m_A u - \int R \, dt = m_A v'_A \tag{13.39}$$

where the integral extends over the period of restitution.

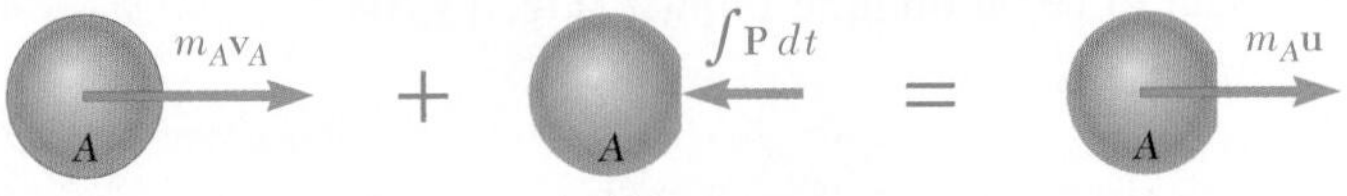

(a) Period of deformation

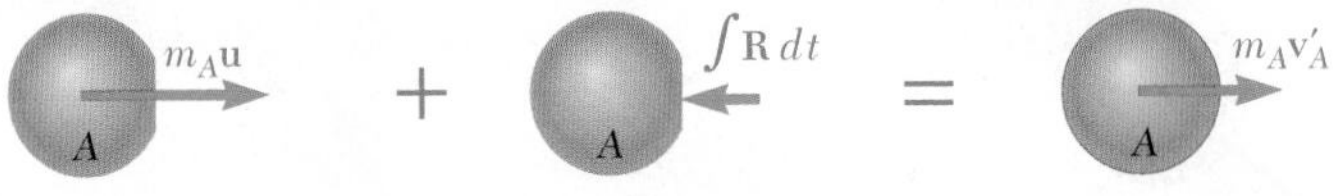

(b) Period of restitution

Fig. 13.22 Impulse-momentum diagram for particle A during (a) the period of deformation, and (b) during the period of restoration.

In general, the force $\mathbf{R}$ exerted on A during the period of restitution differs from the force $\mathbf{P}$ exerted during the period of deformation, and the magnitude $\int R \, dt$ of its impulse is smaller than the magnitude $\int P \, dt$ of the impulse of $\mathbf{P}$. The ratio of the magnitudes of the impulses, corresponding, respectively, to the period of restitution and to the period of deformation, is called the **coefficient of restitution** and is denoted by e. We have

$$e = \frac{\int R \, dt}{\int P \, dt} \tag{13.40}$$

The value of the coefficient e is always between 0 and 1. It depends to a large extent on the two materials involved, but it also varies considerably with the impact velocity and the shape and size of the two colliding bodies.

Solving Eqs. (13.38) and (13.39) for the two impulses and substituting into Eq. (13.40), we obtain

$$e = \frac{u - v'_A}{v_A - u} \tag{13.41}$$

A similar analysis of particle B leads to the relation

$$e = \frac{v'_B - u}{u - v_B} \tag{13.42}$$

Since the quotients in Eqs. (13.41) and (13.42) are equal, they are also equal to the quotient obtained by adding, respectively, their numerators and their denominators. We therefore have

$$e = \frac{(u - v'_A) + (v'_B - u)}{(v_A - u) + (u - v_B)} = \frac{v'_B - v'_A}{v_A - v_B}$$

and

Coefficient of restitution

$$v'_B - v'_A = e(v_A - v_B) \qquad \text{(13.43)}$$

Photo 13.5 The height the tennis ball bounces decreases after each impact because it has a coefficient of restitution less than one and energy is lost with each bounce.

Since $v'_B - v'_A$ represents the relative velocity of the two particles after impact and $v_A - v_B$ represents their relative velocity before impact, formula (13.43) says:

We can obtain the relative velocity of the two particles after impact by multiplying their relative velocity before impact by the coefficient of restitution.

This property is used to determine experimentally the value of the coefficient of restitution of two given materials.

We can now obtain the velocities of the two particles after impact by solving Eqs. (13.37) and (13.43) simultaneously for v'_A and v'_B. Recall that the derivations of Eqs. (13.37) and (13.43) were based on the assumption that particle B is located to the right of A and that both particles are initially moving to the right. If particle B is initially moving to the left, the scalar v_B should be considered negative. The same sign convention holds for the velocities after impact: A positive sign for v'_A indicates that particle A moves to the right after impact, and a negative sign indicates that it moves to the left.

Two particular cases of impact are of special interest.

1. $e = 0$, **Perfectly Plastic Impact**. When $e = 0$, Eq. (13.43) yields $v'_B = v'_A$. There is no period of restitution, and both particles stay together after impact. Substituting $v'_B = v'_A = v'$ into Eq. (13.37), which expresses that the total momentum of the particles is conserved, we have

$$m_A v_A + m_B v_B = (m_A + m_B)v' \qquad \text{(13.44)}$$

We can solve this equation for the common velocity v' of the two particles after impact.

2. $e = 1$, **Perfectly Elastic Impact**. When $e = 1$, Eq. (13.43) reduces to

$$v'_B - v'_A = v_A - v_B \qquad \text{(13.45)}$$

This equation says that the relative velocities before and after impact are equal. This means that the impulses received by each particle during the period of deformation and during the period of restitution are equal. We can obtain the velocities v'_A and v'_B by solving Eqs. (13.37) and (13.45) simultaneously.

It is worth noting that **in the idealized case of a perfectly elastic impact, the total energy of the two particles**, as well as their total momentum, **is conserved**. We can write Eqs. (13.37) and (13.45) as

$$m_A(v_A - v'_A) = m_B(v'_B - v_B) \qquad \text{(13.37')}$$
$$v_A + v'_A = v_B + v'_B \qquad \text{(13.45')}$$

Multiplying Eqs. (13.37') and (13.45') member by member, we have

$$m_A(v_A - v'_A)(v_A + v'_A) = m_B(v'_B - v_B)(v'_B + v_B)$$
$$m_A v_A^2 - m_A(v'_A)^2 = m_B(v'_B)^2 - m_B v_B^2$$

Rearranging the terms in this equation and multiplying by 1/2, we obtain

$$\tfrac{1}{2}m_A v_A^2 + \tfrac{1}{2}m_B v_B^2 = \tfrac{1}{2}m_A(v'_A)^2 + \tfrac{1}{2}m_B(v'_B)^2 \qquad \text{(13.46)}$$

This equation states that the kinetic energy of the particles is conserved.

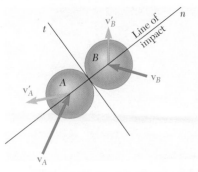

Fig. 13.23 In an oblique central impact, the velocities of the colliding particles are not directed along the line of impact.

Photo 13.6 When pool balls strike each other there is a transfer of momentum.

Note, however, that **in the general case of impact**, i.e., when e is not equal to 1, **the total energy of the particles is not conserved**. This can be shown in any given case by comparing the kinetic energies before and after impact. The lost kinetic energy may be transformed into other forms of energy, such as heat, sound, generation of elastic waves within the two colliding bodies, or permanent deformation of the bodies.

13.4B Oblique Central Impact

Let us now consider the case when the velocities of the two colliding particles are *not* directed along the line of impact (Fig. 13.23). As mentioned earlier, the impact is said to be **oblique**. Since the velocities $\mathbf{v}'_A$ and $\mathbf{v}'_B$ of the particles after impact are unknown in direction as well as in magnitude, their determination requires the use of four independent equations.

We choose as coordinate axes the n axis along the line of impact (i.e., along the common normal to the surfaces in contact) and the t axis along their common tangent. In very special cases where we can assume that the particles are perfectly smooth and frictionless, we observe that the only impulses exerted on the particles during the impact are due to internal forces directed along the line of impact, i.e., along the n axis (Fig. 13.24). This leads to the following results.

1. The component along the t axis of the momentum of each particle, considered separately, is conserved because no impulses act in the t direction; hence the t component of the velocity of each particle remains unchanged. We have

$$(v_A)_t = (v'_A)_t \qquad (v_B)_t = (v'_B)_t \tag{13.47}$$

2. The component along the n axis of the total momentum of the two particles is conserved because the two impulses are equal and opposite to one another. We have

$$m_A(v_A)_n + m_B(v_B)_n = m_A(v'_A)_n + m_B(v'_B)_n \tag{13.48}$$

3. We obtain the component along the n axis of the relative velocity of the two particles after impact by multiplying the n component of their relative velocity before impact by the coefficient of restitution. Indeed, a derivation similar to that given in Sec. 13.4A for direct central impact yields

$$(v'_B)_n - (v'_A)_n = e[(v_A)_n - (v_B)_n] \tag{13.49}$$

We have thus obtained four independent equations that can be solved for the components of the velocities of A and B after impact. This method of solution is illustrated in Sample Prob. 13.20.

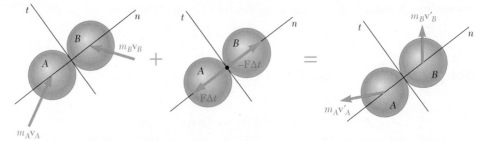

Fig. 13.24 Impulse-momentum diagram for an oblique impact. By including the internal impulses as equal and opposite, you also have the impulse-momentum diagram for each individual object (just ignore the other object).

Our analysis of the oblique central impact of two particles has been based so far on the assumption that both particles move freely before and after the impact. Let us now examine the case when one or both of the colliding particles is constrained in its motion. Consider, for instance, the collision between block A, which is constrained to move on a horizontal surface, and ball B, which is free to move in the plane of the figure (Fig. 13.25). Assuming no friction between the block and the ball or between the block and the horizontal surface, we note that the impulses exerted on the system consist of the impulses of the internal forces $\mathbf{F}$ and $-\mathbf{F}$ directed along the line of impact, i.e., along the n axis, and of the impulse of the external force $\mathbf{F}_{\text{ext}}$ exerted by the horizontal surface on block A and directed along the vertical, as shown in the impulse-momentum diagram (Fig. 13.26).

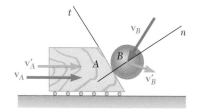

Fig. 13.25 An impact between a block moving on a horizontal surface and a ball moving in the vertical plane is called a "constrained impact."

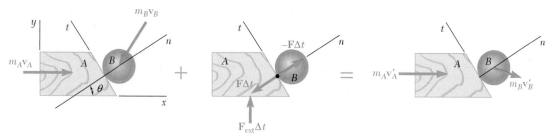

Fig. 13.26 Impulse-momentum diagram for a constrained impact between block A and ball B.

The velocities of block A and ball B immediately after the impact are represented by three unknowns: the magnitude of the velocity $\mathbf{v}'_A$ of block A, which is known to be horizontal, and the magnitude and direction of the velocity $\mathbf{v}'_B$ of ball B. We must therefore write three equations. We do this by using the impulse-momentum diagram and observing the following behavior.

1. The component along the t axis of the momentum of ball B is conserved because no impulses act on the ball in the t direction; hence, the t component of the velocity of ball B remains unchanged. We have

$$(v_B)_t = (v'_B)_t \tag{13.50}$$

2. The component along the horizontal x axis of the total momentum of block A and ball B is conserved because no external impulses act in the x-direction. We write this as

$$m_A v_A + m_B (v_B)_x = m_A v'_A + m_B (v'_B)_x \tag{13.51}$$

3. We obtain the component along the n axis of the relative velocity of block A and ball B after impact by multiplying the n component of their relative velocity before impact by the coefficient of restitution. We again have

$$(v'_B)_n - (v'_A)_n = e[(v_A)_n - (v_B)_n] \tag{13.49}$$

Note, however, that in the case considered here, we cannot establish the validity of Eq. (13.49) through a mere extension of the derivation given in Sec. 13.4A for the direct central impact of two particles moving in a straight line. Indeed, these particles were not subjected to any external impulse, whereas block A in the present analysis is subjected to the impulse exerted by the horizontal surface. To prove that Eq. (13.49) is still valid, we first apply the principle of impulse and momentum to block A

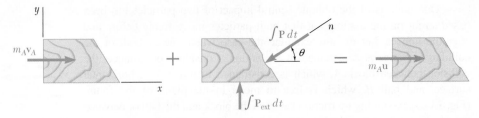

Fig. 13.27 Impulse-momentum diagram for block A.

over the period of deformation (Fig. 13.27). Considering only the horizontal components, we have

$$m_A v_A - (\int P\, dt) \cos \theta = m_A u \tag{13.52}$$

where the integral extends over the period of deformation and $\mathbf{u}$ represents the velocity of block A at the end of that period. Considering now the period of restitution, we have similarly

$$m_A u - (\int R\, dt) \cos \theta = m_A v_A' \tag{13.53}$$

where the integral extends over the period of restitution.

Recalling from Sec. 13.4A the definition of the coefficient of restitution, we have

$$e = \frac{\int R\, dt}{\int P\, dt} \tag{13.40}$$

Solving Eqs. (13.52) and (13.53) for the integrals $\int P\, dt$ and $\int R\, dt$ and substituting into Eq. (13.40), we have, after reductions,

$$e = \frac{u - v_A'}{v_A - u}$$

Then multiplying all velocities by $\cos \theta$ to obtain their projections on the line of impact gives

$$e = \frac{u_n - (v_A')_n}{(v_A)_n - u_n} \tag{13.54}$$

Note that Eq. (13.54) is identical to Eq. (13.41) except for the subscripts n that we use here to indicate that we are considering velocity components along the line of impact. Since the motion of ball B is unconstrained, we can complete the proof of Eq. (13.49) in the same manner as the derivation of Eq. (13.43). Thus, we conclude that the relation in Eq. (13.49) between the components along the line of impact of the relative velocities of two colliding particles remains valid when one of the particles is constrained in its motion. The validity of this relation is easily extended to the case when both particles are constrained in their motion.

13.4C Problems Involving Multiple Principles

You now have at your disposal three different methods for the solution of kinetics problems.

- The direct application of Newton's second law, $\Sigma \mathbf{F} = m\mathbf{a}$.
- The method of work and energy, $T_1 + V_{g_1} + V_{e_1} + U_{1 \to 2}^{NC} = T_2 + V_{g_2} + V_{e_2}$, where $U_{1 \to 2}^{NC}$ is the work of external non-conservative forces such as friction.
- The method of impulse and momentum, $m\mathbf{v}_1 + \mathbf{Imp}_{1 \to 2} = m\mathbf{v}_2$.

To derive maximum benefit from these three methods, you should be able to choose the method best suited for the solution of a given problem. You also should be prepared to solve problems that require you to use multiple principles.

You have already seen that the method of work and energy is in many cases more expeditious than the direct application of Newton's second law. As indicated in Sec. 13.1C, however, the method of work and energy has limitations, and it must sometimes be supplemented by the use of $\Sigma \mathbf{F} = m\mathbf{a}$. This is the case, for example, when you wish to determine an acceleration or a normal force.

For the solution of problems involving no impulsive forces, usually the equation $\Sigma \mathbf{F} = m\mathbf{a}$ yields a solution just as fast as the method of impulse and momentum, and the method of work and energy (if it applies) is more rapid and more convenient. However, in problems involving impact, the method of impulse and momentum is the only practicable method. A solution based on the direct application of $\Sigma \mathbf{F} = m\mathbf{a}$ would be unwieldy, and the method of work and energy cannot be used, because impact (unless perfectly elastic) involves a loss of mechanical energy.

Many problems involve only conservative forces except for a short impact phase during which impulsive forces act. The solution of such problems can be divided into several parts. The part corresponding to the impact phase calls for the use of the method of impulse and momentum and of the relation between relative velocities. The other parts usually can be solved by using the method of work and energy. If the problem involves the determination of a normal force, however, the use of $\Sigma \mathbf{F} = m\mathbf{a}$ is necessary.

Consider, for example, a pendulum A, with a mass m_A and a length l, that is released with no velocity from a position A_1 (Fig. 13.28a). The pendulum swings freely in a vertical plane and hits a second pendulum B, with a mass m_B and the same length l, that is initially at rest. After the impact (with coefficient of restitution e), pendulum B swings through an angle θ that we wish to determine.

The solution of the problem can be divided into three parts:

1. **Pendulum A Swings from A_1 to A_2.** Use the principle of conservation of energy to determine the velocity $(\mathbf{v}_A)_2$ of the pendulum at A_2 (Fig. 13.28b).

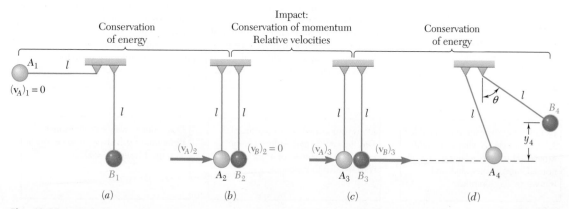

Fig. 13.28 Analyzing an impact between two pendulum bobs by conservation of energy and conservation of momentum.

2. **Pendulum *A* Hits Pendulum *B*.** Use the fact that the total momentum of the two pendulums is conserved, and use the relation between their relative velocities—that is, the coefficient of restitution—to determine the velocities $(\mathbf{v}_A)_3$ and $(\mathbf{v}_B)_3$ of the two pendulums after impact (Fig. 13.28c).

3. **Pendulum *B* Swings from B_3 to B_4.** Apply the principle of conservation of energy to pendulum *B* to determine the maximum elevation y_4 reached by that pendulum (Fig. 13.28d). You can then determine the angle θ by trigonometry.

Note that if you need to determine the tensions in the cords holding the pendulums, the method of solution just described should be supplemented by the use of $\Sigma\mathbf{F} = m\mathbf{a}$. A summary of all the kinetics principles we have discussed so far and some clues as to when to apply them are shown in Fig. 13.29.

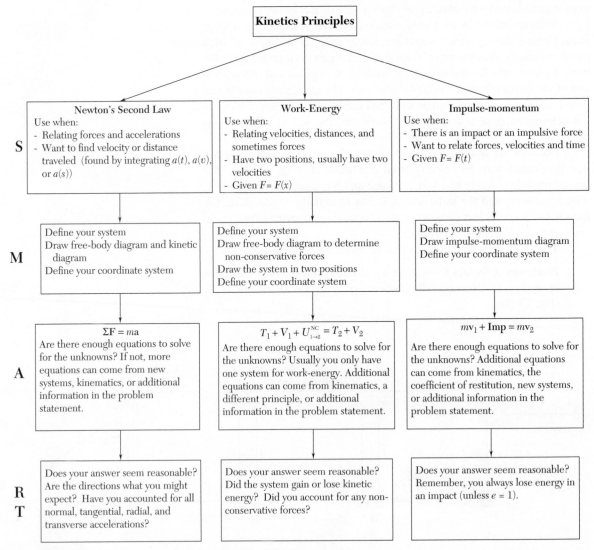

Fig. 13.29 The three kinetics principles using the SMART methodology.

Sample Problem 13.18

A 20-Mg railroad car moving at a speed of 0.5 m/s to the right collides with a 35-Mg car at rest. After the collision, the 35-Mg car moves to the right at a speed of 0.3 m/s. Determine the coefficient of restitution between the two cars.

STRATEGY: Since there is an impact and no external impulses, use the conservation of linear momentum. You will also need to use the equation for the coefficient of restitution.

MODELING: Choose your system to be both railroad cars and model them as particles. The impulse-momentum diagram for this system is shown in Fig. 1. There are no external impulses acting on this system.

Fig. 1 Velocities and linear momenta of the cars before and after impact.

ANALYSIS: The total momentum of the two cars is conserved, so

$$m_A \mathbf{v}_A + m_B \mathbf{v}_B = m_A \mathbf{v}'_A + m_B \mathbf{v}'_B$$

Substituting in the known values gives

$$(20 \text{ Mg})(+0.5 \text{ m/s}) + (35 \text{ Mg})(0) = (20 \text{ Mg})v'_A + (35 \text{ Mg})(+0.3 \text{ m/s})$$
$$v'_A = -0.025 \text{ m/s} \qquad \mathbf{v}'_A = 0.025 \text{ m/s} \leftarrow$$

You can obtain the coefficient of restitution from its definition as

$$e = \frac{v'_B - v'_A}{v_A - v_B} = \frac{+0.3 - (-0.025)}{+0.5 - 0} = \frac{0.325}{0.5}$$

$$e = 0.65 \quad \blacktriangleleft$$

REFLECT and THINK: The railroad cars are constrained to move along the track, so this is a one-dimensional direct central impact. The interaction forces are large, but they last for only a very short time. Mechanical energy is lost during this impact, so you could not have used the conservation of energy.

Sample Problem 13.19

A ball is thrown against a frictionless, vertical wall. Immediately before the ball strikes the wall, its velocity has a magnitude of v and forms an angle of 30° with the horizontal. Knowing that $e = 0.90$, determine the magnitude and direction of the velocity of the ball as it rebounds from the wall.

STRATEGY: An impact occurs, and you are given the coefficient of restitution, so use conservation of momentum and the definition of the coefficient of restitution.

MODELING: Choose your system to be the ball, and model it as a particle. The impulse-momentum diagram for this system is shown in Fig. 1.

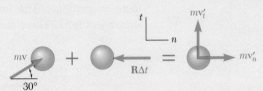

Fig. 1 Impulse-momentum diagram for the ball.

ANALYSIS: Resolve the initial velocity of the ball into components perpendicular and parallel to the wall, as shown in Fig. 2.

$$v_n = v \cos 30° = 0.866v \qquad v_t = v \sin 30° = 0.500v$$

Motion Parallel to the Wall. Since the wall is frictionless, the impulse it exerts on the ball is perpendicular to the wall. Thus, the component of the momentum of the ball parallel to the wall is conserved. You have

$$\mathbf{v}'_t = \mathbf{v}_t = 0.500v \uparrow$$

Motion Perpendicular to the Wall. Since the mass of the wall (and of the earth) is essentially infinite, writing an equation for conservation of the total momentum of the ball and wall would yield no useful information. However, using the equation for coefficient of restitution, you have

$$0 - v'_n = e(v_n - 0)$$
$$v'_n = -0.90(0.866v) = -0.779v \qquad \mathbf{v}'_n = 0.779v \leftarrow$$

Resultant Motion. Adding vectorially the components $\mathbf{v}'_n$ and $\mathbf{v}'_t$ (Fig. 3), you find

$$\mathbf{v}' = 0.926v \searrow 32.7° \quad \blacktriangleleft$$

REFLECT and THINK: Tests similar to this are done to make sure that sporting equipment—such as tennis balls, golf balls, and basketballs—are consistent and fall within certain specifications. Testing modern golf balls and clubs shows that the coefficient of restitution actually decreases with increasing club speed (from about 0.84 at a speed of 90 mph to about 0.80 at club speeds of 130 mph).

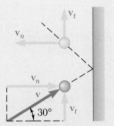

Fig. 2 Components of the initial velocity.

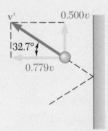

Fig. 3 Finding the magnitude and direction for the final velocity.

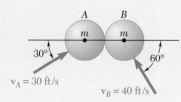

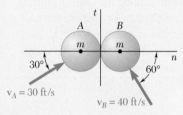

Fig. 1 Initial velocities of A and B and the coordinate system to be used.

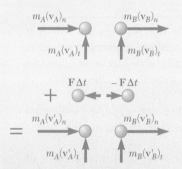

Fig. 2 Impulse-momentum diagram for the system.

Sample Problem 13.20

The magnitudes and directions of the velocities of two identical friction-less balls are shown before they strike each other. Assuming $e = 0.90$, determine the magnitude and direction of the velocity of each ball after the impact.

STRATEGY: Since an impact occurs, use the principle of impulse and momentum. You also need the equation for the coefficient of restitution.

MODELING: Choose your system to be both balls. Assuming they are small and do not rotate, you can model them as particles. Figure 1 shows the normal and tangential directions; Fig. 2 shows the impulse-momentum diagram for this system. The impulsive forces that the balls exert on each other during the impact are directed along a line joining the centers of the balls (the line of impact). Therefore, it is best to resolve the velocities into components directed, respectively, along the line of impact and along the common tangent to the surfaces in contact. Thus,

$$(v_A)_n = v_A \cos 30° = +26.0 \text{ ft/s}$$
$$(v_A)_t = v_A \sin 30° = +15.0 \text{ ft/s}$$
$$(v_B)_n = -v_B \cos 60° = -20.0 \text{ ft/s}$$
$$(v_B)_t = v_B \sin 60° = +34.6 \text{ ft/s}$$

ANALYSIS:

Motion Along the Common Tangent. Considering only the t components, apply the principle of impulse and momentum to each ball *separately*. Since the impulsive forces are directed along the line of impact, the t component of the momentum, and hence the t component of the velocity of each ball, is unchanged. You have

$$(\mathbf{v}'_A)_t = 15.0 \text{ ft/s} \uparrow \qquad (\mathbf{v}'_B)_t = 34.6 \text{ ft/s} \uparrow$$

Motion Along the Line of Impact. In the n direction, consider the two balls as a single system. By Newton's third law, the internal impulses are, respectively, $\mathbf{F} \, \Delta t$ and $-\mathbf{F} \, \Delta t$, so they cancel. Thus, the total momentum of the balls is conserved as

$$m_A(v_A)_n + m_B(v_B)_n = m_A(v'_A)_n + m_B(v'_B)_n$$
$$m(26.0) + m(-20.0) = m(v'_A)_n + m(v'_B)_n$$
$$(v'_A)_n + (v'_B)_n = 6.0 \quad (1)$$

Using the equation for the coefficient of restitution relating the relative velocities, you have

$$(v'_B)_n - (v'_A)_n = e[(v_A)_n - (v_B)_n]$$

You can now substitute the known quantities into this equation. It is important to use the signs correctly when substituting into this equation, e.g. $(\mathbf{v}_B)_n = -20$. This gives

$$(v'_B)_n - (v'_A)_n = (0.90)[26.0 - (-20.0)]$$
$$(v'_B)_n - (v'_A)_n = 41.4 \quad (2)$$

(continued)

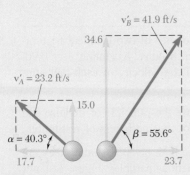

Fig. 3 The velocity components can be resolved into their magnitudes and directions.

$v'_B = 41.9$ ft/s

34.6

$v'_A = 23.2$ ft/s

15.0

$\alpha = 40.3°$

17.7

$\beta = 55.6°$

23.7

Solving Eqs. (1) and (2) simultaneously yields

$$(v'_A)_n = -17.7 \qquad (v'_B)_n = +23.7$$
$$(\mathbf{v}'_A)_n = 17.7 \text{ ft/s} \leftarrow \qquad (\mathbf{v}'_B)_n = 23.7 \text{ ft/s} \rightarrow$$

Resultant Motion. Adding the velocity components of each ball vectorially (Fig. 3), you obtain

$$\mathbf{v}'_A = 23.2 \text{ ft/s} \searrow 40.3° \qquad \mathbf{v}'_B = 41.9 \text{ ft/s} \nearrow 55.6° \blacktriangleleft$$

REFLECT and THINK: Rather than choosing your system to be both balls, you could have applied impulse-momentum along the line of impact for each ball individually. This would have resulted in two equations and one additional unknown, $F\Delta t$. To determine the impulsive force F, you would need to be given the time for the impact, Δt.

Sample Problem 13.21

Ball B is hanging from an inextensible cord BC. An identical ball A is released from rest when it is just touching the cord and acquires a velocity $\mathbf{v}_0$ before striking ball B. Assuming a perfectly elastic impact ($e = 1$) and no friction, determine the velocity of each ball immediately after impact.

STRATEGY: Since an impact occurs, use the impulse-momentum principle. You also need the equation for coefficient of restitution.

MODELING: You have several choices of systems in this problem. If you choose A as your system, you obtain the impulse-momentum diagram shown in Fig. 1. Choosing the system to be both balls results in the impulse-momentum diagram shown in Fig. 2.

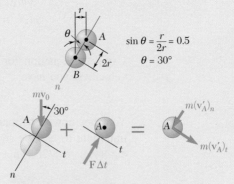

$$\sin \theta = \frac{r}{2r} = 0.5$$
$$\theta = 30°$$

Fig. 1 Impulse-momentum diagram for ball A.

ANALYSIS:

Impulse-Momentum Principle: Ball A. Applying the conservation of momentum of ball A along the common tangent to balls A and B (Fig. 1) gives

$$m\mathbf{v}_A + \mathbf{F}\,\Delta t = m\mathbf{v}_A'$$

$+\searrow t$ components: $\qquad mv_0 \sin 30° + 0 = m(v_A')_t$

$$(v_A')_t = 0.5v_0 \qquad\qquad (1)$$

Impulse-Momentum Principle: Balls A and B. Since ball B is constrained to move in a circle with center C, its velocity $\mathbf{v}_B$ after impact must be horizontal. Applying the conservation of momentum for a system containing both balls (Fig. 2) gives

$$m\mathbf{v}_A + \mathbf{T}\,\Delta t = m\mathbf{v}_A' + m\mathbf{v}_B'$$

$\xrightarrow{+} x$ components: $\qquad 0 = m(v_A')_t \cos 30° - m(v_A')_n \sin 30° - mv_B'$

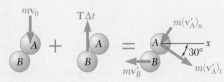

Fig. 2 Impulse-momentum diagram for both balls.

This equation expresses the conservation of total momentum in the x direction. Substituting for $(v_A')_t$ from Eq. (1) and rearranging terms, you have

$$0.5(v_A')_n + v_B' = 0.433v_0 \qquad\qquad (2)$$

Relative Velocities Along the Line of Impact. Since $e = 1$, the equation for the coefficient of restitution gives

$$(v_B')_n - (v_A')_n = (v_A)_n - (v_B)_n$$
$$v_B' \sin 30° - (v_A')_n = v_0 \cos 30° - 0$$
$$0.5v_B' - (v_A')_n = 0.866v_0 \qquad\qquad (3)$$

It is important to note that the coefficient of restitution always uses the components of the velocities along the line of impact; that is, the n-direction. Solving Eqs. (2) and (3) simultaneously, you obtain

$$(v_A')_n = -0.520v_0 \qquad v_B' = 0.693v_0$$

$$\mathbf{v}_B' = 0.693v_0 \leftarrow \quad\blacktriangleleft$$

Recalling Eq. (1), draw a sketch (Fig. 3) and obtain by trigonometry

$$v_A' = 0.721v_0 \qquad \beta = 46.1° \qquad \alpha = 46.1° - 30° = 16.1°$$

$$\mathbf{v}_B' = 0.721v_0 \measuredangle 16.1° \quad\blacktriangleleft$$

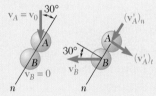

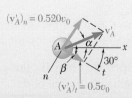

Fig. 3 Diagram to find the magnitude and direction for the final velocity of B.

REFLECT and THINK: Since $e = 1$, the impact between A and B is perfectly elastic. Therefore, rather than using the coefficient of restitution, you could have used the conservation of energy as your final equation.

Sample Problem 13.22

A 30-kg block is dropped from a height of 2 m onto the 10-kg pan of a spring scale. The constant of the spring is $k = 20$ kN/m. Assuming the impact to be perfectly plastic, determine the maximum deflection of the pan.

STRATEGY: This problem has three distinct phases, as shown in Fig. 1. In phase 1, A falls (use the conservation of energy); in phase 2, A hits B (use the conservation of momentum); and in phase 3, A and B move down together (use the conservation of energy).

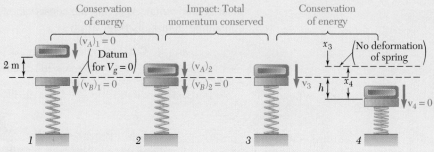

Fig. 1 Three phases of the motion.

MODELING: For each phase of the motion, define a different system. For phase 1, choose A as your system, and for phase 2, define your system as A and B together. For phase 3, your system is A, B, and the spring.

ANALYSIS:

Conservation of Energy A. Block A weighs

$$W_A = (30 \text{ kg})(9.81 \text{ m/s}^2) = 294 \text{ N}.$$

Thus,

$$T_1 = \tfrac{1}{2}m_A(v_A)_1^2 = 0 \qquad V_1 = W_A y = (294 \text{ N})(2 \text{ m}) = 588 \text{ J}$$
$$T_2 = \tfrac{1}{2}m_A(v_A)_2^2 = \tfrac{1}{2}(30 \text{ kg})(v_A)_2^2 \qquad V_2 = 0$$
$$T_1 + V_1 = T_2 + V_2: \qquad 0 + 588 \text{ J} = \tfrac{1}{2}(30 \text{ kg})(v_A)_2^2 + 0$$
$$(v_A)_2 = +6.26 \text{ m/s} \qquad (\mathbf{v}_A)_2 = 6.26 \text{ m/s} \downarrow$$

Impact: Conservation of Momentum for A and B. The impact is perfectly plastic, so $e = 0$; the block and pan move together after the impact.

$$m_A(v_A)_2 + m_B(v_B)_2 = (m_A + m_B)v_3$$
$$(30 \text{ kg})(6.26 \text{ m/s}) + 0 = (30 \text{ kg} + 10 \text{ kg})v_3$$
$$v_3 = +4.70 \text{ m/s} \qquad \mathbf{v}_3 = 4.70 \text{ m/s} \downarrow$$

Conservation of Energy for A, B, and the Spring. Initially, the spring supports the weight W_B of the pan; thus the initial deflection of the spring is

$$x_3 = \frac{W_B}{k} = \frac{(10 \text{kg})(9.81 \text{ m/s}^2)}{20 \times 10^3 \text{ N/m}} = \frac{98.1 \text{ N}}{20 \times 10^3 \text{ N/m}} = 4.91 \times 10^{-3} \text{ m}$$

Denoting the total maximum deflection of the spring by x_4, you have

$$T_3 = \tfrac{1}{2}(m_A + m_B)v_3^2 = \tfrac{1}{2}(30 \text{ kg} + 10 \text{ kg})(4.70 \text{ m/s})^2 = 442 \text{ J}$$
$$V_3 = V_g + V_e = 0 + \tfrac{1}{2}kx_3^2 = \tfrac{1}{2}(20 \times 10^3)(4.91 \times 10^{-3})^2 = 0.241 \text{ J}$$
$$T_4 = 0$$
$$V_4 = V_g + V_e = (W_A + W_B)(-h) + \tfrac{1}{2}kx_4^2 = -(392)h + \tfrac{1}{2}(20 \times 10^3)x_4^2$$

The displacement of the pan is $h = x_4 - x_3$, so the final result is

$$T_3 + V_3 = T_4 + V_4:$$
$$442 + 0.241 = 0 - 392(x_4 - 4.91 \times 10^{-3}) + \tfrac{1}{2}(20 \times 10^3)x_4^2$$
$$x_4 = 0.230 \text{ m} \qquad h = x_4 - x_3 = 0.230 \text{ m} - 4.91 \times 10^{-3} \text{ m}$$
$$h = 0.225 \text{ m} \qquad\qquad h = 225 \text{ mm} \quad ◀$$

REFLECT and THINK: The spring constant for this scale is pretty large, but the block is fairly massive and is dropped from a height of 2 m. From this perspective, the deflection seems reasonable.

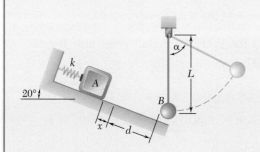

Sample Problem 13.23

A 2-kg block A is pushed up against a spring, compressing it a distance $x = 0.1$ m. The block is then released from rest and slides down the 20° incline until it strikes a 1-kg sphere B that is suspended by a 1-m inextensible rope. The spring constant $k = 800$ N/m, the coefficient of friction between A and the ground is 0.2, block A slides from the unstretched length of the spring a distance $d = 1.5$ m, and the coefficient of restitution between A and B is 0.8. When $\alpha = 40°$, determine (a) the speed of B, (b) the tension in the rope.

STRATEGY: A lot of things are going on in this problem, so you need to break the motion into steps.

Step 1: Block A slides down the incline, so there are two positions. Therefore, use the work–energy principle between position 1 and position 2 to find the velocity of A just before it strikes ball B (Fig. 1).

Step 2: Block A hits B, so an impact occurs. Therefore, use impulse-momentum and the equation for the coefficient of restitution.

Step 3: Ball B is swinging up, so you have two positions (position 2 and position 3 in Fig. 1). You are asked to find the speed at position 3, therefore, use the conservation of energy.

Step 4: To find the tension when $\alpha = 40°$, use Newton's second law with normal and tangential coordinates.

Fig. 1 Three positions of interest for this problem.

(continued)

MODELING: Each step requires a different system. For Step 1, your system is A and the spring. For Step 2, it is A and B. Finally, for Steps 3 and 4, it is B. We model A and B as particles and draw the appropriate figures in the analysis section.

ANALYSIS:

Step 1. Block Slides Down the Incline. The principle of work and energy between where the block is released to the point it strikes B is

$$T_1 + V_{g_1} + V_{e_1} + U^{NC}_{1 \to 2} = T_2 + V_{g_2} + V_{e_2} \tag{1}$$

Work. The only non-conservative force that does work is the friction force. A free-body diagram for A is shown in Fig. 2. Applying Newton's second law gives

$$+\nearrow \Sigma F_y = 0: \qquad N - m_A g \cos \theta = 0 \qquad \text{or}$$
$$N = m_A g \cos \theta = (2 \text{ kg})(9.81 \text{ m/s}^2) \cos 20° = 18.437 \text{ N}$$

and the friction force is

$$F_f = \mu_k N = (0.2)(18.437 \text{ N}) = 3.687 \text{ N}$$

So the work is

$$U^{NC}_{1 \to 2} = -F_f(x + d) = -(3.687 \text{ N})(1.6 \text{ m}) = -5.900 \text{ J}$$

Position 1. Place your datum for V_g at the impact point near B (see Fig. 1). Calculate the initial energy as

$$T_1 = 0, \quad V_{e_1} = \tfrac{1}{2}kx_1^2 = \frac{1}{2}(800)(0.1)^2 = 4.00 \text{ J}$$
$$V_{g_1} = m_A g h_1 = m_A g(x + d)\sin \theta = (2)(9.81)(1.6) \sin 20° = 10.737 \text{ J}$$

Position 2. The energy at position 2 is

$$T_2 = \tfrac{1}{2}m_A v_A^2 = \tfrac{1}{2}(2)v_A^2 = 1.000v_A^2 \qquad V_2 = 0$$

Substituting into Eq. (1) gives $0 + 10.737 \text{ J} + 4.00 \text{ J} - 5.900 \text{ J} = 1.00\, v_A^2 + 0$. Solving for v_A gives $v_A = 2.973$ m/s.

Step 2. Impact. The impulse-momentum diagram for A and B is shown in Fig. 3.

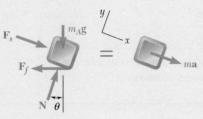

Fig. 2 Free-body diagram and kinetic diagram for block A.

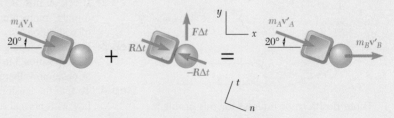

Fig. 3 Impulse-momentum diagram for A and B.

Note that two coordinate systems are defined: n defines the line of impact between the block and ball and y is in the direction of the impulsive force of the rope. Since no impulsive forces act in the horizontal direction, apply impulse-momentum in the x direction. Thus,

$\xrightarrow{+}x$ components: $\qquad m_A v_A \cos\theta + 0 = m_A v'_A \cos\theta + m_B v'_B$ (2)

Coefficient of Restitution.

$$(v'_B)_n - (v'_A)_n = e[(v_A)_n - (v_B)_n] \quad \text{or} \quad v_B \cos\theta - v'_A = e v_A \qquad (3)$$

In Eqs. (2) and (3) you can solve for two unknowns, v'_A and v'_B. This gives

$$v'_A = 1.0382 \text{ m/s}$$
$$v'_B = 3.6356 \text{ m/s}$$

Step 3. Sphere _B_ Rises. The tension does no work, so use the conservation of energy for B between positions 2 and 3. Again, define the datum as shown in Fig 1.

$$T_2 + V_{g_2} + V_{e_2} = T_3 + V_{g_3} + V_{e_3} \qquad (4)$$

Position 2.

$$T_2 = \frac{1}{2}m_B(v'_B)^2, \; V_{g_2} = 0, \; V_{e_2} = 0$$

Position 3.

$$T_3 = \frac{1}{2}m_B v_3^2, \; V_{g_3} = m_B g L(1 - \cos\alpha), V_{e_3} = 0$$

Substituting these into Eq. (4) and solving for v_{B_3} gives

$$v_{B_3} = 2.94 \text{ m/s} \quad \blacktriangleleft$$

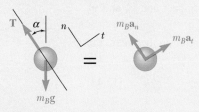

Fig. 4

Step 4. Tension in the Rope. A free-body diagram and kinetic diagram for the sphere at position 3 are shown in Fig. 4. Applying Newton's second law in the normal direction gives

$$+\nwarrow \Sigma F_n = m_B a_n: \qquad T - m_B g \cos\alpha = m_B a_n = m_B \frac{v_{B_3}^2}{L}$$

Solving for T, you find

$$T = 16.14 \text{ N} \quad \blacktriangleleft$$

REFLECT and THINK: You cannot use work–energy from position 1 to position 3 because a loss of energy occurs when A hits B. If the coefficient of friction had been larger, say $\mu_k = 0.4$, you would find that after the impact, B has a speed of 2.10 m/s. Plugging this into Eq. (4) gives an imaginary number for the speed at $\alpha = 40°$, meaning sphere B does not reach this angle.

SOLVING PROBLEMS
ON YOUR OWN

This section deals with the **impact of two smooth bodies**, i.e., with a collision occurring in a very small interval of time. You solved several impact problems by noting that the total momentum of the two bodies is conserved and by expressing the relationship between the relative velocities of the two bodies before and after impact.

1. **As a first step in your solution,** you should select and draw two coordinate axes: the t axis, which is tangent to the surfaces of contact of the two colliding bodies; and the n axis, which is normal to the surfaces of contact and defines the line of impact. In all of the problems in this section, the line of impact passes through the mass centers of the colliding bodies, and the impact is referred to as a **central impact**.

2. **Next draw an impulse-momentum diagram** showing the momenta of the bodies before impact, the impulses exerted on the bodies during impact, and the final momenta of the bodies after impact (Fig. 13.24). Then observe whether the impact is a **direct central impact** or an **oblique central impact**.

3. **Direct central impact** [Sample Prob. 13.18]. This occurs when the velocities of bodies A and B are both directed along the line of impact before impact (Fig. 13.20a).

    a. Conservation of momentum. Since the impulsive forces are internal to the system, the total momentum of A and B is conserved as

$$m_A v_A + m_B v_B = m_A v_A' + m_B v_B' \qquad \textbf{(13.37)}$$

where v_A and v_B denote the velocities of bodies A and B before impact and v_A' and v_B' denote their velocities after impact.

    b. Coefficient of restitution. You can also write the relation between the relative velocities of the two bodies before and after impact as

$$v_B' - v_A' = e(v_A - v_B) \qquad \textbf{(13.43)}$$

where e is the coefficient of restitution between the two bodies.

    Note that Eqs. (13.37) and (13.43) are scalar equations that you can solve for two unknowns. Also, be careful to adopt a consistent sign convention for all velocities.

4. **Oblique central impact** [Sample Prob. 13.20]. This occurs when one or both of the initial velocities of the two bodies is not directed along the line of impact (Fig. 13.20b). Again, these solution steps are only applicable to problems where the impulsive forces in the tangential direction are negligible (e.g., you would not use these to solve Prob. 13.146). To solve problems of this type, you should first resolve the momenta and impulses shown in your diagram into components along the t axis and the n axis.

a. Conservation of momentum. Since the impulsive forces act along the line of impact, i.e., along the n axis, the component along the t axis of the momentum *of each body* is conserved. Therefore, for each body, you can write that the t components of its velocity before and after impact are equal. So,

$$(v_A)_t = (v'_A)_t \qquad (v_B)_t = (v'_B)_t \qquad \textbf{(13.47)}$$

Also, the component along the n axis of the total momentum of the system is conserved as

$$m_A(v_A)_n + m_B(v_B)_n = m_A(v'_A)_n + m_B(v'_B)_n \qquad \textbf{(13.48)}$$

b. Coefficient of restitution. The relation between the relative velocities of the two bodies before and after impact can be written in the n direction only. Hence,

$$(v'_B)_n - (v'_A)_n = e[(v_A)_n - (v_B)_n)] \qquad \textbf{(13.49)}$$

You now have four equations that you can solve for four unknowns. Note that after finding all of the velocities, you can determine the impulse exerted by body A on body B by drawing an impulse-momentum diagram for B alone and equating components in the n direction.

c. When the motion of one of the colliding bodies is constrained, you must include the impulses of the external forces in your diagram [Sample Probs. 13.21 and 13.23]. You will then observe that some of the previous relations do not hold. However, in the example shown in Fig. 13.26, the total momentum of the system is conserved in a direction perpendicular to the external impulse. Also note that, when a body A bounces off a fixed surface B, the only conservation of momentum equation that you can use is the first of Eq. (13.47) [Sample Prob. 13.19].

5. Remember that energy is lost during most impacts. The only exception is for **perfectly elastic** impacts ($e = 1$), where energy is conserved. Thus, in the general case of impact where $e < 1$, mechanical energy is not conserved. Therefore, be careful *not to apply* the principle of conservation of energy through an impact situation. Instead, apply this principle separately to the motions preceding and following the impact [Sample Probs. 13.22 and 13.23].

Problems

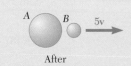

Before   After

Fig. P13.CQ6

CONCEPT QUESTIONS

13.CQ6 A 5-kg ball A strikes a 1-kg ball B that is initially at rest. Is it possible that after the impact A is not moving and B has a speed of $5v$?
a. Yes
b. No
Explain your answer.

IMPULSE-MOMENTUM DIAGRAM PRACTICE PROBLEMS

13.F6 A sphere with a speed v_0 rebounds after striking a frictionless inclined plane as shown. Draw the impulse-momentum diagram that can be used to find the velocity of the sphere after the impact.

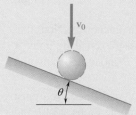

Fig. P13.F6

13.F7 An 80-Mg railroad engine A coasting at 6.5 km/h strikes a 20-Mg flatcar C carrying a 30-Mg load B which can slide along the floor of the car ($\mu_k = 0.25$). The flatcar was at rest with its brakes released. Instead of A and C coupling as expected, it is observed that A rebounds with a speed of 2 km/h after the impact. Draw impulse-momentum diagrams that can be used to determine (*a*) the coefficient of restitution and the speed of the flatcar immediately after impact, (*b*) the time it takes the load to slide to a stop relative to the car.

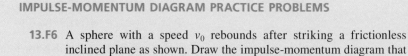

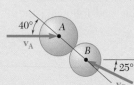

Fig. P13.F8

Fig. P13.F7

13.F8 Two frictionless balls strike each other as shown. The coefficient of restitution between the balls is e. Draw the impulse-momentum diagram that could be used to find the velocities of A and B after the impact.

13.F9 A 10-kg ball A moving horizontally at 12 m/s strikes a 10-kg block B. The coefficient of restitution of the impact is 0.4 and the coefficient of kinetic friction between the block and the inclined surface is 0.5. Draw the impulse-momentum diagram that can be used to determine the speeds of A and B after the impact.

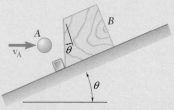

Fig. P13.F9

13.F10 Block A of mass m_A strikes ball B of mass m_B with a speed of v_A as shown. Draw the impulse-momentum diagram that can be used to determine the speeds of A and B after the impact and the impulse during the impact.

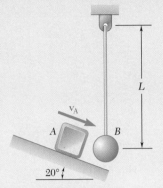

Fig. P13.F10

END-OF-SECTION PROBLEMS

13.155 The coefficient of restitution between the two collars is known to be 0.70. Determine (a) their velocities after impact, (b) the energy loss during impact.

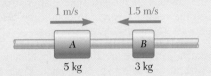

Fig. P13.155

13.156 Collars A and B, of the same mass m, are moving toward each other with identical speeds as shown. Knowing that the coefficient of restitution between the collars is e, determine the energy lost in the impact as a function of m, e, and v.

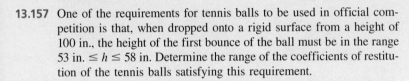

Fig. P13.156

13.157 One of the requirements for tennis balls to be used in official competition is that, when dropped onto a rigid surface from a height of 100 in., the height of the first bounce of the ball must be in the range 53 in. $\leq h \leq$ 58 in. Determine the range of the coefficients of restitution of the tennis balls satisfying this requirement.

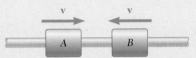

13.158 Two disks sliding on a frictionless horizontal plane with opposite velocities of the same magnitude v_0 hit each other squarely. Disk A is known to have a weight of 6 lb and is observed to have zero velocity after impact. Determine (a) the weight of disk B, knowing that the coefficient of restitution between the two disks is 0.5, (b) the range of possible values of the weight of disk B if the coefficient of restitution between the two disks is unknown.

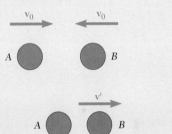

Fig. P13.158

13.159 To apply shock loading to an artillery shell, a 20-kg pendulum A is released from a known height and strikes impactor B at a known velocity $\mathbf{v}_0$. Impactor B then strikes the 1-kg artillery shell C. Knowing the coefficient of restitution between all objects is e, determine the mass of B to maximize the impulse applied to the artillery shell C.

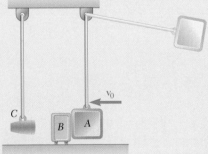

Fig. P13.159

13.160 Packages in an automobile parts supply house are transported to the loading dock by pushing them along on a roller track with very little friction. At the instant shown, packages B and C are at rest, and package A has a velocity of 2 m/s. Knowing that the coefficient of restitution between the packages is 0.3, determine (a) the velocity of package C after A hits B and B hits C, (b) the velocity of A after it hits B for the second time.

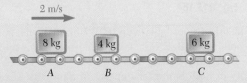

Fig. P13.160

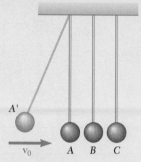

Fig. P13.161

13.161 Three steel spheres of equal mass are suspended from the ceiling by cords of equal length that are spaced at a distance slightly greater than the diameter of the spheres. After being pulled back and released, sphere A hits sphere B, which then hits sphere C. Denoting the coefficient of restitution between the spheres by e and the velocity of A just before it hits B by $\mathbf{v}_0$, determine (a) the velocities of A and B immediately after the first collision, (b) the velocities of B and C immediately after the second collision. (c) Assuming now that n spheres are suspended from the ceiling and that the first sphere is pulled back and released as described here, determine the velocity of the last sphere after it is hit for the first time. (d) Use the result of part c to obtain the velocity of the last sphere when $n = 8$ and $e = 0.9$.

13.162 At an amusement park there are 200-kg bumper cars A, B, and C that have riders with masses of 40 kg, 60 kg, and 35 kg, respectively. Car A is moving to the right with a velocity $\mathbf{v}_A = 2$ m/s and car C has a velocity $\mathbf{v}_B = 1.5$ m/s to the left, but car B is initially at rest. The coefficient of restitution between each car is 0.8. Determine the final velocity of each car, after all impacts, assuming (a) cars A and C hit car B at the same time, (b) car A hits car B before car C does.

Fig. *P13.162* and P13.163

13.163 At an amusement park there are 200-kg bumper cars A, B, and C that have riders with masses of 40 kg, 60 kg, and 35 kg, respectively. Car A is moving to the right with a velocity $\mathbf{v}_A = 2$ m/s when it hits stationary car B. The coefficient of restitution between each car is 0.8. Determine the velocity of car C so that after car B collides with car C the velocity of car B is zero.

13.164 Two identical billiard balls can move freely on a horizontal table. Ball A has a velocity $\mathbf{v}_0$ as shown and hits ball B, which is at rest, at a point C defined by $\theta = 45°$. Knowing that the coefficient of restitution between the two balls is $e = 0.8$ and assuming no friction, determine the velocity of each ball after impact.

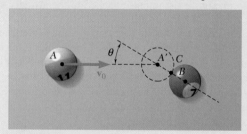

Fig. P13.164

13.165 Two identical pool balls with a 2.37-in. diameter may move freely on a pool table. Ball B is at rest, and ball A has an initial velocity of $\mathbf{v} = v_0\mathbf{i}$. (a) Knowing that $b = 2$ in. and $e = 0.7$, determine the velocity of each ball after impact. (b) Show that if $e = 1$, the final velocities of the balls form a right angle for all values of b.

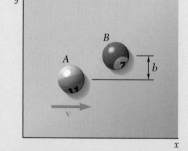

Fig. P13.165

13.166 A 600-g ball A is moving with a velocity of magnitude 6 m/s when it is hit as shown by a 1-kg ball B that has a velocity of magnitude 4 m/s. Knowing that the coefficient of restitution is 0.8 and assuming no friction, determine the velocity of each ball after impact.

13.167 Two identical hockey pucks are moving on a hockey rink at the same speed of 3 m/s and in perpendicular directions when they strike each other as shown. Assuming a coefficient of restitution $e = 0.9$, determine the magnitude and direction of the velocity of each puck after impact.

13.168 The coefficient of restitution is 0.9 between the two 60-mm-diameter billiard balls A and B. Ball A is moving in the direction shown with a velocity of 1 m/s when it strikes ball B, which is at rest. Knowing that after impact B is moving in the x direction, determine (a) the angle θ, (b) the velocity of B after impact.

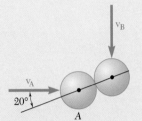

Fig. P13.166

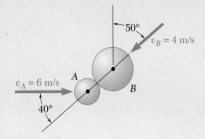

Fig. P13.167

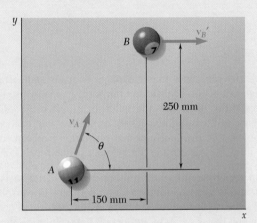

Fig. P13.168

13.169 A boy located at point O of a semicircular wall and the wall itself throws a ball at the wall in a direction forming an angle of 45° with OA. Knowing that after hitting the wall the ball rebounds in a direction parallel to OA, determine the coefficient of restitution between the ball and the wall.

13.170 The Mars Pathfinder spacecraft used large airbags to cushion its impact with the planet's surface when landing. Assuming the spacecraft had an impact velocity of 18.5 m/s at an angle of 45° with respect to the horizontal, the coefficient of restitution is 0.85 and neglecting friction, determine (a) the height of the first bounce, (b) the length of the first bounce. (Acceleration of gravity on Mars = 3.73 m/s².)

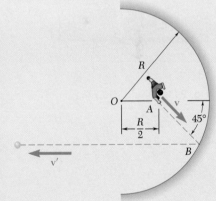

Fig. P13.169

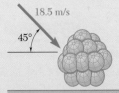

Fig. P13.170

13.171 A girl throws a ball at an inclined wall from a height of 3 ft, hitting the wall at *A* with a horizontal velocity $\mathbf{v}_0$ of magnitude 25 ft/s. Knowing that the coefficient of restitution between the ball and the wall is 0.9 and neglecting friction, determine the distance *d* from the foot of the wall to the point *B* where the ball will hit the ground after bouncing off the wall.

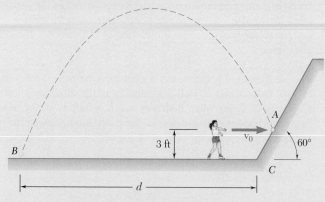

Fig. P13.171

13.172 Rockfalls can cause major damage to roads and infrastructure. To design mitigation bridges and barriers, engineers use the coefficient of restitution to model the behavior of the rocks. Rock *A* falls a distance of 20 m before striking an incline with a slope of $\alpha = 40°$. Knowing that the coefficient of restitution between rock *A* and the incline is 0.2, determine the velocity of the rock after the impact.

13.173 From experimental tests, smaller boulders tend to have a greater coefficient of restitution than larger boulders. Rock *A* falls a distance of 20 meters before striking an incline with a slope of $\alpha = 45°$. Knowing that $h = 30$ m and $d = 20$ m, determine if a boulder will land on the road or beyond the road for a coefficient of restitution of (*a*) $e = 0.2$, (*b*) $e = 0.1$.

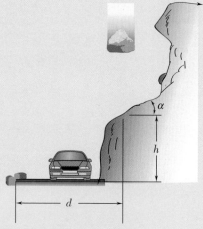

Fig. P13.172 and *P13.173*

13.174 Two cars of the same mass run head-on into each other at *C*. After the collision, the cars skid with their brakes locked and come to a stop in the positions shown in the lower part of the figure. Knowing that the speed of car *A* just before impact was 5 mi/h and that the coefficient of kinetic friction between the pavement and the tires of both cars is 0.30, determine (*a*) the speed of car *B* just before impact, (*b*) the effective coefficient of restitution between the two cars.

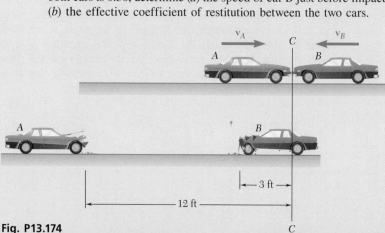

Fig. P13.174

13.175 A 1-kg block B is moving with a velocity $\mathbf{v}_0$ of magnitude $v_0 = 2$ m/s as it hits the 0.5-kg sphere A, which is at rest and hanging from a cord attached at O. Knowing that $\mu_k = 0.6$ between the block and the horizontal surface and $e = 0.8$ between the block and the sphere, determine after impact (a) the maximum height h reached by the sphere, (b) the distance x traveled by the block.

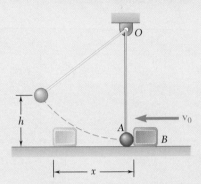

Fig. P13.175

13.176 A 0.25-lb ball thrown with a horizontal velocity $\mathbf{v}_0$ strikes a 1.5-lb plate attached to a vertical wall at a height of 36 in. above the ground. It is observed that after rebounding, the ball hits the ground at a distance of 24 in. from the wall when the plate is rigidly attached to the wall (Fig. 1) and at a distance of 10 in. when a foam-rubber mat is placed between the plate and the wall (Fig. 2). Determine (a) the coefficient of restitution e between the ball and the plate, (b) the initial velocity $\mathbf{v}_0$ of the ball.

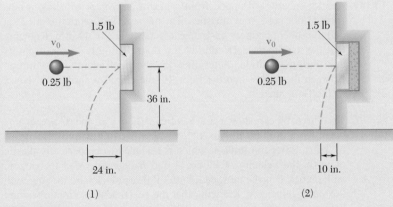

Fig. P13.176

13.177 After having been pushed by an airline employee, an empty 40-kg luggage carrier A hits with a velocity of 5 m/s an identical carrier B containing a 15-kg suitcase equipped with rollers. The impact causes the suitcase to roll into the left wall of carrier B. Knowing that the coefficient of restitution between the two carriers is 0.80 and that the coefficient of restitution between the suitcase and the wall of carrier B is 0.30, determine (a) the velocity of carrier B after the suitcase hits its wall for the first time, (b) the total energy lost in that impact.

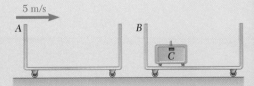

Fig. P13.177

13.178 Blocks A and B each weigh 0.8 lb and block C weighs 2.4 lb. The coefficient of friction between the blocks and the plane is $\mu_k = 0.30$. Initially block A is moving at a speed $v_0 = 15$ ft/s and blocks B and C are at rest (Fig. 1). After A strikes B and B strikes C, all three blocks come to a stop in the positions shown (Fig. 2). Determine (a) the coefficients of restitution between A and B and between B and C, (b) the displacement x of block C.

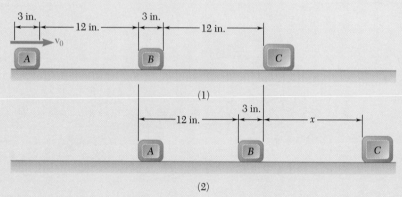

Fig. P13.178

13.179 A 5-kg sphere is dropped from a height of $y = 2$ m to test newly designed spring floors used in gymnastics. The mass of the floor section is 10 kg, and the effective stiffness of the floor is $k = 120$ kN/m. Knowing that the coefficient of restitution between the ball and the platform is 0.6, determine (a) the height h reached by the sphere after rebound, (b) the maximum force in the springs.

13.180 A 5-kg sphere is dropped from a height of $y = 3$ m to test a new spring floor used in gymnastics. The mass of floor section B is 12 kg, and the sphere bounces back upwards a distance of 44 mm. Knowing that the maximum deflection of the floor section is 33 mm from its equilibrium position, determine (a) the coefficient of restitution between the sphere and the floor, (b) the effective spring constant k of the floor section.

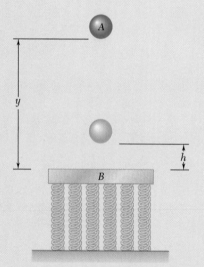

Fig. P13.179 and P13.180

13.181 The three blocks shown are identical. Blocks B and C are at rest when block B is hit by block A, which is moving with a velocity $\mathbf{v}_A$ of 3 ft/s. After the impact, which is assumed to be perfectly plastic ($e = 0$), the velocity of blocks A and B decreases due to friction, while block C picks up speed, until all three blocks are moving with the same velocity $\mathbf{v}$. Knowing that the coefficient of kinetic friction between all surfaces is $\mu_k = 0.20$, determine (a) the time required for the three blocks to reach the same velocity, (b) the total distance traveled by each block during that time.

Fig. P13.181

13.182 Block *A* is released from rest and slides down the frictionless surface of *B* until it hits a bumper on the right end of *B*. Block *A* has a mass of 10 kg and object *B* has a mass of 30 kg and *B* can roll freely on the ground. Determine the velocities of *A* and *B* immediately after impact when (*a*) $e = 0$, (*b*) $e = 0.7$.

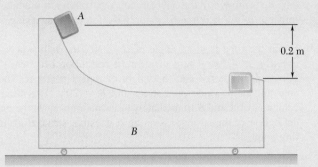

Fig. P13.182

13.183 A 340-g ball *B* is hanging from an inextensible cord attached to a support *C*. A 170-g ball *A* strikes *B* with a velocity $\mathbf{v}_0$ with a magnitude of 1.5 m/s at an angle of 60° with the vertical. Assuming perfectly elastic impact ($e = 1$) and no friction, determine the height *h* reached by ball *B*.

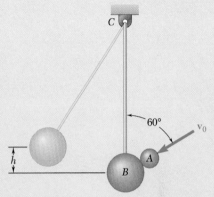

Fig. P13.183

13.184 A test machine that kicks soccer balls has a 5-lb simulated foot attached to the end of a 6-ft long pendulum arm of negligible mass. Knowing that the arm is released from the horizontal position and that the coefficient of restitution between the foot and the 1-lb ball is 0.8, determine the exit velocity of the ball (*a*) if the ball is stationary, (*b*) if the ball is struck when it is rolling towards the foot with a velocity of 10 ft/s.

13.185 Ball *B* is hanging from an inextensible cord. An identical ball *A* is released from rest when it is just touching the cord and drops through the vertical distance $h_A = 8$ in. before striking ball *B*. Assuming $e = 0.9$ and no friction, determine the resulting maximum vertical displacement h_B of the ball *B*.

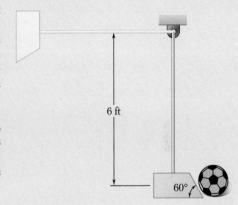

Fig. P13.184

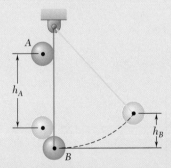

Fig. P13.185

13.186 A 70-g ball *B* dropped from a height $h_0 = 1.5$ m reaches a height $h_2 = 0.25$ m after bouncing twice from identical 210-g plates. Plate *A* rests directly on hard ground, while plate *C* rests on a foam-rubber mat. Determine (*a*) the coefficient of restitution between the ball and the plates, (*b*) the height h_1 of the ball's first bounce.

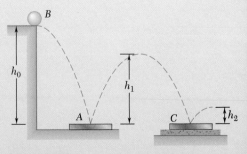

Fig. P13.186

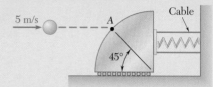

Fig. P13.187

13.187 A 2-kg sphere moving to the right with a velocity of 5 m/s strikes at A, which is on the surface of a 9-kg quarter cylinder that is initially at rest and in contact with a spring with a constant of 20 kN/m. The spring is held by cables, so it is initially compressed 50 mm. Neglecting friction and knowing that the coefficient of restitution is 0.6, determine (a) the velocity of the sphere immediately after impact, (b) the maximum compressive force in the spring.

13.188 When the rope is at an angle of $\alpha = 30°$, the 1-lb sphere A has a speed $v_0 = 4$ ft/s. The coefficient of restitution between A and the 2-lb wedge B is 0.7 and the length of rope $l = 2.6$ ft. The spring constant has a value of 2 lb/in. and $\theta = 20°$. Determine (a) the velocities of A and B immediately after the impact, (b) the maximum deflection of the spring, assuming A does not strike B again before this point.

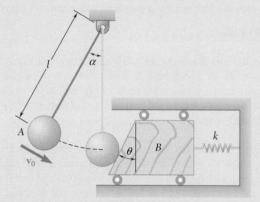

Fig. P13.188 and P13.189

13.189 When the rope is at an angle of $\alpha = 30°$, the 1-kg sphere A has a speed $v_0 = 0.6$ m/s. The coefficient of restitution between A and the 2-kg wedge B is 0.8 and the length of rope $l = 0.9$ m. The spring constant has a value of 1500 N/m and $\theta = 20°$. Determine (a) the velocities of A and B immediately after the impact, (b) the maximum deflection of the spring, assuming A does not strike B again before this point.

Review and Summary

This chapter was devoted to presenting the method of work and energy and the method of impulse and momentum. In the first half of the chapter, we studied the method of work and energy and its application to the analysis of the motion of particles.

Work of a Force

We first considered a force **F** acting on a particle A and defined the **work of F corresponding to the small displacement** $d\mathbf{r}$ [Sec. 13.1] as the quantity

$$dU = \mathbf{F} \cdot d\mathbf{r} \qquad (13.1)$$

or recalling the definition of the scalar product of two vectors, as

$$dU = F \, ds \cos \alpha \qquad (13.1')$$

where α is the angle between **F** and $d\mathbf{r}$ (Fig. 13.30). We obtained the work of **F** during a finite displacement from A_1 to A_2, denoted by $U_{1 \to 2}$, by integrating Eq. (13.1) along the path described by the particle as

$$U_{1 \to 2} = \int_{A_1}^{A_2} \mathbf{F} \cdot d\mathbf{r} \qquad (13.2)$$

For a force defined by its rectangular components, we wrote

$$U_{1 \to 2} = \int_{A_1}^{A_2} (F_x \, dx + F_y \, dy + F_z \, dz) \qquad (13.2'')$$

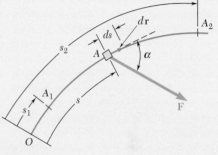

Fig. 13.30

Work of a Weight

We obtain the work of the weight **W** of a body as its center of gravity moves from the elevation y_1 to y_2 (Fig. 13.31) by substituting $F_x = F_z = 0$ and $F_y = -W$ into Eq. (13.2'') and integrating. We found

$$U_{1 \to 2} = -\int_{y_1}^{y_2} W \, dy = Wy_1 - Wy_2 \qquad (13.4)$$

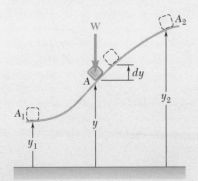

Fig. 13.31

Work of the Force Exerted by a Spring

The work of a force **F** exerted by a spring on a body A during a finite displacement of the body (Fig. 13.32) from A_1 ($x = x_1$) to A_2 ($x = x_2$) was obtained by

$$dU = -F \, dx = -kx \, dx$$

$$U_{1 \to 2} = -\int_{x_1}^{x_2} kx \, dx = \tfrac{1}{2}kx_1^2 - \tfrac{1}{2}kx_2^2 \qquad (13.6)$$

The work of **F** is therefore positive when the spring is returning to its undeformed position.

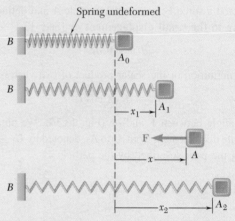

Fig. 13.32

Work of the Gravitational Force

We obtained the **work of the gravitational force F** exerted by a particle of mass M located at O on a particle of mass m as the latter moves from A_1 to A_2 (Fig. 13.33) by recalling from Sec. 12.2C the expression for the magnitude of **F** and writing

$$U_{1 \to 2} = -\int_{r_1}^{r_2} \frac{GMm}{r^2} \, dr = \frac{GMm}{r_2} - \frac{GMm}{r_1} \qquad (13.7)$$

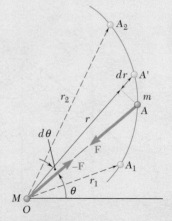

Fig. 13.33

Kinetic Energy of a Particle

We defined the **kinetic energy of a particle** of mass m moving with a velocity $\mathbf{v}$ [Sec. 13.1B] as the scalar quantity

$$T = \tfrac{1}{2}mv^2 \tag{13.9}$$

Principle of Work and Energy

From Newton's second law, we derived the **principle of work and energy**, which states that we can obtain the kinetic energy of a particle at A_2 by adding its kinetic energy at A_1 to the work done during the displacement from A_1 to A_2 by the force $\mathbf{F}$ exerted on the particle as

$$T_1 + U_{1 \to 2} = T_2 \tag{13.11}$$

Method of Work and Energy

The method of work and energy simplifies the solution of many problems dealing with forces, displacements, and velocities, since it does not require the determination of accelerations [Sec. 13.1C]. We also note that it involves only scalar quantities, and we do not need to consider forces that do no work [Sample Probs. 13.1 and 13.4]. However, this method should be supplemented by the direct application of Newton's second law to determine a force normal to the path of the particle [Sample Prob. 13.6].

Power and Mechanical Efficiency

The power developed by a machine and its mechanical efficiency were discussed in Sec. 13.1D. We defined power as the time rate at which work is done by

$$\text{Power} = \frac{dU}{dt} = \mathbf{F} \cdot \mathbf{v} \tag{13.12, 13.13}$$

where $\mathbf{F}$ is the force exerted on the particle and $\mathbf{v}$ is the velocity of the particle [Sample Prob. 13.7]. The **mechanical efficiency**, denoted by η, was expressed as

$$\eta = \frac{\text{power output}}{\text{power input}} \tag{13.15}$$

Conservative Force and Potential Energy

When the work of a force $\mathbf{F}$ is independent of the path followed [Secs. 13.2A and 13.2B], the force $\mathbf{F}$ is said to be a **conservative force**, and its work is equal to minus the change in the potential energy V associated with $\mathbf{F}$

$$U_{1 \to 2} = V_1 - V_2 \tag{13.19'}$$

We obtained the following expressions for the potential energy associated with each of the forces considered earlier.

Force of gravity (weight):

$$V_g = Wy \tag{13.16}$$

Gravitational force:

$$V_g = -\frac{GMm}{r} \tag{13.17}$$

Elastic force exerted by a spring:

$$V_e = \tfrac{1}{2}kx^2 \tag{13.18}$$

Principle of Conservation of Energy

Substituting for $U_{1\rightarrow2}$ from Eq. (13.19′) into Eq. (13.11) and rearranging the terms [Sec. 13.2C], we obtained

$$T_1 + V_1 = T_2 + V_2 \tag{13.24}$$

or

$$T_1 + V_{g_1} + V_{e_1} = T_2 + V_{g_2} + V_{e_2} \tag{13.24′}$$

This is the **principle of conservation of energy**, which states that, when a particle moves under the action of conservative forces, the sum of its kinetic and potential energies remains constant. The application of this principle facilitates the solution of problems involving only conservative forces [Sample Probs. 13.8 and 13.9].

Alternative Expression for the Principle of Work and Energy

Rather than finding the work due to all external forces, you can write an alternative expression for the work-energy principle such that

$$T_1 + V_{g_1} + V_{e_1} + U_{1\rightarrow2}^{\text{NC}} = T_2 + V_{g_2} + V_{e_2} \tag{13.24″}$$

where $U_{1\rightarrow2}^{\text{NC}}$ is the work of external non-conservative forces such as friction [Sample Prob. 13.10].

Motion Under a Gravitational Force

Recalling from Sec. 12.2B that when a particle moves under a central force **F** its angular momentum about the center of force O remains constant, we observed [Sec. 13.D] that, if the central force **F** is also conservative, the principles of conservation of angular momentum and of conservation of energy can be used jointly to analyze the motion of the particle [Sample Prob. 13.11]. Since the gravitational force exerted by the earth on a space vehicle is both central and conservative, this approach was used to study the motion of such vehicles [Sample Prob. 13.12] and was found particularly effective in the case of an **oblique launching**. Considering the initial position P_0 and an arbitrary position P of the vehicle (Fig. 13.34), we have

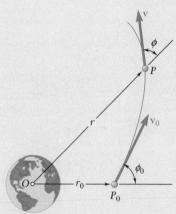

Fig. 13.34

$$(H_O)_0 = H_O: \qquad r_0 m v_0 \sin\phi_0 = rmv \sin\phi \tag{13.25}$$

$$T_0 + V_0 = T + V: \qquad \tfrac{1}{2}mv_0^2 - \frac{GMm}{r_0} = \tfrac{1}{2}mv^2 - \frac{GMm}{r} \tag{13.26}$$

where m is the mass of the vehicle and M the mass of the earth.

Principle of Impulse and Momentum for a Particle

The second half of this chapter was devoted to the method of impulse and momentum and to its application to the solution of various types of problems involving the motion of particles.

We defined the **linear momentum of a particle** [Sec. 13.3A] as the product $m\mathbf{v}$ of the mass m of the particle and its velocity **v**. From Newton's second law, $\mathbf{F} = m\mathbf{a}$, we derived the relation

$$m\mathbf{v}_1 + \int_{t_1}^{t_2} \mathbf{F}\, dt = m\mathbf{v}_2 \tag{13.28}$$

where $m\mathbf{v}_1$ and $m\mathbf{v}_2$ represent the momentum of the particle at a time t_1 and a time t_2, respectively, and where the integral defines the **linear impulse of the force F** during the corresponding time interval. Therefore, we have

$$m\mathbf{v}_1 + \mathbf{Imp}_{1\rightarrow2} = m\mathbf{v}_2 \qquad \textbf{(13.30)}$$

which expresses the principle of impulse and momentum for a particle.

When the particle considered is subjected to several forces, we need to use the sum of the impulses of these forces; we have

$$m\mathbf{v}_1 + \Sigma\mathbf{Imp}_{1\rightarrow2} = m\mathbf{v}_2 \qquad \textbf{(13.32)}$$

Since Eqs. (13.30) and (13.32) involve vector quantities, it is necessary to consider their x and y components separately when applying them to the solution of a given problem [Sample Probs. 13.13 through 13.15].

Impulsive Motion

The method of impulse and momentum is particularly effective in the study of the **impulsive motion** of a particle, when very large forces, called **impulsive forces**, are applied for a very short interval of time Δt, since this method involves the impulses $\mathbf{F}_{avg}\Delta t$ of the forces, rather than the forces themselves [Sec. 13.3B]. Assuming that all non-impulsive forces (e.g., weight) are negligible, we wrote

$$m\mathbf{v}_1 + \Sigma\mathbf{F}_{avg}\Delta t = m\mathbf{v}_2 \qquad \textbf{(13.35)}$$

In the case of the impulsive motion of several particles, we had

$$\Sigma m\mathbf{v}_1 + \Sigma\mathbf{F}_{avg}\Delta t = \Sigma m\mathbf{v}_2 \qquad \textbf{(13.36)}$$

where the second term involves only impulsive, external forces [Sample Probs. 13.16 and 13.17].

In the particular case when the sum of the impulses of the external forces is zero, Eq. (13.36) reduces to $\Sigma m\mathbf{v}_1 = \Sigma m\mathbf{v}_2$; that is, the total momentum of the particles is conserved. For two particles, this reduces to

$$m_A\mathbf{v}_A + m_B\mathbf{v}_B = m_A\mathbf{v}'_A + m_B\mathbf{v}'_B \qquad \textbf{(13.34′)}$$

Direct Central Impact

In Sec. 13.4, we considered the **central impact** of two colliding bodies. In the case of a **direct central impact** [Sec. 13.4A], the two colliding bodies A and B were moving along the **line of impact** with velocities $\mathbf{v}_A$ and $\mathbf{v}_B$, respectively (Fig. 13.35). Two equations could be used to determine their velocities $\mathbf{v}'_A$ and $\mathbf{v}'_B$ after the impact. The first expressed conservation of the total momentum of the two bodies as

$$m_A v_A + m_B v_B = m_A v'_A + m_B v'_B \qquad \textbf{(13.37)}$$

where a positive sign indicates that the corresponding velocity is directed to the right. The second equation, called the coefficient of restitution equation, related the *relative velocities* of the two bodies before and after the impact as

$$v'_B - v'_A = e(v_A - v_B) \qquad \textbf{(13.43)}$$

The constant e is known as the **coefficient of restitution**; its value lies between 0 and 1 and depends in a large measure on the materials involved. When $e = 0$, the impact is said to be **perfectly plastic**; when $e = 1$, it is said to be **perfectly elastic**. Eq. (13.43) is only valid for direct central impact. [Sample Prob. 13.18].

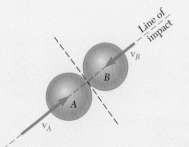

Fig. 13.35

Oblique Central Impact

In the case of an **oblique central impact** [Sec. 13.4B], the velocities of the two colliding smooth bodies before and after the impact were resolved into n components along the line of impact and t components along the common tangent to the surfaces in contact (Fig. 13.36). We observed that the t component of the velocity of each body remained unchanged, while the n components satisfied equations similar to Eqs. (13.37) and (13.43) [Sample Probs. 13.19 and 13.20]. We showed that, although this method was developed for bodies moving freely before and after the impact, it could be extended to the case when one or both of the colliding bodies is constrained in its motion [Sample Prob. 13.21]. When the velocities are not along the line of impact, the coefficient of restitution equation uses the normal component,

$$(v'_B)_n - (v'_A)_n = e[(v_A)_n - (v_B)_n] \tag{13.49}$$

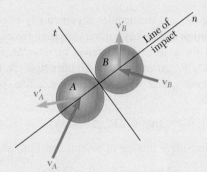

Fig. 13.36

Using the Three Fundamental Methods of Kinetic Analysis

In Sec. 13.4C, we discussed the relative advantages of the three fundamental methods presented in this chapter and the preceding one, namely, Newton's second law, work and energy, and impulse and momentum. We noted that we can combine the method of work and energy and the method of impulse and momentum to solve problems involving a short impact phase during which impulsive forces must be taken into consideration [Sample Probs. 13.22 and 13.23].

Review Problems

13.190 A 32,000-lb airplane lands on an aircraft carrier and is caught by an arresting cable. The cable is inextensible and is paid out at A and B from mechanisms located below deck and consisting of pistons moving in long oil-filled cylinders. Knowing that the piston-cylinder system maintains a constant tension of 85 kips in the cable during the entire landing, determine the landing speed of the airplane if it travels a distance $d = 95$ ft after being caught by the cable.

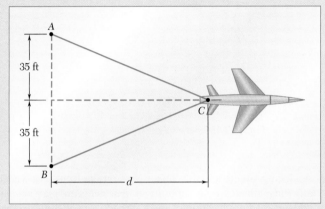

Fig. P13.190

13.191 A 2-oz pellet shot vertically from a spring-loaded pistol on the surface of the earth rises to a height of 300 ft. The same pellet shot from the same pistol on the surface of the moon rises to a height of 1900 ft. Determine the energy dissipated by aerodynamic drag when the pellet is shot on the surface of the earth. (The acceleration of gravity on the surface of the moon is 0.165 times that on the surface of the earth.)

13.192 A satellite describes an elliptic orbit about a planet of mass M. The minimum and maximum values of the distance r from the satellite to the center of the planet are, respectively, r_0 and r_1. Use the principles of conservation of energy and conservation of angular momentum to derive the relation

$$\frac{1}{r_0} + \frac{1}{r_1} = \frac{2GM}{h^2}$$

where h is the angular momentum per unit mass of the satellite and G is the constant of gravitation.

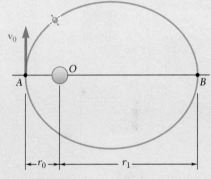

Fig. P13.192

13.193 A 60-g steel sphere attached to a 200-mm cord can swing about point O in a vertical plane. It is subjected to its own weight and to a force $\mathbf{F}$ exerted by a small magnet embedded in the ground. The magnitude of that force expressed in newtons is $F = 3000/r^2$, where r is the distance from the magnet to the sphere expressed in millimeters. Knowing that the sphere is released from rest at A, determine its speed as it passes through point B.

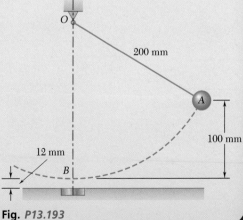

Fig. P13.193

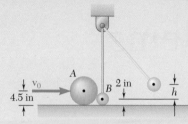

Fig. P13.194

13.194 A 50-lb sphere A with a radius of 4.5 in. is moving with a velocity of magnitude $v_0 = 6$ ft/s. Sphere A strikes a 4.6-lb sphere B that has a radius of 2 in., is hanging from an inextensible cord, and is initially at rest. Knowing that sphere B swings to a maximum height of $h = 0.75$ ft, determine the coefficient of restitution between the two spheres.

13.195 A 300-g block is released from rest after a spring of constant $k = 600$ N/m has been compressed 160 mm. Determine the force exerted by the loop $ABCD$ on the block as the block passes through (a) point A, (b) point B, (c) point C. Assume no friction.

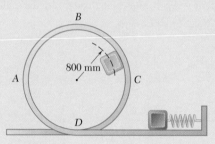

Fig. P13.195

13.196 A kicking-simulation attachment goes on the front of a wheelchair, allowing athletes with mobility impairments to play soccer. The athletes load up the spring shown through a ratchet mechanism that pulls the 2-kg "foot" back to the position 1. They then release the "foot" to impact the 0.45-kg soccer ball that is rolling towards the "foot" with a speed of 2 m/s at an angle $\theta = 30°$, as shown in the figure. The impact occurs with a coefficient of restitution $e = 0.75$ when the foot is at position 2, where the springs are unstretched. Knowing that the effective friction coefficient during rolling is $\mu_k = 0.1$, determine (a) the necessary spring coefficient to make the ball roll 30 m, (b) the direction the ball will travel after it is kicked.

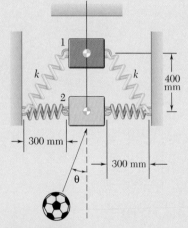

Fig. P13.196

13.197 A 300-g collar A is released from rest, slides down a frictionless rod, and strikes a 900-g collar B that is at rest and supported by a spring of constant 500 N/m. Knowing that the coefficient of restitution between the two collars is 0.9, determine (a) the maximum distance collar A moves up the rod after impact, (b) the maximum distance collar B moves down the rod after impact.

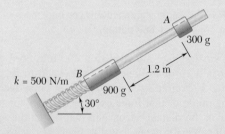

Fig. P13.197

13.198 Blocks A and B are connected by a cord which passes over pulleys and through a collar C. The system is released from rest when $x = 1.7$ m. As block A rises, it strikes collar C with perfectly plastic impact ($e = 0$). After impact, the two blocks and the collar keep moving until they come to a stop and reverse their motion. As A and C move down, C hits the ledge and blocks A and B keep moving until they come to another stop. Determine (*a*) the velocity of the blocks and collar immediately after A hits C, (*b*) the distance the blocks and collar move after the impact before coming to a stop, (*c*) the value of x at the end of one complete cycle.

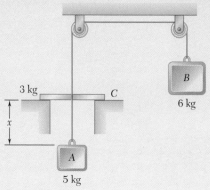

Fig. P13.198

13.199 A 2-kg ball B is traveling horizontally at 10 m/s when it strikes 2-kg ball A. Ball A is initially at rest and is attached to a spring with constant 100 N/m and an unstretched length of 1.2 m. Knowing the coefficient of restitution between A and B is 0.8 and friction between all surfaces is negligible, determine the normal force between A and the ground when it is at the bottom of the hill.

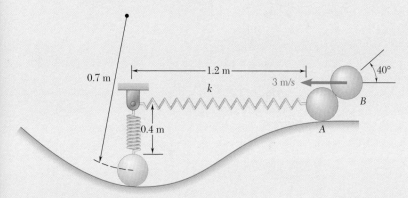

Fig. *P13.199*

13.200 A 2-kg block A is pushed up against a spring compressing it a distance x. The block is then released from rest and slides down the 20° incline until it strikes a 1-kg sphere B which is suspended from a 1-m inextensible rope. The spring constant $k = 800$ N/m, the coefficient of friction between A and the ground is 0.2, the distance A slides from the unstretched length of the spring $d = 1.5$ m, and the coefficient of restitution between A and B is 0.8. Knowing the tension in the rope is 20 N when $\alpha = 30°$, determine the initial compression x of the spring.

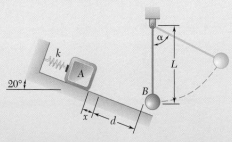

Fig. P13.200

*13.201 The 2-lb ball at A is suspended by an inextensible cord and given an initial horizontal velocity of $\mathbf{v}_0$. If $l = 2$ ft, $x_B = 0.3$ ft, and $y_B = 0.4$ ft, determine the initial velocity $\mathbf{v}_0$ so that the ball will enter the basket. (*Hint:* Use a computer to solve the resulting set of equations.)

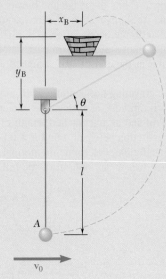

Fig. P13.201

14

Systems of Particles

The thrust for this XR-5M15 prototype engine is produced by gas particles being ejected at a high velocity. The determination of the forces on the test stand is based on the analysis of the motion of a *variable system of particles,* i.e., the motion of a large number of air particles considered together rather than separately.

Objectives

- **Apply** Newton's second law to a system of particles.
- **Calculate** the linear momentum and the angular momentum about a point of a system of particles.
- **Describe** the motion of the center of mass of a system of particles.
- **Determine** the kinetic energy of a system of particles.
- **Analyze** the motion of a system of particles by using the principle of work and energy and the principle of impulse and momentum.
- **Analyze** the motion of steady streams of particles
- **Analyze** systems of particles gaining or losing mass.

Introduction

In this chapter, you will study the motion of **systems of particles**; that is, the motion of a large number of particles considered together. In the first part of the chapter, we examine systems consisting of well-defined particles, like a set of billiard balls or a projectile that fragments into pieces. In the second part, we consider the motion of variable systems; these are systems that are continually gaining or losing particles or doing both at the same time. This could describe the motion of a stream of water or of a rocket during launch.

We start by applying Newton's second law to each particle of the system. We show that the *external forces* acting on the various particles form a system equipollent to the system of $m_i\mathbf{a}_i$ for the various particles. In other words, both systems have the same resultant and the same moment resultant about any given point. We further show that the resultant and moment resultant of the external forces are equal, respectively, to the rate of change of the total linear momentum and to the rate of change of the total angular momentum of the particles of the system.

We then define the *mass center* of a system of particles and describe the motion of that point, along with an analysis of the motion of the particles about their mass center. We discuss the conditions under which the linear momentum and the angular momentum of a system of particles are conserved and apply these results to the solution of various problems.

In Sec. 14.2, we apply the work–energy principle to a system of particles, and then we apply the impulse–momentum principle. We use these ideas to solve several problems of practical interest.

Note that although the derivations given in the first part of this chapter are carried out for a system of independent particles, they remain valid when the particles of the system are rigidly connected, i.e., when they form a rigid body. In fact, these results form the foundation of our discussion of the kinetics of rigid bodies in Chaps. 16 through 18.

In Sec. 14.3, we consider steady streams of particles, such as a stream of water diverted by a fixed vane or the flow of air through a jet engine. We show how to determine the force exerted by the stream on the vane and the thrust developed by the engine. Finally, we analyze systems that gain mass by continually absorbing particles or lose mass by continually expelling particles. Among the various practical applications of this analysis is the determination of the thrust developed by a rocket engine.

14.1 APPLYING NEWTON'S SECOND LAW AND MOMENTUM PRINCIPLES TO SYSTEMS OF PARTICLES

In statics, we studied the effects of forces on particles and on rigid bodies in equilibrium. However, when you consider particles in motion, the situation of particles acting together but not forming a rigid body occurs in several important and practical applications. We analyze this kind of problem by applying Newton's laws to the system. The results are an interesting middle ground between the dynamics of particles and the dynamics of rigid bodies, which we will study next.

14.1A Newton's Second Law for a System of Particles

In order to derive the equations of motion for a system of n particles, let us begin by writing Newton's second law for each individual particle of the system. Consider the particle P_i, where $1 \leq i \leq n$. Let m_i be the mass of P_i and let $\mathbf{a}_i$ be its acceleration with respect to the newtonian frame of reference $Oxyz$. The force exerted on P_i by another particle P_j of the system (Fig. 14.1), called an **internal force**, is denoted by $\mathbf{f}_{ij}$. The resultant of the internal forces exerted on P_i by all the other particles of the system is thus $\sum_{j=1}^{n} \mathbf{f}_{ij}$ (where $\mathbf{f}_{ii}$ has no meaning and is assumed to be equal to zero). On the other hand, denoting the resultant of all the **external forces** acting on P_i by $\mathbf{F}_i$, we write Newton's second law for the particle P_i as

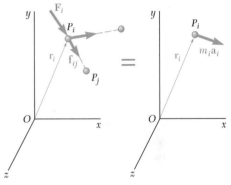

Fig. 14.1 Newton's second law for the ith particle in a system of particles.

$$\mathbf{F}_i + \sum_{j=1}^{n} \mathbf{f}_{ij} = m_i \mathbf{a}_i \qquad \textbf{(14.1)}$$

Denoting the position vector of P_i by $\mathbf{r}_i$ and taking the moments about O of the various terms in Eq. (14.1), we also have

$$\mathbf{r}_i \times \mathbf{F}_i + \sum_{j=1}^{n} (\mathbf{r}_i \times \mathbf{f}_{ij}) = \mathbf{r}_i \times m_i \mathbf{a}_i \qquad \textbf{(14.2)}$$

Repeating this procedure for each particle P_i of the system, we obtain n equations of the type in Eq. (14.1) and n equations of the type in Eq. (14.2), where i takes successively the values 1, 2, . . . , n. Thus, these equations state that the external forces $\mathbf{F}_i$ and the internal forces $\mathbf{f}_{ij}$ acting on the various particles form a system equivalent to the system of

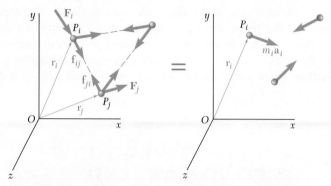

Fig. 14.2 The sum of internal forces equals zero, and the sum of external forces equals the sum of the mass times acceleration for every particle in the system.

the $m_i\mathbf{a}_i$ terms (i.e., one system may be replaced by the other) (Fig. 14.2).

Before proceeding further with our derivation, let us examine the internal forces $\mathbf{f}_{ij}$. These forces occur in pairs as $\mathbf{f}_{ij}$, $\mathbf{f}_{ji}$, where $\mathbf{f}_{ij}$ represents the force exerted by the particle P_j on the particle P_i and $\mathbf{f}_{ji}$ represents the force exerted by P_i on P_j (see Fig. 14.2). Now, according to Newton's third law (Sec. 6.1), as extended by Newton's law of gravitation to particles acting at a distance (Sec. 12.2C), the forces $\mathbf{f}_{ij}$ and $\mathbf{f}_{ji}$ are equal and opposite and have the same line of action. Their sum is therefore $\mathbf{f}_{ij} + \mathbf{f}_{ji} = 0$, and the sum of their moments about O is

$$\mathbf{r}_i \times \mathbf{f}_{ij} + \mathbf{r}_j \times \mathbf{f}_{ji} = \mathbf{r}_i \times (\mathbf{f}_{ij} + \mathbf{f}_{ji}) + (\mathbf{r}_j - \mathbf{r}_i) \times \mathbf{f}_{ji} = 0$$

since the vectors $\mathbf{r}_j - \mathbf{r}_i$ and $\mathbf{f}_{ji}$ in the last term are collinear. Adding all of the internal forces of the system and summing their moments about O, we obtain the equations

$$\sum_{i=1}^{n} \sum_{j=1}^{n} \mathbf{f}_{ij} = 0 \qquad \sum_{i=1}^{n} \sum_{j=1}^{n} (\mathbf{r}_i \times \mathbf{f}_{ij}) = 0 \tag{14.3}$$

These equations state that the resultant and the moment resultant of the internal forces of the system are zero.

Returning now to the n equations (14.1), where $i = 1, 2, \ldots, n$, we sum their left-hand sides and sum their right-hand sides. Taking into account the first of Eqs. (14.3), we obtain

$$\sum_{i=1}^{n} \mathbf{F}_i = \sum_{i=1}^{n} m_i\mathbf{a}_i \tag{14.4}$$

Proceeding similarly with Eq. (14.2) and taking into account the second of Eqs. (14.3), we have

$$\sum_{i=1}^{n} (\mathbf{r}_i \times \mathbf{F}_i) = \sum_{i=1}^{n} (\mathbf{r}_i \times m_i\mathbf{a}_i) \tag{14.5}$$

Equations (14.4) and (14.5) express the fact that the system of the

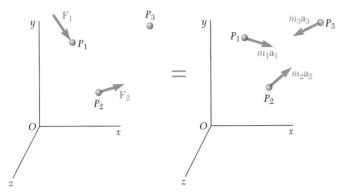

Fig. 14.3 The free-body diagram for a system of particles is equal to the kinetic diagram for a system of particles.

external forces $\mathbf{F}_i$ and the system of $m_i\mathbf{a}_i$ have the same resultant and the same moment resultant. Referring to the definition given in *Statics* Sec. 3.4B for two equipollent systems of vectors, we can therefore state that **the system of the external forces acting on the particles and the system of the** $m_i\mathbf{a}_i$ **terms of the particles are equipollent** (Fig. 14.3). Figure 14.3 basically shows that a free-body diagram for a system of particles is equal to its kinetic diagram.

Equations (14.3) state that the system of internal forces $\mathbf{f}_{ij}$ is equipollent to zero. Note, however, that it does *not* follow that the internal forces have no effect on the individual particles under consideration. Indeed, the gravitational forces that the sun and the planets exert on one another are internal to the solar system and are equipollent to zero. Yet these forces are responsible for the motion of the planets about the sun.

Similarly, it does not follow from Eqs. (14.4) and (14.5) that two systems of external forces that have the same resultant and that the same moment resultant will have the same effect on a given system of particles. Clearly, the systems shown in Figs. 14.4*a* and 14.4*b* have the same resultant and the same moment resultant; yet the first system accelerates particle *A* and leaves particle *B* unaffected, whereas the second system accelerates *B* and does not affect *A*. It is important to recall that when we stated in Sec. 3.4B that two equipollent systems of forces acting on a rigid body are also equivalent, we specifically noted that this property could *not* be extended to a system of forces acting on a set of independent particles such as those considered in this chapter.

In order to avoid any confusion, we use blue equals signs to connect equipollent systems of vectors, such as those shown in Figs. 14.3 and 14.4. These signs indicate that the two systems of vectors have the same resultant and the same moment resultant. We continue to use red equals signs to indicate that two systems of vectors are equivalent, i.e., that one system can actually be replaced by the other (Fig. 14.2).

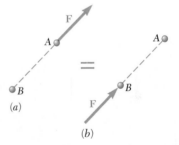

Fig. 14.4 (*a*) A system of resultant force and moment applied to particle *A* is not equivalent to (*b*) the same force and moment applied to particle *B*.

14.1B Linear and Angular Momentum of a System of Particles

We can express Eqs. (14.4) and (14.5) in a more condensed form by introducing the linear and the angular momentum of the system of particles. We define the linear momentum $\mathbf{L}$ of the system of particles as the sum of the linear momenta of the various particles of the system (Sec. 12.1B). Then we have

**Linear momentum,
system of particles**

$$\mathbf{L} = \sum_{i=1}^{n} m_i \mathbf{v}_i \tag{14.6}$$

Defining the angular momentum $\mathbf{H}_O$ about O of the system of particles in a similar way (Sec. 12.2A) gives us

**Angular momentum,
system of particles**

$$\mathbf{H}_O = \sum_{i=1}^{n} (\mathbf{r}_i \times m_i \mathbf{v}_i) \tag{14.7}$$

Differentiating both sides of Eqs. (14.6) and (14.7) with respect to t, we have

$$\dot{\mathbf{L}} = \sum_{i=1}^{n} m_i \dot{\mathbf{v}}_i = \sum_{i=1}^{n} m_i \mathbf{a}_i \tag{14.8}$$

and

$$\dot{\mathbf{H}}_O = \sum_{i=1}^{n} (\dot{\mathbf{r}}_i \times m_i \mathbf{v}_i) + \sum_{i=1}^{n} (\mathbf{r}_i \times m_i \dot{\mathbf{v}}_i)$$

$$= \sum_{i=1}^{n} (\mathbf{v}_i \times m_i \mathbf{v}_i) + \sum_{i=1}^{n} (\mathbf{r}_i \times m_i \mathbf{a}_i)$$

Because the vectors $\mathbf{v}_i$ and $m_i \mathbf{v}_i$ are collinear, this last equation reduces to

$$\dot{\mathbf{H}}_O = \sum_{i=1}^{n} (\mathbf{r}_i \times m_i \mathbf{a}_i) \tag{14.9}$$

Note that the right-hand sides of Eqs. (14.8) and (14.9) are identical to the right-hand sides of Eqs. (14.4) and (14.5), respectively. It follows that the left-hand sides of these equations are also equal. Recall that the left-hand side of Eq. (14.5) represents the sum of the moments $\mathbf{M}_O$ about O of the external forces acting on the particles of the system. So, omitting the subscript i from the sums, we have

$$\Sigma \mathbf{F} = \dot{\mathbf{L}} \tag{14.10}$$
$$\Sigma \mathbf{M}_O = \dot{\mathbf{H}}_O \tag{14.11}$$

These equations state:

> **The resultant and the moment resultant about the fixed point O of the external forces are equal to the rates of change of the linear momentum and of the angular momentum about O, respectively, of the system of particles.**

14.1C Motion of the Mass Center of a System of Particles

We can write Eq. (14.10) in an alternative form by considering the **mass center** of the system of particles. The mass center of the system is the point G defined by the position vector $\bar{\mathbf{r}}$, which satisfies the relation

$$m\bar{\mathbf{r}} = \sum_{i=1}^{n} m_i \mathbf{r}_i \qquad \textbf{(14.12)}$$

where m represents the total mass $m = \sum_{i=1}^{n} m_i$ of the particles. Resolving the position vectors $\bar{\mathbf{r}}$ and $\mathbf{r}_i$ into rectangular components, we obtain the following three scalar equations, which we can use to determine the coordinates $\bar{x}, \bar{y}, \bar{z}$ of the mass center:

$$m\bar{x} = \sum_{i=1}^{n} m_i x_i \qquad m\bar{y} = \sum_{i=1}^{n} m_i y_i \qquad m\bar{z} = \sum_{i=1}^{n} m_i z_i \qquad \textbf{(14.12′)}$$

Since $m_i g$ represents the weight of the particle P_i, and mg is the total weight of the particles, G is also the center of gravity of the system of particles. However, in order to avoid any confusion, we refer to G as the *mass center* of the system of particles when we are discussing properties associated with the *mass* of the particles, and as the *center of gravity* of the system when we consider properties associated with the *weight* of the particles. Particles located outside the gravitational field of the earth, for example, have a mass but no weight. We can then properly refer to their mass center, but obviously not to their center of gravity.[†]

Differentiating both members of Eq. (14.12) with respect to t, we obtain

$$m\dot{\bar{\mathbf{r}}} = \sum_{i=1}^{n} m_i \dot{\mathbf{r}}_i$$

or

$$m\bar{\mathbf{v}} = \sum_{i=1}^{n} m_i \mathbf{v}_i \qquad \textbf{(14.13)}$$

where $\bar{\mathbf{v}}$ represents the velocity of the mass center G of the system of particles. But the right-hand side of Eq. (14.13) is, by definition, the linear momentum $\mathbf{L}$ of the system [see Eq. (14.6)]. We therefore have

$$\mathbf{L} = m\bar{\mathbf{v}} \qquad \textbf{(14.14)}$$

and, differentiating both members with respect to t,

$$\dot{\mathbf{L}} = m\bar{\mathbf{a}} \qquad \textbf{(14.15)}$$

[†]We should also point out that the mass center and the center of gravity of a system of particles do not exactly coincide, since the weights of the particles are directed toward the center of the earth and thus do not truly form a system of parallel forces. For particles on the earth, this difference is extremely small.

where $\bar{\mathbf{a}}$ represents the acceleration of the mass center G. Substituting for $\dot{\mathbf{L}}$ from Eq. (14.15) into Eq. (14.10), we obtain

$$\Sigma\mathbf{F} = m\bar{\mathbf{a}} \qquad\qquad (14.16)$$

which defines the motion of the mass center G of the system of particles.

Note that Eq. (14.16) is identical to the equation we would obtain for a particle of mass m equal to the total mass of the particles of the system, acted upon by all the external forces. We therefore state:

> **The mass center of a system of particles moves as if the entire mass of the system and all of the external forces were concentrated at that point.**

This principle is best illustrated by the motion of an exploding projectile. We know that if air resistance is neglected, we can assume that a projectile will travel along a parabolic path. After it has exploded, the mass center G of the fragments of the projectile will continue to travel along the same path. Indeed, point G must move as if the mass and the weight of all fragments were concentrated at G; it must therefore move as if the projectile had not exploded.

Also note that the preceding derivation does not involve the moments of the external forces. Therefore, *it would be wrong to assume* that the external forces are equipollent to a vector $m\bar{\mathbf{a}}$ attached at the mass center G. In general, this is not the case since, as you will see next, the sum of the moments about G of the external forces is not in general equal to zero.

14.1D Angular Momentum of a System of Particles About Its Mass Center

In some applications (for example, in analyzing the motion of a rigid body), it is convenient to consider the motion of the particles of the system with respect to a centroidal frame of reference $Gx'y'z'$ that translates with respect to the newtonian frame of reference $Oxyz$ (Fig. 14.5). Although a centroidal frame is not, in general, a Newtonian frame of reference, we will show that the fundamental relation in Eq. (14.11) holds when the frame $Oxyz$ is replaced by $Gx'y'z'$.

Let's denote the position vector and the velocity of the particle P_i relative to the moving frame of reference $Gx'y'z'$ by $\mathbf{r}_i'$ and $\mathbf{v}_i'$, respectively. We then define the **angular momentum $\mathbf{H}_G'$** of the system of particles **about the mass center G** as

$$\mathbf{H}_G' = \sum_{i=1}^{n} (\mathbf{r}_i' \times m_i\mathbf{v}_i') \qquad\qquad (14.17)$$

We now differentiate both members of Eq. (14.17) with respect to t. This operation is similar to that performed earlier on Eq. (14.7), so we can write immediately

$$\dot{\mathbf{H}}_G' = \sum_{i=1}^{n} (\mathbf{r}_i' \times m_i\mathbf{a}_i') \qquad\qquad (14.18)$$

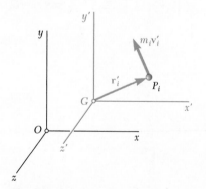

Fig. 14.5 A centroidal frame of reference $Gx'y'z'$ moving in translation with respect to a newtonian frame of reference $Oxyz$.

where $\mathbf{a}_i'$ denotes the acceleration of P_i relative to the moving frame of reference. Referring to Sec. 11.4D, we have

$$\mathbf{a}_i = \overline{\mathbf{a}} + \mathbf{a}_i'$$

where $\mathbf{a}_i$ and $\overline{\mathbf{a}}$ denote, respectively, the accelerations of P_i and G relative to the frame $Oxyz$. Solving for $\mathbf{a}_i'$ and substituting into Eq. (14.18), we have

$$\dot{\mathbf{H}}_G' = \sum_{i=1}^{n} (\mathbf{r}_i' \times m_i \mathbf{a}_i) - \left(\sum_{i=1}^{n} m_i \mathbf{r}_i' \right) \times \overline{\mathbf{a}} \qquad \textbf{(14.19)}$$

However, by Eq. (14.12), the second sum in Eq. (14.19) is equal to $m\overline{\mathbf{r}}'$ and thus to zero, since the position vector $\overline{\mathbf{r}}'$ of G relative to the frame $Gx'y'z'$ is clearly zero. On the other hand, since $\mathbf{a}_i$ represents the acceleration of P_i relative to a newtonian frame, we can use Eq. (14.1) and replace $m_i \mathbf{a}_i$ by the sum of the internal forces $\mathbf{f}_{ij}$ and of the resultant $\mathbf{F}_i$ of the external forces acting on P_i. But a reasoning similar to that used in Sec. 14.1A shows that the moment resultant about G of the internal forces $\mathbf{f}_{ij}$ of the entire system is zero. The first sum in Eq. (14.19) therefore reduces to the resultant moment about G of the external forces acting on the particles of the system, and we have

$$\Sigma \mathbf{M}_G = \dot{\mathbf{H}}_G' \qquad \textbf{(14.20)}$$

This equation states:

The resultant moment about G of the external forces is equal to the rate of change of the angular momentum about G of the system of particles.

Note that in Eq. (14.17) we defined the angular momentum $\mathbf{H}_G'$ as the sum of the moments about G of the momenta of the particles $m_i \mathbf{v}_i'$ *in their motion relative to the centroidal frame of reference $Gx'y'z'$*. We may sometimes want to compute the sum $\mathbf{H}_G$ of the moments about G of the momenta of the particles $m_i \mathbf{v}_i$ *in their absolute motion*, i.e., in their motion as observed from the newtonian frame of reference $Oxyz$ (Fig. 14.6):

$$\mathbf{H}_G = \sum_{i=1}^{n} (\mathbf{r}_i' \times m_i \mathbf{v}_i) \qquad \textbf{(14.21)}$$

Remarkably, the angular momenta $\mathbf{H}_G'$ and $\mathbf{H}_G$ are identically equal. This can be verified by referring to Sec. 11.4D and writing

$$\mathbf{v}_i = \overline{\mathbf{v}} + \mathbf{v}_i' \qquad \textbf{(14.22)}$$

Substituting for $\mathbf{v}_i$ from Eq. (14.22) into Eq. (14.21), we have

$$\mathbf{H}_G = \left(\sum_{i=1}^{n} m_i \mathbf{r}_i' \right) \times \overline{\mathbf{v}} + \sum_{i=1}^{n} (\mathbf{r}_i' \times m_i \mathbf{v}_i')$$

But, as observed earlier, the first sum is equal to zero. Thus, $\mathbf{H}_G$ reduces to the second sum, which by definition is equal to $\mathbf{H}_G'$.[†]

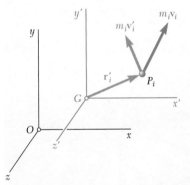

Fig. 14.6 The linear momentum of particle P_i with respect to the centroidal frame ($m_i v_i'$) and with respect to a newtonian frame ($m_i v_i$).

[†]Note that this property is peculiar to the centroidal frame $Gx'y'z'$ and does not, in general, hold for other frames of reference (see Prob. 14.29).

Taking advantage of the property we have just established, we simplify our notation by dropping the prime (′) from Eq. (14.20) and writing

$$\Sigma \mathbf{M}_G = \dot{\mathbf{H}}_G \qquad (14.23)$$

Here we can compute the angular momentum $\mathbf{H}_G$ by taking the moments about G of the momenta of the particles with respect to either the Newtonian frame $Oxyz$ or the centroidal frame $Gx'y'z'$:

$$\mathbf{H}_G = \sum_{i=1}^{n} (\mathbf{r}'_i \times m_i \mathbf{v}_i) = \sum_{i=1}^{n} (\mathbf{r}'_i \times m_i \mathbf{v}'_i) \qquad (14.24)$$

14.1E Conservation of Momentum for a System of Particles

If no external force acts on the particles of a system, the left-hand sides of Eqs. (14.10) and (14.11) are equal to zero. These equations then reduce to $\dot{\mathbf{L}} = 0$ and $\dot{\mathbf{H}}_O = 0$. We conclude that

$$\mathbf{L} = \text{constant} \qquad \mathbf{H}_O = \text{constant} \qquad (14.25)$$

These equations state that the linear momentum of the system of particles and its angular momentum about the fixed point O are conserved.

In some applications, such as problems involving central forces, the moment about a fixed point O of each of the external forces can be zero without any of the forces being zero. In such cases, the second of Eqs. (14.25) still holds; the angular momentum of the system of particles about O is conserved.

We can also apply the concept of conservation of momentum to the analysis of the motion of the mass center G of a system of particles and to the analysis of the motion of the system about G. For example, if the sum of the external forces is zero, the first of Eqs. (14.25) applies. Recalling Eq. (14.14), we have

$$\bar{\mathbf{v}} = \text{constant} \qquad (14.26)$$

This equation says that the mass center G of the system moves in a straight line and at a constant speed. On the other hand, if the sum of the moments about G of the external forces is zero, it follows from Eq. (14.23) that the angular momentum of the system about its mass center is conserved:

$$\mathbf{H}_O = \text{constant} \qquad (14.27)$$

Photo 14.1 No external impulsive forces act on a fireworks as it explodes, so linear and angular momenta of the system are conserved.

Sample Problem 14.1

A 200-kg space vehicle passes through the origin of a newtonian reference frame $Oxyz$ at time $t = 0$ with velocity $\mathbf{v}_0 = (150 \text{ m/s})\mathbf{i}$ relative to the frame. Following the detonation of explosive charges, the vehicle separates into three parts A, B, and C, each with a mass of 100 kg, 60 kg, and 40 kg, respectively. Knowing that at $t = 2.5$ s, the positions of parts A and B are observed to be $A(555, -180, 240)$ and $B(255, 0, -120)$, where the coordinates are expressed in meters. Determine the position of part C at that time.

STRATEGY: There are no external forces, so the linear momentum of the system is conserved. Use kinematics to relate the motion of the center of mass of the spacecraft and the rectangular coordinates of its position.

MODELING and ANALYSIS: The system is the space vehicle. After the explosion, the system is composed of all three parts: A, B, and C. The mass center G of the system moves with the constant velocity $\mathbf{v}_0 = (150 \text{ m/s})\mathbf{i}$. At $t = 2.5$ s, its position is

$$\bar{\mathbf{r}} = \mathbf{v}_0 t = (150 \text{ m/s})\mathbf{i}(2.5\,\text{s}) = (375\,\text{m})\mathbf{i}$$

Recalling Eq. (14.12), you have

$$m\bar{\mathbf{r}} = m_A \mathbf{r}_A + m_B \mathbf{r}_B + m_C \mathbf{r}_C$$
$$(200 \text{ kg})(375 \text{ m})\mathbf{i} = (100 \text{ kg})[(555 \text{ m})\mathbf{i} - (180 \text{ m})\mathbf{j} + (240 \text{ m})\mathbf{k}]$$
$$+ (60 \text{ kg})[(255 \text{ m})\mathbf{i} - (120 \text{ m})\mathbf{k}] + (40 \text{ kg})\mathbf{r}_C$$
$$\mathbf{r}_C = (105 \text{ m})\mathbf{i} + (450 \text{ m})\mathbf{j} - (420 \text{ m})\mathbf{k} \ \blacktriangleleft$$

REFLECT and THINK: This kind of calculation can serve as a model for any situation involving fragmentation of a projectile with no external forces present.

Sample Problem 14.2

A 20-lb projectile is moving with a velocity of 100 ft/s when it explodes into two fragments A and B, weighing 5 lb and 15 lb, respectively. Knowing that immediately after the explosion, fragments A and B travel in directions defined respectively by $\theta_A = 45°$ and $\theta_B = 30°$, determine the velocity of each fragment.

STRATEGY: There are no external forces, so apply the conservation of linear momentum to the system.

(continued)

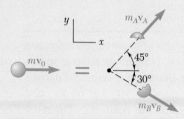

MODELING and ANALYSIS: The system is the projectile. After the explosion, the system is composed of the two fragments. The impulse-momentum diagram for this system is shown in Fig. 1. There are no external impulses acting on this system, so linear momentum is conserved and:

$$m_A\mathbf{v}_A + m_B\mathbf{v}_B = m\mathbf{v}_0$$
$$(5/g)\mathbf{v}_A + (15/g)\mathbf{v}_B = (20/g)\mathbf{v}_0$$

Fig. 1 Impulse-momentum diagram for the projectile.

Applying this equation in the x and y directions gives you two scalar equations. Thus,

$\xrightarrow{+}$ x components: $\qquad 5v_A \cos 45° + 15v_B \cos 30° = 20(100)$

$+\uparrow y$ components: $\qquad 5v_A \sin 45° - 15v_B \sin 30° = 0$

Solving the two equations for v_A and v_B simultaneously gives

$$v_A = 207 \text{ ft/s} \qquad v_B = 97.6 \text{ ft/s}$$

$$\mathbf{v}_A = 207 \text{ ft/s} \measuredangle 45° \qquad \mathbf{v}_B = 97.6 \text{ ft/s} \searrow 30° \ \blacktriangleleft$$

REFLECT and THINK: As you might have predicted, the less massive fragment winds up with a larger magnitude of velocity and departs the original trajectory at a larger angle.

Sample Problem 14.3

A system consists of three particles A, B, and C, with masses $m_A = 1$ kg, $m_B = 2$ kg, and $m_C = 3$ kg. The velocities of the particles expressed in m/s are, respectively, $\mathbf{v}_A = 3\mathbf{i} - 2\mathbf{j} + 4\mathbf{k}$, $\mathbf{v}_B = 4\mathbf{i} + 3\mathbf{j}$, and $\mathbf{v}_C = 2\mathbf{i} + 5\mathbf{j} - 3\mathbf{k}$. Determine (a) the angular momentum $\mathbf{H}_O$ of the system about O, (b) the position vector $\bar{\mathbf{r}}$ of the mass center G of the system, (c) the angular momentum $\mathbf{H}_G$ of the system about G.

STRATEGY: You have a system of particles, so use the definitions of angular momentum and center of mass.

MODELING: Choose the three particles as your system.

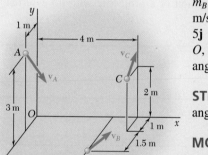

ANALYSIS: The linear momentum of each particle expressed in kg·m/s is

$$m_A\mathbf{v}_A = 3\mathbf{i} - 2\mathbf{j} + 4\mathbf{k}$$
$$m_B\mathbf{v}_B = 8\mathbf{i} + 6\mathbf{j}$$
$$m_C\mathbf{v}_C = 6\mathbf{i} + 15\mathbf{j} - 9\mathbf{k}$$

The position vectors (in meters) are

$$\mathbf{r}_A = 3\mathbf{j} + \mathbf{k} \qquad \mathbf{r}_B = 3\mathbf{i} + 2.5\mathbf{k} \qquad \mathbf{r}_C = 4\mathbf{i} + 2\mathbf{j} + \mathbf{k}$$

a. Angular Momentum About O. Using the definition of angular momentum about O (in $kg \cdot m^2/s$) you find

$$\mathbf{H}_O = \mathbf{r}_A \times (m_A \mathbf{v}_A) + \mathbf{r}_B \times (m_B \mathbf{v}_B) + \mathbf{r}_C \times (m_C \mathbf{v}_C)$$

$$= \begin{vmatrix} \mathbf{i} & \mathbf{j} & \mathbf{k} \\ 0 & 3 & 1 \\ 3 & -2 & 4 \end{vmatrix} + \begin{vmatrix} \mathbf{i} & \mathbf{j} & \mathbf{k} \\ 3 & 0 & 2.5 \\ 8 & 6 & 0 \end{vmatrix} + \begin{vmatrix} \mathbf{i} & \mathbf{j} & \mathbf{k} \\ 4 & 2 & 1 \\ 6 & 15 & -9 \end{vmatrix}$$

$$= (14\mathbf{i} + 3\mathbf{j} - 9\mathbf{k}) + (-15\mathbf{i} + 20\mathbf{j} + 18\mathbf{k}) + (-33\mathbf{i} + 42\mathbf{j} + 48\mathbf{k})$$

$$= 34\mathbf{i} + 65\mathbf{j} + 57\mathbf{k}$$

$$\mathbf{H}_O = -(34 \ kg \cdot m^2/s)\mathbf{i} + (65 \ kg \cdot m^2/s)\mathbf{j} + (57 \ kg \cdot m^2/s)\mathbf{k} \quad \blacktriangleleft$$

b. Mass Center. Using the definition of mass center, you find

$$(m_A + m_B + m_C)\bar{\mathbf{r}} = m_A \mathbf{r}_A + m_B \mathbf{r}_B + m_C \mathbf{r}_C$$

$$6\bar{\mathbf{r}} = (1)(3\mathbf{j} + \mathbf{k}) + (2)(3\mathbf{i} + 2.5\mathbf{k}) + (3)(4\mathbf{i} + 2\mathbf{j} + \mathbf{k})$$

$$\bar{\mathbf{r}} = 3\mathbf{i} + 1.5\mathbf{j} + 1.5\mathbf{k}$$

$$\bar{\mathbf{r}} = (3.00 \ m)\mathbf{i} + (1.500 \ m)\mathbf{j} + (1.500 \ m)\mathbf{k} \quad \blacktriangleleft$$

c. Angular Momentum About G. The angular momentum of the system about G is

$$\mathbf{H}_G = \mathbf{r}'_A \times m_A \mathbf{v}_A + \mathbf{r}'_B \times m_B \mathbf{v}_B + \mathbf{r}'_C \times m_C \mathbf{v}_C$$

where $\mathbf{r}'_A$, $\mathbf{r}'_B$, and $\mathbf{r}'_C$ are the position vectors from the particles to the center of mass; that is

$$\mathbf{r}'_A = \mathbf{r}_A - \bar{\mathbf{r}} = -3\mathbf{i} + 1.5\mathbf{j} - 0.5\mathbf{k}$$

$$\mathbf{r}'_B = \mathbf{r}_B - \bar{\mathbf{r}} = -1.5\mathbf{j} + \mathbf{k}$$

$$\mathbf{r}'_C = \mathbf{r}_C - \bar{\mathbf{r}} = \mathbf{i} + 0.5\mathbf{j} - 0.5\mathbf{k}$$

Therefore, you can calculate the angular momentum as

$$\mathbf{H}_G = \mathbf{r}'_A \times m_A \mathbf{v}_A + \mathbf{r}'_B \times m_B \mathbf{v}_B + \mathbf{r}'_C \times m_C \mathbf{v}_C$$

$$= \begin{vmatrix} \mathbf{i} & \mathbf{j} & \mathbf{k} \\ -3 & 1.5 & -0.5 \\ 3 & -2 & 4 \end{vmatrix} + \begin{vmatrix} \mathbf{i} & \mathbf{j} & \mathbf{k} \\ 0 & -1.5 & 1 \\ 8 & 6 & 0 \end{vmatrix} + \begin{vmatrix} \mathbf{i} & \mathbf{j} & \mathbf{k} \\ 4 & 0.5 & -0.5 \\ 6 & 15 & -9 \end{vmatrix}$$

$$= (5\mathbf{i} + 10.5\mathbf{j} + 1.5\mathbf{k}) + (-6\mathbf{i} + 8\mathbf{j} + 12\mathbf{k}) + (3\mathbf{i} + 6\mathbf{j} + 12\mathbf{k})$$

$$= 2\mathbf{i} + 24.5\mathbf{j} + 25.5\mathbf{k}$$

$$\mathbf{H}_G = (2.00 \ kg \cdot m^2/s)\mathbf{i} + (24.5 \ kg \cdot m^2/s)\mathbf{j} + (25.5 \ kg \cdot m^2/s)\mathbf{k} \quad \blacktriangleleft$$

REFLECT and THINK: You should be able to verify that the answers to this problem satisfy the equations given in Prob. 14.27; that is, $\mathbf{H}_O = \bar{\mathbf{r}} \times m\bar{\mathbf{v}} + \mathbf{H}_G$. Because no impulses act on the system, the linear momentum of the overall system is constant; the location of the center of mass of the system, however, changes with time.

SOLVING PROBLEMS
ON YOUR OWN

This chapter dealt with the motion of **systems of particles** where the motion of a large number of particles is considered together, rather than separately. In this first section, you learned to compute the **linear momentum** and the **angular momentum** of a system of particles. We defined the linear momentum **L** of a system of particles as the sum of the linear momenta of the particles, and we defined the angular momentum $\mathbf{H}_O$ of the system as the sum of the angular momenta of the particles about O:

$$\mathbf{L} = \sum_{i=1}^{n} m_i \mathbf{v}_i \qquad \mathbf{H}_O = \sum_{i=1}^{n} (\mathbf{r}_i \times m_i \mathbf{v}_i) \qquad \textbf{(14.6, 14.7)}$$

In this section, you will be asked to solve several problems of practical interest, either by observing that the linear momentum of a system of particles is conserved or by considering the motion of the mass center of a system of particles.

1. Conservation of the linear momentum of a system of particles. This occurs *when the resultant of the external forces acting on the particles of the system is zero*. You may encounter such a situation in the following types of problems.

   a. Problems involving the rectilinear motion of objects, such as colliding automobiles and railroad cars. After you have checked that the resultant of the external forces is zero, equate the algebraic sums of the initial momenta and final momenta to obtain an equation that you can solve for one unknown.

   b. Problems involving the two-dimensional or three-dimensional motion of objects, such as exploding shells or colliding aircraft, automobiles, or billiard balls. After you have checked that the resultant of the external forces is zero, add the initial momenta of the objects vectorially, add their final momenta vectorially, and equate the two sums to obtain a vector equation expressing that the linear momentum of the system is conserved.

   In the case of two-dimensional motion, you can replace this equation with two scalar equations that you can solve for two unknowns. In the case of three-dimensional motion, you can replace the equation with three scalar equations that you can solve for three unknowns.

2. Motion of the mass center of a system of particles. You saw in Sec. 14.1C that *the mass center of a system of particles moves as if the entire mass of the system and all of the external forces were concentrated at that point.*

   a. In the case of a body exploding while in motion, it follows that the mass center of the resulting fragments moves as the body itself would have moved if the explosion had not occurred. You can solve problems of this type by writing the equation of motion of the mass center of the system in vector form and expressing the position vector of the mass center in terms of the position vectors of the various fragments [Eq. (14.12) and Sample Prob. 14.1]. You can then rewrite the vector equation as two or three scalar equations and solve the equations for an equivalent number of unknowns.

   b. In the case of the collision of several moving bodies, it follows that the motion of the mass center of the various bodies is unaffected by the collision. You can solve problems of this type by writing the equation of motion of the mass center of the system in vector form and expressing its position vector before and after the collision in terms of the position vectors of the relevant bodies [Eq. (14.12)]. You can then rewrite the vector equation as two or three scalar equations and solve these equations for an equivalent number of unknowns.

Problems

14.1 A 30-g bullet is fired with a horizontal velocity of 450 m/s and becomes embedded in block B, which has a mass of 3 kg. After the impact, block B slides on 30-kg carrier C until it impacts the end of the carrier. Knowing the impact between B and C is perfectly plastic and the coefficient of kinetic friction between B and C is 0.2, determine (*a*) the velocity of the bullet and B after the first impact, (*b*) the final velocity of the carrier.

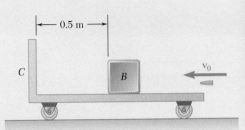

Fig. P14.1

14.2 Two identical 1350-kg automobiles A and B are at rest with their brakes released when B is struck by a 5400-kg truck C that is moving to the left at 8 km/h. A second collision then occurs when B strikes A. Assuming the first collision is perfectly plastic and the second collision is perfectly elastic, determine the velocities of the three vehicles just after the second collision.

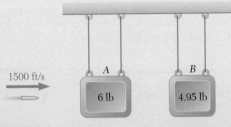

Fig. P14.2

14.3 An airline employee tosses two suitcases with weights of 30 lb and 40 lb, respectively, onto a 50-lb baggage carrier in rapid succession. Knowing that the carrier is initially at rest and that the employee imparts a 9-ft/s horizontal velocity to the 30-lb suitcase and a 6-ft/s horizontal velocity to the 40-lb suitcase, determine the final velocity of the baggage carrier if the first suitcase tossed onto the carrier is (*a*) the 30-lb suitcase, (*b*) the 40-lb suitcase.

Fig. P14.3

14.4 A bullet is fired with a horizontal velocity of 1500 ft/s through a 6-lb block A and becomes embedded in a 4.95-lb block B. Knowing that blocks A and B start moving with velocities of 5 ft/s and 9 ft/s, respectively, determine (*a*) the weight of the bullet, (*b*) its velocity as it travels from block A to block B.

Fig. P14.4

14.5 Two swimmers *A* and *B*, of weight 190 lb and 125 lb, respectively, are at diagonally opposite corners of a floating raft when they realize that the raft has broken away from its anchor. Swimmer *A* immediately starts walking toward *B* at a speed of 2 ft/s relative to the raft. Knowing that the raft weighs 300 lb, determine (*a*) the speed of the raft if *B* does not move, (*b*) the speed with which *B* must walk toward *A* if the raft is not to move.

14.6 A 180-lb man and a 120-lb woman stand side by side at the same end of a 300-lb boat, ready to dive, each with a 16-ft/s velocity relative to the boat. Determine the velocity of the boat after they have both dived, if (*a*) the woman dives first, (*b*) the man dives first.

Fig. P14.6

14.7 A 40-Mg boxcar *A* is moving in a railroad switchyard with a velocity of 9 km/h toward cars *B* and *C*, which are both at rest with their brakes off at a short distance from each other. Car *B* is a 25-Mg flatcar supporting a 30-Mg container, and car *C* is a 35-Mg boxcar. As the cars hit each other they get automatically and tightly coupled. Determine the velocity of car *A* immediately after each of the two couplings, assuming that the container (*a*) does not slide on the flatcar, (*b*) slides after the first coupling but hits a stop before the second coupling occurs, (*c*) slides and hits the stop only after the second coupling has occurred.

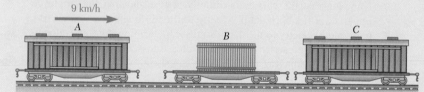

Fig. P14.7

14.8 Two identical cars *A* and *B* are at rest on a loading dock with brakes released. Car *C*, of a slightly different style but of the same weight, has been pushed by dockworkers and hits car *B* with a velocity of 1.5 m/s. Knowing that the coefficient of restitution is 0.8 between *B* and *C* and 0.5 between *A* and *B*, determine the velocity of each car after all collisions have taken place.

Fig. P14.8

14.9 A 20-kg base satellite deploys three sub-satellites, each which has its own thrust capabilities, to perform research on tether propulsion. The masses of sub-satellites A, B, and C are 4 kg, 6 kg, and 8 kg, respectively, and their velocities expressed in m/s are given by $\mathbf{v}_A = 4\mathbf{i} - 2\mathbf{j} + 2\mathbf{k}$, $\mathbf{v}_B = \mathbf{i} + 4\mathbf{j}$, $\mathbf{v}_C = 2\mathbf{i} + 2\mathbf{j} + 4\mathbf{k}$. At the instant shown, what is the angular momentum $\mathbf{H}_O$ of the system about the base satellite?

14.10 For the satellite system of Prob. 14.9, assuming that the velocity of the base satellite is zero, determine (a) the position vector $\bar{\mathbf{r}}$ of the mass center G of the system, (b) the linear momentum $\mathbf{L}$ of the system, (c) the angular momentum $\mathbf{H}_G$ of the system about G. Also, verify that the answers to this problem and to Prob. 14.9 satisfy the equation given in Prob. 14.27.

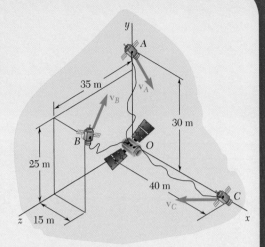

Fig. P14.9 and P14.10

14.11 A system consists of three identical 19.32-lb particles A, B, and C. The velocities of the particles are, respectively, $\mathbf{v}_A = v_A\mathbf{j}$, $\mathbf{v}_B = v_B\mathbf{i}$, and $\mathbf{v}_C = v_C\mathbf{k}$. Knowing that the angular momentum of the system about O expressed in ft·lb·s is $\mathbf{H}_O = -1.2\mathbf{k}$, determine (a) the velocities of the particles, (b) the angular momentum of the system about its mass center G.

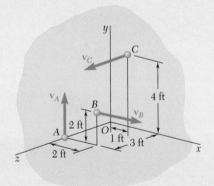

Fig. P14.11 and P14.12

14.12 A system consists of three identical 19.32-lb particles A, B, and C. The velocities of the particles are, respectively, $\mathbf{v}_A = v_A\mathbf{j}$, $\mathbf{v}_B = v_B\mathbf{i}$, and $\mathbf{v}_C = v_C\mathbf{k}$, and the magnitude of the linear momentum $\mathbf{L}$ of the system is 9 lb·s. Knowing that $\mathbf{H}_G = \mathbf{H}_O$, where $\mathbf{H}_G$ is the angular momentum of the system about its mass center G and $\mathbf{H}_O$ is the angular momentum of the system about O, determine (a) the velocities of the particles, (b) the angular momentum of the system about O.

14.13 A system consists of three particles A, B, and C. We know that $m_A = 3$ kg, $m_B = 2$ kg, and $m_C = 4$ kg and that the velocities of the particles expressed in m/s are, respectively, $\mathbf{v}_A = 4\mathbf{i} + 2\mathbf{j} + 2\mathbf{k}$, $\mathbf{v}_B = 4\mathbf{i} + 3\mathbf{j}$, and $\mathbf{v}_C = -2\mathbf{i} + 4\mathbf{j} + 2\mathbf{k}$. Determine the angular momentum $\mathbf{H}_O$ of the system about O.

14.14 For the system of particles of Prob. 14.13, determine (a) the position vector $\bar{\mathbf{r}}$ of the mass center G of the system, (b) the linear momentum $m\bar{\mathbf{v}}$ of the system, (c) the angular momentum $\mathbf{H}_G$ of the system about G. Also verify that the answers to this problem and to Problem 14.13 satisfy the equation given in Prob. 14.27.

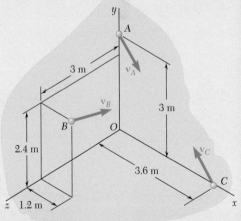

Fig. P14.13

14.15 A 13-kg projectile is passing through the origin O with a velocity $\mathbf{v}_0 = (35 \text{ m/s})\mathbf{i}$ when it explodes into two fragments A and B, of mass 5 kg and 8 kg, respectively. Knowing that 3 s later the position of fragment A is (90 m, 7 m, −14 m), determine the position of fragment B at the same instant. Assume $a_y = -g = -9.81 \text{ m/s}^2$ and neglect air resistance.

14.16 A 300-kg space vehicle traveling with a velocity $\mathbf{v}_0 = (360 \text{ m/s})\mathbf{i}$ passes through the origin O at $t = 0$. Explosive charges then separate the vehicle into three parts A, B, and C, with mass, respectively, 150 kg, 100 kg, and 50 kg. Knowing that at $t = 4$ s, the positions of parts A and B are observed to be A (1170 m, −290 m, –585 m) and B (1975 m, 365 m, 800 m), determine the corresponding position of part C. Neglect the effect of gravity.

14.17 A 2-kg model rocket is launched vertically and reaches an altitude of 70 m with a speed of 30 m/s at the end of powered flight, time $t = 0$. As the rocket approaches its maximum altitude it explodes into two parts of masses $m_A = 0.7$ kg and $m_B = 1.3$ kg. Part A is observed to strike the ground 80 m west of the launch point at $t = 6$ s. Determine the position of part B at that time.

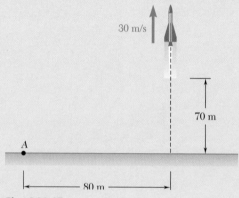

Fig. P14.17

14.18 An 18-kg cannonball and a 12-kg cannonball are chained together and fired horizontally with a velocity of 165 m/s from the top of a 15-m wall. The chain breaks during the flight of the cannonballs and the 12-kg cannonball strikes the ground at $t = 1.5$ s, at a distance of 240 m from the foot of the wall, and 7 m to the right of the line of fire. Determine the position of the other cannonball at that instant. Neglect the resistance of the air.

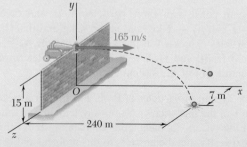

Fig. P14.18

14.19 and 14.20   Car A was traveling east at high speed when it collided at point O with car B, which was traveling north at 45 mi/h. Car C, which was traveling west at 60 mi/h, was 32 ft east and 10 ft north of point O at the time of the collision. Because the pavement was wet, the driver of car C could not prevent his car from sliding into the other two cars, and the three cars, stuck together, kept sliding until they hit the utility pole P. Knowing that the weights of cars A, B, and C are, respectively, 3000 lb, 2600 lb, and 2400 lb, and neglecting the forces exerted on the cars by the wet pavement, solve the problems indicated.

14.19   Knowing that the speed of car A was 75 mi/h and that the time elapsed from the first collision to the stop at P was 2.4 s, determine the coordinates of the utility pole P.

14.20   Knowing that the coordinates of the utility pole are $x_p = 46$ ft and $y_p = 59$ ft, determine (a) the time elapsed from the first collision to the stop at P, (b) the speed of car A.

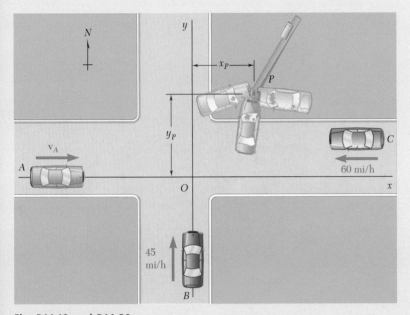

Fig. P14.19 and P14.20

14.21   An expert archer demonstrates his ability by hitting tennis balls thrown by an assistant. A 2-oz tennis ball has a velocity of $(32 \text{ ft/s})\mathbf{i} - (7 \text{ ft/s})\mathbf{j}$ and is 33 ft above the ground when it is hit by a 1.2-oz arrow traveling with a velocity of $(165 \text{ ft/s})\mathbf{j} + (230 \text{ ft/s})\mathbf{k}$ where $\mathbf{j}$ is directed upwards. Determine the position P where the ball and arrow will hit the ground, relative to point O located directly under the point of impact.

14.22   Two spheres, each of mass m, can slide freely on a frictionless, horizontal surface. Sphere A is moving at a speed $v_0 = 16$ ft/s when it strikes sphere B, which is at rest, and the impact causes sphere B to break into two pieces, each of mass $m/2$. Knowing that 0.7 s after the collision one piece reaches point C and 0.9 s after the collision the other piece reaches point D, determine (a) the velocity of sphere A after the collision, (b) the angle θ and the speeds of the two pieces after the collision.

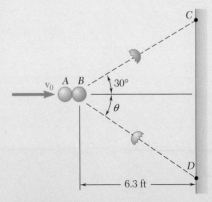

Fig. P14.22

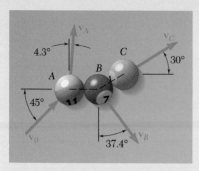

Fig. P14.23

14.23 In a game of pool, ball A is moving with a velocity $\mathbf{v}_0$ when it strikes balls B and C which are at rest and aligned as shown. Knowing that after the collision the three balls move in the directions indicated and that $v_0 = 12$ ft/s and $v_C = 6.29$ ft/s, determine the magnitude of the velocity of (*a*) ball A, (*b*) ball B.

14.24 A 6-kg shell moving with a velocity $\mathbf{v}_0 = (12 \text{ m/s})\mathbf{i} - (9 \text{ m/s})\mathbf{j} - (360 \text{ m/s})\mathbf{k}$ explodes at point D into three fragments A, B, and C of mass, respectively, 3 kg, 2 kg, and 1 kg. Knowing that the fragments hit the vertical wall at the points indicated, determine the speed of each fragment immediately after the explosion. Assume that elevation changes due to gravity may be neglected.

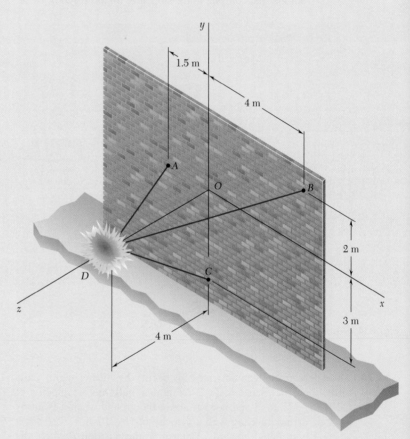

Fig. P14.24 and P14.25

14.25 A 6-kg shell moving with a velocity $\mathbf{v}_0 = (12 \text{ m/s})\mathbf{i} - (9 \text{ m/s})\mathbf{j} - (360 \text{ m/s})\mathbf{k}$ explodes at point D into three fragments A, B, and C of mass, respectively, 2 kg, 1 kg, and 3 kg. Knowing that the fragments hit the vertical wall at the points indicated, determine the speed of each fragment immediately after the explosion. Assume that elevation changes due to gravity may be neglected.

14.26 In a scattering experiment, an alpha particle A is projected with the velocity $\mathbf{u}_0 = -(600 \text{ m/s})\mathbf{i} + (750 \text{ m/s})\mathbf{j} - (800 \text{ m/s})\mathbf{k}$ into a stream of oxygen nuclei moving with a common velocity $\mathbf{v}_0 = (600 \text{ m/s})\mathbf{j}$. After colliding successively with the nuclei B and C, particle A is observed to move along the path defined by the points A_1 (280, 240, 120) and A_2 (360, 320, 160), while nuclei B and C are observed to move along paths defined, respectively, by B_1 (147, 220, 130) and B_2 (114, 290, 120), and by C_1 (240, 232, 90) and C_2 (240, 280, 75). All paths are along straight lines and all coordinates are expressed in millimeters. Knowing that the mass of an oxygen nucleus is four times that of an alpha particle, determine the speed of each of the three particles after the collisions.

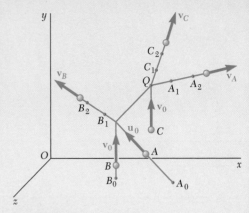

Fig. P14.26

14.27 Derive the relation

$$\mathbf{H}_O = \bar{\mathbf{r}} \times m\bar{\mathbf{v}} + H_G$$

between the angular momenta $\mathbf{H}_O$ and $\mathbf{H}_G$ defined in Eqs. (14.7) and (14.24), respectively. The vectors $\bar{\mathbf{r}}$ and $\bar{\mathbf{v}}$ define, respectively, the position and velocity of the mass center G of the system of particles relative to the newtonian frame of reference $Oxyz$, and m represents the total mass of the system.

14.28 Show that Eq. (14.23) may be derived directly from Eq. (14.11) by substituting for $\mathbf{H}_O$ the expression given in Prob. 14.27.

14.29 Consider the frame of reference $Ax'y'z'$ in translation with respect to the newtonian frame of reference $Oxyz$. We define the angular momentum $\mathbf{H}'_A$ of a system of n particles about A as the sum

$$\mathbf{H}'_A = \sum_{i=1}^{n} \mathbf{r}'_i \times m_i \mathbf{v}'_i \qquad (1)$$

of the moments about A of the momenta $m_i\mathbf{v}'_i$ of the particles in their motion relative to the frame $Ax'y'z'$. Denoting by $\mathbf{H}_A$ the sum

$$\mathbf{H}_A = \sum_{i=1}^{n} \mathbf{r}'_i \times m_i \mathbf{v}_i$$

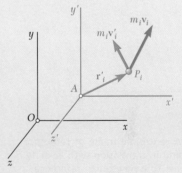

Fig. P14.29

of the moments about A of the momenta $m_i\mathbf{v}_i$ of the particles in their motion relative to the newtonian frame $Oxyz$, show that $\mathbf{H}_A = \mathbf{H}'_A$ at a given instant if, and only if, one of the following conditions is satisfied at that instant: (a) A has zero velocity with respect to the frame $Oxyz$, (b) A coincides with the mass center G of the system, (c) the velocity $\mathbf{v}_A$ relative to $Oxyz$ is directed along the line AG.

14.30 Show that the relation $\Sigma\mathbf{M}_A = \dot{\mathbf{H}}'_A$, where $\mathbf{H}'_A$ is defined by Eq. (1) of Prob. 14.29 and where $\Sigma\mathbf{M}_A$ represents the sum of the moments about A of the external forces acting on the system of particles, is valid if, and only if, one of the following conditions is satisfied: (a) the frame $Ax'y'z'$ is itself a newtonian frame of reference, (b) A coincides with the mass center G, (c) the acceleration $\mathbf{a}_A$ of A relative to $Oxyz$ is directed along the line AG.

14.2   ENERGY AND MOMENTUM METHODS FOR A SYSTEM OF PARTICLES

Solving problems involving a system of particles is often made easier by applying energy and momentum methods, just as it was as for a single particle in Chapter 13. Definitions of terms and statements of the work–energy and impulse-momentum principles are very similar to the single-particle versions, especially when you take into account the mass center of the particles.

14.2A   Kinetic Energy of a System of Particles

We define the kinetic energy T of a system of particles as the sum of the kinetic energies of the various particles of the system. Referring to Sec. 13.1B, we have

Kinetic energy, system of particles

$$T = \frac{1}{2} \sum_{i=1}^{n} m_i v_i^2 \qquad (14.28)$$

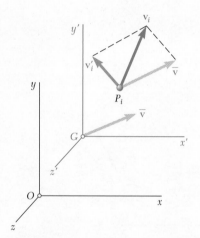

Fig. 14.7 A centroidal frame of reference $Gx'y'z'$ moving in translation with velocity $\bar{\mathbf{v}}$ with respect to a newtonian reference frame $Oxyz$.

Using a Centroidal Frame of Reference.   It is often convenient when computing the kinetic energy of a system comprised of a large number of particles (as in the case of a rigid body) to consider the motion of the mass center G of the system and the motion of the system relative to a moving frame attached to G separately.

Let P_i be a particle of the system, $\mathbf{v}_i$ be its velocity relative to the newtonian frame of reference $Oxyz$, and $\mathbf{v}_i'$ be its velocity relative to the moving frame $Gx'y'z'$ that is in translation with respect to $Oxyz$ (Fig. 14.7). Recall from Sec. 14.1D that

$$\mathbf{v}_i = \bar{\mathbf{v}} + \mathbf{v}_i' \qquad (14.22)$$

where $\bar{\mathbf{v}}$ denotes the velocity of the mass center G relative to the newtonian frame $Oxyz$. Observing that v_i^2 is equal to the scalar product $\mathbf{v}_i \cdot \mathbf{v}_i$, we can express the kinetic energy T of the system relative to the newtonian frame $Oxyz$ as

$$T = \frac{1}{2} \sum_{i=1}^{n} m_i v_i^2 = \frac{1}{2} \sum_{i=1}^{n} (m_i \mathbf{v}_i \cdot \mathbf{v}_i)$$

or, substituting for $\mathbf{v}_i$ from Eq. (14.22),

$$T = \frac{1}{2} \sum_{i=1}^{n} [m_i (\bar{\mathbf{v}} + \mathbf{v}_i') \cdot (\bar{\mathbf{v}} + \mathbf{v}_i')]$$

$$= \frac{1}{2} \left(\sum_{i=1}^{n} m_i \right) \bar{v}^2 + \bar{\mathbf{v}} \cdot \sum_{i=1}^{n} m_i \mathbf{v}_i' + \frac{1}{2} \sum_{i=1}^{n} m_i v_i'^2$$

In this equation, the first sum represents the total mass m of the system. Recalling Eq. (14.13), we note that the second sum is equal to $m\bar{\mathbf{v}}'$ and thus to zero, since $\bar{\mathbf{v}}'$, which represents the velocity of G relative to the frame $Gx'y'z'$, is clearly zero. We therefore have

$$T = \tfrac{1}{2}m\bar{v}^2 + \frac{1}{2}\sum_{i=1}^{n} m_i v_i'^2 \qquad (14.29)$$

This equation states that we can obtain the kinetic energy T of a system of particles **by adding the kinetic energy of the mass center G and the kinetic energy of the system in its motion relative to the frame $Gx'y'z'$.**

14.2B Work-Energy Principle and Conservation of Energy for a System of Particles

We can apply the principle of work and energy to each particle P_i of a system of particles, obtaining for each particle P_i

$$(T_1)_i + (U_{1\rightarrow2})_i = (T_2)_i$$

where $(U_{1\rightarrow2})_i$ represents the work done by the internal forces $\mathbf{f}_{ij}$ and the resultant external force $\mathbf{F}_i$ acting on P_i. Adding the kinetic energies of the various particles of the system and considering the work of all the forces involved, we obtain an expression for the entire system as

Work-energy principle, system of particles

$$T_1 + U_{1\rightarrow2} = T_2 \qquad (14.30)$$

The quantities T_1 and T_2 now represent the kinetic energy of the entire system and can be computed from either Eq. (14.28) or Eq. (14.29). The quantity $U_{1\rightarrow2}$ represents the work of all the forces acting on the particles of the system. Note that although the internal forces $\mathbf{f}_{ij}$ and $\mathbf{f}_{ji}$ are equal and opposite, the work of these forces does not, in general, cancel out, since the particles P_i and P_j on which they act generally undergo different displacements. Therefore, in computing $U_{1\rightarrow2}$, **we must consider the work of the internal forces $\mathbf{f}_{ij}$ as well as the work of the external forces $\mathbf{F}_i$.** An alternative way of writing Eq. (14.30) is

$$T_1 + V_{g_1} + V_{e_1} + U_{1\rightarrow2}^{\text{NC}} = T_2 + V_{g_2} + V_{e_2} \qquad (14.30')$$

where V_g is the gravitational potential energy of the system, V_e is the elastic potential energy, and $U_{1\rightarrow2}^{\text{NC}}$ is the work due to non-conservative forces.

If all of the forces acting on the particles of the system are conservative, we can replace Eq. (14.30) by

Conservation of energy, system of particles

$$T_1 + V_1 = T_2 + V_2 \qquad (14.31)$$

where V represents the potential energy associated with the internal and external forces acting on the particles of the system.

14.2C Impulse-Momentum Principle and Conservation of Momentum for a System of Particles

Integrating Eqs. (14.10) and (14.11) with respect to t from t_1 to t_2, we have

$$\sum \int_{t_1}^{t_2} \mathbf{F}\, dt = \mathbf{L}_2 - \mathbf{L}_1 \tag{14.32}$$

$$\sum \int_{t_1}^{t_2} \mathbf{M}_O\, dt = (\mathbf{H}_O)_2 - (\mathbf{H}_O)_1 \tag{14.33}$$

From the definition of the linear impulse of a force given in Sec. 13.3A, the integrals in Eq. (14.32) represent the linear impulses of the external forces acting on the particles of the system. In a similar way, we shall refer to the integrals in Eq. (14.33) as the **angular impulses** about O of the external forces. Thus, Eq. (14.32) states that the sum of the linear impulses of the external forces acting on the system is equal to the change in linear momentum of the system. Similarly, Eq. (14.33) says that the sum of the angular impulses about O of the external forces is equal to the change in angular momentum about O of the system.

To clarify the physical significance of Eqs. (14.32) and (14.33), we rearrange the terms in these equations, obtaining

$$\mathbf{L}_1 + \sum \int_{t_1}^{t_2} \mathbf{F}\, dt = \mathbf{L}_2 \tag{14.34}$$

$$(\mathbf{H}_O)_1 + \sum \int_{t_1}^{t_2} \mathbf{M}_O\, dt = (\mathbf{H}_O)_2 \tag{14.35}$$

In parts a and c of Fig. 14.8, we have sketched the momenta of the particles of the system at times t_1 and t_2, respectively. In part b, we show terms equal to the sum of the linear impulses of the external forces and the sum of the angular impulses about O of the external forces.

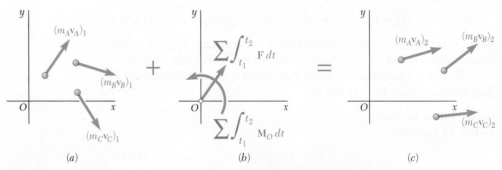

Fig. 14.8 The impulse–momentum diagram for a system of particles contains (a) momenta of particles at time t_1; (b) impulses of the external forces and moments about O; (c) momenta of the particles at time t_2.

For simplicity, we have assumed the particles move in the plane of the figure, but the present discussion remains valid in the case of particles moving in space. Recall from Eq. (14.6) that $\mathbf{L}$, by definition, is the resultant of the momenta $m_i\mathbf{v}_i$. Then Eq. (14.34) says that the resultant of the vectors shown in parts a and b of Fig. 14.8 is equal to the resultant of the vectors shown in part c. Recalling from Eq. (14.7) that $\mathbf{H}_O$ is the angular momentum, we note that Eq. (14.35) similarly says that the angular momentum of the vectors in parts a added to the angular impulses in part b of Fig. 14.8 is equal to the angular momentum of the vectors in part c. Together, Eqs. (14.34) and (14.35) state:

The momenta of the particles at time t_1 and the impulses of the external forces from t_1 to t_2 form a system of vectors equipollent to the system of the momenta of the particles at time t_2.

This is indicated in Fig. 14.8 by the use of blue plus and equal signs.

   If no external force acts on the particles of the system, the integrals in Eqs. (14.34) and (14.35) are zero, and these equations yield

Conservation of linear and angular momentum

$$\mathbf{L}_1 = \mathbf{L}_2 \tag{14.36}$$

$$(\mathbf{H}_O)_1 = (\mathbf{H}_O)_2 \tag{14.37}$$

We thus check the result obtained in Sec. 14.1E: If no external force acts on the particles of a system, the linear momentum and the angular momentum about O of the system of particles are conserved. The system of the initial momenta is equipollent to the system of the final momenta, and it follows that the angular momentum of the system of particles about *any* fixed point is conserved.

Sample Problem 14.4

For the 200-kg space vehicle of Sample Prob. 14.1, it is known that at $t = 2.5$ s, the velocity of part A is $\mathbf{v}_A = (270$ m/s$)\mathbf{i} - (120$ m/s$)\mathbf{j} + (160$ m/s$)\mathbf{k}$, and the velocity of part B is parallel to the xz plane. Determine (a) the velocity of part C, (b) the energy gained during the detonation.

STRATEGY: Since there are no external forces, use the conservation of linear momentum. Although it is not immediately apparent, you will also need to use the conservation of angular momentum to solve this problem.

MODELING and ANALYSIS: Choose the space vehicle as your system. After the explosion, the system is composed of three parts: A, B, and C. Figure 1 shows the momenta of the system before and after the explosion. From the conservation of linear momentum, you have

$$\mathbf{L}_1 = \mathbf{L}_2: \qquad m\mathbf{v}_0 = m_A\mathbf{v}_A + m_B\mathbf{v}_B + m_C\mathbf{v}_C \qquad (1)$$

From conservation of angular momentum about point O you have

$$(\mathbf{H}_O)_1 = (\mathbf{H}_O)_2: \quad 0 = \mathbf{r}_A \times m_A\mathbf{v}_A + \mathbf{r}_B \times m_B\mathbf{v}_B + \mathbf{r}_C \times m_C\mathbf{v}_C \qquad (2)$$

Recall from Sample Prob. 14.1 that $\mathbf{v}_0 = (150$ m/s$)\mathbf{i}$ and

$$m_A = 100 \text{ kg} \qquad m_B = 60 \text{ kg} \qquad m_C = 40 \text{ kg}$$
$$\mathbf{r}_A = (555 \text{ m})\mathbf{i} - (180 \text{ m})\mathbf{j} + (240 \text{ m})\mathbf{k}$$
$$\mathbf{r}_B = (255 \text{ m})\mathbf{i} - (120 \text{ m})\mathbf{k}$$
$$\mathbf{r}_C = (105 \text{ m})\mathbf{i} + (450 \text{ m})\mathbf{j} - (420 \text{ m})\mathbf{k}$$

Then, using the information given in the statement of this problem, rewrite Eqs. (1) and (2) as

$$200(150\mathbf{i}) = 100(270\mathbf{i} - 120\mathbf{j} + 160\mathbf{k}) + 60[(v_B)_x\mathbf{i} + (v_B)_z\mathbf{k}]$$
$$+ 40[(v_C)_x\mathbf{i} + (v_C)_y\mathbf{j} + (v_C)_z\mathbf{k}] \quad (1')$$

$$0 = 100\begin{vmatrix} \mathbf{i} & \mathbf{j} & \mathbf{k} \\ 555 & -180 & 240 \\ 270 & -120 & 160 \end{vmatrix} + 60\begin{vmatrix} \mathbf{i} & \mathbf{j} & \mathbf{k} \\ 255 & 0 & -120 \\ (v_B)_x & 0 & (v_B)_z \end{vmatrix}$$

$$+ 40\begin{vmatrix} \mathbf{i} & \mathbf{j} & \mathbf{k} \\ 105 & 450 & -420 \\ (v_C)_x & (v_C)_y & (v_C)_z \end{vmatrix} \quad (2')$$

Equate the coefficient of $\mathbf{j}$ in Eq. (1') and the coefficients of $\mathbf{i}$ and $\mathbf{k}$ in Eq. (2'). After reductions, you obtain the three scalar equations of

$$(v_C)_y - 300 = 0$$
$$450(v_C)_z + 420(v_C)_y = 0$$
$$105(v_C)_y - 450(v_C)_x - 45\,000 = 0$$

which yield, respectively,

$$(v_C)_y = 300 \qquad (v_C)_z = -280 \qquad (v_C)_x = -30$$

The velocity of part C is thus

$$\mathbf{v}_C = -(30 \text{ m/s})\mathbf{i} + (300 \text{ m/s})\mathbf{j} - (280 \text{ m/s})\mathbf{k} \quad ◀$$

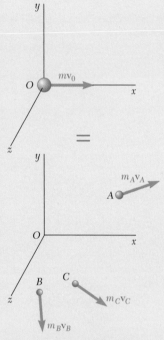

Fig. 1 Impulse-momentum diagram for the system.

Equating the coefficients of the **i** and **k** terms on each side of Eq. (1′) and solving for the unknown components of the velocity of B gives

$$(v_B)_x = 70 \text{ m/s} \qquad (v_B)_z = -80 \text{ m/s}$$

So

$$v_A = \sqrt{(270 \text{ m/s})^2 + (-120 \text{ m/s})^2 + (160 \text{ m/s})^2} = 336.0 \text{ m/s}$$

$$v_B = \sqrt{(70 \text{ m/s})^2 + (0)^2 + (-80 \text{ m/s})^2} = 106.3 \text{ m/s}$$

$$v_C = \sqrt{(-30 \text{ m/s})^2 + (300)^2 + (-280 \text{ m/s})^2} = 411.5 \text{ m/s}$$

The initial kinetic energy is

$$T_1 = \tfrac{1}{2}mv_0^2 = \frac{1}{2}(200 \text{ kg})(150 \text{ m/s})^2 = 2250 \text{ kJ}$$

The final kinetic energy is

$$T_2 = \tfrac{1}{2}m_A v_A^2 + \tfrac{1}{2}m_A v_A^2 + \tfrac{1}{2}m_A v_A^2$$

$$= \frac{1}{2}(100 \text{ kg})(336.0 \text{ m/s})^2 + \frac{1}{2}(60 \text{ kg})(106.3 \text{ m/s})^2 + \frac{1}{2}(40 \text{ kg})(411.5 \text{ m/s})^2$$

$$= 9370 \text{ kJ}$$

So

$$\Delta T = T_2 - T_1 = 9370 \text{ kJ} - 2250 \text{ kJ} \qquad \Delta T = 7120 \text{ kJ} \ \blacktriangleleft$$

REFLECT and THINK: The negative signs for $(v_C)_x$ and $(v_C)_z$ indicate that the velocity is not directed as shown in Fig. 1. We also notice that the directions of the components of $\mathbf{v}_C$ are opposite to those of $\mathbf{v}_A.$ Given the lack of external forces, it seems reasonable to expect a more symmetric spread of velocities in all directions. You should also notice that the explosion added a lot of energy to the system.

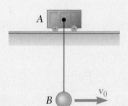

Sample Problem 14.5

Ball B, with a mass of m_B, is suspended from a cord with a length l attached to cart A, with a mass of m_A, that can roll freely on a frictionless horizontal track. If the ball is given an initial horizontal velocity $\mathbf{v}_0$ while the cart is at rest, determine (a) the velocity of B as it reaches its maximum elevation, (b) the maximum vertical distance h through which B will rise. (Assume $v_0^2 < 2gl.$)

STRATEGY: You are asked about the velocity of the system at two different positions, so use the principle of work and energy for the cart–ball system. You will also use the impulse-momentum principle, since momentum is conserved in the x-direction.

(continued)

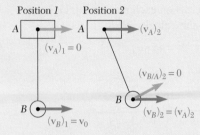

Fig. 1 Velocity vectors at the two positions.

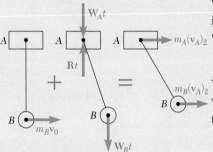

Fig. 2 Impulse–momentum diagram for the system.

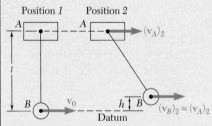

Fig. 3 The system drawn in position 1 and position 2.

MODELING and ANALYSIS: For your system, choose the ball and the cart and model them as particles.

Velocities.

Position 1: $(\mathbf{v}_A)_1 = 0$ $(\mathbf{v}_B)_1 = \mathbf{v}_0$ (1)

Position 2: When ball B reaches its maximum elevation, its velocity $(\mathbf{v}_{B/A})_2$ relative to its support A is zero (Fig. 1). Thus, at that instant, its absolute velocity is

$$(\mathbf{v}_B)_2 = (\mathbf{v}_A)_2 + (\mathbf{v}_{B/A})_2 = (\mathbf{v}_A)_2 \qquad (2)$$

Impulse–Momentum Principle. The external impulses consist of $\mathbf{W}_A t$, $\mathbf{W}_B t$, and $\mathbf{R}t$, where $\mathbf{R}$ is the reaction of the track on the cart. Recalling Eqs. (1) and (2), draw the impulse–momentum diagram (Fig. 2) and write

$$\Sigma m\mathbf{v}_1 + \Sigma\mathbf{Ext\ Imp}_{1\rightarrow 2} = \Sigma m\mathbf{v}_2$$

$\overset{+}{\underset{\rightarrow}{}}x$ components: $m_B v_0 = (m_A + m_B)(v_A)_2$

This expresses that the linear momentum of the system is conserved in the horizontal direction. Solving for $(v_A)_2$, you have

$$(v_A)_2 = \frac{m_B}{m_A + m_B}v_0 \qquad (\mathbf{v}_B)_2 = (\mathbf{v}_A)_2 = \frac{m_B}{m_A + m_B}v_0 \rightarrow \quad \blacktriangleleft$$

Conservation of Energy. The system is shown in Fig. 3 in the two positions. Define your datum at the location of B in position 1 (although you could also choose to place it at A). You can now calculate the kinetic and potential energies in the two positions:

Position 1. *Potential Energy:* $V_1 = m_A gl$
             Kinetic Energy: $T_1 = \frac{1}{2}m_B v_0^2$
Position 2. *Potential Energy:* $V_2 = m_A gl + m_B gh$
             Kinetic Energy: $T_2 = \frac{1}{2}(m_A + m_B)(v_A)_2^2$

Substituting these into the conservation of energy gives

$$T_1 + V_1 = T_2 + V_2: \quad \tfrac{1}{2}m_B v_0^2 + m_A gl = \tfrac{1}{2}(m_A + m_B)(v_A)_2^2 + m_A gl + m_B gh$$

Solving for h, you have

$$h = \frac{v_0^2}{2g} - \frac{m_A + m_B}{m_B}\frac{(v_A)_2^2}{2g}$$

or substituting $(v_A)_2$ from above, you have

$$h = \frac{v_0^2}{2g} - \frac{m_B}{m_A + m_B}\frac{v_0^2}{2g} \qquad h = \frac{m_A}{m_A + m_B}\frac{v_0^2}{2g} \quad \blacktriangleleft$$

REFLECT and THINK: Recalling that $v_0^2 < 2gl$, it follows from the last equation that $h < l$; this verifies that B stays below A, as assumed in the solution. For $m_A \gg m_B$, the answers reduce to $(\mathbf{v}_B)_2 = (\mathbf{v}_A)_2 = 0$ and $h = v_0^2/2g$; B oscillates as a simple pendulum with A fixed. For $m_A \ll m_B$, they reduce to $(\mathbf{v}_B)_2 = (\mathbf{v}_A)_2 = \mathbf{v}_0$ and $h = 0$; A and B move with the same constant velocity $\mathbf{v}_0$.

Sample Problem 14.6

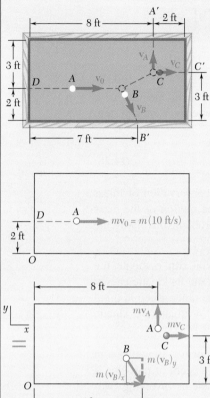

Fig. 1 Impulse–momentum diagram for the system.

In a game of billiards, ball A is given an initial velocity $\mathbf{v}_0$ with a magnitude of $v_0 = 10$ ft/s along line DA parallel to the axis of the table. It hits ball B and then ball C, which are both at rest. Balls A and C hit the sides of the table squarely at points A' and C', respectively, and B hits the side obliquely at B'. Assuming frictionless surfaces and perfectly elastic impacts, determine the velocities $\mathbf{v}_A$, $\mathbf{v}_B$, and $\mathbf{v}_C$ with which the balls hit the sides of the table. (*Remark:* In this sample problem and in several of the problems that follow, we assume the billiard balls are particles moving freely in a horizontal plane, rather than the rolling and sliding spheres they actually are.)

STRATEGY: Since there are no externally applied forces, use the conservation of linear and angular momentum. Because you are told that the impacts are perfectly elastic, you can also use the conservation of energy (but note that in general, energy is lost in an impact).

MODELING and ANALYSIS: Choose the system to be all three billiard balls and model them as particles.

Conservation of Momentum. There is no external force, so the initial momentum $m\mathbf{v}_0$ is equipollent to the system of momenta after the two collisions (and before any of the balls hit the sides of the table). Referring to Fig. 1, you have

$\xrightarrow{+} x$ components: $\qquad m(10 \text{ ft/s}) = m(v_B)_x + mv_C \qquad$ **(1)**

$+\uparrow y$ components: $\qquad\qquad 0 = mv_A - m(v_B)_y \qquad$ **(2)**

$+\curvearrowleft$ moments about O: $\quad -(2 \text{ ft})m(10 \text{ ft/s}) = (8 \text{ ft})mv_A$
$$-(7 \text{ ft})m(v_B)_y - (3 \text{ ft})mv_C \quad \textbf{(3)}$$

Solving the three equations for v_A, $(v_B)_x$, and $(v_B)_y$ in terms of v_C gives

$$v_A = (v_B)_y = 3v_C - 20 \qquad (v_B)_x = 10 - v_C \qquad \textbf{(4)}$$

Conservation of Energy. The surfaces are frictionless and the impacts are perfectly elastic, so the initial kinetic energy $\frac{1}{2}mv_0^2$ is equal to the final kinetic energy of the system:

$$\tfrac{1}{2}mv_0^2 = \tfrac{1}{2}mv_A^2 + \tfrac{1}{2}mv_B^2 + \tfrac{1}{2}mv_C^2$$
$$v_A^2 + (v_B)_x^2 + (v_B)_y^2 + v_C^2 = (10 \text{ ft/s})^2 \qquad \textbf{(5)}$$

Substituting for v_A, $(v_B)_x$, and $(v_B)_y$ from Eq. (4) into Eq. (5), you have

$$2(3v_C - 20)^2 + (10 - v_C)^2 + v_C^2 = 100$$
$$20v_C^2 - 260v_C + 800 = 0$$

Solving for v_C, you find $v_C = 5$ ft/s and $v_C = 8$ ft/s. Since only the second root yields a positive value for v_A after substitution into Eqs. (4), then $v_C = 8$ ft/s and

$$v_A = (v_B)_y = 3(8) - 20 = 4 \text{ ft/s} \qquad (v_B)_x = 10 - 8 = 2 \text{ ft/s}$$
$$\mathbf{v}_A = 4 \text{ ft/s}\uparrow \qquad \mathbf{v}_B = 4.47 \text{ ft/s} \searrow 63.4° \qquad \mathbf{v}_C = 8 \text{ ft/s}\rightarrow \quad \blacktriangleleft$$

REFLECT and THINK: In a real situation, energy would not be conserved, and you would need to know the coefficient of restitution between the balls to solve this problem. We also neglected friction and the rotation of the balls in our analysis, which is often a poor assumption in pool or billiards. We discuss rigid-body impacts in Chapter 17.

SOLVING PROBLEMS
ON YOUR OWN

In Sec. 14.1, we defined the linear momentum and the angular momentum of a system of particles. In this section, we defined the **kinetic energy** T of a system of particles as

$$T = \frac{1}{2}\sum_{i=1}^{n} m_i v_i^2 \qquad \textbf{(14.28)}$$

The solutions of the problems in Sec. 14.1 were based on the conservation of linear momentum of a system of particles or on the observation of the motion of the mass center of a system of particles. In this section, you will solve problems involving the following concepts.

1. Computation of the kinetic energy lost in collisions. You can compute the kinetic energy T_1 of the system of particles before the collisions and its kinetic energy T_2 after the collisions from Eq. (14.28) and subtract one from the other. Keep in mind that although linear momentum and angular momentum are vector quantities, kinetic energy is a *scalar* quantity.

2. Conservation of linear momentum and conservation of energy. As you saw in Sec. 14.1, when the resultant of the external forces acting on a system of particles is zero, the linear momentum of the system is conserved. In problems involving two-dimensional motion, expressing that the initial linear momentum and the final linear momentum of the system are equipollent yields two algebraic equations. Equating the initial total energy of the system of particles (including potential energy as well as kinetic energy) to its final total energy yields an additional equation. Thus, you can write three equations that you can solve for three unknowns [Sample Prob. 14.6]. Note that if the resultant of the external forces is not zero but has a fixed direction, the component of the linear momentum in a direction perpendicular to the resultant is still conserved; the number of equations that you can use is then reduced to two [Sample Prob. 14.5].

3. Conservation of linear and angular momentum. When no external forces act on a system of particles, both the linear momentum of the system and its angular momentum about some arbitrary point are conserved. In the case of three-dimensional motion, this enables you to write as many as six equations, although you may need to solve only some of them to obtain the desired answers [Sample Prob. 14.4]. In the case of two-dimensional motion, you will be able to write three equations that you can solve for three unknowns.

4. Conservation of linear and angular momentum and conservation of energy. In the case of the two-dimensional motion of a system of particles that is not subjected to any external forces, you can obtain two algebraic equations by expressing that the linear momentum of the system is conserved; one equation by writing that the angular momentum of the system about some arbitrary point is conserved; and a fourth equation by expressing that the total energy of the system is conserved. These equations can be solved for four unknowns.

Problems

14.31 Determine the energy lost due to friction and the impacts for Prob. 14.1.

14.32 Assuming that the airline employee of Prob. 14.3 first tosses the 30-lb suitcase on the baggage carrier, determine the energy lost (*a*) as the first suitcase hits the carrier, (*b*) as the second suitcase hits the carrier.

14.33 In Prob. 14.6, determine the work done by the woman and by the man as each dives from the boat, assuming that the woman dives first.

14.34 Determine the energy lost as a result of the series of collisions described in Prob. 14.8.

14.35 Two automobiles *A* and *B*, of mass m_A and m_B, respectively, are traveling in opposite directions when they collide head on. The impact is assumed perfectly plastic, and it is further assumed that the energy absorbed by each automobile is equal to its loss of kinetic energy with respect to a moving frame of reference attached to the mass center of the two-vehicle system. Denoting by E_A and E_B, respectively, the energy absorbed by automobile *A* and by automobile *B*, (*a*) show that $E_A/E_B = m_B/m_A$, that is, the amount of energy absorbed by each vehicle is inversely proportional to its mass, (*b*) compute E_A and E_B, knowing that $m_A = 1600$ kg and $m_B = 900$ kg and that the speeds of *A* and *B* are, respectively, 90 km/h and 60 km/h.

Fig. P14.35

14.36 It is assumed that each of the two automobiles involved in the collision described in Prob. 14.35 had been designed to safely withstand a test in which it crashed into a solid, immovable wall at the speed v_0. The severity of the collision of Prob. 14.35 may then be measured for each vehicle by the ratio of the energy it absorbed in the collision to the energy it absorbed in the test. On that basis, show that the collision described in Prob. 14.35 is $(m_A/m_B)^2$ times more severe for automobile *B* than for automobile *A*.

14.37 Solve Sample Problem 14.5, assuming that cart *A* is given an initial horizontal velocity $\mathbf{v}_0$ while ball *B* is at rest.

14.38 Two hemispheres are held together by a cord which maintains a spring under compression (the spring is not attached to the hemispheres). The potential energy of the compressed spring is 120 J and the assembly has an initial velocity $\mathbf{v}_0$ of magnitude $v_0 = 8$ m/s. Knowing that the cord is severed when $\theta = 30°$, causing the hemispheres to fly apart, determine the resulting velocity of each hemisphere.

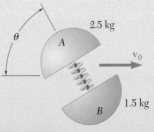

Fig. P14.38

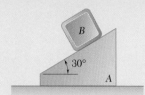

Fig. P14.39

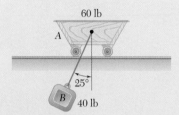

Fig. P14.40

14.39 A 15-lb block B starts from rest and slides on the 25-lb wedge A, which is supported by a horizontal surface. Neglecting friction, determine (a) the velocity of B relative to A after it has slid 3 ft down the inclined surface of the wedge, (b) the corresponding velocity of A.

14.40 A 40-lb block B is suspended from a 6-ft cord attached to a 60-lb cart A, which may roll freely on a frictionless, horizontal track. If the system is released from rest in the position shown, determine the velocities of A and B as B passes directly under A.

14.41 and 14.42 In a game of pool, ball A is moving with a velocity $\mathbf{v}_0$ with a magnitude of $v_0 = 15$ ft/s when it strikes balls B and C, which are at rest and aligned as shown. Knowing that after the collision the three balls move in the directions indicated and assuming frictionless surfaces and perfectly elastic impact (that is, conservation of energy), determine the magnitudes of the velocities $\mathbf{v}_A$, $\mathbf{v}_B$, and $\mathbf{v}_C$.

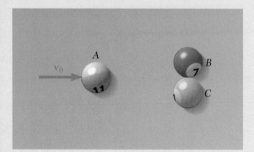

Fig. P14.43

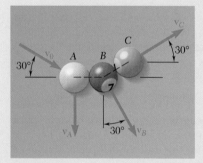

Fig. P14.41

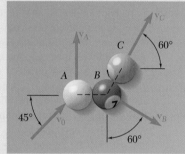

Fig. P14.42

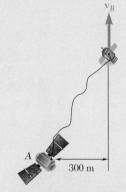

Fig. P14.44

14.43 Three spheres, each with a mass of m, can slide freely on a frictionless, horizontal surface. Spheres A and B are attached to an inextensible, inelastic cord with a length l and are at rest in the position shown when sphere B is struck squarely by sphere C, which is moving with a velocity $\mathbf{v}_0$. Knowing that the cord is taut when sphere B is struck by sphere C and assuming perfectly elastic impact between B and C, and thus the conservation of energy for the entire system, determine the velocity of each sphere immediately after impact.

14.44 In a game of pool, ball A is moving with the velocity $\mathbf{v}_0 = v_0\mathbf{i}$ when it strikes balls B and C, which are at rest side by side. Assuming frictionless surfaces and perfectly elastic impact (i.e., conservation of energy), determine the final velocity of each ball, assuming that the path of A is (a) perfectly centered and that A strikes B and C simultaneously, (b) not perfectly centered and that A strikes B slightly before it strikes C.

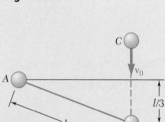

Fig. P14.45

14.45 The 2-kg sub-satellite B has an initial velocity $\mathbf{v}_B = (3 \text{ m/s})\mathbf{j}$. It is connected to the 20-kg base-satellite A by a 500-m space tether. Determine the velocity of the base satellite and sub-satellite immediately after the tether becomes taut (assuming no rebound).

14.46 A 900-lb space vehicle traveling with a velocity $\mathbf{v}_0 = (1500 \text{ ft/s})\mathbf{k}$ passes through the origin O. Explosive charges then separate the vehicle into three parts A, B, and C, with masses of 150 lb, 300 lb, and 450 lb, respectively. Knowing that shortly thereafter the positions of the three parts are, respectively, A(250, 250, 2250), B(600, 1300, 3200), and C(−475, −950, 1900), where the coordinates are expressed in feet, that the velocity of B is $\mathbf{v}_B = (500 \text{ ft/s})\mathbf{i} + (1100 \text{ ft/s})\mathbf{j} + (2100 \text{ ft/s})\mathbf{k}$, and that the x component of the velocity of C is −400 ft/s, determine the velocity of part A.

14.47 Four small disks A, B, C, and D can slide freely on a frictionless horizontal surface. Disks B, C, and D are connected by light rods and are at rest in the position shown when disk B is struck squarely by disk A which is moving to the right with a velocity $\mathbf{v}_0 = (38.5 \text{ ft/s})\mathbf{i}$. The weights of the disks are $W_A = W_B = W_C = 15$ lb, and $W_D = 30$ lb. Knowing that the velocities of the disks immediately after the impact are $\mathbf{v}_A = \mathbf{v}_B = (8.25 \text{ ft/s})\mathbf{i}$, $\mathbf{v}_C = v_C\mathbf{i}$, and $\mathbf{v}_D = v_D\mathbf{i}$, determine (a) the speeds v_C and v_D, (b) the fraction of the initial kinetic energy of the system which is dissipated during the collision.

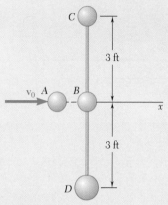

Fig. P14.47

14.48 In the scattering experiment of Prob. 14.26, it is known that the alpha particle is projected from A_0(300, 0, 300) and that it collides with the oxygen nucleus C at Q(240, 200, 100), where all coordinates are expressed in millimeters. Determine the coordinates of point B_0 where the original path of nucleus B intersects the zx plane. (*Hint:* Express that the angular momentum of the three particles about Q is conserved.)

14.49 Three identical small spheres, each of weight 2 lb, can slide freely on a horizontal frictionless surface. Spheres B and C are connected by a light rod and are at rest in the position shown when sphere B is struck squarely by sphere A which is moving to the right with a velocity $\mathbf{v}_0 = (8 \text{ ft/s})\mathbf{i}$. Knowing that $\theta = 45°$ and that the velocities of spheres A and B immediately after the impact are $\mathbf{v}_A = 0$ and $\mathbf{v}_B = (6 \text{ ft/s})\mathbf{i} + (v_B)_y\mathbf{j}$, determine $(v_B)_y$ and the velocity of C immediately after impact.

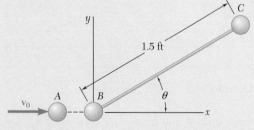

Fig. P14.49

14.50 Three small spheres A, B, and C, each of mass m, are connected to a small ring D of negligible mass by means of three inextensible, inelastic cords of length l. The spheres can slide freely on a frictionless horizontal surface and are rotating initially at a speed v_0 about ring D which is at rest. Suddenly the cord CD breaks. After the other two cords have again become taut, determine (a) the speed of ring D, (b) the relative speed at which spheres A and B rotate about D, (c) the fraction of the original energy of spheres A and B that is dissipated when cords AD and BD again became taut.

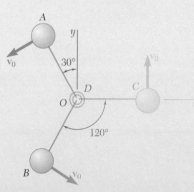

Fig. P14.50

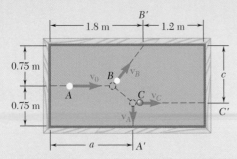

Fig. P14.51

14.51 In a game of billiards, ball A is given an initial velocity $\mathbf{v}_0$ along the longitudinal axis of the table. It hits ball B and then ball C, which are both at rest. Balls A and C are observed to hit the sides of the table squarely at A' and C', respectively, and ball B is observed to hit the side obliquely at B'. Knowing that $v_0 = 4$ m/s, $v_A = 1.92$ m/s, and $a = 1.65$ m, determine (a) the velocities $\mathbf{v}_B$ and $\mathbf{v}_C$ of balls B and C, (b) the point C' where ball C hits the side of the table. Assume frictionless surfaces and perfectly elastic impacts (i.e., conservation of energy).

14.52 For the game of billiards of Prob. 14.51, it is now assumed that $v_0 = 5$ m/s, $v_C = 3.2$ m/s, and $c = 1.22$ m. Determine (a) the velocities $\mathbf{v}_A$ and $\mathbf{v}_B$ of balls A and B, (b) the point A' where ball A hits the side of the table.

14.53 Two small disks A and B, of mass 3 kg and 1.5 kg, respectively, may slide on a horizontal, frictionless surface. They are connected by a cord, 600 mm long, and spin counterclockwise about their mass center G at the rate of 10 rad/s. At $t = 0$, the coordinates of G are $\bar{x}_0 = 0, \bar{y}_0 = 2$ m, and its velocity $\bar{\mathbf{v}}_0 = (1.2 \text{ m/s})\mathbf{i} + (0.96 \text{ m/s})\mathbf{j}$. Shortly thereafter the cord breaks; disk A is then observed to move along a path parallel to the y axis and disk B along a path which intersects the x axis at a distance $b = 7.5$ m from O. Determine (a) the velocities of A and B after the cord breaks, (b) the distance a from the y axis to the path of A.

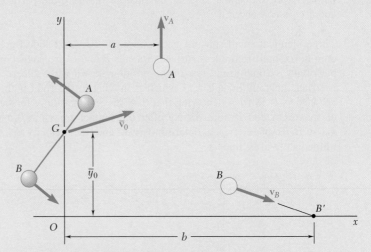

Fig. P14.53 and P14.54

14.54 Two small disks A and B, of mass 2 kg and 1 kg, respectively, may slide on a horizontal and frictionless surface. They are connected by a cord of negligible mass and spin about their mass center G. At $t = 0$, G is moving with the velocity $\bar{\mathbf{v}}_0$ and its coordinates are $\bar{x}_0 = 0, \bar{y}_0 = 1.89$ m. Shortly thereafter, the cord breaks and disk A is observed to move with a velocity $\mathbf{v}_A = (5 \text{ m/s})\mathbf{j}$ in a straight line and at a distance $a = 2.56$ m from the y axis, while B moves with a velocity $\mathbf{v}_B = (7.2 \text{ m/s})\mathbf{i} - (4.6 \text{ m/s})\mathbf{j}$ along a path intersecting the x axis at a distance $b = 7.48$ m from the origin O. Determine (a) the initial velocity $\bar{\mathbf{v}}_0$ of the mass center G of the two disks, (b) the length of the cord initially connecting the two disks, (c) the rate in rad/s at which the disks were spinning about G.

14.55 Three small identical spheres A, B, and C, which can slide on a horizontal, frictionless surface, are attached to three 9-in.-long strings, which are tied to a ring G. Initially the spheres rotate clockwise about the ring with a relative velocity of 2.6 ft/s and the ring moves along the x axis with a velocity $\mathbf{v}_0 = (1.3 \text{ ft/s})\mathbf{i}$. Suddenly the ring breaks and the three spheres move freely in the xy plane with A and B following paths parallel to the y axis at a distance $a = 1.0$ ft from each other and C following a path parallel to the x axis. Determine (*a*) the velocity of each sphere, (*b*) the distance d.

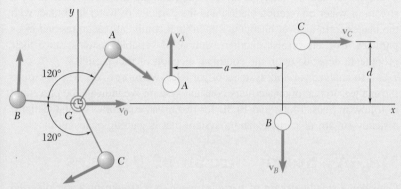

Fig. P14.55 and P14.56

14.56 Three small identical spheres A, B, and C, which can slide on a horizontal, frictionless surface, are attached to three strings of length l which are tied to a ring G. Initially the spheres rotate clockwise about the ring which moves along the x axis with a velocity $\mathbf{v}_0$. Suddenly the ring breaks and the three spheres move freely in the xy plane. Knowing that $\mathbf{v}_A = (3.5 \text{ ft/s})\mathbf{j}$, $\mathbf{v}_C = (6.0 \text{ ft/s})\mathbf{i}$, $a = 16$ in., and $d = 9$ in., determine (*a*) the initial velocity of the ring, (*b*) the length l of the strings, (*c*) the rate in rad/s at which the spheres were rotating about G.

*14.3   VARIABLE SYSTEMS OF PARTICLES

All of the systems considered so far consisted of well-defined particles. These systems did not gain or lose any particles during their motion. In a large number of engineering applications, however, it is necessary to consider **variable systems of particles**, i.e., systems that are continually gaining or losing particles, or doing both at the same time. Consider, for example, a hydraulic turbine. Its analysis involves determining the forces exerted by a stream of water on rotating blades, and the particles of water in contact with the blades form an ever-changing system that continually acquires and loses particles. Rockets furnish another example of variable systems, since their propulsion depends upon the continual ejection of fuel particles.

To analyze variable systems of particles, we must find a way to reduce the analysis to that of an auxiliary constant system. We indicate the procedure to follow in Secs. 14.3A and 14.3B for two broad categories of applications: a steady stream of particles and a system that is gaining or losing mass.

*14.3A   Steady Stream of Particles

Consider a steady stream of particles, such as a stream of water diverted by a fixed vane or a flow of air through a duct or through a blower. In order to determine the resultant of the forces exerted on the particles in contact with the vane, duct, or blower, we isolate these particles and define them to be a system S (Fig. 14.9). Note that S is a variable system of particles, since it continually gains particles flowing in and loses an equal number of particles flowing out. Therefore, the kinetics principles that we have established so far do not apply directly to S.

However, we can easily define an auxiliary system of particles that does remain constant for a short interval of time Δt. Consider at time t the system S *plus* the particles that will enter S during the interval of time Δt (Fig. 14.10a). Next, consider at time $t + \Delta t$ the system S *plus* the particles

Fig. 14.9   A system of particles in a steady stream.

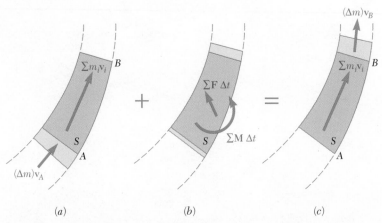

Fig. 14.10   The impulse–momentum diagram for a stream of particles contains (a) momenta of particles entering and in the system S plus (b) impulses during the time interval Δt, and (c) momenta of the particles in and leaving the system.

that have left S during the interval Δt (Fig. 14.10c). Clearly, *the same particles are involved in both cases,* and we can apply the principle of impulse and momentum to those particles. Since the total mass m of the system S remains constant, the particles entering the system and those leaving the system in the time Δt must have the same mass Δm. Suppose we denote the velocities of the particles entering S at A and leaving S at B by $\mathbf{v}_A$ and $\mathbf{v}_B$, respectively. Then we can represent the momentum of the particles entering S by $(\Delta m)\mathbf{v}_A$ (Fig. 14.10a) and the momentum of the particles leaving S by $(\Delta m)\mathbf{v}_B$ (Fig. 14.10c). We also represent the momenta $m_i\mathbf{v}_i$ of the particles forming S and the impulses of the forces exerted on S by appropriate vectors. Then we indicate by blue plus and equals signs that the system of the momenta and impulses in parts a and b of Fig. 14.10 is equipollent to the system of the momenta in part c of the same figure.

The resultant $\Sigma m_i\mathbf{v}_i$ of the momenta of the particles of S is found on both sides of the equals sign and thus can be omitted. We conclude:

The system formed by the momentum $(\Delta m)\mathbf{v}_A$ of the particles entering S in the time Δt and the impulses of the forces exerted on S during that time is equipollent to the momentum $(\Delta m)\mathbf{v}_B$ of the particles leaving S in the same time Δt.

Mathematically, we have

$$(\Delta m)\mathbf{v}_A + \Sigma \mathbf{F}\,\Delta t = (\Delta m)\mathbf{v}_B \qquad \textbf{(14.38)}$$

We can obtain a similar equation by taking the moments of the vectors involved (see Sample Prob. 14.7). Dividing all terms of Eq. (14.38) by Δt and letting Δt approach zero, we obtain at the limit

$$\Sigma \mathbf{F} = \frac{dm}{dt}(\mathbf{v}_B - \mathbf{v}_A) \qquad \textbf{(14.39)}$$

where $\mathbf{v}_B - \mathbf{v}_A$ represents the difference between the *vector* $\mathbf{v}_B$ and the *vector* $\mathbf{v}_A$.

If we use SI units, dm/dt is expressed in kg/s and the velocities in m/s; we check that both sides of Eq. (14.39) are expressed in the same units (newtons). If we use U.S. customary units, dm/dt must be expressed in slugs/s and the velocities in ft/s; we check again that both sides of the equation are expressed in the same units (pounds).[†]

We can use this principle to analyze a large number of engineering applications. Let's look at some of the more common of these applications.

Fluid Stream Diverted by a Vane. If the vane is fixed, we can apply directly the method of analysis given here to find the force $\mathbf{F}$ exerted by the vane on the stream. Note that $\mathbf{F}$ is the only force we need to consider, because the pressure in the stream is constant (atmospheric pressure). The force exerted by the stream on the vane is equal and opposite to $\mathbf{F}$.

[†]It is often convenient to express the mass rate of flow dm/dt as the product ρQ, where ρ is the density of the stream (mass per unit volume) and Q is its volume rate of flow (volume per unit time). If you use SI units, ρ is in kg/m^3 (for instance, $\rho = 1000$ kg/m^3 for water) and Q is in m^3/s. However, if you use U.S. customary units, ρ generally has to be computed from the corresponding specific weight γ (weight per unit volume), $\rho = \gamma/g$. Since γ is expressed in lb/ft^3 (for instance, $\gamma = 62.4$ lb/ft^3 for water), we obtain ρ in slug/ft^3. The volume rate of flow Q is expressed in ft^3/s.

If the vane moves with a constant velocity, the stream is not steady. However, it will appear steady to an observer moving with the vane. We should therefore choose a system of axes moving with the vane. Since this system of axes is not accelerated, we can still use Eq. (14.38), but we must replace $\mathbf{v}_A$ and $\mathbf{v}_B$ by the *relative velocities* of the stream with respect to the vane (see Sample Prob. 14.8).

Fluid Flowing Through a Pipe. We can determine the force exerted by the fluid on a pipe transition, such as a bend or a contraction, by considering the system of particles S in contact with the transition. Since, in general, the pressure in the flow will vary, we should also consider the forces exerted on S by the adjoining portions of the fluid.

Jet Engine. In a jet engine, air enters the front of the engine with no velocity and leaves through the rear with a high velocity. The energy required to accelerate the air particles is obtained by burning fuel. The mass of the burned fuel in the exhaust gases is usually small enough compared with the mass of the air flowing through the engine that it can be neglected. Thus, the analysis of a jet engine reduces to that of an airstream. We can consider this stream as a steady stream if we measure all velocities with respect to the airplane. We assume, therefore, that the airstream enters the engine with a velocity $\mathbf{v}$ of magnitude equal to the speed of the airplane and leaves with a velocity $\mathbf{u}$ equal to the relative velocity of the exhaust gases (Fig. 14.11a). Since the intake and exhaust pressures are nearly atmospheric, the only external force we need to

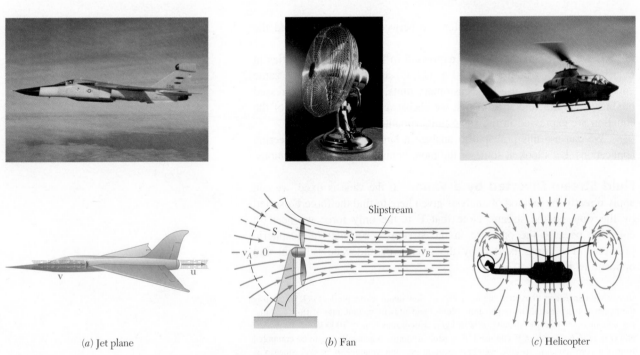

(a) Jet plane   (b) Fan   (c) Helicopter

Fig. 14.11 Applications of a steady stream of particles.

consider is the force exerted by the engine on the airstream. This force is equal and opposite to the thrust.[†]

Fan. Consider the system of particles S shown in Fig. 14.11b. We assume the velocity $\mathbf{v}_A$ of the particles entering the system is equal to zero, and the velocity $\mathbf{v}_B$ of the particles leaving the system is the velocity of the *slipstream*. We can obtain the rate of flow by multiplying v_B by the cross-sectional area of the slipstream. Since the pressure all around S is atmospheric, the only external force acting on S is the thrust of the fan.

Helicopter. Determining the thrust created by the rotating blades of a hovering helicopter is similar to the determination of the thrust of a fan (Fig. 14.11c). We assume the velocity $\mathbf{v}_A$ of the air particles as they approach the blades is zero, and we obtain the rate of flow by multiplying the magnitude of the velocity $\mathbf{v}_B$ of the slipstream by its cross-sectional area.

*14.3B Systems Gaining or Losing Mass

Let us now analyze a different type of variable system of particles; namely, a system that gains mass by continually absorbing particles or loses mass by continually expelling particles. Consider the system S shown in Fig. 14.12. Its mass, equal to m at the instant t, increases by Δm in the time interval Δt. In order to apply the principle of impulse and momentum to this system, we must consider at time t the system S *plus* the particles of mass Δm that S absorbs during the time interval Δt. The velocity of S at time t is denoted by $\mathbf{v}$, the velocity of S at time $t + \Delta t$ is denoted by $\mathbf{v} + \Delta \mathbf{v}$, and the absolute velocity of the particles absorbed is denoted by $\mathbf{v}_a$. Applying the principle of impulse and momentum, we have

$$m\mathbf{v} + (\Delta m)\mathbf{v}_a + \Sigma\mathbf{F}\,\Delta t = (m + \Delta m)(\mathbf{v} + \Delta \mathbf{v}) \qquad \textbf{(14.40)}$$

Solving for the sum $\Sigma\mathbf{F}\,\Delta t$ of the impulses of the external forces acting on S (excluding the forces exerted by the particles being absorbed), we obtain

$$\Sigma\mathbf{F}\,\Delta t = m\Delta\mathbf{v} + \Delta m(\mathbf{v} - \mathbf{v}_a) + (\Delta m)(\Delta\mathbf{v}) \qquad \textbf{(14.41)}$$

Now we introduce the *relative velocity* $\mathbf{u}$ with respect to S of the particles that are absorbed. We have $\mathbf{u} = \mathbf{v}_a - \mathbf{v}$ and note, since $v_a < v$, that the relative velocity $\mathbf{u}$ is directed to the left, as shown in Fig. 14.12. Neglecting the last term in Eq. (14.41), which is of the second order, we have

$$\Sigma\mathbf{F}\,\Delta t = m\,\Delta\mathbf{v} - (\Delta m)\mathbf{u}$$

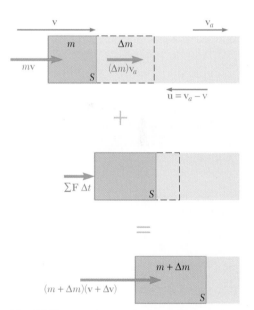

Fig. 14.12 Impulse–momentum diagram for a system that gains mass.

[†]Note that if the airplane is accelerating, we cannot use it as a newtonian frame of reference. However, we can obtain the same result for the thrust by using a reference frame at rest with respect to the atmosphere. In this frame, the air particles enter the engine with no velocity and leave it with a velocity of magnitude $u - v$.

Photo 14.3 As booster rockets are fired, the gas particles they eject provide the thrust required for liftoff.

Now we divide through by Δt and let Δt approach zero. In the limit we obtain[†]

$$\Sigma\mathbf{F} = m\frac{d\mathbf{v}}{dt} - \frac{dm}{dt}\mathbf{u} \qquad (14.42)$$

Rearranging terms and recalling that $d\mathbf{v}/dt = \mathbf{a}$, where $\mathbf{a}$ is the acceleration of the system S, we have

$$\Sigma\mathbf{F} + \frac{dm}{dt}\mathbf{u} = m\mathbf{a} \qquad (14.43)$$

This equation states that the action on S of the particles being absorbed is equivalent to a thrust

$$\mathbf{P} = \frac{dm}{dt}\mathbf{u} \qquad (14.44)$$

that tends to slow down the motion of S, since the relative velocity $\mathbf{u}$ of the particles is directed to the left. If we use SI units, dm/dt is expressed in kg/s, the relative velocity u is in m/s, and the corresponding thrust is in newtons. If we use U.S. customary units, dm/dt must be expressed in slug/s, u in ft/s, and the corresponding thrust in pounds.

We can also use these equations to determine the motion of a system S losing mass. In this case, the rate of change of mass is negative, and the action on S of the particles being expelled is equivalent to a thrust in the direction of $-\mathbf{u}$; that is, in the direction opposite to that in which the particles are being expelled. A *rocket* represents a typical case of a system continually losing mass (see Sample Prob. 14.9).

[†]When the absolute velocity $\mathbf{v}_a$ of the particles absorbed is zero, $\mathbf{u} = -\mathbf{v}$ and formula (14.42) becomes

$$\Sigma\mathbf{F} = \frac{d}{dt}(m\mathbf{v})$$

Comparing this formula with Eq. (12.3) of Sec. 12.1B, we see that this is Newton's second law applied to a system gaining mass, *provided that the particles absorbed are initially at rest*. We can also apply it to a system losing mass, *provided that the velocity of the particles expelled is zero* with respect to the chosen frame of reference.

Sample Problem 14.7

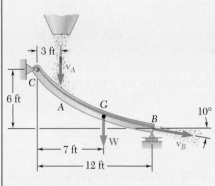

Grain falls from a hopper onto a chute CB at the rate of 240 lb/s. It hits the chute at A with a velocity of 20 ft/s and leaves at B with a velocity of 15 ft/s, forming an angle of 10° with the horizontal. Knowing that the combined weight of the chute and of the grain it supports is a force **W** with a magnitude of 600 lb applied at G, determine the reaction at the roller support B and the components of the reaction at the hinge C.

STRATEGY: Since we have a steady stream of particles, apply the principle of impulse and momentum for the time interval Δt.

MODELING: Choose a system that consists of the chute, the grain it supports, and the amount of grain that hits the chute in the interval Δt. The impulse–momentum diagram for this system is shown in Fig. 1. Since the chute does not move, it has no momentum. Note that the sum $\Sigma m_i\mathbf{v}_i$ of the momenta of the particles supported by the chute is the same at t and $t + \Delta t$ and thus can be omitted.

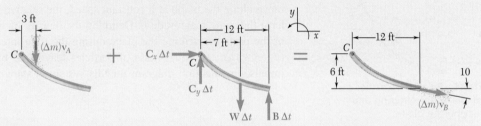

Fig. 1 Impulse–momentum diagram for the system.

ANALYSIS: You can use the impulse–momentum diagram to obtain scalar equations for the x and y directions and for moments about point C.

$\xrightarrow{+} x$ components: $\qquad C_x\,\Delta t = (\Delta m)v_B\cos 10°$ (1)

$+\uparrow y$ components: $\qquad -(\Delta m)v_A + C_y\,\Delta t - W\,\Delta t + B\,\Delta t$
$$= -(\Delta m)v_B\sin 10° \quad (2)$$

$+\uparrow$ moments about C: $\quad -3(\Delta m)v_A - 7(W\,\Delta t) + 12(B\,\Delta t)$
$$= 6(\Delta m)v_B\cos 10° - 12(\Delta m)v_B\sin 10° \quad (3)$$

Using the given data, $W = 600$ lb, $v_A = 20$ ft/s, $v_B = 15$ ft/s, and $\Delta m/\Delta t = 240/32.2 = 7.45$ slug/s, and solving Eq. (3) for B and Eq. (1) for C_x, you obtain

$$12B = 7(600) + 3(7.45)(20) + 6(7.45)(15)(\cos 10° - 2\sin 10°)$$
$$12B = 5075 \qquad B = 423\text{ lb} \qquad\qquad \mathbf{B} = 423\text{ lb}\uparrow \quad \blacktriangleleft$$

$$C_x = (7.45)(15)\cos 10° = 110.1\text{ lb} \qquad \mathbf{C}_x = 110.1\text{ lb}\rightarrow \quad \blacktriangleleft$$

Substituting for B and solving Eq. (2) for C_y, you end up with

$$C_y = 600 - 423 + (7.45)(20 - 15\sin 10°) = 307\text{ lb}$$
$$\mathbf{C}_y = 307\text{ lb}\uparrow \quad \blacktriangleleft$$

REFLECT and THINK: This kind of situation is common in factory and storage settings. Being able to determine the reactions is essential for designing a proper chute that will support the stream safely. We can compare this situation to the case when there is no mass flow, which results in reactions of $B_y = 350$ lb, $C_y = 250$ lb, and $C_x = 0$ lb.

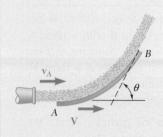

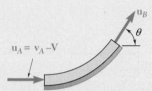

Fig. 1 Relative velocities of the water entering and leaving the blade.

Sample Problem 14.8

A nozzle discharges a stream of water of cross-sectional area A with a velocity $\mathbf{v}_A$. The stream is deflected by a *single* blade that moves to the right with a constant velocity $\mathbf{V}$. Assuming that the water moves along the blade at constant speed, determine (*a*) the components of the force $\mathbf{F}$ exerted by the blade on the stream, (*b*) the velocity $\mathbf{V}$ for which maximum power is developed.

STRATEGY: Since you have a steady stream of particles, apply the principle of impulse and momentum.

MODELING: Choose the system to be the particles in contact with the blade and the particles striking the blade in the time Δt, and use a coordinate system that moves with the blade at a constant velocity $\mathbf{V}$. The particles of water strike the blade with a relative velocity $\mathbf{u}_A = \mathbf{v}_A - \mathbf{V}$ and leave the blade with a relative velocity $\mathbf{u}_B$, as shown in Fig. 1. Since the particles move along the blade at a constant speed, the relative velocities $\mathbf{u}_A$ and $\mathbf{u}_B$ have the same magnitude u. Denoting the density of water by ρ, the mass of the particles striking the blade during the time interval Δt is $\Delta m = A\rho (v_A - V) \Delta t$; an equal mass of particles leaves the blade during Δt. The impulse–momentum diagram for this system is shown in Fig. 2.

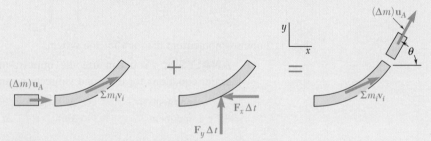

Fig. 2 Impulse–momentum diagram for the system.

ANALYSIS:

a. Components of Force Exerted on Stream.
Recalling that $\mathbf{u}_A$ and $\mathbf{u}_B$ have the same magnitude u and omitting the momentum $\Sigma m_i \mathbf{v}_i$ that appears on both sides, applying the principle of impulse and momentum gives you

$\xrightarrow{+} x$ components: $(\Delta m)u - F_x \Delta t = (\Delta m)u \cos \theta$
$+\uparrow y$ components: $+F_y \Delta t = (\Delta m)u \sin \theta$

Substituting $\Delta m = A\rho (v_A - V) \Delta t$ and $u = v_A - V$, you obtain

$$\mathbf{F}_x = A\rho(v_A - V)^2(1 - \cos \theta) \leftarrow \qquad \mathbf{F}_y = A\rho(v_A - V)^2 \sin \theta \uparrow \quad ◀$$

b. Velocity of Blade for Maximum Power.
You can obtain the power by multiplying the velocity V of the blade by the component F_x of the force exerted by the stream on the blade.

$$\text{Power} = F_x V = A\rho(v_A - V)^2(1 - \cos \theta)V$$

Differentiating the power with respect to V and setting the derivative equal to zero, you have

$$\frac{d(\text{power})}{dV} = A\rho(v_A^2 - 4v_A V + 3V^2)(1 - \cos\theta) = 0$$

$$V = v_A \qquad V = \tfrac{1}{3}v_A \qquad \text{For maximum power } \mathbf{V} = \tfrac{1}{3}v_A \rightarrow \quad \blacktriangleleft$$

REFLECT and THINK: These results are valid only when a *single* blade deflects the stream. Different results appear when a series of blades deflects the stream, as in a Pelton-wheel turbine (see Prob. 14.81).

Sample Problem 14.9

A rocket of initial mass m_0 (including shell and fuel) is fired vertically at time $t = 0$. The fuel is consumed at a constant rate $q = dm/dt$ and is expelled at a constant speed u relative to the rocket. Derive an expression for the magnitude of the velocity of the rocket at time t, neglecting the resistance of the air.

STRATEGY: Since you have a system that is losing mass, apply the principle of impulse and momentum. This gives you an equation you can integrate to obtain the velocity.

MODELING: Choose the rocket shell and its fuel as your system. At time t, the mass of the rocket shell and remaining fuel is $m = m_0 - qt$, and the velocity is $\mathbf{v}$. During the time interval Δt, a mass of fuel $\Delta m = q\,\Delta t$ is expelled with a speed u relative to the rocket. The impulse–momentum diagram for this system is shown in Fig. 1, where $\mathbf{v}_e$ is the absolute velocity of the expelled fuel.

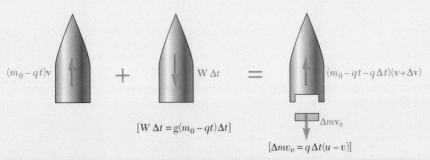

$(m_0 - qt)\mathbf{v}$ $+$ $W\,\Delta t$ $=$ $(m_0 - qt - q\,\Delta t)(\mathbf{v} + \Delta\mathbf{v})$

$[W\,\Delta t = g(m_0 - qt)\,\Delta t]$

$\Delta m\mathbf{v}_e$

$[\Delta m v_e = q\,\Delta t(u - v)]$

Fig. 1 Impulse–momentum diagram for the system.

(continued)

ANALYSIS: Apply the principle of impulse and momentum between time t and time $t + \Delta t$ to find

$$(m_0 - qt)v - g(m_0 - qt)\,\Delta t = (m_0 - qt - q\,\Delta t)(v + \Delta v) - q\,\Delta t(u - v)$$

Divide through by Δt and let Δt approach zero for

$$-g(m_0 - qt) = (m_0 - qt)\frac{dv}{dt} - qu$$

Separating variables and integrating from $t = 0$, $v = 0$ to $t = t$, $v = v$, you have

$$dv = \left(\frac{qu}{m_0 - qt} - g\right)dt$$

$$\int_0^v dv = \int_0^t \left(\frac{qu}{m_0 - qt} - g\right)dt$$

$$v = [-u\ln(m_0 - qt) - gt]_0^t \qquad v = u\ln\frac{m_0}{m_0 - qt} - gt \;\blacktriangleleft$$

REFLECT and THINK: The mass remaining at time t_f, after all of the fuel has been expended, is equal to the mass of the rocket shell $m_s = m_0 - qt_f$, and the maximum velocity attained by the rocket is $v_m = u\ln(m_0/m_s) - gt_f$. Assuming that the fuel is expelled in a relatively short period of time, the term gt_f is small, and we have $v_m \approx u\ln(m_0/m_s)$. In order to escape the gravitational field of Earth, a rocket must reach a velocity of 11.18 km/s. Assuming $u = 2200$ m/s and $v_m = 11.18$ km/s, we obtain $m_0/m_s = 161$. Thus, to project each kilogram of the rocket shell into space, it is necessary to consume more than 161 kg of fuel if we use a propellant yielding $u = 2200$ m/s.

SOLVING PROBLEMS ON YOUR OWN

This section was devoted to the motion of **variable systems of particles**, i.e., systems that are continually *gaining or losing particles* or doing both at the same time. The problems you will be asked to solve will involve (1) **steady streams of particles** and (2) **systems gaining or losing mass**.

1. To solve problems involving a steady stream of particles [Sample Probs. 14.7 and 14.8], consider a portion S of the stream and express mathematically that the system formed by the momentum of the particles entering S at A in the time Δt and that the impulses of the forces exerted on S during that time is equipollent to the momentum of the particles leaving S at B in the same time Δt (Fig. 14.10). Considering only the resultants of the vector systems involved, you can write the vector equation

$$(\Delta m)\mathbf{v}_A + \Sigma\mathbf{F}\,\Delta t = (\Delta m)\mathbf{v}_B \qquad\qquad (14.38)$$

You may also want to consider the angular momentum of the particle systems to obtain an additional equation [Sample Prob. 14.7]. However, many problems can be solved using Eq. (14.38) or the equation obtained by dividing all terms by Δt and letting Δt approach zero,

$$\Sigma\mathbf{F} = \frac{dm}{dt}(\mathbf{v}_B - \mathbf{v}_A) \qquad\qquad (14.39)$$

Here $\mathbf{v}_B - \mathbf{v}_A$ represents a *vector subtraction*, and the mass rate of flow dm/dt can be expressed as the product ρQ of the density ρ of the stream (mass per unit volume) and the volume rate of flow Q (volume per unit time). In U.S. customary units, ρ is expressed as the ratio γ/g, where γ is the specific weight of the stream and g is the acceleration due to gravity.

Typical problems involving a steady stream of particles have been described in Sec. 14.3A. You may be asked to determine the following,

    a. Thrust caused by a diverted flow. Equation (14.39) is applicable, but you will get a better understanding of the problem if you use a solution based on Eq. (14.38).

    b. Reactions at supports of vanes or conveyor belts. First draw a diagram showing on one side of the equals sign the momentum $(\Delta m)\mathbf{v}_A$ of the particles impacting the vane or belt in the time Δt, as well as the impulses of the loads and reactions at the supports during that time. On the other side, show the momentum $(\Delta m)\mathbf{v}_B$ of the particles leaving the vane or belt in the time Δt [Sample Prob. 14.7]. Equating the x components, y components, and moments of the quantities on both sides of the equals sign will yield three scalar equations that you can solve for three unknowns.

    c. Thrust developed by a jet engine, a propeller, or a fan. In most cases, a single unknown is involved, and you can obtain that unknown by solving the scalar equation derived from Eq. (14.38) or Eq. (14.39).

2. To solve problems involving systems gaining mass, consider the system S, which has a mass m and is moving with a velocity $\mathbf{v}$ at time t, and the particles of mass Δm with velocity $\mathbf{v}_a$ that S absorbs in the time interval Δt (Fig. 14.12). You will then express that the total momentum of S and of the particles absorbed, *plus* the impulse of the external forces exerted on S, are equipollent to the momentum of S at time $t + \Delta t$. Noting that the mass of S and its velocity at that time are, respectively, $m + \Delta m$ and $\mathbf{v} + \Delta \mathbf{v}$, you will write the vector equation

$$m\mathbf{v} + (\Delta m)\mathbf{v}_a + \Sigma\mathbf{F}\,\Delta t = (m + \Delta m)(\mathbf{v} + \Delta\mathbf{v}) \tag{14.40}$$

As we showed in Sec. 14.3B, if you introduce the relative velocity $\mathbf{u} = \mathbf{v}_a - \mathbf{v}$ of the particles being absorbed, you obtain the following expression for the resultant of the external forces applied to S as

$$\Sigma\mathbf{F} = m\frac{d\mathbf{v}}{dt} - \frac{dm}{dt}\mathbf{u} \tag{14.42}$$

Furthermore, the action on S of the particles being absorbed is equivalent to a thrust

$$\mathbf{P} = \frac{dm}{dt}\mathbf{u} \tag{14.44}$$

exerted in the direction of the relative velocity of the particles being absorbed.

Examples of systems gaining mass are conveyor belts, moving railroad cars being loaded with gravel or sand, and chains being pulled out of a pile.

3. To solve problems involving systems losing mass, such as rockets and rocket engines, you can use Eqs. (14.40) through (14.44)—provided that you give negative values to the increment of mass Δm and to the rate of change of mass dm/dt [Sample Prob. 14.9]. It follows that the thrust defined by Eq. (14.44) is exerted in a direction opposite to the direction of the relative velocity of the particles being ejected.

Problems

14.57 A stream of water with a density of $\rho = 1000$ kg/m³ is discharged from a nozzle at the rate of 0.06 m³/s. Using Bernoulli's equation, the gage pressure P in the pipe just upstream from the nozzle is $P = 0.5\,\rho(v_2^2 - v_1^2)$. Knowing the nozzle is held to the pipe by six flange bolts, determine the tension in each bolt, neglecting the initial tension caused by the tightening of the nuts.

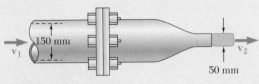

Fig. P14.57

14.58 A jet ski is placed in a channel and is tethered so that it is stationary. Water enters the jet ski with velocity $\mathbf{v}_1$ and exits with velocity $\mathbf{v}_2$. Knowing the inlet area is A_1 and the exit area is A_2, determine the tension in the tether.

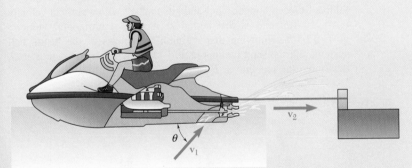

Fig. P14.58

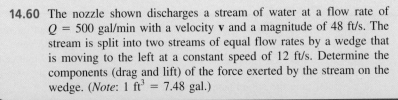

14.59 The nozzle shown discharges a stream of water at a flow rate of $Q = 475$ gal/min with a velocity $\mathbf{v}$ and a magnitude of 60 ft/s. The stream is split into two streams with equal flow rates by a wedge that is kept in a fixed position. Determine the components (drag and lift) of the force exerted by the stream on the wedge. (*Note*: 1 ft³ = 7.48 gal.)

14.60 The nozzle shown discharges a stream of water at a flow rate of $Q = 500$ gal/min with a velocity $\mathbf{v}$ and a magnitude of 48 ft/s. The stream is split into two streams of equal flow rates by a wedge that is moving to the left at a constant speed of 12 ft/s. Determine the components (drag and lift) of the force exerted by the stream on the wedge. (*Note*: 1 ft³ = 7.48 gal.)

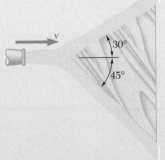

Fig. P14.59 and P14.60

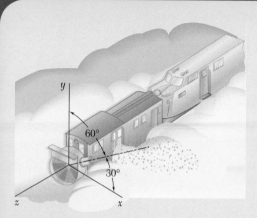

Fig. P14.61

14.61 A rotary power plow is used to remove snow from a level section of railroad track. The plow car is placed ahead of an engine that propels it at a constant speed of 20 km/h. The plow car clears 160 Mg of snow per minute, projecting it in the direction shown with a velocity of 12 m/s relative to the plow car. Neglecting friction, determine (*a*) the force exerted by the engine on the plow car, (*b*) the lateral force exerted by the track on the plow.

14.62 Tree limbs and branches are being fed at *A* at the rate of 5 kg/s into a shredder which spews the resulting wood chips at *C* with a velocity of 20 m/s. Determine the horizontal component of the force exerted by the shredder on the truck hitch at *D*.

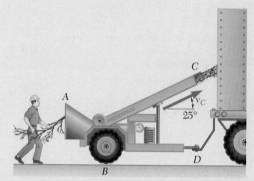

Fig. P14.62

14.63 Sand falls from three hoppers onto a conveyor belt at a rate of 90 lb/s for each hopper. The sand hits the belt with a vertical velocity $v_1 = 10$ ft/s and is discharged at *A* with a horizontal velocity $v_2 = 13$ ft/s. Knowing that the combined mass of the beam, belt system, and the sand it supports is 1300 lb with a mass center at *G*, determine the reaction at *E*.

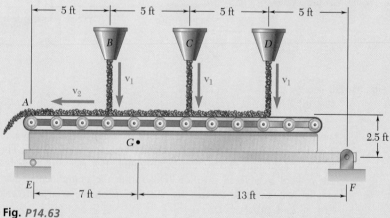

Fig. P14.63

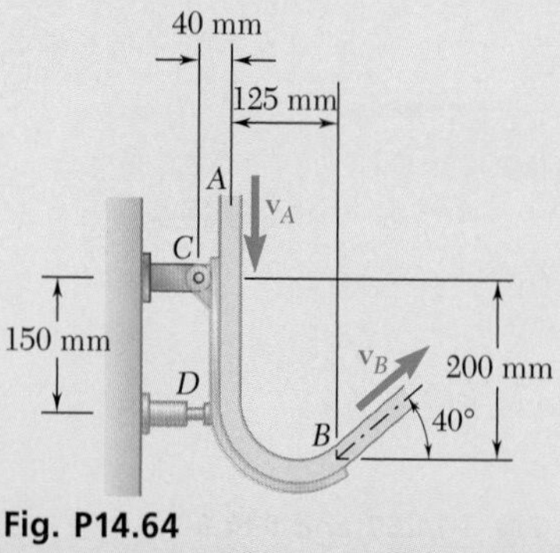

14.64 The stream of water shown flows at a rate of 550 L/min and moves with a velocity of magnitude 18 m/s at both *A* and *B*. The vane is supported by a pin and bracket at *C* and by a load cell at *D* that can exert only a horizontal force. Neglecting the weight of the vane, determine the components of the reactions at *C* and *D*.

Fig. P14.64

14.65 The nozzle shown discharges water at the rate of 40 ft³/min. Knowing that at both A and B the stream of water moves with a velocity of magnitude 75 ft/s and neglecting the weight of the vane, determine the components of the reactions at C and D.

14.66 A stream of water flowing at a rate of 1.2 m³/min and moving with a speed of 30 m/s at both A and B is deflected by a vane welded to a hinged plate. Knowing that the combined mass of the vane and plate is 20 kg with the mass center at point G, determine (a) the angle θ, (b) the reaction at C.

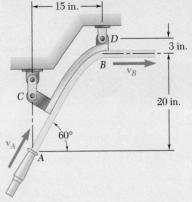

Fig. P14.65

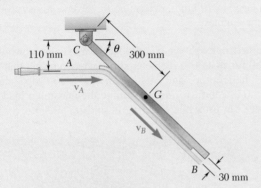

Fig. P14.66 and P14.67

14.67 A stream of water flowing at a rate of 1.2 m³/min and moving with a speed of v at both A and B is deflected by a vane welded to a hinged plate. The combined mass of the vane and plate is 20 kg with the mass center at point G. Knowing that $\theta = 45°$, determine (a) the speed v of the flow, (b) the reaction at C.

14.68 Coal is being discharged from a first conveyor belt at the rate of 120 kg/s. It is received at A by a second belt that discharges it again at B. Knowing that $v_1 = 3$ m/s and $v_2 = 4.25$ m/s and that the second belt assembly and the coal it supports have a total mass of 472 kg, determine the components of the reactions at C and D.

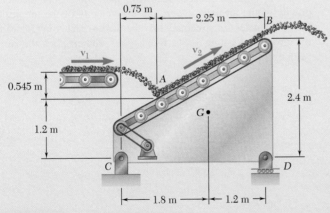

Fig. P14.68

14.69 The total drag due to air friction on a jet airplane traveling at 900 km/h is 35 kN. Knowing that the exhaust velocity is 600 m/s relative to the airplane, determine the mass of air that must pass through the engine per second to maintain the speed of 900 km/h in level flight.

Fig. P14.71

14.70 While cruising in level flight at a speed of 600 mi/h, a jet plane scoops in air at the rate of 200 lb/s and discharges it with a velocity of 2100 ft/s relative to the airplane. Determine the total drag due to air friction on the airplane.

14.71 In order to shorten the distance required for landing, a jet airplane is equipped with movable vanes that partially reverse the direction of the air discharged by each of its engines. Each engine scoops in the air at a rate of 120 kg/s and discharges it with a velocity of 600 m/s relative to the engine. At an instant when the speed of the airplane is 270 km/h, determine the reverse thrust provided by each of the engines.

14.72 The helicopter shown can produce a maximum downward air speed of 80 ft/s in a 30-ft-diameter slipstream. Knowing that the weight of the helicopter and its crew is 3500 lb and assuming $\gamma = 0.076$ lb/ft^3 for air, determine the maximum load that the helicopter can lift while hovering in midair.

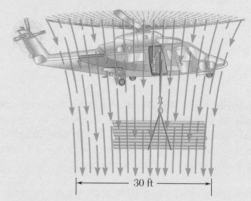

Fig. P14.72

14.73 Prior to takeoff, the pilot of a 3000-kg twin-engine airplane tests the reversible-pitch propellers by increasing the reverse thrust with the brakes at point *B* locked. Knowing that point *G* is the center of gravity of the airplane, determine the velocity of the air in the two 2.2-m-diameter slipstreams when the nose wheel *A* begins to lift off the ground. Assume $\rho = 1.21$ kg/m^3 and neglect the approach velocity of the air.

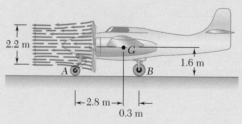

Fig. *P14.73*

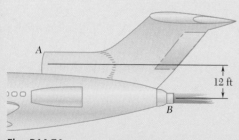

Fig. P14.74

14.74 The jet engine shown scoops in air at *A* at a rate of 200 lb/s and discharges it at *B* with a velocity of 2000 ft/s relative to the airplane. Determine the magnitude and line of action of the propulsive thrust developed by the engine when the speed of the airplane is (*a*) 300 mi/h, (*b*) 600 mi/h.

14.75 A jet airliner is cruising at a speed of 900 km/h with each of its three engines discharging air with a velocity of 800 m/s relative to the plane. Determine the speed of the airliner after it has lost the use of (*a*) one of its engines, (*b*) two of its engines. Assume that the drag due to air friction is proportional to the square of the speed and that the remaining engines keep operating at the same rate.

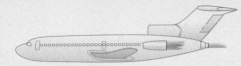

Fig. P14.75

14.76 A 16-Mg jet airplane maintains a constant speed of 774 km/h while climbing at an angle $\alpha = 18°$. The airplane scoops in air at a rate of 300 kg/s and discharges it with a velocity of 665 m/s relative to the airplane. If the pilot changes to a horizontal flight while maintaining the same engine setting, determine (*a*) the initial acceleration of the plane, (*b*) the maximum horizontal speed that will be attained. Assume that the drag due to air friction is proportional to the square of the speed.

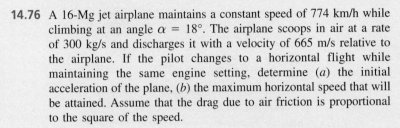

Fig. P14.76

14.77 The propeller of a small airplane has a 2-m-diameter slipstream and produces a thrust of 3600 N when the airplane is at rest on the ground. Assuming $\rho = 1.225$ kg/m^3 for air, determine (*a*) the speed of the air in the slipstream, (*b*) the volume of air passing through the propeller per second, (*c*) the kinetic energy imparted per second to the air in the slipstream.

14.78 The wind turbine generator shown has an output-power rating of 1.5 MW for a wind speed of 36 km/h. For the given wind speed, determine (*a*) the kinetic energy of the air particles entering the 82.5-m-diameter circle per second, (*b*) the efficiency of this energy conversion system. Assume $\rho = 1.21$ kg/m^3 for air.

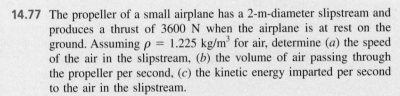

82.5 m

Fig. P14.78 and P14.79

14.79 A wind turbine generator system having a diameter of 82.5 m produces 1.5 MW at a wind speed of 12 m/s. Determine the diameter of blade necessary to produce 10 MW of power assuming the efficiency is the same for both designs and $\rho = 1.21$ kg/m^3 for air.

14.80 While cruising in level flight at a speed of 570 mi/h, a jet airplane scoops in air at a rate of 240 lb/s and discharges it with a velocity of 2200 ft/s relative to the airplane. Determine (*a*) the power actually used to propel the airplane, (*b*) the total power developed by the engine, (*c*) the mechanical efficiency of the airplane.

14.81 In a Pelton-wheel turbine, a stream of water is deflected by a series of blades so that the rate at which water is deflected by the blades is equal to the rate at which water issues from the nozzle ($\Delta m/\Delta t = A\rho v_A$). Using the same notation as in Sample Prob. 14.8, (*a*) determine the velocity **V** of the blades for which maximum power is developed, (*b*) derive an expression for the maximum power, (*c*) derive an expression for the mechanical efficiency.

Fig. P14.81

14.82 A circular reentrant orifice (also called Borda's mouthpiece) of diameter D is placed at a depth h below the surface of a tank. Knowing that the speed of the issuing stream is $v = \sqrt{2gh}$ and assuming that the speed of approach v_1 is zero, show that the diameter of the stream is $d = D/\sqrt{2}$. (*Hint:* Consider the section of water indicated, and note that P is equal to the pressure at a depth h multiplied by the area of the orifice.)

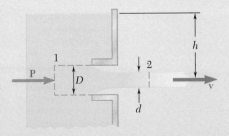

Fig. P14.82

14.83 A railroad car with length L and mass m_0 when empty is moving freely on a horizontal track while being loaded with sand from a stationary chute at a rate $dm/dt = q$. Knowing that the car was approaching the chute at a speed v_0, determine (a) the mass of the car and its load after the car has cleared the chute, (b) the speed of the car at that time.

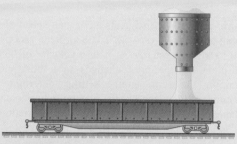

Fig. P14.83

*14.84 The depth of water flowing in a rectangular channel of width b at a speed v_1 and a depth d_1 increases to a depth d_2 at a *hydraulic jump*. Express the rate of flow Q in terms of b, d_1, and d_2.

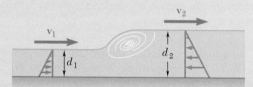

Fig. P14.84

*14.85 Determine the rate of flow in the channel of Prob. 14.84, knowing that $b = 12$ ft, $d_1 = 4$ ft, and $d_2 = 5$ ft.

14.86 A chain of length l and mass m lies in a pile on the floor. If its end A is raised vertically at a constant speed v, express in terms of the length y of chain that is off the floor at any given instant (a) the magnitude of the force **P** applied to A, (b) the reaction of the floor.

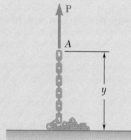

Fig. P14.86

14.87 Solve Prob. 14.86, assuming that the chain is being *lowered* to the floor at a constant speed v.

14.88 The ends of a chain lie in piles at A and C. When released from rest at time $t = 0$, the chain moves over the pulley at B, which has a negligible mass. Denoting by L the length of chain connecting the two piles and neglecting friction, determine the speed v of the chain at time t.

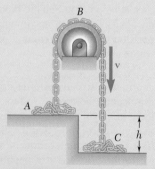

Fig. P14.88

14.89 A toy car is propelled by water that squirts from an internal tank at a constant 6 ft/s relative to the car. The weight of the empty car is 0.4 lb and it holds 2 lb of water. Neglecting other tangential forces, determine the top speed of the car.

Fig. P14.89 and P14.90

14.90 A toy car is propelled by water that squirts from an internal tank. The weight of the empty car is 0.4 lb and it holds 2 lb of water. Knowing the top speed of the car is 8 ft/s, determine the relative velocity of the water that is being ejected.

14.91 The main propulsion system of a space shuttle consists of three identical rocket engines that provide a total thrust of 6 MN. Determine the rate at which the hydrogen-oxygen propellant is burned by each of the three engines, knowing that it is ejected with a relative velocity of 3750 m/s.

14.92 The main propulsion system of a space shuttle consists of three identical rocket engines, each of which burns the hydrogen-oxygen propellant at the rate of 750 lb/s and ejects it with a relative velocity of 12,000 ft/s. Determine the total thrust provided by the three engines.

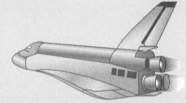

Fig. P14.91 and *P14.92*

14.93 A rocket sled burns fuel at the constant rate of 120 lb/s. The initial weight of the sled is 1800 lb, including 360 lb of fuel. Assume that the track is lubricated and the sled is aerodynamically designed so that air resistance and friction are negligible. (*a*) Derive a formula for the acceleration *a* of the sled as a function of time *t* and the exhaust velocity v_{ex} of the burned fuel relative to the sled. Plot the ratio a/v_{ex} versus time *t* for the range $0 < t < 4$ s, and check the slope of the graph at $t = 0$ and $t = 4$ s using the formula for *a*. (*b*) Determine the ratio of the velocity v_b of the sled at burnout to the exhaust velocity v_{ex}.

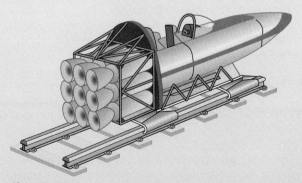

Fig. *P14.93*

Fig. P14.94

14.94 A space vehicle describing a circular orbit about the earth at a speed of 24×10^3 km/h releases at its front end a capsule that has a gross mass of 600 kg, including 400 kg of fuel. If the fuel is consumed at the rate of 18 kg/s and ejected with a relative velocity of 3000 m/s, determine (*a*) the tangential acceleration of the capsule as its engine is fired, (*b*) the maximum speed attained by the capsule.

14.95 A 540-kg spacecraft is mounted on top of a rocket with a mass of 19 Mg, including 17.8 Mg of fuel. Knowing that the fuel is consumed at a rate of 225 kg/s and ejected with a relative velocity of 3600 m/s, determine the maximum speed imparted to the spacecraft if the rocket is fired vertically from the ground.

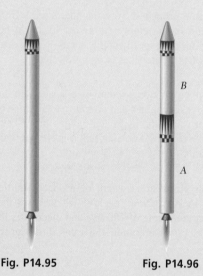

Fig. P14.95   Fig. P14.96

14.96 The rocket used to launch the 540-kg spacecraft of Prob. 14.95 is redesigned to include two stages *A* and *B*, each of mass 9.5 Mg, including 8.9 Mg of fuel. The fuel is again consumed at a rate of 225 kg/s and ejected with a relative velocity of 3600 m/s. Knowing that when stage *A* expels its last particle of fuel, its casing is released and jettisoned, determine (*a*) the speed of the rocket at that instant, (*b*) the maximum speed imparted to the spacecraft.

14.97 The weight of a spacecraft, including fuel, is 11,600 lb when the rocket engines are fired to increase its velocity by 360 ft/s. Knowing that 1000 lb of fuel is consumed, determine the relative velocity of the fuel ejected.

14.98 The rocket engines of a spacecraft are fired to increase its velocity by 450 ft/s. Knowing that 1200 lb of fuel is ejected at a relative velocity of 5400 ft/s, determine the weight of the spacecraft after the firing.

Fig. P14.97 and P14.98

14.99 Determine the distance traveled by the spacecraft of Prob. 14.97 during the rocket engine firing, knowing that its initial speed was 7500 ft/s and the duration of the firing was 60 s.

14.100 A rocket weighs 2600 lb, including 2200 lb of fuel, which is consumed at the rate of 25 lb/s and ejected with a relative velocity of 13,000 ft/s. Knowing that the rocket is fired vertically from the ground, determine (a) its acceleration as it is fired, (b) its acceleration as the last particle of fuel is being consumed, (c) the altitude at which all the fuel has been consumed, (d) the velocity of the rocket at that time.

14.101 Determine the altitude reached by the spacecraft of Prob. 14.95 when all the fuel of its launching rocket has been consumed.

14.102 For the spacecraft and the two-stage launching rocket of Prob. 14.96, determine the altitude at which (a) stage A of the rocket is released, (b) the fuel of both stages has been consumed.

14.103 In a jet airplane, the kinetic energy imparted to the exhaust gases is wasted as far as propelling the airplane is concerned. The useful power is equal to the product of the force available to propel the airplane and the speed of the airplane. If v is the speed of the airplane and u is the relative speed of the expelled gases, show that the mechanical efficiency of the airplane is $\eta = 2v/(u + v)$. Explain why $\eta = 1$ when $u = v$.

14.104 In a rocket, the kinetic energy imparted to the consumed and ejected fuel is wasted as far as propelling the rocket is concerned. The useful power is equal to the product of the force available to propel the rocket and the speed of the rocket. If v is the speed of the rocket and u is the relative speed of the expelled fuel, show that the mechanical efficiency of the rocket is $\eta = 2uv/(u^2 + v^2)$. Explain why $\eta = 1$ when $u = v$.

Review and Summary

In this chapter, we analyzed the motion of **systems of particles**, i.e., the motion of a large number of particles considered together. In the first part of the chapter, we considered systems consisting of well-defined particles, whereas in the second part, we analyzed systems that are continually gaining or losing particles or doing both at the same time.

Newton's Second Law for a System of Particles

We showed that **the system of the external forces acting on the particles and the system of the $m_i \mathbf{a}_i$ terms of the particles are equipollent**; i.e., both systems have the *same resultant* and the *same moment resultant* about O:

$$\sum_{i=1}^{n} \mathbf{F}_i = \sum_{i=1}^{n} m_i \mathbf{a}_i \tag{14.4}$$

$$\sum_{i=1}^{n} (\mathbf{r}_i \times \mathbf{F}_i) = \sum_{i=1}^{n} (\mathbf{r}_i \times m_i \mathbf{a}_i) \tag{14.5}$$

Linear and Angular Momentum of a System of Particles

We defined the *linear momentum* $\mathbf{L}$ and the *angular momentum* $\mathbf{H}_O$ *about point* O of the system of particles [Sec. 14.1B] as

$$\mathbf{L} = \sum_{i=1}^{n} m_i \mathbf{v}_i \qquad \mathbf{H}_O = \sum_{i=1}^{n} (\mathbf{r}_i \times m_i \mathbf{v}_i) \tag{14.6, 14.7}$$

Then we showed that we can replace Eqs. (14.4) and (14.5) with the equations

$$\Sigma \mathbf{F} = \dot{\mathbf{L}} \qquad \Sigma \mathbf{M}_O = \dot{\mathbf{H}}_O \tag{14.10, 14.11}$$

Together, these equations state that **the sum of external forces is equal to the rate of change of the linear momentum, and the sum of the moments about O is equal to the rate of change of the angular momentum about O.**

Motion of the Mass Center of a System of Particles

In Sec. 14.1C, we defined the mass center of a system of particles as the point G whose position vector $\bar{\mathbf{r}}$ satisfies the equation

$$m\bar{\mathbf{r}} = \sum_{i=1}^{n} m_i \mathbf{r}_i \tag{14.12}$$

where m represents the total mass $m = \sum_{i=1}^{n} m_i$ of the particles. Differentiating both sides of Eq. (14.12) twice with respect to t, we obtained the relations

$$\mathbf{L} = m\bar{\mathbf{v}} \qquad \dot{\mathbf{L}} = m\bar{\mathbf{a}} \tag{14.14, 14.15}$$

where $\bar{\mathbf{v}}$ and $\bar{\mathbf{a}}$ represent, respectively, the velocity and the acceleration of the mass center G. Substituting for $\mathbf{L}$ from Eq. (14.15) into Eq. (14.10), we obtained

$$\Sigma \mathbf{F} = m\bar{\mathbf{a}} \tag{14.16}$$

From this, we concluded that **the mass center of a system of particles moves as if the entire mass of the system and all of the external forces were concentrated at that point** [Sample Prob. 14.1].

Angular Momentum of a System of Particles about its Mass Center

In Sec. 14.1D, we considered the motion of the particles of a system with respect to a centroidal frame $Gx'y'z'$ attached to the mass center G of the system and in translation with respect to the newtonian frame $Oxyz$ (Fig. 14.13). We defined the *angular momentum* of the system *about its mass center G* as the sum of the moments about G of the momenta $m_i\mathbf{v}_i'$ of the particles relative to the frame $Gx'y'z'$. We also noted that we can obtain the same earlier result by considering the moments about G of the momenta $m_i\mathbf{v}_i$ of the particles in their absolute motion. We therefore wrote

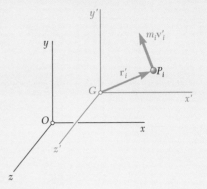

Fig. 14.13

$$\mathbf{H}_G = \sum_{i=1}^{n} (\mathbf{r}_i' \times m_i\mathbf{v}_i) = \sum_{i=1}^{n} (\mathbf{r}_i' \times m_i\mathbf{v}_i') \tag{14.24}$$

and derived the relation

$$\Sigma \mathbf{M}_G = \dot{\mathbf{H}}_G \tag{14.23}$$

This equation states that **the resultant moment about G of the external forces is equal to the rate of change of the angular momentum about G of the system of particles**. As you will see later, this relation is fundamental to the study of the motion of rigid bodies.

Conservation of Momentum

When no external force acts on a system of particles [Sec. 14.1E], it follows from Eqs. (14.10) and (14.11) that the linear momentum $\mathbf{L}$ and the angular momentum $\mathbf{H}_O$ of the system are conserved [Sample Probs. 14.2 and 14.4]. In problems involving central forces, the angular momentum of the system about the center of force O is also conserved.

Kinetic Energy of a System of Particles

The kinetic energy T of a system of particles was defined as the sum of the kinetic energies of the particles [Sec. 14.2A]:

$$T = \frac{1}{2} \sum_{i=1}^{n} m_i v_i^2 \tag{14.28}$$

Using the centroidal frame of reference $Gx'y'z'$ of Fig. 14.13, we noted that we can also obtain the kinetic energy of the system by adding the kinetic energy $\frac{1}{2}m\bar{v}^2$ associated with the motion of the mass center G and the kinetic energy of the system relative to the frame $Gx'y'z'$. Thus,

$$T = \tfrac{1}{2}m\bar{v}^2 + \frac{1}{2} \sum_{i=1}^{n} m_i v_i'^2 \tag{14.29}$$

Principle of Work and Energy

We applied the **principle of work and energy** to a system of particles as well as to individual particles [Sec. 14.2B]. We have

$$T_1 + U_{1\rightarrow 2} = T_2 \tag{14.30}$$

and noted that $U_{1\rightarrow 2}$ represents the work of *all* of the forces acting on the particles of the system—internal as well as external.

Conservation of Energy

If all of the forces acting on the particles of a system are *conservative*, we can determine the potential energy V of the system and write

$$T_1 + V_1 = T_2 + V_2 \tag{14.31}$$

which expresses the **principle of conservation of energy** for a system of particles.

Principle of Impulse and Momentum

We saw in Sec. 14.2C that the **principle of impulse and momentum** for a system of particles can be expressed graphically, as shown in Fig. 14.14. The principle states that the momenta of the particles at time t_1 and the impulses of the external forces from t_1 to t_2 form a system of vectors equipollent to the system of the momenta of the particles at time t_2.

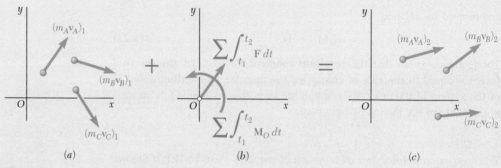

Fig. 14.14

If no external force acts on the particles of the system, the systems of momenta shown in parts a and c of Fig. 14.14 are equipollent, and we have

$$\mathbf{L}_1 = \mathbf{L}_2 \qquad (\mathbf{H}_O)_1 = (\mathbf{H}_O)_2 \tag{14.36, 14.37}$$

Use of Conservation Principles in the Solution of Problems Involving Systems of Particles

We can solve many problems involving the motion of systems of particles by applying simultaneously the principle of impulse and momentum and the principle of conservation of energy [Sample Prob. 14.5] or by expressing that the linear momentum, angular momentum, and energy of the system are conserved [Sample Prob. 14.6].

Steady Stream of Particles

In the second part of the chapter, we considered **variable systems of particles**. First we considered a **steady stream of particles**, such as a stream of water diverted by a fixed vane or the flow of air through a jet engine [Sec. 14.3A]. We applied the principle of impulse and momentum to a system S of particles during a time interval Δt, including the particles that enter the system at A during that time interval and those (of the same mass Δm) that leave the system at B. We concluded that **the system formed by the momentum $(\Delta m)\mathbf{v}_A$ of the particles entering S in the time Δt and the impulses of the forces**

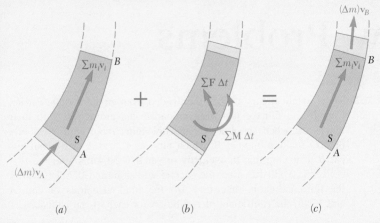

Fig. 14.15

exerted on S during that time is equipollent to the momentum $(\Delta m)\mathbf{v}_B$ of the particles leaving S in the same time Δt (Fig. 14.15). Equating the x components, y components, and moments about a fixed point of the vectors involved, we could obtain as many as three equations that you could solve for the desired unknowns [Sample Probs. 14.7 and 14.8]. From this result, we also derived the expression for the resultant $\Sigma\mathbf{F}$ of the forces exerted on S as

$$\Sigma\mathbf{F} = \frac{dm}{dt}(\mathbf{v}_B - \mathbf{v}_A) \tag{14.39}$$

where $\mathbf{v}_B - \mathbf{v}_A$ represents the difference between the *vectors* $\mathbf{v}_B$ and $\mathbf{v}_A$ and dm/dt is the mass rate of flow of the stream (see footnote, page 951).

Systems Gaining or Losing Mass

We considered next a system of particles gaining mass by continually absorbing particles or losing mass by continually expelling particles [Sec. 14.3B], as in the case of a rocket. We applied the principle of impulse and momentum to the system during a time interval Δt, being careful to include the particles gained or lost during that time interval [Sample Prob. 14.9]. We also noted that the action on a system S of the particles being *absorbed* by S was equivalent to a thrust

$$\mathbf{P} = \frac{dm}{dt}\mathbf{u} \tag{14.44}$$

where dm/dt is the rate at which mass is being absorbed and $\mathbf{u}$ is the velocity of the particles *relative to S*. In the case of particles being *expelled* by S, the rate dm/dt is negative, and the thrust $\mathbf{P}$ is exerted in a direction opposite to that in which the particles are being expelled.

Review Problems

14.105 Three identical cars are being unloaded from an automobile carrier. Cars B and C have just been unloaded and are at rest with their brakes off when car A leaves the unloading ramp with a velocity of 5.76 ft/s and hits car B, which hits car C. Car A then again hits car B. Knowing that the velocity of car B is 5.04 ft/s after the first collision, 0.630 ft/s after the second collision, and 0.709 ft/s after the third collision, determine (a) the final velocities of cars A and C, (b) the coefficient of restitution for each of the collisions.

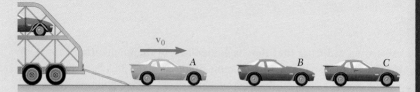

Fig. P14.105

14.106 A 30-g bullet is fired with a velocity of 480 m/s into block A, which has a mass of 5 kg. The coefficient of kinetic friction between block A and cart BC is 0.50. Knowing that the cart has a mass of 4 kg and can roll freely, determine (a) the final velocity of the cart and block, (b) the final position of the block on the cart.

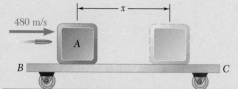

Fig. P14.106

14.107 An 80-Mg railroad engine A coasting at 6.5 km/h strikes a 20-Mg flatcar C carrying a 30-Mg load B that can slide along the floor of the car ($\mu_k = 0.25$). Knowing that the car was at rest with its brakes released and that it automatically coupled with the engine upon impact, determine the velocity of the car (a) immediately after impact, (b) after the load has slid to a stop relative to the car.

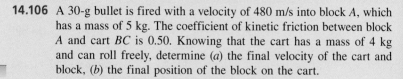

Fig. P14.107

14.108 In a game of pool, ball A is moving with a velocity $\mathbf{v}_0$ when it strikes balls B and C, which are at rest and aligned as shown. Knowing that after the collision the three balls move in the directions indicated and that $v_0 = 12$ ft/s and $v_C = 6.29$ ft/s, determine the magnitude of the velocity of (a) ball A, (b) ball B.

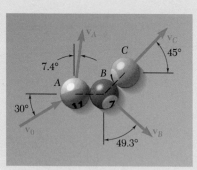

Fig. P14.108

14.109 Mass C, which has a mass of 4 kg, is suspended from a cord attached to cart A, which has a mass of 5 kg and can roll freely on a friction-less horizontal track. A 60-g bullet is fired with a speed $v_0 = 500$ m/s and gets lodged in block C. Determine (a) the velocity of C as it reaches its maximum elevation, (b) the maximum vertical distance h through which C will rise.

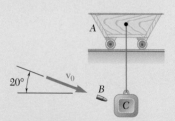

Fig. P14.109

14.110 A 15-lb block B is at rest and a spring of constant $k = 72$ lb/in. is held compressed 3 in. by a cord. After 5-lb block A is placed against the end of the spring, the cord is cut causing A and B to move. Neglecting friction, determine the velocities of blocks A and B immediately after A leaves B.

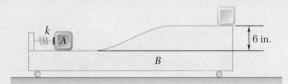

Fig. P14.110

14.111 Car A of mass 1800 kg and car B of mass 1700 kg are at rest on a 20-Mg flatcar which is also at rest. Cars A and B then accelerate and quickly reach constant speeds relative to the flatcar of 2.35 m/s and 1.175 m/s, respectively, before decelerating to a stop at the opposite end of the flatcar. Neglecting friction and rolling resistance, determine the velocity of the flatcar when the cars are moving at constant speeds.

Fig. P14.111

14.112 The nozzle shown discharges water at the rate of 200 gal/min. Knowing that at both B and C the stream of water moves with a velocity of magnitude 100 ft/s, and neglecting the weight of the vane, determine the force-couple system that must be applied at A to hold the vane in place (1 ft^3 = 7.48 gal).

14.113 An airplane with a weight W and a total wing span b flies horizontally at a constant speed v. Use the airplane as a reference frame; that is, consider the airplane to be motionless and the air to flow past it with speed v. Suppose that a cylinder of air with diameter b is deflected downward by the wing (the cross section of the cylinder is the dashed circle in in the figure). Show that the angle through which the cylinder stream is deflected (called the *downwash angle*) is determined by the formula $\sin \theta = 4W/(\pi b^2 \rho v^2)$, where ρ is the mass density of the air.

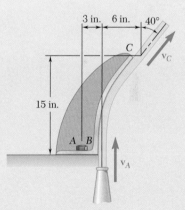

Fig. P14.112

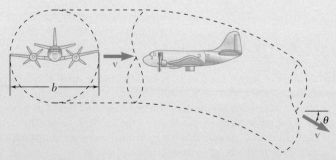

Fig. P14.113

14.114 The final component of a conveyor system receives sand at a rate of 100 kg/s at A and discharges it at B. The sand is moving horizontally at A and B with a velocity of magnitude $v_A = v_B = 4.5$ m/s. Knowing that the combined weight of the component and of the sand it supports is $W = 4$ kN, determine the reactions at C and D.

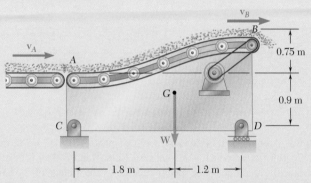

Fig. P14.114

14.115 A garden sprinkler has four rotating arms, each of which consists of two horizontal straight sections of pipe forming an angle of 120° with each other. Each arm discharges water at a rate of 20 L/min with a velocity of 18 m/s relative to the arm. Knowing that the friction between the moving and stationary parts of the sprinkler is equivalent to a couple of magnitude $M = 0.375$ N·m, determine the constant rate at which the sprinkler rotates.

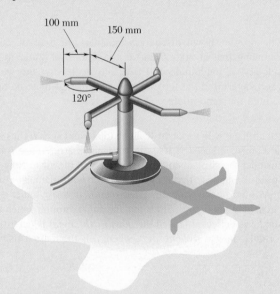

Fig. P14.115

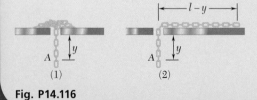

Fig. P14.116

14.116 A chain of length l and mass m falls through a small hole in a plate. Initially, when y is very small, the chain is at rest. In each case shown, determine (a) the acceleration of the first link A as a function of y, (b) the velocity of the chain as the last link passes through the hole. In case 1, assume that the individual links are at rest until they fall through the hole; in case 2, assume that at any instant all links have the same speed. Ignore the effect of friction.

15

Kinematics of Rigid Bodies

This huge crank belongs to a large diesel engine. In this chapter, you will learn to perform the *kinematic* analysis of rigid bodies that undergo *translation*, *fixed axis rotation*, and *general plane motion*.

Objectives

- **Describe** the five basic types of rigid body motion: translation, rotation about a fixed axis, general plane motion, motion about a fixed point, and general motion.
- **Use** angular kinematic relationships involving θ, ω, and α to determine the angular motion of a rigid body.
- **Identify** the directions of terms in the relative velocity and relative acceleration equations.
- **Calculate** the linear velocity and acceleration of any point on a rigid body undergoing translation, fixed axis rotation, or general plane motion.
- **Solve** planar rigid body kinematics problems using the relative velocity and relative acceleration equations.
- **Determine** the instantaneous center of rotation and use it to analyze the planar velocity kinematics of a rigid body.
- When appropriate, **define** a rotating coordinate frame and use it to solve planar and three-dimensional kinematics problems.
- **Determine** the angular velocity and angular acceleration of a body undergoing three-dimensional motion.
- **Calculate** the linear velocity and acceleration of any point on a rigid body undergoing three-dimensional motion.

Introduction

In this chapter, we consider the kinematics of **rigid bodies**. We will investigate the relations between the time, the positions, the velocities, and the accelerations of the various particles forming a rigid body. As you will see, the various types of rigid-body motion can be conveniently grouped as follows:

1. **Translation.** A motion is said to be a translation if any straight line inside the body maintains the same orientation during the motion. In a translation, all of the particles forming the body move along parallel paths. If these paths are straight lines, the motion is called **rectilinear translation** (Fig. 15.1); if the paths are curved lines, the motion is called **curvilinear translation** (Fig. 15.2).

2. **Rotation About a Fixed Axis.** In this motion, the particles forming the rigid body move in parallel planes along circles centered on the same fixed axis (Fig. 15.3). If this axis, called the **axis of rotation**, intersects the rigid body, the particles located on the axis have zero velocity and zero acceleration.

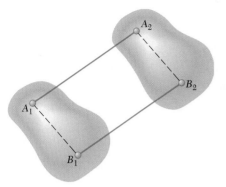

Fig. 15.1 A rigid body in rectilinear translation.

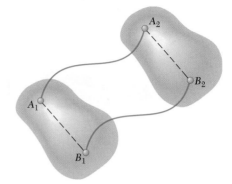

Fig. 15.2 A rigid body in curvilinear translation.

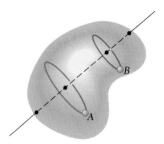

Fig. 15.3 A rigid body rotating about a fixed axis.

Be careful not to confuse rotation with certain types of curvilinear translation. For example, the plate shown in Fig. 15.4*a* is in curvilinear translation, with all of its particles moving along *parallel* circles, whereas the plate shown in Fig. 15.4*b* is in rotation, with all of its particles moving along *concentric* circles. In the first case, any given straight line drawn on the plate maintains the same direction, whereas in the second case, the orientation of the plate changes throughout the rotation. Because each particle moves in a given plane, the rotation of a body about a fixed axis is said to be a **plane motion**.

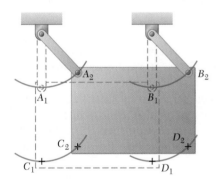

(*a*) Curvilinear translation

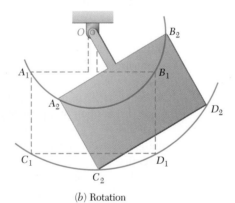

(*b*) Rotation

Fig. 15.4 (*a*) In curvilinear motion, particles move along parallel circles, whereas (*b*) in fixed-axis rotation, particles move along concentric circles.

3. **General Plane Motion.** Many other types of plane motion can occur, i.e., motions in which all the particles of the body move in a single plane. Any plane motion that is neither a rotation nor a translation is referred to as general plane motion. Figure 15.5 shows two examples of general plane motion.

4. **Motion About a Fixed Point.** The three-dimensional motion of a rigid body attached at a fixed point O, such as the motion of a top on a rough floor (Fig. 15.6), is known as motion about a fixed point.

5. **General Motion.** Any motion of a rigid body that does not fall in any of these categories is referred to as a general motion.

After a brief discussion of the motion of translation, we consider the rotation of a rigid body about a fixed axis. We define the *angular velocity* and the *angular acceleration* of a rigid body rotating about a fixed axis,

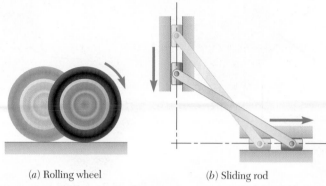

(a) Rolling wheel   (b) Sliding rod

Fig. 15.5 (a) A rolling wheel and (b) a sliding rod are common examples of general plane motion.

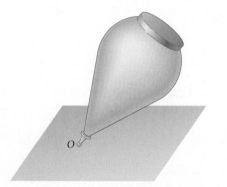

Fig. 15.6 The motion of a spinning top on a rough surface is an example of three-dimensional motion about a fixed point.

and you will see how to express the velocity and acceleration of a given point of the body in terms of its position vector and the angular velocity and angular acceleration of the body.

Afterwards, we study the general plane motion of a rigid body and apply the results to the analysis of mechanisms such as gears, connecting rods, and pin-connected linkages. If we resolve the plane motion of a rigid body into a translation and a rotation, we can then express the velocity of a point B of the body as the sum of the velocity of a reference point A and of the velocity of B relative to a frame of reference translating with A (i.e., moving with A but not rotating). We use the same approach later in Sec. 15.4 to express the acceleration of B in terms of the acceleration of A and of the acceleration of B relative to a frame translating with A. We also present an alternative method for analyzing velocities in plane motion based on the concept of the *instantaneous center of rotation,* and we discuss still another method of analysis based on the use of parametric expressions for the coordinates of a given point.

The motion of a particle relative to a rotating frame of reference and the concept of *Coriolis acceleration* are discussed in Sec. 15.5. We apply the results to the analysis of the plane motion of mechanisms containing parts that slide on each other.

In the remainder of this chapter, we analyze the three-dimensional motion of a rigid body, specifically, the motion of a rigid body with a fixed point and the general motion of a rigid body. We use a fixed frame of reference or a frame of reference in translation to carry out this analysis, then we consider the motion of the body relative to a rotating frame or to a frame in general motion. Again, we use the concept of Coriolis acceleration.

15.1   TRANSLATION AND FIXED AXIS ROTATION

We noted in the introduction that we can resolve a general plane motion into a translation and a rotation. Thus, our first step is to formulate the mathematical descriptions of simple translations and rotations.

15.1A   Translation

Consider a rigid body in translation (either rectilinear or curvilinear translation), and let A and B be any two of its particles (Fig. 15.7a). Denoting the position vectors of A and B with respect to a fixed frame of

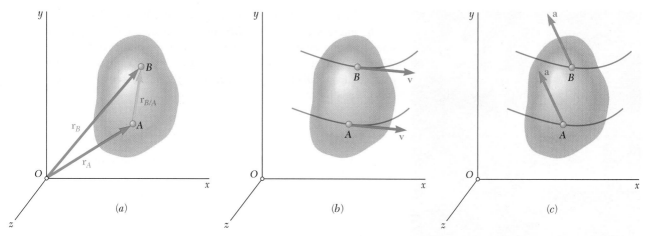

Fig. 15.7 For a rigid body in translation: (*a*) the position vector between any two points is constant in magnitude and direction; (*b*) every point has the same velocity; (*c*) every point has the same acceleration.

reference by $\mathbf{r}_A$ and $\mathbf{r}_B$, respectively, and the vector from A to B by $\mathbf{r}_{B/A}$, we have

$$\mathbf{r}_B = \mathbf{r}_A + \mathbf{r}_{B/A} \tag{15.1}$$

To obtain the relationship between the velocities of A and B, we differentiate this expression with respect to t. Note that, from the very definition of a translation, the vector $\mathbf{r}_{B/A}$ must maintain a constant direction; its magnitude must also be constant, since A and B belong to the same rigid body. Thus, the derivative of $\mathbf{r}_{B/A}$ is zero, and we have

$$\mathbf{v}_B = \mathbf{v}_A \tag{15.2}$$

Differentiating once more, we obtain the relationship between the accelerations of A and B as

$$\mathbf{a}_B = \mathbf{a}_A \tag{15.3}$$

Thus, **when a rigid body is in translation, all the points of the body have the same velocity and the same acceleration at any given instant** (Fig. 15.7*b* and *c*). In the case of curvilinear translation, the velocity and acceleration change in direction as well as in magnitude at every instant. In the case of rectilinear translation, all particles of the body move along parallel straight lines, and their velocity and acceleration keep the same direction during the entire motion.

15.1B Rotation About a Fixed Axis

Consider a rigid body that rotates about a fixed axis AA'. Let P be a point of the body and $\mathbf{r}$ be its position vector with respect to a fixed frame of reference. For convenience, let us assume that the frame is centered at point O on AA' and that the z axis coincides with AA' (Fig. 15.8). Let B be the projection of P on AA'. Since P must remain at a constant distance from B, it describes a circle with a center B and radius $r \sin \phi$, where ϕ denotes the angle formed by $\mathbf{r}$ and AA'.

Photo 15.1 The horizontal linkage of a locomotive undergoes curvilinear translation.

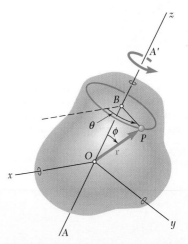

Fig. 15.8 For a rigid body in rotation about a fixed axis, each point of the body moves in a circular path centered on the axis.

Photo 15.2 For the central gear rotating about a fixed axis, the angular velocity and angular acceleration of that gear are vectors directed along the vertical axis of rotation.

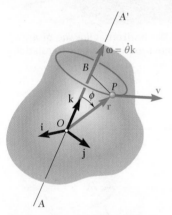

Fig. 15.9 For a rigid body in rotation about a fixed axis, the velocity of a particle is the vector product of the angular velocity of the body and the position vector of the particle.

The position both of P and of the entire body is completely defined by the angle θ that the line BP forms with the zx plane. The angle θ is known as the **angular coordinate** of the body and is defined as positive when viewed as counterclockwise from A'. The angular coordinate is expressed in radians (rad) or, occasionally, in degrees (°) or revolutions (rev). Recall that

$$1 \text{ rev} = 2\pi \text{ rad} = 360°$$

Recall from Sec. 11.4A that the velocity $\mathbf{v} = d\mathbf{r}/dt$ of a particle P is a vector tangent to the path of P and with a magnitude of $v = ds/dt$. The length Δs of the arc described by P when the body rotates through $\Delta\theta$ is

$$\Delta s = (BP)\,\Delta\theta = (r\sin\phi)\,\Delta\theta$$

Then dividing both members by Δt, we obtain in the limit, as Δt approaches zero,

$$v = \frac{ds}{dt} = r\dot{\theta}\sin\phi \tag{15.4}$$

where $\dot{\theta}$ denotes the time derivative of θ. (Note that the angle θ depends on the position of P within the body, but the rate of change $\dot{\theta}$ is itself independent of P.) We conclude that the velocity $\mathbf{v}$ of P is a vector perpendicular to the plane containing AA' and $\mathbf{r}$, and of magnitude v defined by Eq. (15.4). But this is precisely the result we would obtain if we drew a vector $\boldsymbol{\omega} = \dot{\theta}\mathbf{k}$ along AA' and formed the vector product $\boldsymbol{\omega} \times \mathbf{r}$ (Fig. 15.9). We thus have

$$\mathbf{v} = \frac{d\mathbf{r}}{dt} = \boldsymbol{\omega} \times \mathbf{r} \tag{15.5}$$

The vector

$$\boldsymbol{\omega} = \omega\mathbf{k} = \dot{\theta}\mathbf{k} \tag{15.6}$$

is directed along the axis of rotation. It is called the **angular velocity** of the body and is equal in magnitude to the rate of change $\dot{\theta}$ of the angular coordinate. You can obtain the sense of the vector by using the right hand rule (Sec. 3.1E); using your right hand, curl your fingers in the direction of the angular velocity, and your thumb will point in the direction of the vector.[†]

Now we can determine the acceleration $\mathbf{a}$ of particle P. Differentiating Eq. (15.5) and recalling the rule for the differentiation of a vector product (Sec. 11.4B), we have

$$\begin{aligned}
\mathbf{a} = \frac{d\mathbf{v}}{dt} &= \frac{d}{dt}(\boldsymbol{\omega} \times \mathbf{r}) \\
&= \frac{d\boldsymbol{\omega}}{dt} \times \mathbf{r} + \boldsymbol{\omega} \times \frac{d\mathbf{r}}{dt} \\
&= \frac{d\boldsymbol{\omega}}{dt} \times \mathbf{r} + \boldsymbol{\omega} \times \mathbf{v} \tag{15.7}
\end{aligned}$$

[†]We will show in Sec. 15.6 the more general case of a rigid body, rotating simultaneously about axes having different directions, where angular velocities obey the parallelogram law of addition and thus are actually vector quantities.

The vector $d\boldsymbol{\omega}/dt$ is denoted by $\boldsymbol{\alpha}$ and is called the **angular acceleration** of the body. Substituting for $\mathbf{v}$ from Eq. (15.5), we have

$$\mathbf{a} = \boldsymbol{\alpha} \times \mathbf{r} + \boldsymbol{\omega} \times (\boldsymbol{\omega} \times \mathbf{r}) \qquad (15.8)$$

Differentiating Eq. (15.6) and recalling that $\mathbf{k}$ is constant in magnitude and direction, we have

$$\boldsymbol{\alpha} = \alpha\mathbf{k} = \dot{\omega}\mathbf{k} = \ddot{\theta}\mathbf{k} \qquad (15.9)$$

Thus, the angular acceleration of a body rotating about a fixed axis is a vector directed along the axis of rotation and is equal in magnitude to the rate of change $\dot{\omega}$ of the angular velocity.

Returning to Eq. (15.8), we note that the acceleration of P is the sum of two vectors. The first vector is equal to the vector product $\boldsymbol{\alpha} \times \mathbf{r}$; it is tangent to the circle described by P and therefore represents the tangential component of the acceleration. The second vector is equal to the *vector triple product* $\boldsymbol{\omega} \times (\boldsymbol{\omega} \times \mathbf{r})$ obtained by forming the vector product of $\boldsymbol{\omega}$ and $\boldsymbol{\omega} \times \mathbf{r}$. Since $\boldsymbol{\omega} \times \mathbf{r}$ is tangent to the circle described by P, the vector triple product is directed toward the center B of the circle and therefore represents the normal component of the acceleration.

Rotation of a Representative Slab.

We can express the rotation of a rigid body about a fixed axis by examining the motion of a representative slab in a reference plane perpendicular to the axis of rotation. We choose the xy plane as the reference plane and assume that it coincides with the plane of the figure with the z axis pointing out of the page (Fig. 15.10). Recalling from Eq. (15.6) that $\boldsymbol{\omega} = \omega\mathbf{k}$, we note that a positive value of the scalar ω corresponds to a counterclockwise rotation of the representative slab, and a negative value corresponds to a clockwise rotation. Substituting $\omega\mathbf{k}$ for $\boldsymbol{\omega}$ in Eq. (15.5), we express the velocity of any given point P of the slab as

$$\mathbf{v} = \omega\mathbf{k} \times \mathbf{r} \qquad (15.10)$$

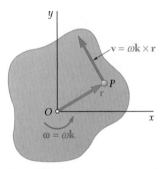

Fig. 15.10 For an object undergoing fixed-axis rotation, the velocity of a point P equals the vector product of the angular velocity vector and the position vector of P. A positive value of the scalar ω corresponds to counterclockwise motion.

Since the vectors $\mathbf{k}$ and $\mathbf{r}$ are mutually perpendicular, the magnitude of the velocity $\mathbf{v}$ is

$$v = r\omega \qquad (15.10')$$

We can obtain its direction by rotating $\mathbf{r}$ through $90°$ in the sense of rotation of the slab.

If we substitute $\omega\mathbf{k}$ into Eq. (15.8), we obtain $\omega\mathbf{k} \times (\omega\mathbf{k} \times \mathbf{r})$, which simplifies to $-\omega^2\mathbf{r}$. This indicates that the direction of the normal acceleration is $-\mathbf{r}$, or toward the center of rotation, which is exactly what we expect. Using this expression and $\boldsymbol{\alpha} = \alpha\mathbf{k}$ in Eq. (15.8), we obtain

$$\mathbf{a} = \alpha\mathbf{k} \times \mathbf{r} - \omega^2\mathbf{r} \qquad (15.11)$$

Resolving $\mathbf{a}$ into tangential and normal components (Fig. 15.11) gives

$$\mathbf{a}_t = \alpha\mathbf{k} \times \mathbf{r} \qquad a_t = r\alpha \qquad (15.11')$$
$$\mathbf{a}_n = -\omega^2\mathbf{r} \qquad a_n = r\omega^2$$

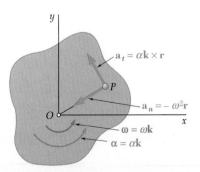

Fig. 15.11 For an object undergoing fixed-axis rotation, the acceleration of a point P has a tangential component that depends on angular acceleration and a normal component that depends on angular velocity.

Photo 15.3 If the lower roll has a constant angular velocity, the speed of the paper being wound onto it increases as the radius of the roll increases.

The tangential component $\mathbf{a}_t$ points in the counterclockwise direction if the scalar α is positive and in the clockwise direction if α is negative. The normal component $\mathbf{a}_n$ always points in the direction opposite to that of $\mathbf{r}$, that is, toward O.

15.1C Equations Defining the Rotation of a Rigid Body About a Fixed Axis

The motion of a rigid body rotating about a fixed axis AA' is said to be *known* when we can express its angular coordinate θ as a known function of t. In practice, however, we can seldom describe the rotation of a rigid body by a relation between θ and t. More often, the conditions of motion are specified by the angular acceleration of the body. For example, α may be given as a function of t, as a function of θ, or as a function of ω. From the relations in Eqs. (15.6) and (15.9), we have

$$\omega = \frac{d\theta}{dt} \tag{15.12}$$

$$\alpha = \frac{d\omega}{dt} = \frac{d^2\theta}{dt^2} \tag{15.13}$$

or solving Eq. (15.12) for dt and substituting into Eq. (15.13), we have

$$\alpha = \omega \frac{d\omega}{d\theta} \tag{15.14}$$

These equations are similar to those obtained in Chap. 11 for the rectilinear motion of a particle, so we can integrate them by following the procedures outlined in Sec. 11.1B.

Two particular cases of rotation occur frequently:

1. *Uniform Rotation.* This case is characterized by the fact that the angular acceleration is zero, therefore the angular velocity is constant and the angular position is given by

$$\theta = \theta_0 + \omega t$$

2. *Uniformly Accelerated Rotation.* In this case, the angular acceleration is constant. We can derive the following formulas relating angular velocity, angular position, and time in a manner similar to that described in Sec. 11.2B. The similarity between the formulas derived here and those obtained for the rectilinear uniformly accelerated motion of a particle is apparent.

$$\omega = \omega_0 + \alpha t$$
$$\theta = \theta_0 + \omega_0 t + \tfrac{1}{2}\alpha t^2 \tag{15.16}$$
$$\omega^2 = \omega_0^2 + 2\alpha(\theta - \theta_0)$$

We emphasize that you can use formula (15.15) only when $\alpha = 0$, and formulas (15.16) only when $\alpha = $ constant. In any other case, you need to use the general Eq. (15.12) through Eq. (15.14).

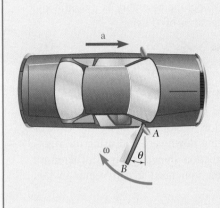

Sample Problem 15.1

A driver starts his car with the door on the passenger's side wide open ($\theta = 0$). As the car moves forward with constant acceleration, the angular acceleration of the door is $\alpha = 2.5 \cos \theta$, where α is in rad/s^2. Determine the angular velocity of the door as it slams shut ($\theta = 90°$).

STRATEGY: You are given the angular acceleration as a function of θ, so use the kinematic relationships between angular acceleration, angular velocity, angular position, and time.

MODELING and ANALYSIS: Model the door as a rigid body. Using the basic kinematic relationship gives

$$\alpha = \frac{d\omega}{dt} = \omega \frac{d\omega}{d\theta} = 2.5 \cos \theta$$

Separating variables gives

$$\omega \, d\omega = 2.5 \cos \theta \, d\theta$$

Integrating, using $\omega = 0$ when $\theta = 0$, you have

$$\int_0^\omega \omega \, d\omega = \int_0^\theta 2.5 \cos \theta \, d\theta$$

$$\frac{1}{2}\omega^2 = 2.5 \sin \theta \Big|_0^{\pi/2} = 2.5$$

$$\omega = 2.24 \text{ rad/s} \curvearrowright \quad \blacktriangleleft$$

REFLECT and THINK: If the angular acceleration of the door had been a constant 2.5 rad/s^2, you would have found $\frac{1}{2}\omega^2 = 2.5|_0^{\pi/2}$ or $\omega = 2.80 \text{ rad/s}$. Since $\alpha = 2.5 \cos \theta$ decreases as θ increases, it makes sense that the answer you found in this case is smaller than the case for constant angular acceleration.

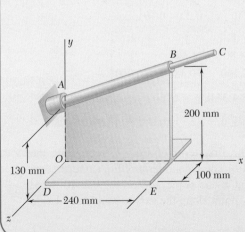

Sample Problem 15.2

The assembly shown rotates about the rod AC. At the instant shown, the assembly has an angular velocity of 5 rad/s that is increasing with an angular acceleration of 25 rad/s^2. Knowing that the y-component of the velocity of corner D is negative at this instant in time, determine the velocity and acceleration of corner E.

STRATEGY: You are interested in determining the velocity and acceleration of a point on a body undergoing fixed-axis rotation, so use rigid body kinematics.

(continued)

MODELING and ANALYSIS: Model the assembly as a rigid body. You can find the velocity and acceleration of E using

$$\mathbf{v}_E = \boldsymbol{\omega} \times \mathbf{r}_{E/B} \tag{1}$$

$$\mathbf{a}_E = \boldsymbol{\alpha} \times \mathbf{r}_{E/B} + \boldsymbol{\omega} \times (\boldsymbol{\omega} \times \mathbf{r}_{E/B}) = \boldsymbol{\alpha} \times \mathbf{r}_{E/B} + \boldsymbol{\omega} \times \mathbf{v}_E \tag{2}$$

To use these equations, you need the angular velocity vector, the angular acceleration vector, and the position vector. The direction of the angular velocity and acceleration vectors are along the axis of rotation. Since the corner D is moving downward and using the right-hand rule, you know $\boldsymbol{\omega}$ is in the direction shown in Fig. 1. Therefore, to write the angular velocity vector, you need a unit vector in this direction. You know that

$$\mathbf{AB} = (0.24 \text{ m})\mathbf{i} + (0.07 \text{ m})\mathbf{j}$$

so the unit vector from A to B is

$$\boldsymbol{\lambda}_{AB} = \frac{(0.24 \text{ m})\mathbf{i} + (0.07 \text{ m})\mathbf{j}}{\sqrt{(0.24 \text{ m})^2 + (0.07 \text{ m})^2}} = 0.960\mathbf{i} + 0.280\mathbf{j}$$

Thus, the angular velocity and angular acceleration are

$$\boldsymbol{\omega} = \omega\boldsymbol{\lambda}_{AB} = (5 \text{ rad/s})(0.960\mathbf{i} + 0.280\mathbf{j}) = (4.80 \text{ rad/s})\mathbf{i} + (1.40 \text{ rad/s})\mathbf{j}$$

$$\boldsymbol{\alpha} = \alpha\boldsymbol{\lambda}_{AB} = (25 \text{ rad/s})(0.960\mathbf{i} + 0.280\mathbf{j}) = (24.0 \text{ rad/s}^2)\mathbf{i} + (7.00 \text{ rad/s}^2)\mathbf{j}$$

The position vector of E with respect to B is

$$\mathbf{r}_{E/B} = (-0.20 \text{ m})\mathbf{j} + (0.10 \text{ m})\mathbf{k}$$

Substituting these expressions into Eqs. (1) and (2) gives

$$\mathbf{v}_E = \boldsymbol{\omega} \times \mathbf{r}_{E/B} = \begin{vmatrix} \mathbf{i} & \mathbf{j} & \mathbf{k} \\ 4.80 & 1.40 & 0 \\ 0 & -0.20 & 0.10 \end{vmatrix} = 0.140\mathbf{i} - 0.480\mathbf{j} - 0.960\mathbf{k}$$

$$\mathbf{v}_E = (0.140 \text{ m/s})\mathbf{i} - (0.480 \text{ m/s})\mathbf{j} - (0.960 \text{ m/s})\mathbf{k} \blacktriangleleft$$

$$\mathbf{a}_E = \boldsymbol{\alpha} \times \mathbf{r}_{E/B} + \boldsymbol{\omega} \times \mathbf{v}_E = \begin{vmatrix} \mathbf{i} & \mathbf{j} & \mathbf{k} \\ 24.0 & 7.00 & 0 \\ 0 & -0.20 & 0.10 \end{vmatrix}$$

$$+ \begin{vmatrix} \mathbf{i} & \mathbf{j} & \mathbf{k} \\ 4.80 & 1.40 & 0 \\ 0.140 & -0.480 & -0.960 \end{vmatrix}$$

$$= 0.70\mathbf{i} - 2.40\mathbf{j} - 4.80\mathbf{k} - 1.344\mathbf{i} + 4.608\mathbf{j} + (-2.304 - 0.196)\mathbf{k}$$

$$\mathbf{a}_E = -(0.644 \text{ m/s}^2)\mathbf{i} + (2.21 \text{ m/s}^2)\mathbf{j} - (7.30 \text{ m/s}^2)\mathbf{k} \blacktriangleleft$$

REFLECT and THINK: The first term of Eq. (2) represents the tangential acceleration of point E. The second term of Eq. (2) represents the normal acceleration of point E and points toward the bar AB. Note that you could have chosen any point along the axis of rotation to define your position vector.

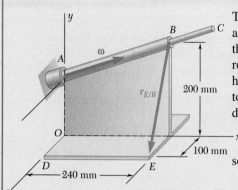

Fig. 1 Direction of the angular velocity and the position vector to point E.

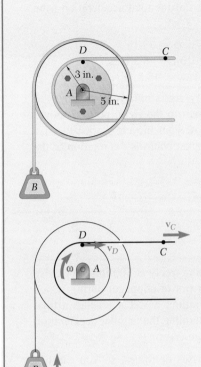

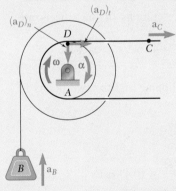

Fig. 1 The velocity of two point on an inextensible cable are equal.

Fig. 2 Acceleration of B, C, and D.

Sample Problem 15.3

Load B is connected to a double pulley by one of the two inextensible cables shown. The motion of the pulley is controlled by cable C, which has a constant acceleration of 9 in./s^2 and an initial velocity of 12 in./s, both directed to the right. Determine (a) the number of revolutions executed by the pulley in 2 s, (b) the velocity and change in position of the load B after 2 s, and (c) the acceleration of point D on the rim of the inner pulley at $t = 0$.

STRATEGY: This is a case of uniformly accelerated rotation, so you can use the kinematic relationships between angular acceleration, angular velocity, angular position, and time. You also need to use the kinematic relationships for the velocity and acceleration of a point on an object undergoing fixed axis rotation.

MODELING and ANALYSIS:

a. Motion of Pulley. You can model the pulley as a rigid body rotating about a fixed axis A. Since the cable is inextensible, the velocity of point D is equal to the velocity of point C (Fig. 1), and the tangential component of the acceleration of D is equal to the acceleration of C (Fig. 2).

$$(\mathbf{v}_D)_0 = (\mathbf{v}_C)_0 = 12 \text{ in./s} \rightarrow \qquad (\mathbf{a}_D)_t = \mathbf{a}_C = 9 \text{ in./s}^2 \rightarrow$$

The distance from D to the center of the pulley is 3 in., so you have

$$\begin{aligned} (v_D)_0 &= r\omega_0 & 12 \text{ in./s} &= (3 \text{ in.})\omega_0 & \omega_0 &= 4 \text{ rad/s} \downarrow \\ (a_D)_t &= r\alpha & 9 \text{ in./s}^2 &= (3 \text{ in.})\alpha & \boldsymbol{\alpha} &= 3 \text{ rad/s}^2 \downarrow \end{aligned}$$

Using the equations of uniformly accelerated motion, for $t = 2$ s you obtain

$$\omega = \omega_0 + \alpha t = 4 \text{ rad/s} + (3 \text{ rad/s}^2)(2 \text{ s}) = 10 \text{ rad/s}$$
$$\boldsymbol{\omega} = 10 \text{ rad/s} \downarrow$$
$$\theta = \omega_0 t + \tfrac{1}{2}\alpha t^2 = (4 \text{ rad/s})(2 \text{ s}) + \tfrac{1}{2}(3 \text{ rad/s}^2)(2 \text{ s})^2 = 14 \text{ rad}$$
$$\theta = 14 \text{ rad} \downarrow$$

$$\text{Number of revolutions} = (14 \text{ rad})\left(\frac{1 \text{ rev}}{2\pi \text{ rad}}\right) = 2.23 \text{ rev} \quad \blacktriangleleft$$

b. Motion of Load B. The motion of load B is the same as a point on the outer rim of the double pulley. Using $r = 5$ in., you have

$$v_B = r\omega = (5 \text{ in.})(10 \text{ rad/s}) = 50 \text{ in./s} \qquad \mathbf{v}_B = 50 \text{ in./s} \uparrow \quad \blacktriangleleft$$
$$\Delta y_B = r\theta = (5 \text{ in.})(14 \text{ rad}) = 70 \text{ in.} \qquad \Delta y_B = 70 \text{ in. upward} \quad \blacktriangleleft$$

c. Acceleration of Point D at $t = 0$. The acceleration of point D has a tangential and a normal component (Fig. 2). The tangential component of the acceleration is

$$(\mathbf{a}_D)_t = \mathbf{a}_C = 9 \text{ in./s}^2 \rightarrow$$

Since, at $t = 0$, $\omega_0 = 4$ rad/s, the normal component of the acceleration is

$$(a_D)_n = r_D \omega_0^2 = (3 \text{ in.})(4 \text{ rad/s})^2 = 48 \text{ in./s}^2 \qquad (\mathbf{a}_D)_n = 48 \text{ in./s}^2 \downarrow$$

(continued)

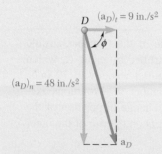

Fig. 3 Vector triangle for resolving the acceleration vector into a magnitude and direction.

You can obtain the magnitude and direction of the total acceleration from Fig. 3.

$$\tan \phi = (48 \text{ in./s}^2)/(9 \text{ in./s}^2) \qquad \phi = 79.4°$$
$$a_D \sin 79.4° = 48 \text{ in./s}^2 \qquad a_D = 48.8 \text{ in./s}^2$$
$$\mathbf{a}_D = 48.8 \text{ in./s}^2 \searrow 79.4° \blacktriangleleft$$

REFLECT and THINK: A double pulley acts similarly to a system of gears; for every 3 inches that point C moves to the right, point B moves 5 inches upward. This is also similar to how your bicycle works; the size ratio of the front chainring to the rear sprocket controls the rotation of the rear tire.

Sample Problem 15.4

Two friction wheels A and B are both rotating freely at 300 rpm clockwise when they are brought into contact. After 6 s of slippage, during which each wheel has a constant angular acceleration, wheel A reaches a final angular velocity of 60 rpm clockwise. Determine the angular acceleration of each wheel during the period of slippage.

STRATEGY: You are not given any masses or forces, so you can use kinematics to solve this problem.

MODELING and ANALYSIS: Model each wheel as a rigid body.

Initial Data. The initial angular velocities of the wheels are $(\omega_A)_0 = (\omega_B)_0 = 300 \text{ rpm} = 31.42 \text{ rad/s}$, both clockwise. After 6 s of slippage, the final angular velocity of A is $\omega_A = 60 \text{ rpm} = 6.28 \text{ rad/s}$ clockwise.

Wheel A. You are told the angular accelerations of the wheels are constant, so

$$\omega_A = (\omega_A)_0 + \alpha_A t: \qquad 6.28 \text{ rad/s} = 31.42 \text{ rad/s} + \alpha_A (6 \text{ s})$$
$$\alpha_A = -4.19 \text{ rad/s}^2 \qquad \boldsymbol{\alpha}_A = 4.19 \text{ rad/s}^2 \curvearrowright \blacktriangleleft$$

Wheel B. At $t = 6$ s, the wheels stop slipping and the two points in contact have the same velocity (Fig. 1). Thus,

$$r_A \omega_A = r_B \omega_B$$

so

$$\omega_B = \frac{r_A \omega_A}{r_B} = \frac{(125 \text{ mm})(6.28 \text{ rad/s})}{(75 \text{ mm})} = 10.47 \text{ rad/s} \curvearrowright$$

The angular acceleration of B is constant, so

$$\omega_B = (\omega_B)_0 + \alpha_B t: \qquad -10.47 \text{ rad/s} = 31.42 \text{ rad/s} + \alpha_B (6 \text{ s})$$
$$\alpha_B = -6.98 \text{ rad/s}^2$$

$$\boldsymbol{\alpha}_B = 6.98 \text{ rad/s}^2 \curvearrowright \blacktriangleleft$$

REFLECT and THINK: The initial angular velocity of B is clockwise, and its final angular velocity is counterclockwise. There must be some time when this wheel has an angular velocity of zero and changes direction from rotating clockwise to rotating counterclockwise.

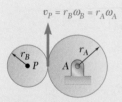

$v_P = r_B \omega_B = r_A \omega_A$

Fig. 1 The wheels will stop slipping when the velocities of the points of contact are equal.

SOLVING PROBLEMS
ON YOUR OWN

In this section, we began the study of the motion of rigid bodies by considering two particular types of motion: **translation** and **rotation about a fixed axis**.

1. Rigid body in translation. At any given instant, all the points of a rigid body in translation have the *same velocity* and the *same acceleration* (Fig. 15.7).

2. Rigid body rotating about a fixed axis. The position of a rigid body rotating about a fixed axis is defined at any given instant by the **angular position** θ, which is usually measured in radians. Selecting the unit vector **k** along the fixed axis in such a way that the rotation of the body appears counterclockwise as seen from the tip of **k**, we define the **angular velocity** ω and the **angular acceleration** α of the body as

$$\omega = \dot{\theta}\mathbf{k} \qquad \alpha = \ddot{\theta}\mathbf{k} \qquad\qquad (15.6, 15.9)$$

In solving problems, keep in mind that the vectors ω and α are both directed along the fixed axis of rotation and that their sense can be obtained by the right-hand rule [Sample Prob. 15.2].

    a. The velocity of a point *P* of a body rotating about a fixed axis is

$$\mathbf{v} = \omega \times \mathbf{r} \qquad\qquad (15.5)$$

where ω is the angular velocity of the body and **r** is the position vector drawn from any point on the axis of rotation to point *P* (Fig. 15.9).

    b. The acceleration of point *P* of a body rotating about a fixed axis is

$$\mathbf{a} = \alpha \times \mathbf{r} + \omega \times (\omega \times \mathbf{r}) \qquad\qquad (15.8)$$

Since vector products are not commutative, *be sure to write the vectors in the order shown* when using either of the above two equations.

3. Rotation of a representative slab. In many problems, you will be able to reduce the analysis of the rotation of a three-dimensional body about a fixed axis to the case of the rotation of a representative slab in a plane perpendicular to the fixed axis. The *z* axis should be directed along the axis of rotation and point out of the page. Thus, the representative slab rotates in the *xy* plane about the origin *O* of the coordinate system (Fig. 15.10).

To solve problems of this type, you should do the following steps.

    a. Draw a diagram of the representative slab showing its dimensions, its angular velocity and angular acceleration, and the vectors representing the velocities and accelerations of the points of the slab.

b. Relate the rotation of the slab and the motion of points of the slab by writing

$$v = r\omega \tag{15.10'}$$

$$a_t = r\alpha \qquad a_n = r\omega^2 \tag{15.11'}$$

Remember that the velocity $\mathbf{v}$ and the component $\mathbf{a}_t$ of the acceleration of a point P of the slab are tangent to the circular path described by P [Sample Probs. 15.3 and 15.4]. You can find the directions of $\mathbf{v}$ and $\mathbf{a}_t$ by rotating the position vector $\mathbf{r}$ through $90°$ in the sense indicated by $\boldsymbol{\omega}$ and $\boldsymbol{\alpha}$, respectively. The normal component $\mathbf{a}_n$ of the acceleration of P is always directed toward the axis of rotation.

4. Equations defining the rotation of a rigid body. Note the similarity between the equations defining the rotation of a rigid body about a fixed axis [Eqs. (15.12) through (15.16)] and those in Chap. 11 defining the rectilinear motion of a particle [Eqs. (11.1) through (11.8)]. All you have to do to obtain the new set of equations is to substitute θ, ω, and α for x, v, and a, respectively, in the equations of Chap. 11 [Sample Prob. 15.1].

Problems

CONCEPT QUESTIONS

15.CQ1 A rectangular plate swings from arms of equal length as shown. What is the magnitude of the angular velocity of the plate?
- **a.** 0 rad/s
- **b.** 1 rad/s
- **c.** 2 rad/s
- **d.** 3 rad/s
- **e.** Need to know the location of the center of gravity.

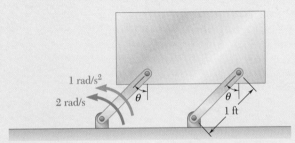

Fig. P15.CQ1

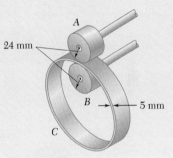

Fig. P15.CQ2

15.CQ2 Knowing that wheel A rotates with a constant angular velocity and that no slipping occurs between ring C and wheel A and wheel B, which of the following statements concerning the angular speeds of the three objects is true?
- **a.** $\omega_a = \omega_b$
- **b.** $\omega_a > \omega_b$
- **c.** $\omega_a < \omega_b$
- **d.** $\omega_a = \omega_c$
- **e.** The contact points between A and C have the same acceleration.

END-OF-SECTION PROBLEMS

15.1 The brake drum is attached to a larger flywheel that is not shown. The motion of the brake drum is defined by the relation $\theta = 36t - 1.6t^2$, where θ is expressed in radians and t in seconds. Determine (*a*) the angular velocity at $t = 2$ s, (*b*) the number of revolutions executed by the brake drum before coming to rest.

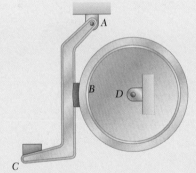

Fig. P15.1

15.2 The motion of an oscillating flywheel is defined by the relation $\theta = \theta_0 e^{-3\pi t} \cos 4\pi t$, where θ is expressed in radians and t in seconds. Knowing that $\theta_0 = 0.5$ rad, determine the angular coordinate, the angular velocity, and the angular acceleration of the flywheel when (*a*) $t = 0$, (*b*) $t = 0.125$ s.

15.3 The motion of an oscillating flywheel is defined by the relation $\theta = \theta_0 e^{-7\pi t/6} \sin 4\pi t$, where θ is expressed in radians and t in seconds. Knowing that $\theta_0 = 0.4$ rad, determine the angular coordinate, the angular velocity, and the angular acceleration of the flywheel when (*a*) $t = 0.125$ s, (*b*) $t = \infty$.

Fig. P15.2 and P15.3

15.4 The rotor of a gas turbine is rotating at a speed of 6900 rpm when the turbine is shut down. It is observed that 4 min is required for the rotor to coast to rest. Assuming uniformly accelerated motion, determine (*a*) the angular acceleration, (*b*) the number of revolutions that the rotor executes before coming to rest.

15.5 A small grinding wheel is attached to the shaft of an electric motor which has a rated speed of 3600 rpm. When the power is turned on, the unit reaches its rated speed in 5 s, and when the power is turned off, the unit coasts to rest in 70 s. Assuming uniformly accelerated motion, determine the number of revolutions that the motor executes (*a*) in reaching its rated speed, (*b*) in coasting to rest.

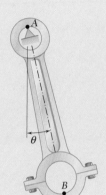

Fig. P15.6

Fig. P15.5

15.6 A connecting rod is supported by a knife-edge at point *A*. For small oscillations the angular acceleration of the connecting rod is governed by the relation $\alpha = -6\theta$ where α is expressed in rad/s^2 and θ in radians. Knowing that the connecting rod is released from rest when $\theta = 20°$, determine (*a*) the maximum angular velocity, (*b*) the angular position when $t = 2$ s.

15.7 When studying whiplash resulting from rear-end collisions, the rotation of the head is of primary interest. An impact test was performed, and it was found that the angular acceleration of the head is defined by the relation $\alpha = 700 \cos \theta + 70 \sin \theta$, where α is expressed in rad/s^2 and θ in radians. Knowing that the head is initially at rest, determine the angular velocity of the head when $\theta = 30°$.

Fig. *P15.7*

15.8 The angular acceleration of an oscillating disk is defined by the relation $\alpha = -k\theta$, where alpha is expressed in rad/s^2 and theta is expressed in radians. Determine (*a*) the value of *k* for which $\omega = 12$ rad/s when $\theta = 0$ and $\theta = 6$ rad when $\omega = 0$, (*b*) the angular velocity of the disk when $\theta = 3$ rad.

15.9 The angular acceleration of a shaft is defined by the relation $\alpha = -0.5\omega$, where α is expressed in rad/s^2 and ω in rad/s. Knowing that at $t = 0$ the angular velocity of the shaft is 30 rad/s, determine (*a*) the number of revolutions the shaft will execute before coming to rest, (*b*) the time required for the shaft to come to rest, (*c*) the time required for the angular velocity of the shaft to reduce to 2 percent of its initial value.

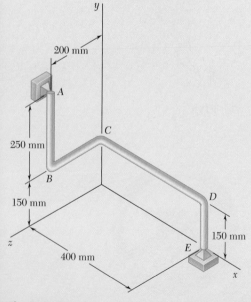

Fig. P15.10

15.10 The bent rod *ABCDE* rotates about a line joining points *A* and *E* with a constant angular velocity of 9 rad/s. Knowing that the rotation is clockwise as viewed from *E*, determine the velocity and acceleration of corner *C*.

15.11 In Prob. 15.10, determine the velocity and acceleration of corner *B*, assuming that the angular velocity is 9 rad/s and increases at the rate of 45 rad/s^2.

15.12 The rectangular block shown rotates about the diagonal OA with a constant angular velocity of 6.76 rad/s. Knowing that the rotation is counterclockwise as viewed from A, determine the velocity and acceleration of point B at the instant shown.

15.13 The rectangular block shown rotates about the diagonal OA with an angular velocity of 3.38 rad/s that is decreasing at the rate of 5.07 rad/s^2. Knowing that the rotation is counterclockwise as viewed from A, determine the velocity and acceleration of point B at the instant shown.

15.14 A circular plate of 120-mm radius is supported by two bearings A and B as shown. The plate rotates about the rod joining A and B with a constant angular velocity of 26 rad/s. Knowing that, at the instant considered, the velocity of point C is directed to the right, determine the velocity and acceleration of point E.

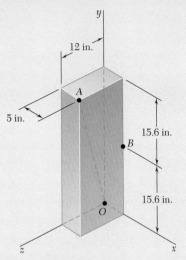

Fig. P15.12 and P15.13

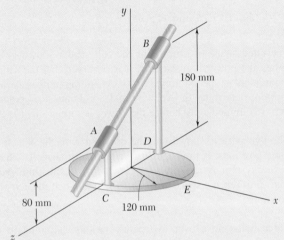

Fig. P15.14

15.15 In Prob. 15.14, determine the velocity and acceleration of point E, assuming that the angular velocity is 26 rad/s and increases at the rate of 65 rad/s^2.

15.16 The earth makes one complete revolution around the sun in 365.24 days. Assuming that the orbit of the earth is circular and has a radius of 93,000,000 mi, determine the velocity and acceleration of the earth.

15.17 The earth makes one complete revolution on its axis in 23 h 56 min. Knowing that the mean radius of the earth is 3960 mi, determine the linear velocity and acceleration of a point on the surface of the earth (*a*) at the equator, (*b*) at Philadelphia, latitude 40° north, (*c*) at the North Pole.

15.18 A series of small machine components being moved by a conveyor belt pass over a 120-mm-radius idler pulley. At the instant shown, the velocity of point A is 300 mm/s to the left and its acceleration is 180 mm/s^2 to the right. Determine (*a*) the angular velocity and angular acceleration of the idler pulley, (*b*) the total acceleration of the machine component at B.

15.19 A series of small machine components being moved by a conveyor belt pass over a 120-mm-radius idler pulley. At the instant shown, the angular velocity of the idler pulley is 4 rad/s clockwise. Determine the angular acceleration of the pulley for which the magnitude of the total acceleration of the machine component at B is 2400 mm/s^2.

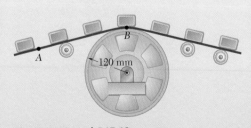

Fig. P15.18 and P15.19

15.20 The belt sander shown is initially at rest. If the driving drum B has a constant angular acceleration of 120 rad/s² counterclockwise, determine the magnitude of the acceleration of the belt at point C when (a) $t = 0.5$ s, (b) $t = 2$ s.

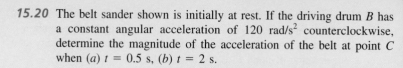
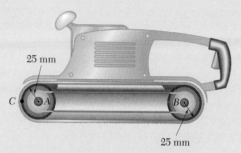

Fig. *P15.20* and *P15.21*

15.21 The rated speed of drum B of the belt sander shown is 2400 rpm. When the power is turned off, it is observed that the sander coasts from its rated speed to rest in 10 s. Assuming uniformly decelerated motion, determine the velocity and acceleration of point C of the belt, (a) immediately before the power is turned off, (b) 9 s later.

15.22 The two pulleys shown may be operated with the V belt in any of three positions. If the angular acceleration of shaft A is 6 rad/s² and if the system is initially at rest, determine the time required for shaft B to reach a speed of 400 rpm with the belt in each of the three positions.

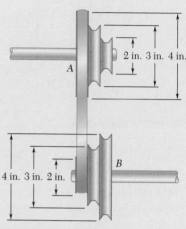

Fig. **P15.22**

15.23 Three belts move over two pulleys without slipping in the speed reduction system shown. At the instant shown, the velocity of point A on the input belt is 2 ft/s to the right, decreasing at the rate of 6 ft/s². Determine, at this instant, (a) the velocity and acceleration of point C on the output belt, (b) the acceleration of point B on the output pulley.

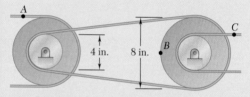

Fig. **P15.23**

15.24 A gear reduction system consists of three gears A, B, and C. Knowing that gear A rotates clockwise with a constant angular velocity $\omega_A = 600$ rpm, determine (a) the angular velocities of gears B and C, (b) the accelerations of the points on gears B and C which are in contact.

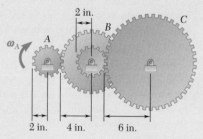

Fig. **P15.24**

15.25 A belt is pulled to the right between cylinders A and B. Knowing that the speed of the belt is a constant 5 ft/s and no slippage occurs, determine (a) the angular velocities of A and B, (b) the accelerations of the points which are in contact with the belt.

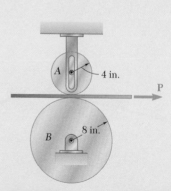

Fig. **P15.25**

15.26 Ring C has an inside radius of 55 mm and an outside radius of 60 mm and is positioned between two wheels A and B, each of 24-mm outside radius. Knowing that wheel A rotates with a constant angular velocity of 300 rpm and that no slipping occurs, determine (a) the angular velocity of ring C and of wheel B, (b) the acceleration of the points on A and B that are in contact with C.

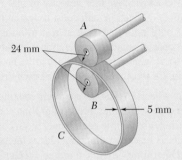

Fig. P15.26

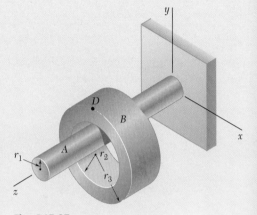

Fig. P15.27

15.27 Ring B has an inside radius r_2 and hangs from the horizontal shaft A as shown. Shaft A rotates with a constant angular velocity of 25 rad/s and no slipping occurs. Knowing that $r_1 = 12$ mm, $r_2 = 30$ mm, and $r_3 = 40$ mm, determine (a) the angular velocity of ring B, (b) the accelerations of the points of shaft A and ring B which are in contact, (c) the magnitude of the acceleration of point D.

15.28 A plastic film moves over two drums. During a 4-s interval the speed of the tape is increased uniformly from $v_0 = 2$ ft/s to $v_1 = 4$ ft/s. Knowing that the tape does not slip on the drums, determine (a) the angular acceleration of drum B, (b) the number of revolutions executed by drum B during the 4-s interval.

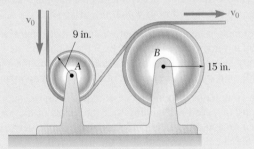

Fig. P15.28

15.29 Cylinder A is moving downward with a velocity of 3 m/s when the brake is suddenly applied to the drum. Knowing that the cylinder moves 6 m downward before coming to rest and assuming uniformly accelerated motion, determine (a) the angular acceleration of the drum, (b) the time required for the cylinder to come to rest.

15.30 The system shown is held at rest by the brake-and-drum system shown. After the brake is partially released at $t = 0$ it is observed that the cylinder moves 5 m in 4.5 s. Assuming uniformly accelerated motion, determine (a) the angular acceleration of the drum, (b) the angular velocity of the drum at $t = 3.5$ s.

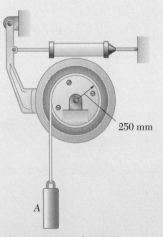

Fig. P15.29 and P15.30

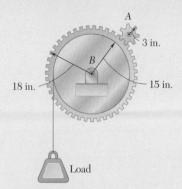

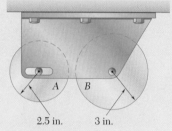

Load

Fig. P15.31

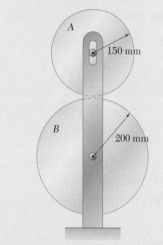

2.5 in.   3 in.

Fig. P15.32 and P15.33

A

150 mm

B

200 mm

Fig. *P15.34* and *P15.35*

15.31 A load is to be raised 20 ft by the hoisting system shown. Assuming gear A is initially at rest, accelerates uniformly to a speed of 120 rpm in 5 s, and then maintains a constant speed of 120 rpm, determine (*a*) the number of revolutions executed by gear A in raising the load, (*b*) the time required to raise the load.

15.32 A simple friction drive consists of two disks A and B. Initially, disk B has a clockwise angular velocity of 500 rpm, and disk A is at rest. It is known that disk B will coast to rest in 60 s. However, rather than waiting until both disks are at rest to bring them together, disk A is given a constant angular acceleration of 3 rad/s^2 counterclockwise. Determine (*a*) at what time the disks can be brought together if they are not to slip, (*b*) the angular velocity of each disk as contact is made.

15.33 Two friction wheels A and B are both rotating freely at 300 rpm counterclockwise when they are brought into contact. After 12 s of slippage, during which time each wheel has a constant angular acceleration, wheel B reaches a final angular velocity of 75 rpm counterclockwise. Determine (*a*) the angular acceleration of each wheel during the period of slippage, (*b*) the time at which the angular velocity of wheel A is equal to zero.

15.34 Two friction disks A and B are to be brought into contact without slipping when the angular velocity of disk A is 240 rpm counterclockwise. Disk A starts from rest at time $t = 0$ and is given a constant angular acceleration with a magnitude α. Disk B starts from rest at time $t = 2$ s and is given a constant clockwise angular acceleration, also with a magnitude α. Determine (*a*) the required angular acceleration magnitude α, (*b*) the time at which the contact occurs.

15.35 Two friction disks A and B are brought into contact when the angular velocity of disk A is 240 rpm counterclockwise and disk B is at rest. A period of slipping follows and disk B makes two revolutions before reaching its final angular velocity. Assuming that the angular acceleration of each disk is constant and inversely proportional to the cube of its radius, determine (*a*) the angular acceleration of each disk, (*b*) the time during which the disks slip.

*15.36 Steel tape is being wound onto a spool that rotates with a constant angular velocity ω_0. Denoting by r the radius of the spool and tape at any given time and by b the thickness of the tape, derive an expression for the acceleration of the tape as it approaches the spool.

*15.37 In a continuous printing process, paper is drawn into the presses at a constant speed v. Denoting by r the radius of the paper roll at any given time and by b the thickness of the paper, derive an expression for the angular acceleration of the paper roll.

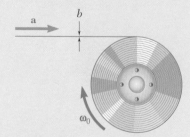

Fig. P15.36

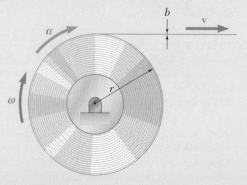

Fig. P15.37

15.2 GENERAL PLANE MOTION: VELOCITY

As indicated in the chapter introduction, general plane motion describes a plane motion that is neither a pure translation nor a pure rotation. As you will presently see, however, **a general plane motion can always be considered as the sum of a translation and a rotation**.

15.2A Analyzing General Plane Motion

As an example of general plane motion, consider a wheel rolling on a straight track (Fig. 15.12). Over some interval of time, two given points A and B will have moved, respectively, from A_1 to A_2 and from B_1 to B_2. However, we could obtain the same result through a translation that would bring A_1 and B_1 into A_2 and B_1' (the line AB remaining vertical), followed by a rotation about A, bringing B into B_2. The original rolling motion differs from the combination of translation and rotation when these motions are taken in succession, but we can duplicate the original motion exactly using a combination of simultaneous translation and rotation.

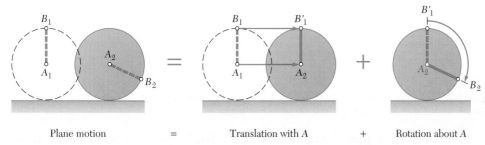

Plane motion   =   Translation with A +   Rotation about A

Fig. 15.12 The general plane motion of a rolling wheel can be analyzed as a combination of translation plus a fixed-axis rotation.

Another example of plane motion is shown in Fig. 15.13, which represents a rod whose ends slide along a horizontal and a vertical track. We can replace this motion using a horizontal translation and a rotation about A (Fig. 15.13a) or using a vertical translation and a rotation about B (Fig. 15.13b).

In the general case of plane motion, we consider a small displacement that brings two particles A and B of a representative rigid body, respectively, from A_1 and B_1 into A_2 and B_2 (Fig. 15.14). We can divide this displacement into two parts: in one, the particles move into A_2 and B_1' while the line AB maintains the same direction; in the other, B moves into B_2 while A remains fixed. The first part of the motion is clearly a translation, and the second part is clearly a rotation about A.

Recall from Sec. 11.4D the definition of the relative motion of a particle with respect to a moving frame of reference—as opposed to its absolute motion with respect to a fixed frame of reference. With that definition in mind, we can restate our results: Given two particles A and B of a rigid body in plane motion, the relative motion of B with respect

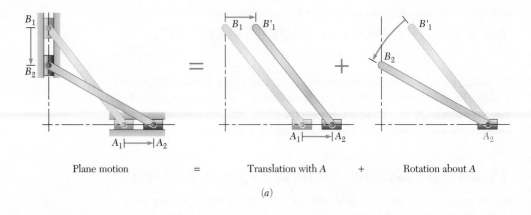

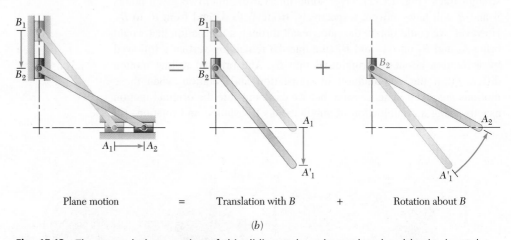

Fig. 15.13 The general plane motion of this sliding rod can be analyzed as (*a*) a horizontal translation plus a fixed-axis rotation about *A* or (*b*) a vertical translation and a fixed-axis rotation about *B*. The results are the same either way.

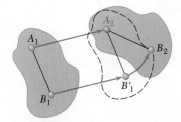

Fig. 15.14 General plane motion is a combination of a translation plus a fixed-axis rotation. To an observer moving with *A* but not rotating, particle *B* appears to travel in a circle centered at *A*.

to a frame attached to *A* and of fixed orientation is a rotation. To an observer moving with *A* but not rotating, particle *B* appears to describe an arc of a circle centered at *A*.

15.2B Absolute and Relative Velocity in Plane Motion

We have just seen that any plane motion of a rigid body can be replaced by a translation of an arbitrary reference point *A* and a simultaneous rotation about *A*. We can obtain the absolute velocity $\mathbf{v}_B$ of a particle *B* of the rigid body from the relative velocity formula derived in Sec. 11.4D, as

$$\mathbf{v}_B = \mathbf{v}_A + \mathbf{v}_{B/A} \qquad (15.17)$$

where the right-hand side represents a vector sum. The velocity $\mathbf{v}_A$ corresponds to the translation of the rigid body with *A*, whereas the relative velocity $\mathbf{v}_{B/A}$ is associated with the rotation of the rigid body about *A* and is measured with respect to axes centered at *A* and of fixed orientation (Fig. 15.15). Denoting the position vector of *B* relative to *A* by $\mathbf{r}_{B/A}$ (which points from *A* to *B*) and the angular velocity of the rigid body with respect to axes of fixed orientation by $\omega\mathbf{k}$, we have from Eqs. (15.10) and (15.10′)

$$\mathbf{v}_{B/A} = \omega\mathbf{k} \times \mathbf{r}_{B/A} \qquad v_{B/A} = r\omega \qquad (15.18)$$

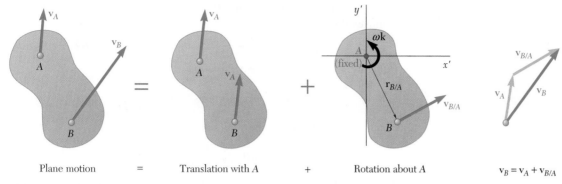

Fig. 15.15 A pictorial representation of the vector equation relating the velocity of two points on a rigid body undergoing general plane motion.

where r is the distance from A to B. Substituting for $\mathbf{v}_{B/A}$ from Eq. (15.18) into Eq. (15.17), we also have

Relative velocity for two points on a rigid body

Photo 15.4 Planetary gear systems are used in applications requiring a large reduction ratio and a high torque-to-weight ratio. The small gears undergo general plane motion.

$$\boxed{\mathbf{v}_B = \mathbf{v}_A + \omega \mathbf{k} \times \mathbf{r}_{B/A}} \qquad (15.17')$$

As an example, let us again consider rod AB of Fig. 15.13. Assuming that we know the velocity $\mathbf{v}_A$ of end A, we propose to find the velocity $\mathbf{v}_B$ of end B and the angular velocity $\boldsymbol{\omega}$ of the rod in terms of the velocity $\mathbf{v}_A$, the length l, and the angle θ. Choosing A as a reference point, the given motion is equivalent to a translation of A and a simultaneous rotation about A (Fig. 15.16). The absolute velocity of B therefore must be equal to the vector sum

$$\mathbf{v}_B = \mathbf{v}_A + \mathbf{v}_{B/A} \qquad (15.17)$$

Note that although we know the direction of $\mathbf{v}_{B/A}$, its magnitude $l\omega$ is unknown. However, this is compensated for by the fact that the direction of $\mathbf{v}_B$ is known. We can therefore complete the vector diagram of Fig. 15.16. Solving for the magnitudes v_B and ω, we obtain

$$v_B = v_A \tan \theta \qquad \omega = \frac{v_{B/A}}{l} = \frac{v_A}{l \cos \theta} \qquad (15.19)$$

Alternatively, we can also solve this problem by using the vector relationship in Eq. (15.17′). Recognizing that point A is constrained to

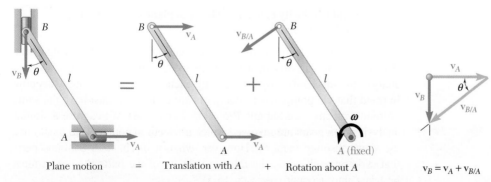

Fig. 15.16 Pictorial representation of Eq. (15.17) for a sliding rod. The relative velocity $\mathbf{v}_{B/A}$ is perpendicular to the line connecting A and B.

move only in the x direction and B moves only in the y direction (assume it moves down), we can write

$$-v_B\mathbf{j} = v_A\mathbf{i} + \omega\mathbf{k} \times(-l \sin \theta\mathbf{i} + l \cos \theta\mathbf{j}) = (v_A - \omega l \cos \theta)\mathbf{i} - \omega l \sin \theta\mathbf{j}$$

Equating components in the x direction, we obtain

$$v_A - \omega l \cos \theta = 0 \qquad \omega = \frac{v_A}{l \cos \theta}$$

Equating components in the y direction, we obtain

$$v_B = \omega l \sin \theta = \left(\frac{v_A}{l \cos \theta}\right)l \sin \theta = v_A \tan \theta$$

These are the same results as we obtained in Eq. 15.19. We obtain the same result by using B as a point of reference. Resolving the given motion into a translation of B and a simultaneous rotation about B (Fig. 15.17), we have the equation

$$\mathbf{v}_A = \mathbf{v}_B + \mathbf{v}_{A/B} = \mathbf{v}_B + \omega\mathbf{k} \times \mathbf{r}_{A/B} \qquad\qquad \textbf{(15.20)}$$

which is represented graphically in Fig. 15.17. Note that $\mathbf{v}_{A/B}$ and $\mathbf{v}_{B/A}$ have the same magnitude $l\omega$ but opposite sense. The sense of the relative velocity depends, therefore, upon the point of reference that we have selected and should be carefully ascertained from the appropriate diagram (Fig. 15.16 or 15.17).

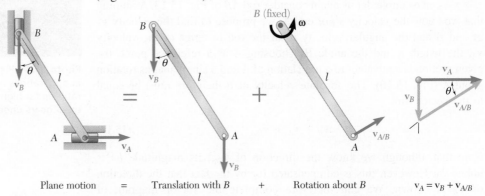

Plane motion   =   Translation with B +   Rotation about B $v_A = v_B + v_{A/B}$

Fig. 15.17   Pictorial representation of Eq. (15.20) for a sliding rod. The relative velocity $\mathbf{v}_{A/B}$ is perpendicular to the line connecting A and B.

Finally, observe that the angular velocity $\boldsymbol{\omega}$ of the rod in its rotation about B is the same as in its rotation about A. It is measured in both cases by the rate of change of the angle θ. This result is quite general; you should therefore bear in mind that

> **The angular velocity $\boldsymbol{\omega}$ of a rigid body in plane motion is independent of the reference point.**

Most mechanisms consist not of one but of *several* moving parts. When the various parts of a mechanism are connected by pins, we can analyze the mechanism by considering each part as a rigid body, keeping in mind that the points where two parts are connected must have the same absolute velocity (see Sample Probs. 15.7 and 15.8). We can use a similar analysis when gears are involved, since the teeth in contact also must have the same absolute velocity. However, when a mechanism contains parts that slide on each other, the relative velocity of the parts in contact must be taken into account (see Sec. 15.5).

Sample Problem 15.5

Collars A and B are pin-connected to bar ABD and can slide along fixed rods. Knowing that at the instant shown the velocity of A is 0.9 m/s to the right, determine (a) the angular velocity of ABD, (b) the velocity of point D.

STRATEGY: Use the kinematic equation that relates the velocity of two points on the same rigid body. Because you know the directions of the velocities of points A and B, choose these two points to relate.

MODELING and ANALYSIS: Model bar ABD as a rigid body. From kinematics you know

$$\mathbf{v}_B = \mathbf{v}_A + \mathbf{v}_{B/A} = \mathbf{v}_A + \boldsymbol{\omega} \times \mathbf{r}_{B/A}$$

Substituting in known values (Fig. 1) and assuming $\boldsymbol{\omega} = \omega\mathbf{k}$ gives you

$$v_B \cos 60°\mathbf{i} + v_B \sin 60°\mathbf{j} = (0.9)\mathbf{i} + $$
$$\omega\mathbf{k} \times [(0.3 \cos 30°)\mathbf{i} + (0.3 \sin 30°)\mathbf{j}]$$
$$0.500 v_B\mathbf{i} + 0.866 v_B\mathbf{j} = (0.9 - 0.15\omega)\mathbf{i} + 0.260\omega\mathbf{j}$$

Equating components,

$$\mathbf{i}: = 0.500 v_B = 0.9 - 0.15\omega$$
$$\mathbf{j}: = 0.866 v_B = 0.260\omega$$

Solving these equations gives you $v_B = 0.900$ m/s and $\omega = 3.00$ rad/s

$$\boldsymbol{\omega} = 3.00 \text{ rad/s} \curvearrowleft \quad \blacktriangleleft$$

Velocity of D. The relationship between the velocities of A and D is

$$\mathbf{v}_D = \mathbf{v}_A + \mathbf{v}_{D/A} = \mathbf{v}_D + \boldsymbol{\omega} \times \mathbf{r}_{D/A}$$

Substituting in values from above gives

$$\mathbf{v}_D = (0.9)\mathbf{i} + 3.00\mathbf{k} \times [(0.6 \cos 30°)\mathbf{i} + (0.6 \sin 30°)\mathbf{j}]$$
$$\mathbf{v}_D = (0.9 - 0.9)\mathbf{i} + 1.559\mathbf{j}$$

$$\mathbf{v}_D = 1.559 \text{ m/s} \uparrow \quad \blacktriangleleft$$

REFLECT and THINK: The velocity of point D is straight up at this instant in time, but as the bar continues to rotate counterclockwise, the direction of the velocity of D will continuously change.

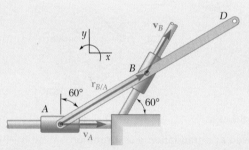

Fig. 1 Position vector and directions of the velocities of A and B.

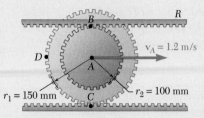

Sample Problem 15.6

The double gear shown rolls on the stationary lower rack; the velocity of its center A is 1.2 m/s directed to the right. Determine (a) the angular velocity of the gear, (b) the velocities of the upper rack R and of point D of the gear.

STRATEGY: The double gear is undergoing general motion, so use rigid body kinematics. Resolve the rolling motion into two component motions: a translation of point A and a rotation about the center A (Fig. 1). In the translation, all points of the gear move with the same velocity $\mathbf{v}_A$. In the rotation, each point P of the gear moves about A with a relative velocity $\mathbf{v}_{P/A} = \omega\mathbf{k} \times \mathbf{r}_{P/A}$, where $\mathbf{r}_{P/A}$ is the position vector of P relative to A.

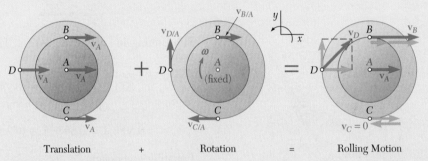

| Translation | + | Rotation | = | Rolling Motion |

Fig. 1 The gear motion can be modeled as a translation plus a rotation.

MODELING and ANALYSIS:

a. Angular Velocity of the Gear. Since the gear rolls on the lower rack, its center A moves through a distance equal to the outer circumference $2\pi r_1$ for each full revolution of the gear. Noting that 1 rev = 2π rad, and that when A moves to the right ($x_A > 0$), the gear rotates clockwise ($\theta < 0$), you have

$$\frac{x_A}{2\pi r_1} = -\frac{\theta}{2\pi} \qquad x_A = -r_1\theta$$

Differentiating with respect to the time t and substituting the known values $v_A = 1.2$ m/s and $r_1 = 150$ mm = 0.150 m, you obtain

$$v_A = -r_1\omega \qquad 1.2 \text{ m/s} = -(0.150 \text{ m})\omega \qquad \omega = -8 \text{ rad/s}$$
$$\boldsymbol{\omega} = \omega\mathbf{k} = -(8 \text{ rad/s})\mathbf{k} \quad \blacktriangleleft$$

where $\mathbf{k}$ is a unit vector pointing out of the page.

b. Velocity of Upper Rack. The velocity of the upper rack is equal to the velocity of point B; you have

$$\mathbf{v}_R = \mathbf{v}_B = \mathbf{v}_A + \mathbf{v}_{B/A} = \mathbf{v}_A + \omega\mathbf{k} \times \mathbf{r}_{B/A}$$
$$= (1.2 \text{ m/s})\mathbf{i} - (8 \text{ rad/s})\mathbf{k} \times (0.100 \text{ m})\mathbf{j}$$
$$= (1.2 \text{ m/s})\mathbf{i} + (0.8 \text{ m/s})\mathbf{i} = (2 \text{ m/s})\mathbf{i}$$
$$\mathbf{v}_R = 2 \text{ m/s} \rightarrow \quad \blacktriangleleft$$

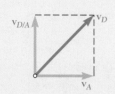

Fig. 2 The two components of the velocity of *D*.

Velocity of Point D. The velocity of point D has two components (Fig. 2):

$$\mathbf{v}_D = \mathbf{v}_A + \mathbf{v}_{D/A} = \mathbf{v}_A + \omega\mathbf{k} \times \mathbf{r}_{D/A}$$
$$= (1.2 \text{ m/s})\mathbf{i} - (8 \text{ rad/s})\mathbf{k} \times (-0.150 \text{ m})\mathbf{i}$$
$$= (1.2 \text{ m/s})\mathbf{i} + (1.2 \text{ m/s})\mathbf{j}$$

$$\mathbf{v}_D = 1.697 \text{ m/s} \measuredangle 45° \quad \blacktriangleleft$$

REFLECT and THINK: The principles involved in this problem are similar to those that you used in Sample Prob. 15.3, but in this problem, point A was free to translate. Point C, since it is in contact with the fixed lower rack, has a velocity of zero. Every point along diameter CAB has a velocity vector directed to the right (Fig. 1) and the magnitude of the velocity increases linearly as the distance from point C increases.

Sample Problem 15.7

In the engine system shown, the crank AB has a constant clockwise angular velocity of 2000 rpm. For the crank position shown, determine (*a*) the angular velocity of the connecting rod BD, (*b*) the velocity of the piston P.

STRATEGY: Connecting rod BD is undergoing general motion, so use rigid-body kinematics. Crank AB is undergoing fixed axis rotation, and piston P is translating. The motion of the piston is the same as the end D of the connecting rod.

MODELING and ANALYSIS:

Motion of Crank AB. The crank AB rotates about point A. Expressing ω_{AB} in rad/s and writing $v_B = r\omega_{AB}$, you have (Fig. 1)

Fig. 1 Crank *AB* is undergoing fixed axis rotation.

$$\omega_{AB} = \left(2000\frac{\text{rev}}{\text{min}}\right)\left(\frac{1 \text{ min}}{60 \text{ s}}\right)\left(\frac{2\pi \text{ rad}}{1 \text{ rev}}\right) = 209.4 \text{ rad/s}$$

$$v_B = (AB)\omega_{AB} = (3 \text{ in.})(209.4 \text{ rad/s}) = 628.3 \text{ in./s}$$
$$\mathbf{v}_B = 628.3 \text{ in./s} \measuredangle 50°$$

Motion of Connecting Rod BD. Consider this as a general plane motion. Using the law of sines, compute the angle β between the connecting rod and the horizontal as

$$\frac{\sin 40°}{8 \text{ in.}} = \frac{\sin \beta}{3 \text{ in.}} \qquad \beta = 13.95°$$

The velocity $\mathbf{v}_D$ of point D where the rod is attached to the piston must be horizontal, while the velocity of point B is equal to the velocity $\mathbf{v}_B$ obtained previously. Expressing the relation between the velocities $\mathbf{v}_D$, $\mathbf{v}_B$, and $\mathbf{v}_{D/B}$, you have

$$\mathbf{v}_D = \mathbf{v}_B + \mathbf{v}_{D/B}$$

(continued)

This equation is shown pictorially in Fig. 2 where the motion of *BD* is resolved into a translation of *B* and a rotation about *B*.

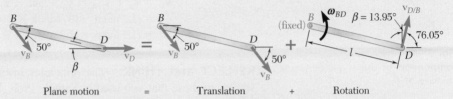

Plane motion   =   Translation   +   Rotation

Fig. 2 The general plane motion of the connecting rod can be modelled as a translation plus a rotation.

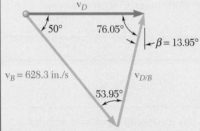

Fig. 3 Vector triangle showing the relationship between the velocities of *B* and *D*.

Draw the vector diagram corresponding to this equation (Fig. 3). Recalling that $\beta = 13.95°$, you can determine the angles of the triangle and write

$$\frac{v_D}{\sin 53.95°} = \frac{v_{D/B}}{\sin 50°} = \frac{628.3 \text{ in./s}}{\sin 76.05°}$$

$$v_{D/B} = 495.9 \text{ in./s} \qquad \mathbf{v}_{D/B} = 495.9 \text{ in./s} \measuredangle 76.05°$$
$$v_D = 523.4 \text{ in./s} = 43.6 \text{ ft/s} \qquad \mathbf{v}_D = 43.6 \text{ ft/s} \rightarrow$$
$$\mathbf{v}_P = \mathbf{v}_D = 43.6 \text{ ft/s} \rightarrow \ \blacktriangleleft$$

Since $v_{D/B} = l\omega_{BD}$, you have

$$495.9 \text{ in./s} = (8 \text{ in.})\omega_{BD} \qquad \omega_{BD} = 62.0 \text{ rad/s} \ \gamma \ \blacktriangleleft$$

REFLECT and THINK: Note that as the crank continues to move clockwise below the center line, the piston changes direction and starts to move to the left. Can you see what happens to the motion of the connecting rod at that point? You can also solve this problem using the vector relationship expressed in Eq. (15.17′); this type of approach is shown in Sample Prob. 15.8.

Sample Problem 15.8

In the position shown, bar *AB* has an angular velocity of 4 rad/s clockwise. Determine the angular velocity of bars *BD* and *DE*.

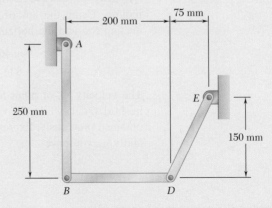

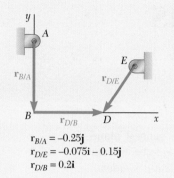

$\mathbf{r}_{B/A} = -0.25\mathbf{j}$
$\mathbf{r}_{D/E} = -0.075\mathbf{i} - 0.15\mathbf{j}$
$\mathbf{r}_{D/B} = 0.2\mathbf{i}$

Fig. 1 Relative position vectors for points B and D.

STRATEGY: The bars *AB* and *DE* are undergoing fixed-axis rotation, whereas bar *BD* is undergoing general plane motion. You will need to use rigid-body kinematics to analyze the motion.

MODELING and ANALYSIS: Model the bars as rigid bodies. The angular velocity of *AB* is given and is equal to $\omega_{AB} = -(4 \text{ rad/s})\mathbf{k}$. You can use vector algebra to relate the velocities of points *B* and *D* on bar *BD* after you find the velocities of *B* and *D* from the connecting bars. Position vectors are defined in Fig. 1.

Bar AB. (Rotation about *A*)

$$\mathbf{v}_B = \omega_{AB} \times \mathbf{r}_{B/A} = (-4\mathbf{k}) \times (-0.25\mathbf{j}) = -(1.00 \text{ m/s})\mathbf{i} \tag{1}$$

Bar ED. (Rotation about *E*) Assuming ω_{DE} is positive, you have

$$\mathbf{v}_D = \omega_{DE}\mathbf{k} \times \mathbf{r}_{D/E} = \omega_{DE}\mathbf{k} \times (-0.075\mathbf{i} - 0.15\mathbf{j}) = 0.15\omega_{DE}\mathbf{i} - 0.075\omega_{DE}\mathbf{j} \tag{2}$$

Bar BD. (Translation with *B* and rotation about *B*.)

$$\mathbf{v}_D = \mathbf{v}_B + \mathbf{v}_{D/B} \tag{3}$$

where you assume ω_{BD} is positive. The relative velocity is

$$\mathbf{v}_{D/B} = \omega_{BD}\mathbf{k} \times \mathbf{r}_{D/B} = \omega_{BD}\mathbf{k} \times 0.2\mathbf{i} = 0.2\omega_{BD}\mathbf{j} \tag{4}$$

Substituting Eqs. (1), (2), and (4) into Eq. (3) gives

$$0.15 \,\omega_{DE}\mathbf{i} - 0.075\omega_{DE}\mathbf{j} = -1.00\mathbf{i} + 0.2\omega_{BD}\mathbf{j}$$

Equating components allows you to solve for the unknown angular velocities:

i: $0.15\omega_{DE} = -1.00$, $\qquad \omega_{DE} = -6.667 \text{ rad/s}$ $\qquad \omega_{DE} = 6.67 \text{ rad/s} \downdownarrows$ ◀

j: $-0.075\omega_{DE} = 0.2\omega_{BD}$ $\qquad \omega_{BD} = \dfrac{-(0.075)(-6.667)}{0.2}$

$$\omega_{BD} = 2.50 \text{ rad/s} \upuparrows \text{ ◀}$$

REFLECT and THINK: The vector algebra approach is very straightforward for problems like this. It makes sense that if *AB* is rotating clockwise, *BD* is rotating counterclockwise and *DE* is rotating clockwise.

SOLVING PROBLEMS
ON YOUR OWN

In this section, you learned how to analyze the velocity of bodies in **general plane motion**. You found that you can always consider a general plane motion to be the sum of the two motions you studied in the Sec. 15.1, namely, *a translation and a rotation*.

To solve a problem involving the velocity of a body in plane motion, you should take the following steps.

1. Whenever possible, determine the velocity of the points of the body where it is connected to another body whose motion is known [Sample Prob. 15.6]. That other body may be an arm or crank rotating with a given angular velocity [Sample Probs. 15.7 and 15.8].

2. Next, draw a diagram to use in your solution (Figs. 15.15 and 15.16) if you are not using the vector algebra approach. This diagram consists of the following diagrams.

   a. Plane motion diagram: Draw a diagram of the body including all dimensions and showing those points for which you know or seek the velocity.

   b. Translation diagram: Select a reference point A for which you know the direction and/or the magnitude of the velocity $\mathbf{v}_A$, and draw a second diagram showing the body in translation with all of its points having the same velocity $\mathbf{v}_A$.

   c. Rotation diagram: Consider point A as a fixed point and draw a diagram showing the body in rotation about A. Show the angular velocity $\boldsymbol{\omega} = \omega\mathbf{k}$ of the body and the relative velocities with respect to A of the other points, such as the velocity $\mathbf{v}_{B/A}$ of B relative to A.

3. Write the relative velocity formula as

$$\mathbf{v}_B = \mathbf{v}_A + \mathbf{v}_{B/A} \tag{15.17}$$

or for plane motion as

$$\mathbf{v}_B = \mathbf{v}_A + \omega\mathbf{k} \times \mathbf{r}_{B/A} \tag{15.17'}$$

You can solve this vector equation analytically by writing the corresponding scalar equations, or you can solve it by using a vector triangle (Fig. 15.16).

4. Use a different reference point to obtain an equivalent solution. For example, if you select point B as the reference point, the relative velocity of point A is

$$\mathbf{v}_A = \mathbf{v}_B + \mathbf{v}_{A/B} = \mathbf{v}_B + \omega\mathbf{k} \times \mathbf{r}_{A/B} \tag{15.20}$$

Note that the relative velocities $\mathbf{v}_{B/A}$ and $\mathbf{v}_{A/B}$ have the same magnitude but opposite sense. Relative velocities, therefore, depend upon the reference point that you select. The angular velocity, however, is independent of the choice of reference point.

5. Write additional relative velocity equations if you are analyzing a multi-body linkage. For problems such as the crankshaft-piston in Sample Prob. 15.7, you may have to write multiple relative velocity equations. In that problem, you can express the velocity of P with respect to B and then the velocity of B with respect to A. Generally, the ends of the linkages will have some type of constraint (e.g., the piston moving only in the x direction).

Problems

15.CQ3 The ball rolls without slipping on the fixed surface as shown. What is the direction of the velocity of point A?
   a. →   **b.** ↗   **c.** ↑   **d.** ↓   **e.** ↘

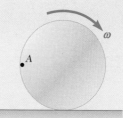

Fig. P15.CQ3

15.CQ4 Three uniform rods—ABC, DCE, and FGH—are connected as shown. Which of the following statements concerning the angular speed of the three objects is true?
   a. $\omega_{ABC} = \omega_{DCE} = \omega_{FGH}$
   b. $\omega_{DCE} > \omega_{ABC} > \omega_{FGH}$
   c. $\omega_{DCE} < \omega_{ABC} < \omega_{FGH}$
   d. $\omega_{ABC} > \omega_{DCE} > \omega_{FGH}$
   e. $\omega_{FGH} = \omega_{DCE} < \omega_{ABC}$

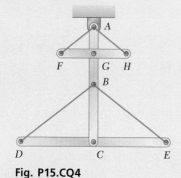

Fig. P15.CQ4

END-OF-SECTION PROBLEMS

15.38 An automobile travels to the right at a constant speed of 48 mi/h. If the diameter of a wheel is 22 in., determine the velocities of points B, C, D, and E on the rim of the wheel.

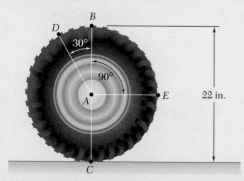

Fig. P15.38

15.39 The motion of rod AB is guided by pins attached at A and B that slide in the slots shown. At the instant shown, $\theta = 40°$ and the pin at B moves upward to the left with a constant velocity of 6 in./s. Determine (a) the angular velocity of the rod, (b) the velocity of the pin at end A.

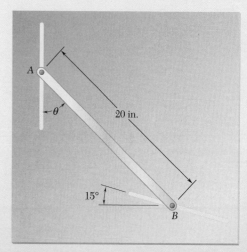

Fig. P15.39

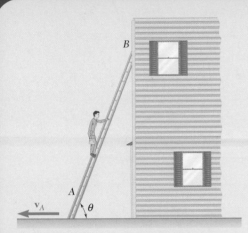

Fig. P15.40

15.40 A painter is halfway up a 10-m ladder when the bottom starts sliding out from under him. Knowing that point A has a velocity $\mathbf{v}_A = 2$ m/s directed to the left when $\theta = 60°$, determine (*a*) the angular velocity of the ladder, (*b*) the velocity of the painter.

15.41 Rod AB can slide freely along the floor and the inclined plane. At the instant shown, the velocity of end A is 1.4 m/s to the left. Determine (*a*) the angular velocity of the rod, (*b*) the velocity of end B of the rod.

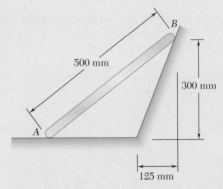

Fig. P15.41 and *P15.42*

15.42 Rod AB can slide freely along the floor and the inclined plane. At the instant shown, the angular velocity of the rod is 4.2 rad/s counterclockwise. Determine (*a*) the velocity of end A of the rod, (*b*) the velocity of end B of the rod.

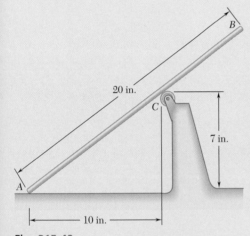

Fig. P15.43

15.43 Rod AB moves over a small wheel at C while end A moves to the right with a constant velocity of 25 in./s. At the instant shown, determine (*a*) the angular velocity of the rod, (*b*) the velocity of end B of the rod.

15.44 The disk shown moves in the xy plane. Knowing that $(v_A)_y = -7$ m/s, $(v_B)_x = -7.4$ m/s, and $(v_C)_x = -1.4$ m/s, determine (*a*) the angular velocity of the disk, (*b*) the velocity of point B.

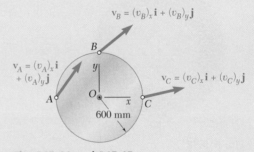

Fig. P15.44 and P15.45

15.45 The disk shown moves in the xy plane. Knowing that $(v_A)_y = -7$ m/s, $(v_B)_x = -7.4$ m/s, and $(v_C)_x = -1.4$ m/s, detemine (*a*) the velocity of point O, (*b*) the point of the disk with zero velocity.

15.46 The plate shown moves in the xy plane. Knowing that $(v_A)_x = 250$ mm/s, $(v_B)_y = -450$ mm/s, and $(v_C)_x = -500$ mm/s, determine (a) the angular velocity of the plate, (b) the velocity of point A.

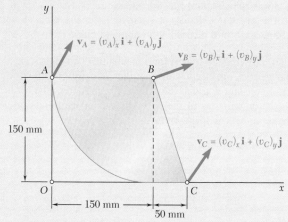

Fig. P15.46

15.47 Velocity sensors are placed on a satellite that is moving only in the xy plane. Knowing that at the instant shown the unidirectional sensors measure $(v_A)_x = 2$ ft/s, $(v_B)_x = -0.333$ ft/s, and $(v_C)_y = -2$ ft/s, determine (a) the angular velocity of the satellite, (b) the velocity of point B.

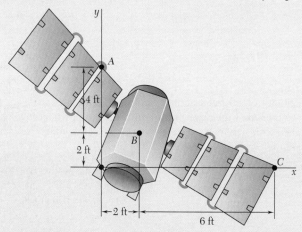

Fig. P15.47

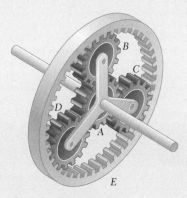

Fig. P15.48 and P15.49

15.48 In the planetary gear system shown, the radius of gears A, B, C, and D is a and the radius of the outer gear E is $3a$. Knowing that the angular velocity of gear A is ω_A clockwise and that the outer gear E is stationary, determine (a) the angular velocity of each planetary gear, (b) the angular velocity of the spider connecting the planetary gears.

15.49 In the planetary gear system shown, the radius of gears A, B, C, and D is 30 mm and the radius of the outer gear E is 90 mm. Knowing that gear E has an angular velocity of 180 rpm clockwise and that the central gear A has an angular velocity of 240 rpm clockwise, determine (a) the angular velocity of each planetary gear, (b) the angular velocity of the spider connecting the planetary gears.

15.50 Arm AB rotates with an angular velocity of 20 rad/s counterclockwise. Knowing that the outer gear C is stationary, determine (a) the angular velocity of gear B, (b) the velocity of the gear tooth located at point D.

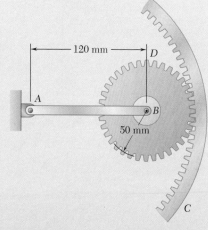

Fig. P15.50

1009

Fig. P15.51

15.51 In the simplified sketch of a ball bearing shown, the diameter of the inner race A is 60 mm and the diameter of each ball is 12 mm. The outer race B is stationary while the inner race has an angular velocity of 3600 rpm. Determine (*a*) the speed of the center of each ball, (*b*) the angular velocity of each ball, (*c*) the number of times per minute each ball describes a complete circle.

15.52 A simplified gear system for a mechanical watch is shown. Knowing that gear A has a constant angular velocity of 1 rev/h and gear C has a constant angular velocity of 1 rpm, determine (*a*) the radius r, (*b*) the magnitudes of the accelerations of the points on gear B that are in contact with gears A and C.

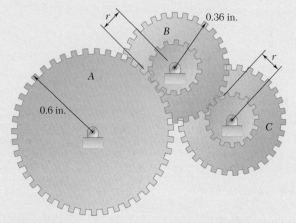

Fig. *P15.52*

15.53 and **15.54** Arm ACB rotates about point C with an angular velocity of 40 rad/s counterclockwise. Two friction disks A and B are pinned at their centers to arm ACB as shown. Knowing that the disks roll without slipping at surfaces of contact, determine the angular velocity of (*a*) disk A, (*b*) disk B.

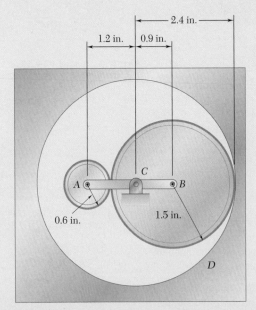

Fig. P15.53

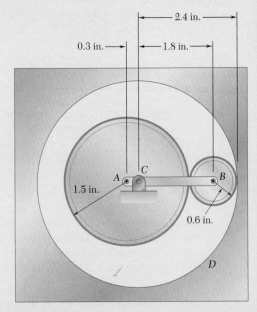

Fig. *P15.54*

15.55 Knowing that at the instant shown the velocity of collar A is 900 mm/s to the left, determine (*a*) the angular velocity of rod ADB, (*b*) the velocity of point B.

15.56 Knowing that at the instant shown the angular velocity of rod DE is 2.4 rad/s clockwise, determine (*a*) the velocity of collar A, (*b*) the velocity of point B.

15.57 Knowing that the disk has a constant angular velocity of 15 rad/s clockwise, determine the angular velocity of bar BD and the velocity of collar D when (*a*) $\theta = 0$, (*b*) $\theta = 90°$, (*c*) $\theta = 180°$.

15.58 The disk has a constant angular velocity of 20 rad/s clockwise. (*a*) Determine the two values of the angle θ for which the velocity of collar D is zero. (*b*) For each of these values of θ, determine the corresponding value of the angular velocity of bar BD.

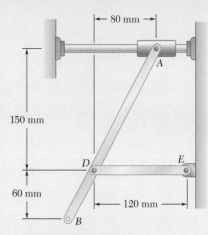

Fig. P15.55 and P15.56

15.59 The test rig shown was developed to perform fatigue testing on fitness trampolines. A motor drives the 9-in.-radius flywheel AB, which is pinned at its center point A, in a counterclockwise direction. The flywheel is attached to slider CD by the 18-in. connecting rod BC. Knowing that the "feet" at D should hit the trampoline twice every second, at the instant when $\theta = 0°$, determine (*a*) the angular velocity of the connecting rod BC, (*b*) the velocity of D, (*c*) the velocity of midpoint CB.

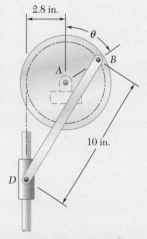

Fig. P15.57 and P15.58

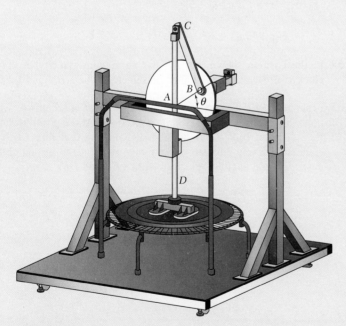

Fig. *P15.59*

15.60 In the eccentric shown, a disk of 2-in. radius revolves about shaft O that is located 0.5 in. from the center A of the disk. The distance between the center A of the disk and the pin at B is 8 in. Knowing that the angular velocity of the disk is 900 rpm clockwise, determine the velocity of the block when $\theta = 30°$.

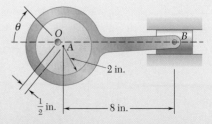

Fig. *P15.60*

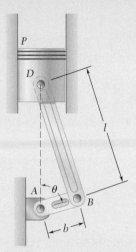

Fig. P15.61 and P15.62

15.61 In the engine system shown, $l = 160$ mm and $b = 60$ mm. Knowing that the crank AB rotates with a constant angular velocity of 1000 rpm clockwise, determine the velocity of the piston P and the angular velocity of the connecting rod when (a) $\theta = 0$, (b) $\theta = 90°$.

15.62 In the engine system shown, $l = 160$ mm and $b = 60$ mm. Knowing that crank AB rotates with a constant angular velocity of 1000 rpm clockwise, determine the velocity of the piston P and the angular velocity of the connecting rod when $\theta = 60°$.

15.63 Knowing that at the instant shown the angular velocity of rod AB is 15 rad/s clockwise, determine (a) the angular velocity of rod BD, (b) the velocity of the midpoint of rod BD.

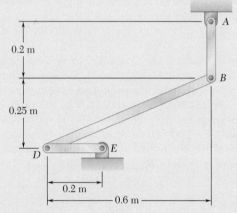

Fig. P15.63

15.64 In the position shown, bar AB has an angular velocity of 4 rad/s clockwise. Determine the angular velocity of bars BD and DE.

15.65 Linkage $DBEF$ is part of a windshield wiper mechanism, where points O, F, and D are fixed pin connections. At the position shown, $\theta = 30°$ and link EB is horizontal. Knowing that link EF has a counterclockwise angular velocity of 4 rad/s at the instant shown, determine the angular velocity of links EB and DB.

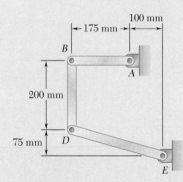

Fig. P15.64

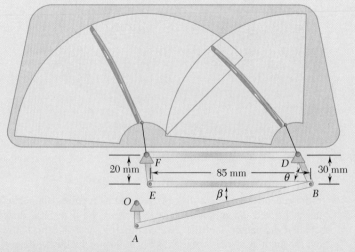

Fig. P15.65

15.66 Roberts linkage is named after Richard Roberts (1789–1864) and can be used to draw a close approximation to a straight line by locating a pen at point F. The distance AB is the same as BF, DF, and DE. Knowing that the angular velocity of bar AB is 5 rad/s clockwise in the position shown, determine (a) the angular velocity of bar DE, (b) the velocity of point F.

15.67 Roberts linkage is named after Richard Roberts (1789–1864) and can be used to draw a close approximation to a straight line by locating a pen at point F. The distance AB is the same as BF, DF, and DE. Knowing that the angular velocity of plate BDF is 2 rad/s counterclockwise when $\theta = 90°$, determine (a) the angular velocities of bars AB and DE, (b) the velocity of point F. When $\theta = 90°$, point F may be assumed to coincide with point E, with negligible error in the velocity analysis.

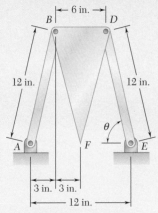

Fig. P15.66 and P15.67

15.68 In the position shown, bar DE has a constant angular velocity of 10 rad/s clockwise. Knowing that $h = 500$ mm, determine (a) the angular velocity of bar FBD, (b) the velocity of point F.

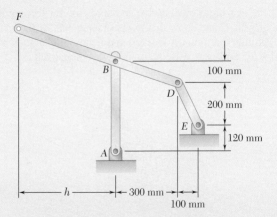

Fig. P15.68 and P15.69

15.69 In the position shown, bar DE has a constant angular velocity of 10 rad/s clockwise. Determine (a) the distance h for which the velocity of point F is vertical, (b) the corresponding velocity of point F.

15.70 Both 6-in.-radius wheels roll without slipping on the horizontal surface. Knowing that the distance AD is 5 in., the distance BE is 4 in., and D has a velocity of 6 in./s to the right, determine the velocity of point E.

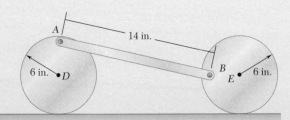

Fig. P15.70

15.71 The 80-mm-radius wheel shown rolls to the left with a velocity of 900 mm/s. Knowing that the distance AD is 50 mm, determine the velocity of the collar and the angular velocity of rod AB when (a) $\beta = 0$, (b) $\beta = 90°$.

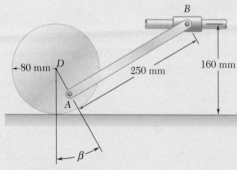

Fig. P15.71

*15.72 For the gearing shown, derive an expression for the angular velocity ω_C of gear C and show that ω_C is independent of the radius of gear B. Assume that point A is fixed and denote the angular velocities of rod ABC and gear A by ω_{ABC} and ω_A, respectively.

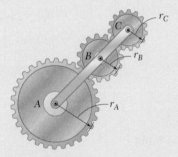

Fig. P15.72

15.3 INSTANTANEOUS CENTER OF ROTATION

Consider the general plane motion of a rigid body. We will show that, at any given instant, the velocities of the various particles of the rigid body are the same as if the body were rotating about an axis perpendicular to the plane of the body, called the **instantaneous axis of rotation**. This axis intersects the plane of the rigid body at a point C, called the **instantaneous center of rotation** of the body or the **instantaneous center of zero velocity**. This gives us an alternative method for solving problems involving the velocities of points on an object in plane motion, and it is sometimes simpler than using the equations in Sec. 15.2.

Recall that we can always replace the plane motion of a rigid body by a translation defined by the motion of an arbitrary reference point A and by a rotation about A. As far as the velocities are concerned, the translation is characterized by the velocity $\mathbf{v}_A$ of the reference point A and the rotation is characterized by the angular velocity $\boldsymbol{\omega}$ of the body (which is independent of the choice of A). Thus, the velocity $\mathbf{v}_A$ of point A and the angular velocity $\boldsymbol{\omega}$ of the rigid body define completely the velocities of all the other particles of the body (Fig. 15.18a).

Photo 15.5 If the tires of this car are rolling without sliding, the instantaneous center of rotation of each tire is the point of contact between the road and the tire.

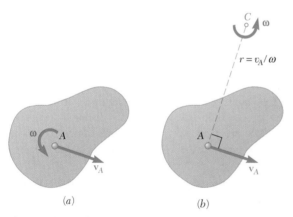

Fig. 15.18 As far as velocities are concerned, at every instant in time the rigid body seems to rotate about a point called the instantaneous center C.

Now let us assume that $\mathbf{v}_A$ and $\boldsymbol{\omega}$ are known and that they are both different from zero. (If $\mathbf{v}_A = 0$, point A is itself the instantaneous center of rotation, and if $\boldsymbol{\omega} = 0$, you have rigid body translation where all of the particles have the same velocity $\mathbf{v}_A$.) We could obtain these velocities by letting the rigid body rotate with the angular velocity $\boldsymbol{\omega}$ about a point C located on the perpendicular to $\mathbf{v}_A$ at a distance $r = v_A/\omega$ from A, as shown in Fig. 15.18b. We check that the velocity of A would be perpendicular to AC and that its magnitude would be $r\omega = (v_A/\omega)\omega = v_A$. Thus, the velocities of all the other particles of the body are the same as originally defined. Therefore, *as far as the velocities are concerned, the rigid body seems to rotate about the instantaneous center C at the instant considered.*

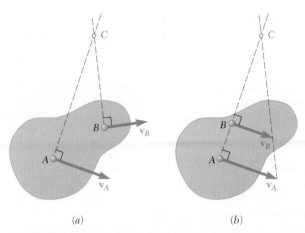

Fig. 15.19 Locating the instantaneous center of rotation C
(a) when you know the directions of the velocities of two points;
(b) when the velocities of two points are perpendicular to line AB.

We can define the position of the instantaneous center in two other ways. If we know the directions of the velocities of two particles A and B of the rigid body and if they are different, we can obtain the instantaneous center C by drawing the perpendicular to $\mathbf{v}_A$ through A and the perpendicular to $\mathbf{v}_B$ through B. The point C is where these two lines intersect (Fig. 15.19a). If the velocities $\mathbf{v}_A$ and $\mathbf{v}_B$ of two particles A and B are perpendicular to line AB and we know their magnitudes, we can find the instantaneous center by intersecting line AB with the line joining the ends of the vectors $\mathbf{v}_A$ and $\mathbf{v}_B$ (Fig. 15.19b). Note that if $\mathbf{v}_A$ and $\mathbf{v}_B$ were parallel in Fig. 15.19a or if $\mathbf{v}_A$ and $\mathbf{v}_B$ had the same magnitude in Fig. 15.19b, the instantaneous center C would be at an infinite distance and $\boldsymbol{\omega}$ would be zero; all points of the rigid body would have the same velocity.

To see how we can use the concept of the instantaneous center of rotation, let us consider again the sliding rod of Sec. 15.2. Drawing the perpendicular to $\mathbf{v}_A$ through A and the perpendicular to $\mathbf{v}_B$ through B (Fig. 15.20), we obtain the instantaneous center C. At the instant

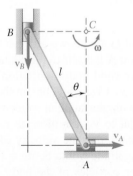

Fig. 15.20 Instantaneous center of rotation C for the sliding rod AB.

considered, the velocities of all the particles of the rod are thus the same as if the rod rotated about C. Now, if we know the magnitude v_A of the velocity of A, we can obtain the magnitude ω of the angular velocity of the rod from

$$\omega = \frac{v_A}{AC} = \frac{v_A}{l\cos\theta}$$

Then we obtain the magnitude of the velocity of B as

$$v_B = (BC)\omega = l\sin\theta\frac{v_A}{l\cos\theta} = v_A\tan\theta$$

Note that we used only *absolute* velocities in the computation.

The instantaneous center of a body in plane motion can be located either on the body or outside the body. If it is located on the rigid body, the particle C coinciding with the instantaneous center at a given instant t must have zero velocity at that instant. However, the instantaneous center of rotation is valid only at a given instant. Thus, particle C of the rigid body that coincides with the instantaneous center at time t generally does not coincide with the instantaneous center at time $t + \Delta t$. Its velocity is zero at time t, but it will probably be different from zero at time $t + \Delta t$. This means, in general, that particle C *does not have zero acceleration* and, therefore, that the accelerations of the various particles of the rigid body cannot be determined as if the body were rotating about C.

As the motion of the rigid body proceeds, the instantaneous center moves in space. However, we just pointed out that the position of the instantaneous center on the body keeps changing. Thus, the instantaneous center describes one curve in space, called the *space centrode*, and another curve on the rigid body, called the *body centrode* (Fig. 15.21). It can be shown that at any instant, these two curves are tangent at C and that as the rigid body moves, the body centrode appears to roll on the space centrode.

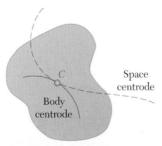

Fig. 15.21 The space centrode and the body centrode are tangent to each other.

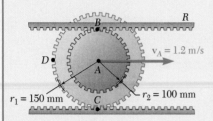

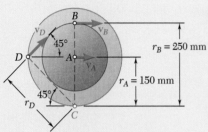

Fig. 1 Distances from the instantaneous center of rotation to A, B, and D.

Sample Problem 15.9

Solve Sample Prob. 15.6 using the method of the instantaneous center of rotation.

STRATEGY: You know the velocity direction of two points on the same rigid body, so you can find an instantaneous center of rotation. Since the gear rolls on the stationary lower rack, the point of contact C of the gear with the rack has no velocity; point C is therefore the instantaneous center of rotation.

MODELING and ANALYSIS:

a. Angular Velocity of the Gear. You can calculate the angular velocity directly from the data in Fig. 1.

$$v_A = r_A\omega \qquad 1.2 \text{ m/s} = (0.150 \text{ m})\,\omega$$

$$\omega = 8 \text{ rad/s} \downdownarrows \quad \blacktriangleleft$$

b. Velocities. As far as velocities are concerned, all points of the gear seem to rotate about the instantaneous center.

Velocity of Upper Rack. Recalling that $v_R = v_B$, you have

$$v_R = v_B = r_B\omega \qquad v_R = (0.250 \text{ m})(8 \text{ rad/s}) = 2 \text{ m/s}$$

$$\mathbf{v}_R = 2 \text{ m/s} \rightarrow \quad \blacktriangleleft$$

Velocity of Point D. Since $r_D = (0.150 \text{ m})\sqrt{2} = 0.2121 \text{ m}$, you obtain

$$v_D = r_D\omega \qquad v_D = (0.2121 \text{ m})(8 \text{ rad/s}) = 1.697 \text{ m/s}$$

$$\mathbf{v}_D = 1.697 \text{ m/s} \measuredangle 45° \quad \blacktriangleleft$$

REFLECT and THINK: The results are the same as in Sample Prob. 15.6, as you would expect, but it took much less computation to get them.

Sample Problem 15.10

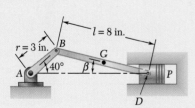

Solve Sample Prob. 15.7 using the method of the instantaneous center of rotation.

STRATEGY: You know the velocity of point B from the motion of the crank (see Sample Prob. 15.7), and you know the direction of the velocity of point D. Therefore, you can find an instantaneous center of rotation.

MODELING and ANALYSIS:

Motion of Crank AB. Referring to Sample Prob. 15.7, you obtain the velocity of point B; $\mathbf{v}_B = 628.3$ in./s $\diagdown 50°$.

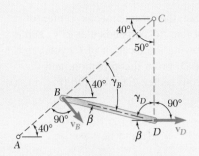

Fig. 1 Instantaneous center of rotation for bar *BD*.

Motion of the Connecting Rod *BD*. First locate the instantaneous center *C* by drawing lines perpendicular to the absolute velocities $\mathbf{v}_B$ and $\mathbf{v}_D$ (Fig. 1). Recalling from Sample Prob. 15.7 that $\beta = 13.95°$ and that $BD = 8$ in., solve the triangle *BCD*.

$$\gamma_B = 40° + \beta = 53.95° \qquad \gamma_D = 90° - \beta = 76.05°$$

$$\frac{BC}{\sin 76.05°} = \frac{CD}{\sin 53.95°} = \frac{8 \text{ in.}}{\sin 50°}$$

$$BC = 10.14 \text{ in.} \qquad CD = 8.44 \text{ in.}$$

Since the connecting rod *BD* seems to rotate about point *C*, you have

$$v_B = (BC)\omega_{BD}$$
$$628.3 \text{ in./s} = (10.14 \text{ in.})\omega_{BD}$$
$$\boldsymbol{\omega}_{BD} = 62.0 \text{ rad/s} \;\ing \quad \blacktriangleleft$$

$$v_D = (CD)\omega_{BD} = (8.44 \text{ in.})(62.0 \text{ rad/s})$$
$$= 523 \text{ in./s} = 43.6 \text{ ft/s}$$
$$\mathbf{v}_P = \mathbf{v}_D = 43.6 \text{ ft/s} \rightarrow \quad \blacktriangleleft$$

REFLECT and THINK: Often, the hardest part of solving a problem using the instantaneous center of rotation is the geometry. Remembering how to use the law of sines or the law of cosines is often helpful.

Sample Problem 15.11

Two 20-in. rods *AB* and *DE* are connected as shown. Point *D* is the midpoint of rod *AB*, and at the instant shown, rod *DE* is horizontal. Knowing that the velocity of point *A* is 1 ft/s downward, determine (*a*) the angular velocity of rod *DE*, (*b*) the velocity of point *E*.

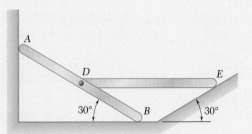

STRATEGY: You know the directions of several points on these objects, so you can use instantaneous centers of rotation to solve this problem.

(continued)

MODELING and ANALYSIS: Locate the instantaneous center of rotation C of bar AB as the intersection of line AC perpendicular to $\mathbf{v}_A$ and line BC perpendicular to $\mathbf{v}_B$ (Fig. 1). Knowing the location of C, you can determine the direction of the velocity of D. From this direction, and the direction of E, you can find the instantaneous center, I, for bar DE (Fig. 1).

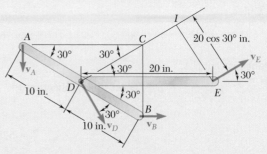

Fig. 1 The instantaneous centers of rotation for bar AB and DE are C and I, respectively.

a. Angular velocity of DE. From geometry, $r_{A/C} = (20 \cos 30°)$ in., so

$$\omega_{AB} = \frac{v_A}{r_{A/C}} = \frac{12 \text{ in./s}}{20 \cos 30° \text{ in.}} = 0.6928 \text{ rad/s} \, \lefttop$$

Now you can find v_D since $r_{D/C} = 10$ in.

$$v_D = \omega_{AB} r_{D/C} = (0.6928 \text{ rad/s})(10 \text{ in.}) = 6.928 \text{ in./s}$$
$$v_D = 6.928 \text{ in./s} \, \measuredangle \, 30°$$

Now, since you know the directions of the velocities of D and E, $\mathbf{v}_E = v_E \measuredangle 30°$, you can find point I, which is the instantaneous center of bar DE. From geometry, $r_{D/I} = 20 \cos 30°$ in., and therefore

$$\omega_{DE} = \frac{v_D}{r_{D/I}} = \frac{6.928 \text{ in./s}}{20 \cos 30° \text{ in.}} = 0.400 \text{ rad/s} \qquad \boldsymbol{\omega}_{DE} = 0.400 \text{ rad/s} \, \lefttop \quad \blacktriangleleft$$

b. Velocity of E. Using this angular velocity, you can easily determine the velocity of E:

$$v_E = \omega_{DE} r_{E/I} = (0.400 \text{ rad/s})(20 \sin 30° \text{ in.}) = 4.00 \text{ in./s}$$

$$\mathbf{v}_E = 0.333 \text{ ft/s} \measuredangle 30° \quad \blacktriangleleft$$

REFLECT and THINK: The direction of ω_{DE} makes intuitive sense; you would expect it to be rotating counterclockwise at the instant shown. You could have also solved this problem using vector equations.

SOLVING PROBLEMS
ON YOUR OWN

In this section, we introduced the **instantaneous center of rotation** in plane motion. This provides us with an alternative way of solving problems involving the *velocities* of the various points of a body in plane motion [Sample Probs. 15.9 through 15.11]. As its name suggests, the instantaneous center of rotation is the point about which you can assume a body is rotating at a given instant; you can use the instantaneous center to determine the velocity of any point on the body at that instant in time.

A. To determine the instantaneous center of rotation of a body in plane motion, you should use one of the following procedures.

1. If you know both the velocity v_A of a point A and the angular velocity ω of the body (Fig. 15.18):

   a. Draw a sketch of the body, showing point A, its velocity $\mathbf{v}_A$, and the angular velocity $\boldsymbol{\omega}$ of the body.

   b. From A draw a line perpendicular to v_A on the side of $\mathbf{v}_A$ from which this velocity is viewed as having *the same sense as $\boldsymbol{\omega}$*.

   c. Locate the instantaneous center C on this line at a distance $r = v_A/\omega$ from point A.

2. If you know the directions of the velocities of two points A and B and they are different (Fig. 15.19a):

   a. Draw a sketch of the body showing points A and B and their velocities $\mathbf{v}_A$ and $\mathbf{v}_B$.

   b. From A and B draw lines perpendicular to v_A and v_B, respectively. The instantaneous center C is located at the point where the two lines intersect.

   c. If you know the velocity of one of the two points, you can determine the angular velocity of the body at that instant in time. For example, if you know $\mathbf{v}_A$, you can write $\omega = v_A/AC$, where AC is the distance from point A to the instantaneous center C.

3. If you know the velocities of two points A and B and both are perpendicular to the line AB (Fig. 15.19b):

   a. Draw a sketch of the body, showing points A and B with their velocities $\mathbf{v}_A$ and $\mathbf{v}_B$ *drawn to scale.*

   b. Draw a line through points A and B, and another line through the tips of the vectors $\mathbf{v}_A$ and $\mathbf{v}_B$. The instantaneous center C is located at the point where the two lines intersect.

c. Obtain the angular velocity of the body by either dividing $\mathbf{v}_A$ by AC or $\mathbf{v}_B$ by BC.

d. If the velocities $\mathbf{v}_A$ and $\mathbf{v}_B$ have the same magnitude, the two lines drawn in part b do not intersect; the instantaneous center C is at an infinite distance. The angular velocity $\boldsymbol{\omega}$ is zero and the body is in translation.

B. Once you have determined the instantaneous center and the angular velocity of a body, you can determine the velocity $\mathbf{v}_P$ of any point P of the body in the following way.

1. Draw a sketch of the body, showing point P, the instantaneous center of rotation C, and the angular velocity $\boldsymbol{\omega}$.

2. Draw a line from P to the instantaneous center C and measure or calculate the distance from P to C.

3. The velocity $\mathbf{v}_P$ is a vector perpendicular to the line PC, of the same sense as $\boldsymbol{\omega}$, and with a magnitude of $v_P = (PC)\omega$.

Finally, keep in mind that the instantaneous center of rotation can be used *only* to determine velocities at a specific instant in time. It cannot be used to determine accelerations.

Problems

CONCEPT QUESTIONS

15.CQ5 The disk rolls without sliding on the fixed horizontal surface. At the instant shown, the instantaneous center of zero velocity for rod AB would be located in which region?
  a. Region 1
  b. Region 2
  c. Region 3
  d. Region 4
  e. Region 5
  f. Region 6

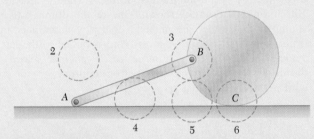

Fig. P15.CQ5

15.CQ6 Bar BDE is pinned to two links, AB and CD. At the instant shown, the angular velocities of link AB, link CD, and bar BDE are ω_{AB}, ω_{CD}, and ω_{BDE}, respectively. Which of the following statements concerning the angular speeds of the three objects is true at this instant?
  a. $\omega_{AB} = \omega_{CD} = \omega_{BDE}$
  b. $\omega_{BDE} > \omega_{AB} > \omega_{CD}$
  c. $\omega_{AB} = \omega_{CD} > \omega_{BDE}$
  d. $\omega_{AB} > \omega_{CD} > \omega_{BDE}$
  e. $\omega_{CD} > \omega_{AB} > \omega_{BDE}$

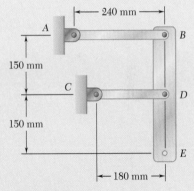

Fig. P15.CQ6

END-OF-SECTION PROBLEMS

15.73 A juggling club is thrown vertically into the air. The center of gravity G of the 20-in. club is located 12 in. from the knob. Knowing that at the instant shown, G has a velocity of 4 ft/s upwards and the club has an angular velocity of 30 rad/s counterclockwise, determine (*a*) the speeds of points A and B, (*b*) the location of the instantaneous center of rotation.

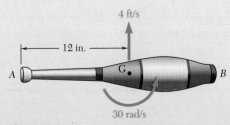

Fig. P15.73

15.74 At the instant shown during deceleration, the velocity of an automobile is 40 ft/s to the right. Knowing that the velocity of the contact point A of the wheel with the ground is 5 ft/s to the right, determine (a) the instantaneous center of rotation of the wheel, (b) the velocity of point B, (c) the velocity of point D.

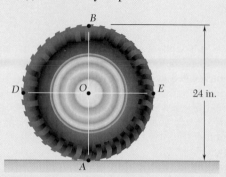

Fig. P15.74

15.75 A helicopter moves horizontally in the x direction at a speed of 120 mi/h. Knowing that the main blades rotate clockwise when viewed from above with an angular velocity of 180 rpm, determine the instantaneous axis of rotation of the main blades.

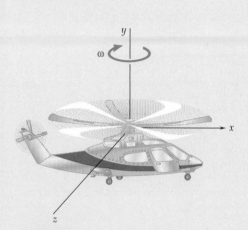

Fig. P15.75

15.76 and 15.77 A 60-mm-radius drum is rigidly attached to a 100-mm-radius drum as shown. One of the drums rolls without sliding on the surface shown, and a cord is wound around the other drum. Knowing that end E of the cord is pulled to the left with a velocity of 120 mm/s, determine (a) the angular velocity of the drums, (b) the velocity of the center of the drums, (c) the length of cord wound or unwound per second.

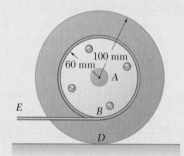

Fig. P15.76

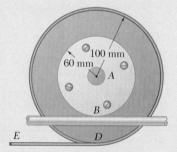

Fig. P15.77

15.78 The spool of tape shown and its frame assembly are pulled upward at a speed $v_A = 750$ mm/s. Knowing that the 80-mm-radius spool has an angular velocity of 15 rad/s clockwise and that at the instant shown the total thickness of the tape on the spool is 20 mm, determine (a) the instantaneous center of rotation of the spool, (b) the velocities of points B and D.

15.79 The spool of tape shown and its frame assembly are pulled upward at a speed $v_A = 100$ mm/s. Knowing that end B of the tape is pulled downward with a velocity of 300 mm/s and that at the instant shown the total thickness of the tape on the spool is 20 mm, determine (a) the instantaneous center of rotation of the spool, (b) the velocity of point D of the spool.

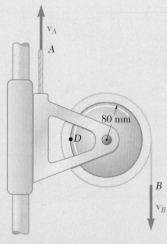

Fig. P15.78 and P15.79

15.80 The arm *ABC* rotates with an angular velocity of 4 rad/s counterclockwise. Knowing that the angular velocity of the intermediate gear *B* is 8 rad/s counterclockwise, determine (*a*) the instantaneous centers of rotation of gears *A* and *C*, (*b*) the angular velocities of gears *A* and *C*.

15.81 The double gear rolls on the stationary left rack *R*. Knowing that the rack on the right has a constant velocity of 2 ft/s, determine (*a*) the angular velocity of the gear, (*b*) the velocities of points *A* and *D*.

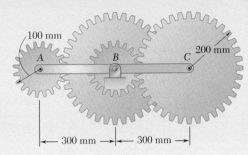

Fig. P15.80

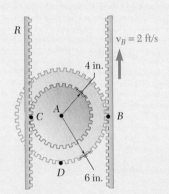

Fig. P15.81

15.82 An overhead door is guided by wheels at *A* and *B* that roll in horizontal and vertical tracks. Knowing that when $\theta = 40°$ the velocity of wheel *B* is 1.5 ft/s upward, determine (*a*) the angular velocity of the door, (*b*) the velocity of end *D* of the door.

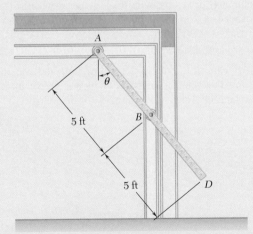

Fig. P15.82

15.83 Rod *ABD* is guided by wheels at *A* and *B* that roll in horizontal and vertical tracks. Knowing that at the instant $\beta = 60°$ and the velocity of wheel *B* is 40 in./s downward, determine (*a*) the angular velocity of the rod, (*b*) the velocity of point *D*.

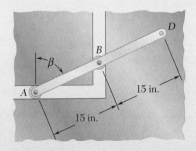

Fig. P15.83

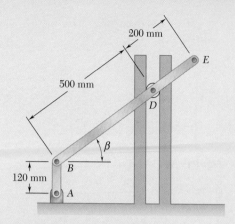

Fig. P15.84 and P15.85

15.84 Rod *BDE* is partially guided by a roller at *D* that moves in a vertical track. Knowing that at the instant shown the angular velocity of crank *AB* is 5 rad/s clockwise and that β = 25°, determine (*a*) the angular velocity of the rod, (*b*) the velocity of point *E*.

15.85 Rod *BDE* is partially guided by a roller at *D* that moves in a vertical track. Knowing that at the instant shown β = 30°, point *E* has a velocity of 2 m/s down and to the right, determine the angular velocities of rod *BDE* and crank *AB*.

15.86 A motor at *O* drives the windshield wiper mechanism so that *OA* has a constant angular velocity of 15 rpm. Knowing that at the instant shown linkage *OA* is vertical, θ = 30°, and β = 15°, determine (*a*) the angular velocity of bar *AB*, (*b*) the velocity of the center of bar *AB*.

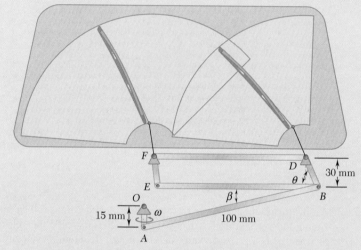

Fig. P15.86 and P15.87

15.87 A motor at *O* drives the windshield wiper mechanism so that point *B* has a speed of 2 m/s. Knowing that at the instant shown linkage *OA* is vertical, θ = 30°, and β = 15°, determine (*a*) the angular velocity of bar *OA*, (*b*) the velocity of the center of bar *AB*.

15.88 Rod *AB* can slide freely along the floor and the inclined plane. Denoting the velocity of point *A* by **v**$_A$, derive an expression for (*a*) the angular velocity of the rod, (*b*) the velocity of end *B*.

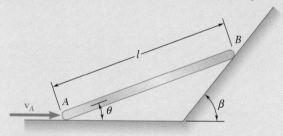

Fig. P15.88

15.89 Small wheels have been attached to the ends of bar *AB* and roll freely along the surfaces shown. Knowing that the velocity of wheel *B* is 7.5 ft/s to the right at the instant shown, determine (*a*) the velocity of end *A* of the bar, (*b*) the angular velocity of the bar, (*c*) the velocity of the midpoint of the bar.

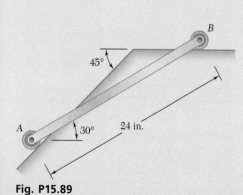

Fig. P15.89

15.90 Two slots have been cut in plate *FG* and the plate has been placed so that the slots fit two fixed pins *A* and *B*. Knowing that at the instant shown the angular velocity of crank *DE* is 6 rad/s clockwise, determine (*a*) the velocity of point *F*, (*b*) the velocity of point *G*.

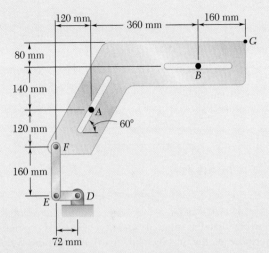

Fig. P15.90

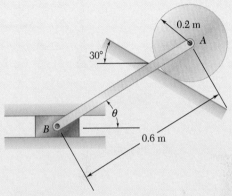

Fig. P15.91

15.91 The disk is released from rest and rolls down the incline. Knowing that the speed of *A* is 1.2 m/s when $\theta = 0°$, determine at that instant (*a*) the angular velocity of the rod, (*b*) the velocity of *B*. (Only portions of the two tracks are shown.)

15.92 The pin at *B* is attached to member *ABD* and can slide freely along the slot cut in the fixed plate. Knowing that at the instant shown the angular velocity of arm *DE* is 3 rad/s clockwise, determine (*a*) the angular velocity of member *ABD*, (*b*) the velocity of point *A*.

15.93 Two identical rods *ABF* and *DBE* are connected by a pin at *B*. Knowing that at the instant shown the velocity of point *D* is 200 mm/s upward, determine the velocity of (*a*) point *E*, (*b*) point *F*.

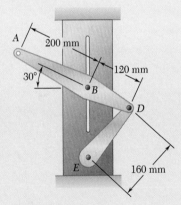

Fig. P15.92

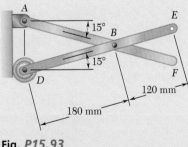

Fig. P15.93

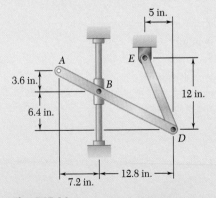

Fig. P15.94

15.94 Arm *ABD* is connected by pins to a collar at *B* and to crank *DE*. Knowing that the velocity of collar *B* is 16 in./s upward, determine (*a*) the angular velocity of arm *ABD*, (*b*) the velocity of point *A*.

15.95 Two 25-in. rods are pin-connected at D as shown. Knowing that B moves to the left with a constant velocity of 24 in./s, determine at the instant shown (a) the angular velocity of each rod, (b) the velocity of E.

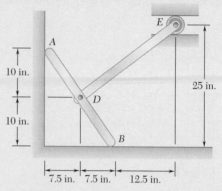

Fig. P15.95

15.96 Two rods ABD and DE are connected to three collars as shown. Knowing that the angular velocity of ABD is 5 rad/s clockwise, determine at the instant shown (a) the angular velocity of DE, (b) the velocity of collar E.

15.97 At the instant shown, the velocity of collar A is 0.4 m/s to the right and the velocity of collar B is 1 m/s to the left. Determine (a) the angular velocity of bar AD, (b) the angular velocity of bar BD, (c) the velocity of point D.

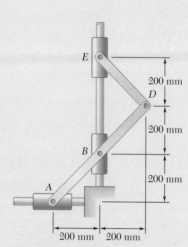

Fig. P15.96

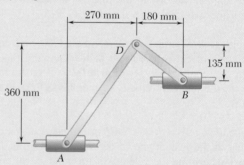

Fig. P15.97

15.98 Two rods AB and DE are connected as shown. Knowing that point D moves to the left with a velocity of 40 in./s, determine (a) the angular velocity of each rod, (b) the velocity of point A.

15.99 Describe the space centrode and the body centrode of rod ABD of Prob. 15.83. (*Hint:* The body centrode need not lie on a physical portion of the rod.)

15.100 Describe the space centrode and the body centrode of the gear of Sample Prob. 15.6 as the gear rolls on the stationary horizontal rack.

15.101 Using the method of Sec. 15.3, solve Prob. 15.60.

15.102 Using the method of Sec. 15.3, solve Prob. 15.64.

15.103 Using the method of Sec. 15.3, solve Prob. 15.65.

15.104 Using the method of Sec. 15.3, solve Prob. 15.38.

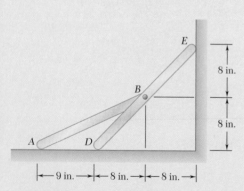

Fig. P15.98

15.4 GENERAL PLANE MOTION: ACCELERATION

We saw in Sec. 15.2A that any plane motion can be replaced by a translation defined by the motion of an arbitrary reference point A and a simultaneous rotation about A. We used this property in Sec. 15.2B to determine the velocity of the various points of a moving rigid body. We now use this same property to determine the acceleration of the points of the body.

15.4A Absolute and Relative Acceleration in Plane Motion

We first recall that the absolute acceleration $\mathbf{a}_B$ of a particle of the rigid body can be obtained from the relative-acceleration formula derived in Sec. 11.4D,

$$\mathbf{a}_B = \mathbf{a}_A + \mathbf{a}_{B/A} \tag{15.21}$$

Photo 15.6 The central gear rotates about a fixed axis and is pin-connected to three bars in general plane motion.

where the right-hand side represents a vector sum. The acceleration $\mathbf{a}_A$ corresponds to the translation of the rigid body with A. The relative acceleration $\mathbf{a}_{B/A}$ is associated with the rotation of the body about A and is measured with respect to axes centered at A and with fixed orientation.

Recall from Sec. 15.1B that we can resolve the relative acceleration $\mathbf{a}_{B/A}$ into two components: a **tangential component** $(\mathbf{a}_{B/A})_t$ perpendicular to the line AB and a **normal component** $(\mathbf{a}_{B/A})_n$ directed toward A (Fig. 15.22). We denote the position vector of B relative to A by $\mathbf{r}_{B/A}$ and the angular velocity and angular acceleration of the rigid body with respect to axes of fixed orientation by $\omega\mathbf{k}$ and $\alpha\mathbf{k}$, respectively. Then we have

$$\begin{aligned} (\mathbf{a}_{B/A})_t &= \alpha\mathbf{k} \times \mathbf{r}_{B/A} & (a_{B/A})_t &= r\alpha \\ (\mathbf{a}_{B/A})_n &= -\omega^2\mathbf{r}_{B/A} & (a_{B/A})_n &= r\omega^2 \end{aligned} \tag{15.22}$$

where r is the distance from A to B. Substituting the expressions obtained for the tangential and normal components of $\mathbf{a}_{B/A}$ into Eq. (15.21), we also have

Relative acceleration for two points on a rigid body

$$\mathbf{a}_B = \mathbf{a}_A + \alpha\mathbf{k} \times \mathbf{r}_{B/A} - \omega^2\mathbf{r}_{B/A} \tag{15.21'}$$

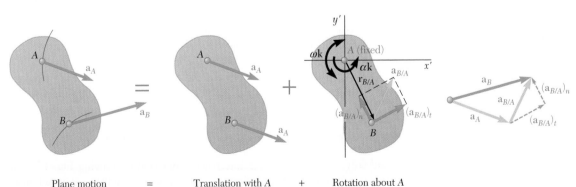

Plane motion   =   Translation with A +   Rotation about A

Fig. 15.22 Pictorial representation of the vector equation relating the acceleration of two points on a rigid body undergoing general plane motion.

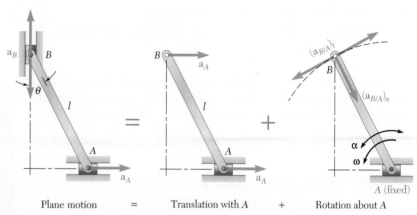

Fig. 15.23 For a sliding rod in general plane motion, the acceleration of point B relative to point A may have a tangential component in either direction perpendicular to the rod. The normal acceleration of B relative to A will always point toward A.

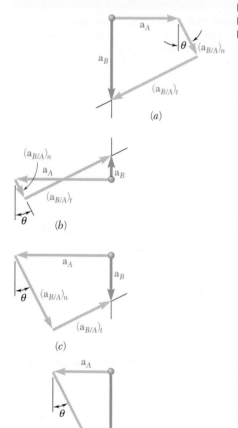

Fig. 15.24 Four possible vector polygons for the acceleration of the sliding rod.

As an example, let us again consider the rod AB whose ends slide along a horizontal and a vertical track (Fig. 15.23). Assuming that we know the velocity $\mathbf{v}_A$ and the acceleration $\mathbf{a}_A$ of A, we propose to determine the acceleration $\mathbf{a}_B$ of B and the angular acceleration $\boldsymbol{\alpha}$ of the rod. Choosing A as a reference point, the given motion is equivalent to a translation with A and a rotation about A. The absolute acceleration of B must be equal to the sum

$$\mathbf{a}_B = \mathbf{a}_A + \mathbf{a}_{B/A}$$
$$= \mathbf{a}_A + (\mathbf{a}_{B/A})_n + (\mathbf{a}_{B/A})_t \tag{15.23}$$

where $(\mathbf{a}_{B/A})_n$ has magnitude $l\omega^2$ and is *directed toward A*, while $(\mathbf{a}_{B/A})_t$ has the magnitude $l\alpha$ and is perpendicular to AB. Note that there is no way to tell whether the tangential component $(\mathbf{a}_{B/A})_t$ is directed to the left or to the right, and therefore, both possible directions for this component are indicated in Fig. 15.23. Similarly, both possible senses for $\mathbf{a}_B$ are indicated, since we do not know whether point B is accelerated upward or downward.

We can illustrate Eq. (15.23) geometrically. Figure 15.24 shows four different vector polygons, depending upon the sense of $\mathbf{a}_A$ and the relative magnitudes of $\mathbf{a}_A$ and $(\mathbf{a}_{B/A})_n$. To determine a_B and α from one of these diagrams, we must know not only a_A and θ but also ω. Therefore, we need to determine the angular velocity of the rod separately, by one of the methods indicated in Secs. 15.2 and 15.3. Then we can obtain the values of a_B and α by considering successively the x and y components of the vectors shown in Fig. 15.24. In the case of polygon a, we are assuming that $\boldsymbol{\alpha}$ is in the counter-clockwise direction and $\mathbf{a}_B$ is down. Therefore, we have

$\xrightarrow{+} x$ components: $\qquad 0 = a_A + l\omega^2 \sin \theta - l\alpha \cos \theta$
$+\uparrow y$ components: $\qquad -a_B = -l\omega^2 \cos \theta - l\alpha \sin \theta$

and solve for a_B and α. An alternative approach to drawing Fig. 15.24 is to use a vector algebra solution; that is, you substitute the vector quantities into (15.21'), take the cross product, and equate components to obtain the two scalar equations shown previously.

Clearly, the determination of accelerations is considerably more involved than the determination of velocities. Yet in the example considered here, the ends A and B of the rod were moving along straight tracks, and the diagrams drawn were relatively simple. If A and B had moved along curved tracks, it would have been necessary to resolve the accelerations $\mathbf{a}_A$ and $\mathbf{a}_B$ into normal and tangential components and the solution of the problem would have involved six different vectors.

When a mechanism consists of several moving parts that are pin-connected, we can analyze the mechanism by considering each part to be a rigid body, keeping in mind that the points at which two parts are connected must have the same absolute acceleration (see Sample Prob. 15.15). In the case of meshed gears (see Sample Prob. 15.13), the tangential components of the accelerations of the teeth in contact are equal, but their normal components are different.

*15.4B Analysis of Plane Motion in Terms of a Parameter

Fig. 15.25 The coordinates of the ends of the rod may be expressed in terms of the parameter θ.

In analyzing some mechanisms, it is possible to express the coordinates x and y of all the significant points of the mechanism by means of simple analytic expressions containing a single parameter. It is sometimes advantageous in such a case to determine the absolute velocity and the absolute acceleration of the various points of the mechanism directly, since we can obtain the components of the velocity and of the acceleration of a given point by differentiating the coordinates x and y of that point.

Let us consider again the rod AB whose ends slide, respectively, in a horizontal and a vertical track (Fig. 15.25). We can express the coordinates x_A and y_B of the ends of the rod in terms of the angle θ that the rod forms with the vertical:

$$x_A = l \sin \theta \qquad y_B = l \cos \theta \qquad \textbf{(15.24)}$$

Differentiating Eqs. (15.24) twice with respect to t, we have

$$v_A = \dot{x}_A = l\dot{\theta} \cos \theta$$
$$a_A = \ddot{x}_A = -l\dot{\theta}^2 \sin \theta + l\ddot{\theta} \cos \theta$$

$$v_B = \dot{y}_B = -l\dot{\theta} \sin \theta$$
$$a_B = \ddot{y}_B = -l\dot{\theta}^2 \cos \theta - l\ddot{\theta} \sin \theta$$

Recalling that $\dot{\theta} = \omega$ and $\ddot{\theta} = \alpha$, we obtain

$$v_A = l\omega \cos \theta \qquad\qquad v_B = -l\omega \sin \theta \qquad \textbf{(15.25)}$$
$$a_A = -l\omega^2 \sin \theta + l\alpha \cos \theta \qquad a_B = -l\omega^2 \cos \theta - l\alpha \sin \theta$$
$$\textbf{(15.26)}$$

Note that a positive sign for v_A or a_A indicates that the velocity $\mathbf{v}_A$ or the acceleration $\mathbf{a}_A$ is directed to the right; a positive sign for v_B or a_B indicates that $\mathbf{v}_B$ or $\mathbf{a}_B$ is directed upward. We can use Eqs. (15.25) to determine, for example, v_B and ω when v_A and θ, are known. Substituting for ω in Eqs. (15.26), we can then determine a_B and α if we know a_A.

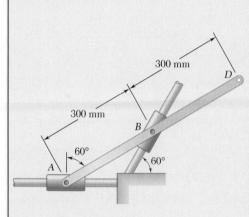

Sample Problem 15.12

Collars A and B are pin-connected to bar ABD and can slide along fixed rods. Knowing that, at the instant shown, the velocity of A is a constant 0.9 m/s to the right, determine the angular acceleration of AB and the acceleration of B.

STRATEGY: Use the kinematic equation that relates the acceleration of two points on the same rigid body. Because you know that the directions of the accelerations of A and B must be along the fixed rods, choose these two points to relate.

MODELING and ANALYSIS: Model bar ABD as a rigid body. From Sample Prob. 15.5, you know $\boldsymbol{\omega} = 3.00$ rad/s $\uparrow$. The accelerations of A and B are related by

$$\mathbf{a}_B = \mathbf{a}_A + \mathbf{a}_{B/A} = \mathbf{a}_A + \boldsymbol{\alpha} \times \mathbf{r}_{B/A} - \omega^2 \mathbf{r}_{B/A}$$

Substituting in known values (Fig. 1) and assuming $\boldsymbol{\alpha} = \alpha\mathbf{k}$ gives

$$a_B\cos 60°\mathbf{i} + a_B\sin 60°\mathbf{j} = 0\mathbf{i} + \alpha\mathbf{k} \times [(0.3\cos 30°)\mathbf{i} + (0.3\sin 30°)\mathbf{j}]$$
$$- 3^2[(0.3\cos 30°)\mathbf{i} + (0.3\sin 30°)\mathbf{j}]$$

$$0.500a_B\mathbf{i} + 0.866a_B\mathbf{j} = (0 - 0.15\alpha - 2.338)\mathbf{i} + (0.260\alpha - 1.350)\mathbf{j}$$

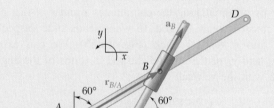

Fig. 1 Position vector and the assumed direction of the acceleration of point B.

Equating components, you have

i: $0.500a_B = -0.15\alpha - 2.338$

j: $0.866a_B = 0.260\alpha - 1.350$

Solving these equations gives $a_B = -3.12$ m/s^2 and $\alpha = -5.20$ rad/s^2.

$$\boldsymbol{\alpha} = 5.20 \text{ rad/s}^2 \downarrow \quad \blacktriangleleft$$
$$\mathbf{a}_B = 3.12 \text{ m/s}^2 \nearrow 60° \quad \blacktriangleleft$$

REFLECT and THINK: Even though A is traveling at a constant speed, bar AB still has an angular acceleration, and B has a linear acceleration. Just because one point on a body is moving at a constant speed doesn't mean the rest of the points on the body also have a constant speed.

Sample Problem 15.13

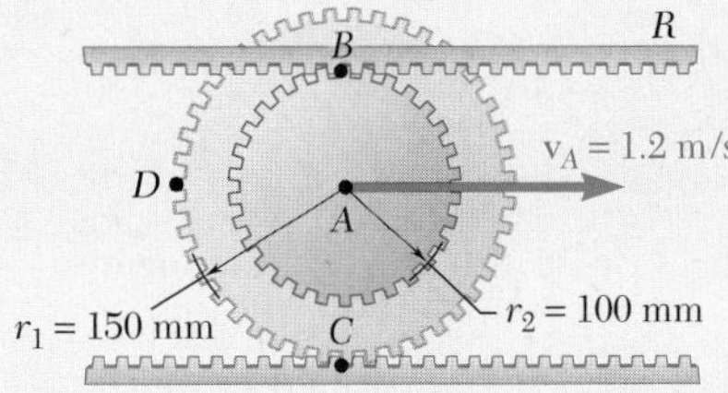

The center of the double gear of Sample Prob. 15.6 has a velocity of 1.2 m/s to the right and an acceleration of 3 m/s^2 to the right. Recalling that the lower rack is stationary, determine (a) the angular acceleration of the gear, (b) the acceleration of points B, C, and D of the gear.

STRATEGY: The double gear is a rigid body undergoing general plane motion, so use acceleration kinematics. You can also differentiate the equation for the gear's velocity and use that to find the gear's acceleration.

MODELING and ANALYSIS:

a. Angular Acceleration of the Gear. In Sample Prob. 15.6, you found that $x_A = -r_1\theta$ and $v_A = -r_1\omega$. Differentiating the second equation with respect to time, you obtain $a_A = -r_1\alpha$.

$$v_A = -r_1\omega \qquad 1.2 \text{ m/s} = -(0.150 \text{ m})\omega \qquad \omega = -8 \text{ rad/s}$$
$$a_A = -r_1\alpha \qquad 3 \text{ m/s}^2 = -(0.150 \text{ m})\alpha \qquad \alpha = -20 \text{ rad/s}^2$$
$$\boldsymbol{\alpha} = \alpha\mathbf{k} = -(20 \text{ rad/s}^2)\mathbf{k} \quad \blacktriangleleft$$

b. Accelerations. The relationship between the acceleration of any two points on a rigid body undergoing general plane motion is

$$\mathbf{a}_B = \mathbf{a}_A + \mathbf{a}_{B/A} = \mathbf{a}_A + (\mathbf{a}_{B/A})_t + (\mathbf{a}_{B/A})_n$$
$$= \mathbf{a}_A + \alpha\mathbf{k} \times \mathbf{r}_{B/A} - \omega^2\mathbf{r}_{B/A} \tag{1}$$

This equation indicates that the rolling motion of the gear can be thought of as a translation with A and a rotation about A (Fig. 1).

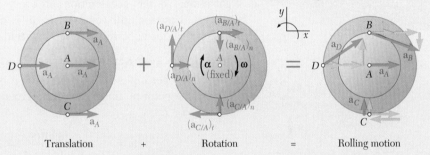

Translation + Rotation = Rolling motion

Fig. 1 A pictorial representation of Eq. 1.

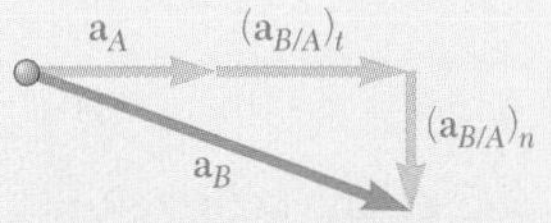

Acceleration of Point B. Substituting values into Eq. (1) gives

$$\mathbf{a}_B = \mathbf{a}_A + \mathbf{a}_{B/A} = \mathbf{a}_A + (\mathbf{a}_{B/A})_t + (\mathbf{a}_{B/A})_n$$
$$= \mathbf{a}_A + \alpha\mathbf{k} \times \mathbf{r}_{B/A} - \omega^2\mathbf{r}_{B/A}$$
$$= (3 \text{ m/s}^2)\mathbf{i} - (20 \text{ rad/s}^2)\mathbf{k} \times (0.100 \text{ m})\mathbf{j} - (8 \text{ rad/s})^2(0.100 \text{ m})\mathbf{j}$$
$$= (3 \text{ m/s}^2)\mathbf{i} + (2 \text{ m/s}^2)\mathbf{i} - (6.40 \text{ m/s}^2)\mathbf{j}$$
$$\mathbf{a}_B = 8.12 \text{ m/s}^2 \searrow 52.0° \quad \blacktriangleleft$$

The vector triangle corresponding to this equation is shown in Fig. 2.

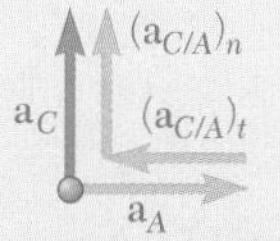

Acceleration of Point C Referring to Fig 3,

$$\mathbf{a}_C = \mathbf{a}_A + \mathbf{a}_{C/A} = \mathbf{a}_A + \alpha\mathbf{k} \times \mathbf{r}_{C/A} - \omega^2\mathbf{r}_{C/A}$$
$$= (3 \text{ m/s}^2)\mathbf{i} - (20 \text{ rad/s}^2)\mathbf{k} \times (-0.150 \text{ m})\mathbf{j} - (8 \text{ rad/s})^2(-0.150 \text{ m})\mathbf{j}$$
$$= (3 \text{ m/s}^2)\mathbf{i} - (3 \text{ m/s}^2)\mathbf{i} + (9.60 \text{ m/s}^2)\mathbf{j}$$
$$\mathbf{a}_C = 9.60 \text{ m/s}^2 \uparrow \quad \blacktriangleleft$$

(continued)

Fig. 2 Vector diagram relating the accelerations of A and B.

Fig. 3 Vector diagram of the equation relating the accelerations of A and C.

$r_1 = 150$ mm $\qquad r_2 = 100$ mm

$v_A = 1.2$ m/s

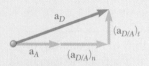

Fig. 4 Vector diagram relating the accelerations of *A* and *D*.

Acceleration of Point *D* (Fig. 4).

$$\mathbf{a}_D = \mathbf{a}_A + \mathbf{a}_{D/A} = \mathbf{a}_A + \alpha\mathbf{k} \times \mathbf{r}_{D/A} - \omega^2\mathbf{r}_{D/A}$$
$$= (3 \text{ m/s}^2)\mathbf{i} - (20 \text{ rad/s}^2)\mathbf{k} \times (-0.150 \text{ m})\mathbf{i} - (8 \text{ rad/s})^2(-0.150 \text{ m})\mathbf{i}$$
$$= (3 \text{ m/s}^2)\mathbf{i} + (3 \text{ m/s}^2)\mathbf{j} + (9.60 \text{ m/s}^2)\mathbf{i}$$
$$\mathbf{a}_D = 12.95 \text{ m/s}^2 \ \measuredangle \ 13.4° \ \blacktriangleleft$$

REFLECT and THINK: It is interesting to note that the *x*-component of acceleration for point *C* is zero since it is in direct contact with the fixed lower rack. It does, however, have a normal acceleration pointed upward. This is also true for a wheel rolling without slip.

Sample Problem 15.14

Two adjacent identical wheels of a train can be modeled as rolling cylinders connected by a horizontal link. The distance between *A* and *D* is 10 in. Assume the wheels roll without sliding on the tracks. Knowing that the train is traveling at a constant 30 mph, determine the acceleration of the center of mass of *DE*.

STRATEGY: The connecting bar *DE* is undergoing curvilinear translation, so the acceleration of every point is identical; that is, $a_G = a_D$. Therefore, all you need to do is determine the acceleration of *D* using the kinematic relationship between *A* and *D*.

MODELING and ANALYSIS: Model the wheels and bar *DE* as rigid bodies. The speed of *A* is $v_A = 30$ mph $= 44$ ft/s. Since the wheel does not slip, the point of contact with the ground, *C* (Fig. 1), has a velocity of zero, so

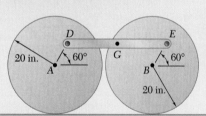

$$\omega = \frac{v_A}{r_{A/C}} = \frac{44 \text{ ft/s}}{(20/12) \text{ ft.}} = 26.4 \text{ rad/s}$$

Acceleration of *D*. The acceleration of *D* is

$$\mathbf{a}_D = \mathbf{a}_A + \mathbf{a}_{D/A} = \mathbf{a}_A + \boldsymbol{\alpha} \times \mathbf{r}_{D/A} - \omega^2\mathbf{r}_{D/A} \quad (1)$$

The train is traveling at a constant speed, so a_A and α are both zero. Substituting known quantities into Eq. (1) gives

$$\mathbf{a}_D = 0 + 0 - (26.4 \text{ rad/s})^2\left[\left(\frac{10}{12}\cos 60° \text{ ft}\right)\mathbf{i} + \left(\frac{10}{12}\sin 60° \text{ ft}\right)\mathbf{j}\right]$$

$$= -(290.4 \text{ ft/s}^2)\mathbf{i} - (503.0 \text{ ft/s}^2)\mathbf{j}$$

$$\mathbf{a}_G = \mathbf{a}_D = -(290 \text{ ft/s}^2)\mathbf{i} - (503 \text{ ft/s}^2)\mathbf{j} \ \blacktriangleleft$$

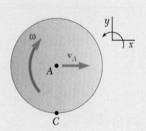

Fig. 1 Velocity and angular velocity of the wheel.

REFLECT and THINK: Instead of using vector algebra, you could have recognized that the direction of $-\omega^2\mathbf{r}_{D/A}$ is directed from *D* to *A*. So the final acceleration of *D* is simply $-\omega^2\mathbf{r}_{D/A} \ \measuredangle 60°$.

Sample Problem 15.15

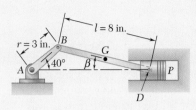

Fig. 1 The acceleration of B is only in the normal direction.

Crank *AB* of the engine system of Sample Prob. 15.7 has a constant clockwise angular velocity of 2000 rpm. For the crank position shown, determine the angular acceleration of the connecting rod *BD* and the acceleration of point *D*.

STRATEGY: The linkage consists of two rigid bodies: crank *AB* is rotating about a fixed axis and connecting rod *BD* is undergoing general plane motion. Therefore, you need to use rigid-body kinematics.

MODELING and ANALYSIS:

Motion of Crank *AB*. Since the crank rotates about *A* with constant $\omega_{AB} = 2000$ rpm $= 209.4$ rad/s, you have $\alpha_{AB} = 0$. The acceleration of *B* is therefore directed toward *A* (Fig. 1) and has the magnitude of

$$a_B = r\omega_{AB}^2 = (\tfrac{3}{12}\text{ ft})(209.4\text{ rad/s})^2 = 10{,}962\text{ ft/s}^2$$
$$\mathbf{a}_B = 10{,}962\text{ ft/s}^2 \; \nearrow\!\!\nwarrow\; 40°$$

Motion of Connecting Rod *BD*. The angular velocity ω_{BD} and the value of β were obtained in Sample Prob. 15.7 using relative velocity equations:

$$\omega_{BD} = 62.0 \text{ rad/s} \uparrow \qquad \beta = 13.95°$$

Resolve the motion of *BD* into a translation with *B* and a rotation about *B* (Fig. 2). Resolve the relative acceleration $\mathbf{a}_{D/B}$ into normal and tangential components:

$$(a_{D/B})_n = (BD)\omega_{BD}^2 = (\tfrac{8}{12}\text{ ft})(62.0\text{ rad/s})^2 = 2563\text{ ft/s}^2$$
$$(\mathbf{a}_{D/B})_n = 2563\text{ ft/s}^2 \; \searrow\; 13.95°$$
$$(a_{D/B})_t = (BD)\alpha_{BD} = (\tfrac{8}{12})\alpha_{BD} = 0.6667\alpha_{BD}$$
$$(\mathbf{a}_{D/B})_t = 0.6667\alpha_{BD} \; \swarrow\!\!\nearrow\; 76.05°$$

Although $(\mathbf{a}_{D/B})_t$ must be perpendicular to *BD*, its sense is not known.

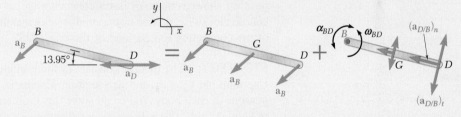

Plane motion   =   Translation   +   Rotation

Fig. 2 General plane motion is a translation plus a rotation.

Noting that the acceleration $\mathbf{a}_D$ must be horizontal, you have

$$\mathbf{a}_D = \mathbf{a}_B + \mathbf{a}_{D/B} = \mathbf{a}_B + (\mathbf{a}_{D/B})_n + (\mathbf{a}_{D/B})_t \qquad (1)$$
$$[a_D \leftrightarrow] = [10{,}962 \; \nearrow\!\!\nwarrow\; 40°] + [2563 \; \searrow\; 13.95°] + [0.6667\alpha_{BD} \; \swarrow\!\!\nearrow\; 76.05°]$$

Equating *x* and *y* components, you obtain the following scalar equations, as

$\xrightarrow{+} x$ components:
$$-a_D = -10{,}962 \cos 40° - 2563 \cos 13.95° + 0.6667\alpha_{BD} \sin 13.95°$$

$+\uparrow y$ components:
$$0 = -10{,}962 \sin 40° + 2563 \sin 13.95° + 0.6667\alpha_{BD} \cos 13.95°$$

(continued)

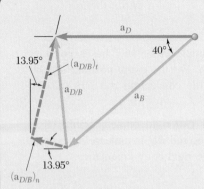

Fig. 3 Vector polygon relating the accelerations of B and D.

Solving the equations simultaneously gives $\alpha_{BD} = +9940$ rad/s² and $a_D = +9290$ ft/s². The positive signs indicate that the senses shown on the vector polygon (Fig. 3) are correct.

$$\alpha_{BD} = 9940 \text{ rad/s}^2 \,\gamma \quad \blacktriangleleft$$
$$\mathbf{a}_D = 9290 \text{ ft/s}^2 \leftarrow \quad \blacktriangleleft$$

REFLECT and THINK: In this solution, you looked at the magnitude and direction of each term in Eq. (1) and then found the x and y components. Alternatively, you could have assumed that a_D was to the left, α_{BD} was positive, and then substituted in the vector quantities to get

$$\mathbf{a}_D = \mathbf{a}_B + \alpha \mathbf{k} \times \mathbf{r}_{D/B} - \omega^2 \mathbf{r}_{D/B}$$

$$-a_D\mathbf{i} = -a_B \cos 40°\mathbf{i} - a_B \sin 40°\mathbf{j} + \alpha_{BD}\mathbf{k} \times (l \cos \beta\mathbf{i} - l \sin \beta\mathbf{j})$$
$$- \omega_{BD}^2 (l \cos \beta\mathbf{i} - l \sin \beta\mathbf{j})$$
$$= -a_B \cos 40°\mathbf{i} - a_B \sin 40°\mathbf{j} + \alpha_{BD} l \cos \beta\mathbf{j} + \alpha_{BD} l \sin \beta\mathbf{i}$$
$$- \omega_{BD}^2 l \cos \beta\mathbf{i} + \omega_{BD}^2 l \sin \beta\mathbf{j}$$

Equating components gives

i: $\quad -a_D = -a_B \cos 40° + \alpha_{BD} l \sin \beta - \omega_{BD}^2 l \cos \beta$

j: $\quad 0 = -a_B \sin 40° + \alpha_{BD} l \cos \beta + \omega_{BD}^2 l \sin \beta$

These are identical to the previous equations if you substitute in the numbers.

Sample Problem 15.16

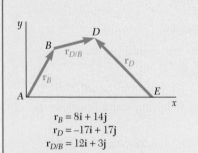

Fig. 1 Position vectors for points B, D, and E.

The linkage *ABDE* moves in the vertical plane. Knowing that, in the position shown, crank *AB* has a constant angular velocity ω_1 of 20 rad/s counterclockwise, determine the angular velocities and angular accelerations of the connecting rod *BD* and of the crank *DE*.

STRATEGY: The linkage consists of three interconnected rigid bodies. Use multiple velocity and acceleration kinematic equations to relate the motions of each body. You could solve this problem with the method used in Sample Prob. 15.15; however, we illustrate a vector approach, choosing position vectors $\mathbf{r}_B$, $\mathbf{r}_D$, and $\mathbf{r}_{D/B}$ as shown in Fig. 1.

MODELING and ANALYSIS:

Velocities. Assuming that the angular velocities of *BD* and *DE* are counterclockwise, you have

$$\boldsymbol{\omega}_{AB} = \omega_{AB}\mathbf{k} = (20 \text{ rad/s})\mathbf{k} \qquad \boldsymbol{\omega}_{BD} = \omega_{BD}\mathbf{k} \qquad \boldsymbol{\omega}_{DE} = \omega_{DE}\mathbf{k}$$

where $\mathbf{k}$ is a unit vector pointing out of the page. We can obtain the velocity of D by relating it to point E, as

$$\mathbf{v}_D = \mathbf{v}_E + \mathbf{v}_{D/E} = 0 + \omega_{AB}\mathbf{k} \times \mathbf{r}_D \tag{1}$$

In the Fig. 1 position vectors:
$$r_B = 8\mathbf{i} + 14\mathbf{j}$$
$$r_D = -17\mathbf{i} + 17\mathbf{j}$$
$$r_{D/B} = 12\mathbf{i} + 3\mathbf{j}$$

We can obtain the velocity of B by relating it to point A, as

$$\mathbf{v}_B = \mathbf{v}_A + \mathbf{v}_{B/A} = 0 + \omega_{AB}\mathbf{k} \times \mathbf{r}_B \qquad (2)$$

The relationship between the velocities of D and B is

$$\mathbf{v}_D = \mathbf{v}_B + \mathbf{v}_{D/B} \qquad (3)$$

Substituting Eqs. (1) and (2) into Eq. (3) and using $\mathbf{v}_{D/B} = \omega_{BD}\mathbf{k} \times \mathbf{r}_{D/B}$ gives

$$\omega_{DE}\mathbf{k} \times \mathbf{r}_D = \omega_{AB}\mathbf{k} \times \mathbf{r}_B + \omega_{BD}\mathbf{k} \times \mathbf{r}_{D/B}$$
$$\omega_{DE}\mathbf{k} \times (-17\mathbf{i} + 17\mathbf{j}) = 20\mathbf{k} \times (8\mathbf{i} + 14\mathbf{j}) + \omega_{BD}\mathbf{k} \times (12\mathbf{i} + 3\mathbf{j})$$
$$-17\omega_{DE}\mathbf{j} - 17\omega_{DE}\mathbf{i} = 160\mathbf{j} - 280\mathbf{i} + 12\omega_{BD}\mathbf{j} - 3\omega_{BD}\mathbf{i}$$

Equating the coefficients of the unit vectors $\mathbf{i}$ and $\mathbf{j}$, the following two scalar equations are

$$-17\omega_{DE} = -280 - 3\omega_{BD}$$
$$-17\omega_{DE} = +160 + 12\omega_{BD}$$

Solving these gives you $\omega_{BD} = -(29.33 \text{ rad/s})\mathbf{k}$ $\omega_{DE} = (11.29 \text{ rad/s})\mathbf{k}$ ◀

Accelerations.   At the instant considered, crank AB has a constant angular velocity, so you have

$$\boldsymbol{\alpha}_{AB} = 0 \qquad \boldsymbol{\alpha}_{BD} = \alpha_{BD}\mathbf{k} \qquad \boldsymbol{\alpha}_{DE} = \alpha_{DE}\mathbf{k}$$
$$\mathbf{a}_D = \mathbf{a}_B + \mathbf{a}_{D/B} \qquad (4)$$

Evaluate each term of Eq. (4) separately:

Bar *DE:* $\begin{aligned}\mathbf{a}_D &= \alpha_{DE}\mathbf{k} \times \mathbf{r}_D - \omega_{DE}^2\mathbf{r}_D \\ &= \alpha_{DE}\mathbf{k} \times (-17\mathbf{i} + 17\mathbf{j}) - (11.29)^2(-17\mathbf{i} + 17\mathbf{j}) \\ &= -17\alpha_{DE}\mathbf{j} - 17\alpha_{DE}\mathbf{i} + 2170\mathbf{i} - 2170\mathbf{j}\end{aligned}$

Bar *AB:* $\begin{aligned}\mathbf{a}_B &= \alpha_{AB}\mathbf{k} \times \mathbf{r}_B - \omega_{AB}^2\mathbf{r}_B = 0 - (20)^2(8\mathbf{i} + 14\mathbf{j}) \\ &= -3200\mathbf{i} - 5600\mathbf{j}\end{aligned}$

Bar *BD:* $\begin{aligned}\mathbf{a}_{D/B} &= \alpha_{BD}\mathbf{k} \times \mathbf{r}_{D/B} - \omega_{BD}^2\mathbf{r}_{D/B} \\ &= \alpha_{BD}\mathbf{k} \times (12\mathbf{i} + 3\mathbf{j}) - (29.33)^2(12\mathbf{i} + 3\mathbf{j}) \\ &= 12\alpha_{BD}\mathbf{j} - 3\alpha_{BD}\mathbf{i} - 10{,}320\mathbf{i} - 2580\mathbf{j}\end{aligned}$

Substituting into Eq. (4) and equating the coefficients of $\mathbf{i}$ and $\mathbf{j}$, you obtain

$$-17\alpha_{DE} + 3\alpha_{BD} = -15{,}690$$
$$-17\alpha_{DE} - 12\alpha_{BD} = -6010$$

Solving these gives you $\alpha_{BD} = -(645 \text{ rad/s}^2)\mathbf{k}$ $\alpha_{DE} = (809 \text{ rad/s}^2)\mathbf{k}$ ◀

REFLECT and THINK: The vector approach is preferred when there are more than two linkages. It is a very methodic approach and is easier to program when simulating mechanism movement over time.

SOLVING PROBLEMS ON YOUR OWN

This section was devoted to determining the accelerations of the points of a rigid body in plane motion. As you did previously for velocities, you will again consider the plane motion of a rigid body as the sum of two motions, namely, a translation and a rotation.

To solve a problem involving accelerations in plane motion, use the following steps.

1. Determine the angular velocity of the body. To find ω, you can either

    a. Consider the motion of the body as the sum of a translation and a rotation, as you did in Sec. 15.2, or

    b. Use the vector approach, as you did in Sec. 15.2, or the instantaneous center of rotation of the body, as you did in Sec. 15.3. However, keep in mind that you cannot use the instantaneous center to determine accelerations.

2. A diagram may be helpful to visualize the kinematics of the rigid bodies. The diagram will include the following diagrams (Fig 15.22):

    a. Plane motion diagram. Draw a sketch of the body, including all dimensions, as well as the angular velocity ω. Show the angular acceleration α with its magnitude and sense if you know them. Also show those points for which you know or seek the accelerations, indicating all that you know about these accelerations.

    b. Translation diagram. Select a reference point A for which you know the direction, the magnitude, or a component of the acceleration $\mathbf{a}_A$. Draw a second diagram showing the body in translation with each point having the same acceleration as point A.

    c. Rotation diagram. Considering point A as a fixed reference point, draw a third diagram showing the body in rotation about A. Indicate the normal and tangential components of the relative accelerations of other points, such as the components $(\mathbf{a}_{B/A})_n$ and $(\mathbf{a}_{B/A})_t$ of the acceleration of point B with respect to point A.

3. Write the relative-acceleration formula relating two points of interest on the body being analyzed

$$\mathbf{a}_B = \mathbf{a}_A + \mathbf{a}_{B/A} \qquad \text{or} \qquad \mathbf{a}_B = \mathbf{a}_A + (\mathbf{a}_{B/A})_n + (\mathbf{a}_{B/A})_t$$

    a. Graphical approach. Select a point for which you know the direction, the magnitude, or a component of the acceleration and draw a vector diagram of the equation [Sample Prob. 15.15]. Starting at the same point, draw all known acceleration components in tip-to-tail fashion for each member of the equation. Complete the diagram by drawing the two remaining vectors in appropriate directions and in such a way that the two sums of vectors end at a common point.

   b. Vector approach. For a single rigid body, it is straightforward to apply

$$\mathbf{a}_B = \mathbf{a}_A + \boldsymbol{\alpha}_{AB} \times \mathbf{r}_{B/A} - \omega_{AB}^2 \mathbf{r}_{B/A}$$

For linkage type problems, you will need to write multiple relative acceleration equations relating the accelerations of points along the linkage [Sample Prob. 15.16].

4. The analysis of plane motion in terms of a parameter completed this section. This method should be used *only* if it is possible to express the coordinates x and y of all significant points of the body in terms of a single parameter (Sec. 15.4B). By differentiating the coordinates x and y of a given point twice with respect to t, you can determine the rectangular components of the absolute velocity and absolute acceleration of that point.

Problems

15.CQ7 A rear-wheel-drive car starts from rest and accelerates to the left so that the tires do not slip on the road. What is the direction of the acceleration of the point on the tire in contact with the road, that is, point A?

a. ← **b.** ↖ **c.** ↑ **d.** ↓ **e.** ↙

Fig. P15.CQ7

END-OF-SECTION PROBLEMS

15.105 A 5-m steel beam is lowered by means of two cables unwinding at the same speed from overhead cranes. As the beam approaches the ground, the crane operators apply brakes to slow the unwinding motion. At the instant considered, the deceleration of the cable attached at B is 2.5 m/s^2, while that of the cable attached at D is 1.5 m/s^2. Determine (a) the angular acceleration of the beam, (b) the acceleration of points A and E.

15.106 For a 5-m steel beam AE, the acceleration of point A is 2 m/s^2 downward and the angular acceleration of the beam is 1.2 rad/s^2 counterclockwise. Knowing that at the instant considered the angular velocity of the beam is zero, determine the acceleration (a) of cable B, (b) of cable D.

15.107 A 900-mm rod rests on a horizontal table. A force **P** applied as shown produces the following accelerations: $\mathbf{a}_A = 3.6$ m/s^2 to the right, $\alpha = 6$ rad/s^2 counterclockwise as viewed from above. Determine the acceleration (a) of point G, (b) of point B.

15.108 In Prob. 15.107, determine the point of the rod that (a) has no acceleration, (b) has an acceleration of 2.4 m/s^2 to the right.

15.109 Knowing that at the instant shown crank BC has a constant angular velocity of 45 rpm clockwise, determine the acceleration (a) of point A, (b) of point D.

15.110 End A of rod AB moves to the right with a constant velocity of 6 ft/s. For the position shown, determine (a) the angular acceleration of rod AB, (b) the acceleration of the midpoint G of rod AB.

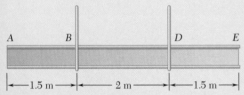

Fig. P15.105 and P15.106

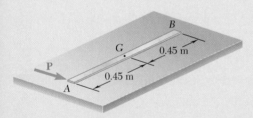

Fig. P15.107 and P15.108

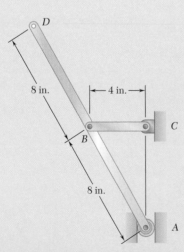

Fig. P15.109

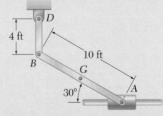

Fig. P15.110

15.111 An automobile travels to the left at a constant speed of 72 km/h. Knowing that the diameter of the wheel is 560 mm, determine the acceleration (*a*) of point *B*, (*b*) of point *C*, (*c*) of point *D*.

15.112 The 18-in.-radius flywheel is rigidly attached to a 1.5-in.-radius shaft that can roll along parallel rails. Knowing that at the instant shown the center of the shaft has a velocity of 1.2 in./s and an acceleration of 0.5 in./s², both directed down to the left, determine the acceleration (*a*) of point *A*, (*b*) of point *B*.

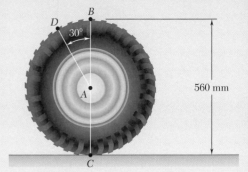

Fig. P15.111

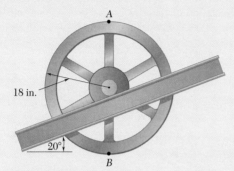

Fig. P15.112

15.113 and 15.114 A 3-in.-radius drum is rigidly attached to a 5-in.-radius drum as shown. One of the drums rolls without sliding on the surface shown, and a cord is wound around the other drum. Knowing that at the instant shown end *D* of the cord has a velocity of 8 in./s and an acceleration of 30 in./s², both directed to the left, determine the accelerations of points *A*, *B*, and *C* of the drums.

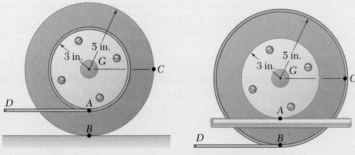

Fig. P15.113   **Fig. P15.114**

15.115 A heavy crate is being moved a short distance using three identical cylinders as rollers. Knowing that at the instant shown the crate has a velocity of 200 mm/s and an acceleration of 400 mm/s², both directed to the right, determine (*a*) the angular acceleration of the center cylinder, (*b*) the acceleration of point *A* on the center cylinder.

Fig. P15.115

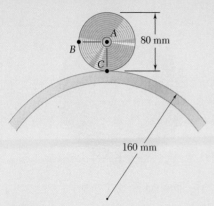

15.116 A wheel rolls without slipping on a fixed cylinder. Knowing that at the instant shown the angular velocity of the wheel is 10 rad/s clockwise and its angular acceleration is 30 rad/s² counterclockwise, determine the acceleration of (*a*) point *A*, (*b*) point *B*, (*c*) point *C*.

15.117 The 100-mm-radius drum rolls without slipping on a portion of a belt that moves downward to the left with a constant velocity of 120 mm/s. Knowing that at a given instant the velocity and acceleration of the center *A* of the drum are as shown, determine the acceleration of point *D*.

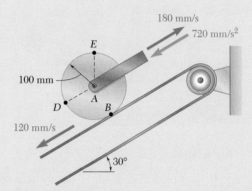

15.118 In the planetary gear system shown, the radius of gears *A*, *B*, *C*, and *D* is 3 in. and the radius of the outer gear *E* is 9 in. Knowing that gear *A* has a constant angular velocity of 150 rpm clockwise and that the outer gear *E* is stationary, determine the magnitude of the acceleration of the tooth of gear *D* that is in contact with (*a*) gear *A*, (*b*) gear *E*.

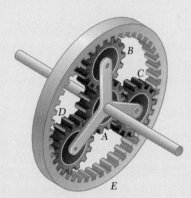

15.119 The 200-mm-radius disk rolls without sliding on the surface shown. Knowing that the distance *BG* is 160 mm and that at the instant shown the disk has an angular velocity of 8 rad/s counterclockwise and an angular acceleration of 2 rad/s² clockwise, determine the acceleration of *A*.

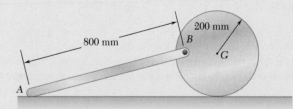

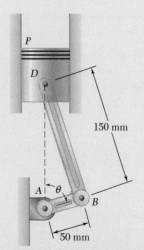

15.120 Knowing that crank *AB* rotates about point *A* with a constant angular velocity of 900 rpm clockwise, determine the acceleration of the piston *P* when *θ* = 60°.

15.121 Knowing that crank *AB* rotates about point *A* with a constant angular velocity of 900 rpm clockwise, determine the acceleration of the piston *P* when *θ* = 120°.

15.122 In the two-cylinder air compressor shown, the connecting rods *BD* and *BE* are each 190 mm long and crank *AB* rotates about the fixed point *A* with a constant angular velocity of 1500 rpm clockwise. Determine the acceleration of each piston when $\theta = 0$.

15.123 The disk shown has a constant angular velocity of 500 rpm counterclockwise. Knowing that rod *BD* is 10 in. long, determine the acceleration of collar *D* when (*a*) $\theta = 90°$, (*b*) $\theta = 180°$.

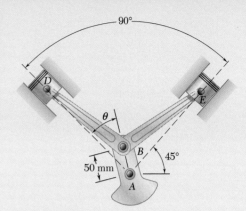

Fig. P15.122

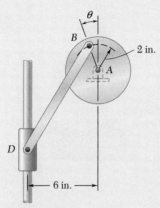

Fig. P15.123

15.124 Arm *AB* has a constant angular velocity of 16 rad/s counterclockwise. At the instant when $\theta = 90°$, determine the acceleration (*a*) of collar *D*, (*b*) of the midpoint *G* of bar *BD*.

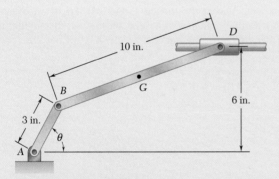

Fig. P15.124 and P15.125

15.125 Arm *AB* has a constant angular velocity of 16 rad/s counterclockwise. At the instant when $\theta = 60°$, determine the acceleration of collar *D*.

15.126 A straight rack rests on a gear of radius $r = 3$ in. and is attached to a block *B* as shown. Knowing that at the instant shown $\theta = 20°$, the angular velocity of gear *D* is 3 rad/s clockwise, and it is speeding up at a rate of 2 rad/s², determine (*a*) the angular acceleration of *AB*, (*b*) the acceleration of block *B*.

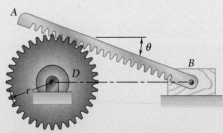

Fig. P15.126

15.127 The elliptical exercise machine has fixed axes of rotation at points *A* and *E*. Knowing that at the instant shown the flywheel *AB* has a constant angular velocity of 6 rad/s clockwise, determine the acceleration of point *D*.

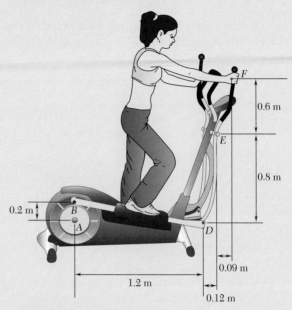

Fig. P15.127 and P15.128

15.128 The elliptical exercise machine has fixed axes of rotation at points *A* and *E*. Knowing that at the instant shown the flywheel *AB* has a constant angular velocity of 6 rad/s clockwise, determine (*a*) the angular acceleration of bar *DEF*, (*b*) the acceleration of point *F*.

15.129 Knowing that at the instant shown bar *AB* has a constant angular velocity of 19 rad/s clockwise, determine (*a*) the angular acceleration of bar *BGD*, (*b*) the angular acceleration of bar *DE*.

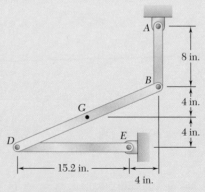

Fig. P15.129 and P15.130

15.130 Knowing that at the instant shown bar *DE* has a constant angular velocity of 18 rad/s clockwise, determine (*a*) the acceleration of point *B*, (*b*) the acceleration of point *G*.

15.131 and 15.132 Knowing that at the instant shown bar *AB* has a constant angular velocity of 4 rad/s clockwise, determine the angular acceleration (*a*) of bar *BD*, (*b*) of bar *DE*.

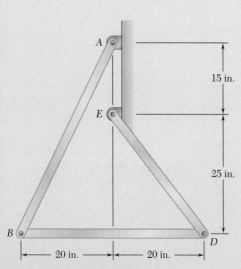

Fig. P15.131 and P15.133

15.133 and 15.134 Knowing that at the instant shown bar *AB* has an angular velocity of 4 rad/s and an angular acceleration of 2 rad/s², both clockwise, determine the angular acceleration (*a*) of bar *BD*, (*b*) of bar *DE* by using the vector approach as is done in Sample Prob. 15.16.

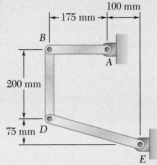

15.135 Roberts linkage is named after Richard Roberts (1789–1864) and can be used to draw a close approximation to a straight line by locating a pen at point *F*. The distance *AB* is the same as *BF*, *DF*, and *DE*. Knowing that at the instant shown, bar *AB* has a constant angular velocity of 4 rad/s clockwise, determine (*a*) the angular acceleration of bar *DE*, (*b*) the acceleration of point *F*.

Fig. P15.132 and P15.134

15.136 For the oil pump rig shown, link *AB* causes the beam *BCE* to oscillate as the crank *OA* revolves. Knowing that *OA* has a radius of 0.6 m and a constant clockwise angular velocity of 20 rpm, determine the velocity and acceleration of point *D* at the instant shown.

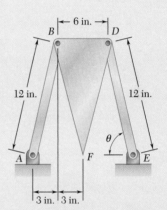

Fig. P15.135

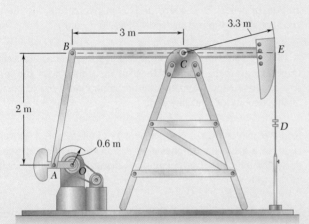

Fig. P15.136

15.137 Denoting by $\mathbf{r}_A$ the position vector of a point *A* of a rigid slab that is in plane motion, show that (*a*) the position vector $\mathbf{r}_C$ of the instantaneous center of rotation is

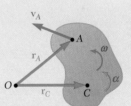

Fig. P15.137

$$\mathbf{r}_C = \mathbf{r}_A + \frac{\boldsymbol{\omega} \times \mathbf{v}_A}{\omega^2}$$

where $\boldsymbol{\omega}$ is the angular velocity of the slab and $\mathbf{v}_A$ is the velocity of point *A*, (*b*) the acceleration of the instantaneous center of rotation is zero if, and only if,

$$\mathbf{a}_A = \frac{\alpha}{\omega} \mathbf{v}_A + \boldsymbol{\omega} \times \mathbf{v}_A$$

where $\boldsymbol{\alpha} = \alpha \mathbf{k}$ is the angular acceleration of the slab.

*15.138 The drive disk of the Scotch crosshead mechanism shown has an angular velocity $\boldsymbol{\omega}$ and an angular acceleration $\boldsymbol{\alpha}$, both directed counterclockwise. Using the method of Sec. 15.4B, derive expressions for the velocity and acceleration of point *B*.

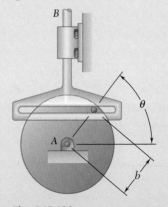

Fig. P15.138

*15.139 The wheels attached to the ends of rod *AB* roll along the surfaces shown. Using the method of Sec. 15.4B, derive an expression for the angular velocity of the rod in terms of v_B, θ, l, and β.

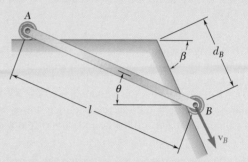

Fig. P15.139 and P15.140

*15.140 The wheels attached to the ends of rod *AB* roll along the surfaces shown. Using the method of Sec. 15.4B and knowing that the acceleration of wheel *B* is zero, derive an expression for the angular acceleration of the rod in terms of v_B, θ, l, and β.

*15.141 A disk of radius *r* rolls to the right with a constant velocity **v**. Denoting by *P* the point of the rim in contact with the ground at $t = 0$, derive expressions for the horizontal and vertical components of the velocity of *P* at any time *t*.

*15.142 Rod *AB* moves over a small wheel at *C* while end *A* moves to the right with a constant velocity $\mathbf{v}_A$. Using the method of Sec. 15.4B, derive expressions for the angular velocity and angular acceleration of the rod.

*15.143 Rod *AB* moves over a small wheel at *C* while end *A* moves to the right with a constant velocity $\mathbf{v}_A$. Using the method of Sec. 15.4B, derive expressions for the horizontal and vertical components of the velocity of point *B*.

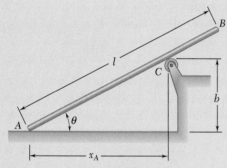

Fig. P15.142 and P15.143

15.144 Crank *AB* rotates with a constant clockwise angular velocity **ω**. Using the method of Sec. 15.4B, derive expressions for the angular velocity of rod *BD* and the velocity of the point on the rod coinciding with point *E* in terms of θ, ω, *b*, and *l*.

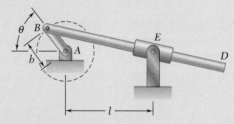

Fig. P15.144 and P15.145

15.145 Crank *AB* rotates with a constant clockwise angular velocity **ω**. Using the method of Sec. 15.4B, derive an expression for the angular acceleration of rod *BD* in terms of θ, ω, *b*, and *l*.

15.146 Solve the engine system from Sample Prob. 15.15 using the methods of Section 15.4B. *Hint:* Define the angle between the horizontal and the crank *AB* as θ and derive the motion in terms of this parameter.

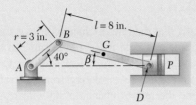

Fig. P15.146

*15.147 The position of rod *AB* is controlled by a disk of radius *r* that is attached to yoke *CD*. Knowing that the yoke moves vertically upward with a constant velocity $\mathbf{v}_0$, derive expressions for the angular velocity and angular acceleration of rod *AB*.

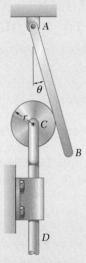

Fig. P15.147

*15.148 A wheel of radius *r* rolls without slipping along the inside of a fixed cylinder of radius *R* with a constant angular velocity $\boldsymbol{\omega}$. Denoting by *P* the point of the wheel in contact with the cylinder at $t = 0$, derive expressions for the horizontal and vertical components of the velocity of *P* at any time *t*. (The curve described by point *P* is a *hypocycloid*.)

*15.149 In Prob. 15.148, show that the path of *P* is a vertical straight line when $r = R/2$. Derive expressions for the corresponding velocity and acceleration of *P* at any time *t*.

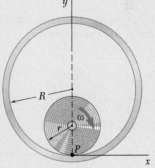

Fig. P15.148

Photo 15.7 A Geneva mechanism is used to convert rotary motion into intermittent motion.

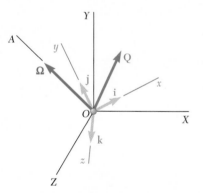

Fig. 15.26 A fixed frame of reference *OXYZ* and a rotating frame *Oxyz* with angular velocity $\boldsymbol{\Omega}$.

15.5 ANALYZING MOTION WITH RESPECT TO A ROTATING FRAME

We saw in Sec. 11.4B that the rate of change of a vector is the same with respect to a fixed frame and with respect to a frame in translation. In this section, we consider the rates of change of a vector $\mathbf{Q}$ with respect to a fixed frame and with respect to a rotating frame of reference.[†] You will see how to determine the rate of change of $\mathbf{Q}$ with respect to one frame of reference when $\mathbf{Q}$ is defined by its components in another frame. This kind of analysis is very useful for designing mechanisms that convert one kind of motion into another, such as continuous rotation into intermittent rotation. It is also helpful when you have, say, an extending linear actuator that is also rotating.

15.5A Rate of Change of a Vector with Respect to a Rotating Frame

Consider two frames of reference centered at *O*: a fixed frame *OXYZ* and a frame *Oxyz* that rotates about the fixed axis *OA*. Let $\boldsymbol{\Omega}$ denote the angular velocity of the frame *Oxyz* at a given instant (Fig. 15.26). Consider now a vector function $\mathbf{Q}(t)$ represented by the vector $\mathbf{Q}$ attached at *O;* as the time *t* varies, both the direction and the magnitude of $\mathbf{Q}$ change. The variation of $\mathbf{Q}$ is viewed differently by an observer using *OXYZ* as a frame of reference and by an observer using *Oxyz*, so we should expect the rate of change of $\mathbf{Q}$ to depend upon the frame of reference that has been selected. Therefore, we denote the rate of change of $\mathbf{Q}$ with respect to the fixed frame *OXYZ* by $(\dot{\mathbf{Q}})_{OXYZ}$ and the rate of change of $\mathbf{Q}$ with respect to the rotating frame *Oxyz* by $(\dot{\mathbf{Q}})_{Oxyz}$. We propose to determine the relation between these two rates of change.

Let us first resolve the vector $\mathbf{Q}$ into components along the *x*, *y*, and *z* axes of the rotating frame. Denoting the corresponding unit vectors by $\mathbf{i}$, $\mathbf{j}$, and $\mathbf{k}$, we have

$$\mathbf{Q} = Q_x\mathbf{i} + Q_y\mathbf{j} + Q_z\mathbf{k} \tag{15.27}$$

Differentiating Eq. (15.27) with respect to *t* and considering the unit vectors $\mathbf{i}$, $\mathbf{j}$, $\mathbf{k}$ to be fixed, we obtain the rate of change of $\mathbf{Q}$ with respect to the rotating frame *Oxyz*, as

$$(\dot{\mathbf{Q}})_{Oxyz} = \dot{Q}_x\mathbf{i} + \dot{Q}_y\mathbf{j} + \dot{Q}_z\mathbf{k} \tag{15.28}$$

To obtain the rate of change of $\mathbf{Q}$ with respect to the fixed frame *OXYZ*, we must consider the unit vectors $\mathbf{i}$, $\mathbf{j}$, $\mathbf{k}$ to be variable when differentiating Eq. (15.27). This gives

$$(\dot{\mathbf{Q}})_{OXYZ} = \dot{Q}_x\mathbf{i} + \dot{Q}_y\mathbf{j} + \dot{Q}_z\mathbf{k} + Q_x\frac{d\mathbf{i}}{dt} + Q_y\frac{d\mathbf{j}}{dt} + Q_z\frac{d\mathbf{k}}{dt} \tag{15.29}$$

From Eq. (15.28), we observe that the sum of the first three terms in the right-hand side of Eq. (15.29) represents the rate of change $(\dot{\mathbf{Q}})_{Oxyz}$.

[†]Recall that the selection of a fixed frame of reference is arbitrary. Any frame may be designated as "fixed"; all others are then considered as moving.

We note, on the other hand, that the rate of change $(\dot{\mathbf{Q}})_{OXYZ}$ would reduce to the last three terms in Eq. (15.29) if vector $\mathbf{Q}$ were fixed within the frame $Oxyz$, since $(\dot{\mathbf{Q}})_{Oxyz}$ would then be zero. But in that case, $(\dot{\mathbf{Q}})_{OXYZ}$ would represent the velocity of a particle located at the tip of $\mathbf{Q}$ and belonging to a body rigidly attached to the frame $Oxyz$. Thus, the last three terms in Eq. (15.29) represent the velocity of that particle. Since the frame $Oxyz$ has an angular velocity $\mathbf{\Omega}$ with respect to $OXYZ$ at the instant considered, we have, by Eq. (15.5),

$$Q_x\frac{d\mathbf{i}}{dt} + Q_y\frac{d\mathbf{j}}{dt} + Q_z\frac{d\mathbf{k}}{dt} = \mathbf{\Omega} \times \mathbf{Q} \qquad \text{(15.30)}$$

Substituting from Eqs. (15.28) and (15.30) into Eq. (15.29), we obtain the fundamental relation

$$(\dot{\mathbf{Q}})_{OXYZ} = (\dot{\mathbf{Q}})_{Oxyz} + \mathbf{\Omega} \times \mathbf{Q} \qquad \text{(15.31)}$$

We conclude that the rate of change of vector $\mathbf{Q}$ with respect to the fixed frame $OXYZ$ consists of two parts: The first part represents the rate of change of $\mathbf{Q}$ with respect to the rotating frame $Oxyz$; the second part, $\mathbf{\Omega} \times \mathbf{Q}$, is induced by the rotation of the frame $Oxyz$.

The use of the relation in Eq. (15.31) simplifies the determination of the rate of change of a vector $\mathbf{Q}$ with respect to a fixed frame of reference $OXYZ$ when vector $\mathbf{Q}$ is defined by its components along the axes of a rotating frame $Oxyz$. In particular, this relation does not require separate computations of the derivatives of the unit vectors defining the orientation of the rotating frame.

15.5B Plane Motion of a Particle Relative to a Rotating Frame

Consider two frames of reference with both centered at O and both in the plane of the figure: a fixed frame OXY and a rotating frame Oxy (Fig. 15.27). Let P be a particle moving in the plane of the figure. The position vector $\mathbf{r}$ of P is the same in both frames, but its rate of change depends upon which frame of reference you select.

The absolute velocity $\mathbf{v}_P$ of the particle is defined as the velocity observed from the fixed frame OXY and is equal to the rate of change $(\dot{\mathbf{r}})_{OXY}$ of $\mathbf{r}$ with respect to that frame. We can, however, express $\mathbf{v}_P$ in terms of the rate of change $(\dot{\mathbf{r}})_{Oxy}$ observed from the rotating frame if we make use of Eq. (15.31). Denoting the angular velocity of the frame Oxy with respect to OXY at the instant considered by $\mathbf{\Omega}$, we have

$$\mathbf{v}_P = (\dot{\mathbf{r}})_{OXY} = \mathbf{\Omega} \times \mathbf{r} + (\dot{\mathbf{r}})_{Oxy} \qquad \text{(15.32)}$$

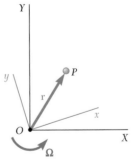

Fig. 15.27 We can express the motion of particle P in either a fixed ($OXYZ$) or a rotating ($Oxyz$) frame of reference.

where $(\dot{\mathbf{r}})_{Oxy}$ defines the velocity of particle P relative to the rotating frame Oxy and is sometimes denoted as $\mathbf{v}_{\text{rel}}$. There also may be instances where point O is not fixed and has a velocity denoted by $\mathbf{v}_O$. Therefore, an alternative way to express Eq. (15.32) is

$$\mathbf{v}_P = \mathbf{v}_O + \mathbf{\Omega} \times \mathbf{r} + \mathbf{v}_{\text{rel}} \qquad \text{(15.32')}$$

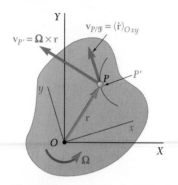

Fig. 15.28 The velocity of a point P is equal to the velocity of a point P' coincident with P but attached to the rotating frame plus the velocity of P with respect to the rotating frame.

The relative velocity, $\mathbf{v}_{\text{rel}}$ or $(\dot{\mathbf{r}})_{Oxy}$, is the velocity of point P with respect to the rotating frame. Denoting the rotating frame by $\mathcal{F}$, another way to represent the velocity $(\dot{\mathbf{r}})_{Oxy}$ of P relative to the rotating frame is $\mathbf{v}_{P/\mathcal{F}}$. Let us imagine that a rigid body has been attached to the rotating frame. Then $\mathbf{v}_{P/\mathcal{F}}$ represents the velocity of P along the path that it describes on that body (Fig. 15.28), and the term $\mathbf{\Omega} \times \mathbf{r}$ in Eq. (15.32) represents the velocity $\mathbf{v}_{P'}$ of the point P' of the rigid body—or rotating frame—that coincides with P at the instant considered. Thus, we have

$$\mathbf{v}_P = \mathbf{v}_{P'} + \mathbf{v}_{P/\mathcal{F}} \tag{15.33}$$

where

$\mathbf{v}_P$ = absolute velocity of particle P

$\mathbf{v}_{P'}$ = velocity of point P' of moving frame $\mathcal{F}$ coinciding with P

$\mathbf{v}_{P/\mathcal{F}}$ = velocity of P relative to moving frame $\mathcal{F}$

We define the absolute acceleration $\mathbf{a}_P$ of the particle as the rate of change of $\mathbf{v}_P$ with respect to the fixed frame OXY. Computing the rates of change with respect to OXY of the terms in Eq. (15.32), we have

$$\mathbf{a}_P = \dot{\mathbf{v}}_P = \dot{\mathbf{\Omega}} \times \mathbf{r} + \mathbf{\Omega} \times \dot{\mathbf{r}} + \frac{d}{dt}[(\dot{\mathbf{r}})_{Oxy}] \tag{15.34}$$

where all derivatives are defined with respect to OXY, except where indicated otherwise. Referring to Eq. (15.31), we note that we can express the last term in Eq. (15.34) as

$$\frac{d}{dt}[(\dot{\mathbf{r}})_{Oxy}] = (\ddot{\mathbf{r}})_{Oxy} + \mathbf{\Omega} \times (\dot{\mathbf{r}})_{Oxy}$$

On the other hand, $\dot{\mathbf{r}}$ represents the velocity $\mathbf{v}_P$ and can be replaced by the right-hand side of Eq. (15.32). After completing these two substitutions into Eq. (15.34), we obtain

$$\mathbf{a}_P = \dot{\mathbf{\Omega}} \times \mathbf{r} + \mathbf{\Omega} \times (\mathbf{\Omega} \times \mathbf{r}) + 2\mathbf{\Omega} \times (\dot{\mathbf{r}})_{Oxy} + (\ddot{\mathbf{r}})_{Oxy} \tag{15.35}$$

As we had for the velocity expression, our reference point O might also be accelerating. For plane motion,

$$\mathbf{a}_P = \mathbf{a}_O + \dot{\mathbf{\Omega}} \times \mathbf{r} - \Omega^2\mathbf{r} + 2\mathbf{\Omega} \times \mathbf{v}_{\text{rel}} + \mathbf{a}_{\text{rel}} \tag{15.35'}$$

where

$\mathbf{a}_O$ = the linear acceleration of point O

$\dot{\mathbf{\Omega}}$ = angular acceleration of the rotating frame

$\mathbf{\Omega}$ = angular velocity of the rotating frame

$\mathbf{r}$ = position vector from the origin to point P

$\mathbf{v}_{\text{rel}}$ = relative velocity of point P with respect to the rotating frame

$\mathbf{a}_{\text{rel}}$ = relative acceleration of point P with respect to the rotating frame

From expression (15.8) obtained in Sec. 15.1B for the acceleration of a particle on a rigid body rotating about a fixed axis, we note that the sum of the first two terms in Eq. (15.35) represents the acceleration $\mathbf{a}_{P'}$ of the point P' of the rotating frame that coincides with P at the instant considered. The last term defines the acceleration $\mathbf{a}_{P/\mathscr{F}}$ of P relative to the rotating frame. If it were not for the third term, which has not been accounted for, we could write a relation similar to Eq. (15.33) for the accelerations, and $\mathbf{a}_P$ could be expressed as the sum of $\mathbf{a}_{P'}$ and $\mathbf{a}_{P/\mathscr{F}}$. However, it is clear that *such a relation would be incorrect* and that we must include the additional term. This term, which we denote by $\mathbf{a}_C$, is called the **Coriolis acceleration**, after the French mathematician Gaspard de Coriolis (1792–1843). We have

$$\mathbf{a}_P = \mathbf{a}_{P'} + \mathbf{a}_{P/\mathscr{F}} + \mathbf{a}_C \tag{15.36}$$

where

$\mathbf{a}_P$ = absolute acceleration of particle P

$\mathbf{a}_{P'}$ = acceleration of point P' of moving frame $\mathscr{F}$ coinciding with P

$\mathbf{a}_{P/\mathscr{F}}$ = acceleration of P relative to moving frame $\mathscr{F}$

$\mathbf{a}_C = 2\boldsymbol{\Omega} \times (\dot{\mathbf{r}})_{Oxy} = 2\boldsymbol{\Omega} \times \mathbf{v}_{P/\mathscr{F}}$

= Coriolis acceleration

Note the difference between Eq. (15.36) and Eq. (15.21). When we wrote

$$\mathbf{a}_B = \mathbf{a}_A + \mathbf{a}_{B/A} \tag{15.21}$$

in Sec. 15.4A, we were expressing the absolute acceleration of point B as the sum of the acceleration $\mathbf{a}_{B/A}$ relative to a frame in translation and the acceleration $\mathbf{a}_A$ of a point of that frame. We are now relating the absolute acceleration of point P to its acceleration $\mathbf{a}_{P/\mathscr{F}}$ relative to a rotating frame $\mathscr{F}$ and to the acceleration $\mathbf{a}_{P'}$ of point P' of that frame, which coincides with P. Equation (15.36) shows that, because the frame is rotating, it is necessary to include an additional term to represent the Coriolis acceleration $\mathbf{a}_C$.

Note that since point P' moves in a circle about the origin O, its acceleration $\mathbf{a}_{P'}$ has, in general, two components: $(\mathbf{a}_{P'})_t$ tangent to the circle and $(\mathbf{a}_{P'})_n$ directed toward O. Similarly, the acceleration $\mathbf{a}_{P/\mathscr{F}}$ generally has two components: $(\mathbf{a}_{P/\mathscr{F}})_t$ tangent to the path that P describes on the rotating rigid body and $(\mathbf{a}_{P/\mathscr{F}})_n$ directed toward the center of curvature of that path. We further note that since the vector $\boldsymbol{\Omega}$ is perpendicular to the plane of motion, and thus to $\mathbf{v}_{P/\mathscr{F}}$, the magnitude of the Coriolis acceleration $\mathbf{a}_C = 2\boldsymbol{\Omega} \times \mathbf{v}_{P/\mathscr{F}}$ is equal to $2\Omega v_{P/\mathscr{F}}$, and its direction can be obtained by rotating the vector $\mathbf{v}_{P/\mathscr{F}}$ through 90° in the sense of rotation of the moving frame (Fig. 15.29). The Coriolis acceleration reduces to zero when either $\boldsymbol{\Omega}$ or $\mathbf{v}_{P/\mathscr{F}}$ is zero.

Consider a collar P that is made to slide at a constant relative speed u along a rod OB rotating at a constant angular velocity $\boldsymbol{\omega}$ about O (Fig. 15.30a). According to formula (15.36), we can obtain the absolute acceleration of P by adding vectorially the acceleration $\mathbf{a}_A$ of the point A of the rod coinciding with P, the relative acceleration $\mathbf{a}_{P/OB}$ of P with respect to the rod, and the Coriolis acceleration $\mathbf{a}_C$.

Since the angular velocity $\boldsymbol{\omega}$ of the rod is constant, $\mathbf{a}_A$ reduces to its normal component $(\mathbf{a}_A)_n$ with a magnitude of $r\omega^2$; and since u is constant, the relative acceleration $\mathbf{a}_{P/OB}$ is zero. According to the definition given previously, the Coriolis acceleration is a vector perpendicular to OB, has a

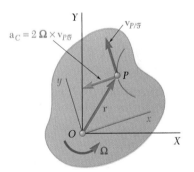

Fig. 15.29 The Coriolis acceleration is perpendicular to the relative velocity of P with respect to the rotating frame.

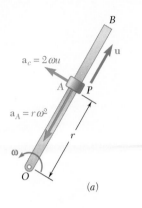

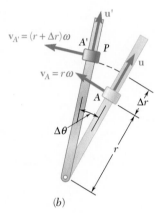

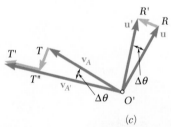

(c)

Fig. 15.30 (*a*) A collar sliding at constant speed along a rotating rod; (*b*) velocities of the collar at two points in time; (*c*) the acceleration components equal the changes in velocity.

magnitude of $2\omega u$, and is directed as shown in Figure 15.30. The acceleration of the collar P consists, therefore, of the two vectors shown in Fig. 15.30*a*. Note that you can check this result by applying the relation in Eq. (11.43).

To understand better the significance of the Coriolis acceleration, let us consider the absolute velocity of P at time t and at time $t + \Delta t$ (Fig. 15.30*b*). We can resolve the velocity at time t into its components $\mathbf{u}$ and $\mathbf{v}_A$; we can resolve the velocity at time $t + \Delta t$ into its components $\mathbf{u}'$ and $\mathbf{v}_{A'}$. Drawing these components from the same origin (Fig. 15.30*c*), we note that the change in velocity during the time Δt can be represented by the sum of three vectors: $\overrightarrow{RR'}$, $\overrightarrow{TT''}$, and $\overrightarrow{T''T'}$. The vector $\overrightarrow{TT''}$ measures the change in direction of the velocity $\mathbf{v}_A$, and the quotient $\overrightarrow{TT''}/\Delta t$ represents the acceleration $\mathbf{a}_A$ when Δt approaches zero. We check that the direction of $\overrightarrow{TT''}$ is that of $\mathbf{a}_A$ when Δt approaches zero and that

$$\lim_{\Delta t \to 0} \frac{TT''}{\Delta t} = \lim_{\Delta t \to 0} v_A \frac{\Delta \theta}{\Delta t} = r\omega\omega = r\omega^2 = a_A$$

The vector $\overrightarrow{RR'}$ measures the change in direction of $\mathbf{u}$ due to the rotation of the rod; the vector $\overrightarrow{T''T'}$ measures the change in magnitude of $\mathbf{v}_A$ due to the motion of P on the rod. The vectors $\overrightarrow{RR'}$ and $\overrightarrow{T''T'}$ result from the *combined effect* of the relative motion of P and of the rotation of the rod; they would vanish if *either* of these two motions stopped. It is easily verified that the sum of these two vectors defines the Coriolis acceleration. Their direction is that of $\mathbf{a}_C$ when Δt approaches zero, and since $RR' = u \, \Delta\theta$ and $T''T' = v_{A'} - v_A = (r + \Delta r) \, \omega - r\omega = \omega \, \Delta r$, we check that a_C is equal to

$$\lim_{\Delta t \to 0} \left(\frac{RR'}{\Delta t} + \frac{T''T'}{\Delta t} \right) = \lim_{\Delta t \to 0} \left(u \frac{\Delta \theta}{\Delta t} + \omega \frac{\Delta r}{\Delta t} \right) = u\omega + \omega u = 2\omega u$$

We can use formulas (15.33) and (15.36) to analyze the motion of mechanisms that contain parts sliding on each other. They make it possible, for example, to relate the absolute and relative motions of sliding pins and collars (see Sample Probs. 15.18 through 15.20). The concept of Coriolis acceleration is also very useful in the study of long-range projectiles and of other objects whose motions are appreciably affected by the rotation of the earth. As we pointed out in Sec. 12.1A, a system of axes attached to the earth does not truly constitute a newtonian frame of reference; such a system of axes actually should be considered rotating. Thus, the formulas derived in this section facilitate the study of the motion of bodies with respect to axes attached to the earth.

Sample Problem 15.17

At the instant shown, the truck is moving forward with a speed of 2 ft/s and is slowing down at a rate of 0.25 ft/s². The length of the boom AB is decreasing at a constant rate of 0.5 ft/s, the angular velocity of the boom is 0.1 rad/s, and the angular acceleration of the boom is 0.02 rad/s², both clockwise. Determine the velocity and acceleration of point B.

STRATEGY: Since you are not given any forces and are asked to find the velocity and acceleration of a point, use rigid-body kinematics. The boom is moving with respect to the truck, so use a rotating reference frame.

MODELING and ANALYSIS: Attach a rotating coordinate system to the boom housing with its origin at A (Fig. 1).

Velocity of B. From Eq. (15.32′), you know

$$\mathbf{v}_B = \mathbf{v}_A + \boldsymbol{\Omega} \times \mathbf{r}_{B/A} + \mathbf{v}_{rel} \tag{1}$$

where $\mathbf{v}_A = (2 \text{ ft/s})\mathbf{i}$, $\mathbf{r}_{B/A} = (20 \cos 30° \text{ ft})\mathbf{i} + (20 \sin 30° \text{ ft})\mathbf{j}$, and $\boldsymbol{\Omega} = (-0.1 \text{ rad/s})\mathbf{k}$. To find the relative velocity, ask yourself what the velocity of B would be, assuming that the rotating coordinate system is not moving. In this case, $\mathbf{v}_{rel} = -(0.5 \cos 30° \text{ ft/s})\mathbf{i} - (0.5 \sin 30° \text{ ft/s})\mathbf{j}$. Substituting into Eq. (1) gives

$$\mathbf{v}_B = 2\mathbf{i} + (-0.1\mathbf{k}) \times (17.32\mathbf{i} + 10\mathbf{j}) - (0.433\mathbf{i} + 0.25\mathbf{j})$$

$$\mathbf{v}_B = (2.57 \text{ ft/s})\mathbf{i} - (1.982 \text{ ft/s})\mathbf{j} \quad \blacktriangleleft$$

Acceleration of B. From Eq. (15.35′), you know

$$\mathbf{a}_B = \mathbf{a}_A + \dot{\boldsymbol{\Omega}} \times \mathbf{r}_{B/A} - \Omega^2 \mathbf{r}_{B/A} + 2\boldsymbol{\Omega} \times \mathbf{v}_{rel} + \mathbf{a}_{rel} \tag{2}$$

where $\mathbf{a}_A = -(0.25 \text{ ft/s}^2)\mathbf{i}$, $\dot{\boldsymbol{\Omega}} = -(0.02 \text{ rad/s}^2)\mathbf{k}$, and $\mathbf{a}_{rel} = 0$. Substituting into Eq. (2) gives

$$\mathbf{a}_B = -0.25\mathbf{i} + (-0.02\mathbf{k}) \times (17.32\mathbf{i} + 10\mathbf{j}) - 0.1^2(17.32\mathbf{i} + 10\mathbf{j})$$
$$+ 2(-0.1\mathbf{k}) \times (-0.433\mathbf{i} - 0.25\mathbf{j}) + 0$$

$$= -0.25\mathbf{i} + (-0.3464\mathbf{j} + 0.2\mathbf{i}) - (0.1732\mathbf{i} + 0.10\mathbf{j}) + (0.0866\mathbf{j} - 0.05\mathbf{i}) + 0$$

$$\mathbf{a}_B = (-0.273 \text{ ft/s}^2)\mathbf{i} - (0.360 \text{ ft/s}^2)\mathbf{j} \quad \blacktriangleleft$$

REFLECT and THINK: The biggest challenge with this problem is interpreting what you are given in the problem statement. After that, it is straightforward to substitute into the governing equations. The last four terms in Eq. (2) are analogous to the polar coordinate expressions we used in Chapter 11. The following terms represent the same physical quantities: $\dot{\boldsymbol{\Omega}} \times \mathbf{r}_{B/A} \rightarrow r\ddot{\theta}$, $-\Omega^2 \mathbf{r}_{B/A} \rightarrow -r\dot{\theta}^2$, $2\boldsymbol{\Omega} \times \mathbf{v}_{rel} \rightarrow 2\dot{r}\dot{\theta}$, and $\mathbf{a}_{rel} \rightarrow \ddot{r}$.

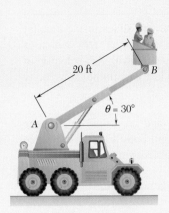

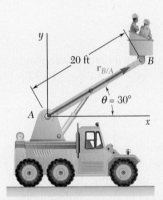

Fig. 1 The rotating coordinate system is attached to the truck at A.

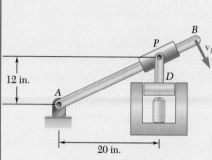

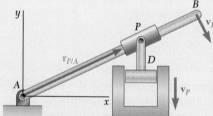

Fig. 1 The rotating coordinate system is attached to arm *AB*.

Sample Problem 15.18

In a can crusher, bar *AB* has a length of 30 in. and slides inside a collar located at point *P*. This collar is attached to plunger *DP*, which is constrained to move vertically. At the instant shown, the velocity of point *B* is a constant 4 ft/s perpendicular to the bar. Determine the velocity and acceleration of the plunger *D*.

STRATEGY: You are not given any forces and are asked to find the velocity and acceleration of a point, so use rigid-body kinematics. Since the collar is moving with respect to the bar, use a rotating reference frame.

MODELING and ANALYSIS: Attach a rotating coordinate system to the bar with its origin at *A* (Fig. 1).

Angular Velocity of *AB*. Rod *AB* is undergoing fixed-axis rotation, so

$$\omega_{AB} = \frac{v_B}{r_{B/A}} = \frac{48 \text{ in./s}}{30 \text{ in.}} = 1.60 \text{ rad/s} \downarrow$$

Velocity of *P*. Points *D* and *P* have the same velocity and acceleration because the plunger is constrained to translate only. From Eq. (15.32′) you know

$$\mathbf{v}_P = \mathbf{v}_A + \mathbf{\Omega} \times \mathbf{r}_{P/A} + \mathbf{v}_{\text{rel}} \tag{1}$$

where $\mathbf{v}_A = 0$, $\mathbf{r}_{P/A} = (20 \text{ in.})\mathbf{i} + (12 \text{ in.})\mathbf{j}$, and $\mathbf{\Omega} = -(1.6 \text{ rad/s})\mathbf{k}$. To find the relative velocity, ask yourself what the velocity of *P* would be assuming that the rotating coordinate system is not moving. In this case, $\mathbf{v}_{\text{rel}} = v_{\text{rel}} \cos\theta \mathbf{i} + v_{\text{rel}} \sin\theta \mathbf{j}$ where $\theta = \tan^{-1}(12/20) = 30.96°$. Substituting into Eq. (1) gives

$$-v_P\mathbf{j} = 0 + (-1.6\mathbf{k}) \times (20\mathbf{i} + 12\mathbf{j}) + (v_{\text{rel}} \cos\theta \, \mathbf{i} + v_{\text{rel}} \sin\theta \, \mathbf{j})$$
$$= -32\mathbf{j} + 19.2\mathbf{i} + 0.8575 v_{\text{rel}}\mathbf{i} + 0.5145 v_{\text{rel}}\mathbf{j}$$

Equating components allows you to solve for the unknown velocities:

i: $\quad 0 = 19.2 + 0.8575 v_{\text{rel}} \quad\quad \longrightarrow \quad\quad v_{\text{rel}} = -22.39 \text{ in./s}$
j: $\quad -v_P = -32 + 0.5145 v_{\text{rel}} \quad \longrightarrow \quad\quad v_P = 43.53 \text{ in./s}$

$$\mathbf{v}_P = 43.53 \text{ in./s} \downarrow \quad \blacktriangleleft$$

Acceleration of *P*. From Eq. (15.35′), you know

$$\mathbf{a}_P = \mathbf{a}_A + \dot{\mathbf{\Omega}} \times \mathbf{r}_{P/A} - \Omega^2\mathbf{r}_{P/A} + 2\mathbf{\Omega} \times \mathbf{v}_{\text{rel}} + \mathbf{a}_{\text{rel}} \tag{2}$$

where $\mathbf{a}_A = 0$, $\dot{\mathbf{\Omega}} = 0$, $\mathbf{a}_{\text{rel}} = a_{\text{rel}} \cos\theta \mathbf{i} + a_{\text{rel}} \sin\theta \mathbf{j}$. Substituting into Eq. (2) gives

$$-a_P\mathbf{j} = 0 + 0 - 1.6^2(20\mathbf{i} + 12\mathbf{j}) + 2(-1.6\mathbf{k}) \times$$
$$(-22.39 \cos\theta\mathbf{i} - 22.39 \sin\theta\mathbf{j}) + (a_{\text{rel}} \cos\theta\mathbf{i} + a_{\text{rel}} \sin\theta\mathbf{j})$$
$$= (-51.2\mathbf{i} - 30.72\mathbf{j}) + (61.44\mathbf{j} - 36.86\mathbf{i}) + (0.8575 a_{\text{rel}}\mathbf{i} + 0.5145 a_{\text{rel}}\mathbf{j})$$

Equating components allows you to solve for the unknown accelerations:

i: $0 = -51.2 - 36.86 + 0.8575a_{rel}$ $\longrightarrow$ $a_{rel} = 102.7$ in./s^2

j: $-a_P = -30.72 + 61.44 + 0.5145a_{rel}$ $\longrightarrow$ $a_P = -83.56$ in./s^2

$$a_P = -83.6 \text{ in./s}^2 \downarrow \quad \blacktriangleleft$$

REFLECT and THINK: You used the same strategy for the telescoping boom in Sample Prob. 15.17 as you did for the sliding collar in this problem. For each case, the point of interest was moving with respect to a coordinate frame attached to a rigid body. The same strategy is used in problems where pins move within slotted bodies (such as the Geneva mechanism in Sample Prob. 15.19).

Sample Problem 15.19

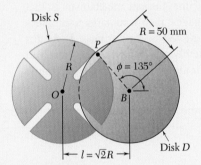

The Geneva mechanism shown is used in many counting instruments and in other applications where an intermittent rotary motion is required. Disk D rotates with a constant counterclockwise angular velocity ω_D of 10 rad/s. A pin P is attached to disk D and slides along one of several slots cut in disk S. It is desirable that the angular velocity of disk S be zero as the pin enters and leaves each slot; in the case of four slots, this occurs if the distance between the centers of the disks is $l = \sqrt{2}R$.

At the instant when $\phi = 150°$, determine (a) the angular velocity of disk S, (b) the velocity of pin P relative to disk S.

STRATEGY: You have two rigid bodies whose motions are related; therefore use rigid-body kinematics. Since point P is moving in a slot, use a rotating reference frame.

MODELING and ANALYSIS:

Using geometry, you can solve triangle OPB, which corresponds to the position $\phi = 150°$ (Fig. 1). Using the law of cosines, you have

$$r^2 = R^2 + l^2 - 2Rl \cos 30° = 0.551R^2 \qquad r = 0.742R = 37.1 \text{ mm}$$

Then, from the law of sines, you have

$$\frac{\sin \beta}{R} = \frac{\sin 30°}{r} \qquad \sin \beta = \frac{\sin 30°}{0.742} \qquad \beta = 42.4°$$

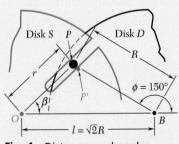

Fig. 1 Distances and angles relating points O, P, and B.

Since pin P is attached to disk D and disk D rotates about point B, the magnitude of the absolute velocity of P is

$$v_P = R\omega_D = (50 \text{ mm})(10 \text{ rad/s}) = 500 \text{ mm/s}$$
$$\mathbf{v}_P = 500 \text{ mm/s} \; \nearrow \; 60°$$

(continued)

Now consider the motion of pin P along the slot in disk S. Denote the point of disk S that coincides with P by P' at the instant considered and select a rotating frame $\mathscr{S}$ attached to disk S. Then from Eq. (15.33), you have

$$\mathbf{v}_P = \mathbf{v}_{P'} + \mathbf{v}_{P/\mathscr{S}} \tag{1}$$

In Eq. (1), $\mathbf{v}_{P'}$ is perpendicular to the radius OP, and $\mathbf{v}_{P/\mathscr{S}}$ is directed along the slot. Draw the velocity triangle corresponding to Eq. (1) (see Fig. 2). From the triangle, you can compute

$$\gamma = 90° - 42.4° - 30° = 17.6°$$
$$v_{P'} = v_P \sin \gamma = (500 \text{ mm/s}) \sin 17.6°$$
$$\mathbf{v}_{P'} = 151.2 \text{ mm/s} \; \measuredangle \; 42.4°$$
$$v_{P/\mathscr{S}} = v_P \cos \gamma = (500 \text{ mm/s}) \cos 17.6°$$
$$\mathbf{v}_{P/S} = \mathbf{v}_{P/\mathscr{S}} = 477 \text{ mm/s} \; \measuredangle \; 42.4° \quad \blacktriangleleft$$

Since $\mathbf{v}_{P'}$ is perpendicular to the radius OP, you have

$$v_{P'} = r\omega_{\mathscr{S}} \qquad 151.2 \text{ mm/s} = (37.1 \text{ mm})\omega_{\mathscr{S}}$$
$$\boldsymbol{\omega}_S = \boldsymbol{\omega}_{\mathscr{S}} = 4.08 \text{ rad/s} \; \downarrow \quad \blacktriangleleft$$

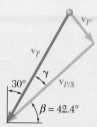

Fig. 2 Vector diagram for the velocity of point P.

REFLECT and THINK: The result of the Geneva mechanism is that disk S rotates ¼ turn each time pin P engages, then it remains motionless while pin P rotates around before entering the next slot. Disk D rotates continuously, but disk S rotates intermittently. An alternative approach to drawing the vector triangle is to use vector algebra, as was done in Sample Prob. 15.18.

Sample Problem 15.20

In the Geneva mechanism of Sample Prob. 15.19, disk D rotates with a constant counterclockwise angular velocity $\boldsymbol{\omega}_D$ of 10 rad/s. At the instant when $\phi = 150°$, determine the angular acceleration of disk S.

STRATEGY: You have two rigid bodies whose motions are related; therefore use rigid-body kinematics. Since point P is moving in a slot, use a rotating reference frame.

MODELING and ANALYSIS: Since you are computing accelerations instead of velocities, you need to use Eq. (15.36), which includes the Coriolis acceleration. You found the angular velocity of the frame $\mathscr{S}$ attached to disk S and the velocity of the pin relative to $\mathscr{S}$ in Sample Prob. 15.19:

$$\omega_{\mathscr{S}} = 4.08 \text{ rad/s} \; \downarrow$$
$$\beta = 42.4° \qquad \mathbf{v}_{P/\mathscr{S}} = 477 \text{ mm/s} \; \measuredangle \; 42.4°$$

Since pin P moves with respect to the rotating frame $\mathscr{S}$, you have

$$\mathbf{a}_P = \mathbf{a}_{P'} + \mathbf{a}_{P/\mathscr{S}} + \mathbf{a}_c \tag{1}$$

Investigate each term of this vector equation separately.

Absolute Acceleration a_P. Since disk D rotates with a constant angular velocity, the absolute acceleration $\mathbf{a}_P$ is directed toward B. This gives

$$a_P = R\omega_D^2 = (500 \text{ mm})(10 \text{ rad/s})^2 = 5000 \text{ mm/s}^2$$
$$\mathbf{a}_P = 5000 \text{ mm/s}^2 \;\searrow\; 30°$$

Acceleration $a_{P'}$ of the Coinciding Point P'. Resolve into normal and tangential components the acceleration $\mathbf{a}_{P'}$ of the point P' of the frame $\mathcal{S}$ that coincides with P at the given instant. (Recall from Sample Prob. 15.19 that $r = 37.1$ mm.)

$$(a_{P'})_n = r\omega_{\mathcal{S}}^2 = (37.1 \text{ mm})(4.08 \text{ rad/s})^2 = 618 \text{ mm/s}^2$$
$$(\mathbf{a}_{P'})_n = 618 \text{ mm/s}^2 \;\nearrow\; 42.4°$$
$$(a_{P'})_t = r\alpha_{\mathcal{S}} = 37.1\alpha_{\mathcal{S}} \qquad (\mathbf{a}_{P'})_t = 37.1\alpha_{\mathcal{S}} \;\nwarrow\; 42.4°$$

Relative Acceleration $a_{P/\mathcal{S}}$. Since the pin P moves in a straight slot cut in disk S, the relative acceleration $\mathbf{a}_{P/\mathcal{S}}$ must be parallel to the slot; i.e., its direction must be $\swarrow$ 42.4°.

Coriolis Acceleration a_C. Rotating the relative velocity $\mathbf{v}_{P/\mathcal{S}}$ through 90° in the sense of $\boldsymbol{\omega}_{\mathcal{S}}$, you obtain the direction of the Coriolis component of the acceleration: $\nwarrow$ 42.4°. You have

$$a_C = 2\omega_{\mathcal{S}} v_{P/\mathcal{S}} = 2(4.08 \text{ rad/s})(477 \text{ mm/s}) = 3890 \text{ mm/s}^2$$
$$\mathbf{a}_C = 3890 \text{ mm/s}^2 \;\nwarrow\; 42.4°$$

Rewrite Eq. (1) and substitute the accelerations found (Fig. 1):

$$\mathbf{a}_P = (\mathbf{a}_{P'})_n + (\mathbf{a}_{P'})_t + \mathbf{a}_{P/\mathcal{S}} + \mathbf{a}_C$$
$$[5000 \;\searrow\; 30°] = [618 \;\nearrow\; 42.4°] + [37.1\alpha_{\mathcal{S}} \;\nwarrow\; 42.4°]$$
$$+ [a_{P/\mathcal{S}} \;\swarrow\; 42.4°] + [3890 \;\nwarrow\; 42.4°]$$

Equating components in a direction perpendicular to the slot,

$$5000 \cos 17.6° = 37.1\alpha_{\mathcal{S}} - 3890$$
$$\alpha_S = \alpha_{\mathcal{S}} = 233 \text{ rad/s}^2 \;\curvearrowright \quad \blacktriangleleft$$

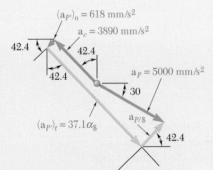

$(a_{P'})_n = 618 \text{ mm/s}^2$

$a_c = 3890 \text{ mm/s}^2$

42.4

42.4

42.4

$a_P = 5000 \text{ mm/s}^2$

30

$(a_{P'})_t = 37.1\alpha_{\mathcal{S}}$

$a_{P/\mathcal{S}}$

42.4

Fig. 1 Vector polygon for the acceleration of point P.

REFLECT and THINK: It seems reasonable that, since disk S starts and stops over the very short time intervals when pin P is engaged in the slots, the disk must have a very large angular acceleration. An alternative approach would have been to use the vector algebra.

SOLVING PROBLEMS
ON YOUR OWN

In this section you studied the rate of change of a vector with respect to a rotating frame and then applied that idea to the analysis of the plane motion of a particle relative to a rotating frame.

1. Rate of change of a vector with respect to a fixed frame and with respect to a rotating frame. Denoting the rate of change of a vector $\mathbf{Q}$ with respect to a fixed frame $OXYZ$ by $(\dot{\mathbf{Q}})_{OXYZ}$ and its rate of change with respect to a rotating frame $Oxyz$ by $(\dot{\mathbf{Q}})_{Oxyz}$, we obtained the fundamental relation

$$(\dot{\mathbf{Q}})_{OXYZ} = (\dot{\mathbf{Q}})_{Oxyz} + \boldsymbol{\Omega} \times \mathbf{Q} \tag{15.31}$$

where $\boldsymbol{\Omega}$ is the angular velocity of the rotating frame.

You can now apply this fundamental relation to the solution of two-dimensional problems.

2. Plane motion of a particle relative to a rotating frame. Using Eq. (15.31) and designating the rotating frame by $\mathscr{F}$, we obtained the following expressions for the velocity and the acceleration of a particle P:

$$\mathbf{v}_P = \mathbf{v}_{P'} + \mathbf{v}_{P/\mathscr{F}} \tag{15.33}$$

or

$$\mathbf{v}_P = \mathbf{v}_O + \boldsymbol{\Omega} \times \mathbf{r} + \mathbf{v}_{\text{rel}} \tag{15.32'}$$

and

$$\mathbf{a}_P = \mathbf{a}_{P'} + \mathbf{a}_{P/\mathscr{F}} + \mathbf{a}_C \tag{15.36}$$

or

$$\mathbf{a}_P = \mathbf{a}_O + \boldsymbol{\alpha} \times \mathbf{r} - \Omega^2 \mathbf{r} + 2\boldsymbol{\Omega} \times \mathbf{v}_{\text{rel}} + \mathbf{a}_{\text{rel}} \tag{15.35'}$$

The notation in Eqs. (15.33) and (15.36) is as follows.

 a. The subscript P refers to the absolute motion of the particle P; that is, to its motion with respect to a fixed or newtonian frame of reference OXY.

 b. The subscript P' refers to the motion of the point P' of the rotating frame $\mathscr{F}$ that coincides with P at the instant considered.

 c. The subscript $P/\mathscr{F}$ refers to the motion of the particle P relative to the rotating frame $\mathscr{F}$.

 d. The term $\mathbf{a}_C$ represents the Coriolis acceleration of point P. Its magnitude is $2\Omega v_{P/\mathscr{F}}$, and its direction is found by rotating $\mathbf{v}_{P/\mathscr{F}}$ through 90° in the sense of rotation of the frame $\mathscr{F}$.

You should keep in mind that you need to take the Coriolis acceleration into account whenever a point has a relative velocity in a rotating frame. The problems you will encounter in this section involve collars that slide on rotating rods, booms that extend from cranes rotating in a vertical plane, etc.

When solving a problem involving a rotating frame, you can either *a*) draw vector diagrams representing Eqs. (15.33) and (15.36), respectively, and use these diagrams to obtain either an analytical or a graphical solution, or *b*) use vector algebra.

Problems

CONCEPT QUESTIONS

15.CQ8 A person walks radially inward on a platform that is rotating counterclockwise about its center. Knowing that the platform has a constant angular velocity ω and the person walks with a constant speed u relative to the platform, what is the direction of the acceleration of the person at the instant shown?

  a. Negative x
  b. Negative y
  c. Negative x and positive y
  d. Positive x and positive y
  e. Negative x and negative y

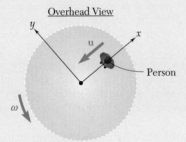

Fig. P15.CQ8

END-OF-SECTION PROBLEMS

15.150 and 15.151 Pin P is attached to the collar shown; the motion of the pin is guided by a slot cut in rod BD and by the collar that slides on rod AE. Knowing that at the instant considered the rods rotate clockwise with constant angular velocities, determine for the given data the velocity of pin P.

  15.150 $\omega_{AE} = 8$ rad/s, $\omega_{BD} = 3$ rad/s
  15.151 $\omega_{AE} = 7$ rad/s, $\omega_{BD} = 4.8$ rad/s

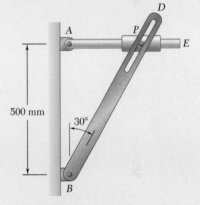

Fig. P15.150 and P15.151

15.152 and 15.153 Two rotating rods are connected by slider block P. The rod attached at A rotates with a constant clockwise angular velocity ω_A. For the given data, determine for the position shown (*a*) the angular velocity of the rod attached at B, (*b*) the relative velocity of slider block P with respect to the rod on which it slides.

  15.152 $b = 8$ in., $\omega_A = 6$ rad/s
  15.153 $b = 300$ mm, $\omega_A = 10$ rad/s

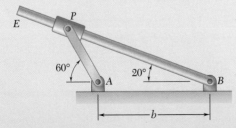

Fig. P15.152

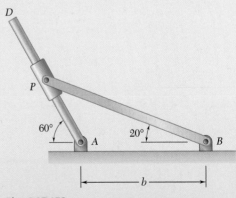

Fig. P15.153

15.154 Pin P is attached to the wheel shown and slides in a slot cut in bar BD. The wheel rolls to the right without slipping with a constant angular velocity of 20 rad/s. Knowing that $x = 480$ mm when $\theta = 0$, determine the angular velocity of the bar and the relative velocity of pin P with respect to the rod when (a) $\theta = 0$, (b) $\theta = 90°$.

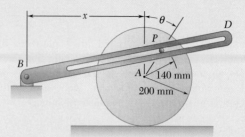

Fig. P15.154

15.155 Knowing that at the instant shown the angular velocity of bar AB is 15 rad/s clockwise and the angular velocity of bar EF is 10 rad/s clockwise, determine (a) the angular velocity of rod DE, (b) the relative velocity of collar B with respect to rod DE.

15.156 Knowing that at the instant shown the angular velocity of rod DE is 10 rad/s clockwise and the angular velocity of bar EF is 15 rad/s counterclockwise, determine (a) the angular velocity of bar AB, (b) the relative velocity of collar B with respect to rod DE.

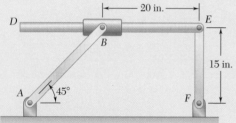

Fig. P15.155 and P15.156

15.157 The motion of pin P is guided by slots cut in rods AD and BE. Knowing that bar AD has a constant angular velocity of 4 rad/s clockwise and bar BE has an angular velocity of 5 rad/s counterclockwise and is slowing down at a rate of 2 rad/s², determine the velocity of P for the position shown.

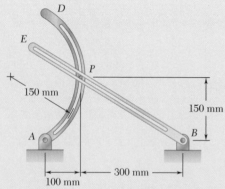

Fig. P15.157

15.158 Four pins slide in four separate slots cut in a circular plate as shown. When the plate is at rest, each pin has a velocity directed as shown and of the same constant magnitude u. If each pin maintains the same velocity relative to the plate when the plate rotates about O with a constant counterclockwise angular velocity ω, determine the acceleration of each pin.

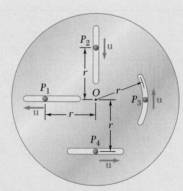

Fig. P15.158

15.159 Solve Prob. 15.158, assuming that the plate rotates about O with a constant clockwise angular velocity ω.

15.160 The cage of a mine elevator moves downward at a constant speed of 12.2 m/s. Determine the magnitude and direction of the Coriolis acceleration of the cage if the elevator is located (a) at the equator, (b) at latitude 40° north, (c) at latitude 40° south.

15.161 Pin *P* is attached to the collar shown; the motion of the pin is guided by a slot cut in bar *BD* and by the collar that slides on rod *AE*. Rod *AE* rotates with a constant angular velocity of 6 rad/s clockwise and the distance from *A* to *P* increases at a constant rate of 8 ft/s. Determine at the instant shown (a) the angular acceleration of bar *BD*, (b) the relative acceleration of pin *P* with respect to bar *BD*.

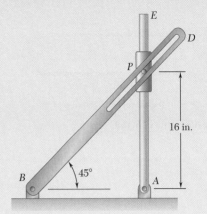

Fig. P15.161

15.162 A rocket sled is tested on a straight track that is built along a meridian. Knowing that the track is located at latitude 40° north, determine the Coriolis acceleration of the sled when it is moving north at a speed of 900 km/h.

15.163 Solve the Geneva mechanism of Sample Prob. 15.20 using vector algebra.

15.164 At the instant shown the length of the boom *AB* is being *decreased* at the constant rate of 0.2 m/s and the boom is being lowered at the constant rate of 0.08 rad/s. Determine (a) the velocity of point *B*, (b) the acceleration of point *B*.

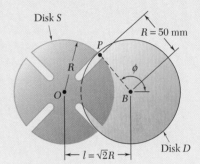

Fig. P15.163

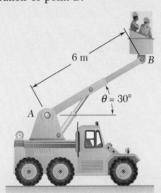

Fig. P15.164 and P15.165

15.165 At the instant shown the length of the boom *AB* is being *increased* at the constant rate of 0.2 m/s and the boom is being lowered at the constant rate of 0.08 rad/s. Determine (a) the velocity of point *B*, (b) the acceleration of point *B*.

15.166 In the automated welding setup shown, the position of the two welding tips *G* and *H* is controlled by the hydraulic cylinder *D* and rod *BC*. The cylinder is bolted to the vertical plate that at the instant shown rotates counterclockwise about *A* with a constant angular velocity of 1.6 rad/s. Knowing that at the same instant the length *EF* of the welding assembly is increasing at the constant rate of 300 mm/s, determine (a) the velocity of tip *H*, (b) the acceleration of tip *H*.

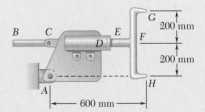

Fig. P15.166 and P15.167

15.167 In the automated welding setup shown, the position of the two welding tips *G* and *H* is controlled by the hydraulic cylinder *D* and rod *BC*. The cylinder is bolted to the vertical plate that at the instant shown rotates counterclockwise about *A* with a constant angular velocity of 1.6 rad/s. Knowing that at the same instant the length *EF* of the welding assembly is increasing at the constant rate of 300 mm/s, determine (a) the velocity of tip *G*, (b) the acceleration of tip *G*.

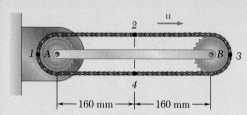

Fig. P15.168 and P15.169

15.168 and 15.169 A chain is looped around two gears of radius 40 mm that can rotate freely with respect to the 320-mm arm *AB*. The chain moves about arm *AB* in a clockwise direction at the constant rate of 80 mm/s relative to the arm. Knowing that in the position shown arm *AB* rotates clockwise about *A* at the constant rate $\omega = 0.75$ rad/s, determine the acceleration of each of the chain links indicated.

    15.168 Links 1 and 2
    15.169 Links 3 and 4

15.170 A basketball player shoots a free throw in such a way that his shoulder can be considered a pin joint at the moment of release as shown. Knowing that at the instant shown the upper arm *SE* has a constant angular velocity of 2 rad/s counterclockwise and the forearm *EW* has a constant clockwise angular velocity of 4 rad/s with respect to *SE*, determine the velocity and acceleration of the wrist *W*.

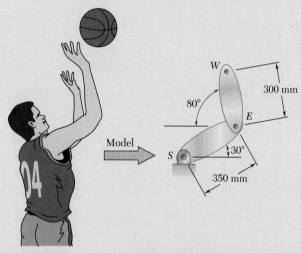

Fig. P15.170

15.171 The human leg can be crudely approximated as two rigid bars (the femur and the tibia) connected with a pin joint. At the instant shown, the velocity of the ankle *A* is zero, the tibia *AK* has an angular velocity of 1.5 rad/s counterclockwise and an angular acceleration of 1 rad/s^2 counterclockwise. Determine the relative angular velocity and relative angular acceleration of the femur *KH* with respect to *AK* so that the velocity and acceleration of *H* are both straight up at this instant.

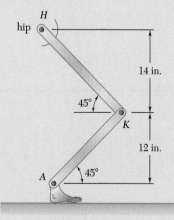

Fig. P15.171

15.172 The collar *P* slides outward at a constant relative speed *u* along rod *AB*, which rotates counterclockwise with a constant angular velocity of 20 rpm. Knowing that $r = 250$ mm when $\theta = 0$ and that the collar reaches *B* when $\theta = 90°$, determine the magnitude of the acceleration of the collar *P* just as it reaches *B*.

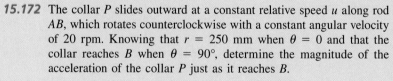

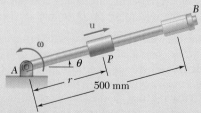

Fig. P15.172

15.173 Pin P slides in a circular slot cut in the plate shown at a constant relative speed $u = 90$ mm/s. Knowing that at the instant shown the plate rotates clockwise about A at the constant rate $\omega = 3$ rad/s, determine the acceleration of the pin if it is located at (a) point A, (b) point B, (c) point C.

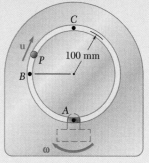

Fig. P15.173

15.174 Rod AD is bent in the shape of an arc of a circle with a radius of $b = 150$ mm. The position of the rod is controlled by pin B that slides in a horizontal slot and also slides along the rod. Knowing that at the instant shown pin B moves to the right at a constant speed of 75 mm/s, determine (a) the angular velocity of the rod, (b) the angular acceleration of the rod.

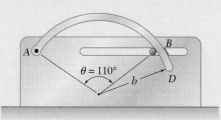

Fig. P15.174

15.175 Solve Prob. 15.174 when $\theta = 90°$.

15.176 Knowing that at the instant shown the rod attached at A has an angular velocity of 5 rad/s counterclockwise and an angular acceleration of 2 rad/s² clockwise, determine the angular velocity and the angular acceleration of the rod attached at B.

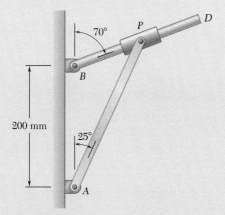

Fig. P15.176

15.177 The Geneva mechanism shown is used to provide an intermittent rotary motion of disk S. Disk D rotates with a constant counterclockwise angular velocity ω_D of 8 rad/s. A pin P is attached to disk D and can slide in one of the six equally spaced slots cut in disk S. It is desirable that the angular velocity of disk S be zero as the pin enters and leaves each of the six slots; this will occur if the distance between the centers of the disks and the radii of the disks are related as shown. Determine the angular velocity and angular acceleration of disk S at the instant when $\phi = 150°$.

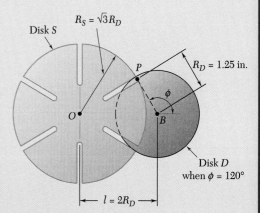

Fig. P15.177

15.178 In Prob. 15.177, determine the angular velocity and angular acceleration of disk S at the instant when $\phi = 135°$.

15.179 At the instant shown bar BC has an angular velocity of 3 rad/s and an angular acceleration of 2 rad/s^2, both counterclockwise; determine the angular acceleration of the plate.

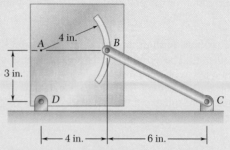

Fig. P15.179 and P15.180

15.180 At the instant shown bar BC has an angular velocity of 3 rad/s and an angular acceleration of 2 rad/s^2, both clockwise; determine the angular acceleration of the plate.

*15.181 Rod AB passes through a collar that is welded to link DE. Knowing that at the instant shown block A moves to the right at a constant speed of 75 in./s, determine (a) the angular velocity of rod AB, (b) the velocity relative to the collar of the point of the rod in contact with the collar, (c) the acceleration of the point of the rod in contact with the collar. (*Hint:* Rod AB and link DE have the same ω and the same α.)

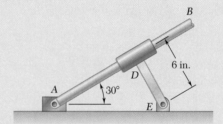

Fig. P15.181

*15.182 Solve Prob. 15.181 assuming block A moves to the left at a constant speed of 75 in./s.

*15.183 In Prob. 15.157, determine the acceleration of pin P.

*15.6 MOTION OF A RIGID BODY IN SPACE

Extending the study of motion in two dimensions to analyzing three-dimensional motion uses most of the same concepts as before, but with some added computational complexity. We introduce these ideas in this section and the next, and we will return to them when discussing kinetics of a rigid body in Chapter 18.

15.6A Motion About a Fixed Point

In Sec. 15.1B, we considered the motion of a rigid body constrained to rotate about a fixed axis. Here we examine the more general case of the three-dimensional motion of a rigid body that has a fixed point O. First, we prove:

> **The most general displacement of a rigid body with a fixed point O is equivalent to a rotation of the body about an axis through O.**

This statement is known as Euler's theorem. We analyze the motion of a sphere with a center O; this analysis can be extended to a rigid body of any shape. Since three points define the position of a solid in space, we let the center O and two points A and B on the surface of the sphere define the position of the sphere and thus the position of the body. Let A_1 and B_1 characterize the position of the sphere at one instant, and let A_2 and B_2 characterize its position at a later instant (Fig. 15.31a). Since the sphere is rigid, the lengths of the arcs of great circles A_1B_1 and A_2B_2 must be equal, but except for this requirement, the positions of A_1, A_2, B_1, and B_2 are arbitrary. We will show that the points A and B can be brought, respectively, from A_1 and B_1 into A_2 and B_2 by a single rotation of the sphere about an axis.

For convenience, and without loss of generality, we select point B so that its initial position coincides with the final position of A; thus, $B_1 = A_2$ (Fig. 15.31b). We draw the arcs of great circles A_1A_2, A_2B_2, and the arcs bisecting, respectively, A_1A_2 and A_2B_2. Let C be the point of intersection of these last two arcs. We complete the construction by drawing A_1C, A_2C, and B_2C. As pointed out above, because of the rigidity of the sphere, $A_1B_1 = A_2B_2$. Since C is by construction equidistant from A_1, A_2, and B_2, we also have $A_1C = A_2C = B_2C$. As a result, the spherical triangles A_1CA_2 and B_1CB_2 are congruent, and the angles A_1CA_2 and B_1CB_2 are equal. Denoting the common value of these angles by θ, we conclude that the sphere can be brought from its initial position into its final position by a single rotation through θ about the axis OC.

It follows that we can consider the motion during a time interval Δt of a rigid body with a fixed point O as a rotation through $\Delta\theta$ about a certain axis. Drawing a vector with a magnitude of $\Delta\theta/\Delta t$ along that axis and letting Δt approach zero, we obtain in the limit the **instantaneous axis of rotation** and the angular velocity $\boldsymbol{\omega}$ of the body at the instant considered (Fig. 15.32). We can then obtain the velocity of a particle P

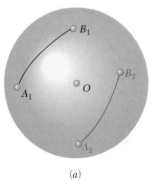

(a)

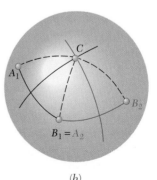

(b)

Fig. 15.31 (a) Positions of two points on a rotating sphere; (b) the sphere can be brought into this new position by a single rotation.

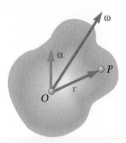

Fig. 15.32 Angular velocity and angular acceleration of a rigid body moving about a fixed point *O*.

of the body, as in Sec. 15.1B, by forming the vector product of $\boldsymbol{\omega}$ and of the position vector $\mathbf{r}$ of the particle:

$$\mathbf{v} = \frac{d\mathbf{r}}{dt} = \boldsymbol{\omega} \times \mathbf{r} \tag{15.37}$$

We obtain the acceleration of the particle by differentiating Eq. (15.37) with respect to *t*. As in Sec. 15.1B, we have

$$\mathbf{a} = \boldsymbol{\alpha} \times \mathbf{r} + \boldsymbol{\omega} \times (\boldsymbol{\omega} \times \mathbf{r}) \tag{15.38}$$

Here we have defined the angular acceleration $\boldsymbol{\alpha}$ as the derivative

$$\boldsymbol{\alpha} = \frac{d\boldsymbol{\omega}}{dt} \tag{15.39}$$

of the angular velocity $\boldsymbol{\omega}$.

In the case of the motion of a rigid body with a fixed point, the direction of $\boldsymbol{\omega}$ and of the instantaneous axis of rotation changes from one instant to the next. The angular acceleration $\boldsymbol{\alpha}$ therefore reflects the change in direction of $\boldsymbol{\omega}$ as well as its change in magnitude. Thus, in general, $\boldsymbol{\alpha}$ *is not directed along the instantaneous axis of rotation.* Although the particles of the body located on the instantaneous axis of rotation have zero velocity at the instant considered, they do not have zero acceleration. Also, the accelerations of the various particles of the body *cannot* be determined as if the body were rotating permanently about the instantaneous axis.

Recalling the definition of the velocity of a particle with position vector $\mathbf{r}$, we note that the angular acceleration $\boldsymbol{\alpha}$, as expressed in Eq. (15.39), represents the velocity of the tip of vector $\boldsymbol{\omega}$. This property may be useful in determining the angular acceleration of a rigid body. For example, it follows that vector $\boldsymbol{\alpha}$ is tangent to the curve described in space by the tip of vector $\boldsymbol{\omega}$.

Note that vector $\boldsymbol{\omega}$ moves within the body, as well as in space. It thus generates two cones called, respectively, the **body cone** and the **space cone** (Fig. 15.33).[†] It can be shown that, at any given instant, the two cones are tangent along the instantaneous axis of rotation and that, as the body moves, the body cone appears to *roll* on the space cone.

Before concluding our analysis of the motion of a rigid body with a fixed point, we should prove that angular velocities are actually vectors. Some quantities, such as the *finite rotations* of a rigid body, have magnitude and direction but do not obey the parallelogram law of addition; these quantities cannot be considered to be vectors. In contrast, angular velocities (and also *infinitesimal rotations*), as we demonstrate presently, do obey the parallelogram law and thus are truly vector quantities.

Consider a rigid body with a fixed point *O* that rotates at a given instant simultaneously about the axes *OA* and *OB* with angular velocities $\boldsymbol{\omega}_1$ and $\boldsymbol{\omega}_2$ (Fig. 15.34*a*). We know that this motion must be equivalent at the instant considered to a single rotation of angular velocity $\boldsymbol{\omega}$. We propose to show that

$$\boldsymbol{\omega} = \boldsymbol{\omega}_1 + \boldsymbol{\omega}_2 \tag{15.40}$$

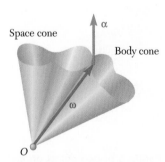

Fig. 15.33 The angular velocity vector generates a body cone and a space cone as it changes direction.

[†]Recall that a cone is, by definition, a surface generated by a straight line passing through a fixed point. In general, the cones considered here are not circular cones.

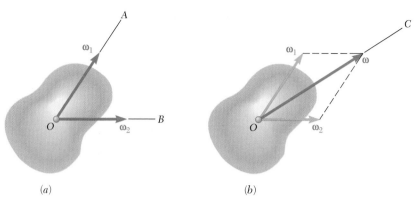

Fig. 15.34 (a) A rigid body rotating about two axes simultaneously; (b) the motion is equivalent to a single rotation with angular velocity equal to the vector sum of the initial angular velocities.

i.e., that we can obtain the resulting angular velocity by adding $\boldsymbol{\omega}_1$ and $\boldsymbol{\omega}_2$ using the parallelogram law (Fig. 15.34b).

Consider a particle P of the body, defined by the position vector $\mathbf{r}$. Denoting the velocity of P when the body rotates about OA only, about OB only, and about both axes simultaneously, by $\mathbf{v}_1$, $\mathbf{v}_2$, and $\mathbf{v}$, respectively, we have

$$\mathbf{v} = \boldsymbol{\omega} \times \mathbf{r} \qquad \mathbf{v}_1 = \boldsymbol{\omega}_1 \times \mathbf{r} \qquad \mathbf{v}_2 = \boldsymbol{\omega}_2 \times \mathbf{r} \qquad \textbf{(15.41)}$$

But the vectorial character of *linear* velocities is well established (since they represent the derivatives of position vectors). We therefore have

$$\mathbf{v} = \mathbf{v}_1 + \mathbf{v}_2$$

where the plus sign indicates vector addition. Substituting from Eq. (15.41), we obtain

$$\boldsymbol{\omega} \times \mathbf{r} = \boldsymbol{\omega}_1 \times \mathbf{r} + \boldsymbol{\omega}_2 \times \mathbf{r}$$
$$\boldsymbol{\omega} \times \mathbf{r} = (\boldsymbol{\omega}_1 + \boldsymbol{\omega}_2) \times \mathbf{r}$$

where the plus sign still indicates vector addition. Since the relation obtained holds for an arbitrary $\mathbf{r}$, we conclude that Eq. (15.40) must be true.

*15.6B General Motion

We now consider the most general motion of a rigid body in space. Let A and B be two particles of the body. Recall from Sec. 11.4D that we can express the velocity of B with respect to the fixed frame of reference $OXYZ$ as

$$\mathbf{v}_B = \mathbf{v}_A + \mathbf{v}_{B/A} \qquad \textbf{(15.42)}$$

where $\mathbf{v}_{B/A}$ is the velocity of B relative to a frame $AX'Y'Z'$ attached to A and of fixed orientation (Fig. 15.35). Since A is fixed in this frame, the motion of the body relative to $AX'Y'Z'$ is the motion of a body with a fixed point. Therefore we can obtain the relative velocity $\mathbf{v}_{B/A}$ from Eq. (15.37) after replacing $\mathbf{r}$ by the position vector $\mathbf{r}_{B/A}$ of B relative to A. Substituting for $\mathbf{v}_{B/A}$ into Eq. (15.42), we have

$$\mathbf{v}_B = \mathbf{v}_A + \boldsymbol{\omega} \times \mathbf{r}_{B/A} \qquad \textbf{(15.43)}$$

where $\boldsymbol{\omega}$ is the angular velocity of the body at the instant considered.

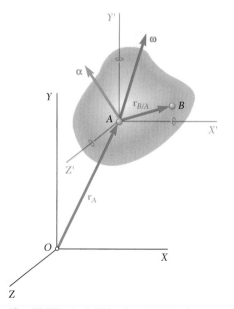

Photo 15.8 You can obtain the angular velocity of a fire truck ladder rotating about its fixed base by adding the angular velocities that correspond to simultaneous rotations about two different axes.

Fig. 15.35 A rigid body moving relative to a fixed reference frame $OXYZ$ and a reference frame attached to the body but with fixed orientation, $OX'Y'Z'$.

We can obtain the acceleration of B by a similar reasoning. We first write

$$\mathbf{a}_B = \mathbf{a}_A + \mathbf{a}_{B/A}$$

and, from Eq. (15.38),

$$\mathbf{a}_B = \mathbf{a}_A + \boldsymbol{\alpha} \times \mathbf{r}_{B/A} + \boldsymbol{\omega} \times (\boldsymbol{\omega} \times \mathbf{r}_{B/A}) \tag{15.44}$$

where $\boldsymbol{\alpha}$ is the angular acceleration of the body at the instant considered.

The angular acceleration $\boldsymbol{\alpha}$ represents the rate of change $(\dot{\boldsymbol{\omega}})_{OXYZ}$ of vector $\boldsymbol{\omega}$ with respect to a fixed frame of reference $OXYZ$ and reflects both a change in magnitude and a change in direction of the angular velocity. When computing $\boldsymbol{\alpha}$, you will usually find it convenient to first compute the rate of change $(\dot{\boldsymbol{\omega}})_{Oxyz}$ of $\boldsymbol{\omega}$ with respect to a rotating frame of reference $Oxyz$ of your choice and use Eq. (15.31) to obtain $\boldsymbol{\alpha}$. You have

$$\boldsymbol{\alpha} = (\dot{\boldsymbol{\omega}})_{OXYZ} = (\dot{\boldsymbol{\omega}})_{Oxyz} + \boldsymbol{\Omega} \times \boldsymbol{\omega}$$

where $\boldsymbol{\Omega}$ is the angular velocity of the rotating frame $Oxyz$ [Sample Prob. 15.21].

Equations (15.43) and (15.44) show that **the most general motion of a rigid body is equivalent, at any given instant, to the sum of a translation** (in which all of the particles of the body have the same velocity and acceleration as a reference particle A) **and of a motion in which particle A is assumed to be fixed.**[†]

By solving Eqs. (15.43) and (15.44) for $\mathbf{v}_A$ and $\mathbf{a}_A$, it can be shown that the motion of the body with respect to a frame attached to B would be characterized by the same vectors $\boldsymbol{\omega}$ and $\boldsymbol{\alpha}$ as its motion relative to $AX'Y'Z'$. Thus, the angular velocity and angular acceleration of a rigid body at a given instant are independent of the choice of reference point. If $AX'Y'Z'$ is a non-rotating frame, you should keep in mind that whether the moving frame is attached to A or to B, it should maintain a fixed orientation; that is, it should remain parallel to the fixed reference frame $OXYZ$ throughout the motion of the rigid body.

In many problems, it will be more convenient to use a moving frame that is allowed to rotate as well as to translate. We discuss the use of such moving frames in Sec. 15.7.

[†]Recall from Sec 15.6A that, in general, vectors $\boldsymbol{\omega}$ and $\boldsymbol{\alpha}$ are not collinear and that the accelerations of the particles of the body in their motion relative to the frame $AX'Y'Z'$ cannot be determined as if the body were rotating permanently about the instantaneous axis through A.

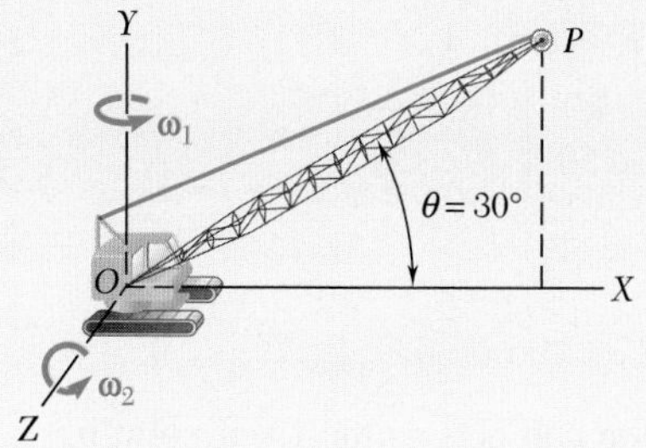

Sample Problem 15.21

The crane shown rotates horizontally with a constant angular velocity $\boldsymbol{\omega}_1$ of 0.30 rad/s. Simultaneously, the boom is being raised with a constant angular velocity $\boldsymbol{\omega}_2$ of 0.50 rad/s relative to the cab. Knowing that the length of the boom OP is $l = 12$ m, determine (a) the angular velocity $\boldsymbol{\omega}$ of the boom, (b) the angular acceleration $\boldsymbol{\alpha}$ of the boom, (c) the velocity $\mathbf{v}$ of the tip of the boom, (d) the acceleration $\mathbf{a}$ of the tip of the boom.

STRATEGY: There are multiple rotational axes, so you need to use the general motion velocity and acceleration kinematic equations. Add the given angular velocities vectorially to find the overall angular velocity of the boom, and differentiate that to find the angular acceleration.

MODELING and ANALYSIS:

a. Angular Velocity of Boom. Add the angular velocity $\boldsymbol{\omega}_1$ of the cab and the angular velocity $\boldsymbol{\omega}_2$ of the boom relative to the cab to obtain the angular velocity $\boldsymbol{\omega}$ of the boom at the instant considered:

$$\boldsymbol{\omega} = \boldsymbol{\omega}_1 + \boldsymbol{\omega}_2 \qquad \boldsymbol{\omega} = (0.30 \text{ rad/s})\mathbf{j} + (0.50 \text{ rad/s})\mathbf{k} \quad \blacktriangleleft$$

b. Angular Acceleration of Boom. Obtain the angular acceleration $\boldsymbol{\alpha}$ of the boom by differentiating $\boldsymbol{\omega}$. Since the vector $\boldsymbol{\omega}_1$ is constant in magnitude and direction, you have

$$\boldsymbol{\alpha} = \dot{\boldsymbol{\omega}} = \dot{\boldsymbol{\omega}}_1 + \dot{\boldsymbol{\omega}}_2 = 0 + \dot{\boldsymbol{\omega}}_2$$

where the rate of change $\dot{\boldsymbol{\omega}}_2$ is to be computed with respect to the fixed frame $OXYZ$. However, it is more convenient to use a frame $Oxyz$ attached to the cab and rotating with it, since the vector $\boldsymbol{\omega}_2$ also rotates with the cab and therefore has zero rate of change with respect to that frame. Using Eq. (15.31) with $\mathbf{Q} = \boldsymbol{\omega}_2$ and $\boldsymbol{\Omega} = \boldsymbol{\omega}_1$, you have

$$(\dot{\mathbf{Q}})_{OXYZ} = (\dot{\mathbf{Q}})_{Oxyz} + \boldsymbol{\Omega} \times \mathbf{Q}$$
$$(\dot{\boldsymbol{\omega}}_2)_{OXYZ} = (\dot{\boldsymbol{\omega}}_2)_{Oxyz} + \boldsymbol{\omega}_1 \times \boldsymbol{\omega}_2$$
$$\boldsymbol{\alpha} = (\dot{\boldsymbol{\omega}}_2)_{OXYZ} = 0 + (0.30 \text{ rad/s})\mathbf{j} \times (0.50 \text{ rad/s})\mathbf{k}$$

$$\boldsymbol{\alpha} = (0.15 \text{ rad/s}^2)\mathbf{i} \quad \blacktriangleleft$$

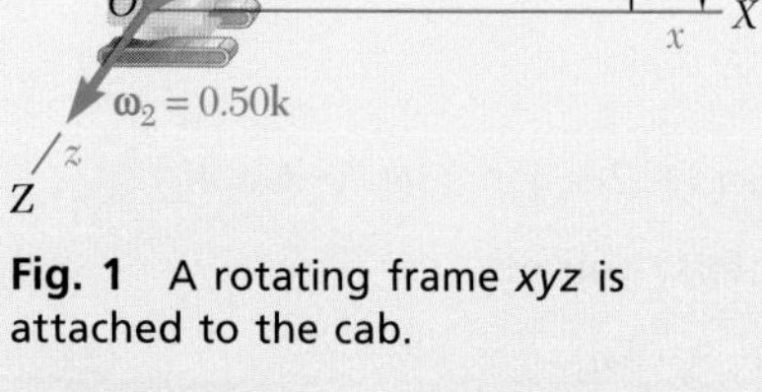

c. Velocity of Tip of Boom. Noting that the position vector of point P is $\mathbf{r} = (10.39 \text{ m})\mathbf{i} + (6 \text{ m})\mathbf{j}$ (Fig. 1) and using the expression found for $\boldsymbol{\omega}$ in part (a), you get

$$\mathbf{v} = \boldsymbol{\omega} \times \mathbf{r} = \begin{vmatrix} \mathbf{i} & \mathbf{j} & \mathbf{k} \\ 0 & 0.30 \text{ rad/s} & 0.50 \text{ rad/s} \\ 10.39 \text{ m} & 6 \text{ m} & 0 \end{vmatrix}$$

$$\mathbf{v} = -(3 \text{ m/s})\mathbf{i} + (5.20 \text{ m/s})\mathbf{j} - (3.12 \text{ m/s})\mathbf{k} \quad \blacktriangleleft$$

Fig. 1 A rotating frame *xyz* is attached to the cab.

(continued)

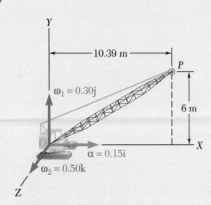

Fig. 2 Angular velocities and accelerations of the boom.

d. Acceleration of Tip of Boom. Recall that $\mathbf{v} = \boldsymbol{\omega} \times \mathbf{r}$. Then, from Fig. 2,

$$\mathbf{a} = \boldsymbol{\alpha} \times \mathbf{r} + \boldsymbol{\omega} \times (\boldsymbol{\omega} \times \mathbf{r}) = \boldsymbol{\alpha} \times \mathbf{r} + \boldsymbol{\omega} \times \mathbf{v}$$

$$\mathbf{a} = \begin{vmatrix} \mathbf{i} & \mathbf{j} & \mathbf{k} \\ 0.15 & 0 & 0 \\ 10.39 & 6 & 0 \end{vmatrix} + \begin{vmatrix} \mathbf{i} & \mathbf{j} & \mathbf{k} \\ 0 & 0.30 & 0.50 \\ -3 & 5.20 & -3.12 \end{vmatrix}$$

$$= 0.90\mathbf{k} - 0.94\mathbf{i} - 2.60\mathbf{i} - 1.50\mathbf{j} + 0.90\mathbf{k}$$

$$\mathbf{a} = -(3.54 \text{ m/s}^2)\mathbf{i} - (1.50 \text{ m/s}^2)\mathbf{j} + (1.80 \text{ m/s}^2)\mathbf{k} \quad \blacktriangleleft$$

REFLECT and THINK: The base of the cab acts as the fixed point of the motion. Even though both components of angular velocity are constant, there is an angular acceleration due to the change in direction of the angular velocity $\boldsymbol{\omega}_2$. The angular velocity vector $\boldsymbol{\omega}_2$ changes due to the rotation of the cab, $\boldsymbol{\omega}_1$.

Sample Problem 15.22

The rod AB has a length of 7 in. and is attached to the disk by a ball-and-socket connection and to the collar B by a clevis. The disk rotates in the yz plane at a constant rate of $\omega_1 = 12$ rad/s, while the collar is free to slide along the horizontal rod CD. For the position $\theta = 0$, determine (a) the velocity of the collar, (b) the angular velocity of the rod.

STRATEGY: Use the velocity and acceleration kinematic equations to relate the velocities of points A and B.

MODELING and ANALYSIS:

a. Velocity of Collar. Since point A is attached to the disk and since collar B moves in a direction parallel to the x axis, you have (Fig. 1)

$$\mathbf{v}_A = \boldsymbol{\omega}_1 \times \mathbf{r}_A = 12\mathbf{i} \times 2\mathbf{k} = -24\mathbf{j} \qquad \mathbf{v}_B = v_B\mathbf{i}$$

Denoting the angular velocity of the rod by $\boldsymbol{\omega}$, you obtain

$$\mathbf{v}_B = \mathbf{v}_A + \mathbf{v}_{B/A} = \mathbf{v}_A + \boldsymbol{\omega} \times \mathbf{r}_{B/A}$$

$$v_B\mathbf{i} = -24\mathbf{j} + \begin{vmatrix} \mathbf{i} & \mathbf{j} & \mathbf{k} \\ \omega_x & \omega_y & \omega_z \\ 6 & 3 & -2 \end{vmatrix}$$

$$v_B\mathbf{i} = -24\mathbf{j} + (-2\omega_y - 3\omega_z)\mathbf{i} + (6\omega_z + 2\omega_x)\mathbf{j} + (3\omega_x - 6\omega_y)\mathbf{k}$$

Equating the coefficients of the unit vectors, you get

$$v_B = \qquad -2\omega_y \quad -3\omega_z \qquad (1)$$
$$24 = 2\omega_x \qquad\qquad +6\omega_z \qquad (2)$$
$$0 = 3\omega_x \quad -6\omega_y \qquad\qquad (3)$$

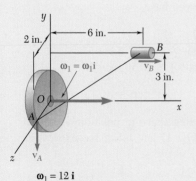

$$\boldsymbol{\omega}_1 = 12\,\mathbf{i}$$
$$\mathbf{r}_A = 2\mathbf{k}$$
$$\mathbf{r}_B = 6\mathbf{i} + 3\mathbf{j}$$
$$\mathbf{r}_{B/A} = 6\mathbf{i} + 3\mathbf{j} - 2\mathbf{k}$$

Fig. 1 Angular velocity of the disk and the direction of the velocities of A and B.

You have three equations and four unknowns in these equations. Fortunately, multiplying Eqs. (1), (2), (3), respectively, by 6, 3, −2 and adding gives you

$$6v_B + 72 = 0 \qquad v_B = -12 \qquad \mathbf{v}_B = -(12 \text{ in./s})\mathbf{i} \blacktriangleleft$$

b. Angular Velocity of Rod *AB*. Note that you cannot determine the angular velocity from Eqs. (1), (2), and (3) because the determinant formed by the coefficients of ω_x, ω_y, and ω_z is zero. You must therefore obtain an additional equation by considering the constraint imposed by the clevis at *B*.

The collar–clevis connection at *B* permits rotation of *AB* about rod *CD* and also about an axis perpendicular to the plane containing *AB* and *CD*. It prevents rotation of *AB* about the axis *EB*, which is perpendicular to *CD* and lies in the plane containing *AB* and *CD* (Fig. 2). Thus, the projection of $\boldsymbol{\omega}$ on $\mathbf{r}_{E/B}$ must be zero, and you have

$$\boldsymbol{\omega} \cdot \mathbf{r}_{E/B} = 0$$

$$(\omega_x\mathbf{i} + \omega_y\mathbf{j} + \omega_z\mathbf{k}) \cdot (-3\mathbf{j} + 2\mathbf{k}) = 0$$

$$-3\omega_y + 2\omega_z = 0 \qquad (4)$$

Solving Eqs. (1) through (4) simultaneously, you obtain

$$v_B = -12 \qquad \omega_x = 3.69 \qquad \omega_y = 1.846 \qquad \omega_z = 2.77$$
$$\boldsymbol{\omega} = (3.69 \text{ rad/s})\mathbf{i} + (1.846 \text{ rad/s})\mathbf{j} + (2.77 \text{ rad/s})\mathbf{k} \blacktriangleleft$$

REFLECT and THINK: Note that the direction of *EB* is that of the vector triple product

$$\mathbf{r}_{B/C} \times (\mathbf{r}_{B/C} \times \mathbf{r}_{B/A})$$

so you could write

$$\boldsymbol{\omega} \cdot [\mathbf{r}_{B/C} \times (\mathbf{r}_{B/C} \times \mathbf{r}_{B/A})] = 0$$

This formulation would be particularly useful if the rod *CD* were not in a convenient direction.

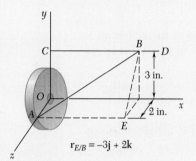

Fig. 2 The collar-clevis prevents rotation about *EB*.

$\mathbf{r}_{E/B} = -3\mathbf{j} + 2\mathbf{k}$

SOLVING PROBLEMS
ON YOUR OWN

In this section, you started the study of the **kinematics of rigid bodies in three dimensions**. You first studied the **motion of a rigid body about a fixed point** and then the **general motion of a rigid body**.

A. Motion of a rigid body about a fixed point. To analyze the motion of a point B of a body rotating about a fixed point O, you may have to take some or all of the following steps.

1. Determine the position vector r connecting the fixed point O to point B.

2. Determine the angular velocity ω of the body with respect to a fixed frame of reference. You can often obtain the angular velocity ω by adding two component angular velocities ω_1 and ω_2 [Sample Prob. 15.21].

3. Compute the velocity of B from the equation

$$\mathbf{v} = \boldsymbol{\omega} \times \mathbf{r} \qquad \qquad \textbf{(15.37)}$$

Your computation is usually easier if you express the vector product as a determinant.

4. Determine the angular acceleration α of the body. The angular acceleration $\boldsymbol{\alpha}$ represents the rate of change $(\dot{\boldsymbol{\omega}})_{OXYZ}$ of the vector $\boldsymbol{\omega}$ with respect to a fixed frame of reference $OXYZ$ and reflects both a change in magnitude and a change in direction of the angular velocity. However, when computing $\boldsymbol{\alpha}$, you may find it convenient to first compute the rate of change $(\dot{\boldsymbol{\omega}})_{Oxyz}$ of $\boldsymbol{\omega}$ with respect to a rotating frame of reference $Oxyz$ of your choice and use Eq. (15.31). You have

$$\boldsymbol{\alpha} = (\dot{\boldsymbol{\omega}})_{OXYZ} = (\dot{\boldsymbol{\omega}})_{Oxyz} + \boldsymbol{\Omega} \times \boldsymbol{\omega}$$

where $\boldsymbol{\Omega}$ is the angular velocity of the rotating frame $Oxyz$ [Sample Prob. 15.21].

5. Compute the acceleration of B by using the equation

$$\mathbf{a} = \boldsymbol{\alpha} \times \mathbf{r} + \boldsymbol{\omega} \times (\boldsymbol{\omega} \times \mathbf{r}) \qquad \qquad \textbf{(15.38)}$$

Note that the vector product $(\boldsymbol{\omega} \times \mathbf{r})$ represents the velocity of point B and was computed in Step 3. Also, the computation of the first vector product in Eq. (15.38) is often simpler if you express this product in determinant form. Remember that, as was the case with the plane motion of a rigid body, the instantaneous axis of rotation *cannot* be used to determine accelerations.

B. General motion of a rigid body. The general motion of a rigid body may be considered as *the sum of a translation and a rotation.* Keep the following in mind:

   a. In the translation part of the motion, all of the points of the body have the *same velocity* $\mathbf{v}_A$ *and the same acceleration* $\mathbf{a}_A$ as point A of the body that has been selected as the reference point.

   b. In the rotation part of the motion, the same reference point A is treated as if it were a *fixed point.*

1. To determine the velocity of a point B of the rigid body when you know the velocity $\mathbf{v}_A$ of the reference point A and the angular velocity $\boldsymbol{\omega}$ of the body, you simply add $\mathbf{v}_A$ to the velocity $\mathbf{v}_{B/A} = \boldsymbol{\omega} \times \mathbf{r}_{B/A}$ of B in its rotation about A:

$$\mathbf{v}_B = \mathbf{v}_A + \boldsymbol{\omega} \times \mathbf{r}_{B/A} \tag{15.43}$$

As indicated earlier, the computation of the vector product is usually simpler if you express this product in determinant form.

You can also use Eq. (15.43) to determine the magnitude of $\mathbf{v}_B$ when its direction is known, even if $\boldsymbol{\omega}$ is not known. Although the corresponding three scalar equations are linearly dependent and the components of $\boldsymbol{\omega}$ are indeterminate, you can eliminate these components and find $\mathbf{v}_A$ by using an appropriate linear combination of the three equations [Sample Prob. 15.22, part (a)]. Alternatively, you can assign an arbitrary value to one of the components of $\boldsymbol{\omega}$ and solve the equations for $\mathbf{v}_A$. However, you must seek an additional equation in order to determine the true values of the components of $\boldsymbol{\omega}$ [Sample Prob. 15.22, part (b)].

2. To determine the acceleration of a point B of the rigid body when you know the acceleration $\mathbf{a}_A$ of the reference point A and the angular acceleration $\boldsymbol{\alpha}$ of the body, you simply add $\mathbf{a}_A$ to the acceleration of B in its rotation about A, as expressed by Eq. (15.38):

$$\mathbf{a}_B = \mathbf{a}_A + \boldsymbol{\alpha} \times \mathbf{r}_{B/A} + \boldsymbol{\omega} \times (\boldsymbol{\omega} \times \mathbf{r}_{B/A}) \tag{15.44}$$

Note that the vector product $(\boldsymbol{\omega} \times \mathbf{r}_{B/A})$ represents the velocity $\mathbf{v}_{B/A}$ of B relative to A and already may have been computed as part of your calculation of $\mathbf{v}_B$.

You can also use the three scalar equations associated with Eq. (15.44) to determine the magnitude of $\mathbf{a}_B$ when its direction is known, even if $\boldsymbol{\omega}$ and $\boldsymbol{\alpha}$ are not known. Although the components of $\boldsymbol{\omega}$ and $\boldsymbol{\alpha}$ are indeterminate, you can assign arbitrary values to one of the components of $\boldsymbol{\omega}$ and to one of the components of $\boldsymbol{\alpha}$ and solve the equations for $\mathbf{a}_B$.

Problems

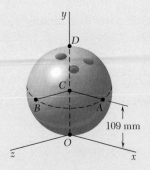

Fig. P15.184 and P15.185

15.184 The bowling ball shown rolls without slipping on the horizontal xz plane with an angular velocity $\boldsymbol{\omega} = \omega_x\mathbf{i} + \omega_y\mathbf{j} + \omega_z\mathbf{k}$. Knowing that $\mathbf{v}_A = (4.8 \text{ m/s})\mathbf{i} - (4.8 \text{ m/s})\mathbf{j} + (3.6 \text{ m/s})\mathbf{k}$ and $\mathbf{v}_D = (9.6 \text{ m/s})\mathbf{i} + (7.2 \text{ m/s})\mathbf{k}$, determine (a) the angular velocity of the bowling ball, (b) the velocity of its center C.

15.185 The bowling ball shown rolls without slipping on the horizontal xz plane with an angular velocity $\boldsymbol{\omega} = \omega_x\mathbf{i} + \omega_y\mathbf{j} + \omega_z\mathbf{k}$. Knowing that $\mathbf{v}_B = (3.6 \text{ m/s})\mathbf{i} - (4.8 \text{ m/s})\mathbf{j} + (4.8 \text{ m/s})\mathbf{k}$ and $\mathbf{v}_D = (7.2 \text{ m/s})\mathbf{i} + (9.6 \text{ m/s})\mathbf{k}$, determine (a) the angular velocity of the bowling ball, (b) the velocity of its center C.

15.186 Plate ABD and rod OB are rigidly connected and rotate about the ball-and-socket joint O with an angular velocity $\boldsymbol{\omega} = \omega_x\mathbf{i} + \omega_y\mathbf{j} + \omega_z\mathbf{k}$. Knowing that $\mathbf{v}_A = (80 \text{ mm/s})\mathbf{i} + (360 \text{ mm/s})\mathbf{j} + (v_A)_z\mathbf{k}$ and $\omega_x = 1.5$ rad/s, determine (a) the angular velocity of the assembly, (b) the velocity of point D.

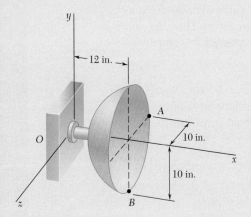

Fig. P15.187

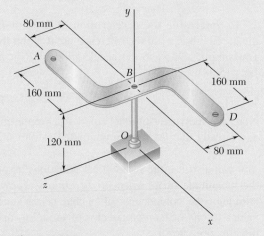

Fig. P15.186

15.187 At the instant considered, the radar antenna shown rotates about the origin of coordinates with an angular velocity $\boldsymbol{\omega} = \omega_x\mathbf{i} + \omega_y\mathbf{j} + \omega_z\mathbf{k}$. Knowing that $(v_A)_y = 15$ in./s, $(v_B)_y = 9$ in./s, and $(v_B)_z = 18$ in./s, determine (a) the angular velocity of the antenna, (b) the velocity of point A.

15.188 The rotor of an electric motor rotates at the constant rate $\omega_1 = 1800$ rpm. Determine the angular acceleration of the rotor as the motor is rotated about the y axis with a constant angular velocity ω_2 of 6 rpm counterclockwise when viewed from the positive y axis.

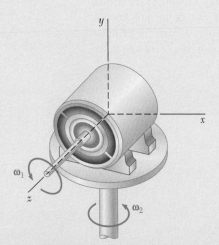

Fig. P15.188

15.189 The disk of a portable sander rotates at the constant rate $\omega_1 = 4400$ rpm as shown. Determine the angular acceleration of the disk as a worker rotates the sander about the z axis with an angular velocity of 0.5 rad/s and an angular acceleration of 2.5 rad/s², both clockwise when viewed from the positive z axis.

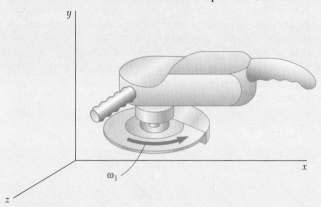

Fig. P15.189

15.190 A flight simulator is used to train pilots on how to recognize spatial disorientation. It has four degrees of freedom, and can rotate around a planetary axis as well as in yaw, pitch, and roll. Knowing that the simulator is rotating around the planetary axis with a constant angular velocity of 20 rpm counterclockwise as seen from above, determine the angular acceleration of the cab if (*a*) the cab has a constant pitch angular velocity of $+3\mathbf{k}$ rad/s, (*b*) the cab has a constant roll angular velocity of $-4\mathbf{i}$ rad/s.

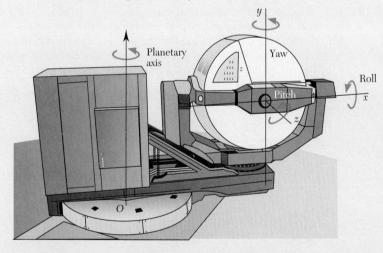

Fig. P15.190

15.191 In the system shown, disk A is free to rotate about the horizontal rod OA. Assuming that disk B is stationary ($\omega_2 = 0$), and that shaft OC rotates with a constant angular velocity $\boldsymbol{\omega}_1$, determine (*a*) the angular velocity of disk A, (*b*) the angular acceleration of disk A.

15.192 In the system shown, disk A is free to rotate about the horizontal rod OA. Assuming that shaft OC and disk B rotate with constant angular velocities $\boldsymbol{\omega}_1$ and $\boldsymbol{\omega}_2$, respectively, both counterclockwise, determine (*a*) the angular velocity of disk A, (*b*) the angular acceleration of disk A.

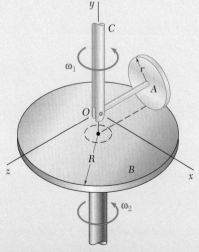

Fig. *P15.191* and *P15.192*

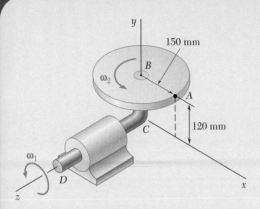

Fig. P15.193

15.193 The L-shaped arm *BCD* rotates about the *z* axis with a constant angular velocity $\omega_1 = 5$ rad/s. Knowing that the 150-mm-radius disk rotates about *BC* with a constant angular velocity $\omega_2 = 4$ rad/s, determine (*a*) the velocity of point *A*, (*b*) the acceleration of point *A*.

15.194 A gun barrel of length *OP* = 4 m is mounted on a turret as shown. To keep the gun aimed at a moving target, the azimuth angle β is being increased at the rate $d\beta/dt = 30°/s$ and the elevation angle γ is being increased at the rate $d\gamma/dt = 10°/s$. For the position $\beta = 90°$ and $\gamma = 30°$, determine (*a*) the angular velocity of the barrel, (*b*) the angular acceleration of the barrel, (*c*) the velocity and acceleration of point *P*.

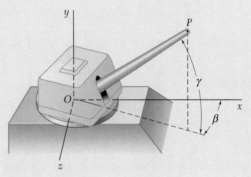

Fig. *P15.194*

15.195 A 3-in.-radius disk spins at the constant rate $\omega_2 = 4$ rad/s about an axis held by a housing attached to a horizontal rod that rotates at the constant rate $\omega_1 = 5$ rad/s. For the position shown, determine (*a*) the angular acceleration of the disk, (*b*) the acceleration of point *P* on the rim of the disk if $\theta = 0$, (*c*) the acceleration of point *P* on the rim of the disk if $\theta = 90°$.

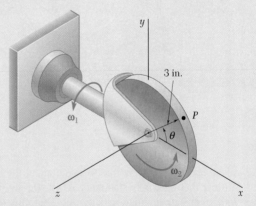

Fig. P15.195 and P15.196

15.196 A 3-in.-radius disk spins at the constant rate $\omega_2 = 4$ rad/s about an axis held by a housing attached to a horizontal rod that rotates at the constant rate $\omega_1 = 5$ rad/s. Knowing that $\theta = 30°$, determine the acceleration of point *P* on the rim of the disk.

15.197 The cone shown rolls on the zx plane with its apex at the origin of coordinates. Denoting by $\boldsymbol{\omega}_1$ the constant angular velocity of the axis OB of the cone about the y axis, determine (a) the rate of spin of the cone about the axis OB, (b) the total angular velocity of the cone, (c) the angular acceleration of the cone.

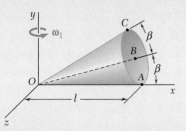

Fig. P15.197

15.198 At the instant shown, the robotic arm ABC is being rotated simultaneously at the constant rate $\omega_1 = 0.15$ rad/s about the y axis, and at the constant rate $\omega_2 = 0.25$ rad/s about the z axis. Knowing that the length of arm ABC is 1 m, determine (a) the angular acceleration of the arm, (b) the velocity of point C, (c) the acceleration of point C.

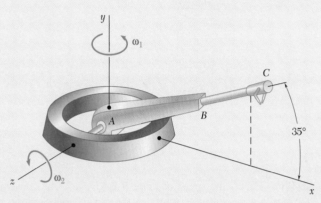

Fig. P15.198

15.199 In the planetary gear system shown, gears A and B are rigidly connected to each other and rotate as a unit about the inclined shaft. Gears C and D rotate with constant angular velocities of 30 rad/s and 20 rad/s, respectively (both counterclockwise when viewed from the right). Choosing the x axis to the right, the y axis upward, and the z axis pointing out of the plane of the figure, determine (a) the common angular velocity of gears A and B, (b) the angular velocity of shaft FH, which is rigidly attached to the inclined shaft.

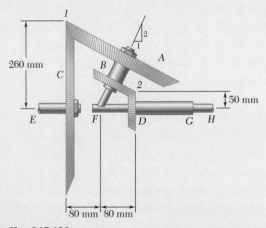

Fig. P15.199

15.200 In Prob. 15.199, determine (a) the common angular acceleration of gears A and B, (b) the acceleration of the tooth of gear A that is in contact with gear C at point 1.

15.201 Several rods are brazed together to form the robotic guide arm shown that is attached to a ball-and-socket joint at O. Rod OA slides in a straight inclined slot while rod OB slides in a slot parallel to the z axis. Knowing that at the instant shown $\mathbf{v}_B = (9 \text{ in./s})\mathbf{k}$, determine (a) the angular velocity of the guide arm, (b) the velocity of point A, (c) the velocity of point C.

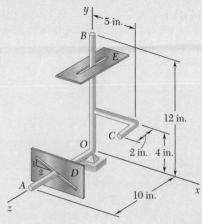

Fig. *P15.201*

15.202 In Prob. 15.201, the speed of point B is known to be constant. For the position shown, determine (a) the angular acceleration of the guide arm, (b) the acceleration of point C.

15.203 Rod AB of length 25 in. is connected by ball-and-socket joints to collars A and B, which slide along the two rods shown. Knowing that collar B moves toward point E at a constant speed of 20 in./s, determine the velocity of collar A as collar B passes through point D.

15.204 Rod AB has a length of 13 in. and is connected by ball-and-socket joints to collars A and B that slide along the two rods shown. Knowing that collar B moves toward point D at a constant speed of 36 in./s, determine the velocity of collar A when $b = 4$ in.

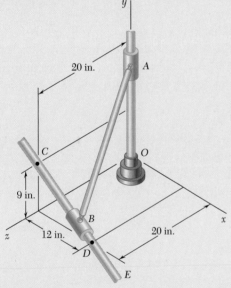

Fig. **P15.203**

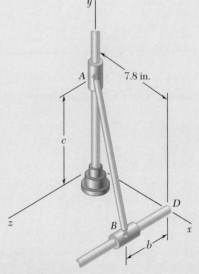

Fig. **P15.204**

15.205 Rod *BC* and *BD* are each 840 mm long and are connected by ball-and-socket joints to collars that may slide on the fixed rods shown. Knowing that collar *B* moves toward *A* at a constant speed of 390 mm/s, determine the velocity of collar *C* for the position shown.

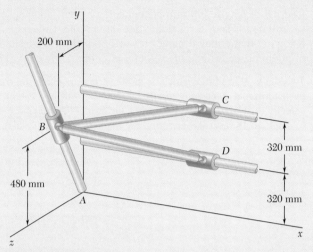

Fig. P15.205

15.206 Rod *AB* is connected by ball-and-socket joints to collar *A* and to the 16-in.-diameter disk *C*. Knowing that disk *C* rotates counterclockwise at the constant rate $\omega_0 = 3$ rad/s in the *zx* plane, determine the velocity of collar *A* for the position shown.

15.207 Rod *AB* of length 29 in. is connected by ball-and-socket joints to the rotating crank *BC* and to the collar *A*. Crank *BC* is of length 8 in. and rotates in the horizontal *xz* plane at the constant rate $\omega_0 = 10$ rad/s. At the instant shown, when crank *BC* is parallel to the *z* axis, determine the velocity of collar *A*.

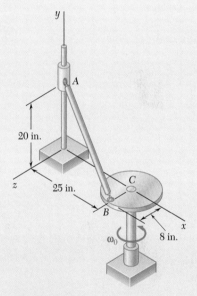

Fig. P15.206

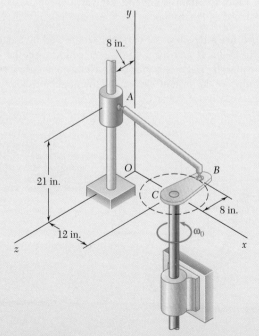

Fig. P15.207

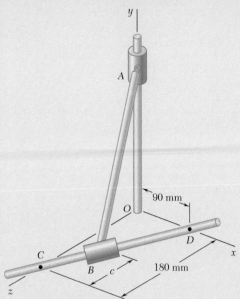

Fig. P15.208 and P15.209

15.208 Rod AB of length 300 mm is connected by ball-and-socket joints to collars A and B, which slide along the two rods shown. Knowing that collar B moves toward point D at a constant speed of 50 mm/s, determine the velocity of collar A when $c = 80$ mm.

15.209 Rod AB of length 300 mm is connected by ball-and-socket joints to collars A and B, which slide along the two rods shown. Knowing that collar B moves toward point D at a constant speed of 50 mm/s, determine the velocity of collar A when $c = 120$ mm.

15.210 Two shafts AC and EG, which lie in the vertical yz plane, are connected by a universal joint at D. Shaft AC rotates with a constant angular velocity ω_1 as shown. At a time when the arm of the crosspiece attached to shaft AC is vertical, determine the angular velocity of shaft EG.

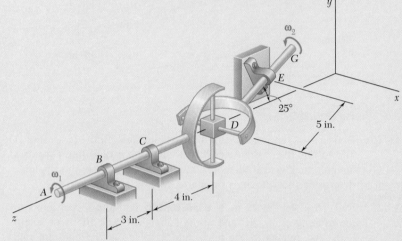

Fig. P15.210

15.211 Solve Prob. 15.210, assuming that the arm of the crosspiece attached to shaft AC is horizontal.

15.212 Rod BC has a length of 42 in. and is connected by a ball-and-socket joint to collar B and by a clevis connection to collar C. Knowing that collar B moves toward A at a constant speed of 19.5 in./s, determine at the instant shown (*a*) the angular velocity of the rod, (*b*) the velocity of collar C.

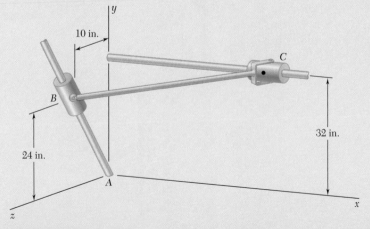

Fig. P15.212

15.213 Rod *AB* has a length of 275 mm and is connected by a ball-and-socket joint to collar *A* and by a clevis connection to collar *B*. Knowing that collar *B* moves down at a constant speed of 1.35 m/s, determine at the instant shown (*a*) the angular velocity of the rod, (*b*) the velocity of collar *A*.

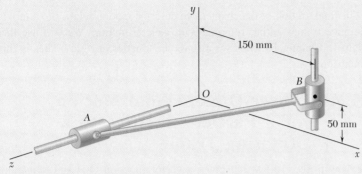

Fig. P15.213

15.214 For the mechanism of the Prob.15.204, determine the acceleration of collar *A*.

15.215 In Prob. 15.205, determine the acceleration of collar *C*.

15.216 In Prob. 15.206, determine the acceleration of collar *A*.

15.217 In Prob. 15.207, determine the acceleration of collar *A*.

15.218 In Prob. 15.208, determine the acceleration of collar *A*.

15.219 In Prob. 15.209, determine the acceleration of collar *A*.

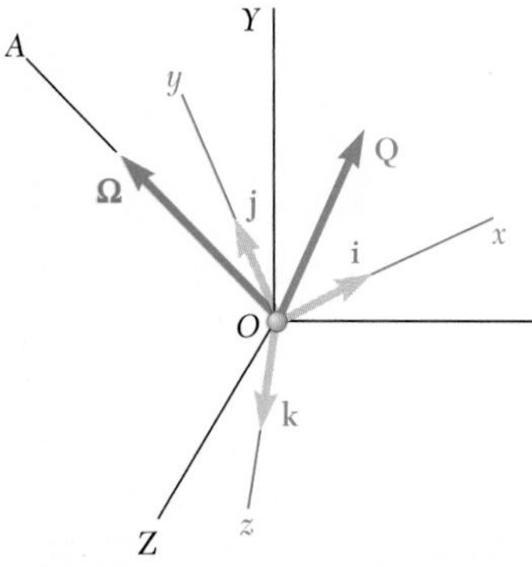

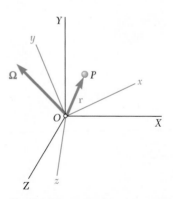

Fig. 15.36 Reference frame *Oxyz* rotating about an instantaneous axis in fixed frame *OXYZ* with angular velocity $\mathbf{\Omega}$.

*15.7 MOTION RELATIVE TO A MOVING REFERENCE FRAME

In this final section of the chapter, we describe motion relative to a moving reference frame—either rotating or in general motion. We will use these results in Chapter 18 when we discuss the kinetics of rigid bodies in three dimensions.

15.7A Three-Dimensional Motion of a Particle Relative to a Rotating Frame

We saw in Sec. 15.5A that given a vector function $\mathbf{Q}(t)$ and two frames of reference centered at *O*—a fixed frame *OXYZ* and a rotating frame *Oxyz*—the rates of change of $\mathbf{Q}$ with respect to the two frames satisfy the relation

$$(\dot{\mathbf{Q}})_{OXYZ} = (\dot{\mathbf{Q}})_{Oxyz} + \mathbf{\Omega} \times \mathbf{Q} \tag{15.31}$$

We had assumed at the time that the frame *Oxyz* was constrained to rotate about a fixed axis *OA*. However, the derivation given in Sec. 15.5A remains valid when the frame *Oxyz* is constrained to have only a fixed point *O*. Under this more general assumption, the axis *OA* represents the *instantaneous* axis of rotation of the frame *Oxyz* (Sec. 15.6A) and the vector $\mathbf{\Omega}$ represents its angular velocity at the instant considered (Fig. 15.36).

Let us now consider the three-dimensional motion of a particle *P* relative to a rotating frame *Oxyz* constrained to have a fixed origin *O*. Let $\mathbf{r}$ be the position vector of *P* at a given instant, and let $\mathbf{\Omega}$ be the angular velocity of the frame *Oxyz* with respect to the fixed frame *OXYZ* at the same instant (Fig. 15.37). The derivations given in Sec. 15.5B for the two-dimensional motion of a particle can be readily extended to the three-dimensional case. Then we can express the absolute velocity $\mathbf{v}_P$ of *P* (i.e., its velocity with respect to the fixed frame *OXYZ*) as

$$\mathbf{v}_P = \mathbf{\Omega} \times \mathbf{r} + (\dot{\mathbf{r}})_{Oxyz} \tag{15.45}$$

Fig. 15.37 A particle *P* moving relative to the rotating frame.

where $(\dot{\mathbf{r}})_{Oxyz}$ is the relative velocity of point *P* with respect to the rotating frame. Sometimes this is also written as $\mathbf{v}_{\text{rel}}$. Denoting the rotating frame *Oxyz* by $\mathscr{F}$, we can write this relation in the alternative form

$$\mathbf{v}_P = \mathbf{v}_{P'} + \mathbf{v}_{P/\mathscr{F}} \tag{15.46}$$

where $\mathbf{v}_P$ = absolute velocity of particle *P*

$\mathbf{v}_{P'}$ = velocity of point *P'* of moving frame $\mathscr{F}$ coinciding with *P*

$\mathbf{v}_{P/\mathscr{F}}$ = velocity of *P* relative to moving frame $\mathscr{F}$

The absolute acceleration $\mathbf{a}_P$ of P can be expressed as

$$\mathbf{a}_P = \dot{\boldsymbol{\Omega}} \times \mathbf{r} + \boldsymbol{\Omega} \times (\boldsymbol{\Omega} \times \mathbf{r}) + 2\boldsymbol{\Omega} \times (\dot{\mathbf{r}})_{Oxyz} + (\ddot{\mathbf{r}})_{Oxyz} \qquad (15.47)$$

An alternative form is

$$\mathbf{a}_P = \mathbf{a}_{P'} + \mathbf{a}_{P/\mathscr{F}} + \mathbf{a}_C \qquad (15.48)$$

where $\mathbf{a}_P$ = absolute acceleration of particle P

$\quad \mathbf{a}_{P'}$ = acceleration of point P' of moving frame $\mathscr{F}$ coinciding with P

$\mathbf{a}_{P/\mathscr{F}}$ = acceleration of P relative to moving frame $\mathscr{F}$

$\quad \mathbf{a}_C = 2\boldsymbol{\Omega} \times (\dot{\mathbf{r}})_{Oxyz} = 2\boldsymbol{\Omega} \times \mathbf{v}_{P/\mathscr{F}}$ = Coriolis acceleration

Note the difference between this equation and Eq. (15.21) of Sec.15.4A, and recall the discussion following Eq. (15.36) of Sec. 15.5B.

Also note that the Coriolis acceleration is perpendicular to the vectors $\boldsymbol{\Omega}$ and $\mathbf{v}_{P/\mathscr{F}}$. However, since these vectors are usually not perpendicular to each other, the magnitude of $\mathbf{a}_C$ is in general *not* equal to $2\Omega v_{P/\mathscr{F}}$—as was the case for the plane motion of a particle. We further note that the Coriolis acceleration reduces to zero when the vectors $\boldsymbol{\Omega}$ and $\mathbf{v}_{P/\mathscr{F}}$ are parallel or when either of them is zero.

Rotating frames of reference are particularly useful in the study of the three-dimensional motion of rigid bodies. If a rigid body has a fixed point O—as was the case for the crane of Sample Prob. 15.21—we can use a frame $Oxyz$ that can rotate. Denoting the angular velocity of the frame $Oxyz$ by $\boldsymbol{\Omega}$, we then resolve the angular velocity $\boldsymbol{\omega}$ of the body into the components $\boldsymbol{\Omega}$ and $\boldsymbol{\omega}_{B/\mathscr{F}}$, where the second component represents the angular velocity of the body relative to the frame $Oxyz$ (see Sample Prob. 15.24). An appropriate choice of a rotating frame often leads to a simpler analysis of the motion of the rigid body than would be possible with axes of fixed orientation. This is especially true in the case of the general three-dimensional motion of a rigid body, i.e., when the rigid body under consideration has no fixed point (see Sample Prob. 15.25).

*15.7B Frame of Reference in General Motion

Consider a fixed frame of reference $OXYZ$ and a frame $Axyz$ that moves in a known, but arbitrary, fashion with respect to $OXYZ$ (Fig. 15.38). Let P be a particle moving in space. The position of P defined at any instant by the vector $\mathbf{r}_P$ in the fixed frame and by the vector $\mathbf{r}_{P/A}$ in the moving frame. Denoting the position vector of A in the fixed frame by $\mathbf{r}_A$, we have

$$\mathbf{r}_P = \mathbf{r}_A + \mathbf{r}_{P/A} \qquad (15.49)$$

We obtain the absolute velocity $\mathbf{v}_P$ of the particle by differentiating, as

$$\mathbf{v}_P = \dot{\mathbf{r}}_P = \dot{\mathbf{r}}_A + \dot{\mathbf{r}}_{P/A} \qquad (15.50)$$

where the derivatives are defined with respect to the fixed frame $OXYZ$. The first term in the right-hand side of Eq. (15.50) thus represents the velocity $\mathbf{v}_A$ of the origin A of the moving axes. Since the rate of change

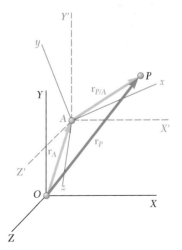

Fig. 15.38 Reference frame *Axyz* moves arbitrarily relative to fixed frame *OXYZ*.

of a vector is the same with respect to both a fixed frame and a frame in translation (See. 11.4B), we can regard the second term as the velocity $\mathbf{v}_{P/A}$ of P relative to the frame $AX'Y'Z'$ with the same orientation as $OXYZ$ and the same origin as $Axyz$. We therefore have

$$\mathbf{v}_P = \mathbf{v}_A + \mathbf{v}_{P/A} \tag{15.51}$$

However, we can obtain the velocity $\mathbf{v}_{P/A}$ of P relative to $AX'Y'Z'$ from Eq. (15.45) by substituting $\mathbf{r}_{P/A}$ for $\mathbf{r}$ in that equation. We get

$$\mathbf{v}_P = \mathbf{v}_A + \mathbf{\Omega} \times \mathbf{r}_{P/A} + (\dot{\mathbf{r}}_{P/A})_{Axyz} \tag{15.52}$$

where $\mathbf{\Omega}$ is the angular velocity of the frame $Axyz$ at the instant considered.

We obtain the absolute acceleration $\mathbf{a}_P$ of the particle by differentiating Eq. (15.51), as

$$\mathbf{a}_P = \dot{\mathbf{v}}_P = \dot{\mathbf{v}}_A + \dot{\mathbf{v}}_{P/A} \tag{15.53}$$

where the derivatives are defined with respect to either of the frames $OXYZ$ or $AX'Y'Z'$. Thus, the first term in the right-hand side of Eq. (15.53) represents the acceleration $\mathbf{a}_A$ of the origin A of the moving axes, and the second term represents the acceleration $\mathbf{a}_{P/A}$ of P relative to the frame $AX'Y'Z'$. We can obtain this acceleration from Eq. (15.47) by substituting $\mathbf{r}_{P/A}$ for $\mathbf{r}$. We therefore have

$$\mathbf{a}_P = \mathbf{a}_A + \dot{\mathbf{\Omega}} \times \mathbf{r}_{P/A} + \mathbf{\Omega} \times (\mathbf{\Omega} \times \mathbf{r}_{P/A})$$
$$+ 2\mathbf{\Omega} \times (\dot{\mathbf{r}}_{P/A})_{Axyz} + (\ddot{\mathbf{r}}_{P/A})_{Axyz} \tag{15.54}$$

Photo 15.9 The motion of air particles in a hurricane can be considered as motion relative to a frame of reference attached to the Earth and rotating with it.

Formulas (15.52) and (15.54) enable us to determine the velocity and acceleration of a given particle with respect to a fixed frame of reference when we know the motion of the particle with respect to a moving frame. These formulas become more significant, and considerably easier to remember, if we note that the sum of the first two terms in Eq. (15.52) represents the velocity of the point P' of the moving frame that coincides with P at the instant considered and that the sum of the first three terms in Eq. (15.54) represents the acceleration of the same point. Thus, relations in Eqs. (15.46) and (15.48) of the preceding section are still valid in the case of a reference frame in general motion, and we have

$$\mathbf{v}_P = \mathbf{v}_{P'} + \mathbf{v}_{P/\mathscr{F}} \tag{15.46}$$
$$\mathbf{a}_P = \mathbf{a}_{P'} + \mathbf{a}_{P/\mathscr{F}} + \mathbf{a}_C \tag{15.48}$$

where the various vectors involved were defined earlier.

Note that if the moving reference frame $\mathscr{F}$ (or $Axyz$) is in translation, the velocity and acceleration of the point P' of the frame that coincides with P become, respectively, equal to the velocity and acceleration of the origin A of the frame. On the other hand, since the frame maintains a fixed orientation, $\mathbf{a}_c$ is zero, and the relations in Eqs. (15.46) and (15.48) reduce, respectively, to the relations in Eqs. 11.32 and 11.33 derived in Sec. 11.4D.

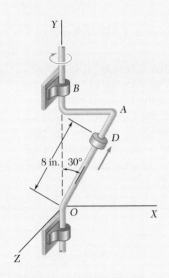

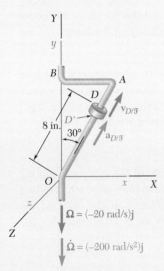

Fig. 1 The rotating coordinate system *xyz* is attached to rod *OAB*.

Sample Problem 15.23

The bent rod *OAB* rotates about the vertical axis *OB*. At the instant considered, its angular velocity and angular acceleration are, respectively, 20 rad/s and 200 rad/s^2, which are both clockwise when viewed from the positive *Y* axis. The collar *D* moves along the rod, and at the instant considered, *OD* = 8 in. The velocity and acceleration of the collar relative to the rod are, respectively, 50 in./s and 600 in./s^2, where both are upward. Determine (*a*) the velocity of the collar, (*b*) the acceleration of the collar.

STRATEGY: Use rigid-body kinematics with a rotating coordinate system since collar *D* is moving relative to the bent rod. Attach the rotating reference frame to the bent rod; then you can calculate its motion relative to the fixed frame and the collar's motion relative to the rotating frame.

MODELING:

Frames of Reference. The angular velocity and angular acceleration of the bent rod (and rotating frame *Oxyz*) relative to the fixed frame *OXYZ* are $\mathbf{\Omega} = (-20 \text{ rad/s})\mathbf{j}$ and $\dot{\mathbf{\Omega}} = (-200 \text{ rad/s}^2)\mathbf{j}$, respectively (Fig. 1). The position vector of *D* is

$$\mathbf{r} = (8 \text{ in.})(\sin 30°\mathbf{i} + \cos 30°\mathbf{j}) = (4 \text{ in.})\mathbf{i} + (6.93 \text{ in.})\mathbf{j}$$

ANALYSIS:

a. Velocity $\mathbf{v}_D$. Denote the point of the rod that coincides with *D* by *D'* and the rotating frame *Oxyz* by $\mathscr{F}$. Then from Eq. (15.46) you have

$$\mathbf{v}_D = \mathbf{v}_{D'} + \mathbf{v}_{D/\mathscr{F}} \tag{1}$$

where

$$\mathbf{v}_{D'} = \mathbf{\Omega} \times \mathbf{r} = (-20 \text{ rad/s})\mathbf{j} \times [(4 \text{ in.})\mathbf{i} + (6.93 \text{ in.})\mathbf{j}] = (80 \text{ in./s})\mathbf{k}$$
$$\mathbf{v}_{D/\mathscr{F}} = (50 \text{ in./s})(\sin 30°\mathbf{i} + \cos 30°\mathbf{j}) = (25 \text{ in./s})\mathbf{i} + (43.3 \text{ in./s})\mathbf{j}$$

Substituting the values obtained for $\mathbf{v}_{D'}$ and $\mathbf{v}_{D/\mathscr{F}}$ into Eq. (1) gives

$$\mathbf{v}_D = (25 \text{ in./s})\mathbf{i} + (43.3 \text{ in./s})\mathbf{j} + (80 \text{ in./s})\mathbf{k} \quad \blacktriangleleft$$

b. Acceleration $\mathbf{a}_D$. From Eq. (15.48), you have

$$\mathbf{a}_D = \mathbf{a}_{D'} + \mathbf{a}_{D/\mathscr{F}} + \mathbf{a}_C \tag{2}$$

where

$$\mathbf{a}_{D'} = \dot{\mathbf{\Omega}} \times \mathbf{r} + \mathbf{\Omega} \times (\mathbf{\Omega} \times \mathbf{r})$$
$$= (-200 \text{ rad/s}^2)\mathbf{j} \times [(4 \text{ in.})\mathbf{i} + (6.93 \text{ in.})\mathbf{j}] - (20 \text{ rad/s})\mathbf{j} \times (80 \text{ in./s})\mathbf{k}$$
$$= +(800 \text{ in./s}^2)\mathbf{k} - (1600 \text{ in./s}^2)\mathbf{i}$$
$$\mathbf{a}_{D/\mathscr{F}} = (600 \text{ in./s}^2)(\sin 30°\mathbf{i} + \cos 30°\mathbf{j}) = (300 \text{ in./s}^2)\mathbf{i} + (520 \text{ in./s}^2)\mathbf{j}$$
$$\mathbf{a}_C = 2\mathbf{\Omega} \times \mathbf{v}_{D/\mathscr{F}}$$
$$= 2(-20 \text{ rad/s})\mathbf{j} \times [(25 \text{ in./s})\mathbf{i} + (43.3 \text{ in./s})\mathbf{j}] = (1000 \text{ in./s}^2)\mathbf{k}$$

(continued)

Substituting the values obtained for $\mathbf{a}_{D'}$, $\mathbf{a}_{D/\mathcal{F}}$, and $\mathbf{a}_c$ into Eq. (2), you obtain

$$\mathbf{a}_D = -(1300 \text{ in./s}^2)\mathbf{i} + (520 \text{ in./s}^2)\mathbf{j} + (1800 \text{ in./s}^2)\mathbf{k} \quad \blacktriangleleft$$

REFLECT and THINK: For this problem, the 800 in./s² $\mathbf{k}$ in the $\mathbf{a}_{D'}$, term corresponds to a tangential acceleration due to $\dot{\boldsymbol{\Omega}}$, while the -1600 in./s²$\mathbf{i}$ corresponds to a normal acceleration toward the axis of rotation. The Coriolis term reflects the fact that the $\mathbf{v}_{D/\mathcal{F}}$ term is changing its direction due to $\boldsymbol{\Omega}$. When solving three-dimensional problems like this, the vector algebra approach is clearly superior to the method discussed in Sample Problem 15.20, since it is very difficult to visualize the direction of the acceleration terms.

Sample Problem 15.24

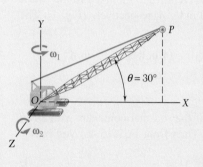

The crane shown rotates with a constant angular velocity $\boldsymbol{\omega}_1$ of 0.30 rad/s. Simultaneously, the boom is being raised with a constant angular velocity $\boldsymbol{\omega}_2$ of 0.50 rad/s relative to the cab. Knowing that the length of the boom OP is $l = 12$ m, determine (a) the velocity of the tip of the boom, (b) the acceleration of the tip of the boom.

STRATEGY: Use rigid body kinematics with a rotating coordinate system because $\boldsymbol{\omega}_2$ is given relative to the cab. Attach a rotating reference frame to the cab; then you can calculate its motion relative to the fixed frame and the motion of the crane tip relative to the rotating frame.

MODELING:

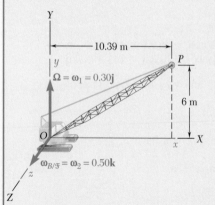

Fig. 1 The rotating coordinate system *xyz* is attached to the cab.

Frames of Reference. The angular velocity of the cab (and rotating frame *Oxyz*) with respect to the fixed frame *OXYZ* is $\boldsymbol{\Omega} = \boldsymbol{\omega}_1 = (0.30 \text{ rad/s})\mathbf{j}$ (Fig. 1). The angular velocity of the boom relative to the cab and the rotating frame *Oxyz* (or $\mathcal{F}$ for short) is $\boldsymbol{\omega}_{B/\mathcal{F}} = \boldsymbol{\omega}_2 = (0.50 \text{ rad/s})\mathbf{k}$.

ANALYSIS:

a. Velocity $\mathbf{v}_P$. From Eq. (15.46), you have

$$\mathbf{v}_P = \mathbf{v}_{P'} + \mathbf{v}_{P/\mathcal{F}} \tag{1}$$

where $\mathbf{v}_{P'}$ is the velocity of the point P' of the rotating frame that coincides with P as

$$\mathbf{v}_{P'} = \boldsymbol{\Omega} \times \mathbf{r} = (0.30 \text{ rad/s})\mathbf{j} \times [(10.39 \text{ m})\mathbf{i} + (6 \text{ m})\mathbf{j}] = -(3.12 \text{ m/s})\mathbf{k}$$

and where $\mathbf{v}_{P/\mathcal{F}}$ is the velocity of P relative to the rotating frame *Oxyz*. However, you know that the angular velocity of the boom relative to *Oxyz* is $\boldsymbol{\omega}_{B/\mathcal{F}} = (0.50 \text{ rad/s})\mathbf{k}$. The velocity of its tip P relative to *Oxyz* is therefore

$$\mathbf{v}_{P/\mathcal{F}} = \boldsymbol{\omega}_{B/\mathcal{F}} \times \mathbf{r} = (0.50 \text{ rad/s})\mathbf{k} \times [(10.39 \text{ m})\mathbf{i} + (6 \text{ m})\mathbf{j}]$$
$$= -(3 \text{ m/s})\mathbf{i} + (5.20 \text{ m/s})\mathbf{j}$$

Substituting the values obtained for $\mathbf{v}_{P'}$ and $\mathbf{v}_{B/\mathcal{F}}$ into Eq. (1), you find

$$\mathbf{v}_P = -(3 \text{ m/s})\mathbf{i} + (5.20 \text{ m/s})\mathbf{j} - (3.12 \text{ m/s})\mathbf{k} \quad \blacktriangleleft$$

b. Acceleration $\mathbf{a}_P$. From Eq. (15.48), you have

$$\mathbf{a}_P = \mathbf{a}_{P'} + \mathbf{a}_{P/\mathcal{F}} + \mathbf{a}_C \quad\quad (2)$$

Since $\mathbf{\Omega}$ and $\mathbf{\omega}_{B/\mathcal{F}}$ are both constant, you obtain

$$\mathbf{a}_{P'} = \mathbf{\Omega} \times (\mathbf{\Omega} \times \mathbf{r}) = (0.30 \text{ rad/s})\mathbf{j} \times (-3.12 \text{ m/s})\mathbf{k} = -(0.94 \text{ m/s}^2)\mathbf{i}$$

$$\mathbf{a}_{P/\mathcal{F}} = \mathbf{\omega}_{B/\mathcal{F}} \times (\mathbf{\omega}_{B/\mathcal{F}} \times \mathbf{r})$$
$$= (0.50 \text{ rad/s})\mathbf{k} \times [-(3 \text{ m/s})\mathbf{i} + (5.20 \text{ m/s})\mathbf{j}]$$
$$= -(1.50 \text{ m/s}^2)\mathbf{j} - (2.60 \text{ m/s}^2)\mathbf{i}$$

$$\mathbf{a}_C = 2\mathbf{\Omega} \times \mathbf{v}_{P/\mathcal{F}}$$
$$= 2(0.30 \text{ rad/s})\mathbf{j} \times [-(3 \text{ m/s})\mathbf{i} + (5.20 \text{ m/s})\mathbf{j}] = (1.80 \text{ m/s}^2)\mathbf{k}$$

Substituting for $\mathbf{a}_{P'}$, $\mathbf{a}_{P/\mathcal{F}}$, and $\mathbf{a}_C$ into Eq. (2), you find

$$\mathbf{a}_P = -(3.54 \text{ m/s}^2)\mathbf{i} - (1.50 \text{ m/s}^2)\mathbf{j} + (1.80 \text{ m/s}^2)\mathbf{k} \quad \blacktriangleleft$$

REFLECT and THINK: You also could have attached your reference frame to rotate with the boom:

$$\mathbf{\Omega}_B = \mathbf{\omega}_{\mathcal{F}} + \mathbf{\omega}_{B/\mathcal{F}} = (0.30 \text{ rad/s})\mathbf{j} + (0.50 \text{ rad/s})\mathbf{k}$$

and used Eq. 15.52 for

$$\mathbf{v}_P = \mathbf{\Omega}_B \times \mathbf{r} = [(0.30 \text{ rad/s})\mathbf{j} + (0.5 \text{ rad/s})\mathbf{k}] \times [(10.39 \text{ m})\mathbf{i} + (6 \text{ m})\mathbf{j}]$$
$$= -(3.0 \text{ m/s})\mathbf{i} + (5.20 \text{ m/s})\mathbf{j} - (3.12 \text{ m/s})\mathbf{k}$$

which is the same answer you found previously. Similarly, you could use Eq. 15.54 to solve for the acceleration. If the crane were moving forward, you would just add its translational velocity and acceleration to that due to the rotations.

Sample Problem 15.25

Disk D has a radius R and is pinned to end A of the arm OA. OA has a length L and is located in the plane of the disk. The arm rotates about a vertical axis through O at the constant rate ω_1, and the disk rotates about A at the constant rate ω_2. Determine (a) the velocity of point P located directly above A, (b) the acceleration of P, (c) the angular velocity and angular acceleration of the disk.

STRATEGY: Use rigid-body kinematics with a rotating coordinate system, since disk D is moving relative to the arm OA.

MODELING:

Frames of Reference. Attach a moving frame $Axyz$ to arm OA. Its angular velocity with respect to the fixed frame $OXYZ$ is therefore $\mathbf{\Omega} = \omega_1\mathbf{j}$

(continued)

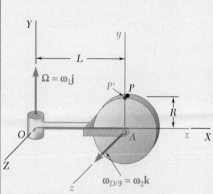

Fig. 1 The rotating coordinate system *xyz* is attached to arm *OA* at point *A*.

(Fig. 1). The angular velocity of disk D relative to the moving frame $Axyz$ (or $\mathscr{F}$ for short) is $\boldsymbol{\omega}_{D/\mathscr{F}} = \omega_2\mathbf{k}$. The position vector of P relative to O is $\mathbf{r} = L\mathbf{i} + R\mathbf{j}$, and its position vector relative to A is $\mathbf{r}_{P/A} = R\mathbf{j}$.

ANALYSIS:

a. Velocity $\mathbf{v}_P$. Denote by P' the point of the moving frame that coincides with P. Then from Eq. (15.46), you have

$$\mathbf{v}_P = \mathbf{v}_{P'} + \mathbf{v}_{P/\mathscr{F}} \tag{1}$$

where $\mathbf{v}_{P'} = \boldsymbol{\Omega} \times \mathbf{r} = \omega_1\mathbf{j} \times (L\mathbf{i} + R\mathbf{j}) = -\omega_1 L\mathbf{k}$

$$\mathbf{v}_{P/\mathscr{F}} = \boldsymbol{\omega}_{D/\mathscr{F}} \times \mathbf{r}_{P/A} = \omega_2\mathbf{k} \times R\mathbf{j} = -\omega_2 R\mathbf{i}$$

Substituting the values obtained for $\mathbf{v}_{P'}$ and $\mathbf{v}_{D/\mathscr{F}}$ into Eq. (1), you obtain

$$\mathbf{v}_P = -\omega_2 R\mathbf{i} - \omega_1 L\mathbf{k} \quad \blacktriangleleft$$

b. Acceleration $\mathbf{a}_P$. From Eq. (15.48), you have

$$\mathbf{a}_P = \mathbf{a}_{P'} + \mathbf{a}_{P/\mathscr{F}} + \mathbf{a}_C \tag{2}$$

Since $\boldsymbol{\Omega}$ and $\boldsymbol{\omega}_{D/\mathscr{F}}$ are both constant, you obtain

$$\mathbf{a}_{P'} = \boldsymbol{\Omega} \times (\boldsymbol{\Omega} \times \mathbf{r}) = \omega_1\mathbf{j} \times (-\omega_1 L\mathbf{k}) = -\omega_1^2 L\mathbf{i}$$
$$\mathbf{a}_{P/\mathscr{F}} = \boldsymbol{\omega}_{D/\mathscr{F}} \times (\boldsymbol{\omega}_{D/\mathscr{F}} \times \mathbf{r}_{P/A}) = \omega_2\mathbf{k} \times (-\omega_2 R\mathbf{i}) = -\omega_2^2 R\mathbf{j}$$
$$\mathbf{a}_C = 2\boldsymbol{\Omega} \times \mathbf{v}_{P/\mathscr{F}} = 2\omega_1\mathbf{j} \times (-\omega_2 R\mathbf{i}) = 2\omega_1\omega_2 R\mathbf{k}$$

Substituting these values into Eq. (2), you find

$$\mathbf{a}_P = -\omega_1^2 L\mathbf{i} - \omega_2^2 R\mathbf{j} + 2\omega_1\omega_2 R\mathbf{k} \quad \blacktriangleleft$$

c. Angular Velocity and Angular Acceleration of Disk.

$$\boldsymbol{\omega} = \boldsymbol{\Omega} + \boldsymbol{\omega}_{D/\mathscr{F}} \qquad \boldsymbol{\omega} = \omega_1\mathbf{j} + \omega_2\mathbf{k} \quad \blacktriangleleft$$

Using Eq. (15.31) with $\boldsymbol{\Omega} = \boldsymbol{\omega}$, you obtain

$$\boldsymbol{\alpha} = (\dot{\boldsymbol{\omega}})_{OXYZ} = (\dot{\boldsymbol{\omega}})_{Axyz} + \boldsymbol{\Omega} \times \boldsymbol{\omega}$$
$$= 0 + \omega_1\mathbf{j} \times (\omega_1\mathbf{j} + \omega_2\mathbf{k})$$

$$\boldsymbol{\alpha} = \omega_1\omega_2\mathbf{i} \quad \blacktriangleleft$$

REFLECT and THINK: Knowing the absolute angular velocity of the disk is equal to $\omega_1\mathbf{j} + \omega_2\mathbf{k}$, you could have determined the velocity of P by attaching the rotating axes to the disk and using Eq. 15.52,

$$\mathbf{v}_P = \mathbf{v}_A + \boldsymbol{\Omega}_D \times \mathbf{r}_{P/A} + \mathbf{v}_{P/A} = \omega_1\mathbf{j} \times L\mathbf{i} + (\omega_1\mathbf{j} + \omega_2\mathbf{k}) \times R\mathbf{j} + \mathbf{0}$$
$$= -\omega_1 L\mathbf{k} - \omega_2 R\mathbf{i}$$

which is the same answer we found earlier. Similarly,

$$\mathbf{a}_P = \mathbf{a}_A + \dot{\boldsymbol{\Omega}}_D \times \mathbf{r}_{P/A} + \boldsymbol{\Omega}_D \times (\boldsymbol{\Omega}_D \times \mathbf{r}_{P/A}) + 2\boldsymbol{\Omega}_D \times \dot{\mathbf{r}}_{P/A} + \ddot{\mathbf{r}}_{P/A}$$
$$= -\omega_1^2 L\mathbf{i} + \omega_1\omega_2\mathbf{i} \times R\mathbf{j} + (\omega_1\mathbf{j} + \omega_2\mathbf{k}) \times [(\omega_1\mathbf{j} + \omega_2\mathbf{k}) \times R\mathbf{j}] + 0 + 0$$
$$= -\omega_1^2 L\mathbf{i} + \omega_1\omega_2 R\mathbf{k} + (\omega_1\mathbf{j} + \omega_2\mathbf{k}) \times (-\omega_2 R\mathbf{i})$$
$$= -\omega_1^2 L\mathbf{i} - \omega_2^2 R\mathbf{j} + 2\omega_1\omega_2 R\mathbf{k}$$

which, again, is the same answer shown previously.

SOLVING PROBLEMS
ON YOUR OWN

I n this section, we concluded our presentation of the kinematics of rigid bodies by showing you how to use an auxiliary frame of reference $\mathcal{F}$ to analyze the three-dimensional motion of a rigid body. This auxiliary frame may be a *rotating frame* with a fixed origin O or it may be a *frame in general motion.*

A. Using a rotating frame of reference. As you approach a problem involving the use of a rotating frame $\mathcal{F}$, you should take the following steps.

1. Select the rotating frame $\mathcal{F}$ that you wish to use and draw the corresponding coordinate axes x, y, and z from the fixed point O.

2. Determine the angular velocity $\boldsymbol{\Omega}$ **of the frame** $\mathcal{F}$ with respect to a fixed frame $OXYZ$. In most cases, you will have selected a frame that is attached to some rotating element of the system; $\boldsymbol{\Omega}$ is then the angular velocity of that element.

3. Designate as P' **the point of the rotating frame** $\mathcal{F}$ that coincides with the point P of interest at the instant you are considering. Determine the velocity $\mathbf{v}_{P'}$ and the acceleration $\mathbf{a}_{P'}$ of point P'. Since P' is part of $\mathcal{F}$ and has the same position vector $\mathbf{r}$ as P, you will find

$$\mathbf{v}_{P'} = \boldsymbol{\Omega} \times \mathbf{r} \quad \text{and} \quad \mathbf{a}_{P'} = \boldsymbol{\alpha} \times \mathbf{r} + \boldsymbol{\Omega} \times (\boldsymbol{\Omega} \times \mathbf{r})$$

where $\boldsymbol{\alpha}$ is the angular acceleration of $\mathcal{F}$.

4. Determine the velocity and acceleration of point P with respect to the frame $\mathcal{F}$. As you are trying to determine $\mathbf{v}_{P/\mathcal{F}}$ and $\mathbf{a}_{P/\mathcal{F}}$, you will find it useful to visualize the motion of P on frame $\mathcal{F}$ when the frame is not rotating. If P is a point of a rigid body $\mathcal{B}$ that has an angular velocity $\boldsymbol{\omega}_{\mathcal{B}}$ and an angular acceleration $\boldsymbol{\alpha}_{\mathcal{B}}$ relative to $\mathcal{F}$ [Sample Prob. 15.24], you will find that

$$\mathbf{v}_{P/\mathcal{F}} = \boldsymbol{\omega}_{\mathcal{B}} \times \mathbf{r} \quad \text{and} \quad \mathbf{a}_{P/\mathcal{F}} = \boldsymbol{\alpha}_{\mathcal{B}} \times \mathbf{r} + \boldsymbol{\omega}_{\mathcal{B}} \times (\boldsymbol{\omega}_{\mathcal{B}} \times \mathbf{r})$$

5. Determine the Coriolis acceleration. Considering the angular velocity $\boldsymbol{\Omega}$ of frame $\mathcal{F}$ and the velocity $\mathbf{v}_{P/\mathcal{F}}$ of point P relative to that frame, which was computed in Step 4, you have

$$\mathbf{a}_C = 2\boldsymbol{\Omega} \times \mathbf{v}_{P/\mathcal{F}}$$

6. The velocity and the acceleration of P with respect to the fixed frame $OXYZ$ can now be obtained by adding the expressions you have determined:

$$\mathbf{v}_P = \mathbf{v}_{P'} + \mathbf{v}_{P/\mathcal{F}} \tag{15.46}$$

$$\mathbf{a}_P = \mathbf{a}_{P'} + \mathbf{a}_{P/\mathcal{F}} + \mathbf{a}_C \tag{15.48}$$

(continued)

B. Using a frame of reference in general motion. The steps that you will take differ only slightly from those listed under part A. They consist of the following:

1. Select the frame $\mathcal{F}$ that you wish to use and a reference point A in that frame from which you will draw the coordinate axes, x, y, and z, defining that frame. Consider the motion of the frame as the sum of a **translation with A and a rotation about** A.

2. Determine the velocity v_A of point A and the angular velocity Ω of the frame. In most cases, you will have selected a frame that is attached to some element of the system; Ω is then the angular velocity of that element.

3. Designate as P' the point of frame $\mathcal{F}$ that coincides with the point P of interest at the instant you are considering, and determine the velocity $\mathbf{v}_{P'}$ and the acceleration $\mathbf{a}_{P'}$ of that point. In some cases, you can do this by visualizing the motion of P if that point were prevented from moving with respect to $\mathcal{F}$ [Sample Prob. 15.25]. A more general approach is to recall that the motion of P' is the sum of a translation with the reference point A and a rotation about A. You can obtain the velocity $\mathbf{v}_{P'}$ and the acceleration $\mathbf{a}_{P'}$ of P', therefore, by adding $\mathbf{v}_A$ and $\mathbf{a}_A$, respectively, to the expressions found in part A, Step 3, and replacing the position vector $\mathbf{r}$ by the vector $\mathbf{r}_{P/A}$ drawn from A to P:

$$\mathbf{v}_{P'} = \mathbf{v}_A + \boldsymbol{\Omega} \times \mathbf{r}_{P/A} \qquad \mathbf{a}_{P'} = \mathbf{a}_A + \boldsymbol{\alpha} \times \mathbf{r}_{P/A} + \boldsymbol{\Omega} \times (\boldsymbol{\Omega} \times \mathbf{r}_{P/A})$$

4, 5, and 6 are the same as in part A of this summary, except that the vector $\mathbf{r}$ should again be replaced by $\mathbf{r}_{P/A}$. Thus, Eqs. (15.46) and (15.48) can still be used to obtain the velocity and the acceleration of P with respect to the fixed frame of reference $OXYZ$.

C. Alternative approach using a frame of reference in general motion. As shown in the sample problems, you can also use Eqs. (15.52) and (15.54) to determine the velocity and acceleration of point P, respectively.

$$\mathbf{v}_P = \mathbf{v}_A + \boldsymbol{\Omega} \times \mathbf{r}_{P/A} + (\dot{\mathbf{r}}_{P/A})_{Axyz} \tag{15.52}$$

$$\mathbf{a}_P = \mathbf{a}_A + \dot{\boldsymbol{\Omega}} \times \mathbf{r}_{P/A} + \boldsymbol{\Omega} \times (\boldsymbol{\Omega} \times \mathbf{r}_{P/A})$$
$$+ 2\boldsymbol{\Omega} \times (\dot{\mathbf{r}}_{P/A})_{Axyz} + (\ddot{\mathbf{r}}_{P/A})_{Axyz} \tag{15.54}$$

You first need to determine a reference point A and attach your rotating frame of reference at that point; generally this is attached to a specific part of the object under consideration (e.g., the cab or boom of a crane). Define the angular velocity of the frame as $\boldsymbol{\Omega}$ and the angular acceleration of the frame as $\dot{\boldsymbol{\Omega}}$. The terms $(\dot{\mathbf{r}}_{P/A})_{Axyz}$ and $(\ddot{\mathbf{r}}_{P/A})_{Axyz}$ represent the velocity and acceleration of point P relative to the rotating frame of reference A_{xyz}.

Problems

15.220 A flight simulator is used to train pilots on how to recognize spatial disorientation. It has four degrees of freedom and can rotate around a planetary axis as well as in yaw, pitch, and roll. The pilot is seated so that her head B is located at $\mathbf{r} = 2\mathbf{i} + 1\mathbf{j}$ ft with respect to the center of the cab A. Knowing that the cab is rotating about the planetary axis with a constant angular velocity of 20 rpm counterclockwise as seen from above, and pitches with a constant angular velocity of $+3\mathbf{k}$ rad/s, determine (*a*) the velocity of the pilot's head, (*b*) the angular acceleration of the cab, (*c*) the acceleration of the pilot's head.

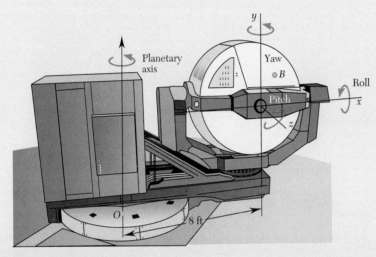

Fig. P15.220 and P15.221

15.221 A flight simulator is used to train pilots on how to recognize spatial disorientation. It has four degrees of freedom and can rotate around a planetary axis as well as in yaw, pitch, and roll. The pilot is seated so that her head B is located at $r = 2\mathbf{i} + 1\mathbf{j}$ ft with respect to the center of the cab A. The cab is rotating about the planetary axis with an angular velocity of 20 rpm counterclockwise as seen from above and this is increasing by 1 rad/s². Knowing that the cab rolls with a constant angular velocity of $-4\mathbf{i}$ rad/s, determine (*a*) the velocity of the pilot's head, (*b*) the angular acceleration of the cab, (*c*) the acceleration of the pilot's head.

15.222 and 15.223 The rectangular plate shown rotates at the constant rate $\omega_2 = 12$ rad/s with respect to arm AE, which itself rotates at the constant rate $\omega_1 = 9$ rad/s about the Z axis. For the position shown, determine the velocity and acceleration of the point of the plate indicated.

    15.222   Corner B
    15.223   Corner C

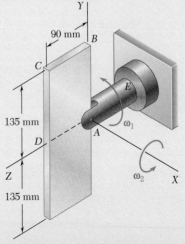

Fig. P15.222 and P15.223

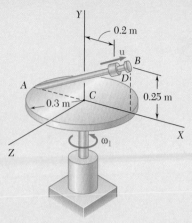

15.224 Rod AB is welded to the 0.3-m-radius plate that rotates at the constant rate $\omega_1 = 6$ rad/s. Knowing that collar D moves toward end B of the rod at a constant speed $u = 1.3$ m/s, determine, for the position shown, (a) the velocity of D, (b) the acceleration of D.

15.225 The bent rod shown rotates at the constant rate of $\omega_1 = 5$ rad/s and collar C moves toward point B at a constant relative speed of $u = 39$ in./s. Knowing that collar C is halfway between points B and D at the instant shown, determine its velocity and acceleration.

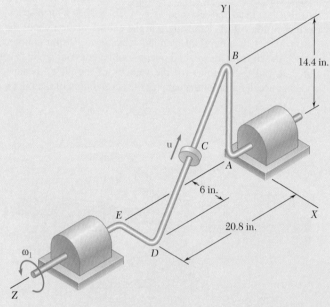

Fig. P15.225

15.226 The bent pipe shown rotates at the constant rate $\omega_1 = 10$ rad/s. Knowing that a ball bearing D moves in portion BC of the pipe toward end C at a constant relative speed $u = 2$ ft/s, determine at the instant shown (a) the velocity of D, (b) the acceleration of D.

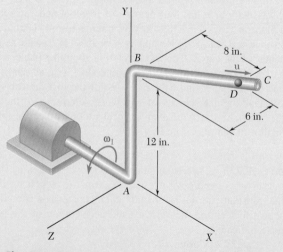

Fig. P15.226

15.227 The circular plate shown rotates about its vertical diameter at the constant rate $\omega_1 = 10$ rad/s. Knowing that in the position shown the disk lies in the XY plane and point D of strap CD moves upward at a constant relative speed $u = 1.5$ m/s, determine (a) the velocity of D, (b) the acceleration of D.

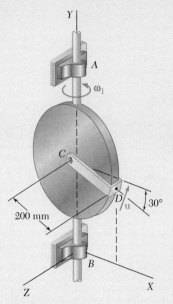

Fig. P15.227

15.228 Manufactured items are spray-painted as they pass through the automated work station shown. Knowing that the bent pipe ACE rotates at the constant rate $\omega_1 = 0.4$ rad/s and that at point D the paint moves through the pipe at a constant relative speed $u = 150$ mm/s, determine, for the position shown, (a) the velocity of the paint at D, (b) the acceleration of the paint at D.

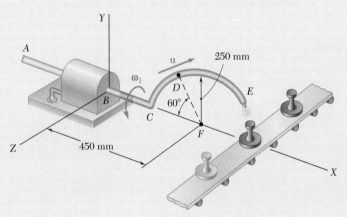

Fig. P15.228

15.229 Solve Prob. 15.227, assuming that at the instant shown the angular velocity ω_1 of the plate is 10 rad/s and is decreasing at the rate of 25 rad/s², while the relative speed u of point D of strap CD is 1.5 m/s and is decreasing at the rate of 3 m/s².

15.230 Solve Prob. 15.225, assuming that at the instant shown the angular velocity ω_1 of the rod is 5 rad/s and is increasing at the rate of 10 rad/s² while the relative speed u of the collar C is 39 in./s and is decreasing at the rate of 260 in./s².

15.231 Using the method of Sec. 15.7A, solve Prob. 15.192.

15.232 Using the method of Sec. 15.7A, solve Prob. 15.196.

15.233 Using the method of Sec. 15.7A, solve Prob. 15.198.

15.234 The 400-mm bar AB is made to rotate at the constant rate of $\omega_2 = d\theta/dt = 8$ rad/s with respect to the frame CD that rotates at the constant rate of $\omega_1 = 12$ rad/s about the Y axis. Knowing that $\theta = 60°$ at the instant shown, determine the velocity and acceleration of point A.

15.235 The 400-mm bar AB is made to rotate at the rate $\omega_2 = d\theta/dt$ with respect to the frame CD that rotates at the rate ω_1 about the Y axis. At the instant shown, $\omega_1 = 12$ rad/s, $d\omega_1/dt = -16$ rad/s², $\omega_2 = 8$ rad/s, $d\omega_2/dt = 10$ rad/s², and $\theta = 60°$. Determine the velocity and acceleration of point A at this instant.

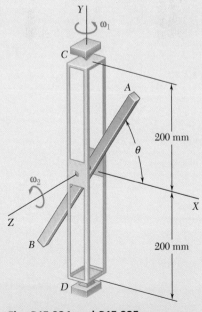

Fig. P15.234 and P15.235

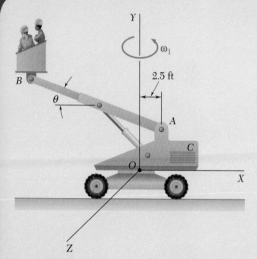

15.236 The arm AB of length 16 ft is used to provide an elevated platform for construction workers. In the position shown, arm AB is being raised at the constant rate $d\theta/dt = 0.25$ rad/s; simultaneously, the unit is being rotated about the Y axis at the constant rate $\omega_1 = 0.15$ rad/s. Knowing that $\theta = 20°$, determine the velocity and acceleration of point B.

15.237 The remote manipulator system (RMS) shown is used to deploy payloads from the cargo bay of space shuttles. At the instant shown, the whole RMS is rotating at the constant rate $\omega_1 = 0.03$ rad/s about the axis AB. At the same time, portion BCD rotates as a rigid body at the constant rate $\omega_2 = d\beta/dt = 0.04$ rad/s about an axis through B parallel to the X axis. Knowing that $\beta = 30°$, determine (a) the angular acceleration of BCD, (b) the velocity of D, (c) the acceleration of D.

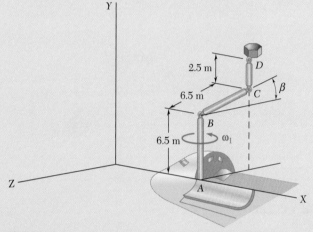

Fig. P15.237

15.238 A disk with a radius of 120 mm rotates at the constant rate of $\omega_2 = 5$ rad/s with respect to the arm AB that rotates at the constant rate of $\omega_1 = 3$ rad/s. For the position shown, determine the velocity and acceleration of point C.

15.239 The crane shown rotates at the constant rate $\omega_1 = 0.25$ rad/s; simultaneously, the telescoping boom is being lowered at the constant rate $\omega_2 = 0.40$ rad/s. Knowing that at the instant shown the length of the boom is 20 ft and is increasing at the constant rate $u = 1.5$ ft/s, determine the velocity and acceleration of point B.

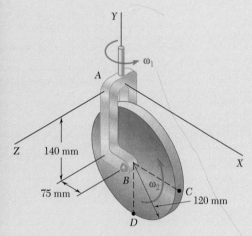

Fig. P15.238

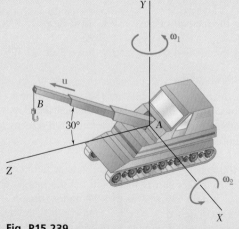

Fig. P15.239

15.240 The vertical plate shown is welded to arm *EFG*, and the entire unit rotates at the constant rate $\omega_1 = 1.6$ rad/s about the *Y* axis. At the same time, a continuous link belt moves around the perimeter of the plate at a constant speed $u = 4.5$ in./s. For the position shown, determine the acceleration of the link of the belt located (*a*) at point *A*, (*b*) at point *B*.

15.241 The vertical plate shown is welded to arm *EFG*, and the entire unit rotates at the constant rate $\omega_1 = 1.6$ rad/s about the *Y* axis. At the same time, a continuous link belt moves around the perimeter of the plate at a constant speed $u = 4.5$ in./s. For the position shown, determine the acceleration of the link of the belt located (*a*) at point *C*, (*b*) at point *D*.

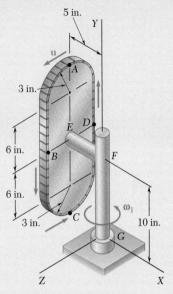

Fig. P15.240 and P15.241

15.242 A disk of 180-mm radius rotates at the constant rate $\omega_2 = 12$ rad/s with respect to arm *CD*, which itself rotates at the constant rate $\omega_1 = 8$ rad/s about the *Y* axis. Determine at the instant shown the velocity and acceleration of point *A* on the rim of the disk.

15.243 A disk of 180-mm radius rotates at the constant rate $\omega_2 = 12$ rad/s with respect to arm *CD*, which itself rotates at the constant rate $\omega_1 = 8$ rad/s about the *Y* axis. Determine at the instant shown the velocity and acceleration of point *B* on the rim of the disk.

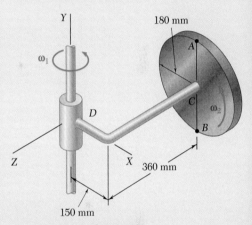

Fig. P15.242 and P15.243

15.244 A square plate of side $2r$ is welded to a vertical shaft that rotates with a constant angular velocity ω_1. At the same time, rod *AB* of length r rotates about the center of the plate with a constant angular velocity ω_2 with respect to the plate. For the position of the plate shown, determine the acceleration of end *B* of the rod if (*a*) $\theta = 0$, (*b*) $\theta = 90°$, (*c*) $\theta = 180°$.

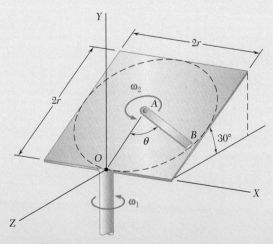

Fig. P15.244

15.245 Two disks, each of 130-mm radius, are welded to the 500-mm rod CD. The rod-and-disks unit rotates at the constant rate $\omega_2 = 3$ rad/s with respect to arm AB. Knowing that at the instant shown $\omega_1 = 4$ rad/s, determine the velocity and acceleration of (*a*) point E, (*b*) point F.

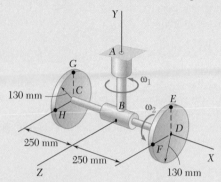

Fig. P15.245

15.246 In Prob. 15.245, determine the velocity and acceleration of (*a*) point G, (*b*) point H.

15.247 The position of the stylus tip A is controlled by the robot shown. In the position shown, the stylus moves at a constant speed $u = 180$ mm/s relative to the solenoid BC. At the same time, arm CD rotates at the constant rate $\omega_2 = 1.6$ rad/s with respect to component DEG. Knowing that the entire robot rotates about the X axis at the constant rate $\omega_1 = 1.2$ rad/s, determine (*a*) the velocity of A, (*b*) the acceleration of A.

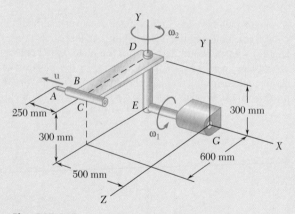

Fig. P15.247

Review and Summary

This chapter was devoted to the study of the kinematics of rigid bodies.

Rigid Body in Translation

We first considered the **translation** of a rigid body [Sec. 15.1A] and observed that in such a motion **all points of the body have the same velocity and the same acceleration at any given instant**.

Rigid Body in Rotation About a Fixed Axis

We next considered the **rotation** of a rigid body about a fixed axis [Sec. 15.1B]. The position of the body is defined by the angle θ that the line BP, drawn from the axis of rotation to a point P of the body, forms with a fixed plane (Fig. 15.39). We found that the magnitude of the velocity of P is

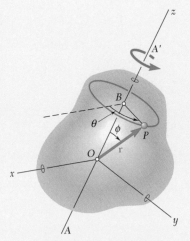

Fig. 15.39

$$v = \frac{ds}{dt} = r\dot{\theta} \sin \phi \qquad (15.4)$$

where $\dot{\theta}$ is the time derivative of θ. We then expressed the velocity of P as

$$\mathbf{v} = \frac{d\mathbf{r}}{dt} = \boldsymbol{\omega} \times \mathbf{r} \qquad (15.5)$$

where the vector

$$\boldsymbol{\omega} = \omega\mathbf{k} = \dot{\theta}\mathbf{k} \qquad (15.6)$$

is directed along the fixed axis of rotation and represents the angular velocity of the body.

Denoting the derivative $d\boldsymbol{\omega}/dt$ of the angular velocity by $\boldsymbol{\alpha}$, we expressed the acceleration of P as

$$\mathbf{a} = \boldsymbol{\alpha} \times \mathbf{r} + \boldsymbol{\omega} \times (\boldsymbol{\omega} \times \mathbf{r}) \qquad (15.8)$$

Differentiating Eq. (15.6) and recalling that $\mathbf{k}$ is constant in magnitude and direction, we found that

$$\boldsymbol{\alpha} = \alpha\mathbf{k} = \dot{\omega}\mathbf{k} = \ddot{\theta}\mathbf{k} \qquad (15.9)$$

The vector $\boldsymbol{\alpha}$ represents the angular acceleration of the body and is directed along the fixed axis of rotation [Sample Prob. 15.2].

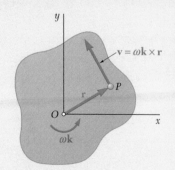

Fig. 15.40

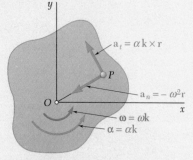

Fig. 15.41

Rotation of a Representative Slab:

Tangential and Normal Components

Next we considered the motion of a representative slab located in a plane perpendicular to the axis of rotation of the body (Fig. 15.40). Since the angular velocity is perpendicular to the slab, we expressed the velocity of a point P of the slab as

$$\mathbf{v} = \omega\mathbf{k} \times \mathbf{r} \tag{15.10}$$

where $\mathbf{v}$ is contained in the plane of the slab. Substituting $\boldsymbol{\omega} = \omega\mathbf{k}$ and $\boldsymbol{\alpha} = \alpha\mathbf{k}$ into Eq. (15.8), we found that we could resolve the acceleration of P into tangential and normal components (Fig. 15.41) respectively equal to

$$\mathbf{a}_t = \alpha\mathbf{k} \times \mathbf{r} \qquad a_t = r\alpha$$
$$\mathbf{a}_n = -\omega^2\mathbf{r} \qquad a_n = r\omega^2 \tag{15.11'}$$

Angular Velocity and Angular Acceleration of a Rotating Rigid Body

Recalling Eqs. (15.6) and (15.9), we obtained the following expressions for the *angular velocity* and the *angular acceleration* of the rigid body [Sec. 15.1C]:

$$\omega = \frac{d\theta}{dt} \tag{15.12}$$

$$\alpha = \frac{d\omega}{dt} = \frac{d^2\theta}{dt^2} \tag{15.13}$$

or

$$\alpha = \omega\frac{d\omega}{d\theta} \tag{15.14}$$

We noted that these expressions are similar to those obtained in Chap. 11 for the rectilinear motion of a particle.

Two particular cases of rotation are frequently encountered: *uniform rotation* and *uniformly accelerated rotation*. You can solve problems involving either of these motions by using equations similar to those used in Sec. 11.2 for the uniform rectilinear motion and the uniformly accelerated rectilinear motion of a particle, but where x, v, and a are replaced by θ, ω, and α, respectively [Sample Prob. 15.1].

Velocities in Plane Motion

We can consider the **most general plane motion** of a rigid body as the **sum of a translation and a rotation** [Sec. 15.2A]. For example, the body shown in Fig. 15.42 can be assumed to translate with point A, while simultaneously rotating about A. It follows [Sec. 15.2B] that the velocity of any point B of the rigid body can be expressed as

$$\mathbf{v}_B = \mathbf{v}_A + \mathbf{v}_{B/A} \tag{15.17}$$

where $\mathbf{v}_A$ is the velocity of A and $\mathbf{v}_{B/A}$ is the relative velocity of B with respect to A or, more precisely, with respect to axes $x'y'$ translating with A. Denoting the position vector of B relative to A by $\mathbf{r}_{B/A}$, we found that

$$\mathbf{v}_{B/A} = \omega\mathbf{k} \times \mathbf{r}_{B/A} \qquad v_{B/A} = r\omega \tag{15.18}$$

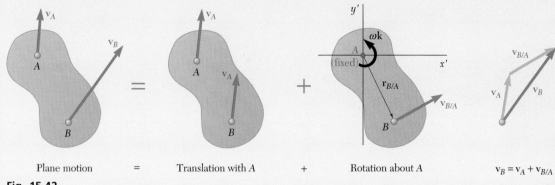

| Plane motion | = | Translation with A | + | Rotation about A | $v_B = v_A + v_{B/A}$ |

Fig. 15.42

The fundamental equation (15.17) relating the absolute velocities of points A and B and the relative velocity of B with respect to A was expressed in the form of a vector diagram, which can be used to solve problems involving the motion of various types of mechanisms [Sample Probs. 15.6 and 15.7].

Instantaneous Center of Rotation

We presented another approach to the solution of problems involving the velocities of the points of a rigid body in plane motion in Sec. 15.3 and used it in Sample Probs. 15.9 through 15.11. It is based on the determination of the **instantaneous center of rotation** C of the rigid body (Fig. 15.43).

Accelerations in Plane Motion

In Sec. 15.4A, we used the fact that any plane motion of a rigid body can be considered the sum of a translation of the body with a reference point A and a rotation about A. Knowing this, we can find the absolute acceleration of A by adding the relative acceleration of B with respect to A to the absolute acceleration of B.

$$\mathbf{a}_B = \mathbf{a}_A + \mathbf{a}_{B/A} \tag{15.21}$$

where $\mathbf{a}_{B/A}$ consisted of a *normal component* $(\mathbf{a}_{B/A})_n$ with a magnitude $r\omega^2$ directed toward A and a *tangential component* $(\mathbf{a}_{B/A})_t$ with a magnitude $r\alpha$

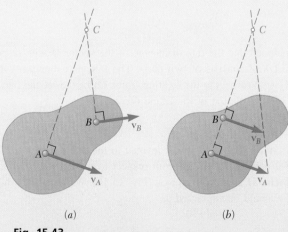

(a)   (b)

Fig. 15.43

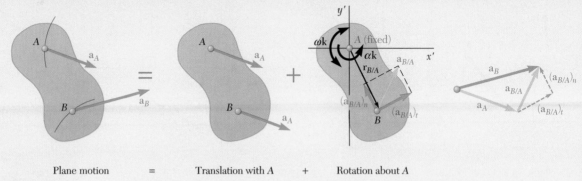

Fig. 15.44

Plane motion = Translation with A + Rotation about A

perpendicular to the line *AB* (Fig. 15.44). We expressed the fundamental relation in Eq. (15.21) in terms of vector diagrams or vector equations and used them to determine the accelerations of given points of various mechanisms [Sample Probs. 15.12 through 15.16]. We noted that we cannot use the instantaneous center of rotation *C* considered in Sec. 15.3 for the determination of accelerations, since point *C*, in general, does *not* have zero acceleration.

Coordinates Expressed in Terms of a Parameter

In the case of certain mechanisms, it is possible to express the coordinates *x* and *y* of all significant points of the mechanism by means of simple analytic expressions containing a *single parameter*. We can obtain the components of the absolute velocity and acceleration of a given point by differentiating twice with respect to the time *t* the coordinates *x* and *y* of that point [Sec. 15.4B].

Rate of Change of a Vector with Respect to a Rotating Frame

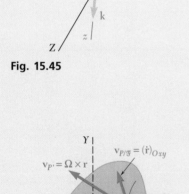

Fig. 15.45

The rate of change of a vector is the same with respect to a fixed frame of reference and with respect to a frame in translation, but the rate of change of a vector with respect to a rotating frame is different. Therefore, in order to study the motion of a particle relative to a rotating frame, we first had to compare the rates of change of a general vector **Q** with respect to a fixed frame *OXYZ* and with respect to a frame *Oxyz* rotating with an angular velocity **Ω** [Sec. 15.5A] (Fig. 15.45). We obtained the fundamental relation

$$(\dot{\mathbf{Q}})_{OXYZ} = (\dot{\mathbf{Q}})_{Oxyz} + \mathbf{\Omega} \times \mathbf{Q} \tag{15.31}$$

and we concluded that the rate of change of the vector **Q** with respect to the fixed frame *OXYZ* consists of two parts: The first part represents the rate of change of **Q** with respect to the rotating frame *Oxyz*; the second part, **Ω** × **Q**, is induced by the rotation of the frame *Oxyz*.

Plane Motion of a Particle Relative to a Rotating Frame

The next section [Sec. 15.5B] was devoted to the two-dimensional kinematic analysis of a particle *P* moving with respect to a frame $\mathscr{F}$ rotating with an angular velocity **Ω** about a fixed axis (Fig. 15.46). We found that the absolute velocity of *P* could be expressed as

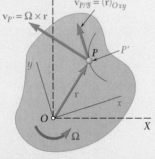

Fig. 15.46

$$\mathbf{v}_P = \mathbf{v}_{P'} + \mathbf{v}_{P/\mathscr{F}} \tag{15.33}$$

or

$$\mathbf{v}_P = \mathbf{v}_O + \boldsymbol{\Omega} \times \mathbf{r} + \mathbf{v}_{\text{rel}} \qquad \textbf{(15.32$'$)}$$

where $\mathbf{v}_P$ = absolute velocity of particle P

   $\mathbf{v}_{P'} = \mathbf{v}_O + \boldsymbol{\Omega} \times \mathbf{r}$ = velocity of point P' of moving frame $\mathcal{F}$ coinciding with P

   $\mathbf{v}_{P/\mathcal{F}} = \mathbf{v}_{\text{rel}}$ = velocity of P relative to moving frame $\mathcal{F}$

We noted that we obtain the same expression for $\mathbf{v}_P$ if the frame is in translation rather than in rotation. However, when the frame is in rotation, the expression for the acceleration of P contains an additional term $\mathbf{a}_c$ called the **Coriolis acceleration**. We have

$$\mathbf{a}_P = \mathbf{a}_{P'} + \mathbf{a}_{P/\mathcal{F}} + \mathbf{a}_C \qquad \textbf{(15.36)}$$

or

$$\mathbf{a}_P = \mathbf{a}_O + \dot{\boldsymbol{\Omega}} \times \mathbf{r} - \Omega^2 \mathbf{r} + 2\boldsymbol{\Omega} \times \mathbf{v}_{\text{rel}} + \mathbf{a}_{\text{rel}}$$

where $\mathbf{a}_P$ = absolute acceleration of particle P

   $\mathbf{a}_{P'} = \mathbf{a}_O + \dot{\boldsymbol{\Omega}} \times \mathbf{r} - \Omega^2 \mathbf{r}$ = acceleration of point P' of moving frame $\mathcal{F}$ coinciding with P

   $\mathbf{a}_{P/\mathcal{F}} = \mathbf{a}_{\text{rel}}$ = acceleration of P relative to moving frame $\mathcal{F}$

   $\mathbf{a}_C = 2\boldsymbol{\Omega} \times (\dot{\mathbf{r}})_{Oxyz} = 2\boldsymbol{\Omega} \times \mathbf{v}_{P/\mathcal{F}} = 2\boldsymbol{\Omega} \times \mathbf{v}_{\text{rel}}$ = Coriolis acceleration

Since $\boldsymbol{\Omega}$ and $\mathbf{v}_{P/\mathcal{F}}$ are perpendicular to each other in the case of plane motion, the Coriolis acceleration has a magnitude $a_C = 2\Omega v_{P/\mathcal{F}}$ and points in the direction obtained by rotating the vector $\mathbf{v}_{P/\mathcal{F}}$ through 90° in the sense of rotation of the moving frame. We can use formulas (15.33) and (15.36) to analyze the motion of mechanisms that contain parts sliding on each other [Sample Probs. 15.17 through 15.20].

Motion of a Rigid Body with a Fixed Point

In the last part of this chapter, we studied the kinematics of rigid bodies in three dimensions. We first considered the motion of a rigid body with a fixed point [Sec. 15.6A]. After proving that the most general displacement of a rigid body with a fixed point O is equivalent to a rotation of the body about an axis through O, we were able to define the angular velocity $\boldsymbol{\omega}$ and the **instantaneous axis of rotation** of the body at a given instant. The velocity of a point P of the body (Fig. 15.47) again could be expressed as

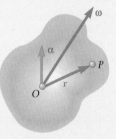

Fig. 15.47

$$\mathbf{v} = \frac{d\mathbf{r}}{dt} = \boldsymbol{\omega} \times \mathbf{r} \qquad \textbf{(15.37)}$$

Differentiating this expression gave

$$\mathbf{a} = \boldsymbol{\alpha} \times \mathbf{r} + \boldsymbol{\omega} \times (\boldsymbol{\omega} \times \mathbf{r}) \qquad \textbf{(15.38)}$$

However, since the direction of $\boldsymbol{\omega}$ changes from one instant to the next, the angular acceleration $\boldsymbol{\alpha}$ is, in general, not directed along the instantaneous axis of rotation [Sample Prob. 15.21].

General Motion in Space

We showed in Sec. 15.6B that **the most general motion of a rigid body in space is equivalent, at any given instant, to the sum of a translation and a rotation**. Considering two particles A and B of the body, we found that

$$\mathbf{v}_B = \mathbf{v}_A + \mathbf{v}_{B/A} \tag{15.42}$$

where $\mathbf{v}_{B/A}$ is the velocity of B relative to a frame $AX'Y'Z'$ attached to A and of fixed orientation (Fig. 15.48). Denoting by $\mathbf{r}_{B/A}$ the position vector of B relative to A, we have

$$\mathbf{v}_B = \mathbf{v}_A + \boldsymbol{\omega} \times \mathbf{r}_{B/A} \tag{15.43}$$

where $\boldsymbol{\omega}$ is the angular velocity of the body at the instant considered [Sample Prob. 15.22]. We obtained the acceleration of B using a similar reasoning. We first wrote

$$\mathbf{a}_B = \mathbf{a}_A + \mathbf{a}_{B/A}$$

and, recalling Eq. (15.38),

$$\mathbf{a}_B = \mathbf{a}_A + \boldsymbol{\alpha} \times \mathbf{r}_{B/A} + \boldsymbol{\omega} \times (\boldsymbol{\omega} \times \mathbf{r}_{B/A}) \tag{15.44}$$

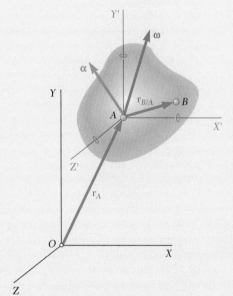

Fig. 15.48

Three-Dimensional Motion of a Particle Relative to a Rotating Frame

In the final section of this chapter, we considered the three-dimensional motion of a particle P relative to a frame $Oxyz$ rotating with an angular velocity $\boldsymbol{\Omega}$ with respect to a fixed frame $OXYZ$ (Fig. 15.49). In Sec. 15.7A, we expressed the absolute velocity $\mathbf{v}_P$ of P as

$$\mathbf{v}_P = \boldsymbol{\Omega} \times \mathbf{r} + (\dot{\mathbf{r}})_{Oxyz} \tag{15.45}$$

or alternatively as

$$\mathbf{v}_P = \mathbf{v}_{P'} + \mathbf{v}_{P/\mathcal{F}} \tag{15.46}$$

where $\mathbf{v}_P$ = absolute velocity of particle P
$\mathbf{v}_{P'}$ = velocity of point P' of moving frame $\mathcal{F}$ coinciding with P
$\mathbf{v}_{P/\mathcal{F}}$ = velocity of P relative to moving frame $\mathcal{F}$

The absolute acceleration $\mathbf{a}_P$ of P can be expressed as

$$\mathbf{a}_P = \dot{\boldsymbol{\Omega}} \times \mathbf{r} + \boldsymbol{\Omega} \times (\boldsymbol{\Omega} \times \mathbf{r}) + 2\boldsymbol{\Omega} \times (\dot{\mathbf{r}})_{Oxyz} + (\ddot{\mathbf{r}})_{Oxyz} \tag{15.47}$$

or alternatively

$$\mathbf{a}_P = \mathbf{a}_{P'} + \mathbf{a}_{P/\mathcal{F}} + \mathbf{a}_C \tag{15.48}$$

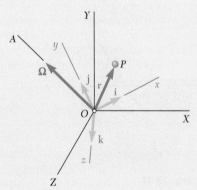

Fig. 15.49

where $\mathbf{a}_P$ = absolute acceleration of particle P

$\quad\mathbf{a}_{P'}$ = acceleration of point P' of moving frame $\mathscr{F}$ coinciding with P

$\quad\mathbf{a}_{P/\mathscr{F}}$ = acceleration of P relative to moving frame $\mathscr{F}$

$\quad\mathbf{a}_C = 2\mathbf{\Omega} \times (\dot{\mathbf{r}})_{Oxyz} = 2\mathbf{\Omega} \times \mathbf{v}_{P/\mathscr{F}}$ = Coriolis acceleration

We noted that the magnitude a_c of the Coriolis acceleration is not equal to $2\Omega v_{P/\mathscr{F}}$ [Sample Prob. 15.23] except in the special case when $\mathbf{\Omega}$ and $\mathbf{v}_{P/\mathscr{F}}$ are perpendicular to each other. Additionally, we usually will have to use Eq. 15.31 to determine the angular acceleration $\dot{\mathbf{\Omega}}$ of the rotating frame.

Frame of Reference in General Motion

We also observed [Sec. 15.7B] that Eqs. (15.46) and (15.48) remain valid when the frame $Axyz$ moves in a known—but arbitrary—fashion with respect to the fixed frame $OXYZ$ (Fig. 15.50), provided that the motion of A is included in the terms $\mathbf{v}_{P'}$ and $\mathbf{a}_{P'}$ representing the absolute velocity and acceleration of the coinciding point P'. We obtained

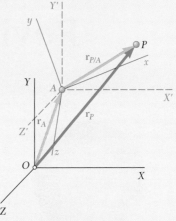

Fig. 15.50

$$\mathbf{v}_P = \mathbf{v}_A + \mathbf{\Omega} \times \mathbf{r}_{P/A} + (\dot{\mathbf{r}}_{P/A})_{Axyz} \qquad \textbf{(15.52)}$$

and

$$\mathbf{a}_P = \mathbf{a}_A + \dot{\mathbf{\Omega}} \times \mathbf{r}_{P/A} + \mathbf{\Omega} \times (\mathbf{\Omega} \times \mathbf{r}_{P/A})$$
$$+ 2\mathbf{\Omega} \times (\dot{\mathbf{r}}_{P/A})_{Axyz} + (\ddot{\mathbf{r}}_{P/A})_{Axyz} \qquad \textbf{(15.54)}$$

Rotating frames of reference are particularly useful in the study of the three-dimensional motion of rigid bodies. Indeed, in many cases, an appropriate choice of the rotating frame leads to a simpler analysis of the motion of the rigid body than would be possible with axes of fixed orientation [Sample Probs. 15.24 and 15.25].

Review Problems

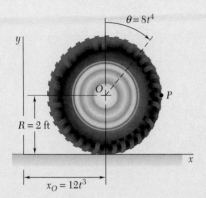

Fig. P15.248

15.248 A wheel moves in the xy plane in such a way that the location of its center is given by the equations $x_O = 12t^3$ and $y_O = R = 2$, where x_O and y_O are measured in feet and t is measured in seconds. The angular displacement of a radial line measured from a vertical reference line is $\theta = 8t^4$, where θ is measured in radians. Determine the velocity of point P located on the horizontal diameter of the wheel at $t = 1$ s.

15.249 Two blocks and a pulley are connected by inextensible cords as shown. The relative velocity of block A with respect to block B is 2.5 ft/s to the left at time $t = 0$ and 1.25 ft/s to the left when $t = 0.25$ s. Knowing that the angular acceleration of the pulley is constant, find (a) the relative acceleration of block A with respect to block B, (b) the distance block A moves relative to block B during the interval $0 \le t \le 0.25$ s.

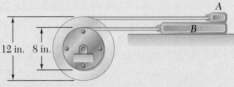

Fig. P15.249

15.250 A baseball pitching machine is designed to deliver a baseball with a ball speed of 70 mph and a ball rotation of 300 rpm clockwise. Knowing that there is no slipping between the wheels and the baseball during the ball launch, determine the angular velocities of wheels A and B.

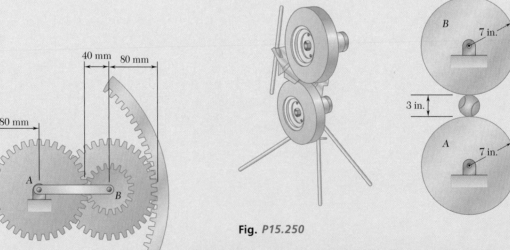

Fig. P15.250

Fig. P15.251

15.251 Knowing that inner gear A is stationary and outer gear C starts from rest and has a constant angular acceleration of 4 rad/s² clockwise, determine at $t = 5$ s (a) the angular velocity of arm AB, (b) the angular velocity of gear B, (c) the acceleration of the point on gear B that is in contact with gear A.

15.252 Knowing that at the instant shown bar AB has an angular velocity of 10 rad/s clockwise and it is slowing down at a rate of 2 rad/s², determine the angular accelerations of bar BD and bar DE.

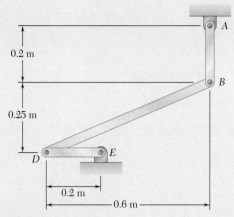

Fig. P15.252

15.253 Knowing that at the instant shown rod AB has zero angular acceleration and an angular velocity of 15 rad/s counterclockwise, determine (*a*) the angular acceleration of arm DE, (*b*) the acceleration of point D.

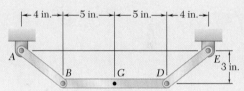

Fig. P15.253

15.254 Rod AB is attached to a collar at A and is fitted with a wheel at B that has a radius $r = 15$ mm. Knowing that when $\theta = 60°$ the collar has a velocity of 250 mm/s upward and it is slowing down at a rate of 150 mm/s², determine (*a*) the angular acceleration of rod AB, (*b*) the angular acceleration of the wheel.

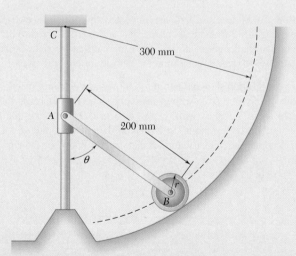

Fig. P15.254

15.255 Water flows through a curved pipe AB that rotates with a constant clockwise angular velocity of 90 rpm. If the velocity of the water relative to the pipe is 8 m/s, determine the total acceleration of a particle of water at point P.

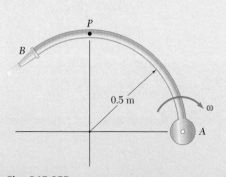

Fig. P15.255

15.256 A disk of 0.15-m radius rotates at the constant rate ω_2 with respect to plate BC, which itself rotates at the constant rate ω_1 about the y axis. Knowing that $\omega_1 = \omega_2 = 3$ rad/s, determine, for the position shown, the velocity and acceleration (a) of point D, (b) of point F.

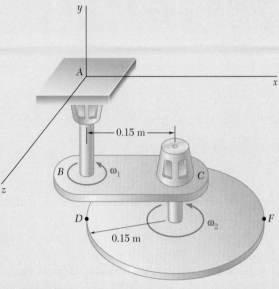

Fig. P15.256

15.257 Two rods AE and BD pass through holes drilled into a hexagonal block. (The holes are drilled in different planes so that the rods will not touch each other.) Knowing that rod AE has an angular velocity of 20 rad/s clockwise and an angular acceleration of 4 rad/s² counterclockwise when $\theta = 90°$, determine (a) the relative velocity of the block with respect to each rod, (b) the relative acceleration of the block with respect to each rod.

15.258 Rod BC of length 24 in. is connected by ball-and-socket joints to a rotating arm AB and to a collar C that slides on the fixed rod DE. Knowing that the length of arm AB is 4 in. and that it rotates at the constant rate $\omega_1 = 10$ rad/s, determine the velocity of collar C when $\theta = 0$.

15.259 In the position shown the thin rod moves at a constant speed $u = 3$ in./s out of the tube BC. At the same time, tube BC rotates at the constant rate $\omega_2 = 1.5$ rad/s with respect to arm CD. Knowing that the entire assembly rotates about the X axis at the constant rate $\omega_1 = 1.2$ rad/s, determine the velocity and acceleration of end A of the rod.

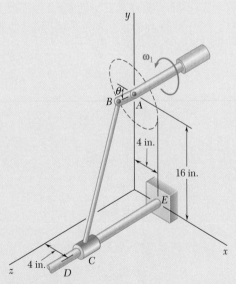

Fig. P15.257

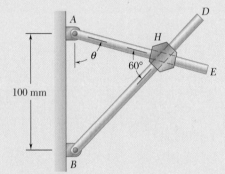

Fig. P15.258

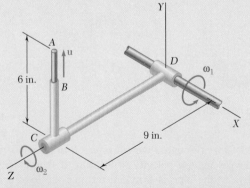

Fig. P15.259

16

Plane Motion of Rigid Bodies: Forces and Accelerations

The blades of the wind turbines shown in this picture are subjected to large forces and moments during motion. In this chapter, you will learn to analyze the motion of a rigid body by considering the motion of its mass center, the motion relative to its mass center, and the external forces acting on it.

Objectives

- **Discuss** how the mass and mass moment of inertia affect the linear and angular accelerations of a rigid body.

- **Model** physical systems involving rigid bodies by drawing correct free-body diagrams and kinetic diagrams.

- Using rigid-body kinetics principles, **determine** whether a body slips or tips and if a wheel rolls with or without slip.

- **Apply** appropriate kinetic equations and kinematics relationships to solve kinetics problems for a rigid body undergoing translation, centroidal rotation, or general plane motion.

- **Analyze** systems of connected rigid bodies using appropriate kinetic and kinematic equations.

- **Analyze** constrained motion of rigid bodies, including fixed-axis rotation and rolling disks and wheels.

Introduction

In this chapter and in Chaps. 17 and 18, you will study the **kinetics of rigid bodies**; i.e., the relations between the forces acting on a rigid body, the shape and mass of the body, and the motion produced. You studied similar relations in Chaps. 12 and 13, assuming then that you could consider the body as a particle, with its mass concentrated in one point and all forces acting at that point. Now you have to take into account the shape of the body, as well as the exact location of the points of application of the forces. You also will be concerned not only with the motion of the body as a whole but with the motion of the body about its mass center.

Our approach will be to consider rigid bodies as made up of large numbers of particles and to use the results obtained in Chap. 14 for the motion of systems of particles. Specifically, we use two equations from Chap. 14: Eq. (14.16), $\Sigma \mathbf{F} = m\bar{\mathbf{a}}$, which relates the resultant of the external forces and the acceleration of the mass center G of the system of particles, and Eq. (14.23), $\Sigma \mathbf{M}_G = \dot{\mathbf{H}}_G$, which relates the resultant moment of the external forces and the angular momentum of the system of particles about G.

Except for Sec. 16.1A, which applies to the most general case of the motion of a rigid body, the results derived in this chapter are limited in two ways: (1) They are restricted to the *plane motion* of rigid bodies, i.e., where all motion occurs in a single two-dimensional reference plane. (2) The rigid bodies considered consist only of plane rigid bodies and of bodies that are symmetrical with respect to a reference plane (or more generally, bodies that have a principal centroidal axis of inertia perpendicular to a reference plane). The study of the plane motion of nonsymmetrical three-dimensional bodies and, more generally, the motion of rigid bodies in three-dimensional space will be postponed until Chap. 18.

In Sec. 16.1B, we define the angular momentum of a rigid body in plane motion and show that the rate of change of the angular momentum $\dot{\mathbf{H}}_G$ about the mass center is equal to the product $\bar{I}\boldsymbol{\alpha}$ of the centroidal mass moment of inertia $\bar{I}$ and the angular acceleration $\boldsymbol{\alpha}$ of the body. We then prove that the external forces acting on a rigid body are equivalent to a vector $m\bar{\mathbf{a}}$ attached at the mass center and a couple of moment $\bar{I}\boldsymbol{\alpha}$.

We also derive the principle of transmissibility using only the parallelogram law and Newton's laws of motion, allowing us to remove this principle from the list of axioms (*Statics*, Sec. 1.2) required for the study of the statics and dynamics of rigid bodies. We then discuss the use of the free-body diagram and kinetic diagram in the solution of all problems involving the plane motion of rigid bodies.

We consider the plane motion of connected rigid bodies in Sec. 16.1F, which will prepare you to solve a variety of problems involving the translation, centroidal rotation, and unconstrained motion of rigid bodies. In the remaining part of this chapter, we present the solutions of problems involving noncentroidal rotation, rolling motion, and other partially constrained plane motions of rigid bodies.

16.1 KINETICS OF A RIGID BODY

As we saw in Chapter 15, we can generally consider the motion of a rigid body to be a combination of translation of the body and rotation about its mass center. We use this same idea to analyze the relationship between forces and moments acting on a rigid body and the body's linear and angular acceleration.

16.1A Equations of Motion for a Rigid Body

Consider a rigid body acted upon by several external forces $\mathbf{F}_1, \mathbf{F}_2, \mathbf{F}_3, \ldots$ (Fig. 16.1). We can assume that the body is made of a large number n of particles of mass Δm_i ($i = 1, 2, \ldots, n$) and apply the results obtained in Chap. 14 for a system of particles (Fig. 16.2). Consider first the motion of the mass center G of the body with respect to the newtonian frame of reference $Oxyz$. From Eq. (14.16), we have

Translational equation of motion

$$\Sigma\mathbf{F} = m\bar{\mathbf{a}} \tag{16.1}$$

where m is the mass of the body and $\bar{\mathbf{a}}$ is the acceleration of the mass center G. Turning now to the motion of the body relative to the centroidal frame of reference $Gx'y'z'$, from Eq. (14.23), we have

Rotational equation of motion

$$\Sigma\mathbf{M}_G = \dot{\mathbf{H}}_G \tag{16.2}$$

where $\dot{\mathbf{H}}_G$ represents the rate of change of $\mathbf{H}_G$, which is the angular momentum about G of the system of particles forming the rigid body.

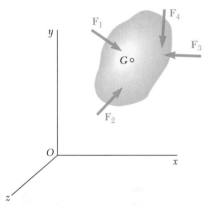

Fig. 16.1 A rigid body acted on by several external forces.

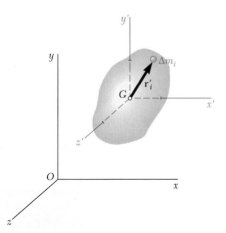

Fig. 16.2 A particle of a rigid body in relation to the mass center G.

In the following discussion, we refer to $\mathbf{H}_G$ simply as the **angular momentum of the rigid body about its mass center** G. Together, Eqs. (16.1) and (16.2) express that

> **The system of the external forces and moments is equipollent to the system consisting of the vector $m\bar{\mathbf{a}}$ attached at G and the couple of moment $\dot{\mathbf{H}}_G$ (Fig. 16.3).**

As you will see in Chap. 18, Eqs. (16.1) and (16.2) apply to general three-dimensional motion of a rigid body. In the rest of this chapter, however, we limit our analysis to the **plane motion** of rigid bodies, i.e., to a motion in which each particle remains within a fixed reference plane. We also assume that the rigid bodies considered consist only of plane rigid bodies and of bodies that are symmetrical with respect to the plane of motion. Further study of the plane motion of nonsymmetrical three-dimensional bodies and of the motion of rigid bodies in three-dimensional space will be postponed until Chap. 18.

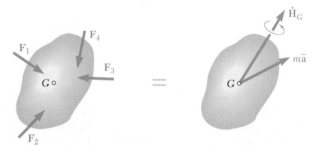

Fig. 16.3 A system of external forces is equipollent to an inertial vector $m\bar{\mathbf{a}}$ and a couple of moment $\dot{\mathbf{H}}_G$ acting at the mass center.

Photo 16.1 The system of external forces acting on the man and wakeboard includes the weights, the tension in the tow rope, and the forces exerted by the water and the air.

16.1B Angular Momentum of a Rigid Body in Plane Motion

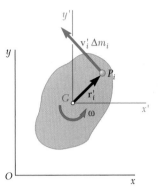

Consider a rigid body in plane motion. Assume that the body is made of a large number n of particles P_i with a mass $\Delta m_i.$ Then from Eq. (14.24) of Sec. 14.1D, we can compute the angular momentum $\mathbf{H}_G$ of the rigid body about its mass center G by taking the moments about G of the momenta of the particles of the body with respect to either of the frames Oxy or $Gx'y'$ (Fig. 16.4). Choosing the second option gives

$$\mathbf{H}_G = \sum_{i=1}^{n} (\mathbf{r}'_i \times \mathbf{v}'_i \,\Delta m_i) \qquad (16.3)$$

Fig. 16.4 The angular momentum about G of a particle of a rigid body is $\mathbf{r}'_i \times \mathbf{v}'_i \,\Delta m_i.$

where $\mathbf{r}'_i$ and $\mathbf{v}'_i \Delta m_i$ denote, respectively, the position vector and the linear momentum of the particle P_i relative to the centroidal frame of reference $Gx'y'$. However, since the particle is part of the rigid body, we have $\mathbf{v}'_i = \boldsymbol{\omega} \times \mathbf{r}'_i$, where $\boldsymbol{\omega}$ is the angular velocity of the body at the instant considered. We have

$$\mathbf{H}_G = \sum_{i=1}^{n} [\mathbf{r}'_i \times (\boldsymbol{\omega} \times \mathbf{r}'_i) \,\Delta m_i]$$

Referring to Fig. 16.4, we easily verify that this expression represents a vector of the same direction as $\boldsymbol{\omega}$ (i.e., perpendicular to the body) and with a magnitude of $\omega \Sigma r_i'^2 \,\Delta m_i$ Recalling that the sum $\Sigma r_i'^2 \,\Delta m_i$ represents the moment of inertia $\bar{I}$ of the rigid body about a centroidal axis perpendicular to the body, we conclude that the angular momentum $\mathbf{H}_G$ of the rigid body about its mass center is

**Angular momentum of a
rigid body about G**

$$\mathbf{H}_G = \bar{I}\boldsymbol{\omega} \qquad (16.4)$$

Differentiating both sides of Eq. (16.4), we obtain

**Rate of change of angular
momentum about G**

$$\dot{\mathbf{H}}_G = \bar{I}\dot{\boldsymbol{\omega}} = \bar{I}\boldsymbol{\alpha} \qquad (16.5)$$

Thus, the rate of change of the angular momentum of the rigid body is represented by a vector in the same direction as $\boldsymbol{\alpha}$ (i.e., perpendicular to the body) with a magnitude $\bar{I}\alpha$.

Keep in mind that the results obtained in this section have been derived for a rigid body in plane motion. As you will see in Chap. 18, they remain valid in the case of the plane motion of rigid bodies that are symmetrical with respect to a reference plane (or, more generally, bodies that have a principal centroidal axis of inertia perpendicular to a reference plane). However, they do not apply in the case of nonsymmetrical bodies or in the case of three-dimensional motion.

Photo 16.2 The hard disk and pick-up arm of a computer hard drive undergo fixed-axis rotation.

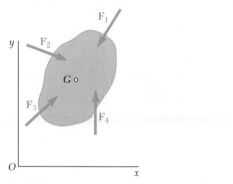

Fig. 16.5 A rigid body acted upon by several external forces in the plane of the body.

16.1C Plane Motion of a Rigid Body

Consider a rigid body with a mass m moving under the action of several external forces $\mathbf{F}_1, \mathbf{F}_2, \mathbf{F}_3, \ldots$ contained in the plane of the body (Fig. 16.5). Substituting $\dot{\mathbf{H}}_G$ from Eq. (16.5) into Eq. (16.2) and writing the fundamental equations of motion from Eqs. (16.1) and (16.2) in scalar form, we have

$$\Sigma F_x = m\bar{a}_x \quad \Sigma F_y = m\bar{a}_y \quad \Sigma M_G = \bar{I}\alpha \tag{16.6}$$

Equations (16.6) show that we can obtain the acceleration of the mass center G of the rigid body and its angular acceleration $\boldsymbol{\alpha}$ once we have determined the resultant of the external forces acting on the body and their moment resultant about G. Given appropriate initial conditions, we can then obtain the coordinates $\bar{x}$ and $\bar{y}$ of the mass center and the angular coordinate θ of the body by integration at any instant t. Thus,

The motion of the rigid body is completely defined by the resultant force and resultant moment about G acting on the body.

Since the motion of a rigid body depends only upon the resultant and resultant moment of the external forces acting on it, it follows that **two systems of forces that are equipollent** (i.e., that have the same resultant and the same moment resultant) **are also equivalent**. That is, they have exactly the same effect on a given rigid body.

Consider in particular the system of external forces acting on a rigid body (Fig. 16.6a) and the system of inertial terms associated with the particles forming the rigid body (Fig. 16.6b). We showed in Sec. 14.1A that the two systems thus defined are equipollent. But since the particles we are considering now form a rigid body, it follows from the discussion above that the two systems are also equivalent. We can thus state that

The external forces acting on a rigid body are equivalent to the inertial terms of the various particles forming the body.

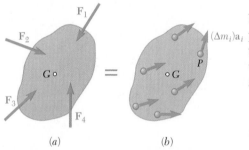

Fig. 16.6 The external forces acting on the rigid body are equivalent to the inertial terms of the particles of the body.

The fact that the system of external forces is equivalent to the system of inertial terms has been emphasized by the use of red equal signs in Fig. 16.6 and also in Fig. 16.7. Here, using results obtained earlier in this section, we replaced the inertial terms by a vector $m\bar{\mathbf{a}}$ attached at the mass center G of the rigid body and the rotational inertial term $\bar{I}\boldsymbol{\alpha}$.

Let's look at three examples of rigid-body plane motion.

Translation. In the case of a body in translation, the angular acceleration of the body is equal to zero and its inertial terms reduce to the vector $m\bar{\mathbf{a}}$ attached at G (Fig. 16.8). Thus, the resultant of the external forces acting on a rigid body in translation passes through the mass center of the body and is equal to $m\bar{\mathbf{a}}$.

Centroidal Rotation. When a rigid body, or more generally, a body symmetrical with respect to a reference plane, rotates about a fixed axis perpendicular to the reference plane and passing through its mass center G, we say that the body is in *centroidal rotation*. Since the acceleration $\bar{\mathbf{a}}$ is identically equal to zero, the inertial terms of the body reduce to the couple $\bar{I}\boldsymbol{\alpha}$ (Fig. 16.9). Thus, the external forces acting on a body in centroidal rotation are equivalent to the rotational inertia $\bar{I}\boldsymbol{\alpha}$.

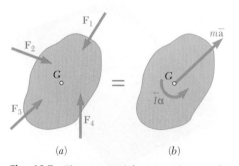

Fig. 16.7 The external forces acting on the rigid body are also equivalent to a vector $m\bar{\mathbf{a}}$ attached to the mass center G and a rotational inertia $\bar{I}\boldsymbol{\alpha}$.

General Plane Motion. Comparing Fig. 16.7 with Figs. 16.8 and 16.9, we observe that, from the point of view of *kinetics,* the most general plane motion of a rigid body symmetrical with respect to a reference plane can be replaced by the sum of a translation and a centroidal rotation. Note that this statement is more restrictive than the similar statement made earlier from the point of view of *kinematics* (Sec. 15.2A), since we now require that the mass center of the body be selected as the reference point.

Referring to Eqs. (16.6), we observe that the first two equations are identical with the equations of motion of a particle of mass m acted upon by the given forces $\mathbf{F}_1$, $\mathbf{F}_2$, $\mathbf{F}_3$, ... We thus check that

> **The mass center G of a rigid body in plane motion moves as if the entire mass of the body were concentrated at that point, and as if all the external forces act on it.**

Recall that we already obtained this result in Sec. 14.1C in the general case of a system of particles with the particles being not necessarily rigidly connected. We also note, as we did earlier, that the system of the external forces does not, in general, reduce to a single vector $m\bar{\mathbf{a}}$ attached at G. Therefore, in the general case of the plane motion of a rigid body, **the resultant of the external forces acting on the body does not pass through the mass center of the body.**

Finally, note that the last of Eqs. (16.6) would still be valid if the rigid body, while subjected to the same applied forces, were constrained to rotate about a fixed axis through G. Thus, **a rigid body in plane motion rotates about its mass center as if this point were fixed.**

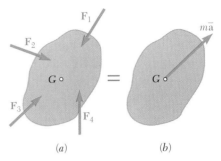

Fig. 16.8 A rigid body in translation has a vector $m\bar{\mathbf{a}}$ attached to the mass center G but no rotational inertia $\bar{I}\alpha$.

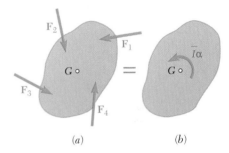

Fig. 16.9 A rigid body in centroidal rotation has a rotational inertia $\bar{I}\alpha$ but no $m\bar{\mathbf{a}}$.

*16.1D A Remark on the Axioms of the Mechanics of Rigid Bodies

The fact that two equipollent systems of external forces acting on a rigid body are also equivalent—i.e., have the same effect on that rigid body—has already been established in *Statics,* Sec. 3.4B. However, there we derived it from the *principle of transmissibility,* which is one of the axioms used in our study of the statics of rigid bodies. We have not used this axiom in the present chapter because Newton's second and third laws of motion make its use unnecessary in the study of the dynamics of rigid bodies.

In fact, we can now derive the principle of transmissibility from the other axioms used in the study of mechanics. This principle stated, without proof (Sec. 3.1B), that the conditions of equilibrium or motion of a rigid body remain unchanged if a force $\mathbf{F}$ acting at a given point of the rigid body is replaced by a force $\mathbf{F}'$ of the same magnitude and same direction—but acting at a different point—provided that the two forces have the same line of action. But since $\mathbf{F}$ and $\mathbf{F}'$ have the same moment about any given point, it is clear that they form two equipollent systems of external forces. Thus, we may now *prove,* as a result of what we established in the preceding section, that $\mathbf{F}$ and $\mathbf{F}'$ have the same effect on the rigid body (see Fig. 3.3 repeated here).

We can therefore remove the principle of transmissibility from the list of axioms required for the study of the mechanics of rigid bodies. These axioms are reduced to the parallelogram law of addition of vectors and to Newton's laws of motion.

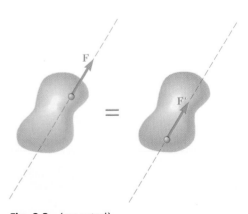

Fig. 3.3 (repeated)

16.1E Solution of Problems Involving the Motion of a Rigid Body

We saw in Sec. 16.1C that when a rigid body is in plane motion a fundamental relation exists between the forces $\mathbf{F}_1$, $\mathbf{F}_2$, $\mathbf{F}_3$, . . . , acting on the body, the acceleration $\bar{\mathbf{a}}$ of its mass center, and the angular acceleration $\boldsymbol{\alpha}$ of the body. This relation is represented in Fig. 16.7 in the form of a **free-body diagram and a kinetic diagram**. We can use these diagrams to determine the acceleration $\bar{\mathbf{a}}$ and the angular acceleration $\boldsymbol{\alpha}$ produced by a given system of forces acting on a rigid body or, conversely, to determine the forces that produce a given motion of the rigid body.

We can use the three algebraic equations of Eq. (16.6) to solve problems of plane motion.[†] However, our experience in statics suggests that the solution of many problems involving rigid bodies can be simplified by an appropriate choice of the point about which we compute the moments of the forces. It is therefore preferable to remember the relation between the forces and the accelerations in the pictorial form shown in Fig. 16.7 and to derive from this fundamental relation the component or moment equations that best fit the solution of the problem under consideration.

Drawing a free-body diagram for rigid bodies follows the same basic steps as we discussed in Chapter 12. For rigid bodies, however, it is important to draw your forces at their location of action, since you will be summing moments about specific points. Labeling different dimensions on your free-body diagram is particularly helpful when summing these moments.

The kinetic diagram is also slightly modified from Chap. 12. The translational inertial term $m\bar{\mathbf{a}}$ is always located at the center of mass of the body. We are now concerned with the rotational inertia of the body, so we include an additional term on our kinetic diagram, $\bar{I}\boldsymbol{\alpha}$. This is also located at the center of mass of the body.

We can apply the steps from Chap. 12 to the pendulum shown in Fig. 16.10, where a moment M is applied to the bar. These steps include:

1. Isolating the **body**
2. Defining the **axes**
3. Replacing constraints with **support forces**
4. Adding **applied forces and moments**, as well as **body forces** to the diagram
5. Labeling the free-body diagram with **dimensions**

For the kinetic diagram, we typically draw the translational inertial term in component form (e.g., $m\bar{a}_x$ and $m\bar{a}_y$) at the center of mass of the body and add the rotational inertial term $\bar{I}\alpha$. Using these steps gives you the free-body diagram and kinetic diagram shown in Fig. 16.11.

We use the pendulum shown in Fig. 16.10 to illustrate an alternative form of the moment equation. It is straightforward to apply Eq. 16.6 to this problem, where the sum of moments about the center of mass results in

$$+\!\!\uparrow\!\Sigma M_G = \bar{I}\alpha: \qquad M - P_y\!\left(\frac{L}{2}\right) = \bar{I}\alpha$$

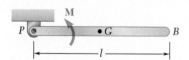

Fig. 16.10 A pendulum with mass m, length l, and an applied moment **M**.

[†]Recall that the last of Eq. (16.6) is valid only in the case of the plane motion of a rigid body symmetrical with respect to a reference plane. In all other cases, you need to use the methods of Chap. 18.

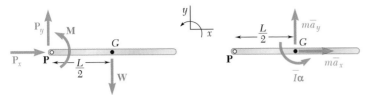

Fig. 16.11 Free-body diagram and kinetic diagram for a pendulum with an external moment applied.

Alternatively, we could choose an arbitrary point P about which to sum moments. If we choose P to be at the left end of the rod, then we also have to sum the moments about P due to the inertial terms. In this case, we obtain

$$+\curvearrowleft \Sigma M_P = \bar{I}\alpha + m\bar{a}d_\perp: \qquad M - W\left(\frac{L}{2}\right) = \bar{I}\alpha + m\bar{a}_y\left(\frac{L}{2}\right) + m\bar{a}_x(0)$$

where $d_\perp$ is the perpendicular distance from point P to the line of action of the resultant acceleration vector $\bar{\mathbf{a}}$. As in statics, you can also determine the moment about a point P by using vector products, as

$$m\bar{a}d_\perp = \mathbf{r}_{G/P} \times m\bar{\mathbf{a}}$$

where $\mathbf{r}_{G/P}$ is the vector from point P to the center of mass of the body. Therefore, we can also write Eq (16.6) as

$$\Sigma F_x = m\bar{a}_x \qquad \Sigma F_y = m\bar{a}_y$$
$$\Sigma M_G = \bar{I}\alpha \text{ or } \Sigma M_P = \bar{I}\alpha + m\bar{a}d_\perp \text{ or } \Sigma M_P = \bar{I}\alpha + \mathbf{r}_{G/P} \times m\bar{\mathbf{a}} \quad \textbf{(16.6')}$$

The use of a free-body diagram and a kinetic diagram, showing vectorially the relationship between the forces applied on the rigid body and the resulting linear and angular accelerations, presents considerable advantages over the blind application of formulas (16.6). We can summarize these advantages as follows.

1. The use of a pictorial representation provides a much clearer understanding of the effect of the forces on the motion of the body.
2. This approach makes it possible to divide the solution of a dynamics problem into two parts: In the first part, the analysis of the kinematic and kinetic characteristics of the problem leads to the free-body diagram and the kinetic diagram of Fig. 16.7; in the second part, you can use the diagrams to analyze the various forces and vectors involved.

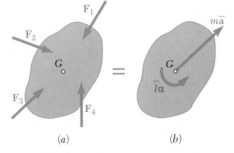

Fig. 16.7 (repeated)

3. A unified approach is provided for the analysis of the plane motion of a rigid body, regardless of the particular type of motion involved. Although the kinematics of the various motions considered may vary from one case to the other, the approach to the kinetics of the motion is consistently the same. In every case, you draw a diagram showing the external forces, the vector $m\bar{\mathbf{a}}$ associated with the motion of G, and the couple $\bar{I}\alpha$ associated with the rotation of the body about G.
4. The resolution of the plane motion of a rigid body into a translation and a centroidal rotation, which we use here, is a basic concept that can be applied effectively throughout the study of mechanics. We will use it again in Chap. 17 with both the method of work and energy and the method of impulse and momentum.

5. As you will see in Chap. 18, we can extend this approach to the study of the general three-dimensional motion of a rigid body. The motion of the body is again resolved into a translation and a rotation about the mass center, and we use free-body diagrams and kinetic diagrams to indicate the relationship between the external forces and the rates of change of the linear and angular momenta of the body.

16.1F Systems of Rigid Bodies

The method just described also can be used in problems involving the plane motion of several connected rigid bodies. For each part of the system, you draw a diagram similar to Fig. 16.7. You can obtain the equations of motion from these diagrams and solve them simultaneously.

In some cases, as in Sample Prob. 16.4, you can draw a single diagram for the entire system. This diagram should include all of the external forces as well as the vectors $m\bar{\mathbf{a}}$ and the couples $\bar{I}\boldsymbol{\alpha}$ associated with the various parts of the system. However, you can omit internal forces, such as the forces exerted by connecting cables, because they occur in pairs of equal and opposite forces and are thus equipollent to zero. The equations obtained by expressing that the system of external forces is equipollent to the system of inertial terms can be solved for the remaining unknowns (note that we cannot speak of equivalent systems because we are not dealing with a single rigid body). For systems involving multiple rigid bodies, the general equation of motion is written as

$$\Sigma \mathbf{F} = \Sigma m_i \bar{\mathbf{a}}_i \quad \text{and} \quad \sum \mathbf{M}_P = \dot{\mathbf{H}}_P$$

where

$$\dot{\mathbf{H}}_P = \Sigma \bar{I}_i \boldsymbol{\alpha}_i + \Sigma m_i \bar{\mathbf{a}}_i (d_\perp)_i = \Sigma \bar{I}_i \boldsymbol{\alpha}_i + \Sigma [(\mathbf{r}_{G/P})_i \times m_i \bar{\mathbf{a}}_i]$$

Historically, sometimes these equations have been written as

$$\Sigma \mathbf{F} = \Sigma \mathbf{F}_{\text{eff}} \quad \text{and} \quad \Sigma \mathbf{M}_P = \Sigma (\mathbf{M}_P)_{\text{eff}}$$

where the left-hand sides of these equations come from the free-body diagram and the right-hand sides come from the kinetic diagram. We have chosen not to use this notation because the terms on the right-hand side are due to the inertial terms and not due to external forces and moments.

It is not possible to include more than one rigid body in your system in problems involving more than three unknowns, since only three equations of motion are available when a single diagram is used. We will not elaborate upon this point, since the discussion involved would be completely similar to that given in Sec. 6.3B in the case of the equilibrium of a system of rigid bodies.

Photo 16.3 The forklift and moving load can be analyzed as a system of two connected rigid bodies in plane motion.

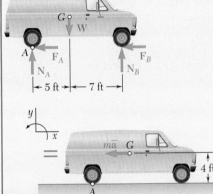

Fig. 1 Free-body diagram and kinetic diagram for the van.

Sample Problem 16.1

When the forward speed of the van shown is 30 ft/s, the brakes are suddenly applied, causing all four wheels to stop rotating. The van skids to rest in 20 ft. Determine the magnitude of the normal reaction and of the friction force at each wheel as the van skids to rest.

STRATEGY: You are given enough information to determine the acceleration and you want to find forces, so use Newton's second law. The motion described is pure translation, so the angular acceleration is zero.

MODELING: Choose the van to be your system and model it as a rigid body. A free-body diagram and a kinetic diagram for this system are shown in Fig. 1. The external forces consist of the weight **W** of the truck and of the normal reactions and friction forces at the wheels. The vectors **N**$_A$ and **F**$_A$ represent the sum of the reactions at the rear wheels, while **N**$_B$ and **F**$_B$ represent the sum of the reactions at the front wheels. Since the truck is in translation, $\alpha = 0$ and the inertial terms reduce to the vector $m\bar{a}$ attached at G.

ANALYSIS:

Kinematics of Motion. Choose the positive sense to the right and use the equations of uniformly accelerated motion. You have

$$\bar{v}_0 = +30 \text{ ft/s} \qquad \bar{v}^2 = \bar{v}_0^2 + 2\bar{a}\bar{x} \qquad 0 = (30)^2 + 2\bar{a}(20)$$
$$\bar{a} = -22.5 \text{ ft/s}^2 \qquad \mathbf{\bar{a}} = 22.5 \text{ ft/s}^2 \leftarrow$$

Equations of Motion. You can obtain three equations of motion by expressing that the system of the external forces from your free-body diagram is equivalent to the inertial terms from your kinetic diagram. Applying Newton's second law in the x and y directions gives

$$+\uparrow\Sigma F_y = m\bar{a}_y: \qquad N_A + N_B - W = 0 \qquad (1)$$

$$\xrightarrow{+}\Sigma F_x = m\bar{a}_x: \qquad -(F_A + F_B) = -m\bar{a} \qquad (2)$$

Taking moments about any point gives you a third equation. For moments about point A, you find

$$+\curvearrowleft\Sigma M_A = \bar{I}\alpha + m\bar{a}d_\perp: \qquad - W(5 \text{ ft}) + N_B(12 \text{ ft}) = m\bar{a}(4 \text{ ft}) \qquad (3)$$

In these three equations you have five unknowns, N_A, N_B, F_A, F_B, and $\bar{a}$. Since $F_A = \mu_k N_A$ and $F_B = \mu_k N_B$, where μ_k is the coefficient of kinetic friction, you have from Eq. (1)

$$F_A + F_B = \mu_k(N_A + N_B) = \mu_k W$$

Substituting into Eq. (2) and using $m = W/g$ gives

$$-\mu_k W = -\frac{W}{32.2 \text{ ft/s}^2}\bar{a} = -\frac{W}{32.2 \text{ ft/s}^2}(22.5 \text{ ft/s}^2)$$

(continued)

or $\mu_k = 0.699$. Solving Eq. (3) for N_B gives you $N_B = 0.640W$. Substituting this into Eq. (1), you find $N_A = 0.350W$. The friction forces are easily determined once you know the normal forces $F_A = \mu_k N_A = (0.699)(0.350W) = 0.245W$ and $F_B = \mu_k N_B = (0.699)(0.650W) = 0.454W$.

Reactions at Each Wheel. Recall that the values computed here represent the sum of the reactions at the two front wheels or the two rear wheels. You obtain the magnitude of the reactions at each wheel by writing

$$N_{\text{front}} = \tfrac{1}{2}N_B = 0.325W \qquad N_{\text{rear}} = \tfrac{1}{2}N_A = 0.175W \blacktriangleleft$$
$$F_{\text{front}} = \tfrac{1}{2}F_B = 0.227W \qquad F_{\text{rear}} = \tfrac{1}{2}F_A = 0.122W \blacktriangleleft$$

REFLECT and THINK: Note that even though the angular acceleration of the van is zero, the sum of the moments about point A is not equal to zero, since from the kinetic diagram, $m\bar{a}$ produces a moment about A. Rather than taking moments about point A, you also could have chosen to take moments about the center of mass, G. In this case, the sum of the moments would have been equal to zero. You only get three independent equations for a rigid body in plane motion: ΣF_x, ΣF_y, and one moment equation.

Sample Problem 16.2

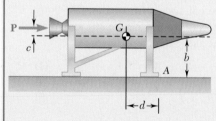

A sled is jet-propelled along a straight track by a force P that increases linearly with time according to $P = kt$, where k is a constant. The coefficient of sliding friction between the sled runners and the track is μ_k, the coefficient of static friction is μ_s, and the mass of the sled is m. Determine (a) the time at which the tip of the rocket begins to rotate downward, (b) the acceleration of the sled at this instant. Neglect loss of mass due to fuel consumption and assume that the sled will slide before it tips.

STRATEGY: Since you are given a force, use Newton's second law to find the acceleration required for the rocket to begin rotating forward. You can then find the time using $P = kt$.

MODELING: Choose the sled as your system and model it as a rigid body. The rocket force must overcome the static friction force before it begins moving. Define this time to be t_0. Figure 1 shows a free-body diagram when the motion is impending. In this case, both of the friction forces are set equal to the maximum allowable friction force $\mu_s N$. Free-body and kinetic diagrams for when the sled is about to tip are shown in Fig. 2. Just as the sled starts to tip, the normal force on the rear of the sled goes to zero.

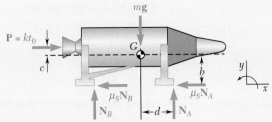

Fig. 1 Free-body diagram when motion is impending.

ANALYSIS: Using Fig. 1 and applying Newton's second law in the y- and x-directions gives

$$+\uparrow\Sigma F_y = m\bar{a}_y: \quad N_A + N_B - mg = 0 \quad \text{or} \quad N_A + N_B = mg$$

$$\xrightarrow{+}\Sigma F_x = m\bar{a}_x: \quad kt_0 - (\mu_s N_A + \mu_s N_B) = 0$$

or

$$kt_0 = \mu_s(N_A + N_B) = \mu_s mg \tag{1}$$

Now that you know when the sled begins to slide, you can determine the time it will start to tip using Fig. 2.

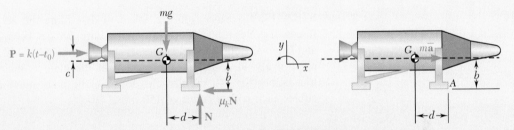

Fig. 2 Free-body diagram and kinetic diagram for the sled after it begins to move.

From this diagram, you can apply Newton's second law in the x and y directions and sum moments about any point. If you choose to take moments about G, you find

$$\xrightarrow{+}\Sigma F_x = m\bar{a}_x: \quad k(t - t_0) - \mu_k N = m\bar{a} \tag{2}$$

$$+\uparrow\Sigma F_y = m\bar{a}_y: \quad N - mg = 0 \tag{3}$$

$$+\curvearrowleft \Sigma M_G = \bar{I}\alpha: \quad Nd - \mu_k Nb - k(t - t_0)c = 0 \tag{4}$$

Solving Eqs. (1), (2), (3), and (4) for $t_0, t, N, \bar{a}$, you find $N = mg$, $t_0 = \mu_s mg/k$ and

$$t = \frac{mg(d + c\mu_s - b\mu_k)}{kc} \quad \blacktriangleleft$$

$$\bar{a} = \frac{g(d - c\mu_k - b\mu_k)}{c} \quad \blacktriangleleft$$

(continued)

REFLECT and THINK: Rather than taking moments about G, you could have chosen any other point. For example, for moments about A, you have

$$+\uparrow\Sigma M_A = \bar{I}\alpha + m\bar{a}d: \quad mgd - k(t - t_0)(b + c) = -m\bar{a}b$$

Using this equation rather than Eq. (4) will give you the same answer. To check the assumption that the sled slides before it tips, you would need to use Fig. 1 and show that both N_A and N_B are positive for the given value of $P = kt_0$.

Sample Problem 16.3

The thin plate $ABCD$ has a mass of 8 kg and is held in the position shown by the wire BH and two links AE and DF. Neglecting the mass of the links, determine immediately after wire BH has been cut (a) the acceleration of the plate, (b) the force in each link.

STRATEGY: Since you are asked to determine the acceleration and forces, use Newton's second law. After wire BH has been cut, corners A and D move along parallel circles, each with a radius of 150 mm centered, respectively, at E and F. The motion of the plate is thus a curvilinear translation (Fig. 1); the particles forming the plate move along parallel circles each with a radius of 150 mm.

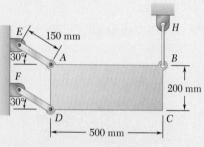

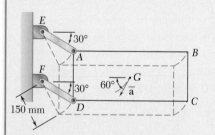

Fig. 1 Curvilinear translation of the plate.

MODELING: Choose the plate to be your system and model it as a rigid body. To draw the kinetic diagram, you need to consider the kinematics of the motion. At the instant wire BH is cut, the velocity of the plate is zero. Thus, the acceleration of the mass center G of the plate is tangent to the circular path described by G (Fig. 1). The free-body diagram and kinetic diagram for this system are shown in Fig. 2. The external forces consist of the weight $\mathbf{W}$ and the forces $\mathbf{F}_{AE}$ and $\mathbf{F}_{DF}$ exerted by the links. Since the plate is in translation, the kinetic diagram is the vector $m\bar{\mathbf{a}}$ attached at G and directed along the t axis.

ANALYSIS:

a. Acceleration of the Plate.

$$+\nearrow\Sigma F_t = m\bar{a}_t:$$

$$W \cos 30° = m\bar{a}$$

$$mg \cos 30° = m\bar{a}$$

$$\bar{a} = g \cos 30° = (9.81 \text{ m/s}^2) \cos 30° \qquad (1)$$

$$\bar{\mathbf{a}} = 8.50 \text{ m/s}^2 \; \nearrow \; 60° \quad \blacktriangleleft$$

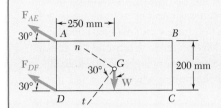

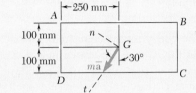

Fig. 2 Free-body diagram and kinetic diagram for the plate.

b. Forces in Links *AE* and *DF*.

$+\nwarrow\Sigma F_n = m\bar{a}_n$: $F_{AE} + F_{DF} - W\sin 30° = 0$ (2)

$+\downarrow\Sigma M_G = \bar{I}\alpha$:

$(F_{AE}\sin 30°)(250\text{ mm}) - (F_{AE}\cos 30°)(100\text{ mm})$
$\qquad\qquad + (F_{DF}\sin 30°)(250\text{ mm}) + (F_{DF}\cos 30°)(100\text{ mm}) = 0$

$$38.4F_{AE} + 211.6F_{DF} = 0$$

$$F_{DF} = -0.1815F_{AE}\qquad\qquad (3)$$

Substituting F_{DF} from Eq. (3) into Eq. (2), you have

$$F_{AE} - 0.1815F_{AE} - W\sin 30° = 0$$
$$F_{AE} = 0.6109W$$
$$F_{DF} = -0.1815(0.6109W) = -0.1109W$$

Noting that $W = mg = (8\text{ kg})(9.81\text{ m/s}^2) = 78.48\text{ N}$, you have

$$F_{AE} = 0.6109(78.48\text{ N})\qquad F_{AE} = 47.9\text{ N } T \;\blacktriangleleft$$
$$F_{DF} = -0.1109(78.48\text{ N})\qquad F_{DF} = 8.70\text{ N } C \;\blacktriangleleft$$

where bar *AE* is in tension and bar *DF* is in compression.

REFLECT and THINK: If *AE* and *DF* had been cables rather than links, the answers you just determined indicate that *DF* would have gone slack (i.e., you can't push on a rope), since the analysis showed that it would be in compression. Therefore, the plate would not be undergoing curvilinear translation, but it would have been undergoing general plane motion. It is important to note that that there is always more than one way to solve problems like this, since you can choose to take moments about any point you wish. In this case, you took them about *G*, but you could have also chosen to take them about *A* or *D*.

Sample Problem 16.4

A pulley weighing 12 lb and having a radius of gyration of 8 in. is connected to two blocks as shown. Assuming no axle friction, determine the angular acceleration of the pulley and the acceleration of each block.

STRATEGY: Since you want to determine accelerations and are given the weights, use Newton's second law.

MODELING: Choose the pulley and the two blocks as a single system. The pulley moves in pure rotation and each block moves in pure translation.

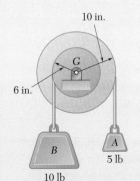

(continued)

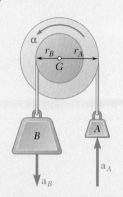

Fig. 1 Acceleration directions assuming a CCW angular acceleration.

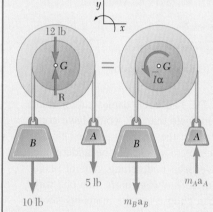

Fig. 2 Free-body diagram and kinetic diagram for the system.

Sense of Motion. Although you can assume an arbitrary sense of motion as shown in Fig. 1 (since no friction forces are involved) and later check it by the sign of the answer, you may prefer to determine the actual sense of rotation of the pulley. First determine the weight of block B, W_B', required to maintain the equilibrium of the pulley when it is acted upon by the 5-lb block A.

$$+\gamma\Sigma M_G = 0: \qquad W_B'(6 \text{ in.}) - (5 \text{ lb})(10 \text{ in.}) = 0 \qquad W_B' = 8.33 \text{ lb}$$

Since block B actually weighs 10 lb, the pulley rotates counterclockwise. The free-body and kinetic diagrams for this system are shown in Fig. 2. The forces external to the system consist of the weights of the pulley and the two blocks and of the reaction at G (Fig. 2). The forces exerted by the cables on the pulley and on the blocks are internal to the system and cancel out. Since the motion of the pulley is a centroidal rotation and the motion of each block is a translation, the inertial terms reduce to the couple $\bar{I}\alpha$ and the two vectors $m\mathbf{a}_A$ and $m\mathbf{a}_B$.

ANALYSIS:

Kinematics of Motion. Assuming α is counterclockwise and noting that $a_A = r_A\alpha$ and $a_B = r_B\alpha$, you obtain

$$\mathbf{a}_A = (\tfrac{10}{12} \text{ ft})\alpha \uparrow \qquad \mathbf{a}_B = (\tfrac{6}{12} \text{ ft})\alpha \downarrow$$

Equations of Motion. The centroidal moment of inertia of the pulley is

$$\bar{I} = m\bar{k}^2 = \frac{W}{g}\bar{k}^2 = \frac{12 \text{ lb}}{32.2 \text{ ft/s}^2}(\tfrac{8}{12} \text{ ft})^2 = 0.1656 \text{ lb·ft·s}^2$$

Since the system of external forces is equivalent to the system of inertial terms, you have

$$+\gamma\Sigma M_G = \dot{H}_G:$$

$$(10 \text{ lb})(\tfrac{6}{12} \text{ ft}) - (5 \text{ lb})(\tfrac{10}{12} \text{ ft}) = +\bar{I}\alpha + m_Ba_B(\tfrac{6}{12} \text{ ft}) + m_Aa_A(\tfrac{10}{12} \text{ ft})$$

$$(10)(\tfrac{6}{12}) - (5)(\tfrac{10}{12}) = 0.1656\alpha + \tfrac{10}{32.2}(\tfrac{6}{12}\alpha)(\tfrac{6}{12}) + \tfrac{5}{32.2}(\tfrac{10}{12}\alpha)(\tfrac{10}{12})$$

$$\alpha = +2.374 \text{ rad/s}^2 \qquad\qquad \alpha = 2.37 \text{ rad/s}^2 \gamma \quad \blacktriangleleft$$

$$a_A = r_A\alpha = (\tfrac{10}{12} \text{ ft})(2.374 \text{ rad/s}^2) \qquad \mathbf{a}_A = 1.978 \text{ ft/s}^2 \uparrow \quad \blacktriangleleft$$

$$a_B = r_B\alpha = (\tfrac{6}{12} \text{ ft})(2.374 \text{ rad/s}^2) \qquad \mathbf{a}_B = 1.187 \text{ ft/s}^2 \downarrow \quad \blacktriangleleft$$

REFLECT and THINK: You could also solve this problem by considering the pulley and each block as separate systems, but you would have more resulting equations. You would have to use this approach if you wanted to know the forces in the cables.

Fig. 1 Assumed directions for the angular acceleration and the acceleration of the center of mass.

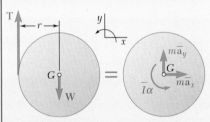

Fig. 2 Free-body diagram and kinetic diagram for the disk.

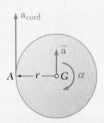

Fig. 3 Acceleration of points A and G on the disk.

Sample Problem 16.5

A cord is wrapped around a homogeneous disk with a radius of $r = 0.5$ m and a mass of $m = 15$ kg. If the cord is pulled upward with a force **T** of magnitude 180 N, determine (a) the acceleration of the center of the disk, (b) the angular acceleration of the disk, (c) the acceleration of the cord.

STRATEGY: Since you have forces and are interested in determining accelerations, use Newton's second law.

MODELING: Choose the disk and the cord as your system. Assume that the components $\bar{\mathbf{a}}_x$ and $\bar{\mathbf{a}}_y$ of the acceleration of the center are directed, respectively, to the right and upward and that the angular acceleration of the disk is counterclockwise (Fig. 1). A free-body diagram and kinetic diagram for this system are shown in Fig. 2. The external forces acting on the disk consist of its weight **W** and the force **T** exerted by the cord.

ANALYSIS:

Equations of Motion. Applying Newton's second law in the x and y directions gives

$$\xrightarrow{+}\Sigma F_x = m\bar{a}_x: \qquad\qquad 0 = m\bar{a}_x \qquad\qquad \bar{\mathbf{a}}_x = \mathbf{0} \quad \blacktriangleleft$$
$$+\uparrow\Sigma F_y = m\bar{a}_y: \qquad\qquad T - W = m\bar{a}_y$$
$$\bar{a}_y = \frac{T - W}{m}$$

Since $T = 180$ N, $m = 15$ kg, and $W = (15\ \text{kg})(9.81\ \text{m/s}^2) = 147.1$ N, you have

$$\bar{a}_y = \frac{180\ \text{N} - 147.1\ \text{N}}{15\ \text{kg}} = +2.19\ \text{m/s}^2 \qquad \bar{\mathbf{a}}_y = 2.19\ \text{m/s}^2\uparrow \quad \blacktriangleleft$$

Now taking moments about the center of gravity, you get

$$+\upharpoonleft\Sigma M_G = \bar{I}\alpha: \qquad\qquad -Tr = \bar{I}\alpha$$
$$-Tr = (\tfrac{1}{2}mr^2)\alpha$$
$$\alpha = -\frac{2T}{mr} = -\frac{2(180\ \text{N})}{(15\ \text{kg})(0.5\ \text{m})} = -48.0\ \text{rad/s}^2$$
$$\boldsymbol{\alpha} = 48.0\ \text{rad/s}^2 \downdownarrows \quad \blacktriangleleft$$

Acceleration of Cord. The acceleration of the cord is equal to the tangential component of the acceleration of point A on the disk, so you have (Fig. 3)

$$\mathbf{a}_{\text{cord}} = (\mathbf{a}_A)_t = \bar{\mathbf{a}} + (\mathbf{a}_{A/G})_t$$
$$= [2.19\ \text{m/s}^2\uparrow] + [(0.5\ \text{m})(48\ \text{rad/s}^2)\uparrow]$$
$$\mathbf{a}_{\text{cord}} = 26.2\ \text{m/s}^2\uparrow \quad \blacktriangleleft$$

REFLECT and THINK: The angular acceleration is clockwise, as we would expect. A similar analysis would apply in many practical situations, such as pulling wire off a spool or paper off a roll. In such cases, you would need to be sure that the tension pulling on the disk is not larger than the tensile strength of the material.

Sample Problem 16.6

A uniform sphere with mass m and radius r is projected along a rough horizontal surface with a linear velocity $\bar{\mathbf{v}}_0$ and no angular velocity. Denoting the coefficient of kinetic friction between the sphere and the floor by μ_k, determine (a) the time t_1 at which the sphere starts rolling without sliding, (b) the linear velocity and angular velocity of the sphere at time t_1.

STRATEGY: Since you have forces acting on the sphere, use Newton's second law. To relate the acceleration to the velocity, you need to use the basic kinematic relationships. The sphere starts out rotating and sliding; it stops sliding when the instantaneous point of contact with the ground has a velocity of zero.

MODELING: Choose the sphere as your system and model it as a rigid body. The assumed positive directions for the acceleration of the mass center and the angular acceleration are shown in Fig. 1. Free-body and kinetic diagrams for this system are shown in Fig. 2. Since the point of the sphere in contact with the surface is sliding to the right, the friction force **F** is directed to the left. While the sphere is sliding, the magnitude of the friction force is $F = \mu_k N$.

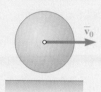

Fig. 1 Assumed directions for the angular and linear acceleration of the sphere.

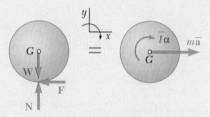

Fig. 2 Free-body diagram and kinetic diagram for the sphere.

ANALYSIS:

Equations of Motion. Applying Newton's second law in the x and y directions gives

$$+\uparrow \Sigma F_y = m\bar{a}_y: \qquad N - W = 0$$
$$N = W = mg \qquad F = \mu_k N = \mu_k mg$$

$$\xrightarrow{+} \Sigma F_x = m\bar{a}_x: \qquad -F = m\bar{a} \qquad -\mu_k mg = m\bar{a} \qquad \bar{a} = -\mu_k g$$

Now taking moments about the center of gravity, you get

$$+\downarrow \Sigma M_G = \bar{I}\alpha: \qquad Fr = \bar{I}\alpha$$

Noting that $\bar{I} = \frac{2}{5}mr^2$ and substituting the given value for F, you have

$$(\mu_k mg)r = \frac{2}{5}mr^2\alpha \qquad \alpha = \frac{5}{2}\frac{\mu_k g}{r}$$

Kinematics of Motion.   As long as the sphere both rotates and slides, its linear and angular accelerations are constant. Therefore, you can use the constant-acceleration equations to relate these accelerations to the linear velocity and angular velocity.

$$t = 0, \bar{v} = \bar{v}_0 \qquad \bar{v} = \bar{v}_0 + \bar{a}t = \bar{v}_0 - \mu_k g t \tag{1}$$

$$t = 0, \omega_0 = 0 \qquad \omega = \omega_0 + \alpha t = 0 + \left(\frac{5}{2}\frac{\mu_k g}{r}\right)t \tag{2}$$

The sphere starts rolling without sliding when the velocity $\mathbf{v}_C$ of the point of contact C is zero (Fig. 3). At that time, $t = t_1$, point C becomes the instantaneous center of rotation, and you have

$$\bar{v}_1 = r\omega_1 \tag{3}$$

Substituting in Eq. (3) the values obtained for v_1 and ω_1 by making $t = t_1$ in Eqs. (1) and (2), respectively, you obtain

$$\bar{v}_0 - \mu_k g t_1 = r\left(\frac{5}{2}\frac{\mu_k g}{r} t_1\right) \qquad t_1 = \frac{2}{7}\frac{\bar{v}_0}{\mu_k g} \;\blacktriangleleft$$

Substituting for t_1 into Eq. (2), you have

$$\omega_1 = \frac{5}{2}\frac{\mu_k g}{r} t_1 = \frac{5}{2}\frac{\mu_k g}{r}\left(\frac{2}{7}\frac{\bar{v}_0}{\mu_k g}\right) \qquad \omega_1 = \frac{5}{7}\frac{\bar{v}_0}{r} \qquad \boldsymbol{\omega}_1 = \frac{5}{7}\frac{\bar{v}_0}{r} \;\downarrow \;\blacktriangleleft$$

$$\bar{v}_1 = r\omega_1 = r\left(\frac{5}{7}\frac{\bar{v}_0}{r}\right) \qquad \bar{v}_1 = \tfrac{5}{7}\bar{v}_0 \qquad \mathbf{v}_1 = \tfrac{5}{7}\bar{v}_0 \;\rightarrow \;\blacktriangleleft$$

REFLECT and THINK: Notice we chose a different coordinate system then we usually do, with the positive rotation going clockwise. This means that you will not be able to use vector algebra solutions since it is not a right-handed coordinate system.

You could use this type of analysis to determine how long it takes a bowling ball to begin to roll without slip or to see how the coefficient of friction affects this motion. Instead of taking moments about the center of gravity, you could have chosen to take moments about point C, in which case your third equation would have been $\Sigma M_C = \dot{H}_C \longrightarrow 0 = m\bar{a}r + \bar{I}\alpha$.

Fig. 3 The point of contact has zero velocity when the sphere starts rolling.

SOLVING PROBLEMS
ON YOUR OWN

This chapter deals with the *plane motion* of **rigid bodies**, and in this first section, we considered rigid bodies that are free to move under the action of applied forces.

1. **Free-body diagram and kinetic diagram.** After choosing a system, your first step in the solution of a problem is to draw a *free-body diagram and a kinetic diagram*.

    a. A free-body diagram shows **the forces exerted on the body**, including the applied forces and moments, the reactions at the supports, and the weight of the body.

    b. A kinetic diagram shows the **inertial terms**: vector $m\bar{\mathbf{a}}$ and the couple $\bar{I}\boldsymbol{\alpha}$.

2. **Using your free-body diagram and kinetic diagram, generate the equations of motion for the system.** Drawing good free-body and kinetic diagrams will allow you to *sum components in any direction and to sum moments about any point*. For a single body, you can obtain a maximum of three independent equations (two translational and one moment) that can be used to help analyze the system. Noting that the external forces and moments are equivalent to the inertial terms, we wrote

$$\Sigma F_x = m\bar{a}_x \quad \Sigma F_y = m\bar{a}_y$$

$$\Sigma M_G = \bar{I}\alpha \ \text{ or } \ \Sigma M_P = \bar{I}\alpha + m\bar{a}d_\perp \ \text{ or } \ \Sigma M_P = \bar{I}\alpha + \mathbf{r}_{G/P} \times m\bar{\mathbf{a}} \quad \textbf{(16.6′)}$$

where G is the center of mass of the body, P is any arbitrary point, and $d_\perp$ is the perpendicular distance between point P and the line of action of the acceleration of the center of mass.

3. **Apply kinematic relationships.** Often, you will have more than three unknowns and will need to generate additional equations. You can usually do this by applying kinematic relationships, such as $a_n = r\omega^2$ and $a_t = r\alpha$ or for a rigid body undergoing fixed-axis rotation or the more general expression relating the acceleration of two points on a rigid body, as

$$\mathbf{a}_B = \mathbf{a}_A + \alpha\mathbf{k} \times \mathbf{r}_{B/A} - \omega^2\mathbf{r}_{B/A} \quad \textbf{(15.21′)}$$

4. **Plane motion of a rigid body.** The problems that you will be asked to solve will fall into one of the following categories.

    a. Rigid body in translation. For a body in translation, the angular acceleration is zero. The kinetic diagram, therefore, is simply the vector $m\bar{\mathbf{a}}$ applied at the mass center [Sample Probs. 16.1 through 16.3].

b. Rigid body in centroidal rotation. For a body in centroidal rotation, the linear acceleration of the mass center is zero. Therefore, the kinetic diagram is simply the couple $\bar{I}\alpha$ [Sample Prob. 16.4].

c. Rigid body in general plane motion. You can consider the general plane motion of a rigid body to be the sum of a translation and a centroidal rotation. The kinetic diagram contains the vector $m\bar{a}$ and the couple $\bar{I}\alpha$ [Sample Probs. 16.5 and 16.6].

5. Plane motion of a system of rigid bodies. You first should draw a free-body diagram and a kinetic diagram that includes all of the rigid bodies of the system. A vector $m\bar{a}$ and a couple $\bar{I}\alpha$ are attached to each body. However, the forces exerted on each other by the various bodies of the system can be omitted, since they occur in pairs of equal and opposite forces.

a. If no more than three unknowns are involved, you can use the free-body and kinetic diagrams to sum components in any direction and sum moments about any point, obtaining equations that can be solved for the desired unknowns [Sample Prob. 16.4].

b. If more than three unknowns are involved, you must choose a new system, use kinematics, or use additional information in the problem statement to find additional equations.

Problems

CONCEPT QUESTIONS

16.CQ1 Two pendulums, A and B, with the masses and lengths shown are released from rest. Which system has a larger mass moment of inertia about its pivot point?

a. A

b. B

c. They are the same.

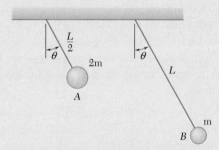

Fig. P16.CQ1 and P16.CQ2

16.CQ2 Two pendulums, A and B, with the masses and lengths shown are released from rest. Which system has a larger angular acceleration immediately after release?

a. A

b. B

c. They are the same.

16.CQ3 Two solid cylinders, A and B, have the same mass m and the radii $2r$ and r, respectively. Each is accelerated from rest with a force applied as shown. In order to impart identical angular accelerations to both cylinders, what is the relationship between F_1 and F_2?

a. $F_1 = 0.5F_2$

b. $F_1 = F_2$

c. $F_1 = 2F_2$

d. $F_1 = 4F_2$

e. $F_1 = 8F_2$

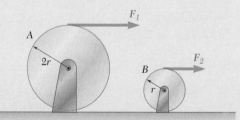

Fig. P16.CQ3

FREE-BODY PRACTICE PROBLEMS

16.F1 A 6-ft board is placed in a truck with one end resting against a block secured to the floor and the other leaning against a vertical partition. Draw the FBD and KD necessary to determine the maximum allowable acceleration of the truck if the board is to remain in the position shown.

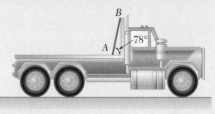

Fig. P16.F1

16.F2 A uniform circular plate of mass 3 kg is attached to two links *AC* and *BD* of the same length. Knowing that the plate is released from rest in the position shown, in which lines joining *G* to *A* and *B* are, respectively, horizontal and vertical, draw the FBD and KD for the plate.

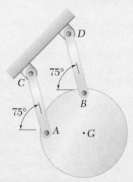

Fig. P16.F2

16.F3 Two uniform disks and two cylinders are assembled as indicated. Disk *A* weighs 20 lb and disk *B* weighs 12 lb. Knowing that the system is released from rest, draw the FBD and KD for the whole system.

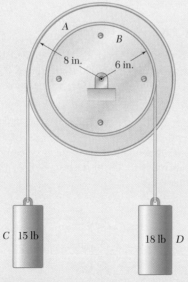

Fig. P16.F3

16.F4 The 400-lb crate shown is lowered by means of two overhead cranes. Knowing the tension in each cable, draw the FBD and KD that can be used to determine the angular acceleration of the crate and the acceleration of the center of gravity.

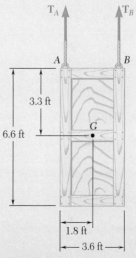

Fig. P16.F4

END-OF-SECTION PROBLEMS

16.1 A 60-lb uniform thin panel is placed in a truck with end A resting on a rough horizontal surface and end B supported by a smooth vertical surface. Knowing that the deceleration of the truck is 12 ft/s^2, determine (a) the reactions at ends A and B, (b) the minimum required coefficient of static friction at end A.

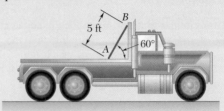

Fig. P16.1 and P16.2

16.2 A 60-lb uniform thin panel is placed in a truck with end A resting on a rough horizontal surface and end B supported by a smooth vertical surface. Knowing that the panel remains in the position shown, determine (a) the maximum allowable acceleration of the truck, (b) the corresponding minimum required coefficient of static friction at end A.

16.3 Knowing that the coefficient of static friction between the tires and the road is 0.80 for the automobile shown, determine the maximum possible acceleration on a level road, assuming (a) four-wheel drive, (b) rear-wheel drive, (c) front-wheel drive.

Fig. P16.3

16.4 The motion of the 2.5-kg rod AB is guided by two small wheels that roll freely in horizontal slots. If a force **P** of magnitude 8 N is applied at B, determine (a) the acceleration of the rod, (b) the reactions at A and B.

16.5 A uniform rod BC of mass 4 kg is connected to a collar A by a 250-mm cord AB. Neglecting the mass of the collar and cord, determine (a) the smallest constant acceleration $\mathbf{a}_A$ for which the cord and the rod will lie in a straight line, (b) the corresponding tension in the cord.

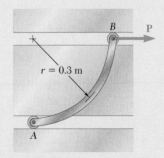

Fig. P16.4

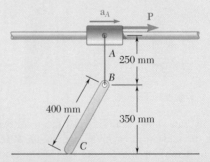

Fig. P16.5

16.6 A 2000-kg truck is being used to lift a 400-kg boulder *B* that is on a 50-kg pallet *A*. Knowing the acceleration of the rear-wheel-drive truck is 1 m/s², determine (*a*) the reaction at each of the front wheels, (*b*) the force between the boulder and the pallet.

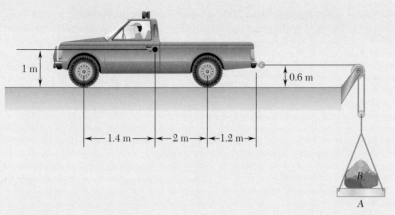

1 m

0.6 m

1.4 m — 2 m — 1.2 m

B

A

Fig. P16.6

16.7 The support bracket shown is used to transport a cylindrical can from one elevation to another. Knowing that $\mu_s = 0.25$ between the can and the bracket, determine (*a*) the magnitude of the upward acceleration **a** for which the can will slide on the bracket, (*b*) the smallest ratio *h/d* for which the can will tip before it slides.

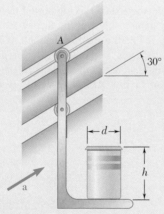

A

30°

d

h

a

Fig. *P16.7*

16.8 Solve Prob. 16.7, assuming that the acceleration **a** of the bracket is directed downward.

16.9 A 20-kg cabinet is mounted on casters that allow it to move freely ($\mu = 0$) on the floor. If a 100-N force is applied as shown, determine (*a*) the acceleration of the cabinet, (*b*) the range of values of *h* for which the cabinet will not tip.

16.10 Solve Prob. 16.9, assuming that the casters are locked and slide on the rough floor ($\mu_k = 0.25$).

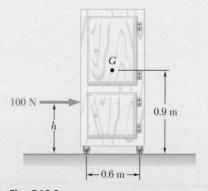

G

100 N

h

0.9 m

0.6 m

Fig. P16.9

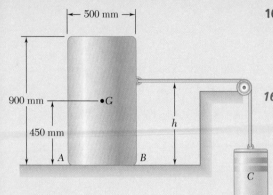

Fig. P16.11

16.11 A completely filled barrel and its contents have a combined mass of 90 kg. A cylinder C is connected to the barrel at a height $h = 550$ mm as shown. Knowing $\mu_s = 0.40$ and $\mu_k = 0.35$, determine the maximum mass of C so the barrel will not tip.

16.12 A 40-kg vase has a 200-mm-diameter base and is being moved using a 100-kg utility cart as shown. The cart moves freely ($\mu = 0$) on the ground. Knowing the coefficient of static friction between the vase and the cart is $\mu_s = 0.4$, determine the maximum force $\mathbf{F}$ that can be applied if the vase is not to slide or tip.

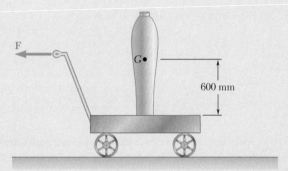

Fig. P16.12

16.13 The retractable shelf shown is supported by two identical linkage-and-spring systems; only one of the systems is shown. A 20-kg machine is placed on the shelf so that half of its weight is supported by the system shown. If the springs are removed and the system is released from rest, determine (a) the acceleration of the machine, (b) the tension in link AB. Neglect the weight of the shelf and links.

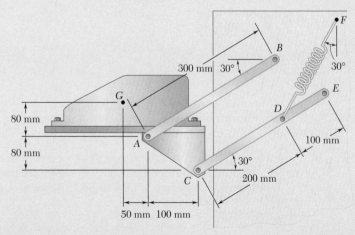

Fig. P16.13

16.14 Bars *AB* and *BE*, each with a mass of 4 kg, are welded together and are pin-connected to two links *AC* and *BD*. Knowing that the assembly is released from rest in the position shown and neglecting the masses of the links, determine (*a*) the acceleration of the assembly, (*b*) the forces in the links.

16.15 At the instant shown, the tensions in the vertical ropes *AB* and *DE* are 300 N and 200 N, respectively. Knowing that the mass of the uniform bar *BE* is 5 kg, determine, at this instant, (*a*) the force **P**, (*b*) the magnitude of the angular velocity of each rope, (*c*) the angular acceleration of each rope.

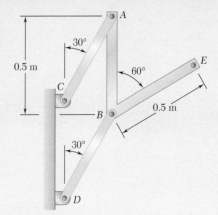

Fig. P16.14

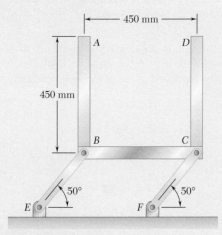

Fig. P16.15

16.16 Three bars, each of mass 3 kg, are welded together and pin-connected to two links *BE* and *CF*. Neglecting the weight of the links, determine the force in each link immediately after the system is released from rest.

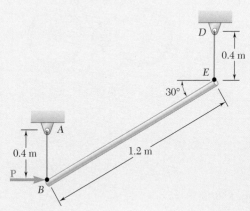

Fig. P16.16

16.17 Members *ACE* and *DCB* are each 600 mm long and are connected by a pin at *C*. The mass center of the 10-kg member *AB* is located at *G*. Determine (*a*) the acceleration of *AB* immediately after the system has been released from rest in the position shown, (*b*) the corresponding force exerted by roller *A* on member *AB*. Neglect the weight of members *ACE* and *DCB*.

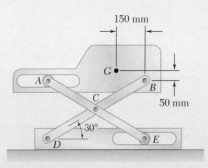

Fig. P16.17

1133

16.18 A prototype rotating bicycle rack is designed to save space at a train station. The combined weight of platform *BD* and the bicycle is 40 lbs and is centered 1 ft above the midpoint of the platform. The motor at *A* causes the support beam *AB* to have an angular velocity of 10 rpm and zero angular acceleration at $\theta = 30°$. At this instant, determine the vertical components of the forces exerted on platform *BD* by the pins at *B* and *D*.

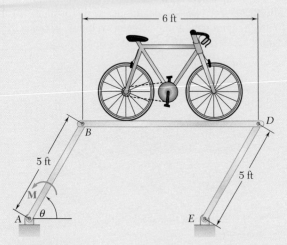

Fig. P16.18

16.19 The triangular weldment *ABC* is guided by two pins that slide freely in parallel curved slots of radius 6 in. cut in a vertical plate. The weldment weighs 16 lb and its mass center is located at point *G*. Knowing that at the instant shown the velocity of each pin is 30 in./s downward along the slots, determine (*a*) the acceleration of the weldment, (*b*) the reactions at *A* and *B*.

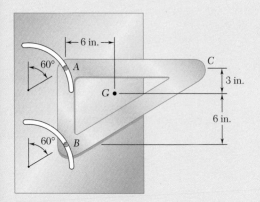

Fig. P16.19

16.20 The coefficients of friction between the 30-lb block and the 5-lb platform *BD* are $\mu_s = 0.50$ and $\mu_k = 0.40$. Determine the accelerations of the block and of the platform immediately after wire *AB* has been cut.

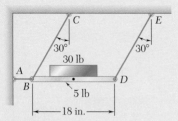

Fig. P16.20

16.21 Draw the shear and bending-moment diagrams for the vertical rod *AB* of Prob. 16.16.

*16.22 Draw the shear and bending-moment diagrams for each of the bars *AB* and *BE* of Prob. 16.14.

16.23 For a rigid body in translation, show that the system of the inertial terms consists of vectors $(\Delta m_i)\bar{\mathbf{a}}$ attached to the various particles of the body, where $\bar{\mathbf{a}}$ is the acceleration of the mass center G of the body. Further show, by computing their sum and the sum of their moments about G, that the inertial terms reduce to a single vector $m\bar{\mathbf{a}}$ attached at G.

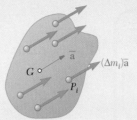

Fig. P16.23

16.24 For a rigid body in centroidal rotation, show that the system of the inertial terms consists of vectors $-(\Delta m_i)\omega^2\mathbf{r}_i'$ and $(\Delta m_i)(\boldsymbol{\alpha} \times \mathbf{r}_i')$ attached to the various particles P_i of the body, where $\boldsymbol{\omega}$ and $\boldsymbol{\alpha}$ are the angular velocity and angular acceleration of the body, and where $\mathbf{r}_i'$ denotes the position vector of the particle P_i relative to the mass center G of the body. Further show, by computing their sum and the sum of their moments about G, that the inertial terms reduce to a couple $\bar{I}\boldsymbol{\alpha}$.

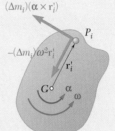

Fig. P16.24

16.25 It takes 10 min for a 2.4-Mg flywheel to coast to rest from an angular velocity of 300 rpm. Knowing that the radius of gyration of the flywheel is 1 m, determine the average magnitude of the couple due to kinetic friction in the bearing.

16.26 The rotor of an electric motor has an angular velocity of 3600 rpm when the load and power are cut off. The 120-lb rotor, which has a centroidal radius of gyration of 9 in., then coasts to rest. Knowing that kinetic friction results in a couple of magnitude 2.5 lb·ft exerted on the rotor, determine the number of revolutions that the rotor executes before coming to rest.

16.27 The 8-in.-radius brake drum is attached to a larger flywheel that is not shown. The total mass moment of inertia of the drum and the flywheel is 14 lb·ft·s² and the coefficient of kinetic friction between the drum and the brake shoe is 0.35. Knowing that the angular velocity of the flywheel is 360 rpm counterclockwise when a force **P** of magnitude 75 lb is applied to the pedal C, determine the number of revolutions executed by the flywheel before it comes to rest.

16.28 Solve Prob. 16.27, assuming that the initial angular velocity of the flywheel is 360 rpm clockwise.

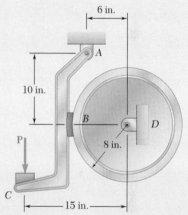

Fig. P16.27

16.29 The 100-mm-radius brake drum is attached to a flywheel that is not shown. The drum and flywheel together have a mass of 300 kg and a radius of gyration of 600 mm. The coefficient of kinetic friction between the brake band and the drum is 0.30. Knowing that a force **P** of magnitude 50 N is applied at A when the angular velocity is 180 rpm counterclockwise, determine the time required to stop the flywheel when $a = 200$ mm and $b = 160$ mm.

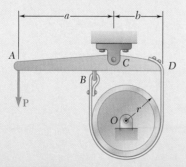

Fig. P16.29

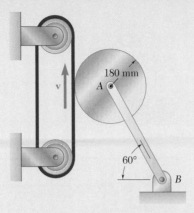

Fig. P16.30

16.30 The 180-mm-radius disk is at rest when it is placed in contact with a belt moving at a constant speed. Neglecting the weight of the link *AB* and knowing that the coefficient of kinetic friction between the disk and the belt is 0.40, determine the angular acceleration of the disk while slipping occurs.

16.31 Solve Prob. 16.30, assuming that the direction of motion of the belt is reversed.

16.32 In order to determine the mass moment of inertia of a flywheel of radius 600 mm, a 12-kg block is attached to a wire that is wrapped around the flywheel. The block is released and is observed to fall 3 m in 4.6 s. To eliminate bearing friction from the computation, a second block of mass 24 kg is used and is observed to fall 3 m in 3.1 s. Assuming that the moment of the couple due to friction remains constant, determine the mass moment of inertia of the flywheel.

Fig. *P16.32* and P16.33

16.33 The flywheel shown has a radius of 20 in., a weight of 250 lb, and a radius of gyration of 15 in. A 30-lb block *A* is attached to a wire that is wrapped around the flywheel, and the system is released from rest. Neglecting the effect of friction, determine (*a*) the acceleration of block *A*, (*b*) the speed of block *A* after it has moved 5 ft.

16.34 Each of the double pulleys shown has a mass moment of inertia of 15 lb·ft·s² and is initially at rest. The outside radius is 18 in., and the inner radius is 9 in. Determine (*a*) the angular acceleration of each pulley, (*b*) the angular velocity of each pulley after point *A* on the cord has moved 10 ft.

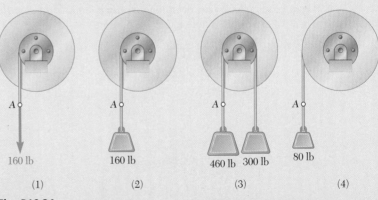

Fig. P16.34

16.35 Each of the gears A and B has a mass of 9 kg and has a radius of gyration of 200 mm; gear C has a mass of 3 kg and has a radius of gyration of 75 mm. If a couple $\mathbf{M}$ of constant magnitude 5 N·m is applied to gear C, determine (a) the angular acceleration of gear A, (b) the tangential force that gear C exerts on gear A.

16.36 Solve Prob. 16.35, assuming that the couple $\mathbf{M}$ is applied to disk A.

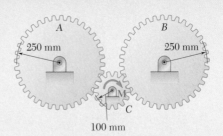
Fig. P16.35

16.37 Gear A weighs 1 lb and has a radius of gyration of 1.3 in.; gear B weighs 6 lb and has a radius of gyration of 3 in.; gear C weighs 9 lb and has a radius of gyration of 4.3 in. Knowing a couple $\mathbf{M}$ of constant magnitude of 40 lb·in. is applied to gear A, determine (a) the angular acceleration of gear C, (b) the tangential force that gear B exerts on gear C.

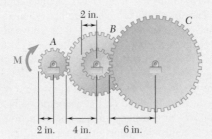
Fig. P16.37

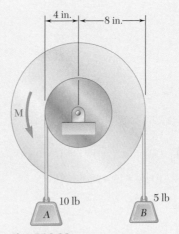
Fig. P16.38

16.38 The 25-lb double pulley shown is at rest and in equilibrium when a constant 3.5 lb·ft couple $\mathbf{M}$ is applied. Neglecting the effect of friction and knowing that the radius of gyration of the double pulley is 6 in., determine (a) the angular acceleration of the double pulley, (b) the tension in each rope.

16.39 A belt of negligible mass passes between cylinders A and B and is pulled to the right with a force $\mathbf{P}$. Cylinders A and B weigh, respectively, 5 and 20 lb. The shaft of cylinder A is free to slide in a vertical slot and the coefficients of friction between the belt and each of the cylinders are $\mu_s = 0.50$ and $\mu_k = 0.40$. For $P = 3.6$ lb, determine (a) whether slipping occurs between the belt and either cylinder, (b) the angular acceleration of each cylinder.

16.40 Solve Prob. 16.39 for $P = 2.00$ lb.

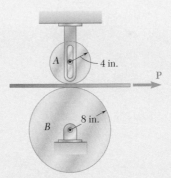
Fig. P16.39

16.41 Disk A has a mass of 6 kg and an initial angular velocity of 360 rpm clockwise; disk B has a mass of 3 kg and is initially at rest. The disks are brought together by applying a horizontal force of magnitude 20 N to the axle of disk A. Knowing that $\mu_k = 0.15$ between the disks and neglecting bearing friction, determine (a) the angular acceleration of each disk, (b) the final angular velocity of each disk.

16.42 Solve Prob. 16.41, assuming that initially disk A is at rest and disk B has an angular velocity of 360 rpm clockwise.

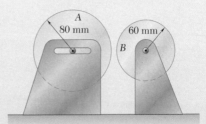
Fig. P16.41

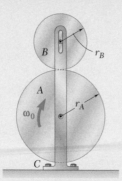

Fig. P16.43 and P16.44

16.43 Disk A has a mass $m_A = 4$ kg, a radius $r_A = 300$ mm, and an initial angular velocity $\omega_0 = 300$ rpm clockwise. Disk B has a mass $m_B = 1.6$ kg, a radius $r_B = 180$ mm, and is at rest when it is brought into contact with disk A. Knowing that $\mu_k = 0.35$ between the disks and neglecting bearing friction, determine (a) the angular acceleration of each disk, (b) the reaction at the support C.

16.44 Disk B is at rest when it is brought into contact with disk A, which has an initial angular velocity ω_0. (a) Show that the final angular velocities of the disks are independent of the coefficient of friction μ_k between the disks as long as $\mu_k \neq 0$. (b) Express the final angular velocity of disk A in terms of ω_0 and the ratio of the masses of the two disks m_A/m_B.

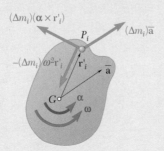

Fig. P16.45

16.45 Cylinder A has an initial angular velocity of 720 rpm clockwise, and cylinders B and C are initially at rest. Disks A and B each weigh 5 lb and have radius $r = 4$ in. Disk C weighs 20 lb and has a radius of 8 in. The disks are brought together when C is placed gently onto A and B. Knowing that $\mu_k = 0.25$ between A and C and no slipping occurs between B and C, determine (a) the angular acceleration of each disk, (b) the final angular velocity of each disk.

16.46 Show that the system of the inertial terms for a rigid body in plane motion reduces to a single vector, and express the distance from the mass center G of the body to the line of action of this vector in terms of the centroidal radius of gyration $\bar{k}$ of the body, the magnitude $\bar{a}$ of the acceleration of G, and the angular acceleration α.

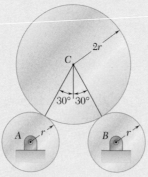

Fig. P16.47

16.47 For a rigid body in plane motion, show that the system of the inertial terms consists of vectors $(\Delta m_i)\bar{\mathbf{a}}$, $-(\Delta m_i)\omega^2 \mathbf{r}'_i$, and $(\Delta m_i)(\boldsymbol{\alpha} \times \mathbf{r}'_i)$ attached to the various particles P_i of the body, where $\bar{\mathbf{a}}$ is the acceleration of the mass center G of the body, $\boldsymbol{\omega}$ is the angular velocity of the body, $\boldsymbol{\alpha}$ is its angular acceleration, and $\mathbf{r}'_i$ denotes the position vector of the particle P_i, relative to G. Further show, by computing their sum and the sum of their moments about G, that the inertial terms reduce to a vector $m\bar{\mathbf{a}}$ attached at G and a couple $\bar{I}\boldsymbol{\alpha}$.

16.48 A uniform slender rod AB rests on a frictionless horizontal surface, and a force $\mathbf{P}$ of magnitude 0.25 lb is applied at A in a direction perpendicular to the rod. Knowing that the rod weighs 1.75 lb, determine (a) the acceleration of point A, (b) the acceleration of point B, (c) the location of the point on the bar that has zero acceleration.

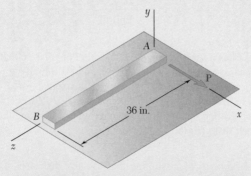

Fig. P16.48

16.49 (*a*) In Prob. 16.48, determine the point of the rod *AB* at which the force **P** should be applied if the acceleration of point *B* is to be zero. (*b*) Knowing that $P = 0.25$ lb, determine the corresponding acceleration of point *A*.

16.50 and 16.51 A force **P** with a magnitude of 3 N is applied to a tape wrapped around the body indicated. Knowing that the body rests on a frictionless horizontal surface, determine the acceleration of (*a*) point *A*, (*b*) point *B*.

    16.50 A thin hoop of mass 2.4 kg.
    16.51 A uniform disk of mass 2.4 kg.

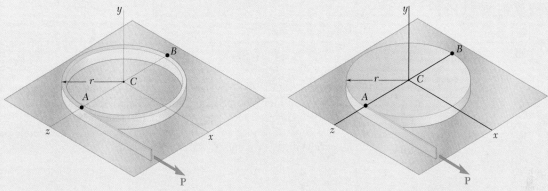

Fig. P16.50   **Fig. P16.51**

16.52 A 250-lb satellite has a radius of gyration of 24 in. with respect to the *y* axis and is symmetrical with respect to the *zx* plane. Its orientation is changed by firing four small rockets *A*, *B*, *C*, and *D*, each of which produces a 4-lb thrust **T** directed as shown. Determine the angular acceleration of the satellite and the acceleration of its mass center *G* (*a*) when all four rockets are fired, (*b*) when all rockets except *D* are fired.

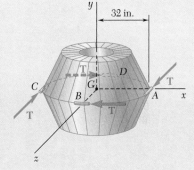

Fig. P16.52

16.53 A rectangular plate of mass 5 kg is suspended from four vertical wires, and a force **P** of magnitude 6 N is applied to corner *C* as shown. Immediately after **P** is applied, determine the acceleration of (*a*) the midpoint of edge *BC*, (*b*) corner *B*.

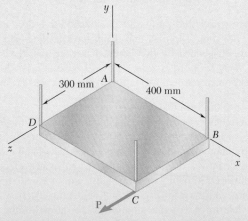

Fig. P16.53

16.54 A uniform semicircular plate with a mass of 6 kg is suspended from three vertical wires at points A, B, and C, and a force **P** with a magnitude of 5 N is applied to point B. Immediately after **P** is applied, determine the acceleration of (a) the mass center of the plate, (b) point C.

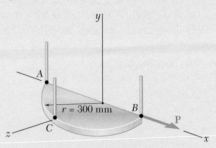

Fig. P16.54

16.55 A drum with a 200-mm radius is attached to a disk with a radius of $r_A = 150$ mm. The disk and drum have a combined mass of 5 kg and a combined radius of gyration of 120 mm and are suspended by two cords. Knowing that $T_A = 35$ N and $T_B = 25$ N, determine the accelerations of points A and B on the cords.

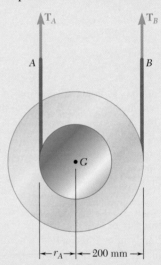

Fig. P16.55 and P16.56

16.56 A drum with a 200-mm radius is attached to a disk with a radius of $r_A = 140$ mm. The disk and drum have a combined mass of 5 kg and are suspended by two cords. Knowing that the acceleration of point B on the cord is zero, $T_A = 40$ N, and $T_B = 20$ N, determine the combined radius of gyration of the disk and drum.

16.57 The 12-lb uniform disk shown has a radius of $r = 3.2$ in. and rotates counterclockwise. Its center C is constrained to move in a slot cut in the vertical member AB, and a 11-lb horizontal force **P** is applied at B to maintain contact at D between the disk and the vertical wall. The disk moves downward under the influence of gravity and the friction at D. Knowing that the coefficient of kinetic friction between the disk and the wall is 0.12 and neglecting friction in the vertical slot, determine (a) the angular acceleration of the disk, (b) the acceleration of the center C of the disk.

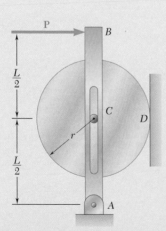

Fig. P16.57

16.58 The steel roll shown has a mass of 1200 kg, a centroidal radius of gyration of 150 mm, and is lifted by two cables looped around its shaft. Knowing that for each cable $T_A = 3100$ N and $T_B = 3300$ N, determine (*a*) the angular acceleration of the roll, (*b*) the acceleration of its mass center.

16.59 The steel roll shown has a mass of 1200 kg, has a centroidal radius of gyration of 150 mm, and is lifted by two cables looped around its shaft. Knowing that at the instant shown the acceleration of the roll is 150 mm/s² downward and that for each cable $T_A = 3000$ N, determine (*a*) the corresponding tension T_B, (*b*) the angular acceleration of the roll.

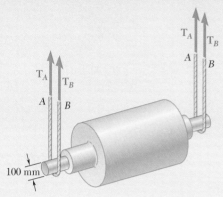

Fig. P16.58 and P16.59

16.60 and 16.61 A 15-ft beam weighing 500 lb is lowered by means of two cables unwinding from overhead cranes. As the beam approaches the ground, the crane operators apply brakes to slow the unwinding motion. Knowing that the deceleration of cable A is 20 ft/s² and the deceleration of cable B is 2 ft/s², determine the tension in each cable.

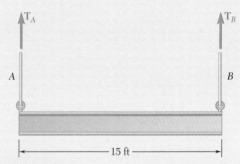

Fig. P16.60

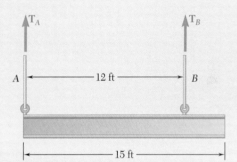

Fig. *P16.61*

16.62 Two uniform cylinders, each of weight $W = 14$ lb and radius $r = 5$ in., are connected by a belt as shown. If the system is released from rest, determine (*a*) the angular acceleration of each cylinder, (*b*) the tension in the portion of belt connecting the two cylinders, (*c*) the velocity of the center of the cylinder A after it has moved through 3 ft.

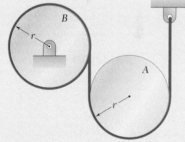

16.63 and 16.64 A beam AB with a mass *m* and of uniform cross section is suspended from two springs as shown. If spring 2 breaks, determine at that instant (*a*) the angular acceleration of the beam, (*b*) the acceleration of point A, (*c*) the acceleration of point B.

Fig. *P16.62*

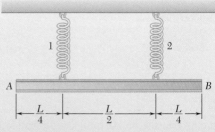

Fig. P16.63

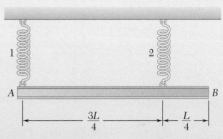

Fig. P16.64

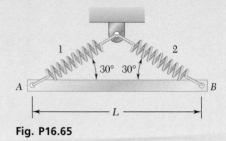

Fig. P16.65

16.65 A uniform slender bar AB with a mass m is suspended from two springs as shown. If spring 2 breaks, determine at that instant (a) the angular acceleration of the bar, (b) the acceleration of point A, (c) the acceleration of point B.

16.66 through 16.68 A thin plate of the shape indicated and of mass m is suspended from two springs as shown. If spring 2 breaks, determine the acceleration at that instant (a) of point A, (b) of point B.

  16.66 A square plate of side b
  16.67 A thin hoop of diameter b
  16.68 A rectangular plate of height b and width a

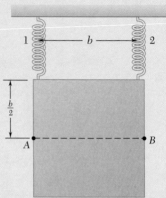

Fig. P16.66

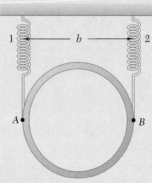

Fig. P16.67

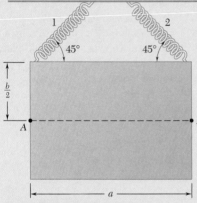

Fig. P16.68

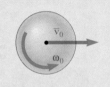

Fig. P16.69

16.69 A sphere of radius r and mass m is projected along a rough horizontal surface with the initial velocities indicated. If the final velocity of the sphere is to be zero, express, in terms of v_0, r, and μ_k, (a) the required magnitude of $\boldsymbol{\omega}_0$, (b) the time t_1 required for the sphere to come to rest, (c) the distance the sphere will move before coming to rest.

16.70 Solve Prob. 16.69, assuming that the sphere is replaced by a uniform thin hoop of radius r and mass m.

16.71 A bowler projects an 8-in.-diameter ball weighing 12 lb along an alley with a forward velocity $\mathbf{v}_0$ of 15 ft/s and a backspin $\boldsymbol{\omega}_0$ of 9 rad/s. Knowing that the coefficient of kinetic friction between the ball and the alley is 0.10, determine (a) the time t_1 at which the ball will start rolling without sliding, (b) the speed of the ball at time t_1, (c) the distance the ball will have traveled at time t_1.

Fig. P16.71

16.72 Solve Prob. 16.71, assuming that the bowler projects the ball with the same forward velocity but with a backspin of 18 rad/s.

16.73 A uniform sphere of radius r and mass m is placed with no initial velocity on a belt that moves to the right with a constant velocity $\mathbf{v}_1$. Denoting by μ_k the coefficient of kinetic friction between the sphere and the belt, determine (a) the time t_1 at which the sphere will start rolling without sliding, (b) the linear and angular velocities of the sphere at time t_1.

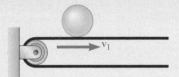

Fig. P16.73

16.74 A sphere of radius r and mass m has a linear velocity $\mathbf{v}_0$ directed to the left and no angular velocity as it is placed on a belt moving to the right with a constant velocity $\mathbf{v}_1$. If after first sliding on the belt the sphere is to have no linear velocity relative to the ground as it starts rolling on the belt without sliding, determine in terms of v_1 and the coefficient of kinetic friction μ_k between the sphere and the belt (a) the required value of v_0, (b) the time t_1 at which the sphere will start rolling on the belt, (c) the distance the sphere will have moved relative to the ground at time t_1.

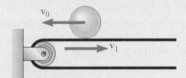

Fig. P16.74

16.2 CONSTRAINED PLANE MOTION

Most engineering applications deal with rigid bodies that are moving under given constraints. For example, cranks must rotate about a fixed axis, wheels must roll without sliding, and connecting rods must describe certain prescribed motions. In all such cases, definite relations exist between the components of the acceleration $\bar{\mathbf{a}}$ of the mass center G of the body considered and its angular acceleration $\boldsymbol{\alpha}$. The corresponding motion is said to be a **constrained motion**.

As discussed in the previous section, we draw our free-body and kinetic diagrams (Fig. 16.13) and then write the equations of motion. The solution of a problem involving a constrained plane motion also calls for a *kinematic analysis* of the problem. Consider, for example, a slender rod AB with a length l and a mass m, where the extremities are connected to blocks of negligible mass that slide along horizontal and vertical friction-less tracks. The rod is pulled by a force $\mathbf{P}$ applied at A (Fig. 16.12). We know from Sec. 15.4A that we can determine the acceleration $\bar{\mathbf{a}}$ of the mass center G of the rod at any given instant from the position of the rod, its angular velocity, and its angular acceleration at that instant. Suppose, for example, that we know the values of θ, ω, and α at a given instant, and we wish to determine the corresponding value of the force $\mathbf{P}$ as well as the reactions at A and B. We should first *determine the components $\bar{a}_x$ and $\bar{a}_y$ of the acceleration of the mass center G* using the method in Sec. 15.4A. We next solve our equations of motion using the expressions obtained for $\bar{a}_x$ and $\bar{a}_y$. We can then find the unknown forces $\mathbf{P}$, $\mathbf{N}_A$, and $\mathbf{N}_B$ by solving the appropriate equations.

Suppose now that we know the applied force $\mathbf{P}$, the angle θ, and the angular velocity ω of the rod at a given instant and that we wish to find the angular acceleration α of the rod and the components $\bar{a}_x$ and $\bar{a}_y$ of the acceleration of its mass center at that instant, as well as the reactions at A and B. The preliminary kinematic study of the problem will aim *to express the components $\bar{a}_x$ and $\bar{a}_y$ of the acceleration of G in terms of the angular acceleration α of the rod*. This is done by first expressing the acceleration of a suitable reference point such as A in terms of the angular acceleration α. We can then determine the components $\bar{a}_x$ and $\bar{a}_y$ of the acceleration of G in terms of α and carry these expressions into Fig. 16.13. We can then derive three equations in terms of α, N_A, and N_B and solve for the three unknowns (see Sample Prob. 16.12).

When a mechanism consists of several moving parts, we can use the approach just described with each part of the mechanism. The procedure required to determine the various unknowns is then similar to the procedure followed in the case of the equilibrium of a system of connected rigid bodies (Sec. 6.3B).

Earlier, we analyzed two particular cases of constrained plane motion: translation of a rigid body, in which the angular acceleration of the body is constrained to be zero, and centroidal rotation, in which the acceleration $\bar{\mathbf{a}}$ of the mass center of the body is constrained to be zero. Two other particular cases of constrained plane motion are of special interest: *noncentroidal rotation* of a rigid body and *rolling motion* of a disk or wheel. We can analyze these two cases using one of the general

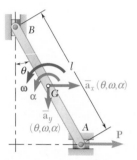

Fig. 16.12 Kinematic variables for a constrained rod pulled to the right.

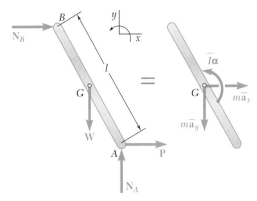

Fig. 16.13 Free-body diagram and kinetic diagram for the rod in Fig. 16.12.

methods described previously. However, in view of the range of their applications, they deserve a few special comments.

Noncentroidal Rotation. The motion of a rigid body constrained to rotate about a fixed axis that does not pass through its mass center is called **noncentroidal rotation**. The mass center G of the body moves along a circle with a radius $\bar{r}$ centered at point O, where the axis of rotation intersects the plane of reference (Fig. 16.14). Denoting the angular velocity and the angular acceleration of the line OG by $\boldsymbol{\omega}$ and $\boldsymbol{\alpha}$, respectively, we obtain the following expressions for the tangential and normal components of the acceleration of G:

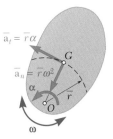

Fig. 16.14 For noncentroidal fixed-axis rotation, the center of mass has a tangential and a normal component of acceleration.

$$\bar{a}_t = \bar{r}\alpha \qquad \bar{a}_n = \bar{r}\omega^2 \tag{16.7}$$

Since line OG belongs to the body, its angular velocity $\boldsymbol{\omega}$ and its angular acceleration $\boldsymbol{\alpha}$ also represent the angular velocity and the angular acceleration of the body. Equations (16.7) define, therefore, the kinematic relation between the motion of the mass center G and the motion of the body about G.

We obtain an interesting relation by equating the moments about the fixed point O of the forces and vectors shown, respectively, in Fig. 16.15a and b. We have

$$+\!\uparrow\! \Sigma M_O = \bar{I}\alpha + (m\bar{r}\alpha)\bar{r} = (\bar{I} + m\bar{r}^2)\alpha$$

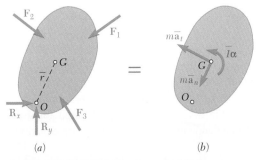

(a)   (b)

Fig. 16.15 Free-body diagram and kinetic diagram for the rigid body in Fig. 16.14.

But according to the parallel-axis theorem, we have $\bar{I} + m\bar{r}^2 = I_O$, where I_O denotes the moment of inertia of the rigid body about the fixed axis. We therefore obtain

**Moments about
a fixed axis**

$$\Sigma M_O = I_O\alpha \tag{16.8}$$

Although formula (16.8) expresses an important relation between the sum of the moments of the external forces about the fixed point O and the product $I_O\alpha$, we will still need to apply Eq. (16.1) to find the forces at O.

A particular case of noncentroidal rotation is of special interest—the case of *uniform rotation,* in which the angular velocity $\boldsymbol{\omega}$ is constant. Since $\boldsymbol{\alpha}$ is zero, the inertia couple in Fig. 16.15 vanishes, and the inertia vector reduces to its normal component. This component (also called *centrifugal force* in layman's terms) represents the tendency of the rigid body to break away from the axis of rotation.

Rolling Motion.   Another important case of plane motion is the motion of a disk or wheel rolling on a plane surface. If the disk is constrained to roll without sliding, the acceleration $\bar{\mathbf{a}}$ of its mass center G and its angular acceleration $\boldsymbol{\alpha}$ are not independent. Assuming that the disk is balanced so that its mass center and its geometric center coincide, the distance $\bar{x}$ traveled by G during a rotation θ of the disk is $\bar{x} = r\theta$, where r is the radius of the disk. Differentiating this relation twice, we have

$$\bar{a} = r\alpha \tag{16.9}$$

Recall that the system of the inertial terms in plane motion reduces to a vector $m\bar{\mathbf{a}}$ and a couple $\bar{I}\boldsymbol{\alpha}$. We find that, in the particular case of the rolling motion of a balanced disk, these terms reduce to a vector of magnitude $mr\alpha$ attached at G and to a couple with a magnitude of $\bar{I}\alpha$. We may thus say that the external forces are equivalent to the vector and couple shown in Fig. 16.16.

When a disk **rolls without sliding**, there is no relative motion between the point of the disk in contact with the ground and the ground itself. Thus, as far as the computation of the friction force $\mathbf{F}$ is concerned, a rolling disk can be compared with a block at rest on a surface. The magnitude F of the friction force can have any value, as long as this value does not exceed the maximum value $F_m = \mu_s N$, where μ_s is the coefficient of static friction and N is the magnitude of the normal force. In the case of a rolling disk, the magnitude F of the friction force therefore should be determined independently of N by solving the equation obtained from Fig. 16.16.

When *sliding is impending,* the friction force reaches its maximum value $F_m = \mu_s N$ and can be obtained after solving for N.

Photo 16.4   As the ball hits the bowling alley, it first spins and slides, then rolls without sliding.

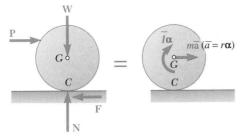

Fig. 16.16 Free-body diagram and kinetic diagram for a disk rolling without slipping on a fixed surface.

When the disk *rotates and slides* at the same time, a relative motion exists between the point of the disk in contact with the ground and the ground itself. The force of friction has the magnitude $F_k = \mu_k N$, where μ_k is the coefficient of kinetic friction. In this case, however, the motion of the mass center G of the disk and the rotation of the disk about G are independent, and $\bar{a}$ is not equal to $r\alpha$.

We can summarize these three different cases as

Rolling, no sliding $F \le \mu_s N$ $\bar{a} = r\alpha$

Rolling, sliding impending $F = \mu_s N$ $\bar{a} = r\alpha$

Rotating and sliding $F = \mu_k N$ $\bar{a}$ and α independent

If you do not know whether or not a disk slides, you should first assume that the disk rolls without sliding. You will then be able to solve your system of equations by assuming that $\bar{a} = r\alpha$. If F is found to be smaller than or equal to $\mu_s N$, the assumption is proved correct. If F is found to be larger than $\mu_s N$, the assumption is incorrect, and you should start the problem again, assuming rotating, sliding, and that $F = \mu_k N$.

When a disk is *unbalanced,* i.e., when its mass center G does not coincide with its geometric center O, the relation in Eq. (16.9) does not hold between $\bar{a}$ and α. However, a similar relation holds between the magnitude a_O of the acceleration of the geometric center and the angular acceleration α of an unbalanced disk that rolls without sliding. We have

$$a_O = r\alpha \tag{16.10}$$

To determine $\bar{a}$ in terms of the angular acceleration α and the angular velocity ω of the disk, we can use the relative-acceleration formula, as

$$\begin{aligned}\bar{\mathbf{a}} = \bar{\mathbf{a}}_G &= \mathbf{a}_O + \mathbf{a}_{G/O} \\ &= \mathbf{a}_O + (\mathbf{a}_{G/O})_t + (\mathbf{a}_{G/O})_n\end{aligned} \tag{16.11}$$

where the three component accelerations have the directions indicated in Fig. 16.17 and the magnitudes $a_O = r\alpha$, $(a_{G/O})_t = (OG)\alpha$, and $(a_{G/O})_n = (OG)\omega^2$. These terms also can be solved using the relationship between two points on a rigid body undergoing plane motion:

$$\bar{\mathbf{a}} = \mathbf{a}_O + \boldsymbol{\alpha} \times \mathbf{r}_{G/O} - \omega^2 \mathbf{r}_{G/O} \tag{16.12}$$

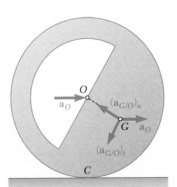

Fig. 16.17 Accelerations of the geometric center O and center of mass G for a rolling unbalanced disk.

Sample Problem 16.7

The portion *AOB* of a mechanism consists of a 400-mm steel rod *OB* welded to a gear *E* with a radius of 120 mm that can rotate about a horizontal shaft *O*. It is actuated by a gear *D* and, at the instant shown, has a clockwise angular velocity of 8 rad/s and a counterclockwise angular acceleration of 40 rad/s². Knowing that rod *OB* has a mass of 3 kg and gear *E* has a mass of 4 kg and a radius of gyration of 85 mm, determine (*a*) the tangential force exerted by gear *D* on gear *E*, (*b*) the components of the reaction at point *O* on the shaft.

STRATEGY: Since you are asked to determine forces, use Newton's second law.

MODELING: For your system, choose the single object that consists of the steel rod *OB* and the gear *E*. Since these two objects are welded together, they have the same angular velocity and angular acceleration. Rather than finding the center of mass for this object, use the center of mass for gear *E* and for rod *OB* separately in your kinetic diagram. Therefore, first determine the components of the acceleration of the mass center G_{OB} of the rod (Fig. 1) as

$$(\bar{a}_{OB})_t = \bar{r}\alpha = (0.200 \text{ m})(40 \text{ rad/s}^2) = 8 \text{ m/s}^2$$
$$(\bar{a}_{OB})_n = \bar{r}\omega^2 = (0.200 \text{ m})(8 \text{ rad/s})^2 = 12.8 \text{ m/s}^2$$

A free-body diagram and kinetic diagram for the system are shown in Fig. 2. The inertial terms on your kinetic diagram include a couple $\bar{I}_E\alpha$ (since gear *E* is in centroidal rotation), a couple $\bar{I}_{OB}\alpha$, and two vector components $m_{OB}(\bar{a}_{OB})_n$ and $m_{OB}(\bar{a}_{OB})_t$ at the mass center of *OB*.

ANALYSIS:

Preliminary Calculations: The magnitudes of the weights are

$$W_E = m_E g = (4 \text{ kg})(9.81 \text{ m/s}^2) = 39.2 \text{ N}$$
$$W_{OB} = m_{OB}g = (3 \text{ kg})(9.81 \text{ m/s}^2) = 29.4 \text{ N}$$

Since you know the accelerations, you can compute the magnitudes of the components and couples on your kinetic diagram, as

$$\bar{I}_E\alpha = m_E\bar{k}_E^2\alpha = (4 \text{ kg})(0.085 \text{ m})^2(40 \text{ rad/s}^2) = 1.156 \text{ N·m}$$
$$m_{OB}(\bar{a}_{OB})_t = (3 \text{ kg})(8 \text{ m/s}^2) = 24.0 \text{ N}$$
$$m_{OB}(\bar{a}_{OB})_n = (3 \text{ kg})(12.8 \text{ m/s}^2) = 38.4 \text{ N}$$
$$\bar{I}_{OB}\alpha = (\tfrac{1}{12}m_{OB}L^2)\alpha = \tfrac{1}{12}(3 \text{ kg})(0.400 \text{ m})^2(40 \text{ rad/s}^2) = 1.600 \text{ N·m}$$

Equations of Motion. Setting the system of the external forces shown in your free-body diagram equal to the inertia terms in your kinetic diagram, you obtain the following equations, which you can solve as

$+\mathord{\uparrow}\Sigma M_O = \dot{H}_O$:

$$F(0.120 \text{ m}) = \bar{I}_E\alpha + m_{OB}(\bar{a}_{OB})_t(0.200 \text{ m}) + \bar{I}_{OB}\alpha$$
$$F(0.120 \text{ m}) = 1.156 \text{ N·m} + (24.0 \text{ N})(0.200 \text{ m}) + 1.600 \text{ N·m}$$

$$F = 63.0 \text{ N} \qquad \mathbf{F} = 63.0 \text{ N}\downarrow \quad \blacktriangleleft$$

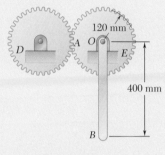

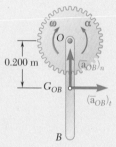

Fig. 1 Acceleration of the center of gravity of the bar.

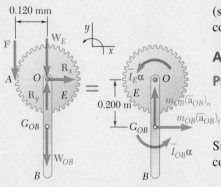

Fig. 2 Free-body diagram and kinetic diagram for the system.

$$\overset{+}{\rightarrow}\Sigma F_x = \Sigma m\bar{a}_x: \qquad R_x = m_{OB}(\bar{a}_{OB})_t$$
$$R_x = 24.0 \text{ N} \qquad \mathbf{R}_x = 24.0 \text{ N} \rightarrow \ \blacktriangleleft$$

$$+\uparrow\Sigma F_y = \Sigma m\bar{a}_y: \qquad R_y - F - W_E - W_{OB} = m_{OB}(\bar{a}_{OB})_n$$
$$R_y - 63.0 \text{ N} - 39.2 \text{ N} - 29.4 \text{ N} = 38.4 \text{ N}$$

$$R_y = 170.0 \text{ N} \qquad \mathbf{R}_y = 170.0 \text{ N} \uparrow \ \blacktriangleleft$$

REFLECT and THINK: When you drew your kinetic diagram, you put your inertia terms at the center of mass for the gear and for the rod. Alternatively, you could have found the center of mass for the system and put the vectors $\bar{I}_{AOB}\alpha$, $m_{AOB}\bar{a}_x$ and $m_{AOB}\bar{a}_y$ on the diagram. Finally, you could have found an overall I_O for the combined gear and rod and used Eq. 16.8 to solve for force F.

Sample Problem 16.8

A 6 × 8 in. rectangular plate weighing 60 lb is suspended from two pins A and B. If pin B is suddenly removed, determine (a) the angular acceleration of the plate, (b) the components of the reaction at pin A immediately after pin B has been removed.

STRATEGY: You are asked to determine forces and the angular acceleration of the plate, so use Newton's second law.

MODELING: Choose the plate to be your system and model it as a rigid body. Observe that as the plate rotates about point A, its mass center G describes a circle with a radius $\bar{r}$ and its center at A (Fig. 1). The free-body diagram and kinetic diagram for this system are shown in Fig. 2. The plate is released from rest ($\omega = 0$), so the normal component of the acceleration of G is zero. The magnitude of the acceleration $\bar{a}$ of the mass center G is thus $\bar{a} = \bar{r}\alpha$.

ANALYSIS:

a. Angular Acceleration. Using your free-body diagram and kinetic diagram, you can take moments about A to find

$$+\downarrow\Sigma M_A = \bar{I}\alpha + m\bar{a}d_\perp: \qquad W\bar{x} = \bar{I}\alpha + (m\bar{a})\bar{r}$$

Since $\bar{a} = \bar{r}\alpha$, you have

$$W\bar{x} = \bar{I}\alpha + (m\bar{r}\alpha)\bar{r} \qquad \alpha = \dfrac{W\bar{x}}{\dfrac{W}{g}\bar{r}^2 + \bar{I}} \qquad (1)$$

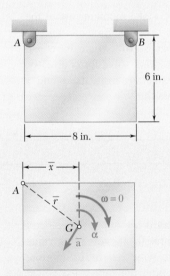

Fig. 1 The plate travels in a circle about A.

6 in.

8 in.

$\bar{x}$

A

$\bar{r}$

$\omega = 0$

G

α

$\bar{a}$

(continued)

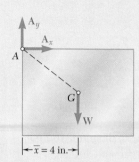

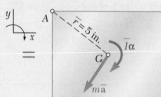

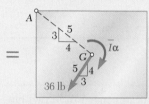

Fig. 2 Free-body diagram and kinetic diagram for the plate.

The centroidal moment of inertia of the plate is

$$\bar{I} = \frac{m}{12}(a^2 + b^2) = \frac{60 \text{ lb}}{12(32.2 \text{ ft/s}^2)}[(\tfrac{8}{12}\text{ ft})^2 + (\tfrac{6}{12}\text{ ft})^2]$$

$$= 0.1078 \text{ lb·ft·s}^2$$

Substituting this value of $\bar{I}$ together with $W = 60$ lb, $\bar{r} = \tfrac{5}{12}$ ft, and $\bar{x} = \tfrac{4}{12}$ ft into Eq. (1), you obtain

$$\alpha = +46.4 \text{ rad/s}^2 \qquad \boldsymbol{\alpha} = 46.4 \text{ rad/s}^2 \text{⤸} \quad \blacktriangleleft$$

b. Reaction at A. Using the computed value of α, determine the magnitude of the vector $m\bar{a}$ attached at G as

$$m\bar{a} = m\bar{r}\alpha = \frac{60 \text{ lb}}{32.2 \text{ ft/s}^2}(\tfrac{5}{12}\text{ ft})(46.4 \text{ rad/s}^2) = 36.0 \text{ lb}$$

Applying Newton's second law in the x and y directions gives

$$\overset{+}{\rightarrow} \Sigma F_x = m\bar{a}_x: \qquad A_x = -\tfrac{3}{5}(36 \text{ lb})$$
$$= -21.6 \text{ lb} \qquad\qquad \mathbf{A}_x = 21.6 \text{ lb} \leftarrow \quad \blacktriangleleft$$

$$+\uparrow \Sigma F_y = m\bar{a}_y: \qquad A_y - 60 \text{ lb} = -\tfrac{4}{5}(36 \text{ lb})$$
$$A_y = +31.2 \text{ lb} \qquad\qquad \mathbf{A}_y = 31.2 \text{ lb} \uparrow \quad \blacktriangleleft$$

REFLECT and THINK: If you had chosen to take moments about the center of gravity rather than point A, the two reaction forces A_x and A_y would have been in the resulting equation; that is, you would have had one equation and three unknowns, and you could not solve for α directly. Therefore, you would also need to use the equations from the x and y directions to solve for the three unknowns. Note that for convenience, we used a non-right handed coordinate system.

Sample Problem 16.9

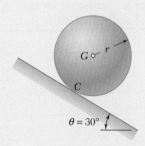

A sphere with a radius r and a weight W is released with no initial velocity on an incline and rolls without slipping. Determine (a) the minimum value of the coefficient of static friction compatible with the rolling motion, (b) the velocity of the center G of the sphere after the sphere has rolled 10 ft, (c) the velocity of G if the sphere were to move 10 ft down a frictionless 30° incline.

STRATEGY: Use Newton's second law to determine the acceleration of the center of gravity. Then determine the velocity from kinematics.

MODELING: Choose the sphere to be your system and model it as a rigid body. Recall that for rolling motion, the instantaneous point of contact has a velocity of zero, which leads to $\bar{a} = r\alpha$ (Fig. 1). A free-body diagram and kinetic diagram for this system are shown in Fig. 2. The external forces $\mathbf{W}$, $\mathbf{N}$, and $\mathbf{F}$ form a system equivalent to the inertial terms represented by the vector $m\bar{a}$ and the couple $\bar{I}\alpha$.

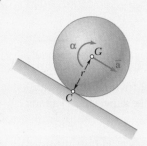

Fig. 1 The acceleration of G down the incline.

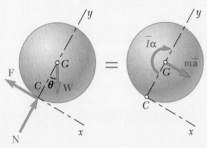

Fig. 2 Free-body diagram and kinetic diagram for the sphere.

ANALYSIS:

a. Minimum μ_s for Rolling Motion. Since the sphere rolls without sliding, you have $\bar{a} = r\alpha$ and can sum moments about C:

$$+\downarrow\Sigma M_C = \bar{I}\alpha + m\bar{a}d_\perp: \qquad (W\sin\theta)r = \bar{I}\alpha + (m\bar{a})r$$

$$(W\sin\theta)r = \bar{I}\alpha + (mr\alpha)r$$

Noting that $m = W/g$ and $\bar{I} = \frac{2}{5}mr^2$, you have

$$(W\sin\theta)r = \frac{2}{5}\frac{W}{g}r^2\alpha + \left(\frac{W}{g}r\alpha\right)r \qquad \alpha = +\frac{5g\sin\theta}{7r}$$

$$\bar{a} = r\alpha = \frac{5g\sin\theta}{7} = \frac{5(32.2\text{ ft/s}^2)\sin 30°}{7} = 11.50\text{ ft/s}^2$$

Applying Newton's second law in the x and y directions gives

$$+\searrow\Sigma F_x = m\bar{a}_x: \qquad W\sin\theta - F = m\bar{a}$$

$$W\sin\theta - F = \frac{W}{g}\frac{5g\sin\theta}{7}$$

$$F = +\frac{2}{7}W\sin\theta = \frac{2}{7}W\sin 30° \qquad \mathbf{F} = 0.143W \searrow 30°$$

$$+\nearrow\Sigma F_y = m\bar{a}_y: \qquad N - W\cos\theta = 0$$

$$N = W\cos\theta = 0.866W \qquad \mathbf{N} = 0.866W \measuredangle 60°$$

$$\mu_s = \frac{F}{N} = \frac{0.143W}{0.866W} \qquad\qquad \mu_s = 0.165 \quad\blacktriangleleft$$

b. Velocity of Rolling Sphere. This is a case of uniformly accelerated motion, so

$$\bar{v}_0 = 0 \qquad \bar{a} = 11.50\text{ ft/s}^2 \qquad \bar{x} = 10\text{ ft} \qquad \bar{x}_0 = 0$$

$$\bar{v}^2 = \bar{v}_0^2 + 2\bar{a}(\bar{x} - \bar{x}_0) \qquad \bar{v}^2 = 0 + 2(11.50\text{ ft/s}^2)(10\text{ ft})$$

$$\bar{v} = 15.17\text{ ft/s} \qquad \bar{\mathbf{v}} = 15.17\text{ ft/s} \searrow 30° \quad\blacktriangleleft$$

c. Velocity of Sliding Sphere. Now assuming no friction, you have $F = 0$ and obtain

$$+\downarrow\Sigma M_G = \bar{I}\alpha: \qquad 0 = \bar{I}\alpha \qquad \alpha = 0$$

$$+\searrow\Sigma F_x = m\bar{a}_x: \qquad W\sin 30° = m\bar{a} \qquad 0.50W = \frac{W}{g}\bar{a}$$

$$\bar{a} = +16.1\text{ ft/s}^2 \qquad \bar{\mathbf{a}} = 16.1\text{ ft/s}^2 \searrow 30°$$

Substituting $\bar{a} = 16.1\text{ ft/s}^2$ into the equations for uniformly accelerated motion, you obtain

$$\bar{v}^2 = \bar{v}_0^2 + 2\bar{a}(\bar{x} - \bar{x}_0) \qquad \bar{v}^2 = 0 + 2(16.1\text{ ft/s}^2)(10\text{ ft})$$

$$\bar{v} = 17.94\text{ ft/s} \qquad\qquad \bar{\mathbf{v}} = 17.94\text{ ft/s} \searrow 30° \quad\blacktriangleleft$$

REFLECT and THINK: Note that the sphere moving down a frictionless surface has a higher velocity than the rolling sphere, as you would expect. It is also interesting to note that the expression you obtained for the acceleration of the center of mass, that is, $\bar{a} = 5g\sin\theta/7$, is independent of the radius of the sphere and the mass of the sphere. This means that any two solid spheres, as long they are rolling without sliding, have the same linear acceleration.

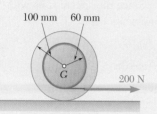

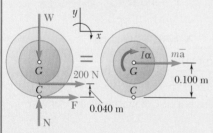

Fig. 1 Linear and angular acceleration of the wheel.

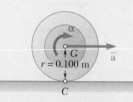

Fig. 2 Free-body diagram and kinetic diagram for the wheel assuming the friction force is to the right.

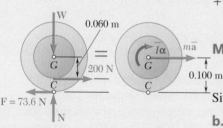

Fig. 3 Free-body diagram and kinetic diagram for the wheel when it is sliding and rotating.

Sample Problem 16.10

A cord is wrapped around the inner drum of a wheel and pulled horizontally with a force of 200 N. The wheel has a mass of 50 kg and a radius of gyration of 70 mm. Knowing that the coefficients of friction are $\mu_s = 0.20$ and $\mu_k = 0.15$, determine the acceleration of G and the angular acceleration of the wheel.

STRATEGY: Since you have forces acting on the wheel and are interested in accelerations, use Newton's second law. Assume the wheel rolls without sliding and compare the friction force needed with the maximum possible friction force. If the force needed exceeds the force available, redo the problem assuming rotation and sliding.

MODELING: Choose the wheel as your system and model it as a rigid body. The acceleration of G is to the right and the angular acceleration is clockwise (Fig. 1). The free-body and kinetic diagrams for this system are shown in (Fig. 2).

ANALYSIS:

a. Assume Rolling without Sliding. In this case, you have

$$\bar{a} = r\alpha = (0.100 \text{ m})\alpha$$

The moment of inertia of the wheel is

$$\bar{I} = m\bar{k}^2 = (50 \text{ kg})(0.070 \text{ m})^2 = 0.245 \text{ kg·m}^2$$

Equations of Motion. Setting the system of external forces in your free-body diagram equal to the system of inertial terms in your kinetic diagram, you obtain

$$+\!\!\downarrow\!\Sigma M_C = \bar{I}\alpha + m\bar{a}d_\perp: \quad (200 \text{ N})(0.040 \text{ m}) = \bar{I}\alpha + (m\bar{a})(0.100 \text{ m})$$
$$8.00 \text{ N·m} = (0.245 \text{ kg·m}^2)\alpha + (50 \text{ kg})(0.100 \text{ m})\alpha(0.100 \text{ m})$$
$$\alpha = +10.74 \text{ rad/s}^2$$
$$\bar{a} = r\alpha = (0.100 \text{ m})(10.74 \text{ rad/s}^2) = 1.074 \text{ m/s}^2$$

$$\overset{+}{\rightarrow} \Sigma F_x = m\bar{a}_x: \quad F + 200 \text{ N} = m\bar{a}$$
$$F + 200 \text{ N} = (50 \text{ kg})(1.074 \text{ m/s}^2)$$
$$F = -146.3 \text{ N} \qquad\qquad \mathbf{F = 146.3 \text{ N}} \leftarrow$$

$$+\!\!\uparrow\!\Sigma F_y = m\bar{a}_y:$$
$$N - W = 0 \quad N - W = mg = (50 \text{ kg})(9.81 \text{ m/s}^2) = 490.5 \text{ N}$$
$$\mathbf{N = 490.5 \text{ N}} \uparrow$$

Maximum Possible Friction Force.

$$F_{\max} = \mu_s N = 0.20(490.5 \text{ N}) = 98.1 \text{ N}$$

Since $F > F_{\max}$, the assumed motion is impossible.

b. Rotating and Sliding. Since the wheel must rotate and slide at the same time, we draw new free-body and kinetic diagrams (Fig. 3), where $\bar{\mathbf{a}}$ and $\boldsymbol{\alpha}$ are independent and

$$F = F_k = \mu_k N = 0.15(490.5 \text{ N}) = 73.6 \text{ N}$$

From the computation of part (a), you found that **F** is directed to the left. You can obtain and solve the following equations of motion as

$$\xrightarrow{+}\Sigma F_x = m\bar{a}_x: \quad 200\text{ N} - 73.6\text{ N} = (50\text{ kg})\bar{a}$$
$$\bar{a} = +2.53\text{ m/s}^2 \qquad \mathbf{\bar{a}} = 2.53\text{ m/s}^2 \rightarrow \quad \triangleleft$$

$$+\!\!\downarrow\Sigma M_G = \bar{I}\alpha:$$
$$(73.6\text{ N})(0.100\text{ m}) - (200\text{ N})(0.060\text{ m}) = (0.245\text{ kg}\cdot\text{m}^2)\alpha$$
$$\alpha = -18.94\text{ rad/s}^2 \qquad \mathbf{\alpha} = 18.94\text{ rad/s}^2 \text{↖} \quad \triangleleft$$

REFLECT and THINK: The wheel has larger linear and angular accelerations under conditions of rotating while sliding than when rolling without sliding.

Sample Problem 16.11

Overhead cranes are often used to move large containers in shipyards. A simplified model of a 60,000-lb container and crane is shown. The uniform container is at rest when the connection at B fails. Determine the tension in the cable connecting the pulley to the container at A.

STRATEGY: Since you are asked to find a tension, use Newton's second law.

MODELING: Start by choosing the container to be your system. After the connection at B fails, the only external forces acting on the container are the tension in the cable at A and the weight. A free-body diagram and kinetic diagram for this system immediately after the connection at B fails are shown in Fig. 1. Since the container is undergoing general plane motion, in the kinetic diagram you can represent the acceleration of the center of mass as having a vertical and a horizontal component.

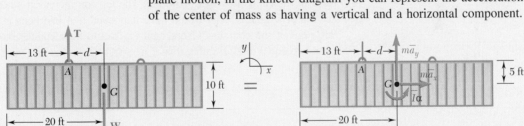

Fig. 1 Free-body diagram and kinetic diagram for the container.

ANALYSIS: Using Fig. 1 and applying Newton's second law in the x-direction and y-direction and summing moments about point G gives you

$$\xrightarrow{+}\Sigma F_x = m\bar{a}_x: \qquad 0 = m\bar{a}_x \tag{1}$$

$$+\!\!\uparrow\Sigma F_y = m\bar{a}_y: \qquad T - W = m\bar{a}_y \tag{2}$$

$$+\!\!\curvearrowleft\Sigma M_G = \bar{I}\alpha: \qquad -Td = \bar{I}\alpha \tag{3}$$

(continued)

where

$$d = 7 \text{ ft and}$$

$$m = \frac{W}{g} = \frac{60{,}000 \text{ lb}}{32.2 \text{ ft/s}^2} = 1863 \text{ lb·s}^2/\text{ft}$$

$$\bar{I} = \tfrac{1}{12}m(b^2 + c^2) = \tfrac{1}{12}(1836 \text{ lb·s}^2/\text{ft})\big[(40 \text{ ft})^2 + (10 \text{ ft})^2\big] = 26{,}400 \text{ lb·ft·s}^2$$

In Eqs. (1) through (3), you have four unknowns: T, $\bar{a}_x$, $\bar{a}_y$, and α. You can use kinematics to obtain additional equations. You want to relate the acceleration of the center of mass to that of another point on the container. At the instant the cable breaks, the angular velocity of the cable is zero, so point A has no normal acceleration, but it has an acceleration perpendicular to the cable. The accelerations of A and G are related by

$$\mathbf{a}_G = \mathbf{a}_A + \mathbf{a}_{G/A} = \mathbf{a}_A + \boldsymbol{\alpha} \times \mathbf{r}_{G/A} - \omega^2 \mathbf{r}_{G/A}$$

Substituting in known values and letting $\boldsymbol{\omega} = 0$ and $\boldsymbol{\alpha} = \alpha\mathbf{k}$ gives you

$$\bar{a}_x\mathbf{i} + \bar{a}_y\mathbf{j} = a_A\mathbf{i} + \alpha\mathbf{k} \times [d\mathbf{i} - 5\mathbf{j}] - 0 = a_A\mathbf{i} + (d\alpha)\mathbf{j} + (5\alpha)\mathbf{i}$$

Equating components gives

i: $\bar{a}_x = a_A + 5\alpha$ \hfill (4)

j: $\bar{a}_y = d\alpha$ \hfill (5)

Solving Eqs. (1–5) for T, $\bar{a}_y$, $\bar{a}_x$, a_A, and α gives you $T = 44{,}580$ lb, $\bar{a}_y = -8.275 \text{ ft/s}^2$, $\bar{a}_x = 0$, $a_A = 5.911 \text{ ft/s}^2$, and $\alpha = -1.182 \text{ rad/s}^2$.

$$\mathbf{T} = 44{,}600 \text{ lb} \uparrow \quad \blacktriangleleft$$

REFLECT and THINK: You don't need all five equations to solve for the required unknowns; that is, you could have chosen to just use Eqs. (2), (3), and (5). The acceleration of the center of gravity is only in the vertical direction at the instant the cable breaks. When the container was at rest, the force in the cable at A was 30,000 lb. The tension increased when the connection at B failed. What would have happened if A had been at the upper left edge of the container? Your analysis would be identical except that d would be equal to 20 ft rather than 7 ft. Substituting this into your equations and solving gives you $T = 15{,}690$ lb, which is less than 30,000 lb.

Sample Problem 16.12

The ends of a 4-ft rod weighing 50 lb can move freely and with no friction along two straight tracks as shown. If the rod is released from rest at the position shown, determine (*a*) the angular acceleration of the rod, (*b*) the reactions at A and B.

STRATEGY: Since you are asked to determine forces and accelerations, use Newton's second law. The motion is constrained, so the acceleration of G must be related to the angular acceleration $\boldsymbol{\alpha}$. To obtain this relation, first determine the magnitude of the acceleration $\mathbf{a}_A$ of point A in terms of α.

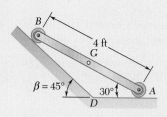

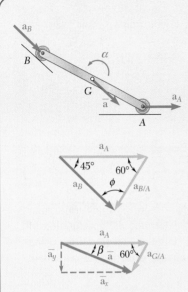

Fig. 1 Vector diagrams for accelerations of points on the rod.

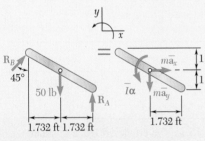

Fig. 2 Free-body diagram and kinetic diagram for the rod assuming a downward acceleration.

MODELING and ANALYSIS: Choose the rod to be your system and model it as a rigid body. Before drawing the kinetic diagram, you need to relate the acceleration of G to the angular acceleration of the rod. You can do this using kinematics.

Kinematics of Motion.   Assume that $\boldsymbol{\alpha}$ is directed counterclockwise. Noting that $a_{B/A} = 4\alpha$, you have (Fig. 1)

$$\mathbf{a}_B = \mathbf{a}_A + \mathbf{a}_{B/A}$$
$$[a_B \searrow 45°] = [a_A \rightarrow] + [4\alpha \nearrow 60°]$$

Noting that $\phi = 75°$ and using the law of sines, you obtain

$$a_A = 5.46\alpha \qquad a_B = 4.90\alpha$$

Now you can find the acceleration of G from

$$\overline{\mathbf{a}} = \mathbf{a}_G = \mathbf{a}_A + \mathbf{a}_{G/A}$$
$$\overline{\mathbf{a}} = [5.46\alpha \rightarrow] + [2\alpha \nearrow 60°]$$

Resolving $\overline{\mathbf{a}}$ into x and y components, you obtain

$$\overline{a}_x = 5.46\alpha - 2\alpha \cos 60° = 4.46\alpha \qquad \overline{\mathbf{a}}_x = 4.46\alpha \rightarrow$$
$$\overline{a}_y = -2\alpha \sin 60° = -1.732\alpha \qquad \overline{\mathbf{a}}_y = 1.732\alpha \downarrow$$

Kinetics of Motion.   Draw a free-body-diagram and kinetic diagram for your system (Fig. 2). Compute the following magnitudes.

$$\overline{I} = \tfrac{1}{12}ml^2 = \frac{1}{12}\frac{50 \text{ lb}}{32.2 \text{ ft/s}^2}(4 \text{ ft})^2 = 2.07 \text{ lb·ft·s}^2 \qquad \overline{I}\alpha = 2.07\alpha$$

$$m\overline{a}_x = \frac{50}{32.2}(4.46\alpha) = 6.93\alpha \qquad m\overline{a}_y = -\frac{50}{32.2}(1.732\alpha) = -2.69\alpha$$

Equations of Motion.

$+\curvearrowleft \Sigma M_B = \overline{I}\alpha + m\overline{a}d_\perp$:

$$R_A(4 \cos 30° \text{ ft}) - W(2 \cos 30° \text{ ft}) = \overline{I}\alpha + (m\overline{a}_x)(2 \sin 30° \text{ ft})$$
$$- (m\overline{a}_y)(2 \cos 30° \text{ ft})$$

$$R_A(3.464) - (50 \text{ lb})(1.732) = 2.07\alpha + (6.93\alpha)(1.000)$$
$$- (2.69\alpha)(1.732) \tag{1}$$

$\xrightarrow{+} \Sigma F_x = m\overline{a}_x$: $R_B \sin 45° = 6.93\alpha$ (2)

$+\uparrow \Sigma F_y = m\overline{a}_y$: $R_A + R_B \cos 45° - 50 = -2.69\alpha$ (3)

Solving these equations gives

$$\boldsymbol{\alpha} = 2.30 \text{ rad/s}^2 \curvearrowleft \quad \blacktriangleleft$$
$$\mathbf{R}_B = 22.5 \text{ lb} \measuredangle 45° \quad \blacktriangleleft$$
$$\mathbf{R}_A = 27.9 \text{ lb} \uparrow \quad \blacktriangleleft$$

REFLECT and THINK: For the kinematics, you could have used the vector algebra approach rather than the method demonstrated in this example. Using the vector algebra approach, you can write

$$\mathbf{a}_B = \mathbf{a}_A + \alpha\mathbf{k} \times \mathbf{r}_{B/A} - \omega^2\mathbf{r}_{B/A}$$

(continued)

Substituting the directions assumed in Fig. 1, you find

$$\frac{a_B}{\sqrt{2}}\mathbf{i} - \frac{a_B}{\sqrt{2}}\mathbf{j} = a_A\mathbf{i} + \alpha\mathbf{k} \times (-3.464\mathbf{i} + 2\mathbf{j}) + 0$$
$$= a_A\mathbf{i} + (-3.464\alpha\mathbf{j} - 2\alpha\mathbf{i})$$

Equating components gives

i: $\quad \dfrac{a_B}{\sqrt{2}} = a_A - 2\alpha$

j: $\quad \dfrac{a_B}{\sqrt{2}} = -3.46\alpha$

Solving these, you find $a_B = 4.90\alpha$ and $a_A = 5.46\alpha$, which are similar to the approach shown previously. You can determine the acceleration of the center of gravity in terms of the angular acceleration using $\mathbf{a}_G = \mathbf{a}_A + \alpha\mathbf{k} \times r_{G/A} - \omega^2\mathbf{r}_{G/A}$. Substituting the directions assumed in Fig. 1, you find

$$\bar{a}_x\mathbf{i} + \bar{a}_y\mathbf{j} = a_A\mathbf{i} + \alpha\mathbf{k} \times (-1.732\mathbf{i} + 1\mathbf{j}) + 0 = a_A\mathbf{i} + (-1.732\alpha\mathbf{j} - 1\alpha\mathbf{i})$$

Equating components gives

i: $\quad \bar{a}_x = a_A - 1\alpha = 4.46\alpha$

j: $\quad \bar{a}_y = -1.732\alpha$

These are identical to the answers determined previously.

Sample Problem 16.13

In the engine system from Sample Prob. 15.15, the crank AB has a constant clockwise angular velocity of 2000 rpm. Knowing that the connecting rod BD weighs 4 lb and the piston P weighs 5 lb, determine the forces on the connecting rod at B and D. Assume the center of mass of BD is at its geometric center and it can be treated as a uniform, slender rod.

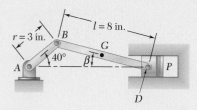

STRATEGY: Since you are asked to find forces at the instant shown, use Newton's second law.

MODELING: Since you want to determine the forces at B and D, start by choosing the connecting rod BD as your system. The pin forces at B and D are represented by horizontal and vertical components, and since the rod is undergoing general plane motion, you can represent the acceleration of the center of mass in the kinetic diagram as having a vertical and a horizontal component. The free-body and kinetic diagrams for this system are shown in Fig. 1, where $\ell = 8$ in. $= 0.6667$ ft and $\beta = 13.95°$.

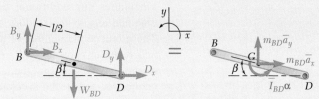

Fig. 1

ANALYSIS: Using Fig. 1, applying Newton's second law in the x-direction and y-direction, and summing moments about point G gives

$$\xrightarrow{+} \Sigma F_x = m\bar{a}_x: \qquad B_x + D_x = m_{BD}\bar{a}_x \tag{1}$$

$$+\uparrow \Sigma F_y = m\bar{a}_y: \qquad B_y + D_y - W_{BD} = m_{BD}\bar{a}_y \tag{2}$$

$$+\!\!\upharpoonleft \Sigma M_G = \bar{I}\alpha: \quad -B_y(\ell/2)\cos\beta - B_x(\ell/2)\sin\beta + D_y(\ell/2)\cos\beta$$
$$+ D_x(\ell/2)\sin\beta = \bar{I}_{BD}\alpha_{BD} \tag{3}$$

where

$$m_{BD} = \frac{W_{BD}}{g} = \frac{4 \text{ lb}}{32.2 \text{ ft/s}^2} = 0.1242 \text{ lb·s}^2/\text{ft}$$

$$\bar{I}_{BD} = \tfrac{1}{12}m_{BD}\ell^2 = \tfrac{1}{12}(0.1242 \text{ lb·s}^2/\text{ft})(0.6667 \text{ ft})^2 = 0.004601 \text{ lb·ft·s}^2$$

In Eqs. (1) through (3), you have seven unknowns: B_x, B_y, D_x, D_y, $\bar{a}_x$, $\bar{a}_y$, and α_{BD}. Therefore, you need more equations. You can get them from kinematics or by choosing another system. Choose the piston to be your system, model it as a particle, and draw its free-body and kinetic diagrams (Fig. 2).

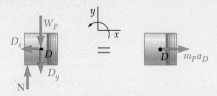

Fig. 2 Free-body diagram and kinetic diagram for the piston.

Note that you must draw D_x and D_y in the opposite directions to what you drew for the connecting rod. Using Fig. 2 and applying Newton's second law in the x-direction and y-direction gives

$$\xrightarrow{+} \Sigma F_x = m\bar{a}_x: \qquad -D_x = m_P a_D \tag{4}$$

$$+\uparrow \Sigma F_y = m\bar{a}_y: \qquad -D_y + N - W_P = 0 \tag{5}$$

where

$$m_P = \frac{W_P}{g} = \frac{5 \text{ lb}}{32.2 \text{ ft/s}^2} = 0.1553 \text{ lb·s}^2/\text{ft}$$

You now have five equations and nine unknowns: N, B_x, B_y, D_x, D_y, $\bar{a}_x$, $\bar{a}_y$, α_{BD}, and a_D. You could choose crank AB as another system, but since this will introduce three additional unknowns (the reactions at A and the driving torque) and you are not provided its mass, you should turn to kinematics for additional equations. From Sample Prob. 15.15, you obtained $\omega_{BD} = 62.0$ rad/s↰, $a_D = 9290$ ft/s²←, and $\alpha_{BD} = 9940$ rad/s² ↰. These reduce the number of unknowns by two, so you have five equations and seven unknowns: N, B_x, B_y, D_x, D_y, $\bar{a}_x$, and $\bar{a}_y$. You can find two more equations by relating the acceleration of the center of mass of the connecting rod to the acceleration of D,

$$\mathbf{a}_G = \mathbf{a}_D + \mathbf{a}_{G/D} = \mathbf{a}_D + \boldsymbol{\alpha} \times \mathbf{r}_{G/D} - \omega_{BD}^2 \mathbf{r}_{G/D}$$

Substituting in known and assumed values (Fig. 1) $\mathbf{a}_D = a_D\mathbf{i}$, where $a_D = -9290$ ft/s², and $\boldsymbol{\alpha}_{BD} = \alpha_{BD}\mathbf{k}$, where $\alpha_{BD} = 9940$ rad/s², gives

$$\bar{a}_x\mathbf{i} + \bar{a}_y\mathbf{j} = a_D\mathbf{i} + \alpha_{BD}\mathbf{k} \times \left[-\tfrac{\ell}{2}\cos\beta\mathbf{i} + \tfrac{\ell}{2}\sin\beta\mathbf{j}\right] - \omega_{BD}^2\left[-\tfrac{\ell}{2}\cos\beta\mathbf{i} + \tfrac{\ell}{2}\sin\beta\mathbf{j}\right]$$

$$= a_D\mathbf{i} - \alpha_{BD}\tfrac{\ell}{2}\cos\beta\mathbf{j} - \alpha_{BD}\tfrac{\ell}{2}\sin\beta\mathbf{i} + \omega_{BD}^2\tfrac{\ell}{2}\cos\beta\mathbf{i} - \omega_{BD}^2\tfrac{\ell}{2}\sin\beta\mathbf{j}$$

Equating components, you have

$$\mathbf{i}:\ \bar{a}_x = a_D - \alpha_{BD}\tfrac{\ell}{2}\sin\beta + \omega_{BD}^2\tfrac{\ell}{2}\cos\beta \tag{6}$$

$$\mathbf{j}:\ \bar{a}_y = -\alpha_{BD}\tfrac{\ell}{2}\cos\beta - \omega_{BD}^2\tfrac{\ell}{2}\sin\beta \tag{7}$$

You now have seven equations and seven unknowns. Substituting in numerical values and solving these equations using your calculator or software such as MathCad, Maple, Matlab, or Mathematica gives you $B_x = -2541$ lb, $B_y = 207.2$ lb, $D_x = 1442$ lb, $D_y = -641$ lb, $N = -636$ lb, $\bar{a}_x = -8845$ ft/s², and $\bar{a}_y = -3524$ ft/s².

$$B_x = 2541\ \text{lb} \leftarrow \qquad B_y = 207\ \text{lb} \uparrow \quad ◄$$

$$D_x = 1442\ \text{lb} \rightarrow \qquad D_y = 641\ \text{lb} \downarrow \quad ◄$$

REFLECT and THINK: The calculated forces are much larger than the weight of the piston and the connecting rod. This problem required multiple systems and rigid body kinematics to solve, most of which was done in Sample Prob. 15.15. In problems like this, it is a good practice to focus on the problem formulation and to keep track of equations and unknowns. Once you have enough equations to solve for all the unknowns, using a computer or calculator to solve the resulting equations is often the easiest approach.

SOLVING PROBLEMS ON YOUR OWN

In this section, we considered the **plane motion of rigid bodies under constraints**. We found that the types of constraints involved in engineering problems vary widely. For example, a rigid body may be constrained to rotate about a fixed axis or to roll on a given surface, or it may be pin-connected to collars or to other bodies.

1. Your solution of a problem involving the constrained motion of a rigid body consists, in general, of three steps. First, you should model your system by drawing the free-body diagram and the kinetic diagram. Second, use these diagrams to write out your equations of motion. Finally, you will generally need to consider the *kinematics of the motion* to have enough equations to solve the problem. Sometimes it is helpful to examine the kinematics first to help you draw the kinetic diagram and choose an appropriate coordinate system.

2. Free-body diagram and kinetic diagram. Your first step in the solution of a problem is to draw a *free-body diagram and a kinetic diagram*.

    a. A free-body diagram shows ***the forces exerted on the body***, including the applied forces, the reactions at the supports, and the weight of the body.

    b. *A kinetic diagram* shows the **inertial terms**: vector $m\bar{\mathbf{a}}$ and couple $\bar{I}\boldsymbol{\alpha}$.

3. Using your free-body diagram and kinetic diagram, generate the equations of motion for the system. Drawing good free-body and kinetic diagrams will allow you to *sum components in any direction and to sum moments about any point*. For a single body, you can obtain a maximum of three independent equations (two translational and one moment) that can be used to help analyze the system.

$$\Sigma F_x = m\bar{a}_x \qquad \Sigma F_y = m\bar{a}_y$$

$$\Sigma M_G = \bar{I}\alpha \quad \text{or} \quad \Sigma M_O = I_O\alpha \quad \text{or} \quad \Sigma M_P = \bar{I}\alpha + m\bar{a}d_\perp \quad \text{or} \quad \Sigma M_P = \bar{I}\alpha + r_{G/P} \times m\bar{\mathbf{a}}$$

where G is the center of mass of the body, O is a fixed axis of rotation, P is any arbitrary point, and $d_\perp$ is the perpendicular distance between point P and the line of action of the acceleration of the center of mass.

4. The kinematic analysis of the motion uses the methods you learned in Chap. 15. Due to the constraints, linear and angular accelerations are related. You should establish relationships among the accelerations (angular as well as linear), and your goal should be to express all accelerations in terms of a single unknown acceleration.

    a. For a body in noncentroidal rotation about a fixed axis, the components of the acceleration of the mass center are $\bar{a}_t = \bar{r}\alpha$ and $\bar{a}_n = \bar{r}\omega^2$, where ω is generally known [Sample Probs. 16.7 and 16.8].

(continued)

b. For a rolling disk or wheel, the acceleration of the geometric center is $\bar{a} = r\alpha$ [Sample Prob. 16.9].

c. For a body in general plane motion, your best course of action if neither $\bar{a}$ nor α is known or readily obtainable is to express $\bar{a}$ in terms of α [Sample Probs. 16.10 through 16.13]. This can be done by relating the acceleration of the center of mass to some reference point:

$$\bar{\mathbf{a}} = \mathbf{a}_A + \alpha\mathbf{k} \times \mathbf{r}_{G/A} - \omega^2\mathbf{r}_{G/A}$$

5. When solving problems involving rolling disks or wheels, keep in mind the following situations.

a. If sliding is impending, the friction force exerted on the rolling body has reached its maximum value, so $F_m = \mu_s N$, where N is the normal force exerted on the body and μ_s is the coefficient of static friction between the surfaces of contact.

b. If sliding is not impending, the friction force F can have any value smaller than F_m and therefore should be considered an independent unknown. After you have determined F, be sure to check that it is smaller than F_m; if it is not, the body does not roll but rotates and slides as described in the next paragraph.

c. If the body rotates and slides at the same time, then the body is not rolling, and the acceleration $\bar{a}$ of the mass center is *independent* of the angular acceleration α of the body: $\bar{a} \neq r\alpha$. On the other hand, the friction force has a well-defined value, $F = \mu_k N$, where μ_k is the coefficient of kinetic friction between the surfaces of contact.

d. For an unbalanced rolling disk or wheel, the relation $\bar{a} = r\alpha$ between the acceleration $\bar{a}$ of the mass center G and the angular acceleration α of the disk or wheel does not hold any more. However, a similar relation holds between the acceleration a_O of the geometric center O and the angular acceleration α of the disk or wheel: $a_O = r\alpha$. This relation can be used to express $\bar{a}$ in terms of α and ω (Fig. 16.17).

6. For a system of connected rigid bodies, the goal of your kinematic analysis should be to determine all the accelerations from the given data or to express them all in terms of a single unknown. For systems with several degrees of freedom, you will need to use as many unknowns as there are degrees of freedom.

Your kinetic analysis will sometimes be carried out by drawing a free-body diagram and a kinetic diagram for the entire system. If you only have three unknowns, this is usually the best approach. In most cases, however, it will be necessary to analyze each rigid body separately in order to obtain enough equations to solve for all the unknown quantities in the problem.

Problems

16.CQ4 A cord is attached to a spool when a force **P** is applied to the cord as shown. Assuming the spool rolls without slipping, what direction does the spool move for each case?

    Case 1: **a.** left   **b.** right **c.** It would not move.

    Case 2: **a.** left   **b.** right **c.** It would not move.

    Case 3: **a.** left   **b.** right **c.** It would not move.

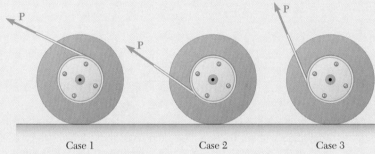

      Case 1   Case 2   Case 3

Fig. P16.CQ4 and P16.CQ5

16.CQ5 A cord is attached to a spool when a force **P** is applied to the cord as shown. Assuming the spool rolls without slipping, in what direction does the friction force act for each case?

    Case 2: **a.** left   **b.** right **c.** The friction force would be zero.

    Case 3: **a.** left   **b.** right **c.** The friction force would be zero.

16.CQ6 A front-wheel-drive car starts from rest and accelerates to the right. Knowing that the tires do not slip on the road, what is the direction of the friction force the road applies to the front tires?

a. left

b. right

c. The friction force is zero.

16.CQ7 A front-wheel-drive car starts from rest and accelerates to the right. Knowing that the tires do not slip on the road, what is the direction of the friction force the road applies to the rear tires?

a. left

b. right

c. The friction force is zero.

6 in.

8 in.

A

Fig. P16.F5

16.F5 A uniform 6 × 8-in. rectangular plate of mass m is pinned at A. Knowing the angular velocity of the plate at the instant shown is ω, draw the FBD and KD.

16.F6 Two identical 4-lb slender rods AB and BC are connected by a pin at B and by the cord AC. The assembly rotates in a vertical plane under the combined effect of gravity and a couple $\mathbf{M}$ applied to rod AB. Knowing that in the position shown the angular velocity of the assembly is ω, draw the FBD and KD that can be used to determine the angular acceleration of the assembly.

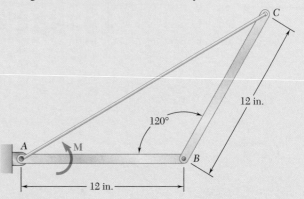

Fig. P16.F6

16.F7 The 4-lb uniform rod AB is attached to collars of negligible mass that slide without friction along the fixed rods shown. Rod AB is at rest in the position $\theta = 25°$ when a horizontal force $\mathbf{P}$ is applied to collar A causing it to start moving to the left. Draw the FBD and KD for the rod.

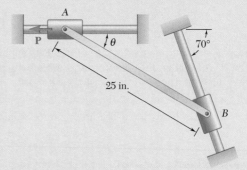

Fig. P16.F7

16.F8 A uniform disk of mass $m = 4$ kg and radius $r = 150$ mm is supported by a belt $ABCD$ that is bolted to the disk at B and C. If the belt suddenly breaks at a point located between A and B, draw the FBD and KD for the disk immediately after the break.

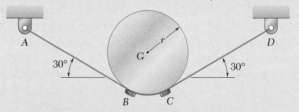

Fig. P16.F8

END-OF-SECTION PROBLEMS

16.75 Show that the couple $\bar{I}\alpha$ of Fig. 16.15 can be eliminated by attaching the vectors $m\bar{\mathbf{a}}_t$ and $m\bar{\mathbf{a}}_n$ at a point P called the *center of percussion*, located on line OG at a distance $GP = \bar{k}^2/\bar{r}$ from the mass center of the body.

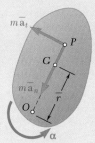

Fig. P16.75

16.76 A uniform slender rod of length $L = 900$ mm and mass $m = 4$ kg is suspended from a hinge at C. A horizontal force **P** of magnitude 75 N is applied at end B. Knowing that $\bar{r} = 225$ mm, determine (*a*) the angular acceleration of the rod, (*b*) the components of the reaction at C.

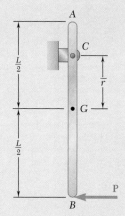

Fig. P16.76

16.77 In Prob. 16.76, determine (*a*) the distance $\bar{r}$ for which the horizontal component of the reaction at C is zero, (*b*) the corresponding angular acceleration of the rod.

16.78 A uniform slender rod of length $L = 36$ in. and weight $W = 4$ lb hangs freely from a hinge at A. If a force **P** of magnitude 1.5 lb is applied at B horizontally to the left ($h = L$), determine (*a*) the angular acceleration of the rod, (*b*) the components of the reaction at A.

16.79 In Prob. 16.78, determine (*a*) the distance h for which the horizontal component of the reaction at A is zero, (*b*) the corresponding angular acceleration of the rod.

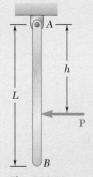

Fig. P16.78

16.80 An athlete performs a leg extension on a machine using a 20-kg mass at A located 400 mm away from the knee joint at center O. Biomechanical studies show that the patella tendon inserts at B, which is 100 mm below point O and 20 mm from the center line of the tibia (see figure). The mass of the lower leg and foot is 5 kg, the center of gravity of this segment is 300 mm from the knee, and the radius of gyration about the knee is 350 mm. Knowing that the leg is moving at a constant angular velocity of 30 degrees per second when $\theta = 60°$, determine (*a*) the force **F** in the patella tendon, (*b*) the magnitude of the joint force at the knee joint center O.

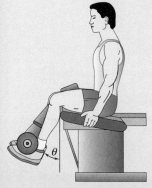

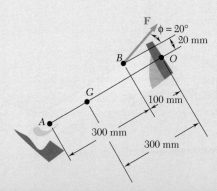

Fig. P16.80

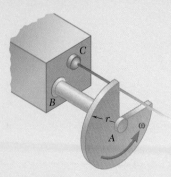

Fig. P16.81

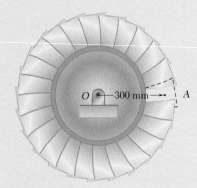

$O \quad \bullet \text{—} 300 \text{ mm} \text{—} \cdot \quad A$

Fig. P16.83

Fig. P16.84

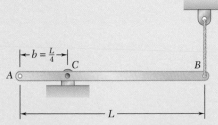

Fig. P16.85

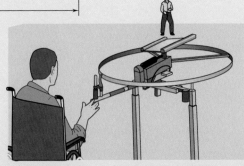

Fig. P16.86

16.81 The shutter shown was formed by removing one quarter of a disk of 0.75-in. radius and is used to interrupt a beam of light emanating from a lens at *C*. Knowing that the shutter weighs 0.125 lb and rotates at the constant rate of 24 cycles per second, determine the magnitude of the force exerted by the shutter on the shaft at *A*.

16.82 A 6-in.-diameter hole is cut as shown in a thin disk of 15-in. diameter. The disk rotates in a horizontal plane about its geometric center *A* at the constant rate of 480 rpm. Knowing that the disk has a mass of 60 lb after the hole has been cut, determine the horizontal component of the force exerted by the shaft on the disk at *A*.

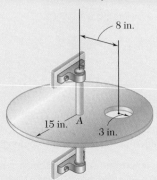

Fig. P16.82

16.83 A turbine disk of mass 26 kg rotates at a constant rate of 9600 rpm. Knowing that the mass center of the disk coincides with the center of rotation *O*, determine the reaction at *O* immediately after a single blade at *A*, of mass 45 g, becomes loose and is thrown off.

16.84 and 16.85 A uniform rod of length *L* and mass *m* is supported as shown. If the cable attached at end *B* suddenly breaks, determine (*a*) the acceleration of end *B*, (*b*) the reaction at the pin support.

16.86 An adapted launcher uses a torsional spring about point *O* to help people with mobility impairments throw a Frisbee®. Just after the Frisbee leaves the arm, the angular velocity of the throwing arm is 200 rad/s and its acceleration is 10 rad/s²; both are counterclockwise. The rotation point *O* is located 1 in. from the two sides. Assume that you can model the 2-lb throwing arm as a uniform rectangle. Just after the Frisbee leaves the arm, determine (*a*) the moment about *O* caused by the spring, (*b*) the forces on the pin at *O*.

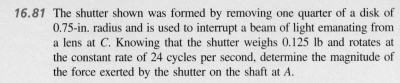

16.87 A 1.5-kg slender rod is welded to a 5-kg uniform disk as shown. The assembly swings freely about C in a vertical plane. Knowing that in the position shown the assembly has an angular velocity of 10 rad/s clockwise, determine (a) the angular acceleration of the assembly, (b) the components of the reaction at C.

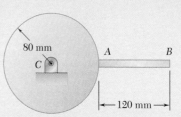

Fig. P16.87

16.88 Two identical 4-lb slender rods AB and BC are connected by a pin at B and by the cord AC. The assembly rotates in a vertical plane under the combined effect of gravity and a 6 lb·ft couple M applied to rod AB. Knowing that in the position shown the angular velocity of the assembly is zero, determine (a) the angular acceleration of the assembly, (b) the tension in cord AC.

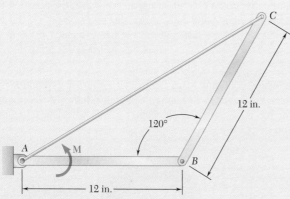

Fig. P16.88

16.89 The object ABC consists of two slender rods welded together at point B. Rod AB has a weight of 2 lb and bar BC has a weight of 4 lb. Knowing the magnitude of the angular velocity of ABC is 10 rad/s when $\theta = 0°$, determine the components of the reaction at point C at this location.

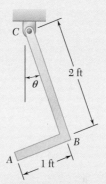

Fig. P16.89

16.90 A 3.5-kg slender rod AB and a 2-kg slender rod BC are connected by a pin at B and by the cord AC. The assembly can rotate in a vertical plane under the combined effect of gravity and a couple M applied to rod BC. Knowing that in the position shown the angular velocity of the assembly is zero and the tension in cord AC is equal to 25 N, determine (a) the angular acceleration of the assembly, (b) the magnitude of the couple M.

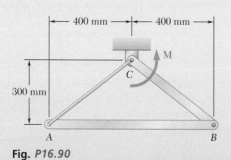

Fig. P16.90

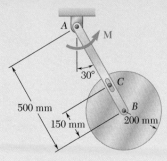

Fig. P16.91

16.91 A 9-kg uniform disk is attached to the 5-kg slender rod AB by means of frictionless pins at B and C. The assembly rotates in a vertical plane under the combined effect of gravity and of a couple **M** that is applied to rod AB. Knowing that at the instant shown the assembly has an angular velocity of 6 rad/s and an angular acceleration of 25 rad/s², both counterclockwise, determine (a) the couple **M**, (b) the force exerted by pin C on member AB.

16.92 Derive the equation $\Sigma M_C = I_C \alpha$ for the rolling disk of Fig. 16.16, where ΣM_C represents the sum of the moments of the external forces about the instantaneous center C, and I_C is the moment of inertia of the disk about C.

16.93 Show that in the case of an unbalanced disk, the equation derived in Prob. 16.92 is valid only when the mass center G, the geometric center O, and the instantaneous center C happen to lie in a straight line.

16.94 A wheel of radius r and centroidal radius of gyration $\bar{k}$ is released from rest on the incline and rolls without sliding. Derive an expression for the acceleration of the center of the wheel in terms of r, $\bar{k}$, β, and g.

Fig. P16.94

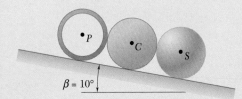

Fig. P16.95

16.95 A homogeneous sphere S, a uniform cylinder C, and a thin pipe P are in contact when they are released from rest on the incline shown. Knowing that all three objects roll without slipping, determine, after 4 s of motion, the distance between (a) the pipe and the cylinder, (b) the cylinder and the sphere.

16.96 A 40-kg flywheel of radius R = 0.5 m is rigidly attached to a shaft of radius r = 0.05 m that can roll along parallel rails. A cord is attached as shown and pulled with a force **P** of magnitude 150 N. Knowing the centroidal radius of gyration is $\bar{k} = 0.4$ m, determine (a) the angular acceleration of the flywheel, (b) the velocity of the center of gravity after 5 s.

16.97 A 40-kg flywheel of radius R = 0.5 m is rigidly attached to a shaft of radius r = 0.05 m that can roll along parallel rails. A cord is attached as shown and pulled with a force **P**. Knowing the centroidal radius of gyration is $\bar{k} = 0.4$ m and the coefficient of static friction is $\mu_s = 0.4$, determine the largest magnitude of force **P** for which no slipping will occur.

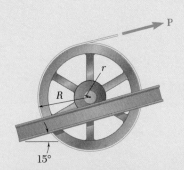

Fig. P16.96 and P16.97

16.98 through 16.101 A drum of 60-mm radius is attached to a disk of 120-mm radius. The disk and drum have a total mass of 6 kg and a combined radius of gyration of 90 mm. A cord is attached as shown and pulled with a force **P** of magnitude 20 N. Knowing that the disk rolls without sliding, determine (*a*) the angular acceleration of the disk and the acceleration of *G*, (*b*) the minimum value of the coefficient of static friction compatible with this motion.

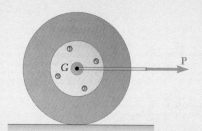

Fig. P16.98 and P16.102

16.102 through 16.105 A drum of 4-in. radius is attached to a disk of 8-in. radius. The disk and drum have a combined weight of 10 lb and a combined radius of gyration of 6 in. A cord is attached as shown and pulled with a force **P** of magnitude 5 lb. Knowing that the coefficients of static and kinetic friction are $\mu_s = 0.25$ and $\mu_k = 0.20$, respectively, determine (*a*) whether or not the disk slides, (*b*) the angular acceleration of the disk and the acceleration of *G*.

Fig. P16.99 and P16.103

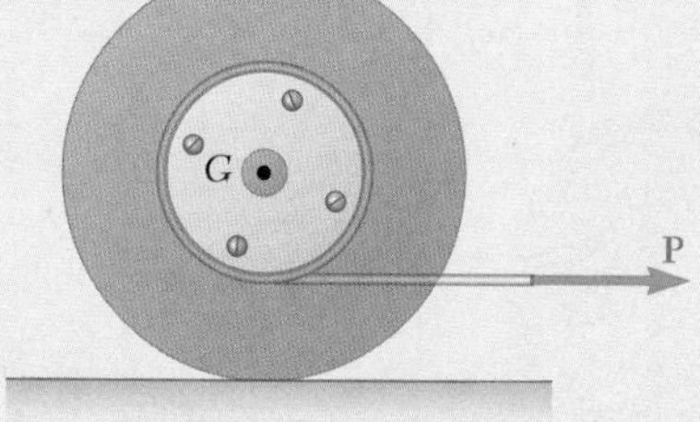

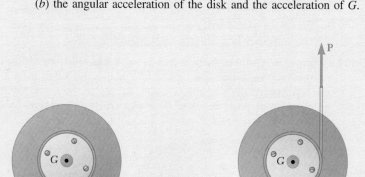

Fig. P16.100 and P16.104

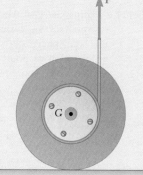

Fig. P16.101 and P16.105

16.106 and 16.107 A 12-in.-radius cylinder of weight 16 lb rests on a 6-lb carriage. The system is at rest when a force **P** of magnitude 4 lb is applied. Knowing that the cylinder rolls without sliding on the carriage and neglecting the mass of the wheels of the carriage, determine (*a*) the acceleration of the carriage, (*b*) the acceleration of point *A*, (*c*) the distance the cylinder has rolled with respect to the carriage after 0.5 s.

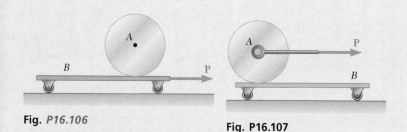

Fig. P16.106

Fig. P16.107

16.108 Gear *C* has a mass of 5 kg and a centroidal radius of gyration of 75 mm. The uniform bar *AB* has a mass of 3 kg and gear *D* is stationary. If the system is released from rest in the position shown, determine (*a*) the angular acceleration of gear *C*, (*b*) the acceleration of point *B*.

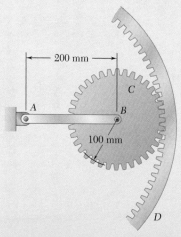

Fig. P16.108

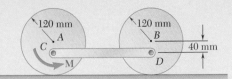

Fig. P16.109

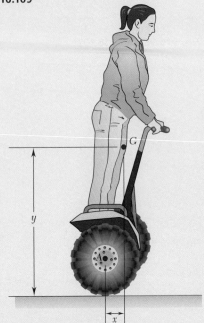

Fig. *P16.110*

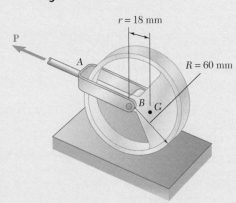

Fig. P16.113

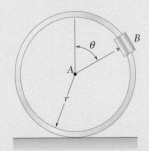

Fig. P16.114 and P16.115

16.109 Two uniform disks A and B, each with a mass of 2 kg, are connected by a 2.5-kg rod CD as shown. A counterclockwise couple **M** of moment 2.25 N·m is applied to disk A. Knowing that the disks roll without sliding, determine (a) the acceleration of the center of each disk, (b) the horizontal component of the force exerted on disk B by pin D.

16.110 A single-axis personal transport device starts from rest with the rider leaning slightly forward. Together, the two wheels weigh 25 lbs, and each has a radius of 10 in. The mass moment of inertia of the wheels about the axle is 0.15 slug·ft². The combined weight of the rest of the device and the rider (excluding the wheels) is 200 lbs, and the center of gravity G of this weight is located at $x = 4$ in. in front of axle A and $y = 36$ in. above the ground. An initial clockwise torque **M** is applied by the motor to the wheels. Knowing that the coefficients of static and kinetic friction are 0.7 and 0.6, respectively, determine (a) the torque **M** that will keep the rider in the same angular position, (b) the corresponding linear acceleration of the rider.

16.111 A hemisphere of weight W and radius r is released from rest in the position shown. Determine (a) the minimum value of μ_s for which the hemisphere starts to roll without sliding, (b) the corresponding acceleration of point B. [*Hint:* Note that $OG = \frac{3}{8}r$ and that, by the parallel-axis theorem, $\bar{I} = \frac{2}{5}mr^2 - m(OG)^2$.]

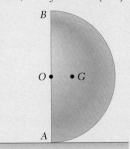

Fig. P16.111

16.112 Solve Prob. 16.111, considering a half cylinder instead of a hemisphere. [*Hint:* Note that $OG = 4r/3\pi$ and that, by the parallel-axis theorem, $\bar{I} = \frac{1}{2}mr^2 - m(OG)^2$.]

16.113 The center of gravity G of a 1.5-kg unbalanced tracking wheel is located at a distance $r = 18$ mm from its geometric center B. The radius of the wheel is $R = 60$ mm and its centroidal radius of gyration is 44 mm. At the instant shown, the center B of the wheel has a velocity of 0.35 m/s and an acceleration of 1.2 m/s², both directed to the left. Knowing that the wheel rolls without sliding and neglecting the mass of the driving yoke AB, determine the horizontal force **P** applied to the yoke.

16.114 A small clamp of mass m_B is attached at B to a hoop of mass m_h. The system is released from rest when $\theta = 90°$ and rolls without sliding. Knowing that $m_h = 3m_B$, determine (a) the angular acceleration of the hoop, (b) the horizontal and vertical components of the acceleration of B.

16.115 A small clamp of mass m_B is attached at B to a hoop of mass m_h. Knowing that the system is released from rest and rolls without sliding, derive an expression for the angular acceleration of the hoop in terms of m_B, m_h, r, and θ.

16.116 A 4-lb bar is attached to a 10-lb uniform cylinder by a square pin, P, as shown. Knowing that $r = 16$ in., $h = 8$ in., $\theta = 20°$, $L = 20$ in., and $\omega = 2$ rad/s at the instant shown, determine the reactions at P at this instant assuming that the cylinder rolls without sliding down the incline.

16.117 The uniform rod AB with a mass m and a length of $2L$ is attached to collars of negligible mass that slide without friction along fixed rods. If the rod is released from rest in the position shown, derive an expression for (a) the angular acceleration of the rod, (b) the reaction at A.

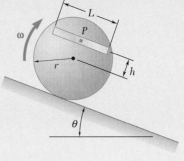

Fig. P16.116

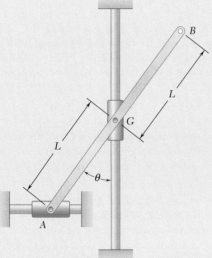

Fig. P16.117 and P16.118

16.118 The 10-lb-uniform rod AB has a total length of $2L = 2$ ft and is attached to collars of negligible mass that slide without friction along fixed rods. If rod AB is released from rest when $\theta = 30°$, determine immediately after release (a) the angular acceleration of the rod, (b) the reaction at A.

16.119 A 40-lb ladder rests against a wall when the bottom begins to slide out. The ladder is 30 ft long and the coefficient of kinetic friction between the ladder and all surfaces is 0.2. For $\theta = 40°$, determine (a) the angular acceleration of the ladder, (b) the forces at A and B.

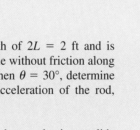

Fig. P16.119

16.120 A beam AB of length L and mass m is supported by two cables as shown. If cable BD breaks, determine at that instant the tension in the remaining cable as a function of its initial angular orientation θ.

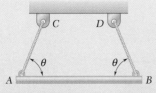

Fig. P16.120

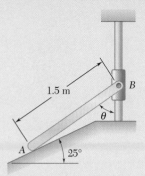

Fig. P16.121 and *P16.122*

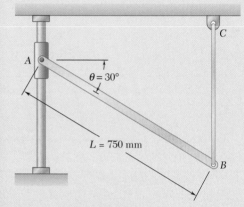

Fig. *P16.123*

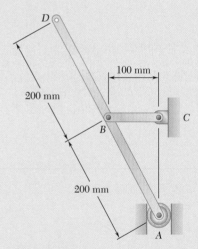

Fig. P16.124

16.121 End *A* of the 6-kg uniform rod *AB* rests on the inclined surface, while end *B* is attached to a collar of negligible mass that can slide along the vertical rod shown. Knowing that the rod is released from rest when $\theta = 35°$ and neglecting the effect of friction, determine immediately after release (*a*) the angular acceleration of the rod, (*b*) the reaction at *B*.

16.122 End *A* of the 6-kg uniform rod *AB* rests on the inclined surface, while end *B* is attached to a collar of negligible mass that can slide along the vertical rod shown. When the rod is at rest, a vertical force **P** is applied at *B*, causing end *B* of the rod to start moving upward with an acceleration of 4 m/s². Knowing that $\theta = 35°$, determine the force **P**.

16.123 End *A* of the 8-kg uniform rod *AB* is attached to a collar that can slide without friction on a vertical rod. End *B* of the rod is attached to a vertical cable *BC*. If the rod is released from rest in the position shown, determine immediately after release (*a*) the angular acceleration of the rod, (*b*) the reaction at *A*.

16.124 The 4-kg uniform rod *ABD* is attached to the crank *BC* and is fitted with a small wheel that can roll without friction along a vertical slot. Knowing that at the instant shown crank *BC* rotates with an angular velocity of 6 rad/s clockwise and an angular acceleration of 15 rad/s² counterclockwise, determine the reaction at *A*.

16.125 The 3-lb uniform rod *BD* is connected to crank *AB* and to a collar of negligible weight. A couple (not shown) is applied to crank *AB*, causing it to rotate with an angular velocity of 12 rad/s counterclockwise and an angular acceleration of 80 rad/s² clockwise at the instant shown. Neglecting the effect of friction, determine the reaction at *D*.

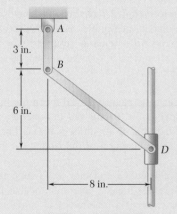

Fig. P16.125 and P16.126

16.126 The 3-lb uniform rod *BD* is connected to crank *AB* and to a collar of negligible weight. A couple (not shown) is applied to crank *AB* causing it to rotate. At the instant shown, crank *AB* has an angular velocity of 12 rad/s and an angular acceleration of 80 rad/s²; both are counterclockwise. Neglecting the effect of friction, determine the reaction at *D*.

16.127 The test rig shown was developed to perform fatigue testing on fitness trampolines. A motor drives the 200-mm radius flywheel AB, which is pinned at its center point A, in a counterclockwise direction with a constant angular velocity of 120 rpm. The flywheel is attached to slider CD by the 400-mm connecting rod BC. The mass of the connecting rod BC is 5 kg, and the mass of the link CD and foot is 2 kg. At the instant when $\theta = 0°$ and the foot is just above the trampoline, determine the force exerted by pin C on rod BC.

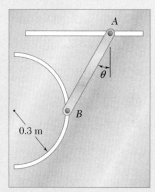

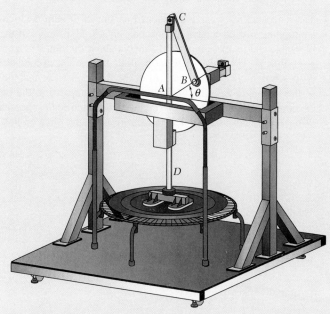

Fig. P16.127

Fig. P16.129

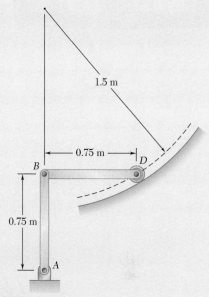

Fig. *P16.130*

16.128 Solve Prob. 16.127 for $\theta = 90°$.

16.129 The 4-kg uniform slender bar BD is attached to bar AB and a wheel of negligible mass that rolls on a circular surface. Knowing that at the instant shown bar AB has an angular velocity of 6 rad/s and no angular acceleration, determine the reaction at point D.

16.130 The motion of the uniform slender rod of length $L = 0.5$ m and mass $m = 3$ kg is guided by pins at A and B that slide freely in frictionless slots, circular and horizontal, cut into a vertical plate as shown. Knowing that at the instant shown the rod has an angular velocity of 3 rad/s counterclockwise and $\theta = 30°$, determine the reactions at points A and B.

16.131 At the instant shown, the 20-ft-long, uniform 100-lb pole ABC has an angular velocity of 1 rad/s counterclockwise and point C is sliding to the right. A 120-lb horizontal force $\mathbf{P}$ acts at B. Knowing the coefficient of kinetic friction between the pole and the ground is 0.3, determine at this instant (*a*) the acceleration of the center of gravity, (*b*) the normal force between the pole and the ground.

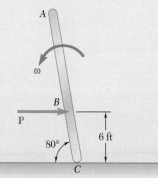

Fig. P16.131

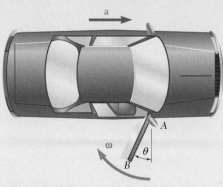

Fig. P16.132

16.132 A driver starts his car with the door on the passenger's side wide open ($\theta = 0$). The 80-lb door has a centroidal radius of gyration $\bar{k} = 12.5$ in., and its mass center is located at a distance $r = 22$ in. from its vertical axis of rotation. Knowing that the driver maintains a constant acceleration of 6 ft/s^2, determine the angular velocity of the door as it slams shut ($\theta = 90°$).

16.133 For the car of Prob. 16.132, determine the smallest constant acceleration that the driver can maintain if the door is to close and latch, knowing that as the door hits the frame its angular velocity must be at least 2 rad/s for the latching mechanism to operate.

16.134 The hatchback of a car is positioned as shown to help determine the appropriate size for a damping mechanism AB. The weight of the door is 40 lbs, and its mass moment of inertia about the center of gravity G is 15 lb·ft·s^2. The linkage $DEFH$ controls the motion of the hatch and is shown in more detail in part (b) of the figure. Assume that the mass of the links DE, EF, and FH are negligible, compared to the mass of the door. With AB removed, determine (a) the initial angular acceleration of the 40-lb door as it is released from rest, (b) the force on link FH.

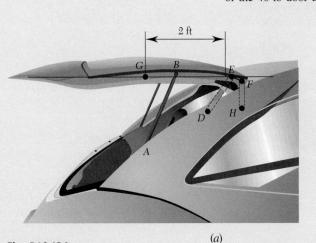

(a)

Fig. P16.134

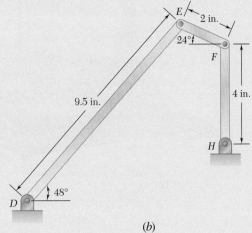

(b)

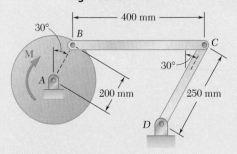

Fig. P16.135 and P16.136

*16.135 The 6-kg rod BC connects a 10-kg disk centered at A to a 5-kg rod CD. The motion of the system is controlled by the couple **M** applied to disk A. Knowing that at the instant shown disk A has an angular velocity of 36 rad/s clockwise and no angular acceleration, determine (a) the couple **M**, (b) the components of the force exerted at C on rod BC.

*16.136 The 6-kg rod BC connects a 10-kg disk centered at A to a 5-kg rod CD. The motion of the system is controlled by the couple **M** applied to disk A. Knowing that at the instant shown disk A has an angular velocity of 36 rad/s clockwise and an angular acceleration of 150 rad/s^2 counterclockwise, determine (a) the couple **M**, (b) the components of the force exerted at C on rod BC.

16.137 In the engine system shown $l = 250$ mm and $b = 100$ mm. The connecting rod BD is assumed to be a 1.2-kg uniform slender rod and is attached to the 1.8-kg piston P. During a test of the system, crank AB is made to rotate with a constant angular velocity of 600 rpm clockwise with no force applied to the face of the piston. Determine the forces exerted on the connecting rod at B and D when $\theta = 180°$. (Neglect the effect of the weight of the rod.)

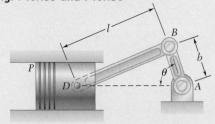

Fig. P16.137

16.138 Solve Prob. 16.137 when $\theta = 90°$.

16.139 The 4-lb uniform slender rod AB, the 8-lb uniform slender rod BF, and the 4-lb uniform thin sleeve CE are connected as shown and move without friction in a vertical plane. The motion of the linkage is controlled by the couple $\mathbf{M}$ applied to rod AB. Knowing that at the instant shown the angular velocity of rod AB is 15 rad/s and the magnitude of the couple $\mathbf{M}$ is 5 ft·lb, determine (a) the angular acceleration of rod AB, (b) the reaction at point D.

16.140 The 4-lb uniform slender rod AB, the 8-lb uniform slender rod BF, and the 4-lb uniform thin sleeve CE are connected as shown and move without friction in a vertical plane. The motion of the linkage is controlled by the couple $\mathbf{M}$ applied to rod AB. Knowing that at the instant shown the angular velocity of rod AB is 30 rad/s and the angular acceleration of rod AB is 96 rad/s² clockwise, determine (a) the magnitude of the couple $\mathbf{M}$, (b) the reaction at point D.

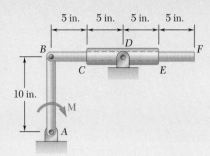

Fig. P16.139 and P16.140

16.141 Two rotating rods in the vertical plane are connected by a slider block P of negligible mass. The rod attached at A has a weight of 1.6 lb and a length of 8 in. Rod BP weighs 2 lb and is 10 in. long and the friction between block P and AE is negligible. The motion of the system is controlled by a couple $\mathbf{M}$ applied to rod BP. Knowing that rod BP has a constant angular velocity of 20 rad/s clockwise, determine (a) the couple $\mathbf{M}$, (b) the components of the force exerted on AE by block P.

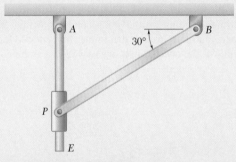

Fig. P16.141 and P16.142

16.142 Two rotating rods in the vertical plane are connected by a slider block P of negligible mass. The rod attached at A has a mass of 0.8 kg and a length of 160 mm. Rod BP has a mass of 1 kg and is 200 mm long and the friction between block P and AE is negligible. The motion of the system is controlled by a couple $\mathbf{M}$ applied to bar BP. Knowing that at the instant shown rod BP has an angular velocity of 20 rad/s clockwise and an angular acceleration of 80 rad/s² clockwise, determine (a) the couple $\mathbf{M}$, (b) the components of the force exerted on AE by block P.

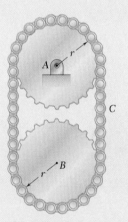

Fig. P16.143

* 16.143 Two disks, each with a mass m and a radius r, are connected as shown by a continuous chain belt of negligible mass. If a pin at point C of the chain belt is suddenly removed, determine (a) the angular acceleration of each disk, (b) the tension in the left-hand portion of the belt, (c) the acceleration of the center of disk B.

* 16.144 A uniform slender bar AB of mass m is suspended as shown from a uniform disk of the same mass m. Neglecting the effect of friction, determine the accelerations of points A and B immediately after a horizontal force $\mathbf{P}$ has been applied at B.

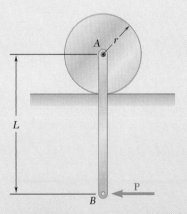

Fig. P16.144

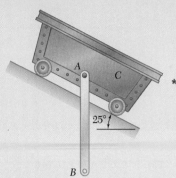

Fig. P16.145

16.145 A uniform rod *AB*, of mass 15 kg and length 1 m, is attached to the 20-kg cart *C*. Neglecting friction, determine immediately after the system has been released from rest, (*a*) the acceleration of the cart, (*b*) the angular acceleration of the rod.

* 16.146 The uniform slender 2-kg bar *BD* is attached to the uniform 6-kg disk by a pin at *B* and released from rest in the position shown. Assuming that the disk rolls without slipping, determine (*a*) the initial reaction at the contact point *A*, (*b*) the corresponding smallest allowable value of the coefficient of static friction.

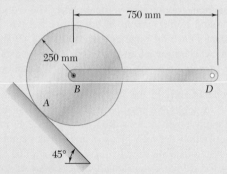

Fig. P16.146

*16.147 and *16.148 The 6-lb cylinder *B* and the 4-lb wedge *A* are held at rest in the position shown by cord *C*. Assuming that the cylinder rolls without sliding on the wedge and neglecting friction between the wedge and the ground, determine, immediately after cord *C* has been cut, (*a*) the acceleration of the wedge, (*b*) the angular acceleration of the cylinder.

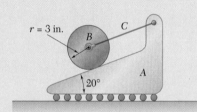

Fig. P16.147

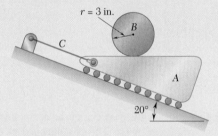

Fig. P16.148

*16.149 Each of the 3-kg bars *AB* and *BC* is of length *L* = 500 mm. A horizontal force **P** of magnitude 20 N is applied to bar *BC* as shown. Knowing that *b* = *L* (**P** is applied at *C*), determine the angular acceleration of each bar.

*16.150 Each of the 3-kg bars *AB* and *BC* is of length *L* = 500 mm. A horizontal force **P** of magnitude 20 N is applied to bar *BC*. For the position shown, determine (*a*) the distance *b* for which the bars move as if they formed a single rigid body, (*b*) the corresponding angular acceleration of the bars.

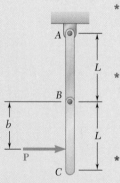

Fig. P16.149 and P16.150

* 16.151 (*a*) Determine the magnitude and the location of the maximum bending moment in the rod of Prob. 16.78. (*b*) Show that the answer to part *a* is independent of the weight of the rod.

*16.152 Draw the shear and bending-moment diagrams for the rod of Prob. 16.84 immediately after the cable at *B* breaks.

Review and Summary

In this chapter, we studied the **kinetics of rigid bodies**, i.e., the relations between the forces acting on a rigid body, the shape and mass of the body, and the motion produced. Except for the first two sections, which apply to the most general case of the motion of a rigid body, our analysis was restricted to the **plane motion of rigid bodies** and rigid bodies symmetrical with respect to the plane of motion. We will study the plane motion of nonsymmetrical rigid bodies and the motion of rigid bodies in three-dimensional space in Chap. 18.

Fundamental Equations of Motion for a Rigid Body

We first recalled [Sec. 16.1A] the two fundamental equations derived in Chap. 14 for the motion of a system of particles and observed that they apply in the most general case of the motion of a rigid body. The first equation defines the motion of the mass center G of the body; we have

$$\Sigma \mathbf{F} = m\bar{\mathbf{a}} \qquad (16.1)$$

where m is the mass of the body and $\bar{\mathbf{a}}$ is the acceleration of G. The second equation is related to the motion of the body relative to a centroidal frame of reference; we have

$$\Sigma \mathbf{M}_G = \dot{\mathbf{H}}_G \qquad (16.2)$$

where $\dot{\mathbf{H}}_G$ is the rate of change of the angular momentum $\mathbf{H}_G$ of the body about its mass center G. Together, Eqs. (16.1) and (16.2) state that **the system of the external forces is equipollent to the system consisting of the vector $m\bar{\mathbf{a}}$ attached at G and the couple of moment $\dot{\mathbf{H}}_G$** (Fig. 16.18).

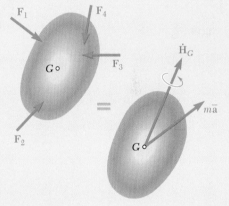

Fig. 16.18

Angular Momentum in Plane Motion

Restricting our analysis at this point and for the rest of the chapter to the plane motion of rigid bodies and rigid bodies symmetrical with respect to the plane of motion, we showed [Sec. 16.1B] that the angular momentum of the body could be expressed as

$$\mathbf{H}_G = \bar{I}\boldsymbol{\omega} \qquad (16.4)$$

where $\bar{I}$ is the moment of inertia of the body about a centroidal axis perpendicular to the reference plane and $\boldsymbol{\omega}$ is the angular velocity of the body. Differentiating both sides of Eq. (16.4), we obtained

$$\dot{\mathbf{H}}_G = \bar{I}\dot{\boldsymbol{\omega}} = \bar{I}\boldsymbol{\alpha} \qquad (16.5)$$

which shows that, in the restricted case considered here, we can represent the rate of change of the angular momentum of the rigid body by a vector of the same direction as $\boldsymbol{\alpha}$ (i.e., perpendicular to the plane of reference) and of magnitude $\bar{I}\alpha$.

Equations for the Plane Motion of a Rigid Body

It follows from [Sec. 16.1E] that the plane motion of a rigid body or of a rigid body symmetrical with respect to the reference plane is defined by the three scalar equations. You will have one equation for the x-direction, one for the y-direction, and one moment equation, as

$$\Sigma F_x = m\bar{a}_x \qquad \Sigma F_y = m\bar{a}_y$$

$$\Sigma M_G = \bar{I}\alpha \quad \text{or} \quad \Sigma M_O = I_O\alpha \quad \text{or} \quad \Sigma M_P = \bar{I}\alpha + m\bar{a}d_\perp \quad \text{or}$$

$$\Sigma M_P = \bar{I}\alpha + \mathbf{r}_{G/P} \times m\bar{\mathbf{a}}$$

where G is the center of mass of the body, O is a fixed axis of rotation, P is any arbitrary point, and $d_\perp$ is the perpendicular distance between point P and the line of action of the acceleration of the center of mass.

Plane Motion of a Rigid Body

It further follows that *the external forces acting on the rigid body are actually **equivalent** to the inertial terms of the various particles forming the body.* This statement can be represented by a free-body diagram and kinetic diagram as shown in Fig. 16.19, where the inertial terms have been represented by a vector $m\bar{\mathbf{a}}$ attached at G and a couple $\bar{I}\boldsymbol{\alpha}$. In the particular case of a rigid body in *translation*, the inertial terms shown in part b of this figure reduce to the single vector $m\bar{\mathbf{a}}$, whereas in the particular case of a rigid body in *centroidal rotation*, they reduce to the single couple $\bar{I}\boldsymbol{\alpha}$. In any other case of plane motion, both the vector $m\bar{\mathbf{a}}$ and the couple $\bar{I}\boldsymbol{\alpha}$ should be included.

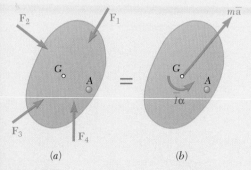

(a)   (b)

Fig. 16.19

Free-Body Diagram and Kinetic Diagram

Any kinetics problem involving the plane motion of a rigid body may be solved by drawing a **free-body diagram and kinetic diagram** similar to that of Fig. 16.19 [Sec. 16.1E]. You can then obtain three equations of motion (see previous equations) by equating the x components, y components, and moments about a chosen point (such as G or some arbitrary point P) of the forces and vectors involved [Sample Probs. 16.1 through 16.5].

Connected Rigid Bodies

We can also use the method described previously to solve problems involving the plane motion of several connected rigid bodies [Sec. 16.1F]. You draw a free-body diagram and kinetic diagram for each system and solve the equations of motion simultaneously. In some cases, however, you can include multiple objects in your system and draw a single diagram for the entire system, including all of the external forces as well as the vectors $m\bar{\mathbf{a}}$ and the couples $\bar{I}\alpha$ associated with the various parts of the system [Sample Prob. 16.4].

Constrained Plane Motion

In the second section of this chapter, we were concerned with rigid bodies *moving under given constraints* [Sec. 16.2]. Although the kinetic analysis of the constrained plane motion of a rigid body is the same as before, it must be supplemented by a *kinematic analysis* that aims to express the components $\bar{a}_x$ and $\bar{a}_y$ of the acceleration of the mass center G of the body in terms of its angular acceleration α. This often involves using analyses that we examined in Ch. 15, including the relationship between two points on a body undergoing general plane motion:

$$\bar{\mathbf{a}} = \mathbf{a}_A + \alpha\mathbf{k} \times \mathbf{r}_{G/A} - \omega^2\mathbf{r}_{G/A}$$

Problems solved in this way included the noncentroidal rotation of rods and plates [Sample Probs. 16.7 and 16.8], the rolling motion of spheres and wheels [Sample Probs. 16.9 and 16.10], the general plane motion of a body with no fixed point [Sample Probs. 16.11 and 16.12], and the plane motion of various types of linkages [Sample Prob. 16.13].

Review Problems

16.153 A cyclist is riding a bicycle at a speed of 20 mph on a horizontal road. The distance between the axles is 42 in., and the mass center of the cyclist and the bicycle is located 26 in. behind the front axle and 40 in. above the ground. If the cyclist applies the brakes only on the front wheel, determine the shortest distance in which he can stop without being thrown over the front wheel.

16.154 The forklift truck shown weighs 2250 lb and is used to lift a crate of weight $W = 2500$ lb. The truck is moving to the left at a speed of 10 ft/s when the brakes are applied on all four wheels. Knowing that the coefficient of static friction between the crate and the fork lift is 0.30, determine the smallest distance in which the truck can be brought to a stop if the crate is not to slide and if the truck is not to tip forward.

Fig. P16.154

16.155 The total mass of the Baja car and driver, including the wheels, is 250 kg. Each pair of 58-cm radius wheels and the axle has a total mass of 20 kg and a mass moment of inertia of 2.9 kg·m². The center of gravity of the driver and Baja body (not including the wheels) is located $x = 0.70$ m from the rear axle A and $y = 0.55$ m from the ground. The wheelbase is $L = 1.60$ m. If the engine exerts a torque of 500 N·m on the rear axle, what is the car's acceleration?

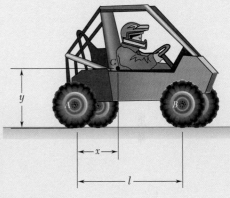

Fig. P16.155

16.156 Identical cylinders of mass m and radius r are pushed by a series of moving arms. Assuming the coefficient of friction between all surfaces to be $\mu < 1$ and denoting by a the magnitude of the acceleration of the arms, derive an expression for (a) the maximum allowable value of a if each cylinder is to roll without sliding, (b) the minimum allowable value of a if each cylinder is to move to the right without rotating.

Fig. P16.156

16.157 The uniform rod AB of weight W is released from rest when $\beta = 70°$. Assuming that the friction force between end A and the surface is large enough to prevent sliding, determine immediately after release (*a*) the angular acceleration of the rod, (*b*) the normal reaction at A, (*c*) the friction force at A.

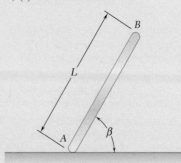

Fig. P16.157 and P16.158

16.158 The uniform rod AB of weight W is released from rest when $\beta = 70°$. Assuming that the friction force is zero between end A and the surface, determine immediately after release (*a*) the angular acceleration of the rod, (*b*) the acceleration of the mass center of the rod, (*c*) the reaction at A.

16.159 A bar of mass $m = 5$ kg is held as shown between four disks, each of mass $m' = 2$ kg and radius $r = 75$ mm. Knowing that the normal forces on the disks are sufficient to prevent any slipping, for each of the cases shown determine the acceleration of the bar immediately after it has been released from rest.

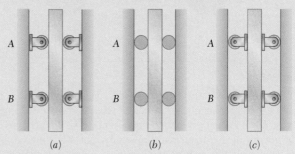

Fig. *P16.159*

16.160 A uniform plate of mass m is suspended in each of the ways shown. For each case determine immediately after the connection B has been released (*a*) the angular acceleration of the plate, (*b*) the acceleration of its mass center.

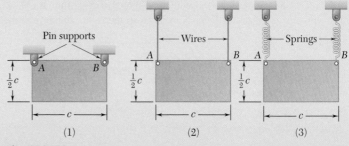

Fig. P16.160

16.161 A cylinder with a circular hole is rolling without slipping on a fixed curved surface as shown. The cylinder would have a weight of 16 lb without the hole, but with the hole it has a weight of 15 lb. Knowing that at the instant shown the disk has an angular velocity of 5 rad/s clockwise, determine (*a*) the angular acceleration of the disk, (*b*) the components of the reaction force between the cylinder and the ground at this instant.

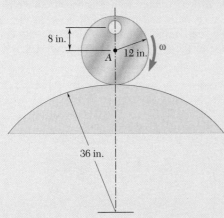

Fig. P16.161

16.162 The motion of a square plate of side 150 mm and mass 2.5 kg is guided by pins at corners *A* and *B* that slide in slots cut in a vertical wall. Immediately after the plate is released from rest in the position shown, determine (*a*) the angular acceleration of the plate, (*b*) the reaction at corner *A*.

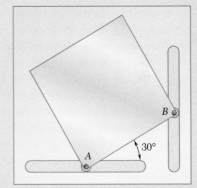

Fig. P16.162

16.163 The motion of a square plate of side 150 mm and mass 2.5 kg is guided by a pin at corner *A* that slides in a horizontal slot cut in a vertical wall. Immediately after the plate is released from rest in the position shown, determine (*a*) the angular acceleration of the plate, (*b*) the reaction at corner *A*.

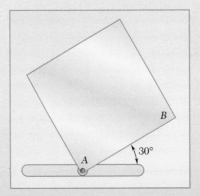

Fig. P16.163

16.164 The Geneva mechanism shown is used to provide an intermittent rotary motion of disk S. Disk D weighs 2 lb and has a radius of gyration of 0.9 in., and disk S weighs 6 lb and has a radius of gyration of 1.5 in. The motion of the system is controlled by a couple **M** applied to disk D. A pin P is attached to disk D and can slide in one of the six equally spaced slots cut in disk S. It is desirable that the angular velocity of disk S be zero as the pin enters and leaves each of the six slots; this will occur if the distance between the centers of the disks and the radii of the disks are related as shown. Knowing disk D rotates with a constant counterclockwise angular velocity of 8 rad/s and the friction between the slot and pin P is negligible, determine when $\phi = 150°$ (a) the couple **M**, (b) the magnitude of the force pin P applies to disk S.

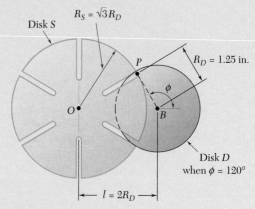

Fig. P16.164

1180

17

Plane Motion of Rigid Bodies: Energy and Momentum Methods

In this chapter the energy and momentum methods will be added to the tools available for your study of the motion of rigid bodies. We can analyze the transfer between potential and kinetic energy as the gymnast goes from a high position to a lower one, and we can use conservation of angular momentum to examine how changes in the gymnast's body position affect his angular velocity.

Objectives

- **Calculate** the work done by a force or a moment on a rigid body.
- **Calculate** the kinetic energy of a rigid body in plane motion.
- **Solve** rigid body kinetics problems using the principle of work and energy.
- **Solve** rigid body kinetics problems using conservation of energy.
- **Calculate** the power of a mechanical system of rigid bodies.
- **Draw** complete and accurate impulse–momentum diagrams for problems involving rigid bodies.
- **Solve** rigid body kinetics problems using the principles of linear impulse and momentum and of angular impulse and momentum.
- **Solve** rigid body kinetics problems using conservation of angular momentum.
- **Solve** rigid body problems involving eccentric impact by using the principle of impulse and momentum and the coefficient of restitution.

Introduction

In this chapter, we return to the method of work and energy and the method of impulse and momentum that were introduced in Chapter 13 in the context of particle kinetics. Here we use them to analyze the plane motion of rigid bodies and of systems of rigid bodies.

We consider the method of work and energy first. We define the work of a force and of a couple, and we obtain an expression for the kinetic energy of a rigid body in plane motion. Then we use the principle of work and energy to solve problems involving displacements and velocities. We also apply the principle of conservation of energy to solve a variety of engineering problems.

In the second section, we apply the principle of impulse and momentum to solve problems involving velocities and time. We also discuss the concept of conservation of angular momentum for rigid bodies in plane motion.

In the last section of this chapter, we consider problems involving the eccentric impact of rigid bodies. As we did in Chap. 13, where we analyzed the impact of particles, we use the coefficient of restitution between colliding bodies, together with the principle of impulse and momentum, to solve impact problems. We will show that the method used is applicable not only when the colliding bodies move freely after the impact but also when the bodies are partially constrained in their motion.

17.1 ENERGY METHODS FOR A RIGID BODY

We now use the principle of work and energy to analyze the plane motion of rigid bodies. As we pointed out in Chap. 13, the method of work and energy is particularly well adapted to solving problems involving velocities and displacements. Its main advantage is that the work of forces and the kinetic energy of particles are scalar quantities.

17.1A Principle of Work and Energy

To apply the principle of work and energy to the motion of a rigid body, we again assume that the rigid body is made up of a large number n of particles of mass Δm_i. From Eq. (14.30) of Sec. 14.2B, we have

Principle of work and energy, rigid body

$$T_1 + U_{1 \to 2} = T_2 \tag{17.1}$$

where $T_1, T_2 =$ the initial and final values of total kinetic energy of particles forming the rigid body

$U_{1 \to 2} =$ work of all forces acting on various particles of the body

Just as we did in Chap. 13, we can express the work done by nonconservative forces as $U_{1 \to 2}^{NC}$, and we can define potential energy terms for conservative forces. Then we can express Eq. (17.1) as

$$T_1 + V_{g_1} + V_{e_1} + U_{1 \to 2}^{NC} = T_2 + V_{g_2} + V_{e_2} \tag{17.1'}$$

where V_{g_1} and V_{g_2} are the initial and final gravitational potential energy of the center of mass of the rigid body with respect to a reference point or datum, and V_{e_1} and V_{e_2} are the initial and final values of the elastic energy associated with springs in the system.

We obtain the total kinetic energy

$$T = \frac{1}{2} \sum_{i=1}^{n} \Delta m_i v_i^2 \tag{17.2}$$

Photo 17.1 The work done by friction reduces the kinetic energy of the wheel.

by adding positive scalar quantities, so it is itself a positive scalar quantity. You will see later how to determine T for various types of motion of a rigid body.

The expression $U_{1 \to 2}$ in Eq. (17.1) represents the work of all the forces acting on the various particles of the body, whether these forces are internal or external. However, the total work of the internal forces holding together the particles of a rigid body is zero. To see this, consider two particles A and B of a rigid body and the two equal and opposite forces $\mathbf{F}$ and $-\mathbf{F}$ they exert on each other (Fig. 17.1). Although, in general, small displacements $d\mathbf{r}$ and $d\mathbf{r}'$ of the two particles are different, the components of these displacements along AB must be equal; otherwise, the particles would not remain at the same distance from each other and the body would not be rigid. Therefore, the work of $\mathbf{F}$ is equal in magnitude and opposite

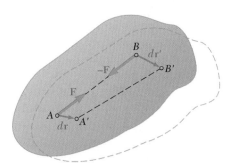

Fig. 17.1 The total work of the internal forces acting on the particles of a rigid body is zero.

in sign to the work of $-\mathbf{F}$, and their sum is zero. Thus, the total work of the internal forces acting on the particles of a rigid body is zero, and the expression $U_{1\rightarrow2}$ in Eq. (17.1) reduces to the work of the external forces acting on the body during the displacement considered.

17.1B   Work of Forces Acting on a Rigid Body

We saw in Sec. 13.1A that the work of a force $\mathbf{F}$ during a displacement of its point of application from A_1 to A_2 is

Work of a force
$$U_{1\rightarrow2} = \int_{A_1}^{A_2} \mathbf{F}\cdot d\mathbf{r} \qquad \textbf{(17.3)}$$

or

$$U_{1\rightarrow2} = \int_{s_1}^{s_2} (F\cos\alpha)\,ds \qquad \textbf{(17.3')}$$

where F is the magnitude of the force, α is the angle it forms with the direction of motion of its point of application A, and s is the variable of integration that measures the distance traveled by A along its path.

In computing the work of the external forces acting on a rigid body, it is often convenient to determine the work of a couple without considering the work of each of the two forces forming the couple separately. Consider the two forces $\mathbf{F}$ and $-\mathbf{F}$ forming a couple of moment $\mathbf{M}$ and acting on a rigid body (Fig. 17.2). Any small displacement of the rigid body bringing A and B, respectively, into A' and B'' can be divided into two parts: in one part, points A and B undergo equal displacements $d\mathbf{r}_1$; in the other part, A' remains fixed, while B' moves into B'' through a displacement $d\mathbf{r}_2$ with a magnitude of $ds_2 = r\,d\theta$. In the first part of the motion, the work of $\mathbf{F}$ is equal in magnitude and opposite in sign to the work of $-\mathbf{F}$, and their sum is zero. In the second part of the motion, only force $\mathbf{F}$ works, and its work is $dU = F\,ds_2 = Fr\,d\theta$. But the product Fr is equal to the magnitude M of the moment of the couple. Thus, the work of a couple of moment $\mathbf{M}$ acting on a rigid body is

$$dU = M\,d\theta \qquad \textbf{(17.4)}$$

where $d\theta$ is the small angle through which the body rotates and is expressed in radians. (We again note that work should be expressed in units obtained by multiplying units of force by units of length.) To obtain the work of the couple during a finite rotation of the rigid body, we integrate both members of Eq. (17.4) from the initial value θ_1 of the angle θ to its final value θ_2.

$$U_{1\rightarrow2} = \int_{\theta_1}^{\theta_2} M\,d\theta \qquad \textbf{(17.5)}$$

When the moment $\mathbf{M}$ of the couple is constant, formula (17.5) reduces to

$$U_{1\rightarrow2} = \mathrm{M}(\theta_2 - \theta_1) \qquad \textbf{(17.6)}$$

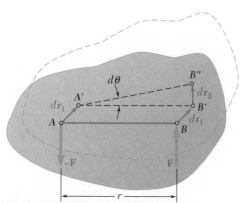

Fig. 17.2 The work of a couple acting on a rigid body equals the integral of the moment $\mathbf{M}$ of the couple with respect to the angular displacement of the body.

We pointed out in Sec. 13.1A that some forces encountered in problems of kinetics *do no work*. These include forces applied to fixed points or acting in a direction perpendicular to the displacement of their point of application. Among these forces are the reaction at a frictionless pin when the body rotates about the pin; the reaction at a frictionless surface when the body in contact moves along the surface; and the weight of a body when its center of gravity moves horizontally. We can now add that

When a rigid body rolls without sliding on a fixed surface, the friction force F at the point of contact C does no work.

The velocity $\mathbf{v}_C$ of the point of contact C is zero, and the work of the friction force $\mathbf{F}$ during a small displacement of the rigid body is

$$dU = F\,ds_C = F(v_C\,dt) = 0$$

17.1C Kinetic Energy of a Rigid Body in Plane Motion

Consider a rigid body with a mass m in plane motion. Recall from Sec. 14.2A that, if the absolute velocity $\mathbf{v}_i$ of each particle P_i of the body is expressed as the sum of the velocity $\bar{\mathbf{v}}$ of the mass center G of the body and of the velocity $\mathbf{v}'_i$ of the particle relative to a frame $Gx'y'$ attached to G and of fixed orientation (Fig. 17.3), we can express the kinetic energy of the system of particles forming the rigid body in the form

$$T = \tfrac{1}{2}m\bar{v}^2 + \frac{1}{2}\sum_{i=1}^{n}\Delta m_i v'^2_i \tag{17.7}$$

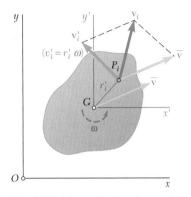

Fig. 17.3 The velocity of a particle P_i is the vector sum of the velocity of the mass center G and the tangential velocity $r'_i\omega$ due to rotation about G.

As you can see in Fig. 17.3, v'_i of particle P_i is equal to the product $r'_i\omega$, where r'_i is the distance from G to P_i and ω is the angular velocity of the body at the instant considered. Substituting into Eq. (17.7), we have

$$T = \tfrac{1}{2}m\bar{v}^2 + \frac{1}{2}\left(\sum_{i=1}^{n} r'^2_i\,\Delta m_i\right)\omega^2 \tag{17.8}$$

The sum represents the moment of inertia $\bar{I}$ of the body about the axis through G, so we have

**Kinetic energy of
a rigid body**

$$T = \tfrac{1}{2}m\bar{v}^2 + \tfrac{1}{2}\bar{I}\omega^2 \tag{17.9}$$

Note that, in the particular case of a body in translation ($\omega = 0$), this expression reduces to $\tfrac{1}{2}m\bar{v}^2$, whereas in the case of a centroidal rotation ($\bar{v} = 0$), it reduces to $\tfrac{1}{2}\bar{I}\omega^2$. We conclude that we can separate the kinetic energy of a rigid body in plane motion into two parts: (1) the kinetic energy $\tfrac{1}{2}m\bar{v}^2$ associated with the motion of the mass center G of the body and (2) the kinetic energy $\tfrac{1}{2}\bar{I}\omega^2$ associated with the rotation of the body about G.

Noncentroidal Rotation. The relation in Eq. (17.9) is valid for any type of plane motion, so we can use it to express the kinetic energy

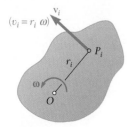

Fig. 17.4 For noncentroidal rotation, the velocity of a particle P_i is the tangential velocity $r_i\omega$ due to rotation about O.

of a rigid body rotating with an angular velocity $\boldsymbol{\omega}$ about a fixed axis through O (Fig. 17.4). In that case, however, we can express the kinetic energy of the body more directly by noting that the speed v_i of particle P_i is equal to $r_i\omega$, where r_i is the distance from the fixed axis to P_i and ω is the angular velocity of the body at the instant considered. Substituting into Eq. (17.2), we have

$$T = \frac{1}{2}\sum_{i=1}^{n} \Delta m_i (r_i\omega)^2 = \frac{1}{2}\left(\sum_{i=1}^{n} r_i^2\,\Delta m_i \right)\omega^2$$

The last sum represents the moment of inertia I_O of the body about the fixed axis through O, so this equation reduces to

$$T = \tfrac{1}{2}I_O\omega^2 \tag{17.10}$$

Note that these results are not limited to the motion of plane rigid bodies or to the motion of bodies that are symmetrical with respect to the reference plane—we can apply them to the study of the plane motion of any rigid body regardless of its shape. However, remember that Eq. (17.9) is applicable to any plane motion, whereas Eq. (17.10) is applicable only in cases involving rotating about a fixed axis.

17.1D   Systems of Rigid Bodies

When a problem involves several rigid bodies, we usually analyze all of the bodies together as a system instead of analyzing each individual rigid body separately. Adding the kinetic energies of all the rigid bodies and considering the work of all the forces involved, we can write the equation of work and energy for the entire system. We have

$$T_1 + U_{1\rightarrow2} = T_2 \tag{17.11}$$

where T represents the arithmetic sum of the kinetic energies of the rigid bodies forming the system (all terms are positive) and $U_{1\rightarrow2}$ represents the work of all the forces acting on the various bodies—whether these forces are *internal* or *external* from the point of view of the system as a whole.

The method of work and energy is particularly useful in solving problems involving pin-connected members, blocks and pulleys connected by inextensible cords, and meshed gears. In all of these cases, the internal forces occur in pairs of equal and opposite forces, and the points of application of the forces in each pair *move through equal distances* during a small displacement of the system. As a result, the work of the internal forces is zero, and $U_{1\rightarrow2}$ reduces to the work of the *forces external to the system.*

17.1E   Conservation of Energy

We saw in Sec. 13.2A that the work of conservative forces, such as the weight of a body or the force exerted by a spring, can be expressed as a change in potential energy. When a rigid body, or a system of rigid bodies, moves under the action of conservative forces, we can express the principle of work and energy in a modified form. Substituting for $U_{1\to2}$ from Eq. (13.19′) into Eq. (17.1), we have

**Conservation of
energy, rigid body**

$$T_1 + V_1 = T_2 + V_2 \qquad\qquad \textbf{(17.12)}$$

In Ch. 13, we discussed two types of potential energy: gravitational potential energy, V_g, and elastic potential energy, V_e. Therefore, another way to write Eq. (17.12) is

$$T_1 + V_{g_1} + V_{e_1} = T_2 + V_{g_2} + V_{e_2} \qquad\qquad \textbf{(17.12′)}$$

Formulas (17.12) and (17.12′) indicate that when a rigid body, or a system of rigid bodies, moves under the action of conservative forces, **the sum of the kinetic energy and of the potential energy of the system remains constant**. Note that, in the case of the plane motion of a rigid body, the kinetic energy of the body should include both the *translational* term $\frac{1}{2}m\bar{v}^2$ and the *rotational* term $\frac{1}{2}\bar{I}\omega^2$.

As an example of applying the principle of conservation of energy, let us consider a slender rod AB with a length l and a mass m, whose ends are connected to blocks of negligible mass sliding along horizontal and vertical tracks. We assume that the rod is released with no initial velocity from a horizontal position (Fig. 17.5a), and we wish to determine its angular velocity after it has rotated through an angle θ (Fig. 17.5b).

Since the initial velocity is zero, we have $T_1 = 0$. Measuring the potential energy from the level of the horizontal track, we have $V_1 = 0$. After the rod has rotated through θ, the center of gravity G of the rod is at a distance $\frac{1}{2}l\sin\theta$ below the reference level, and we have

$$V_2 = -\tfrac{1}{2}Wl \sin\theta = -\tfrac{1}{2}mgl \sin\theta$$

In this position, the instantaneous center of the rod is located at C and $CG = \frac{1}{2}l$, so $\bar{v}_2 = \frac{1}{2}l\omega$, and we obtain

$$T_2 = \tfrac{1}{2}m\bar{v}_2^2 + \tfrac{1}{2}\bar{I}\omega_2^2 = \tfrac{1}{2}m(\tfrac{1}{2}l\omega)^2 + \tfrac{1}{2}(\tfrac{1}{12}ml^2)\omega^2$$

$$= \frac{1}{2}\frac{ml^2}{3}\omega^2$$

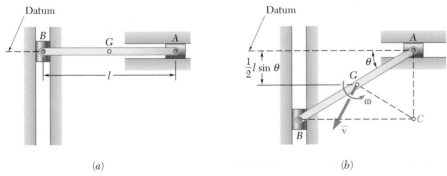

(a)   (b)

Fig. 17.5 (a) Rod AB in position 1 with the datum defined as shown. (b) Rod AB in position 2 with an instantaneous center C.

Applying the principle of conservation of energy gives

$$T_1 + V_1 = T_2 + V_2$$

$$0 = \frac{1}{2}\frac{ml^2}{3}\omega^2 - \tfrac{1}{2}mgl\sin\theta$$

$$\omega = \left(\frac{3g}{l}\sin\theta\right)^{1/2}$$

The advantages of the method of work and energy, as well as its shortcomings, were indicated in Sec. 13.1C. Here we should add that we need to supplement the method of work and energy by the application of Newton's second law when we need to determine reactions at fixed axles, rollers, or sliding blocks. For example, in order to compute the reactions at the ends A and B of the rod of Fig. 17.5b, we need to draw a free-body diagram and a kinetic diagram to show that the system of the external forces applied to the rod is equivalent to both the vector $m\bar{\mathbf{a}}$ and the couple $\bar{I}\boldsymbol{\alpha}$. However, we first need to determine the angular velocity $\boldsymbol{\omega}$ of the rod using the method of work and energy before solving the equations of motion for the reactions. The complete analysis of the motion of the rod and of the forces exerted on the rod requires, therefore, the combined use of the method of work and energy and of the principle of equivalence of the external forces and moments and inertial terms.

17.1F   Power

We defined **power** in Sec. 13.1D as the time rate at which work is done. In the case of a body acted upon by a force $\mathbf{F}$ and moving with a velocity $\mathbf{v}$, we expressed the power as

$$\text{Power} = \frac{dU}{dt} = \mathbf{F}\cdot\mathbf{v} \tag{13.13}$$

In the case of a rigid body rotating with an angular velocity $\boldsymbol{\omega}$ and acted upon by a couple of moment $\mathbf{M}$ parallel to the axis of rotation, we have, by Eq. (17.4),

$$\text{Power} = \frac{dU}{dt} = \frac{M\,d\theta}{dt} = M\omega \tag{17.13}$$

The various units used to measure power, such as the watt and the horse-power, were defined in Sec. 13.1D.

Sample Problem 17.1

1.25 ft

240 lb

A 240-lb block is suspended from an inextensible cable that is wrapped around a drum with a 1.25-ft radius that is rigidly attached to a flywheel. The drum and flywheel have a combined centroidal moment of inertia of $\bar{I} = 10.5$ lb·ft·s². At the instant shown, the velocity of the block is 6 ft/s directed downward. Knowing that the bearing at A is poorly lubricated so that the bearing friction is equivalent to a couple **M** of magnitude 60 lb·ft, determine the velocity of the block after it has moved 4 ft downward.

STRATEGY: Since you have two positions and are interested in determining the velocity of the block, use the principle of work and energy.

MODELING: Consider the system formed by the flywheel and the block. Since the cable is inextensible, the work done by the internal forces exerted by the cable cancels out to zero. The initial and final positions of the system and the external forces acting on the system are shown in Fig. 1.

ANALYSIS: Apply the principle of work and energy

$$T_1 + U_{1\to2} = T_2 \tag{1}$$

Kinetic Energy. You need to calculate the initial and final kinetic energy and the work.

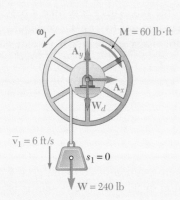

ω_1 M = 60 lb·ft

A_y

A_x

W_d

$\bar{v}_1 = 6$ ft/s $s_1 = 0$

W = 240 lb

Position 1.

Block: $\qquad\qquad \bar{v}_1 = 6$ ft/s

Flywheel: $\qquad\qquad w_1 = \dfrac{\bar{v}_1}{r} = \dfrac{6 \text{ ft/s}}{1.25 \text{ ft}} = 4.80$ rad/s

$$T_1 = \tfrac{1}{2}m\bar{v}_1^2 + \tfrac{1}{2}\bar{I}\omega_1^2$$

$$= \frac{1}{2}\frac{240 \text{ lb}}{32.2 \text{ ft/s}^2}(6 \text{ ft/s})^2 + \tfrac{1}{2}(10.5 \text{ lb·ft·s}^2)(4.80 \text{ rad/s})^2$$

$$= 255 \text{ ft·lb}$$

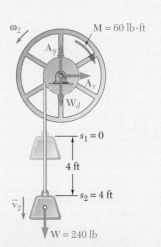

ω_2 M = 60 lb·ft

A_y

A_x

W_d

$s_1 = 0$

4 ft

$s_2 = 4$ ft

$\bar{v}_2$

W = 240 lb

Position 2. Noting that $\omega_2 = \bar{v}_2/1.25$, you have

$$T_2 = \tfrac{1}{2}m\bar{v}_2^2 + \tfrac{1}{2}\bar{I}\omega_2^2$$

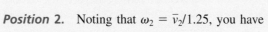
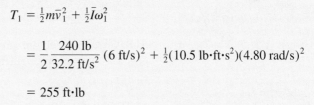

$$= \frac{1}{2}\frac{240}{32.2}(\bar{v}_2)^2 + (\tfrac{1}{2})(10.5)\left(\frac{\bar{v}_2}{1.25}\right)^2 = 7.09\bar{v}_2^2$$

Fig. 1 Free body diagram of the system in positions 1 and 2.

(continued)

Work.   During the motion, only the weight $\mathbf{W}$ of the block and the friction couple $\mathbf{M}$ do work. Note that $\mathbf{W}$ does positive work, and the friction couple $\mathbf{M}$ does negative work. The total work done is

$$s_1 = 0 \qquad s_2 = 4 \text{ ft}$$

$$\theta_1 = 0 \qquad \theta_2 = \frac{s_2}{r} = \frac{4 \text{ ft}}{1.25 \text{ ft}} = 3.20 \text{ rad}$$

$$U_{1 \to 2} = W(s_2 - s_1) - M(\theta_2 - \theta_1)$$

$$= (240 \text{ lb})(4 \text{ ft}) - (60 \text{ lb·ft})(3.20 \text{ rad})$$

$$= 768 \text{ ft·lb}$$

Substituting these expressions into Eq. (1) gives

$$T_1 + U_{1 \to 2} = T_2$$
$$255 \text{ ft·lb} + 768 \text{ ft·lb} = 7.09\bar{v}_2^2$$
$$\bar{v}_2 = 12.01 \text{ ft/s} \qquad \bar{\mathbf{v}}_2 = 12.01 \text{ ft/s} \downarrow \quad \blacktriangleleft$$

REFLECT and THINK: The speed of the block increases as it falls, but much more slowly than if it were in free fall. This seems like a reasonable result. Rather than calculating the work done by gravity, you could have also treated the effect of the weight using gravitational potential energy, V_g.

Sample Problem 17.2

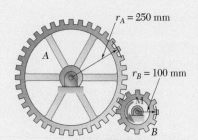

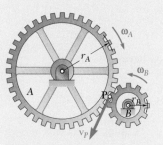

Fig. 1 The point of contact has the same velocity on each gear.

Gear A has a mass of 10 kg and a radius of gyration of 200 mm; gear B has a mass of 3 kg and a radius of gyration of 80 mm. The system is at rest when a couple $\mathbf{M}$ of magnitude 6 N·m is applied to gear B. Neglecting friction, determine (a) the number of revolutions executed by gear B before its angular velocity reaches 600 rpm, (b) the tangential force that gear B exerts on gear A.

STRATEGY: You are given a couple and are asked to determine the position at a given angular velocity, so use the principle of work and energy.

MODELING: For part (a), choose the system to be both gears and model each as a rigid body. In part (b), you are asked to determine an internal force, so you need to choose gear A as your system.

ANALYSIS:

Kinematics.   The velocity of the point of contact, P, is the same for both gears (Fig. 1), so you have

$$v_P = r_A \omega_A = r_B \omega_B \qquad \omega_A = \omega_B \frac{r_B}{r_A} = \omega_B \frac{100 \text{ mm}}{250 \text{ mm}} = 0.40 \omega_B$$

Calculations.   For $\omega_B = 600$ rpm, you have

$$\omega_B = 62.8 \text{ rad/s} \qquad \omega_A = 0.40\omega_B = 25.1 \text{ rad/s}$$
$$\bar{I}_A = m_A\bar{k}_A^2 = (10 \text{ kg})(0.200 \text{ m})^2 = 0.400 \text{ kg·m}^2$$
$$\bar{I}_B = m_B\bar{k}_B^2 = (3 \text{ kg})(0.080 \text{ m})^2 = 0.0192 \text{ kg·m}^2$$

Principle of Work and Energy: Apply the principle of work and energy

$$T_1 + U_{1\rightarrow 2} = T_2 \tag{1}$$

You need to calculate the initial and final kinetic energy and the work.

Kinetic Energy.   The system is initially at rest, so $T_1 = 0$. Adding the kinetic energies of the two gears when $\omega_B = 600$ rpm gives

$$T_2 = \tfrac{1}{2}\bar{I}_A\omega_A^2 + \tfrac{1}{2}\bar{I}_B\omega_B^2$$
$$= \tfrac{1}{2}(0.400 \text{ kg·m}^2)(25.1 \text{ rad/s})^2 + \tfrac{1}{2}(0.0192 \text{ kg·m}^2)(62.8 \text{ rad/s})^2$$
$$= 163.9 \text{ J}$$

Work.   Denote the angular displacement of gear B by θ_B. Then

$$U_{1\rightarrow 2} = M\theta_B = (6 \text{ N·m})(\theta_B \text{ rad}) = (6\theta_B) \text{ J}$$

Substituting these terms into Eq. (1) gives you

$$0 + (6\theta_B) \text{ J} = 163.9 \text{ J}$$
$$\theta_B = 27.32 \text{ rad} \qquad \theta_B = 4.35 \text{ rev} \quad \blacktriangleleft$$

Motion of Gear A.

Kinetic Energy.   Initially, gear A is at rest, so $T_1 = 0$. When $\omega_B = 600$ rpm, the kinetic energy of gear A is

$$T_2 = \tfrac{1}{2}\bar{I}_A\omega_A^2 = \tfrac{1}{2}(0.400 \text{ kg·m}^2)(25.1 \text{ rad/s})^2 = 126.0 \text{ J}$$

Work.   The forces acting on gear A are shown in Fig. 2. The tangential force **F** does work equal to the product of its magnitude and of the length $\theta_A r_A$ of the arc described by the point of contact. Since $\theta_A r_A = \theta_B r_B$, you have

$$U_{1\rightarrow 2} = F(\theta_B r_B) = F(27.3 \text{ rad})(0.100 \text{ m}) = F(2.73 \text{ m})$$

Substituting these values into work and energy gives

$$T_1 + U_{1\rightarrow 2} = T_2$$
$$0 + F(2.73 \text{ m}) = 126.0 \text{ J}$$
$$F = +46.2 \text{ N} \qquad \mathbf{F} = 46.2 \text{ N}\swarrow \quad \blacktriangleleft$$

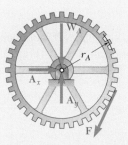

Fig. 2 Free-body diagram for gear *A*.

REFLECT and THINK: When the system was both gears, the tangential force between the gears did not appear in the work and energy equation, since it was internal to the system and therefore did no work. If you want to determine an internal force, you need to define a system where the force of interest is an external force. This problem, like most problems, also could have been solved using Newton's second law and kinematic relationships.

Sample Problem 17.3

A sphere, a cylinder, and a hoop, each having the same mass and the same radius, are released from rest on an incline. Determine the velocity of each body after it has rolled through a distance corresponding to a change in elevation h.

STRATEGY: You are given two positions, want to find the velocities, and the friction force $\mathbf{F}$ in rolling motion does no work, so use the conservation of energy. First solve the problem in general terms, and then find the results for each body. Denote the mass by m, the centroidal moment of inertia by $\bar{I}$, and the radius by r.

MODELING: Choose the rolling object as your system and model it as a rigid body. Since each body rolls, the instantaneous center of rotation is located at C (Fig. 1). Free-body diagrams of at the two locations are shown in Fig. 2.

ANALYSIS:

Conservation of Energy.

$$T_1 + V_{g_1} + V_{e_1} = T_2 + V_{g_2} + V_{e_2} \tag{1}$$

Potential Energy.

Because there is no spring in the system, $V_{e_1} = V_{e_2} = 0$. If you place your datum at the center of mass of the system when it is at position 2, you have $V_{g_2} = 0$ and $V_{g_1} = mgh$.

Kinetic Energy.

$$T_1 = 0$$
$$T_2 = \tfrac{1}{2}m\bar{v}^2 + \tfrac{1}{2}\bar{I}\omega^2$$

Kinematics. You need to relate $\bar{v}$ and ω using kinematics. Since each body rolls, the instantaneous center of rotation is located at C (Fig. 1), which gives

$$\omega = \frac{\bar{v}}{r}$$

Substituting this into T_2 gives

$$T_2 = \tfrac{1}{2}m\bar{v}^2 + \tfrac{1}{2}\bar{I}\left(\frac{\bar{v}}{r}\right)^2 = \tfrac{1}{2}\left(m + \frac{\bar{I}}{r^2}\right)\bar{v}^2$$

Substituting these energy expressions into Eq. (1) gives

$$0 + mgh + 0 = \tfrac{1}{2}\left(m + \frac{\bar{I}}{r^2}\right)\bar{v}^2 + 0 + 0$$

Solving for the speed at position 2, you find

$$\bar{v}^2 = \frac{2gh}{1 + \bar{I}/mr^2}$$

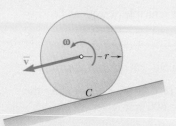

Fig. 1 Angular velocity and the velocity of the center of mass of the rolling object.

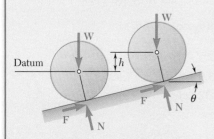

Fig. 2 Free-body diagrams of the system in positions 1 and 2.

Velocities of Sphere, Cylinder, and Hoop. Introducing the particular expressions for $\bar{I}$, you obtain

Sphere: $\qquad\qquad\qquad\qquad \bar{I} = \frac{2}{5} mr^2 \qquad\qquad \bar{v} = 0.845\sqrt{2gh}$ ◀

Cylinder: $\qquad\qquad\qquad\quad \bar{I} = \frac{1}{2} mr^2 \qquad\qquad \bar{v} = 0.816\sqrt{2gh}$ ◀

Hoop: $\qquad\qquad\qquad\qquad \bar{I} = mr^2 \qquad\qquad\quad \bar{v} = 0.707\sqrt{2gh}$ ◀

REFLECT and THINK: Comparing the results, we note that the velocity of the body is independent of both its mass and radius. However, the velocity does depend upon the quotient of $\bar{I}/mr^2 = \bar{k}^2/r^2$, which measures the ratio of the rotational kinetic energy to the translational kinetic energy. Thus the hoop, which has the largest $\bar{k}$ for a given radius r, attains the smallest velocity, whereas the sliding block, which does not rotate, attains the largest velocity.

Let us compare the results with the velocity attained by a frictionless block sliding through the same distance. The solution is identical to the previous solution except that $\omega = 0$; we find $\bar{v} = \sqrt{2gh}$. So, all the rolling objects are slower than one moving down a frictionless surface.

Sample Problem 17.4

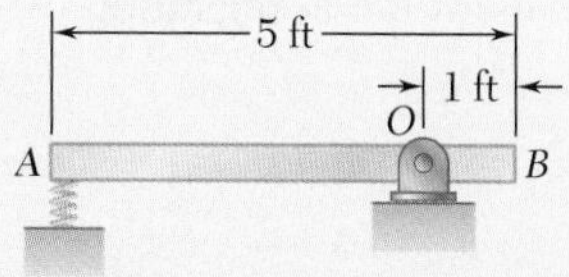

A 30-lb slender rod AB is 5 ft long and is pivoted about a point O that is 1 ft from end B. The other end is pressed against a vertical spring with a constant of $k = 1800$ lb/in. until the spring is compressed 1 in. The rod is then in a horizontal position. If the rod is released from this position, determine its angular velocity and the reaction at the pivot O as the rod passes through a vertical position.

STRATEGY: Since you are given two positions, want to find the velocities, and no external forces do work, use the conservation of energy. To determine the reactions at position 2, use a free-body diagram and a kinetic diagram.

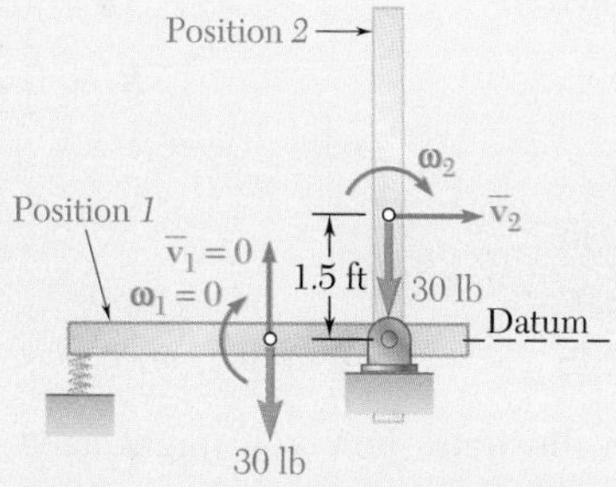

MODELING: Choose the rod and the spring as your system and model the rod as a rigid body. Denote the initial position as position 1 and the vertical position as position 2 (Fig. 1). Choose your datum to be at position 1.

ANALYSIS:

Conservation of Energy.

$$T_1 + V_{g_1} + V_{e_1} = T_2 + V_{g_2} + V_{e_2} \qquad (1)$$

You need to calculate the energy at position 1 and position 2.

Fig. 1 The rod in positions 1 and 2.

(continued)

Position 1.

Potential Energy. The spring is compressed 1 in., so you have $x_1 = 1$ in. The elastic potential energy is

$$V_{e_1} = \tfrac{1}{2}kx_1^2 = \tfrac{1}{2}(1800 \text{ lb/in.})(1 \text{ in.})^2 = 900 \text{ in·lb} = 75 \text{ ft·lb}$$

Since the datum is at position 1, you have $V_{g_1} = 0$.

Kinetic Energy. The velocity in position 1 is zero, so you have $T_1 = 0$.

Position 2.

Potential Energy. The elongation of the spring is zero, so you have $V_{e_2} = 0$. Since the center of gravity of the rod is now 1.5 ft above the datum, you have

$$V_{g_2} = mgy = (30 \text{ lb})(1.5 \text{ ft}) = 45 \text{ ft·lb}$$

Kinetic Energy. Denote the angular velocity of the rod in position 2 by $\boldsymbol{\omega}_2$. The rod rotates about O, so you have $\bar{v}_2 = \bar{r}\omega_2 = 1.5\omega_2$ and

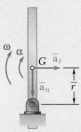

Fig. 2 The acceleration of the center of mass and the angular velocity and acceleration of the rod.

$$\bar{I} = \tfrac{1}{12}ml^2 = \frac{1}{12}\frac{30 \text{ lb}}{32.2 \text{ ft/s}^2}(5 \text{ ft})^2 = 1.941 \text{ lb·ft·s}^2$$

$$T_2 = \tfrac{1}{2}m\bar{v}_2^2 + \tfrac{1}{2}\bar{I}\omega_2^2 = \frac{1}{2}\frac{30}{32.2}(1.5\omega_2)^2 + \tfrac{1}{2}(1.941)\omega_2^2 = 2.019\omega_2^2$$

Substituting these expressions into Eq. (1) give

$$0 + 0 + 75 \text{ ft·lb} = 2.019\omega_2^2 + 45 \text{ ft·lb} + 0 \qquad \omega_2 = 3.86 \text{ rad/s} \!\downarrow \quad \blacktriangleleft$$

Reaction. Since $\omega_2 = 3.86$ rad/s, the components of the acceleration of G as the rod passes through position 2 are (Fig. 2)

$$\bar{a}_n = \bar{r}\omega_2^2 = (1.5 \text{ ft})(3.86 \text{ rad/s})^2 = 22.3 \text{ ft/s}^2 \qquad \mathbf{\bar{a}}_n = 22.3 \text{ ft/s}^2 \downarrow$$

$$\bar{a}_t = \bar{r}\alpha \qquad\qquad\qquad\qquad\qquad \mathbf{\bar{a}}_t = \bar{r}\alpha \ \rightarrow$$

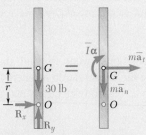

Fig. 3 Free-body diagram and kinetic diagram for the rod.

Draw free-body and kinetic diagrams (Fig. 3) to express that the system of external forces is equivalent to the vector of components $m\mathbf{\bar{a}}_t$ and $m\mathbf{\bar{a}}_n$ attached at G and the couple $\bar{I}\alpha$.

$$+\!\downarrow\!\Sigma M_O = \bar{I}\alpha + m\bar{a}d_\perp: \qquad 0 = \bar{I}\alpha + m(\bar{r}\alpha)\bar{r} \qquad\qquad \alpha = 0$$

$$\overset{+}{\rightarrow}\Sigma F_x = m\bar{a}_x: \qquad\qquad R_x = m(\bar{r}\alpha) \qquad\qquad\qquad R_x = 0$$

$$+\uparrow\Sigma F_y = m\bar{a}_y: \qquad R_y - 30 \text{ lb} = -m\bar{a}_n$$

$$R_y - 30 \text{ lb} = -\frac{30 \text{ lb}}{32.2 \text{ ft/s}^2}(22.3 \text{ ft/s}^2)$$

$$R_y = +9.22 \text{ lb} \qquad\qquad \mathbf{R} = 9.22 \text{ lb} \uparrow \quad \blacktriangleleft$$

REFLECT and THINK: This problem illustrates how you might need to supplement the conservation of energy with Newton's second law. What if the spring constant had been smaller, say 180 lb/in.? You would have found $V_{e_1} = 7.5$ ft·lb and then solved Eq. (1) to obtain $\omega_2^2 = -18.57$. This is clearly impossible and means that the rod would not make it to position 2 as assumed.

Sample Problem 17.5

A large box with a mass m and a flat bottom rests on two identical homogeneous cylindrical rollers, where each has radius r and a mass half that of the crate. The system is released from rest on a plane that is inclined at angle ϕ to the horizontal. Determine the speed of the box at the instant when the rollers have turned through an angle θ. Neglect rolling resistance and assume that the rollers do not slide.

STRATEGY: You are interested in the velocity after the rollers have moved a specified distance, $r\theta$, and the friction force in rolling motion does no work, so use the conservation of energy.

MODELING: Choose the box and the two cylindrical rollers as your system and model them as rigid bodies. In order for you to draw the system in its initial and final positions, you need to know how far each mass travels. You can determine this by using the instantaneous center of rotation. The rollers do not slide, so the instantaneous center of rotation for each roller is located at the point of contact, C, between the roller and the ground (Fig. 1). Using this instantaneous center of velocity, you know $v_B = 2\omega r$ and $v_R = \omega r$. Therefore, the box moves down a distance $2h$ when the rollers move a distance h (Fig. 2). Since you have three masses in the system (the two rollers and the box), you may define an individual datum for each mass to simplify the calculation of the gravitational potential energy.

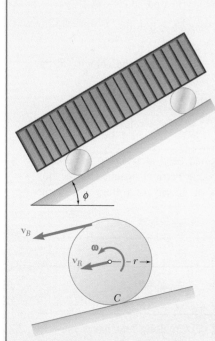

Fig. 1 Velocity of various points on the roller.

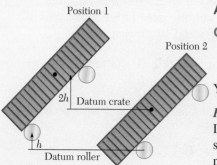

Fig. 2 The system in positions 1 and 2.

ANALYSIS:

Conservation of Energy.

$$T_1 + V_{g_1} + V_{e_1} = T_2 + V_{g_2} + V_{e_2} \tag{1}$$

You need to calculate the energy at position 1 and position 2.

Potential Energy. Because there is no spring in the system, $V_{e_1} = V_{e_2} = 0$. If you place your datum at the center of mass of each object when the system is at position 2, you have $V_{g_2} = 0$. The vertical distance a roller moves is $h = r\theta\sin\phi$, so

$$V_{g_1} = mg(2h) + 2(\tfrac{m}{2})g(h) = 3mgh = 3mgr\theta\sin\phi$$

Kinetic Energy. The velocity in position 1 is zero, so $T_1 = 0$.

At position 2,

$$T_2 = 2(\tfrac{1}{2}m_R v_R^2 + \tfrac{1}{2}\bar{I}\omega^2) + \tfrac{1}{2}mv_B^2$$

where m_R is the mass of the roller and $\bar{I}$ is the mass moment of inertia of the roller about its center of gravity. Substituting $v_B = 2\omega r$, $v_R = \omega r$, and $\bar{I} = \tfrac{1}{2}m_R r^2 = \tfrac{1}{2}(\tfrac{m}{2})r^2 = \tfrac{1}{4}mr^2$ into T_2 gives

$$T_2 = \tfrac{11}{4}mr^2\omega^2$$

(continued)

Substituting these expressions into Eq. (1), you find

$$0 + 3mgr\theta\sin\phi + 0 = \tfrac{11}{4}mr^2\omega^2 + 0 + 0$$

Solving for the angular velocity, $\omega = \sqrt{\dfrac{12g\theta\sin\phi}{11r}}$, so the velocity of the box at $v_B = 2\omega r$ is

$$v_B = 4\sqrt{\frac{3}{11}gr\theta\sin\phi} \quad \blacktriangleleft$$

REFLECT and THINK: If the rollers had been attached to the box by brackets, they would have traveled the same vertical distance as the box and the change in height of the centers of gravity of rollers and of the box would have been equal.

Sample Problem 17.6

Each of the two slender rods shown is 0.75 m long and has a mass of 6 kg. If the system is released from rest with $\beta = 60°$, determine (a) the angular velocity of rod AB when $\beta = 20°$, (b) the velocity of point D at the same instant.

STRATEGY: You have two positions and are interested in velocities, so use the conservation of energy. You will also need to use kinematics to relate the velocity terms in the kinetic energy expression.

MODELING: Choose the system to be both bars and model them as rigid bodies.

ANALYSIS: To illustrate that the order in which you solve a problem doesn't matter, let's start with kinematics.

Kinematics of Motion When $\beta = 20°$. Since $\mathbf{v}_B$ is perpendicular to the rod AB and $\mathbf{v}_D$ is horizontal, the instantaneous center of rotation of rod BD is located at C (Fig. 1). From the geometry of the figure, you obtain

$$BC = 0.75 \text{ m} \qquad CD = 2(0.75 \text{ m}) \sin 20° = 0.513 \text{ m}$$

Apply the law of cosines to triangle CDE, where E is located at the mass center of rod BD. You find $EC = 0.522$ m. Denoting the angular velocity of rod AB by ω, you have (Fig. 2)

$$\bar{v}_{AB} = (0.375 \text{ m})\omega \qquad\qquad \bar{\mathbf{v}}_{AB} = 0.375\omega \searrow$$
$$v_B = (0.75 \text{ m})\omega \qquad\qquad v_B = 0.75\omega \searrow$$

Since rod BD seems to rotate about point C, you have

$$v_B = (BC)\omega_{BD} \qquad (0.75 \text{ m})\omega = (0.75 \text{ m})\omega_{BD} \qquad \omega_{BD} = \omega \nwarrow$$
$$\bar{v}_{BD} = (EC)\omega_{BD} = (0.522 \text{ m})\omega \qquad \bar{\mathbf{v}}_{BD} = 0.522\omega \searrow$$

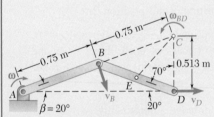

Fig. 1 Instantaneous center of rotation C for bar BD.

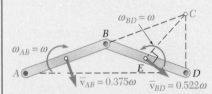

Fig. 2 Velocities of the center of masses of AB and BD in terms of ω.

Conservation of Energy. Since there are no springs in the system

$$T_1 + V_{g_1} = T_2 + V_{g_2}$$

You first need to determine the energy at the two positions.

Position 1.

Potential Energy. Choose the datum as shown in Fig. 3, and observe that $W = (6 \text{ kg})(9.81 \text{ m/s}^2) = 58.86 \text{ N}$. Then you have

$$V_{g_1} = 2W\bar{y}_1 = 2(58.86 \text{ N})(0.325 \text{ m}) = 38.26 \text{ J}$$

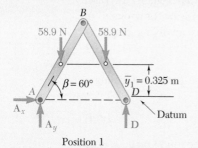

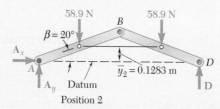

Fig. 3 Free-body diagram and distance from the datum in position 1.

Fig. 4 Free-body diagram and distance from the datum in position 2.

Kinetic Energy. Initially, the system is at rest, so $T_1 = 0$.

Position 2.

Potential Energy. Compute the new height of the mass centers of the rods to be $\bar{y}_2 = 0.75\sin(20) = 0.1283 \text{ m}$ (Fig. 4).

$$V_{g_2} = 2W\bar{y}_2 = 2(58.86 \text{ N})(0.1283 \text{ m}) = 15.10 \text{ J}$$

Kinetic Energy.

$$I_{AB} = \bar{I}_{BD} = \tfrac{1}{12}ml^2 = \tfrac{1}{12}(6 \text{ kg})(0.75 \text{ m})^2 = 0.281 \text{ kg·m}^2$$
$$T_2 = \tfrac{1}{2}m\bar{v}_{AB}^2 + \tfrac{1}{2}\bar{I}_{AB}\omega_{AB}^2 + \tfrac{1}{2}m\bar{v}_{BD}^2 + \tfrac{1}{2}\bar{I}_{BD}\omega_{BD}^2$$
$$= \tfrac{1}{2}(6)(0.375\omega)^2 + \tfrac{1}{2}(0.281)\omega^2 + \tfrac{1}{2}(6)(0.522\omega)^2 + \tfrac{1}{2}(0.281)\omega^2$$
$$= 1.520\omega^2$$

Conservation of Energy. Now you can write

$$T_1 + V_{g_1} = T_2 + V_{g_2}$$
$$0 + 38.26 \text{ J} = 1.520\omega^2 + 15.10 \text{ J}$$
$$\omega = 3.90 \text{ rad/s} \qquad \omega_{AB} = 3.90 \text{ rad/s} \; \downdownarrow \quad \blacktriangleleft$$

Velocity of Point D.

$$v_D = (CD)\omega = (0.513 \text{ m})(3.90 \text{ rad/s}) = 2.00 \text{ m/s}$$
$$\mathbf{v}_D = 2.00 \text{ m/s} \rightarrow \quad \blacktriangleleft$$

REFLECT and THINK: The only step in which you need to use forces is when calculating the gravitational potential energy in each position. However, it is good engineering practice to show the complete free-body diagram in each case to identify which, if any, forces do work. Rather than use the instantaneous center of rotation, you could have also used vector algebra to relate the velocities of the various objects.

SOLVING PROBLEMS
ON YOUR OWN

In this section, we introduced energy methods to determine the velocity of rigid bodies for various positions during their motion. As you saw previously in Chap. 13, energy methods are particularly useful for problems involving displacements and velocities.

1. **The method of work and energy**, when applied to all of the particles forming a rigid body, yields the equation

$$T_1 + U_{1 \rightarrow 2} = T_2 \tag{17.1}$$

where T_1 and T_2 are, respectively, the initial and final values of the total kinetic energy of the particles forming the body and $U_{1 \rightarrow 2}$ is the work done by the external forces exerted on the rigid body. If we express the work done by nonconservative forces as $U_{1 \rightarrow 2}^{NC}$ and define the potential energy terms for conservative forces, we can express Eq. (17.1) as

$$T_1 + V_{g_1} + V_{e_1} + U_{1 \rightarrow 2}^{NC} = T_2 + V_{g_2} + V_{e_2} \tag{17.1'}$$

where V_{g_1} and V_{g_2} are the initial and final gravitational potential energy of the center of mass of the rigid body and V_{e_1} and V_{e_2} are the initial and final values of the elastic energy associated with springs in the system. Recall that, for a linear spring, $V_e = \frac{1}{2}kx^2$, where x is the deflection of the spring from its unstretched length. For a single rigid body, $V_g = mgy$, where y is the elevation of the center of mass from a reference plane or datum.

   a. **Work of forces and couples.** To the expression for the work of a force (Chap. 13), we added the expression for the work of a couple and wrote

$$U_{1 \rightarrow 2} = \int_{A_1}^{A_2} \mathbf{F} \cdot d\mathbf{r} \qquad U_{1 \rightarrow 2} = \int_{\theta_1}^{\theta_2} M d\theta \tag{17.3, 17.5}$$

When the moment of a couple is constant, the work of the couple is

$$U_{1 \rightarrow 2} = M(\theta_2 - \theta_1) \tag{17.6}$$

where θ_1 and θ_2 are expressed in radians [Sample Probs. 17.1 and 17.2].

   b. **The kinetic energy of a rigid body in plane motion** was found by considering the motion of the body as the sum of a translation with its mass center and a rotation about the mass center. So

$$T = \tfrac{1}{2} m\bar{v}^2 + \tfrac{1}{2}\bar{I}\omega^2 \tag{17.9}$$

where $\bar{v}$ is the velocity of the mass center and ω is the angular velocity of the body [Sample Probs. 17.3 and 17.4]. You will generally need to use kinematics to relate $\bar{v}$ and ω.

2. For a system of rigid bodies we again used the equation

$$T_1 + U_{1 \to 2} = T_2 \tag{17.1}$$

where T is the sum of the kinetic energies of the bodies forming the system and U is the work done by *all the forces acting on the bodies*—internal as well as external. Your computations will be simplified if you keep the following ideas in mind.

a. The forces exerted on each other by pin-connected members or by meshed gears are equal and opposite, and since they have the same point of application, they undergo equal small displacements. Therefore, *their total work is zero* and can be omitted from your calculations [Sample Prob. 17.2].

b. The forces exerted by an inextensible cord on the two bodies it connects have the same magnitude and their points of application move through equal distances, but the work of one force is positive and the work of the other is negative. Therefore, *their total work is zero* and again can be omitted from your calculations [Sample Prob. 17.1].

c. The forces exerted by a spring on the two bodies it connects also have the same magnitude, but their points of application generally move through different distances. Therefore, *their total work is usually not zero* and should be taken into account in your calculations. The easiest way to handle springs, therefore, is to use elastic potential energy.

3. The principle of conservation of energy can be expressed as

$$T_1 + V_1 = T_2 + V_2 \tag{17.12}$$

where V represents the potential energy of the system. If you prefer to write this equation in terms of gravitational potential energy, V_g, and elastic potential energy, V_e, you get

$$T_1 + V_{g_1} + V_{e_1} = T_2 + V_{g_2} + V_{e_2} \tag{17.12'}$$

You can use this principle when a body or a system of bodies is acted upon by conservative forces, such as the force exerted by a spring or the force of gravity [Sample Probs. 17.4 through 17.6].

4. The last part of this section was devoted to power, which is the time rate at which work is done. For a body acted upon by a couple of moment $\mathbf{M}$, the power can be expressed as

$$\text{Power} = M\omega \tag{17.13}$$

where ω is the angular velocity of the body, expressed in rad/s. As you did in Chap. 13, you should express power either in watts or in horsepower (1 hp = 550 ft·lb/s).

Problems

CONCEPT QUESTIONS

17.CQ1 A round object of mass m and radius r is released from rest at the top of a curved surface and rolls without slipping until it leaves the surface with a horizontal velocity as shown. Will a solid sphere, a solid cylinder, or a hoop travel the greatest distance x?

   a. Solid sphere
   b. Solid cylinder
   c. Hoop
   d. They will all travel the same distance.

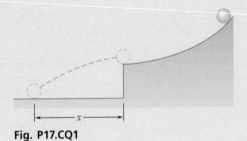

Fig. P17.CQ1

17.CQ2 A solid steel sphere A of radius r and mass m is released from rest and rolls without slipping down an incline as shown. After traveling a distance d, the sphere has a speed v. If a solid steel sphere of radius $2r$ is released from rest on the same incline, what will its speed be after rolling a distance d?

   a. $0.25v$
   b. $0.5v$
   c. v
   d. $2v$
   e. $4v$

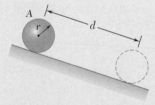

Fig. P17.CQ2

17.CQ3 Slender bar A is rigidly connected to a massless rod BC in Case 1 and two massless cords in Case 2 as shown. The vertical thickness of bar A is negligible compared to L. In both cases A is released from rest at an angle $\theta = \theta_0$. When $\theta = 0°$, which system will have the larger kinetic energy?

a. Case 1
b. Case 2
c. The kinetic energy will be the same.

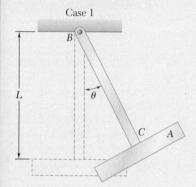

Fig. P17.CQ3 and P17.CQ5

17.CQ4 In Prob. 17.CQ3, how will the speeds of the centers of gravity compare for the two cases when $\theta = 0°$?

a. Case 1 will be larger.
b. Case 2 will be larger.
c. The speeds will be the same.

17.CQ5 Slender bar A is rigidly connected to a massless rod BC in Case 1 and two massless cords in Case 2 as shown. The vertical thickness of bar A is not negligible compared to L. In both cases A is released from rest at an angle $\theta = \theta_0$. When $\theta = 0°$, which system will have the largest kinetic energy?

a. Case 1
b. Case 2
c. The kinetic energy will be the same.

END-OF-SECTION PROBLEMS

17.1 A 200-kg flywheel is at rest when a constant 300 N·m couple is applied. After executing 560 revolutions, the flywheel reaches its rated speed of 2400 rpm. Knowing that the radius of gyration of the flywheel is 400 mm, determine the average magnitude of the couple due to kinetic friction in the bearing.

17.2 The rotor of an electric motor has an angular velocity of 3600 rpm when the load and power are cut off. The 110-lb rotor, which has a centroidal radius of gyration of 9 in., then coasts to rest. Knowing that the kinetic friction of the rotor produces a couple with a magnitude of 2.5 lb·ft, determine the number of revolutions that the rotor executes before coming to rest.

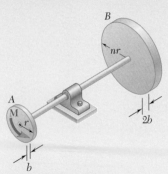

Fig. P17.3 and P17.4

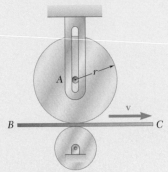

Fig. P17.7

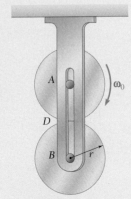

Fig. P17.8

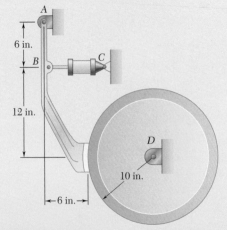

Fig. P17.9

17.3 Two uniform disks of the same material are attached to a shaft as shown. Disk A has a weight of 10 lb and a radius of $r = 6$ in. Disk B is twice as thick as disk A. Knowing that a couple $\mathbf{M}$ with a magnitude of 22 lb·ft is applied to disk A when the system is at rest, determine the radius nr of disk B if the angular velocity of the system is to be 480 rpm after five revolutions.

17.4 Two disks of the same material are attached to a shaft as shown. Disk A has a radius r and a thickness b, while disk B has a radius nr and a thickness $2b$. A couple $\mathbf{M}$ with a constant magnitude is applied when the system is at rest and is removed after the system has executed two revolutions. Determine the value of n that results in the largest final speed for a point on the rim of disk B.

17.5 The flywheel of a small punch rotates at 300 rpm. It is known that 1800 ft·lb of work must be done each time a hole is punched. It is desired that the speed of the flywheel after one punching be not less than 90 percent of the original speed of 300 rpm. (*a*) Determine the required moment of inertia of the flywheel. (*b*) If a constant 25-lb·ft couple is applied to the shaft of the flywheel, determine the number of revolutions that must occur between each punching, knowing that the initial velocity is to be 300 rpm at the start of each punching.

17.6 The flywheel of a punching machine has a mass of 300 kg and a radius of gyration of 600 mm. Each punching operation requires 2500 J of work. (*a*) Knowing that the speed of the flywheel is 300 rpm just before a punching, determine the speed immediately after the punching. (*b*) If a constant 25-N·m couple is applied to the shaft of the flywheel, determine the number of revolutions executed before the speed is again 300 rpm.

17.7 Disk A, of weight 10 lb and radius $r = 6$ in., is at rest when it is placed in contact with belt BC, which moves to the right with a constant speed $v = 40$ ft/s. Knowing that $\mu_k = 0.20$ between the disk and the belt, determine the number of revolutions executed by the disk before it attains a constant angular velocity.

17.8 The uniform 4-kg cylinder A with a radius of $r = 150$ mm has an angular velocity of $\omega_0 = 50$ rad/s when it is brought into contact with an identical cylinder B that is at rest. The coefficient of kinetic friction at the contact point D is μ_k. After a period of slipping, the cylinders attain constant angular velocities of equal magnitude and opposite direction at the same time. Knowing that cylinder A executes three revolutions before it attains a constant angular velocity and cylinder B executes one revolution before it attains a constant angular velocity, determine (*a*) the final angular velocity of each cylinder, (*b*) the coefficient of kinetic friction μ_k.

17.9 The 10-in.-radius brake drum is attached to a larger flywheel which is not shown. The total mass moment of inertia of the flywheel and drum is 16 lb·ft·s² and the coefficient of kinetic friction between the drum and the brake shoe is 0.40. Knowing that the initial angular velocity is 240 rpm clockwise, determine the force that must be exerted by the hydraulic cylinder if the system is to stop in 75 revolutions.

17.10 Solve Prob. 17.9, assuming that the initial angular velocity of the flywheel is 240 rpm counterclockwise.

17.11 Each of the gears A and B has a mass of 2.4 kg and a radius of gyration of 60 mm, while gear C has a mass of 12 kg and a radius of gyration of 150 mm. A couple $\mathbf{M}$ of constant magnitude 10 N·m is applied to gear C. Determine (*a*) the number of revolutions of gear C required for its angular velocity to increase from 100 to 450 rpm, (*b*) the corresponding tangential force acting on gear A.

17.12 Solve Prob. 17.11, assuming that the 10-N·m couple is applied to gear B.

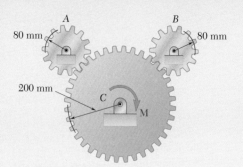

Fig. P17.11

17.13 The gear train shown consists of four gears of the same thickness and of the same material; two gears are of radius r, and the other two are of radius nr. The system is at rest when the couple $\mathbf{M}_0$ is applied to shaft C. Denoting by I_0 the moment of inertia of a gear of radius r, determine the angular velocity of shaft A if the couple $\mathbf{M}_0$ is applied for one revolution of shaft C.

17.14 The double pulley shown has a mass of 15 kg and a centroidal radius of gyration of 160 mm. Cylinder A and block B are attached to cords that are wrapped on the pulleys as shown. The coefficient of kinetic friction between block B and the surface is 0.2. Knowing that the system is at rest in the position shown when a constant force $\mathbf{P} = 200$ N is applied to cylinder A, determine (*a*) the velocity of cylinder A as it strikes the ground, (*b*) the total distance that block B moves before coming to rest.

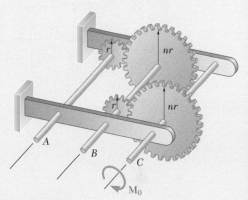

Fig. P17.13

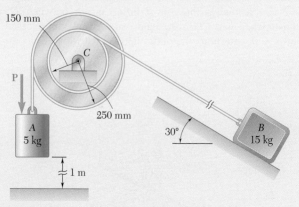

Fig. P17.14

17.15 Gear A has a mass of 1 kg and a radius of gyration of 30 mm; gear B has a mass of 4 kg and a radius of gyration of 75 mm; gear C has a mass of 9 kg and a radius of gyration of 100 mm. The system is at rest when a couple $\mathbf{M}_0$ of constant magnitude 4 N·m is applied to gear C. Assuming that no slipping occurs between the gears, determine the number of revolutions required for disk A to reach an angular velocity of 300 rpm.

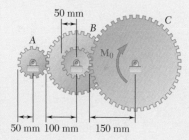

Fig. P17.15

17.16 A slender rod of length l and weight W is pivoted at one end as shown. It is released from rest in a horizontal position and swings freely. (*a*) Determine the angular velocity of the rod as it passes through a vertical position and determine the corresponding reaction at the pivot. (*b*) Solve part *a* for $W = 1.8$ lb and $l = 3$ ft.

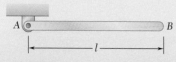

Fig. P17.16

17.17 A slender rod of length l is pivoted about a point C located at a distance b from its center G. It is released from rest in a horizontal position and swings freely. Determine (a) the distance b for which the angular velocity of the rod as it passes through a vertical position is maximum, (b) the corresponding values of its angular velocity and of the reaction at C.

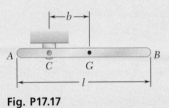

Fig. P17.17

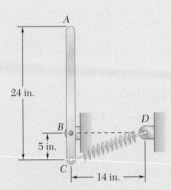

Fig. P17.18

17.18 A slender 9-lb rod can rotate in a vertical plane about a pivot at B. A spring of constant $k = 30$ lb/ft and of unstretched length 6 in. is attached to the rod as shown. Knowing that the rod is released from rest in the position shown, determine its angular velocity after it has rotated through 90°.

17.19 An adapted golf device attaches to a wheelchair to help people with mobility impairments play putt-putt. The stationary frame OD is attached to the wheelchair, and a club holder OB is attached to the pin at O. Holder OB is 6 in. long and weighs 8 oz, and the distance between O and D is $x = 1$ ft. The putter shaft has a length of $L = 36$ in. and weighs 10 oz, while the putter head at A weighs 12 oz. Knowing that the 1-lb/in. spring between D and B is unstretched when $\theta = 90°$ and that the putter is released from rest at $\theta = 0$, determine the putter head speed when it hits the golf ball.

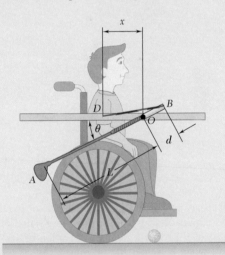

Fig. P17.19

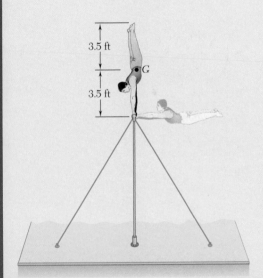

Fig. P17.20

17.20 A 160-lb gymnast is executing a series of full-circle swings on the horizontal bar. In the position shown, he has a small and negligible clockwise angular velocity and will maintain his body straight and rigid as he swings downward. Assuming that during the swing the centroidal radius of gyration of his body is 1.5 ft, determine his angular velocity and the force exerted on his hands after he has rotated through (a) 90°, (b) 180°.

17.21 A collar with a mass of 1 kg is rigidly attached at a distance $d = 300$ mm from the end of a uniform slender rod AB. The rod has a mass of 3 kg and is of length $L = 600$ mm. Knowing that the rod is released from rest in the position shown, determine the angular velocity of the rod after it has rotated through 90°.

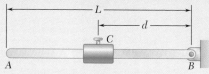
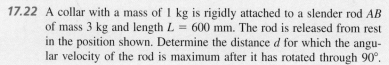

Fig. P17.21 and P17.22

17.22 A collar with a mass of 1 kg is rigidly attached to a slender rod AB of mass 3 kg and length $L = 600$ mm. The rod is released from rest in the position shown. Determine the distance d for which the angular velocity of the rod is maximum after it has rotated through 90°.

17.23 Two identical slender rods AB and BC are welded together to form an L-shaped assembly. The assembly is pressed against a spring at D and released from the position shown. Knowing that the maximum angle of rotation of the assembly in its subsequent motion is 90° counterclockwise, determine the magnitude of the angular velocity of the assembly as it passes through the position where rod AB forms an angle of 30° with the horizontal.

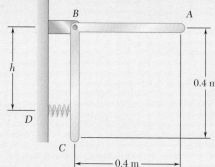

Fig. P17.23

17.24 The 30-kg turbine disk has a centroidal radius of gyration of 175 mm and is rotating clockwise at a constant rate of 60 rpm when a small blade of weight 0.5 N at point A becomes loose and is thrown off. Neglecting friction, determine the change in the angular velocity of the turbine disk after it has rotated through (*a*) 90°, (*b*) 270°.

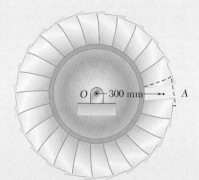

Fig. P17.24

17.25 A rope is wrapped around a cylinder of radius r and mass m as shown. Knowing that the cylinder is released from rest, determine the velocity of the center of the cylinder after it has moved downward a distance s.

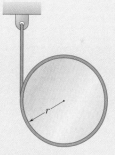

Fig. P17.25

17.26 Solve Prob. 17.25, assuming that the cylinder is replaced by a thin-walled pipe of radius r and mass m.

17.27 Greek engineers had the unenviable task of moving large columns from the quarries to the city. One engineer, Chersiphron, tried several different techniques to do this. One method was to cut pivot holes into the ends of the stone and then use oxen to pull the column. The 4-ft diameter column weighs 12,000 lbs, and the team of oxen generates a constant pull force of 1500 lbs on the center of the cylinder G. Knowing that the column starts from rest and rolls without slipping, determine (*a*) the velocity of its center G after it has moved 5 ft, (*b*) the minimum static coefficient of friction that will keep it from slipping.

Fig. P17.27

17.28 A small sphere of mass m and radius r is released from rest at A and rolls without sliding on the curved surface to point B where it leaves the surface with a horizontal velocity. Knowing that $a = 1.5$ m and $b = 1.2$ m, determine (a) the speed of the sphere as it strikes the ground at C, (b) the corresponding distance c.

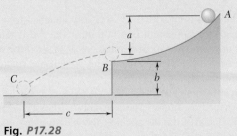

Fig. P17.28

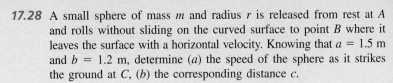

Fig. P17.29

17.29 The mass center G of a 3-kg wheel of radius $R = 180$ mm is located at a distance $r = 60$ mm from its geometric center C. The centroidal radius of gyration of the wheel is $\bar{k} = 90$ mm. As the wheel rolls without sliding, its angular velocity is observed to vary. Knowing that $\omega = 8$ rad/s in the position shown, determine (a) the angular velocity of the wheel when the mass center G is directly above the geometric center C, (b) the reaction at the horizontal surface at the same instant.

17.30 A half-cylinder with mass m and radius r is released from rest in the position shown. Knowing that the half-cylinder rolls without sliding, determine (a) its angular velocity after it has rolled through 90°, (b) the reaction at the horizontal surface at the same instant. [*Hint*: Note that $GO = 4r/3\pi$ and that, by the parallel-axis theorem, $\bar{I} = \frac{1}{2}mr^2 - m(GO)^2$.]

Fig. P17.30

17.31 A sphere of mass m and radius r rolls without slipping inside a curved surface of radius R. Knowing that the sphere is released from rest in the position shown, derive an expression for (a) the linear velocity of the sphere as it passes through B, (b) the magnitude of the vertical reaction at that instant.

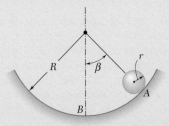

Fig. P17.31

17.32 Two uniform cylinders, each of weight $W = 14$ lb and radius $r = 5$ in., are connected by a belt as shown. Knowing that at the instant shown the angular velocity of cylinder B is 30 rad/s clockwise, determine (*a*) the distance through which cylinder A will rise before the angular velocity of cylinder B is reduced to 5 rad/s, (*b*) the tension in the portion of belt connecting the two cylinders.

17.33 Two uniform cylinders, each of weight $W = 14$ lb and radius $r = 5$ in., are connected by a belt as shown. If the system is released from rest, determine (*a*) the velocity of the center of cylinder A after it has moved through 3 ft, (*b*) the tension in the portion of belt connecting the two cylinders.

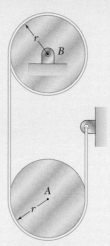

Fig. P17.32 and P17.33

17.34 A bar of mass $m = 5$ kg is held as shown between four disks each of mass $m' = 2$ kg and radius $r = 75$ mm. Knowing that the forces exerted on the disks are sufficient to prevent slipping and that the bar is released from rest, for each of the cases shown, determine the velocity of the bar after it has moved through the distance h.

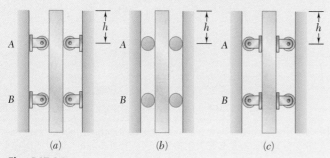

(a)   (b)   (c)

Fig. P17.34

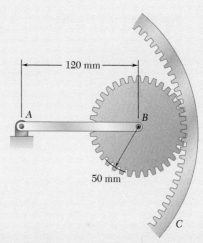

Fig. P17.35

17.35 The 1.5-kg uniform slender bar AB is connected to the 3-kg gear B that meshes with the stationary outer gear C. The centroidal radius of gyration of gear B is 30 mm. Knowing that the system is released from rest in the position shown, determine (*a*) the angular velocity of the bar as it passes through the vertical position, (*b*) the corresponding angular velocity of gear B.

17.36 The motion of the uniform rod AB is guided by small wheels of negligible mass that roll on the surface shown. If the rod is released from rest when $\theta = 0$, determine the velocities of A and B when $\theta = 30°$.

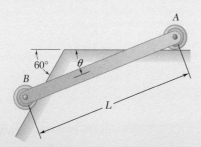

Fig. P17.36

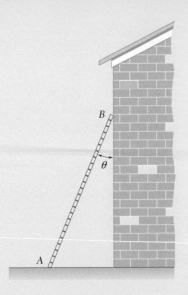

Fig. P17.37 and P17.38

17.37 A 5-m-long ladder has a mass of 15 kg and is placed against a house at an angle $\theta = 20°$. Knowing that the ladder is released from rest, determine the angular velocity of the ladder and the velocity of end A when $\theta = 45°$. Assume the ladder can slide freely on the horizontal ground and on the vertical wall.

17.38 A long ladder of length l, mass m, and centroidal mass moment of inertia $\bar{I}$ is placed against a house at an angle $\theta = \theta_0$. Knowing that the ladder is released from rest, determine the angular velocity of the ladder when $\theta = \theta_2$. Assume the ladder can slide freely on the horizontal ground and on the vertical wall.

17.39 The ends of a 9-lb rod AB are constrained to move along slots cut in a vertical plate as shown. A spring of constant $k = 3$ lb/in. is attached to end A in such a way that its tension is zero when $\theta = 0$. If the rod is released from rest when $\theta = 50°$, determine the angular velocity of the rod and the velocity of end B when $\theta = 0$.

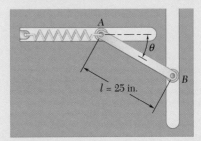

Fig. P17.39

17.40 The mechanism shown is one of two identical mechanisms attached to the two sides of a 200-lb uniform rectangular door. Edge ABC of the door is guided by wheels of negligible mass that roll in horizontal and vertical tracks. A spring with a constant of $k = 40$ lb/ft is attached to wheel B. Knowing that the door is released from rest in the position $\theta = 30°$ with the spring unstretched, determine the velocity of wheel A just as the door reaches the vertical position.

17.41 The mechanism shown is one of two identical mechanisms attached to the two sides of a 200-lb uniform rectangular door. Edge ABC of the door is guided by wheels of negligible mass that roll in horizontal and vertical tracks. A spring with a constant k is attached to wheel B in such a way that its tension is zero when $\theta = 30°$, Knowing that the door is released from rest in the position $\theta = 45°$ and reaches the vertical position with an angular velocity of 0.6 rad/s, determine the spring constant k.

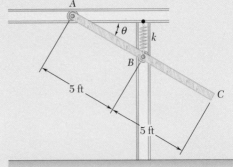

Fig. P17.40 and *P17.41*

17.42 Each of the two rods shown is of length $L = 1$ m and has a mass of 5 kg. Point D is connected to a spring of constant $k = 20$ N/m and is constrained to move along a vertical slot. Knowing that the system is released from rest when rod BD is horizontal and the spring connected to point D is initially unstretched, determine the velocity of point D when it is directly to the right of point A.

17.43 The 4-kg rod AB is attached to a collar of negligible mass at A and to a flywheel at B. The flywheel has a mass of 16 kg and a radius of gyration of 180 mm. Knowing that in the position shown the angular velocity of the flywheel is 60 rpm clockwise, determine the velocity of the flywheel when point B is directly below C.

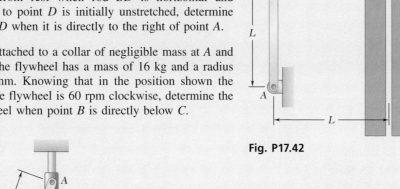

Fig. P17.42

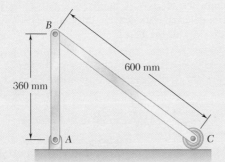

Fig. P17.43 and P17.44

17.44 If in Prob. 17.43 the angular velocity of the flywheel is to be the same in the position shown and when point B is directly above C, determine the required value of its angular velocity in the position shown.

17.45 The uniform rods AB and BC weigh 2.4 kg and 4 kg, respectively, and the small wheel at C is of negligible weight. If the wheel is moved slightly to the right and then released, determine the velocity of pin B after rod AB has rotated through 90°.

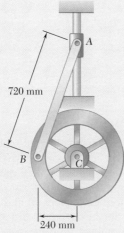

Fig. P17.45 and P17.46

17.46 The uniform rods AB and BC weigh 2.4 kg and 4 kg, respectively, and the small wheel at C is of negligible weight. Knowing that in the position shown the velocity of wheel C is 2 m/s to the right, determine the velocity of pin B after rod AB has rotated through 90°.

17.47 The 80-mm-radius gear shown has a mass of 5 kg and a centroidal radius of gyration of 60 mm. The 4-kg rod AB is attached to the center of the gear and to a pin at B that slides freely in a vertical slot. Knowing that the system is released from rest when $\theta = 60°$, determine the velocity of the center of the gear when $\theta = 20°$.

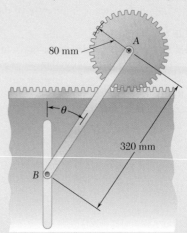

Fig. P17.47

17.48 Knowing that the maximum allowable couple that can be applied to a shaft is 15.5 kip·in., determine the maximum horsepower that can be transmitted by the shaft at (*a*) 180 rpm, (*b*) 480 rpm.

17.49 Three shafts and four gears are used to form a gear train which will transmit 7.5 kW from the motor at A to a machine tool at F. (Bearings for the shafts are omitted from the sketch.) Knowing that the frequency of the motor is 30 Hz, determine the magnitude of the couple that is applied to shaft (*a*) AB, (*b*) CD, (*c*) EF.

17.50 The shaft-disk-belt arrangement shown is used to transmit 2.4 kW from point A to point D. Knowing that the maximum allowable couples that can be applied to shafts AB and CD are 25 N·m and 80 N·m, respectively, determine the required minimum speed of shaft AB.

17.51 The drive belt on a vintage sander transmits ½ hp to a pulley that has a diameter of $d = 4$ in. Knowing that the pulley rotates at 1450 rpm, determine the tension difference $T_1 - T_2$ between the tight and slack sides of the belt.

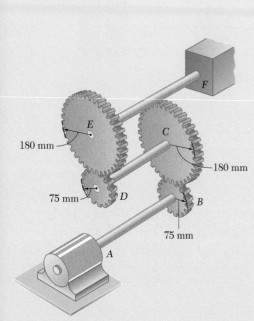

Fig. P17.49

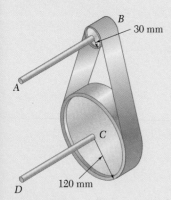

Fig. P17.50

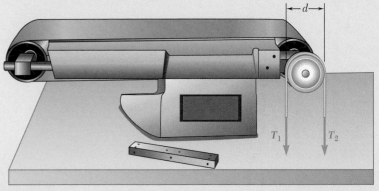

Fig. P17.51

17.2   MOMENTUM METHODS FOR A RIGID BODY

We now apply the principle of impulse and momentum to the plane motion of rigid bodies and of systems of rigid bodies. As we pointed out in Chap. 13, the method of impulse and momentum is particularly well adapted to the solution of problems involving time and velocities. Moreover, the principle of impulse and momentum provides the only practicable method for the solution of problems involving impulsive motion or impact (Sec. 17.3).

17.2A   Principle of Impulse and Momentum

Consider again a rigid body made of a large number of particles P_i. Recall from Sec. 14.2C that impulse–momentum diagrams are a pictorial representation of the principle of impulse and momentum. They show (*a*) the system formed by the momenta of the particles at time t_1 and (*b*) the system of the impulses of the external forces applied from t_1 to t_2 are together equipollent to (*c*) the system formed by the momenta of the particles at time t_2 (Fig. 17.6). We can consider the vectors associated with a rigid body to be sliding vectors, so it follows (*Statics*, Sec. 3.4B) that the systems of vectors shown in Fig. 17.6 are not only equipollent, but they are truly *equivalent*. In other words, the vectors on the left-hand side of the equal sign can be transformed into the vectors on the right-hand side through the use of the fundamental operations listed in Sec. 3.3B. We therefore have

> **Syst Momenta₁ + Syst Ext Imp₁→₂ = Syst Momenta₂**   **(17.14)**

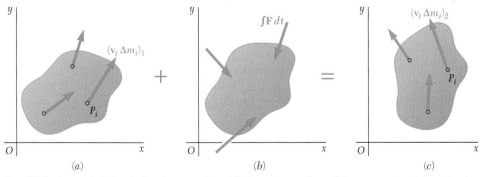

Fig. 17.6   For a rigid body in plane motion: (*a*) the system of particle momenta at time t_1 plus (*b*) the system of impulses of the external forces from time t_1 to t_2 is equivalent to (*c*) the system of particle momenta at time t_2.

The momenta $\mathbf{v}_i \, \Delta m_i$ of the particles can be reduced to a vector attached at G that is equal to their sum

$$\mathbf{L} = \sum_{i=1}^{n} \mathbf{v}_i \, \Delta m_i$$

Photo 17.2 A Charpy impact test is used to determine the amount of energy absorbed by a material during impact. To determine the amount of energy absorbed, the final gravitational potential energy of the arm is subtracted from its initial gravitational potential energy.

and a couple of moment equal to the sum of their moments about G, as

$$\mathbf{H}_G = \sum_{i=1}^{n} \mathbf{r}'_i \times \mathbf{v}_i \, \Delta m_i$$

Recall from Sec. 14.1B that $\mathbf{L}$ and $\mathbf{H}_G$ define, respectively, the linear momentum and the angular momentum about G of the system of particles forming the rigid body. Also note from Eq. (14.14) that $\mathbf{L} = m\bar{\mathbf{v}}$. On the other hand, by restricting the present analysis to the plane motion of a rigid body or of a rigid body symmetrical with respect to the reference plane, we recall from Eq. (16.4) that $\mathbf{H}_G = \bar{I}\boldsymbol{\omega}$. We thus conclude that the system of the momenta $\mathbf{v}_i \, \Delta m_i$ is equivalent to the **linear momentum vector** $m\bar{\mathbf{v}}$ attached at G and to the **angular momentum couple** $\bar{I}\boldsymbol{\omega}$ (Fig. 17.7).

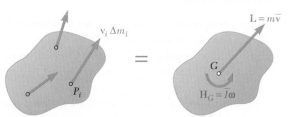

Fig. 17.7 The system of momenta of a rigid body is equivalent to a linear momentum vector attached at G and an angular momentum couple.

The system of momenta reduces to the vector $m\bar{\mathbf{v}}$ in the particular case of a translation ($\boldsymbol{\omega} = 0$) and to the couple $\bar{I}\boldsymbol{\omega}$ in the particular case of a centroidal rotation ($\bar{\mathbf{v}} = 0$). Thus, we verify once more that the plane motion of a rigid body that is symmetrical with respect to the reference plane can be resolved into a translation with the mass center G and a rotation about G.

Replacing the system of momenta in Fig. 17.6a and c by the equivalent linear momentum vector and angular momentum couple, we obtain the three diagrams shown in Fig. 17.8. This impulse–momentum diagram is a visual representation of the fundamental relation in Eq. (17.14) in the case of the plane motion of a rigid body or of a rigid body symmetrical with respect to the reference plane.

We can derive three equations of motion from Fig. 17.8. Two equations come from summing and equating the x and y *components* of the momenta

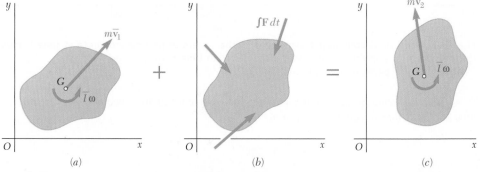

Fig. 17.8 An impulse–momentum diagram is used for applying the principle of impulse and momentum.

and impulses. The third equation is obtained by summing and equating the *moments* of these vectors *about any given point.* We can choose the coordinate axes to be fixed in space or allowed to move with the mass center of the body while maintaining a fixed direction. In either case, the point about which moments are taken should keep the same position relative to the coordinate axes during the interval of time considered. If you choose to sum moments about a point P, Eq. (17.14) can be expressed as

$$\bar{I}\omega_1 + m\bar{v}_1 d_\perp + \sum \int_{t_1}^{t_2} M_P dt = \bar{I}\omega_2 + m\bar{v}_2 d_\perp \qquad \textbf{(17.14′)}$$

where $d_\perp$ is the perpendicular distance from point P to the line of action of the linear velocity of G. If you choose to sum moments about the center of gravity of the body, then Eq. (17.14′) reduces to

$$\bar{I}\omega_1 + \sum \int_{t_1}^{t_2} M_G dt = \bar{I}\omega_2 \qquad \textbf{(17.14″)}$$

In deriving the three equations of motion for a rigid body, you should take care to avoid adding linear and angular momenta indiscriminately. Remember that $m\bar{v}_x$ and $m\bar{v}_y$ represent the *components of a vector,* namely, the linear momentum vector $m\bar{\mathbf{v}}$, whereas $\bar{I}\omega$ represents the *magnitude of a couple,* namely, the angular momentum couple $\bar{I}\omega$. Thus, you should add the quantity $\bar{I}\omega$ only to the *moment* of the linear momentum $m\bar{\mathbf{v}}$—never to this vector itself nor to its components. All angular momentum quantities involved then will be expressed in the same units, namely N·m·s or lb·ft·s.

Noncentroidal Rotation.

In this particular case of plane motion, the magnitude of the velocity of the mass center of the body is $\bar{v} = \bar{r}\omega$, where $\bar{r}$ represents the distance from the mass center to the fixed axis of rotation and $\boldsymbol{\omega}$ represents the angular velocity of the body at the instant considered. The magnitude of the momentum vector attached at G is thus $m\bar{v} = m\bar{r}\omega$. Summing the moments about O of the momentum vector and momentum couple (Fig. 17.9) and using the parallel-axis theorem for moments of inertia, we find that the angular momentum $\mathbf{H}_O$ of the body about fixed axis O has the magnitude[†]

$$\bar{I}\omega + (m\bar{r}\omega)\bar{r} = (\bar{I} + m\bar{r}^2)\omega = I_O \omega \qquad \textbf{(17.15)}$$

Equating the moments about O of the momenta and impulses in Eq. (17.14), we have

$$I_O \omega_1 + \sum \int_{t_1}^{t_2} M_O \, dt = I_O \omega_2 \qquad \textbf{(17.16)}$$

In the general case of plane motion of a rigid body symmetrical with respect to the reference plane, you can use Eq. (17.16) with respect to the instantaneous axis of rotation under certain conditions. We recommend, however, that all problems of plane motion be solved by the general method described earlier in this section.

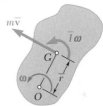

Fig. 17.9 The linear and angular momenta for a noncentroidal rotation.

[†]Note that the sum $\mathbf{H}_P$ of the moments about an arbitrary point P of the momenta of the particles of a rigid body is, in general, not equal to $I_P\boldsymbol{\omega}$ (see Prob. 17.67).

17.2B Systems of Rigid Bodies

We can analyze the motion of several rigid bodies by applying the principle of impulse and momentum to each body separately (Sample Prob. 17.7). However, in solving problems involving no more than three unknowns (including the impulses of unknown reactions), it is often convenient to apply the principle of impulse and momentum to the system as a whole.

To do this, first draw impulse–momentum diagrams for the entire system of bodies. For each moving part of the system, the diagrams of momenta should include a linear momentum vector and a momentum couple. You can omit impulses of forces internal to the system from the diagram showing the impulses, since they occur in pairs of equal and opposite vectors. Summing and equating successively the x components, y components, and moments of all vectors involved, you obtain three relations expressing that the momenta at time t_1 and the impulses of the external forces form a system equipollent to the system of the momenta at time t_2. Again, you should take care not to add linear and angular momenta indiscriminately; check each equation to make sure that consistent units are used. This approach has been used in Sample Probs. 17.9 through 17.13.

17.2C Conservation of Angular Momentum

When no external force acts on a rigid body or a system of rigid bodies, the impulses of the external forces are zero and the system of the momenta at time t_1 is equipollent to the system of the momenta at time t_2. Summing and equating successively the x components, y components, and moments of the momenta at times t_1 and t_2, we conclude that the total linear momentum of the system is conserved in any direction and that its total angular momentum is conserved about any point.

In many engineering applications, however, the linear momentum is not conserved, yet the angular momentum $\mathbf{H}_P$ of the system about a given point P is conserved. That is,

$$(\mathbf{H}_P)_1 = (\mathbf{H}_P)_2 \tag{17.17}$$

Such cases occur when the lines of action of all external forces pass through P or, more generally, when the sum of the angular impulses of the external forces about P is zero.

You can solve problems involving the **conservation of angular momentum** about a point P using the general method of impulse and momentum, i.e., by drawing impulse-momentum diagrams as described earlier. You then obtain Eq. (17.17) by summing and equating moments about P (Sample Prob. 17.9). As you will see in Sample Prob. 17.11, you can obtain two additional equations by summing and equating the x and y components of the linear momentum; then you can use these equations to determine two unknown linear impulses, such as the impulses of the reaction components at a fixed point.

Photo 17.3 A figure skater at the beginning and at the end of a spin. By using the principle of conservation of angular momentum you will find that her angular velocity is much higher at the end of the spin.

Sample Problem 17.7

Gear A has a mass of 10 kg and a radius of gyration of 200 mm, and gear B has a mass of 3 kg and a radius of gyration of 80 mm. The system is at rest when a couple **M** with a magnitude of 6 N·m is applied to gear B. (These gears were considered in Sample Prob. 17.2.) Neglecting friction, determine (a) the time required for the angular velocity of gear B to reach 600 rpm, (b) the tangential force that gear B exerts on gear A.

$r_A = 250$ mm

$r_B = 100$ mm

STRATEGY: Since you are given an angular velocity and are asked for time, use the principle of impulse and momentum.

MODELING: You are asked to find the internal tangential force, so you need two systems for this problem; that is, gear A and gear B. Model the gears as rigid bodies. Since all forces and couples are constant, you can obtain the impulses by multiplying the forces and moments by the unknown time t.

ANALYSIS: Recall from Sample Prob. 17.2 that the centroidal moments of inertia and the final angular velocities are

$$\bar{I}_A = 0.400 \text{ kg·m}^2 \qquad \bar{I}_B = 0.0192 \text{ kg·m}^2$$

$$(\omega_A)_2 = 25.1 \text{ rad/s} \qquad (\omega_B)_2 = 62.8 \text{ rad/s}$$

Principle of Impulse and Momentum for Gear A. The impulse-momentum diagram (Fig. 1) for gear A shows the initial momenta, impulses, and final momenta.

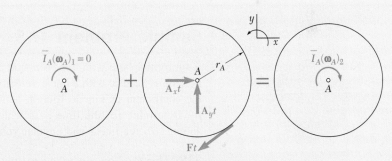

Fig. 1 Impulse–momentum diagram for gear A.

$$\text{Syst Momenta}_1 + \text{Syst Ext Imp}_{1 \rightarrow 2} = \text{Syst Momenta}_2$$

$+\!\uparrow$moments about A: $\qquad\qquad 0 - Ftr_A = -\bar{I}_A(\omega_A)_2$

$$Ft(0.250 \text{ m}) = (0.400 \text{ kg·m}^2)(25.1 \text{ rad/s})$$

$$Ft = 40.2 \text{ N·s}$$

(continued)

Principle of Impulse and Momentum for Gear B. Draw a separate impulse–momentum diagram for gear B (Fig. 2).

$\bar{I}_B(\omega_B)_1 = 0$ Ft $\bar{I}_B(\omega_B)_2$

$B_x t$ B Mt

$B_y t$

Fig. 2 Impulse–momentum diagram for gear B.

Syst Momenta$_1$ + Syst Ext Imp$_{1\to2}$ = Syst Momenta$_2$

$+\!\!\gamma$ moments about B: $0 + Mt - Ftr_B = \bar{I}_B(\omega_B)_2$

$+(6\text{ N·m})t - (40.2\text{ N·s})(0.100\text{ m}) = (0.0192\text{ kg·m}^2)(62.8\text{ rad/s})$

$$t = 0.871 \text{ s} \quad \blacktriangleleft$$

Recall that $Ft = 40.2$ N·m, so you have

$$F(0.871\text{ s}) = 40.2\text{ N·s} \qquad F = +46.2\text{ N}$$

Thus, the force exerted by gear B on gear A is

$$\mathbf{F} = 46.2\text{ N} \swarrow \quad \blacktriangleleft$$

REFLECT and THINK: This is the same answer obtained in Sample Prob. 17.2 by the method of work and energy, as you would expect. The difference is that in Sample Prob. 17.2, you were asked to find the number of revolutions, and in this problem, you were asked to find the time. What you are asked to find will often determine the best approach to use when solving a problem.

Sample Problem 17.8

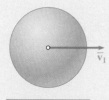

A uniform sphere with a mass m and a radius r is projected along a rough horizontal surface with a linear velocity $\bar{\mathbf{v}}_1$ and no angular velocity. Denote the coefficient of kinetic friction between the sphere and the surface by μ_k. Determine (*a*) the time t_2 at which the sphere starts rolling without sliding, (*b*) the linear and angular velocities of the sphere at time t_2.

STRATEGY: You are asked to find the time, so use the principle of impulse and momentum. You can apply this principle to the sphere from the time $t_1 = 0$ when it is placed on the surface until the time $t_2 = t$ when it starts rolling without sliding.

MODELING: Choose the sphere as your system and model it as a rigid body. While the sphere is sliding relative to the surface, it is acted upon by the normal force $\mathbf{N}$, the friction force $\mathbf{F}$, and its weight $\mathbf{W}$ with a magnitude of $W = mg$. An impulse-momentum diagram for this system is shown in Fig. 1.

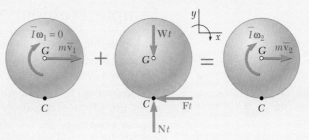

Fig. 1 Impulse–momentum diagram for the sphere.

ANALYSIS:

Principle of Impulse and Momentum. Apply the principle of impulse and momentum for this system between time t_1 and t_2

$$\textbf{Syst Momenta}_1 + \textbf{Syst Ext Imp}_{1\rightarrow2} = \textbf{Syst Momenta}_2$$

$+\uparrow y$ components:
$$Nt - Wt = 0 \tag{1}$$

$\xrightarrow{+} x$ components:
$$m\bar{v}_1 - Ft = m\bar{v}_2 \tag{2}$$

$+\downarrow$ moments about G:
$$Ftr = \bar{I}\omega_2 \tag{3}$$

From Eq. (1) you obtain $N = W = mg$. During the entire time interval considered, sliding occurs at point C, and $F = \mu_k N = \mu_k mg$. Substituting this expression for F into Eq. (2), you have

$$m\bar{v}_1 - \mu_k mgt = m\bar{v}_2 \qquad \bar{v}_2 = \bar{v}_1 - \mu_k gt \tag{4}$$

Substituting $F = \mu_k mg$ and $\bar{I} = \tfrac{2}{5}mr^2$ into Eq. (3) gives

$$\mu_k mgtr = \tfrac{2}{5}mr^2\omega_2 \qquad \omega_2 = \frac{5}{2}\frac{\mu_k g}{r}t \tag{5}$$

The sphere starts rolling without sliding when the velocity $\mathbf{v}_C$ of the point of contact is zero. At that time, point C becomes the instantaneous center of rotation, and you have $\bar{v}_2 = r\omega_2$. Substituting Eqs. (4) and (5) into this equation, you obtain

$$\bar{v}_1 - \mu_k gt = r\left(\frac{5}{2}\frac{\mu_k g}{r}t\right) \qquad\qquad t = \frac{2}{7}\frac{\bar{v}_1}{\mu_k g} \quad \blacktriangleleft$$

Substituting this expression for t into Eq. (5), you have

$$\omega_2 = \frac{5}{2}\frac{\mu_k g}{r}\left(\frac{2}{7}\frac{\bar{v}_1}{\mu_k g}\right) \qquad \omega_2 = \frac{5}{7}\frac{\bar{v}_1}{r} \qquad \omega_2 = \frac{5}{7}\frac{\bar{v}_1}{r} \downarrow \quad \blacktriangleleft$$

$$\bar{v}_2 = r\omega_2 \qquad\qquad \bar{v}_2 = r\left(\frac{5}{7}\frac{v_1}{r}\right) \qquad \mathbf{\bar{v}}_2 = \tfrac{5}{7}\bar{v}_1 \rightarrow \quad \blacktriangleleft$$

REFLECT and THINK: This is the same answer obtained in Sample Prob. 16.6 by first dealing directly with force and acceleration and then applying kinematic relationships.

Sample Problem 17.9

Two solid spheres with a radius 3 in. and weighing 2 lb each are mounted at A and B on the horizontal rod $A'B'$ that rotates freely about a vertical axis with a counterclockwise angular velocity of 6 rad/s. The spheres are held in position by a cord, which is suddenly cut. The centroidal moment of inertia of the rod and pivot is $\bar{I}_R = 0.25$ lb·ft·s². Determine (a) the angular velocity of the rod after the spheres have moved to positions A' and B', (b) the energy lost due to the plastic impact of the spheres and the stops at A' and B'.

STRATEGY: You can first use the principle of impulse and momentum to find the angular velocity of the rod and then use the definition of kinetic energy to determine the change in energy.

MODELING: Choose the two solid spheres and the horizontal rod as your system and model these as rigid bodies. The impulse–momentum diagram for this system is shown in Fig. 1.

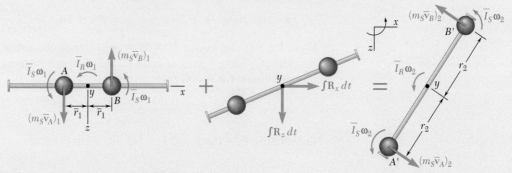

Fig. 1 Impulse–momentum diagram for the system.

ANALYSIS:

a. Principle of Impulse and Momentum.   Apply the principle of impulse and momentum for this system between time t_1 (when the spheres are at r_1) and t_2 (when the spheres are at r_1)

$$\textbf{Syst Momenta}_1 + \textbf{Syst Ext Imp}_{1\rightarrow2} = \textbf{Syst Momenta}_2$$

The external forces consist of the weights and the reaction at the pivot, which have no moment about the y axis. Noting that the rod is undergoing centroidal rotation and $\bar{v}_A = \bar{v}_B = \bar{r}\omega$, you can equate moments about the y axis as

$$2(m_S\bar{r}_1\omega_1)\bar{r}_1 + 2\bar{I}_S\omega_1 + \bar{I}_R\omega_1 = 2(m_S\bar{r}_2\omega_2)\bar{r}_2 + 2\bar{I}_S\omega_2 + \bar{I}_R\omega_2$$

$$(2m_S\bar{r}_1^2 + 2\bar{I}_S + \bar{I}_R)\omega_1 = (2m_S\bar{r}_2^2 + 2\bar{I}_S + \bar{I}_R)\omega_2 \tag{1}$$

This states that the angular momentum of the system about the y axis is conserved. You can now compute

$$\bar{I}_S = \tfrac{2}{5}m_S a^2 = \tfrac{2}{5}(2 \text{ lb}/32.2 \text{ ft/s}^2)(\tfrac{3}{12} \text{ ft})^2 = 0.00155 \text{ lb·ft·s}^2$$

$$m_S\bar{r}_1^2 = (2/32.2)(\tfrac{5}{12})^2 = 0.0108 \qquad m_S\bar{r}_2^2 = (2/32.2)(\tfrac{25}{12})^2 = 0.2696$$

Substituting these values, along with $\bar{I}_R = 0.25$ lb·ft·s^2 and $\omega_1 = 6$ rad/s, into Eq. (1) gives

$$0.275(6 \text{ rad/s}) = 0.792\omega_2 \quad \boldsymbol{\omega_2 = 2.08 \text{ rad/s}} \, \gamma \quad \blacktriangleleft$$

b. Energy Lost. The kinetic energy of the system at any instant is

$$T = 2(\tfrac{1}{2}m_S\bar{v}^2 + \tfrac{1}{2}\bar{I}_S\omega^2) + \tfrac{1}{2}\bar{I}_R\omega^2 = \tfrac{1}{2}(2m_S\bar{r}^2 + 2\bar{I}_S + \bar{I}_R)\omega^2$$

Using the numerical values found here, you have

$$T_1 = \tfrac{1}{2}(0.275)(6)^2 = 4.95 \text{ ft·lb} \qquad T_2 = \tfrac{1}{2}(0.792)(2.08)^2 = 1.713 \text{ ft·lb}$$

$$\Delta T = T_2 - T_1 = 1.71 - 4.95 \qquad \Delta T = -3.24 \text{ ft·lb} \quad \blacktriangleleft$$

REFLECT and THINK: As expected, when the spheres move outward, the angular velocity of the system decreases. This is similar to an ice skater who throws her arms outward to reduce her angular speed.

Sample Problem 17.10

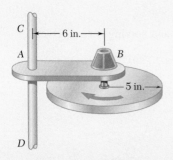

A 10-lb uniform disk is attached to the shaft of a motor mounted on arm *AB* that is free to rotate about the vertical axle *CD*. The arm-and-motor unit has a moment of inertia of 0.032 lb·ft·s^2 about axle *CD*. Knowing that the system is initially at rest, determine the angular velocities of the arm and of the disk when the motor reaches a speed of 360 rpm.

STRATEGY: Since you have two times—when the system starts from rest and when the motor has reached a speed of 360 rpm—use the conservation of angular momentum. You cannot use the conservation of energy because the motor converts electrical energy into mechanical energy.

MODELING: Choose the arm *AB*, the motor, and the disk to be your system and model them as rigid bodies. The impulse–momentum diagram for this system is shown in Fig. 1.

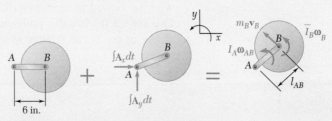

Fig. 1 Impulse–momentum diagram for the system.

(continued)

Moments of Inertia. The mass moment of inertia of the arm and motor about the axle is $I_A = 0.032 \, \text{lb} \cdot \text{ft} \cdot \text{s}^2$, and the mass moment of inertia of disk B about its center of mass is

$$\bar{I}_B = \frac{1}{2}\frac{W}{g}r^2 = \frac{1}{2}\left(\frac{10}{32.2}\right)\left(\frac{5}{12}\right)^2 = 0.02696 \, \text{lb} \cdot \text{ft} \cdot \text{s}^2$$

ANALYSIS:

Principle of Impulse and Momentum. Apply the principle of impulse and momentum for this system between time t_1 (when the system is at rest) and t_2 (when the motor has an angular velocity of 360 rpm)

$$\textbf{Syst Momenta}_1 + \textbf{Syst Ext Imp}_{1\rightarrow 2} = \textbf{Syst Momenta}_2$$

Taking moments about A gives

$+\upharpoonleft$ moments about A: $0 + 0 = (m_B v_B)l_{AB} + I_A \omega_{AB} + \bar{I}_B \omega_B$ (1)

Kinematics. You can relate the velocity of B to the angular velocity of AB using

$$v_B = l_{AB}\omega_{AB} = \tfrac{6}{12}\omega_{AB}$$ (2)

The velocity of the motor is $\omega_M = 360 \, \text{rpm} = 12\pi \, \text{rad/s}$, which is the angular velocity of the disk relative to the arm. Thus,

$$\omega_B = \omega_{AB} + \omega_M$$ (3)

Substituting Eqs. (2) and (3) into Eq. (1) and solving for ω_{AB} gives

$$(m_B l_{AB}^2 + I_A)\omega_{AB} + \bar{I}_B(\omega_{AB} + \omega_M) = 0$$

$$\left[\left(\frac{10}{32.2}\right)\left(\frac{6}{12}\right)^2 + 0.032\right]\omega_{AB} + 0.02696(\omega_{AB} + 12\pi) = 0$$

$$\omega_{AB} = -7.44 \, \text{rad/s}$$

$$\boldsymbol{\omega}_{AB} = 71.0 \, \text{rpm} \downarrow \quad \blacktriangleleft$$

The angular velocity of the disk is

$$\omega_B = -7.44 + 12\pi = 30.26 \, \text{rad/s}$$

$$\boldsymbol{\omega}_B = 289 \, \text{rpm} \upharpoonleft \quad \blacktriangleleft$$

REFLECT and THINK: When the motor spins the disk counterclockwise (as viewed from above), the arm AB rotates in a clockwise direction. One key to solving this problem is recognizing that the angular velocity of the motor is the relative angular velocity of the disk with respect to the bar.

SOLVING PROBLEMS
ON YOUR OWN

In this section, we described how to use the method of impulse and momentum to solve problems involving the plane motion of rigid bodies. As you found out previously in Chap. 13, this method is most effective when used in the solution of problems involving velocities and time.

1. The principle of impulse and momentum for the plane motion of a rigid body is expressed by the vector equation:

$$\textbf{Syst Momenta}_1 + \textbf{Syst Ext Imp}_{1\to2} = \textbf{Syst Momenta}_2 \qquad (17.14)$$

where **Syst Momenta** represents the system of the momenta of the particles forming the rigid body and **Syst Ext Imp** represents the system of all the external impulses exerted during the motion.

   a. The system of the momenta of a rigid body is equivalent to a linear momentum vector $m\bar{\textbf{v}}$ attached at the mass center of the body and an angular momentum couple about the center of mass $\bar{I}\boldsymbol{\omega}$ (Fig. 17.7).

   b. You should draw an impulse–momentum diagram for the rigid body to express the vector equation (17.14) graphically. Your diagram should consist of three sketches of the body representing, respectively, the initial momenta, the impulses of the external forces, and the final momenta. This shows that the system of the initial momenta and the system of the impulses of the external forces are together equivalent to the system of the final momenta (Fig. 17.8).

   c. By using the impulse–momentum diagram, you can sum components in any direction and sum moments about any point. For a single rigid body, if you choose to sum moments about an arbitrary point P, you can write Eq. (17.14) as

$$\bar{I}\omega_1 + m\bar{v}_1d_\perp + \sum \int_{t_1}^{t_2} M_P dt = \bar{I}\omega_2 + m\bar{v}_2 d_\perp \qquad (17.14')$$

where $d_\perp$ is the perpendicular distance from point P to the line of action of the linear velocity of G. If you choose to sum moments about the center of gravity of the body, Eq. (17.14′) reduces to

$$\bar{I}\omega_1 + \sum \int_{t_1}^{t_2} M_G dt = \bar{I}\omega_2 \qquad (17.14'')$$

If you choose to sum moments about a fixed point O, Eq. (17.14′) reduces to

$$I_O\omega_1 + \sum \int_{t_1}^{t_2} M_O \, dt = I_O\omega_2 \qquad (17.16)$$

where I_O is the mass moment of inertia about point O. In most cases, you will be able to select and solve an equation that involves only one unknown.

2. **In problems involving a system of rigid bodies,** you can apply the principle of impulse and momentum to the system as a whole. Since internal forces occur in equal and opposite pairs, they should not be part of your solution [Sample Probs. 17.9 and 17.10].

3. **Conservation of angular momentum about a given axis** occurs when, for a system of rigid bodies, *the sum of the moments of the external impulses about that axis is zero.* You can indeed easily observe from the impulse–momentum diagram that the initial and final angular momenta of the system about that axis are equal and, thus, that the angular momentum of the system about the given axis is conserved. You can then sum the angular momenta of the various bodies of the system and the moments of their linear momenta about that axis to obtain an equation that you can solve for one unknown [Sample Prob. 17.9 and 17.10].

Problems

CONCEPT QUESTIONS

17.CQ6 Slender bar A is rigidly connected to a massless rod BC in Case 1 and two massless cords in Case 2 as shown. The vertical thickness of bar A is negligible compared to L. If bullet D strikes A with a speed v_0 and becomes embedded in it, how will the speeds of the center of gravity of A immediately after the impact compare for the two cases?

  a. Case 1 will be larger.
  b. Case 2 will be larger.
  c. The speeds will be the same.

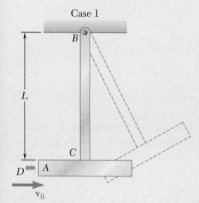
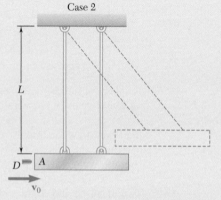

Fig. P17.CQ6

17.CQ7 A 1-m-long uniform slender bar AB has an angular velocity of 12 rad/s and its center of gravity has a velocity of 2 m/s as shown. About which point is the angular momentum of A smallest at this instant?

  a. P_1
  b. P_2
  c. P_3
  d. P_4
  e. It is the same about all the points.

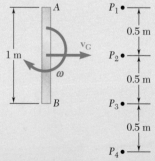

Fig. P17.CQ7

17.F1 The 350-kg flywheel of a small hoisting engine has a radius of gyration of 600 mm. If the power is cut off when the angular velocity of the flywheel is 100 rpm clockwise, draw an impulse–momentum diagram that can be used to determine the time required for the system to come to rest.

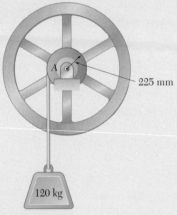

225 mm

120 kg

Fig. P17.F1

Fig. P17.F2

17.F2 A sphere of radius r and mass m is placed on a horizontal floor with no linear velocity but with a clockwise angular velocity ω_0. Denoting by μ_k the coefficient of kinetic friction between the sphere and the floor, draw the impulse–momentum diagram that can be used to determine the time t_1 at which the sphere will start rolling without sliding.

17.F3 Two panels A and B are attached with hinges to a rectangular plate and held by a wire as shown. The plate and the panels are made of the same material and have the same thickness. The entire assembly is rotating with an angular velocity ω_0 when the wire breaks. Draw the impulse–momentum diagram that is needed to determine the angular velocity of the assembly after the panels have come to rest against the plate.

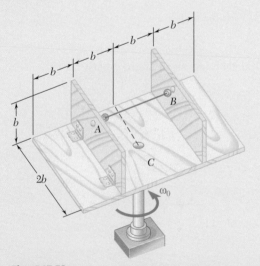

Fig. P17.F3

17.52 The rotor of an electric motor has a mass of 25 kg, and it is observed that 4.2 min is required for the rotor to coast to rest from an angular velocity of 3600 rpm. Knowing that kinetic friction produces a couple of magnitude 1.2 N·m, determine the centroidal radius of gyration for the rotor.

17.53 A small grinding wheel is attached to the shaft of an electric motor that has a rated speed of 3600 rpm. When the power is turned off, the unit coasts to rest in 70 s. The grinding wheel and rotor have a combined weight of 6 lb and a combined radius of gyration of 2 in. Determine the average magnitude of the couple due to kinetic friction in the bearings of the motor.

Fig. P17.53

17.54 A bolt located 50 mm from the center of an automobile wheel is tightened by applying the couple shown for 0.10 s. Assuming that the wheel is free to rotate and is initially at rest, determine the resulting angular velocity of the wheel. The wheel has a mass of 19 kg and has a radius of gyration of 250 mm.

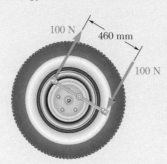

100 N
460 mm
100 N

Fig. P17.54

17.55 A uniform 144-lb cube is attached to a uniform 136-lb circular shaft as shown, and a couple **M** with a constant magnitude is applied to the shaft when the system is at rest. Knowing that $r = 4$ in., $L = 12$ in., and the angular velocity of the system is 960 rpm after 4 s, determine the magnitude of the couple **M**.

17.56 A uniform 75-kg cube is attached to a uniform 70-kg circular shaft as shown, and a couple **M** with a constant magnitude of 20 N·m is applied to the shaft. Knowing that $r = 100$ mm and $L = 300$ mm, determine the time required for the angular velocity of the system to increase from 1000 rpm to 2000 rpm.

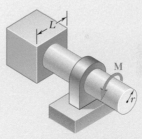

Fig. P17.55 and P17.56

17.57 A disk of constant thickness, initially at rest, is placed in contact with a belt that moves with a constant velocity **v**. Denoting by μ_k the coefficient of kinetic friction between the disk and the belt, derive an expression for the time required for the disk to reach a constant angular velocity.

17.58 Disk A, of weight 5 lb and radius $r = 3$ in., is at rest when it is placed in contact with a belt that moves at a constant speed $v = 50$ ft/s. Knowing that $\mu_k = 0.20$ between the disk and the belt, determine the time required for the disk to reach a constant angular velocity.

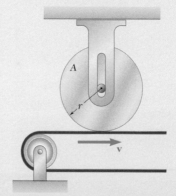

Fig. P17.57 and P17.58

17.59 A cylinder of radius r and weight W with an initial counterclockwise angular velocity ω_0 is placed in the corner formed by the floor and a vertical wall. Denoting by μ_k the coefficient of kinetic friction between the cylinder and the wall and the floor, derive an expression for the time required for the cylinder to come to rest.

Fig. P17.59

17.60 Each of the double pulleys shown has a centroidal mass moment of inertia of 0.25 kg·m², an inner radius of 100 mm, and an outer radius of 150 mm. Neglecting bearing friction, determine (a) the velocity of the cylinder 3 s after the system is released from rest, (b) the tension in the cord connecting the pulleys.

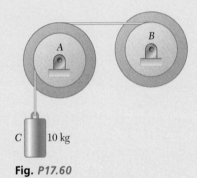

Fig. P17.60

17.61 Each of the gears A and B has a mass of 675 g and a radius of gyration of 40 mm, while gear C has a mass of 3.6 kg and a radius of gyration of 100 mm. Assume that kinetic friction in the bearings of gears A, B, and C produces couples of constant magnitude 0.15 N·m, 0.15 N·m, and 0.3 N·m, respectively. Knowing that the initial angular velocity of gear C is 2000 rpm, determine the time required for the system to come to rest.

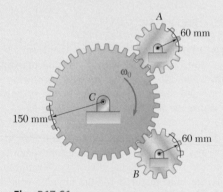

Fig. P17.61

17.62 Disk B has an initial angular velocity ω_0 when it is brought into contact with disk A, which is at rest. Show that the final angular velocity of disk B depends only on ω_0 and the ratio of the masses m_A and m_B of the two disks.

17.63 The 7.5-lb disk A has a radius $r_A = 6$ in. and is initially at rest. The 10-lb disk B has a radius $r_B = 8$ in. and an angular velocity ω_0 of 900 rpm when it is brought into contact with disk A. Neglecting friction in the bearings, determine (a) the final angular velocity of each disk, (b) the total impulse of the friction force exerted on disk A.

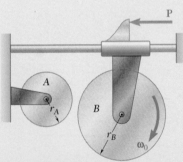

Fig. P17.62 and P17.63

17.64 A tape moves over the two drums shown. Drum A weighs 1.4 lb and has a radius of gyration of 0.75 in., while drum B weighs 3.5 lb and has a radius of gyration of 1.25 in. In the lower portion of the tape the tension is constant and equal to $T_A = 0.75$ lb. Knowing that the tape is initially at rest, determine (a) the required constant tension T_B if the velocity of the tape is to be $v = 10$ ft/s after 0.24 s, (b) the corresponding tension in the portion of the tape between the drums.

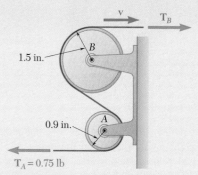

1.5 in.

0.9 in.

$T_A = 0.75$ lb

Fig. P17.64

17.65 Show that the system of momenta for a rigid body in plane motion reduces to a single vector, and express the distance from the mass center G to the line of action of this vector in terms of the centroidal radius of gyration $\bar{k}$ of the body, the magnitude $\bar{v}$ of the velocity of G, and the angular velocity $\boldsymbol{\omega}$.

17.66 Show that, when a rigid body rotates about a fixed axis through O perpendicular to the body, the system of the momenta of its particles is equivalent to a single vector of magnitude $m\bar{r}\omega$, perpendicular to the line OG, and applied to a point P on this line, called the *center of percussion*, at a distance $GP = \bar{k}^2/\bar{r}$ from the mass center of the body.

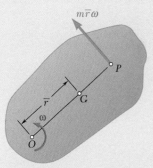

$m\bar{r}\omega$

Fig. P17.66

17.67 Show that the sum $\mathbf{H}_A$ of the moments about a point A of the momenta of the particles of a rigid body in plane motion is equal to $I_A\boldsymbol{\omega}$, where $\boldsymbol{\omega}$ is the angular velocity of the body at the instant considered and I_A the moment of inertia of the body about A, if and only if one of the following conditions is satisfied: (a) A is the mass center of the body, (b) A is the instantaneous center of rotation, (c) the velocity of A is directed along a line joining point A and the mass center G.

17.68 Consider a rigid body initially at rest and subjected to an impulsive force $\mathbf{F}$ contained in the plane of the body. We define the *center of percussion* P as the point of intersection of the line of action of $\mathbf{F}$ with the perpendicular drawn from G. (a) Show that the instantaneous center of rotation C of the body is located on line GP at a distance $GC = \bar{k}^2/GP$ on the opposite side of G. (b) Show that if the center of percussion were located at C, the instantaneous center of rotation would be located at P.

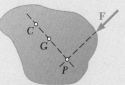

Fig. P17.68

17.69 A flywheel is rigidly attached to a 1.5-in.-radius shaft that rolls without sliding along parallel rails. Knowing that after being released from rest the system attains a speed of 6 in./s in 30 s, determine the centroidal radius of gyration of the system.

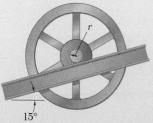

r

15°

Fig. P17.69

17.70 A wheel of radius r and centroidal radius of gyration $\bar{k}$ is released from rest on the incline shown at time $t = 0$. Assuming that the wheel rolls without sliding, determine (a) the velocity of its center at time t, (b) the coefficient of static friction required to prevent slipping.

Fig. P17.70

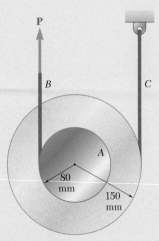

Fig. P17.71

17.71 The double pulley shown has a mass of 3 kg and a radius of gyration of 100 mm. Knowing that when the pulley is at rest, a force **P** of magnitude 24 N is applied to cord B, determine (a) the velocity of the center of the pulley after 1.5 s, (b) the tension in cord C.

17.72 and *17.73* A 9-in.-radius cylinder of weight 18 lb rests on a 6-lb carriage. The system is at rest when a force **P** of magnitude 2.5 lb is applied as shown for 1.2 s. Knowing that the cylinder rolls without sliding on the carriage and neglecting the mass of the wheels of the carriage, determine the resulting velocity of (a) the carriage, (b) the center of the cylinder.

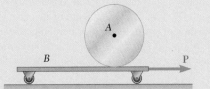

Fig. P17.72

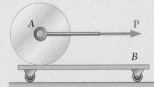

Fig. *P17.73*

17.74 Two uniform cylinders, each of mass $m = 6$ kg and radius $r = 125$ mm, are connected by a belt as shown. If the system is released from rest when $t = 0$, determine (a) the velocity of the center of cylinder B at $t = 3$ s, (b) the tension in the portion of belt connecting the two cylinders.

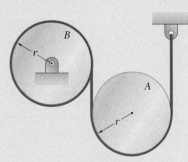

Fig. P17.74 and P17.75

17.75 Two uniform cylinders, each of mass $m = 6$ kg and radius $r = 125$ mm, are connected by a belt as shown. Knowing that at the instant shown the angular velocity of cylinder A is 30 rad/s counterclockwise, determine (a) the time required for the angular velocity of cylinder A to be reduced to 5 rad/s, (b) the tension in the portion of belt connecting the two cylinders.

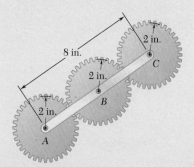

Fig. *P17.76*

17.76 In the gear arrangement shown, gears A and C are attached to rod ABC, that is free to rotate about B, while the inner gear B is fixed. Knowing that the system is at rest, determine the magnitude of the couple **M** that must be applied to rod ABC, if 2.5 s later the angular velocity of the rod is to be 240 rpm clockwise. Gears A and C weigh 2.5 lb each and may be considered as disks of radius 2 in.; rod ABC weighs 4 lb.

17.77 A sphere of radius r and mass m is projected along a rough horizontal surface with the initial velocities shown. If the final velocity of the sphere is to be zero, express (a) the required magnitude of $\boldsymbol{\omega}_0$ in terms of v_0 and r, (b) the time required for the sphere to come to rest in terms of v_0 and the coefficient of kinetic friction μ_k.

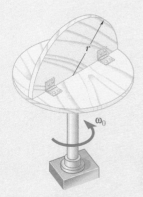

Fig. P17.77

17.78 A bowler projects an 8.5-in.-diameter ball weighing 16 lb along an alley with a forward velocity $\mathbf{v}_0$ of 25 ft/s and a backspin ω_0 of 9 rad/s. Knowing that the coefficient of kinetic friction between the ball and the alley is 0.10, determine (a) the time t_1 at which the ball will start rolling without sliding, (b) the speed of the ball at time t_1.

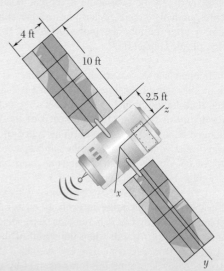

Fig. P17.78

17.79 A semicircular panel with a radius r is attached with hinges to a circular plate with a radius r and initially is held in the vertical position as shown. The plate and the panel are made of the same material and have the same thickness. Knowing that the entire assembly is rotating freely with an initial angular velocity of $\boldsymbol{\omega}_0$, determine the angular velocity of the assembly after the panel has been released and comes to rest against the plate.

Fig. P17.79

17.80. A satellite has a total weight (on Earth) of 250 lbs, and each of the solar panels weighs 15 lbs. The body of the satellite has mass moment of inertia about the z-axis of 6 slug-ft^2, and the panels can be modeled as flat plates. The satellite spins with a rate of 10 rpm about the z-axis when the solar panels are positioned in the xy plane. Determine the spin rate about z after a motor on the satellite has rotated both panels to be positioned in the yz plane (as shown in the figure).

Fig. P17.80

17.81 Two 10-lb disks and a small motor are mounted on a 15-lb rectangular platform that is free to rotate about a central vertical spindle. The normal operating speed of the motor is 180 rpm. If the motor is started when the system is at rest, determine the angular velocity of all elements of the system after the motor has attained its normal operating speed. Neglect the mass of the motor and of the belt.

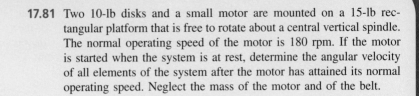

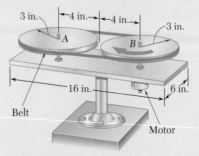

Fig. P17.81

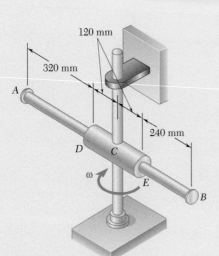

Fig. P17.82

17.82 A 3-kg rod of length 800 mm can slide freely in the 240-mm cylinder *DE*, which in turn can rotate freely in a horizontal plane. In the position shown, the assembly is rotating with an angular velocity of magnitude $\omega = 40$ rad/s and end *B* of the rod is moving toward the cylinder at a speed of 75 mm/s relative to the cylinder. Knowing that the centroidal mass moment of inertia of the cylinder about a vertical axis is 0.025 kg·m^2 and neglecting the effect of friction, determine the angular velocity of the assembly as end *B* of the rod strikes end *E* of the cylinder.

17.83 A 1.6-kg tube *AB* can slide freely on rod *DE*, which in turn can rotate freely in a horizontal plane. Initially the assembly is rotating with an angular velocity of magnitude $\omega = 5$ rad/s and the tube is held in position by a cord. The moment of inertia of the rod and bracket about the vertical axis of rotation is 0.30 kg·m^2 and the centroidal moment of inertia of the tube about a vertical axis is 0.0025 kg·m^2. If the cord suddenly breaks, determine (*a*) the angular velocity of the assembly after the tube has moved to end *E*, (*b*) the energy lost during the plastic impact at *E*.

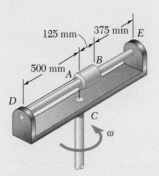

Fig. P17.83

17.84 In the helicopter shown, a vertical tail propeller is used to prevent rotation of the cab as the speed of the main blades is changed. Assuming that the tail propeller is not operating, determine the final angular velocity of the cab after the speed of the main blades has been changed from 180 to 240 rpm. (The speed of the main blades is measured relative to the cab, and the cab has a centroidal moment of inertia of 650 lb·ft·s^2. Each of the four main blades is assumed to be a slender 14-ft rod weighing 55 lb.)

17.85 Assuming that the tail propeller in Prob. 17.84 is operating and that the angular velocity of the cab remains zero, determine the final horizontal velocity of the cab when the speed of the main blades is changed from 180 to 240 rpm. The cab weighs 1250 lb and is initially at rest. Also, determine the force exerted by the tail propeller if the change in speed takes place uniformly in 12 s.

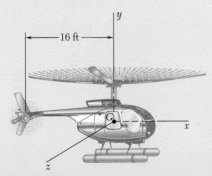

Fig. P17.84

17.86 The circular platform A is fitted with a rim of 200-mm inner radius and can rotate freely about the vertical shaft. It is known that the platform-rim unit has a mass of 5 kg and a radius of gyration of 175 mm with respect to the shaft. At a time when the platform is rotating with an angular velocity of 50 rpm, a 3-kg disk B of radius 80 mm is placed on the platform with no velocity. Knowing that disk B then slides until it comes to rest relative to the platform against the rim, determine the final angular velocity of the platform.

17.87 The 30-kg uniform disk A and the bar BC are at rest and the 5-kg uniform disk D has an initial angular velocity of ω_1 with a magnitude of 440 rpm when the compressed spring is released and disk D contacts disk A. The system rotates freely about the vertical spindle BE. After a period of slippage, disk D rolls without slipping. Knowing that the magnitude of the final angular velocity of disk D is 176 rpm, determine the final angular velocities of bar BC and disk A. Neglect the mass of bar BC.

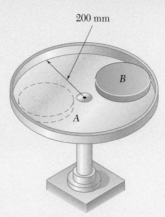

Fig. P17.86

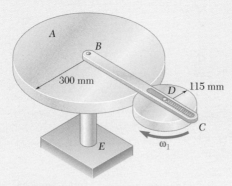

Fig. P17.87

17.88 The 4-kg rod AB can slide freely inside the 6-kg tube. The rod was entirely within the tube ($x = 0$) and released with no initial velocity relative to the tube when the angular velocity of the assembly was 5 rad/s. Neglecting the effect of friction, determine the speed of the rod relative to the tube when $x = 400$ mm.

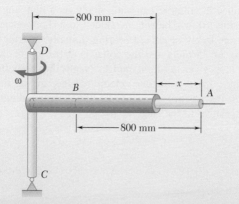

Fig. P17.88

1231

17.89 A 1.8-kg collar A and a 0.7-kg collar B can slide without friction on a frame, consisting of the horizontal rod OE and the vertical rod CD, which is free to rotate about its vertical axis of symmetry. The two collars are connected by a cord running over a pulley that is attached to the frame at O. At the instant shown, the velocity $\mathbf{v}_A$ of collar A has a magnitude of 2.1 m/s and a stop prevents collar B from moving. The stop is suddenly removed and collar A moves toward E. As it reaches a distance of 0.12 m from O, the magnitude of its velocity is observed to be 2.5 m/s. Determine at that instant the magnitude of the angular velocity of the frame and the moment of inertia of the frame and pulley system about CD.

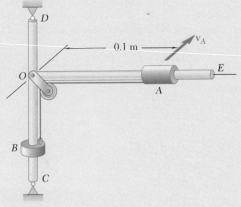

Fig. P17.89

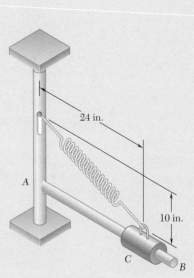

Fig. P17.90

17.90 A 6-lb collar C is attached to a spring and can slide on rod AB, which in turn can rotate in a horizontal plane. The mass moment of inertia of rod AB with respect to end A is 0.35 lb·ft·s². The spring has a constant $k = 15$ lb/in. and an undeformed length of 10 in. At the instant shown, the velocity of the collar relative to the rod is zero and the assembly is rotating with an angular velocity of 12 rad/s. Neglecting the effect of friction, determine (a) the angular velocity of the assembly as the collar passes through a point located 7.5 in. from end A of the rod, (b) the corresponding velocity of the collar relative to the rod.

17.91 A small 4-lb collar C can slide freely on a thin ring of weight 6 lb and radius 10 in. The ring is welded to a short vertical shaft, which can rotate freely in a fixed bearing. Initially, the ring has an angular velocity of 35 rad/s and the collar is at the top of the ring ($\theta = 0$) when it is given a slight nudge. Neglecting the effect of friction, determine (a) the angular velocity of the ring as the collar passes through the position $\theta = 90°$, (b) the corresponding velocity of the collar relative to the ring.

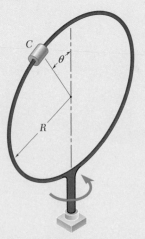

Fig. P17.91

17.92 Rod *AB* has a weight of 6 lb and is attached to a 10-lb cart *C*. Knowing that the system is released from rest in the position shown and neglecting friction, determine (*a*) the velocity of point *B* as rod *AB* passes through a vertical position, (*b*) the corresponding velocity of the cart *C*.

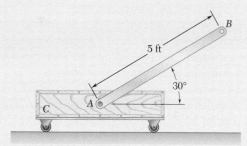

Fig. *P17.92*

17.93 In Prob. 17.82, determine the velocity of rod *AB* relative to cylinder *DE* as end *B* of the rod strikes end *E* of the cylinder.

17.94 In Prob. 17.83, determine the velocity of the tube relative to the rod as the tube strikes end *E* of the assembly.

17.95 The 6-lb steel cylinder *A* of radius *r* and the 10-lb wooden cart *B* are at rest in the position shown when the cylinder is given a slight nudge, causing it to roll without sliding along the top surface of the cart. Neglecting friction between the cart and the ground, determine the velocity of the cart as the cylinder passes through the lowest point of the surface at *C*.

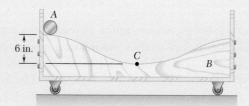

Fig. P17.95

17.3 ECCENTRIC IMPACT

You saw in Chap. 13 that the method of impulse and momentum is the only practicable method for solving problems involving the impulsive motion of a particle. Now you will see that problems involving the impulsive motion of a rigid body are particularly well suited to a solution using the method of impulse and momentum. Since the time interval considered in the computation of linear impulses and angular impulses is very short, we can assume the bodies involved occupy the same position during that time interval, making the computation quite simple.

In Sec. 13.4, we described how to solve problems of **central impact**, i.e., problems in which the mass centers of the two colliding bodies are located on the line of impact. We now analyze the **eccentric impact** of two rigid bodies.

Consider two colliding bodies and denote the velocities of the two points of contact A and B before impact by $\mathbf{v}_A$ and $\mathbf{v}_B$ (Fig. 17.10a). Under

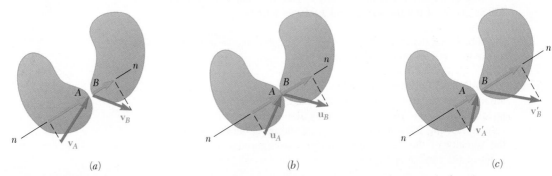

(a)   (b)   (c)

Fig. 17.10 When two rigid bodies collide, (a) the velocities of the points of contact before impact (b) change during the period of deformation and (c) change again during the period of restitution.

Photo 17.4 A swinging bat applies an impulsive force on contact with the ball. You can use the principle of impulse and momentum to determine the final velocities of the ball and bat.

the impact, the two bodies *deform,* and at the end of the period of deformation, the velocities $\mathbf{u}_A$ and $\mathbf{u}_B$ of A and B have equal components along the line of impact nn (Fig. 17.10b). A period of *restitution* then takes place, at the end of which points A and B have velocities of $\mathbf{v}'_A$ and $\mathbf{v}'_B$ (Fig. 17.10c). Assuming that the bodies are frictionless, we find that the forces they exert on each other are directed along the line of impact. We denote the magnitude of the impulse of one of these forces during the period of deformation by $\int Pdt$ and the magnitude of its impulse during the period of restitution by $\int Rdt$. Recall that we define the coefficient of restitution e as the ratio of

$$e = \frac{\int R\, dt}{\int P\, dt} \tag{17.18}$$

We propose to show that the relation established in Sec. 13.4 between the relative velocities of two particles before and after impact also holds between the components along the line of impact of the relative velocities of the two points of contact A and B. That is, we want to show that

$$(v'_B)_n - (v'_A)_n = e[(v_A)_n - (v_B)_n] \tag{17.19}$$

First, we assume that the motion of each of the two colliding bodies of Fig. 17.10 is unconstrained. Thus, the only impulsive forces exerted on the bodies during the impact are applied at A and B, respectively. Consider the body to which point A belongs and draw the impulse–momentum diagram corresponding to the period of deformation (Fig. 17.11). We

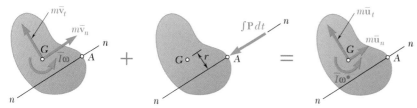

Fig. 17.11 An impulse–momentum diagram for a body undergoing an eccentric impact during the period of deformation.

denote the velocity of the mass center at the beginning and at the end of the period of deformation by $\bar{\mathbf{v}}$ and $\bar{\mathbf{u}}$, respectively, and we denote the angular velocity of the body at the same instants by $\boldsymbol{\omega}$ and $\boldsymbol{\omega}^*$. Summing and equating the components of the momenta and impulses along the line of impact nn, we have

$$m\bar{v}_n - \int P \, dt = m\bar{u}_n \tag{17.20}$$

Summing and equating the moments about G of the momenta and impulses, we also have

$$\bar{I}\omega - r\int P \, dt = \bar{I}\omega^* \tag{17.21}$$

where r represents the perpendicular distance from G to the line of impact. Considering now the period of restitution, we obtain in a similar way

$$m\bar{u}_n - \int R \, dt = m\bar{v}'_n \tag{17.22}$$
$$\bar{I}\omega^* - r\int R \, dt = \bar{I}\omega' \tag{17.23}$$

where $\bar{\mathbf{v}}'$ and $\boldsymbol{\omega}'$ represent, respectively, the velocity of the mass center and the angular velocity of the body after impact. First solving Eqs. (17.20) and (17.22) for the two impulses and substituting into Eq. (17.18) and then solving Eqs. (17.21) and (17.23) for the same two impulses and substituting again into Eq. (17.18), we obtain the two alternative expressions for the coefficient of restitution as

$$e = \frac{\bar{u}_n - \bar{v}'_n}{\bar{v}_n - \bar{u}_n} \qquad e = \frac{\omega^* - \omega'}{\omega - \omega^*} \tag{17.24}$$

Multiplying the numerator and denominator of the second expression for e by r and adding them, respectively, to the numerator and denominator of the first expression, we have

$$e = \frac{\bar{u}_n + r\omega^* - (\bar{v}'_n + r\omega')}{\bar{v}_n + r\omega - (\bar{u}_n + r\omega^*)} \tag{17.25}$$

Observe that $\bar{v}_n + r\omega$ represents the component $(v_A)_n$ along nn of the velocity of the point of contact A and that, similarly, $\bar{u}_n + r\omega^*$ and $\bar{v}'_n + r\omega'$ represent, respectively, the components $(u_A)_n$ and $(v'_A)_n$. Thus, we have

$$e = \frac{(u_A)_n - (v'_A)_n}{(v_A)_n - (u_A)_n} \tag{17.26}$$

The analysis of the motion of the second body leads to a similar expression for e in terms of the components along nn of the successive velocities of point B. Recalling that $(u_A)_n = (u_B)_n$, and eliminating these two velocity components by a manipulation similar to the one used in Sec. 13.4, we obtain the relation in Eq. (17.19).

If one or both of the colliding bodies is constrained to rotate about a fixed point O—as in the case of a compound pendulum (Fig. 17.12a)—an impulsive reaction is exerted at O (Fig. 17.12b). Let us verify that, although

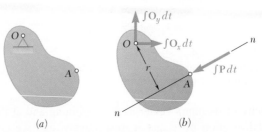

Fig. 17.12 (*a*) A rigid body constrained to rotate about a fixed point O; (*b*) impulsive reaction at O resulting from an eccentric impact.

their derivation must be modified, Eqs. (17.26) and (17.19) remain valid. Applying formula (17.16) to the period of deformation and to the period of restitution, we have

$$I_O\omega - r\!\int\! P\,dt = I_O\omega^* \tag{17.27}$$

$$I_O\omega^* - r\!\int\! R\,dt = I_O\omega' \tag{17.28}$$

where r represents the perpendicular distance from the fixed point O to the line of impact. We solve Eqs. (17.27) and (17.28) for the two impulses and substitute them into Eq. (17.18). Noting that $r\omega$, $r\omega^*$, and $r\omega'$ represent the components along nn of the successive velocities of point A, we obtain

$$e = \frac{\omega^* - \omega'}{\omega - \omega^*} = \frac{r\omega^* - r\omega'}{r\omega - r\omega^*} = \frac{(u_A)_n - (v'_A)_n}{(v_A)_n - (u_A)_n}$$

This verifies that Eq. (17.26) still holds. Thus, Eq. (17.19) remains valid when one or both of the colliding bodies is constrained to rotate about a fixed point O.

In order to determine the velocities of the two colliding bodies after impact, we need to use the relation in Eq. (17.19) in conjunction with one or several other equations obtained by applying the principle of impulse and momentum (Sample Prob. 17.11 and 17.13).

Sample Problem 17.11

A 0.05-lb bullet B is fired with a horizontal velocity of 1500 ft/s into the side of a 20-lb square panel suspended from a hinge at A. Knowing that the panel is initially at rest, determine (a) the angular velocity of the panel immediately after the bullet becomes embedded, (b) the impulsive reaction at A, assuming that the bullet becomes embedded in 0.0006 s.

STRATEGY: Since you have an impact, use the principle of impulse and momentum.

MODELING: Choose your system to be the bullet and the panel, where you model the bullet as a particle and the panel as a rigid body. The impulse-momentum diagram for this system is shown in Fig. 1. Since the time interval $\Delta t = 0.0006$ s is very short, you can neglect all nonimpulsive forces and consider only the external impulses $\mathbf{A}_x \Delta t$ and $\mathbf{A}_y \Delta t$.

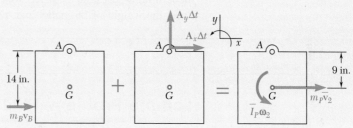

Fig. 1 Impulse–momentum diagram for the system. The bullet is neglected at time 2.

ANALYSIS:

Principle of Impulse and Momentum.

$$\text{Syst Momenta}_1 + \text{Syst Ext Imp}_{1\to2} = \text{Syst Momenta}_2$$

$+\curvearrowleft$moments about A: $\quad m_B v_B(\tfrac{14}{12}\text{ ft}) + 0 = m_P \bar{v}_2(\tfrac{9}{12}\text{ ft}) + \bar{I}_P\omega_2$ **(1)**

$\xrightarrow{+}x$ components: $\quad m_B v_B + A_x \Delta t = m_P \bar{v}_2$ **(2)**

$+\uparrow y$ components: $\quad 0 + A_y \Delta t = 0$ **(3)**

Note that the weight of the bullet is negligible compared to the weight of the panel, so we did not include it on the right-hand side of Eq. (1). The centroidal mass moment of inertia of the square panel is

$$\bar{I}_P = \tfrac{1}{6}m_P b^2 = \frac{1}{6}\left(\frac{20\text{ lb}}{32.2}\right)(\tfrac{18}{12}\text{ ft})^2 = 0.2329\text{ lb·ft·s}^2$$

Substituting this value as well as the given data into Eq. (1) and noting that from kinematics, you know

$$\bar{v}_2 = (\tfrac{9}{12}\text{ ft})\omega_2$$

(continued)

So you have

$$\left(\frac{0.05}{32.2}\right)(1500)(\tfrac{14}{12}) = 0.2329\omega_2 + \left(\frac{20}{32.2}\right)(\tfrac{9}{12}\omega_2)(\tfrac{9}{12})$$

$$\omega_2 = 4.67 \text{ rad/s} \qquad\qquad \omega_2 = 4.67 \text{ rad/s}\curvearrowleft \quad \blacktriangleleft$$

$$\bar{v}_2 = (\tfrac{9}{12}\text{ ft})\omega_2 = (\tfrac{9}{12}\text{ ft})(4.67\,\text{rad/s}) = 3.50 \text{ ft/s}$$

Substituting $\bar{v}_2 = 3.50$ ft/s, $\Delta t = 0.0006$ s, and the given data into Eq. (2) gives you

$$\left(\frac{0.05}{32.2}\right)(1500) + A_x(0.0006) = \left(\frac{20}{32.2}\right)(3.50)$$

$$A_x = -259 \text{ lb} \qquad\qquad \mathbf{A}_x = 259 \text{ lb} \leftarrow \quad \blacktriangleleft$$

From Eq. (3), you find $A_y = 0$.

$$\mathbf{A}_y = 0 \quad \blacktriangleleft$$

REFLECT and THINK: The speed of the bullet is in the range of a modern high-performance rifle. Notice that the reaction at A is over 5000 times the weight of the bullet and over 10 times the weight of the plate.

Sample Problem 17.12

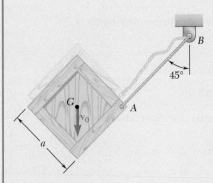

A uniformly loaded square crate is falling freely with a velocity $\mathbf{v}_0$ when cable AB suddenly becomes taut. Assuming that the impact is perfectly plastic, determine the angular velocity of the crate and the velocity of its mass center immediately after the cable becomes taut.

STRATEGY: Since impact occurs, use the principle of impulse and momentum.

MODELING: Choose the crate as your system and model it as a rigid body. The impulse–momentum diagram for this system is shown in Fig. 1. The mass moment of inertia of the plate about G is $\bar{I} = \tfrac{1}{6}ma^2$.

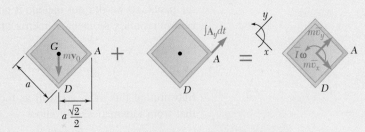

Syst Momenta$_1$ + Syst Ext Imp$_{1\to2}$ = Syst Momenta$_2$

Fig. 1 Impulse–momentum diagram for the crate.

ANALYSIS:

Principle of Impulse and Momentum. Applying the impulse–momentum principle in the x-direction and taking moments about A gives

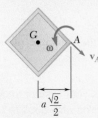

Fig. 2 Velocity of point A.

$$+\!\!\upharpoonleft \text{ moments about } A: \qquad mv_0 a \frac{\sqrt{2}}{2} + 0 = \bar{I}\omega + m\bar{v}_x \frac{a}{2} - m\bar{v}_y \frac{a}{2} \qquad (1)$$

$$\overset{+}{\searrow}\ x \text{ components:} \qquad mv_0 \frac{\sqrt{2}}{2} + 0 = m\bar{v}_x \qquad (2)$$

There are three unknowns in these two equations, ω, $\bar{v}_x$, and $\bar{v}_y$. For additional equations, you can use kinematics. Since you are told the impact is perfectly plastic, point A has a velocity perpendicular to the rope (Fig. 2). Therefore, you can relate the acceleration of A to that of G, as

$$\bar{\mathbf{v}} = \mathbf{v}_G = \mathbf{v}_A + \mathbf{v}_{G/A}$$

$$= \left[v_A \searrow 45° \right] + \left[a\frac{\sqrt{2}}{2}\omega \downarrow \right]$$

Equating components in the x and y directions, you find

$$\overset{+}{\searrow}\ x\text{-components:} \qquad \bar{v}_x = v_A + a\frac{\sqrt{2}}{2}\omega \frac{\sqrt{2}}{2} = v_A + \frac{a\omega}{2} \qquad (3)$$

$$\overset{+}{\nearrow}\ y\text{-components:} \qquad \bar{v}_y = -a\frac{\sqrt{2}}{2}\omega \frac{\sqrt{2}}{2} = -\frac{a\omega}{2} \qquad (4)$$

You now have four equations and four unknowns. Solving these gives

$$\omega = \frac{3\sqrt{2}}{5}\frac{v_0}{a} \qquad \bar{v}_x = \frac{\sqrt{2}}{2}v_0 \qquad \bar{v}_y = -\frac{3\sqrt{2}}{10}v_0 \qquad v_A = \frac{\sqrt{2}}{5}v_0$$

So

$$\omega = 0.849 \frac{v_0}{d} \upharpoonleft \quad \blacktriangleleft$$

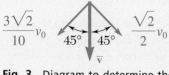

Fig. 3 Diagram to determine the magnitude and direction of $\bar{v}$

Resolving the velocity of the center of mass into a magnitude and direction using Fig. 3 gives you

$$\bar{\mathbf{v}} = 0.825v_0 \searrow 76.0° \quad \blacktriangleleft$$

REFLECT and THINK: If the impact had not been plastic, point A would have rebounded and the rope would have become slack. To solve the problem in this case, you would have needed to use the equation for the coefficient of restitution.

Sample Problem 17.13

A 2-kg sphere moving horizontally to the right with an initial velocity of 5 m/s strikes the lower end of an 8-kg rigid rod AB. The rod is suspended from a hinge at A and is initially at rest. Knowing that the coefficient of restitution between the rod and the sphere is 0.80, determine the angular velocity of the rod and the velocity of the sphere immediately after the impact.

STRATEGY: Since you have an impact, use the principle of impulse and momentum.

MODELING: Choose the sphere and the rod as your system; model the sphere as a particle and the rod as a rigid body. You also need to use the coefficient of restitution equation. The impulse–momentum diagram for this system is shown in Fig. 1. Note that the only impulsive force external to the system is the impulsive reaction at A.

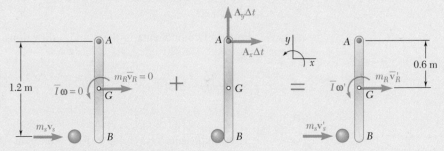

Fig. 1 Impulse–momentum diagram for the system.

ANALYSIS:

Principle of Impulse and Momentum.

$$\textbf{Syst Momenta}_1 + \textbf{Syst Ext Imp}_{1\to2} = \textbf{Syst Momenta}_2$$

$+\uparrow$moments about A:

$$m_s v_s(1.2 \text{ m}) = m_s v_s'(1.2 \text{ m}) + m_R \bar{v}_R'(0.6 \text{ m}) + \bar{I}\omega' \tag{1}$$

In this case, the mass of the sphere is not negligible compared to the rod, so we must include it on the right-hand side of Eq. (1). Since the rod rotates about A, from kinematics, you know $\bar{v}_R' = \bar{r}\omega' = (0.6 \text{ m})\omega'$. Also,

$$\bar{I} = \tfrac{1}{12}mL^2 = \tfrac{1}{12}(8 \text{ kg})(1.2 \text{ m})^2 = 0.96 \text{ kg·m}^2$$

Substituting these values and the given data into Eq. (1), you obtain

$$(2 \text{ kg})(5 \text{ m/s})(1.2 \text{ m}) = (2 \text{ kg})v_s'(1.2 \text{ m}) + (8 \text{ kg})(0.6 \text{ m})\omega'(0.6 \text{ m})$$
$$+ (0.96 \text{ kg·m}^2)\omega'$$
$$12 = 2.4v_s' + 3.84\omega' \tag{2}$$

Coefficient of Restitution. Choosing positive to the right, you have

$$v'_B - v'_s = e(v_s - v_B)$$

Substituting $v_s = 5$ m/s, $v_B = 0$, and $e = 0.80$ gives

$$v'_B - v'_s = 0.8(5 \text{ m/s} - 0) \tag{3}$$

Again noting that the rod rotates about A, you have

$$v'_B = (1.2 \text{ m}) \, \omega' \tag{4}$$

Solving Eqs. (2) to (4) simultaneously, you obtain

$$\omega' = 3.21 \text{ rad/s} \qquad \omega' = 3.21 \text{ rad/s} \; \text{↱} \; ◀$$
$$v'_s = -0.143 \text{ m/s} \qquad v'_s = 0.143 \text{ m/s} \leftarrow \; ◀$$

REFLECT and THINK: The negative value for the velocity of the sphere after impact means that it bounces back to the left. Given the masses of the sphere and the rod, this seems reasonable.

Sample Problem 17.14

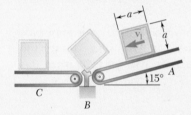

A square package of side a and mass m moves down a conveyor belt A with a constant velocity $\bar{\mathbf{v}}_1$. At the end of the conveyor belt, the corner of the package strikes a rigid support at B. Assuming that the impact at B is perfectly plastic, derive an expression for the smallest magnitude of the velocity $\bar{\mathbf{v}}_1$ for which the package will rotate about B and reach conveyor belt C.

STRATEGY: Because you have an impact, use the principle of impulse and momentum for when the package strikes the rigid support at B, and then apply the conservation of energy for the rotation of the package about the support B after the impact.

MODELING: Choose the package to be your system and model it as a rigid body. The impulse–momentum diagram for this system is shown in Fig. 1. Note that the only impulsive force external to the package is the impulsive reaction at B.

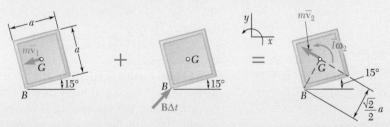

Fig. 1 Impulse–momentum diagram for the crate.

(continued)

ANALYSIS:

Principle of Impulse and Momentum.

$$\textbf{Syst Momenta}_1 + \textbf{Syst Ext Imp}_{1\rightarrow2} = \textbf{Syst Momenta}_2$$

$+\,\uparrow$ moments about B: $(m\bar{v}_1)(\tfrac{1}{2}a) + 0 = (m\bar{v}_2)(\tfrac{1}{2}\sqrt{2}a) + \bar{I}\omega_2$ (1)

Position 2

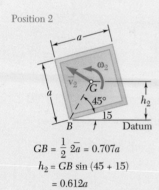

$GB = \tfrac{1}{2}\overline{2a} = 0.707a$

$h_2 = GB \sin(45 + 15)$

     $= 0.612a$

Since the package rotates about B, from kinematics you have $\bar{v}_2 = (GB)\omega_2 = \tfrac{1}{2}\sqrt{2}a\omega_2$.

  Substitute this expression, together with $\bar{I} = \tfrac{1}{6}ma^2$, into Eq. (1) for

$$(m\bar{v}_1)(\tfrac{1}{2}a) = m(\tfrac{1}{2}\sqrt{2}a\omega_2)(\tfrac{1}{2}\sqrt{2}a) + \tfrac{1}{6}ma^2\omega_2 \quad \bar{v}_1 = \tfrac{4}{3}a\omega_2 \quad (2)$$

Conservation of Energy.
Apply the principle of conservation of energy between position 2 and position 3 (Fig. 2) as

$$T_2 + V_2 = T_3 + V_3 \tag{3}$$

You need to determine the energy at these two positions.

Position 3

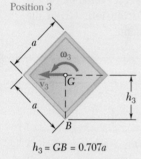

$h_3 = GB = 0.707a$

Fig. 2 The crate in positions 2 and 3.

Position 2. $V_2 = Wh_2$. Since $\bar{v}_2 = \tfrac{1}{2}\sqrt{2}a\omega_2$, you have

$$T_2 = \tfrac{1}{2}m\bar{v}_2^2 + \tfrac{1}{2}\bar{I}\omega_2^2 = \tfrac{1}{2}m(\tfrac{1}{2}\sqrt{2}a\omega_2)^2 + \tfrac{1}{2}(\tfrac{1}{6}ma^2)\omega_2^2 = \tfrac{1}{3}ma^2\omega_2^2$$

Position 3.   The package must reach conveyor belt C, so it must pass through position 3 where G is directly above B. Also, since you wish to determine the smallest velocity for which the package will reach this position, choose $\bar{v}_3 = \omega_3 = 0$. Therefore, $T_3 = 0$ and $V_3 = Wh_3$.

  Substituting these into Eq. (3)

$$\tfrac{1}{3}ma^2\omega_2^2 + Wh_2 = 0 + Wh_3$$

$$\omega_2^2 = \frac{3W}{ma^2}(h_3 - h_2) = \frac{3g}{a^2}(h_3 - h_2) \tag{4}$$

Substituting the computed values of h_2 and h_3 into Eq. (4), you obtain

$$\omega_2^2 = \frac{3g}{a^2}(0.707a - 0.612a) = \frac{3g}{a^2}(0.095a) \quad \omega_2 = \sqrt{0.285g/a}$$

$$\bar{v}_1 = \tfrac{4}{3}a\omega_2 = \tfrac{4}{3}a\sqrt{0.285g/a} \qquad\qquad \bar{v}_1 = 0.712\sqrt{ga} \;\blacktriangleleft$$

REFLECT and THINK: The combination of energy and momentum methods is typical of many design analyses. If you had been interested in determining the reaction at B immediately after the impact or at some other point in the motion, you would have needed to draw a free-body diagram and a kinetic diagram and apply Newton's second law.

Sample Problem 17.15

A soccer ball tester consists of a 15-kg slender rod AB with a 1.1-kg simulated foot located at A and a torsional spring located at pin B. The torsional spring has a spring constant of $k_t = 910$ N·m and is unstretched when AB is vertical. The length of AB is 0.9 m, and you can assume that the foot can be modeled as a point mass. Knowing that the velocity of the 0.45-kg soccer ball is 30 ft/s after impact, determine (*a*) the coefficient of restitution between the simulated foot and the ball, (*b*) the impulse at B during the impact.

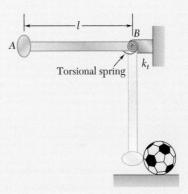

STRATEGY: This problem can be broken into two distinct stages of motion. In stage 1, the arm moves downward under the influence of gravity and the torsional spring. You can use the conservation of energy for this stage. In stage 2, the foot hits the ball, and you need to use both the principle of impulse and momentum and the coefficient of restitution.

MODELING: Each stage requires a different system. For stage 1, your system is rod AB, foot B, and the torsional spring. In stage 2, your system is rod AB, foot B, and the soccer ball. The appropriate diagrams are drawn in the analysis section. You can model AB as a slender rod, so its mass moment of inertia is

$$\bar{I}_{AB} = \tfrac{1}{12}m_{AB}l^2 = \tfrac{1}{12}(15 \text{ kg})(0.9 \text{ m})^2 = 1.0125 \text{ kg·m}^2$$

ANALYSIS:

Rod *AB* Moves Down. Apply the principle of conservation of energy

$$T_1 + V_{g_1} + V_{e_1} = T_2 + V_{g_2} + V_{e_2} \tag{1}$$

(continued)

Position 1. The system starts from rest, so $T_1 = 0$. Using the datum defined in Fig. 1, you know $V_{g_1} = 0$, and since the spring is unstretched at position 2, you find

$$V_{e_1} = \tfrac{1}{2}k_t\theta^2 = \tfrac{1}{2}(910 \text{ N·m})(\tfrac{\pi}{2})^2 = 1123 \text{ J}$$

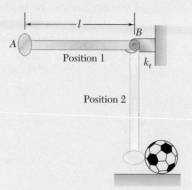

Fig. 1 The rod in positions 1 and 2.

Position 2. The elastic potential energy is $V_{e_2} = 0$, and the gravitational potential energy is

$$V_{g_2} = -m_{AB}g\frac{l}{2} - m_A gl = -(15 \text{ kg})(9.81 \text{ m/s}^2)(0.45 \text{ m}) - (1.1 \text{ kg})(9.81 \text{ m/s}^2)(0.9 \text{ m})$$
$$= -75.93 \text{ J}$$

The kinetic energy is

$$T_2 = \tfrac{1}{2}m_A v_A^2 + \tfrac{1}{2}m_{AB}v_G^2 + \tfrac{1}{2}\bar{I}_{AB}\omega^2$$

You can relate the velocity of the foot and the velocity of the center of gravity of the rod to the angular velocity of AB by recognizing that AB is undergoing fixed-axis rotation. Therefore, $v_G = \omega\frac{l}{2}$ and $v_A = \omega l$. Substituting these into the expression for T_2 and putting in values gives

$$T_2 = \frac{1}{2}\left(m_A l^2 + m_{AB}\left(\frac{l}{2}\right)^2 + \bar{I}_{AB}\right)\omega^2 = 2.4705\omega^2$$

Substituting these energy terms into Eq. (1) gives

$$0 + 0 + 1123 = 2.4705\omega^2 - 75.93 + 0$$

Solving for the angular velocity, you find $\omega = 22.03$ rad/s. Knowing ω, you can calculate the velocities $v_G = 9.912$ m/s and $v_A = 19.824$ m/s.

Foot *A* Impacts the Soccer Ball. Impulse–momentum diagrams for the impact on the ball are shown in Fig. 2.

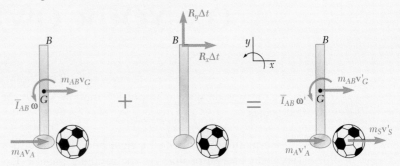

Syst Momenta₁ + Syst Ext Imp₁→₂ = Syst Momenta₂

Fig. 2

Taking moments about *B* gives you

+↰ moments about *B*:

$$m_A v_A l + m_{AB} v_G \frac{l}{2} + I_{AB}\omega + 0 = m_A v_A' l + m_{AB} v_G' \frac{l}{2} + \bar{I}_{AB}\omega' + m_S v_S' l \qquad (2)$$

The equation for the coefficient of restitution is

$$v_S' - v_A' = e(v_A - 0) \qquad (3)$$

where $v_S' = 30$ m/s. From kinematics, you know $v_A' = \omega'l$ and $v_G' = \omega'(l/2)$. Using these kinematic equations and Eqs. (2) and (3), you can solve for the unknown quantities

$$v_A' = 17.61 \text{ m/s} \quad v_G' = 8.81 \text{ m/s} \quad \omega' = 19.57 \text{ rad/s} \quad e = 0.625$$

$$e = 0.625 \quad \blacktriangleleft$$

Impulses During Impact. Applying impulse–momentum in the *x*- and *y*-directions gives

$\xrightarrow{+}$ *x*-components: $\quad m_{AB} v_G + m_A v_A + R_x \Delta t = m_{AB} v_G' + m_A v_A' + m_S v_S' \qquad$ (4)

+↑ *y*-components: $\qquad\qquad\qquad 0 + R_y \Delta t = 0 \qquad$ (5)

Solving these equations, you find $R_x \Delta t = -5.53$ N and $R_y \Delta t = 0$.

$$\mathbf{R}\Delta t = 5.53 \text{ N} \leftarrow \quad \blacktriangleleft$$

REFLECT and THINK: This coefficient of restitution seems reasonable. As you decrease the pressure in the ball, you would expect the coefficient of restitution to decrease; therefore, the distance the ball travels will decrease. If you had been asked to determine the reactions at *B* after the impact, you would need to draw a free-body diagram and kinetic diagram for your system and apply Newton's second law.

SOLVING PROBLEMS
ON YOUR OWN

This section was devoted to **impulsive motion** and to the **eccentric impact** of rigid bodies.

1. Impulsive motion occurs when a rigid body is subjected to a very large force **F** for a very short interval of time Δt; the resulting impulse $\mathbf{F}_{\text{avg}} \Delta t$ is both finite and different from zero. Such forces are referred to as **impulsive forces** and arise whenever an impact occurs between two rigid bodies. Forces for which the impulse is negligible are referred to as **nonimpulsive forces**. As discussed in Chap. 13, you can assume the following forces to be nonimpulsive: the weight of a body, the force exerted by a spring, and any other force that is known to be small by comparison with the impulsive forces. Unknown reactions, however, cannot be assumed to be nonimpulsive.

2. Eccentric impact of rigid bodies. When two bodies collide, the velocity components along the line of impact of the points of contact A and B before and after impact satisfy

$$(v'_B)_n - (v'_A)_n = e[(v_A)_n - (v_B)_n] \qquad \textbf{(17.19)}$$

where the left-hand side is the *relative velocity after the impact* and the right-hand side is the product of the coefficient of restitution and the *relative velocity before the impact.*

This equation expresses the same relation between the velocity components of the points of contact before and after an impact that you used for particles in Chap. 13.

3. To solve a problem involving an impact you should use the *method of impulse and momentum* and take the following steps.

    a. Draw an impulse–momentum diagram of the system showing the momenta immediately before impact plus the impulses of the external forces acting during the impact; this sum is equivalent to the momenta immediately after impact.

    b. Write the governing equations for the angular momentum about some point. Depending on the problem type (especially when you want to find support impulsive reactions), you may also need to write the equations for linear momentum [Sample Prob. 17.11].

    c. In the case of an impact in which $e > 0$, the number of unknowns will be greater than the number of equations that you can write by summing components and moments. You should supplement the equations obtained from the impulse–momentum diagram with the coefficient of restitution from Eq. (17.19) that relates the relative velocities of the points of contact before and after impact [Sample Prob. 17.13 and 17.15].

    d. During an impact, you must use the method of impulse and momentum. However, *before and after the impact* you can, if necessary, use some of the other methods of solution that you have learned, such as the conservation of energy [Sample Prob. 17.14 and 17.15] or Newton's second law.

Problems

IMPULSE–MOMENTUM DIAGRAM PRACTICE PROBLEMS

17.F4 A uniform slender rod AB of mass m is at rest on a frictionless horizontal surface when hook C engages a small pin at A. Knowing that the hook is pulled upward with a constant velocity $\mathbf{v}_0$, draw the impulse-momentum diagram that is needed to determine the impulse exerted on the rod at A and B. Assume that the velocity of the hook is unchanged and that the impact is perfectly plastic.

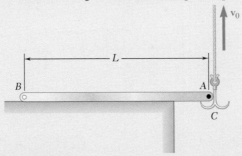

Fig. P17.F4

17.F5 A uniform slender rod AB of length L is falling freely with a velocity $\mathbf{v}_0$ when cord AC suddenly becomes taut. Assuming that the impact is perfectly plastic, draw the impulse–momentum diagram that is needed to determine the angular velocity of the rod and the velocity of its mass center immediately after the cord becomes taut.

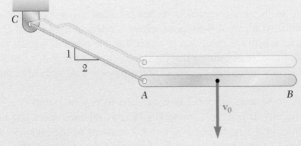

Fig. P17.F5

17.F6 A slender rod CDE of length L and mass m is attached to a pin support at its midpoint D. A second and identical rod AB is rotating about a pin support at A with an angular velocity $\boldsymbol{\omega}_1$ when its end B strikes end C of rod CDE. The coefficient of restitution between the rods is e. Draw the impulse–momentum diagrams that are needed to determine the angular velocity of each rod immediately after the impact.

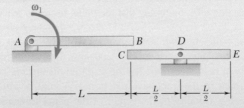

Fig. P17.F6

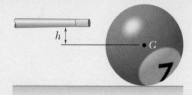

Fig. P17.96

END-OF-SECTION PROBLEMS

17.96 At what height h above its center G should a billiard ball of radius r be struck horizontally by a cue if the ball is to start rolling without sliding?

17.97 A bullet weighing 0.08 lb is fired with a horizontal velocity of 1800 ft/s into the lower end of a slender 15-lb bar of length $L = 30$ in. Knowing that $h = 12$ in. and that the bar is initially at rest, determine (*a*) the angular velocity of the bar immediately after the bullet becomes embedded, (*b*) the impulsive reaction at C, assuming that the bullet becomes embedded in 0.001 s.

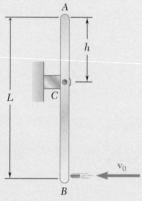

Fig. P17.97

17.98 In Prob. 17.97, determine (*a*) the required distance h if the impulsive reaction at C is to be zero, (*b*) the corresponding angular velocity of the bar immediately after the bullet becomes embedded.

17.99 A 16-lb wooden panel is suspended from a pin support at A and is initially at rest. A 4-lb metal sphere is released from rest at B and falls into a hemispherical cup C attached to the panel at a point located on its top edge. Assuming that the impact is perfectly plastic, determine the velocity of the mass center G of the panel immediately after the impact.

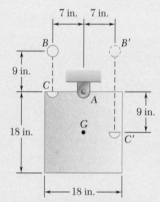

Fig. P17.99 and P17.100

17.100 A 16-lb wooden panel is suspended from a pin support at A and is initially at rest. A 4-lb metal sphere is released from rest at B' and falls into a hemispherical cup C' attached to the panel at the same level as the mass center G. Assuming that the impact is perfectly plastic, determine the velocity of the mass center G of the panel immediately after the impact.

17.101 A 45-g bullet is fired with a velocity of 400 m/s at $\theta = 30°$ into a 9-kg square panel of side $b = 200$ mm. Knowing that $h = 150$ mm and that the panel is initially at rest, determine (*a*) the velocity of the center of the panel immediately after the bullet becomes embedded, (*b*) the impulsive reaction at A, assuming that the bullet becomes embedded in 2 ms.

17.102 A 45-g bullet is fired with a velocity of 400 m/s at $\theta = 5°$ into a 9-kg square panel of side $b = 200$ mm. Knowing that the panel is initially at rest, determine (*a*) the required distance h if the horizontal component of the impulsive reaction at A is to be zero, (*b*) the corresponding velocity of the center of the panel immediately after the bullet becomes embedded.

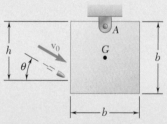

Fig. P17.101 and P17.102

17.103 Two uniform rods, each of mass m, form the L-shaped rigid body ABC, which is initially at rest on the frictionless horizontal surface when hook D of the carriage E engages a small pin at C. Knowing that the carriage is pulled to the right with a constant velocity $\mathbf{v}_0$, determine immediately after the impact (a) the angular velocity of the body, (b) the velocity of corner B. Assume that the velocity of the carriage is unchanged and that the impact is perfectly plastic.

17.104 The uniform slender rod AB of weight 5 lb and length 30 in. forms an angle $\beta = 30°$ with the vertical as it strikes the smooth corner shown with a vertical velocity $\mathbf{v}_1$ of magnitude 8 ft/s and no angular velocity. Assuming that the impact is perfectly plastic, determine the angular velocity of the rod immediately after the impact.

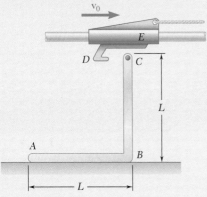

Fig. P17.103

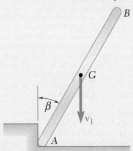

Fig. P17.104

17.105 A bullet weighing 0.08 lb is fired with a horizontal velocity of 1800 ft/s into the 15-lb wooden rod AB of length $L = 30$ in. The rod, which is initially at rest, is suspended by a cord of length $L = 30$ in. Determine the distance h for which, immediately after the bullet becomes embedded, the instantaneous center of rotation of the rod is point C.

17.106 A prototype of an adapted bowling device is a simple ramp that attaches to a wheelchair. The bowling ball has a mass moment of inertia about its center of gravity of cmr^2, where c is a unitless constant, r is the radius, and m is its mass. The athlete nudges the ball slightly from a height of h, and the ball rolls down the ramp without sliding. It hits the bowling lane, and after slipping for a short distance, it begins to roll again. Assuming that the ball does not bounce as it hits the lane, determine the angular velocity and velocity of the mass center of the ball after it has resumed rolling.

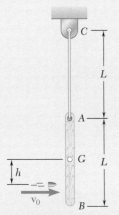

Fig. P17.105

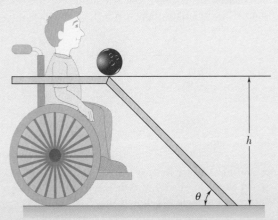

Fig. P17.106

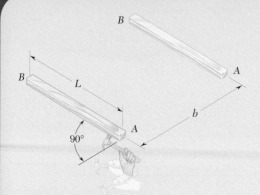

Fig. P17.107

17.107 A uniform slender rod AB is at rest on a frictionless horizontal table when end A of the rod is struck by a hammer that delivers an impulse that is perpendicular to the rod. In the subsequent motion, determine the distance b through which the rod will move each time it completes a full revolution.

17.108 A bullet of mass m is fired with a horizontal velocity $\mathbf{v}_0$ and at a height $h = \frac{1}{2}R$ into a wooden disk of much larger mass M and radius R. The disk rests on a horizontal plane and the coefficient of friction between the disk and the plane is finite. (*a*) Determine the linear velocity $\bar{\mathbf{v}}_1$ and the angular velocity $\boldsymbol{\omega}_1$ of the disk immediately after the bullet has penetrated the disk. (*b*) Describe the ensuing motion of the disk and determine its linear velocity after the motion has become uniform.

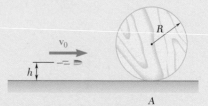

Fig. P17.108 and P17.109

17.109 Determine the height h at which the bullet of Prob. 17.108 should be fired (*a*) if the disk is to roll without sliding immediately after impact, (*b*) if the disk is to slide without rolling immediately after impact.

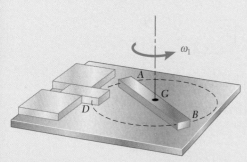

Fig. P17.110

17.110 A uniform slender bar of length $L = 200$ mm and mass $m = 0.5$ kg is supported by a frictionless horizontal table. Initially the bar is spinning about its mass center G with a constant angular speed $\omega_1 = 6$ rad/s. Suddenly latch D is moved to the right and is struck by end A of the bar. Knowing that the coefficient of restitution between A and D is $e = 0.6$, determine the angular velocity of the bar and the velocity of its mass center immediately after the impact.

17.111 A uniform slender rod of length L is dropped onto rigid supports at A and B. Since support B is slightly lower than support A, the rod strikes A with a velocity $\bar{\mathbf{v}}_1$ before it strikes B. Assuming perfectly elastic impact at both A and B, determine the angular velocity of the rod and the velocity of its mass center immediately after the rod (*a*) strikes support A, (*b*) strikes support B, (*c*) again strikes support A.

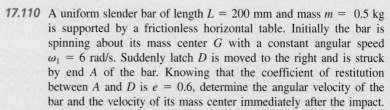

Fig. P17.111

17.112 A uniform slender rod AB has a mass m, a length L, and is falling freely with a velocity $\mathbf{v}_0$ when end B strikes a smooth inclined surface as shown. Assuming that the impact is perfectly elastic, determine the angular velocity of the rod and the velocity of its mass center immediately after the impact.

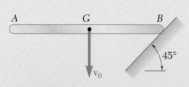

Fig. P17.112

1250

17.113 The slender rod AB of length $L = 1$ m forms an angle $\beta = 30°$ with the vertical as it strikes the frictionless surface shown with a vertical velocity $\bar{\mathbf{v}}_1 = 2$ m/s and no angular velocity. Knowing that the coefficient of restitution between the rod and the ground is $e = 0.8$, determine the angular velocity of the rod immediately after the impact.

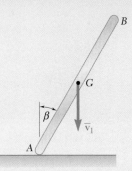

Fig. P17.113

17.114 The trapeze/lanyard air drop (t/LAD) launch is a proposed innovative method for airborne launch of a payload-carrying rocket. The release sequence involves several steps as shown in (1) where the payload rocket is shown at various instances during the launch. To investigate the first step of this process, where the rocket body drops freely from the carrier aircraft until the 2-m lanyard stops the vertical motion of B, a trial rocket is tested as shown in (2). The rocket can be considered a uniform 1×7-m rectangle with a mass of 4000 kg. Knowing that the rocket is released from rest and falls vertically 2 m before the lanyard becomes taut, determine the angular velocity of the rocket immediately after the lanyard is taut.

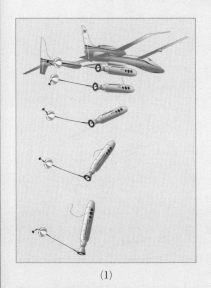

(1)

(2)

Fig. P17.114

17.115 The uniform rectangular block shown is moving along a frictionless surface with a velocity $\bar{\mathbf{v}}_1$ when it strikes a small obstruction at B. Assuming that the impact between corner A and obstruction B is perfectly plastic, determine the magnitude of the velocity $\bar{\mathbf{v}}_1$ for which the maximum angle θ through which the block will rotate will be 30°.

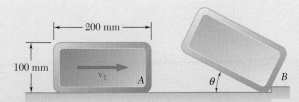

Fig. P17.115

17.119 A 1-oz bullet is fired with a horizontal velocity of 750 mi/h into the 18-lb wooden beam AB. The beam is suspended from a collar of negligible mass that can slide along a horizontal rod. Neglecting friction between the collar and the rod, determine the maximum angle of rotation of the beam during its subsequent motion.

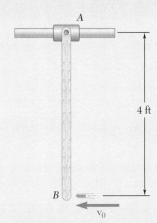

Fig. P17.119

17.120 For the beam of Prob. 17.119, determine the velocity of the 1-oz bullet for which the maximum angle of rotation of the beam will be 90°.

17.121 The plank CDE has a mass of 15 kg and rests on a small pivot at D. The 55-kg gymnast A is standing on the plank at C when the 70-kg gymnast B jumps from a height of 2.5 m and strikes the plank at E. Assuming perfectly plastic impact and that gymnast A is standing absolutely straight, determine the height to which gymnast A will rise.

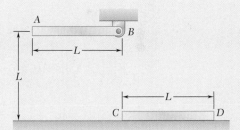

Fig. P17.121

17.122 Solve Prob. 17.121, assuming that the gymnasts change places so that gymnast A jumps onto the plank while gymnast B stands at C.

17.123 A slender rod AB is released from rest in the position shown. It swings down to a vertical position and strikes a second and identical rod CD that is resting on a frictionless surface. Assuming that the coefficient of restitution between the rods is 0.4, determine the velocity of rod CD immediately after the impact.

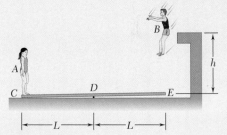

Fig. P17.123 and P17.124

17.124 A slender rod AB is released from rest in the position shown. It swings down to a vertical position and strikes a second and identical rod CD that is resting on a frictionless surface. Assuming that the impact between the rods is perfectly elastic, determine the velocity of rod CD immediately after the impact.

17.125 Block A has a mass m and is attached to a cord that is wrapped around a uniform disk with a mass M. The block is released from rest and falls through a distance h before the cord becomes taut. Derive expressions for the velocity of the block and the angular velocity of the disk immediately after the impact. Assume that the impact is (a) perfectly plastic, (b) perfectly elastic.

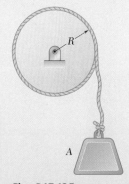

Fig. *P17.125*

17.126 A 2-kg solid sphere of radius $r = 40$ mm is dropped from a height $h = 200$ mm and lands on a uniform slender plank AB of mass 4 kg and length $L = 500$ mm that is held by two inextensible cords. Knowing that the impact is perfectly plastic and that the sphere remains attached to the plank at a distance $a = 40$ mm from the left end, determine the velocity of the sphere immediately after impact. Neglect the thickness of the plank.

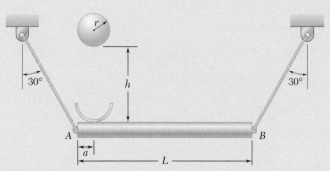

Fig. P17.126

17.127 and 17.128 Member ABC has a mass of 2.4 kg and is attached to a pin support at B. An 800-g sphere D strikes the end of member ABC with a vertical velocity $\mathbf{v}_1$ of 3 m/s. Knowing that $L = 750$ mm and that the coefficient of restitution between the sphere and member ABC is 0.5, determine immediately after the impact (a) the angular velocity of member ABC, (b) the velocity of the sphere.

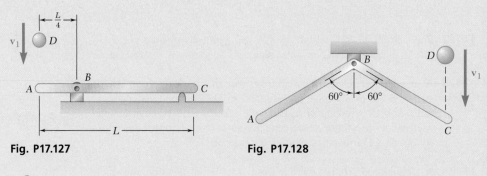

Fig. P17.127   Fig. P17.128

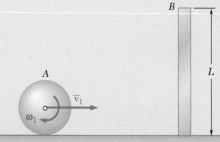

Fig. P17.129

17.129 Sphere A of mass $m_A = 2$ kg and radius $r = 40$ mm rolls without slipping with a velocity $\overline{\mathbf{v}}_1 = 2$ m/s on a horizontal surface when it hits squarely a uniform slender bar B of mass $m_B = 0.5$ kg and length $L = 100$ mm that is standing on end and is at rest. Denoting by μ_k the coefficient of kinetic friction between the sphere and the horizontal surface, neglecting friction between the sphere and the bar, and knowing the coefficient of restitution between A and B is 0.1, determine the angular velocities of the sphere and the bar immediately after the impact.

17.130 A large 3-lb sphere with a radius $r = 3$ in. is thrown into a light basket at the end of a thin, uniform rod weighing 2 lb and length $L = 10$ in. as shown. Immediately before the impact, the angular velocity of the rod is 3 rad/s counterclockwise and the velocity of the sphere is 2 ft/s down. Assume the sphere sticks in the basket. Determine after the impact (a) the angular velocity of the bar and sphere, (b) the components of the reactions at A.

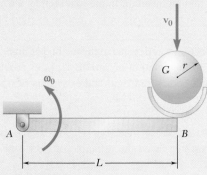

Fig. P17.130

17.131 A small rubber ball of radius r is thrown against a rough floor with a velocity $\bar{\mathbf{v}}_A$ of magnitude $\mathbf{v}_0$ and a backspin $\boldsymbol{\omega}_A$ of magnitude ω_0. It is observed that the ball bounces from A to B, then from B to A, then from A to B, etc. Assuming perfectly elastic impact, determine the required magnitude ω_0 of the backspin in terms of $\bar{\mathbf{v}}_0$ and r.

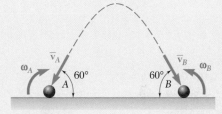

Fig. P17.131

17.132 Sphere A of mass m and radius r rolls without slipping with a velocity $\bar{\mathbf{v}}_1$ on a horizontal surface when it hits squarely an identical sphere B that is at rest. Denoting by μ_k the coefficient of kinetic friction between the spheres and the surface, neglecting friction between the spheres, and assuming perfectly elastic impact, determine (a) the linear and angular velocities of each sphere immediately after the impact, (b) the velocity of each sphere after it has started rolling uniformly.

Fig. P17.132

17.133 In a game of pool, ball A is rolling without slipping with a velocity $\bar{\mathbf{v}}_0$ as it hits obliquely ball B, which is at rest. Denoting by r the radius of each ball and by μ_k the coefficient of kinetic friction between a ball and the table, and assuming perfectly elastic impact, determine (a) the linear and angular velocity of each ball immediately after the impact, (b) the velocity of ball B after it has started rolling uniformly.

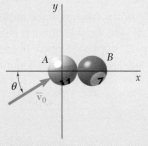

Fig. P17.133

17.134 Each of the bars AB and BC is of length $L = 400$ mm and mass $m = 1.2$ kg. Determine the angular velocity of each bar immediately after the impulse $\mathbf{Q}\Delta t = (1.5$ N·s$)\mathbf{i}$ is applied at C.

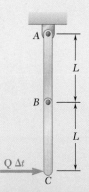

Fig. P17.134

Review and Summary

In this chapter, we again considered the method of work and energy and the method of impulse and momentum. In the first section, we applied the method of work and energy to the analysis of the motion of rigid bodies and systems of rigid bodies.

The second section was devoted to the method of impulse and momentum and its application to the solution of various types of problems involving the plane motion of rigid bodies and rigid bodies symmetrical with respect to the reference plane.

Principle of Work and Energy for a Rigid Body

In Sec. 17.1, we first expressed the principle of work and energy for a rigid body in the form

$$T_1 + U_{1 \to 2} = T_2 \tag{17.1}$$

where T_1 and T_2 represent the initial and final values of the kinetic energy of the rigid body and $U_{1 \to 2}$ represents the work of the external forces acting on it. If we express the work done by nonconservative forces as $U_{1 \to 2}^{NC}$ and define potential energy terms for conservative forces, we can express Eq. (17.1) as

$$T_1 + V_{g_1} + V_{e_1} + U_{1 \to 2}^{NC} = T_2 + V_{g_2} + V_{e_2} \tag{17.1'}$$

where V_{g_1} and V_{g_2} are the initial and final gravitational potential energy of the center of mass of the rigid body and V_{e_1} and V_{e_2} are the initial and final values of the elastic energy associated with springs in the system, respectively.

Work of a Force or a Couple

In Sec. 17.1B, we recalled the expression found in Chap. 13 for the work of a force $\mathbf{F}$ applied at a point A, namely

$$U_{1 \to 2} = \int_{A_1}^{A_2} \mathbf{F} \cdot d\mathbf{r} \tag{17.3}$$

or

$$U_{1 \to 2} = \int_{s_1}^{s_2} (F \cos \alpha)\, ds \tag{17.3'}$$

where F is the magnitude of the force, α is the angle it forms with the direction of motion of A, and s is the variable of integration measuring the distance traveled by A along its path. We also derived the expression for the work of a couple of moment $\mathbf{M}$ applied to a rigid body during a rotation in θ of the rigid body as

$$U_{1 \to 2} = \int_{\theta_1}^{\theta_2} M\, d\theta \tag{17.5}$$

Kinetic Energy in Plane Motion

We then derived an expression for the kinetic energy of a rigid body in plane motion [Sec. 17.1C]:

$$T = \tfrac{1}{2} m \bar{v}^2 + \tfrac{1}{2} \bar{I} \omega^2 \tag{17.9}$$

where $\bar{v}$ is the speed of the mass center G of the body, ω is the angular speed of the body, and $\bar{I}$ is its moment of inertia about an axis through G perpendicular to the plane of reference (Fig. 17.13) [Sample Prob. 17.3]. We noted that the kinetic energy of a rigid body in plane motion can be separated into two parts: (1) the kinetic energy $\frac{1}{2}m\bar{v}^2$ associated with the motion of the mass center G of the body and (2) the kinetic energy $\frac{1}{2}\bar{I}\omega^2$ associated with the rotation of the body about G. You will generally need to use kinematics to relate $\bar{v}$ and ω.

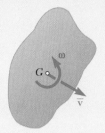

Fig. 17.13

Kinetic Energy in Rotation About a Fixed Axis

For a rigid body rotating about a fixed axis through O with an angular velocity ω, we had

$$T = \tfrac{1}{2}I_O\omega^2 \qquad\qquad (17.10)$$

where I_O is the moment of inertia of the body about the fixed axis. We noted that this result is not limited to the rotation of plane rigid bodies or of bodies symmetrical with respect to the reference plane, but it also is valid regardless of the shape of the body or of the location of the axis of rotation.

Systems of Rigid Bodies

Equation (17.1) can be applied to the motion of systems of rigid bodies [Sec. 17.1D] as long as all the forces acting on the various bodies involved—internal as well as external to the system—are included in the computation of $U_{1\to2}$. However, in the case of systems consisting of pin-connected members or blocks and pulleys connected by inextensible cords or meshed gears, the points of application of the internal forces move through equal distances and the work of these forces cancels out [Sample Probs. 17.1, 17.2, and 17.6].

Conservation of Energy

When a rigid body or a system of rigid bodies moves under the action of conservative forces, the principle of work and energy can be expressed in the form

$$T_1 + V_1 = T_2 + V_2 \qquad\qquad (17.12)$$

or

$$T_1 + V_{g_1} + V_{e_1} = T_2 + V_{g_2} + V_{e_2} \qquad\qquad (17.12')$$

This is referred to as the *principle of conservation of energy* [Sec. 17.1E]. We can use this principle to solve problems involving conservative forces such as the force of gravity or the force exerted by a spring [Sample Probs. 17.4 through 17.6]. However, if we need to determine a reaction, we must supplement the principle of conservation of energy by using Newton's second law [Sample Prob. 17.4].

Power

In Sec. 17.1F, we extended the concept of power to a rotating body subjected to a couple as

$$\text{Power} = \frac{dU}{dt} = \frac{M\,d\theta}{dt} = M\omega \qquad\qquad (17.13)$$

where M is the magnitude of the couple and ω is the magnitude of the angular velocity of the body.

Principle of Impulse and Momentum for a Rigid Body

In Sec. 17.2, we applied the principle of impulse and momentum as had been derived in Sec. 14.2C for a system of particles to the motion of a rigid body [Sec. 17.2A]. We have

$$\textbf{Syst Momenta}_1 + \textbf{Syst Ext Imp}_{1\to2} = \textbf{Syst Momenta}_2 \quad \textbf{(17.14)}$$

Next we showed that, for a rigid body symmetrical with respect to the reference plane, the system of the momenta of the particles forming the body is equivalent to a vector $m\bar{\mathbf{v}}$ attached at the mass center G of the body and a couple $\bar{I}\boldsymbol{\omega}$ (Fig. 17.14). The vector $m\bar{\mathbf{v}}$ is associated with the translation of the body with G and represents the *linear momentum* of the body, whereas the couple $\bar{I}\boldsymbol{\omega}$ corresponds to the rotation of the body about G and represents the *angular momentum* of the body about an axis through G.

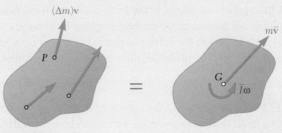

Fig. 17.14

We can express Eq. (17.14) graphically using an impulse–momentum diagram, as shown in Fig. 17.15. This diagram represents the system of the

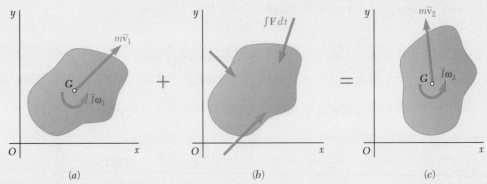

(a)   (b)   (c)

Fig. 17.15

initial momenta of the body, the impulses of the external forces acting on the body, and the system of the final momenta of the body, respectively. We can choose to sum moments about an arbitrary point P using

$$\bar{I}\omega_1 + m\bar{v}_1 d_\perp + \sum \int_{t_1}^{t_2} M_P dt = \bar{I}\omega_2 + m\bar{v}_2 d_\perp \quad \textbf{(17.14$'$)}$$

the center of mass G using

$$\bar{I}\omega_1 + \sum \int_{t_1}^{t_2} M_G dt = \bar{I}\omega_2 \quad \textbf{(17.14$''$)}$$

or a fixed axis of rotation O using

$$I_O\omega_1 + \sum \int_{t_1}^{t_2} M_O dt = I_O\omega_2 \quad \textbf{(17.16)}$$

Using one of these expressions and the *x and y components* of the linear impulse–momentum equation, we obtain three equations of motion that we can solve for the desired unknowns [Sample Probs. 17.7 and 17.8].

In problems dealing with several connected rigid bodies [Sec. 17.2B], we can consider each body separately [Sample Prob. 17.7], or if no more than three unknowns are involved, we can apply the principle of impulse and momentum to the entire system, considering the impulses of the external forces only [Sample Prob. 17.9].

Conservation of Angular Momentum

When the lines of action of all the external forces acting on a system of rigid bodies pass through a given point O, the angular momentum of the system about O is conserved [Sec. 17.2C]. We suggested that problems involving conservation of angular momentum be solved by the general method described previously [Sample Prob. 17.9 and 17.10].

Impulsive Motion

Section 17.3 was devoted to the **impulsive motion** and the **eccentric impact** of rigid bodies. We recalled that the method of impulse and momentum is the only practicable method for the solution of problems involving impulsive motion and that the computation of impulses in such problems is particularly simple [Sample Prob. 17.11 and 17.12].

Eccentric Impact

We also recalled that the eccentric impact of two rigid bodies is defined as an impact in which the mass centers of the colliding bodies are *not* located on the line of impact. We showed that, in such a situation, a relation similar to that derived in Chap. 13 for the central impact of two particles and involving the coefficient of restitution e still holds, but *the velocities of points A and B where contact occurs during the impact should be used*. We have

$$(v'_B)_n - (v'_A)_n = e[(v_A)_n - (v_B)_n] \tag{17.19}$$

where $(v_A)_n$ and $(v_B)_n$ are the components along the line of impact of the velocities of A and B before the impact, and $(v'_A)_n$ and $(v'_B)_n$ are their components after the impact (Fig. 17.16). Equation (17.19) applies not only when the colliding bodies move freely after the impact but also when the bodies are partially constrained in their motion. You should use it in conjunction with one or several other equations obtained by applying the principle of impulse and momentum [Sample Prob. 17.13]. We also considered problems where the method of impulse and momentum and the method of work and energy can be combined [Sample Prob. 17.14].

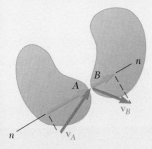

(*a*) Before impact

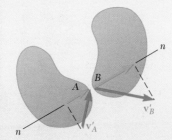

(*b*) After impact

Fig. 17.16

Review Problems

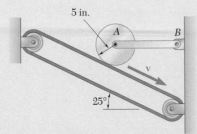

5 in.

A

B

v

25°

Fig. P17.135

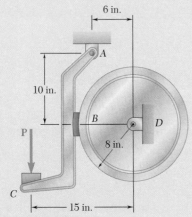

6 in.

A

10 in.

B

D

8 in.

P

C

15 in.

Fig. P17.136

17.135 A uniform disk of constant thickness and initially at rest is placed in contact with the belt shown, which moves at a constant speed $v = 80$ ft/s. Knowing that the coefficient of kinetic friction between the disk and the belt is 0.15, determine (*a*) the number of revolutions executed by the disk before it reaches a constant angular velocity, (*b*) the time required for the disk to reach that constant angular velocity.

17.136 The 8-in.-radius brake drum is attached to a larger flywheel that is not shown. The total mass moment of inertia of the flywheel and drum is 14 lb·ft·s² and the coefficient of kinetic friction between the drum and the brake shoe is 0.35. Knowing that the initial angular velocity of the flywheel is 360 rpm counterclockwise, determine the vertical force **P** that must be applied to the pedal *C* if the system is to stop in 100 revolutions.

17.137 Charpy impact test pendulums are used to determine the amount of energy a test specimen absorbs during an impact (see ASTM Standard E23). The hammer weighs 71.2 lbs and has a mass moment of inertia about its center of gravity G_H of 20.9 slug·in². The arm weighs 19.5 lbs and has a mass moment of inertia about its own center of gravity G_A of 47.1 slug·in². The pendulum is released from rest from an initial position of $\theta = 39°$. Knowing that the friction at pin *O* is negligible, determine (*a*) the impact speed when the hammer hits the test specimen, (*b*) the force on the pin *O* just before the hammer hits the test specimen, (*c*) the amount of energy that the test specimen absorbs if the hammer swings up to a maximum of $\phi = 70°$ after the impact.

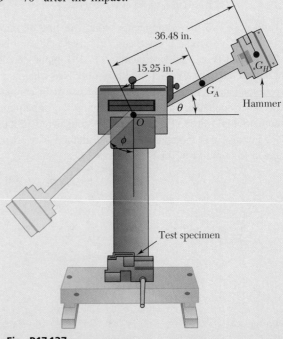

36.48 in.

15.25 in.

G_A

G_H

Hammer

θ

O

ϕ

Test specimen

Fig. P17.137

17.138 The gear shown has a radius $R = 150$ mm and a radius of gyration $\bar{k} = 125$ mm. The gear is rolling without sliding with a velocity $\bar{\mathbf{v}}_1$ of magnitude 3 m/s when it strikes a step of height $h = 75$ mm. Because the edge of the step engages the gear teeth, no slipping occurs between the gear and the step. Assuming perfectly plastic impact, determine (a) the angular velocity of the gear immediately after the impact, (b) the angular velocity of the gear after it has rotated to the top of the step.

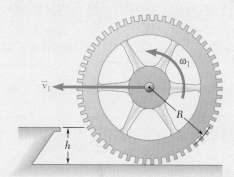

Fig. P17.138

17.139 A uniform slender rod is placed at corner B and is given a slight clockwise motion. Assuming that the corner is sharp and becomes slightly embedded in the end of the rod so that the coefficient of static friction at B is very large, determine (a) the angle β through which the rod will have rotated when it loses contact with the corner, (b) the corresponding velocity of end A.

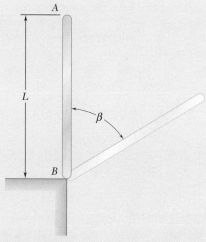

Fig. P17.139

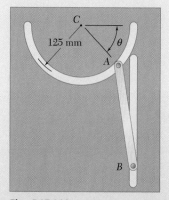

Fig. P17.140

17.140 The motion of the slender 250-mm rod AB is guided by pins at A and B that slide freely in slots cut in a vertical plate as shown. Knowing that the rod has a mass of 2 kg and is released from rest when $\theta = 0$, determine the reactions at A and B when $\theta = 90°$.

17.141 A baseball attachment that helps people with mobility impairments play T-ball and baseball is powered by a spring that is unstretched at position 2. The spring is attached to a cord that is fastened to point B on the 75-mm radius pulley. The pulley is fixed at point O, rotates backwards to the cocked position at θ, and the rope wraps around the pulley and stretches the spring with a stiffness of $k = 2000$ N/m. The combined mass moment of inertia of all the rotating components about point O is 0.40 kg·m^2. The swing is timed perfectly to strike a 145-gram baseball travelling with a speed of $v_0 = 10$ m/s at a distance of $h = 0.7$ m away from point O. Knowing that the coefficient of restitution between the bat and ball is 0.59, determine the velocity of the baseball immediately after the impact. Assume that the ball is travelling primarily in the horizontal plane and that its spin is negligible.

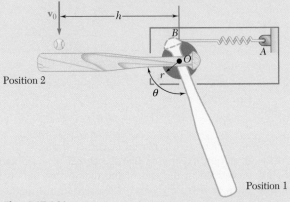

Position 2

Position 1

Fig. P17.141

17.142 Two panels A and B are attached with hinges to a rectangular plate and held by a wire as shown. The plate and the panels are made of the same material and have the same thickness. The entire assembly is rotating with an angular velocity ω_0 when the wire breaks. Determine the angular velocity of the assembly after the panels have come to rest against the plate.

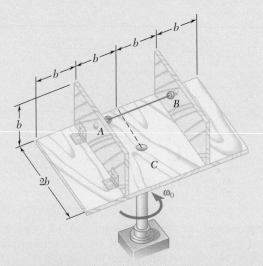

Fig. P17.142

17.143 Disks A and B are made of the same material, are of the same thickness, and can rotate freely about the vertical shaft. Disk B is at rest when it is dropped onto disk A, which is rotating with an angular velocity of 500 rpm. Knowing that disk A has a mass of 8 kg, determine (a) the final angular velocity of the disks, (b) the change in kinetic energy of the system.

17.144 A square block of mass m is falling with a velocity $\bar{v}_1$ when it strikes a small obstruction at B. Knowing that the coefficient of restitution for the impact between corner A and the obstruction B is $e = 0.5$, determine immediately after the impact (a) the angular velocity of the block, (b) the velocity of its mass center G.

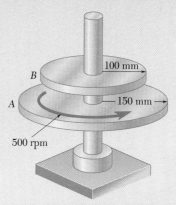

Fig. P17.143

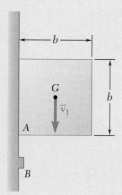

Fig. *P17.144*

17.145 A 3-kg bar AB is attached by a pin at D to a 4-kg square plate, which can rotate freely about a vertical axis. Knowing that the angular velocity of the plate is 120 rpm when the bar is vertical, determine (a) the angular velocity of the plate after the bar has swung into a horizontal position and has come to rest against pin C, (b) the energy lost during the plastic impact at C.

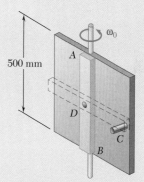

Fig. P17.145

17.146 A 1.8-lb javelin DE impacts a 10-lb slender rod ABC with a horizontal velocity of $v_0 = 30$ ft/s as shown. Knowing that the javelin becomes embedded into the end of the rod at point C and does not penetrate very far into it, determine immediately after the impact (a) the angular velocity of the rod ABC, (b) the components of the reaction at B. Assume the javelin and the rod move as a single rigid body after the impact.

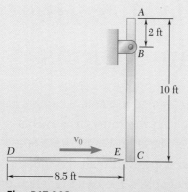

Fig. *P17.146*

18

Kinetics of Rigid Bodies in Three Dimensions

While the general principles that you learned in earlier chapters can be used again to solve problems involving the three-dimensional motion of rigid bodies, the solution of these problems requires a new approach and is considerably more involved than the solution of two-dimensional problems. One example is the determination of the forces acting on the robotic arm of the spacecraft.

Objectives

- **Calculate** the angular momentum and kinetic energy of a rigid body undergoing general three-dimensional motion.
- **Define** the inertia tensor, products of inertia, and principal axes of inertia.
- **Apply** the principle of impulse and momentum to solve three-dimensional rigid body kinetics problems.
- **Solve** three-dimensional rigid body kinetics problems, including fixed point rotation, fixed axis rotation, and gyroscopic motion.
- **Describe** the relationship between applied moment, precession, and spin of a gyroscope undergoing steady precession.
- **Analyze** the motion of a rotating axisymmetric body under no external forces.

Introduction

In Chaps. 16 and 17, we were concerned with the plane motion of rigid bodies and of systems of rigid bodies. In Chap. 16 and in the second half of Chap. 17 (impulse and momentum), our study was further restricted to the motion of plane rigid bodies and of bodies symmetrical with respect to the reference plane. However, many of the fundamental results obtained in these two chapters remain valid in the case of the motion of a rigid body in three dimensions. For example, the two fundamental equations

$$\Sigma \mathbf{F} = m\bar{\mathbf{a}} \tag{18.1}$$

$$\Sigma \mathbf{M}_G = \dot{\mathbf{H}}_G \tag{18.2}$$

on which we based the analysis of the plane motion of a rigid body remain valid in the most general case of motion of a rigid body. As indicated in Sec. 16.1, these equations express that the system of external forces is equipollent to the system consisting of the vector $m\bar{\mathbf{a}}$ attached at G and the couple of moment $\dot{\mathbf{H}}_G$ (Fig. 18.1).

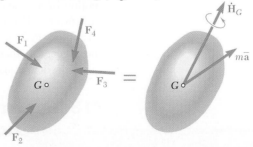

Fig. 18.1 The external forces acting on the rigid body are equipollent to a vector $m\bar{\mathbf{a}}$ attached to the mass center G and a rotational inertia vector $\dot{\mathbf{H}}_G$.

The relation $\mathbf{H}_G = \bar{I}\boldsymbol{\omega}$ enabled us to determine the angular momentum of a rigid body and played an important part in the solution of problems involving the plane motion of rigid bodies and bodies symmetrical with respect to the reference plane. However, this equation ceases to be valid in the case of nonsymmetrical bodies or three-dimensional motion. Thus, we need to develop a more general method for computing the angular momentum $\mathbf{H}_G$ of a rigid body in three dimensions.

Similarly, the main feature of the impulse–momentum method discussed in Sec. 17.2A is the reduction of the momenta of the particles of a rigid body to a linear momentum vector $m\bar{\mathbf{v}}$ attached at the mass center G of the body and an angular momentum couple $\mathbf{H}_G$. This method remains valid in the more general case, but we must discard the relation $\mathbf{H}_G = \bar{I}\boldsymbol{\omega}$ and replace it with a more general relation before we can apply this method to the three-dimensional motion of a rigid body (Sec. 18.1B).

Also note that the work–energy principle and the principle of conservation of energy still apply in the case of the motion of a rigid body in three dimensions. However, we need to replace the expression obtained in Sec. 17.1C for the kinetic energy of a rigid body in plane motion with a new expression for a rigid body in three-dimensional motion.

In the second part of this chapter, you will learn to determine the rate of change $\dot{\mathbf{H}}_G$ of the angular momentum $\mathbf{H}_G$ of a three-dimensional rigid body using a rotating frame of reference where the moments and products of inertia of the body remain constant. Then you can express Eqs. (18.1) and (18.2) in the form of free-body and kinetic diagrams that you can use to solve various problems involving the three-dimensional motion of rigid bodies (Sec. 18.2).

The last part of this chapter (Sec. 18.3) is devoted to the study of the motion of gyroscopes or, more generally, of an axisymmetric body with a fixed point located on its axis of symmetry. We first consider the particular case of the steady precession of a gyroscope and then analyze the motion of an axisymmetric body subjected to no force except its own weight.

18.1   ENERGY AND MOMENTUM OF A RIGID BODY

All of the methods you studied in earlier chapters for analyzing the plane motion of a rigid body have corresponding versions for motion in three dimensions. However, some of the formulas for determining kinetic quantities such as energy and angular momentum need to be replaced by more general equations. In this section, we examine some of the basic quantities and equations needed for the study of motion in space.

*18.1A   Angular Momentum of a Rigid Body in Three Dimensions

In this section you will see how to determine the angular momentum $\mathbf{H}_G$ of a body about its mass center G from the angular velocity $\boldsymbol{\omega}$ of the body in the case of three-dimensional motion.

According to Eq. (14.24), we can express the angular momentum of the body about G as

$$\mathbf{H}_G = \sum_{i=1}^{n} (\mathbf{r}'_i \times \mathbf{v}'_i \, \Delta m_i) \tag{18.3}$$

where $\mathbf{r}'_i$ and $\mathbf{v}'_i$ denote, respectively, the position vector and the velocity of the particle P_i with a mass Δm_i that is relative to the centroidal frame $Gxyz$ (Fig. 18.2). However, $\mathbf{v}'_i = \boldsymbol{\omega} \times \mathbf{r}'_i$, where $\boldsymbol{\omega}$ is the angular velocity of the body at the instant considered. Substituting into Eq. (18.3), we have

$$\mathbf{H}_G = \sum_{i=1}^{n} [\mathbf{r}'_i \times (\boldsymbol{\omega} \times \mathbf{r}'_i) \, \Delta m_i]$$

From the rule for determining the rectangular components of a vector product (*Statics*, Sec. 3.1D, or Appendix A), we obtain the following expression for the x-component of the angular momentum as

$$
\begin{aligned}
H_x &= \sum_{i=1}^{n} [y_i(\boldsymbol{\omega} \times \mathbf{r}'_i)_z - z_i(\boldsymbol{\omega} \times \mathbf{r}'_i)_y] \, \Delta m_i \\
&= \sum_{i=1}^{n} [y_i(\omega_x y_i - \omega_y x_i) - z_i(\omega_z x_i - \omega_x z_i)] \, \Delta m_i \\
&= \omega_x \sum_i (y_i^2 + z_i^2) \, \Delta m_i - \omega_y \sum_i x_i y_i \, \Delta m_i - \omega_z \sum_i z_i x_i \, \Delta m_i
\end{aligned}
$$

Replacing the sums by integrals in this expression and in the two similar expressions obtained for H_y and H_z, we have

$$
\begin{aligned}
H_x &= \omega_x \!\int (y^2 + z^2)\, dm - \omega_y \!\int xy\, dm - \omega_z \!\int zx\, dm \\
H_y &= -\omega_x \!\int xy\, dm + \omega_y \!\int (z^2 + x^2)\, dm - \omega_z \!\int yz\, dm \\
H_z &= -\omega_x \!\int zx\, dm - \omega_y \!\int yz\, dm + \omega_z \!\int (x^2 + y^2)\, dm
\end{aligned}
\tag{18.4}
$$

Note that the integrals containing squares represent the *centroidal mass moments of inertia* of the body about the x, y, and z axes, respectively (*Statics*, Sec. 9.5A, or Appendix B). That is,

$$
\bar{I}_x = \int (y^2 + z^2)\, dm \qquad \bar{I}_y = \int (z^2 + x^2)\, dm
$$
$$
\bar{I}_z = \int (x^2 + y^2)\, dm
\tag{18.5}
$$

Similarly, the integrals containing products of coordinates represent the *centroidal mass products of inertia* of the body (Sec. 9.6A); we have

$$
\bar{I}_{xy} = \int xy\, dm \qquad \bar{I}_{yz} = \int yz\, dm \qquad \bar{I}_{zx} = \int zx\, dm \tag{18.6}
$$

Substituting from Eqs. (18.5) and (18.6) into Eq. (18.4), we obtain the components of the angular momentum $\mathbf{H}_G$ of the body about its mass center as

Angular momentum about mass center

$$
\begin{aligned}
H_x &= +\bar{I}_x\, \omega_x - \bar{I}_{xy}\omega_y - \bar{I}_{xz}\omega_z \\
H_y &= -\bar{I}_{yx}\omega_x + \bar{I}_y\, \omega_y - \bar{I}_{yz}\omega_z \\
H_z &= -\bar{I}_{zx}\omega_x - \bar{I}_{zy}\omega_y + \bar{I}_z\, \omega_z
\end{aligned}
\tag{18.7}
$$

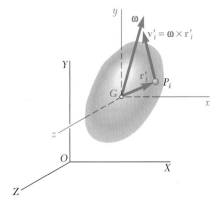

Fig. 18.2 The velocity of particle P_i is needed to derive the angular momentum of a rigid body in three dimensions.

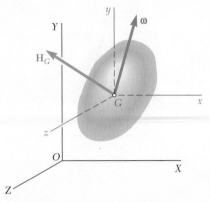

Fig. 18.3 In general, the angular momentum and the angular velocity are not in the same direction.

The relations in Eq. (18.7) show that the operation transforming the vector **ω** into the vector $\mathbf{H}_G$ (Fig. 18.3) is characterized by the array of moments and products of inertia as

Inertia tensor

$$\begin{pmatrix} \bar{I}_x & -\bar{I}_{xy} & -\bar{I}_{xz} \\ -\bar{I}_{yx} & \bar{I}_y & -\bar{I}_{yz} \\ -\bar{I}_{zx} & -\bar{I}_{zy} & \bar{I}_z \end{pmatrix} \tag{18.8}$$

The array in Eq. (18.8) defines the **inertia tensor** of the body at its mass center G.[†] We obtain a new array of moments and products of inertia if we use a different system of axes. The angular momentum $\mathbf{H}_G$ corresponding to a given angular velocity **ω** is independent of the choice of the coordinate axes.

As we showed in *Statics*, Sec. 9.6, or in Appendix B, it is always possible to select a system of axes $Gx'y'z'$, called *principal axes of inertia*, with respect to which all the products of inertia of a given body are zero. The array of Eq. (18.8) then takes the diagonalized form as

$$\begin{pmatrix} \bar{I}_{x'} & 0 & 0 \\ 0 & \bar{I}_{y'} & 0 \\ 0 & 0 & \bar{I}_{z'} \end{pmatrix} \tag{18.9}$$

where $\bar{I}_{x'}, \bar{I}_{y'}, \bar{I}_{z'}$ represent the *principal centroidal moments of inertia* of the body, and the relations in Eq. (18.7) reduce to

$$H_{x'} = \bar{I}_{x'}\omega_{x'} \qquad H_{y'} = \bar{I}_{y'}\omega_{y'} \qquad H_{z'} = \bar{I}_{z'}\omega_{z'} \tag{18.10}$$

Note that if the three principal centroidal moments of inertia $\bar{I}_{x'}, \bar{I}_{y'}, \bar{I}_{z'}$ are equal, the components $H_{x'}, H_{y'}, H_{z'}$ of the angular momentum about G are proportional to the components $\omega_{x'}, \omega_{y'}, \omega_{z'}$ of the angular velocity, and the vectors $\mathbf{H}_G$ and **ω** are collinear. In general, however, the principal moments of inertia are different, and the vectors $\mathbf{H}_G$ and **ω** have different directions except when two of the three components of **ω** happen to be zero, i.e., when **ω** is directed along one of the coordinate axes. Thus,

> **The angular momentum $\mathbf{H}_G$ of a rigid body and its angular velocity ω have the same direction if, and only if, ω is directed along a principal axis of inertia.[‡]**

This condition is satisfied in the case of the plane motion of a rigid body that is symmetrical with respect to the reference plane, so in Secs. 16.1 and 17.2, we were able to represent the angular momentum $\mathbf{H}_G$ of such a body by the vector $\bar{I}\boldsymbol{\omega}$. We must realize, however, that this result cannot be extended to the case of the plane motion of a nonsymmetrical body or to

---

[†]Setting $\bar{I}_x = I_{11}$, $\bar{I}_y = I_{22}$, $\bar{I}_z = I_{33}$, and $-\bar{I}_{xy} = I_{12}$, $-\bar{I}_{xz} = I_{13}$, etc., we can write the inertia tensor of Eq. (18.8) in the standard form

$$\begin{pmatrix} I_{11} & I_{12} & I_{13} \\ I_{21} & I_{22} & I_{23} \\ I_{31} & I_{32} & I_{33} \end{pmatrix}$$

[‡]In the particular case when $\bar{I}_{x'} = \bar{I}_{y'} = \bar{I}_{z'}$, any line through G can be considered to be a principal axis of inertia, and the vectors $\mathbf{H}_G$ and **ω** are always collinear.

the case of the three-dimensional motion of a rigid body. Except when $\boldsymbol{\omega}$ happens to be directed along a principal axis of inertia, the angular momentum and angular velocity of a rigid body have different directions, and you must use the relation in Eq. (18.7) or (18.10) to determine $\mathbf{H}_G$ from $\boldsymbol{\omega}$.

Reduction of the Momenta of the Particles of a Rigid Body to a Momentum Vector and a Couple at G.

We saw in Sec. 17.2A that we can reduce the system formed by the momenta of the various particles of a rigid body to a vector $\mathbf{L}$ that is attached at the mass center G of the body, representing the linear momentum of the body, and to a couple $\mathbf{H}_G$, representing the angular momentum of the body about G (Fig. 18.4). We are now in a position to determine the vector $\mathbf{L}$ and the couple $\mathbf{H}_G$ in the most general case of three-dimensional motion of a rigid body. As in the case of the two-dimensional motion considered earlier, the linear momentum $\mathbf{L}$ of the body is equal to the product $m\bar{\mathbf{v}}$ of its mass m and velocity $\bar{\mathbf{v}}$ of its mass center G. However, we can no longer obtain the angular momentum $\mathbf{H}_G$ by simply multiplying the angular velocity $\boldsymbol{\omega}$ of the body by the scalar $\bar{I}$. Instead, we obtain it from the components of $\boldsymbol{\omega}$ and from the centroidal moments and products of inertia of the body through the use of Eq. (18.7) or (18.10).

We should also note that once we have determined the linear momentum $m\bar{\mathbf{v}}$ and the angular momentum $\mathbf{H}_G$ of a rigid body, we can obtain its angular momentum $\mathbf{H}_O$ about any given point O by adding the moments about O of vector $m\bar{\mathbf{v}}$ and of couple $\mathbf{H}_G$. We have

$$\mathbf{H}_O = \bar{\mathbf{r}} \times m\bar{\mathbf{v}} + \mathbf{H}_G \qquad (18.11)$$

Angular Momentum of a Rigid Body Constrained to Rotate about a Fixed Point.

In the particular case of a rigid body constrained to rotate in three-dimensional space about a fixed point O (Fig. 18.5a), it is sometimes convenient to determine the angular momentum $\mathbf{H}_O$ of the body about O. Although we could obtain $\mathbf{H}_O$ by first computing $\mathbf{H}_G$ as indicated previously and then using Eq. (18.11), it is often advantageous to determine $\mathbf{H}_O$ directly from the angular velocity $\boldsymbol{\omega}$ of the body and its moments and products of inertia with respect to a frame $Oxyz$ centered at O. From Eq. (14.7), we have

$$\mathbf{H}_O = \sum_{i=1}^{n} (\mathbf{r}_i \times \mathbf{v}_i \, \Delta m_i) \qquad (18.12)$$

Photo 18.1 The design of a robotic welder for an automobile assembly line requires a three-dimensional study of both kinematics and kinetics.

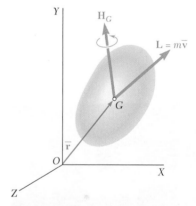

Fig. 18.4 A momentum vector attached to the mass center of a rigid body and the angular momentum of the body about its mass center.

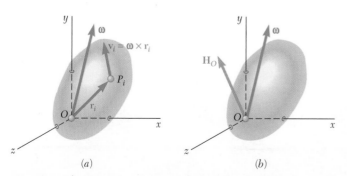

Fig. 18.5 (a) The velocity of particle P_i of a rigid body rotating with angular velocity $\boldsymbol{\omega}$; (b) angular velocity and angular momentum of a rigid body.

where $\mathbf{r}_i$ and $\mathbf{v}_i$ denote, respectively, the position vector and the velocity of particle P_i with respect to the fixed frame $Oxyz$. Substituting $\mathbf{v}_i = \boldsymbol{\omega} \times \mathbf{r}_i$ and after making manipulations similar to those used in the earlier part of this section, we find that the components of the angular momentum $\mathbf{H}_O$ (Fig. 18.5b) are given by the relations

**Angular momentum about
a fixed point O**

$$
\begin{aligned}
H_x &= +I_x\,\omega_x - I_{xy}\omega_y - I_{xz}\omega_z \\
H_y &= -I_{yx}\omega_x + I_y\,\omega_y - I_{yz}\omega_z \\
H_z &= -I_{zx}\omega_x - I_{zy}\omega_y + I_z\,\omega_z
\end{aligned}
\tag{18.13}
$$

where we compute the moments of inertia I_x, I_y, I_z and the products of inertia I_{xy}, I_{yz}, I_{zx} with respect to the frame $Oxyz$ centered at the fixed point O.

*18.1B Applying the Principle of Impulse and Momentum to the Three-Dimensional Motion of a Rigid Body

Before we can apply the fundamental equation (18.2) to the solution of problems involving the three-dimensional motion of a rigid body, we must be able to compute the derivative of the vector $\mathbf{H}_G$. We show how to do this in Sec. 18.2A. However, we can use the results obtained already to solve problems using the impulse–momentum method.

Recall that the system formed by the momenta of the particles of a rigid body reduces to a linear momentum vector $m\bar{\mathbf{v}}$ attached at the mass center G of the body and an angular momentum couple $\mathbf{H}_G$. We can represent the fundamental relation

$$\textbf{Syst Momenta}_1 + \textbf{Syst Ext Imp}_{1\rightarrow2} = \textbf{Syst Momenta}_2 \tag{17.14}$$

graphically by means of the impulse–momentum diagram shown in Fig. 18.6. To solve a given problem, we can use this diagram to write appropriate component and moment equations, keeping in mind that the components of the angular momentum $\mathbf{H}_G$ are related to the components of the angular velocity $\boldsymbol{\omega}$ by Eqs. (18.7).

Photo 18.2 As a result of the impulsive force applied by the bowling ball, a pin acquires both linear momentum and angular momentum.

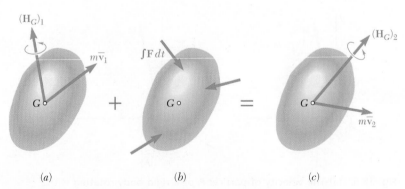

(a) (b) (c)

Fig. 18.6 Impulse–momentum diagram for applying the principle of impulse and momentum to the motion of a rigid body in space.

In solving problems dealing with the motion of a body rotating about a fixed point O, it will be convenient to eliminate the impulse of the reaction at O by writing an equation involving the moments of the momenta and impulses about O. Recall that you can obtain the angular momentum $\mathbf{H}_O$ of the body about the fixed point O either directly from Eqs. (18.13) or by first computing its linear momentum $m\overline{\mathbf{v}}$ and its angular momentum $\mathbf{H}_G$ and then using Eq. (18.11).

*18.1C Kinetic Energy of a Rigid Body in Three Dimensions

Consider a rigid body with a mass m in three-dimensional motion. Recall from Sec. 14.2A that, if we express the absolute velocity $\mathbf{v}_i$ of each particle P_i of the body as the sum of velocity $\overline{\mathbf{v}}$ of the mass center G of the body and velocity $\mathbf{v}'_i$ of the particle relative to a frame $Gxyz$ attached to G and of fixed orientation (Fig. 18.7), we can write the kinetic energy of the system of particles forming the rigid body as

$$T = \tfrac{1}{2}m\overline{v}^2 + \frac{1}{2}\sum_{i=1}^{n}\Delta m_i v_i'^2 \tag{18.14}$$

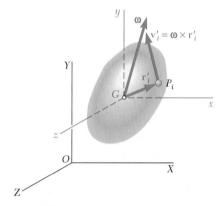

Fig. 18.7 The relative velocity of a particle P_i with respect to the mass center is $\boldsymbol{\omega} \times \mathbf{r}'_i$.

Here the last term represents the kinetic energy T' of the body relative to the centroidal frame $Gxyz$. Since $v'_i = |\mathbf{v}'_i| = |\boldsymbol{\omega} \times \mathbf{r}'_i|$, we have

$$T' = \frac{1}{2}\sum_{i=1}^{n}\Delta m_i v_i'^2 = \frac{1}{2}\sum_{i=1}^{n}|\boldsymbol{\omega} \times \mathbf{r}'_i|^2\,\Delta m_i$$

Expressing the square in terms of the rectangular components of the vector product and replacing the sums by integrals, we have

$$T' = \tfrac{1}{2}\int [(\omega_x y - \omega_y x)^2 + (\omega_y z - \omega_z y)^2 + (\omega_z x - \omega_x z)^2]\,dm$$
$$= \tfrac{1}{2}[\omega_x^2\int(y^2 + z^2)\,dm + \omega_y^2\int(z^2 + x^2)\,dm + \omega_z^2\int(x^2 + y^2)\,dm$$
$$- 2\omega_x\omega_y\int xy\,dm - 2\omega_y\omega_z\int yz\,dm - 2\omega_z\omega_x\int zx\,dm]$$

or recalling the relations of Eqs. (18.5) and (18.6), we have

$$T' = \tfrac{1}{2}(\overline{I}_x\omega_x^2 + \overline{I}_y\omega_y^2 + \overline{I}_z\omega_z^2 - 2\overline{I}_{xy}\omega_x\omega_y - 2\overline{I}_{yz}\omega_y\omega_z - 2\overline{I}_{zx}\omega_z\omega_x) \tag{18.15}$$

Substituting Eq. (18.15) for the kinetic energy of the body relative to centroidal axes into Eq. (18.14), we obtain

Kinetic energy of a rigid body

$$T = \tfrac{1}{2}m\bar{v}^2 + \tfrac{1}{2}(\bar{I}_x\omega_x^2 + \bar{I}_y\omega_y^2 + \bar{I}_z\omega_z^2 - 2\bar{I}_{xy}\omega_x\omega_y \\ - 2\bar{I}_{yz}\omega_y\omega_z - 2\bar{I}_{zx}\omega_z\omega_x)$$

(18.16)

If we choose the axes of coordinates so that they coincide with the principal axes x', y', z' of the body at the instant considered, this relation reduces to

$$T = \tfrac{1}{2}m\bar{v}^2 + \tfrac{1}{2}(\bar{I}_{x'}\omega_{x'}^2 + \bar{I}_{y'}\omega_{y'}^2 + \bar{I}_{z'}\omega_{z'}^2)$$

(18.17)

where $\bar{\mathbf{v}}$ = velocity of mass center

$\boldsymbol{\omega}$ = angular velocity

m = mass of rigid body

$\bar{I}_{x'}, \bar{I}_{y'}, \bar{I}_{z'}$ = principal centroidal moments of inertia

These results enable us to apply the principles of work and energy (Sec. 17.1A) and the conservation of energy (Sec. 17.1E) to the three-dimensional motion of a rigid body.

Kinetic Energy of a Rigid Body with a Fixed Point.

In the particular case of a rigid body rotating in three-dimensional space about a fixed point O, we can express the kinetic energy of the body in terms of its moments and products of inertia with respect to axes attached at O (Fig. 18.8). Recalling the definition of the kinetic energy of a system of particles and substituting $v_i = |\mathbf{v}_i| = |\boldsymbol{\omega} \times \mathbf{r}_i|$, we have

$$T = \frac{1}{2}\sum_{i=1}^{n}\Delta m_i v_i^2 = \frac{1}{2}\sum_{i=1}^{n}|\boldsymbol{\omega} \times \mathbf{r}_i|^2\,\Delta m_i$$

(18.18)

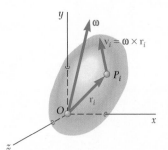

Fig. 18.8 The velocity of every particle P_i of a rigid body undergoing fixed axis rotation is $\boldsymbol{\omega} \times \mathbf{r}_i$.

Manipulations similar to those used to derive Eq. (18.15) yield

$$T = \tfrac{1}{2}(I_x\omega_x^2 + I_y\omega_y^2 + I_z\omega_z^2 - 2I_{xy}\omega_x\omega_y - 2I_{yz}\omega_y\omega_z - 2I_{zx}\omega_z\omega_x)$$

(18.19)

or if we choose the principal axes x', y', z' of the body at the origin O as coordinate axes, we have

$$T = \tfrac{1}{2}(I_{x'}\omega_{x'}^2 + I_{y'}\omega_{y'}^2 + I_{z'}\omega_{z'}^2)$$

(18.20)

Sample Problem 18.1

A rectangular plate with a mass m is suspended from two wires at A and B and is hit at D in a direction perpendicular to the plate. Denoting the impulse applied at D by $\mathbf{F}\,\Delta t$, determine immediately after the impact (a) the velocity of the mass center G, (b) the angular velocity of the plate.

STRATEGY: Since you have an impulse applied to the plate, use the principle of impulse and momentum.

MODELING: Choose the plate to be your system and model it as a rigid body undergoing three-dimensional motion.

ANALYSIS: Assume that the wires remain taut. Therefore, the components $\bar{v}_y$ of $\bar{\mathbf{v}}$ and ω_z of $\boldsymbol{\omega}$ are zero after the impact. Then you have

$$\bar{\mathbf{v}} = \bar{v}_x\mathbf{i} + \bar{v}_z\mathbf{k} \qquad \boldsymbol{\omega} = \omega_x\mathbf{i} + \omega_y\mathbf{j}$$

The x, y, z axes are principal axes of inertia, so you have

$$\mathbf{H}_G = \bar{I}_x\omega_x\mathbf{i} + \bar{I}_y\omega_y\mathbf{j} \qquad \mathbf{H}_G = \tfrac{1}{12}mb^2\omega_x\mathbf{i} + \tfrac{1}{12}ma^2\omega_y\mathbf{j} \qquad (1)$$

Principle of Impulse and Momentum. Since the initial momenta are zero, the system of the impulses must be equivalent to the system of the final momenta (Fig. 1).

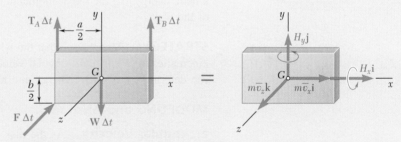

Fig. 1 Impulse–momentum diagram for the plate.

a. Velocity of Mass Center. Equate the components of the impulses and momenta in the x and z directions as

x components: $\qquad\qquad 0 = m\bar{v}_x \qquad \bar{v}_x = 0$

z components: $\qquad\quad -F\,\Delta t = m\bar{v}_z \qquad \bar{v}_z = -F\,\Delta t/m$

$$\bar{\mathbf{v}} = \bar{v}_x\mathbf{i} + \bar{v}_z\mathbf{k} \qquad \bar{\mathbf{v}} = -(F\,\Delta t/m)\mathbf{k} \quad \blacktriangleleft$$

b. Angular Velocity. Equate the moments of the impulses and momenta about the x and y axes as

About x axis: $\qquad\qquad \tfrac{1}{2}bF\,\Delta t = H_x$

About y axis: $\qquad\qquad -\tfrac{1}{2}aF\,\Delta t = H_y$

$$\mathbf{H}_G = H_x\mathbf{i} + H_y\mathbf{j} \qquad \mathbf{H}_G = \tfrac{1}{2}bF\,\Delta t\mathbf{i} - \tfrac{1}{2}aF\,\Delta t\mathbf{j} \qquad (2)$$

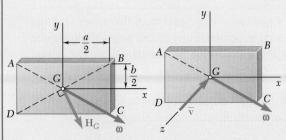

Fig. 2 Directions of the angular velocity, angular momentum, and velocity of G immediately after the impulse.

Comparing Eqs. (1) and (2), you can conclude that

$$\omega_x = 6F\,\Delta t/mb \qquad \omega_y = -6F\,\Delta t/ma$$
$$\boldsymbol{\omega} = \omega_x\mathbf{i} + \omega_y\mathbf{j} \qquad \boldsymbol{\omega} = (6F\,\Delta t/mab)(a\mathbf{i} - b\mathbf{j}) \quad \blacktriangleleft$$

Note that $\boldsymbol{\omega}$ is directed along the diagonal AC (Fig. 2).

REFLECT and THINK: Equating the y components of the impulses and momenta and their moments about the z axis, you can obtain two additional equations that yield $T_A = T_B = \frac{1}{2}W$. This verifies that the wires remain taut and that the initial assumption was correct. If the impulse was at G, this would reduce to a two-dimensional problem.

Sample Problem 18.2

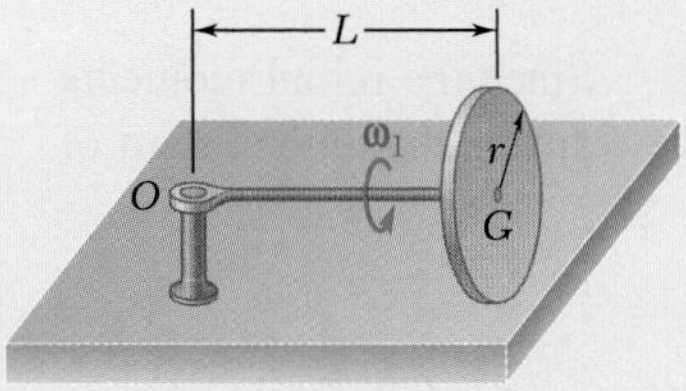

A homogeneous disk of radius r and mass m is mounted on an axle OG of length L and negligible mass. The axle is pivoted at the fixed point O, and the disk is constrained to roll on a horizontal floor. The disk rotates counterclockwise at the rate ω_1 about the axle OG. Determine (a) the angular velocity of the disk, (b) its angular momentum about O, (c) its kinetic energy, (d) the linear momentum and angular momentum about G of the disk.

STRATEGY: Recognizing that the wheel rolls without slip, you can use kinematics to calculate the angular velocity of the bar around O. Then you can determine the kinetic energy and momenta of the system.

MODELING and ANALYSIS:

a. Angular Velocity. As the disk rotates about the axle OG, it also rotates with the axle about the y axis at a rate of ω_2 clockwise (Fig. 1). The total angular velocity of the disk is therefore

$$\boldsymbol{\omega} = \omega_1\mathbf{i} - \omega_2\mathbf{j} \tag{1}$$

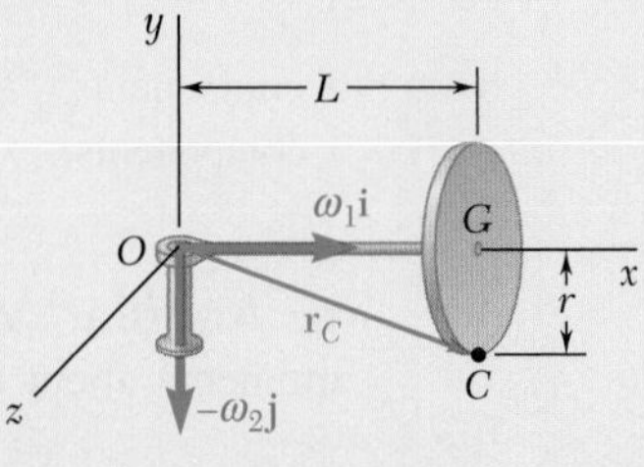

Fig. 1 Angular velocity of the system.

(continued)

The disk is rolling, so set the velocity of C to zero to determine ω_2 as

$$\mathbf{v}_C = \boldsymbol{\omega} \times \mathbf{r}_C = 0$$
$$(\omega_1\mathbf{i} - \omega_2\mathbf{j}) \times (L\mathbf{i} - r\mathbf{j}) = 0$$
$$(L\omega_2 - r\omega_1)\mathbf{k} = 0 \qquad \omega_2 = r\omega_1/L$$

Substituting into Eq. (1) for ω_2 gives

$$\boldsymbol{\omega} = \omega_1\mathbf{i} - (r\omega_1/L)\mathbf{j} \quad \blacktriangleleft$$

b. Angular Momentum about O.

Assuming the axle to be part of the disk, you can consider the disk to have a fixed point at O. Since the x, y, and z axes are principal axes of inertia for the disk, you have

$$H_x = I_x\omega_x = (\tfrac{1}{2}mr^2)\omega_1$$
$$H_y = I_y\omega_y = (mL^2 + \tfrac{1}{4}mr^2)(-r\omega_1/L)$$
$$H_z = I_z\omega_z = (mL^2 + \tfrac{1}{4}mr^2)0 = 0$$
$$\mathbf{H}_O = \tfrac{1}{2}mr^2\omega_1\mathbf{i} - m(L^2 + \tfrac{1}{4}r^2)(r\omega_1/L)\mathbf{j} \quad \blacktriangleleft$$

c. Kinetic Energy.

Using the values obtained for the moments of inertia and the components of $\boldsymbol{\omega}$, you have

$$T = \tfrac{1}{2}(I_x\omega_x^2 + I_y\omega_y^2 + I_z\omega_z^2) = \tfrac{1}{2}[\tfrac{1}{2}mr^2\omega_1^2 + m(L^2 + \tfrac{1}{4}r^2)(-r\omega_1/L)^2]$$
$$T = \tfrac{1}{8}mr^2\left(6 + \frac{r^2}{L^2}\right)\omega_1^2 \quad \blacktriangleleft$$

d. Linear Momentum and Angular Momentum about G.

The linear momentum vector $m\overline{\mathbf{v}}$ and the angular momentum couple $\mathbf{H}_G$ are (Fig.2)

$$m\overline{\mathbf{v}} = mr\omega_1\mathbf{k} \quad \blacktriangleleft$$

and

$$\mathbf{H}_G = \overline{I}_{x'}\omega_x\mathbf{i} + \overline{I}_{y'}\omega_y\mathbf{j} + \overline{I}_{z'}\omega_z\mathbf{k} = \tfrac{1}{2}mr^2\omega_1\mathbf{i} + \tfrac{1}{4}mr^2(-r\omega_1/L)\mathbf{j}$$
$$\mathbf{H}_G = \tfrac{1}{2}mr^2\omega_1\left(\mathbf{i} - \frac{r}{2L}\mathbf{j}\right) \quad \blacktriangleleft$$

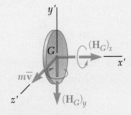

Fig. 2 Linear and angular momenta for the system.

REFLECT and THINK: If the mass of the axle was not negligible and it was instead modeled as a slender rod with a mass M_{axle}, it would also contribute to the kinetic energy $T_{\text{axle}} = \tfrac{1}{2}(\tfrac{1}{3}M_{\text{axle}}L^2)\omega_2^2$ and to the angular momentum $\mathbf{H}_{\text{axle}} = -(\tfrac{1}{3}M_{\text{axle}}L^2)\omega_2\mathbf{j}$ of the system.

SOLVING PROBLEMS
ON YOUR OWN

In this section, you saw how to compute the **angular momentum of a rigid body in three dimensions** and to apply the principle of impulse and momentum to the three-dimensional motion of a rigid body. You also learned how to compute the **kinetic energy of a rigid body in three dimensions**. It is important for you to keep in mind that, except for very special situations, the angular momentum of a rigid body in three dimensions *cannot* be expressed as the product $\bar{I}\boldsymbol{\omega}$ and, therefore, does not have the same direction as the angular velocity $\boldsymbol{\omega}$ (Fig. 18.3).

1. To compute the angular momentum $\mathbf{H}_G$ of a rigid body about its mass center G, you must first determine the angular velocity $\boldsymbol{\omega}$ of the body with respect to a system of axes centered at G and of fixed orientation. Since you will be asked in the problems to determine the angular momentum of the body *at a given instant only,* select the system of axes that will be most convenient for your computations.

    a. If the principal axes of inertia of the body at G are known, use these axes as coordinate axes x', y', and z', since the corresponding products of inertia of the body are equal to zero. Resolve $\boldsymbol{\omega}$ into components $\omega_{x'}, \omega_{y'},$ and $\omega_{z'}$ along these axes and compute the principal moments of inertia as $\bar{I}_{x'}, \bar{I}_{y'}, \bar{I}_{z'}$. The corresponding components of the angular momentum $\mathbf{H}_G$ are

$$H_{x'} = \bar{I}_{x'}\omega_{x'} \qquad H_{y'} = \bar{I}_{y'}\omega_{y'} \qquad H_{z'} = \bar{I}_{z'}\omega_{z'} \qquad \textbf{(18.10)}$$

    b. If the principal axes of inertia of the body at G are not known, you must use Eqs. (18.7) to determine the components of the angular momentum $\mathbf{H}_G$. These equations require prior computation of the products of inertia of the body as well as prior computation of its moments of inertia with respect to the selected axes.

    c. The magnitude and direction cosines of $\mathbf{H}_G$ are obtained from formulas similar to those used in *Statics* [Sec. 2.4A]. We have

$$H_G = \sqrt{H_x^2 + H_y^2 + H_z^2}$$

$$\cos\theta_x = \frac{H_x}{H_G} \qquad \cos\theta_y = \frac{H_y}{H_G} \qquad \cos\theta_z = \frac{H_z}{H_G}$$

    d. Once you have determined $\mathbf{H}_G$, you can obtain the angular momentum of the body about any given point O by observing from Fig. (18.4) that

$$\mathbf{H}_O = \bar{\mathbf{r}} \times m\bar{\mathbf{v}} + \mathbf{H}_G \qquad \textbf{(18.11)}$$

where $\bar{\mathbf{r}}$ is the position vector of G relative to O and $m\bar{\mathbf{v}}$ is the linear momentum of the body.

2. To compute the angular momentum $\mathbf{H}_O$ of a rigid body with a fixed point O, follow the procedure described in paragraph 1, except that you should now use axes centered at the fixed point O. Alternatively, you can use Eq. 18.11.

    a. If you know the principal axes of inertia of the body at O, resolve $\boldsymbol{\omega}$ into components along these axes [Sample Prob. 18.2]. Obtain the corresponding components of the angular momentum $\mathbf{H}_G$ from equations similar to Eqs. (18.10).

b. If you do not know the principal axes of inertia of the body at O, you must compute the products as well as the moments of inertia of the body with respect to the axes that you have selected. Then use Eqs. (18.13) to determine the components of the angular momentum $\mathbf{H}_O$.

3. To apply the principle of impulse and momentum to the solution of a problem involving the three-dimensional motion of a rigid body, use the same vector equation that you used for plane motion in Chap. 17:

$$\textbf{Syst Momenta}_1 + \textbf{Syst Ext Imp}_{1 \rightarrow 2} = \textbf{Syst Momenta}_2 \qquad (17.14)$$

where the initial and final systems of momenta are each represented by a *linear-momentum vector* $m\overline{\mathbf{v}}$ and an *angular-momentum couple* $\mathbf{H}_G$. Now, however, these vector-and-couple systems should be represented in three dimensions, as shown in Fig. 18.6, and $\mathbf{H}_G$ should be determined as explained in paragraph 1.

a. In problems involving the application of a known impulse to a rigid body, draw the impulse–momentum diagram corresponding to Eq. (17.14). Equating the components of the vectors involved, you can determine the final linear momentum $m\overline{\mathbf{v}}$ of the body and, thus, the corresponding velocity $\overline{\mathbf{v}}$ of its mass center. Equating moments about G, you can determine the final angular momentum $\mathbf{H}_G$ of the body. Then substitute the values obtained for the components of $\mathbf{H}_G$ into Eq. (18.10) or (18.7) and solve for the corresponding values of the components of the angular velocity $\boldsymbol{\omega}$ of the body [Sample Prob. 18.1].

b. In problems involving unknown impulses, draw the impulse–momentum diagram corresponding to Eq. (17.14) and write equations that do not involve the unknown impulses. You can obtain such equations by equating moments about the point or line of impact.

4. To compute the kinetic energy of a rigid body with a fixed point O, resolve the angular velocity $\boldsymbol{\omega}$ into components along axes of your choice and compute the moments and products of inertia of the body with respect to these axes. As was the case for the computation of the angular momentum, use the principal axes of inertia x', y', and z' if you can easily determine them. The products of inertia are then zero [Sample Prob. 18.2], and the expression for the kinetic energy reduces to

$$T = \tfrac{1}{2}(I_{x'}\omega_{x'}^2 + I_{y'}\omega_{y'}^2 + I_{z'}\omega_{x'}^2) \qquad (18.20)$$

If you must use axes other than the principal axes of inertia, express the kinetic energy of the body as shown in Eq. (18.19).

5. To compute the kinetic energy of a rigid body in general motion, consider the motion as the sum of a *translation with the mass center G and a rotation about G.* The kinetic energy associated with the translation is $\tfrac{1}{2}m\overline{v}^2$. If you can use principal axes of inertia, express the kinetic energy associated with the rotation about G in the form used in Eq. (18.20). The total kinetic energy of the rigid body is then

$$T = \tfrac{1}{2}m\overline{v}^2 + \tfrac{1}{2}(\overline{I}_{x'}\omega_{x'}^2 + \overline{I}_{y'}\omega_{y'}^2 + \overline{I}_{z'}\omega_{z'}^2) \qquad (18.17)$$

If you must use axes other than the principal axes of inertia to determine the kinetic energy associated with the rotation about G, express the total kinetic energy of the body as shown in Eq. (18.16).

Problems

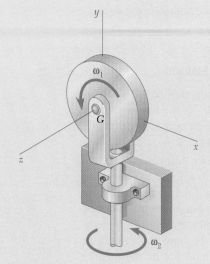

Fig. P18.1

18.1 A thin, homogeneous disk of mass m and radius r spins at the constant rate ω_1 about an axle held by a fork-ended vertical rod that rotates at the constant rate ω_2. Determine the angular momentum $\mathbf{H}_G$ of the disk about its mass center G.

18.2 A thin rectangular plate of weight 15 lb rotates about its vertical diagonal AB with an angular velocity $\boldsymbol{\omega}$. Knowing that the z axis is perpendicular to the plate and that $\boldsymbol{\omega}$ is constant and equal to 5 rad/s, determine the angular momentum of the plate about its mass center G.

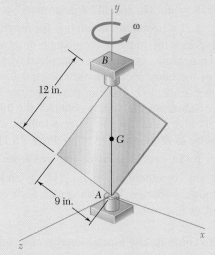

Fig. P18.2

18.3 Two uniform rods AB and CE, each of weight 3 lb and length 2 ft, are welded to each other at their midpoints. Knowing that this assembly has an angular velocity of constant magnitude $\omega = 12$ rad/s, determine the magnitude and direction of the angular momentum $\mathbf{H}_D$ of the assembly about D.

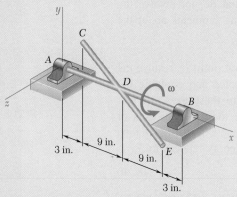

Fig. P18.3

18.4 A homogeneous disk of weight $W = 6$ lb rotates at the constant rate $\omega_1 = 16$ rad/s with respect to arm ABC, which is welded to a shaft DCE rotating at the constant rate $\omega_2 = 8$ rad/s. Determine the angular momentum $\mathbf{H}_A$ of the disk about its center A.

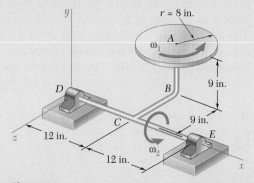

Fig. P18.4

18.5 A thin disk of mass $m = 4$ kg rotates at the constant rate $\omega_2 = 15$ rad/s with respect to arm ABC, which itself rotates at the constant rate $\omega_1 = 5$ rad/s about the y axis. Determine the angular momentum of the disk about its center C.

18.6 A solid rectangular parallelepiped of mass m has a square base of side a and a length $2a$. Knowing that it rotates at the constant rate ω about its diagonal AC' and that its rotation is observed from A as counterclockwise, determine (a) the magnitude of the angular momentum $\mathbf{H}_G$ of the parallelepiped about its mass center G, (b) the angle that $\mathbf{H}_G$ forms with the diagonal AC'.

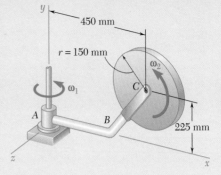

Fig. P18.5

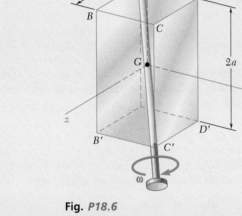

Fig. P18.6

18.7 Solve Prob. 18.6, assuming that the solid rectangular parallelepiped has been replaced by a hollow one consisting of six thin metal plates welded together.

18.8 A thin homogeneous disk with a mass m and radius r is mounted on the horizontal axle AB. The plane of the disk forms an angle of $\beta = 20°$ with the vertical. Knowing that the axle rotates with an angular velocity $\boldsymbol{\omega}$, determine the angle θ formed by the axle and the angular momentum of the disk about G.

18.9 Determine the angular momentum $\mathbf{H}_D$ of the disk of Prob. 18.4 about point D.

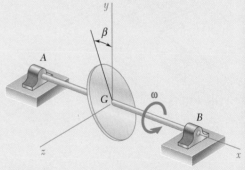

Fig. P18.8

18.10 Determine the angular momentum of the disk of Prob. 18.5 about point A.

18.11 Determine the angular momentum $\mathbf{H}_O$ of the disk of Sample Prob. 18.2 from the expressions obtained for its linear momentum $m\bar{\mathbf{v}}$ and its angular momentum $\mathbf{H}_G$, using Eqs. (18.11). Verify that the result obtained is the same as that obtained by direct computation.

18.12 The 100-kg projectile shown has a radius of gyration of 100 mm about its axis of symmetry Gx and a radius of gyration of 250 mm about the transverse axis Gy. Its angular velocity $\boldsymbol{\omega}$ can be resolved into two components; one component, directed along Gx, measures the *rate of spin* of the projectile, while the other component, directed along GD, measures its *rate of precession*. Knowing that $\theta = 6°$ and that the angular momentum of the projectile about its mass center G is $\mathbf{H}_G = (500 \text{ g·m}^2/\text{s})\mathbf{i} - (10 \text{ g·m}^2/\text{s})\mathbf{j}$, determine (a) the rate of spin, (b) the rate of precession.

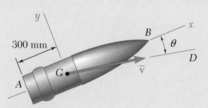

Fig. P18.12

18.13 Determine the angular momentum $\mathbf{H}_A$ of the projectile of Prob. 18.12 about the center A of its base, knowing that its mass center G has a velocity $\overline{\mathbf{v}}$ of 750 m/s. Give your answer in terms of components respectively parallel to the x and y axes shown and to a third axis z pointing toward you.

18.14 (a) Show that the angular momentum $\mathbf{H}_B$ of a rigid body about point B can be obtained by adding to the angular momentum $\mathbf{H}_A$ of that body about point A the vector product of the vector $\mathbf{r}_{A/B}$ drawn from B to A and the linear momentum $m\overline{\mathbf{v}}$ of the body:

$$\mathbf{H}_B = \mathbf{H}_A + \mathbf{r}_{A/B} \times m\overline{\mathbf{v}}$$

(b) Further show that when a rigid body rotates about a fixed axis, its angular momentum is the same about any two points A and B located on the fixed axis ($\mathbf{H}_A = \mathbf{H}_B$) if, and only if, the mass center G of the body is located on the fixed axis.

18.15 Two L-shaped arms each have a mass of 5 kg and are welded at the one-third points of the 600-mm shaft AB to form the assembly shown. Knowing that the assembly rotates at the constant rate of 360 rpm, determine (a) the angular momentum $\mathbf{H}_A$ of the assembly about point A, (b) the angle formed by $\mathbf{H}_A$ and AB.

18.16 For the assembly of Prob. 18.15, determine (a) the angular momentum $\mathbf{H}_B$ of the assembly about point B, (b) the angle formed by $\mathbf{H}_B$ and BA.

18.17 A 10-lb rod of uniform cross section is used to form the shaft shown. Knowing that the shaft rotates with a constant angular velocity $\boldsymbol{\omega}$ of magnitude 12 rad/s, determine (a) the angular momentum $\mathbf{H}_G$ of the shaft about its mass center G, (b) the angle formed by $\mathbf{H}_G$ and the axis AB.

18.18 Determine the angular momentum of the shaft of Prob. 18.17 about (a) point A, (b) point B.

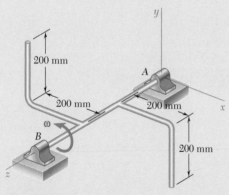

Fig. P18.15

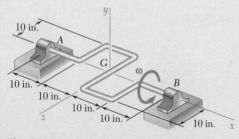

Fig. P18.17

18.19 Two triangular plates each have a mass of 8 kg and are welded to a vertical shaft AB. Knowing that the system rotates at the constant rate of $\omega = 6$ rad/s, determine its angular momentum about G.

18.20 The assembly shown consists of two pieces of sheet aluminum with a uniform thickness and total mass of 1.6 kg welded to a light axle supported by bearings A and B. Knowing that the assembly rotates with an angular velocity of constant magnitude $\omega = 20$ rad/s, determine the angular momentum $\mathbf{H}_G$ of the assembly about point G.

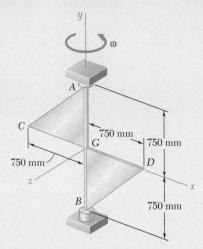

Fig. P18.19

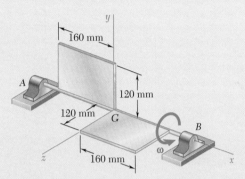

Fig. P18.20

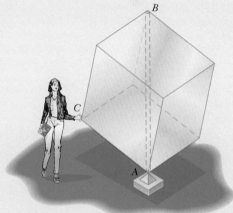

18.21 One of the sculptures displayed on a university campus consists of a hollow cube made of six aluminum sheets, each 1.5×1.5 m, welded together and reinforced with internal braces of negligible weight. The cube is mounted on a fixed base at A and can rotate freely about its vertical diagonal AB. As she passes by this display on the way to a class in mechanics, an engineering student grabs corner C of the cube and pushes it for 1.2 s in a direction perpendicular to the plane ABC with an average force of 50 N. Having observed that it takes 5 s for the cube to complete one full revolution, she flips out her calculator and proceeds to determine the mass of the cube. What is the result of her calculation? (*Hint:* The perpendicular distance from the diagonal joining two vertices of a cube to any of its other six vertices can be obtained by multiplying the side of the cube by $\sqrt{2/3}$.)

Fig. P18.21

18.22 If the aluminum cube of Prob. 18.21 were replaced by a cube of the same size, made of six plywood sheets with mass 8 kg each, how long would it take for that cube to complete one full revolution if the student pushed its corner C in the same way that she pushed the corner of the aluminum cube?

18.23 A uniform rod of total mass m is bent into the shape shown and is suspended by a wire attached at B. The bent rod is hit at D in a direction perpendicular to the plane containing the rod (in the negative z direction). Denoting the corresponding impulse by $\mathbf{F} \, \Delta t$, determine (*a*) the velocity of the mass center of the rod, (*b*) the angular velocity of the rod.

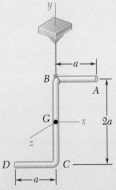

Fig. P18.23

18.24 Solve Prob. 18.23, assuming that the bent rod is hit at C.

18.25 Three slender rods, each of mass m and length $2a$, are welded together to form the assembly shown. The assembly is hit at A in a vertical downward direction. Denoting the corresponding impulse by $\mathbf{F}\,\Delta t$, determine immediately after the impact (a) the velocity of the mass center G, (b) the angular velocity of the rod.

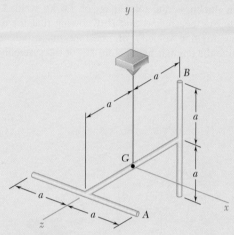

Fig. P18.25

18.26 Solve Prob. 18.25, assuming that the assembly is hit at B in the negative x direction.

18.27 Two circular plates, each of mass 4 kg, are rigidly connected by a rod AB of negligible mass and are suspended from point A as shown. Knowing that an impulse $\mathbf{F}\,\Delta t = -(2.4 \text{ N·s})\mathbf{k}$ is applied at point D, determine (a) the velocity of the mass center G of the assembly, (b) the angular velocity of the assembly.

18.28 Two circular plates, each of mass 4 kg, are rigidly connected by a rod AB of negligible mass and are suspended from point A as shown. Knowing that an impulse $\mathbf{F}\,\Delta t = (2.4 \text{ N·s})\mathbf{j}$ is applied at point D, determine (a) the velocity of the mass center G of the assembly, (b) the angular velocity of the assembly.

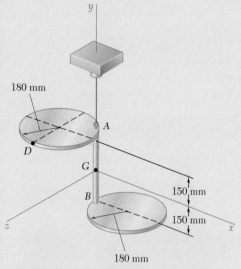

Fig. P18.27 and P18.28

18.29 A circular plate of mass m is falling with a velocity $\overline{\mathbf{v}}_0$ and no angular velocity when its edge C strikes an obstruction. A line passing the origin and parallel to the line CG makes a 45° angle with the x-axis. Assuming the impact to be perfectly plastic ($e = 0$), determine the angular velocity of the plate immediately after the impact.

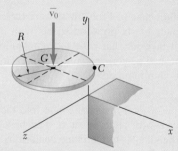

Fig. P18.29

18.30 For the plate of Prob. 18.29, determine (a) the velocity of its mass center G immediately after the impact, (b) the impulse exerted on the plate by the obstruction during the impact.

18.31 A square plate of side a and mass m supported by a ball-and-socket joint at A is rotating about the y axis with a constant angular velocity $\boldsymbol{\omega} = \omega_0\mathbf{j}$ when an obstruction is suddenly introduced at B in the xy plane. Assuming the impact at B to be perfectly plastic ($e = 0$), determine immediately after the impact (a) the angular velocity of the plate, (b) the velocity of its mass center G.

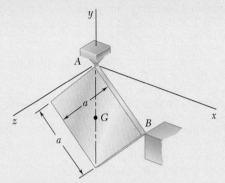

Fig. P18.31

18.32 Determine the impulse exerted on the plate of Prob. 18.31 during the impact by (a) the obstruction at B, (b) the support at A.

18.33 The coordinate axes shown represent the principal centroidal axes of inertia of a 3000-lb space probe whose radii of gyration are $k_x = 1.375$ ft, $k_y = 1.425$ ft, and $k_z = 1.250$ ft. The probe has no angular velocity when a 5-oz meteorite strikes one of its solar panels at point A with a velocity $\mathbf{v}_0 = (2400 \text{ ft/s})\mathbf{i} - (3000 \text{ ft/s})\mathbf{j} + (3200 \text{ ft/s})\mathbf{k}$ relative to the probe. Knowing that the meteorite emerges on the other side of the panel with no change in the direction of its velocity, but with a speed reduced by 20 percent, determine the final angular velocity of the probe.

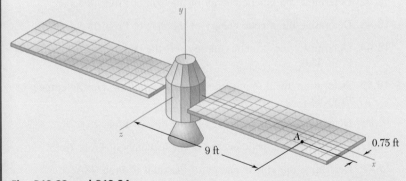

Fig. P18.33 and P18.34

18.34 The coordinate axes shown represent the principal centroidal axes of inertia of a 3000-lb space probe whose radii of gyration are $k_x = 1.375$ ft, $k_y = 1.425$ ft, and $k_z = 1.250$ ft. The probe has no angular velocity when a 5-oz meteorite strikes one of its solar panels at point A and emerges on the other side of the panel with no change in the direction of its velocity, but with a speed reduced by 25 percent. Knowing that the final angular velocity of the probe is $\boldsymbol{\omega} = (0.05 \text{ rad/s})\mathbf{i} - (0.12 \text{ rad/s})\mathbf{j} + \omega_z\mathbf{k}$ and that the x-component of the resulting change in the velocity of the mass center of the probe is -0.675 in./s, determine (a) the component ω_z of the final angular velocity of the probe, (b) the relative velocity $\mathbf{v}_0$ with which the meteorite strikes the panel.

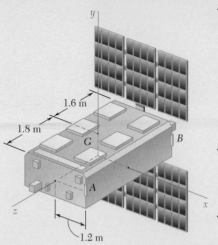

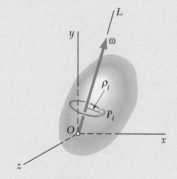

Fig. P18.35

Fig. P18.38

18.35 A 1200-kg satellite designed to study the sun has an angular velocity of $\boldsymbol{\omega}_0 = (0.050 \text{ rad/s})\mathbf{i} + (0.075 \text{ rad/s})\mathbf{k}$ when two small jets are activated at A and B in a direction parallel to the y axis. Knowing that the coordinate axes are principal centroidal axes, that the radii of gyration of the satellite are $\bar{k}_x = 1.120$ m, $\bar{k}_y = 1.200$ m, and $\bar{k}_z = 0.900$ m, and that each jet produces a 50-N thrust, determine (a) the required operating time of each jet if the angular velocity of the satellite is to be reduced to zero, (b) the resulting change in the velocity of the mass center G.

18.36 If jet A in Prob. 18.35 is inoperative, determine (a) the required operating time of jet B to reduce the x-component of the angular velocity of the satellite to zero, (b) the resulting final angular velocity, (c) the resulting change in the velocity of the mass center G.

18.37 Denoting, respectively, by $\boldsymbol{\omega}$, $\mathbf{H}_O$, and T the angular velocity, the angular momentum, and the kinetic energy of a rigid body with a fixed point O, (a) prove that $\mathbf{H}_O \cdot \boldsymbol{\omega} = 2T$; (b) show that the angle θ between $\boldsymbol{\omega}$ and $\mathbf{H}_O$ will always be acute.

18.38 Show that the kinetic energy of a rigid body with a fixed point O can be expressed as $T = \frac{1}{2}I_{OL}\omega^2$, where $\boldsymbol{\omega}$ is the instantaneous angular velocity of the body and I_{OL} is its moment of inertia about the line of action OL of $\boldsymbol{\omega}$. Derive this expression (a) from Eqs. (9.46) (or Eq. B.19 in the Appendix) and (18.19), (b) by considering T as the sum of the kinetic energies of particles P_i describing circles of radius ρ_i about line OL.

18.39 Determine the kinetic energy of the disk of Prob. 18.1.

18.40 Determine the kinetic energy of the plate of Prob. 18.2.

18.41 Determine the kinetic energy of the assembly of Prob. 18.3.

18.42 Determine the kinetic energy of the disk of Prob. 18.4.

18.43 Determine the kinetic energy of the disk of Prob. 18.5.

18.44 Determine the kinetic energy of the solid parallelepiped of Prob. 18.6.

18.45 Determine the kinetic energy of the hollow parallelepiped of Prob. 18.7.

18.46 Determine the kinetic energy of the disk of Prob. 18.8.

18.47 Determine the kinetic energy of the assembly of Prob. 18.15.

18.48 Determine the kinetic energy of the shaft of Prob. 18.17.

18.49 Determine the kinetic energy of the assembly of Prob. 18.19.

18.50 Determine the kinetic energy imparted to the cube of Prob. 18.21.

18.51 Determine the kinetic energy lost when edge C of the plate of Prob. 18.29 hits the obstruction.

18.52 Determine the kinetic energy lost when the plate of Prob. 18.31 hits the obstruction at B.

18.53 Determine the kinetic energy of the space probe of Prob. 18.33 in its motion about its mass center after its collision with the meteorite.

18.54 Determine the kinetic energy of the space probe of Prob. 18.34 in its motion about its mass center after its collision with the meteorite.

*18.2 MOTION OF A RIGID BODY IN THREE DIMENSIONS

As indicated in Sec. 18.1A, the fundamental equations

$$\Sigma \mathbf{F} = m\bar{\mathbf{a}} \qquad \textbf{(18.1)}$$

$$\Sigma \mathbf{M}_G = \dot{\mathbf{H}}_G \qquad \textbf{(18.2)}$$

remain valid in the most general case of the motion of a rigid body. Before we could apply Eq. (18.2) to the three-dimensional motion of a rigid body, however, it was necessary to derive Eqs. (18.7), which relate the components of the angular momentum $\mathbf{H}_G$ and those of the angular velocity $\boldsymbol{\omega}$. It still remains for us to find an effective and convenient way to compute the components of the derivative $\dot{\mathbf{H}}_G$ of the angular momentum. In this section, we do that first and then show how we can use the results to analyze motion of a rigid body in space.

18.2A Rate of Change of Angular Momentum

The notation $\mathbf{H}_G$ represents the angular momentum of a rigid body in its motion relative to centroidal axes $GX'Y'Z'$ with a fixed orientation (Fig. 18.9). Since $\dot{\mathbf{H}}_G$ represents the rate of change of $\mathbf{H}_G$ with respect to the same axes, it would seem natural to use components of $\boldsymbol{\omega}$ and $\mathbf{H}_G$ along the axes X', Y', Z' in writing the relations of Eq. (18.7). However, since the body rotates, its moments and products of inertia change continually, and it would be necessary to determine their values as functions of time. It is therefore more convenient to use axes x, y, z attached to the body, ensuring that its moments and products of inertia maintain the same values during the motion. The angular velocity $\boldsymbol{\omega}$, however, still should be *defined* with respect to the frame $GX'Y'Z'$ with a fixed orientation. We can then *resolve* the vector $\boldsymbol{\omega}$ into components along the rotating $x, y,$ and z axes. Applying the relations of Eq. (18.7), we obtain the *components* of vector $\mathbf{H}_G$ along the rotating axes. Vector $\mathbf{H}_G$, however, represents the angular momentum about G of the body *in its motion relative to the frame $GX'Y'Z'$*.

Differentiating the components of the angular momentum in Eq. (18.7) with respect to t, we define the rate of change of vector $\mathbf{H}_G$ with respect to the rotating frame $Gxyz$ as

$$(\dot{\mathbf{H}}_G)_{Gxyz} = \dot{H}_x\mathbf{i} + \dot{H}_y\mathbf{j} + \dot{H}_z\mathbf{k} \qquad \textbf{(18.21)}$$

where $\mathbf{i}$, $\mathbf{j}$, and $\mathbf{k}$ are the unit vectors along the rotating axes. Recall from Sec. 15.5A that the rate of change $\dot{\mathbf{H}}_G$ of vector $\mathbf{H}_G$ with respect to the frame $GX'Y'Z'$ is found by adding $(\dot{\mathbf{H}}_G)_{Gxyz}$ to the vector product $\boldsymbol{\Omega} \times \mathbf{H}_G$, where $\boldsymbol{\Omega}$ denotes the angular velocity of the rotating frame. That is,

$$\dot{\mathbf{H}}_G = (\dot{\mathbf{H}}_G)_{Gxyz} + \boldsymbol{\Omega} \times \mathbf{H}_G \qquad \textbf{(18.22)}$$

where $\mathbf{H}_G$ = angular momentum of the body with respect to frame $GX'Y'Z'$ with a fixed orientation

$(\dot{\mathbf{H}}_G)_{Gxyz}$ = rate of change of $\mathbf{H}_G$ with respect to rotating frame $Gxyz$ to be computed from the relations in Eqs. (18.7) and (18.21)

$\boldsymbol{\Omega}$ = angular velocity of rotating frame $Gxyz$

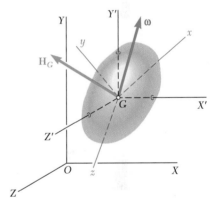

Fig. 18.9 Angular velocity and angular momentum of a rigid body with centroidal axes $X'Y'Z'$ of fixed orientation and centroidal axes xyz attached to the body.

Substituting for $\dot{\mathbf{H}}_G$ from Eq. (18.22) into Eq. (18.2), we have

$$\Sigma\mathbf{M}_G = (\dot{\mathbf{H}}_G)_{Gxyz} + \boldsymbol{\Omega} \times \mathbf{H}_G \tag{18.23}$$

If the rotating frame is attached to the body as we have assumed in this discussion, its angular velocity $\boldsymbol{\Omega}$ is identically equal to the angular velocity $\boldsymbol{\omega}$ of the body. In many applications, however, it is advantageous to use a frame of reference that is not actually attached to the body but rotates in an independent manner. For example, if the body considered is axisymmetric, as in Sample Prob. 18.5 or Sec. 18.3, it is possible to select a frame of reference where the moments and products of inertia of the body remain constant, but which rotate less than the body itself. As a result, it is possible to obtain simpler expressions for the angular velocity $\boldsymbol{\omega}$ and the angular momentum $\mathbf{H}_G$ of the body than we could have obtained if the frame of reference had actually been attached to the body. It is clear that in such cases the angular velocity $\boldsymbol{\Omega}$ of the rotating frame and the angular velocity $\boldsymbol{\omega}$ of the body are different.

*18.2B   Euler's Equations of Motion

If we choose the x, y, and z axes to coincide with the principal axes of inertia of the body, we can use the simplified relations in Eq. (18.10) to determine the components of the angular momentum $\mathbf{H}_G$. Omitting the primes from the subscripts, we have

$$\mathbf{H}_G = \bar{I}_x\omega_x\mathbf{i} + \bar{I}_y\omega_y\mathbf{j} + \bar{I}_z\omega_z\mathbf{k} \tag{18.24}$$

where $\bar{I}_x$, $\bar{I}_y$, and $\bar{I}_z$ denote the principal centroidal moments of inertia of the body. Substituting for $\mathbf{H}_G$ from Eq. (18.24) into Eq. (18.23) and setting $\boldsymbol{\Omega} = \boldsymbol{\omega}$, we obtain the three scalar equations:

**Euler's equations
of motion**

$$\begin{aligned}
\Sigma M_x &= \bar{I}_x\dot{\omega}_x - (\bar{I}_y - \bar{I}_z)\omega_y\omega_z \\
\Sigma M_y &= \bar{I}_y\dot{\omega}_y - (\bar{I}_z - \bar{I}_x)\omega_z\omega_x \\
\Sigma M_z &= \bar{I}_z\dot{\omega}_z - (\bar{I}_x - \bar{I}_y)\omega_x\omega_y
\end{aligned} \tag{18.25}$$

We can use these equations, called **Euler's equations of motion** after the Swiss mathematician Leonhard Euler (1707–1783), to analyze the motion of a rigid body about its mass center. In the following sections, however, we will use Eq. (18.23) in preference to Eqs. (18.25), since Eq. (18.23) is more general, and the compact vectorial form in which it is expressed is easier to remember.

Writing Eq. (18.1) in scalar form, we obtain the three additional equations of

$$\Sigma F_x = m\bar{a}_x \qquad \Sigma F_y = m\bar{a}_y \qquad \Sigma F_z = m\bar{a}_z \tag{18.26}$$

Together with Euler's equations, these form a system of six differential equations. Given appropriate initial conditions, these differential equations have a unique solution. Thus, the motion of a rigid body in three dimensions is completely defined by the resultant and the moment resultant of the external forces acting on it. This result is a generalization of a similar result obtained in Sec. 16.1C in the case of the plane motion of a rigid body. It follows that in three as well as in two dimensions, two systems of forces that are equipollent are also equivalent; that is, they have the same effect on a given rigid body.

Considering in particular the system of the external forces acting on a rigid body (Fig. 18.10a) and the system of the inertial terms associated with the particles forming the rigid body (Fig. 18.10b), we can state that the two systems—which were shown in Sec. 14.1A to be equipollent—are also equivalent. Replacing the inertia terms in Fig. 18.10b by $m\bar{\mathbf{a}}$ and $\dot{\mathbf{H}}_G$, we can verify that the system of the external forces acting on a rigid body in three-dimensional motion is equivalent to the system consisting of the vector $m\bar{\mathbf{a}}$ attached at the mass center G of the body and the couple of moment $\dot{\mathbf{H}}_G$ (Fig. 18.11), where we obtain $\dot{\mathbf{H}}_G$ from the relations in Eqs. (18.7) and (18.22). Note that the equivalence of the systems of vectors shown in Figs. 18.10 and 18.11 has been indicated by *red* equals signs. You can solve problems involving the three-dimensional motion of a rigid body by considering the free-body diagram and kinetic diagram represented in Fig. 18.11 and by writing appropriate scalar equations relating the components or moments of the external forces and the inertial terms (see Sample Prob. 18.3).

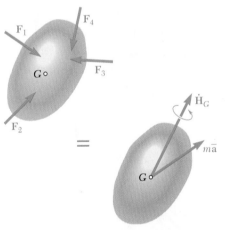

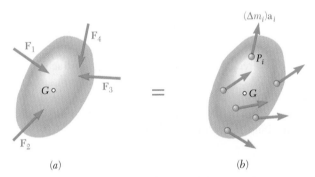

Fig. 18.10 (a) The system of external forces acting on a rigid body is equivalent to (b) the system of inertia terms associated with the particles of the rigid body.

Fig. 18.11 The free-body diagram and kinetic diagram show that the system of external forces is equivalent to the system consisting of the vectors $m\bar{\mathbf{a}}$ attached at the mass center G and $\dot{\mathbf{H}}_G$.

*18.2C Motion of a Rigid Body About a Fixed Point

If we want to analyze the motion of a rigid body constrained to rotate about a fixed point O, it is useful to write an equation involving the moments about O of the external forces and of the inertial terms, since this equation contains the unknown reaction at O. Although we can obtain such an equation from Fig. 18.11, it may be more convenient to write it by considering the rate of change of the angular momentum $\mathbf{H}_O$ of the body about the fixed point O (Fig. 18.12). Recalling Eq. (14.11), we have

$$\Sigma\mathbf{M}_O = \dot{\mathbf{H}}_O \qquad (18.27)$$

where $\dot{\mathbf{H}}_O$ denotes the rate of change of the vector $\mathbf{H}_O$ with respect to the fixed frame $OXYZ$. A derivation similar to that used in Sec. 18.2A enables

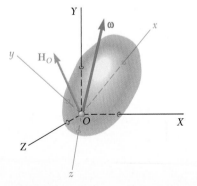

Fig. 18.12 Angular velocity and angular momentum of a rigid body rotating about a fixed point.

Photo 18.3 The revolving radio telescope is an example of a structure constrained to rotate about a fixed point.

us to relate $\dot{\mathbf{H}}_O$ to the rate of change $(\dot{\mathbf{H}}_O)_{Oxyz}$ of $\mathbf{H}_O$ with respect to the rotating frame $Oxyz$. Substitution into Eq. (18.27) leads to

$$\Sigma \mathbf{M}_O = (\dot{\mathbf{H}}_O)_{Oxyz} + \mathbf{\Omega} \times \mathbf{H}_O \qquad (18.28)$$

where $\Sigma \mathbf{M}_O$ = sum of moments about O of forces applied to the rigid body

$\mathbf{H}_O$ = angular momentum of the body with respect to fixed frame $OXYZ$

$(\dot{\mathbf{H}}_O)_{Oxyz}$ = rate of change of $\mathbf{H}_O$ with respect to rotating frame $Oxyz$ to be computed from relations in Eq. (18.13)

$\mathbf{\Omega}$ = angular velocity of rotating frame $Oxyz$

If the rotating frame is attached to the body, its angular velocity $\mathbf{\Omega}$ is identically equal to the angular velocity $\boldsymbol{\omega}$ of the body. However, as indicated in the last paragraph of Sec. 18.2A, in many applications it is advantageous to use a frame of reference that is not actually attached to the body but rotates in an independent manner.

*18.2D   Rotation of a Rigid Body About a Fixed Axis

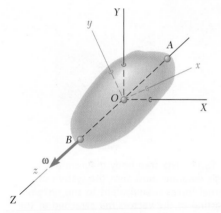

Fig. 18.13 Angular velocity of a rigid body rotating about a fixed axis AB.

We can use Eq. (18.28) to analyze the motion of a rigid body constrained to rotate about a fixed axis AB (Fig. 18.13). First, we note that the angular velocity of the body with respect to the fixed frame $OXYZ$ is represented by the vector $\boldsymbol{\omega}$ directed along the axis of rotation. Attaching the moving frame of reference $Oxyz$ to the body, with the z axis along AB, we have $\boldsymbol{\omega} = \omega \mathbf{k}$. Substituting $\omega_x = 0$, $\omega_y = 0$, and $\omega_z = \omega$ into the relations of Eq. (18.13), we obtain the components along the rotating axes of the angular momentum $\mathbf{H}_O$ of the body about O as

$$H_x = -I_{xz}\omega \qquad H_y = -I_{yz}\omega \qquad H_z = I_z\omega$$

Since the frame $Oxyz$ is attached to the body, we have $\mathbf{\Omega} = \boldsymbol{\omega}$, and Eq. (18.28) yields

$$\Sigma \mathbf{M}_O = (\dot{\mathbf{H}}_O)_{Oxyz} + \boldsymbol{\omega} \times \mathbf{H}_O$$
$$= (-I_{xz}\mathbf{i} - I_{yz}\mathbf{j} + I_z\mathbf{k})\dot{\omega} + \omega\mathbf{k} \times (-I_{xz}\mathbf{i} - I_{yz}\mathbf{j} + I_z\mathbf{k})\omega$$
$$= (-I_{xz}\mathbf{i} - I_{yz}\mathbf{j} + I_z\mathbf{k})\alpha + (-I_{xz}\mathbf{j} + I_{yz}\mathbf{i})\omega^2$$

We can express this result by the three scalar equations

$$\Sigma M_x = -I_{xz}\alpha + I_{yz}\omega^2$$
$$\Sigma M_y = -I_{yz}\alpha - I_{xz}\omega^2 \qquad (18.29)$$
$$\Sigma M_z = I_z\alpha$$

When the forces and moments applied to the body are known, you can obtain the angular acceleration α from the last of Eqs. (18.29). You can then determine the angular velocity ω by integration and substitute the values obtained for α and ω into the first two of Eqs. (18.29). You can then use these equations plus the three equations (18.26) that define the motion of the mass center of the body to determine the reactions at the bearings A and B.

It is possible to select axes other than those shown in Fig. 18.13 to analyze the rotation of a rigid body about a fixed axis. In many cases, the principal axes of inertia of the body will be more advantageous. It is therefore a good idea to revert to Eq. (18.28) and select the system of axes that best fits the problem under consideration.

If the rotating body is symmetrical with respect to the xy plane, the products of inertia I_{xz} and I_{yz} are equal to zero. Then Eqs. (18.29) reduce to

$$\Sigma M_x = 0 \qquad \Sigma M_y = 0 \qquad \Sigma M_z = I_z\alpha \qquad \textbf{(18.30)}$$

which is in agreement with the results obtained in Chap. 16. If, on the other hand, the products of inertia I_{xz} and I_{yz} are different from zero, the sum of the moments of the external forces about the x and y axes are also different from zero, even when the body rotates at a constant rate ω. Indeed, in this case, Eqs. (18.29) yield

$$\Sigma M_x = I_{yz}\omega^2 \qquad \Sigma M_y = -I_{xz}\omega^2 \qquad \Sigma M_z = 0 \qquad \textbf{(18.31)}$$

This last observation leads us to discuss the **balancing of rotating shafts.** Consider, for instance, the crankshaft shown in Fig. 18.14a that is symmetrical about its mass center G. We first observe that, when the crankshaft is at rest, it exerts no lateral thrust on its supports, since its center of gravity G is located directly above A. The shaft is said to be *statically balanced.* The reaction at A is often referred to as a *static reaction* and is vertical, and its magnitude is equal to the weight W of the shaft. Let us now assume that the shaft rotates with a constant angular velocity $\boldsymbol{\omega}$. Attaching our frame of reference to the shaft with its origin at G, the z axis along AB, and the y axis in the plane of symmetry of the shaft (Fig. 18.14b), we note that I_{xz} is zero and that I_{yz} is positive. According to Eqs. (18.31), there is an inertial term $I_{yz}\omega^2\mathbf{i}$. Summing the moments about G in the x direction and applying Eq. (18.31), we have

$$\mathbf{A}_y = \frac{I_{yz}\omega^2}{l}\mathbf{j} \qquad \mathbf{B} = -\frac{I_{yz}\omega^2}{l}\mathbf{j} \qquad \textbf{(18.32)}$$

Since the bearing reactions are proportional to ω^2, the shaft has a tendency to tear away from its bearings when rotating at high speeds. Moreover, since the bearing reactions $\mathbf{A}_y$ and $\mathbf{B}$, which are called *dynamic reactions,* are contained in the yz plane, they rotate with the shaft and cause the structure supporting it to vibrate. These undesirable effects can be avoided if, by rearranging the distribution of mass around the shaft or by adding corrective masses, we let I_{yz} become equal to zero. Then the dynamic reactions $\mathbf{A}_y$ and $\mathbf{B}$ vanish and the reactions at the bearings reduce to the static reaction $\mathbf{A}_z$—the direction of which is fixed. The shaft is then **dynamically as well as statically balanced.**

Photo 18.4 The rotating automobile crankshaft causes static and dynamic reactions on its bearings. The crankshaft can be designed to minimize dynamic imbalances and reduce these reaction forces.

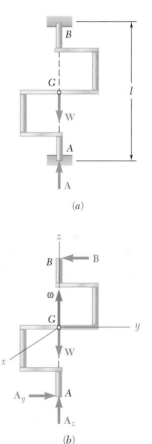

Fig. 18.14 (a) The crankshaft at rest is statically balanced; (b) the crankshaft rotating with constant angular velocity may or may not be dynamically balanced.

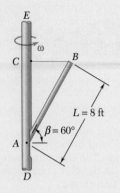

Sample Problem 18.3

A slender rod AB with a length of $L = 8$ ft and a weight of $W = 40$ lb is pinned at A to a vertical axle DE that rotates with a constant angular velocity $\boldsymbol{\omega}$ of 15 rad/s. The rod is maintained in position by means of a horizontal wire BC attached to the axle and to end B of the rod. Determine the tension in the wire and the reaction at A.

STRATEGY: Since you have a rigid body that is not symmetrical with respect to the plane of motion, you need to use the three-dimensional form of Newton's second law.

MODELING: Choose the rod AB as your system. The angular velocity is shown in Fig. 1, and the free-body and kinetic diagrams consisting of the vector $m\bar{\mathbf{a}}$ attached at G and the couple $\dot{\mathbf{H}}_G$ are shown in Fig. 2.

ANALYSIS: Since G describes a horizontal circle with a radius of $\bar{r} = \frac{1}{2}L\cos\beta$ and BG rotates at the constant rate ω (Fig. 1), you have

$$\bar{\mathbf{a}} = \mathbf{a}_n = -\bar{r}\omega^2\mathbf{I} = -(\tfrac{1}{2}L\cos\beta)\omega^2\mathbf{I} = -(450\ \text{ft/s}^2)\mathbf{I}$$

$$m\bar{\mathbf{a}} = \frac{40}{g}(-450\mathbf{I}) = -(559\ \text{lb})\mathbf{I}$$

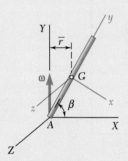

Fig. 1 Angular velocity of the rod.

Determination of $\dot{\mathbf{H}}_G$. First compute the angular momentum $\mathbf{H}_G$. Using the principal centroidal axes of inertia x, y, and z, you have

$$\bar{I}_x = \tfrac{1}{12}mL^2 \qquad \bar{I}_y = 0 \qquad \bar{I}_z = \tfrac{1}{12}mL^2$$

$$\omega_x = -\omega\cos\beta \qquad \omega_y = \omega\sin\beta \qquad \omega_z = 0$$

$$\mathbf{H}_G = \bar{I}_x\omega_x\mathbf{i} + \bar{I}_y\omega_y\mathbf{j} + \bar{I}_z\omega_z\mathbf{k}$$

$$\mathbf{H}_G = -\tfrac{1}{12}mL^2\omega\cos\beta\,\mathbf{i}$$

Obtain the rate of change $\dot{\mathbf{H}}_G$ of $\mathbf{H}_G$ with respect to axes of fixed orientation from Eq. (18.22). Observe that the rate of change $(\dot{\mathbf{H}}_G)_{Gxyz}$ of $\mathbf{H}_G$ with respect to the rotating frame $Gxyz$ is zero and that the angular velocity $\boldsymbol{\Omega}$ of that frame is equal to the angular velocity $\boldsymbol{\omega}$ of the rod. Thus, you have

$$\dot{\mathbf{H}}_G = (\dot{\mathbf{H}}_G)_{Gxyz} + \boldsymbol{\omega} \times \mathbf{H}_G$$

$$\dot{\mathbf{H}}_G = 0 + (-\omega\cos\beta\,\mathbf{i} + \omega\sin\beta\,\mathbf{j}) \times (-\tfrac{1}{12}mL^2\omega\cos\beta\,\mathbf{i})$$

$$\dot{\mathbf{H}}_G = \tfrac{1}{12}mL^2\omega^2\sin\beta\cos\beta\,\mathbf{k} = (645\ \text{lb·ft})\,\mathbf{k}$$

Equations of Motion. The system of the external forces is equivalent to the inertia terms (Fig. 2). This gives

$$\Sigma\mathbf{M}_A = \dot{\mathbf{H}}_A = \mathbf{r} \times m\bar{\mathbf{a}} + \dot{\mathbf{H}}_G:$$

$$6.93\mathbf{J} \times (-T\mathbf{I}) + 2\mathbf{I} \times (-40\mathbf{J}) = 3.46\mathbf{J} \times (-559\mathbf{I}) + 645\mathbf{K}$$

$$(6.93T - 80)\mathbf{K} = (1934 + 645)\mathbf{K} \qquad T = 384\ \text{lb} \blacktriangleleft$$

$$\Sigma\mathbf{F} = m\bar{\mathbf{a}}: \qquad A_X\mathbf{I} + A_Y\mathbf{J} + A_Z\mathbf{K} - 384\mathbf{I} - 40\mathbf{J} = -559\mathbf{I}$$

$$\mathbf{A} = -(175\ \text{lb})\mathbf{I} + (40\ \text{lb})\mathbf{J} \blacktriangleleft$$

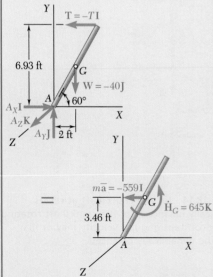

Fig. 2 Free-body diagram and kinetic diagram of the rod.

REFLECT and THINK: You could have obtained the value of T from $\mathbf{H}_A$ and Eq. (18.28). Even though the rod rotates with a constant angular velocity, the asymmetry of the rod causes a moment about the z axis. Note that we calculated the inertial term $\dot{\mathbf{H}}_A$ by adding $\mathbf{r} \times m\bar{\mathbf{a}}$ and the couple $\dot{\mathbf{H}}_G$.

Sample Problem 18.4

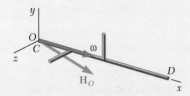

Two 100-mm rods A and B each have a mass of 300 g and are welded to shaft CD that is supported by bearings at C and D. If a couple $\mathbf{M}$ with a magnitude of 6 N·m is applied to the shaft, determine the components of the dynamic reactions at C and D at the instant when the shaft has reached an angular velocity of 1200 rpm. Neglect the moment of inertia of the shaft itself.

STRATEGY: Use the three-dimensional form of Newton's second law in the form of Eq. (18.28) for the case of rotation about a fixed axis, where $\Omega = \omega$.

MODELING: Choose the shaft and the two rods as your system. The angular momentum and angular velocity are shown in Fig. 1, and a free-body diagram is shown in Fig. 2.

Fig. 1 Angular momentum and angular velocity of the system.

ANALYSIS:

Angular Momentum About O. Attach the frame of reference $Oxyz$ to the body and note that the axes chosen are not principal axes of inertia for the body. Since the body rotates about the x axis, you know that $\omega_x = \omega$ and $\omega_y = \omega_z = 0$ (Fig. 1). Substituting into Eqs. (18.13), you have

$$H_x = I_x\omega \qquad H_y = -I_{xy}\omega \qquad H_z = -I_{xz}\omega$$
$$\mathbf{H}_O = (I_x\mathbf{i} - I_{xy}\mathbf{j} - I_{xz}\mathbf{k})\omega$$

Moments of the External Forces About O. Since the frame of reference rotates with the angular velocity $\boldsymbol{\omega}$ and the only angular acceleration term is $\alpha_x = \alpha$, Eq. (18.28) gives

$$\Sigma\mathbf{M}_O = (\dot{\mathbf{H}}_O)_{Oxyz} + \boldsymbol{\omega} \times \mathbf{H}_O$$
$$= (I_x\mathbf{i} - I_{xy}\mathbf{j} - I_{xz}\mathbf{k})\alpha + \omega\mathbf{i} \times (I_x\mathbf{i} - I_{xy}\mathbf{j} - I_{xz}\mathbf{k})\omega$$
$$= I_x\alpha\mathbf{i} - (I_{xy}\alpha - I_{xz}\omega^2)\mathbf{j} - (I_{xz}\alpha + I_{xy}\omega^2)\mathbf{k} \qquad (1)$$

Dynamic Reaction at D. The external forces consist of the weights of the shaft and rods, the couple $\mathbf{M}$, the static reactions at C and D, and the dynamic reactions at C and D. Since the weights and static reactions are balanced, the external forces reduce to the couple $\mathbf{M}$ and the dynamic reactions $\mathbf{C}$ and $\mathbf{D}$, as shown in Fig. 2. Taking moments about O, you have

$$\Sigma\mathbf{M}_O = L\mathbf{i} \times (D_y\mathbf{j} + D_z\mathbf{k}) + M\mathbf{i} = M\mathbf{i} - D_zL\mathbf{j} + D_yL\mathbf{k} \qquad (2)$$

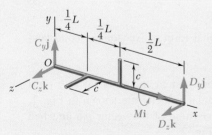

Fig. 2 Free-body diagram for the system.

Equating the coefficients of the unit vector $\mathbf{i}$ in Eqs. (1) and (2) gives

$$M = I_x\alpha \qquad M = 2(\tfrac{1}{3}mc^2)\alpha \qquad \alpha = 3M/2mc^2$$

Equating the coefficients of $\mathbf{k}$ and $\mathbf{j}$ in Eqs. (1) and (2) provides

$$D_y = -(I_{xz}\alpha + I_{xy}\omega^2)/L \qquad D_z = (I_{xy}\alpha - I_{xz}\omega^2)/L \qquad (3)$$

Using the parallel-axis theorem and noting that the product of inertia of each rod is zero with respect to their own centroidal axes, you obtain

$$I_{xy} = \Sigma m \bar{x} \bar{y} = m(\tfrac{1}{2}L)(\tfrac{1}{2}c) = \tfrac{1}{4}mLc$$
$$I_{xz} = \Sigma m \bar{x} \bar{z} = m(\tfrac{1}{4}L)(\tfrac{1}{2}c) = \tfrac{1}{8}mLc$$

Substituting into Eq. (3) the values found for I_{xy}, I_{xz}, and α gives

$$D_y = -\tfrac{3}{16}(M/c) - \tfrac{1}{4}mc\omega^2 \qquad D_z = \tfrac{3}{8}(M/c) - \tfrac{1}{8}mc\omega^2$$

Substituting $\omega = 1200$ rpm $= 125.7$ rad/s, $c = 0.100$ m, $M = 6$ N·m, and $m = 0.300$ kg, you have

$$D_y = -129.8 \text{ N} \qquad D_z = -36.8 \text{ N} \quad \blacktriangleleft$$

Dynamic Reaction at C. Using a frame of reference attached at D, you obtain equations similar to Eqs. (3) that yield

$$C_y = -152.2 \text{ N} \qquad C_z = -155.2 \text{ N} \quad \triangle$$

REFLECT and THINK: The dynamic forces are larger at C than at D. Rod A is closer to this end of the bar, so you would expect it to affect this end more than the other end. Note that two small 300-g rods end up causing forces of over 150 N. You often have to account for these large forces when designing mechanical systems involving rotary equipment (e.g., automobiles, turbines, mills).

Sample Problem 18.5

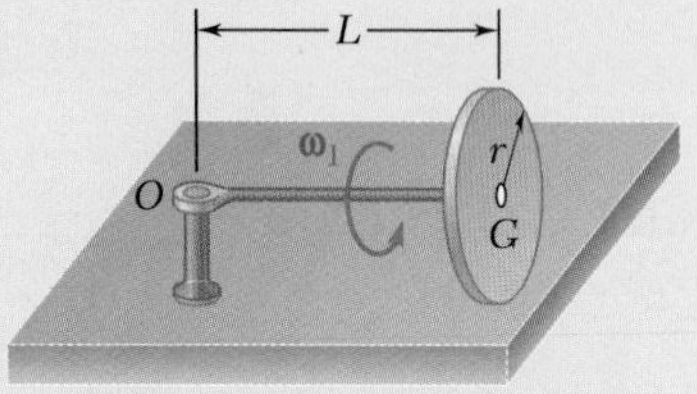

A homogeneous disk with radius r and mass m is mounted on an axle OG with length L and a negligible mass. The axle is pivoted at the fixed point O, and the disk is constrained to roll on a horizontal surface. The disk rotates counterclockwise at the constant rate ω_1 about the axle. Determine (a) the force (assumed vertical) exerted by the floor on the disk, (b) the reaction at the pivot O.

STRATEGY: Use the three-dimensional form of Newton's second law; that is, Eqs (18.1) and (18.2).

MODELING: Choose the disk as your system and model it as a rigid body. The angular momentum and angular velocity are shown in Fig. 1, and free-body and kinetic diagrams consisting of the vector $m\bar{\mathbf{a}}$ attached at G and the couple $\dot{\mathbf{H}}_G$ are shown in Fig. 2.

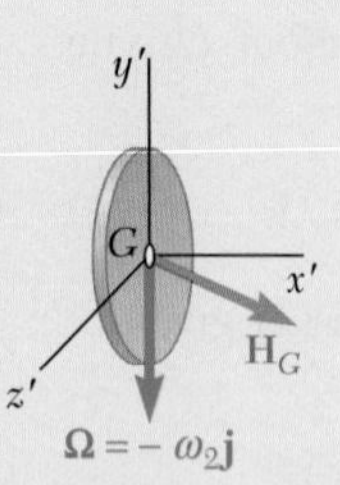

ANALYSIS: From Sample Prob. 18.2, the axle rotates about the y axis at the rate $\omega_2 = r\omega_1/L$, so you have

$$m\bar{\mathbf{a}} = -mL\omega_2^2 \mathbf{i} = -mL(r\omega_1/L)^2 \mathbf{i} = -(mr^2\omega_1^2/L)\mathbf{i} \tag{1}$$

Fig. 1 The angular momentum and angular velocity of the disk.

(continued)

Determination of $\dot{\mathbf{H}}_G$. Allow the x, y, z axes to rotate with the bar OG but not with the disk; the x', y', z' axes rotate with both the bar and the disk. Recall from Sample Prob. 18.2 that the angular momentum of the disk about G is

$$\mathbf{H}_G = \tfrac{1}{2}mr^2\omega_1\left(\mathbf{i} - \frac{r}{2L}\mathbf{j}\right)$$

where $\mathbf{H}_G$ is resolved into components along the rotating axes $x', y', z'; x'$ is along OG; and y' is vertical at the instant shown (Fig. 1). Obtain the rate of change $\dot{\mathbf{H}}_G$ of $\mathbf{H}_G$ with respect to axes of fixed orientation from Eq. (18.22). Note that the rate of change $(\dot{\mathbf{H}}_G)_{Gx'y'z'}$ of $\mathbf{H}_G$ with respect to the rotating frame is zero and that the angular velocity $\boldsymbol{\Omega}$ of that frame is

$$\boldsymbol{\Omega} = -\omega_2\mathbf{j} = -\frac{r\omega_1}{L}\mathbf{j}$$

Then you have

$$
\begin{aligned}
\dot{\mathbf{H}}_G &= (\dot{\mathbf{H}}_G)_{Gx'y'z'} + \boldsymbol{\Omega} \times \mathbf{H}_G \\
&= 0 - \frac{r\omega_1}{L}\mathbf{j} \times \tfrac{1}{2}mr^2\omega_1\left(\mathbf{i} - \frac{r}{2L}\mathbf{j}\right) \\
&= \tfrac{1}{2}mr^2(r/L)\omega_1^2\mathbf{k}
\end{aligned}
\tag{2}
$$

Equations of Motion. The system of the external forces is equivalent to the system of the inertial terms (Fig. 2), so you have

$$\Sigma\mathbf{M}_O = \dot{\mathbf{H}}_G: \qquad L\mathbf{i} \times (N\mathbf{j} - W\mathbf{j}) = \dot{\mathbf{H}}_G$$

$$(N - W)L\mathbf{k} = \tfrac{1}{2}mr^2(r/L)\omega_1^2\mathbf{k}$$

$$N = W + \tfrac{1}{2}mr(r/L)^2\omega_1^2 \qquad \mathbf{N} = [W + \tfrac{1}{2}mr(r/L)^2\omega_1^2]\mathbf{j} \quad (3) \quad \blacktriangleleft$$

$$\Sigma\mathbf{F} = m\overline{\mathbf{a}}: \qquad \mathbf{R} + N\mathbf{j} - W\mathbf{j} = m\overline{\mathbf{a}}$$

Substituting for N from Eq. (3), for $m\overline{\mathbf{a}}$ from Eq. (1), and solving for $\mathbf{R}$, you have

$$\mathbf{R} = -(mr^2\omega_1^2/L)\mathbf{i} - \tfrac{1}{2}mr(r/L)^2\omega_1^2\mathbf{j}$$

$$\mathbf{R} = -\frac{mr^2\omega_1^2}{L}\left(\mathbf{i} + \frac{r}{2L}\mathbf{j}\right) \quad \blacktriangleleft$$

REFLECT and THINK: This is a case where the coordinate system attached to the rotating object has its own angular velocity. The change in direction of the angular momentum of the disk ends up increasing the normal force.

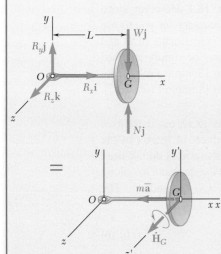

Fig. 2 Free-body diagram and kinetic diagram for the system.

SOLVING PROBLEMS ON YOUR OWN

In this section, you were asked to solve problems involving the three-dimensional motion of rigid bodies. The method you used is basically the same one you used in Chap. 16 in your study of the plane motion of rigid bodies. You made free-body and kinetic diagrams showing that the system of the external forces is equivalent to the system of the inertia terms. You equated sums of components and sums of moments on both sides of this equation. Now, however, the system of the inertia terms is represented by the vector $m\bar{\mathbf{a}}$ and a couple vector $\dot{\mathbf{H}}_G$, which are explained next in paragraphs 1 and 2.

To solve a problem involving the three-dimensional motion of a rigid body, you should take the following steps.

1. **Determine the angular momentum $\mathbf{H}_G$ of the body about its mass center G** from its angular velocity $\boldsymbol{\omega}$ with respect to a frame of reference $GX'Y'Z'$ of fixed orientation. This is an operation you learned to perform in Sect. 18.1. However, since the configuration of the body is changing with time, it is now necessary for you to use an auxiliary system of axes $Gx'y'z'$ (Fig. 18.9) to compute the components of $\boldsymbol{\omega}$ and the moments and products of inertia of the body. These axes may be rigidly attached to the body, in which case their angular velocity is equal to $\boldsymbol{\omega}$ [Sample Probs. 18.3 and 18.4], or they may have an angular velocity $\boldsymbol{\Omega}$ of their own [Sample Prob. 18.5].

   Recall the following ideas from the preceding section.

   a. If you know the principal axes of inertia of the body at G, use these axes as coordinate axes x', y', and z', since the corresponding products of inertia of the body are equal to zero. (Note that if the body is axisymmetric, these axes do not need to be rigidly attached to the body.) Resolve $\boldsymbol{\omega}$ into components $\omega_{x'}$, $\omega_{y'}$, and $\omega_{z'}$ along these axes and compute the principal moments of inertia $\bar{I}_{x'}$, $\bar{I}_{y'}$, and $\bar{I}_{z'}$. The corresponding components of the angular momentum $\mathbf{H}_G$ are

   $$H_{x'} = \bar{I}_{x'}\omega_{x'} \qquad H_{y'} = \bar{I}_{y'}\omega_{y'} \qquad H_{z'} = \bar{I}_{z'}\omega_{z'} \tag{18.10}$$

   b. If you do not know the principal axes of inertia of the body at G, you must use Eqs. (18.7) to determine the components of the angular momentum $\mathbf{H}_G$. These equations require your prior computation of the products of inertia of the body—as well as of its moments of inertia—with respect to the selected axes.

2. Compute the rate of change $\dot{\mathbf{H}}_G$ of the angular momentum $\mathbf{H}_G$ with respect to the frame $GX'Y'Z'$. Note that this frame has a *fixed orientation*, whereas the frame $Gx'y'z'$ you used when calculating the components of the vector $\boldsymbol{\omega}$ was a *rotating frame*. (Review the discussion in Sec. 15.5A of the rate of change of a vector with respect to a rotating frame.) Recalling Eq. (15.31), you can express the rate of change $\dot{\mathbf{H}}_G$ as

$$\dot{\mathbf{H}}_G = (\dot{\mathbf{H}}_G)_{Gx'y'z'} + \boldsymbol{\Omega} \times \mathbf{H}_G \tag{18.22}$$

The first term in the right-hand side of Eq. (18.22) represents the rate of change of $\mathbf{H}_G$ with respect to the rotating frame $Gx'y'z'$. This term drops out if $\boldsymbol{\omega}$—and thus $\mathbf{H}_G$—remains constant in both magnitude and direction when viewed from that frame. On the other hand, if any of the time derivatives $\dot{\omega}_{x'}$, $\dot{\omega}_{y'}$, or $\dot{\omega}_{z'}$ is different from zero, $(\dot{\mathbf{H}}_G)_{Gx'y'z'}$ is also different from zero, and you should determine its components by differentiating Eqs. (18.10) with respect to t. Finally, we remind you that if the rotating frame is rigidly attached to the body, its angular velocity is the same as that of the body, and $\boldsymbol{\Omega}$ can be replaced by $\boldsymbol{\omega}$.

3. Draw the free-body and kinetic diagrams for the rigid body showing that the system of the external forces exerted on the body is equivalent to the vector $m\bar{\mathbf{a}}$ applied at G and the couple vector $\dot{\mathbf{H}}_G$ (Fig. 18.11). By equating components in any direction and moments about any point, you can write as many as six independent scalar equations of motion [Sample Probs. 18.3 and 18.5].

4. When solving problems involving the motion of a rigid body about a fixed point O, you may find it convenient to use the following equation that was derived in Sec. 18.2C and eliminates the components of the reaction at the support O. So

$$\Sigma\mathbf{M}_O = (\dot{\mathbf{H}}_O)_{Oxyz} + \boldsymbol{\Omega} \times \mathbf{H}_O \tag{18.28}$$

Here the first term on the right-hand side represents the rate of change of $\mathbf{H}_O$ with respect to the rotating frame $Oxyz$, and $\boldsymbol{\Omega}$ is the angular velocity of that frame.

5. When determining the reactions at the bearings of a rotating shaft, use Eq. (18.28) and take the following steps.

   a. Place the fixed point O at one of the two bearings supporting the shaft and attach the rotating frame $Oxyz$ to the shaft with one of the axes directed along it. Assuming, for instance, that the x axis has been aligned with the shaft, you will have $\boldsymbol{\Omega} = \boldsymbol{\omega} = \omega\mathbf{i}$ [Sample Prob. 18.4].

   b. Since the selected axes are usually not the principal axes of inertia at O, you must compute the products of inertia of the shaft—as well as its moments of inertia—with respect to these axes and use Eqs. (18.13) to determine $\mathbf{H}_O$. Assuming again that the x axis has been aligned with the shaft, Eqs. (18.13) reduce to

$$H_x = I_x\omega \qquad H_y = -I_{yx}\omega \qquad H_z = -I_{zx}\omega \tag{18.13'}$$

These equations show that $\mathbf{H}_O$ is not directed along the shaft.

(continued)

c. To obtain $\dot{\mathbf{H}}_O$, **substitute these expressions into Eq. (18.28), and let** $\boldsymbol{\Omega} = \boldsymbol{\omega} = \omega\mathbf{i}$. If the angular velocity of the shaft is constant, the first term in the right-hand side of the equation drops out. However, if the shaft has an angular acceleration $\boldsymbol{\alpha} = \alpha\mathbf{i}$, the first term is not zero and must be determined by differentiating the expressions in Eq. (18.13′) with respect to t. The result will be equations similar to Eqs. (18.13′) with ω replaced by α. The result also can be expressed by the three scalar equations of Eq. (18.29).

d. Since point O **coincides with one of the bearings,** you can solve the three scalar equations corresponding to Eq. (18.28) for the components of the dynamic reaction at the other bearing. If the mass center G of the shaft is located on the line joining the two bearings, the inertial term $m\bar{\mathbf{a}}$ is zero. Drawing the free-body diagram and kinetic diagram of the shaft, you then observe that the components of the dynamic reaction at the first bearing must be equal and opposite to those you have just determined. If G is not located on the line joining the two bearings, you can determine the reaction at the first bearing by placing the fixed point O at the second bearing and repeating the earlier procedure [Sample Prob. 18.4]; or you can obtain additional equations of motion from the free-body and kinetic diagrams of the shaft, making sure to first determine and include the inertial term $m\bar{\mathbf{a}}$ applied at G.

e. Most problems call for the determination of the "dynamic reactions" at the bearings, that is, for the additional forces exerted by the bearings on the shaft when the shaft is rotating. When determining dynamic reactions, ignore the effect of static loads, such as the weight of the shaft.

Problems

18.55 Determine the rate of change $\dot{\mathbf{H}}_G$ of the angular momentum $\mathbf{H}_G$ of the disk of Prob. 18.1.

18.56 Determine the rate of change $\dot{\mathbf{H}}_G$ of the angular momentum $\mathbf{H}_G$ of the plate of Prob. 18.2.

18.57 Determine the rate of change $\dot{\mathbf{H}}_D$ of the angular momentum $\mathbf{H}_D$ of the assembly of Prob. 18.3.

18.58 Determine the rate of change $\dot{\mathbf{H}}_A$ of the angular momentum $\mathbf{H}_A$ of the disk of Prob. 18.4.

18.59 Determine the rate of change $\dot{\mathbf{H}}_C$ of the angular momentum $\mathbf{H}_C$ of the disk of Prob. 18.5.

18.60 Determine the rate of change $\dot{\mathbf{H}}_G$ of the angular momentum $\mathbf{H}_G$ of the disk of Prob. 18.8 for an arbitrary value of β, knowing that its angular velocity $\boldsymbol{\omega}$ remains constant.

18.61 Determine the rate of change $\dot{\mathbf{H}}_D$ of the angular momentum $\mathbf{H}_D$ of the assembly of Prob. 18.3, assuming that at the instant considered the assembly has an angular velocity $\boldsymbol{\omega} = (12 \text{ rad/s})\mathbf{i}$ and an angular acceleration $\boldsymbol{\alpha} = -(96 \text{ rad/s}^2)\mathbf{i}$.

18.62 Determine the rate of change $\dot{\mathbf{H}}_D$ of the angular momentum $\mathbf{H}_D$ of the assembly of Prob. 18.3, assuming that at the instant considered the assembly has an angular velocity $\boldsymbol{\omega} = (12 \text{ rad/s})\mathbf{i}$ and an angular acceleration $\boldsymbol{\alpha} = (96 \text{ rad/s}^2)\mathbf{i}$.

18.63 A thin, homogeneous square of mass m and side a is welded to a vertical shaft AB with which it forms an angle of 45°. Knowing that the shaft rotates with an angular velocity $\boldsymbol{\omega} = \omega\mathbf{j}$ and an angular acceleration $\boldsymbol{\alpha} = \alpha\mathbf{j}$, determine the rate of change $\dot{\mathbf{H}}_A$ of the angular momentum $\mathbf{H}_A$ of the plate assembly.

18.64 Determine the rate of change $\dot{\mathbf{H}}_G$ of the angular momentum $\mathbf{H}_G$ of the disk of Prob. 18.8 for an arbitrary value of β, knowing that the disk has an angular velocity $\boldsymbol{\omega} = \omega\mathbf{i}$ and an angular acceleration $\boldsymbol{\alpha} = \alpha\mathbf{i}$.

18.65 A slender, uniform rod AB of mass m and a vertical shaft CD, each of length $2b$, are welded together at their midpoints G. Knowing that the shaft rotates at the constant rate ω, determine the dynamic reactions at C and D.

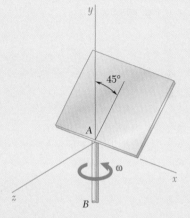

Fig. P18.63

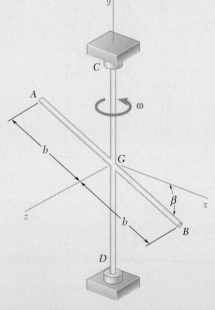

Fig. P18.65

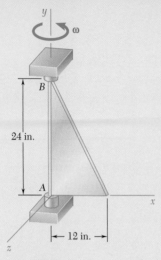

24 in.

12 in.

Fig. P18.66

18.66 A thin, homogeneous triangular plate of weight 10 lb is welded to a light, vertical axle supported by bearings at A and B. Knowing that the plate rotates at the constant rate $\omega = 8$ rad/s, determine the dynamic reactions at A and B.

18.67 The assembly shown consists of pieces of sheet aluminum of uniform thickness and of total weight 2.7 lb welded to a light axle supported by bearings at A and B. Knowing that the assembly rotates at the constant rate $\omega = 240$ rpm, determine the dynamic reactions at A and B.

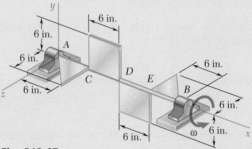

6 in.
6 in.
6 in.
6 in.
6 in.
6 in.
6 in.
6 in.

Fig. P18.67

18.68 The 8-kg shaft shown has a uniform cross section. Knowing that the shaft rotates at the constant rate $\omega = 12$ rad/s, determine the dynamic reactions at A and B.

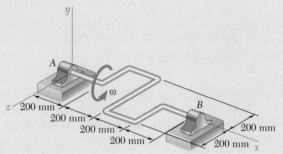

200 mm
200 mm
200 mm
200 mm
200 mm
200 mm

Fig. P18.68

18.69 After attaching the 18-kg wheel shown to a balancing machine and making it spin at the rate of 15 rev/s, a mechanic has found that to balance the wheel both statically and dynamically, he should use two corrective masses, a 170-g mass placed at B and a 56-g mass placed at D. Using a right-handed frame of reference rotating with the wheel (with the z axis perpendicular to the plane of the figure), determine before the corrective masses have been attached (a) the distance from the axis of rotation to the mass center of the wheel and the products of inertia I_{xy} and I_{zx}, (b) the force-couple system at C equivalent to the forces exerted by the wheel on the machine.

18.70 When the 18-kg wheel shown is attached to a balancing machine and made to spin at a rate of 12.5 rev/s, it is found that the forces exerted by the wheel on the machine are equivalent to a force-couple system consisting of a force $\mathbf{F} = (160 \text{ N})\mathbf{j}$ applied at C and a couple $\mathbf{M}_C = (14.7 \text{ N·m})\mathbf{k}$, where the unit vectors form a triad that rotates with the wheel. (a) Determine the distance from the axis of rotation to the mass center of the wheel and the products of inertia I_{xy} and I_{zx}. (b) If only two corrective masses are to be used to balance the wheel statically and dynamically, what should these masses be and at which of the points A, B, D, or E should they be placed?

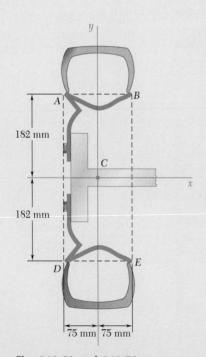

182 mm

182 mm

75 mm   75 mm

Fig. P18.69 and P18.70

18.71 Knowing that the assembly of Prob. 18.65 is initially at rest ($\omega = 0$) when a couple of moment $\mathbf{M}_0 = M_0\mathbf{j}$ is applied to shaft CD, determine (*a*) the resulting angular acceleration of the assembly, (*b*) the dynamic reactions at C and D immediately after the couple is applied.

18.72 Knowing that the plate of Prob. 18.66 is initially at rest ($\omega = 0$) when a couple of moment $\mathbf{M}_0 = (0.75 \text{ ft·lb})\mathbf{j}$ is applied to it, determine (*a*) the resulting angular acceleration of the plate, (*b*) the dynamic reactions A and B immediately after the couple has been applied.

18.73 The assembly of Prob. 18.67 is initially at rest ($\omega = 0$) when a couple $\mathbf{M}_0$ is applied to axle AB. Knowing that the resulting angular acceleration of the assembly is $\boldsymbol{\alpha} = (150 \text{ rad/s}^2)\mathbf{i}$, determine (*a*) the couple $\mathbf{M}_0$, (*b*) the dynamic reactions at A and B immediately after the couple is applied.

18.74 The shaft of Prob. 18.68 is initially at rest ($\omega = 0$) when a couple $\mathbf{M}_0$ is applied to it. Knowing that the resulting angular acceleration of the shaft is $\boldsymbol{\alpha} = (20 \text{ rad/s}^2)\mathbf{i}$, determine (*a*) the couple $\mathbf{M}_0$, (*b*) the dynamic reactions at A and B immediately after the couple is applied.

18.75 The assembly shown weighs 12 lb and consists of 4 thin 16-in.-diameter semicircular aluminum plates welded to a light 40-in.-long shaft AB. The assembly is at rest ($\omega = 0$) at time $t = 0$ when a couple $\mathbf{M}_0$ is applied to it as shown, causing the assembly to complete one full revolution in 2 s. Determine (*a*) the couple $\mathbf{M}_0$, (*b*) the dynamic reactions at A and B at $t = 0$.

18.76 For the assembly of Prob. 18.75, determine the dynamic reactions at A and B at $t = 2$ s.

18.77 The sheet-metal component shown is of uniform thickness and has a mass of 600 g. It is attached to a light axle supported by bearings at A and B located 150 mm apart. The component is at rest when it is subjected to a couple $\mathbf{M}_0$ as shown. If the resulting angular acceleration is $\boldsymbol{\alpha} = (12 \text{ rad/s}^2)\mathbf{k}$, determine (*a*) the couple $\mathbf{M}_0$, (*b*) the dynamic reactions A and B immediately after the couple has been applied.

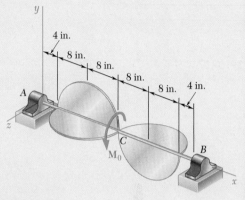

Fig. P18.75

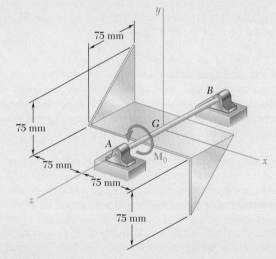

Fig. *P18.77*

18.78 For the sheet-metal component of Prob. 18.77, determine (*a*) the angular velocity of the component 0.6 s after the couple $\mathbf{M}_0$ has been applied to it, (*b*) the magnitude of the dynamic reactions at A and B at that time.

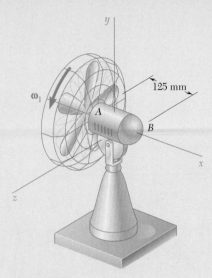

18.79 The blade of an oscillating fan and the rotor of its motor have a total mass of 300 g and a combined radius of gyration of 75 mm. They are supported by bearings at A and B, 125 mm apart, and rotate at the rate $\omega_1 = 1800$ rpm. Determine the dynamic reactions at A and B when the motor casing has an angular velocity $\boldsymbol{\omega}_2 = (0.6 \text{ rad/s})\mathbf{j}$.

18.80 The blade of a portable saw and the rotor of its motor have a total weight of 2.5 lb and a combined radius of gyration of 1.5 in. Knowing that the blade rotates as shown at the rate $\omega_1 = 1500$ rpm, determine the magnitude and direction of the couple $\mathbf{M}$ that a worker must exert on the handle of the saw to rotate it with a constant angular velocity $\boldsymbol{\omega}_2 = -(2.4 \text{ rad/s})\mathbf{j}$.

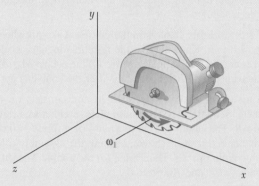

Fig. *P18.80*

18.81 The flywheel of an automobile engine, which is rigidly attached to the crankshaft, is equivalent to a 400-mm-diameter, 15-mm-thick steel plate. Determine the magnitude of the couple exerted by the flywheel on the horizontal crankshaft as the automobile travels around an unbanked curve of 200-m radius at a speed of 90 km/h, with the flywheel rotating at 2700 rpm. Assume the automobile to have (*a*) a rear-wheel drive with the engine mounted longitudinally, (*b*) a front-wheel drive with the engine mounted transversely. (Density of steel = 7860 kg/m^3.)

18.82 Each wheel of an automobile has a mass of 22 kg, a diameter of 575 mm, and a radius of gyration of 225 mm. The automobile travels around an unbanked curve of radius 150 m at a speed of 95 km/h. Knowing that the transverse distance between the wheels is 1.5 m, determine the additional normal force exerted by the ground on each outside wheel due to the motion of the car.

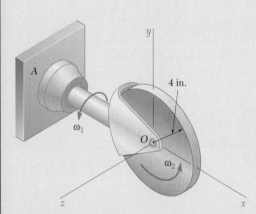

Fig. *P18.83*

18.83 The uniform, thin 5-lb disk spins at a constant rate $\omega_2 = 6$ rad/s about an axis held by a housing attached to a horizontal rod that rotates at the constant rate $\omega_1 = 3$ rad/s. Determine the couple that represents the dynamic reaction at the support A.

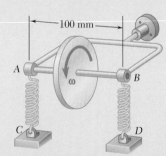

Fig. P18.84

18.84 The essential structure of a certain type of aircraft turn indicator is shown. Each spring has a constant of 500 N/m, and the 200-g uniform disk of 40-mm radius spins at the rate of 10 000 rpm. The springs are stretched and exert equal vertical forces on yoke AB when the airplane is traveling in a straight path. Determine the angle through which the yoke will rotate when the pilot executes a horizontal turn of 750-m radius to the right at a speed of 800 km/h. Indicate whether point A will move up or down.

18.85 A slender rod is bent to form a square frame of side 6 in. The frame is attached by a collar at A to a vertical shaft that rotates with a constant angular velocity ω. Determine the value of ω for which line AB forms an angle $\beta = 48°$ with the horizontal x axis.

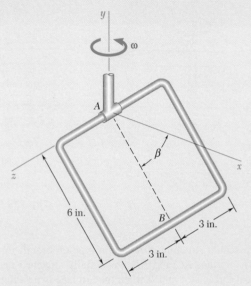

Fig. P18.85

18.86 A uniform square plate with side $a = 225$ mm is hinged at points A and B to a clevis that rotates with a constant angular velocity ω about a vertical axis. Determine (a) the constant angle β that the plate forms with the horizontal x axis when $\omega = 12$ rad/s, (b) the largest value of ω for which the plate remains vertical ($\beta = 90°$).

18.87 A uniform square plate with side $a = 300$ mm is hinged at points A and B to a clevis that rotates with a constant angular velocity ω about a vertical axis. Determine (a) the value of ω for which the plate forms a constant angle $\beta = 60°$ with the horizontal x axis, (b) the largest value of ω for which the plate remains vertical ($\beta = 90°$).

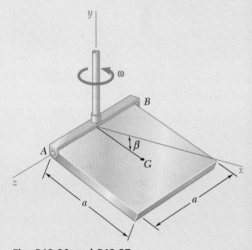

Fig. P18.86 and P18.87

18.88 The 2-lb gear A is constrained to roll on the fixed gear B but is free to rotate about axle AD. Axle AD has a length of 20 in., a negligible weight, and is connected by a clevis to the vertical shaft DE that rotates as shown with a constant angular velocity ω_1. Assuming that gear A can be approximated by a thin disk with a radius of 4 in., determine the largest allowable value of ω_1 if gear A is not to lose contact with gear B.

18.89 Determine the force $\mathbf{F}$ exerted by gear B on gear A of Prob. 18.88 when shaft DE rotates with the constant angular speed of $\omega_1 = 4$ rad/s. (*Hint:* The force $\mathbf{F}$ must be perpendicular to the line drawn from D to C.)

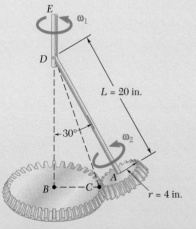

Fig. P18.88

18.90 and 18.91 The slender rod *AB* is attached by a clevis to arm *BCD* that rotates with a constant angular velocity $\boldsymbol{\omega}$ about the centerline of its vertical portion *CD*. Determine the magnitude of the angular velocity $\boldsymbol{\omega}$.

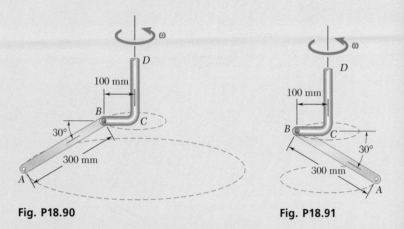

Fig. P18.90   **Fig. P18.91**

18.92 The essential structure of a certain type of aircraft turn indicator is shown. Springs *AC* and *BD* are initially stretched and exert equal vertical forces at *A* and *B* when the airplane is traveling in a straight path. Each spring has a constant of 600 N/m and the uniform disk has a mass of 250 g and spins at the rate of 12 000 rpm. Determine the angle through which the yoke will rotate when the pilot executes a horizontal turn of 800-m radius to the right at a speed of 720 km/h. Indicate whether point *A* will move up or down.

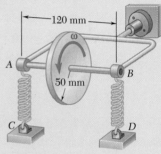

Fig. P18.92

18.93 The 10-oz disk shown spins at the rate $\omega_1 = 750$ rpm, while axle *AB* rotates as shown with an angular velocity $\boldsymbol{\omega}_2$ of magnitude 6 rad/s. Determine the dynamic reactions at *A* and *B*.

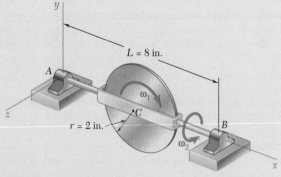

Fig. P18.93 and P18.94

18.94 The 10-oz disk shown spins at the rate $\omega_1 = 750$ rpm, while axle *AB* rotates as shown with an angular velocity $\boldsymbol{\omega}_2$. Determine the maximum allowable magnitude of $\boldsymbol{\omega}_2$ if the dynamic reactions at *A* and *B* are not to exceed 0.25 lb each.

18.95 Two disks each have a mass of 5 kg and a radius 300 mm. They spin as shown at the rate of $\omega_1 = 1200$ rpm about a rod AB of negligible mass that rotates about the horizontal z axis at the rate of $\omega_2 = 60$ rpm. (*a*) Determine the dynamic reactions at points C and D. (*b*) Solve part (*a*) assuming that the direction of spin of disk A is reversed.

18.96 Two disks each have a mass of 5 kg and a radius of 300 mm. They spin as shown at the rate of $\omega_1 = 1200$ rpm about a rod AB of negligible mass that rotates about the horizontal z axis at the rate ω_2. Determine the maximum allowable value of ω_2 if the magnitudes of the dynamic reactions at points C and D are not to exceed 350 N each.

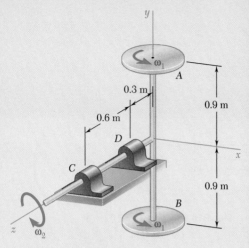

Fig. P18.95 and P18.96

18.97 A stationary horizontal plate is attached to the ceiling by means of a fixed vertical tube. A wheel of radius a and mass m is mounted on a light axle AC that is attached by means of a clevis at A to a rod AB fitted inside the vertical tube. The rod AB is made to rotate with a constant angular velocity Ω causing the wheel to roll on the lower face of the stationary plate. Determine the minimum angular velocity Ω for which contact is maintained between the wheel and the plate. Consider the particular cases (*a*) when the mass of the wheel is concentrated in the rim, (*b*) when the wheel is equivalent to a thin disk of radius a.

18.98 Assuming that the wheel of Prob. 18.97 weighs 8 lb, has a radius $a = 4$ in., and a radius of gyration of 3 in., and that $R = 20$ in., determine the force exerted by the plate on the wheel when $\Omega = 25$ rad/s.

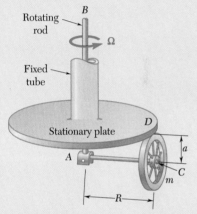

Fig. P18.97

18.99 A thin disk of mass $m = 4$ kg rotates with an angular velocity ω_2 with respect to arm ABC, which itself rotates with an angular velocity ω_1 about the y axis. Knowing that $\omega_1 = 5$ rad/s and $\omega_2 = 15$ rad/s and that both are constant, determine the force-couple system representing the dynamic reaction at the support at A.

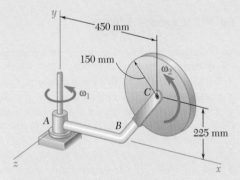

Fig. P18.99

18.100 An experimental Fresnel-lens solar-energy concentrator can rotate about the horizontal axis AB that passes through its mass center G. It is supported at A and B by a steel framework that can rotate about the vertical y axis. The concentrator has a mass of 30 Mg, a radius of gyration of 12 m about its axis of symmetry CD, and a radius of gyration of 10 m about any transverse axis through G. Knowing that the angular velocities ω_1 and ω_2 have constant magnitudes equal to 0.20 rad/s and 0.25 rad/s, respectively, determine for the position $\theta = 60°$ (*a*) the forces exerted on the concentrator at A and B, (*b*) the couple $M_2\mathbf{k}$ applied to the concentrator at that instant.

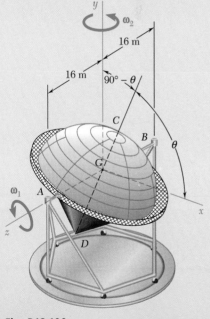

Fig. P18.100

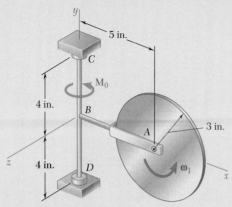

Fig. P18.101 and P18.102

18.101 A 6-lb homogeneous disk of radius 3 in. spins as shown at the constant rate $\omega_1 = 60$ rad/s. The disk is supported by the fork-ended rod AB, which is welded to the vertical shaft CBD. The system is at rest when a couple $\mathbf{M}_0 = (0.25 \text{ ft·lb})\mathbf{j}$ is applied to the shaft for 2 s and then removed. Determine the dynamic reactions at C and D after the couple has been removed.

18.102 A 6-lb homogeneous disk of radius 3 in. spins as shown at the constant rate $\omega_1 = 60$ rad/s. The disk is supported by the fork-ended rod AB, which is welded to the vertical shaft CBD. The system is at rest when a couple $\mathbf{M}_0$ is applied as shown to the shaft for 3 s and then removed. Knowing that the maximum angular velocity reached by the shaft is 18 rad/s, determine (a) the couple $\mathbf{M}_0$, (b) the dynamic reactions at C and D after the couple has been removed.

18.103 A 2.5-kg homogeneous disk of radius 80 mm rotates with an angular velocity ω_1 with respect to arm ABC, which is welded to a shaft DCE rotating as shown at the constant rate $\omega_2 = 12$ rad/s. Friction in the bearing at A causes ω_1 to decrease at the rate of 15 rad/s². Determine the dynamic reactions at D and E at a time when ω_1 has decreased to 50 rad/s.

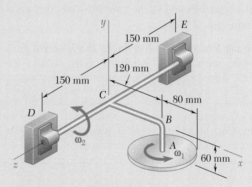

Fig. P18.103 and P18.104

18.104 A 2.5-kg homogeneous disk of radius 80 mm rotates at the constant rate $\omega_1 = 50$ rad/s with respect to arm ABC, which is welded to a shaft DCE. Knowing that at the instant shown, shaft DCE has an angular velocity $\omega_2 = (12 \text{ rad/s})\mathbf{k}$ and an angular acceleration $\alpha_2 = (8 \text{ rad/s}^2)\mathbf{k}$, determine (a) the couple that must be applied to shaft DCE to produce that acceleration, (b) the corresponding dynamic reactions at D and E.

18.105 For the disk of Prob. 18.99, determine (a) the couple $M_1\mathbf{j}$ that should be applied to arm ABC to give it an angular acceleration $\alpha_1 = -(7.5 \text{ rad/s}^2)\mathbf{j}$ when $\omega_1 = 5$ rad/s, knowing that the disk rotates at the constant rate $\omega_2 = 15$ rad/s, (b) the force-couple system representing the dynamic reaction at A at that instant. Assume that ABC has a negligible mass.

***18.106** A slender homogeneous rod AB of mass m and length L is made to rotate at a constant rate ω_2 about the horizontal z axis, while frame CD is made to rotate at the constant rate ω_1 about the y axis. Express as a function of the angle θ (a) the couple $\mathbf{M}_1$ required to maintain the rotation of the frame, (b) the couple $\mathbf{M}_2$ required to maintain the rotation of the rod, (c) the dynamic reactions at the supports C and D.

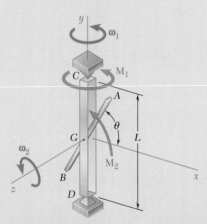

Fig. P18.106

*18.3 MOTION OF A GYROSCOPE

A **gyroscope** consists essentially of a rotor that can spin freely about its geometric axis. When mounted in a Cardan's suspension (Fig. 18.15), a gyroscope can assume any orientation, but its mass center must remain fixed in space. Because a gyroscope can measure its orientation in space and maintain that orientation, it has become an indispensable part of modern navigational equipment. In this section, we examine the motion of a gyroscope as a practical example of analyzing the motion of a rigid body in three dimensions.

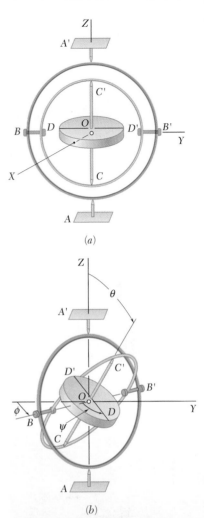

18.3A Eulerian Angles

In order to define the position of a gyroscope at a given instant, let us select a fixed frame of reference $OXYZ$ with the origin O located at the mass center of the gyroscope and the Z axis directed along the line defined by the bearings A and A' of the outer gimbal. We consider a reference position of the gyroscope in which the two gimbals and a given diameter DD' of the rotor are located in the fixed YZ plane (Fig. 18.15a). The gyroscope can be brought from this reference position into any arbitrary position (Fig. 18.15b) by means of the following steps.

1. A rotation of the outer gimbal through an angle ϕ about the axis AA'.
2. A rotation of the inner gimbal through θ about BB'.
3. A rotation of the rotor through ψ about CC'.

The angles ϕ, θ, and ψ are called the **Eulerian angles**; they completely characterize the position of the gyroscope at any given instant. Their derivatives $\dot{\phi}$, $\dot{\theta}$, and $\dot{\psi}$ define, respectively, the rate of **precession**, the rate of **nutation**, and the rate of **spin** of the gyroscope at the instant considered. Precession is the revolution of the axis BB' about the Z-axis, and nutation is the back-and-forth motion of CC' as the object precesses.

In order to compute the components of the angular velocity and of the angular momentum of the gyroscope, we will use a rotating system of axes $Oxyz$ attached to the inner gimbal, with the y axis along BB' and the z axis along CC' (Fig. 18.16). These axes are principal axes of inertia for the gyroscope. Although they follow it in its precession and nutation, they do not spin with $\dot{\psi}$; for that reason, they are more convenient to use than axes actually attached to the gyroscope. The angular velocity $\boldsymbol{\omega}$ of the gyroscope with respect to the fixed frame of reference $OXYZ$ now can be expressed as the sum of three partial angular velocities that correspond to the precession, the nutation, and the spin of the gyroscope, respectively. Denoting the unit vectors along the rotating axes by $\mathbf{i}$, $\mathbf{j}$, and $\mathbf{k}$ and the unit vector along the fixed Z axis by $\mathbf{K}$, we have

$$\boldsymbol{\omega} = \dot{\phi}\mathbf{K} + \dot{\theta}\mathbf{j} + \dot{\psi}\mathbf{k} \tag{18.33}$$

Since the vector components obtained for $\boldsymbol{\omega}$ in Eq. (18.33) are not orthogonal (Fig. 18.16), we resolve the unit vector $\mathbf{K}$ into components along the x and z axes; we obtain

$$\mathbf{K} = -\sin\theta\,\mathbf{i} + \cos\theta\,\mathbf{k} \tag{18.34}$$

Fig. 18.15 (a) Reference position of a gyroscope; (b) arbitrary position of the gyroscope by rotation through the three Eulerian angles.

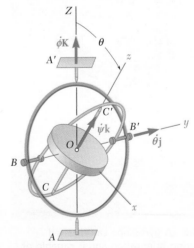

Fig. 18.16 Precession $\dot{\phi}$, nutation $\dot{\theta}$, and spin $\dot{\psi}$ of a gyroscope.

Then, substituting for **K** into Eq. (18.33), we have

$$\boldsymbol{\omega} = -\dot{\phi}\sin\theta\,\mathbf{i} + \dot{\theta}\,\mathbf{j} + (\dot{\psi} + \dot{\phi}\cos\theta)\mathbf{k} \qquad (18.35)$$

The coordinate axes are principal axes of inertia, so we can obtain the components of the angular momentum $\mathbf{H}_O$ by multiplying the components of $\boldsymbol{\omega}$ by the moments of inertia of the rotor about the x, y, and z axes, respectively. Denoting the moment of inertia of the rotor about its spin axis by I, its moment of inertia about a transverse axis through O by I', and neglecting the mass of the gimbals, we have

$$\mathbf{H}_O = -I'\dot{\phi}\sin\theta\,\mathbf{i} + I'\dot{\theta}\,\mathbf{j} + I(\dot{\psi} + \dot{\phi}\cos\theta)\mathbf{k} \qquad (18.36)$$

Recalling that the rotating axes are attached to the inner gimbal and thus do not spin with $\dot{\psi}$, we express their angular velocity as the sum

$$\boldsymbol{\Omega} = \dot{\phi}\mathbf{K} + \dot{\theta}\mathbf{j} \qquad (18.37)$$

or substituting for **K** from Eq. (18.34), we have

$$\boldsymbol{\Omega} = -\dot{\phi}\sin\theta\,\mathbf{i} + \dot{\theta}\,\mathbf{j} + \dot{\phi}\cos\theta\,\mathbf{k} \qquad (18.38)$$

Substituting for $\mathbf{H}_O$ and $\boldsymbol{\Omega}$ from Eqs. (18.36) and (18.38) into the equation gives

$$\Sigma\mathbf{M}_O = (\dot{\mathbf{H}}_O)_{Oxyz} + \boldsymbol{\Omega} \times \mathbf{H}_O \qquad (18.28)$$

We now obtain the three differential equations

$$\Sigma M_x = -I'(\ddot{\phi}\sin\theta + 2\dot{\theta}\dot{\phi}\cos\theta) + I\dot{\theta}(\dot{\psi} + \dot{\phi}\cos\theta)$$
$$\Sigma M_y = I'(\ddot{\theta} - \dot{\phi}^2\sin\theta\cos\theta) + I\dot{\phi}\sin\theta(\dot{\psi} + \dot{\phi}\cos\theta) \qquad (18.39)$$
$$\Sigma M_z = I\frac{d}{dt}(\dot{\psi} + \dot{\phi}\cos\theta)$$

Equations (18.39) define the motion of a gyroscope subjected to a given system of forces when the mass of its gimbals is neglected. We can also use them to define the motion of an **axisymmetric body** (or body of revolution) attached at a point on its axis of symmetry and to define the motion of an axisymmetric body about its mass center. The gimbals of the gyroscope helped us visualize the Eulerian angles, but it is clear that we can use these angles to define the position of any rigid body with respect to axes centered at a point of the body—regardless of the way in which the body is actually supported.

Because Eqs. (18.39) are nonlinear, it is not possible to express the Eulerian angles ϕ, θ, and ψ as analytical functions of time t in general, and you may need to use numerical methods of solution. However, as you will see in the rest of this section, several particular cases of interest can be analyzed easily.

Photo 18.5 A gyroscope can be used for measuring orientation and is capable of maintaining the same absolute direction in space.

*18.3B Steady Precession of a Gyroscope

Let us now investigate the particular case of gyroscopic motion in which the angle θ, the rate of precession $\dot{\phi}$, and the rate of spin $\dot{\psi}$ remain constant. We propose to determine the forces that must be applied to the gyroscope to maintain this motion, which is known as the **steady precession** of a gyroscope.

Instead of applying the general equations (18.39), we determine the sum of the moments of the required forces by computing the rate of change of the angular momentum of the gyroscope in the particular case considered. We first note that the angular velocity $\boldsymbol{\omega}$ of the gyroscope, its angular momentum $\mathbf{H}_O$, and the angular velocity $\boldsymbol{\Omega}$ of the rotating frame of reference (Fig. 18.17) reduce, respectively, to

$$\boldsymbol{\omega} = -\dot{\phi}\sin\theta\,\mathbf{i} + \omega_z\mathbf{k} \qquad (18.40)$$

$$\mathbf{H}_O = -I'\dot{\phi}\sin\theta\,\mathbf{i} + I\omega_z\mathbf{k} \qquad (18.41)$$

$$\boldsymbol{\Omega} = -\dot{\phi}\sin\theta\,\mathbf{i} + \dot{\phi}\cos\theta\,\mathbf{k} \qquad (18.42)$$

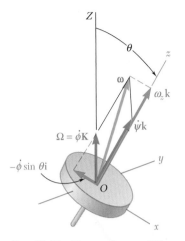

Fig. 18.17 Kinematic quantities used to determine the steady rate of precession of a gyroscope.

where $\omega_z = \dot{\psi} + \dot{\phi}\cos\theta$ is the rectangular component along the spin axis of the total angular velocity of the gyroscope.

Since θ, ϕ, and $\dot{\psi}$ are constant, the vector $\mathbf{H}_O$ is constant in magnitude and direction with respect to the rotating frame of reference. Therefore its rate of change $(\dot{\mathbf{H}}_O)_{Oxyz}$ with respect to that frame is zero. Thus, Eq. (18.28) reduces to

$$\Sigma\mathbf{M}_O = \boldsymbol{\Omega} \times \mathbf{H}_O \qquad (18.43)$$

which yields, after substitutions from Eqs. (18.41) and (18.42),

$$\Sigma\mathbf{M}_O = (I\omega_z - I'\dot{\phi}\cos\theta)\dot{\phi}\sin\theta\,\mathbf{j} \qquad (18.44)$$

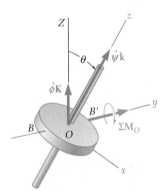

Fig. 18.18 To maintain a gyroscope in steady precession, a couple must be applied about an axis perpendicular to the precession and spin axis.

The mass center of the gyroscope is fixed in space, so using Eq. (18.1), we have $\Sigma\mathbf{F} = 0$. Thus, the forces that must be applied to the gyroscope to maintain its steady precession reduce to a couple of moment equal to the right-hand side of Eq. (18.44). Note that *this couple should be applied about an axis perpendicular to the precession axis and to the spin axis of the gyroscope* (Fig. 18.18).

In the particular case when the precession axis and the spin axis are at a right angle to each other, we have $\theta = 90°$, and Eq. (18.44) reduces to

$$\Sigma\mathbf{M}_O = I\dot{\psi}\dot{\phi}\,\mathbf{j} \qquad (18.45)$$

Thus, if we apply a couple $\mathbf{M}_O$ to the gyroscope about an axis perpendicular to its axis of spin, the gyroscope precesses about an axis perpendicular to both the spin axis and the couple axis. The sense of the precession is such that the vectors representing the spin, the couple, and the precession, respectively, form a right-handed triad (Fig. 18.19). The relationship of this triad also can be represented by writing Eq. (18.45) as the vector equation of

$$\Sigma\mathbf{M}_O = \dot{\boldsymbol{\phi}} \times I\dot{\boldsymbol{\psi}} \qquad (18.45')$$

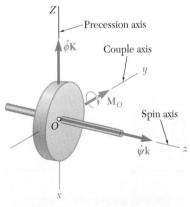

Fig. 18.19 A right-handed triad of the spin, couple, and precession axes.

Because of the relatively large couples required to change the orientation of their axles, gyroscopes are used as stabilizers in torpedoes and ships. Spinning bullets and shells remain tangent to their trajectory because of gyroscopic action. Also, a bicycle is easier to keep balanced at high speeds because of the stabilizing effect of its spinning wheels. However, gyroscopic action is not always welcome; it must be taken into account in the design of bearings supporting rotating shafts subjected to forced precession. The reactions exerted on an airplane by its propellers, which changes the direction of flight, also must be taken into consideration and compensated for whenever possible.

*18.3C Motion of an Axisymmetric Body Under no Force

We can now analyze the motion of an axisymmetric body about its mass center under no force except its own weight. Examples of such motion are furnished by projectiles (if air resistance is neglected) and by satellites and space vehicles after the burnout of their launching rockets.

The sum of the moments of the external forces about the mass center G of the body is zero, so Eq. (18.2) yields $\dot{\mathbf{H}}_G = 0$. It follows that the angular momentum $\mathbf{H}_G$ of the body about G is constant. Thus, the direction of $\mathbf{H}_G$ is fixed in space and can be used to define the Z axis, or axis of precession (Fig. 18.20). Let us select a rotating system of axes $Gxyz$ with the z axis along the axis of symmetry of the body, the x axis in the plane defined by the Z and z axes, and the y axis pointing away from you (Fig. 18.21). This gives us

$$H_x = -H_G \sin \theta \qquad H_y = 0 \qquad H_z = H_G \cos \theta \qquad \text{(18.46)}$$

where θ represents the angle formed by the Z and z axes and H_G denotes the constant magnitude of the angular momentum of the body about G. Since the x, y, and z axes are principal axes of inertia for the body considered, we have

$$H_x = I' \omega_x \qquad H_y = I' \omega_y \qquad H_z = I \omega_z \qquad \text{(18.47)}$$

where I denotes the moment of inertia of the body about its axis of symmetry and I' denotes its moment of inertia about a transverse axis through G. It follows from Eqs. (18.46) and (18.47) that

$$\omega_x = -\frac{H_G \sin \theta}{I'} \qquad \omega_y = 0 \qquad \omega_z = \frac{H_G \cos \theta}{I} \qquad \text{(18.48)}$$

The second of these relations shows that the angular velocity $\boldsymbol{\omega}$ has no component along the y axis, i.e., along an axis perpendicular to the Z–z plane. Thus, the angle θ formed by the Z and z axes remains constant and *the body is in steady precession about the Z axis.*

Dividing the first of the relations in Eqs. (18.48) by the third, and observing from Fig. 18.21 that $-\omega_x/\omega_z = \tan \gamma$, we obtain the following relation between the angles γ and θ that the vectors $\boldsymbol{\omega}$ and $\mathbf{H}_G$, respectively, form with the axis of symmetry of the body.

$$\tan \gamma = \frac{I}{I'} \tan \theta \qquad \text{(18.49)}$$

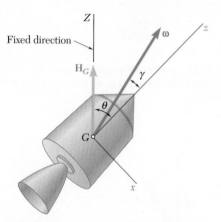

Fig. 18.20 For an axisymmetric body under no force other than its own weight, the angular momentum has a constant direction.

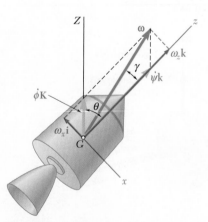

Fig. 18.21 The angular velocity of an axisymmetric body expressed in terms of body-fixed coordinates xyz.

Two particular cases of motion of an axisymmetric body under no force involve no precession.

1. If the body is set to spin about its axis of symmetry, we have $\omega_x = 0$ and, by Eq. (18.47), $H_x = 0$. Thus, the vectors $\boldsymbol{\omega}$ and $\mathbf{H}_G$ have the same orientation, and the body keeps spinning about its axis of symmetry (Fig. 18.22a).

2. If the body is set to spin about a transverse axis, we have $\omega_z = 0$ and, by Eq. (18.47), $H_z = 0$. Again $\boldsymbol{\omega}$ and $\mathbf{H}_G$ have the same orientation, and the body keeps spinning about the given transverse axis (Fig. 18.22b).

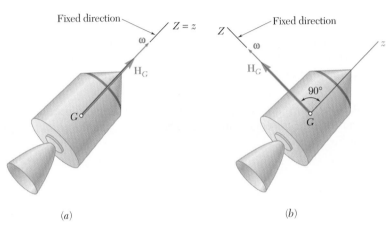

(a)   (b)

Fig. 18.22 (a) A body spinning about its axis of symmetry; (b) a body spinning about a transverse axis.

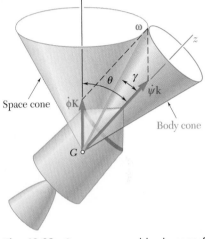

Fig. 18.23 Space cone and body cone for an elongated body ($I < I'$) in direct precession.

Considering now the general case represented in Fig. 18.21, recall from Sec. 15.6A that we can represent the motion of a body about a fixed point—or about its mass center—by the motion of a body cone rolling on a space cone. In the case of steady precession, the two cones are circular, since the angles γ and $\theta - \gamma$ that the angular velocity $\boldsymbol{\omega}$ forms, respectively, with the axis of symmetry of the body and with the precession axis are constant. Two cases should be distinguished.

1. $I < I'$. This is the case of an elongated body, such as the space vehicle of Fig. 18.23. From Eq. (18.49), we have $\gamma < \theta$. The vector $\boldsymbol{\omega}$ lies inside the angle ZGz; the space cone and the body cone are tangent externally; and the spin and the precession are both observed as counterclockwise from the positive z axis. The precession is said to be *direct*.

2. $I > I'$. This is the case of a flattened body, such as the satellite of Fig. 18.24. From Eq. (18.49), we have $\gamma > \theta$. Since the vector $\boldsymbol{\omega}$ must lie outside the angle ZGz, the vector $\dot{\psi}\mathbf{k}$ has a sense opposite to that of the z axis; the space cone is inside the body cone; and the precession and the spin have opposite senses. The precession is said to be *retrograde*.

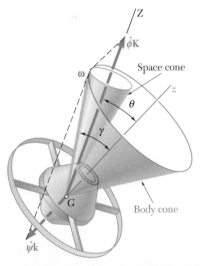

Fig. 18.24 Space cone and body cone for a flattened body ($I > I'$) in retrograde precession.

Sample Problem 18.6

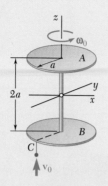

A space satellite with mass m can be modeled as two thin disks of equal mass. The disks have a radius of $a = 800$ mm and are rigidly connected by a light rod with a length of $2a$. Initially, the satellite is spinning freely about its axis of symmetry at the rate $\omega_0 = 60$ rpm. A meteorite with a mass of $m_0 = m/1000$ is traveling with a velocity $\mathbf{v}_0$ of 2000 m/s relative to the satellite, strikes the satellite, and becomes embedded at C. Determine (a) the angular velocity of the satellite immediately after impact, (b) the precession axis of the ensuing motion, (c) the rates of precession and spin of the ensuing motion.

STRATEGY: Since an impact occurs, use the principle of impulse and momentum. Then you can use the relations in this section to determine the gyroscopic motion of the satellite.

MODELING: Choose the meteorite and the satellite as your system. The linear and angular momenta of the system before and after the impact are shown in Fig. 1.

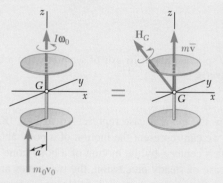

Fig. 1 Momenta before and after the impact.

ANALYSIS:

Moments of Inertia. Note that the axes shown are principal axes of inertia for the satellite. Thus, you have

$$I = I_z = \tfrac{1}{2}ma^2 \qquad I' = I_x = I_y = 2[\tfrac{1}{4}(\tfrac{1}{2}m)a^2 + (\tfrac{1}{2}m)a^2] = \tfrac{5}{4}ma^2$$

Principle of Impulse and Momentum. Since no external force acts on the system, the momenta before and after impact are equal (Fig. 1). Taking moments about G, you have

$$-a\mathbf{j} \times m_0 v_0 \mathbf{k} + I\omega_0 \mathbf{k} = \mathbf{H}_G$$
$$\mathbf{H}_G = -m_0 v_0 a\mathbf{i} + I\omega_0 \mathbf{k} \tag{1}$$

Angular Velocity After Impact. Substitute the values obtained for the components of $\mathbf{H}_G$ in Eq. (1) and for the moments of inertia into

$$H_x = I_x\omega_x \qquad H_y = I_y\omega_y \qquad H_z = I_z\omega_z$$

The result is

$$-m_0v_0a = I'\omega_x = \tfrac{5}{4}ma^2\omega_x \qquad 0 = I'\omega_y \qquad I\omega_0 = I\omega_z$$

$$\omega_x = -\frac{4}{5}\frac{m_0v_0}{ma} \qquad \omega_y = 0 \qquad \omega_z = \omega_0 \qquad (2)$$

For the satellite considered, you have $\omega_0 = 60$ rpm $= 6.283$ rad/s, $m_0/m = 1/1000$, $a = 0.800$ m, and $v_0 = 2000$ m/s. You obtain

$$\omega_x = -2 \text{ rad/s} \qquad \omega_y = 0 \qquad \omega_z = 6.283 \text{ rad/s}$$

$$\omega = \sqrt{\omega_x^2 + \omega_z^2} = 6.594 \text{ rad/s} \qquad \tan\gamma = \frac{-\omega_x}{\omega_z} = +0.3183$$

$$\omega = 63.0 \text{ rpm} \qquad \gamma = 17.7° \quad \blacktriangleleft$$

Precession Axis. In free motion, the direction of the angular momentum $\mathbf{H}_G$ is fixed in space, so the satellite precesses about this direction. The angle θ formed by the precession axis and the z axis is (Fig. 2)

$$\tan\theta = \frac{-H_x}{H_z} = \frac{m_0v_0a}{I\omega_0} = \frac{2m_0v_0}{ma\omega_0} = 0.796 \qquad \theta = 38.5° \quad \blacktriangleleft$$

Rates of Precession and Spin. Sketch the space and body cones for the free motion of the satellite (Fig. 3). Using the law of sines, compute the rates of precession and spin.

$$\frac{\omega}{\sin\theta} = \frac{\dot\phi}{\sin\gamma} = \frac{\dot\psi}{\sin(\theta - \gamma)}$$

$$\dot\phi = 30.8 \text{ rpm} \qquad \dot\psi = 35.9 \text{ rpm} \quad \blacktriangleleft$$

REFLECT and THINK: If you applied the principle of impulse and momentum in the z-direction, you would find that $P\Delta t = m\bar{v}$ where $P\Delta t$ is the impulse the meteorite applies to the satellite. In this problem, we were interested in the three-dimensional rotation of the satellite and modeled it as a rigid body. In Chap. 12, we were concerned with the orbits of satellites over the earth and modeled the satellite as a particle. As engineers, how we model a system depends on what type of problem we are trying to solve.

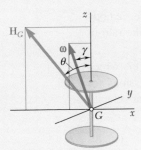

Fig. 2 Angles between the z-axis and the angular velocity and the angular momentum.

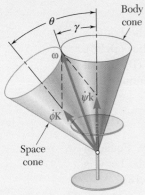

Fig. 3 Space and body cones for the satellite.

SOLVING PROBLEMS
ON YOUR OWN

In this section, we analyzed the motion of **gyroscopes** and of other **axisymmetric bodies** with a fixed point O. In order to define the position of these bodies at any given instant, we introduced the three **Eulerian angles** ϕ, θ, and ψ (Fig. 18.15) and noted that their time derivatives define, respectively, the rate of **precession**, the rate of **nutation**, and the rate of **spin** (Fig. 18.16). The problems you encountered in this section fall into one of the following categories.

1. Steady precession. This is the motion of a gyroscope or other axisymmetric body with a fixed point located on its axis of symmetry in which the angle θ, the rate of precession $\dot{\phi}$, and the rate of spin $\dot{\psi}$ all remain constant.

    a. Using the rotating frame of reference $Oxyz$ shown in Fig. 18.17, which *precesses* with the body, *but does not spin* with it, we obtained the expressions for the angular velocity $\boldsymbol{\omega}$ of the body, its angular momentum $\mathbf{H}_O$, and the angular velocity $\boldsymbol{\Omega}$ of the frame $Oxyz$ as

$$\boldsymbol{\omega} = -\dot{\phi}\sin\theta\,\mathbf{i} + \omega_z\mathbf{k} \tag{18.40}$$

$$\mathbf{H}_O = -I'\dot{\phi}\sin\theta\,\mathbf{i} + I\,\omega_z\mathbf{k} \tag{18.41}$$

$$\boldsymbol{\Omega} = -\dot{\phi}\sin\theta\,\mathbf{i} + \dot{\phi}\cos\theta\,\mathbf{k} \tag{18.42}$$

where I = moment of inertia of body about its axis of symmetry
     I' = moment of inertia of body about a transverse axis through O
   ω_z = rectangular component of $\boldsymbol{\omega}$ along z axis = $\dot{\psi} + \dot{\phi}\cos\theta$

    b. The sum of the moments about O **of the forces applied to the body** is equal to the rate of change of its angular momentum, as expressed by Eq. (18.28). But, since θ and the rates of change $\dot{\phi}$ and $\dot{\psi}$ are constant, it follows from Eq. (18.41) that $\mathbf{H}_O$ remains constant in magnitude and direction when viewed from the frame $Oxyz$. Thus, its rate of change is zero with respect to that frame, and you have

$$\Sigma\mathbf{M}_O = \boldsymbol{\Omega} \times \mathbf{H}_O \tag{18.43}$$

where $\boldsymbol{\Omega}$ and $\mathbf{H}_O$ are defined by Eqs. (18.42) and (18.41), respectively. Equation (18.43) shows that the moment resultant at O of the forces applied to the body is perpendicular to both the axis of precession and the axis of spin (Fig. 18.18).

    c. Keep in mind that the method described applies not only to gyroscopes, where the fixed point O coincides with the mass center G, but also to any axisymmetric body with a fixed point O located on its axis of symmetry. This method, therefore, can be used to analyze the *steady precession of a top* on a rough floor.

    d. When an axisymmetric body has no fixed point but is in steady precession about its mass center G, you should draw a free-body diagram and a kinetic diagram showing that the system of the external forces exerted on the body (including the body's weight) is equivalent to the vector $m\bar{\mathbf{a}}$ applied at G and the couple vector $\dot{\mathbf{H}}_G$.

You can use Eqs. (18.40) through (18.42), replacing $\mathbf{H}_O$ with $\mathbf{H}_G$, and express the moment of the couple as

$$\dot{\mathbf{H}}_G = \boldsymbol{\Omega} \times \mathbf{H}_G$$

You can then use the free-body and kinetic diagrams to write as many as six independent scalar equations.

2. **Motion of an axisymmetric body under no force, except its own weight.** We have $\Sigma \mathbf{M}_G = 0$ and thus $\dot{\mathbf{H}}_G = 0$; it follows that the angular momentum $\mathbf{H}_G$ is constant in magnitude and direction (Sec. 18.3C). The body is in **steady precession** with the precession axis GZ directed along $\mathbf{H}_G$ (Fig. 18.20). Using the rotating frame $Gxyz$ and denoting by γ the angle that $\boldsymbol{\omega}$ forms with the spin axis Gz (Fig. 18.21), we obtained the relation between γ and the angle θ formed by the precession and spin axes as

$$\tan \gamma = \frac{I}{I'} \tan \theta \qquad (18.49)$$

The precession is said to be *direct* if $I < I'$ (Fig. 18.23) and *retrograde* if $I > I'$ (Fig. 18.24).

    a. In many problems dealing with the motion of an axisymmetric body under no force, you will be asked to determine the precession axis and the rates of precession and spin of the body when given the magnitude of its angular velocity $\boldsymbol{\omega}$ and the angle γ that it forms with the axis of symmetry Gz (Fig. 18.21). From Eq. (18.49), determine the angle θ that the precession axis GZ forms with Gz and resolve $\boldsymbol{\omega}$ into its two oblique components $\dot{\phi}\mathbf{K}$ and $\dot{\psi}\mathbf{k}$. Using the law of sines, you then can determine the rate of precession $\dot{\phi}$ and the rate of spin $\dot{\psi}$.

    b. In other problems, the body is subjected to a given impulse and you will first determine the resulting angular momentum $\mathbf{H}_G$. Using Eqs. (18.10), you can calculate the rectangular components of the angular velocity $\boldsymbol{\omega}$, its magnitude ω, and the angle γ that it forms with the axis of symmetry. You then determine the precession axis and the rates of precession and spin as described previously [Sample Prob. 18.6].

3. **General motion of an axisymmetric body with a fixed point O located on its axis of symmetry, and subjected only to its own weight.** This is a motion in which the angle θ is allowed to vary. At any given instant you should take into account the rate of precession $\dot{\phi}$, the rate of spin $\dot{\psi}$, and the rate of nutation $\dot{\theta}$—none of which will remain constant. An example of such a motion is the motion of a top, which is discussed in Probs. 18.137 and 18.138. The rotating frame of reference $Oxyz$ that you will use is still the one shown in Fig. 18.18, but this frame now rotates about the y axis at the rate $\dot{\theta}$. Equations (18.40), (18.41), and (18.42), therefore, should be replaced by

$$\boldsymbol{\omega} = -\dot{\phi} \sin \theta \, \mathbf{i} + \dot{\theta} \, \mathbf{j} + (\dot{\psi} + \dot{\phi} \cos \theta) \, \mathbf{k} \qquad (18.40')$$

$$\mathbf{H}_O = -I'\dot{\phi} \sin \theta \, \mathbf{i} + I'\dot{\theta} \, \mathbf{j} + I(\dot{\psi} + \dot{\phi} \cos \theta) \, \mathbf{k} \qquad (18.41')$$

$$\boldsymbol{\Omega} = -\dot{\phi} \sin \theta \, \mathbf{i} + \dot{\theta} \, \mathbf{j} + \dot{\phi} \cos \theta \, \mathbf{k} \qquad (18.42')$$

Since substituting these expressions into Eq. (18.44) would lead to nonlinear differential equations, it is preferable, whenever feasible, to apply the following conservation principles.

a. Conservation of energy. Denoting the distance between the fixed point O and the mass center G of the body by c and the total energy by E, you have

$$T + V = E: \qquad \tfrac{1}{2}(I'\omega_x^2 + I'\omega_y^2 + I\omega_z^2) + mgc \cos \theta = E$$

Then substitute the expressions obtained in Eq. (18.40') for the components of $\boldsymbol{\omega}$. Note that c is positive or negative depending upon the position of G relative to O. Also, $c = 0$ if G coincides with O; the kinetic energy is then conserved.

b. Conservation of the angular momentum about the axis of precession. Since the support at O is located on the Z axis and the weight of the body and the Z axis are both vertical and thus are parallel to each other, it follows that $\Sigma M_Z = 0$. Thus, H_Z remains constant. We can express this by writing that the scalar product $\mathbf{K} \cdot \mathbf{H}_O$ is constant, where $\mathbf{K}$ is the unit vector along the Z axis.

c. Conservation of the angular momentum about the axis of spin. Since the support at O and the center of gravity G are both located on the z axis, it follows that $\Sigma M_z = 0$ and, thus, that H_z remains constant. Thus, the coefficient of the unit vector $\mathbf{k}$ in Eq. (18.41') is constant. Note that this last conservation principle cannot be applied when the body is restrained from spinning about its axis of symmetry, but in that case, the only variables are θ and ϕ.

Problems

18.107 A uniform thin disk with a 6-in. diameter is attached to the end of a rod AB of negligible mass that is supported by a ball-and-socket joint at point A. Knowing that the disk is observed to precess about the vertical axis AC at the constant rate of 36 rpm in the sense indicated and that its axis of symmetry AB forms an angle $\beta = 60°$ with AC, determine the rate at which the disk spins about rod AB.

18.108 A uniform thin disk with a 6-in. diameter is attached to the end of a rod AB of negligible mass that is supported by a ball-and-socket joint at point A. Knowing that the disk is spinning about its axis of symmetry AB at the rate of 2100 rpm in the sense indicated and that AB forms an angle $\beta = 45°$ with the vertical axis AC, determine the two possible rates of steady precession of the disk about the axis AC.

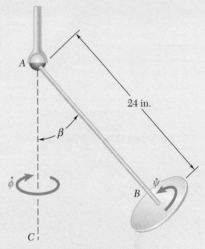

Fig. P18.107 and *P18.108*

18.109 The 85-g top shown is supported at the fixed point O. The radii of gyration of the top with respect to its axis of symmetry and with respect to a transverse axis through O are 21 mm and 45 mm, respectively. Knowing that $c = 37.5$ mm and that the rate of spin of the top about its axis of symmetry is 1800 rpm, determine the two possible rates of steady precession corresponding to $\theta = 30°$.

18.110 The top shown is supported at the fixed point O and its moments of inertia about its axis of symmetry and about a transverse axis through O are denoted, respectively, by I and I'. (a) Show that the condition for steady precession of the top is

$$(I\omega_z - I'\dot{\phi} \cos \theta)\, \dot{\phi} = Wc$$

where $\dot{\phi}$ is the rate of precession and ω_z is the rectangular component of the angular velocity along the axis of symmetry of the top. (b) Show that if the rate of spin $\dot{\psi}$ of the top is very large compared with its rate of precession $\dot{\phi}$, the condition for steady precession is $I\dot{\psi}\dot{\phi} \approx Wc$. (c) Determine the percentage error introduced when this last relation is used to approximate the slower of the two rates of precession obtained for the top of Prob. 18.109.

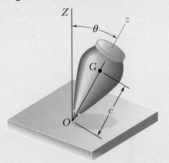

Fig. P18.109 and *P18.110*

18.111 A solid aluminum sphere of radius 4 in. is welded to the end of a 10-in.-long rod AB of negligible mass that is supported by a ball-and-socket joint at A. Knowing that the sphere is observed to precess about a vertical axis at the constant rate of 60 rpm in the sense indicated and that rod AB forms an angle $\beta = 20°$ with the vertical, determine the rate of spin of the sphere about line AB.

18.112 A solid aluminum sphere of radius 4 in. is welded to the end of a 10-in.-long rod AB of negligible mass that is supported by a ball-and-socket joint at A. Knowing that the sphere spins as shown about line AB at the rate of 600 rpm, determine the angle β for which the sphere will precess about a vertical axis at the constant rate of 60 rpm in the sense indicated.

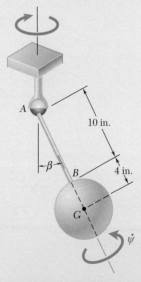

Fig. P18.111 and P18.112

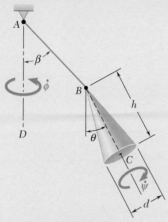

Fig. P18.113 and P18.114

18.113 A homogeneous cone with a height h and a base with a diameter $d < h$ is attached as shown to a cord AB. The cone spins about its axis BC at the constant rate $\dot{\psi}$ and precesses about the vertical through A at the constant rate $\dot{\phi}$. Determine the angle β for which the axis BC of the cone is aligned with cord AB ($\theta = \beta$).

18.114 A homogeneous cone with a height of $h = 12$ in. and a base with a diameter of $d = 6$ in. is attached as shown to a cord AB. Knowing that the angles that cord AB and the axis BC of the cone form with the vertical are, respectively, $\beta = 45°$ and $\theta = 30°$ and that the cone precesses at the constant rate $\dot{\phi} = 8$ rad/s in the sense indicated, determine (a) the rate of spin $\dot{\psi}$ of the cone about its axis BC, (b) the length of cord AB.

18.115 A solid cube of side $c = 80$ mm is attached as shown to cord AB. It is observed to spin at the rate $\dot{\psi} = 40$ rad/s about its diagonal BC and to precess at the constant rate $\dot{\phi} = 5$ rad/s about the vertical axis AD. Knowing that $\beta = 30°$, determine the angle θ that the diagonal BC forms with the vertical. (*Hint:* The moment of inertia of a cube about an axis through its center is independent of the orientation of that axis.)

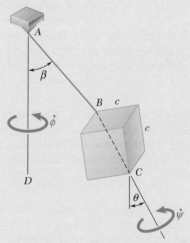

Fig. P18.115 and P18.116

18.116 A solid cube of side $c = 120$ mm is attached as shown to a cord AB of length 240 mm. The cube spins about its diagonal BC and precesses about the vertical axis AD. Knowing that $\theta = 25°$ and $\beta = 40°$, determine (a) the rate of spin of the cube, (b) its rate of precession. (See hint of Prob. 18.115.)

18.117 A high-speed photographic record shows that a certain projectile was fired with a horizontal velocity $\bar{\mathbf{v}}$ of 2000 ft/s and with its axis of symmetry forming an angle $\beta = 3°$ with the horizontal. The rate of spin $\dot{\psi}$ of the projectile was 6000 rpm, and the atmospheric drag was equivalent to a force $\mathbf{D}$ of 25 lb acting at the center of pressure C_P located at a distance $c = 6$ in. from G. (a) Knowing that the projectile has a weight of 45 lb and a radius of gyration of 2 in. with respect to its axis of symmetry, determine its approximate rate of steady precession. (b) If it is further known that the radius of gyration of the projectile with respect to a transverse axis through G is 8 in., determine the exact values of the two possible rates of precession.

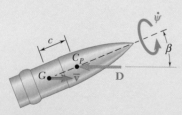

Fig. P18.117

18.118 If the earth were a sphere, the gravitational attraction of the sun, moon, and planets would at all times be equivalent to a single force **R** acting at the mass center of the earth. However, the earth is actually an oblate spheroid and the gravitational system acting on the earth is equivalent to a force **R** and a couple **M**. Knowing that the effect of the couple **M** is to cause the axis of the earth to precess about the axis GA at the rate of one revolution in 25 800 years, determine the average magnitude of the couple **M** applied to the earth. Assume that the average density of the earth is 5.51 g/cm^3, that the average radius of the earth is 6370 km, and that $\bar{I} = \frac{2}{5}mR^2$. (*Note:* This forced precession is known as the precession of the equinoxes and is not to be confused with the free precession discussed in Prob. 18.123.)

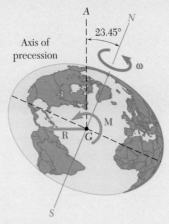

Fig. P18.118

18.119 Show that for an axisymmetric body under no force, the rates of precession and spin can be expressed, respectively, as

$$\dot{\phi} = \frac{H_G}{I'}$$

and

$$\dot{\psi} = \frac{H_G \cos \theta (I' - I)}{II'}$$

where H_G is the constant value of the angular momentum of the body.

18.120 (*a*) Show that for an axisymmetric body under no force, the rate of precession can be expressed as

$$\dot{\phi} = \frac{I\omega_2}{I' \cos \theta}$$

where ω_2 is the rectangular component of **ω** along the axis of symmetry of the body. (*b*) Use this result to check that the condition (18.44) for steady precession is satisfied by an axisymmetric body under no force.

18.121 Show that the angular velocity vector **ω** of an axisymmetric body under no force is observed from the body itself to rotate about the axis of symmetry at the constant rate

$$n = \frac{I' - I}{I'} \omega_2$$

where ω_2 is the rectangular component of **ω** along the axis of symmetry of the body.

18.122 For an axisymmetric body under no force, prove (*a*) that the rate of retrograde precession can never be less than twice the rate of spin of the body about its axis of symmetry, (*b*) that in Fig. 18.24 the axis of symmetry of the body can never lie within the space cone.

18.123 Using the relation given in Prob. 18.121, determine the period of precession of the north pole of the earth about the axis of symmetry of the earth. The earth may be approximated by an oblate spheroid of axial moment of inertia I and of transverse moment of inertia $I' = 0.99671I$. (*Note:* Actual observations show a period of precession of the north pole of about 432.5 mean solar days; the difference between the observed and computed periods is due to the fact that the earth is not a perfectly rigid body. The free precession considered here should not be confused with the much slower precession of the equinoxes, which is a forced precession. See Prob. 18.118.)

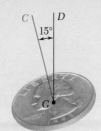

Fig. P18.124

18.124 A coin is tossed into the air. It is observed to spin at the rate of 600 rpm about an axis GC perpendicular to the coin and to precess about the vertical direction GD. Knowing that GC forms an angle of $15°$ with GD, determine (a) the angle that the angular velocity $\boldsymbol{\omega}$ of the coin forms with GD, (b) the rate of precession of the coin about GD.

18.125 The angular velocity vector of a football that has just been kicked is horizontal, and its axis of symmetry OC is oriented as shown. Knowing that the magnitude of the angular velocity is 200 rpm and that the ratio of the axis and transverse moments of inertia is $I/I' = \frac{1}{3}$, determine (a) the orientation of the axis of precession OA, (b) the rates of precession and spin.

Fig. P18.125

18.126 A space station consists of two sections A and B of equal masses that are rigidly connected. Each section is dynamically equivalent to a homogeneous cylinder with a length of 15 m and a radius of 3 m. Knowing that the station is precessing about the fixed direction GD at the constant rate of 2 rev/h, determine the rate of spin of the station about its axis of symmetry CC'.

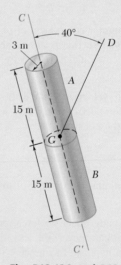

Fig. P18.126 and P18.127

18.127 If the connection between sections A and B of the space station of Prob. 18.126 is severed when the station is oriented as shown and if the two sections are gently pushed apart along their common axis of symmetry, determine (a) the angle between the spin axis and the new precession axis of section A, (b) the rate of precession of section A, (c) its rate of spin.

18.128 Solve Sample Prob. 18.6, assuming that the meteorite strikes the satellite at C with a velocity $\mathbf{v}_0 = (2000 \text{ m/s})\mathbf{i}$.

18.129 An 800-lb geostationary satellite is spinning with an angular velocity $\omega_0 = (1.5 \text{ rad/s})\mathbf{j}$ when it is hit at B by a 6-oz meteorite traveling with a velocity $\mathbf{v}_0 = -(1600 \text{ ft/s})\mathbf{i} + (1300 \text{ ft/s})\mathbf{j} + (4000 \text{ ft/s})\mathbf{k}$ relative to the satellite. Knowing that $b = 20$ in. and that the radii of gyration of the satellite are $\bar{k}_x = \bar{k}_z = 28.8$ in. and $\bar{k}_y = 32.4$ in., determine the precession axis and the rates of precession and spin of the satellite after the impact.

18.130 Solve Prob. 18.129, assuming that the meteorite hits the satellite at A instead of B.

18.131 A homogeneous rectangular plate of mass m and sides c and $2c$ is held at A and B by a fork-ended shaft of negligible mass that is supported by a bearing at C. The plate is free to rotate about AB, and the shaft is free to rotate about a horizontal axis through C. Knowing that, initially, $\theta_0 = 40°$, $\dot\theta_0 = 0$, and $\dot\phi_0 = 10$ rad/s, determine for the ensuing motion (a) the range of values of θ, (b) the minimum value of $\dot\phi$, (c) the maximum value of $\dot\theta$.

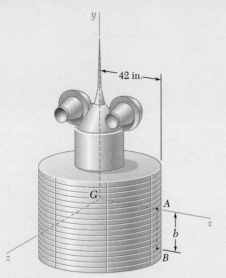

Fig. P18.129

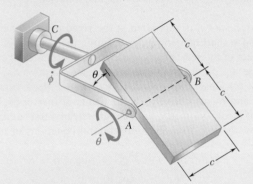

Fig. P18.131 and P18.132

18.132 A homogeneous rectangular plate of mass m and sides c and $2c$ is held at A and B by a fork-ended shaft of negligible mass that is supported by a bearing at C. The plate is free to rotate about AB, and the shaft is free to rotate about a horizontal axis through C. Initially the plate lies in the plane of the fork ($\theta_0 = 0$) and the shaft has an angular velocity $\dot\phi_0 = 10$ rad/s. If the plate is slightly disturbed, determine for the ensuing motion (a) the minimum value of $\dot\phi$, (b) the maximum value of $\dot\theta$.

18.133 A homogeneous square plate with a mass m and side c is held at points A and B by a frame of negligible mass that is supported by bearings at points C and D. The plate is free to rotate about AB, and the frame is free to rotate about the vertical CD. Knowing that, initially, $\theta_0 = 45°$, $\dot\theta_0 = 0$, and $\dot\phi_0 = 8$ rad/s, determine for the ensuing motion (a) the range of values of θ, (b) the minimum value of $\dot\phi$, (c) the maximum value of $\dot\theta$.

18.134 A homogeneous square plate with a mass m and side c is held at points A and B by a frame of negligible mass that is supported by bearings at points C and D. The plate is free to rotate about AB, and the frame is free to rotate about the vertical CD. Initially, the plate lies in the plane of the frame ($\theta_0 = 90°$), and the frame has an angular velocity of $\dot\phi = 8$ rad/s. If the plate is slightly disturbed, determine for the ensuing motion (a) the minimum value of $\dot\phi$, (b) the maximum value of $\dot\theta$.

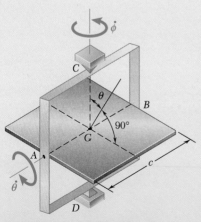

Fig. P18.133 and P18.134

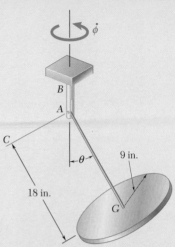

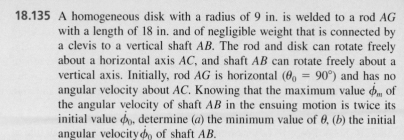

18.135 A homogeneous disk with a radius of 9 in. is welded to a rod AG with a length of 18 in. and of negligible weight that is connected by a clevis to a vertical shaft AB. The rod and disk can rotate freely about a horizontal axis AC, and shaft AB can rotate freely about a vertical axis. Initially, rod AG is horizontal ($\theta_0 = 90°$) and has no angular velocity about AC. Knowing that the maximum value $\dot\phi_m$ of the angular velocity of shaft AB in the ensuing motion is twice its initial value $\dot\phi_0$, determine (a) the minimum value of θ, (b) the initial angular velocity $\dot\phi_0$ of shaft AB.

18.136 A homogeneous disk with a radius of 9 in. is welded to a rod AG with a length of 18 in. and of negligible weight that is connected by a clevis to a vertical shaft AB. The rod and disk can rotate freely about a horizontal axis AC, and shaft AB can rotate freely about a vertical axis. Initially, rod AG is horizontal ($\theta_0 = 90°$) and has no angular velocity about AC. Knowing that the smallest value of θ in the ensuing motion is 30°, determine (a) the initial angular velocity of shaft AB, (b) its maximum angular velocity.

Fig. P18.135 and P18.136

*18.137 The top shown is supported at the fixed point O. Denoting by ϕ, θ, and ψ the Eulerian angles defining the position of the top with respect to a fixed frame of reference, consider the general motion of the top in which all Eulerian angles vary.

(a) Observing that $\Sigma M_Z = 0$ and $\Sigma M_z = 0$, and denoting by I and I', respectively, the moments of inertia of the top about its axis of symmetry and about a transverse axis through O, derive the two first-order differential equations of motion

$$I'\dot\phi \sin^2\theta + I(\dot\psi + \dot\phi \cos\theta)\cos\theta = \alpha \tag{1}$$

$$I(\dot\psi + \dot\phi \cos\theta) = \beta \tag{2}$$

where α and β are constants depending upon the initial conditions. These equations express that the angular momentum of the top is conserved about both the Z and z axes, i.e., that the rectangular component of $\mathbf{H}_O$ along each of these axes is constant.

(b) Use Eqs. (1) and (2) to show that the rectangular component ω_z of the angular velocity of the top is constant and that the rate of precession $\dot\phi$ depends upon the value of the angle of nutation θ.

*18.138 (a) Applying the principle of conservation of energy, derive a third differential equation for the general motion of the top of Prob. 18.137.

(b) Eliminating the derivatives $\dot\phi$ and $\dot\psi$ from the equation obtained and from the two equations of Prob. 18.137, show that the rate of nutation $\dot\theta$ is defined by the differential equation $\dot\theta^2 = f(\theta)$, where

$$f(\theta) = \frac{1}{I'}\left(2E - \frac{\beta^2}{I} - 2mgc\cos\theta\right) - \left(\frac{\alpha - \beta\cos\theta}{I'\sin\theta}\right)^2 \tag{1}$$

(c) Further show, by introducing the auxiliary variable $x = \cos\theta$, that the maximum and minimum values of θ can be obtained by solving for x the cubic equation

$$\left(2E - \frac{\beta^2}{I} - 2mgcx\right)(1 - x^2) - \frac{1}{I'}(\alpha - \beta x)^2 = 0 \tag{2}$$

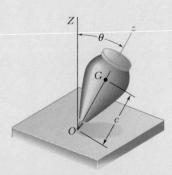

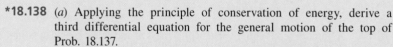

Fig. P18.137 and P18.138

*18.139 A solid cone of height 180 mm with a circular base of radius 60 mm is supported by a ball and socket at A. The cone is released from the position $\theta_0 = 30°$ with a rate of spin $\dot{\psi}_0 = 300$ rad/s, a rate of precession $\dot{\phi}_0 = 20$ rad/s, and a zero rate of nutation. Determine (a) the maximum value of θ in the ensuing motion, (b) the corresponding values of the rates of spin and precession. [*Hint:* Use Eq. (2) of Prob. 18.138; you can either solve this equation numerically or reduce it to a quadratic equation, since one of its roots is known.]

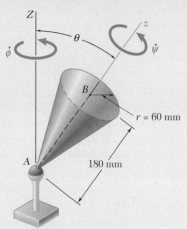

*18.140 A solid cone of height 180 mm with a circular base of radius 60 mm is supported by a ball and socket at A. The cone is released from the position $\theta_0 = 30°$ with a rate of spin $\dot{\psi}_0 = 300$ rad/s, a rate of precession $\dot{\phi}_0 = -4$ rad/s, and a zero rate of nutation. Determine (a) the maximum value of θ in the ensuing motion, (b) the corresponding values of the rates of spin and precession, (c) the value of θ for which the sense of the precession is reversed. (See hint of Prob. 18.139.)

Fig. P18.139 and P18.140

*18.141 A homogeneous sphere of mass m and radius a is welded to a rod AB of negligible mass, which is held by a ball-and-socket support at A. The sphere is released in the position $\beta = 0$ with a rate of precession $\dot{\phi} = \sqrt{17\,g/11a}$ with no spin or nutation. Determine the largest value of β in the ensuing motion.

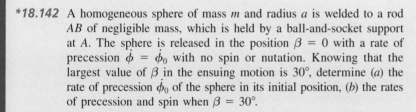

*18.142 A homogeneous sphere of mass m and radius a is welded to a rod AB of negligible mass, which is held by a ball-and-socket support at A. The sphere is released in the position $\beta = 0$ with a rate of precession $\dot{\phi} = \dot{\phi}_0$ with no spin or nutation. Knowing that the largest value of β in the ensuing motion is 30°, determine (a) the rate of precession $\dot{\phi}_0$ of the sphere in its initial position, (b) the rates of precession and spin when $\beta = 30°$.

Fig. P18.141 and P18.142

*18.143 Consider a rigid body of arbitrary shape that is attached at its mass center O and subjected to no force other than its weight and the reaction of the support at O.

(a) Prove that the angular momentum $\mathbf{H}_O$ of the body about the fixed point O is constant in magnitude and direction, that the kinetic energy T of the body is constant, and that the projection along $\mathbf{H}_O$ of the angular velocity $\boldsymbol{\omega}$ of the body is constant.

(b) Show that the tip of the vector $\boldsymbol{\omega}$ describes a curve on a fixed plane in space (called the *invariable plane*), which is perpendicular to $\mathbf{H}_O$ and at a distance $2T/H_O$ from O.

(c) Show that with respect to a frame of reference attached to the body and coinciding with its principal axes of inertia, the tip of the vector $\boldsymbol{\omega}$ appears to describe a curve on an ellipsoid of equation

$$I_x\omega_x^2 + I_y\omega_y^2 + I_z\omega_z^2 = 2T = \text{constant}$$

The ellipsoid (called the *Poinsot ellipsoid*) is rigidly attached to the body and is of the same shape as the ellipsoid of inertia, but of a different size.

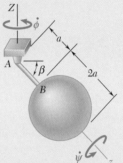

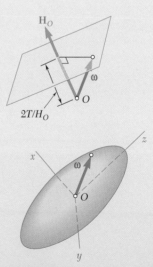

Fig. P18.143

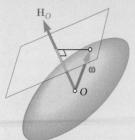

Fig. P18.144

*18.144 Referring to Prob. 18.143, (*a*) prove that the Poinsot ellipsoid is tangent to the invariable plane, (*b*) show that the motion of the rigid body must be such that the Poinsot ellipsoid appears to roll on the invariable plane. [*Hint:* In part *a*, show that the normal to the Poinsot ellipsoid at the tip of **ω** is parallel to **H**$_O$. It is recalled that the direction of the normal to a surface of equation $F(x, y, z) =$ constant at a point P is the same as that of **grad** F at point P.]

*18.145 Using the results obtained in Probs. 18.143 and 18.144, show that for an axisymmetric body attached at its mass center O and under no force other than its weight and the reaction at O, the Poinsot ellipsoid is an ellipsoid of revolution and the space and body cones are both circular and are tangent to each other. Further show that (*a*) the two cones are tangent externally, and the precession is direct, when $I < I'$, where I and I' denote, respectively, the axial and transverse moment of inertia of the body, (*b*) the space cone is inside the body cone, and the precession is retrograde, when $I > I'$.

*18.146 Refer to Probs. 18.143 and 18.144.

(*a*) Show that the curve (called *polhode*) described by the tip of the vector **ω** with respect to a frame of reference coinciding with the principal axes of inertia of the rigid body is defined by the equations

$$I_x\omega_x^2 + I_y\omega_y^2 + I_z\omega_z^2 = 2T = \text{constant} \tag{1}$$
$$I_x^2\omega_x^2 + I_y^2\omega_y^2 + I_z^2\omega_z^2 = H_O^2 = \text{constant} \tag{2}$$

and that this curve can, therefore, be obtained by intersecting the Poinsot ellipsoid with the ellipsoid defined by Eq. (2).

(*b*) Further show, assuming $I_x > I_y > I_z$, that the polhodes obtained for various values of H_O have the shapes indicated in the figure.

(*c*) Using the result obtained in part *b*, show that a rigid body under no force can rotate about a fixed centroidal axis if, and only if, that axis coincides with one of the principal axes of inertia of the body, and that the motion will be stable if the axis of rotation coincides with the major or minor axis of the Poinsot ellipsoid (z or x axis in the figure) and unstable if it coincides with the intermediate axis (y axis).

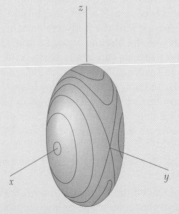

Fig. P18.146

Review and Summary

This chapter was devoted to the kinetic analysis of the motion of rigid bodies in three dimensions.

Fundamental Equations of Motion for a Rigid Body

We first noted that the two fundamental equations derived in Chap. 14 for the motion of a system of particles,

$$\Sigma \mathbf{F} = m\bar{\mathbf{a}} \tag{18.1}$$

$$\Sigma \mathbf{M}_G = \dot{\mathbf{H}}_G \tag{18.2}$$

provide the foundation of our analysis, just as they did in Chap. 16 in the case of the plane motion of rigid bodies. The computation of the angular momentum $\mathbf{H}_G$ of the body and of its derivative $\dot{\mathbf{H}}_G$, however, are now considerably more involved.

Angular Momentum of a Rigid Body in Three Dimensions

In Sec. 18.1A, we saw that we can express the rectangular components of the angular momentum $\mathbf{H}_G$ of a rigid body in terms of the components of its angular velocity $\boldsymbol{\omega}$ and of its centroidal moments and products of inertia:

$$\begin{aligned}
H_x &= +\bar{I}_x\,\omega_x - \bar{I}_{xy}\omega_y - \bar{I}_{xz}\omega_z \\
H_y &= -\bar{I}_{yx}\omega_x + \bar{I}_y\,\omega_y - \bar{I}_{yz}\omega_z \\
H_z &= -\bar{I}_{zx}\omega_x - \bar{I}_{zy}\omega_y + \bar{I}_z\,\omega_z
\end{aligned} \tag{18.7}$$

If we use **principal axes of inertia** $Gx'y'z'$, these relations reduce to

$$H_{x'} = \bar{I}_{x'}\omega_{x'} \qquad H_{y'} = \bar{I}_{y'}\omega_{y'} \qquad H_{z'} = \bar{I}_{z'}\omega_{z'} \tag{18.10}$$

We observed that, in general, *the angular momentum $\mathbf{H}_G$ and the angular velocity $\boldsymbol{\omega}$ do not have the same direction* (Fig. 18.25). They do, however, have the same direction if $\boldsymbol{\omega}$ is directed along one of the principal axes of inertia of the body.

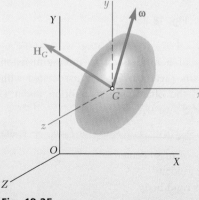

Fig. 18.25

Angular Momentum About a Given Point

Recalling that the system of the momenta of the particles forming a rigid body can be reduced to the vector $m\bar{\mathbf{v}}$ attached at G and the couple $\mathbf{H}_G$ (Fig. 18.26), we noted that, once we have determined the linear momentum $m\bar{\mathbf{v}}$ and the angular momentum $\mathbf{H}_G$ of a rigid body, we can obtain the angular momentum $\mathbf{H}_O$ of the body about any given point O from

$$\mathbf{H}_O = \bar{\mathbf{r}} \times m\bar{\mathbf{v}} + \mathbf{H}_G \qquad (18.11)$$

Rigid Body with a Fixed Point

In the particular case of a rigid body *constrained to rotate about a fixed point O*, we can obtain the components of the angular momentum $\mathbf{H}_O$ of the body about O directly from the components of its angular velocity and from its moments and products of inertia with respect to axes through O.

$$\begin{aligned} H_x &= +I_x\,\omega_x - I_{xy}\omega_y - I_{xz}\omega_z \\ H_y &= -I_{yx}\omega_x + I_y\,\omega_y - I_{yz}\omega_z \\ H_z &= -I_{zx}\omega_x - I_{zy}\omega_y + I_z\,\omega_z \end{aligned} \qquad (18.13)$$

Principle of Impulse and Momentum

The *principle of impulse and momentum* for a rigid body in three-dimensional motion [Sec. 18.1B] is expressed by the same fundamental formula that was used in Chap. 17 for a rigid body in plane motion as

$$\textbf{Syst Momenta}_1 + \textbf{Syst Ext Imp}_{1\to2} = \textbf{Syst Momenta}_2 \qquad (17.14)$$

However, the systems of the initial and final momenta should now be represented as shown in Fig. 18.26, and $\mathbf{H}_G$ should be computed from the relations in Eqs. (18.7) or (18.10) [Sample Probs. 18.1 and 18.2].

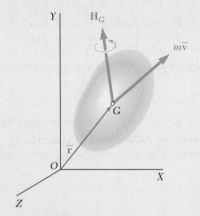

Fig. 18.26

Kinetic Energy of a Rigid Body in Three Dimensions

The kinetic energy of a rigid body in three-dimensional motion can be divided into two parts [Sec. 18.1C]: one associated with the motion of its mass center G and the other with its motion about G. Using principal centroidal axes x', y', z', we wrote

$$T = \tfrac{1}{2}m\bar{v}^2 + \tfrac{1}{2}(\bar{I}_{x'}\omega_{x'}^2 + \bar{I}_{y'}\omega_{y'}^2 + \bar{I}_{z'}\omega_{z'}^2) \qquad (18.17)$$

where $\bar{\mathbf{v}}$ = velocity of mass center

$\quad\boldsymbol{\omega}$ = angular velocity

$\quad m$ = mass of rigid body

$\bar{I}_{x'}, \bar{I}_{y'}, \bar{I}_{z'}$ = principal centroidal moments of inertia

We also noted that, in the case of a rigid body *constrained to rotate about a fixed point O*, we can express the kinetic energy of the body as

$$T = \tfrac{1}{2}(I_{x'}\omega_{x'}^2 + I_{y'}\omega_{y'}^2 + I_{z'}\omega_{z'}^2) \qquad \textbf{(18.20)}$$

where the x', y', and z' axes are the principal axes of inertia of the body about O. These results make it possible to extend the application of the principle of work and energy and the principle of conservation of energy to the three-dimensional motion of a rigid body.

Using a Rotating Frame to Write the Equations of Motion of a Rigid Body in Space

Section 18.2 was devoted to applying the fundamental equations

$$\Sigma \mathbf{F} = m\overline{\mathbf{a}} \qquad \textbf{(18.1)}$$

$$\Sigma \mathbf{M}_G = \dot{\mathbf{H}}_G \qquad \textbf{(18.2)}$$

to the motion of a rigid body in three dimensions. We first recalled [Sec. 18.2A] that $\mathbf{H}_G$ represents the angular momentum of the body relative to a centroidal frame $GX'Y'Z'$ of fixed orientation (Fig. 18.27) and that $\dot{\mathbf{H}}_G$ in Eq. (18.2)

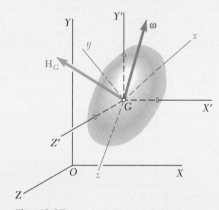

Fig. 18.27

represents the rate of change of $\mathbf{H}_G$ with respect to that frame. We noted that, as the body rotates, its moments and products of inertia with respect to the frame $GX'Y'Z'$ change continually. Therefore, it is more convenient to use a rotating frame $Gxyz$ when resolving $\boldsymbol{\omega}$ into components and computing the moments and products of inertia that are used to determine $\mathbf{H}_G$ from Eqs. (18.7) or (18.10). However, since $\dot{\mathbf{H}}_G$ in Eq. (18.2) represents the rate of change of $\mathbf{H}_G$ with respect to the frame $GX'Y'Z'$ of fixed orientation, we must use the method of Sec. 15.5A to determine its value. Recalling Eq. (15.31), we wrote

$$\dot{\mathbf{H}}_G = (\dot{\mathbf{H}}_G)_{Gxyz} + \boldsymbol{\Omega} \times \mathbf{H}_G \qquad \textbf{(18.22)}$$

where $\mathbf{H}_G$ = angular momentum of body with respect to frame $GX'Y'Z'$ of fixed orientation

$(\dot{\mathbf{H}}_G)_{Gxyz}$ = rate of change of $\mathbf{H}_G$ with respect to rotating frame $Gxyz$ to be computed from relations in Eq. (18.7)

$\boldsymbol{\Omega}$ = angular velocity of the rotating frame $Gxyz$

Substituting for $\dot{\mathbf{H}}_G$ from Eq. (18.22) into Eq. (18.2), we obtained

$$\Sigma \mathbf{M}_G = (\dot{\mathbf{H}}_G)_{Gxyz} + \boldsymbol{\Omega} \times \mathbf{H}_G \qquad \textbf{(18.23)}$$

If the rotating frame is actually attached to the body, its angular velocity $\boldsymbol{\Omega}$ is identically equal to the angular velocity $\boldsymbol{\omega}$ of the body. In many applications, however, it is advantageous to use a frame of reference that is not attached to the body but rotates in an independent manner [Sample Prob. 18.5].

Euler's Equations of Motion

Setting $\boldsymbol{\Omega} = \boldsymbol{\omega}$ in Eq. (18.23), using principal axes, and writing this equation in scalar form, we obtained **Euler's equations of motion** [Sec. 18.2B]. Then we extended Newton's second law to the three-dimensional motion of a rigid body, showing that the system of the external forces acting on the rigid body is not only equipollent but actually *equivalent* to the inertial terms of the body represented by the vector $m\bar{\mathbf{a}}$ and the couple $\dot{\mathbf{H}}_G$ (Fig. 18.28). You can solve problems involving the three-dimensional motion of a rigid body by considering the free-body and kinetic diagrams represented in Fig. 18.28 and writing appropriate scalar equations relating the components or moments of the external forces and inertial terms [Sample Probs. 18.3 and 18.5].

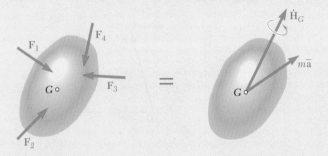

Fig. 18.28

Rigid Body with a Fixed Point

In the case of a rigid body *constrained to rotate about a fixed point O*, we can use an alternative method of solution involving the moments of the forces and the rate of change of the angular momentum about point O. We wrote [Sec. 18.2C]

$$\Sigma\mathbf{M}_O = (\dot{\mathbf{H}}_O)_{Oxyz} + \boldsymbol{\Omega} \times \mathbf{H}_O \qquad (18.28)$$

You can use this approach to solve some types of problems involving the rotation of a rigid body about a fixed axis [Sec. 18.2D]; for example, an unbalanced rotating shaft [Sample Prob. 18.4].

Motion of a Gyroscope

In Section 18.3, we considered the motion of **gyroscopes** and other *axisymmetric bodies*. We introduced the **Eulerian angles** ϕ, θ, and ψ to define the position of a gyroscope (Fig. 18.29), and we observed that their derivatives $\dot{\phi}$, $\dot{\theta}$, and $\dot{\psi}$ represent, respectively, the rates of **precession**, **nutation**, and **spin** of the gyroscope [Sec. 18.3A]. Expressing the angular velocity $\boldsymbol{\omega}$ in terms of these derivatives, we wrote

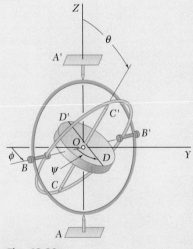

Fig. 18.29

$$\boldsymbol{\omega} = -\dot{\phi}\sin\theta\,\mathbf{i} + \dot{\theta}\,\mathbf{j} + (\dot{\psi} + \dot{\phi}\cos\theta)\mathbf{k} \qquad (18.35)$$

where the unit vectors are associated with a frame $Oxyz$ attached to the inner gimbal of the gyroscope (Fig. 18.30). These vectors rotate, therefore, with the angular velocity

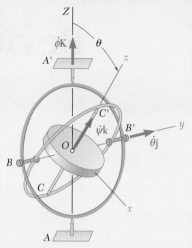

$$\mathbf{\Omega} = -\dot{\phi} \sin \theta \mathbf{i} + \dot{\theta} \mathbf{j} + \dot{\phi} \cos \theta \mathbf{k} \qquad (18.38)$$

Denoting the moment of inertia of the gyroscope with respect to its spin axis z by I and its moment of inertia with respect to a transverse axis through O by I', we wrote

$$\mathbf{H}_O = -I'\dot{\phi} \sin \theta \mathbf{i} + I'\dot{\theta} \mathbf{j} + I(\dot{\psi} + \dot{\phi} \cos \theta)\mathbf{k} \qquad (18.36)$$

Substituting for $\mathbf{H}_O$ and $\mathbf{\Omega}$ into Eq. (18.28) led us to the differential equations defining the motion of a gyroscope.

Fig. 18.30

Steady Precession

In the particular case of the **steady precession** of a gyroscope [Sec. 18.3B], the angle θ, the rate of precession $\dot{\phi}$, and the rate of spin $\dot{\psi}$ remain constant. We saw that such a motion is possible only if the moments of the external forces about O satisfy the relation

$$\Sigma \mathbf{M}_O = (I\omega_z - I'\dot{\phi} \cos \theta)\dot{\phi} \sin \theta \mathbf{j} \qquad (18.44)$$

i.e., if the external forces reduce to a couple of moment equal to the right-hand side of Eq. (18.44) and applied about an axis perpendicular to the precession axis and to the spin axis (Fig. 18.31). This chapter ended with a discussion of the motion of an axisymmetric body spinning and precessing under no force [Sec. 18.3C; Sample Prob. 18.6].

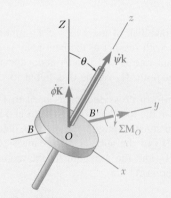

Fig. 18.31

Review Problems

18.147 Three 25-lb rotor disks are attached to a shaft that rotates at 720 rpm. Disk A is attached eccentrically so that its mass center is $\frac{1}{4}$ in. from the axis of rotation, while disks B and C are attached so that their mass centers coincide with the axis of rotation. Where should 2-lb weights be bolted to disks B and C to balance the system dynamically?

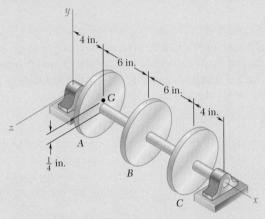

Fig. P18.147

18.148 A homogeneous disk of mass $m = 5$ kg rotates at the constant rate $\omega_1 = 8$ rad/s with respect to the bent axle ABC, which itself rotates at the constant rate $\omega_2 = 3$ rad/s about the y axis. Determine the angular momentum $\mathbf{H}_C$ of the disk about its center C.

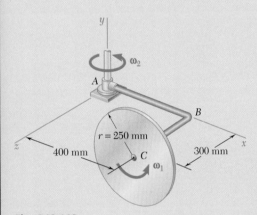

Fig. P18.148

18.149 A rod of uniform cross section is used to form the shaft shown. Denoting by m the total mass of the shaft and knowing that the shaft rotates with a constant angular velocity $\boldsymbol{\omega}$, determine (a) the angular momentum $\mathbf{H}_G$ of the shaft about its mass center G, (b) the angle formed by $\mathbf{H}_G$ and the axis AB, (c) the angular momentum of the shaft about point A.

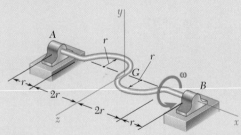

Fig. P18.149

18.150 A uniform rod of mass m and length $5a$ is bent into the shape shown and is suspended from a wire attached at point B. Knowing that the rod is hit at point A in the negative y direction and denoting the corresponding impulse by $-(F\ \Delta t)\mathbf{j}$, determine immediately after the impact (a) the velocity of the mass center G, (b) the angular velocity of the rod.

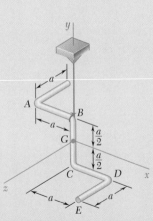

Fig. P18.150

18.151 A four-bladed airplane propeller has a mass of 160 kg and a radius of gyration of 800 mm. Knowing that the propeller rotates at 1600 rpm as the airplane is traveling in a circular path of 600-m radius at 540 km/h, determine the magnitude of the couple exerted by the propeller on its shaft due to the rotation of the airplane.

18.152 A 2.4-kg piece of sheet steel with dimensions 160×640 mm was bent to form the component shown. The component is at rest ($\omega = 0$) when a couple $\mathbf{M}_0 = (0.8 \text{ N·m})\mathbf{k}$ is applied to it. Determine (a) the angular acceleration of the component, (b) the dynamic reactions at A and B immediately after the couple is applied.

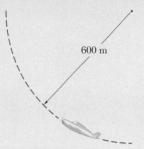

Fig. P18.151

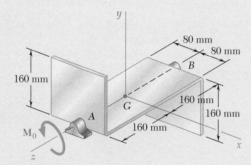

Fig. *P18.152*

18.153 A homogeneous disk of weight $W = 6$ lb rotates at the constant rate $\omega_1 = 16$ rad/s with respect to arm ABC, which is welded to a shaft DCE rotating at the constant rate $\omega_2 = 8$ rad/s. Determine the dynamic reactions at D and E.

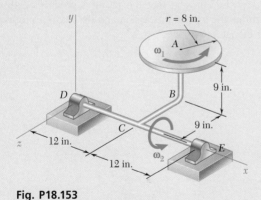

Fig. P18.153

18.154 A 48-kg advertising panel of length $2a = 2.4$ m and width $2b = 1.6$ m is kept rotating at a constant rate ω_1 about its horizontal axis by a small electric motor attached at A to frame ACB. This frame itself is kept rotating at a constant rate ω_2 about a vertical axis by a second motor attached at C to the column CD. Knowing that the panel and the frame complete a full revolution in 6 s and 12 s, respectively, express, as a function of the angle θ, the dynamic reaction exerted on column CD by its support at D.

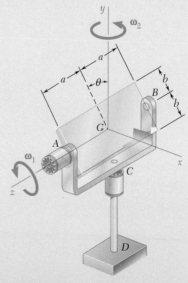

Fig. P18.154

18.155 A 2500-kg satellite is 2.4 m high and has octagonal bases of sides 1.2 m. The coordinate axes shown are the principal centroidal axes of inertia of the satellite, and its radii of gyration are $k_x = k_z = 0.90$ m and $k_y = 0.98$ m. The satellite is equipped with a main 500-N thruster E and four 20-N thrusters A, B, C, and D that can expel fuel in the positive y direction. The satellite is spinning at the rate of 36 rev/h about its axis of symmetry Gy, which maintains a fixed direction in space, when thrusters A and B are activated for 2 s. Determine (*a*) the precession axis of the satellite, (*b*) its rate of precession, (*c*) its rate of spin.

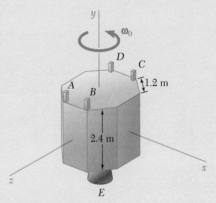

Fig. P18.155

18.156 The space capsule has no angular velocity when the jet at A is activated for 1 s in a direction parallel to the x axis. Knowing that the capsule has a mass of 1000 kg, that its radii of gyration are $\bar{k}_z = \bar{k}_y = 1.00$ m and $\bar{k}_z = 1.25$ m, and that the jet at A produces a thrust of 50 N, determine the axis of precession and the rates of precession and spin after the jet has stopped.

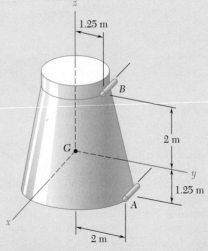

Fig. P18.156

18.157 A homogeneous disk of mass m is connected at A and B to a fork-ended shaft of negligible mass that is supported by a bearing at C. The disk is free to rotate about its horizontal diameter AB and the shaft is free to rotate about a vertical axis through C. Initially the disk lies in a vertical plane ($\theta_0 = 90°$) and the shaft has an angular velocity $\dot{\phi}_0 = 8$ rad/s. If the disk is slightly disturbed, determine for the ensuing motion (a) the minimum value of $\dot{\phi}$, (b) the maximum value of $\dot{\theta}$.

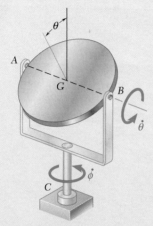

Fig. P18.157

18.158 The essential features of the gyrocompass are shown. The rotor spins at the rate $\dot{\psi}$ about an axis mounted in a single gimbal, which may rotate freely about the vertical axis AB. The angle formed by the axis of the rotor and the plane of the meridian is denoted by θ, and the latitude of the position on the earth is denoted by λ. We note that line OC is parallel to the axis of the earth, and we denote by $\boldsymbol{\omega}_e$ the angular velocity of the earth about its axis.

(a) Show that the equations of motion of the gyrocompass are

$$I'\ddot{\theta} + I\omega_z\omega_e \cos \lambda \sin \theta - I'\omega_e^2 \cos^2 \lambda \sin \theta \cos \theta = 0$$

$$I\dot{\omega}_z = 0$$

where ω_z is the rectangular component of the total angular velocity $\boldsymbol{\omega}$ along the axis of the rotor, and I and I' are the moments of inertia of the rotor with respect to its axis of symmetry and a transverse axis through O, respectively.

(b) Neglecting the term containing ω_e^2, show that for small values of θ, we have

$$\ddot{\theta} + \frac{I\omega_z\omega_e \cos \lambda}{I'}\theta = 0$$

and that the axis of the gyrocompass oscillates about the north–south direction.

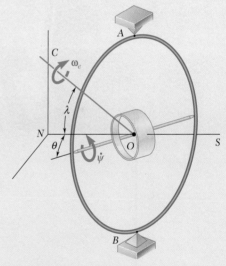

Fig. P18.158

19

Mechanical Vibrations

The Wind Damper inside of a building helps protect against typhoons and earthquakes by reducing the effects of wind and vibrations on the building. Mechanical systems may undergo *free vibrations* or they may be subject to *forced vibrations*. The vibrations are *damped* when there is energy dissipation and *undamped* otherwise. This chapter is an introduction to many fundamental concepts in vibration analysis.

Introduction

Objectives

- **Define, compare, and contrast** simple harmonic motion, undamped free and forced vibrations, and damped free and forced vibrations.
- Using Newton's second law, **determine** the differential equation of motion of a particle or a rigid body undergoing vibratory motion.
- Using the conservation of energy, **determine** the differential equation of motion of a particle or a rigid body undergoing vibratory motion.
- **Calculate** the natural circular frequency, period, and natural frequency for a system undergoing simple harmonic motion.
- **Calculate** the maximum amplitude and the magnification factor for a body undergoing forced vibrations.
- **Compare and contrast** the vibration responses of underdamped, critically damped, and overdamped systems.

Introduction

A **mechanical vibration** is the motion of a particle or body that oscillates about a position of equilibrium. Most vibrations in machines and structures are undesirable because of the increased stresses and energy losses that accompany them. Appropriate design therefore aims to eliminate or reduce vibrations as much as possible. The analysis of vibrations has become increasingly important in recent years owing to the current trend toward higher-speed machines and lighter structures. There is every reason to expect that this trend will continue and that an even greater need for vibration analysis will develop in the future.

The analysis of vibrations is a very extensive subject to which entire texts have been devoted. Our present study is limited to the simplest types of vibrations—namely, the vibrations of a body or a system of bodies with one degree of freedom.

A mechanical vibration generally results when a system is displaced from a position of stable equilibrium. The system tends to return to this position under the action of restoring forces (either elastic forces, as in the case of a mass attached to a spring, or gravitational forces, as in the case of a pendulum). But the system generally reaches its original position with an acquired velocity that carries it beyond that position. Since the process can be repeated indefinitely, the system keeps moving back and forth across its position of equilibrium. The time interval required for the system to complete a full cycle of motion is called the **period** of the vibration. The number of cycles per unit time defines the **frequency**, and the maximum displacement of the system from its position of equilibrium is called the **amplitude** of the vibration.

When the motion is maintained by the restoring forces only, the vibration is said to be a **free vibration**. When a periodic force is applied to the system, the resulting motion is described as a **forced vibration**. If we can neglect the effects of friction, the vibrations are said to be **undamped**. However, all vibrations are actually **damped** to some degree. If a free vibration is only slightly damped, its amplitude slowly decreases until, after a certain time, the motion comes to a stop. But if damping is large enough to prevent any true vibration, the system then slowly regains its original position. A damped forced vibration is maintained as long as the periodic force that produces the vibration is applied. The amplitude of the vibration, however, is affected by the magnitude of the damping forces.

In this chapter, we first examine vibrations without damping, studying vibrations of particles, rigid bodies, and forced vibrations. Then we will look at damped vibrations, including both free and forced vibrations.

19.1 VIBRATIONS WITHOUT DAMPING

The first step in analyzing vibrations is to formulate an equation of motion for the simple case of a particle in free vibration. We will modify this equation as we consider more complicated situations, such as damped and forced vibrations.

19.1A Simple Harmonic Motion and Free Vibrations of Particles

Consider a body with a mass m attached to a spring with a constant k (Fig. 19.1a). At the moment, we are concerned only with the motion of its mass center, so we will refer to this body as a particle. When the particle is in static equilibrium, the forces acting on it are its weight $\mathbf{W}$ and the force $\mathbf{T}$ exerted by the spring, which has a magnitude $T = k\delta_{st}$, where δ_{st} denotes the static elongation of the spring from its unstretched length. We therefore have

$$W = k\delta_{st}$$

Suppose now that the particle is displaced through a distance x_m from its equilibrium position and released with no initial velocity. If we have chosen x_m to be smaller than δ_{st}, the particle moves back and forth through its equilibrium position; a vibration with an amplitude x_m is generated. Note that we can also produce a vibration by imparting an initial velocity to the particle when it is in its equilibrium position $x = 0$ or, more generally, by starting the particle from any given position $x = x_0$ with a given initial velocity $\mathbf{v}_0$.

To analyze the vibration, let us consider the particle in a position P at some arbitrary time t (Fig. 19.1b). Denoting the displacement OP measured from the equilibrium position O (positive downward) by x, we note that the forces acting on the particle are its weight $\mathbf{W}$ and the force $\mathbf{T}$ exerted by the spring. In this position, the spring force has a magnitude $T = k(\delta_{st} + x)$. Recalling that $W = k\delta_{st}$, we find that the magnitude of the resultant $\mathbf{F}$ of the two forces (positive downward) is

$$F = W - k(\delta_{st} + x) = -kx \tag{19.1}$$

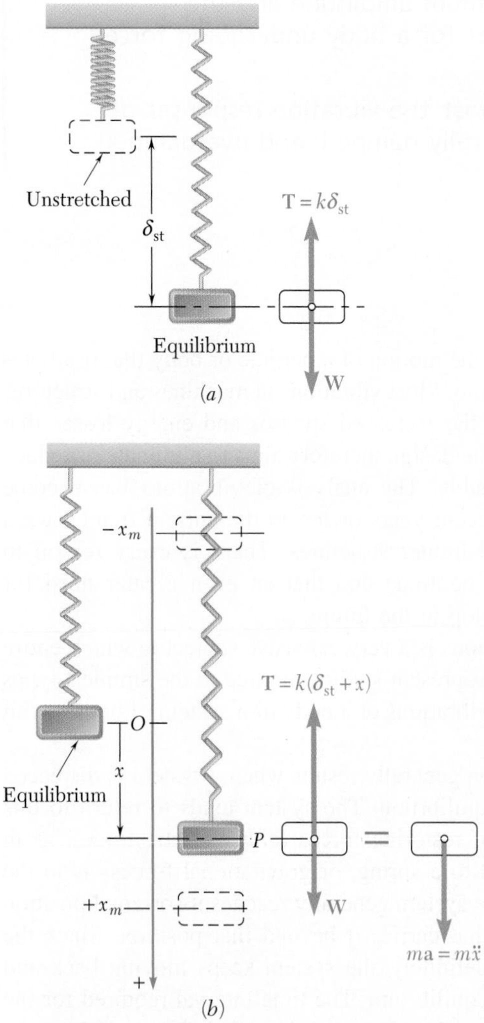

Fig. 19.1 (a) At the equilibrium position, the spring force is equal to the weight; (b) the block at position P with its free-body diagram and kinetic diagram.

Thus, the resultant of the forces exerted on the particle is proportional to the displacement OP **measured from the equilibrium position**. Recalling the sign convention, we note that **F** is always directed *toward* the equilibrium position O. Substituting for F into the fundamental equation $F = ma$ and recalling that a is the second derivative $\ddot{x}$ of x with respect to t, we have

Equation of motion for simple harmonic motion

$$m\ddot{x} + kx = 0 \tag{19.2}$$

Note that we use the same sign convention for the acceleration $\ddot{x}$ and for the displacement x, namely, positive downward. By measuring the displacement from the static equilibrium point, we get a homogeneous differential equation; that is, the right-hand side is equal to zero.

The motion defined by Eq. (19.2) is called **simple harmonic motion**. It is characterized by the fact that **the acceleration is proportional to the displacement and in the opposite direction**. We can verify that each of the functions

$$x_1 = \sin(\sqrt{k/m}\ t) \quad \text{and} \quad x_2 = \cos(\sqrt{k/m}\ t)$$

satisfies Eq. (19.2). These functions, therefore, constitute two *particular solutions* of the differential equation (19.2). We can obtain the *general solution* of Eq. (19.2) by multiplying each of the particular solutions by an arbitrary constant and adding. Thus, the general solution is

$$x = C_1 x_1 + C_2 x_2 = C_1 \sin\left(\sqrt{\frac{k}{m}}\ t\right) + C_2 \cos\left(\sqrt{\frac{k}{m}}\ t\right) \tag{19.3}$$

Note that x is a **periodic function** of the time t and therefore represents a vibration of the particle P. The coefficient of t in the expression we have obtained is referred to as the **natural circular frequency** of the vibration and is denoted by ω_n. We have

$$\text{Natural circular frequency} = \omega_n = \sqrt{\frac{k}{m}} \tag{19.4}$$

Substituting for $\sqrt{k/m}$ into Eq. (19.3) gives

$$x = C_1 \sin \omega_n t + C_2 \cos \omega_n t \tag{19.5}$$

This is the general solution of the differential equation

$$\ddot{x} + \omega_n^2 x = 0 \tag{19.6}$$

that we can obtain from Eq. (19.2) by dividing both terms by m and observing that $k/m = \omega_n^2$. Differentiating both sides of Eq. (19.5) twice with respect to t, we obtain the expressions for the velocity and the acceleration at time t as

$$v = \dot{x} = C_1 \omega_n \cos \omega_n t - C_2 \omega_n \sin \omega_n t \tag{19.7}$$

$$a = \ddot{x} = -C_1 \omega_n^2 \sin \omega_n t - C_2 \omega_n^2 \cos \omega_n t \tag{19.8}$$

The values of the constants C_1 and C_2 depend upon the *initial conditions* of the motion. For example, we have $C_1 = 0$ if the particle is displaced from its equilibrium position and released at $t = 0$ with no initial velocity. Also, we have $C_2 = 0$ if the particle starts from O at $t = 0$ with a given initial velocity. In general, substituting $t = 0$ and the initial values x_0 and v_0 of the displacement and the velocity into Eqs. (19.5) and (19.7), we find that $C_1 = v_0/\omega_n$ and $C_2 = x_0$.

We can write these expressions for the displacement, velocity, and acceleration of a particle in a more compact form if we note that Eq. (19.5) says that the displacement $x = OP$ is the sum of the x components of two vectors $\mathbf{C}_1$ and $\mathbf{C}_2$, respectively, with magnitudes of C_1 and C_2 that are directed as shown in Fig. 19.2a. As t varies, both vectors rotate clockwise; we also note that the magnitude of their resultant $\overrightarrow{OQ}$ is equal to the maximum displacement x_m. Thus, we can obtain the simple harmonic motion of P along the x axis by projecting on this axis the motion of a point Q describing an *auxiliary circle* of radius x_m with a constant angular velocity ω_n. This explains the name natural *circular* frequency given to ω_n. Denoting the angle formed by the vectors $\overrightarrow{OQ}$ and $\mathbf{C}_1$ by ϕ, we have

$$OP = OQ \sin(\omega_n t + \phi) \tag{19.9}$$

This leads to new expressions for the displacement, velocity, and acceleration of P:

$$x = x_m \sin(\omega_n t + \phi) \tag{19.10}$$

$$v = \dot{x} = x_m \omega_n \cos(\omega_n t + \phi) \tag{19.11}$$

$$a = \ddot{x} = -x_m \omega_n^2 \sin(\omega_n t + \phi) \tag{19.12}$$

The displacement-time curve is represented by a sine curve (Fig. 19.2b); the maximum value x_m of the displacement is called the **amplitude** of the vibration, and the angle ϕ that defines the initial position of Q on the circle is called the **phase angle**. As we can see from Fig. 19.2, a full cycle occurs every 2π rad. The corresponding value of t is denoted by τ_n. This is called the **period** of the free vibration and is measured in seconds. We have

$$\text{Period} = \tau_n = \frac{2\pi}{\omega_n} \tag{19.13}$$

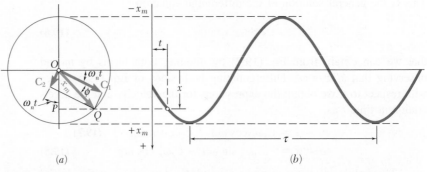

Fig. 19.2 (a) Auxiliary circle of simple harmonic motion: the resultant OQ rotates at constant angular velocity ω_n; (b) the graph of displacement versus time is a sine curve.

The number of cycles described per unit of time is denoted by f_n and is known as the **natural frequency** of the vibration. We have

$$\text{Natural frequency} = f_n = \frac{1}{\tau_n} = \frac{\omega_n}{2\pi} \qquad (19.14)$$

The unit of frequency is called a *hertz* (Hz). It also follows from Eq. (19.14) that a frequency of 1 s^{-1} or 1 Hz corresponds to a circular frequency of 2π rad/s. In problems involving angular velocities expressed in revolutions per minute (rpm), we have 1 rpm $= \frac{1}{60}$ s$^{-1} = \frac{1}{60}$ Hz, or 1 rpm $= (2\pi/60)$ rad/s.

Recall that we defined ω_n in Eq. (19.4) in terms of the constant k of the spring and the mass m of the particle. Thus, the period and the frequency are independent both of the initial conditions and of the amplitude of the vibration. Also, τ_n and f_n depend on the *mass* rather than on the *weight* of the particle and thus are independent of the value of g.

We can represent the velocity-time and acceleration-time curves using sine curves of the same period as the displacement-time curve—but with different amplitudes and different phase angles. From Eqs. (19.11) and (19.12), the maximum values of the magnitudes of the velocity and acceleration are

$$v_m = x_m \omega_n \qquad a_m = x_m \omega_n^2 \qquad (19.15)$$

The point Q describes the auxiliary circle with a radius x_m at the constant angular velocity ω_n, so its velocity and acceleration are equal, respectively, to the expressions of Eq. (19.15). Recalling Eqs. (19.11) and (19.12), we can find the velocity and acceleration of P at any instant by projecting vectors of magnitudes $v_m = x_m \omega_n$ and $a_m = x_m \omega_n^2$ on the x axis. These two vectors represent the velocity and acceleration of Q, respectively, at the same instant (Fig. 19.3).

These results are not limited to the solution of the problem of a mass attached to a spring. We can use them to analyze the rectilinear motion of a particle whenever the resultant **F** of the forces acting on the particle is proportional to the displacement x and directed toward O. In such a case, we can write the fundamental equation of motion $F = ma$ in the form of Eq. (19.6), which is characteristic of a simple harmonic motion. Observing that the coefficient of x must be equal to ω_n^2, we can easily determine the natural circular frequency ω_n of the motion. Substituting the value obtained for ω_n into Eqs. (19.13) and (19.14), we then obtain the period τ_n and the natural frequency f_n of the motion.

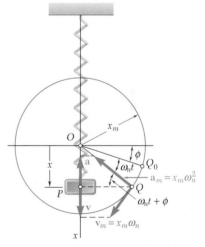

Fig. 19.3 Auxiliary circle of simple harmonic motion showing the maximum values of velocity and acceleration.

19.1B   Simple Pendulum (Approximate Solution)

Many of the vibrations encountered in engineering applications can be represented using simple harmonic motion. Many others can be *approximated* by a simple harmonic motion—provided that their amplitude remains small. Consider, for example, a **simple pendulum** consisting of a bob with

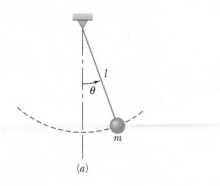

(a)

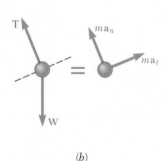

(b)

Fig. 19.4 (a) A simple pendulum consists of a bob of mass m at the end of a cord of length l; (b) free-body diagram and kinetic diagram of the simple pendulum.

a mass m attached to a cord of length l that can oscillate in a vertical plane (Fig. 19.4a). At a given time t, the cord forms an angle θ with the vertical. The forces acting on the bob are its weight $\mathbf{W}$ and the force $\mathbf{T}$ exerted by the cord (Fig. 19.4b). Resolving the vector $m\mathbf{a}$ into tangential and normal components, with $m\mathbf{a}_t$ directed to the right (i.e., in the direction corresponding to increasing values of θ), and observing that $a_t = l\alpha = l\ddot{\theta}$, we have

$$\Sigma F_t = ma_t: \qquad -W \sin\theta = ml\ddot{\theta}$$

Noting that $W = mg$ and dividing through by ml, we obtain

$$\ddot{\theta} + \frac{g}{l}\sin\theta = 0 \tag{19.16}$$

For oscillations of small amplitude, we can replace $\sin\theta$ by θ, which is expressed in radians, obtaining

$$\ddot{\theta} + \frac{g}{l}\theta = 0 \tag{19.17}$$

Comparison with Eq. (19.6) shows that the differential equation (19.17) is that of a simple harmonic motion with a natural circular frequency ω_n equal to $(g/l)^{1/2}$. Thus, we can express the general solution of Eq. (19.17) as

$$\theta = \theta_m \sin(\omega_n t + \phi)$$

where θ_m is the amplitude of the oscillations and ϕ is a phase angle. Substituting the value obtained for ω_n into Eq. (19.13), we get the expression for the period of the small oscillations of a pendulum of length l as

$$\tau_n = \frac{2\pi}{\omega_n} = 2\pi\sqrt{\frac{l}{g}} \tag{19.18}$$

*19.1C   Simple Pendulum (Exact Solution)

Formula (19.18) is only approximate. To obtain an exact expression for the period of the oscillations of a simple pendulum, we must return to Eq. (19.16). Multiplying both terms by $2\dot{\theta}$ and integrating from an initial position corresponding to the maximum deflection (that is, $\theta = \theta_m$ and $\dot{\theta} = 0$, we have

$$\left(\frac{d\theta}{dt}\right)^2 = \frac{2g}{l}(\cos\theta - \cos\theta_m)$$

We replace $\theta\cos\theta$ by $1 - 2\sin^2(\theta/2)$ and $\cos\theta_m$ by a similar expression, solve for dt, and integrate over a quarter period from $t = 0$, $\theta = 0$ to $t = \tau_n/4$, $\theta = \theta_m$. This gives

$$\tau_n = 2\sqrt{\frac{l}{g}}\int_0^{\theta_m} \frac{d\theta}{\sqrt{\sin^2(\theta_m/2) - \sin^2(\theta/2)}}$$

The integral on the right-hand side is known as an *elliptic integral*; it cannot be expressed in terms of the usual algebraic or trigonometric functions. However, setting

$$\sin(\theta/2) = \sin(\theta_m/2)\sin\phi$$

we can write

$$\tau_n = 4\sqrt{\frac{l}{g}}\int_0^{\pi/2}\frac{d\phi}{\sqrt{1 - \sin^2(\theta_m/2)\sin^2\phi}} \tag{19.19}$$

We can calculate this integral, commonly denoted by K, by using a numerical method of integration. It also can be found using computer programs such as Maple, Mathematica, or Matlab or in *tables of elliptic integrals* for various values of $\theta_m/2$.[†]

In order to compare this result with that of the preceding section, we write Eq. (19.19) in the form

$$\tau_n = \frac{2K}{\pi}\left(2\pi\sqrt{\frac{l}{g}}\right) \tag{19.20}$$

Formula (19.20) shows that we can obtain the actual value of the period of a simple pendulum by multiplying the approximate value given in Eq. (19.18) by the correction factor $2K/\pi$. Values of the correction factor are given in Table 19.1 for various values of the amplitude θ_m. Note that for ordinary engineering computations, the correction factor can be omitted as long as the amplitude does not exceed 10°.

Table 19.1 Correction Factor for the Period of a Simple Pendulum

θ_m	0°	10°	20°	30°	60°	90°	120°	150°	180°
K	1.571	1.574	1.583	1.598	1.686	1.854	2.157	2.768	∞
$2K/\pi$	1.000	1.002	1.008	1.017	1.073	1.180	1.373	1.762	∞

[†]See, for example, *Standard Mathematical Tables and Formulae*, CRC Press, Cleveland, Ohio.

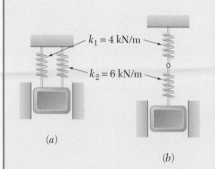

(a)

(b)

Sample Problem 19.1

A 50-kg block moves between vertical guides as shown. The block is pulled 40 mm down from its equilibrium position and released. For each spring arrangement, determine the period of the vibration, the maximum velocity of the block, and the maximum acceleration of the block.

STRATEGY: You first need to calculate the equivalent spring constant for each arrangement of the springs. Then you can use the information in this section to determine the motion.

MODELING and ANALYSIS:

a. Springs Attached in Parallel. First determine the constant k of a single spring equivalent to the two springs *by finding the magnitude of the force* $\mathbf{P}$ required to cause a given deflection δ (Fig. 1). Since for a deflection δ the magnitudes of the forces exerted by the springs are, respectively, $k_1\delta$ and $k_2\delta$, you have

$$P = k_1\delta + k_2\delta = (k_1 + k_2)\delta$$

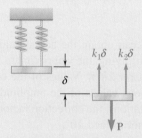

Fig. 1 Springs in parallel elongated a distanced δ.

Thus, the constant k of the single equivalent spring is

$$k = \frac{P}{\delta} = k_1 + k_2 = 4 \text{ kN/m} + 6 \text{ kN/m} = 10 \text{ kN/m} = 10^4 \text{ N/m}$$

Since $m = 50$ kg, Eq. (19.4) yields

Period of Vibration:

$$\omega_n^2 = \frac{k}{m} = \frac{10^4 \text{ N/m}}{50 \text{ kg}} \qquad \omega_n = 14.14 \text{ rad/s}$$

$$\tau_n = 2\pi/\omega_n \qquad \qquad \tau_n = 0.444 \text{ s} \quad \blacktriangleleft$$

Maximum Velocity:

$$v_m = x_m\omega_n = (0.040 \text{ m})(14.14 \text{ rad/s})$$

$$v_m = 0.566 \text{ m/s} \qquad \mathbf{v}_m = 0.566 \text{ m/s} \updownarrow \quad \blacktriangleleft$$

Maximum Acceleration:

$$a_m = x_m\omega_n^2 = (0.040 \text{ m})(14.14 \text{ rad/s})^2$$

$$a_m = 8.00 \text{ m/s}^2 \qquad \mathbf{a}_m = 8.00 \text{ m/s}^2 \updownarrow \quad \blacktriangleleft$$

b. Springs Attached in Series. In this case, determine the constant k of a single spring equivalent to the two springs *by finding the total elongation* δ *of the springs under a given static load* **P** (Fig. 2).

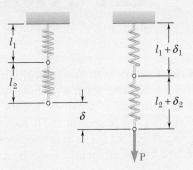

Fig. 2 Springs in series elongated a distance δ.

To facilitate the computation, you can use an arbitrary static load with a magnitude of $P = 12$ kN (this number is chosen since it has four and six as divisors). You obtain

$$\delta = \delta_1 + \delta_2 = \frac{P}{k_1} + \frac{P}{k_2} = \frac{12 \text{ kN}}{4 \text{ kN/m}} + \frac{12 \text{ kN}}{6 \text{ kN/m}} = 5 \text{ m}$$

$$k = \frac{P}{\delta} = \frac{12 \text{ kN}}{5 \text{ m}} = 2.4 \text{ kN/m} = 2400 \text{ N/m}$$

Period of Vibration:

$$\omega_n^2 = \frac{k}{m} = \frac{2400 \text{ N/m}}{50 \text{ kg}} \qquad \omega_n = 6.93 \text{ rad/s}$$

$$\tau_n = \frac{2\pi}{\omega_n} \qquad \tau_n = 0.907 \text{ s} \quad \blacktriangleleft$$

Maximum Velocity:

$$v_m = x_m\omega_n = (0.040 \text{ m})(6.93 \text{ rad/s})$$
$$v_m = 0.277 \text{ m/s} \qquad \mathbf{v}_m = 0.277 \text{ m/s} \updownarrow \quad \blacktriangleleft$$

Maximum Acceleration:

$$a_m = x_m\omega_n^2 = (0.040 \text{ m})(6.93 \text{ rad/s})^2$$
$$a_m = 1.920 \text{ m/s}^2 \qquad \mathbf{a}_m = 1.920 \text{ m/s}^2 \updownarrow \quad \blacktriangleleft$$

REFLECT and THINK: The problem did not ask you to determine the expression for combining springs in series, but from this analysis, it is clear that $\delta = \frac{P}{k_1} + \frac{P}{k_2} = \frac{P}{k}$ or $\frac{1}{k} = \frac{1}{k_1} + \frac{1}{k_2}$. Thus, for springs in series, $\frac{1}{k} = \frac{1}{k_1} + \frac{1}{k_2}$, and for springs in parallel, $k = k_1 + k_2$.

SOLVING PROBLEMS
ON YOUR OWN

This chapter deals with **mechanical vibrations**, i.e., with the motion of a particle or body oscillating about a position of equilibrium. In this first section, we saw that a **free vibration** of a particle occurs when the particle is subjected to a force proportional to its displacement and in the opposite direction, such as the force exerted by a spring (Fig. 19.1). The resulting motion, called **simple harmonic motion**, is characterized by the differential equation as

$$m\ddot{x} + kx = 0 \tag{19.2}$$

where x is the displacement of the particle from the equilibrium point, $\ddot{x}$ is its acceleration, m is its mass, and k is the constant of the spring. We found the solution of this differential equation to be

$$x = x_m \sin(\omega_n t + \phi) \tag{19.10}$$

where x_m = amplitude of the vibration
$\omega_n = \sqrt{k/m}$ = natural circular frequency (rad/s)
ϕ = phase angle (rad)

We defined the **period** of the vibration as the time $\tau_n = 2\pi/\omega_n$ needed for the particle to complete one cycle. The **natural frequency** is the number of cycles per second, $f_n = 1/\tau_n = \omega_n/2\pi$, expressed in Hz or s^{-1}. Differentiating Eq. (19.10) twice yields the velocity and the acceleration of the particle at any time. We found the maximum values of the velocity and acceleration to be

$$v_m = x_m \omega_n \qquad a_m = x_m \omega_n^2 \tag{19.15}$$

To determine the parameters in Eq. (19.10), you can follow these steps:

1. Draw a free-body diagram showing the forces exerted on the particle when the particle is at a distance x from its position of equilibrium. The resultant of these forces is proportional to x, and its direction is opposite to the positive direction of x [Eq. (19.1)].

2. Write the differential equation of motion by equating $m\ddot{x}$ to the resultant of the forces found in Step 1. Note that once you have chosen a positive direction for x, you should use the same sign convention for the acceleration $\ddot{x}$. After transposition, you will obtain an equation of the form of Eq. (19.2).

3. Determine the natural circular frequency ω_n by dividing the coefficient of x by the coefficient of $\ddot{x}$ in this equation and taking the square root of the result. Make sure that ω_n is expressed in rad/s.

4. Determine the amplitude x_m **and the phase angle** ϕ by substituting the value obtained for ω_n and the initial values of x and $\ddot{x}$ into Eq. (19.10) and the equation obtained by differentiating Eq. (19.10) with respect to t.

You can now use Eq. (19.10) and the two equations obtained by differentiating Eq. (19.10) twice with respect to t to find the displacement, velocity, and acceleration of the particle at any time. Equations (19.15) yield the maximum velocity v_m and the maximum acceleration a_m.

5. For the small oscillations of a simple pendulum, the angle θ that the cord of the pendulum forms with the vertical satisfies the differential equation

$$\ddot{\theta} + \frac{g}{l}\theta = 0 \tag{19.17}$$

where l is the length of the cord and θ is expressed in radians [Sec. 19.1B]. This equation defines again a simple harmonic motion, and its solution is of the same form as Eq. (19.10) as

$$\theta = \theta_m \sin{(\omega_n t + \phi)}$$

where the natural circular frequency $\omega_n = \sqrt{g/l}$ is expressed in rad/s. The determination of the various constants in this expression is carried out in a manner similar to that described previously. Remember that the velocity of the bob is tangent to the path and that its magnitude is $v = l\dot{\theta}$, whereas the acceleration of the bob has a tangential component $\mathbf{a}_t$ with a magnitude of $a_t = l\ddot{\theta}$ and a normal component $\mathbf{a}_n$ directed toward the center of the path and with a magnitude of $a_n = l\dot{\theta}^2$.

Problems

19.1 A particle moves in simple harmonic motion. Knowing that the maximum velocity is 200 mm/s and the maximum acceleration is 4 m/s², determine the amplitude and frequency of the motion.

19.2 A particle moves in simple harmonic motion. Knowing that the amplitude is 15 in. and the maximum acceleration is 15 ft/s², determine the maximum velocity of the particle and the frequency of its motion.

19.3 Determine the amplitude and maximum acceleration of a particle that moves in simple harmonic motion with a maximum velocity of 4 ft/s and a frequency of 6 Hz.

19.4 A 32-kg block is attached to a spring and can move without friction in a slot as shown. The block is in its equilibrium position when it is struck by a hammer that imparts to the block an initial velocity of 250 mm/s. Determine (a) the period and frequency of the resulting motion, (b) the amplitude of the motion and the maximum acceleration of the block.

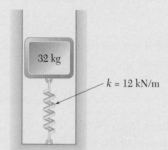

32 kg

$k = 12$ kN/m

Fig. P19.4

19.5 A 12-kg block is supported by the spring shown. If the block is moved vertically downward from its equilibrium position and released, determine (a) the period and frequency of the resulting motion, (b) the maximum velocity and acceleration of the block if the amplitude of its motion is 50 mm.

5 kN/m

12 kg

Fig. P19.5

19.6 An instrument package A is bolted to a shaker table as shown. The table moves vertically in simple harmonic motion at the same frequency as the variable-speed motor that drives it. The package is to be tested at a peak acceleration of 150 ft/s². Knowing that the amplitude of the shaker table is 2.3 in., determine (a) the required speed of the motor in rpm, (b) the maximum velocity of the table.

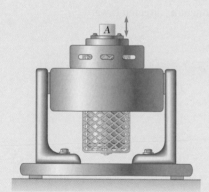

A

Fig. P19.6

19.7 A simple pendulum consisting of a bob attached to a cord oscillates in a vertical plane with a period of 1.3 s. Assuming simple harmonic motion and knowing that the maximum velocity of the bob is 0.4 m/s, determine (a) the amplitude of the motion in degrees, (b) the maximum tangential acceleration of the bob.

19.8 A simple pendulum consisting of a bob attached to a cord of length $l = 800$ mm oscillates in a vertical plane. Assuming simple harmonic motion and knowing that the bob is released from rest when $\theta = 6°$, determine (a) the frequency of oscillation, (b) the maximum velocity of the bob.

l

θ

m

Fig. P19.7 and P19.8

19.9 A 10-lb block A rests on a 40-lb plate B that is attached to an unstretched spring with a constant of $k = 60$ lb/ft. Plate B is slowly moved 2.4 in. to the left and released from rest. Assuming that block A does not slip on the plate, determine (a) the amplitude and frequency of the resulting motion, (b) the corresponding smallest allowable value of the coefficient of static friction.

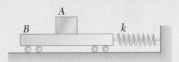

Fig. P19.9

19.10 A 5-kg fragile glass vase is surrounded by packing material in a cardboard box of negligible weight. The packing material has negligible damping and a force-deflection relationship as shown. Knowing that the box is dropped from a height of 1 m and the impact with the ground is perfectly plastic, determine (a) the amplitude of vibration for the vase, (b) the maximum acceleration the vase experiences in g's.

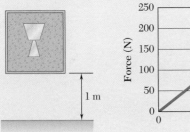
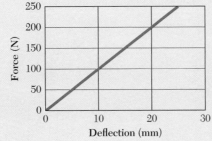

Fig. P19.10

19.11 A 3-lb block is supported as shown by a spring of constant $k = 2$ lb/in. that can act in tension or compression. The block is in its equilibrium position when it is struck from below by a hammer that imparts to the block an upward velocity of 90 in./s. Determine (a) the time required for the block to move 3 in. upward, (b) the corresponding velocity and acceleration of the block.

Fig. P19.11

19.12 In Prob. 19.11, determine the position, velocity, and acceleration of the block 0.90 s after it has been struck by the hammer.

19.13 The bob of a simple pendulum of length $l = 40$ in. is released from rest when $\theta = 5°$. Assuming simple harmonic motion, determine 1.6 s after release (a) the angle θ, (b) the magnitudes of the velocity and acceleration of the bob.

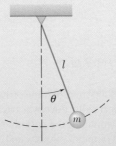

Fig. P19.13

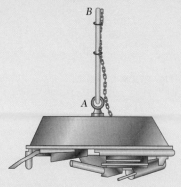

Fig. P19.14

19.14 A 150-kg electromagnet is at rest and is holding 100 kg of scrap steel when the current is turned off and the steel is dropped. Knowing that the cable and the supporting crane have a total stiffness equivalent to a spring of constant 200 kN/m, determine (*a*) the frequency, the amplitude, and the maximum velocity of the resulting motion, (*b*) the minimum tension that will occur in the cable during the motion, (*c*) the velocity of the magnet 0.03 s after the current is turned off.

19.15 A 5-kg collar *C* is released from rest in the position shown and slides without friction on a vertical rod until it hits a spring with a constant of $k = 720$ N/m that it compresses. The velocity of the collar is reduced to zero, and the collar reverses the direction of its motion and returns to its initial position. The cycle is then repeated. Determine (*a*) the period of the motion of the collar, (*b*) the velocity of the collar 0.4 s after it was released. (*Note:* This is a periodic motion, but it is not simple harmonic motion.)

Fig. *P19.15*

19.16 A small bob is attached to a cord of length 1.2 m and is released from rest when $\theta_A = 5°$. Knowing that $d = 0.6$ m, determine (*a*) the time required for the bob to return to point *A*, (*b*) the amplitude θ_C.

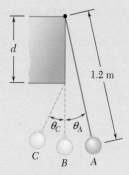

Fig. *P19.16*

19.17 A 25-kg block is supported by the spring arrangement shown. If the block is moved vertically downward from its equilibrium position and released, determine (*a*) the period and frequency of the resulting motion, (*b*) the maximum velocity and acceleration of the block if the amplitude of the motion is 30 mm.

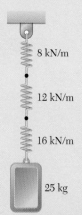

Fig. P19.17

19.18 A 11-lb block is attached to the lower end of a spring whose upper end is fixed and vibrates with a period of 7.2 s. Knowing that the constant k of a spring is inversely proportional to its length (e.g., if you cut a 10 lb/in. spring in half, the remaining two springs each have a spring constant of 20 lb/in.), determine the period of a 7-lb block that is attached to the center of the same spring if the upper and lower ends of the spring are fixed.

19.19 Block A has a mass m and is supported by the spring arrangement as shown. Knowing that the mass of the pulley is negligible and that the block is moved vertically downward from its equilibrium position and released, determine the frequency of the motion.

Fig. P19.19

19.20 A 13.6-kg block is supported by the spring arrangement shown. If the block is moved from its equilibrium position 44 mm vertically downward and released, determine (a) the period and frequency of the resulting motion, (b) the maximum velocity and acceleration of the block.

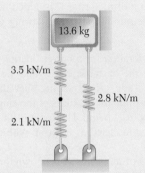

Fig. P19.20

19.21 and 19.22 A 50-kg block is supported by the spring arrangement shown. The block is moved vertically downward from its equilibrium position and released. Knowing that the amplitude of the resulting motion is 60 mm, determine (a) the period and frequency of the motion, (b) the maximum velocity and maximum acceleration of the block.

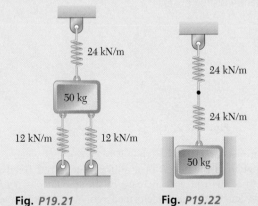

Fig. P19.21   **Fig. P19.22**

19.23 Two springs with constants k_1 and k_2 are connected in series to a block A that vibrates in simple harmonic motion with a period of 5 s. When the same two springs are connected in parallel to the same block, the block vibrates with a period of 2 s. Determine the ratio k_1/k_2 of the two spring constants.

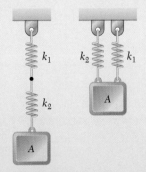

Fig. P19.23

19.24 The period of vibration of the system shown is observed to be 0.8 s. If block A is removed, the period is observed to be 0.7 s. Determine (a) the mass of block C, (b) the period of vibration when both blocks A and B have been removed.

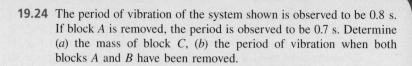

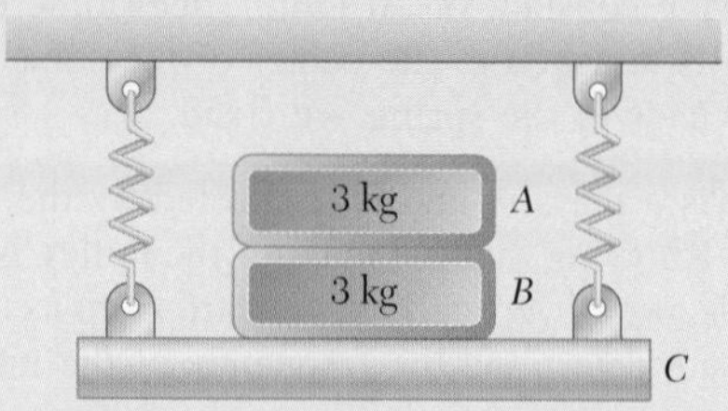

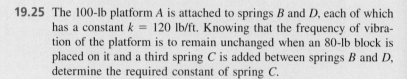

3 kg A

3 kg B

C

Fig. P19.24

19.25 The 100-lb platform A is attached to springs B and D, each of which has a constant $k = 120$ lb/ft. Knowing that the frequency of vibration of the platform is to remain unchanged when an 80-lb block is placed on it and a third spring C is added between springs B and D, determine the required constant of spring C.

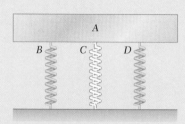

Fig. P19.25

19.26 The period of vibration for a barrel floating in salt water is found to be 0.58 s when the barrel is empty and 1.8 s when it is filled with 55 gallons of crude oil. Knowing that the density of the oil is 900 kg/m^3, determine (a) the mass of the empty barrel, (b) the density of the salt water, ρ_{sw}. [*Hint:* The force of the water on the bottom of the barrel can be modeled as a spring with constant $k = \rho_{sw}gA$.]

572 mm

Fig. P19.26

19.27 From mechanics of materials, it is known that for a simply supported beam of uniform cross section, a static load $\mathbf{P}$ applied at the center will cause a deflection of $\delta_A = PL^3/48EI$, where L is the length of the beam, E is the modulus of elasticity, and I is the moment of inertia of the cross-sectional area of the beam. Knowing that $L = 15$ ft, $E = 30 \times 10^6$ psi, and $I = 2 \times 10^{-3}$ ft^4, determine (a) the equivalent spring constant of the beam, (b) the frequency of vibration of a 1500-lb block attached to the center of the beam. Neglect the mass of the beam and assume that the load remains in contact with the beam.

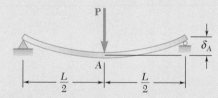

Fig. *P19.27*

19.28 From mechanics of materials it is known that when a static load $\mathbf{P}$ is applied at the end B of a uniform metal rod fixed at end A, the length of the rod will increase by an amount $\delta = PL/AE$, where L is the length of the undeformed rod, A is its cross-sectional area, and E is the modulus of elasticity of the metal. Knowing that $L = 450$ mm and $E = 200$ GPa and that the diameter of the rod is 8 mm, and neglecting the mass of the rod, determine (a) the equivalent spring constant of the rod, (b) the frequency of the vertical vibrations of a block of mass $m = 8$ kg attached to end B of the same rod.

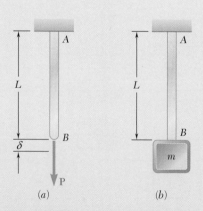

Fig. P19.28

19.29 Denoting by δ_{st} the static deflection of a beam under a given load, show that the frequency of vibration of the load is

$$f = \frac{1}{2\pi}\sqrt{\frac{g}{\delta_{st}}}$$

Neglect the mass of the beam, and assume that the load remains in contact with the beam.

19.30 A 40-mm deflection of the second floor of a building is measured directly under a newly installed 3500-kg piece of rotating machinery that has a slightly unbalanced rotor. Assuming that the deflection of the floor is proportional to the load it supports, determine (*a*) the equivalent spring constant of the floor system, (*b*) the speed in rpm of the rotating machinery that should be avoided if it is not to coincide with the natural frequency of the floor-machinery system.

19.31 If $h = 700$ mm and $d = 500$ mm and each spring has a constant $k = 600$ N/m, determine the mass m for which the period of small oscillations is (*a*) 0.50 s, (*b*) infinite. Neglect the mass of the rod and assume that each spring can act in either tension or compression.

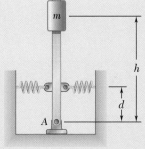

Fig. P19.31

19.32 The force–deflection equation for a nonlinear spring fixed at one end is $F = 1.5x^{1/2}$ where F is the force, expressed in newtons, applied at the other end and x is the deflection expressed in meters. (*a*) Determine the deflection x_0 if a 4-oz block is suspended from the spring and is at rest. (*b*) Assuming that the slope of the force–deflection curve at the point corresponding to this loading can be used as an equivalent spring constant, determine the frequency of vibration of the block if it is given a very small downward displacement from its equilibrium position and released.

*19.33** Expanding the integrand in Eq. (19.19) of Sec. 19.1C into a series of even powers of $\sin \phi$ and integrating, show that the period of a simple pendulum of length l may be approximated by the formula

$$\tau = 2\pi\sqrt{\frac{l}{g}}\left(1 + \tfrac{1}{4}\sin^2\frac{\theta_m}{2}\right)$$

where θ_m is the amplitude of the oscillations.

*19.34** Using the formula given in Prob. 19.33, determine the amplitude θ_m for which the period of a simple pendulum is $\frac{1}{2}$ percent longer than the period of the same pendulum for small oscillations.

*19.35** Using the data of Table 19.1, determine the period of a simple pendulum of length $l = 750$ mm (*a*) for small oscillations, (*b*) for oscillations of amplitude $\theta_m = 60°$, (*c*) for oscillations of amplitude $\theta_m = 90°$.

*19.36** Using the data of Table 19.1, determine the length in inches of a simple pendulum that oscillates with a period of 2 s and an amplitude of 90°.

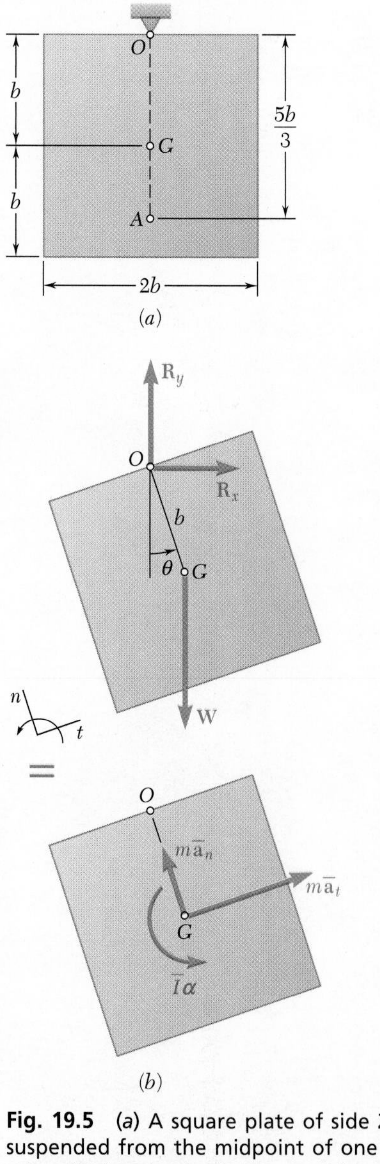

Fig. 19.5 (a) A square plate of side 2b suspended from the midpoint of one of its sides; (b) free-body diagram and kinetic diagram for the plate.

19.2 FREE VIBRATIONS OF RIGID BODIES

The analysis of the vibrations of a rigid body (or of a system of rigid bodies) possessing a single degree of freedom is similar to the analysis of the vibrations of a particle. We choose an appropriate variable, such as a distance x or an angle θ, to define the position of the body or system of bodies and write an equation relating this variable and its second derivative with respect to t. If the equation is of the same form as Eq. (19.6), i.e., if we have

$$\ddot{x} + \omega_n^2 x = 0 \qquad \text{or} \qquad \ddot{\theta} + \omega_n^2 \theta = 0 \qquad (19.21)$$

the vibration is a simple harmonic motion. We can obtain the period and natural frequency of the vibration by identifying ω_n and substituting its value into Eqs. (19.13) and (19.14).

In general, a simple way to obtain one of Eqs. (19.21) is to use Newton's second law. To do this, first draw free-body and kinetic diagrams for the system displaced in the positive direction. The acceleration in your kinetic diagram needs to be in the same positive direction you defined for the displacement. From your drawn diagrams, it is straightforward to write the appropriate equation of motion. Recall that the goal should be the determination of the coefficient of the variable x or θ—*not* the determination of the variable itself or of the derivative $\ddot{x}$ or $\ddot{\theta}$. Setting this coefficient equal to ω_n^2, we obtain the natural circular frequency ω_n from which we can determine τ_n and f_n.

This method can be used to analyze vibrations that are truly represented by a simple harmonic motion or by vibrations of small amplitude that can be *approximated* by a simple harmonic motion. As an example, let us determine the period of the small oscillations of a square plate with a side $2b$ that is suspended from the midpoint O of one of its sides (Fig. 19.5a). We consider the plate in an arbitrary position defined by the angle θ that the line OG forms with the vertical. Then we draw free-body and kinetic diagrams to express that the weight $\mathbf{W}$ of the plate and the components $\mathbf{R}_x$ and $\mathbf{R}_y$ of the reaction at O are equivalent to the vectors $m\mathbf{a}_t$ and $m\mathbf{a}_n$ and to the couple $\bar{I}\boldsymbol{\alpha}$ (Fig. 19.5b). Since the angular velocity and angular acceleration of the plate are equal to $\dot{\theta}$ and $\ddot{\theta}$, respectively, the magnitudes of the two vectors $m\mathbf{a}_t$ and $m\mathbf{a}_n$ are $mb\ddot{\theta}$ and $mb\dot{\theta}^2$, respectively, and the moment of the couple is $\bar{I}\ddot{\theta}$. In previous applications of this method (Chap. 16), we tried whenever possible to assume the correct sense for the acceleration. Here, however, we must assume the same positive sense for θ and $\ddot{\theta}$ in order to obtain an equation of the form in Eq. (19.21). Consequently, we assume the angular acceleration $\ddot{\theta}$ is positive counterclockwise—even though this assumption is obviously unrealistic. Equating moments about O, we have

$$-W(b \sin \theta) = (mb\ddot{\theta})b + \bar{I}\ddot{\theta}$$

Noting that

$$\bar{I} = \tfrac{1}{12}m[(2b)^2 + (2b)^2] = \tfrac{2}{3}mb^2 \text{ and } W = mg$$

we obtain

$$\ddot{\theta} + \frac{3}{5}\frac{g}{b}\sin\theta = 0 \qquad \textbf{(19.22)}$$

For oscillations of small amplitude, we can replace $\sin\theta$ by θ, expressed in radians, which gives

$$\ddot{\theta} + \frac{3}{5}\frac{g}{b}\theta = 0 \qquad \textbf{(19.23)}$$

Comparison with Eq. (19.21) shows that this equation is that of a simple harmonic motion and that the natural circular frequency ω_n of the oscillations is equal to $(3g/5b)^{1/2}$. Substituting into Eq. (19.13), we find that the period of the oscillations is

$$\tau_n = \frac{2\pi}{\omega_n} = 2\pi\sqrt{\frac{5b}{3g}} \qquad \textbf{(19.24)}$$

This result is valid only for oscillations of small amplitude. A more accurate description of the motion of the plate is obtained by comparing Eqs. (19.16) and (19.22). Note that the two equations are identical if we choose l equal to $5b/3$. This means that the plate oscillates as a simple pendulum with a length of $l = 5b/3$, and we can use the results of Sec. 19.1C to correct the value of the period given in Eq. (19.24). Point A of the plate located on line OG at a distance $l = 5b/3$ from O is defined as the **center of oscillation** corresponding to O (Fig. 19.5a).

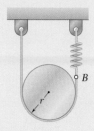

Sample Problem 19.2

A cylinder with weight W and radius r is suspended from a looped cord as shown. One end of the cord is attached directly to a rigid support, and the other end is attached to a spring with a constant k. Determine the period and natural frequency of the vibrations of the cylinder.

STRATEGY: First choose a coordinate to describe the motion, and then use Newton's second law to determine the equations of motion.

MODELING: Choose the cylinder to be your system, and model it as a rigid body. The system of external forces acting on the cylinder consists of the weight $\mathbf{W}$ and of the forces $\mathbf{T}_1$ and $\mathbf{T}_2$ exerted by the cord. Draw free-body and kinetic diagrams (Fig. 1) to express that this system is equivalent to the system represented by the vector $m\bar{a}$ attached at G and the couple $\bar{I}\alpha$.

ANALYSIS:

Kinematics of Motion. Express the linear displacement and the acceleration of the cylinder in terms of the angular displacement θ. Choosing the positive sense clockwise and measuring the displacements from the equilibrium position (Fig. 2), you have

$$\bar{x} = r\theta \qquad \delta = 2\bar{x} = 2r\theta$$
$$\boldsymbol{\alpha} = \ddot{\theta}\!\downarrow \qquad \bar{a} = r\alpha = r\ddot{\theta} \qquad \bar{\mathbf{a}} = r\ddot{\theta}\downarrow \qquad \textbf{(1)}$$

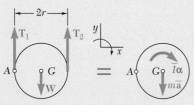

Fig. 1 Free-body diagram and kinetic diagram for the cylinder.

Equations of Motion. Newton's second law gives you (Fig. 1)

$$+\!\!\downarrow\!\Sigma M_A = m\bar{a}d_\perp + \bar{I}\alpha: \qquad Wr - T_2(2r) = m\bar{a}r + \bar{I}\alpha \qquad \textbf{(2)}$$

When the cylinder is in its position of equilibrium, the tension in the cord is $T_0 = \frac{1}{2}W$. Note that for an angular displacement θ, the magnitude of $\mathbf{T}_2$ is

$$T_2 = T_0 + k\delta = \tfrac{1}{2}W + k\delta = \tfrac{1}{2}W + k(2r\theta) \qquad \textbf{(3)}$$

Substituting from Eqs. (1) and (3) into Eq. (2) and recalling that $\bar{I} = \frac{1}{2}mr^2$, you have

$$Wr - (\tfrac{1}{2}W + 2kr\theta)(2r) = m(r\ddot{\theta})r + \tfrac{1}{2}mr^2\ddot{\theta}$$
$$\ddot{\theta} + \frac{8}{3}\frac{k}{m}\theta = 0$$

The motion is simple harmonic, and you have

$$\omega_n^2 = \frac{8}{3}\frac{k}{m} \qquad \omega_n = \sqrt{\frac{8}{3}\frac{k}{m}}$$

$$\tau_n = \frac{2\pi}{\omega_n} \qquad \tau_n = 2\pi\sqrt{\frac{3}{8}\frac{m}{k}} \qquad \blacktriangleleft$$

$$f_n = \frac{\omega_n}{2\pi} \qquad f_n = \frac{1}{2\pi}\sqrt{\frac{8}{3}\frac{k}{m}} \qquad \blacktriangleleft$$

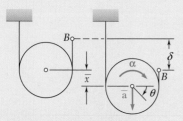

Fig. 2 Linear and angular displacements and linear and angular accelerations of the cylinder.

REFLECT and THINK: If the cylinder had been smooth, it would not have rotated when displaced downward. Also note that the answers you obtained are independent of r.

Sample Problem 19.3

A circular disk weighs 20 lb, has a radius of 8 in., and is suspended from a wire as shown. The disk is rotated (thus twisting the wire) and then released; the period of the torsional vibration is observed to be 1.13 s. A gear is then suspended from the same wire, and the period of torsional vibration for the gear is observed to be 1.93 s. Assuming that the moment of the couple exerted by the wire is proportional to the angle of twist, determine (*a*) the torsional spring constant of the wire, (*b*) the centroidal moment of inertia of the gear, (*c*) the maximum angular velocity reached by the gear if it is rotated through 90° and released.

STRATEGY: Use Newton's second law to obtain the equation of motion. From this, you can find the circular natural frequency in terms of the torsional spring constant and the centroidal moment of inertia. You can determine the torsional spring constant for the wire from the analysis of the disk. Then you can use that to describe the motion of the gear.

MODELING: Choose the disk (or gear) as your system, and model it as a rigid body. The kinematic variables are shown in Fig. 1, and the free-body and kinetic diagrams are shown in Fig. 2.

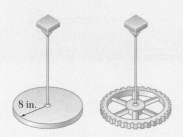

Fig. 1 Angular displacement and acceleration for the disk (or gear).

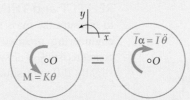

Fig. 2 Free-body diagram and kinetic diagram for the disk (or gear).

ANALYSIS:

a. Vibration of Disk. Denoting the angular displacement of the disk by θ (Fig. 1), you can express that the magnitude of the couple exerted by the wire is $M = K\theta$, where K is the torsional spring constant of the wire. Applying Newton's second law, you have

$$+\!\!\uparrow\Sigma M_O = \bar{I}\alpha: \qquad\qquad +K\theta = -\bar{I}\ddot{\theta}$$

$$\ddot{\theta} + \frac{K}{\bar{I}}\theta = 0$$

The motion is simple harmonic, so you have

$$\omega_n^2 = \frac{K}{\bar{I}} \qquad \tau_n = \frac{2\pi}{\omega_n} \qquad \tau_n = 2\pi\sqrt{\frac{\bar{I}}{K}} \qquad\qquad \textbf{(1)}$$

For the disk,

$$\tau_n = 1.13 \text{ s} \qquad \bar{I} = \tfrac{1}{2}mr^2 = \frac{1}{2}\left(\frac{20 \text{ lb}}{32.2 \text{ ft/s}^2}\right)\left(\frac{8}{12} \text{ ft}\right)^2 = 0.138 \text{ lb·ft·s}^2$$

(continued)

Substituting into Eq. (1), you obtain

$$1.13 = 2\pi\sqrt{\frac{0.138}{K}} \qquad K = 4.27 \text{ lb·ft/rad} \quad \blacktriangleleft$$

b. Vibration of Gear. The period of vibration of the gear is 1.93 s and $K = 4.27$ lb·ft/rad, so Eq. (1) yields

$$1.93 = 2\pi\sqrt{\frac{\bar{I}}{4.27}} \qquad \bar{I}_{\text{gear}} = 0.403 \text{ lb·ft·s}^2 \quad \blacktriangleleft$$

c. Maximum Angular Velocity of Gear. Because it is simple harmonic motion, you have

$$\theta = \theta_m \sin \omega_n t \qquad \omega = \theta_m \omega_n \cos \omega_n t \qquad \omega_m = \theta_m \omega_n$$

Recalling that $\theta_m = 90° = 1.571$ rad and $\tau = 1.93$ s, you have

$$\omega_m = \theta_m \omega_n = \theta_m\left(\frac{2\pi}{\tau}\right) = (1.571 \text{ rad})\left(\frac{2\pi}{1.93 \text{ s}}\right)$$

$$\omega_m = 5.11 \text{ rad/s} \quad \blacktriangleleft$$

REFLECT and THINK: A torsional spring is often used experimentally to measure the mass moment of inertia of different objects. It is common engineering practice to use one situation to determine the dynamic characteristics of a system and then to use those parameters to analyze a slightly different situation.

SOLVING PROBLEMS
ON YOUR OWN

In this section, you saw that a rigid body, or a system of rigid bodies, whose position can be defined by a single coordinate x or θ, executes a simple harmonic motion if the differential equation obtained by applying Newton's second law is of the form

$$\ddot{x} + \omega_n^2 x = 0 \qquad \text{or} \qquad \ddot{\theta} + \omega_n^2 \theta = 0 \tag{19.21}$$

Your goal should be to determine ω_n, from which you can obtain the period τ_n and the natural frequency f_n. Taking into account the initial conditions, you can then write an equation of the form

$$x = x_m \sin(\omega_n t + \phi) \tag{19.10}$$

where you should replace x by θ if a rotation is involved. To solve the problems in this section, you should follow these steps:

1. Choose a coordinate that measures the displacement of the body from its equilibrium position. You will find that many of the problems in this section involve the rotation of a body about a fixed axis and that the angle measuring the rotation of the body from its equilibrium position is the most convenient coordinate to use. In problems involving the general plane motion of a body, where a coordinate x (and possibly a coordinate y) is used to define the position of the mass center G of the body and a coordinate θ is used to measure its rotation about G, kinematic relations will allow you to express x (and y) in terms of θ [Sample Prob. 19.2].

2. Draw a free-body diagram and a kinetic diagram to express that the system of the external forces is equivalent to the vector $m\bar{\mathbf{a}}$ and the couple $\bar{I}\boldsymbol{\alpha}$ where $\bar{a} = \ddot{x}$ and $\alpha = \ddot{\theta}$. Be sure that each applied force or couple is drawn in a direction consistent with the assumed displacement and that the senses of $\bar{\mathbf{a}}$ and $\boldsymbol{\alpha}$ are those in which the coordinates x and θ are increasing.

3. Write the differential equations of motion by equating the sums of the components of the external forces and the inertial terms in the x and y directions and the sums of their moments about a given point. If necessary, use the kinematic relations developed in Step 1 to obtain equations involving only the coordinate θ. If θ is a small angle, replace $\sin \theta$ by θ and $\cos \theta$ by 1 if these functions appear in your equations. Eliminating any unknown reactions, you will obtain an equation of the type of Eqs. (19.21). Note that, in problems involving a body rotating about a fixed axis, you can immediately obtain such an equation by equating the moments of the external forces and inertial terms about the fixed axis.

(continued)

4. Comparing the equation you have obtained with one of Eqs. (19.21), you can identify ω_n^2 and thus determine the natural circular frequency ω_n. Remember that the object of your analysis is *not to solve* the differential equation you have obtained *but to identify* ω_n^2.

5. Determine the amplitude and the phase angle ϕ by substituting the value obtained for ω_n and the initial values of the coordinate and its first derivative into Eq. (19.10) and the equation obtained by differentiating Eq. (19.10) with respect to t. From Eq. (19.10) and the two equations obtained by differentiating Eq. (19.10) twice with respect to t and using the kinematic relations developed in Step 1, you will be able to determine the position, velocity, and acceleration of any point of the body at any given time.

6. In problems involving torsional vibrations, the torsional spring constant K is expressed in N·m/rad or lb·ft/rad. The product of K and the angle of twist θ, where θ is expressed in radians, yields the moment of the restoring couple, which should be equated to the inertial terms in the system. [Sample Prob. 19.3].

Problems

19.37 The uniform rod shown has mass 6 kg and is attached to a spring of constant $k = 700$ N/m. If end B of the rod is depressed 10 mm and released, determine (a) the period of vibration, (b) the maximum velocity of end B.

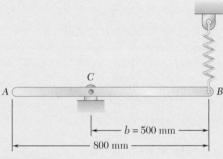

Fig. P19.37

19.38 A belt is placed around the rim of a 500-lb flywheel and attached as shown to two springs, each of constant $k = 85$ lb/in. If end C of the belt is pulled 1.5 in. down and released, the period of vibration of the flywheel is observed to be 0.5 s. Knowing that the initial tension in the belt is sufficient to prevent slipping, determine (a) the maximum angular velocity of the flywheel, (b) the centroidal radius of gyration of the flywheel.

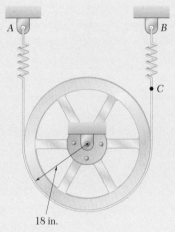

18 in.

Fig. P19.38

19.39 A 6-kg uniform cylinder can roll without sliding on a horizontal surface and is attached by a pin at point C to the 4-kg horizontal bar AB. The bar is attached to two springs, each having a constant of $k = 5$ kN/m, as shown. Knowing that the bar is moved 12 mm to the right of the equilibrium position and released, determine (a) the period of vibration of the system, (b) the magnitude of the maximum velocity of bar AB.

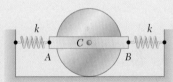

Fig. P19.39 and *P19.40*

19.40 A 6-kg uniform cylinder is assumed to roll without sliding on a horizontal surface and is attached by a pin at point C to the 4-kg horizontal bar AB. The bar is attached to two springs, each having a constant of $k = 3.5$ kN/m, as shown. Knowing that the coefficient of static friction between the cylinder and the surface is 0.5, determine the maximum amplitude of the motion of point C that is compatible with the assumption of rolling.

19.41 A 15-lb slender rod AB is riveted to a 12-lb uniform disk as shown. A belt is attached to the rim of the disk and to a spring that holds the rod at rest in the position shown. If end A of the rod is moved 0.75 in. down and released, determine (a) the period of vibration, (b) the maximum velocity of end A.

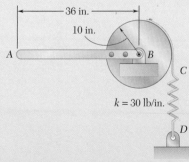

Fig. P19.41

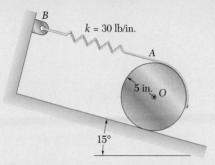

Fig. P19.42

19.42 A 30-lb uniform cylinder can roll without sliding on a 15° incline. A belt is attached to the rim of the cylinder, and a spring holds the cylinder at rest in the position shown. If the center of the cylinder is moved 2 in. down the incline and released, determine (a) the period of vibration, (b) the maximum acceleration of the center of the cylinder.

19.43 A square plate of mass m is held by eight springs, each of constant k. Knowing that each spring can act in either tension or compression, determine the frequency of the resulting vibration if (a) the plate is given a small vertical displacement and released, (b) the plate is rotated through a small angle about G and released.

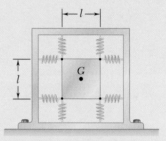

Fig. P19.43

19.44 Two small weights w are attached at A and B to the rim of a uniform disk of radius r and weight W. Denoting by τ_0 the period of small oscillations when $\beta = 0$, determine the angle β for which the period of small oscillations is $2\tau_0$.

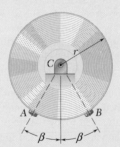

Fig. P19.44 and P19.45

19.45 Two 40-g weights are attached at A and B to the rim of a 1.5-kg uniform disk of radius r = 100 mm. Determine the frequency of small oscillations when $\beta = 60°$.

19.46 A three-blade wind turbine used for research is supported on a shaft so that it is free to rotate about O. One technique to determine the centroidal mass moment of inertia of an object is to place a known weight at a known distance from the axis of rotation and to measure the frequency of oscillations after releasing it from rest with a small initial angle. In this case, a weight of $W_{add} = 50$ lb is attached to one of the blades at a distance R = 20 ft from the axis of rotation. Knowing that when the blade with the added weight is displaced slightly from the vertical axis, and the system is found to have a period of 7.6 s, determine the centroidal mass moment of inertia of the three-blade rotor.

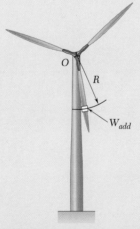

Fig. P19.46

19.47 A connecting rod is supported by a knife-edge at point A; the period of its small oscillations is observed to be 0.87 s. The rod is then inverted and supported by a knife edge at point B and the period of its small oscillations is observed to be 0.78 s. Knowing that $r_a + r_b = 10$ in., determine (*a*) the location of the mass center G, (*b*) the centroidal radius of gyration $\bar{k}$.

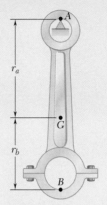

Fig. P19.47

19.48 A semicircular hole is cut in a uniform square plate that is attached to a frictionless pin at its geometric center O. Determine (*a*) the period of small oscillations of the plate, (*b*) the length of a simple pendulum that has the same period.

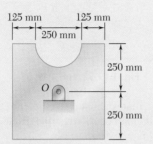

Fig. P19.48

19.49 A uniform disk of radius $r = 250$ mm is attached at A to a 650-mm rod AB of negligible mass that can rotate freely in a vertical plane about B. Determine the period of small oscillations (*a*) if the disk is free to rotate in a bearing at A, (*b*) if the rod is riveted to the disk at A.

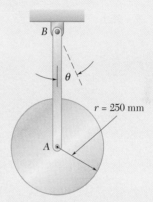

Fig. P19.49

19.50 A small collar of mass 1 kg is rigidly attached to a 3-kg uniform rod of length $L = 750$ mm. Determine (*a*) the distance d to maximize the frequency of oscillation when the rod is given a small initial displacement, (*b*) the corresponding period of oscillation.

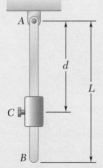

Fig. P19.50

19.51 A thin homogeneous wire is bent into the shape of an isosceles triangle of sides b, b, and $1.6b$. Determine the period of small oscillations if the wire (*a*) is suspended from point A as shown, (*b*) is suspended from point B.

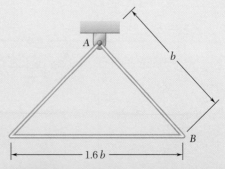

Fig. P19.51

19.52 A *compound pendulum* is defined as a rigid body that oscillates about a fixed point O, called the center of suspension. Show that the period of oscillation of a compound pendulum is equal to the period of a simple pendulum of length OA, where the distance from A to the mass center G is $GA = \bar{k}^2/\bar{r}$. Point A is defined as the center of oscillation and coincides with the center of percussion defined in Prob. 17.66.

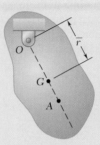

Fig. P19.52 and P19.53

19.53 A rigid slab oscillates about a fixed point O. Show that the smallest period of oscillation occurs when the distance $\bar{r}$ from point O to the mass center G is equal to $\bar{k}$.

19.54 Show that if the compound pendulum of Prob. 19.52 is suspended from A instead of O, the period of oscillation is the same as before and the new center of oscillation is located at O.

19.55 The 8-kg uniform bar AB is hinged at C and is attached at A to a spring of constant $k = 500$ N/m. If end A is given a small displacement and released, determine (a) the frequency of small oscillations, (b) the smallest value of the spring constant k for which oscillations will occur.

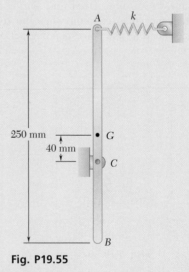

Fig. P19.55

19.56 Two uniform rods each have a mass m and length l and are welded together to form an L-shaped assembly. The assembly is constrained by two springs, each with a constant k, and is in equilibrium in a vertical plane in the position shown. Determine the frequency of small oscillations of the system.

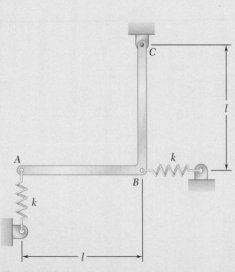

Fig. P19.56

19.57 A uniform disk with radius r and mass m can roll without slipping on a cylindrical surface and is attached to bar ABC with a length L and negligible mass. The bar is attached at point A to a spring with a constant k and can rotate freely about point B in the vertical plane. Knowing that end A is given a small displacement and released, determine the frequency of the resulting vibration in terms of m, L, k, and g.

19.58 A 1300-kg sports car has a center of gravity G located a distance h above a line connecting the front and rear axles. The car is suspended from cables that are attached to the front and rear axles as shown. Knowing that the periods of oscillation are 4.04 s when $L = 4$ m and 3.54 s when $L = 3$ m, determine h and the centroidal radius of gyration.

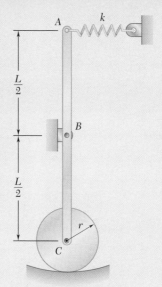

Fig. P19.57

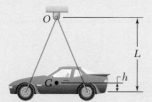

Fig. P19.58

19.59 A 6-lb slender rod is suspended from a steel wire that is known to have a torsional spring constant $K = 1.5$ ft·lb/rad. If the rod is rotated through 180° about the vertical and released, determine (a) the period of oscillation, (b) the maximum velocity of end A of the rod.

19.60 A uniform disk of radius $r = 250$ mm is attached at A to a 650-mm rod AB of negligible mass that can rotate freely in a vertical plane about B. If the rod is displaced 2° from the position shown and released, determine the magnitude of the maximum velocity of point A, assuming that the disk is (a) free to rotate in a bearing at A, (b) riveted to the rod at A.

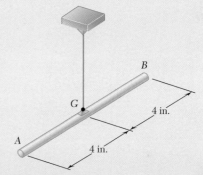

Fig. P19.59

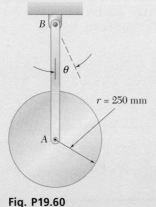

Fig. P19.60

19.61 Two uniform rods, each of mass m and length l, are welded together to form the T-shaped assembly shown. Determine the frequency of small oscillations of the assembly.

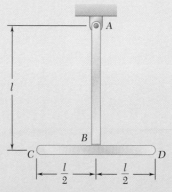

Fig. P19.61

1361

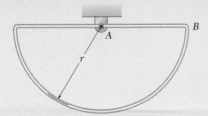

Fig. P19.62

19.62 A homogeneous wire bent to form the figure shown is attached to a pin support at A. Knowing that $r = 220$ mm and that point B is pushed down 20 mm and released, determine the magnitude of the velocity of B, 8 s later.

19.63 A horizontal platform P is held by several rigid bars that are connected to a vertical wire. The period of oscillation of the platform is found to be 2.2 s when the platform is empty and 3.8 s when an object A of unknown moment of inertia is placed on the platform with its mass center directly above the center of the plate. Knowing that the wire has a torsional constant $K = 27$ N·m/rad, determine the centroidal moment of inertia of object A.

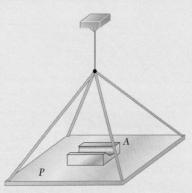

Fig. P19.63

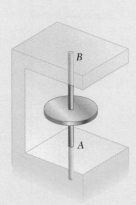

Fig. P19.64

19.64 A uniform disk of radius $r = 120$ mm is welded at its center to two elastic rods of equal length with fixed ends at A and B. Knowing that the disk rotates through an 8° angle when a 500-mN·m couple is applied to the disk and that it oscillates with a period of 1.3 s when the couple is removed, determine (*a*) the mass of the disk, (*b*) the period of vibration if one of the rods is removed.

19.65 A 5-kg uniform rod CD of length $l = 0.7$ m is welded at C to two elastic rods, which have fixed ends at A and B and are known to have a combined torsional spring constant $K = 24$ N·m/rad. Determine the period of small oscillations, if the equilibrium position of CD is (*a*) vertical as shown, (*b*) horizontal.

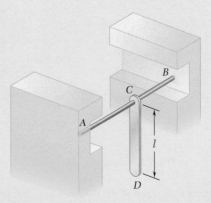

Fig. P19.65

19.66 A uniform equilateral triangular plate with a side b is suspended from three vertical wires of the same length l. Determine the period of small oscillations of the plate when (a) it is rotated through a small angle about a vertical axis through its mass center G, (b) it is given a small horizontal displacement in a direction perpendicular to AB.

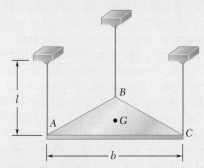

Fig. P19.66

19.67 A period of 6.00 s is observed for the angular oscillations of a 4-oz gyroscope rotor suspended from a wire as shown. Knowing that a period of 3.80 s is obtained when a 1.25-in.-diameter steel sphere is suspended in the same fashion, determine the centroidal radius of gyration of the rotor. (Specific weight of steel = 490 lb/ft^3.)

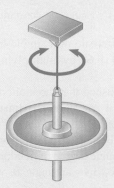

Fig. P19.67

19.68 The centroidal radius of gyration $\bar{k}_y$ of an airplane is determined by suspending the airplane by two 12-ft-long cables as shown. The airplane is rotated through a small angle about the vertical through G and then released. Knowing that the observed period of oscillation is 3.3 s, determine the centroidal radius of gyration $\bar{k}_y$.

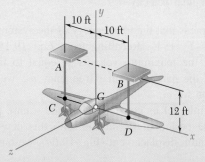

Fig. P19.68

19.3   APPLYING THE PRINCIPLE OF CONSERVATION OF ENERGY

Conservation of energy provides an alternative method to determine the natural frequency of a system. Usually, velocity kinematics are easier than acceleration kinematics, so using energy is sometimes easier than using Newton's second law directly. We saw in Sec. 19.1A that, when a particle with mass m is in simple harmonic motion, the resultant $\mathbf{F}$ of the forces exerted on the particle has a magnitude proportional to the displacement x measured from the position of equilibrium O and is directed toward O; we have $F = -kx$. Referring to Sec. 13.2A, we note that $\mathbf{F}$ is a *conservative force* and that the corresponding potential energy is $V = \frac{1}{2}kx^2$, where V is assumed equal to zero in the equilibrium position $x = 0$. The velocity of the particle is equal to $\dot{x}$, so its kinetic energy is $T = \frac{1}{2}m\dot{x}^2$. We can state that the total energy of the particle is conserved by writing

$$T + V = \text{constant} \qquad \tfrac{1}{2}m\dot{x}^2 + \tfrac{1}{2}kx^2 = \text{constant}$$

Dividing through by $m/2$ and recalling from Sec. 19.1A that $k/m = \omega_n^2$, where ω_n is the natural circular frequency of the vibration, we have

$$\dot{x}^2 + \omega_n^2 x^2 = \text{constant} \qquad\qquad \textbf{(19.25)}$$

Equation (19.25) is characteristic of a simple harmonic motion, since we can obtain it from Eq. (19.6) by multiplying both terms by $2\dot{x}$ and integrating.

Once we have established that the motion of the system is a simple harmonic motion or that it can be approximated by a simple harmonic motion, the principle of conservation of energy provides a convenient way for determining the period of vibration of a rigid body or of a system of rigid bodies possessing a single degree of freedom. Choosing an appropriate variable, such as a distance x or an angle θ, we consider two particular positions of the system:

1. **The displacement of the system is maximum.** We have $T_1 = 0$, and we can express V_1 in terms of the amplitude x_m or θ_m (choosing $V = 0$ in the equilibrium position).
2. **The system passes through its equilibrium position.** We have $V_2 = 0$, and we can express T_2 in terms of the maximum velocity $\dot{x}_m$ or the maximum angular velocity $\dot{\theta}_m$.

We then express that the total energy of the system is conserved and write $T_1 + V_1 = T_2 + V_2$. Recalling from Eq. (19.15) that for simple harmonic motion the maximum velocity is equal to the product of the amplitude and of the natural circular frequency ω_n, we find that we can solve this equation for ω_n.

As an example, let us consider again the square plate of Sec. 19.2 and determine the period of its motion with this new approach. In the position of maximum displacement (Fig. 19.6a), we have

$$T_1 = 0 \qquad V_1 = W(b - b \cos \theta_m) = Wb(1 - \cos \theta_m)$$

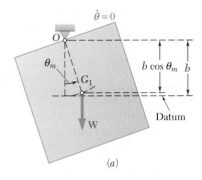

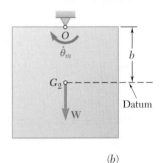

Fig. 19.6 A square plate: (a) in the position of maximum displacement; (b) as it passes through its equilibrium position.

or since $1 - \cos\theta_m = 2\sin^2(\theta_m/2) \approx 2(\theta_m/2)^2 = \theta_m^2/2$ for oscillations of small amplitude,

$$T_1 = 0 \qquad V_1 = \tfrac{1}{2}Wb\theta_m^2 \qquad \textbf{(19.26)}$$

As the plate passes through its position of equilibrium (Fig. 19.6b), its velocity is maximum, and we have

$$T_2 = \tfrac{1}{2}m\bar{v}_m^2 + \tfrac{1}{2}\bar{I}\omega_m^2 = \tfrac{1}{2}mb^2\dot{\theta}_m^2 + \tfrac{1}{2}\bar{I}\dot{\theta}_m^2 \qquad V_2 = 0$$

or recalling from Sec. 19.2 that $\bar{I} = \dfrac{2}{3}mb^2$,

$$T_2 = \tfrac{1}{2}(\tfrac{5}{3}mb^2)\dot{\theta}_m^2 \qquad V_2 = 0 \qquad \textbf{(19.27)}$$

Substituting from Eqs. (19.26) and (19.27) into $T_1 + V_1 = T_2 + V_2$ and noting that the maximum velocity $\dot{\theta}_m$ is equal to the product $\theta_m\omega_n$, we have

$$\tfrac{1}{2}Wb\theta_m^2 = \tfrac{1}{2}(\tfrac{5}{3}mb^2)\theta_m^2\omega_n^2 \qquad \textbf{(19.28)}$$

This give us $\omega_n^2 = 3g/5b$ and

$$\tau_n = \frac{2\pi}{\omega_n} = 2\pi\sqrt{\frac{5b}{3g}} \qquad \textbf{(19.29)}$$

as obtained earlier in Sec. 19.2.

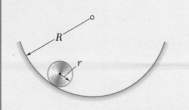

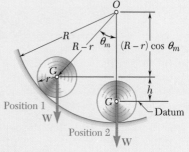

Fig. 1 The cylinder in positions 1 and 2.

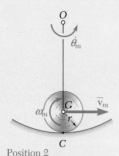

Fig. 2 Kinematic quantities to describe the motion of the disk.

Sample Problem 19.4

Determine the period of small oscillations of a cylinder with radius r that rolls without slipping inside a curved surface with radius R.

STRATEGY: Since the cylinder rolls without slipping, you can apply the principle of conservation of energy between position 1, where $\theta = \theta_m$, and position 2, where $\theta = 0$.

MODELING: Chose the cylinder to be your system and model it as a rigid body. Denote the angle that line OG forms with the vertical by θ (Fig. 1).

ANALYSIS: Position 1.

Kinetic Energy. The velocity of the cylinder is zero, so $T_1 = 0$.

Potential Energy. Choose a datum as shown in Fig. 1 and denote the weight of the cylinder by W. Then you have

$$V_1 = Wh = W(R - r)(1 - \cos \theta)$$

For small oscillations, $(1 - \cos\theta) = 2\sin^2(\theta/2) \approx \theta^2/2$, so you have

$$V_1 = W(R - r)\frac{\theta_m^2}{2}$$

Position 2. Denote the angular velocity of line OG as the cylinder passes through position 2 by $\dot{\theta}_m$, and observe that point C is the instantaneous center of rotation of the cylinder (Fig. 2). Then

$$\bar{v}_m = (R - r)\dot{\theta}_m \qquad \omega_m = \frac{\bar{v}_m}{r} = \frac{R - r}{r}\dot{\theta}_m$$

Kinetic Energy.

$$\begin{aligned} T_2 &= \tfrac{1}{2}m\bar{v}_m^2 + \tfrac{1}{2}\bar{I}\omega_m^2 \\ &= \tfrac{1}{2}m(R - r)^2\dot{\theta}_m^2 + \tfrac{1}{2}(\tfrac{1}{2}mr^2)\left(\frac{R - r}{r}\right)^2\dot{\theta}_m^2 \\ &= \tfrac{3}{4}m(R - r)^2\dot{\theta}_m^2 \end{aligned}$$

Potential Energy.

$$V_2 = 0$$

Conservation of Energy.

$$T_1 + V_1 = T_2 + V_2$$

$$0 + W(R - r)\frac{\theta_m^2}{2} = \tfrac{3}{4}m(R - r)^2\dot{\theta}_m^2 + 0$$

Since $\dot{\theta}_m = \omega_n\theta_m$ and $W = mg$, you have

$$mg(R - r)\frac{\theta_m^2}{2} = \tfrac{3}{4}m(R - r)^2(\omega_n\theta_m)^2 \qquad \omega_n^2 = \frac{2}{3}\frac{g}{R - r}$$

$$\tau_n = \frac{2\pi}{\omega_n} \qquad\qquad \tau_n = 2\pi\sqrt{\frac{3}{2}\frac{R - r}{g}} \quad \blacktriangleleft$$

REFLECT and THINK: This answer makes sense, because as the radius R increases, the period also increases. In the limit as R goes to infinity, the period also goes to infinity, that is, the system would not oscillate. This is the case of a cylinder on a horizontal surface. The small angle approximation, $(1 - \cos\theta) = 2\sin^2(\theta/2) \approx \theta^2/2$, is often used in problems like this one.

SOLVING PROBLEMS
ON YOUR OWN

In the problems that follow, you will be asked to use the *principle of conservation of energy* to determine the period or natural frequency of the simple harmonic motion of a particle or rigid body. Assuming that you choose an angle θ to define the position of the system (with $\theta = 0$ in the equilibrium position), as you will in most of the problems in this section, you will express that the total energy of the system is conserved using $T_1 + V_1 = T_2 + V_2$ between position 1 of maximum displacement $(\theta_1 = \theta_m, \dot{\theta}_1 = 0)$ and position 2 of maximum velocity $(\dot{\theta}_2 = \dot{\theta}_m, \theta_2 = 0)$. It follows that T_1 and V_2 are both zero, and the energy equation reduces to $V_1 = T_2$, where V_1 and T_2 are homogeneous quadratic expressions in θ_m and $\dot{\theta}_m$, respectively. Recalling that for a simple harmonic motion, $\dot{\theta}_m = \theta_m \omega_n$, and substituting this product into the energy equation, after reduction you obtain an equation that you can solve for ω_n^2. Once you have determined the natural circular frequency ω_n, you can obtain the period τ_n and the natural frequency f_n of the vibration.

The steps that you should take are as follows:

1. Calculate the potential energy V_1 of the system in its position of maximum displacement. Draw a sketch of the system in its position of maximum displacement and express the potential energy of all the forces involved in terms of the maximum displacement x_m or θ_m.

   a. The potential energy associated with the weight W of a body is $V_g = Wy$, where y is the elevation of the center of gravity G of the body above its equilibrium position. If the problem you are solving involves the oscillation of a rigid body about a horizontal axis through a point O located at a distance b from G (Fig. 19.6), express y in terms of the angle θ that the line OG forms with the vertical: $y = b(1 - \cos \theta)$. For small values of θ, you can replace this expression with $y = \frac{1}{2}b\theta^2$ [Sample Prob. 19.4]. Therefore, when θ reaches its maximum value θ_m and for oscillations of small amplitude, you can express V_g as

$$V_g = \frac{1}{2}Wb\theta_m^2$$

Note that *if G is located above O* in its equilibrium position (instead of below O, as we have assumed), the vertical displacement y is negative and should be approximated as $y = -\frac{1}{2}b\theta^2$, which results in a negative value for V_g. In the absence of other forces, the equilibrium position is unstable, and the system does not oscillate. (See, for instance, Prob. 19.89.)

(continued)

b. **The potential energy associated with the elastic force exerted by a spring** is $V_e = \frac{1}{2}kx^2$, where k is the constant of the spring and x is its deflection. In problems involving the rotation of a body about an axis, you generally have $x = a\theta$, where a is the distance from the axis of rotation to the point of the body where the spring is attached and θ is the angle of rotation. Therefore, when x reaches its maximum value x_m and θ reaches its maximum value θ_m, you can express V_e as

$$V_e = \tfrac{1}{2}kx_m^2 = \tfrac{1}{2}ka^2\theta_m^2$$

c. **The potential energy V_1 of the system in its position of maximum displacement** is obtained by adding the various potential energies that you have computed. It is equal to the product of a constant and θ_m^2.

2. Calculate the kinetic energy T_2 of the system in its position of maximum velocity. Note that this position is also the equilibrium position of the system.

a. **If the system consists of a single rigid body,** the kinetic energy T_2 of the system is the sum of the kinetic energy associated with the motion of the mass center G of the body and the kinetic energy associated with the rotation of the body about G. Therefore, you can write

$$T_2 = \tfrac{1}{2}m\bar{v}_m^2 + \tfrac{1}{2}\bar{I}\omega_m^2$$

Assuming that the position of the body has been defined by an angle θ, express $\bar{v}_m$ and ω_m in terms of the rate of change $\dot{\theta}_m$ of θ as the body passes through its equilibrium position. The kinetic energy of the body is thus expressed as the product of a constant and $\dot{\theta}_m^2$. Note that if θ measures the rotation of the body about its mass center, as was the case for the plate of Fig. 19.6, then $\omega_m = \dot{\theta}_m$. In other cases, however, the kinematics of the motion should be used to derive a relation between ω_m and $\dot{\theta}_m$ [Sample Prob. 19.4].

b. **If the system consists of several rigid bodies,** repeat the previous computation for each of the bodies using the same coordinate θ and add the results.

3. Equate the potential energy V_1 of the system to its kinetic energy T_2,

$$V_1 = T_2$$

and recalling the first of Eqs. (19.15), replace $\dot{\theta}_m$ in the right-hand term with the product of the amplitude θ_m and the circular frequency ω_n. Since both terms now contain the factor θ_m^2, you can cancel this factor and solve the resulting equation for the circular frequency ω_n.

Problems

19.69 Two blocks each have a mass 1.5 kg and are attached to links that are pin-connected to bar BC as shown. The masses of the links and bar are negligible, and the blocks can slide without friction. Block D is attached to a spring of constant $k = 720$ N/m. Knowing that block A is at rest when it is struck horizontally with a mallet and given an initial velocity of 250 mm/s, determine the magnitude of the maximum displacement of block D during the resulting motion.

19.70 Two small spheres, A and C, each have a mass m and are attached to rod AB that is supported by a pin and bracket at B and by a spring CD with constant k. Knowing that the mass of the rod is negligible and that the system is in equilibrium when the rod is horizontal, determine the frequency of the small oscillations of the system.

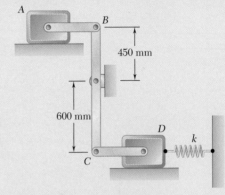

Fig. P19.69

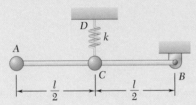

Fig. P19.70

19.71 A 14-oz sphere A and a 10-oz sphere C are attached to the ends of a rod AC of negligible weight that can rotate in a vertical plane about an axis at B. Determine the period of small oscillations of the rod.

19.72 Determine the period of small oscillations of a small particle that moves without friction inside a cylindrical surface of radius R.

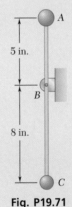

Fig. P19.71

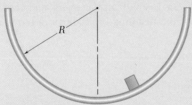

Fig. P19.72

19.73 The inner rim of an 85-lb flywheel is placed on a knife edge, and the period of its small oscillations is found to be 1.26 s. Determine the centroidal moment of inertia of the flywheel.

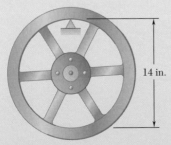

Fig. P19.73

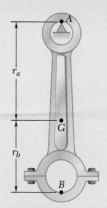

Fig. P19.74

19.74 A connecting rod is supported by a knife edge at point A; the period of its small oscillations is observed to be 1.03 s. Knowing that the distance r_a is 6 in., determine the centroidal radius of gyration of the connecting rod.

19.75 A uniform rod AB can rotate in a vertical plane about a horizontal axis at C located at a distance c above the mass center G of the rod. For small oscillations determine the value of c for which the frequency of the motion will be maximum.

Fig. P19.75

19.76 A homogeneous wire of length $2l$ is bent as shown and allowed to oscillate about a frictionless pin at B. Denoting by τ_0 the period of small oscillations when $\beta = 0$, determine the angle β for which the period of small oscillations is $2\tau_0$.

19.77 A uniform disk of radius r and mass m can roll without slipping on a cylindrical surface and is attached to bar ABC of length L and negligible mass. The bar is attached to a spring of constant k and can rotate freely in the vertical plane about point B. Knowing that end A is given a small displacement and released, determine the frequency of the resulting oscillations in terms of m, L, k, and g.

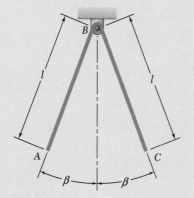

Fig. P19.76

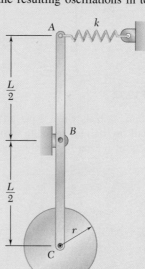

Fig. P19.77

19.78 Two uniform rods, each of weight $W = 1.2$ lb and length $l = 8$ in., are welded together to form the assembly shown. Knowing that the constant of each spring is $k = 0.6$ lb/in. and that end A is given a small displacement and released, determine the frequency of the resulting motion.

19.79 A 15-lb uniform cylinder can roll without sliding on an incline and is attached to a spring AB as shown. If the center of the cylinder is moved 0.4 in. down the incline and released, determine (a) the period of vibration, (b) the maximum velocity of the center of the cylinder.

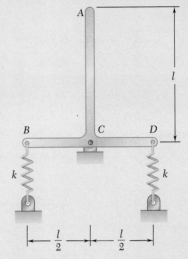

Fig. P19.78

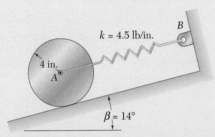

Fig. P19.79

19.80 A 3-kg slender rod AB is bolted to a 5-kg uniform disk. A spring of constant 280 N/m is attached to the disk and is unstretched in the position shown. If end B of the rod is given a small displacement and released, determine the period of vibration of the system.

19.81 A slender 10-kg bar AB with a length of $l = 0.6$ m is connected to two collars of negligible weight. Collar A is attached to a spring with a constant of $k = 1.5$ kN/m and can slide on a horizontal rod, while collar B can slide freely on a vertical rod. Knowing that the system is in equilibrium when bar AB is vertical and that collar A is given a small displacement and released, determine the period of the resulting vibrations.

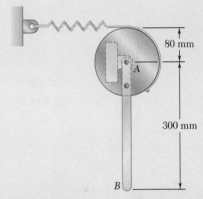

Fig. P19.80

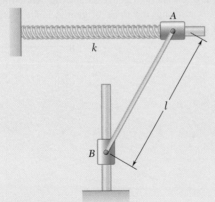

Fig. P19.81 and P19.82

19.82 A slender 5-kg bar AB with a length of $l = 0.6$ m is connected to two collars, each of mass 2.5 kg. Collar A is attached to a spring with a constant of $k = 1.5$ kN/m and can slide on a horizontal rod, while collar B can slide freely on a vertical rod. Knowing that the system is in equilibrium when bar AB is vertical and that collar A is given a small displacement and released, determine the period of the resulting vibrations.

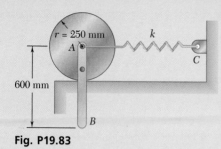

Fig. P19.83

19.83 An 800-g rod *AB* is bolted to a 1.2-kg disk. A spring of constant *k* = 12 N/m is attached to the center of the disk at *A* and to the wall at *C*. Knowing that the disk rolls without sliding, determine the period of small oscillations of the system.

19.84 Three identical 3.6-kg uniform slender bars are connected by pins as shown and can move in a vertical plane. Knowing that bar *BC* is given a small displacement and released, determine the period of vibration of the system.

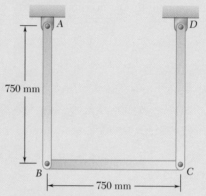

Fig. P19.84

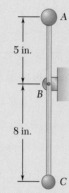

Fig. P19.85

19.85 A 14-oz sphere *A* and a 10-oz sphere *C* are attached to the ends of a 20-oz rod *AC* that can rotate in a vertical plane about an axis at *B*. Determine the period of small oscillations of the rod.

19.86 A 10-lb uniform rod *CD* is welded at *C* to a shaft of negligible mass that is welded to the centers of two 20-lb uniform disks *A* and *B*. Knowing that the disks roll without sliding, determine the period of small oscillations of the system.

19.87 **and** **19.88** Two uniform rods *AB* and *CD*, each of length *l* and mass *m*, are attached to gears as shown. Knowing that the mass of gear *C* is *m* and that the mass of gear *A* is 4*m*, determine the period of small oscillations of the system.

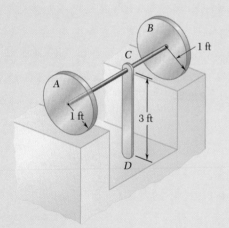

Fig. P19.86

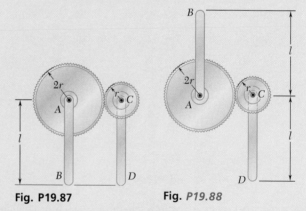

Fig. P19.87   **Fig. *P19.88***

19.89 An inverted pendulum consisting of a rigid bar *ABC* of length *l* and mass *m* is supported by a pin and bracket at *C*. A spring of constant *k* is attached to the bar at *B* and is undeformed when the bar is in the vertical position shown. Determine (*a*) the frequency of small oscillations, (*b*) the smallest value of *a* for which these oscillations will occur.

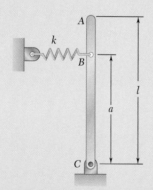

Fig. P19.89

19.90 Two 12-lb uniform disks are attached to the 20-lb rod *AB* as shown. Knowing that the constant of the spring is 30 lb/in. and that the disks roll without sliding, determine the frequency of vibration of the system.

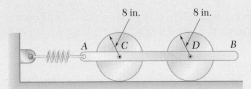

Fig. P19.90

19.91 The 20-lb rod *AB* is attached to two 8-lb disks as shown. Knowing that the disks roll without sliding, determine the frequency of small oscillations of the system.

19.92 A half section of a uniform cylinder of radius *r* and mass *m* rests on two casters *A* and *B*, each of which is a uniform cylinder of radius *r*/4 and mass *m*/8. Knowing that the half cylinder is rotated through a small angle and released and that no slipping occurs, determine the frequency of small oscillations.

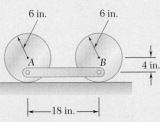

Fig. P19.91

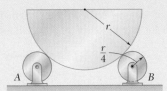

Fig. P19.92

19.93 The motion of the uniform rod *AB* is guided by the cord *BC* and by the small roller at *A*. Determine the frequency of oscillation when the end *B* of the rod is given a small horizontal displacement and released.

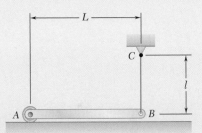

Fig. P19.93

19.94 A uniform rod of length *L* is supported by a ball-and-socket joint at *A* and by a vertical wire *CD*. Derive an expression for the period of oscillation of the rod if end *B* is given a small horizontal displacement and then released.

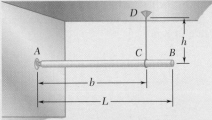

Fig. P19.94

19.95 A section of uniform pipe is suspended from two vertical cables attached at *A* and *B*. Determine the frequency of oscillation when the pipe is given a small rotation about the centroidal axis *OO'* and released.

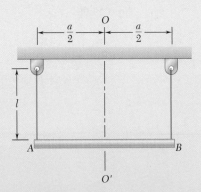

Fig. P19.95

19.96 Three collars each have a mass m and are connected by pins to bars AC and BC, each having length l and negligible mass. Collars A and B can slide without friction on a horizontal rod and are connected by a spring of constant k. Collar C can slide without friction on a vertical rod and the system is in equilibrium in the position shown. Knowing that collar C is given a small displacement and released, determine the frequency of the resulting motion of the system.

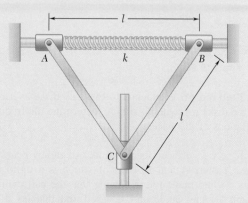

Fig. P19.96

*19.97 A thin plate of length l rests on a half cylinder of radius r. Derive an expression for the period of small oscillations of the plate.

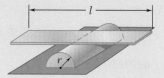

Fig. P19.97

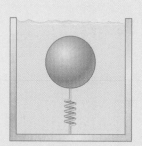

Fig. P19.98

*19.98 As a submerged body moves through a fluid, the particles of the fluid flow around the body and thus acquire kinetic energy. In the case of a sphere moving in an ideal fluid, the total kinetic energy acquired by the fluid is $\frac{1}{4}\rho V v^2$, where ρ is the mass density of the fluid, V is the volume of the sphere, and v is the velocity of the sphere. Consider a 500-g hollow spherical shell of radius 80 mm that is held submerged in a tank of water by a spring of constant 500 N/m. (*a*) Neglecting fluid friction, determine the period of vibration of the shell when it is displaced vertically and then released. (*b*) Solve part *a*, assuming that the tank is accelerated upward at the constant rate of 8 m/s^2.

19.4 FORCED VIBRATIONS

From the point of view of engineering applications, the most important vibrations are the **forced vibrations** of a system. These vibrations occur when a system is subjected to a periodic force or when it is elastically connected to a support that has an alternating motion.

Consider first the case of a body of mass m suspended from a spring and subjected to a periodic force $\mathbf{P}$ with a magnitude of $P = P_m \sin \omega_f t$, where ω_f is the circular frequency of $\mathbf{P}$ and is referred to as the **forced circular frequency** of the motion (Fig. 19.7). This force may be an actual external force applied to the body, or it may be a result of the rotation of some unbalanced part of the body (see Sample Prob. 19.5). Denoting the displacement of the body measured from its equilibrium position by x, the equation of motion is obtained from the free-body diagram and kinetic diagram in Fig. 19.7 as

$$+\downarrow \Sigma F = ma: \qquad P_m \sin \omega_f t + W - k(\delta_{st} + x) = m\ddot{x}$$

Recalling that $W = k\delta_{st}$, we have

$$m\ddot{x} + kx = P_m \sin \omega_f t \qquad (19.30)$$

Next we consider the case of a body with a mass m suspended from a spring attached to a moving support whose displacement δ is equal to $\delta_m \sin \omega_f t$ (Fig. 19.8). Measuring the displacement x of the body from the position of static equilibrium corresponding to $\omega_f t = 0$, we find that the total elongation of the spring at time t is $\delta_{st} + x - \delta_m \sin \omega_f t$. The equation of motion is thus

$$+\downarrow \Sigma F = ma: \qquad W - k(\delta_{st} + x - \delta_m \sin \omega_f t) = m\ddot{x}$$

Again recalling that $W = k\delta_{st}$, we have

$$m\ddot{x} + kx = k\delta_m \sin \omega_f t \qquad (19.31)$$

Note that Eqs. (19.30) and (19.31) are of the same form and that a solution of the first equation will satisfy the second if we set $P_m = k\delta_m$.

A differential equation such as Eq. (19.30) or (19.31), possessing a right-hand side different from zero, is said to be *nonhomogeneous*. We can obtain its general solution by adding a particular solution of the given equation to the general solution of the corresponding *homogeneous* equation (with the right-hand side equal to zero). We can obtain a *particular solution* of Eq. (19.30) or (19.31) by trying a solution of the form

$$x_{part} = x_m \sin \omega_f t \qquad (19.32)$$

Substituting x_{part} for x into Eq. (19.30), we find

$$-m\omega_f^2 x_m \sin \omega_f t + kx_m \sin \omega_f t = P_m \sin \omega_f t$$

We can solve this equation for the amplitude as

$$x_m = \frac{P_m}{k - m\omega_f^2}$$

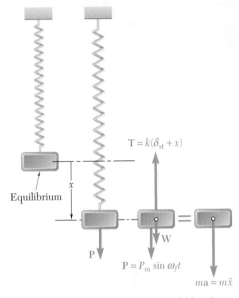

Fig. 19.7 Free-body diagram and kinetic diagram of a block suspended from a spring and subjected to a periodic force.

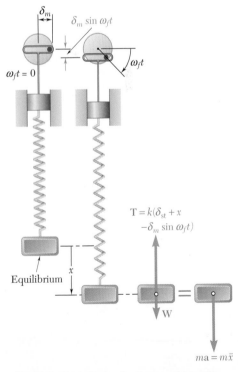

Fig. 19.8 Free-body diagram and kinetic diagram of a block suspended from a spring attached to a harmonically moving support.

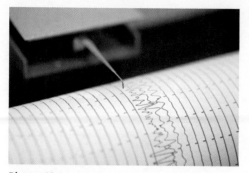

Photo 19.1 A seismometer operates by measuring the amount of electrical energy needed to keep a mass centered in the housing in the presence of strong ground vibration.

Recall from Eq. (19.4) that $k/m = \omega_n^2$, where ω_n is the natural circular frequency of the system. Then we have

$$x_m = \frac{P_m/k}{1 - (\omega_f/\omega_n)^2} \qquad (19.33)$$

If we define the frequency ratio, r, as $r = \omega_f/\omega_n$, we can write this equation as

$$x_m = \frac{P_m/k}{1 - r^2}$$

In a similar way, substituting from Eq. (19.32) into Eq. (19.31), we obtain

$$x_m = \frac{\delta_m}{1 - (\omega_f/\omega_n)^2} \qquad (19.33')$$

or

$$x_m = \frac{\delta_m}{1 - r^2}$$

The homogeneous equation corresponding to Eq. (19.30) or (19.31) is Eq. (19.2), which defines the free vibration of the body. We found its general solution, called the *complementary function,* in Sec. 19.1A:

$$x_{\text{comp}} = C_1 \sin \omega_n t + C_2 \cos \omega_n t \qquad (19.34)$$

Adding the particular solution of Eq. (19.32) to the complementary function of Eq. (19.34), we obtain the **general solution** of Eqs. (19.30) and (19.31) as

$$x = C_1 \sin \omega_n t + C_2 \cos \omega_n t + x_m \sin \omega_f t \qquad (19.35)$$

Note that this vibration consists of two superposed vibrations. The first two terms in Eq. (19.35) represent a free vibration of the system. The frequency of this vibration is the *natural frequency* of the system, which depends only upon the constant k of the spring and the mass m of the body, and the constants C_1 and C_2 can be determined from the initial conditions. This free vibration is also called a **transient** vibration, since in actual practice, it is soon damped out by friction forces (Sec. 19.5B).

The last term in Eq. (19.35) represents the **steady-state** vibration produced and maintained by the impressed force or impressed support movement. Its frequency is the **forced frequency** imposed by this force or movement, and its amplitude x_m, defined by Eq. (19.33) or (19.33′), depends upon the **frequency ratio** $r = \omega_f/\omega_n$. Dividing the amplitude x_m of the steady-state vibration by P_m/k in the case of a periodic force, or by δ_m in the case of an oscillating support, we obtain the **magnification factor**. From Eqs. (19.33) and (19.33′), we obtain

$$\text{Magnification factor} = \frac{x_m}{P_m/k} = \frac{x_m}{\delta_m} = \frac{1}{1 - (\omega_f/\omega_n)^2} \qquad (19.36)$$

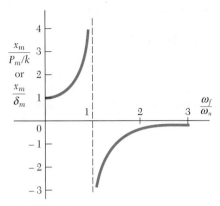

Fig. 19.9 For an undamped system, the magnification factor becomes infinite at a forcing frequency equal to the natural frequency.

In Fig. 19.9, we have plotted the magnification factor against the frequency ratio ω_f/ω_n. Note that when $\omega_f = \omega_n$, the amplitude of the forced vibration becomes infinite. The impressed force or impressed support movement is said to be in **resonance** with the given system. Actually, the amplitude of the vibration remains finite because of damping forces (Sec. 19.5B); nevertheless, such a situation should be avoided, and the forced frequency should not be chosen too close to the natural frequency of the system. Also note that for $\omega_f < \omega_n$, the coefficient of sin $\omega_f t$ in Eq. (19.35) is positive, whereas for $\omega_f > \omega_n$, this coefficient is negative. In the first case, the forced vibration is *in phase* with the impressed force or impressed support movement, while in the second case, it is 180° *out of phase.*

Finally, observe that we can obtain the velocity and acceleration of the steady-state vibration by differentiating the last term of Eq. (19.35) twice with respect to t. The maximum values are given by expressions similar to those of Eqs. (19.15) of Sec. 19.1A, except that these expressions now involve the amplitude and the circular frequency of the forced vibration:

$$v_m = x_m \omega_f \qquad a_m = x_m \omega_f^2 \qquad \textbf{(19.37)}$$

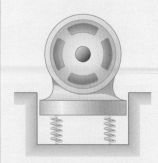

Sample Problem 19.5

A motor weighing 350 lb is supported by four springs, each having a constant of 750 lb/in. The unbalance of the rotor is equivalent to a weight of 1 oz located 6 in. from the axis of rotation. The motor is constrained to move vertically. Determine (a) the speed in rpm at which resonance will occur, (b) the amplitude of the vibration of the motor at a speed of 1200 rpm.

STRATEGY: You can determine the resonance speed directly from the given data since you know $\omega_n = \sqrt{k/m}$. To find the vibration amplitude at a speed of 1200 rpm, you can use Eq. (19.33).

MODELING: Choose the motor to be your system, and model it as a single degree-of-freedom particle undergoing forced oscillation.

ANALYSIS:

a. Resonance Speed. The resonance speed is equal to the natural circular frequency ω_n (in rpm) of the free vibration of the motor. The mass of the motor, M, and the equivalent constant of the supporting springs are

$$M = \frac{350 \text{ lb}}{32.2 \text{ ft/s}^2} = 10.87 \text{ lb·s}^2/\text{ft}$$

$$k = 4(750 \text{ lb/in.}) = 3000 \text{ lb/in.} = 36,000 \text{ lb/ft}$$

$$\omega_n = \sqrt{\frac{k}{M}} = \sqrt{\frac{36,000}{10.87}} = 57.5 \text{ rad/s} = 549 \text{ rpm}$$

$$\text{Resonance speed} = 549 \text{ rpm} \blacktriangleleft$$

b. Amplitude of Vibration at 1200 rpm. The angular velocity of the motor and the mass m of the equivalent 1-oz weight are

$$\omega = 1200 \text{ rpm} = 125.7 \text{ rad/s}$$

$$m = (1 \text{ oz})\frac{1 \text{ lb}}{16 \text{ oz}} \frac{1}{32.2 \text{ ft/s}^2} = 0.001941 \text{ lb·s}^2/\text{ft}$$

To find the equivalent of an applied force, you can draw a free-body diagram and kinetic diagram (Fig. 1).

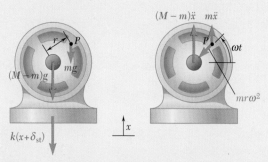

Fig. 1 Free-body diagram and kinetic diagram for the system.

Applying Newton's second law in the vertical direction gives

$$-(M - m)g - mg - k(x + \delta_{st}) = (M - m)\ddot{x} + m\ddot{x} - mr\omega^2 \sin\omega t$$

Recognizing that $Mg = k\delta_{st}$, this equation simplifies to

$$M\ddot{x} + kx = mr\omega^2 \sin \omega t$$

Thus, the rotating unbalanced mass is equivalent to an applied force

$$P_m = mr\omega^2 = (0.001941 \text{ lb·s}^2/\text{ft})(\tfrac{6}{12} \text{ ft})(125.7 \text{ rad/s})^2 = 15.33 \text{ lb}$$

The static deflection that would be caused by a constant load P_m is

$$\frac{P_m}{k} = \frac{15.33 \text{ lb}}{3000 \text{ lb/in.}} = 0.00511 \text{ in.}$$

The forced circular frequency ω_f of the motion is the angular velocity of the motor,

$$\omega_f = \omega = 125.7 \text{ rad/s}$$

Substituting the values of P_m/k, ω_f, and ω_n into Eq. (19.33), we obtain

$$x_m = \frac{P_m/k}{1 - (\omega_f/\omega_n)^2} = \frac{0.00511 \text{ in.}}{1 - (125.7/57.5)^2} = -0.001352 \text{ in.}$$

$$x_m = 0.001352 \text{ in. (out of phase)} \quad \blacktriangleleft$$

REFLECT and THINK: In problems involving an unbalanced mass, the result of the imbalance is equivalent to an applied force of $P_m = mr\omega^2$. In this problem, since $\omega_f > \omega_n$, the vibration is 180° out of phase with the force due to the unbalance of the rotor. For example, when the unbalanced mass is directly below the axis of rotation, the position of the motor is $x_m = 0.001352$ in. above the position of equilibrium.

SOLVING PROBLEMS
ON YOUR OWN

In this section, we analyzed the **forced vibrations** of a mechanical system. These vibrations occur either when the system is subjected to a periodic force **P** (Fig. 19.7) or when it is elastically connected to a support that has an alternating motion (Fig. 19.8). In the first case, the motion of the system is defined by the differential equation

$$m\ddot{x} + kx = P_m \sin \omega_f t \tag{19.30}$$

where the right-hand side represents the magnitude of the force **P** at a given instant. In the second case, the motion is defined by the differential equation

$$m\ddot{x} + kx = k\delta_m \sin \omega_f t \tag{19.31}$$

where the right-hand side is the product of the spring constant k and the displacement of the support at a given instant.

You will be concerned only with the **steady-state** motion of the system, which is defined by a *particular solution* of Eqs. (19.30) and (19.31), of the form

$$x_{\text{part}} = x_m \sin \omega_f t \tag{19.32}$$

1. If the forced vibration is caused by a periodic force P with an amplitude P_m and circular frequency ω_f, the amplitude of the vibration is

$$x_m = \frac{P_m/k}{1 - (\omega_f/\omega_n)^2} \tag{19.33}$$

where ω_n is the *natural circular frequency* of the system, $\omega_n = \sqrt{k/m}$, and k is the spring constant. Note that the circular frequency of the vibration is ω_f and that the amplitude x_m does not depend upon the initial conditions. For $\omega_f = \omega_n$, the denominator in Eq. (19.33) is zero, and x_m is infinite (Fig. 19.9); the impressed force **P** is said to be in **resonance** with the system. Also, for $\omega_f < \omega_n$, x_m is positive and the vibration is *in phase* with **P**, whereas for $\omega_f > \omega_n$, x_m is negative and the vibration is *out of phase*.

 a. In the problems that follow, you may be asked to determine one of the parameters in Eq. (19.33) when the others are known. We suggest that you keep Fig. 19.9 in front of you when solving these problems. For example, if you are asked to find the frequency at which the amplitude of a forced vibration has a given value, but you do not know whether the vibration is in or out of phase with respect to the impressed force, you should note from Fig. 19.9 that there can be two frequencies satisfying this requirement. One frequency corresponds to a positive value of x_m and to a vibration in phase with the impressed force, and the other corresponds to a negative value of x_m and to a vibration out of phase with the impressed force.

b. Once you have obtained the amplitude x_m of the motion of a component of the system from Eq. (19.33), you can use Eqs. (19.37) to determine the maximum values of the velocity and acceleration of that component:

$$v_m = x_m\omega_f \qquad a_m = x_m\omega_f^2 \tag{19.37}$$

c. When the impressed force P is due to the unbalance of the rotor of a motor, its maximum value is $P_m = mr\omega_f^2$, where m is the mass of the rotor, r is the distance between its mass center and the axis of rotation, and ω_f is equal to the angular velocity ω of the rotor expressed in rad/s [Sample Prob. 19.5].

2. If the forced vibration is caused by the simple harmonic motion of a support with an amplitude δ_m and a circular frequency ω_f, the amplitude of the vibration is

$$x_m = \frac{\delta_m}{1 - (\omega_f/\omega_n)^2} \tag{19.33'}$$

where ω_n is the *natural circular frequency* of the system and $\omega_n = \sqrt{k/m}$. Again, note that the circular frequency of the vibration is ω_f and that the amplitude x_m does not depend upon the initial conditions.

a. Be sure to read our comments in paragraphs 1, 1a, and 1b, since they apply equally well to a vibration caused by the motion of a support.

b. If the maximum acceleration a_m **of the support is specified,** rather than its maximum displacement δ_m, remember that, since the motion of the support is a simple harmonic motion, you can use the relation $a_m = \delta_m\omega_f^2$ to determine δ_m; then substitute this value into Eq. (19.33').

Problems

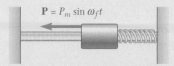

Fig. P19.99, P19.100 and P19.101

19.99 A 4-kg collar can slide on a frictionless horizontal rod and is attached to a spring with a constant of 450 N/m. It is acted upon by a periodic force with a magnitude of $P = P_m \sin \omega_f t$, where $P_m = 13$ N. Determine the amplitude of the motion of the collar if (a) $\omega_f = 5$ rad/s, (b) $\omega_f = 10$ rad/s.

19.100 A 4-kg collar can slide on a frictionless horizontal rod and is attached to a spring with constant k. It is acted upon by a periodic force of magnitude $P = P_m \sin \omega_f t$, where $P_m = 9$ N and $\omega_f = 5$ rad/s. Determine the value of the spring constant k knowing that the motion of the collar has an amplitude of 150 mm and is (a) in phase with the applied force, (b) out of phase with the applied force.

19.101 A collar with mass m that slides on a frictionless horizontal rod is attached to a spring with constant k and is acted upon by a periodic force with a magnitude of $P = P_m \sin \omega_f t$. Determine the range of values of ω_f for which the amplitude of the vibration exceeds three times the static deflection caused by a constant force with a magnitude of P_m.

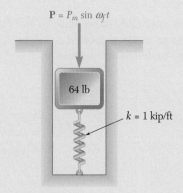

Fig. P19.102

19.102 A 64-lb block is attached to a spring with a constant of $k = 1$ kip/ft and can move without friction in a vertical slot as shown. It is acted upon by a periodic force with a magnitude of $P = P_m \sin \omega_f t$, where $\omega_f = 10$ rad/s. Knowing that the amplitude of the motion is 0.75 in., determine P_m.

19.103 A small 20-kg block A is attached to the rod BC of negligible mass that is supported at B by a pin and bracket and at C by a spring of constant $k = 2$ kN/m. The system can move in a vertical plane and is in equilibrium when the rod is horizontal. The rod is acted upon at C by a periodic force $\mathbf{P}$ of magnitude $P = P_m \sin \omega_f t$, where $P_m = 6$ N. Knowing that $b = 200$ mm, determine the range of values of ω_f for which the amplitude of vibration of block A exceeds 3.5 mm.

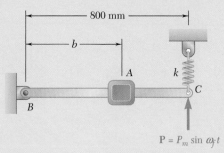

Fig. P19.103

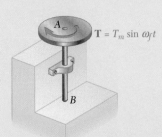

Fig. P19.104

19.104 An 8-kg uniform disk of radius 200 mm is welded to a vertical shaft with a fixed end at B. The disk rotates through an angle of 3° when a static couple of magnitude 50 N·m is applied to it. If the disk is acted upon by a periodic torsional couple of magnitude $T = T_m \sin \omega_f t$, where $T_m = 60$ N·m, determine the range of values of ω_f for which the amplitude of the vibration is less than the angle of rotation caused by a static couple of magnitude T_m.

19.105 An 18-lb block A slides in a vertical frictionless slot and is connected to a moving support B by means of a spring AB of constant $k = 10$ lb/in. Knowing that the displacement of the support is $\delta = \delta_m \sin \omega_f t$, where $\delta_m = 6$ in., determine the range of values of ω_f for which the amplitude of the fluctuating force exerted by the spring on the block is less than 30 lb.

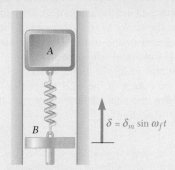

Fig. P19.105

19.106 A beam ABC is supported by a pin connection at A and by rollers at B. A 120-kg block placed on the end of the beam causes a static deflection of 15 mm at C. Assuming that the support at A undergoes a vertical periodic displacement $\delta = \delta_m \sin \omega_f t$, where $\delta_m = 10$ mm and $\omega_f = 18$ rad/s, and the support at B does not move, determine the maximum acceleration of the block at C. Neglect the weight of the beam and assume that the block does not leave the beam.

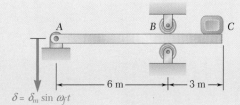

Fig. P19.106

19.107 A small 2-kg sphere B is attached to the bar AB of negligible mass that is supported at A by a pin and bracket and connected at C to a moving support D by means of a spring of constant $k = 3.6$ kN/m. Knowing that support D undergoes a vertical displacement $\delta = \delta_m \sin \omega_f t$, where $\delta_m = 3$ mm and $\omega_f = 15$ rad/s, determine (a) the magnitude of the maximum angular velocity of bar AB, (b) the magnitude of the maximum acceleration of sphere B.

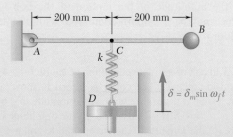

Fig. P19.107

19.108 The crude-oil pumping rig shown is driven at 20 rpm. The inside diameter of the well pipe is 2 in., and the diameter of the pump rod is 0.75 in. The length of the pump rod and the length of the column of oil lifted during the stroke are essentially the same, and equal to 6000 ft. During the downward stroke, a valve at the lower end of the pump rod opens to let a quantity of oil into the well pipe, and the column of oil is then lifted to obtain a discharge into the connecting pipeline. Thus, the amount of oil pumped in a given time depends upon the stroke of the lower end of the pump rod. Knowing that the upper end of the rod at D is essentially sinusoidal with a stroke of 45 in. and the specific weight of crude oil is 56.2 lb/ft^3, determine (a) the output of the well in ft^3/min if the shaft is rigid, (b) the output of the well in ft^3/min if the stiffness of the rod is 2210 N/m, the equivalent mass of the oil and shaft is 290 kg, and damping is negligible.

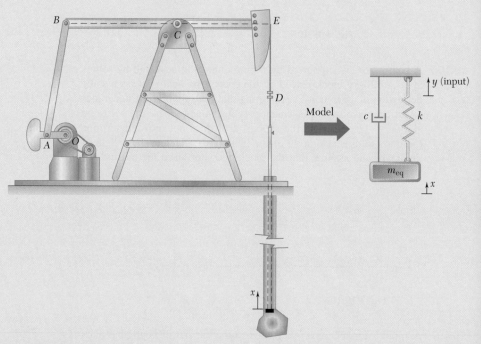

Fig. P19.108

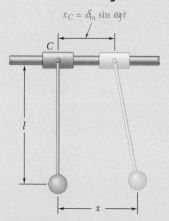

$x_C = \delta_m \sin \omega_f t$

Fig. P19.109 and P19.110

19.109 A simple pendulum of length l is suspended from collar C that is forced to move horizontally according to the relation $x_C = \delta_m \sin \omega_f t$. Determine the range of values of ω_f for which the amplitude of the motion of the bob is less than δ_m. (Assume that δ_m is small compared with the length l of the pendulum.)

19.110 The 2.75-lb bob of a simple pendulum of length $l = 24$ in. is suspended from a 3-lb collar C. The collar is forced to move according to the relation $x_C = \delta_m \sin \omega_f t$, with an amplitude $\delta_m = 0.4$ in. and a frequency $f_f = 0.5$ Hz. Determine (a) the amplitude of the motion of the bob, (b) the force that must be applied to collar C to maintain the motion.

19.111 An 18-lb block A slides in a vertical frictionless slot and is connected to a moving support B by means of a spring AB of constant $k = 8$ lb/ft. Knowing that the acceleration of the support is $a = a_m \sin \omega_f t$, where $a_m = 5$ ft/s^2 and $\omega_f = 6$ rad/s, determine (a) the maximum displacement of block A, (b) the amplitude of the fluctuating force exerted by the spring on the block.

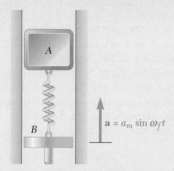

Fig. P19.111

19.112 A variable-speed motor is rigidly attached to a beam BC. When the speed of the motor is less than 600 rpm or more than 1200 rpm, a small object placed at A is observed to remain in contact with the beam. For speeds between 600 and 1200 rpm, the object is observed to "dance" and actually to lose contact with the beam. Determine the speed at which resonance will occur.

Fig. P19.112

19.113 A motor of mass M is supported by springs with an equivalent spring constant k. The unbalance of its rotor is equivalent to a mass m located at a distance r from the axis of rotation. Show that when the angular velocity of the motor is ω_f, the amplitude x_m of the motion of the motor is

$$x_m = \frac{r(m/M)(\omega_f/\omega_n)^2}{1 - (\omega_f/\omega_n)^2}$$

where $\omega_n = \sqrt{k/M}$.

19.114 As the rotational speed of a spring-supported 100-kg motor is increased, the amplitude of the vibration due to the unbalance of its 15-kg rotor first increases and then decreases. It is observed that as very high speeds are reached, the amplitude of the vibration approaches 3.3 mm. Determine the distance between the mass center of the rotor and its axis of rotation. (*Hint:* Use the formula derived in Prob. 19.113.)

19.115 A motor of weight 40 lb is supported by four springs, each of constant 225 lb/in. The motor is constrained to move vertically, and the amplitude of its motion is observed to be 0.05 in. at a speed of 1200 rpm. Knowing that the weight of the rotor is 9 lb, determine the distance between the mass center of the rotor and the axis of the shaft.

Fig. P19.115

19.116 A motor weighing 400 lb is supported by springs having a total constant of 1200 lb/in. The unbalance of the rotor is equivalent to a 1-oz weight located 8 in. from the axis of rotation. Determine the range of allowable values of the motor speed if the amplitude of the vibration is not to exceed 0.06 in.

Fig. P19.117

19.117 A 180-kg motor is bolted to a light horizontal beam. The unbalance of its rotor is equivalent to a 28-g mass located 150 mm from the axis of rotation, and the static deflection of the beam due to the weight of the motor is 12 mm. The amplitude of the vibration due to the unbalance can be decreased by adding a plate to the base of the motor. If the amplitude of vibration is to be less than 60 μm for motor speeds above 300 rpm, determine the required mass of the plate.

19.118 The unbalance of the rotor of a 400-lb motor is equivalent to a 3-oz weight located 6 in. from the axis of rotation. In order to limit to 0.2 lb the amplitude of the fluctuating force exerted on the foundation when the motor is run at speeds of 100 rpm and above, a pad is to be placed between the motor and the foundation. Determine (*a*) the maximum allowable spring constant k of the pad, (*b*) the corresponding amplitude of the fluctuating force exerted on the foundation when the motor is run at 200 rpm.

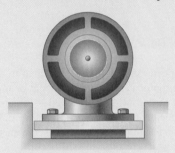

Fig. *P19.118*

19.119 A counter-rotating eccentric mass exciter consisting of two rotating 100-g masses describing circles of radius r at the same speed but in opposite senses is placed on a machine element to induce a steady-state vibration of the element. The total mass of the system is 300 kg, the constant of each spring is $k = 600$ kN/m, and the rotational speed of the exciter is 1200 rpm. Knowing that the amplitude of the total fluctuating force exerted on the foundation is 160 N, determine the radius r.

Fig. P19.119

19.120 A 360-lb motor is supported by springs of total constant 12.5 kips/ft. The unbalance of the rotor is equivalent to a 0.9-oz weight located 7.5 in. from the axis of rotation. Determine the range of speeds of the motor for which the amplitude of the fluctuating force exerted on the foundation is less than 5 lb.

19.121 Figures (1) and (2) show how springs can be used to support a block in two different situations. In Fig. (1), they help decrease the amplitude of the fluctuating force transmitted by the block to the foundation. In Fig. (2), they help decrease the amplitude of the fluctuating displacement transmitted by the foundation to the block. The ratio of the transmitted force to the impressed force or the ratio of the transmitted displacement to the impressed displacement is called the *transmissibility*. Derive an equation for the transmissibility for each situation. Give your answer in terms of the ratio ω_f/ω_n of the frequency ω_f of the impressed force or impressed displacement to the natural frequency ω_n of the spring-mass system. Show that in order to cause any reduction in transmissibility, the ratio ω_f/ω_n must be greater than $\sqrt{2}$.

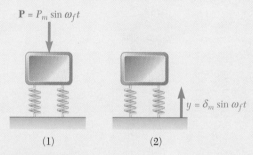

(1)   (2)

Fig. P19.121

19.122 A vibrometer used to measure the amplitude of vibrations consists essentially of a box containing a mass-spring system with a known natural frequency of 120 Hz. The box is rigidly attached to a surface that is moving according to the equation $y = \delta_m \sin \omega_f t$. If the amplitude z_m of the motion of the mass relative to the box is used as a measure of the amplitude δ_m of the vibration of the surface, determine (*a*) the percent error when the frequency of the vibration is 600 Hz, (*b*) the frequency at which the error is zero.

Fig. P19.122 and P19.123

19.123 A certain accelerometer consists essentially of a box containing a mass-spring system with a known natural frequency of 2200 Hz. The box is rigidly attached to a surface that is moving according to the equation $y = \delta_m \sin \omega_f t$. If the amplitude z_m of the motion of the mass relative to the box times a scale factor ω_n^2 is used as a measure of the maximum acceleration $\alpha_m = \delta_m \omega_f^2$ of the vibrating surface, determine the percent error when the frequency of the vibration is 600 Hz.

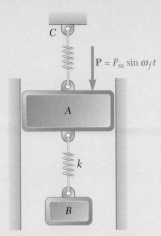

Fig. P19.124

19.124 Block A can move without friction in the slot as shown and is acted upon by a vertical periodic force of magnitude $P = P_m \sin \omega_f t$, where $\omega_f = 2$ rad/s and $P_m = 20$ N. A spring of constant k is attached to the bottom of block A and to a 22-kg block B. Determine (a) the value of the constant k that will prevent a steady-state vibration of block A, (b) the corresponding amplitude of the vibration of block B.

19.125 A 60-lb disk is attached with an eccentricity $e = 0.006$ in. to the midpoint of a vertical shaft AB that revolves at a constant angular velocity ω_f. Knowing that the spring constant k for horizontal movement of the disk is 40,000 lb/ft, determine (a) the angular velocity ω_f at which resonance will occur, (b) the deflection r of the shaft when $\omega_f = 1200$ rpm.

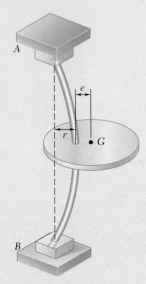

Fig. P19.125

19.126 A small trailer and its load have a total mass of 250 kg. The trailer is supported by two springs, each of constant 10 kN/m, and is pulled over a road, the surface of which can be approximated by a sine curve with an amplitude of 40 mm and a wavelength of 5 m (i.e., the distance between successive crests is 5 m and the vertical distance from crest to trough is 80 mm). Determine (a) the speed at which resonance will occur, (b) the amplitude of the vibration of the trailer at a speed of 50 km/h.

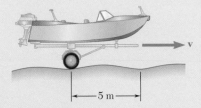

Fig. P19.126

19.5 DAMPED VIBRATIONS

The vibrating systems considered in the first part of this chapter were assumed free of damping. Actually, all vibrations are damped to some degree by friction forces. These forces can be caused by *dry friction,* or *Coulomb friction,* between rigid bodies; by *fluid friction* when a rigid body moves in a fluid; or by *internal friction* between the molecules of a seemingly elastic body. A type of damping of special interest is the *viscous damping* caused by fluid friction at low and moderate speeds. We will first consider free vibrations with viscous damping and then examine the effect of viscous damping on forced vibrations.

*19.5A Damped Free Vibrations

Viscous damping is characterized by the fact that the friction force is *directly proportional and opposite in direction to the velocity* of the moving body. As an example, let us again consider a body with mass m suspended from a spring of constant k, assuming that the body is attached to the plunger of a dashpot (Fig. 19.10). The magnitude of the friction force exerted on the plunger by the surrounding fluid is equal to $c\dot{x}$, where the constant c, expressed in N·s/m or lb·s/ft and known as the *coefficient of viscous damping*, depends upon the physical properties of the fluid and the construction of the dashpot. Examining the free-body and kinetic diagrams, the equation of motion is

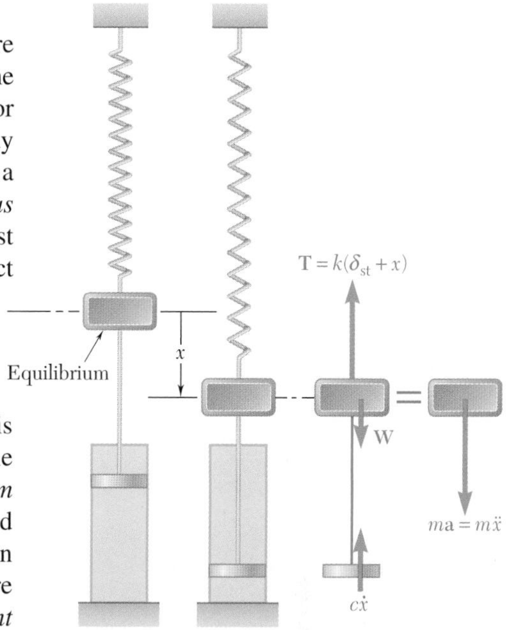

Fig. 19.10 Free-body diagram and kinetic diagram of a spring-mass-damper system.

$$+\downarrow \Sigma F = ma: \qquad W - k(\delta_{st} + x) - c\dot{x} = m\ddot{x}$$

Recalling that $W = k\delta_{st}$, we have

$$m\ddot{x} + c\dot{x} + kx = 0 \tag{19.38}$$

If we substitute $x = e^{\lambda t}$ into Eq. (19.38) and divide through by $e^{\lambda t}$, we obtain

Characteristic equation

$$m\lambda^2 + c\lambda + k = 0 \tag{19.39}$$

and obtain the roots

$$\lambda = -\frac{c}{2m} \pm \sqrt{\left(\frac{c}{2m}\right)^2 - \frac{k}{m}} \tag{19.40}$$

Defining the *critical damping coefficient* c_c as the value of c that makes the radical in Eq. (19.40) equal to zero, we have

$$\left(\frac{c_c}{2m}\right)^2 - \frac{k}{m} = 0 \qquad c_c = 2m\sqrt{\frac{k}{m}} = 2m\omega_n \tag{19.41}$$

where ω_n is the natural circular frequency of the system in the absence of damping. We can distinguish three different cases of damping, depending upon the value of the coefficient c.

1. **Overdamped:** $c > c_c$. The roots λ_1 and λ_2 of the characteristic equation (19.39) are real and distinct, and the general solution of the differential equation (19.38) is

$$x = C_1 e^{\lambda_1 t} + C_2 e^{\lambda_2 t} \tag{19.42}$$

This solution corresponds to a nonvibratory motion. Since λ_1 and λ_2 are both negative, x approaches zero as t increases indefinitely. However, the system actually regains its equilibrium position after a finite time.

2. **Critically damped:** $c = c_c$. The characteristic equation has a double root $\lambda = -c_c/2m = -\omega_n$, and the general solution of Eq. (19.38) is

$$x = (C_1 + C_2 t)e^{-\omega_n t} \tag{19.43}$$

This motion is again nonvibratory. Critically damped systems are of special interest in engineering applications because they regain their equilibrium position in the shortest possible time without oscillation.

3. **Underdamped:** $c < c_c$. The roots of Eq. (19.39) are complex and conjugate, and the general solution of Eq. (19.38) is of the form

$$x = e^{-(c/2m)t}(C_1 \sin \omega_d t + C_2 \cos \omega_d t) \tag{19.44}$$

where ω_d is defined by the relation

$$\omega_d^2 = \frac{k}{m} - \left(\frac{c}{2m}\right)^2$$

Substituting $k/m = \omega_n^2$ and recalling Eq. (19.41), we have

$$\omega_d = \omega_n \sqrt{1 - \left(\frac{c}{c_c}\right)^2} \tag{19.45}$$

where the constant c/c_c is known as the **damping factor** or the **damping ratio**. This quantity is often denoted by ζ. Even though the motion does not actually repeat itself, the constant ω_d is commonly referred to as the *damped circular frequency*. In terms of the damping ratio, the damped circular frequency is

$$\omega_d = \omega_n \sqrt{1 - \zeta^2} \tag{19.45'}$$

A substitution similar to the one used in Sec. 19.1A enables us to write the general solution of Eq. (19.38) in the form

$$x = x_0 e^{-(c/2m)t} \sin(\omega_d t + \phi) \tag{19.46}$$

or

$$x = x_0 e^{-\zeta \omega_n t} \sin(\omega_d t + \phi) \tag{19.46'}$$

The motion defined by Eq. (19.46) is vibratory with diminishing amplitude (Fig. 19.11). The time interval $\tau_d = 2\pi/\omega_d$ separating two successive points where the curve defined by Eq. (19.46) touches one of the limiting curves shown in Fig. 19.11 is commonly referred to as the *period of the damped vibration*. Recalling Eq. (19.45), we observe that $\omega_d < \omega_n$ and, thus, that τ_d is larger than the period of vibration τ_n of the corresponding undamped system.

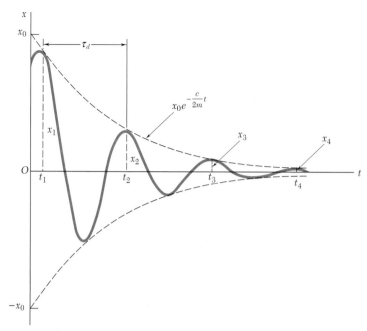

Fig. 19.11 The free response of a viscously damped system decays exponentially and oscillates with a frequency ω_d.

*19.5B Damped Forced Vibrations

If the system considered in the preceding section is subjected to a periodic force **P** of magnitude $P = P_m \sin \omega_f t$, the equation of motion becomes

$$m\ddot{x} + c\dot{x} + kx = P_m \sin \omega_f t \tag{19.47}$$

We can obtain the general solution of Eq. (19.47) by adding a particular solution of Eq. (19.47) to the complementary function or general solution of the homogeneous equation (19.38). The complementary function is given by Eq. (19.42), (19.43), or (19.44), depending upon the type of damping considered. It represents a *transient* motion that is eventually damped out.

Our interest in this section is centered on the steady-state vibration represented by a particular solution of Eq. (19.47) of the form

$$x_{\text{part}} = x_m \sin (\omega_f t - \phi) \tag{19.48}$$

Substituting x_{part} for x into Eq. (19.47), we obtain

$$-m\omega_f^2 x_m \sin (\omega_f t - \phi) + c\omega_f x_m \cos (\omega_f t - \phi) + kx_m \sin (\omega_f t - \phi)$$
$$= P_m \sin \omega_f t$$

Making $\omega_f t - \phi$ successively equal to 0 and to $\pi/2$ gives

$$c\omega_f x_m = P_m \sin \phi \tag{19.49}$$
$$(k - m\omega_f^2)x_m = P_m \cos \phi \tag{19.50}$$

Photo 19.2 The automobile suspension shown consists essentially of a spring and a shock absorber, which will cause the body of the car to undergo *damped forced vibrations* when the car is driven over an uneven road.

Photo 19.3 This truck is experiencing damped forced vibration in the vehicle dynamics test.

Squaring both sides of Eqs. (19.49) and (19.50) and adding, we have

$$[(k - m\omega_f^2)^2 + (c\omega_f)^2] x_m^2 = P_m^2 \tag{19.51}$$

Solving Eq. (19.51) for x_m and dividing Eqs. (19.49) and (19.50) by the result, we obtain, respectively,

$$x_m = \frac{P_m}{\sqrt{(k - m\omega_f^2)^2 + (c\omega_f)^2}} \qquad \tan\phi = \frac{c\omega_f}{k - m\omega_f^2} \tag{19.52}$$

Recalling from Eq. (19.4) that $k/m = \omega_n^2$, where ω_n is the circular frequency of the undamped free vibration, and from Eq. (19.41) that $2m\omega_n = c_c$, where c_c is the critical damping coefficient of the system, we have

$$\frac{x_m}{P_m/k} = \frac{x_m}{\delta_m} = \frac{1}{\sqrt{[1 - (\omega_f/\omega_n)^2]^2 + [2(c/c_c)(\omega_f/\omega_n)]^2}} \tag{19.53}$$

$$\tan\phi = \frac{2(c/c_c)(\omega_f/\omega_n)}{1 - (\omega_f/\omega_n)^2} \tag{19.54}$$

Defining the frequency ratio $r = \omega_f/\omega_n$, we can write the steady-state response of a viscously damped system in terms of the frequency ratio and the damping ratio as

$$\frac{x_m}{P_m/k} = \frac{x_m}{\delta_{st}} = \frac{1}{\sqrt{(1 - r^2)^2 + (2\zeta r)^2}} \tag{19.53'}$$

$$\tan\phi = \frac{2\zeta r}{1 - r^2} \tag{19.54'}$$

We can use these equations to determine the amplitude of the steady-state vibration produced by an impressed force of magnitude $P = P_m \sin \omega_f t$ or by an impressed support movement $\delta = \delta_m \sin \omega_f t$. Using these same parameters, Eq. (19.54) defines the *phase difference* ϕ between the impressed force or impressed support movement and the resulting steady-state vibration of the damped system. The magnification factor has been plotted against the frequency ratio in Fig. 19.12 for various values of the damping ratio. Note that we can keep the amplitude of a forced vibration small by choosing a large coefficient of viscous damping c or by keeping the natural and forced frequencies far apart.

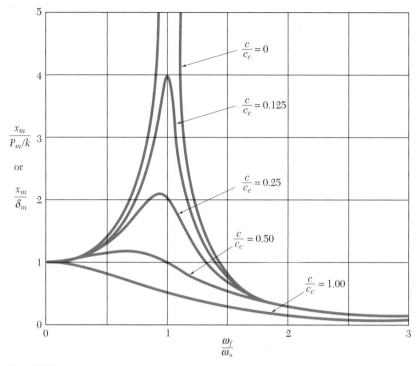

Fig. 19.12 Graph of magnification factor as a function of frequency ratio for several values of the damping ratio.

*19.5C Electrical Analogs

Oscillating electrical circuits are characterized by differential equations of the same type as those just discussed. Their analysis is therefore similar to that of a mechanical system, and the results obtained for a given vibrating system can be readily extended to the equivalent circuit. Conversely, any result obtained for an electrical circuit also applies to the corresponding mechanical system.

Consider an electrical circuit consisting of an inductor of inductance L, a resistor of resistance R, and a capacitor of capacitance C, connected in series with a source of alternating voltage $E = E_m \sin \omega_f t$ (Fig. 19.13). Elementary circuit theory[†] says that if i denotes the current in the circuit and q denotes the electric charge on the capacitor, the drop in potential is $L(di/dt)$ across the inductor, Ri across the resistor, and q/C across the capacitor. The algebraic sum of the applied voltage and of the drops in potential around the circuit loop must be zero, so we have

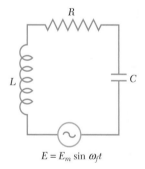

$E = E_m \sin \omega_f t$

Fig. 19.13 An electrical circuit with inductance L, resistance R, capacitance C, and a source of alternating voltage E.

$$E_m \sin \omega_f t - L\frac{di}{dt} - Ri - \frac{q}{C} = 0 \qquad \textbf{(19.55)}$$

[†]See C. R. Paul, S. A. Nasar, and L. E. Unnewehr, *Introduction to Electrical Engineering*, 2nd ed., McGraw-Hill, New York, 1992.

Rearranging the terms and recalling that at any instant the current i is equal to the rate of change $\dot{q}$ of the charge q, we have

$$L\ddot{q} + R\dot{q} + \frac{1}{C}q = E_m \sin \omega_f t \qquad (19.56)$$

We can verify that Eq. (19.56), which defines the oscillations of the electrical circuit of Fig. 19.13, is of the same type as Eq. (19.47), which characterizes the damped forced vibrations of the mechanical system of Fig. 19.10. By comparing the two equations, we can construct a table of the analogous mechanical and electrical expressions.

Table 19.2 can be used to extend the results obtained earlier for various mechanical systems to their electrical analogs. For instance, we can determine the amplitude i_m of the current in the circuit of Fig. 19.13 by noting that it corresponds to the maximum value v_m of the velocity in the analogous mechanical system. Recalling from the first of Eqs. (19.37) that $v_m = x_m\omega_f$, substituting for x_m from Eq. (19.52), and replacing the constants of the mechanical system by the corresponding electrical expressions, we have

$$i_m = \frac{\omega_f E_m}{\sqrt{\left(\dfrac{1}{C} - L\omega_f^2\right)^2 + (R\omega_f)^2}}$$

$$i_m = \frac{E_m}{\sqrt{R^2 + \left(L\omega_f - \dfrac{1}{C\omega_f}\right)^2}} \qquad (19.57)$$

The radical term in this expression is known as the *impedance* of the electrical circuit.

The analogy between mechanical systems and electrical circuits holds for transient as well as steady-state oscillations. The oscillations of the circuit shown in Fig. 19.14, for instance, are analogous to the damped free vibrations of the system of Fig. 19.10. As far as the initial conditions are concerned, we should note that closing the switch S when the charge on the capacitor is $q = q_0$ is equivalent to releasing the mass of the mechanical system with no initial velocity from the position $x = x_0$. Also note that, if a battery of constant voltage E is introduced in the electrical circuit of Fig. 19.14, closing the switch S is equivalent to suddenly applying a force of constant magnitude P to the mass of the mechanical system of Fig. 19.10.

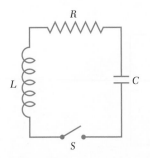

Fig. 19.14 An *LRC* circuit with switch *S*.

Table 19.2 Characteristics of a Mechanical System and of Its Electrical Analog

Mechanical System		Electrical Circuit	
m	Mass	L	Inductance
c	Coefficient of viscous damping	R	Resistance
k	Spring constant	$1/C$	Reciprocal of capacitance
x	Displacement	q	Charge
v	Velocity	i	Current
P	Applied force	E	Applied voltage

This discussion would be of questionable value if its only result were to make it possible for mechanics students to analyze electrical circuits without learning the elements of circuit theory. We hope that this discussion will instead encourage students to apply to the solution of problems in mechanical vibrations the mathematical techniques they may learn in later courses in circuit theory. The chief value of the concept of electrical analogs, however, resides in its application to *experimental methods* for determining the characteristics of a given mechanical system. Indeed, an electrical circuit is much more easily constructed than is a mechanical model, and the fact that we can modify its characteristics by varying the inductance, resistance, or capacitance of its various components makes the use of the electrical analog particularly convenient.

To determine the electrical analog of a given mechanical system, we focus our attention on each moving mass in the system and observe which springs, dashpots, or external forces are applied directly to it. We can then construct an equivalent electrical loop to match each of these mechanical units; the various loops obtained in this way will together form the desired circuit. Consider, for instance, the mechanical system of Fig. 19.15. The mass m_1 is acted upon by two springs with constants k_1 and k_2 and by two dashpots characterized by the coefficients of viscous damping c_1 and c_2. The electrical circuit should therefore include a loop consisting of an inductor of inductance L_1 proportional to m_1; of two capacitors of capacitance C_1 and C_2 inversely proportional to k_1 and k_2, respectively; and of two resistors of resistance R_1 and R_2, proportional to c_1 and c_2, respectively. Since the mass m_2 is acted upon by the spring k_2 and the dashpot c_2, as well as by the force $P = P_m \sin \omega_f t$, the circuit should also include a loop containing the capacitor C_2, the resistor R_2, the new inductor L_2, and the voltage source $E = E_m \sin \omega_f t$ (Fig. 19.16).

To check that the mechanical system of Fig. 19.15 and the electrical circuit of Fig. 19.16 actually satisfy the same differential equations, we first derive the equations of motion for m_1 and m_2. Denoting the displacements of m_1 and m_2 from their equilibrium positions by x_1 and x_2, respectively, we observe that the elongation of the spring k_1 (measured from the equilibrium position) is equal to x_1, while the elongation of the spring k_2 is equal to the relative displacement $x_2 - x_1$ of m_2 with respect to m_1. The equations of motion for m_1 and m_2 are therefore

$$m_1 \ddot{x}_1 + c_1 \dot{x}_1 + c_2(\dot{x}_1 - \dot{x}_2) + k_1 x_1 + k_2(x_1 - x_2) = 0 \qquad \textbf{(19.58)}$$

$$m_2 \ddot{x}_2 + c_2(\dot{x}_2 - \dot{x}_1) + k_2(x_2 - x_1) = P_m \sin \omega_f t \qquad \textbf{(19.59)}$$

Now consider the electrical circuit of Fig. 19.16; we denote the current in the first and second loops by i_1 and i_2, respectively, and by q_1 and q_2 the integrals $\int i_1\, dt$ and $\int i_2\, dt$. Noting that the charge on the capacitor C_1 is q_1 and the charge on C_2 is $q_1 - q_2$, we can state that the sum of the potential differences in each loop is zero and obtain

$$L_1 \ddot{q}_1 + R_1 \dot{q}_1 + R_2(\dot{q}_1 - \dot{q}_2) + \frac{q_1}{C_1} + \frac{q_1 - q_2}{C_2} = 0 \qquad \textbf{(19.60)}$$

$$L_2 \ddot{q}_2 + R_2(\dot{q}_2 - \dot{q}_1) + \frac{q_2 - q_1}{C_2} = E_m \sin \omega_f t \qquad \textbf{(19.61)}$$

We easily check that Eqs. (19.60) and (19.61) reduce to Eqs. (19.58) and (19.59), respectively, after performing the substitutions indicated in Table 19.2.

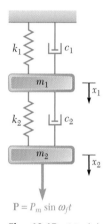

Fig. 19.15 Model of a two-degree-of-freedom harmonically excited system.

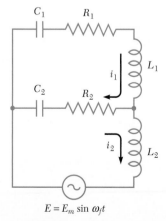

Fig. 19.16 An electrical circuit analogous to the mechanical system in Fig. 19.15.

SOLVING PROBLEMS
ON YOUR OWN

In this section, we developed a more realistic model of a vibrating system by including the effect of the **viscous damping** caused by fluid friction. We represented viscous damping in Fig. 19.10 by the force exerted on the moving body by a plunger moving in a dashpot. This force is equal in magnitude to $c\dot{x}$, where the constant c, expressed in N·s/m or lb·s/ft, is known as the *coefficient of viscous damping*. Keep in mind that the same sign convention should be used for x, $\dot{x}$, and $\ddot{x}$

1. Damped free vibrations. The differential equation defining this motion was found to be

$$m\ddot{x} + c\dot{x} + kx = 0 \tag{19.38}$$

To obtain the solution of this equation, calculate the *critical damping coefficient* c_c, using the formula

$$c_c = 2m\sqrt{k/m} = 2m\omega_n \tag{19.41}$$

where ω_n is the natural circular frequency of the undamped system.

    a. If $c > c_c$ (**overdamped**), the solution of Eq. (19.38) is

$$x = C_1 e^{\lambda_1 t} + C_2 e^{\lambda_2 t} \tag{19.42}$$

where

$$\lambda_{1,2} = -\frac{c}{2m} \pm \sqrt{\left(\frac{c}{2m}\right)^2 - \frac{k}{m}} \tag{19.40}$$

and where the constants C_1 and C_2 can be determined from the initial conditions $x(0)$ and $\dot{x}(0)$. This solution corresponds to a nonvibratory motion.

    b. If $c = c_c$ (**critically damped**), the solution of Eq. (19.38) is

$$x = (C_1 + C_2 t)e^{-\omega_n t} \tag{19.43}$$

which also corresponds to a nonvibratory motion. Critically damped systems are of special interest in engineering applications because they regain their equilibrium position in the shortest possible time without oscillation.

    c. If $c < c_c$ (**underdamped**), the solution of Eq. (19.38) is

$$x = x_0 e^{-(c/2m)t} \sin(\omega_d t + \phi) \tag{19.46}$$

or in terms of the damping ratio $\zeta = c/c_{\text{cr}}$,

$$x = x_0 e^{-\zeta \omega_n t} \sin(\omega_d t + \phi) \tag{19.46'}$$

where

$$\omega_d = \omega_n\sqrt{1 - \left(\frac{c}{c_c}\right)^2} \qquad \text{(19.45)}$$

or

$$\omega_d = \omega_n\sqrt{1 - \zeta^2} \qquad \text{(19.45')}$$

and where x_0 and ϕ can be determined from the initial conditions $x(0)$ and $\dot{x}(0)$. This solution corresponds to oscillations of decreasing amplitude and of period $\tau_d = 2\pi/\omega_d$ (Fig. 19.11).

2. Damped forced vibrations. These vibrations occur when a system with viscous damping is subjected to a periodic force $\mathbf{P}$ with a magnitude of $P = P_m \sin \omega_f t$ or when it is elastically connected to a support with an alternating motion of $\delta = \delta_m \sin \omega_f t$. In the first case, the motion is defined by the differential equation

$$m\ddot{x} + c\dot{x} + kx = P_m \sin \omega_f t \qquad \text{(19.47)}$$

and in the second case, by a similar equation obtained by replacing P_m with $k\delta_m$. You will be concerned only with the *steady-state* motion of the system, which is defined by a *particular solution* of these equations of the form

$$x_{\text{part}} = x_m \sin (\omega_f t - \phi) \qquad \text{(19.48)}$$

where

$$\frac{x_m}{P_m/k} = \frac{x_m}{\delta_m} = \frac{1}{\sqrt{[1 - (\omega_f/\omega_n)^2]^2 + [2(c/c_c)(\omega_f/\omega_n)]^2}} \qquad \text{(19.53)}$$

and

$$\tan \phi = \frac{2(c/c_c)(\omega_f/\omega_n)}{1 - (\omega_f/\omega_n)^2} \qquad \text{(19.54)}$$

The expression given in Eq. (19.53) is referred to as the *magnification factor* and has been plotted against the frequency ratio ω_f/ω_n in Fig. 19.12 for various values of the damping ratio c/c_c. Eqs. (19.53) and (19.54) can be written in terms of the damping ratio ζ and frequency ratio r as shown in Eqs. (19.53') and (19.54'). In the problems that follow, you may be asked to determine one of the parameters in Eqs. (19.53) and (19.54) when the others are known.

Problems

19.127 Show that in the case of heavy damping ($c > c_c$), a body never passes through its position of equilibrium O if it is (*a*) released with no initial velocity from an arbitrary position, (*b*) started from O with an arbitrary initial velocity.

19.128 Show that in the case of heavy damping ($c > c_c$), a body released from an arbitrary position with an arbitrary initial velocity cannot pass more than once through its equilibrium position.

19.129 In the case of light damping ($c < c_c$), the displacements x_1, x_2, x_3, shown in Fig. 19.11 may be assumed equal to the maximum displacements. Show that the ratio of any two successive maximum displacements x_n and x_{n+1} is a constant and that the natural logarithm of this ratio, called the *logarithmic decrement*, is

$$\ln \frac{x_n}{x_{n+1}} = \frac{2\pi(c/c_c)}{\sqrt{1 - (c/c_c)^2}}$$

19.130 In practice, it is often difficult to determine the logarithmic decrement of a system with light damping defined in Prob. 19.129 by measuring two successive maximum displacements. Show that the logarithmic decrement can also be expressed as $(1/k) \ln(x_n/x_{n+k})$, where k is the number of cycles between readings of the maximum displacement.

19.131 In an underdamped system ($c < c_c$), the period of vibration is commonly defined as the time interval $\tau_d = 2\pi/\omega_d$ corresponding to two successive points where the displacement–time curve touches one of the limiting curves shown in Fig. 19.11. Show that the interval of time (*a*) between a maximum positive displacement and the following maximum negative displacement is $\frac{1}{2}\tau_d$, (*b*) between two successive zero displacements is $\frac{1}{2}\tau_d$, (*c*) between a maximum positive displacement and the following zero displacement is greater than $\frac{1}{4}\tau_d$.

19.132 A loaded railroad car weighing 30,000 lb is rolling at a constant velocity $\mathbf{v}_0$ when it couples with a spring and dashpot bumper system (Fig. 1). The recorded displacement–time curve of the loaded railroad car after coupling is as shown (Fig. 2). Determine (*a*) the damping constant, (*b*) the spring constant. (*Hint:* Use the definition of logarithmic decrement given in 19.129.)

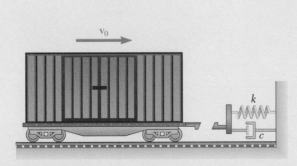

(1)

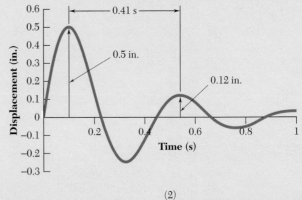

(2)

Fig. P19.132

B.43 through B.46 A section of sheet steel 2 mm thick is cut and bent into the machine component shown. Knowing that the density of steel is 7850 kg/m³, determine the mass products of inertia I_{xy}, I_{yz}, and I_{zx} of the component.

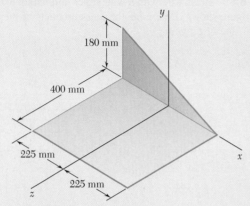

Fig. B.43

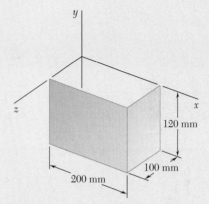

Fig. B.44

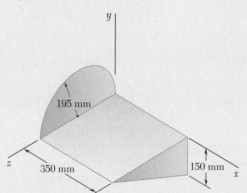

Fig. B.45

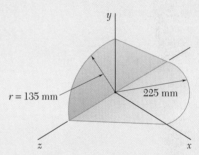

Fig. B.46

B.47 The figure shown is formed of 1.5-mm-diameter aluminum wire. Knowing that the density of aluminum is 2800 kg/m³, determine the mass products of inertia I_{xy}, I_{yz}, and I_{zx} of the wire figure.

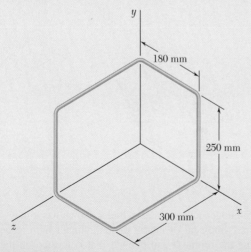

Fig. B.47

A37

19.133 A torsional pendulum has a centroidal mass moment of inertia of 0.3 kg·m^2 and when given an initial twist and released is found to have a frequency of oscillation of 200 rpm. Knowing that when this pendulum is immersed in oil and when given the same initial condition it is found to have a frequency of oscillation of 180 rpm, determine the damping constant for the oil.

19.134 The barrel of a field gun weighs 1500 lb and is returned into firing position after recoil by a recuperator of constant c = 1100 lb·s/ft. Determine (a) the constant k that should be used for the recuperator to return the barrel into firing position in the shortest possible time without any oscillation, (b) the time needed for the barrel to move back two-thirds of the way from its maximum-recoil position to its firing position.

19.135 A 2-kg block is supported by a spring with a constant of k = 128 N/m and a dashpot with a coefficient of viscous damping of c = 0.6 N·s/m. The block is in equilibrium when it is struck from below by a hammer that imparts to the block an upward velocity of 0.4 m/s. Determine (a) the logarithmic decrement, (b) the maximum upward displacement of the block from equilibrium after two cycles.

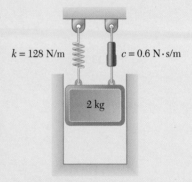

k = 128 N/m c = 0.6 N·s/m

2 kg

Fig. P19.135

19.136 A 4-kg block A is dropped from a height of 800 mm onto a 9-kg block B that is at rest. Block B is supported by a spring of constant k = 1500 N/m and is attached to a dashpot of damping coefficient c = 230 N·s/m. Knowing that there is no rebound, determine the maximum distance the blocks will move after the impact.

19.137 A 0.9-kg block B is connected by a cord to a 2.4-kg block A that is suspended as shown from two springs, each with a constant of k = 180 N/m, and a dashpot with a damping coefficient of c = 7.5 N·s/m. Knowing that the system is at rest when the cord connecting A and B is cut, determine the minimum tension that will occur in each spring during the resulting motion.

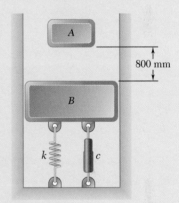

A

800 mm

B

k c

Fig. P19.136

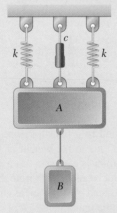

k c k

A

B

Fig. P19.137 and P19.138

19.138 A 0.9-kg block B is connected by a cord to a 2.4-kg block A that is suspended as shown from two springs, each with a constant of k = 180 N/m, and a dashpot with a damping coefficient of c = 60 N·s/m. Knowing that the system is at rest when the cord connecting A and B is cut, determine the velocity of block A after 0.1 s.

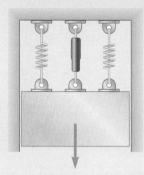

Fig. P19.139

19.139 A machine element weighing 800 lb is supported by two springs, each having a constant of 200 lb/in. A periodic force of maximum value 30 lb is applied to the element with a frequency of 2.5 cycles per second. Knowing that the coefficient of damping is 8 lb·s/in., determine the amplitude of the steady-state vibration of the element.

19.140 In Prob. 19.139, determine the required value of the coefficient of damping if the amplitude of the steady-state vibration of the element is to be 0.15 in.

19.141 In the case of the forced vibration of a system, determine the range of values of the damping factor c/c_c for which the magnification factor will always decrease as the frequency ratio ω_f/ω_n increases.

19.142 Show that for a small value of the damping factor c/c_c, the maximum amplitude of a forced vibration occurs when $\omega_f \approx \omega_n$ and that the corresponding value of the magnification factor is $\frac{1}{2}(c/c_c)$.

19.143 A counter-rotating eccentric mass exciter consisting of two rotating 14-oz weights describing circles of 6-in. radius at the same speed but in opposite senses is placed on a machine element to induce a steady-state vibration of the element and to determine some of the dynamic characteristics of the element. At a speed of 1200 rpm, a stroboscope shows the eccentric masses to be exactly under their respective axes of rotation and the element to be passing through its position of static equilibrium. Knowing that the amplitude of the motion of the element at that speed is 0.6 in. and that the total weight of the system is 300 lb, determine (*a*) the combined spring constant *k*, (*b*) the damping factor c/c_c.

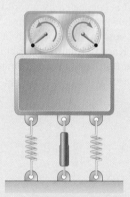

Fig. P19.143

19.144 A 36-lb motor is bolted to a light horizontal beam that has a static deflection of 0.075 in. due to the weight of the motor. Knowing that the unbalance of the rotor is equivalent to a weight of 0.64 oz located 6.25 in. from the axis of rotation, determine the amplitude of the vibration of the motor at a speed of 900 rpm, assuming (*a*) that no damping is present, (*b*) that the damping factor c/c_c is equal to 0.055.

Fig. P19.144 and P19.145

19.145 A 45-kg motor is bolted to a light horizontal beam that has a static deflection of 6 mm due to the weight of the motor. The unbalance of the motor is equivalent to a mass of 110 g located 75 mm from the axis of rotation. Knowing that the amplitude of the vibration of the motor is 0.25 mm at a speed of 300 rpm, determine (*a*) the damping factor c/c_c, (*b*) the coefficient of damping *c*.

19.146 The unbalance of the rotor of a 180-kg motor is equivalent to a mass of 85 g located 150 mm from the axis of rotation. The pad that is placed between the motor and the foundation is equivalent to a spring with a constant of $k = 7.5$ kN/m in parallel with a dashpot with constant c. Knowing that the magnitude of the maximum acceleration of the motor is 9 mm/s² at a speed of 100 rpm, determine the damping factor c/c_c.

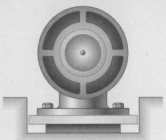

19.147 A machine element is supported by springs and is connected to a dashpot as shown. Show that if a periodic force of magnitude $P = P_m \sin \omega_f t$ is applied to the element, the amplitude of the fluctuating force transmitted to the foundation is

Fig. P19.146

$$F_m = P_m \sqrt{\frac{1 + [2(c/c_c)(\omega_f/\omega_n)]^2}{[1 - (\omega_f/\omega_n)^2]^2 + [2(c/c_c)(\omega_f/\omega_n)]^2}}$$

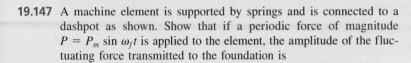

19.148 A 91-kg machine element supported by four springs, each of constant $k = 175$ N/m, is subjected to a periodic force of frequency 0.8 Hz and amplitude 89 N. Determine the amplitude of the fluctuating force transmitted to the foundation if (a) a dashpot with a coefficient of damping $c = 365$ N·s/m is connected to the machine element and to the ground, (b) the dashpot is removed.

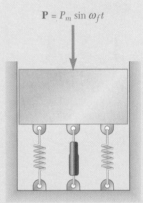

19.149 A simplified model of a washing machine is shown. A bundle of wet clothes forms a weight w_b of 20 lb in the machine and causes a rotating unbalance. The rotating weight is 40 lb (including w_b) and the radius of the washer basket e is 9 in. Knowing the washer has an equivalent spring constant $k = 70$ lb/ft and damping ratio $\zeta = c/c_c = 0.05$ and during the spin cycle the drum rotates at 250 rpm, determine the amplitude of the motion and the magnitude of the force transmitted to the sides of the washing machine.

Fig. P19.147 and P19.148

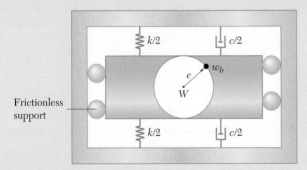

Fig. P19.149

*19.150 For a steady-state vibration with damping under a harmonic force, show that the mechanical energy dissipated per cycle by the dashpot is $E = \pi c x_m^2 \omega_f$, where c is the coefficient of damping, x_m is the amplitude of the motion, and ω_f is the circular frequency of the harmonic force.

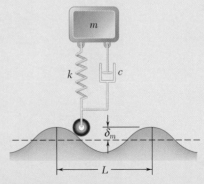

*19.151 The suspension of an automobile can be approximated by the simplified spring-and-dashpot system shown. (a) Write the differential equation defining the vertical displacement of the mass m when the system moves at a speed v over a road with a sinusoidal cross section of amplitude δ_m and wave length L. (b) Derive an expression for the amplitude of the vertical displacement of the mass m.

Fig. P19.151

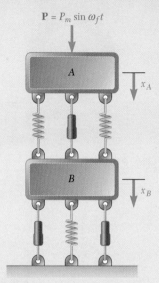

$P = P_m \sin \omega_f t$

A

x_A

B

x_B

Fig. P19.152

*19.152 Two blocks A and B, each of mass m, are supported as shown by three springs of the same constant k. Blocks A and B are connected by a dashpot and block B is connected to the ground by two dashpots, each dashpot having the same coefficient of damping c. Block A is subjected to a force of magnitude $P = P_m \sin \omega_f t$. Write the differential equations defining the displacements x_A and x_B of the two blocks from their equilibrium positions.

19.153 Express in terms of L, C, and E the range of values of the resistance R for which oscillations will take place in the circuit shown when switch S is closed.

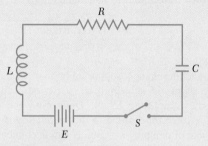

Fig. P19.153

19.154 Consider the circuit of Prob. 19.153 when the capacitor C is removed. If switch S is closed at time t = 0, determine (a) the final value of the current in the circuit, (b) the time t at which the current will have reached $(1 - 1/e)$ times its final value. (The desired value of t is known as the *time constant* of the circuit.)

19.155 and 19.156 Draw the electrical analogue of the mechanical system shown. (*Hint:* Draw the loops corresponding to the free bodies m and A.)

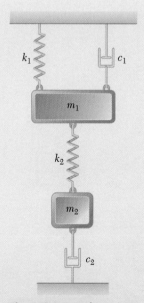

Fig. P19.156 and *P19.158*

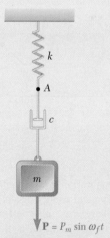

$P = P_m \sin \omega_f t$

Fig. P19.155 and P19.157

19.157 and *19.158* Write the differential equations defining (a) the displacements of the mass m and of the point A, (b) the charges on the capacitors of the electrical analogue.

Review and Summary

This chapter was devoted to the study of **mechanical vibrations**, i.e., to the analysis of the motion of particles and rigid bodies oscillating about a position of equilibrium. In the first part of the chapter [Secs. 19.1 through 19.4], we considered *vibrations without damping,* while the second part was devoted to *damped vibrations* [Sec. 19.5].

Free Vibrations of a Particle

In Sec. 19.1, we considered the **free vibrations of a particle**, i.e., the motion of a particle P subjected to a restoring force proportional to the displacement of the particle—such as the force exerted by a spring. If the displacement x of the particle P is measured from its equilibrium position O (Fig. 19.17), the resultant **F** of the forces acting on P (including its weight) has a magnitude kx and is directed toward O. Applying Newton's second law $F = ma$ and recalling that $a = \ddot{x}$, we wrote the differential equation

$$m\ddot{x} + kx = 0 \tag{19.2}$$

or, setting $\omega_n^2 = k/m$,

$$\ddot{x} + \omega_n^2 x = 0 \tag{19.6}$$

The motion defined by this equation is called **simple harmonic motion**.

The solution of Eq. (19.6), which represents the displacement of the particle P, was expressed as

$$x = x_m \sin(\omega_n t + \phi) \tag{19.10}$$

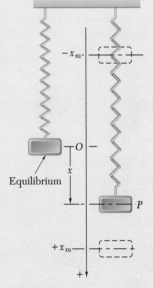

Fig. 19.17

where x_m = amplitude of the vibration
$\omega_n = \sqrt{k/m}$ = natural circular frequency
ϕ = phase angle

The **period of the vibration** (i.e., the time required for a full cycle) and its **natural frequency** (i.e., the number of cycles per second) were expressed as

$$\text{Period} = \tau_n = \frac{2\pi}{\omega_n} \tag{19.13}$$

$$\text{Natural frequency} = f_n = \frac{1}{\tau_n} = \frac{\omega_n}{2\pi} \tag{19.14}$$

We obtained the velocity and acceleration of the particle by differentiating Eq. (19.10), and their maximum values were found to be

$$v_m = x_m \omega_n \qquad a_m = x_m \omega_n^2 \tag{19.15}$$

Since all of the above parameters depend directly upon the natural circular frequency ω_n and thus upon the ratio k/m, it is essential in any given problem to calculate the value of the constant k. This can be done by determining the

relation between the restoring force and the corresponding displacement of the particle [Sample Prob. 19.1].

It was also shown that we can represent the oscillatory motion of particle P by the projection on the x axis of the motion of a point Q describing an auxiliary circle of radius x_m with the constant angular velocity ω_n (Fig. 19.18). Then we can obtain the instantaneous values of the velocity and acceleration of P by projecting on the x axis the vectors $\mathbf{v}_m$ and $\mathbf{a}_m$ representing, respectively, the velocity and acceleration of Q.

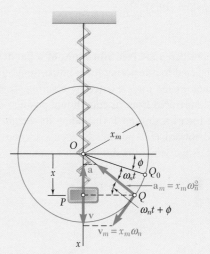

Fig. 19.18

Simple Pendulum

Although the motion of a **simple pendulum** is not truly a simple harmonic motion, we can use the formulas given previously with $\omega_n^2 = g/l$ to calculate the period and natural frequency of the *small oscillations* of a simple pendulum [Sec. 19.1B]. Large-amplitude oscillations of a simple pendulum were discussed in Sec. 19.1C.

Free Vibrations of a Rigid Body

We can analyze the **free vibrations of a rigid body** by choosing an appropriate variable, such as a distance x or an angle θ, to define the position of the body. We then draw a free-body diagram and kinetic diagram to express the equivalence of the external forces and inertial terms and write an equation relating the selected variable and its second derivative [Sec. 19.2]. If the equation obtained is of the form

$$\ddot{x} + \omega_n^2 x = 0 \qquad \text{or} \qquad \ddot{\theta} + \omega_n^2 \theta = 0 \qquad \textbf{(19.21)}$$

the vibration considered is a simple harmonic motion, and its period and natural frequency can be obtained *by identifying* ω_n and substituting its value into Eqs. (19.13) and (19.14) [Sample Probs. 19.2 and 19.3].

Using the Principle of Conservation of Energy

We can use the *principle of conservation of energy* as an alternative method for determining the period and natural frequency of the simple harmonic motion of a particle or rigid body [Sec. 19.3]. Choosing again an appropriate

variable, such as θ, to define the position of the system, we express that the total energy of the system is conserved, using $T_1 + V_1 = T_2 + V_2$, between the position of maximum displacement $(\theta_1 = \theta_m)$ and the position of maximum velocity $(\dot{\theta}_2 = \dot{\theta}_m)$. If the motion considered is simple harmonic, the two sides of the equation obtained consist of homogeneous quadratic expressions in θ_m and $\dot{\theta}_m$, respectively. Substituting $\dot{\theta}_m = \theta_m \omega_n$ in this equation, we can factor out θ_m^2 and solve for the circular frequency ω_n [Sample Prob. 19.4]. It is important to note that if the motion can be approximated only by a simple harmonic motion, such as for the small oscillations of a body under gravity, we must approximate the potential energy by a quadratic expression in θ_m [Sample Prob. 19.4].

Forced Vibrations

In Sec. 19.4, we considered the **forced vibrations** of a mechanical system. These vibrations occur when the system is subjected to a periodic force (Fig. 19.19) or when it is elastically connected to a support that has an alternating motion (Fig. 19.20). Denoting the forced circular frequency by ω_f, we found that in the first case, the motion of the system was defined by the differential equation

$$m\ddot{x} + kx = P_m \sin \omega_f t \qquad (19.30)$$

and that in the second case, it was defined by the differential equation

$$m\ddot{x} + kx = k\delta_m \sin \omega_f t \qquad (19.31)$$

We can obtain the general solution of these equations by adding a particular solution of the form

$$x_{\text{part}} = x_m \sin \omega_f t \qquad (19.32)$$

to the general solution of the corresponding homogeneous equation. The particular solution of Eq. (19.32) represents a **steady-state vibration** of the

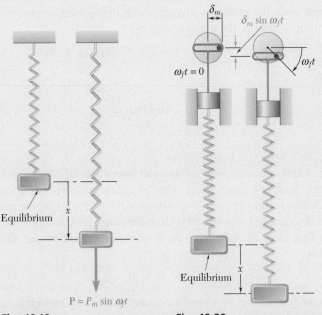

Fig. 19.19

Fig. 19.20

system, whereas the solution of the homogeneous equation represents a **transient free vibration** that can generally be neglected.

Dividing the amplitude x_m of the steady-state vibration by P_m/k in the case of a periodic force or by δ_m in the case of an oscillating support, we defined the **magnification factor** of the vibration and found that

$$\text{Magnification factor} = \frac{x_m}{P_m/k} = \frac{x_m}{\delta_m} = \frac{1}{1 - (\omega_f/\omega_n)^2} \qquad (19.36)$$

According to Eq. (19.36), the amplitude x_m of the forced vibration becomes infinite when $\omega_f = \omega_n$, i.e., when the forced frequency is equal to the natural frequency of the system. The impressed force or impressed support movement is then said to be in **resonance** with the system [Sample Prob. 19.5]. (Actually, the amplitude of the vibration remains finite, due to damping forces.)

Damped Free Vibrations

In Sec. 19.5, we considered the **damped vibrations** of a mechanical system. First, we analyzed the damped free vibrations of a system with **viscous damping** [Sec. 19.5A]. We found that the motion of such a system was defined by the differential equation

$$m\ddot{x} + c\dot{x} + kx = 0 \qquad (19.38)$$

where c is a constant called the *coefficient of viscous damping*. Defining the *critical damping coefficient* c_c as

$$c_c = 2m\sqrt{\frac{k}{m}} = 2m\omega_n \qquad (19.41)$$

where ω_n is the natural circular frequency of the system in the absence of damping, we distinguished three different cases of damping, namely, (1) *overdamped*, when $c > c_c$; (2) *critically damped*, when $c = c_c$; and (3) *underdamped*, when $c < c_c$. In the first two cases, the system when disturbed tends to regain its equilibrium position without any oscillation. In the third case, the motion is vibratory with diminishing amplitude. For an underdamped system, the transient response is

$$x = x_0 e^{-(c/2m)t} \sin(\omega_d t + \phi) \qquad (19.46)$$

where

$$\omega_d = \omega_n \sqrt{1 - \left(\frac{c}{c_c}\right)^2} \qquad (19.45)$$

Damped Forced Vibrations

In Sec. 19.5B, we considered the **damped forced vibrations** of a mechanical system. These vibrations occur when a system with viscous damping is subjected to a periodic force **P** of magnitude $P = P_m \sin \omega_f t$ or when it is elastically connected to a support with an alternating motion of $\delta = \delta_m \sin \omega_f t$. In the first case, the motion of the system was defined by the differential equation

$$m\ddot{x} + c\dot{x} + kx = P_m \sin \omega_f t \qquad (19.47)$$

and in the second case, by a similar equation obtained by replacing P_m by $k\delta_m$ in (19.47).

The *steady-state vibration* of the system is represented by a particular solution of Eq. (19.47) of the form

$$x_{\text{part}} = x_m \sin{(\omega_f t - \phi)} \qquad \textbf{(19.48)}$$

Dividing the amplitude x_m of the steady-state vibration by P_m/k in the case of a periodic force or by δ_m in the case of an oscillating support, we obtained the following expression for the magnification factor as

$$\frac{x_m}{P_m/k} = \frac{x_m}{\delta_m} = \frac{1}{\sqrt{[1 - (\omega_f/\omega_n)^2]^2 + [2(c/c_c)(\omega_f/\omega_n)]^2}} \qquad \textbf{(19.53)}$$

or

$$\frac{x_m}{P_m/k} = \frac{x_m}{\delta_{st}} = \frac{1}{\sqrt{(1 - r^2)^2 + (2\zeta r)^2}}$$

where $\omega_n = \sqrt{k/m}$ = natural circular frequency of undamped system
$c_c = 2m\omega_n$ = critical damping coefficient
$c/c_c = \zeta$ = damping ratio
$r = \omega/\omega_n$ = frequency ratio

We also found that the *phase difference* ϕ between the impressed force or support movement and the resulting steady-state vibration of the damped system was defined by the relation

$$\tan \phi = \frac{2(c/c_c)(\omega_f/\omega_n)}{1 - (\omega_f/\omega_n)^2} \qquad \textbf{(19.54)}$$

or

$$\tan \phi = \frac{2\zeta r}{1 - r^2} \qquad \textbf{(19.54')}$$

Electrical Analogs

This chapter ended with a discussion of *electrical analogs* [Sec. 19.5C] in which we showed that the vibrations of mechanical systems and the oscillations of electrical circuits are defined by the same differential equations. Electrical analogs of mechanical systems therefore can be used to study or predict the behavior of these systems.

Review Problems

Fig. P19.159

19.159 An automobile wheel-and-tire assembly of total weight 47 lb is attached to a mounting plate of negligible weight that is suspended from a steel wire. The torsional spring constant of the wire is known to be $K = 0.40$ lb·in/rad. The wheel is rotated through 90° about the vertical and then released. Knowing that the period of oscillation is observed to be 30 s, determine the centroidal mass moment of inertia and the centroidal radius of gyration of the wheel-and-tire assembly.

19.160 The period of vibration of the system shown is observed to be 0.6 s. After cylinder B has been removed, the period is observed to be 0.5 s. Determine (*a*) the weight of cylinder A, (*b*) the constant of the spring.

3 lb B

Fig. P19.160

19.161 Disks A and B weigh 30 lb and 12 lb, respectively, and a small 5-lb block C is attached to the rim of disk B. Assuming that no slipping occurs between the disks, determine the period of small oscillations of the system.

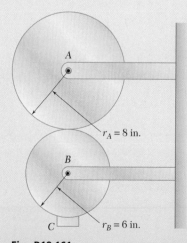

$r_A = 8$ in.

$r_B = 6$ in.

Fig. P19.161

19.162 The block shown is depressed 1.2 in. from its equilibrium position and released. Knowing that after 10 cycles the maximum displacement of the block is 0.5 in., determine (*a*) the damping factor c/c, (*b*) the value of the coefficient of viscous damping. (*Hint:* See Problems 19.129 and 19.130.)

19.163 An 0.8-lb ball is connected to a paddle by means of an elastic cord *AB* of constant $k = 5$ lb/ft. Knowing that the paddle is moved vertically according to the relation $\delta = \delta_m \sin \omega_f t$, where $\delta_m = 8$ in., determine the maximum allowable circular frequency ω_f if the cord is not to become slack.

Fig. P19.162

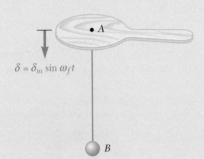

$\delta = \delta_m \sin \omega_f t$

Fig. P19.163

19.164 A 3-kg slender rod *AB* is bolted to a 5-kg uniform disk. A dashpot with a damping coefficient of $c = 9$ N·s/m is attached to the disk as shown. Determine (*a*) the differential equation of motion for small oscillations, (*b*) the damping factor c/c_c.

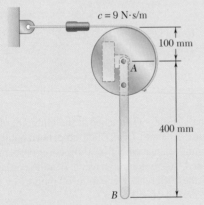

$c = 9$ N·s/m

100 mm

A

400 mm

B

Fig. P19.164

19.165 A 4-lb uniform rod is supported by a pin at *O* and a spring at *A* and is connected to a dashpot at *B*. Determine (*a*) the differential equation of motion for small oscillations, (*b*) the angle that the rod will form with the horizontal 5 s after end *B* has been pushed 0.9 in. down and released.

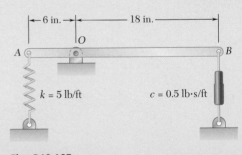

A

$\leftarrow$ 6 in. $\rightarrow$ $\leftarrow$ 18 in. $\rightarrow$

O

B

$k = 5$ lb/ft

$c = 0.5$ lb·s/ft

Fig. P19.165

Fig. P19.166

19.166 A 400-kg motor supported by four springs, each of constant 150 kN/m, and a dashpot of constant $c = 6500$ N·s/m is constrained to move vertically. Knowing that the unbalance of the rotor is equivalent to a 23-g mass located at a distance of 100 mm from the axis of rotation, determine for a speed of 800 rpm (a) the amplitude of the fluctuating force transmitted to the foundation, (b) the amplitude of the vertical motion of the motor.

19.167 The compressor shown has a mass of 250 kg and operates at 2000 rpm. At this operating condition, the force transmitted to the ground is excessively high and is found to be $mr\omega_f^2$, where mr is the unbalance and ω_f is the forcing frequency. To fix this problem, it is proposed to isolate the compressor by mounting it on a square concrete block separated from the rest of the floor as shown. The density of concrete is 2400 kg/m^3 and the spring constant for the soil is found to be 80×10^6 N/m. The geometry of the compressor leads to choosing a block that is 1.5 m by 1.5 m. Determine the depth h that will reduce the force transmitted to the ground by 75 percent.

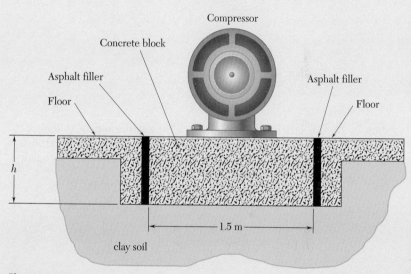

Fig. P19.167

19.168 A small ball of mass m attached at the midpoint of a tightly stretched elastic cord of length l can slide on a horizontal plane. The ball is given a small displacement in a direction perpendicular to the cord and released. Assuming the tension T in the cord to remain constant, (a) write the differential equation of motion of the ball, (b) determine the period of vibration.

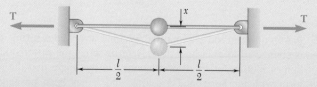

Fig. P19.168

19.169 A certain vibrometer used to measure vibration amplitudes consists essentially of a box containing a slender rod to which a mass m is attached; the natural frequency of the mass–rod system is known to be 5 Hz. When the box is rigidly attached to the casing of a motor rotating at 600 rpm, the mass is observed to vibrate with an amplitude of 0.06 in. relative to the box. Determine the amplitude of the vertical motion of the motor.

Fig. P19.169

19.170 If either a simple or a compound pendulum is used to determine experimentally the acceleration of gravity g, difficulties are encountered. In the case of the simple pendulum, the string is not truly weightless, while in the case of the compound pendulum, the exact location of the mass center is difficult to establish. In the case of a compound pendulum, the difficulty can be eliminated by using a reversible, or Kater, pendulum. Two knife edges A and B are placed so that they are obviously not at the same distance from the mass center G, and the distance l is measured with great precision. The position of a counterweight D is then adjusted so that the period of oscillation τ is the same when either knife edge is used. Show that the period τ obtained is equal to that of a true simple pendulum of length l and that $g = 4\pi^2 l/\tau^2$.

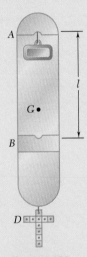

Fig. P19.170

A Some Useful Definitions and Properties of Vector Algebra

The following definitions and properties of vector algebra were discussed fully in Chaps. 2 and 3 of *Vector Mechanics for Engineers: Statics.* They are summarized here for the convenience of the reader, with references to the appropriate sections of the *Statics* volume. Equation and illustration numbers are those used in the original presentation.

A.1 Addition of Vectors (Secs. 2.1B and 2.1C)

Vectors are defined as *mathematical expressions possessing magnitude and direction, which add according to the parallelogram law.* Thus the sum of two vectors **P** and **Q** is obtained by attaching the two vectors to the same point A and constructing a parallelogram, using **P** and **Q** as two sides of the parallelogram (Fig. A.2). The diagonal that passes through A represents the sum of the vectors **P** and **Q**, and this sum is denoted by **P** + **Q**. Vector addition is *associative* and *commutative.*

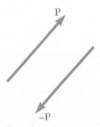

Fig. A.1

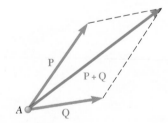

Fig. A.2

The *negative vector* of a given vector **P** is defined as a vector having the same magnitude P and a direction opposite to that of **P** (Fig. A.1); the negative of the vector **P** is denoted by $-\mathbf{P}$. Clearly, we have

$$\mathbf{P} + (-\mathbf{P}) = 0$$

A1

A.2   Product of a Scalar and a Vector (Sec. 2.1C)

The product $k\mathbf{P}$ of a scalar k and a vector $\mathbf{P}$ is defined as a vector having the same direction as $\mathbf{P}$ (if k is positive), or a direction opposite to that of $\mathbf{P}$ (if k is negative), and a magnitude equal to the product of the magnitude P and the absolute value of k (Fig. A.3).

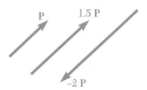

Fig. A.3

A.3   Unit Vectors. Resolution of a Vector into Rectangular Components (Secs. 2.2A and 2.4A)

The vectors $\mathbf{i}$, $\mathbf{j}$, and $\mathbf{k}$, called *unit vectors,* are defined as vectors of magnitude 1, directed, respectively, along the positive x, y, and z axes (Fig. A.4).

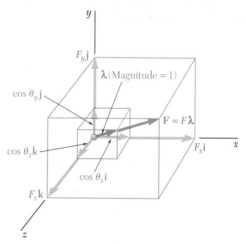

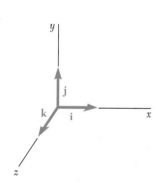

Fig. A.4

Fig. A.5

Denoting by F_x, F_y, and F_z, the scalar components of a vector $\mathbf{F}$, we have (Fig. A.5)

$$\mathbf{F} = F_x\mathbf{i} + F_y\mathbf{j} + F_z\mathbf{k} \tag{2.20}$$

In the particular case of a unit vector $\boldsymbol{\lambda}$ directed along a line forming angles θ_x, θ_y, and θ_z with the coordinate axes, we have

$$\boldsymbol{\lambda} = \cos\theta_x\mathbf{i} + \cos\theta_y\mathbf{j} + \cos\theta_z\mathbf{k} \tag{2.22}$$

A.4   Vector Product of Two Vectors (Secs. 3.1C and 3.1D)

The vector product, or *cross product,* of two vectors $\mathbf{P}$ and $\mathbf{Q}$ is defined as the vector

$$\mathbf{V} = \mathbf{P} \times \mathbf{Q}$$

which satisfies the following conditions:

1. The line of action of **V** is perpendicular to the plane containing **P** and **Q** (Fig. A.6).
2. The magnitude of **V** is the product of the magnitudes of **P** and **Q** and of the sine of the angle θ formed by **P** and **Q** (the measure of which will always be 180° or less); we thus have

$$V = PQ \sin \theta \tag{3.1}$$

3. The direction of **V** is obtained from the *right-hand rule*. Close your right hand and hold it so that your fingers are curled in the same sense as the rotation through θ which brings the vector **P** in line with the vector **Q**; your thumb will then indicate the direction of the vector **V** (Fig. A.6*b*). Note that if **P** and **Q** do not have a common point of application, they should first be redrawn from the same point. The three vectors **P**, **Q**, and **V**—taken in that order—are said to form a *right-handed triad*.

Vector products are *distributive* but *not commutative*. We have

$$\mathbf{Q} \times \mathbf{P} = -(\mathbf{P} \times \mathbf{Q}) \tag{3.4}$$

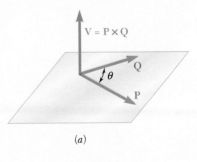

(a)

(b)

Fig. A.6

Vector Products of Unit Vectors. It follows from the definition of the vector product of two vectors that

$$
\begin{array}{lll}
\mathbf{i} \times \mathbf{i} = 0 & \mathbf{j} \times \mathbf{i} = -\mathbf{k} & \mathbf{k} \times \mathbf{i} = \mathbf{j} \\
\mathbf{i} \times \mathbf{j} = \mathbf{k} & \mathbf{j} \times \mathbf{j} = 0 & \mathbf{k} \times \mathbf{j} = -\mathbf{i} \\
\mathbf{i} \times \mathbf{k} = -\mathbf{j} & \mathbf{j} \times \mathbf{k} = \mathbf{i} & \mathbf{k} \times \mathbf{k} = 0
\end{array}
\tag{3.7}
$$

Rectangular Components of Vector Product. Resolving the vectors **P** and **Q** into rectangular components, we obtain the following expressions for the components of their vector product **V**:

$$
\begin{aligned}
V_x &= P_y Q_z - P_z Q_y \\
V_y &= P_z Q_x - P_x Q_z \\
V_z &= P_x Q_y - P_y Q_x
\end{aligned}
\tag{3.9}
$$

In determinant form, we have

$$
\mathbf{V} = \mathbf{P} \times \mathbf{Q} = \begin{vmatrix} \mathbf{i} & \mathbf{j} & \mathbf{k} \\ P_x & P_y & P_z \\ Q_x & Q_y & Q_z \end{vmatrix} \tag{3.10}
$$

A.5 Moment of a Force About a Point (Secs. 3.1E and 3.1F)

The moment of a force **F** (or, more generally, of a vector **F**) about a point O is defined as the vector product

$$\mathbf{M}_O = \mathbf{r} \times \mathbf{F} \tag{3.11}$$

where **r** denotes the *position vector* of the point of application A of **F** (Fig. A.7*a*).

According to the definition of the vector product of two vectors given in Sec. A.4, the moment $\mathbf{M}_O$ must be perpendicular to the plane containing O and the force **F**. Its magnitude is equal to

$$M_O = rF \sin \theta = Fd \tag{3.12}$$

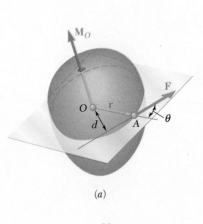

(a)

(b)

Fig. A.7

where d is the perpendicular distance from O to the line of action of $\mathbf{F}$, and its sense is defined by the sense of the rotation which would bring the vector $\mathbf{r}$ in line with the vector $\mathbf{F}$; this rotation should be viewed as *counterclockwise* by an observer located at the tip of $\mathbf{M}_O$. Another way of defining the sense of $\mathbf{M}_O$ is furnished by a variation of the *right-hand rule:* Close your right hand and hold it so that your fingers are curled in the sense of the rotation that $\mathbf{F}$ would impart to the rigid body about a fixed axis directed along the line of action of $\mathbf{M}_O$; your thumb will indicate the sense of the moment $\mathbf{M}_O$ (Fig. A.7*b*).

Rectangular Components of the Moment of a Force.   Denoting by x, y, and z the coordinates of the point of application A of $\mathbf{F}$, we obtain the following expressions for the components of the moment $\mathbf{M}_O$ of $\mathbf{F}$:

$$\begin{aligned} M_x &= yF_z - zF_y \\ M_y &= zF_x - xF_z \\ M_z &= xF_y - yF_x \end{aligned} \qquad (3.18)$$

In determinant form, we have

$$\mathbf{M}_O = \mathbf{r} \times \mathbf{F} = \begin{vmatrix} \mathbf{i} & \mathbf{j} & \mathbf{k} \\ x & y & z \\ F_x & F_y & F_z \end{vmatrix} \qquad (3.19)$$

To compute the moment $\mathbf{M}_B$ about an arbitrary point B of a force $\mathbf{F}$ applied at A, we must use the vector $\mathbf{r}_{A/B} = \mathbf{r}_A - \mathbf{r}_B$ drawn from B to A instead of the vector $\mathbf{r}$. We write

$$\mathbf{M}_B = \mathbf{r}_{A/B} \times \mathbf{F} = (\mathbf{r}_A - \mathbf{r}_B) \times \mathbf{F} \qquad (3.20)$$

or, using the determinant form,

$$\mathbf{M}_B = \begin{vmatrix} \mathbf{i} & \mathbf{j} & \mathbf{k} \\ x_{A/B} & y_{A/B} & z_{A/B} \\ F_x & F_y & F_z \end{vmatrix} \qquad (3.21)$$

where $x_{A/B}$, $y_{A/B}$, $z_{A/B}$ are the components of the vector $\mathbf{r}_{A/B}$:

$$x_{A/B} = x_A - x_B \qquad y_{A/B} = y_A - y_B \qquad z_{A/B} = z_A - z_B$$

A.6   Scalar Product of Two Vectors (Sec. 3.2A)

The scalar product, or *dot product,* of two vectors $\mathbf{P}$ and $\mathbf{Q}$ is defined as the product of the magnitudes of $\mathbf{P}$ and $\mathbf{Q}$ and of the cosine of the angle θ formed by $\mathbf{P}$ and $\mathbf{Q}$ (Fig. A.8). The scalar product of $\mathbf{P}$ and $\mathbf{Q}$ is denoted by $\mathbf{P}\cdot\mathbf{Q}$. We write

$$\mathbf{P}\cdot\mathbf{Q} = PQ \cos \theta \qquad (3.24)$$

Scalar products are *commutative* and *distributive.*

Scalar Products of Unit Vectors.   It follows from the definition of the scalar product of two vectors that

$$\begin{aligned} \mathbf{i}\cdot\mathbf{i} &= 1 & \mathbf{j}\cdot\mathbf{j} &= 1 & \mathbf{k}\cdot\mathbf{k} &= 1 \\ \mathbf{i}\cdot\mathbf{j} &= 0 & \mathbf{j}\cdot\mathbf{k} &= 0 & \mathbf{k}\cdot\mathbf{i} &= 0 \end{aligned} \qquad (3.27)$$

Fig. A.8

Scalar Product Expressed in Terms of Rectangular Components.   Resolving the vectors $\mathbf{P}$ and $\mathbf{Q}$ into rectangular components, we obtain

$$\mathbf{P}\cdot\mathbf{Q} = P_xQ_x + P_yQ_y + P_zQ_z \tag{3.28}$$

Angle Formed by Two Vectors.   It follows from (3.24) and (3.27) that

$$\cos\theta = \frac{\mathbf{P}\cdot\mathbf{Q}}{PQ} = \frac{P_xQ_x + P_yQ_y + P_zQ_z}{PQ} \tag{3.30}$$

Projection of a Vector on a Given Axis.   The projection of a vector $\mathbf{P}$ on the axis OL defined by the unit vector $\boldsymbol{\lambda}$ (Fig. A.9) is

$$P_{OL} = OA = \mathbf{P}\cdot\boldsymbol{\lambda} \tag{3.34}$$

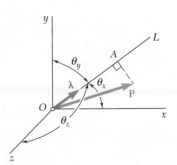

Fig. A.9

A.7   Mixed Triple Product of Three Vectors (Sec. 3.2B)

The mixed triple product of the three vectors $\mathbf{S}$, $\mathbf{P}$, and $\mathbf{Q}$ is defined as the scalar expression

$$\mathbf{S}\cdot(\mathbf{P} \times \mathbf{Q}) \tag{3.36}$$

obtained by forming the scalar product of $\mathbf{S}$ with the vector product of $\mathbf{P}$ and $\mathbf{Q}$. Mixed triple products are invariant under *circular permutations* but change sign under any other permutation:

$$\mathbf{S}\cdot(\mathbf{P} \times \mathbf{Q}) = \mathbf{P}\cdot(\mathbf{Q} \times \mathbf{S}) = \mathbf{Q}\cdot(\mathbf{S} \times \mathbf{P})$$
$$= -\mathbf{S}\cdot(\mathbf{Q} \times \mathbf{P}) = -\mathbf{P}\cdot(\mathbf{S} \times \mathbf{Q}) = -\mathbf{Q}\cdot(\mathbf{P} \times \mathbf{S}) \tag{3.37}$$

Mixed Triple Product Expressed in Terms of Rectangular Components.   The mixed triple product of $\mathbf{S}$, $\mathbf{P}$, and $\mathbf{Q}$ may be expressed in the form of a determinant

$$\mathbf{S}\cdot(\mathbf{P} \times \mathbf{Q}) = \begin{vmatrix} S_x & S_y & S_z \\ P_x & P_y & P_z \\ Q_x & Q_y & Q_z \end{vmatrix} \tag{3.39}$$

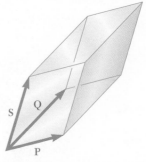

Fig. A.10

The mixed triple product $\mathbf{S}\cdot(\mathbf{P} \times \mathbf{Q})$ measures the volume of the parallelepiped having the vectors $\mathbf{S}$, $\mathbf{P}$, and $\mathbf{Q}$ for sides (Fig. A.10).

A.8   Moment of a Force About a Given Axis (Sec. 3.2C)

The moment M_{OL} of a force $\mathbf{F}$ (or, more generally, of a vector $\mathbf{F}$) about an axis OL is defined as the projection OC on the axis OL of the moment $\mathbf{M}_O$ of $\mathbf{F}$ about O (Fig. A.11). Denoting by $\boldsymbol{\lambda}$ the unit vector along OL, we have

$$M_{OL} = \boldsymbol{\lambda}\cdot\mathbf{M}_O = \boldsymbol{\lambda}\cdot(\mathbf{r} \times \mathbf{F}) \tag{3.40}$$

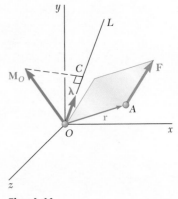

Fig. A.11

or, in determinant form,

$$M_{OL} = \begin{vmatrix} \lambda_x & \lambda_y & \lambda_z \\ x & y & z \\ F_x & F_y & F_z \end{vmatrix} \qquad \textbf{(3.41)}$$

where λ_x, λ_y, λ_z, = direction cosines of axis OL
          x, y, z = coordinates of point of application of $\mathbf{F}$
     F_x, F_y, F_z = components of force $\mathbf{F}$

The moments of the force $\mathbf{F}$ about the three coordinate axes are given by the expressions (3.18) obtained earlier for the rectangular components of the moment $\mathbf{M}_O$ of $\mathbf{F}$ about O:

$$\begin{aligned} M_x &= yF_z - zF_y \\ M_y &= zF_x - xF_z \\ M_z &= xF_y - yF_x \end{aligned} \qquad \textbf{(3.18)}$$

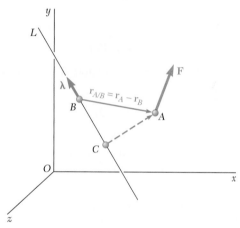

Fig. A.12

More generally, the moment of a force $\mathbf{F}$ applied at A about an axis which does not pass through the origin is obtained by choosing an arbitrary point B on the axis (Fig. A.12) and determining the projection on the axis BL of the moment $\mathbf{M}_B$ of $\mathbf{F}$ about B. We write

$$M_{BL} = \boldsymbol{\lambda} \cdot \mathbf{M}_B = \boldsymbol{\lambda} \cdot (\mathbf{r}_{A/B} \times \mathbf{F}) \qquad \textbf{(3.43)}$$

where $\mathbf{r}_{A/B} = \mathbf{r}_A - \mathbf{r}_B$ represents the vector drawn from B to A. Expressing M_{BL} in the form of a determinant, we have

$$M_{BL} = \begin{vmatrix} \lambda_x & \lambda_y & \lambda_z \\ x_{A/B} & y_{A/B} & z_{A/B} \\ F_x & F_y & F_z \end{vmatrix} \qquad \textbf{(3.44)}$$

where λ_x, λ_y, λ_z = direction cosines of axis BL
      $x_{A/B} = x_A - x_B$, $y_{A/B} = y_A - y_B$, $z_{A/B} = z_A - z_B$
   F_x, F_y, F_z = components of force $\mathbf{F}$

It should be noted that the result obtained is independent of the choice of the point B on the given axis; the same result would have been obtained if point C had been chosen instead of B.

APPENDIX

B Mass Moment of Inertia

B.1 Moment of Inertia of a Simple Mass

Consider a small mass Δm mounted on a rod of negligible mass that can rotate freely about an axis AA' (Fig. B.1a). If we apply a couple to the system, the rod and mass (assumed to be initially at rest) start rotating about AA'. We will study the details of this motion later in dynamics. At present, we wish to indicate only that the time required for the system to reach a given speed of rotation is proportional to the mass Δm and to the square of the distance r. The product $r^2 \Delta m$ thus provides a measure of the **inertia** of the system; i.e., a measure of the resistance the system offers when we try to set it in motion. For this reason, the product $r^2 \Delta m$ is called the **moment of inertia** of the mass Δm with respect to axis AA'.

Now suppose a body of mass m is to be rotated about an axis AA' (Fig. B.1b). Dividing the body into elements of mass Δm_1, Δm_2, etc., we find that the body's resistance to being rotated is measured by the sum $r_1^2 \Delta m_1 + r_2^2 \Delta m_2 + \ldots$. This sum defines the moment of inertia of the body with respect to axis AA'. Increasing the number of elements, we find that the moment of inertia is equal, in the limit, to the integral

Mass moment of inertia

$$I = \int r^2 \, dm \qquad \text{(B.1)}$$

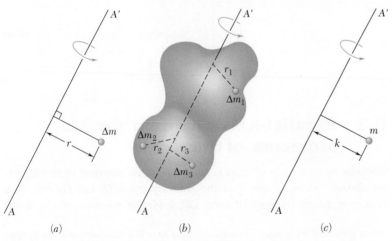

(a)　　　　　(b)　　　　　(c)

Fig. B.1 (a) An element of mass Δm at a distance r from an axis AA'; (b) the moment of inertia of a rigid body is the sum of the moments of inertia of many small masses; (c) the moment of inertia is unchanged if all the mass is concentrated at a point at a distance from the axis equal to the radius of gyration.

We define the **radius of gyration** k of the body with respect to axis AA' by the relation

**Radius of gyration
of a mass**

$$I = k^2 m \quad \text{or} \quad k = \sqrt{\frac{I}{m}} \tag{B.2}$$

The radius of gyration k represents the distance at which the entire mass of the body should be concentrated if its moment of inertia with respect to AA' is to remain unchanged (Fig. B.1c). Whether it stays in its original shape (Fig. B.1b) or is concentrated as shown in Fig. B.1c, the mass m reacts in the same way to a rotation (or *gyration*) about AA'.

If SI units are used, the radius of gyration k is expressed in meters and the mass m in kilograms, so the unit for the moment of inertia of a mass is kg·m². If U.S. customary units are used, the radius of gyration is expressed in feet and the mass in slugs (i.e., in lb·s²/ft), so the derived unit for the moment of inertia of a mass is lb·ft·s².[†]

We can express the moment of inertia of a body with respect to a coordinate axis in terms of the coordinates x, y, z of the element of mass dm (Fig. B.2). Noting, for example, that the square of the distance r from the element dm to the y axis is $z^2 + x^2$, the moment of inertia of the body with respect to the y axis is

$$I_y = \int r^2 \, dm = \int (z^2 + x^2) \, dm$$

We obtain similar expressions for the moments of inertia with respect to the x and z axes.

**Moments of inertia with
respect to coordinate axes**

$$\begin{aligned} I_x &= \int (y^2 + z^2) \, dm \\ I_y &= \int (z^2 + x^2) \, dm \\ I_z &= \int (x^2 + y^2) \, dm \end{aligned} \tag{B.3}$$

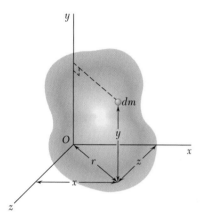

Fig. B.2 An element of mass dm in an x, y, z coordinate system.

B.2 Parallel-Axis Theorem for Mass Moments of Inertia

Consider again a body of mass m and let $Oxyz$ be a system of rectangular coordinates whose origin is at the arbitrary point O. Let $Gx'y'z'$ be a system of parallel centroidal axes; i.e., a system whose origin is at the

Photo B.1 The rotational behavior of this crankshaft depends upon its mass moment of inertia with respect to its axis of rotation.

[†]When converting the moment of inertia of a mass from U.S. customary units to SI units, keep in mind that the base unit (pound) used in the derived unit (lb·ft·s²) is a unit of force (*not* of mass). Therefore, it should be converted into newtons. We have

$$1 \text{ lb·ft·s}^2 = (4.45 \text{ N})(0.3048 \text{ m})(1 \text{ s})^2 = 1.356 \text{ N·m·s}^2$$

or since $1 \text{ N} = 1 \text{ kg·m/s}^2$

$$1 \text{ lb·ft·s}^2 = 1.356 \text{ kg·m}^2$$

center of gravity G of the body and whose axes x', y', z' are parallel to the x, y, and z axes, respectively (Fig. B.3). (Note that we use the term centroidal here to define axes passing through the center of gravity G of the body, regardless of whether or not G coincides with the centroid of the volume of the body.) We denote by $\bar{x}, \bar{y}, \bar{z}$ the coordinates of G with respect to $Oxyz$. Then we have the following relations between the coordinates x, y, z of the element dm with respect to $Oxyz$ and its coordinates x', y', z' with respect to the centroidal axes $Gx'y'z'$:

$$x = x' + \bar{x} \qquad y = y' + \bar{y} \qquad z = z' + \bar{z} \qquad \text{(B.4)}$$

Referring to Eqs. (B.3), we can express the moment of inertia of the body with respect to the x axis as

$$I_x = \int (y^2 + z^2)\, dm = \int [(y' + \bar{y})^2 + (z' + \bar{z})^2]\, dm$$
$$= \int (y'^2 + z'^2)\, dm + 2\bar{y}\int y'\, dm + 2\bar{z}\int z'\, dm + (\bar{y}^2 + \bar{z}^2)\int dm$$

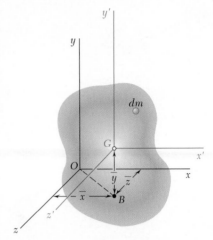

Fig. B.3 A body of mass m with an arbitrary rectangular coordinate system at O and a parallel centroidal coordinate system at G. Also shown is an element of mass dm.

The first integral in this expression represents the moment of inertia $\bar{I}_{x'}$ of the body with respect to the centroidal axis x'. The second and third integrals represent the first moment of the body with respect to the $z'x'$ and $x'y'$ planes, respectively, and since both planes contain G, these two integrals are zero. The last integral is equal to the total mass m of the body. Therefore, we have

$$I_x = \bar{I}_{x'} + m(\bar{y}^2 + \bar{z}^2) \qquad \text{(B.5)}$$

Similarly,

$$I_y = \bar{I}_{y'} + m(\bar{z}^2 + \bar{x}^2) \qquad I_z = \bar{I}_{z'} + m(\bar{x}^2 + \bar{y}^2) \qquad \text{(B.5')}$$

We easily verify from Fig. B.3 that the sum $\bar{z}^2 + \bar{x}^2$ represents the square of the distance OB between the y and y' axes. Similarly, $\bar{y}^2 + \bar{z}^2$ and $\bar{x}^2 + \bar{y}^2$ represent the squares of the distance between the x and x' axes and the z and z' axes, respectively. We denote the distance between an arbitrary axis AA' and a parallel centroidal axis BB' by d (Fig. B.4). Then the general relation between the moment of inertia I of the body with respect to AA' and its moment of inertia [&*obar*{I*N*[1%0]}&] with respect to BB', known as the parallel-axis theorem for mass moments of inertia, is

Parallel-axis theorem for mass moments of inertia

$$I = \bar{I} + md^2 \qquad \text{(B.6)}$$

Expressing the moments of inertia in terms of the corresponding radii of gyration, we can also write

$$k^2 = \bar{k}^2 + d^2 \qquad \text{(B.7)}$$

where k and $\bar{k}$ represent the radii of gyration of the body about AA' and BB', respectively.

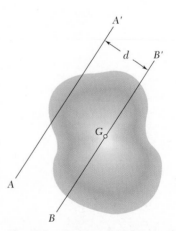

Fig. B.4 We use d to denote the distance between an arbitrary axis AA' and a parallel centroidal axis BB'.

B.3 Moments of Inertia of Thin Plates

Now imagine a thin plate of uniform thickness t, made of a homogeneous material of density ρ (density = mass per unit volume). The mass moment of inertia of the plate with respect to an axis AA' *contained in the plane* of the plate (Fig. B.5a) is

$$I_{AA', \text{mass}} = \int r^2 \, dm$$

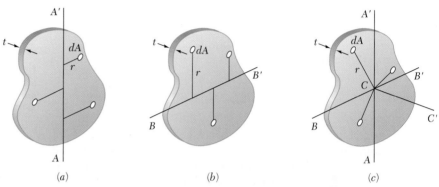

Fig. B.5 (a) A thin plate with an axis AA' in the plane of the plate; (b) an axis BB' in the plane of the plate and perpendicular to AA'; (c) an axis CC' perpendicular to the plate and passing through the intersection of AA' and BB'.

Since $dm = \rho t \, dA$, we have

$$I_{AA', \text{mass}} = \rho t \int r^2 \, dA$$

However, r represents the distance of the element of area dA to the axis AA'. Therefore, the integral is equal to the moment of inertia of the area of the plate with respect to AA'.

$$I_{AA', \text{mass}} = \rho t I_{AA', \text{area}} \tag{B.8}$$

Similarly, for an axis BB' that is contained in the plane of the plate and is perpendicular to AA' (Fig. B.5b), we have

$$I_{BB', \text{mass}} = \rho t I_{BB', \text{area}} \tag{B.9}$$

Consider now the axis CC', which is *perpendicular* to the plate and passes through the point of intersection C of AA' and BB' (Fig. B.5c). This time we have

$$I_{CC', \text{mass}} = \rho t J_{C, \text{area}} \tag{B.10}$$

where J_C is the polar moment of inertia of the area of the plate with respect to point C.

Recall the relation $J_C = I_{AA'} + I_{BB'}$ between the polar and rectangular moments of inertia of an area. We can use this to write the relation between the mass moments of inertia of a thin plate as

$$I_{CC'} = I_{AA'} + I_{BB'} \tag{B.11}$$

Rectangular Plate. In the case of a rectangular plate of sides a and b (Fig. B.6), we obtain the mass moments of inertia with respect to axes through the center of gravity of the plate as

$$I_{AA', \text{mass}} = \rho t I_{AA', \text{area}} = \rho t(\tfrac{1}{12}a^3 b)$$
$$I_{BB', \text{mass}} = \rho t I_{BB', \text{area}} = \rho t(\tfrac{1}{12}ab^3)$$

Since the product ρabt is equal to the mass m of the plate, we can also write the mass moments of inertia of a thin rectangular plate as

$$I_{AA'} = \tfrac{1}{12}ma^2 \qquad I_{BB'} = \tfrac{1}{12}mb^2 \tag{B.12}$$
$$I_{CC'} = I_{AA'} + I_{BB'} = \tfrac{1}{12}m(a^2 + b^2) \tag{B.13}$$

Circular Plate. In the case of a circular plate, or disk, of radius r (Fig. B.7), Eq. (B.8) becomes

$$I_{AA', \text{mass}} = \rho t I_{AA', \text{area}} = \rho t(\tfrac{1}{4}\pi r^4)$$

In this case, the product $\rho \pi r^2 t$ is equal to the mass m of the plate, and $I_{AA'} = I_{BB'}$. Therefore, we can write the mass moments of inertia of a circular plate as

$$I_{AA'} = I_{BB'} = \tfrac{1}{4}mr^2 \tag{B.14}$$
$$I_{CC'} = I_{AA'} + I_{BB'} = \tfrac{1}{2}mr^2 \tag{B.15}$$

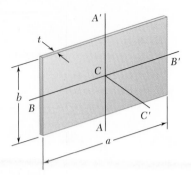

Fig. B.6 A thin rectangular plate of sides a and b.

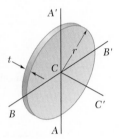

Fig. B.7 A thin circular plate of radius r.

B.4 Determining the Moment of Inertia of a Three-Dimensional Body by Integration

We obtain the moment of inertia of a three-dimensional body by evaluating the integral $I = \int r^2 \, dm$. If the body is made of a homogeneous material with a density ρ, the element of mass dm is equal to $\rho \, dV$, and we have $I = \rho \int r^2 \, dV$. This integral depends only upon the shape of the body. Thus, in order to compute the moment of inertia of a three-dimensional body, it is generally necessary to perform a triple, or at least a double, integration.

However, if the body possesses two planes of symmetry, it is usually possible to determine the body's moment of inertia with a single integration. We do this by choosing as the element of mass dm a thin slab that is perpendicular to the planes of symmetry. In the case of bodies of revolution, for example, the element of mass is a thin disk (Fig. B.8). Using formula (B.15), we can express the moment of inertia of the disk with respect to the axis of revolution as indicated in Fig. B.8. Its moment of inertia with respect to each of the other two coordinate axes is obtained by using formula (B.14) and the parallel-axis theorem. Integration of these expressions yields the desired moment of inertia of the body.

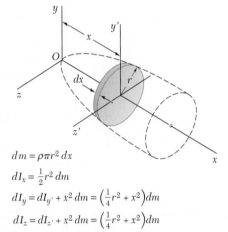

$$dm = \rho \pi r^2 \, dx$$
$$dI_x = \tfrac{1}{2}r^2 \, dm$$
$$dI_y = dI_{y'} + x^2 \, dm = \left(\tfrac{1}{4}r^2 + x^2\right)dm$$
$$dI_z = dI_{z'} + x^2 \, dm = \left(\tfrac{1}{4}r^2 + x^2\right)dm$$

Fig. B.8 Using a thin disk to determine the moment of inertia of a body of revolution.

B.5 Moments of Inertia of Composite Bodies

Figure B.9 lists the moments of inertia of a few common shapes. For a body consisting of several of these simple shapes in combination, you can obtain the moment of inertia of the body with respect to a given axis by first computing the moments of inertia of its component parts about the desired axis and then adding them together. As was the case for areas, the radius of gyration of a composite body *cannot* be obtained by adding the radii of gyration of its component parts.

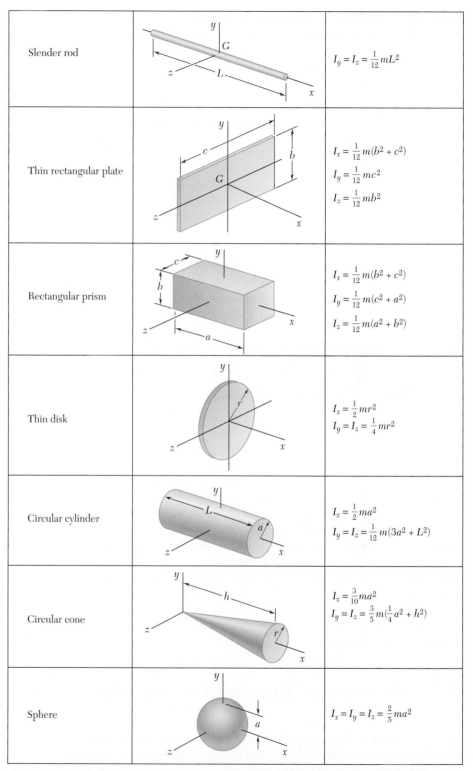

Slender rod		$I_y = I_z = \frac{1}{12}mL^2$
Thin rectangular plate		$I_x = \frac{1}{12}m(b^2 + c^2)$ $I_y = \frac{1}{12}mc^2$ $I_z = \frac{1}{12}mb^2$
Rectangular prism		$I_x = \frac{1}{12}m(b^2 + c^2)$ $I_y = \frac{1}{12}m(c^2 + a^2)$ $I_z = \frac{1}{12}m(a^2 + b^2)$
Thin disk		$I_x = \frac{1}{2}mr^2$ $I_y = I_z = \frac{1}{4}mr^2$
Circular cylinder		$I_x = \frac{1}{2}ma^2$ $I_y = I_z = \frac{1}{12}m(3a^2 + L^2)$
Circular cone		$I_x = \frac{3}{10}ma^2$ $I_y = I_z = \frac{3}{5}m(\frac{1}{4}a^2 + h^2)$
Sphere		$I_x = I_y = I_z = \frac{2}{5}ma^2$

Fig. B.9 Mass moments of inertia of common geometric shapes.

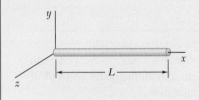

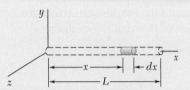

Fig. 1 Differential element of mass.

Sample Problem B.1

Determine the moment of inertia of a slender rod of length L and mass m with respect to an axis that is perpendicular to the rod and passes through one end.

STRATEGY: Approximating the rod as a one-dimensional body enables you to solve the problem by a single integration.

MODELING and ANALYSIS: Choose the differential element of mass shown in Fig. 1 and express it as a mass per unit length.

$$dm = \frac{m}{L}dx$$

$$I_y = \int x^2\, dm = \int_0^L x^2 \frac{m}{L}dx = \left[\frac{m}{L}\frac{x^3}{3}\right]_0^L \quad I_y = \tfrac{1}{3}mL^2 \quad \blacktriangleleft$$

REFLECT and THINK: This problem could also have been solved by starting with the moment of inertia for a slender rod with respect to its centroid, as given in Fig. B.9, and using the parallel-axis theorem to obtain the moment of inertia with respect to an end of the rod.

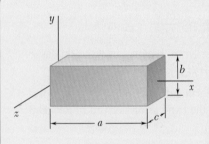

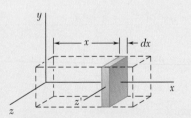

Fig. 1 Differential element of mass.

Sample Problem B.2

For the homogeneous rectangular prism shown, determine the moment of inertia with respect to the z axis.

STRATEGY: You can approach this problem by choosing a differential element of mass perpendicular to the long axis of the prism; find its moment of inertia with respect to a centroidal axis parallel to the z axis; and then apply the parallel-axis theorem.

MODELING and ANALYSIS: Choose as the differential element of mass the thin slab shown in Fig. 1. Then

$$dm = \rho bc\, dx$$

Referring to Sec. B.3, the moment of inertia of the element with respect to the z' axis is

$$dI_{z'} = \tfrac{1}{12}b^2\, dm$$

Applying the parallel-axis theorem, you can obtain the mass moment of inertia of the slab with respect to the z axis.

$$dI_z = dI_{z'} + x^2\, dm = \tfrac{1}{12}b^2\, dm + x^2\, dm = (\tfrac{1}{12}b^2 + x^2)\rho bc\, dx$$

Integrating from $x = 0$ to $x = a$ gives you

$$I_z = \int dI_z = \int_0^a (\tfrac{1}{12}b^2 + x^2)\rho bc\, dx = \rho abc(\tfrac{1}{12}b^2 + \tfrac{1}{3}a^2)$$

Since the total mass of the prism is $m = \rho abc$, you can write

$$I_z = m(\tfrac{1}{12}b^2 + \tfrac{1}{3}a^2) \qquad I_z = \tfrac{1}{12}m(4a^2 + b^2) \quad \blacktriangleleft$$

REFLECT and THINK: Note that if the prism is thin, b is small compared to a, and the expression for I_z reduces to $\tfrac{1}{3}ma^2$, which is the result obtained in Sample Prob. B.1 when $L = a$.

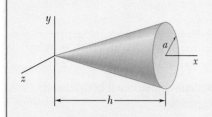

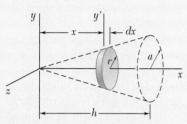

Fig. 1 Differential element of mass.

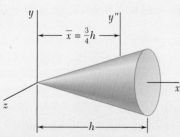

Fig. 2 Centroid of a right circular cone.

Sample Problem B.3

Determine the moment of inertia of a right circular cone with respect to (a) its longitudinal axis, (b) an axis through the apex of the cone and perpendicular to its longitudinal axis, (c) an axis through the centroid of the cone and perpendicular to its longitudinal axis.

STRATEGY: For parts (a) and (b), choose a differential element of mass in the form of a thin circular disk perpendicular to the longitudinal axis of the cone. You can solve part (c) by an application of the parallel-axis theorem.

MODELING and ANALYSIS: Choose the differential element of mass shown in Fig. 1. Express the radius and mass of this disk as

$$r = a\frac{x}{h} \qquad dm = \rho\pi r^2\,dx = \rho\pi\frac{a^2}{h^2}x^2\,dx$$

a. Moment of Inertia I_x. Using the expression derived in Sec. B.3 for a thin disk, compute the mass moment of inertia of the differential element with respect to the x axis.

$$dI_x = \tfrac{1}{2}r^2\,dm = \tfrac{1}{2}\Big(a\frac{x}{h}\Big)^2\Big(\rho\pi\frac{a^2}{h^2}x^2\,dx\Big) = \tfrac{1}{2}\rho\pi\frac{a^4}{h^4}x^4\,dx$$

Integrating from $x = 0$ to $x = h$ gives you

$$I_x = \int dI_x = \int_0^h \tfrac{1}{2}\rho\pi\frac{a^4}{h^4}x^4\,dx = \tfrac{1}{2}\rho\pi\frac{a^4}{h^4}\frac{h^5}{5} = \tfrac{1}{10}\rho\pi a^4 h$$

Since the total mass of the cone is $m = \tfrac{1}{3}\rho\pi a^2 h$, you can write this as

$$I_x = \tfrac{1}{10}\rho\pi a^4 h = \tfrac{3}{10}a^2(\tfrac{1}{3}\rho\pi a^2 h) = \tfrac{3}{10}ma^2 \qquad I_x = \tfrac{3}{10}ma^2 \quad \blacktriangleleft$$

b. Moment of Inertia I_y. Use the same differential element. Applying the parallel-axis theorem and using the expression derived in Sec. B.3 for a thin disk, you have

$$dI_y = dI_{y'} + x^2\,dm = \tfrac{1}{4}r^2\,dm + x^2\,dm = (\tfrac{1}{4}r^2 + x^2)\,dm$$

Substituting the expressions for r and dm into this equation yields

$$dI_y = \Big(\frac{1}{4}\frac{a^2}{h^2}x^2 + x^2\Big)\Big(\rho\pi\frac{a^2}{h^2}x^2\,dx\Big) = \rho\pi\frac{a^2}{h^2}\Big(\frac{a^2}{4h^2} + 1\Big)x^4\,dx$$

$$I_y = \int dI_y = \int_0^h \rho\pi\frac{a^2}{h^2}\Big(\frac{a^2}{4h^2} + 1\Big)x^4\,dx = \rho\pi\frac{a^2}{h^2}\Big(\frac{a^2}{4h^2} + 1\Big)\frac{h^5}{5}$$

Introducing the total mass of the cone m, you can rewrite I_y as

$$I_y = \tfrac{3}{5}(\tfrac{1}{4}a^2 + h^2)\tfrac{1}{3}\rho\pi a^2 h \qquad I_y = \tfrac{3}{5}m(\tfrac{1}{4}a^2 + h^2) \quad \blacktriangleleft$$

c. Moment of Inertia $\bar{I}_{y''}$. Apply the parallel-axis theorem to obtain

$$I_y = \bar{I}_{y''} + m\bar{x}^2$$

Solve for $\bar{I}_{y''}$ and recall from Fig. 5.21 that $\bar{x} = \tfrac{3}{4}h$ (Fig. 2). The result is

$$\bar{I}_{y''} = I_y - m\bar{x}^2 = \tfrac{3}{5}m(\tfrac{1}{4}a^2 + h^2) - m(\tfrac{3}{4}h)^2$$

$$\bar{I}_{y''} = \tfrac{3}{20}m(a^2 + \tfrac{1}{4}h^2) \quad \blacktriangleleft$$

REFLECT and THINK: The parallel-axis theorem for masses can be just as useful as the version for areas. Don't forget to use the reference figures for centroids of volumes when needed.

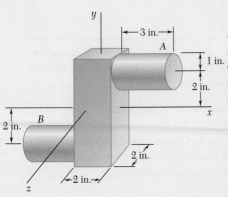

Fig. 1 Geometry of each component.

Sample Problem B.4

A steel forging consists of a $6 \times 2 \times 2$-in. rectangular prism and two cylinders with a diameter of 2 in. and length of 3 in. as shown. Determine the moments of inertia of the forging with respect to the coordinate axes. The specific weight of steel is 490 lb/ft^3.

STRATEGY: Compute the moments of inertia of each component from Fig. B.9 using the parallel-axis theorem when necessary. Note that all lengths should be expressed in feet to be consistent with the units for the given specific weight.

MODELING and ANALYSIS:

Computation of Masses.
Prism

$$V = (2 \text{ in.})(2 \text{ in.})(6 \text{ in.}) = 24 \text{ in}^3$$

$$W = \frac{(24 \text{ in}^3)(490 \text{ lb/ft}^3)}{1728 \text{ in}^3/\text{ft}^3} = 6.81 \text{ lb}$$

$$m = \frac{6.81 \text{ lb}}{32.2 \text{ ft/s}^2} = 0.211 \text{ lb·s}^2/\text{ft}$$

Each Cylinder

$$V = \pi(1 \text{ in.})^2(3 \text{ in.}) = 9.42 \text{ in}^3$$

$$W = \frac{(9.42 \text{ in}^3)(490 \text{ lb/ft}^3)}{1728 \text{ in}^3/\text{ft}^3} = 2.67 \text{ lb}$$

$$m = \frac{2.67 \text{ lb}}{32.2 \text{ ft/s}^2} = 0.0829 \text{ lb·s}^2/\text{ft}$$

Moments of Inertia (Fig. 1).
Prism

$$I_x = I_z = \tfrac{1}{12}(0.211 \text{ lb·s}^2/\text{ft})[(\tfrac{6}{12} \text{ ft})^2 + (\tfrac{2}{12} \text{ ft})^2] = 4.88 \times 10^{-3} \text{ lb·ft·s}^2$$

$$I_y = \tfrac{1}{12}(0.211 \text{ lb·s}^2/\text{ft})[(\tfrac{2}{12} \text{ ft})^2 + (\tfrac{2}{12} \text{ ft})^2] = 0.977 \times 10^{-3} \text{ lb·ft·s}^2$$

Each Cylinder

$$I_x = \tfrac{1}{2}ma^2 + m\bar{y}^2 = \tfrac{1}{2}(0.0829 \text{ lb·s}^2/\text{ft})(\tfrac{1}{12} \text{ ft})^2$$
$$+ (0.0829 \text{ lb·s}^2/\text{ft})(\tfrac{2}{12} \text{ ft})^2 = 2.59 \times 10^{-3} \text{ lb·ft·s}^2$$

$$I_y = \tfrac{1}{12}m(3a^2 + L^2) = m\bar{x}^2 = \tfrac{1}{12}(0.0829 \text{ lb·s}^2/\text{ft})[3(\tfrac{1}{12} \text{ ft})^2 + (\tfrac{3}{12} \text{ ft})^2]$$
$$+ (0.0829 \text{ lb·s}^2/\text{ft})(\tfrac{2.5}{12} \text{ ft})^2 = 4.17 \times 10^{-3} \text{ lb·ft·s}^2$$

$$I_z = \tfrac{1}{12}m(3a^2 + L^2) + m(\bar{x}^2 + \bar{y}^2) = \tfrac{1}{12}(0.0829 \text{ lb·s}^2/\text{ft})[3(\tfrac{1}{12} \text{ ft})^2 + (\tfrac{3}{12} \text{ ft})^2]$$
$$+ (0.0829 \text{ lb·s}^2/\text{ft})[(\tfrac{2.5}{12} \text{ ft})^2 + (\tfrac{2}{12} \text{ ft})^2] = 6.48 \times 10^{-3} \text{ lb·ft·s}^2$$

Entire Body. Adding the values obtained for the prism and two cylinders, you have

$$I_x = 4.88 \times 10^{-3} + 2(2.59 \times 10^{-3}) \qquad I_x = 10.06 \times 10^{-3} \text{ lb·ft·s}^2 \quad \blacktriangleleft$$

$$I_y = 0.977 \times 10^{-3} + 2(4.17 \times 10^{-3}) \qquad I_y = 9.32 \times 10^{-3} \text{ lb·ft·s}^2 \quad \blacktriangleleft$$

$$I_z = 4.88 \times 10^{-3} + 2(6.48 \times 10^{-3}) \qquad I_z = 17.84 \times 10^{-3} \text{ lb·ft·s}^2 \quad \blacktriangleleft$$

REFLECT and THINK: The results indicate this forging has more resistance to rotation about the z axis (largest moment of inertia) than about the x or y axes. This makes intuitive sense, because more of the mass is farther from the z axis than from the x or y axes.

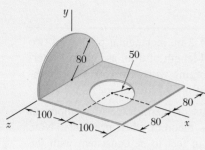

80   50

100

100   80   80

Dimensions in mm

Sample Problem B.5

A thin steel plate that is 4 mm thick is cut and bent to form the machine part shown. The density of the steel is 7850 kg/m^3. Determine the moments of inertia of the machine part with respect to the coordinate axes.

STRATEGY: The machine part consists of a semicircular plate and a rectangular plate from which a circular plate has been removed (Fig. 1). After calculating the moments of inertia for each part, add those of the semicircular plate and the rectangular plate, then subtract those of the circular plate to determine the moments of inertia for the entire machine part.

MODELING and ANALYSIS:

Computation of Masses. *Semicircular Plate*

$$V_1 = \tfrac{1}{2}\pi r^2 t = \tfrac{1}{2}\pi (0.08 \text{ m})^2 (0.004 \text{ m}) = 40.21 \times 10^{-6} \text{ m}^3$$
$$m_1 = \rho V_1 = (7.85 \times 10^3 \text{ kg/m}^3)(40.21 \times 10^{-6} \text{ m}^3) = 0.3156 \text{ kg}$$

Rectangular Plate

$$V_2 = (0.200 \text{ m})(0.160 \text{ m})(0.004 \text{ m}) = 128 \times 10^{-6} \text{ m}^3$$
$$m_2 = \rho V_2 = (7.85 \times 10^3 \text{ kg/m}^3)(128 \times 10^{-6} \text{ m}^3) = 1.005 \text{ kg}$$

Circular Plate

$$V_3 = \pi a^2 t = \pi (0.050 \text{ m})^2 (0.004 \text{ m}) = 31.42 \times 10^{-6} \text{ m}^3$$
$$m_3 = \rho V_3 = (7.85 \times 10^3 \text{ kg/m}^3)(31.42 \times 10^{-6} \text{ m}^3) = 0.2466 \text{ kg}$$

Moments of Inertia. Compute the moments of inertia of each component, using the method presented in Sec. B.3.

Semicircular Plate. Observe from Fig. B.9 that, for a circular plate of mass m and radius r,

$$I_x = \tfrac{1}{2}mr^2 \qquad I_y = I_z = \tfrac{1}{4}mr^2$$

Because of symmetry, halve these values for a semicircular plate. Thus,

$$I_x = \tfrac{1}{2}(\tfrac{1}{2}mr^2) \qquad I_y = I_z = \tfrac{1}{2}(\tfrac{1}{4}mr^2)$$

Since the mass of the semicircular plate is $m_1 = \tfrac{1}{2}m$, you have

$$I_x = \tfrac{1}{2}m_1 r^2 = \tfrac{1}{2}(0.3156 \text{ kg})(0.08 \text{ m})^2 = 1.010 \times 10^{-3} \text{ kg·m}^2$$
$$I_y = I_z = \tfrac{1}{4}(\tfrac{1}{2}mr^2) = \tfrac{1}{4}m_1 r^2 = \tfrac{1}{4}(0.3156 \text{ kg})(0.08 \text{ m})^2 = 0.505 \times 10^{-3} \text{ kg·m}^2$$

Rectangular Plate

$$I_x = \tfrac{1}{12}m_2 c^2 = \tfrac{1}{12}(1.005 \text{ kg})(0.16 \text{ m})^2 = 2.144 \times 10^{-3} \text{ kg·m}^2$$
$$I_z = \tfrac{1}{3}m_2 b^2 = \tfrac{1}{3}(1.005 \text{ kg})(0.2 \text{ m})^2 = 13.400 \times 10^{-3} \text{ kg·m}^2$$
$$I_y = I_x + I_z = (2.144 + 13.400)(10^{-3}) = 15.544 \times 10^{-3} \text{ kg·m}^2$$

Circular Plate

$$I_x = \tfrac{1}{4}m_3 a^2 = \tfrac{1}{4}(0.2466 \text{ kg})(0.05 \text{ m})^2 = 0.154 \times 10^{-3} \text{ kg·m}^2$$
$$I_y = \tfrac{1}{2}m_3 a^2 + m_3 d^2$$
$$= \tfrac{1}{2}(0.2466 \text{ kg})(0.05 \text{ m})^2 + (0.2466 \text{ kg})(0.1 \text{ m})^2 = 2.774 \times 10^{-3} \text{ kg·m}^2$$
$$I_z = \tfrac{1}{4}m_3 a^2 + m_3 d^2 = \tfrac{1}{4}(0.2466 \text{ kg})(0.05 \text{ m})^2 + (0.2466 \text{ kg})(0.1 \text{ m})^2$$
$$= 2.620 \times 10^{-3} \text{ kg·m}^2$$

Entire Machine Part

$$I_x = (1.010 + 2.144 - 0.154)(10^{-3}) \text{ kg·m}^2 \qquad I_x = 3.00 \times 10^{-3} \text{ kg·m}^2 \blacktriangleleft$$
$$I_y = (0.505 + 15.544 - 2.774)(10^{-3}) \text{ kg·m}^2 \quad I_y = 13.28 \times 10^{-3} \text{ kg·m}^2 \blacktriangleleft$$
$$I_z = (0.505 + 13.400 - 2.620)(10^{-3}) \text{ kg·m}^2 \quad I_z = 11.29 \times 10^{-3} \text{ kg·m}^2 \blacktriangleleft$$

$r = 0.08$ m

+

$b = 0.2$ m $c = 0.16$ m

−

$a = 0.05$ m

$d = 0.1$ m

Fig. 1 Modeling the machine part as a combination of simple geometric shapes.

SOLVING PROBLEMS
ON YOUR OWN

In this section, we introduced the **mass moment of inertia** and the **radius of gyration** of a three-dimensional body with respect to a given axis [Eqs. (B.1) and (B.2)]. We also derived a **parallel-axis theorem** for use with mass moments of inertia and discussed the computation of the mass moments of inertia of thin plates and three-dimensional bodies.

1. **Computing mass moments of inertia.** You can calculate the mass moment of inertia I of a body with respect to a given axis directly from the definition given in Eq. (B.1) for simple shapes [Sample Prob. B.1]. In most cases, however, it is necessary to divide the body into thin slabs, compute the moment of inertia of a typical slab with respect to the given axis—using the parallel-axis theorem if necessary—and integrate the resulting expression.

2. **Applying the parallel-axis theorem.** In Sec. B.2, we derived the parallel-axis theorem for mass moments of inertia as

$$I = \bar{I} + md^2 \tag{B.6}$$

This theorem states that the moment of inertia I of a body of mass m with respect to a given axis is equal to the sum of the moment of inertia $\bar{I}$ of that body with respect to a parallel centroidal axis and the product md^2, where d is the distance between the two axes. When you calculate the moment of inertia of a three-dimensional body with respect to one of the coordinate axes, you can replace d^2 by the sum of the squares of distances measured along the other two coordinate axes [Eqs. (B.5) and (B.5′)].

3. **Avoiding unit-related errors.** To avoid errors, you must be consistent in your use of units. Thus, all lengths should be expressed in meters or feet, as appropriate, and for problems using U.S. customary units, masses should be given in lb·s^2/ft. In addition, we strongly recommend that you include units as you perform your calculations [Sample Probs. B.4 and B.5].

4. **Calculating the mass moment of inertia of thin plates.** We showed in Sec. B.3 that you can obtain the mass moment of inertia of a thin plate with respect to a given axis by multiplying the corresponding moment of inertia of the area of the plate by the density ρ and the thickness t of the plate [Eqs. (B.8) through (B.10)]. Note that, since the axis CC' in Fig. B.5c is perpendicular to the plate, $I_{CC',\text{mass}}$ is associated with the *polar* moment of inertia $J_{C,\text{area}}$.

Instead of calculating the moment of inertia of a thin plate with respect to a specified axis directly, you may sometimes find it convenient to first compute its moment of inertia with respect to an axis parallel to the specified axis and to then apply the parallel-axis theorem. Furthermore, to determine the moment of inertia of a thin plate with respect to an axis perpendicular to the plate, you may wish to first determine its moments of inertia with respect to two perpendicular in-plane axes and to then use Eq. (B.11). Finally, remember that the mass of a plate consists of area A, thickness t, and density ρ, as $m = \rho t A$.

(continued)

5. Determining the moment of inertia of a body by direct single integration. We discussed in Sec. B.4 and illustrated in Sample Probs. B.2 and B.3 how you can use single integration to compute the moment of inertia of a body that can be divided into a series of thin, parallel slabs. For such cases, you will often need to express the mass of the body in terms of the body's density and dimensions. Assuming that the body has been divided, as in the sample problems, into thin slabs perpendicular to the x axis, you will need to express the dimensions of each slab as functions of the variable x.

   a. In the special case of a body of revolution, the elemental slab is a thin disk, and you can use the equations given in Fig. B.8 to determine the moments of inertia of the body [Sample Prob. B.3].

   b. In the general case, when the body is not a solid of revolution, the differential element is not a disk but a thin slab of a different shape. You cannot use the equations of Fig. B.8 in this case. See, for example, Sample Prob. B.2, where the element was a thin, rectangular slab. For more complex configurations, you may want to use one or more of the following equations, which are based on Eqs. (B.5) and (B.5′) of Sec. B.2.

$$dI_x = dI_{x'} + (\bar{y}_{el}^2 + \bar{z}_{el}^2)\, dm$$
$$dI_y = dI_{y'} + (\bar{z}_{el}^2 + \bar{x}_{el}^2)\, dm$$
$$dI_z = dI_{z'} + (\bar{x}_{el}^2 + \bar{y}_{el}^2)\, dm$$

Here, the primes denote the centroidal axes of each elemental slab and $\bar{x}_{el}$, $\bar{y}_{el}$, and $\bar{z}_{el}$ represent the coordinates of its centroid. Determine the centroidal moments of inertia of the slab in the manner described earlier for a thin plate: calculate the corresponding moments of inertia of the area of the slab, and multiply the result by the density ρ and the thickness t of the slab. Also, assuming that the body has been divided into thin slabs perpendicular to the x axis, remember that you can obtain $dI_{x'}$ by adding $dI_{y'}$ and $dI_{z'}$ instead of computing it directly. Finally, using the geometry of the body, express the result obtained in terms of the single variable x, and integrate in x.

6. Computing the moment of inertia of a composite body. As stated in Sec. B.5, the moment of inertia of a composite body with respect to a specified axis is equal to the sum of the moments of its components with respect to that axis. Sample Probs. B.4 and B.5 illustrate the appropriate method of solution. Also remember that the moment of inertia of a component is negative only if the component is *removed* (as in the case of a hole).

Although the composite-body problems in this section are relatively straightforward, you will have to work carefully to avoid computational errors. In addition, if some of the moments of inertia that you need are not given in Fig. B.9, you will have to derive your own formulas, using the techniques described in this section.

Problems

B.1 A thin plate with a mass m is cut in the shape of an equilateral triangle of side a. Determine the mass moment of inertia of the plate with respect to (*a*) the centroidal axes AA' and BB', (*b*) the centroidal axis CC' that is perpendicular to the plate.

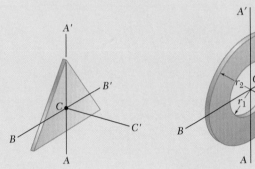

Fig. B.1   **Fig. B.2**

B.2 A ring with a mass m is cut from a thin uniform plate. Determine the mass moment of inertia of the ring with respect to (*a*) the axis AA', (*b*) the centroidal axis CC' that is perpendicular to the plane of the ring.

B.3 A thin, semielliptical plate has a mass m. Determine the mass moment of inertia of the plate with respect to (*a*) the centroidal axis BB', (*b*) the centroidal axis CC' that is perpendicular to the plate.

B.4 The parabolic spandrel shown was cut from a thin, uniform plate. Denoting the mass of the spandrel by m, determine its mass moment of inertia with respect to (*a*) the axis BB', (*b*) the axis DD' that is perpendicular to the spandrel. (*Hint:* See Sample Prob. 9.3.)

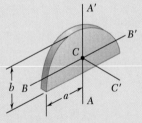

Fig. B.3

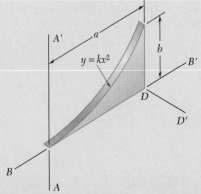

Fig. B.4

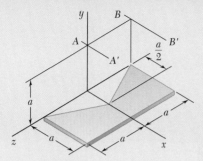

Fig. B.5 and B.6

B.5 A piece of thin, uniform sheet metal is cut to form the machine component shown. Denoting the mass of the component by m, determine its mass moment of inertia with respect to (*a*) the x axis, (*b*) the y axis.

B.6 A piece of thin, uniform sheet metal is cut to form the machine component shown. Denoting the mass of the component by m, determine its mass moment of inertia with respect to (*a*) the axis AA', (*b*) the axis BB', where the AA' and BB' axes are parallel to the x axis and lie in a plane parallel to and at a distance a above the xz plane.

B.7 A thin plate with a mass m has the trapezoidal shape shown. Determine the mass moment of inertia of the plate with respect to (*a*) the x axis, (*b*) the y axis.

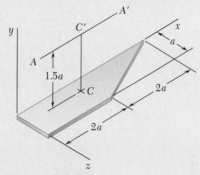

Fig. B.7 and B.8

B.8 A thin plate with a mass m has the trapezoidal shape shown. Determine the mass moment of inertia of the plate with respect to (*a*) the centroidal axis CC' that is perpendicular to the plate, (*b*) the axis AA' that is parallel to the x axis and is located at a distance $1.5a$ from the plate.

B.9 Determine by direct integration the mass moment of inertia with respect to the z axis of the right circular cylinder shown, assuming that it has a uniform density and a mass m.

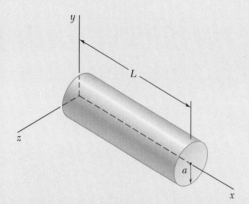

Fig. B.9

B.10 The area shown is revolved about the x axis to form a homogeneous solid of revolution of mass m. Using direct integration, express the mass moment of inertia of the solid with respect to the x axis in terms of m and h.

B.11 The area shown is revolved about the x axis to form a homogeneous solid of revolution of mass m. Determine by direct integration the mass moment of inertia of the solid with respect to (*a*) the x axis, (*b*) the y axis. Express your answers in terms of m and the dimensions of the solid.

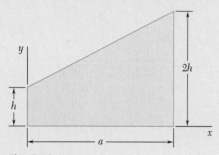

Fig. B.10

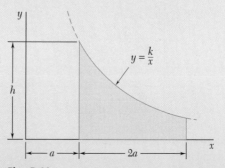

Fig. B.11

B.12 Determine by direct integration the mass moment of inertia with respect to the x axis of the tetrahedron shown, assuming that it has a uniform density and a mass m.

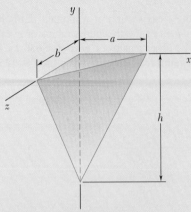

Fig. B.12 and *B.13*

B.13 Determine by direct integration the mass moment of inertia with respect to the y axis of the tetrahedron shown, assuming that it has a uniform density and a mass m.

B.14 Determine by direct integration the mass moment of inertia and the radius of gyration with respect to the x axis of the paraboloid shown, assuming that it has a uniform density and a mass m.

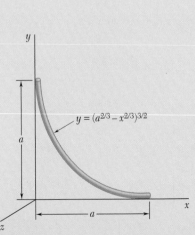

Fig. B.14

B.15 A thin, rectangular plate with a mass m is welded to a vertical shaft AB as shown. Knowing that the plate forms an angle θ with the y axis, determine by direct integration the mass moment of inertia of the plate with respect to (a) the y axis, (b) the z axis.

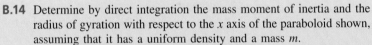

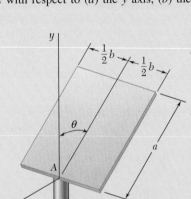

Fig. B.15

***B.16** A thin steel wire is bent into the shape shown. Denoting the mass per unit length of the wire by m', determine by direct integration the mass moment of inertia of the wire with respect to each of the coordinate axes.

Fig. B.16

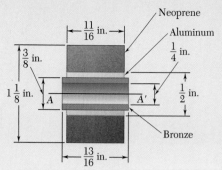

Fig. B.17

B.17 Shown is the cross section of an idler roller. Determine its mass moment of inertia and its radius of gyration with respect to the axis AA'. (The specific weight of bronze is 0.310 lb/in^3; of aluminum, 0.100 lb/in^3; and of neoprene, 0.0452 lb/in^3.)

B.18 Shown is the cross section of a molded flat-belt pulley. Determine its mass moment of inertia and its radius of gyration with respect to the axis AA'. (The density of brass is 8650 kg/m^3, and the density of the fiber-reinforced polycarbonate used is 1250 kg/m^3.)

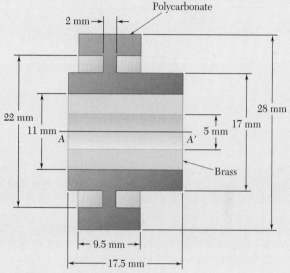

Fig. B.18

B.19 The machine part shown is formed by machining a conical surface into a circular cylinder. For $b = \frac{1}{2}h$, determine the mass moment of inertia and the radius of gyration of the machine part with respect to the y axis.

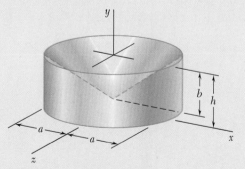

Fig. B.19

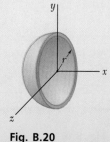

Fig. B.20

B.20 Knowing that the thin hemispherical shell shown has a mass m and thickness t, determine the mass moment of inertia and the radius of gyration of the shell with respect to the x axis. (*Hint:* Consider the shell as formed by removing a hemisphere of radius r from a hemisphere of radius $r + t$; then neglect the terms containing t^2 and t^3 and keep those terms containing t.)

B.21 A square hole is centered in and extends through the aluminum machine component shown. Determine (a) the value of a for which the mass moment of inertia of the component is maximum with respect to the axis AA' that bisects the top surface of the hole, (b) the corresponding values of the mass moment of inertia and the radius of gyration with respect to the axis AA'. (The specific weight of aluminum is 0.100 lb/in^3.)

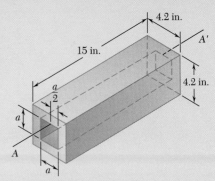

Fig. B.21

B.22 The cups and the arms of an anemometer are fabricated from a material of density ρ. Knowing that the mass moment of inertia of a thin, hemispherical shell with a mass m and thickness t with respect to its centroidal axis GG' is $5ma^2/12$, determine (a) the mass moment of inertia of the anemometer with respect to the axis AA', (b) the ratio of a to l for which the centroidal moment of inertia of the cups is equal to 1 percent of the moment of inertia of the cups with respect to the axis AA'.

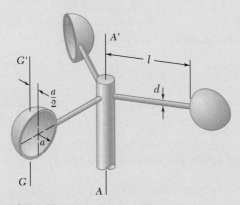

Fig. B.22

B.23 After a period of use, one of the blades of a shredder has been worn to the shape shown and is of mass 0.18 kg. Knowing that the mass moments of inertia of the blade with respect to the AA' and BB' axes are 0.320 g·m^2 and 0.680 g·m^2, respectively, determine (a) the location of the centroidal axis GG', (b) the radius of gyration with respect to axis GG'.

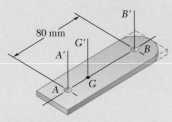

Fig. B.23

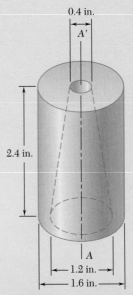

Fig. B.24

B.24 Determine the mass moment of inertia of the 0.9-lb machine component shown with respect to the axis AA'.

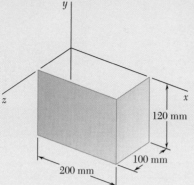

Fig. B.25

B.25 and B.26 A 2-mm-thick piece of sheet steel is cut and bent into the machine component shown. Knowing that the density of steel is 7850 kg/m^3, determine the mass moment of inertia of the component with respect to each of the coordinate axes.

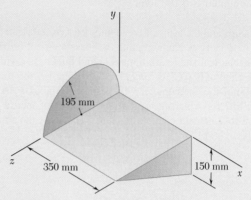

Fig. B.26

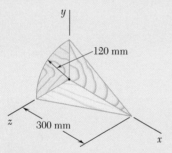

Fig. *B.27*

B.27 A subassembly for a model airplane is fabricated from three pieces of 1.5-mm plywood. Neglecting the mass of the adhesive used to assemble the three pieces, determine the mass moment of inertia of the subassembly with respect to each of the coordinate axes. (The density of the plywood is 780 kg/m^3.)

B.28 A section of sheet steel 0.03 in. thick is cut and bent into the sheet metal machine component shown. Determine the mass moment of inertia of the component with respect to each of the coordinate axes. (The specific weight of steel is 490 lb/ft^3.)

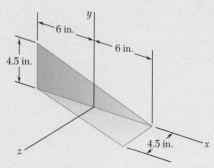

Fig. B.28

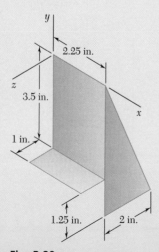

Fig. B.29

B.29 A framing anchor is formed of 0.05-in.-thick galvanized steel. Determine the mass moment of inertia of the anchor with respect to each of the coordinate axes. (The specific weight of galvanized steel is 470 lb/ft^3.)

*B.30 A farmer constructs a trough by welding a rectangular piece of 2-mm-thick sheet steel to half of a steel drum. Knowing that the density of steel is 7850 kg/m^3 and that the thickness of the walls of the drum is 1.8 mm, determine the mass moment of inertia of the trough with respect to each of the coordinate axes. Neglect the mass of the welds.

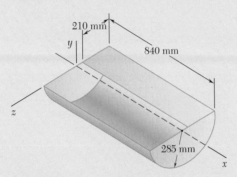

Fig. B.30

B.31 The machine element shown is fabricated from steel. Determine the mass moment of inertia of the assembly with respect to (a) the x axis, (b) the y axis, (c) the z axis. (The density of steel is 7850 kg/m^3.)

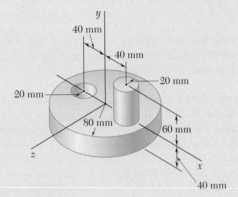

Fig. B.31

B.32 Determine the mass moments of inertia and the radii of gyration of the steel machine element shown with respect to the x and y axes. (The density of steel is 7850 kg/m^3.)

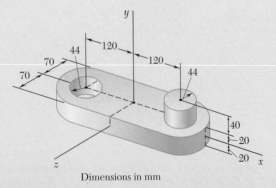

Dimensions in mm

Fig. B.32

B.33 Determine the mass moment of inertia of the steel machine element shown with respect to the *x* axis. (The specific weight of steel is 490 lb/ft³.)

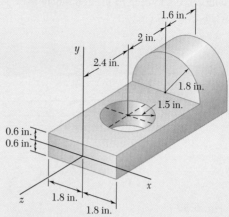

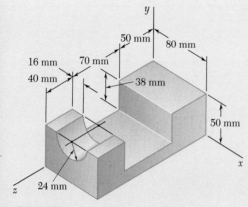

Fig. B.35

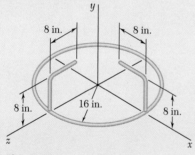

Fig. B.36

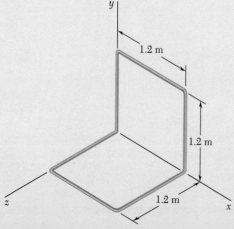

Fig. B.38

Fig. **B.33** and *B.34*

B.34 Determine the mass moment of inertia of the steel machine element shown with respect to the *y* axis. (The specific weight of steel is 490 lb/ft³.)

B.35 Determine the mass moment of inertia of the steel fixture shown with respect to (*a*) the *x* axis, (*b*) the *y* axis, (*c*) the *z* axis. (The density of steel is 7850 kg/m³.)

B.36 Aluminum wire with a weight per unit length of 0.033 lb/ft is used to form the circle and the straight members of the figure shown. Determine the mass moment of inertia of the assembly with respect to each of the coordinate axes.

B.37 The figure shown is formed of ⅛-in.-diameter steel wire. Knowing that the specific weight of the steel is 490 lb/ft³, determine the mass moment of inertia of the wire with respect to each of the coordinate axes.

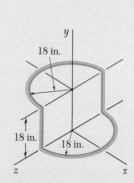

Fig. **B.37**

B.38 A homogeneous wire with a mass per unit length of 0.056 kg/m is used to form the figure shown. Determine the mass moment of inertia of the wire with respect to each of the coordinate axes.

*B.6 Mass Products of Inertia

In this section, you will see how to determine the moment of inertia of a body with respect to an arbitrary axis OL through the origin (Fig. B.10) if its moments of inertia with respect to the three coordinate axes, as well as certain other quantities defined here, have already been determined.

The moment of inertia I_{OL} of the body with respect to OL is equal to $\int p^2\,dm$, where p denotes the perpendicular distance from the element of mass dm to the axis OL. If we denote the unit vector along OL by $\boldsymbol{\lambda}$ and the position vector of the element dm by $\mathbf{r}$, the perpendicular distance p is equal to $r\sin\theta$, which is the magnitude of the vector product $\boldsymbol{\lambda}\times\mathbf{r}$. We therefore have

$$I_{OL} = \int p^2\,dm = \int |\boldsymbol{\lambda}\times\mathbf{r}|^2\,dm \qquad (B.16)$$

Expressing $|\boldsymbol{\lambda}\times\mathbf{r}|^2$ in terms of the rectangular components of the vector product, we have

$$I_{OL} = \int [(\lambda_x y - \lambda_y x)^2 + (\lambda_y z - \lambda_z y)^2 + (\lambda_z x - \lambda_x z)^2]\,dm$$

Here, the components λ_x, λ_y, λ_z of the unit vector $\boldsymbol{\lambda}$ represent the direction cosines of the axis OL, and the components x, y, z of $\mathbf{r}$ represent the coordinates of the element of mass dm. Expanding the squares and rearranging the terms, we obtain

$$I_{OL} = \lambda_x^2 \int (y^2 + z^2)\,dm + \lambda_y^2 \int (z^2 + x^2)\,dm + \lambda_z^2 \int (x^2 + y^2)\,dm$$

$$- 2\lambda_x\lambda_y \int xy\,dm - 2\lambda_y\lambda_z \int yz\,dm - 2\lambda_z\lambda_x \int zx\,dm \qquad (B.17)$$

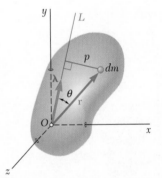

Fig. B.10 An element of mass dm of a body and its perpendicular distance to an arbitrary axis OL through the origin.

Referring to Eqs. (B.3), note that the first three integrals in Eq. (B.17) represent, respectively, the moments of inertia I_x, I_y, and I_z of the body with respect to the coordinate axes. The last three integrals in Eq. (B.17), which involve products of coordinates, are called the **products of inertia** of the body with respect to the x and y axes, the y and z axes, and the z and x axes, respectively.

Mass products of inertia

$$I_{xy} = \int xy\,dm \qquad I_{yz} = \int yz\,dm \qquad I_{zx} = \int zx\,dm \qquad (B.18)$$

Rewriting Eq. (B.17) in terms of the integrals defined in Eqs. (B.3) and (B.18), we have

$$I_{OL} = I_x\lambda_x^2 + I_y\lambda_y^2 + I_z\lambda_z^2 - 2I_{xy}\lambda_x\lambda_y - 2I_{yz}\lambda_y\lambda_z - 2I_{zx}\lambda_z\lambda_x \qquad (B.19)$$

The definition of the products of inertia of a mass given in Eqs. (B.18) is an extension of the definition of the product of inertia of an area. Mass products of inertia reduce to zero under the same conditions of symmetry as do products of inertia of areas, and the parallel-axis theorem for mass products of inertia is expressed by relations similar to the formula derived

for the product of inertia of an area. Substituting the expressions for x, y, and z given in Eqs. (B.4) into Eqs. (B.18), we find that

**Parallel-axis theorem for
mass products of inertia**

$$
\begin{aligned}
I_{xy} &= \bar{I}_{x'y'} + m\bar{x}\bar{y} \\
I_{yz} &= \bar{I}_{y'z'} + m\bar{y}\bar{z} \\
I_{zx} &= \bar{I}_{z'x'} + m\bar{z}\bar{x}
\end{aligned}
\tag{B.20}
$$

Here $\bar{x}, \bar{y}, \bar{z}$ are the coordinates of the center of gravity G of the body and $\bar{I}_{x'y'}, \bar{I}_{y'z'}, \bar{I}_{z'x'}$ denote the products of inertia of the body with respect to the centroidal axes x', y', and z' (see Fig. B.3).

*B.7   Principal Axes and Principal Moments of Inertia

Let us assume that we have determined the moment of inertia of the body considered in the preceding section with respect to a large number of axes OL through the fixed point O. Suppose that we plot a point Q on each axis OL at a distance $OQ = 1/\sqrt{I_{OL}}$ from O. The locus of the points Q forms a surface (Fig. B.11). We can obtain the equation of that surface by substituting $1/(OQ)^2$ for I_{OL} in Eq. (B.19) and then multiplying both sides of the equation by $(OQ)^2$. Observing that

$$
(OQ)\lambda_x = x \qquad (OQ)\,\lambda_y = y \qquad (OQ)\,\lambda_z = z
$$

where x, y, z denote the rectangular coordinates of Q, we have

$$
I_x x^2 + I_y y^2 + I_z z^2 - 2I_{xy}xy - 2I_{yz}yz - 2I_{zx}zx = 1
\tag{B.21}
$$

This is the equation of a *quadric surface*. Since the moment of inertia I_{OL} is different from zero for every axis OL, no point Q can be at an infinite distance from O. Thus, the quadric surface obtained is an *ellipsoid*. This ellipsoid, which defines the moment of inertia of the body with respect to any axis through O, is known as the **ellipsoid of inertia** of the body at O.

Observe that, if we rotate the axes in Fig. B.11, the coefficients of the equation defining the ellipsoid change, since they are equal to the moments and products of inertia of the body with respect to the rotated coordinate axes. However, the *ellipsoid itself remains unaffected*, since its shape depends only upon the distribution of mass in the given body. Suppose that we choose as coordinate axes the principal axes x', y', and z' of the ellipsoid of inertia (Fig. B.12). The equation of the ellipsoid with respect to these coordinate axes is known to be of the form

$$
I_{x'}x'^2 + I_{y'}y'^2 + I_{z'}z'^2 = 1
\tag{B.22}
$$

which does not contain any products of the coordinates. Comparing Eqs. (B.21) and (B.22), we observe that the products of inertia of the body with respect to the x', y', and z' axes must be zero. The x', y', and z' axes

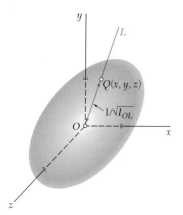

Fig. B.11 The elllipsoid of inertia defines the moment of inertia of a body with respect to any axis through O.

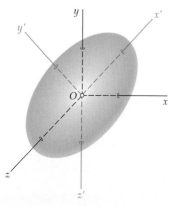

Fig. B.12 Principal axes of inertia x', y', z' of the body at O.

are known as the **principal axes of inertia** of the body at O, and the coefficients $I_{x'}$, $I_{y'}$, and $I_{z'}$ are referred to as the **principal moments of inertia** of the body at O. Note that, given a body of arbitrary shape and a point O, it is always possible to find principal axes of inertia of the body at O; that is, axes with respect to which the products of inertia of the body are zero. Indeed, whatever the shape of the body, the moments and products of inertia of the body with respect to the x, y, and z axes through O define an ellipsoid, and this ellipsoid has principal axes that, by definition, are the principal axes of inertia of the body at O.

If the principal axes of inertia x', y', and z' are used as coordinate axes, the expression in Eq. (B.19) for the moment of inertia of a body with respect to an arbitrary axis through O reduces to

$$I_{OL} = I_{x'}\lambda_{x'}^2 + I_{y'}\lambda_{y'}^2 + I_{z'}\lambda_{z'}^2 \qquad \textbf{(B.23)}$$

The determination of the principal axes of inertia of a body of arbitrary shape is somewhat involved and is discussed in the next section. In many cases, however, these axes can be spotted immediately. Consider, for instance, the homogeneous cone of elliptical base shown in Fig. B.13; this cone possesses two mutually perpendicular planes of symmetry OAA' and OBB'. From the definition of Eq. (B.18), we observe that if we choose the $x'y'$ and $y'z'$ planes to coincide with the two planes of symmetry, all of the products of inertia are zero. The x', y', and z' axes selected in this way are therefore the principal axes of inertia of the cone at O. In the case of the homogeneous regular tetrahedron $OABC$ shown in Fig. B.14, the line joining the corner O to the center D of the opposite face is a principal axis of inertia at O, and any line through O perpendicular to OD is also a principal axis of inertia at O. This property is apparent if we observe that rotating the tetrahedron through 120° about OD leaves its shape and mass distribution unchanged. It follows that the ellipsoid of inertia at O also remains unchanged under this rotation. The ellipsoid, therefore, is a body of revolution whose axis of revolution is OD, and the line OD, as well as any perpendicular line through O, must be a principal axis of the ellipsoid.

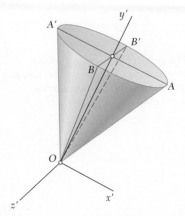

Fig. B.13 A homogeneous cone with elliptical base has two mutually perpendicular planes of symmetry.

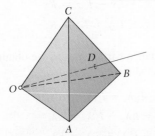

Fig. B.14 A line drawn from a corner to the center of the opposite face of a homogeneous regular tetrahedron is a principal axis, since each 120° rotation of the body about this axis leaves its shape and mass distribution unchanged.

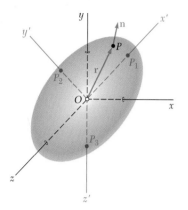

Fig. B.15 The principal axes intersect an ellipsoid of inertia at points where the radius vectors are collinear with the unit normal vectors to the surface.

*B.8 Principal Axes and Moments of Inertia for a Body of Arbitrary Shape

The method of analysis described in this section extends the analysis in the preceding section. However, generally speaking, you should use it only when the body under consideration has no obvious property of symmetry.

Consider the ellipsoid of inertia of a body at a given point O (Fig. B.15). Let $\mathbf{r}$ be the radius vector of a point P on the surface of the ellipsoid, and let $\mathbf{n}$ be the unit vector along the normal to that surface at P. We observe that the only points where $\mathbf{r}$ and $\mathbf{n}$ are collinear are points P_1, P_2, and P_3, where the principal axes intersect the visible portion of the surface of the ellipsoid (along with the corresponding points on the other side of the ellipsoid).

Recall from calculus that the direction of the normal to a surface of equation $f(x, y, z) = 0$ at a point $P(x, y, z)$ is defined by the gradient ∇f of the function f at that point. To obtain the points where the principal axes intersect the surface of the ellipsoid of inertia, we must therefore express that $\mathbf{r}$ and ∇f are collinear,

$$\nabla f = (2K)\mathbf{r} \tag{B.24}$$

where K is a constant, $\mathbf{r} = x\mathbf{i} + y\mathbf{j} + z\mathbf{k}$, and

$$\nabla f = \frac{\partial f}{\partial x}\mathbf{i} + \frac{\partial f}{\partial y}\mathbf{j} + \frac{\partial f}{\partial z}\mathbf{k}$$

Recalling Eq. (B.21), we note that the function $f(x, y, z)$ corresponding to the ellipsoid of inertia is

$$f(x, y, z) = I_x x^2 + I_y y^2 + I_z z^2 - 2I_{xy}xy - 2I_{yz}yz - 2I_{zx}zx - 1$$

Substituting for $\mathbf{r}$ and ∇f into Eq. (B.24) and equating the coefficients of the unit vectors, we obtain

$$\begin{aligned} I_x x \quad - I_{xy}y - I_{zx}z &= Kx \\ -I_{xy}x + I_y y - I_{yz}z &= Ky \\ -I_{zx}x - I_{yz}y + I_z z &= Kz \end{aligned} \tag{B.25}$$

Dividing each term by the distance r from O to P, we obtain similar equations involving the direction cosines λ_x, λ_y, and λ_z:

$$\begin{aligned} I_x\lambda_x - I_{xy}\lambda_y - I_{zx}\lambda_z &= K\lambda_x \\ -I_{xy}\lambda_x + I_y\lambda_y - I_{yz}\lambda_z &= K\lambda_y \\ -I_{zx}\lambda_x - I_{yz}\lambda_y + I_z\lambda_z &= K\lambda_z \end{aligned} \tag{B.26}$$

Transposing the right-hand members leads to the homogeneous linear equations, as

$$(I_x - K)\lambda_x - I_{xy}\lambda_y - I_{zx}\lambda_z = 0$$
$$-I_{xy}\lambda_x + (I_y - K)\lambda_y - I_{yz}\lambda_z = 0 \qquad \textbf{(B.27)}$$
$$-I_{zx}\lambda_x - I_{yz}\lambda_y + (I_z - K)\lambda_z = 0$$

For this system of equations to have a solution different from $\lambda_x = \lambda_y = \lambda_z = 0$, its discriminant must be zero. Thus,

$$\begin{vmatrix} I_x - K & -I_{xy} & -I_{zx} \\ -I_{xy} & I_y - K & -I_{yz} \\ -I_{zx} & -I_{yz} & I_z - K \end{vmatrix} = 0 \qquad \textbf{(B.28)}$$

Expanding this determinant and changing signs, we have

$$K^3 - (I_x + I_y + I_z)K^2 + (I_x I_y + I_y I_z + I_z I_x - I_{xy}^2 - I_{yz}^2 - I_{zx}^2)K$$
$$- (I_x I_y I_z - I_x I_{yz}^2 - I_y I_{zx}^2 - I_z I_{xy}^2 - 2I_{xy} I_{yz} I_{zx}) = 0 \qquad \textbf{(B.29)}$$

This is a cubic equation in K, which yields three real, positive roots: K_1, K_2, and K_3.

To obtain the direction cosines of the principal axis corresponding to the root K_1, we substitute K_1 for K in Eqs. (B.27). Since these equations are now linearly dependent, only two of them may be used to determine λ_x, λ_y, and λ_z. We can obtain an additional equation, however, by recalling from Statics that the direction cosines must satisfy the relation

$$\lambda_x^2 + \lambda_y^2 + \lambda_z^2 = 1 \qquad \textbf{(B.30)}$$

Repeating this procedure with K_2 and K_3, we obtain the direction cosines of the other two principal axes.

We now show that *the roots K_1, K_2, and K_3 of Eq. (B.29) are the principal moments of inertia of the given body.* Let us substitute for K in Eqs. (B.26) the root K_1, and for λ_x, λ_y, and λ_z the corresponding values $(\lambda_x)_1$, $(\lambda_y)_1$, and $(\lambda_z)_1$ of the direction cosines; the three equations are satisfied. We now multiply by $(\lambda_x)_1$, $(\lambda_y)_1$, and $(\lambda_z)_1$, respectively, each term in the first, second, and third equation and add the equations obtained in this way. The result is

$$I_x^2(\lambda_x)_1^2 + I_y^2(\lambda_y)_1^2 + I_z^2(\lambda_z)_1^2 - 2I_{xy}(\lambda_x)_1(\lambda_y)_1$$
$$- 2I_{yz}(\lambda_y)_1(\lambda_z)_1 - 2I_{zx}(\lambda_z)_1(\lambda_x)_1 = K_1[(\lambda_x)_1^2 + (\lambda_y)_1^2 + (\lambda_z)_1^2]$$

Recalling Eq. (B.19), we observe that the left-hand side of this equation represents the moment of inertia of the body with respect to the principal axis corresponding to K_1; it is thus the principal moment of inertia corresponding to that root. On the other hand, recalling Eq. (B.30), we note that the right-hand member reduces to K_1. Thus, K_1 itself is the principal moment of inertia. In the same fashion, we can show that K_2 and K_3 are the other two principal moments of inertia of the body.

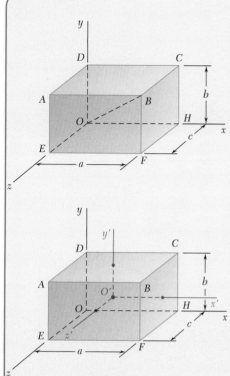

Fig. 1 Centroidal axes for the rectangular prism.

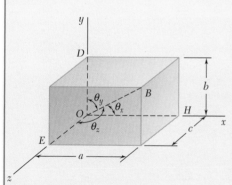

Fig. 2 Direction angles for *OB*.

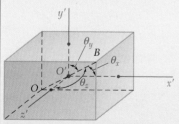

Fig. 3 Line *OB* passes through the centroid *O'*.

Sample Problem B.6

Consider a rectangular prism with a mass of m and sides a, b, *and* c. Determine (*a*) the moments and products of inertia of the prism with respect to the coordinate axes shown, (*b*) its moment of inertia with respect to the diagonal *OB*.

STRATEGY: For part (*a*), you can introduce centroidal axes and apply the parallel-axis theorem. For part (*b*), determine the direction cosines of line *OB* from the given geometry and use either Eq. (B.19) or (B.20).

MODELING and ANALYSIS: a. Moments and Products of Inertia with Respect to the Coordinate Axes.

Moments of Inertia. Introduce the centroidal axes x', y', and z' with respect to which the moments of inertia are given in Fig. B.9, and then apply the parallel-axis theorem (Fig. 1). Thus,

$$I_x = \bar{I}_{x'} + m(\bar{y}^2 + \bar{z}^2) = \tfrac{1}{12}m(b^2 + c^2) + m(\tfrac{1}{4}b^2 + \tfrac{1}{4}c^2)$$
$$I_x = \tfrac{1}{3}m(b^2 + c^2) \quad \blacktriangleleft$$

Similarly,

$$I_y = \tfrac{1}{3}m(c^2 + a^2) \qquad I_z = \tfrac{1}{3}m(a^2 + b^2) \quad \blacktriangleleft$$

Products of Inertia. Because of symmetry, the products of inertia with respect to the centroidal axes x', y', and z' are zero, and these axes are principal axes of inertia. Using the parallel-axis theorem, you have

$$I_{xy} = \bar{I}_{x'y'} + m\bar{x}\bar{y} = 0 + m(\tfrac{1}{2}a)(\tfrac{1}{2}b) \qquad I_{xy} = \tfrac{1}{4}mab \quad \blacktriangleleft$$

Similarly,

$$I_{yz} = \tfrac{1}{4}mbc \qquad I_{zx} = \tfrac{1}{4}mca \quad \blacktriangleleft$$

b. Moment of Inertia with Respect to *OB*. Recall Eq. (B.19):

$$I_{OB} = I_x\lambda_x^2 + I_y\lambda_y^2 + I_z\lambda_z^2 - 2I_{xy}\lambda_x\lambda_y - 2I_{yz}\lambda_y\lambda_z - 2I_{zx}\lambda_z\lambda_x$$

where the direction cosines of *OB* are (Fig. 2)

$$\lambda_x = \cos\theta_x = \frac{OH}{OB} = \frac{a}{(a^2 + b^2 + c^2)^{1/2}}$$

$$\lambda_y = \frac{b}{(a^2 + b^2 + c^2)^{1/2}} \qquad \lambda_z = \frac{c}{(a^2 + b^2 + c^2)^{1/2}}$$

Substituting the values obtained in part (*a*) for the moments and products of inertia and for the direction cosines into the equation for I_{OB}, you obtain

$$I_{OB} = \frac{1}{a^2 + b^2 + c^2}\left[\tfrac{1}{3}m(b^2 + c^2)a^2 + \tfrac{1}{3}m(c^2 + a^2)b^2 + \tfrac{1}{3}m(a^2 + b^2)c^2\right.$$
$$\left. -\tfrac{1}{2}ma^2b^2 - \tfrac{1}{2}mb^2c^2 - \tfrac{1}{2}mc^2a^2\right]$$

$$I_{OB} = \frac{m}{6}\frac{a^2b^2 + b^2c^2 + c^2a^2}{a^2 + b^2 + c^2} \quad \blacktriangleleft$$

REFLECT and THINK: You can also obtain the moment of inertia I_{OB} directly from the principal moments of inertia $\bar{I}_{x'}$, $\bar{I}_{y'}$, and $\bar{I}_{z'}$, since the line *OB* passes through the centroid *O'*. Since the x', y', and z' axes are principal axes of inertia (Fig. 3), use Eq. (B.23) to write

$$I_{OB} = \bar{I}_{x'}\lambda_x^2 + \bar{I}_{y'}\lambda_y^2 + \bar{I}_{z'}\lambda_z^2$$

$$= \frac{1}{a^2 + b^2 + c^2}\left[\frac{m}{12}(b^2 + c^2)a^2 + \frac{m}{12}(c^2 + a^2)b^2 + \frac{m}{12}(a^2 + b^2)c^2\right]$$

$$I_{OB} = \frac{m}{6}\frac{a^2b^2 + b^2c^2 + c^2a^2}{a^2 + b^2 + c^2} \quad \blacktriangleleft$$

Sample Problem B.7

If $a = 3c$ and $b = 2c$ for the rectangular prism of Sample Prob. B.6, determine (a) the principal moments of inertia at the origin O, (b) the principal axes of inertia at O.

STRATEGY: Substituting the data into the results from Sample Prob. B.6 gives you values you can use with Eq. (B.29) to determine the principal moments of inertia. You can then use these values to set up a system of equations for finding the direction cosines of the principal axes.

MODELING and ANALYSIS:

a. Principal Moments of Inertia at the Origin O.
Substituting $a = 3c$ and $b = 2c$ into the solution to Sample Prob. B.6 gives you

$$I_x = \tfrac{5}{3}mc^2 \qquad I_y = \tfrac{10}{3}mc^2 \qquad I_z = \tfrac{13}{3}mc^2$$
$$I_{xy} = \tfrac{3}{2}mc^2 \qquad I_{yz} = \tfrac{1}{2}mc^2 \qquad I_{zx} = \tfrac{3}{4}mc^2$$

Substituting the values of the moments and products of inertia into Eq. (B.29) and collecting terms yields

$$K^3 - (\tfrac{28}{3}mc^2)K^2 + (\tfrac{3479}{144}m^2c^4)K - \tfrac{589}{54}m^3c^6 = 0$$

Now solve for the roots of this equation; from the discussion in Sec. B.8, it follows that these roots are the principal moments of inertia of the body at the origin.

$$K_1 = 0.568867mc^2 \qquad K_2 = 4.20885mc^2 \qquad K_3 = 4.55562mc^2$$
$$K_1 = 0.569mc^2 \qquad K_2 = 4.21mc^2 \qquad K_3 = 4.56mc^2 \qquad \blacktriangleleft$$

b. Principal Axes of Inertia at O.
To determine the direction of a principal axis of inertia, first substitute the corresponding value of K into two of the equations (B.27). The resulting equations, together with Eq. (B.30), constitute a system of three equations from which you can determine the direction cosines of the corresponding principal axis. Thus, for the first principal moment of inertia K_1, you have

$$(\tfrac{5}{3} - 0.568867)mc^2(\lambda_x)_1 - \tfrac{3}{2}mc^2(\lambda_y)_1 - \tfrac{3}{4}mc^2(\lambda_z)_1 = 0$$
$$-\tfrac{3}{2}mc^2(\lambda_x)_1 + (\tfrac{10}{3} - 0.568867)\,mc^2(\lambda_y)_1 - \tfrac{1}{2}mc^2(\lambda_z)_1 = 0$$
$$(\lambda_x)_1^2 + (\lambda_y)_1^2 + (\lambda_z)_1^2 = 1$$

Solving yields

$$(\lambda_x)_1 = 0.836600 \qquad (\lambda_y)_1 = 0.496001 \qquad (\lambda_z)_1 = 0.232557$$

The angles that the first principal axis of inertia forms with the coordinate axes are then

$$(\theta_x)_1 = 33.2° \qquad (\theta_y)_1 = 60.3° \qquad (\theta_z)_1 = 76.6° \qquad \blacktriangleleft$$

Using the same set of equations successively with K_2 and K_3, you can find that the angles associated with the second and third principal moments of inertia at the origin are, respectively,

$$(\theta_x)_2 = 57.8° \qquad (\theta_y)_2 = 146.6° \qquad (\theta_z)_2 = 98.0° \qquad \blacktriangleleft$$

and

$$(\theta_x)_3 = 82.8° \qquad (\theta_y)_3 = 76.1° \qquad (\theta_z)_3 = 164.3° \qquad \blacktriangleleft$$

SOLVING PROBLEMS
ON YOUR OWN

In this section, we defined the **mass products of inertia** I_{xy}, I_{yz}, and I_{zx} of a body and showed you how to determine the moments of inertia of that body with respect to an arbitrary axis passing through the origin O. You also saw how to determine at the origin O the **principal axes of inertia** of a body and the corresponding **principal moments of inertia**.

1. Determining the mass products of inertia of a composite body. You can express the mass products of inertia of a composite body with respect to the coordinate axes as the sums of the products of inertia of its component parts with respect to those axes. For each component part, use the parallel-axis theorem to write Eqs. (B.20)

$$I_{xy} = \bar{I}_{x'y'} + m\bar{x}\bar{y} \qquad I_{yz} = \bar{I}_{y'z'} + m\bar{y}\bar{z} \qquad I_{zx} = \bar{I}_{z'x'} + m\bar{z}\bar{x}$$

Here the primes denote the centroidal axes of each component part, and $\bar{x}$, $\bar{y}$, and $\bar{z}$ represent the coordinates of its center of gravity. Keep in mind that the mass products of inertia can be positive, negative, or zero, and be sure to take into account the signs of $\bar{x}$, $\bar{y}$, and $\bar{z}$.

a. From the properties of symmetry of a component part, you can deduce that two or all three of its centroidal mass products of inertia are zero. For instance, you can verify for a thin plate parallel to the xy plane, a wire lying in a plane parallel to the xy plane, a body with a plane of symmetry parallel to the xy plane, and a body with an axis of symmetry parallel to the z axis that the products of inertia $\bar{I}_{y'z'}$ and $\bar{I}_{z'x'}$ are zero.

For rectangular, circular, or semicircular plates with axes of symmetry parallel to the coordinate axes, straight wires parallel to a coordinate axis, circular and semicircular wires with axes of symmetry parallel to the coordinate axes, and rectangular prisms with axes of symmetry parallel to the coordinate axes, the products of inertia $\bar{I}_{x'y'}$, $\bar{I}_{y'z'}$, and $\bar{I}_{z'x'}$ are all zero.

b. Mass products of inertia that are different from zero can be computed from Eqs. (B.18). Although, in general, you need a triple integration to determine a mass product of inertia, you can use a single integration if you can divide the given body into a series of thin, parallel slabs. The computations are then similar to those discussed in the preceding section for moments of inertia.

2. Computing the moment of inertia of a body with respect to an arbitrary axis OL. In Sec. B.6, we derived an expression for the moment of inertia I_{OL} that was given in Eq. (B.19). Before computing I_{OL}, you must first determine the mass moments and products of inertia of the body with respect to the given coordinate axes, as well as the direction cosines of the unit vector $\boldsymbol{\lambda}$ along OL.

3. Calculating the principal moments of inertia of a body and determining its principal axes of inertia. You saw in Sec. B.7 that it is always possible to find an orientation of the coordinate axes for which the mass products of inertia are zero. These axes are referred to as the **principal axes of inertia**, and the corresponding moments of inertia are known as the **principal moments of inertia** of the body. In many cases, you can determine the principal axes of inertia of a body from its properties of symmetry. The procedure required to determine the principal moments and principal axes of inertia of a body with no obvious property of symmetry was discussed in Sec. B.8 and was illustrated in Sample Prob. B.7. It consists of the following steps.

a. Expand the determinant in Eq. (B.28) and solve the resulting cubic equation. You can obtain the solution by trial and error or (preferably) with an advanced scientific calculator or appropriate computer software. The roots K_1, K_2, and K_3 of this equation are the principal moments of inertia of the body.

b. To determine the direction of the principal axis corresponding to K_1, substitute this value for K in two of the equations (B.27) and solve these equations, together with Eq. (B.30), for the direction cosines of the principal axis corresponding to K_1.

c. Repeat this procedure with K_2 and K_3 to determine the directions of the other two principal axes. As a check of your computations, you may wish to verify that the scalar product of any two of the unit vectors along the three axes you have obtained is zero and, thus, that these axes are perpendicular to each other.

Problems

B.39 Determine the mass products of inertia I_{xy}, I_{yz}, and I_{zx} of the steel fixture shown. (The density of steel is 7850 kg/m³.)

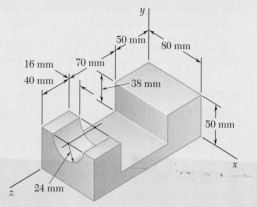

Fig. B.39

B.40 Determine the mass products of inertia I_{xy}, I_{yz}, and I_{zx} of the steel machine element shown. (The density of steel is 7850 kg/m³.)

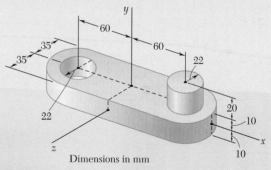

Dimensions in mm

Fig. B.40

B.41 and B.42 Determine the mass products of inertia I_{xy}, I_{yz}, and I_{zx} of the cast aluminum machine component shown. (The specific weight of aluminum is 0.100 lb/in³.)

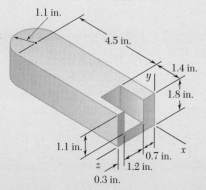

Fig. B.41

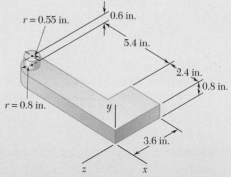

Fig. B.42

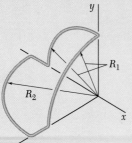

Fig. B.48

B.48 Thin aluminum wire of uniform diameter is used to form the figure shown. Denoting the mass per unit length of the wire by m', determine the mass products of inertia I_{xy}, I_{yz}, and I_{zx} of the wire figure.

B.49 and B.50 Brass wire with a weight per unit length w is used to form the figure shown. Determine the mass products of inertia I_{xy}, I_{yz}, and I_{zx} of the wire figure.

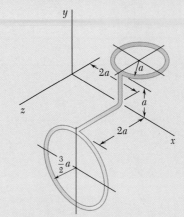

Fig. B.49

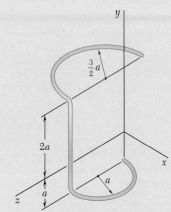

Fig. B.50

B.51 Complete the derivation of Eqs. (B.20) that expresses the parallel-axis theorem for mass products of inertia.

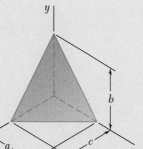

Fig. B.52

B.52 For the homogeneous tetrahedron of mass m shown, (a) determine by direct integration the mass product of inertia I_{zx}, (b) deduce I_{yz} and I_{xy} from the result obtained in part a.

B.53 The homogeneous circular cone shown has a mass m. Determine the mass moment of inertia of the cone with respect to the line joining the origin O and point A.

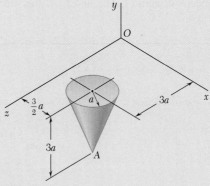

Fig. B.53

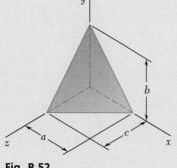

Fig. B.54

B.54 The homogeneous circular cylinder shown has a mass m. Determine the mass moment of inertia of the cylinder with respect to the line joining the origin O and point A that is located on the perimeter of the top surface of the cylinder.

B.55 Shown is the machine element of Prob. B.31. Determine its mass moment of inertia with respect to the line joining the origin O and point A.

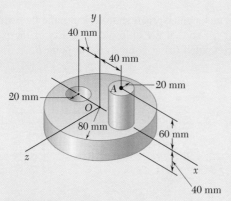

Fig. B.55

B.56 Determine the mass moment of inertia of the steel fixture of Probs. B.35 and B.39 with respect to the axis through the origin that forms equal angles with the x, y, and z axes.

B.57 The thin, bent plate shown is of uniform density and weight W. Determine its mass moment of inertia with respect to the line joining the origin O and point A.

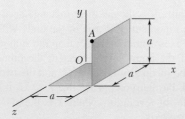

Fig. B.57

B.58 A piece of sheet steel with thickness t and specific weight γ is cut and bent into the machine component shown. Determine the mass moment of inertia of the component with respect to the line joining the origin O and point A.

B.59 Determine the mass moment of inertia of the machine component of Probs. B.26 and B.45 with respect to the axis through the origin characterized by the unit vector $\boldsymbol{\lambda} = (-4\mathbf{i} + 8\mathbf{j} + \mathbf{k})/9$.

B.60 through B.62 For the wire figure of the problem indicated, determine the mass moment of inertia of the figure with respect to the axis through the origin characterized by the unit vector $\boldsymbol{\lambda} = (-3\mathbf{i} - 6\mathbf{j} + 2\mathbf{k})/7$.

    B.60   Prob. B.38
    B.61   Prob. B.37
    B.62   Prob. B.36

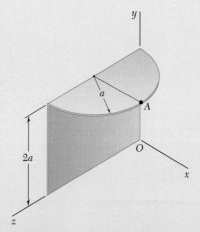

Fig. B.58

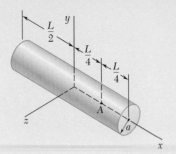

Fig. B.63

B.63 For the homogeneous circular cylinder shown with radius a and length L, determine the value of the ratio a/L for which the ellipsoid of inertia of the cylinder is a sphere when computed (*a*) at the centroid of the cylinder, (*b*) at point A.

B.64 For the rectangular prism shown, determine the values of the ratios b/a and c/a so that the ellipsoid of inertia of the prism is a sphere when computed (*a*) at point A, (*b*) at point B.

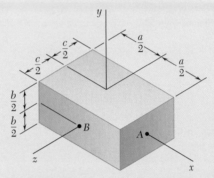

Fig. B.64

B.65 For the right circular cone of Sample Prob. B.3, determine the value of the ratio a/h for which the ellipsoid of inertia of the cone is a sphere when computed (*a*) at the apex of the cone, (*b*) at the center of the base of the cone.

B.66 Given an arbitrary body and three rectangular axes x, y, and z, prove that the mass moment of inertia of the body with respect to any one of the three axes cannot be larger than the sum of the mass moments of inertia of the body with respect to the other two axes. That is, prove that the inequality $I_x \leq I_y + I_z$ and the two similar inequalities are satisfied. Furthermore, prove that $I_y \geq \frac{1}{2}I_x$ if the body is a homogeneous solid of revolution, where x is the axis of revolution and y is a transverse axis.

B.67 Consider a cube with mass m and side a. (*a*) Show that the ellipsoid of inertia at the center of the cube is a sphere, and use this property to determine the moment of inertia of the cube with respect to one of its diagonals. (*b*) Show that the ellipsoid of inertia at one of the corners of the cube is an ellipsoid of revolution, and determine the principal moments of inertia of the cube at that point.

B.68 Given a homogeneous body of mass m and of arbitrary shape and three rectangular axes x, y, and z with origin at O, prove that the sum $I_x + I_y + I_z$ of the mass moments of inertia of the body cannot be smaller than the similar sum computed for a sphere of the same mass and the same material centered at O. Furthermore, using the result of Prob. B.66, prove that, if the body is a solid of revolution where x is the axis of revolution, its mass moment of inertia I_y about a transverse axis y cannot be smaller than $3ma^2/10$, where a is the radius of the sphere of the same mass and the same material.

*B.69 The homogeneous circular cylinder shown has a mass m, and the diameter OB of its top surface forms 45° angles with the x and z axes. (a) Determine the principal mass moments of inertia of the cylinder at the origin O. (b) Compute the angles that the principal axes of inertia at O form with the coordinate axes. (c) Sketch the cylinder, and show the orientation of the principal axes of inertia relative to the x, y, and z axes.

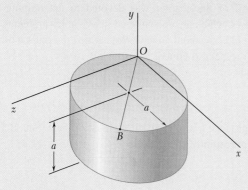

Fig. B.69

B.70 through *B.74* For the component described in the problem indicated, determine (a) the principal mass moments of inertia at the origin, (b) the principal axes of inertia at the origin. Sketch the body and show the orientation of the principal axes of inertia relative to the x, y, and z axes.

 *B.70 Prob. B.55
 *B.71 Probs. B.35 and B.39
 *B.72 Prob. B.57
 *B.73 Prob. B.58
 *B.74 Probs. B.38 and B.60

Review and Summary

Mass Moments of Inertia

The second half of the chapter was devoted to determining **moments of inertia of masses**, which are encountered in dynamics problems involving the rotation of a rigid body about an axis. We defined the mass moment of inertia of a body with respect to an axis AA' (Fig. B.16) as

$$I = \int r^2 \, dm \tag{B.1}$$

where r is the distance from AA' to the element of mass [Sec. B.1]. We defined the **radius of gyration** of the body as

$$k = \sqrt{\frac{I}{m}} \tag{B.2}$$

The moments of inertia of a body with respect to the coordinate axes were expressed as

$$I_x = \int (y^2 + z^2) \, dm$$

$$I_y = \int (z^2 + x^2) \, dm \tag{B.3}$$

$$I_z = \int (x^2 + y^2) \, dm$$

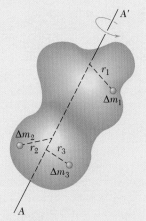

Fig. B.16

Parallel-Axis Theorem

We saw that the **parallel-axis theorem** also applies to mass moments of inertia [Sec. B.2]. Thus, the moment of inertia I of a body with respect to an arbitrary axis AA' (Fig. B.17) can be expressed as

$$I = \bar{I} + md^2 \tag{B.6}$$

where $\bar{I}$ is the moment of inertia of the body with respect to the centroidal axis BB' that is parallel to the axis AA', m is the mass of the body, and d is the distance between the two axes.

Moments of Inertia of Thin Plates

We can readily obtain the moments of inertia of thin plates from the moments of inertia of their areas [Sec. B.3]. We found that for a rectangular plate the moments of inertia with respect to the axes shown (Fig. B.18) are

$$I_{AA'} = \tfrac{1}{12}ma^2 \qquad I_{BB'} = \tfrac{1}{12}mb^2 \tag{B.12}$$

$$I_{CC'} = I_{AA'} + I_{BB'} = \tfrac{1}{12}m(a^2 + b^2) \tag{B.13}$$

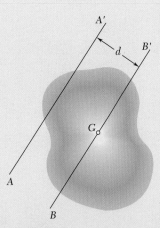

Fig. B.17

whereas for a circular plate (Fig. B.19), they are

$$I_{AA'} = I_{BB'} = \tfrac{1}{4}mr^2 \qquad\qquad \textbf{(B.14)}$$

$$I_{CC'} = I_{AA'} + I_{BB'} = \tfrac{1}{2}mr^2 \qquad\qquad \textbf{(B.15)}$$

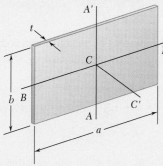

Fig. B.18

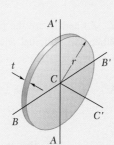

Fig. B.19

Composite Bodies

When a body possesses two planes of symmetry, it is usually possible to use a single integration to determine its moment of inertia with respect to a given axis by selecting the element of mass dm to be a thin plate [Sample Probs. B.2 and B.3]. On the other hand, when a body consists of several common geometric shapes, we can obtain its moment of inertia with respect to a given axis by using the formulas given in Fig. B.9 together with the parallel-axis theorem [Sample Probs. B.4 and B.5].

Moment of Inertia with Respect to an Arbitrary Axis

In the last section of the chapter, we described how to determine the moment of inertia of a body with respect to an arbitrary axis OL that is drawn through the origin O [Sec. B.6]. We denoted the components of the unit vector $\boldsymbol{\lambda}$ along OL by λ_x, λ_y, and λ_z (Fig. B.20) and introduced the **products of inertia as**

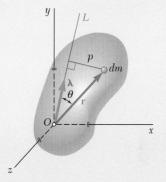

Fig. B.20

$$I_{xy} = \int xy\,dm \qquad I_{yz} = \int yz\,dm \qquad I_{zx} = \int zx\,dm \qquad \textbf{(B.18)}$$

We found that the moment of inertia of the body with respect to OL could be expressed as

$$I_{OL} = I_x\lambda_x^2 + I_y\lambda_y^2 + I_z\lambda_z^2 - 2I_{xy}\lambda_x\lambda_y - 2I_{yz}\lambda_y\lambda_z - 2I_{zx}\lambda_z\lambda_x \qquad \textbf{(B.19)}$$

Ellipsoid of Inertia

By plotting a point Q along each axis OL at a distance $OQ = 1/\sqrt{I_{OL}}$ from O [Sec. B.7], we obtained the surface of an ellipsoid, known as the **ellipsoid of inertia** of the body at point O.

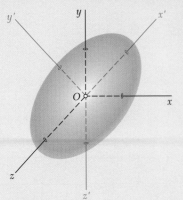

Fig. B.21

Principal Axes and Principal Moments of Inertia

The principal axes x', y', and z' of this ellipsoid (Fig. B.21) are the **principal axes of inertia** of the body; that is, the products of inertia $I_{x'y'}$, $I_{y'z'}$, and $I_{z'x'}$ of the body with respect to these axes are all zero. In many situations, you can deduce the principal axes of inertia of a body from its properties of symmetry. Choosing these axes to be the coordinate axes, we can then express I_{OL} as

$$I_{OL} = I_{x'}\lambda_{x'}^2 + I_{y'}\lambda_{y'}^2 + I_{z'}\lambda_{z'}^2 \tag{B.23}$$

where $I_{x'}$, $I_{y'}$, and $I_{z'}$ are the **principal moments of inertia** of the body at O.

When the principal axes of inertia cannot be obtained by observation [Sec. B.7], it is necessary to solve the cubic equation

$$K^3 - (I_x + I_y + I_z)K^2 + (I_xI_y + I_yI_z + I_zI_x - I_{xy}^2 - I_{yz}^2 - I_{zx}^2)K$$
$$- (I_xI_yI_z - I_xI_{yz}^2 - I_yI_{zx}^2 - I_zI_{xy}^2 - 2I_{xy}I_{yz}I_{zx}) = 0 \tag{B.29}$$

We found [Sec. B.8] that the roots K_1, K_2, and K_3 of this equation are the principal moments of inertia of the given body. The direction cosines $(\lambda_x)_1$, $(\lambda_y)_1$, and $(\lambda_z)_1$ of the principal axis corresponding to the principal moment of inertia K_1 are then determined by substituting K_1 into Eqs. (B.27) and by solving two of these equations and Eq. (B.30) simultaneously. The same procedure is then repeated using K_2 and K_3 to determine the direction cosines of the other two principal axes [Sample Prob. B.7].

C Fundamentals of Engineering Examination

Engineers are required to be licensed when their work directly affects the public health, safety, and welfare. The intent is to ensure that engineers have met minimum qualifications involving competence, ability, experience, and character. The licensing process involves an initial exam, called the *Fundamentals of Engineering Examination;* professional experience; and a second exam, called the *Principles and Practice of Engineering.* Those who successfully complete these requirements are licensed as a *Professional Engineer.* The exams are developed under the auspices of the *National Council of Examiners for Engineering and Surveying.*

The first exam, the *Fundamentals of Engineering Examination,* can be taken just before or after graduation from a four-year accredited engineering program. The exam stresses subject material in a typical undergraduate engineering program, including statics. The topics included in the exam cover much of the material in this book. The following is a list of the main topic areas, with references to the appropriate sections in this book. Also included are problems that can be solved to review this material.

Kinematics (11.1–11.2; 11.4–11.5; 15.1–15.4)
Problems: 11.3, 11.10, 11.34, 11.35, 11.97, 11.102, 15.4, 15.6, 15.28, 15.39, 15.61, 15.63, 15.82, 15.111, 15.112

Force, Mass, and Acceleration (12.1; 16.1–16.2)
Problems: 12.5, 12.6, 12.11, 12.23, 12.36, 12.44, 12.45, 12.50, 16.1, 16.3, 16.9, 16.26, 16.27, 16.50, 16.60, 16.63, 16.78, 16.84

Work and Energy (13.1–13.2; 13.8; 17.1)
Problems: 13.3, 13.6, 13.13, 13.17, 13.40, 13.42, 13.47, 13.64, 13.66, 13.68, 17.1, 17.2, 17.16, 17.20

Impulse and Momentum (13.3–13.4; 17.2–17.3)
Problems: 13.119, 13.120, 13.129, 13.134, 13.146, 13.155, 13.163, 13.169, 17.53, 17.58, 17.70, 17.72, 17.96, 17.97, 17.104

Vibration (19.1; 19.2–19.4)
Problems: 19.1, 19.2, 19.11, 19.18, 19.23, 19.28, 19.50, 19.55, 19.64, 19.79, 19.99, 19.101, 19.105, 19.116

Friction (Problems involving friction occur in each of the above subjects.)

Answers to Problems

CHAPTER 11

11.1 97.5 ft, 49.5 ft/s, 17 ft/s².

11.2 1.000s, 15.00 ft, −6.00 ft/s²; 2.00 s, 14.00 ft, 6.00 ft/s².

11.3 (a) 102.9 mm, −35.6 mm/s, −11.40 mm/s².
(b) −36.1 mm/s, 72.1 mm/s².

11.4 (a) 0 mm, 960 mm/s →, 9220 mm/s² or 9.22 m/s² ←.
(b) 14.16 mm ←, 87.9 mm/s →, 3110 mm/s² or 3.11 m/s² →.

11.5 0.667 s, 0.259m, −8.56 m/s.

11.7 (a) 0.586 s and 3.414 s. (b) 0 m. (c) 3.656 m.

11.9 (a) 77.5 ft/s. (b) 7.75 s.

11.10 1.427 ft/s, 0.363 ft.

11.11 $x(t) = t^4/108 + 10t + 24$ m.
$v(t) = t^3/27 + 10$ m/s.

11.12 (a) 6.00 m/s⁴. (b) $a = 6t^2$, $v = 2t^3 − 8$, $x = t^4/2 − 8t + 8$.

11.15 800 m/s² ↑.

11.16 (a) $−2.43 \times 10^6$ ft/s². (b) 1.366×10^{-3} s.

11.17 (a) 5.89 ft/s. (b) 1.772 ft.

11.18 167.1 mm/s² ↑, 15.19 m/s² ↑.

11.21 (a) 2.52 m²/s², 4.70 m/s.

11.22 (a) 10.00 ft. (b) 1.833 ft/s, 0.440 ft/s².

11.23 (a) 42.0 ft. (b) 12.86 ft/s.

11.24 (a) 29.3 m/s. (b) 0.947 s.

11.25 (a) 4.76 mm/s. (b) 0.171 s.

11.26 1.995 m/s².

11.27 (a) −0.0525 m/s². (b) 6.17 s.

11.28 (a) 7.15 mi. (b) $−275 \times 10^{-6}$ ft/s². (c) 49.9 min.

11.31 (a) $2.36\, v_0T$, $\pi\, v_0/T$. (b) $0.363\, v_0$.

11.32 $r + \dfrac{d_{max}}{2}\cos\theta, -v_{max}\sin\theta, -\dfrac{d_{max}}{2}\ddot\theta\sin\theta - \dfrac{2v_{max}^2}{d_{max}}\cos\theta.$

11.33 (a) 2.0 m/s². (b) 60.0 m/s.

11.34 (a) −0.417 m/s². (b) 18.00 km/h.

11.35 (a) 6.0 s. (b) 180.0 ft.

11.36 (a) 252 ft/s. (b) 1076 ft.

11.39 11.60 s, 50.4 m.

11.40 (a) 1.563 m/s². (b) 3.13 m/s².

11.41 (a) −3.20 ft/s² and 3.72 ft/s². (b) 3.41 s before A reaches the exchange zone.

11.42 (a) 15.05 s, 734 ft from the ititial point of A.
(b) A: 42.5 mi/h. B: 23.7 mi/h.

11.43 (a) $\mathbf{a}_A = 0.767$ ft/s² ←, $\mathbf{a}_B = 0.834$ ft/s² →.
(b) 20.7 s. (c) 51.8 mi/h.

11.44 (a) 1.330 s. (b) 4.68 m below the man.

11.47 (a) 8.00 m/s ↑. (b) 4.00 m/s ↑. (c) 12.00 m/s ↑. (d) 8.00 m/s ↑.

11.48 (a) $\mathbf{a}_E = 2.40$ ft/s² ↑, $\mathbf{a}_C = 4.80$ ft/s² ↓.
(b) 12.00 ft/s ↑.

11.49 (a) 0.125 m/s↑. (b) 0.5154 m/s ∠ 14°.

11.50 (a) 18 ft/s² ←, 6 ft/s² ↑. (b) 9 ft/s ←, 2.25 ft ←.

11.51 (a) 200 mm/s →. (b) 600 mm/s →.
(c) 200 mm/s ←. (d) 400 mm/s →.

11.52 (a) $\mathbf{a}_A = 13.33$ mm/s² ←, $\mathbf{a}_B = 20.0$ mm/s² ←.
(b) 13.33 mm/s² →. (c) 70.0 mm/s →. 440 mm →.

11.55 (a) 2.5 s. (b) 7.5 in.↓.

11.56 (a) 1.000 s. (b) 3.00 in. ↓.

11.57 (a) $\mathbf{a}_A = 345$ mm/s² ↓, $\mathbf{a}_B = 240$ mm/s² ↑.
(b) $(\mathbf{v}_A)_0 = 43.3$ mm/s ↑, $(v_C)_0 = 130.0$ mm/s →.
(c) 728 mm →.

11.58 (a) 10.00 mm/s →. (b) $\mathbf{a}_A = 2.00$ mm/s² ↑,
$\mathbf{a}_C = 6.00$ mm/s² →. (c) 175.0 mm ↑.

11.61 88 ft.

11.62 (b) 5.83 s.

11.63 (a) 10 s to 26 s, $a = −5.00$ m/s²;
41 s to 46 s, $a = 3.00$ m/s²; otherwise $a = 0$.
(b) 1383 m. (c) 9.00 s, 49.5 s.

11.64 (a) Same as Prob. 11.63. (b) 420 m. (c) 10.69 s, 40.0 s.

11.65 (a) 162 ft. (b) 18 s and 30 s.

11.66 (a) 44.8 s. (b) 103.3 m/s².

11.69 (a) 0.600 s. (b) 0.200 m/s, 2.84 m.

11.70 (a) 60.0 m/s, 1194 m. (b) 59.3 m/s.

11.71 (a) A: 52.2 s, B: 52.0 s. (b) 1.879 m.

11.72 9.39 s.

11.73 8.54 s, 58.3 mi/h.

11.74 77.5 ft.

11.75 5.67 s.

11.78 (a) 18.00 s. (b) 178.8 m. (c) 34.7 km/h.

11.79 (a) 5.01 min. (b) 19.18 mi/h.

11.80 (a) 2.00 s. (b) 1.200 ft/s, 0.600 ft/s.

11.83 (a) 2.96 s. (b) 224 ft.

11.84 (a) 163.0 in/s². (b) 114.3 in/s².

11.85 (a) 15.49 s. (b) 4.65 m/s. (c) 2.90 m/s, 8.50 m.

11.89 (a) 6.28 m/s ⦨ 37.2°. (b) 7.49 m.

11.90 (a) 67.1 mm/s ⦛ 63.4°, 256 mm/s² ⦨ 69.4°.
(b) 8.29 mm/s ⦛ 36.2°, 336 mm/s² ⦨ 86.6°.

11.91 (a) $(−12.57$ in/s$)\mathbf{i}$, $(−39.5$ in/s$^2)\mathbf{j}$. (b) $y = x^2/8 − 1$.

11.92 (a) max: 15.00 ft/s, min: 5.00 ft/s
(b) min: $t = 2\pi N$ s, $x = 20\pi N$ ft, $y = 5$ ft, $v_x = 5$ ft/s,
$v_y = 0$, $\theta = 0$.
max: $t = (2N + 1)\,\pi$ s, $x = 20\pi(N + 1)$ ft, $y = 15$ ft,
$v_x = 15$ ft/s, $v_y = 0$, $\theta = 0$.

11.95 $\sqrt{R^2(1 + w_n^2 t^2) + c^2}$, $Rw_n\sqrt{4 + w_n^2 t^2}$.

11.97 1140 ft.

11.98 (a) 2.94 s. (b) 84.9 m. (c) 10.62 m.

11.99 (a) 115.3 km/h $\leq v_0 \leq$ 148.0 km/h.
(b) $h = 0.788$ m, $\alpha = 6.66°$; $h = 1.068$ m, $\alpha = 4.05°$.

11.100 15.38 ft/s $< v_0 <$ 35.0 ft/s.

11.102 (a) Meets max. height requirement. (b) 0.937 m.

11.103 (a) Ball clears the net. (b) 7.01 m from the net.

11.105 22.9 ft/s.

11.106 16.20 m/s $< v_0 <$ 21.0 m/s.

11.107 (a) 29.8 ft/s. (b) 29.6 ft/s.

11.108 37.7 m/s $< v_0 <$ 44.3 m/s.

11.111 (a) 10.38°. (b) 9.74°.

11.112 (a) 4.17°. (b) 285 m. (c) 15.89 s.

11.113 (a) 14.66°. (b) 0.1074 s.

11.114 (a) 4.98 m. (b) 23.8°.

11.117 17.80 ft/s ⦡ 50.9°.

11.118 $\mathbf{v}_A = 125$ mm/s ↑, $\mathbf{v}_B = 75$ mm/s ↓, $\mathbf{v}_C = 175$ mm/s ↓.

11.119 (a) 91.0 ft/s ⦢ 47.0°. (b) 364 ft ⦢ 47.0°. (c) 293 ft.

11.120 3.20 km/h ⭦ 17.8°.

11.123 (a) 4 ft/s ↑. (b) 6 ft/s² ↓.

11.124 (a) 8.53 in/s ⭨ 54.1°. (b) 6.40 in/s ⭨ 54.1°.

11.125 (a) 0.979 m. (b) 12.55 m/s ⭦ 86.5°.

11.126 (a) 0.835 mm/s² ⭨ 75°. (b) 8.35 mm/s ⭨ 75°.

11.127 (a) 5.18 ft/s ⭨ 15°. (b) 1.232 ft/s ⭨ 15°.

11.128 10.54 ft/s ⭧ 81.3°.

11.129 5.96 m/s ⭦ 82.8°.

11.131 15.79 km/h ⭦ 26.0°.

11.133 500 m.

11.134 97.6 km/h.

11.135 12.13 m/s.

11.136 (a) 0.407 ft/s². (b) 0.0333 ft/s². (c) 0.00593 ft/s².

11.137 8.56 s.

11.138 (a) 10.20 mm/s². (b) 25.2 s.

11.139 (a) 178.9 m. (b) 1.118 m/s².

11.141 (a) 189.5 km/h ⭦ 54.0°. (b) 21.8 m/s² ⭦ 5.3°.

11.143 (a) $1.047\mathbf{i} - 33.726\mathbf{j}$ m/s². (b) $-47.55\mathbf{i} - 8.64\mathbf{j}$ m/s.

11.144 1467.9 m.

11.145 (a) 281 m. (b) 209 m.

11.146 (a) 27.6 m. (b) 34.0 m.

11.147 (a) 0.634 m. (b) 9.07 m.

11.149 (a) 14.48 m/s. (b) 21.3 m.

11.151 $(R^2 + c^2)/2w_n R$.

11.152 2.50 ft.

11.153 149.8 Gm.

11.154 1425 Gm.

11.155 16 200 mi/h.

11.156 7740 mi/h.

11.159 1.606 h.

11.161 (a) $(1.624\ \text{in/s})\mathbf{e}_r - (15.56\ \text{in/s})\mathbf{e}_\theta$
(b) $(-49.9\ \text{in/s}^2)\mathbf{e}_r + (-9.74\ \text{in/s}^2)\mathbf{e}_\theta$
(c) $(-3.25\ \text{in/s}^2)\mathbf{e}_r$.

11.162 (a) $3\pi b\mathbf{e}_\theta$ and $-4\pi^2 b\mathbf{e}_r$, 14.48. (b) $\theta = 2N\pi$, $N = 0, 1, 2, \ldots$.

11.163 13.280 m/s ∠ 27.08°, 0.2437 m/s² ⭦ 30.00°.

11.164 (b) 1.787 m/s².

11.165 (a) $\mathbf{v} = bk\mathbf{e}_\theta$, $\mathbf{a} = -(bk^2/2)\mathbf{e}_r$.
(b) $\mathbf{v} = 2bk\mathbf{e}_r + 2bk\mathbf{e}_\theta$, $\mathbf{a} = 2bk^2\mathbf{e}_r + 4bk^2\mathbf{e}_\theta$.

11.166 (a) $a = 4b\dot\theta^2$. (b) directed toward point A.

11.169 $\dot r = 370$ ft/s, $\ddot r = 57.9$ ft/s², $\dot\theta = -0.0924$ rad/s, $\ddot\theta = 0.0315$ rad/s².

11.170 (a) $\dot r = -dw/2$, $\dot\theta = w/2$. (b) $\ddot r = -\sqrt{3}\,dw^2/4$, $\ddot\theta = 0$.

11.171 185.7 km/h.

11.172 61.8 mi/h, 49.7°.

11.175 $be^{\frac{1}{2}\theta^2}\theta(\theta^2 + 4)^{\frac{1}{2}}\omega^2$.

11.176 $\dfrac{b}{\theta^4}(36 + 4\theta^2 + \theta^4)^{\frac{1}{2}}\omega^2$.

11.177 $v = 2\pi\sqrt{A^2 + n^2 B^2 \cos^2 2\pi nt}$,
$a = 4\pi^2\sqrt{A^2 + n^4 B^2 \sin^2 2\pi nt}$.

11.179 (a) $v = \sqrt{A^2 + B^2}$, $a = \sqrt{(1 + 16\pi^2)A^2 + B^2}$.
(b) $v = 2\pi A$, $a = 4\pi^2 A$.

11.180 $\tan^{-1}[R(2 + w_n^2 t^2)/c\sqrt{4 + w_n^2 t^2}]$.

11.181 (a) $\theta_x = 90°$, $\theta_y = 123.7°$, $\theta_z = 33.7°$.
(b) $\theta_x = 103.4°$, $\theta_y = 134.3°$, $\theta_z = 47.4°$.

11.182 (a) 1.00 s and 4.00 s. (b) 1.500 m, 24.5 m.

11.183 (a) 9.6 s. (b) 543.0 m.

11.185 (a) 111.4 km/h ∠ 10.50°. (b) 2.96 km.

11.187 (a) $\mathbf{a}_B = 2.00$ in/s² ↑, $\mathbf{a}_C = 3.00$ in/s² ↓. (b) 0.667 s.
(c) 0.667 in. ↑.

11.188 (a) 38.1 m/s, 20.4 m. (b) 41.1 m/s, 29.6 m.

11.189 (a) 3.21 ft/s² ⭦ 22.4°. (b) 6.43 ft/s² ⭦ 22.4°.

11.190 $1.097\mathbf{e}_t + 19.71\mathbf{e}_n$ m/s².

11.191 (a) 23.4 ft/s. (b) 103.2 ft.

CHAPTER 12

12.1 (a) 844 lb. (b) 26.2 slugs.

12.2 (a) 0°: 4.987 lb, 45°: 5.000 lb, 90°: 5.013 lb.
(b) 5.000 lb. (c) 0.1554 lb·s²/ft.

12.3 2.84×10^6 kg·m/s.

12.5 0.242 mi.

12.6 (a) 1449 ft. (b) 10.0 s.

12.7 (a) 18.84 s. (b) 36.14 m.

12.8 (a) 110.5 km/h. (b) 85.6 km/h. (c) 69.9 km/h.

12.9 (a) 40.1 m. (b) 47.0 m.

12.10 (a) 2.22 s. (b) 3.32 m.

12.11 51.0 m.

12.12 (a) 234 m. (b) 3.33 kN (tension).

12.15 (a) (1): 10.73 ft/s² ↓, (2): 16.10 ft/s² ↓, (3): 0.749 ft/s² ↓.
(b) (1): 14.65 ft/s ↓, (2): 17.94 ft/s ↓, (3): 3.87 ft/s ↓.
(c) (1): 1.864 s, (2): 1.242 s, (3): 26.7 s.

12.16 $\mathbf{a}_A = 0.997$ ft/s² ∠ 15°, $\mathbf{a}_B = 1.619$ ft/s² ∠ 15°.

12.17 (a) 765 lb. (b) 1016 lb.

12.18 (a) 0.986 m/s² ⭨ 25°. (b) 51.7 N.

12.19 (a) 1.794 m/s² ⭨ 25°. (b) 58.2 N.

12.20 (a) 16.19 kN. (b) 2.45 m/s².

12.23 $\mathbf{a}_1 = 19.53$ m/s² ∠ 65°, $\mathbf{a}_2 = 4.24$ m/s² ⭧ 65°.

12.24 1.598 km.

12.25 (a) 335 m. (b) 73.6 mm/s ↓.

12.27 $\sqrt{k/m}\left(\sqrt{l^2 + x_0^2} - l\right)$.

12.28 (a) 10.00 N. (b) 103.1 N.

12.29 (a) 8.94 ft/s² ←, 18.06 lb.
(b) 12.38 ft/s² ←, 15.38 lb. (c) Same as (b).

12.30 20.26 kg.

12.31 (a) 2.43 lb. (b) $\mathbf{a}_A = 3.14$ ft/s² →,
$\mathbf{a}_B = 0.881$ m/s² →, $\mathbf{a}_C = 5.41$ m/s² ↓.

12.34 0.0740 m/s² ∠ 20°, 137.2 N.

12.35 (a) 5.94 m/s² ⭦ 75.6°. (b) 3.74 m/s ⭦ 20°.

12.36 (a) 49.9°. (b) 6.85 N.

12.37 (a) 80.4 N. (b) 2.30 m/s.

12.38 (a) 22.55 s. (b) 6.379°.

12.39 3.47 m/s.

12.40 3.01 m/s ≤ v ≤ 3.85 m/s.

12.42 9.00 ft/s < v_C < 12.31 ft/s.

12.43 2.42 ft/s < v < 13.85 ft/s.

12.44 (a) 122.2 lb. (b) 145.6 lb.

12.45 (a) 668 ft. (b) 120.0 lb ↑.

12.46 434 N.

12.47 (a) 4.63 m/s². (b) 1.962 m/s². (c) 0.1842 m.s².

12.48 77.23 rpm.

12.49 (a) 2.91 N. (b) 13.09°.

12.50 1126 N ⭨ 25.6°.

12.51 (a) 12.19 m/s. (b) 2290 N.

12.53 (a) 0.1858 W. (b) 10.28°.

12.55 7.67 m/s.

12.56 (a) 12.00 m/s. (b) 2.05×10^{-3} N.

12.57 0.236.

12.58 3.71 m.

12.61 0.400.

12.62 (a) 0.1834. (b) left: 10.39°, right 169.6°.

12.63 (a) 2.98 ft/s. (b) left: 19.29°, right 160.7°.

12.64 0°, 180°, and 69.6°.

12.65 (a) no sliding, 0.611 N $\measuredangle$ 75°. (b) sliding, 0.957 N $\measuredangle$ 40°.
12.66 (a) 289.1 lb.
12.67 −2.17 lb and 64.9 lb.
12.68 2.00 s.
12.69 (a) 7.47 N $\measuredangle$ 45°. (b) 6.94 m/s² $\measuredangle$ 45°.
12.71 (a) 126.6 N. (b) 5.48 m/s² →. (c) 4.75 m/s² ↓.
12.72 (a) 142.7 N. (b) 6.18 m/s² →. (c) 4.10 m/s² ↓.
12.74 $v_r = v_0 \sin 2\theta / \sqrt{\cos 2\theta}$, $v_\theta = v_0 \sqrt{\cos 2\theta}$.
12.77 (a) 0. (b) $8m\, v_0^2 / r_0$.
12.78 413×10^{21} lb·s²/ft.
12.79 383×10^3 km, 238×10^3 mi.
12.80 (a) 35 800 km, 22 200 mi. (b) 3.07 km/s, 10.09×10^3 ft/s.
12.81 (b) 24.8 m/s².
12.82 (a) 1.998×10^{30} kg. (b) 276 m/s².
12.85 (a) 1684 N. (b) 2510 km. (c) 1.620 m/s².
12.86 (a) 1551 m/s. (b) −15.8 m/s.
12.87 2.64 km/s.
12.88 (a) 5280 ft/s. (b) 8000 ft/s.
12.89 (a) 5.12×10^3 ft/s. (b) 97.0 ft/s.
12.90 (a) $(a_A)_r = (a_A)_\theta = 0$. (b) 38.4 m/s². (c) 0.800 m/s.
12.91 (a) $(a_B)_r = (a_B)_\theta = 0$. (b) 61.4 ft/s². (c) 2.98 ft/s.
12.100 (a) 10.13 km/s. (b) 2.97 km/s.
12.101 1.147.
12.103 $\sqrt{2} / (2 + \alpha)$.
12.104 (a) 1.637×10^3 m/s. (b) 725 m/s. (c) 0.333.
12.107 (a) 52.4×10^3 ft/s. (b) A: 1318 ft/s, B′: 3900 ft/s.
12.108 5.31×10^9 km.
12.109 91.8×10^3 yr.
12.112 4.95 h.
12.113 50 min 55 s.
12.114 $\cos^{-1}[(1 - n\beta^2)/(1 - \beta^2)]$.
12.115 (a) 4.00 km/s. (b) 0.684.
12.124 (a) 20.5 ft/s² $\measuredangle$ 30°. (b) 17.75 ft/s² →.
12.125 (a) 1.088 ft/s² ←. (b) 233 lb.
12.126 (a) 5.79 m/s². (b) 2.45 m/s². (c) 0.230 m/s².
12.127 18.4 kN $\measuredangle$ 31.97°.
12.128 (a) 0.454, down. (b) 0.1796 down. (c) 0.218, up.
12.129 (a) 539 N. (b) 47.1 m.
12.132 54.0°.
12.133 (a) 0.500 m, 0. (b) 0.270 m, −84.1 N.

CHAPTER 13

13.1 6.17 GJ.
13.2 (a) 140.1 ft·lb, 140.1 ft. (b) 140.1 ft·lb, 850 ft.
13.5 10.51 ft/s.
13.6 9.53 ft.
13.7 (a) 112.2 km/h. (b) 91.6 km/h.
13.8 (a) 17.54 m/s. (b) 0.893.
13.9 (a) 8.70 m. (b) 4.94 m/s $\measuredangle$ 15°.
13.11 6.71 m.
13.12 (a) 2.90 m/s. (b) 0.893 m.
13.15 (a) 57.8 m. (b) 154 N →.
13.16 (a) 7.41 kN. (b) 5.56 kN (tension).
13.17 (a) 124.1 ft. (b) A to B: 19.38 kips (tension);
           B to C: 8.62 kips (tension).
13.18 (a) 279 ft. (b) A to B: 19.38 kips (compression);
           B to C: 8.62 kips (compression).
13.19 (a) 46.0 ft·lb.
           (b) A: 19.76 lb; B: 12.10 lb.
13.20 (a) 7.43 ft/s. (b) 0.800 ft.
13.23 (a) 1.218 m/s ←. (b) 91.0 N.

13.24 1.190 m/s.
13.25 (a) 3.96 m/s. (b) 5.60 m/s.
13.26 (a) 3.29 m/s. (b) 1.533 m.
13.27 (a) 3.29 m/s. (b) 1.472 m.
13.28 (a) 8.83 lb/in. (b) 5.13 in.
13.29 (a) 0.159. (b) 5.92 ft/s.
13.32 $0.759 \sqrt{pAa/m}$.
13.33 (a) 13.43 ft. (b) 386 ft/s².
13.34 A: 5.37 in.; B: 7.21 in.
13.36 (a) 10.39 km/s. (b) 11.14 km/s. (c) 11.18 km/s.
13.37 (a) 0.0316%. (b) 25.4%.
13.38 364 m.
13.39 14.00°.
13.40 (a) $\sqrt{3gl}$. (b) $\sqrt{2gl}$.
13.41 41.8°.
13.44 2.30 m/s.
13.45 (a) 27.4°. (b) 3.81 ft.
13.46 (a) 57.2 kW. (b) 269 kW.
13.47 (a) 2.75 kW. (b) 3.35 kW.
13.48 14.80 kN.
13.51 (a) 14.95 kW. (b) 45.4 kW.
13.52 (a) 17.75 kW. (b) 46.7 kW.
13.54 (a) 8.00 hp. (b) 7.91 hp.
13.55 (a) $k_1 k_2 / (k_1 + k_2)$. (b) $k_1 + k_2$.
13.57 (a) 5.12 m/s. (b) 4.20 m/s.
13.58 49.0 ft/s.
13.59 23.1 ft/s.
13.62 (a) 533 lb/ft. (b) 37.0 ft.
13.64 (a) 2.48 m/s ←. (b) 1.732 m/s ↑.
13.65 (a) 2.92 m/s. (b) (−33.9 N)**i** + (33.3 N)**j**.
13.66 (a) 43.5°. (b) 8.02 ft/s ↓.
13.68 0.269 m.
13.69 0.1744 m.
13.70 731 N.
13.71 (max) 5520 N at D; (min) 731 N just above B.
13.72 14.34 ft/s, 13.77 lb ↑.
13.74 Loop 1: (a) 25.1 ft/s. (b) 1.500 lb ←.
           Loop 2: (a) 24.1 ft/s. (b) 1.000 lb.
13.76 Loop 1: (a) $\sqrt{5gr}$. (b) 3 W →.
           Loop 2: (a) $\sqrt{4gr}$. (b) 2 W →.
13.77 0.488 m.
13.78 3/5l.
13.80 $V = -\ln xyz$.
13.81 (a) $(k - 1)a^2/2$, not conservative. (b) 0, conservative.
13.82 (a) $P_x = x/R$, $P_y = y/R$, $P_z = z/R$, where
           $R = (x^2 + y^2 + z^2)^{1/2}$.
           (b) $U_{OABD} = -\Delta V_{OD} = a\sqrt{3}$.
13.85 (a) 62.5 MJ/kg. (b) 11.18 km/s.
13.86 (a) 9.56 km/s. (b) 2.39 km/s.
13.87 (a) 50.1×10^9 ft·lb. (b) 115.9×10^9 ft·lb.
13.88 (a) 1.918×10^6 ft·lb/lb. (b) 10.51×10^6 ft·lb/lb.
13.89 25.1 Mm/h.
13.90 6.48 km/s.
13.93 $v_r = \pm 3.87$ m/s, $v_\theta = 1.000$ m/s.
13.94 (a) 0.720 m. (b) 0.834 m/s.
13.95 3.77 in, $(28.04$ ft/s$)\mathbf{e}_r + (7.96$ ft/s$)\mathbf{e}_r$.
13.96 (a) 14.36 ft/s. (b) 1.225 ft.
13.97 (a) 4.14 ft/s. (b) 16.58 ft/s.
13.100 27.6×10^3 km/h.
13.101 (a) 7960 ft/s. (b) 4820 ft/s.
13.102 (a) 16 800 ft/s. (b) 32 700 ft/s.
13.103 14.20 km/s.

13.106 (a) 7.35 km/s. (b) 45.0°.

13.107 68.9°.

13.108 $r_{max} = r_0(1 + \sin \alpha)$, $r_{min} = (1 - \sin \alpha)r_0$.

13.109 3450 m/s.

13.110 (a) 11.32×10^3 ft/s. (b) 13.68×10^3 ft/s.

13.111 30.9×10^3 ft/s, 58.9°.

13.115 (b) $v_{esc}\sqrt{\alpha/(1 + \alpha)} < v_0 < v_{esc}\sqrt{(1 + \alpha)/(2 + \alpha)}$.

13.119 4 min 19 s.

13.120 (a) 3.64 s. (b) 27.3 s.

13.121 17.86 lb.

13.123 6.26 s.

13.124 (a) 2280 lb. (b) 3.00 s.

13.125 0.278.

13.126 (a) 18.16 s. (b) 1.94 km.

13.129 (a) 14.78 s. (b) 693 lb (tension).

13.130 (a) 29.6 s. (b) 2500 lb (tension).

13.131 (a) 5.28 s. (b) 17.05 kN (compression).

13.132 (a) 0.549 s. (b) 56.8 N.

13.134 (a) 3730 lb. (b) 7450 lb.

13.136 223 MPa.

13.138 15.36 mi/h.

13.139 76.9 lb.

13.140 1.449 kips.

13.141 6.21 W.

13.142 2.68 kN.

13.145 (a) 1.67 mi/h ←. (b) 0.190 s.

13.146 (a) car A. (b) 115.2 km/h.

13.147 65.0 kN.

13.148 (a) 9.32 ft·lb, 0.932 lb·s.
(b) 7.99 ft·lb, 0.799 lb·s.

13.149 497 ft/s.

13.150 (a) 2.80 ft/s ←. (b) 0.229 ft/s ←.

13.151 (a) 1.694 m/s ↓. (b) 0.1619 J.

13.152 (a) 778.9 m/s. (b) 4.65 J. (c) 19.74 N.

13.155 (a) $v_A = 0.594$ m/s ←, $v_B = 1.156$ m/s →. (b) 2.99 J.

13.156 $(1 - e^2)mv^2$.

13.157 $0.728 \le e \le 0.762$.

13.158 (a) 3.00 lb. (b) 2.00 lb $\le W_B \le$ 6.00 lb.

13.161 (a) $v_0(1 - e)/2$ and $v_0(1+e)/2$. (b) $v_0(1 - e)^2/4$ and $v_0(1 + e)^2/4$.
(c) $v_0(1 + e)^{n-1}/2^{n-1}$. (d) $0.698v_0$.

13.163 0.294 m/s ←.

13.164 $\mathbf{v}'_A = 0.711\ v_0 \measuredangle 39.3°$, $\mathbf{v}'_B = 0.636\ v_0 \measuredangle 45°$.

13.165 (a) $0.848v_0 \measuredangle 27.0°$. (b) $0.456v_0 \measuredangle 57.6°$.

13.166 $\mathbf{v}'_A = 6.37$ m/s $\measuredangle 77.2°$, $\mathbf{v}'_B = 1.802$ m/s $\measuredangle 40°$.

13.167 $\mathbf{v}'_A = 1.322$ m/s $\measuredangle 70.9°$, $\mathbf{v}'_B = 3.85$ m/s $\measuredangle 27.0°$.

13.168 (a) 70.2°. (b) 0.322 m/s.

13.169 0.837.

13.172 13.09 m/s $\measuredangle 26.6°$.

13.174 (a) 20.6 mi/h. (b) 0.203.

13.175 (a) 0.294 m. (b) 54.4 mm.

13.176 (a) 0.324. (b) 14.30 ft/s.

13.177 (a) 2.90 m/s. (b) 100.5 J.

13.179 (a) 8.89 mm. (b) 3758 N.

13.180 (a) 0.588. (b) 148.7 kN/m.

13.182 (a) $\mathbf{v}'_A = 0$, $\mathbf{v}'_B = 0$.
(b) $\mathbf{v}'_A = 1.201$ m/s ←, $\mathbf{v}'_B = 0.400$ m/s →.

13.183 45.5 mm.

13.184 (a) 26.65 ft/s $\measuredangle 30°$.
(b) 31.93 ft/s $\measuredangle 39.0°$.

13.185 3.47 in.

13.186 (a) 0.923. (b) 1.278 m.

13.188 (a) $\mathbf{v}'_A = 2.36$ ft/s $\measuredangle 83.8°$, $\mathbf{v}'_B = 3.23$ ft/s →. (b) 1.97 in.

13.190 102.6 mi/h.

13.191 1.688 ft·lb.

13.194 0.283.

13.195 (a) 13.31 N →. (b) 4.49 N ↓. (c) 13.31 N ←.

13.197 (a) 217 mm. (b) 69.1 mm.

13.198 (a) $v'_A = v'_B = v'_C = 1.368$ m/s. (b) 0.668 m. (c) 1.049 m.

13.200 0.107 m.

CHAPTER 14

14.1 (a) 4.46 m/s ←. (b) 0.409 m/s ←.

14.2 10.67 km/h ←, 4.27 km/h ←, and 4.27 km/h ←.

14.3 (a) 4.25 ft/s →. (b) 4.25 ft/s →.

14.4 (a) 0.800 oz. (b) 900 ft/s →.

14.7 (a) 3.79 km/h →, 2.77 km/h →.
(b) 5.54 km/h →, 2.77 km/h →.
(c) 5.54 km/h →, 3.60 km/h →.

14.8 $v_A = 1.013$ m/s ←, $v_B = 0.338$ m/s ←, $v_C = 0.150$ m/s ←.

14.9 $-(600\ \text{kg·m}^2/\text{s})\mathbf{i} - (1070.0\ \text{kg·m}^2/\text{s})\mathbf{j} + (370.0\ \text{kg·m}^2/\text{s})\mathbf{k}$

14.10 (a) $(22.78\ \text{m})\mathbf{i} + (15.00\ \text{m})\mathbf{j} + (11.67\ \text{m})\mathbf{k}$.
(b) $(38.0\ \text{kg·m/s})\mathbf{i} + (32.0\ \text{kg·m/s})\mathbf{j} + (40.0\ \text{kg·m/s})\mathbf{k}$.
(c) $-(826.67\ \text{kg·m}^2/\text{s})\mathbf{i} - (602.22\ \text{kg·m}^2/\text{s})\mathbf{j} + (211.11\ \text{kg·m}^2/\text{s})\mathbf{k}$.

14.11 (a) $\mathbf{v}_A = (4.00\ \text{ft/s})\mathbf{j}$, $\mathbf{v}_B = (1.000\ \text{ft/s})\mathbf{i}$, $v_C = (3.00\ \text{ft/s})\mathbf{k}$.
(b) $(1.20\ \text{ft·lb·s})\mathbf{i} + (0.60\ \text{ft·lb·s})\mathbf{j} - (2.40\ \text{ft·lb·s})\mathbf{k}$.

14.12 (a) $\mathbf{v}_A = (10.00\ \text{ft/s})\mathbf{j}$, $\mathbf{v}_B = (5.00\ \text{ft/s})\mathbf{i}$, $v_C = (10.00\ \text{ft/s})\mathbf{k}$.
(b) $(6.00\ \text{ft·lb·s})\mathbf{i} + (3.00\ \text{ft·lb·s})\mathbf{j} - (6.00\ \text{ft·lb·s})\mathbf{k}$.

14.15 $(114.4\ \text{m})\mathbf{j} - (76.1\ \text{m})\mathbf{j} + (8.75\ \text{m})\mathbf{k}$.

14.16 $(1180\ \text{m})\mathbf{i} + (140\ \text{m})\mathbf{j} + (155\ \text{m})\mathbf{k}$.

14.19 $x = 45.2$ ft, $y = 54.5$ ft.

14.20 (a) 2.00 s. (b) 92.8 mi/h.

14.21 $(81.5\ \text{ft})\mathbf{i} + (351\ \text{ft})\mathbf{k}$.

14.22 (a) 8.00 ft/s →. (b) 36.6°, $v_C = 10.39$ ft/s, $v_D = 8.72$ ft/s.

14.24 $v_A = 431$ m/s, $v_B = 395$ m/s, $v_C = 528$ m/s.

14.25 $v_A = 646$ m/s, $v_B = 789$ m/s, $v_C = 176$ m/s.

14.26 $v_A = 919$ m/s, $v_B = 717$ m/s, $v_C = 619$ m/s.

14.31 friction: 2.97 J, first impact: 3007 J, second impact: 24.3 J.

14.32 (a) 23.6 ft·lb. (b) 2.85 ft·lb.

14.33 (woman) 382 ft·lb, (man) 447 ft·lb.

14.35 (b) $E_A = 180.0$ kJ, $E_B = 320$ kJ.

14.37 (a) $\mathbf{v}_B = \dfrac{m_A v_0}{m_A + m_B} \rightarrow$. (b) $h = \dfrac{m_A}{m_A + m_B}\dfrac{v_0^2}{2g}$.

14.38 $\mathbf{v}_A = 4.11$ m/s $\measuredangle 46.9°$, $\mathbf{v}_B = 17.39$ m/s $\measuredangle 16.7°$.

14.39 (a) $\mathbf{v}_{B/A} = 11.59$ ft/s $\measuredangle 30°$. (b) $\mathbf{v}_A = 3.76$ ft/s →.

14.40 $\mathbf{v}_A = 3.11$ ft/s ←, $\mathbf{v}_B = 4.66$ ft/s →.

14.41 $v_A = 7.50$ ft/s, $v_B = 6.50$ ft/s, $v_C = 11.25$ ft/s.

14.42 $v_A = 10.61$ ft/s, $v_B = 9.19$ ft/s, $v_C = 5.30$ ft/s.

14.45 $v_A = 0.218$ m/s $\measuredangle 53.1°$ and $v_B = 1.813$ m/s $\measuredangle 43.8°$.

14.46 $(200\ \text{ft/s})\mathbf{i} + (172\ \text{ft/s})\mathbf{j} + (1560\ \text{ft/s})\mathbf{k}$.

14.47 (a) $v_C = 11.00$ ft/s, $v_D = 5.50$ ft/s. (b) 0.786.

14.48 $x = 181.7$ mm, $y = 0$, $z = 139.4$ mm.

14.51 (a) $\mathbf{v}_B = 2.40$ m/s $\measuredangle 53.1°$, $\mathbf{v}_C = 2.56$ m/s →, (b) $c = 1.059$ m.

14.52 (a) $\mathbf{v}_A = 2.40$ m/s ↓, $\mathbf{v}_B = 3.00$ m/s $\measuredangle 53.1°$, (b) $a = 1.864$ m.

14.55 (a) $\mathbf{v}_A = 2.25$ ft ↑, $\mathbf{v}_B = 2.25$ ft/s ↓, $\mathbf{v}_c = 3.90$ ft/s →. (b) 11.1 in.

14.56 (a) 2.00 ft/s →. (b) 0.760 ft. (c) 5.29 rad/s ↓.

14.57 1086.5 N.

14.58 $\rho A_2\ v_2^2 - \rho A_1\ v_1^2 \cos \theta$.

14.59 drag $= 26.3$ lb →, lift $= 12.74$ lb↑.

14.60 drag $= 34.6$ lb →, lift $= 16.76$ lb↑.

14.61 (a) 14.8 kN. (b) 27.7 kN.

14.62 90.6 N ←.

14.64 $D_x = 329$ N, $D_y = 0$, $C_x = -203$ N, $C_y = 271$ N.

14.66 (a) $\theta = 35.4°$. (b) 187.3 N $\measuredangle 53.8°$.

14.67 (a) 26.0 m/s. (b) 230 N $\nearrow$ 48.4°.
14.68 $C_x = 90.0$ N, $C_y = 2360$ N, $D_x = 0$, $D_y = 2900$ N.
14.69 100 kg/s.
14.70 7580 lb.
14.71 33.6 kN ←.
14.72 7180 lb.
14.74 (a) 9690 lb, 3.38 ft. (b) 6960 lb, 9.43 ft.
14.76 (a) 3.03 m/s² $\measuredangle$ 18°. (b) 922 km/h.
14.77 (a) 30.6 m/s. (b) 96.1 m³/s. (c) 55 100 N·m/s.
14.78 (a) 3.23 MW. (b) 0.464.
14.79 213 m.
14.80 (a) 15 450 hp. (b) 28 060 hp. (c) 0.551.
14.83 (a) $m_0 e^{qL/m_0v_0}$. (b) $v_0 e^{-qL/m_0v_0}$.
14.86 (a) $m(v^2 + gy)/l$. (b) $\mathbf{R} = mg(1 - y/l)\uparrow$.
14.87 (a) mgy/l. (b) $m[g(l - y) + v^2]/l \uparrow$.
14.88 $\sqrt{gh} \tan h(\sqrt{gh}\, t/L)$.
14.89 10.10 ft/s.
14.90 4.75 ft/s.
14.91 533 kg/s.
14.94 (a) 90.0 m/s². (b) 35.9×10^3 km/h.
14.95 7930 m/s.
14.96 (a) 1800 m/s. (b) 9240 m/s.
14.99 87.2 mi.
14.100 (a) 92.8 ft/s² ↑. (b) 780 ft/s² ↑. (c) 119.3 mi. (d) 14660 mi/h.
14.101 186.8 km/h.
14.102 (a) 31.2 km. (b) 197.5 km.
14.106 (a) 1.595 m/s. (b) 0.370 m.
14.107 (a) 5.20 km/h →. (b) 4.00 km/h →.
14.108 (a) 6.05 ft/s. (b) 6.81 ft/s.
14.110 $\mathbf{v}_A = 15.38$ ft/s →, $\mathbf{v}_B = 5.13$ ft/s ←.
14.112 $\mathbf{A}_x = 55.5$ lb →, $\mathbf{A}_y = 20.2$ lb ↓, $\mathbf{m}_A = 41.4$ lb·ft $\downarrow$.
14.114 $\mathbf{D} = 2.29$ kN ↑, $\mathbf{C} = 1.712$ kN ↑.
14.115 414 rpm.
14.116 Case 1: (a) 0.333 g ↓. (b) $0.817\sqrt{gl}$.
Case 2: (a) $gy/l \downarrow$. (b) $\sqrt{gl}$.

CHAPTER 15

15.1 (a) 29.6 rad/s. (b) 32.2 rev.
15.2 (a) 0.50 rad, −4.71 rad/s, −34.50 rad/s². (b) 0, −1.934 rad/s, 36.46 rad/s².
15.3 (a) 0.253 rad, −0.927 rad/s, −36.55 rad/s². (b) 0, 0, 0.
15.4 (a) −3.01 rad/s². (b) 13 800 rev.
15.5 (a) 150 rev. (b) 2100 rev.
15.6 (a) 0.855 rad/s. (b) 3.71°.
15.9 (a) 9.55 rev. (b) ∞. (c) 7.82 s.
15.10 −(0.450 m/s)**i** − (1.200 m/s)**j** + (1.500 m/s)**k**, (12.60 m/s²)**i** + (7.65 m/s²)**j** + (9.90 m/s²)**k**.
15.11 (0.750 m/s)**i** + (1.500 m/s)**k**, (12.75 m/s²)**i** + (11.25 m/s²)**j** + (3.00 m/s²)**k**.
15.12 −(37.4 in/s)**i** + (12.00 in/s)**j** − (15.60 in/s)**k**, −(126.1 in/s²)**i** − (74.3 in/s²)**j** + (246 in/s²)**k**.
15.13 −(18.72 in/s)**i** + (6.00 in/s)**j** − (7.80 in/s)**k**, −(3.46 in/s²)**i** − (27.6 in/s²)**j** + (73.1 in/s²)**k**.
15.16 66 700 mi/h, 19.47×10^{-3} ft/s².
15.17 (a) 1525 ft/s, 0.1112 ft/s². (b) 1168 ft/s, 0.0852 ft/s². (c) 0, 0.
15.18 (a) 2.50 rad/s ↖, 1.500 rad/s² $\downarrow$. (b) 771 mm/s² $\searrow$ 76.5°.
15.19 12.00 rad/s² ↖ or 12.00 rad/s² $\downarrow$.
15.22 left: 3.49 s; middle: 6.98 s; right: 13.96 s.
15.23 (a) 0.500 ft/s →, 1.500 ft/s² ←. (b) 4.24 ft/s² $\searrow$ 45°.

15.24 (a) 300 rpm ↖, 100 rpm $\downarrow$. (b) $\mathbf{a}_B = 1974$ in/s² ←, $\mathbf{a}_C = 658$ in/s² →.
15.25 (a) A: 15.00 rad/s ↖; B: 7.50 rad/s $\downarrow$. (b) A: 75.0 ft/s² ↑; B: 37.5 ft/s² ↓.
15.26 (a) C: 120 rpm; B: 275 rpm. (b) A: 23.7 m/s² ↑; B: 19.90 m/s² ↓.
15.27 (a) 10.00 rad/s. (b) A: 7.50 m/s²; B: 3.00 m/s² ↓. (c) 4.00 m/s² ↓.
15.28 (a) 0.400 rad/s² $\downarrow$. (b) 1.528 rev.
15.29 (a) 3.00 rad/s² $\downarrow$. (b) 4.00 s.
15.30 (a) 1.975 rad/s² ↖. (b) 6.91 rad/s ↖.
15.31 (a) 15.28 rev. (b) 10.14 s.
15.32 (a) 15.52 s. (b) $\omega_A = 445$ rpm ↖, $\omega_B = 371$ rpm $\downarrow$.
15.33 (a) $\boldsymbol{\alpha}_A = 3.40$ rad/s² $\downarrow$, $\boldsymbol{\alpha}_B = 1.963$ rad/s² $\downarrow$. (b) 9.23 s.
15.36 $b\omega_0^2/2\pi \rightarrow$.
15.37 $bv^2/2\pi r^3 \downarrow$.
15.38 $\mathbf{v}_B = 140.8$ ft/s →, $\mathbf{v}_C = 0$, $\mathbf{v}_0 = 136.0$ ft/s $\measuredangle$ 15°, $\mathbf{v}_E = 99.6$ ft/s $\searrow$ 45°.
15.39 (a) 0.378 rad/s $\downarrow$. (b) 6.42 in/s ↑.
15.40 (a) 0.231 rad/s $\downarrow$. (b) −(1.00 m/s)**i** − (0.577 m/s)**j**.
15.41 (a) 3.00 rad/s $\downarrow$. (b) 1.30 m/s $\nearrow$ 67.4°.
15.44 (a) 10.00 rad/s ↖. (b) −(7.40 m/s)**i** − (1.00 m/s)**j**.
15.45 (a) −(1.40 m/s)**i** − (1.00 m/s)**j**. (b) $x = 100.0$ mm, $y = -140.0$ mm.
15.47 (a) 0.583 rad/s $\downarrow$. (b) 1.537 ft/s $\searrow$ 77.48°.
15.48 (a) $\boldsymbol{\omega}_B = \boldsymbol{\omega}_C = \boldsymbol{\omega}_D = \frac{1}{2}\boldsymbol{\omega}_A$ ↖. (b) $\boldsymbol{\omega}_S = 0.25\,\boldsymbol{\omega}_A$ $\downarrow$.
15.49 (a) $\boldsymbol{\omega}_B = \boldsymbol{\omega}_C = \boldsymbol{\omega}_D = 150$ rpm $\downarrow$. (b) $\boldsymbol{\omega}_S = 195$ rpm $\downarrow$.
15.50 (a) 48.0 rad/s $\downarrow$. (b) 3.39 m/s $\measuredangle$ 45°.
15.51 (a) 5.65 m/s ↑. (b) 9000 rpm, (c) 1500.
15.53 (a) 200 rad/s ↖. (b) 24.0 rad/s $\downarrow$.
15.55 (a) (6.00 rad/s)**k** or 6.00 rad/s ↖. (b) (360 mm/s)**i** −(672 mm/s)**j** or 762 mm/s $\searrow$ 61.8°.
15.56 (a) 540 mm/s →. (b) 457 mm/s $\searrow$ 61.8°.
15.57 (a) 4.38 rad/s $\downarrow$, 12.25 in/s ↑. (b) 0, 42.0 in/s ↓. (c) 4.38 rad/s ↖, 12.25 in/s ↓.
15.58 (a) 22.9° and 192.6°. (b) 5.60 rad/s $\downarrow$ and 5.60 rad/s ↖.
15.61 (a) $\mathbf{v}_P = 0$, $\boldsymbol{\omega}_{BD} = 39.3$ rad/s ↖. (b) $\mathbf{v}_P = 6.28$ m/s ↓, $\boldsymbol{\omega}_{BD} = 0$.
15.62 $\mathbf{v}_P = 6.52$ m/s $\downarrow$, $\boldsymbol{\omega}_{BD} = 20.8$ rad/s ↖.
15.63 (a) 12.00 rad/s ↖. (b) 3.90 m/s $\nearrow$ 67.4°.
15.64 $\boldsymbol{\omega}_{DE} = 2.55$ rad/s $\downarrow$, $\boldsymbol{\omega}_{BD} = 0.955$ rad/s ↖.
15.65 $\boldsymbol{\omega}_{BD} = 4.00$ rad/s ↖, $\boldsymbol{\omega}_{EB} = 0.600$ rad/s ↖.
15.68 (a) 3.33 rad/s ↖. (b) 2.00 m/s $\searrow$ 56.3°.
15.69 (a) 1.500 m. (b) 5.00 m/s ↓.
15.70 14.76 in/s →.
15.71 (a) 338 mm/s ←, 0. (b) 710 mm/s ←, 2.37 rad/s $\downarrow$.
15.72 $(1 - r_A/r_C)\omega_{ABC}$.
15.74 (a) 1.714 in. below A. (b) 75.0 ft/s →. (c) 53.2 ft/s $\measuredangle$ 41.2°.
15.75 $x = 0$, $z = 9.34$ ft.
15.76 (a) 3.00 rad/s ↖. (b) 300 mm/s ←. (c) 180.0 mm/s (wound).
15.77 (a) 3.00 rad/s $\downarrow$. (b) 180 mm/s →. (c) 300 mm/s (unwound).
15.78 (a) 50 mm to the right of the axle. (b) $\mathbf{v}_B = 750$ mm/s ↓, $\mathbf{v}_D = 1.950$ m/s ↑.
15.79 (a) 25 mm to the right of 0. (b) 420 mm/s ↑.
15.80 (a) A: 300 mm to the left of A. C: 600 mm to the left of C. (b) $\boldsymbol{\omega}_A = 4.00$ rad/s $\downarrow$, $\boldsymbol{\omega}_C = 2.00$ rad/s ↖.
15.82 (a) 0.467 rad/s ↖. (b) 3.49 ft/s $\measuredangle$ 59.2°.
15.83 (a) 3.08 rad/s ↖. (b) 83.3 ft/s $\searrow$ 73.9°.
15.86 (a) 0.122 rad/s ↖. (b) 22.76 mm/s $\measuredangle$ 15°.
15.87 (a) 0.133 rad/s ↖. (b) 18.22 mm/s $\measuredangle$ 15°.
15.88 (a) $(v_A/l) \sin \beta/\cos (\beta - \theta)$. (b) $v_A \cos \theta/\cos(\beta - \theta)$.

15.89 (*a*) 6.72 ft/s $\measuredangle$ 45°. (*b*) 2.75 rad/s $\downarrow$. (*c*) 6.57 ft/s $\measuredangle$ 21.2°.

15.90 (*a*) 0.900 rad/s $\downarrow$. (*b*) 411 mm/s $\searangle$ 20.5°.

15.91 (*a*) 1.00 rad/s $\downarrow$. (*b*) 1.04 m/s →.

15.94 (*a*) 1.58 rad/s $\downarrow$. (*b*) 28.0 in/s $\measuredangle$ 78.3°.

15.95 (*a*) $\boldsymbol{\omega}_{AB}$ = 1.200 rad/s $\downarrow$, $\boldsymbol{\omega}_{DE}$ = 0.450 rad/s $\downarrow$.
   (*b*) 5.25 in/s ←.

15.96 (*a*) 5.00 rad/s $\nwarrow$. (*b*) 3.00 m/s $\downarrow$.

15.97 (*a*) 2.49 rad/s $\nwarrow$. (*b*) 3.73 rad/s $\downarrow$. (*c*) 0.835 m/s $\searrow$ 53.6°.

15.98 (*a*) $\boldsymbol{\omega}_{AB}$ = 1.177 rad/s $\downarrow$, $\boldsymbol{\omega}_{DE}$ = 2.50 rad/s $\downarrow$.
   (*b*) 29.4 in/s ←.

15.99 Space centrode: quarter circle, r = 15 in, centered at *O*. Body
   centrode: semicircle, r = 7.5 in., centered midway between
   A and *B*.

15.100 Space centrode: lower rack.
   Body centrode: circumference of gear.

15.102 $\boldsymbol{\omega}_{BD}$ = 0.955 rad/s $\downarrow$, $\boldsymbol{\omega}_{DE}$ = 2.55 rad/s $\nwarrow$.

15.103 $\boldsymbol{\omega}_{BD}$ = 4.000 rad/s $\nwarrow$, $\boldsymbol{\omega}_{EB}$ = 0.600 rad/s $\nwarrow$.

15.105 (*a*) 0.50 rad/s² $\downarrow$. (*b*) $\mathbf{a}_A$ = 3.25 m/s² $\uparrow$, $\mathbf{a}_E$ = 0.75 m/s² $\uparrow$.

15.106 (*a*) 0.20 m/s² $\downarrow$. (*b*) 2.20 m/s² $\uparrow$.

15.107 (*a*) 0.900 m/s² →. (*b*) 1.800 m/s² ←.

15.108 (*a*) 0.600 m from *A*. (*b*) 0.200 m from *A*.

15.109 (*a*) 51.3 in/s² $\downarrow$. (*b*) 184.9 in/s² $\measuredangle$ 16.1°.

15.110 (*a*) 1.039 rad/s² $\downarrow$. (*b*) (2.60 ft/s²)**i** + (4.50 ft/s²)**j** or
   5.20 ft/s² $\measuredangle$ 60°.

15.111 (*a*) 1430 m/s² $\downarrow$. (*b*) 1430 m/s² $\uparrow$, (*c*) 1430 m/s² $\searangle$ 60°.

15.112 (*a*) 13.35 in/s² $\nearrow$ 61.0°. (*b*) 12.62 in/s² $\measuredangle$ 64.0°.

15.113 $\mathbf{a}_A$ = 56.6 in/s² $\searrow$ 58.0°, $\mathbf{a}_B$ = 80.0 in/s² $\uparrow$,
   $\mathbf{a}_C$ = 172.2 in/s² $\searrow$ 25.8°.

15.114 $\mathbf{a}_A$ = 48.0 in/s² $\uparrow$, $\mathbf{a}_B$ = 85.4 in/s² $\searrow$ 69.4°.
   $\mathbf{a}_C$ = 82.8 in/s² $\nearrow$ 65.0°.

15.115 (*a*) 2.00 rad/s² $\downarrow$. (*b*) 0.224 m/s² $\searangle$ 63.4°.

15.118 (*a*) 92.5 in/s². (*b*) 278 in/s².

15.120 148.3 m/s² $\downarrow$.

15.121 296 m/s² $\uparrow$.

15.122 $\mathbf{a}_D$ = 1558 m/s² $\searangle$ 45°. $\mathbf{a}_E$ = 337 m/s² $\measuredangle$ 45°.

15.124 (*a*) 242 in/s² ←. (*b*) 403 in/s² $\nearrow$ 72.5°.

15.125 694 in/s² ←.

15.127 2.10 m/s² $\measuredangle$ 47.1°.

15.128 (*a*) 1.47 rad/s² $\nwarrow$. (*b*) 1.575 m/s² $\nearrow$ 47.1°.

15.129 (*a*) 228 rad/s² $\nwarrow$. (*b*) 92.0 m/s² $\downarrow$.

15.130 (*a*) 138.1 ft/s² $\searrow$ 78.6°. (*b*) 203 ft/s² $\measuredangle$ 19.5°.

15.132 (*a*) 4.18 rad/s² $\downarrow$. (*b*) 2.43 rad/s² $\downarrow$.

15.133 (*a*) 8.15 rad/s² $\nwarrow$. (*b*) 0.896 rad/s² $\downarrow$.

15.134 (*a*) 3.70 rad/s² $\downarrow$. (*b*) 3.70 rad/s² $\downarrow$.

15.136 $\mathbf{v}_D$ = 1.382 m/s $\downarrow$. $\mathbf{a}_D$ = 0.695 m/s² $\downarrow$.

15.138 $\mathbf{v}_B$ = $b\omega$ cos θ, a_B = $b\alpha$ cos θ − $b\omega^2$ sin θ.

15.139 v_B sin β/l cos θ.

15.140 (v_B sin β/l)² (sin θ/cos³ θ).

15.141 v_x = v[1 − cos (vt/r)]. v_y = v sin (vt/r).

15.142 $\boldsymbol{\omega}$ = $bv_A(b^2 + x_A^2)$ $\nwarrow$, $\boldsymbol{\alpha}$ = $2bx_A\,v_A^2/(b^2 + x_A^2)^2$ $\nwarrow$.

15.143 v_{B_x} = v_A − l$b^2 v_A/(b^2 + x_A^2)^{3/2}$ →, $(v_B)_y$ = l$b\,x_A v_A/(b^2 + x_A^2)^{3/2}$ $\uparrow$.

15.144 $\boldsymbol{\omega}_{BD}$ = $b\omega(b + l$ cos $\theta)/(l^2 + b^2 + 2bl$ cos $\theta)$ $\downarrow$,
   $\mathbf{v}_E$ = $bl\omega$ sin $\theta/(l^2 + b^2 + 2bl$ cos $\theta)$ $\searangle$
   tan^{-1}[(b sin $\theta/(l + b$ cos $\theta)$]

15.145 $bl\omega^2(l^2 − b^2)$ sin $\theta/(l^2 + b^2 + 2bl$ cos $\theta)^2$ $\nwarrow$.

15.147 $\boldsymbol{\omega}$ = v_0 sin² θ/r cos θ $\nwarrow$, $\boldsymbol{\alpha}$ = $(v_0/r)^2$ (1 + cos² θ) tan³ θ $\nwarrow$.

15.148 $(v_\rho)_x$ = $r\omega\left[\cos\dfrac{r\omega t}{R − r} − \cos \omega t\right]$,

   $(v_\rho)_y$ = $r\omega\left[\sin\dfrac{r\omega t}{R − r} + \sin \omega t\right]$.

15.149 Path is the y axis. $\mathbf{v}$ = ($R\omega$ sin ωt)**j**,
   $\mathbf{a}$ = ($R\omega^2$ cos ωt)**j**.

15.150 2.40 m/s $\searangle$ 73.9°.

15.151 2.87 m/s $\searangle$ 44.8°.

15.152 (*a*) 1.815 rad/s $\downarrow$. (*b*) 16.42 in/s $\searangle$ 20°.

15.153 (*a*) 5.16 rad/s $\downarrow$. (*b*) 1.399 m/s $\searangle$ 60°.

15.154 (*a*) 3.81 rad/s $\downarrow$, 6.53 m/s $\measuredangle$ 16.26°.
   (*b*) 3.00 rad/s $\downarrow$, 4.00 m/s →.

15.155 (*a*) 11.25 rad/s $\nwarrow$. (*b*) 75.0 in/s →.

15.160 (*a*) 1.78 × 10^{-3} m/s² west. (*b*) 1.36 × 10^{-3} m/s² west.
   (*c*) 1.36 × 10^{-3} m/s² west.

15.161 (*a*) 54 rad/s² $\downarrow$. (*b*) 33.9 ft/s² $\measuredangle$ 45°.

15.162 0.0234 m/s² west.

15.164 (*a*) 0.520 m/s $\searangle$ 82.6°. (*b*) 50.0 mm/s² $\searangle$ 9.8°.

15.165 (*a*) 0.520 m/s $\searangle$ 37.4°. (*b*) 50.0 mm/s² $\nearrow$ 69.8°.

15.166 (*a*) 1006 mm/s $\measuredangle$ 72.6°. (*b*) 1811 mm/s² $\measuredangle$ 32.0°.

15.167 (*a*) 1018 mm/s $\searangle$70.5°. (*b*) 1537 mm/s² $\nearrow$ 2.4°.

15.168 (1) 303 mm/s² →; (2) 168.5 mm/s² $\nearrow$ 57.7°.

15.169 (3) 483 mm/s² ←; (4) 168.5 mm/s² $\searangle$ 57.7°.

15.170 0.750 m/s $\measuredangle$ 71.3°, 2.13 m/s² $\nearrow$ 61.9°.

15.171 2.79 rad/s $\downarrow$, 2.13 rad/s² $\downarrow$.

15.174 (*a*) 0.436 rad/s $\nwarrow$. (*b*) 0.271 rad/s² $\nwarrow$.

15.175 (*a*) 0.354 rad/s $\nwarrow$. (*b*) 0.125 rad/s² $\nwarrow$.

15.176 7.86 rad/s $\nwarrow$, 81.1 rad/s² $\nwarrow$.

15.177 3.81 rad/s $\downarrow$, 81.4 rad/s² $\downarrow$.

15.178 1.526 rad/s $\downarrow$, 57.6 rad/s² $\downarrow$.

15.181 (*a*) 3.61 rad/s $\nwarrow$. (*b*) 86.6 in/s $\measuredangle$ 30°. (*c*) 563 in/s² $\nearrow$ 46.1°.

15.182 (*a*) 3.61 rad/s $\downarrow$. (*b*) 86.6 in/s $\measuredangle$ 30°. (*c*) 563 in/s² $\nearrow$ 46.1°.

15.183 51.5 m/s² $\searangle$ 44.4°.

15.184 (*a*) (33.0 rad/s)**i** − (44.0 rad/s)**k**. (*b*) (4.80 m/s)**i** + (3.60 m/s)**k**.

15.185 (*a*) (44.0 rad/s)**i** − (33.0 rad/s)**k**. (*b*) (3.60 m/s)**i** + (4.80 m/s)**k**.

15.186 (*a*) (1.5 rad/s)**i** − (3.5 rad/s)**j** − (3.0 rad/s)**k**.
   (*b*) (640 mm/s)**i** − (360 mm/s)**j** + (740 mm/s)**k**.

15.187 (*a*) (0.60 rad/s)**i** − (2.00 rad/s)**j** + (0.75 rad/s)**k**. (*b*) (20.0 in/s)**i**
   + (15.0 in/s)**j** + (24.0 in/s)**k**.

15.188 (118.4 rad/s²)**i**.

15.189 (230 rad/s²)**i** − (2.5 rad/s²)**k**.

15.190 (*a*) (6.28 rad/s²)**i**. (*b*) (8.38 rad/s²)**k**.

15.193 (*a*) −(0.600 m/s)**i** + (0.750 m/s)**j** − (0.600 m/s)**k**.
   (*b*) −(6.15 m/s)**i** − (3.00 m/s)**j**.

15.195 (*a*) −(20.0 rad/s²)**j**. (*b*) −(4.00 ft/s²)**i** + (10.00 ft/s²)**k**.
   (*c*) −(10.25 ft/s²)**j**.

15.196 −(3.46 ft/s²)**i** − (5.13 ft/s²)**j** + (8.66 ft/s²)**k**.

15.197 (*a*) ω_1 / sinβ. (*b*) ω_1 / tan β**i**. (*c*) ω_1^2 / tan β**k**.

15.198 (*a*) (0.0375 rad/s)**i**.
   (*b*) −(0.1434 m/s)**i** + (0.204 m/s)**j** − (0.1228 m/s)**k**.
   (*c*) −(0.696 m/s²)**i** − (0.0358 m/s²)**j** + (0.0430 m/s²)**k**.

15.199 (*a*) (28.4 rad/s)**i** + (5.24 rad/s)**j**. (*b*) (25.8 rad/s)**i**.

15.200 (*a*) (135.1 rad/s²)**k**. (*b*) (5.77 m/s²)**i**. − (232 m/s²)**j**.

15.203 −(33.3 in/s)**j**.

15.204 (15.0 in/s)**j**.

15.205 −(34.5 mm/s)**i**.

15.206 −(30.0 in/s)**j**.

15.207 (45.7 in/s)**j**.

15.210 (ω_2/cos 25°) (−sin 25°**i** + cos 25°)**k**.

15.211 (ω_1 cos 25°) (−sin 25°**i** + cos 25°**k**).

15.212 (*a*) (1.463 rad/s)**i** + (0.1052 rad/s)**j** + (0.0841 rad/s)**k**.
   (*b*) −(1.725 in/s)**i**.

15.213 (*a*) −(4.15 rad/s)**i** + (0.615 rad/s)**j** − (2.77 rad/s)**k**.
   (*b*) (0.30 m/s)**k**.

15.216 −(45.0 in/s²)**j**.

15.217 (205 in/s²)**j**.

15.218 −(9.51 mm/s²)**j**.

15.219 −(8.76 mm/s²)**j**.

15.220 (a) $(-3.00 \text{ ft/s})\mathbf{i} + (6.00 \text{ ft/s})\mathbf{j} - (20.94 \text{ ft/s})\mathbf{k}$.
(b) $(6.28 \text{ rad/s}^2)\mathbf{i}$.
(c) $(-62.87 \text{ ft/s}^2)\mathbf{i} - (9.00 \text{ ft/s}^2)\mathbf{j} + (12.57 \text{ ft/s}^2)\mathbf{k}$.

15.221 (a) $-(24.94 \text{ ft/s})\mathbf{k}$. (b) $(1.00 \text{ rad/s}^2)\mathbf{j} + (8.38 \text{ rad/s}^2)\mathbf{k}$.
(c) $(-60.62 \text{ ft/s}^2)\mathbf{i} - (16.00 \text{ ft/s}^2)\mathbf{j} - (10.00 \text{ ft/s}^2)\mathbf{k}$.

15.222 (a) $-(1.215 \text{ m/s})\mathbf{i} + (1.620 \text{ m/s})\mathbf{k}$. (b) $-(30.4 \text{ m/s}^2)\mathbf{j}$.

15.223 (a) $-(1.215 \text{ m/s})\mathbf{i} - (1.080 \text{ m/s})\mathbf{j} + (1.620 \text{ m/s})\mathbf{k}$.
(b) $(19.44 \text{ m/s}^2)\mathbf{i} - (30.4 \text{ m/s}^2)\mathbf{j} - (12.96 \text{ m/s}^2)\mathbf{k}$.

15.224 (a) $(1.200 \text{ m/s})\mathbf{i} + (0.500 \text{ m/s})\mathbf{j} - (1.200 \text{ m/s})\mathbf{k}$.
(b) $-(7.20 \text{ m/s}^2)\mathbf{i} - (14.40 \text{ m/s}^2)\mathbf{k}$.

15.227 (a) $(0.750 \text{ m/s})\mathbf{i} + (1.299 \text{ m/s})\mathbf{j} - (1.732 \text{ m/s})\mathbf{k}$.
(b) $(27.1 \text{ m/s}^2)\mathbf{i} + (5.63 \text{ m/s}^2)\mathbf{j} - (15.00 \text{ m/s}^2)\mathbf{k}$.

15.228 (a) $(129.9 \text{ mm/s})\mathbf{i} + (75.0 \text{ mm/s})\mathbf{j} + (86.6 \text{ mm/s})\mathbf{k}$.
(b) $(45.0 \text{ mm/s}^2)\mathbf{i} - (112.6 \text{ mm/s}^2)\mathbf{j} + (60.0 \text{ mm/s}^2)\mathbf{k}$.

15.230 $\mathbf{v}_C = -(45.0 \text{ in/s})\mathbf{i} + (36.6 \text{ in/s})\mathbf{j} - (31.2 \text{ in/s})\mathbf{k}$,
$\mathbf{a}_C = -(303 \text{ in/s}^2)\mathbf{i} - (384 \text{ in/s}^2)\mathbf{j} + (208 \text{ in/s}^2)\mathbf{k}$.

15.231 (a) $\omega_1 + (R/r)(\omega_1 - \omega_2)\mathbf{k}$. (b) $\omega_1(\omega_1 - \omega_2)(R/r)\mathbf{j}$.

15.232 $-(41.6 \text{ in/s}^2)\mathbf{i} - (61.5 \text{ in/s}^2)\mathbf{j} + (103.9 \text{ in/s}^2)\mathbf{k}$.

15.233 (a) $(0.0375 \text{ rad/s}^2)\mathbf{i}$.
(b) $-(0.143 \text{ m/s})\mathbf{i} + (0.205 \text{ m/s})\mathbf{j} - (0.123 \text{ m/s})\mathbf{k}$.
(c) $-(0.0696 \text{ m/s}^2)\mathbf{i} - (0.0358 \text{ m/s}^2)\mathbf{j} + (0.0430 \text{ m/s}^2)\mathbf{k}$.

15.234 $\mathbf{v}_A = -(1.39 \text{ m/s})\mathbf{i} + (0.80 \text{ m/s})\mathbf{j} - (1.20 \text{ m/s})\mathbf{k}$,
$\mathbf{a}_A = -(20.8 \text{ m/s}^2)\mathbf{i} - (11.09 \text{ m/s}^2)\mathbf{j} + (33.3 \text{ m/s}^2)\mathbf{k}$.

15.235 $\mathbf{v}_A = -(1.39 \text{ m/s})\mathbf{i} + (0.80 \text{ m/s})\mathbf{j} - (1.20 \text{ m/s})\mathbf{k}$,
$\mathbf{a}_A = -(22.5 \text{ m/s}^2)\mathbf{i} - (10.09 \text{ m/s}^2)\mathbf{j} + (34.9 \text{ m/s}^2)\mathbf{k}$.

15.236. (a) $-(1.37 \text{ ft/s})\mathbf{i} + (3.76 \text{ ft/s})\mathbf{j} + (1.88 \text{ ft/s})\mathbf{k}$.
(b) $(1.22 \text{ ft/s}^2)\mathbf{i} - (0.342 \text{ ft/s}^2)\mathbf{j} - (0.410 \text{ ft/s}^2)\mathbf{k}$.

15.239 (a) $(4.33 \text{ ft/s})\mathbf{i} - (6.18 \text{ ft/s})\mathbf{j} + (5.30 \text{ ft/s})\mathbf{k}$.
(b) $(2.65 \text{ ft/s}^2)\mathbf{i} - (2.64 \text{ ft/s}^2)\mathbf{j} - (3.25 \text{ ft/s}^2)\mathbf{k}$.

15.240 (a) $(27.2 \text{ in/s})\mathbf{i} - (6.75 \text{ in/s})\mathbf{j}$.
(b) $(12.80 \text{ in/s}^2)\mathbf{i} - (7.68 \text{ in/s}^2)\mathbf{k}$.

15.241 (a) $-(1.600 \text{ in/s})\mathbf{i} + (6.75 \text{ in/s})\mathbf{j}$.
(b) $(12.80 \text{ in/s}^2)\mathbf{i} + (7.68 \text{ in/s}^2)\mathbf{k}$.

15.242 $-(5.04 \text{ m/s})\mathbf{i} - (1.200 \text{ m/s})\mathbf{k}$.
$-(9.60 \text{ m/s}^2)\mathbf{i} - (25.9 \text{ m/s}^2)\mathbf{j} + (57.6 \text{ m/s}^2)\mathbf{k}$.

15.243 $-(0.720 \text{ m/s})\mathbf{i} - (1.200 \text{ m/s})\mathbf{k}$,
$-(9.60 \text{ m/s}^2)\mathbf{i} + (25.9 \text{ m/s}^2)\mathbf{j} - (11.52 \text{ m/s}^2)\mathbf{k}$.

15.244 (a) $r\omega_2^2 \sin 30°\mathbf{j} - (r\omega_2^2 \cos 30° + 2r\omega_1\omega_2)\mathbf{k}$.
(b) $-r(\omega_1^2 + \omega_2^2 + 2\omega_1\omega_2 \cos 30°)\mathbf{i} + r\omega_1^2 \cos 30°\mathbf{k}$.
(c) $-r\omega_2^2 \sin 30°\mathbf{j} + r(2\omega_1^2 \cos 30° + \omega_2^2 \cos 30° + 2\omega_1\omega_2)\mathbf{k}$.

15.245 (a) $(0.610 \text{ m/s})\mathbf{k}, -(0.880 \text{ m/s}^2)\mathbf{i} + (1.170 \text{ m/s}^2)\mathbf{j}$.
(b) $(5.20 \text{ m/s})\mathbf{i} - (0.390 \text{ m/s})\mathbf{j} - (1.000 \text{ m/s})\mathbf{k}$,
$-(4.00 \text{ m/s}^2)\mathbf{i} - (3.25 \text{ m/s}^2)\mathbf{k}$.

15.248 $(36.0 \text{ ft/s})\mathbf{i} - (64.0 \text{ ft/s})\mathbf{j}$.

15.249 (a) $5.00 \text{ ft/s}^2 \rightarrow$. (b) $5.63 \text{ in} \leftarrow$.

15.252 $\boldsymbol{\alpha}_{BD} = 306 \text{ rad/s}^2 \;\nwarrow, \boldsymbol{\alpha}_{DE} = 737 \text{ rad/s}^2 \;\nwarrow$.

15.253 (a) $1080 \text{ rad/s}^2 \;\downarrow$. (b) $460 \text{ ft/s}^2 \;\searrow\; 64.9°$.

15.255 $49.4 \text{ m/s}^2 \;\nwarrow\; 26.0°$.

15.256 (a) $(0.450 \text{ m/s})\mathbf{k}, (4.05 \text{ m/s}^2)\mathbf{i}$. (b) $-(1.350 \text{ m/s})\mathbf{k}, -(6.75 \text{ m/s}^2)\mathbf{i}$.

15.258 $(40.0 \text{ in/s})\mathbf{k}$.

15.259 $(9.00 \text{ in/s})\mathbf{i} - (7.80 \text{ in/s})\mathbf{j} + (7.20 \text{ in/s})\mathbf{k}$,
$(9.00 \text{ in/s})\mathbf{i} - (22.1 \text{ in/s})\mathbf{j} - (5.76 \text{ in/s})\mathbf{k}$.

CHAPTER 16

16.1 (a) $R_A = 60.31 \text{ lb} \;\measuredangle\; 84.2°$ and $N_B = 28.5 \text{ lb} \leftarrow$.
(b) $\mu = 0.1023$.

16.2 (a) $18.59 \text{ ft/s}^2 \rightarrow$. (b) 0.577.

16.3 (a) 25.8 ft/s^2. (b) 12.27 ft/s^2. (c) 13.32 ft/s^2.

16.4 (a) 3.20 m/s^2. (b) $\mathbf{A} = 3.82 \text{ N} \uparrow, \mathbf{B} = 20.7 \text{ N} \uparrow$.

16.5 (a) 4.09 m/s^2. (b) 42.5 N.

16.6 (a) $5270 \text{ N} \uparrow$. (b) 4120 N.

16.9 (a) $5.00 \text{ m/s}^2 \rightarrow$. (b) $0.311 \text{ m} \leq h \leq 1.489 \text{ m}$.

16.10 (a) $2.55 \text{ m/s}^2 \rightarrow$. (b) $h \leq 1.047 \text{ m}$.

16.11 195.9 kg.

16.12 229 N.

16.14 (a) $4.91 \text{ m/s}^2 \;\nwarrow\; 30°$. (b) $F_A = 0, F_B = 68.0 \text{ N compression}$.

16.15 (a) $173.2 \text{ N} \rightarrow$. (b) 15.02 rad/s. (c) $86.6 \text{ rad/s}^2 \;\nwarrow$.

16.18 $B_y = 16.48 \text{ lb}$ and $D_y = 17.62 \text{ lb}$.

16.19 (a) $30.6 \text{ ft/s}^2 \;\nwarrow\; 84.1°$. (b) $\mathbf{A} = 0.505 \text{ lb} \;\measuredangle\; 30°$,
$\mathbf{B} = 1.285 \text{ lb} \;\measuredangle\; 30°$.

16.20 Block: $17.01 \text{ ft/s}^2 \;\nwarrow\; 58.5°$; platform: $31.3 \text{ ft/s}^2 \;\nwarrow\; 30°$.

16.25 125.7 N-m.

16.26 9480 rev.

16.27 93.5 rev.

16.28 107.6 rev.

16.29 74.5 s.

16.30 $20.4 \text{ rad/s}^2 \;\downarrow$.

16.31 $32.7 \text{ rad/s}^2 \;\nwarrow$.

16.33 (a) $5.66 \text{ ft/s}^2 \;\downarrow$. (b) $7.52 \text{ ft/s} \;\downarrow$.

16.34 (1): (a) $8.00 \text{ rad/s}^2 \;\nwarrow$. (b) $14.61 \text{ rad/s} \;\nwarrow$.
(2): (a) $6.74 \text{ rad/s}^2 \;\nwarrow$. (b) $13.41 \text{ rad/s} \;\nwarrow$.
(3): (a) $4.24 \text{ rad/s}^2 \;\nwarrow$. (b) $10.64 \text{ rad/s} \;\nwarrow$.
(4): (a) $5.83 \text{ rad/s}^2 \;\nwarrow$. (b) $8.82 \text{ rad/s} \;\nwarrow$.

16.36 (a) $6.06 \text{ rad/s}^2 \;\downarrow$. (b) $11.28 \text{ N} \;\nearrow$.

16.39 (a) No slipping on A; slipping on B.
(b) $\boldsymbol{\alpha}_A = 61.8 \text{ rad/s}^2 \;\nwarrow; \boldsymbol{\alpha}_B = 9.66 \text{ rad/s}^2 \;\downarrow$.

16.40 (a) No slipping at either cylinder.
(b) $\boldsymbol{\alpha}_A = 15.46 \text{ rad/s}^2 \;\nwarrow, \boldsymbol{\alpha}_B = 7.73 \text{ rad/s}^2 \;\downarrow$.

16.41 (a) $\boldsymbol{\alpha}_A = 12.50 \text{ rad/s}^2 \;\nwarrow, \boldsymbol{\alpha}_B = 33.3 \text{ rad/s}^2 \;\nwarrow$.
(b) $\boldsymbol{\omega}_A = 240 \text{ rpm} \;\downarrow, \boldsymbol{\omega}_B = 320 \text{ rpm} \;\nwarrow$.

16.42 (a) $\boldsymbol{\alpha}_A = 12.50 \text{ rad/s}^2 \;\nwarrow, \boldsymbol{\alpha}_B = 33.3 \text{ rpm} \;\nwarrow$.
(b) $\boldsymbol{\omega}_A = 90.0 \text{ rpm} \;\nwarrow, \boldsymbol{\omega}_B = 120.0 \text{ rpm} \;\downarrow$.

16.43 (a) $\boldsymbol{\alpha}_A = 9.16 \text{ rad/s}^2 \;\nwarrow, \boldsymbol{\alpha}_B = 38.2 \text{ rad/s}^2 \;\nwarrow$.
(b) $\mathbf{C} = 54.9 \text{ N} \uparrow, \mathbf{M}_C = 2.64 \text{ N·m} \;\nwarrow$.

16.44 (b) $\omega_0/(1 + m_B/m_A) \;\downarrow$.

16.48 (a) $18.40 \text{ ft/s}^2 \rightarrow$. (b) $9.20 \text{ ft/s}^2 \leftarrow$. (c) $z = 24.0 \text{ in}$.

16.49 (a) $12.0 \text{ in. from end } A$. (b) $9.20 \text{ ft/s}^2 \rightarrow$.

16.50 (a) $2.50 \text{ m/s}^2 \rightarrow$. (b) 0.

16.51 (a) $3.75 \text{ m/s}^2 \rightarrow$. (b) $1.25 \text{ m/s}^2 \leftarrow$.

16.52 (a) $0, -1.374 \text{ rad/s}^2 \mathbf{j}$. (b) $-(0.515 \text{ ft/s}^2)\mathbf{i}, -1.030 \text{ rad/s}^2 \mathbf{j}$.

16.55 $\mathbf{a}_A = 2.71 \text{ m/s}^2 \uparrow$ and $\mathbf{a}_B = 1.496 \text{ m/s}^2 \uparrow$.

16.56 170.9 mm.

16.57 (a) $53.1 \text{ rad/s}^2 \;\downarrow$. (b) $\mathbf{a} = 39.3 \text{ ft/s}^2 \;\downarrow$.

16.58 (a) $0.741 \text{ rad/s}^2 \;\nwarrow$. (b) 0.857 m/s^2.

16.59 (a) 2800 N. (b) $15.11 \text{ rad/s}^2 \;\downarrow$.

16.60 $T_A = 359 \text{ lb}, T_B = 312 \text{ lb}$.

16.63 (a) $\dfrac{3g}{2L} \;\downarrow$. (b) $\dfrac{g}{4} \uparrow$. (c) $\dfrac{5g}{4} \;\downarrow$.

16.64 (a) $\dfrac{2g}{L} \;\downarrow$. (b) $\dfrac{g}{3} \uparrow$. (c) $\dfrac{5g}{3} \;\downarrow$.

16.65 (a) $\dfrac{3g}{L} \;\downarrow$. (b) $1.323 g \;\measuredangle\; 49.1°$. (c) $2.18 g \;\nwarrow\; 66.6°$.

16.66 (a) $0.25 g \uparrow$. (b) $5 g/4 \;\downarrow$.

16.67 (a) 0. (b) $g \downarrow$.

16.69 (a) $5v_0/2r \;\nwarrow$. (b) $v_0/\mu_k g$. (c) $v_0^2/2\mu_k g$.

16.70 (a) $v_0/r \;\nwarrow$. (b) $v_0/\mu_k g$. (c) $v_0^2/2\mu_k g$.

16.71 (a) 1.597 s. (b) 9.86 ft/s. (c) 19.85 ft.

16.72 (a) 1.863 s. (b) 9.00 ft/s. (c) 22.4 ft.

16.76 (a) $107.1 \text{ rad/s}^2 \;\downarrow$. (b) $21.4 \text{ N} \leftarrow, 39.2 \text{ N} \uparrow$.

16.77 (a) 150 mm. (b) $125 \text{ rad/s}^2 \;\downarrow$.

16.78 (a) $12.08 \text{ rad/s}^2 \;\downarrow$. (b) $0.750 \text{ lb} \leftarrow, 4.00 \text{ lb} \uparrow$.

16.79 (a) $8.05 \text{ rad/s}^2 \;\downarrow$. (b) 24.0 in.

16.80 (a) 1522.9 N. (b) 1341.8 N.

16.83 13.64 kN →.

16.84 (a) 1.5 g ↓. (b) 0.25 mg ↑.

16.85 (a) 9g/7. (b) 4mg/7 ↑.

16.86 (a) 0.6727 ft·lb. (b) 1999.2 lb.

16.87 (a) 43.6 rad/s². (b) 21.0 N ←, 54.6 N ↑.

16.88 (a) 3.72 rad/s² ↓. (b) 1.462 lb.

16.94 $r^2 g \sin \beta / (r^2 + \bar{k}^2)$.

16.95 (a) 2.27 m (7.46 ft). (b) 0.649 m (2.13 ft).

16.98 (a) 17.78 rad/s² ↓, 2.13 m/s² →. (b) 0.122.

16.99 (a) 26.7 rad/s² ↓, 3.20 m/s² →. (b) 0.0136.

16.102 (a) no sliding. (b) 15.46 rad/s² ↓, 10.30 ft/s².

16.103 (a) no sliding. (b) 23.2 rad/s² ↓, 15.46 ft/s².

16.104 (a) slides. (b) 4.29 rad/s² ↑, 9.66 ft/s² →.

16.105 (a) slides. (b) 12.88 rad/s² ↑, 3.22 ft/s² ←.

16.107 (a) 6.63 ft/s² →. (b) 3.79 ft/s² →. (c) 0.355 ft →.

16.108 (a) 72.4 rad/s² ↑. (b) 7.24 m/s² ↓.

16.109 (a) 2.64 m/s² ←. (b) 11.87 N ←.

16.111 (a) 0.298. (b) 0.536 g →.

16.112 (a) 0.322. (b) 0.566 g →.

16.113 8.26 N ←.

16.114 (a) 0.125 g/r ↓. (b) 0.125 g →, 0.125 g ↓.

16.115 $m_B g \sin \theta / [2r \{m_h + m_B (1 + \cos \theta)\}]$.

16.116 3.43 lb ∡ 70.5°, 0.1550 ft·lb ↓.

16.117 (a) $\dfrac{g}{L}\left[\dfrac{\sin \theta}{\frac{1}{3} + \sin^2\theta}\right]$ ↓. (b) $\dfrac{mg}{1 + 3\sin^2\theta}$ ↑.

16.118 (a) 27.6 rad/s² ↓. (b) 5.714 lb ↑.

16.119 (a) 0.510 rad/s² ↓. (b) $\mathbf{F}_A$ = 31.80 lb ∡ 78.7° and $\mathbf{F}_B$ = 13.79 lb ↘11.3°.

16.120 mg sin θ/(1 + 3 sin θ).

16.121 (a) 6.26 rad/s² ↓. (b) 13.22 N ←.

16.124 6.40 N ←.

16.125 7.10 lb →.

16.126 5.51 lb →.

16.127 67.62 N ↗56.0°.

16.128 75.13 N ↑.

16.129 25.9 N ↘ 60°.

16.131 (a) 37.8 ft/s² ↘ 26.1°. (b) 48.4 lb ↑.

16.134 (a) 4.36 rad/s² ↑. (b) 31.36 lb↑.

16.135 (a) 36.3 N·m ↑. (b) 231 N ←, 524 N ↑.

16.136 (a) 82.3 N·m ↑. (b) 147.2 N ←, 479 N ↑.

16.137 $\mathbf{B}$ = 805 N ←, $\mathbf{D}$ = 426 N →.

16.138 $\mathbf{B}$ = 525 N ↗ 38.1°, $\mathbf{D}$ = 322 N ↘ 15.7°.

16.139 (a) 24.8 rad/s² ↓. (b) 29.5 lb↑.

16.140 (a) 19.3 ft·lb ↓. (b) 81.9 lb↑.

16.143 (a) $\boldsymbol{\alpha}_A = \dfrac{2}{5}\dfrac{g}{r}$ ↑ and $\boldsymbol{\alpha}_B = \dfrac{2}{5}\dfrac{g}{r}$ ↓. (b) $\dfrac{1}{5}mg$. (c) $\dfrac{4}{5}g$ ↓.

16.146 (a) 50.2 N ∡ 60.3°. (b) 0.273.

16.148 (a) 17.03 ft/s² ↘ 20°. (b) 42.7 rad/s² ↑.

16.151 M_{max} = 10.39 lb·in. located 20.8 in. below A.

16.153 20.6 ft.

16.154 17.34 ft.

16.156 (a) 2μg/(1 + 3μ). (b) 1.000 g.

16.157 (a) 0.513 g/L ↓. (b) 0.912 mg ↑. (c) 0.241 mg →.

16.158 (a) 1.519 g/L ↓. (b) 0.260 g ↓. (c) 0.740 mg ↑.

16.160 (1): (a) 1.200 g/c ↓. (b) 0.671 g ↗ 63.4°.
(2): (a) 24 g/17c ↓, (b) 12 g/17 ↓.
(3): (a) 2.40 g/c ↓, (b) 0.500 g ↓.

16.162 (a) 51.2 rad/s² ↓. (b) 21.0 N ↑.

16.163 (a) 59.8 rad/s² ↓. (b) 20.4 N ↑.

CHAPTER 17

17.1 12.77 N·m.

17.2 8690 rev.

17.3 9.60 in.

17.4 0.841.

17.5 (a) 19.20 lb·ft·s². (b) 11.46 rev.

17.6 (a) 293 rpm. (b) 15.92 rev.

17.7 19.77 rev.

17.10 109.4 lb →.

17.11 (a) 6.35 rev. (b) 7.14 N.

17.12 (a) 2.54 rev. (b) 17.86 N.

17.13 $\omega_A = \dfrac{2n}{n^2 + 1}\sqrt{\dfrac{\pi M_0}{\bar{I}_0}}$.

17.16 (a) $\sqrt{3g/L}$, 2.50 W ↑. (b) 5.67 rad/s ↓, 4.50 lb ↑.

17.17 (a) 0.289 l. (b) $1.861\sqrt{g/l}$, 2.00 mg ↑.

17.18 11.52 rad/s ↑.

17.19 16.23 ft/s.

17.20 (a) 3.94 rad/s ↓, 271 lb ↘ 5.25°.
(b) 5.58 rad/s ↓, 701 lb ↑.

17.23 7.09 rad/s.

17.24 (a) −0.250 rpm. (b) 0.249 rpm.

17.25 $\sqrt{4gs/3}$.

17.26 $\sqrt{gs}$.

17.27 (a) 5.18 ft/s. (b) 0.042.

17.29 (a) 5.00 rad/s. (b) 24.9 N ↑.

17.30 (a) $1.142\sqrt{\dfrac{g}{r}}$ ↓. (b) 1.553 mg ↑.

17.31 (a) $[10g\,(R - r)\,(1 - \cos \beta)/7]^{1/2}$.
(b) mg(17 − 10 cos β)/7.

17.32 (a) 2.06 ft. (b) 4.00 lb.

17.33 (a) 7.43 ft/s ↓. (b) 4.00 lb.

17.35 (a) 11.57 rad/s ↓. (b) 27.8 rad/s ↑.

17.36 $\mathbf{v}_A = 0.775\sqrt{gl}$ ←, $\mathbf{v}_B = 0.775\sqrt{gl}$ ↗ 60°.

17.37 1.170 rad/s ↓, 5.07 m/s ←.

17.38 $[3g\,(\cos \theta_0 - \cos \theta_2)/L]^{1/2}$ ↓.

17.39 3.71 rad/s ↑, 7.74 ft/s ↑.

17.40 15.54 ft/s →.

17.42 2.69 m/s ↓.

17.43 84.7 rpm ↓.

17.44 110.8 rpm ↓.

17.45 3.25 m/s ↓.

17.46 4.43 m/s ↓.

17.47 0.770 m/s ←.

17.48 (a) 44.3 hp. (b) 118.1 hp.

17.49 (a) 39.8 N·m. (b) 95.5 N·m. (c) 229 N·m.

17.50 1146 rpm.

17.51 10.87 lb.

17.52 179.1 mm.

17.53 0.335 lb·in.

17.54 3.87 rad/s.

17.55 24.6 ft·lb.

17.58 3.88 s.

17.59 $(1 + \mu_k^2)\,r\omega_0 / [2\mu_k(1 + \mu_k)g]$.

17.62 $\omega_0 / (1 + m_A/m_B)$.

17.63 (a) $\boldsymbol{\omega}_A$ = 686 rpm ↑, $\boldsymbol{\omega}_B$ = 514 rpm ↓. (b) 4.18 lb·s ↑.

17.64 (a) 5.15 lb. (b) 2.01 lb.

17.65 $\mathbf{X}$ = mv, $d = \bar{k}^2 \omega / \bar{v}$.

17.69 2.79 ft.

17.70 (a) $r^2 gt \sin \beta / (r^2 + \bar{k}^2)$ ↘ β.
(b) $u_s \geq \bar{k}^2 \tan \beta / (r^2 + \bar{k}^2)$.

17.71 (*a*) 2.55 m/s ↑. (*b*) 10.53 N.
17.72 (*a*) 8.05 ft/s →. (*b*) 2.68 ft/s →.
17.74 (*a*) 8.41 m/s ↓. (*b*) 16.82 N.
17.75 (*a*) 0.557 s. (*b*) 16.82 N.
17.77 (*a*) $2.50\,\bar{v}_0/r$. (*b*) $\bar{v}_0/\mu_k g$.
17.78 (*a*) 2.50 s. (*b*) 16.95 ft/s.
17.79 $\dfrac{5}{6}\omega_0$.
17.80 10.19 rpm.
17.81 *A* and *B*: 159.1 rpm ↓; platform 20.9 rpm ↑.
17.82 18.07 rad/s.
17.83 (*a*) 2.54 rad/s. (*b*) 1.902 J.
17.86 37.2 rpm.
17.87 $\boldsymbol{\omega}_{BC}$ = 36.6 rpm ↓ and $\boldsymbol{\omega}_A$ = 16.87 rpm ↑.
17.88 2.51 m/s.
17.89 18.83 rad/s, 0.0508 kg·m².
17.90 (*a*) 31.1 rad/s. (*b*) 18.13 ft/s.
17.91 (*a*) 15.00 rad/s. (*b*) 20.5 ft/s.
17.94 1.542 m/s.
17.95 2.01 ft/s ←.
17.96 0.400 *r*.
17.97 (*a*) 24.4 rad/s ↓. (*b*) 1545 lb →.
17.98 (*a*) 10.00 in. (*b*) 22.6 rad/s ↓.
17.101 (*a*) 2.16 m/s →. (*b*) 4.87 kN ∠ 66.9°.
17.102 (*a*) 158.0 mm. (*b*) 1.992 m/s →.
17.103 (*a*) $0.90\,v_0/L$ ↓. (*b*) $0.10\,v_0$ →.
17.104 2.40 rad/s ↓.
17.105 1.667 in.
17.106 $\boldsymbol{\omega} = \dfrac{\sqrt{2gh}(c + \cos\theta)}{r(1 + c)^{\frac{3}{2}}}$ ↓ and $\mathbf{v} = \dfrac{\sqrt{2gh}(c + \cos\theta)}{(1 + c)^{\frac{3}{2}}}$ →.
17.107 $\dfrac{\pi}{3}L$.
17.108 (*a*) mv_0/M →. (*b*) mv_0/MR ↑.
17.109 (*a*) 1.500 *R*. (*b*) 1.000 *R*.
17.112 $\boldsymbol{\omega} = 2.4\dfrac{v_0}{L}$ ↑ and $\bar{\mathbf{v}} = 0.721v_0$ ⬈ 56.3°.
17.115 2.38 m/s.
17.116 4.867 rad/s ↑.
17.117 (*a*) $0.437\sqrt{g/L}$. (*b*) 5.12°.
17.118 (*a*) $0.250\,\omega_0$ ↓. (*b*) 0.9375. (*c*) 1.50°.
17.119 48.7°.
17.120 1887 ft/s.
17.121 725 mm.
17.122 447 mm.
17.123 $0.606\sqrt{gL}$ →.
17.124 $0.866\sqrt{gL}$ →.
17.127 (*a*) 3.00 rad/s ↑. (*b*) 0.938 m/s ↑.
17.128 (*a*) 2.60 rad/s ↓. (*b*) 1.635 m/s ⬎ 53.4°.
17.131 $1.250\,v_0/r$.
17.132 (*a*) $\mathbf{v}_A = 0$, $\boldsymbol{\omega}_A = v_1/r$ ↓, $\mathbf{v}_B = v_1$ →, $\boldsymbol{\omega}_B = 0$.
(*b*) $\mathbf{v}'_A = 0.286\,v_1$ →, $\mathbf{v}'_B = 0.514\,v_1$ →.
17.133 (*a*) $\mathbf{v}_A = (v_0 \sin\theta)\mathbf{j}$, $\mathbf{v}_B = (v_0 \cos\theta)\mathbf{i}$, $\boldsymbol{\omega}_A = (v_0/r)(-\sin\theta\mathbf{i} + \cos\theta\mathbf{j})$, $\boldsymbol{\omega}_B = 0$.
(*b*) $0.714\,v_0 \cos\theta\mathbf{i}$.
17.134 $\boldsymbol{\omega}_{AB}$ = 2.68 rad/s ↓, $\boldsymbol{\omega}_{BC}$ = 13.39 rad/s ↓.
17.135 (*a*) 106.7 rev. (*b*) 6.98 s.
17.136 70.1 lb ↓.
17.137 (*a*) 18.22 ft/s. (*b*) 359.7 lb ↑. (*c*) 234.2 ft·lbs.
17.139 (*a*) 53.1°. (*b*) $1.095\sqrt{gL}$ ⬎ 53.1°.
17.140 **A** = 100.1 N ↑, **B** = 43.9 N →.

17.142 $0.778\,\omega_0$.
17.143 (*a*) 418 rpm. (*b*) −20.4 J.
17.145 (*a*) 68.6 rpm. (*b*) 2.82 J.

CHAPTER 18

18.1 $0.250\,mr^2 \omega_2\mathbf{j} + 0.500\,mr^2 \omega_1\mathbf{k}$.
18.2 −(0.0408 slug·ft²/s)**i** + (0.1398 slug·ft²/s)**j**.
18.3 0.247 slug·ft²/s, θ_x = 48.6°, θ_y = 41.4°, θ_z = 90°.
18.5 (0.1125 kg·m²/s)**j** + (0.675 kg·m²/s)**k**.
18.7 $0.432\,ma^2\omega$, 20.2°.
18.8 9.7°.
18.9 (1.843 lb·ft·s)**i** − (0.455 lb·ft·s)**j** + (1.118 lb·ft·s)**k**.
18.10 −(2.03 kg·m²/s)**i** + (4.16 kg·m²/s)**j** + (0.675.03 kg·m²/s))**k**.
18.11 $0.500\,mr^2\omega_1\mathbf{i} - m(L^2 + 0.250\,r^2)\,(r\omega_1/L)\mathbf{j}$.
18.12 (*a*) 0.485 rad/s. (*b*) 0.01531 rad/s.
18.15 (*a*) (5.65 kg·m²/s)**i** − (1.885 kg·m²/s)**j** + (12.57 kg·m²/s)**k**.
(*b*) 25.4°.
18.16 (*a*) (5.65 kg·m²/s)**i** − (1.885 kg·m²/s)**j** + (12.57 kg·m²/s)**k**.
(*b*) 154.6°.
18.17 (*a*) (1.078 lb·s·ft)**i** − (0.647 lb·s·ft)**k**. (*b*) 31.0°.
18.18 (*a*) (1.078 lb·s·ft)**i** − (0.647 lb·s·ft)**k**.
(*b*) (1.078 lb·s·ft)**i** − (0.647 lb·s·ft)**k**.
18.21 93.6 kg.
18.22 2.57 s.
18.25 (*a*) 0. (*b*) $(3F\Delta t/ma)\,(\mathbf{i} - 4\mathbf{k})$.
18.26 (*a*) $-(F\Delta t/m)\mathbf{i}$. (*b*) $(3\,F\Delta t/8ma)\,(\mathbf{j} + 4\mathbf{k})$.
18.27 (*a*) −(0.300 m/s)**i**. (*b*) −(0.962 rad/s)**i** − (0.577 rad/s)**j**.
18.28 (*a*) (0.300 m/s)**j**.
(*b*) −(3.46 rad/s)**i** + (1.923 rad/s)**j** − (0.857 rad/s)**k**.
18.31 (*a*) $0.1250\,\omega_0\,(-\mathbf{i} + \mathbf{j})$. (*b*) $0.0884\,a\omega_0\mathbf{k}$.
18.32 (*a*) $0.1031\,ma\omega_0\mathbf{k}$. (*b*) $-0.01473\,ma\omega_0\mathbf{k}$.
18.33 (0.0248 rad/s)**i** − (0.277 rad/s)**j** −(0.360 rad/s)**k**.
18.34 (*a*) −0.726 rad/s.
(*b*) −(2160 ft/s)**i** − (4860 ft/s)**j** + (860 ft/s)**k**.
18.35 (*a*) t_A = 0.129 s, t_B = 1.086 s. (*b*) −(50.6 mm/s)**j**.
18.36 (*a*) 0.941 s. (*b*) (0.0169 rad/s)**j**. (*c*) −(39.2 mm/s)**j**.
18.39 $0.1250\,mr^2\,(\omega_2^2 + 2\omega_1^2)$.
18.40 0.349 ft·lb.
18.41 0.978 ft·lb.
18.42 12.67 ft·lb.
18.43 15.47 J.
18.44 $0.1250\,ma^2\omega^2$.
18.45 $0.203\,ma^2\omega^2$.
18.47 237 J.
18.49 27.0 J.
18.50 46.2 J.
18.51 $0.1000\,m\bar{v}_0^2$.
18.53 16.75 ft·lb.
18.54 39.9 ft·lb.
18.55 $0.500\,mr^2\,\omega_1\omega_2\mathbf{i}$.
18.56 (0.204 ft·lb)**k**.
18.57 (2.22 ft·lb)**k**.
18.58 (5.30 lb·ft)**k**.
18.59 (3.38 N·m)**i**.
18.60 $\dfrac{1}{4}mr^2\omega^2 \sin\beta\cos\beta\mathbf{k}$.
18.64 $\dfrac{1}{4}mr^2\alpha\sin\beta\cos\beta\mathbf{j} + \dfrac{1}{4}mr^2\omega^2\sin\beta\cos\beta\mathbf{k}$.
18.65 **C** = $0.1667\,mb\omega^2 \sin\beta\cos\beta\mathbf{i}$.
D = $-0.1667\,mb\omega^2 \sin\beta\cos\beta\mathbf{i}$.

18.66 $\mathbf{A} = -(4.97 \text{ lb})\mathbf{i}$, $\mathbf{B} = -(1.656 \text{ lb})\mathbf{i}$.

18.67 $\mathbf{A} = -(1.103 \text{ lb})\mathbf{j} - (0.920 \text{ lb})\mathbf{k}$.
$\mathbf{B} = (1.103 \text{ lb})\mathbf{j} + (0.920 \text{ lb})\mathbf{k}$.

18.68 $\mathbf{A} = (14.4 \text{ N})\mathbf{k}$, $\mathbf{B} = -(14.4 \text{ N})\mathbf{k}$.

18.71 (a) $3M_0/mb^2 \cos^2 \beta$. (b) $\mathbf{C} = -\mathbf{D} = (M_0 \tan \beta/2b)\mathbf{k}$.

18.72 (a) $(14.49 \text{ rad/s}^2)\mathbf{j}$. (b) $\mathbf{A} = -(1.125 \text{ lb})\mathbf{k}$, $\mathbf{B} = -(0.375 \text{ lb})\mathbf{k}$.

18.73 (a) $(0.873 \text{ lb·ft})\mathbf{i}$. (b) $\mathbf{A} = -\mathbf{B} = -(0.218 \text{ lb})\mathbf{j} + (0.262 \text{ lb})\mathbf{k}$.

18.74 (a) $(2.67 \text{ N·m})\mathbf{i}$. (b) $\mathbf{A} = -\mathbf{B} = (2.00 \text{ N})\mathbf{j}$.

18.75 (a) $(0.1301 \text{ lb·ft})\mathbf{i}$. (b) $\mathbf{A} = -\mathbf{B} = -(0.0331 \text{ lb})\mathbf{i} + (0.0331 \text{ lb})\mathbf{j}$.

18.76 $\mathbf{A} = -\mathbf{B} = -(0.449 \text{ lb})\mathbf{j} - (0.383 \text{ lb})\mathbf{k}$.

18.79 $\mathbf{A} = -\mathbf{B} = (1.527 \text{ N})\mathbf{j}$.

18.81 (a) 10.47 N·m. (b) 10.47 N·m.

18.82 24.0 N ↑.

18.84 1.138°; up.

18.85 10.20 rad/s.

18.86 (a) 27.0°. (b) 8.09 rad/s.

18.87 (a) 7.53 rad/s. (b) 7.00 rad/s.

18.88 4.84 rad/s.

18.90 7.89 rad/s.

18.91 15.24 rad/s.

18.93 $\mathbf{A} = -\mathbf{B} = (0.1906 \text{ lb})\mathbf{k}$.

18.94 7.87 rad/s.

18.95 (a) $\mathbf{C} = -(592 \text{ N})\mathbf{j}$ and $\mathbf{D} = (592 \text{ N})\mathbf{j}$. (b) $\mathbf{C} = \mathbf{D} = 0$.

18.96 35.5 rpm.

18.99 $-(45.0 \text{ N})\mathbf{i}$, $(3.38 \text{ N·m})\mathbf{i} + (10.13 \text{ N·m})\mathbf{k}$.

18.100 (a) $\mathbf{A} = (1.786 \text{ kN})\mathbf{i} + (143.5 \text{ kN})\mathbf{j}$, $\mathbf{B} = -(1.786 \text{ kN})\mathbf{i} + (150.8 \text{ kN})\mathbf{j}$, (b) $-(35.7 \text{ kN·m})\mathbf{k}$.

18.101 $\mathbf{C} = -(7.81 \text{ lb})\mathbf{i} + (7.43 \text{ lb})\mathbf{k}$, $\mathbf{D} = -(7.81 \text{ lb})\mathbf{i} - (7.43 \text{ lb})\mathbf{k}$.

18.102 $\mathbf{C} = -(12.58 \text{ lb})\mathbf{i} + (9.43 \text{ lb})\mathbf{k}$, $\mathbf{D} = -(12.58 \text{ lb})\mathbf{i} - (9.43 \text{ lb})\mathbf{k}$.

18.103 $\mathbf{D} = -(22.0 \text{ N})\mathbf{i} + (26.8 \text{ N})\mathbf{j}$, $\mathbf{E} = -(21.2 \text{ N})\mathbf{i} - (5.20 \text{ N})\mathbf{j}$.

18.104 (a) $(0.392 \text{ N·m})\mathbf{k}$. (b) $\mathbf{D} = -(21.0 \text{ N})\mathbf{i} + (28.0 \text{ N})\mathbf{j}$, $\mathbf{E} = -(21.0 \text{ N})\mathbf{i} - (4.00 \text{ N})\mathbf{j}$.

18.107 2930 rpm.

18.109 45.9 rpm, 533 rpm.

18.111 442 rpm.

18.112 68.1°.

18.113 $\cos^{-1}\left[\dfrac{2d^2\dot{\psi}}{(d^2 + h^2)\dot{\phi}}\right]$.

18.114 (a) 128.3 rad/s. (b) 2.17 in.

18.115 23.7°.

18.116 (a) 52.7 rad/s. (b) 6.44 rad/s.

18.117 (a) 4.89 rpm. (b) 4.96 rpm, 396 rpm.

18.124 (a) 13.19°. (b) 1242 rpm (retrograde)

18.126 24.8 rev/h.

18.127 (a) 12.85°. (b) 5.78 rev/h. (c) 20.7 rev/h.

18.128 (a) 109.4 rpm, $\gamma_x = 90°$, $\gamma_y = 100.05°$, $\gamma_z = 10.05°$. (b) $\theta_x = 90°$, $\theta_y = 113.9°$, $\theta_z = 23.9°$. (c) precession: 47.1 rpm; spin: 64.6 rpm.

18.130 (a) $\theta_x = 90.0°$, $\theta_y = 26.0°$, $\theta_z = 64.0°$. (b) precession, 0.847 rad/s (retrograde); spin: 0.1593 rad/s.

18.131 (a) $40.0° < \theta < 140.0°$. (b) 5.31 rad/s. (c) 5.58 rad/s.

18.132 (a) 2.00 rad/s. (b) 8.94 rad/s.

18.135 (a) 41.2°. (b) 5.52 rad/s.

18.136 (a) 4.23 rad/s. (b) 12.50 rad/s.

18.139 (a) 47.0°. (b) precession: 15.25 rad/s; spin: 307 rad/s.

18.140 (a) 76.3°. (b) precession: 9.62 rad/s; spin: 294 rad/s. (c) 36.5°.

18.148 $(0.234 \text{ kg·m}^2/\text{s})\mathbf{j} + (1.250 \text{ kg·m}^2/\text{s})\mathbf{k}$.

18.150 (a) 0. (b) $(F\Delta t/ma) (2.50\mathbf{i} - 1.454\mathbf{j} + 2.19\mathbf{k})$.

18.151 4.29 kN·m.

18.153 $\mathbf{D} = -(7.12 \text{ lb})\mathbf{j} + (4.47 \text{ lb})\mathbf{k}$, $\mathbf{E} = -(1.822 \text{ lb})\mathbf{j} + (4.47 \text{ lb})\mathbf{k}$.

18.154 $\mathbf{D} = 0$; $\mathbf{M}_D = (11.23 \text{ N·m}) \cos^2 \theta\mathbf{i} + (11.23 \text{ N·m}) \sin \theta \cos \theta\mathbf{j} - (2.81 \text{ N·m}) \sin \theta \cos \theta\mathbf{k}$.

18.155 (a) $\theta_x = 52.5°$, $\theta_y = 37.5°$, $\theta_z = 90°$. (b) 53.8 rev/h. (c) 6.68 rev/h.

18.156 axis: 32.0°, precession: 1.126 rpm, and spin: 0.344 rpm.

18.157 (a) 4.00 rad/s. (b) 5.66 rad/s.

CHAPTER 19

19.1 10 mm, 3.18 Hz.

19.2 4.33 ft/s, 0.551 Hz.

19.3 1.273 in., 150.8 ft/s^2.

19.4 (a) 0.324 s, 3.08 Hz. (b) 12.91 mm, 4.84 m/s^2.

19.5 (a) 0.308 s, 3.25 Hz. (b) 1.021 m/s, 20.8 m/s^2.

19.6 (a) 267 rpm. (b) 5.36 ft/s.

19.7 (a) 11.29°. (b) 1.933 m/s^2.

19.8 (a) 0.557 Hz. (b) 293 mm/s.

19.11 (a) 0.0352 s. (b) 6.34 ft/s ↑, 64.4 ft/s^2 ↓.

19.12 0.445 ft ↑, 2.27 ft/s ↓, 114.7 ft/s^2 ↓.

19.13 (a) 1.288°. (b) 0.874 ft/s, 0.760 ft/s^2.

19.14 (a) 4.91 mm, 5.81 Hz, 0.1791 m/s. (b) 491 N, (c) 0.1592 m/s ↑.

19.17 (a) 0.517 s, 1.934 Hz. (b) 0.365 m/s, 4.43 m/s^2.

19.18 2.87 s.

19.19 $\sqrt{\dfrac{k}{2m}}$.

19.20 (a) 0.361 s, 2.77 Hz. (b) 0.765 m/s, 13.30 m/s^2.

19.23 4.

19.24 (a) 6.80 kg. (b) 0.583 s.

19.25 192 lb/ft.

19.26 (a) 21.7 kg. (b) 1011 kg/m^3.

19.28 (a) 22.3 MN/m. (b) 266 Hz.

19.30 (a) 858 N/mm. (b) 149.5 rpm.

19.31 (a) 3.56 kg. (b) 43.7 kg.

19.34 16.26°.

19.35 (a) 1.737 s. (b) 1.864 s. (c) 2.05 s.

19.36 28.1 in.

19.37 (a) 0.293 s. (b) 0.215 m/s.

19.38 (a) 1.047 rad/s. (b) 16.42 in.

19.39 (a) 0.227 s. (b) 333 mm/s.

19.41 (a) 0.491 s. (b) 9.60 in/s.

19.42 (a) 0.1957 s. (b) 171.7 ft/s^2.

19.44 75.5°.

19.45 0.346 Hz.

19.48 (a) 2.79 s. (b) 1.933 m.

19.49 (a) 1.617 s. (b) 1.676 s.

19.50 (a) 227 mm. (b) 1.352 s.

19.51 (a) $6.33\sqrt{\dfrac{b}{g}}$. (b) $6.67\sqrt{\dfrac{b}{g}}$.

19.55 (a) 2.21 Hz. (b) 115.3 N/m.

19.56 $\dfrac{1}{2\pi}\sqrt{\dfrac{6k}{5m} + \dfrac{9g}{10l}}$ Hz.

19.57 $\dfrac{1}{2\pi}\sqrt{\dfrac{2k}{3m} + \dfrac{4g}{3L}}$ Hz.

19.59 (a) 0.426 s. (b) 15.44 ft/s.

19.60 (a) 88.1 mm/s. (b) 85.1 mm/s.

19.62 82.2 mm/s ↑.

19.63 6.57 kg·m^2.

19.64 (a) 21.3 kg. (b) 1.836 s.

19.67 0.672 in.

19.68 8.60 ft.

19.69 19.02 mm.

19.70 $\dfrac{1}{2\pi}\sqrt{\dfrac{k}{5m}}$ Hz.

19.71 3.18 s.

19.72 $6.28\sqrt{R/g}$.

19.75 $l/\sqrt{12}$.

19.76 75.5°.

19.77 $0.159\sqrt{(2k/3m)+(4g/3L)}$.

19.78 2.10 Hz.

19.79 (a) 0.715 s. (b) 0.293 ft/s.

19.80 0.821 s.

19.83 1.327 s.

19.85 1.834 s.

19.86 2.39 s.

19.87 $2\pi\sqrt{(12r^2+2l^2)/3gl}$.

19.89 (a) $\sqrt{(6ka^2-3mgl)}/(2\pi)$. (b) $\sqrt{mgl/2k}$.

19.90 2.29 Hz.

19.91 0.911 Hz.

19.92 $0.1312\sqrt{g/r}$.

19.95 $0.276\sqrt{g/l}$.

19.96 $\dfrac{1}{2\pi}\sqrt{\dfrac{12k}{7m}+\dfrac{8g}{7\sqrt{3l}}}$ Hz.

19.97 $1.814l/\sqrt{gr}$.

19.98 0.352 s.

19.99 (a) 37.1 mm. (b) 260 mm.

19.100 (a) 160.0 N/m. (b) 40.0 N/m.

19.101 $\sqrt{\dfrac{2k}{3m}}<\omega_f<\sqrt{\dfrac{4k}{3m}}$.

19.102 50.1 lb.

19.105 $\omega_f<8.46$ rad/s, (no out-of-phase solution).

19.106 3.21 m/s^2.

19.107 (a) 0.450 rad/s. (b) 2.70 m/s^2.

19.109 $\omega_f>\sqrt{2g/l}$.

19.110 (a) 1.034 in. (b) $-0.1033\sin\pi t$ (lb).

19.112 651 rpm.

19.114 22.0 mm.

19.115 0.0999 in.

19.116 $\omega_f<322$ rpm.

19.117 39.1 kg.

19.119 149.3 mm.

19.121 Force transmissibility: $1/(1-\omega_f^2/\omega_n^2)$,
Displacement transmissiblity: $1/(1-\omega_f^2/\omega_n^2)$.

19.122 (a) 4.17%. (b) 84.9 Hz.

19.123 8.04%.

19.125 (a) 1399 rpm. (b) 0.01670 in.

19.132 (a) 6.49 kip · s/ft. (b) 230 kips/ft.

19.133 5.48 N·m·s.

19.134 (a) 6490 lb/ft. (b) 0.1939 s.

19.135 (a) 0.118. (b) 38.4 mm.

19.136 56.9 mm.

19.137 8.82 N.

19.138 106.5 mm/s↑.

19.139 0.1791 in.

19.140 10.61 lb·s/in.

19.141 ≥ 0.707.

19.143 (a) 147 kip/ft. (b) 0.0292.

19.144 0.0162 in.

19.145 (a) 0.127. (b) 462 N·s/m.

19.146 0.487.

19.148 (a) 71.8 N. (b) 39.0 N.

19.149 (a) 4.90 in. (b) 30.3 lb.

19.151 (a) $m\ddot{x}+c\dot{x}+kx=(k\sin\omega_f t+c\omega_f\cos\omega_f t)\delta_m$.
(b) $x=x_m\sin(\omega_f t-\varphi+\psi)$, where
$x_m=\delta_m\sqrt{k^2+(c\omega_f)^2}/\sqrt{(k-m\omega_f^2)^2+(c\omega_f)^2}$,
$\tan\varphi=c\omega_f/(k-m\omega_f^2)$, $\tan\psi=c\omega_f/k$.

19.153 $R<2\sqrt{L/C}$.

19.154 (a) E/R. (b) L/R.

19.157 (a) $c(\dot{x}_A-\dot{x}_m)+kx_A=0$
$m\ddot{x}_m+c(\dot{x}_m-\dot{x}_A)=P_m\sin\omega_f t$
(b) $R(\dot{q}_A-\dot{q}_m)+(1/C)q_A=0$
$L\ddot{q}_m+R(\dot{q}_m-\dot{q}_A)=E_m\sin\omega_f t$

19.159 0.760 lb·s^2· ft, 8.66 in.

19.160 (a) 6.82 lb. (b) 33.4 lb/ft.

19.161 1.785 s.

19.162 (a) 0.0139. (b) 0.0417 lb·s/ft.

19.165 (a) $0.07246\ddot{\theta}+0.3375\dot{\theta}+1.25\theta=0$.
(b) -19.05×10^{-6} degrees.

19.168 (a) $m\ddot{x}+2T(2x/l)=0$. (b) $\pi\sqrt{ml/T}$.

19.169 0.045 in.

Photo Credits

Index